REVIEWS in MINERALOGY and GEOCHEMISTRY

Volume 79   2015

I0031517

Arsenic
Environmental Geochemistry, Mineralogy, and Microbiology

EDITORS

Robert J. Bowell
SRK Consulting, Cardiff, United Kingdom

Charles N. Alpers
USGS, California, U.S.A.

Heather E. Jamieson
Queen's University, Ontario, Canada

D. Kirk Nordstrom
USGS, Colorado, U.S.A.

Juraj Majzlan
Friedrich-Schiller-Universitat, Jena, Germany

Front Cover: Mimetite, 17-23 Floors, 9th Level, Jersey Vein, Bunker Hill Mine, Kellogg, Idaho, U.S.A. Field of view 3 cm. Crystal drawing of mimetite crystal structure.

Back Cover: (*top left*) Erythrite crystals, Bou Azer District, Tazenakht, Ouarzazate Province, Souss-Massa-Draâ Region, Morocco. Field of view 2 cm. (*top right*) Arthurite, scorodite and pharmacosiderite, 150 level Copper Stope, Majuba Mine, Pershing County, Nevada, U.S.A. Field of view 1.2 cm. (*bottom left*) Shultenite (white) in matrix of Cuprian adamite (emerald green) and olivenite (dark green) nest of crystals in tennantite matrix, Tsumeb Mine, Tsumeb, Otjikoto region, Namibia. Field of view 2.5 cm (*bottom right*) Liroconite, Wheal Gorland, St Day, Cornwall, United Kingdom. Field of view 3 cm.

Series Editor: **Jodi J. Rosso**
MINERALOGICAL SOCIETY of AMERICA
GEOCHEMICAL SOCIETY

DE GRUYTER

Reviews in Mineralogy and Geochemistry, Volume 79

Arsenic: Environmental Geochemistry, Mineralogy, and Microbiology

ISSN 1529-6466
ISBN 978-0-939950-94-2

COPYRIGHT 2015

THE MINERALOGICAL SOCIETY OF AMERICA
3635 CONCORDE PARKWAY, SUITE 500
CHANTILLY, VIRGINIA, 20151-1125, U.S.A.
www.degruyter.com

Arsenic
Environmental Geochemistry, Mineralogy, and Microbiology

79   *Reviews in Mineralogy and Geochemistry*   **79**

FROM THE SERIES EDITOR

This volume, edited by Rob Bowell, Charlie Alpers, Heather Jamieson, Kirk Nordstrom, and Juro Majzlan, presents a comprehensive review of the topics covered at the "Environmental Geochemistry, Mineralogy, and Microbiology of Arsenic" short course that followed the 24th Annual V.M. Goldschmidt Conference and held at the Miners Foundry, Nevada City, CA (June 15-16, 2014). This project has been a labor of love for the organizers. Changing short-course venues, changing chapter authorship, and other unexpected obstacles, challenged the organizers and threatened to undermine the completion of this volume. It may have taken several years to bring to fulfillment, but Rob and the other editors overcame the obstacles and successfully compiled the volume that you are now reading.

L. Frank Baum, in *The Marvelous Land of Oz*, wrote "Everything has to come to an end, sometime." With the completion of this volume, I will be stepping down from my position as the Series Editor of RiMG. It has been a wonderful experience to shepherd this series and work with the Mineralogical Society of America, the Geochemical Society, and so many editors and authors over the past 14 years. Thank you!

All supplemental materials associated with this volume can be found at the MSA website. Errata will be posted there as well.

Jodi J. Rosso, Series Editor
Richland, Washington
August 2014

PREFACE

"Put this in any liquid thing you will
And drink it off; and, if you had the strength
Of twenty men, it would dispatch you straight."

Apothecary. Act V Scene 1 (*Romeo and Juliet,* by William Shakespeare)

Arsenic is perhaps history's favorite poison, often termed the "King of Poisons" and the "Poison of Kings" and thought to be the demise of fiction's most famous ill-fated lovers. The toxic nature of arsenic has been known for millennia with the mineral realgar (AsS), originally named "arsenikon" by Theophrastus in 300 B.C.E. meaning literally "potent." For centuries it has been used as rat poison and as an important component of bactericides and wood preservatives. Arsenic is believed to be the cause of death to Napoleon Bonaparte who

1529-6466/14/0079-0000$00.00   http://dx.doi.org/10.2138/rmg.2014.79.0

was exposed to wallpaper colored green from aceto-arsenite of copper (Aldersey-Williams 2011). The use of arsenic as a poison has been featured widely in literature, film, theatre, and television. Its use as a pesticide made it well known in the nineteenth century and it was exploited by Sir Arthur Conan Doyle in the Sherlock Holmes novel *The Golden Pince-Nez* (Conan-Doyle 1903). The dark comedy *Arsenic and Old Lace* is a prime example of arsenic in popular culture, being first a play but becoming famous as a movie.

Arsenic has figured prominently not only in fiction but in historical crimes as well (Kumar 2010). A high profile case of the mid-nineteenth century involved a hydrotherapist, Dr. Thomas Smethurst, who was accused of using arsenic to poison a woman he had befriended (Wharton 2010). Based on analytical evidence from a renowned toxicologist, Alfred Swaine Taylor, a death sentence was imposed, however Taylor had to confess that his apparatus was contaminated. The verdict was overturned after public opinion was voiced against it and a plea for clemency was made to Queen Victoria.

In recent years, arsenic has been recognized as a widespread, low-level, natural ground-water contaminant in many parts of the world, particularly in places such as West Bengal and Bangladesh, where it has given rise to chronic human-health issues. Long-term exposure to arsenic has been shown to cause skin lesions, blackfoot disease, and cancer of the skin, bladder, and lungs, and is also associated with developmental effects, cardiovascular disease, neurotoxicity, and diabetes (WHO 2012). Arsenate's toxicity is caused by its close chemical similarities to phosphate; it uses a phosphate transport system to enter cells. Arsenic occurs in many geological environments including sedimentary basins, and is particularly associated with geothermal waters and hydrothermal ore deposits. It is often a useful indicator of prox-imity to economic concentrations of metals such as gold, copper, and tin, where it occurs in hydrothermally altered wall rocks surrounding the zones of economic mineralization. Arsenic is commonly a persistent problem in metal mining and there has been significant effort to man-age and treat mine waste to mitigate its environmental impacts.

This volume compiles and reviews current information on arsenic from a variety of perspectives, including mineralogy, geochemistry, microbiology, toxicology, and environmental engineering. The first chapter (Bowell et al. 2014) presents an overview of arsenic geochemical cycles and is followed by a chapter on the paragenesis and crystal chemistry of arsenic minerals (chapter 2; Majzlan et al. 2014). The next chapters deal with an assessment of arsenic in natural waters (chapter 3; Campbell and Nordstrom 2014) and a review of thermodynamics of arsenic species (chapter 4; Nordstrom et al. 2014). The next two chapters deal with analytical measurement and assessment starting with measuring arsenic speciation in solids using x-ray absorption spectroscopy (chapter 5; Foster and Kim 2014). Chapter 6 (Leybourne and Johannesson 2014) presents a review on the measurement of arsenic speciation in environmental media: sampling, preservation, and analysis. In chapter 7 (Amend et al. 2014) there is a review of microbial arsenic metabolism and reaction energetics. This is followed by an overview of arsenic toxicity and human health issues (chapter 8; Mitchell 2014) and an assessment of methods used to characterize arsenic bioavailability and bioaccessibility (chapter 9; Basta and Jurasz 2014). This leads into chapter 10 (Craw and Bowell 2014), which describes the characterization of arsenic in mine waste with some examples from New Zealand, followed by a chapter on the management and treatment of arsenic in mining environments (chapter 11; Bowell and Craw 2014). The final three chapters are in-depth case studies of the geochemistry and mineralogy of legacy arsenic contamination in different historical mining environments: the Giant gold mine in Canada (chapter 12; Jamieson 2014), the Sierra Nevada Foothills gold belt of California (chapter 13; Alpers et al. 2014), and finally, the hydrogeochemistry of arsenic in the Tsumeb polymetallic mine in Namibia (chapter 14; Bowell 2014).

We thank all the authors for their comprehensive and timely efforts and for their cooperation with our requests for scheduling, consistency of format, and nomenclature. Special thanks are due to the numerous colleagues who provided peer reviews that substantially improved all chapters. This volume would not have been possible without the wisdom and patience of the series editor, Dr. Jodi Rosso. Finally we thank our families for their support and understanding during the last several months.

R.J. Bowell, SRK Consulting, Churchill House, Cardiff, U.K.
C.N. Alpers, U.S. Geological Survey, Sacramento, California, U.S.A.
H.E. Jamieson, Queen's University, Kingston, Ontario, Canada
D.K. Nordstrom, U.S. Geological Survey, Boulder, Colorado, U.S.A.
J. Majzlan, Friedrich-Schiller-Universität, Jena, Germany

REFERENCES

Aldersey-Williams H (2011) Periodic Tales: The curious lives of the elements. Penguin Books, London 428 p

Alpers CN, Myers PA, Millsap D, Regnier TB (2014) Arsenic associated with historical gold mining in the Sierra Nevada foothills: case study and field trip guide for Empire Mine State Historic Park, California. Rev Mineral Geochem 79:553-587

Amend JP, Saltikov C, Lu G-S, Hernandez J (2014) Microbial arsenic metabolism and reaction energetics. Rev Mineral Geochem 79:391-433

Basta NT, Juhasz A (2014) Using in vivo bioavailability and/or in vitro gastrointestinal bioaccessibility testing to adjust human exposure to arsenic from soil ingestion. Rev Mineral Geochem 79:451-472

Bowell RJ (2014) Hydrogeochemistry of the Tsumeb deposit: implications for arsenate mineral stability. Rev Mineral Geochem 79:589-627

Bowell RJ, Alpers CN, Jamieson HE, Nordstrom DK, Majzlan J (2014) The environmental geochemistry of arsenic: an overview. Rev Mineral Geochem 79:1-16

Bowell RJ, Craw D (2014) The management of arsenic in the mining industry. Rev Mineral Geochem 79:507-532

Campbell KM, Nordstrom DK (2014) Arsenic speciation and sorption in natural environments. Rev Mineral Geochem 79:185-216

Conan-Doyle A (1903) The adventure of the golden pince-nez. *In*: The Return of Sherlock Holmes, Strand Magazine, London. October 1903-December 1904:597-615.

Craw D, Bowell RJ (2014) The characterization of arsenic in mine waste. Rev Mineral Geochem 79:473-506

Foster AL, Kim CS (2014) Arsenic speciation in solids using X-ray absorption spectroscopy. Rev Mineral Geochem 79:257-369

Jamieson HE (2014) The legacy of arsenic contamination from mining and processing refractory gold ore at Giant Mine, Yellowknife, Northwest Territories, Canada. Rev Mineral Geochem 79:533-551

Kumar NSA (2010) The science of Sherlock Holmes. Science Reporter, March 2010, 8-14 *http://nopr.niscair. res.in/bitstream/123456789/7512/1/SR%2047(3)%208-14.pdf*

Leybourne MI, Johannesson KH, Asfaw A (2014) Measuring arsenic speciation in environmental media: sampling, preservation, and analysis. Rev Mineral Geochem 79:371-390

Majzlan J, Drahota P, Filippi M (2014) Parageneses and crystal chemistry of arsenic minerals. Rev Mineral Geochem 79:17-184

Mitchell VL (2014) Health risks associated with chronic exposures to arsenic in the environment. Rev Mineral Geochem 79:435-449

Nordstrom DK, Majzlan J, Königsberger E (2014) Thermodynamic properties for arsenic minerals and aqueous species. Rev Mineral Geochem 79:217-255

Wharton JC (2010) The Arsenic Century: How Victorian Britain Was Poisoned at Home, Work, & Play, Oxford University Press, 412 p

WHO (2012) Arsenic. World Health Organization, Fact Sheet 372. *http://www.who.int/mediacentre/factsheets/ fs372/en/*

Arsenic
Environmental Geochemistry, Mineralogy, and Microbiology

79   *Reviews in Mineralogy and Geochemistry*   **79**

TABLE OF CONTENTS

3   Arsenic Speciation and Sorption in Natural Environments

Kate M. Campbell, D. Kirk Nordstrom

4   Thermodynamic Properties for Arsenic Minerals and Aqueous Species

D. Kirk Nordstrom, Juraj Majzlan,
Erich Königsberger

5   Arsenic Speciation in Solids Using X-ray Absorption Spectroscopy

Andrea L. Foster, Christopher S. Kim

6 Measuring Arsenic Speciation in Environmental Media: Sampling, Preservation, and Analysis

Matthew I. Leybourne, Karen H. Johannesson
Alemayehu Asfaw

7 Microbial Arsenic Metabolism and Reaction Energetics

Jan P. Amend, Chad Saltikov
Guang-Sin Lu. Jaime Hernandez

8   Health Risks Associated with Chronic Exposures to Arsenic in the Environment

Valerie L. Mitchell

9   Using *In Vivo* Bioavailability and/or *In Vitro* Gastrointestinal Bioaccessibility Testing to Adjust Human Exposure to Arsenic from Soil Ingestion

Nicholas T. Basta, Albert Juhasz

10   The Characterization of Arsenic in Mine Waste

Dave Craw, Robert J. Bowell

11   The Management of Arsenic in the Mining Industry

Robert J. Bowell, Dave Craw

12   The Legacy of Arsenic Contamination from Mining and Processing Refractory Gold Ore at Giant Mine, Yellowknife, Northwest Territories, Canada

Heather E. Jamieson

13   Arsenic Associated with Historical Gold Mining in the Sierra Nevada Foothills: Case Study and Field Trip Guide for Empire Mine State Historic Park, California

Charles N. Alpers, Perry A. Myers,
Daniel Millsap, Tamsen Burlak Regnier

14   Hydrogeochemistry of the Tsumeb Deposit: Implications for Arsenate Mineral Stability

Robert J. Bowell

Reviews in Mineralogy & Geochemistry
Vol. 79 pp. 1-16, 2014
Copyright © Mineralogical Society of America

1

The Environmental Geochemistry of Arsenic
— An Overview —

Robert J. Bowell

*SRK Consulting, Churchill House
Cardiff CF10 2HH, United Kingdom*

rbowell@srk.co.uk

Charles N. Alpers

*U.S. Geological Survey, Placer Hall, 6000 J Street
Sacramento, California 95819, U.S.A.*

cnalpers@usgs.gov

Heather E. Jamieson

*Department of Geological Sciences & Geological Engineering
Miller Hall, Queen's University
Kingston, Ontario K7L 3N6, Canada*

jamieson@queensu.ca

D. Kirk Nordstrom

*U.S. Geological Survey, 3215 Marine St., Suite 127
Boulder, Colorado 80303, U.S.A.*

dkn@usgs.gov

Juraj Majzlan

*Friedrich-Schiller-Universität
Jena, Germany*

Juraj.Majzlan@uni-jena.de

INTRODUCTION

Arsenic is one of the most prevalent toxic elements in the environment. The toxicity, mobility, and fate of arsenic in the environment are determined by a complex series of controls dependent on mineralogy, chemical speciation, and biological processes. The element was first described by Theophrastus in 300 B.C. and named arsenikon (also arrhenicon; Caley and Richards 1956) referring to its "potent" nature, although it was originally considered an alternative form of sulfur (Boyle and Jonasson 1973). Arsenikon is believed to be derived from the earlier Persian, *zarnik* (online etymology dictionary, *http://www.etymonline.com/index. php?term=arsenic*). It was not until the thirteenth century that an alchemist, Albertus Magnus, was able to isolate the element from orpiment, an arsenic sulfide (As_2S_3). The complex chemistry required to do this led to arsenic being considered a "bastard metal" or what we now call a "metalloid," having properties of both metals and non-metals. As a chemical element, arsenic is widely distributed in nature and can be concentrated in many different ways. In the Earth's crust, arsenic is concentrated by magmatic and hydrothermal processes and has been

1529-6466/14/0079-0001$05.00   http://dx.doi.org/10.2138/rmg.2014.79.1

used as a "pathfinder" for metallic ore deposits, particularly gold, tin, copper, and tungsten (Boyle and Jonasson 1973; Cohen and Bowell 2014). It has for centuries been considered a potent toxin, is a common poison in actual and fictional crimes, and has led to significant impacts on human health in many areas of the world (Cullen 2008; Wharton 2010).

ARSENIC TOXICITY IN DRINKING WATER

The potential issues associated with elevated As concentrations in water supplies have led to a large body of published research in the last few years related to:

- arsenic impacts in the environment (Chappell et al. 1994, 1999, 2001, 2003; Nriagu 1994a,b; Abernathy et al. 1997; Nordic Ministers Council 1999; Frankenberger 2002; Naidu et al. 2006; Garelick and Jones 2009)

- advances in arsenic chemistry and microbiology (O'Day et al. 2005; Henke 2009; Santini and Ward 2012; Zhu et al. 2014)

- arsenic in groundwater and drinking water (NRC 1977, 1999, 2001; Anwar 2000; Bianchelli 2003; Welch and Stollenwerk 2003; Bhattacharya et al. 2007; Meliker 2007; Aphuja 2008; Bundschuh et al. 2005, 2009; Sorlini and Collivignarellli 2011)

- the health effects of arsenic (Nriagu et al. 1994b; Murphy and Guo 2003; Le and Weinfeld 2004; Parker and Parker 2004; Meharg 2005; Cullen 2008; Ravenscroft et al. 2009; Jean et al. 2010; Wharton 2010; Chen and Chiou 2011; Ng et al. 2012)

- improved methods of arsenic analysis (Le 2001; Clifford et al. 2004; Francesconi and Kuehnelt 2004; Samanta and Clifford 2006).

Based on the mounting evidence for the acute and chronic toxicity of As, the WHO recommended a more stringent drinking water limit for total As which was provisionally reduced in 1993 from 50 µg L^{-1} to 10 µg L^{-1} (NRC 1999). The recommended value, however, is still based largely on analytical capability (NRC 2001). If the standard basis for risk assessment applied to industrial chemicals was applied to As, the maximum permissible concentration would be lower based on toxicology and water consumption. The recommendations are based on an average body weight and water intake per day. Those with hard manual work in tropical regions surpass the average daily water intake by a factor of 2-3 and for them, the limit would still have to be decreased (Chakraborti et al. 2010).

Although many national authorities are bringing limits in line with the WHO guideline value, many developing countries still operate at the 50 µg L^{-1} standard, in part because of lack of adequate testing facilities for lower concentrations. Despite the substantial body of literature, research on As is still continuing and many recent papers have focused on groundwater with little interconnection to understanding the fundamental source(s) of As, its variable speciation in both solid and aqueous form, and its interaction with the biosphere. The purpose of this short course volume is to provide a summary of the current state of knowledge in these areas.

ARSENIC MINERALOGY AND PRIMARY OCCURRENCE

Arsenic is mobilized in the environment through a combination of natural processes such as weathering reactions, biological activity, and volcanic emissions, as well as through a range of anthropogenic activities. It has only one stable isotope (^{75}As) and is the 47th most abundant natural element. The average crustal abundance is 2.5 mg kg^{-1} although it is even more abundant in the upper continental crust (5.7 mg kg^{-1}; Hu and Gao 2008) and generally more abundant in marine shales and mudstones (Tourtelot 1964), with high concentrations associated with hydrothermal ore deposits, coal, and lignite deposits (Table 1).

Table 1. Range of arsenic concentrations in the environment.

Rock/soil type	Average As (mg kg^{-1})	Range As (mg kg^{-1})	Refs.
Ultrabasic	1.5	0.03 – 15.8	[1]
Granite	1.3	0.2 – 15	[1]
Andesite	2.7	0.5 – 5.8	[1]
Basalt	2.3	0.18 – 113	[1]
Slate/phyllite	18	0.5 – 143	[1]
Mudstone/marine shale	3 – 15	<490	[1]
Hornfels	5.5	0.7 – 11	[1]
Sandstone	4.1	0.6 – 120	[1]
Limestone	2.6	0.1 – 20.1	[1]
Phosphorite	21	0.4 – 188	[1]
Coal		0.3 – 35,000	[1]
Alluvial sands (Bangladesh)	2.9	1 – 6.2	[1]
Alluvial muds (Bangladesh)	6.5	2.7 – 14.7	[1]
River bed (Bangladesh)		1.2 – 5.9	[1]
Tropical soils (Ghana)	0.3	0.2 – 1.2	[2]
Tropical baseline soils, gold deposit (Ghana)		2 – 35,600	[2]
Great Basin Alluvium (Nevada, USA)		13.6 – 54	[3]
Loess Silt (Argentina)		5.4 – 18	[1]
Mine-contaminated soil (Cornwall)	1,800	4 – 9,000	[4]
Gold mine waste (California, USA)		10.1 – 15,300	[5][6]
Mine-contaminated sediment (USA)	342	80 – 1,104	[1]
Mine-contaminated reservoir sediment (California, USA)		54 – 301	[7]
Soil, sulfide deposit	126	2 – 8,000	[1]
Glacial till, Canada	9.2	1.9 – 170	[1]
Sewage sludge	9.8	2.4 – 39.6	[1]

References: [1] Smedley and Kinniburgh (2002); [2] Bowell (1994); [3] Theodore et al. (2003); [4] Bowell et al. (2013); [5] MFG (2009); [6] Alpers et al. (2014, this volume); [7] Savage et al. (2000)

As of July 2014, there are 568 known minerals for which arsenic is a critical component. These include elemental arsenic, arsenides, sulfides, oxides, arsenates, mixed-anion arsenates, and arsenites (IMA 2014). High arsenic concentrations are also found in many oxide minerals and hydrous metal oxides, either as part of their periodic structure or as sorbed and occluded species. Iron oxides are particularly well known to accumulate As up to concentrations of several weight per cent. Arsenic (as As(III) or As(V)) can substitute for P(V), Si(IV), Al(III), Fe(III), and Ti(IV) in various mineral structures and is therefore present in many rock-forming minerals, albeit at much lower concentrations. The element is primarily concentrated in sulfide minerals where it can occur as an arsenide or sulfarsenide anion bound to transition metals (e.g., löllingite, FeAs$_2$; arsenopyrite, FeAsS) or more rarely in minerals where arsenic forms nominally a cation (e.g., realgar, AsS).

Despite the large number of known As minerals, the largest reservoir of arsenic in crustal rocks is probably pyrite (Nordstrom 2000) which contains trace to minor contents (up to 16.5 wt% in synthetic marcasite and 10 wt% in natural pyrite; Reich et al. 2005; Neumann et al. 2013; Simon et al. 2013) of this element. Besides being an important constituent of ore bodies, pyrite is also formed in low-temperature sedimentary environments under reducing conditions.

Authigenic pyrite is present in the sediments of many rivers, lakes, oceans, and aquifers and plays a very important role in the present-day geochemical cycles of various elements. Through a series of intermediate phases, pyrite commonly forms in zones of intense reduction such as around buried and decomposing organic matter or in microenvironments where the sulfate-reducing bacteria generate appreciable amounts of sulfide. It is sometimes present in a characteristic form as framboidal pyrite (Roberts 1982; Wilkin and Barnes 1997; Alpers et al. 2002). During the formation of this pyrite, it is likely that As will also be incorporated. Pyrite is not stable in aerobic systems and oxidizes to hydrous iron oxides with the release of large amounts of sulfate, acidity, and associated trace constituents, including As (Nordstrom and Alpers 1999). The presence of pyrite as a minor constituent in sulfide-rich coals is ultimately responsible for the production of "acid rain" and coal mine associated acid mine drainage (AMD), and for the presence of As problems around coal mines and areas of intensive coal burning (e.g., Tourtelot 1964; Finkelman et al. 1999, 2002; Finkelman 2004).

ARSENIC IN THE WEATHERING ENVIRONMENT

Arsenic in secondary minerals and soils

Arsenic behavior is typical of many chalcophile elements in that it is released by sulfide oxidation, modified by various biogeochemical processes, and attenuated by adsorption and co-precipitation with Fe-minerals, clays, and organic matter. It can form a large number of secondary As minerals including native arsenic, arsenates, and in rare cases arsenites (Drahota and Filippi 2009) such as in the Tsumeb deposit of Namibia (Bowell 2014, this volume). The attenuation and concentration of As in surface soils can be a useful indicator of sulfide mineral deposits (Boyle and Jonasson 1973). For example in the northern part of Nevada, As concentrations in soils define zoned anomalies or "chalcophile corridors" around major bedrock gold deposits (Theodore et al. 2003; Fig. 1).

District-scale As geochemical trends in the vicinity of major Carlin-type deposits in alluvium, surface rocks, and stream sediments define "corridors" of anomalous values with a northwest-trending lobate pattern that mimics the distribution of the major gold deposits. Here As occurs in concentrations up to 54 mg kg^{-1} and reflects the concentration of As in pyrite that occurs within the gold-bearing zones (Thompson et al. 2002). Arsenic concentration in exposed surface rocks (up to 90 mg kg^{-1}) correlates to As in derivative stream sediments. The application of X-ray absorption near-edge spectra (XANES) of selected light-density minerals from the stream-sediment samples indicated As is associated with Al-bearing phases, such as gibbsite, amorphous Al oxyhydroxides, or aluminosilicate clay minerals as As(V) (Theodore et al. 2003). This association occurred through chemical weathering of mineralized rock fragments and migration of As in groundwater with fixation in the supergene environment as As(V).

Arsenic in water

The geochemical behavior of As in the surficial environment is characteristic of oxyanion-forming metalloids in that it is mobile not only at the pH values typically found in groundwaters (pH 6.5-8.5) but also under both oxidizing and reducing conditions. Arsenic can occur in the environment in several oxidation states although in most natural groundwaters, it occurs either as the trivalent arsenite [As(III)] or pentavalent arsenate [As(V)]. Organic arsenic forms may be produced by biological activity, mostly in surface waters, but are rarely quantitatively important except in biological tissues where they may be dominant (for example arsenobetaine; Cullen and Reimer 1989). Arsenic is generally present as an oxysalt or oxyanion in oxic environments. In anaerobic soils, on the other hand, it is typically found combined with sulfur. Thus, in uncontaminated aerobic sediments and soils arsenate (As^{V}O$_4^{3-}$) is the predominant species, whereas in anaerobic sediments and soils arsenite (As^{III}O$_3^{3-}$) is the dominant species (Nordstrom and Archer 2003; Campbell and Nordstrom 2014, this volume).

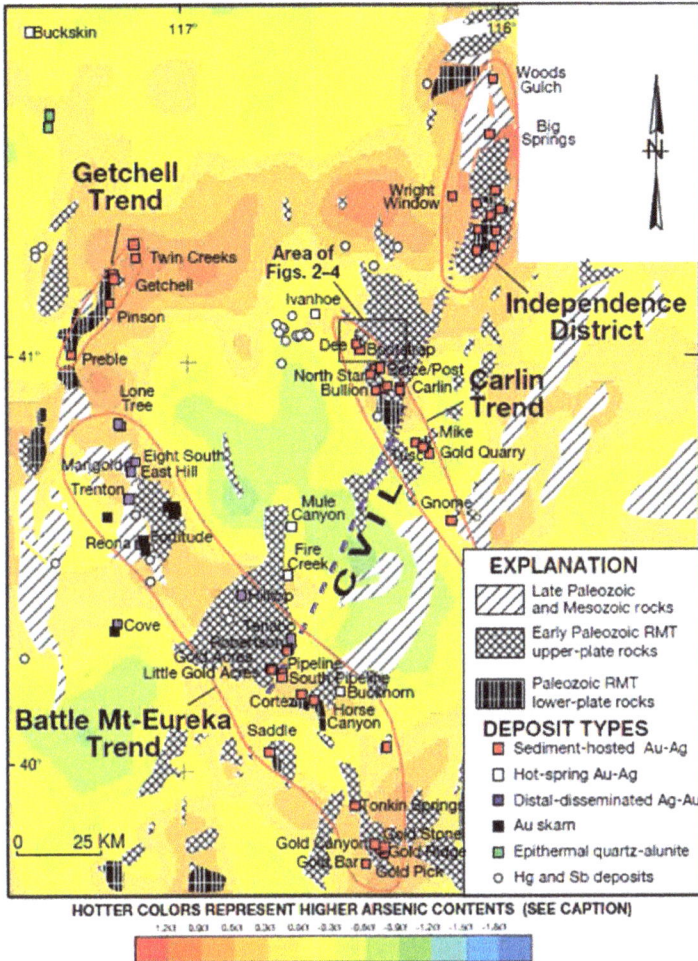

Figure 1. (*for color see Plate 1*) Index map showing relationship of alluvium As concentrations in north-ernmost Carlin trend to bedrock, Crescent Valley-Independence Lineament (CVIL), and major mineral deposits in north-central Nevada (Theodore et al. 2003). Distribution of arsenic contents in sediment samples is in normalized log scale. Reported As contents in approximately 4,300 sediment samples are gridded (1,000-m-wide cells) and filtered (z = 5,000 m) resulting in contours showing standard deviations from the mean of log-transformed metal concentrations. [Reproduced with permission of the Society of Economic Geologists from Theodore et al. (2003), *Econ Geol*, Vol. 98, Fig. 1, p. 288.]

Methylated forms of As can also form in surficial environments and are also pH sensitive. Over the natural range of Eh and pH in soils, both As(III) and As(V) can occur in a range of stable aqueous and solid forms. Commonly both forms occur together in waters and some minerals due to redox disequilibrium. The equilibrium constants for selected As species in aqueous solution are given in Table 2.

Thus, As dispersion in water can be extensive and concentrations can vary considerably in naturally occurring waters (Welch and Stollenwerk 2003) as shown in Table 3. Typically rainwater carries few trace elements except in highly industrialized environments where local conditions can impact rainwater chemistry (Hem 1985). River-water As content is likewise

Table 2. Aqueous speciation of arsenic (from Lewis et al. 1976; Nordstrom et al. 2014, this volume)

Arsenic acid	
$H_3AsO_4 \rightleftharpoons H_2AsO_4^- + H^+$	$pK_a = 2.25$
$H_2AsO_4^- \rightleftharpoons HAsO_4^{2-} + H^+$	$pK_{a2} = 6.98$
$HAsO_4^{2-} \rightleftharpoons AsO_4^{3-} + H^+$	$pK_{a3} = 11.58$
Arsenous acid	
$H_3AsO_3 \rightleftharpoons H_2AsO_3^- + H^+$	$pK_a = 9.24$
Monomethylarsonic acid	
$CH_3AsO(OH)_2 \rightleftharpoons CH_3As(OH)O_2^- + H^+$	$pK_a = 4.19$
$CH_3As(OH)O_2^- \rightleftharpoons CH_3AsO_3^{2-} + H^+$	$pK_{a2} = 8.77$
Dimethylarsinic acid	
$CH_3AsO(OH) \rightleftharpoons (CH_3)_2AsO_2^- + H^+$	$pK_a = 6.14$

Table 3. Range of arsenic concentrations in natural waters.

Type	Range As ($\mu g\ L^{-1}$)	Refs.
Terrestrial rain water	0.013 – 0.032	[1]
Seawater (deep Pacific/Atlantic)	1 – 1.8	[1]
River water	<1 – 12,400	[1]
Lake, mine pit lake	<1 – 508	[1]
Lake water	<0.2 – 0.42	[1]
Groundwater (UK)	<0.5 – 10	[1]
Groundwater (Bangladesh)	<0.5 – 2,500	[1]
Groundwater (West Bengal)	<0.5 – 3,200	[1]
Groundwater (Antofagasta, Chile)	100 – 1,000	[1]
Groundwater, mine impacted (Coeur d'Alene, Idaho, USA)	<1400	[1]
Groundwater, mine impacted (Northern Bavaria, Germany)	<10 – 150	[1]
Yellowstone geothermal water (Wyoming, USA)	160 – 10,000	[2]
Waiotapu geothermal (New Zealand)	710 – 6,500	[2]
Wairakei geothermal (New Zealand)	230 – 3,000	[2]
El Tatio geothermal (Chile)	45,000 – 50,000	[3]
Salton Sea (USA)	30 – 12,000	[1]

References: [1] Smedley and Kinniburgh (2002); [2] Webster and Nordstrom (2003); [3] Ellis and Mahon (1977)

a product of local conditions in which point-source discharges, such as a mine, can impact local river-water chemistry adversely but the area impacted is typically limited and dilution tends to reduce the impact with distance. In addition, such changes can also occur in response to localized changes over time in the physical and chemical conditions of the receiving environment. For example the Mokrsko stream in central Czech Republic is a neutral oxic stream that drains a natural As-Au anomaly with As concentrations decreasing away from this point source and also shows diel and seasonal variations (Drahota et al. 2006, 2009, 2013).

Groundwater arsenic concentrations can vary significantly (Table 3). Apart from volcanic and geothermal inputs and anthropogenic impacts including mining-influenced water (Smedley and Kinniburgh 2002; Webster and Nordstrom 2003), some large aquifers, demonstrate natural concentrations above 50 µg L^{-1} (Table 3). These aquifers have been reported from Bangladesh, West Bengal, Chile, Argentina, China, Mexico, Vietnam, and parts of Canada and the U.S.A. (Smedley and Kinniburgh 2002; McGuigen et al. 2010; Chappells et al. 2014). Although the presence of As concentrations above 50 µg L^{-1} is not uncommon, it is not typical. The conditions controlling these elevated As levels are complex and relate to bedrock type (although not always), past and present hydrogeology, and geochemical environment. The most studied area of natural high As groundwater is in Bangladesh and West Bengal where high As in alluvial and deltaic aquifers has resulted in a significant human-health impact. Here, more than a quarter of all shallow drinking-water wells contain As above 50 µg L^{-1} (Smedley and Kinniburgh 2002).

The impacted aquifers are generally shallow (100-150 m deep) and developed in Holocene age micaceous sands, silts and clays. The sediments were most likely derived from upland Himalayan catchments and West Bengal basement complex. The aquifers are capped by a layer of clay or silt that restricts the ingress of atmospheric oxygen and, together with organic matter in the sediments, has produced reducing conditions that favor the mobilization of As. The As is believed to be derived mainly by desorption and reduction of arsenate from rapidly buried Fe oxides.

Deeper aquifers in Bangladesh tend to show lower concentrations of As. The differences between the two aquifers may relate to variations in the total As reservoir available in host sediments, oxidation state of As, and the speciation of As in the sediments. In addition, groundwater recharge and flushing of the aquifers in the Bengal basin are also contributing factors. Older, deeper sediments have most likely been subject to prolonged periods of groundwater flow assisted by greater hydraulic driving forces during the Pleistocene (Smedley and Kinniburgh 2002). Older sediments are often oxidized and have sufficient hydrated ferric oxides to immobilize As through sorption.

ANTHROPOGENIC ARSENIC CONTAMINATION

Human activity contributes to the mobilization of As through mining and mineral processing, combustion of fossil fuels, and the use of As in pesticides, herbicides, crop desiccants, wood preservatives, and as an additive to livestock feed, particularly for poultry and swine. It was once common practice to dip sheep and cattle in As-rich solutions to rid them of parasites and other pests. Examination of these solutions led to the discovery of arsenate-reducing and arsenite-oxidizing microorganisms. Although the use of arsenical products such as pesticides and herbicides has decreased significantly in the last few decades, their use for wood preservation is still common in developing countries and for municipal use (Matschullat 2000, 2011). The impact on the environment of the use of arsenical compounds, at least locally, will remain for many years.

These anthropogenic outputs can be observed on local- to continent-scale mapping of soil and sediment geochemistry (e.g., Reimann et al. 2009). For example in the Kola region of Russia, the dispersion of As from smelting operations generates localized hot spots of As in the vicinity of the Norilsk Smelter (Fig. 2). Elevated As is also seen in the gold-mining region near Kittilä, Finland, especially in the C-horizon soils (Fig. 2).

Perhaps the most commonly identified source and concern for As in the environment and its toxicity are the impacts related to mining (Craw and Bowell 2014, this volume). Under the extremely acidic conditions of acid mine drainage (AMD), high concentrations of arsenic (up to 850 mg L^{-1}) have been reported (Nordstrom and Alpers 1999). The highest values of

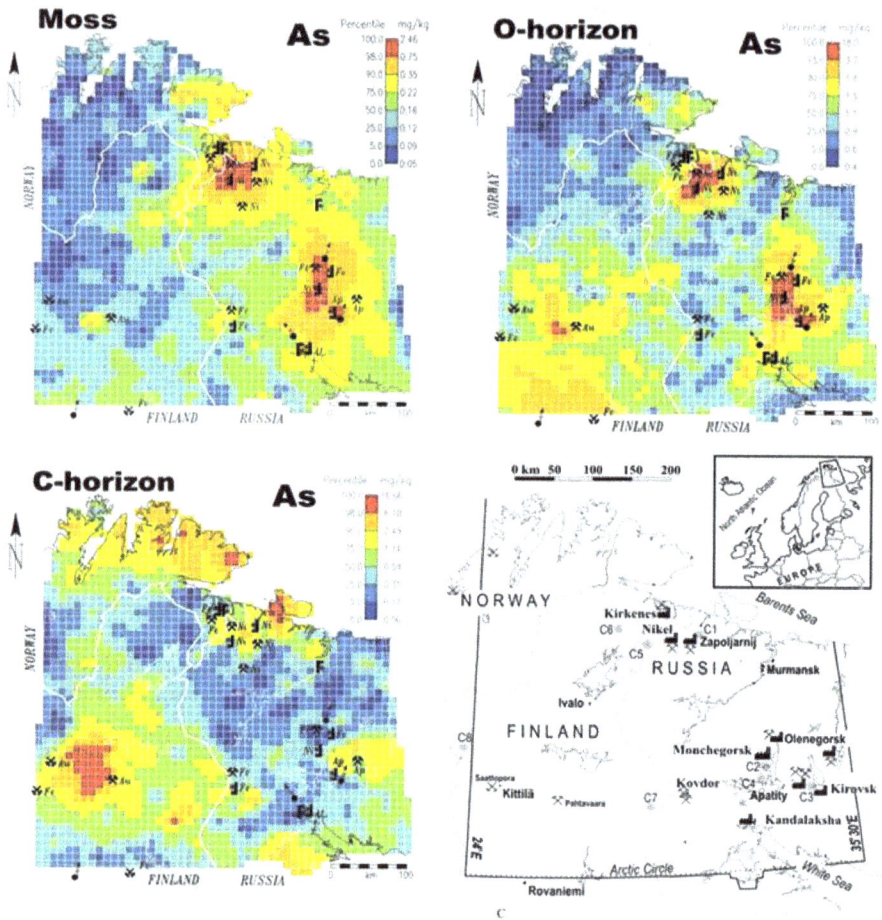

Figure 2. *(for color see Plate 2)* Regional distribution of As in moss and O- and C-horizon soil samples in the Kola Project area. [Reproduced with permission of Elsevier from Reimann et al. (2009) *Appl Geochem*, Vol. 24, Fig. 9, p. 1158.]

dissolved As reported in the literature (130,000 mg L^{-1}) were found at the Jáchymov mine in Czech Republic in highly acidic waters with a pH around 0 (Majzlan et al. 2014). Very-high As concentrations (4,000 mg L^{-1}) have also been measured in pH-neutral waters associated with arsenic trioxide produced from ore roasting (Jamieson 2014, this volume). Extensive study has been made of methods to mitigate the impacts of mining on the environment with respect to As pollution including management of future potential discharges (Riveros et al. 2001; Bowell and Craw 2014, this volume).

ARSENIC IN THE BIOSPHERE

Arsenic is also present in the biosphere and can be transferred through the food chain. Although the element is not an essential nutrient it can be taken up by pathways mimicking that of beneficial nutrients, for example arsenate via the phosphate transporters and arsenite via the aquaglyceroporin channels (Bhattacharjee et al. 2008; Zhao et al. 2010; Yang et al. 2012). The total arsenic content of terrestrial plants is estimated to be 1.8×10^5 tons, approximately four orders of magnitude less than that in soil (Matschullat 2011). This difference reflects generally limited As bioaccumulation because of the low bioavailability of As in soil. There are exceptions, however, such as ferns that can accumulate more than 1,000 mg kg^{-1} As and rice species that have been reported as being hyperaccumulators (Ma et al. 2001; Srivastava et al. 2006; Zhu et al. 2014).

In the tissue of living organisms, As occurs as As(III) or As(V) with As(III) predominant in reduced environments. Biological transformation can also lead to stabilization of As(III) in oxic environments as methylated As or arsenosugar compounds (Zhu et al. 2014).

In marine organisms, As is commonly present as arsenobetaine (AB) and arsenosugars (Francesconi and Edmonds 1996). Arsenic (III) is predominant in reduced environments, although it can occur in oxic environments as a result of biological transformation and redox disequilibrium (Zhu et al. 2014). In most living organisms, arsenite is predominant due to the prevailing low redox conditions. The activity of microbial methylation reactions is well understood. In the presence of microorganisms, methylation of the arsenic oxyanion may occur to form monomethylarsonic acid (MMAA), dimethylarsinic acid (DMAA), trimethylarsinic acid (TMAA), and dimethylarsine (DMA) (Zhu et al. 2014). The reduction of arsenate to arsenite may have been an important process in primeval biological cycles (van Lis et al. 2013; Zhu et al. 2014).

In general, the thermodynamically most stable aqueous As species over the general groundwater pH range of 4-8 is $H_2AsO_4^-$. Under reducing conditions, H_3AsO_3 will be the most stable aqueous arsenic species in the absence of complexing ions and methylating organisms. However, the rate of change in the oxidation state of As is not rapid unless microbially mediated, and microbial catalysis can change the distribution of redox species substantially from equilibrium (Zhu et al. 2014). Although some organisms can fully methylate As over a wide Eh-pH range, others are more specific in the As species with which they can react. These processes are also pH dependent and consequently pH variations affect the distribution of organic as well as inorganic As species (Amend et al. 2014, this volume).

BIOGEOCHEMICAL CYCLING OF ARSENIC

The majority of As in Earth's crust-ocean-atmosphere system is present in the lithosphere (Fig. 3). The size of the lithospheric pool is approximately five orders of magnitude larger than that in the ocean (Matschullat 2011). Weathering of rocks, geothermal and volcanic activities, mining, and smelting release As from the lithosphere to the terrestrial and oceanic environments (Fig. 3). The biosphere reservoir appears to be fairly well characterized (e.g., Ma et al. 2011; Zhu et al. 2014). However the detection of hyperaccumulating plant species and highly variable results still present challenges to providing accurate predictions of As budgets. The shortest-term reservoir, the atmosphere, presents more of an issue. Limited data exist for As in the atmosphere and even less for the anthroposphere, so related estimates of global flux are not very reliable. Due to the very low concentrations observed in atmospheric media compared with soils, the impact of airborne pollution may be difficult to detect except in extreme examples (e.g., Reimann et al. 2009; Jamieson 2014, this volume).

The concentrations of As in natural waters vary by more than four orders of magnitude depending on the source of As, the amount available and the local geochemical environment (Fig. 3). Under natural conditions, the greatest range and the highest concentrations of As are found in groundwaters as a result of the strong influence of water-rock interactions and the greater tendency in aquifers for the physical and geochemical conditions to be favorable for As mobilization and accumulation. The reported ranges are therefore extreme and unrepresentative of natural waters as a whole. Concentrations are commonly higher when riverine inputs are affected by industrial or mining effluent or by geothermal water. Unlike some other trace

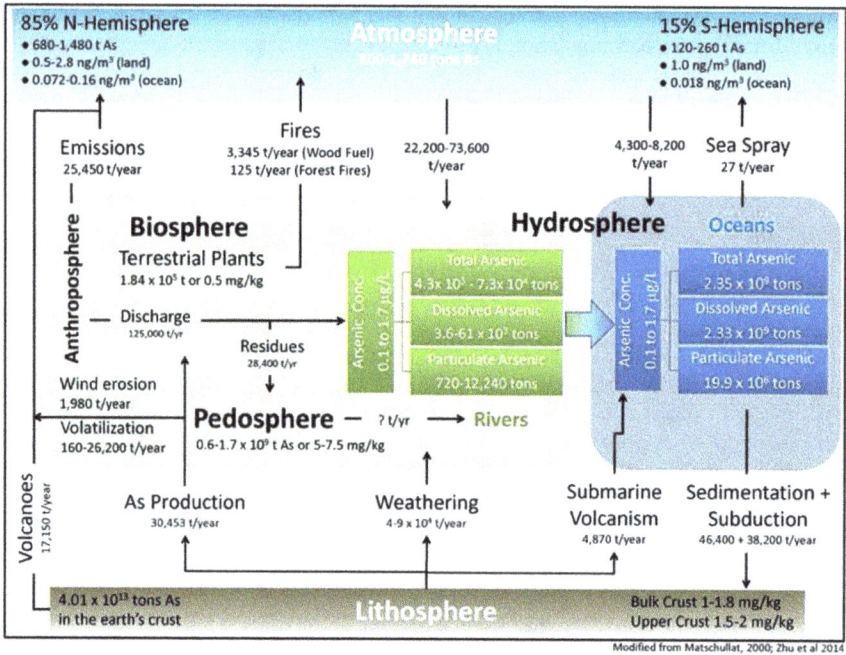

Figure 3. (*for color see Plate 3*) Global arsenic cycle. Redrawn and simplified from Matschullat (2000), Zhu et al (2014).

elements such as boron, saline intrusion of seawater into an aquifer is unlikely to lead to a significant increase of arsenic in the affected groundwater. In lake and river waters, As(V) is generally the predominant species (Pettine et al. 1992), though significant diel and seasonal variations in speciation as well as absolute concentration have been found (Gammons et al. 2007). Concentrations and relative proportions of As(V) and As(III) vary according to changes in input sources, redox conditions and biological activity. The presence of As(III) may be maintained in oxic waters by biological reduction of As(V), particularly during summer months (Lloyd and Oremland 2006; Santini and Ward 2012). Higher relative proportions of As(III) have been found in rivers close to inputs of As(III)-dominated industrial effluent and in waters with a component of geothermal water (Webster and Nordstrom 2003; Morin and Calas 2006).

Proportions of As(III) and As(V) are particularly variable in stratified lakes where redox gradients can be large and seasonally variable (e.g., Kuhn and Sigg 1993). As with estuarine waters, distinct changes in As speciation occur in lake profiles as a result of redox changes. For example, in the stratified, hypersaline and hyperalkaline Mono Lake (California, U.S.A.), there is a predominance of As(V) in the upper oxic layer and of As(III) in the lower reducing layer (Maeda 1994; Oremland et al. 2000). Rapid oxidation of As(III) occurs during the early stages of lake turnover as a result of microbial activity (Oremland et al. 2000). This event precedes Fe(II) oxidation although the speciation of As in lakes does not always follow that expected from thermodynamic considerations.

Welch et al. (1988) found that the Eh calculated from the As(V)-As(III) couple neither agreed with that from the Fe(II)-Fe(III) and other redox couples nor with the measured Eh. Therefore, the As redox couple is not reliable as a redox indicator except in a qualitative manner. Kempton et al. (1990) showed that only the Fe(II/III) and the $Fe(CN)_6^{3-}/Fe(CN)_6^{4-}$ redox couples respond quantitatively to a platinum electrode in a Nernstian equilibrium manner.

Therefore, redox species have to be measured; they cannot be calculated reliably from Eh measurements. Much of this redox disequilibrium has been attributed to the role of micro-organisms in the cycling of As (Zhu et al. 2014).

As a molecular analogue of phosphate, arsenate uses a phosphate transport system to enter cells. Once inside, it inhibits the phosphorylation of ADP and thereby the synthesis of ATP, leading to its toxic legacy. Arsenate can also substitute for phosphate in various biomolecules, thus disrupting key pathways, including glycolysis. Arsenite is even more toxic than arsenate and enters the cell much like glycerol molecules (Meng 2004). Arsenite binds with glutathione, a key enzyme in mammalian metabolism, inhibiting its function and it binds to thiolates in cysteine residues, disrupting the function of many proteins (Mukhopadhyay et al. 2002).

Arsenic levels in edible plants are generally low, even in crops grown on contaminated land with lower levels in Fe-rich and clay-rich soils and higher levels in plants grown on sandy or organic-rich soils such as aridisols, alluvium, or peat (Abrahams and Thornton 1987). The degree of uptake is variable from species to species. Unlike marine and freshwater organisms grown in contact with sediments, the As level in plants remains below that of the associated soils. In plants, roots show higher As levels than stems, leaves, or fruit and lower plants and grasses have a greater uptake than higher-order plants. For example in SW England, crops growing directly on As-contaminated spoil rarely exceed 1 mg kg^{-1} As while grass growing on identical substrate can have levels in excess of 2000 mg kg^{-1} As. In urban areas, similar grass species growing on soil with 20 mg kg^{-1} As were found to have a maximum of 3 mg kg^{-1} As dry weight (Abrahams and Thornton 1987).

Arsenic uptake by plants tends only to be significant in alkaline soils or where extremely high As levels (>10,000 mg kg^{-1} As) occur in the substrate (Table 4). In the Bau area of Sar-awak, Malaysia, high As levels in the soil (15-50,000 mg kg^{-1} As) coupled with an alkaline soil pH (7.5-8.4) has led to high As uptake in bamboo species (up to 4,500 mg kg^{-1} in roots and 2,650 mg kg^{-1} in leaves). In the Kutna Hora district, Czech Republic, a strong As uptake from mine-im-pacted soils and root vegetables has been demonstrated in laboratory tests (Száková et al. 2010). The highest As concentrations occur around areas where the soil is comprised of lime (used in the kilns) and mine spoil.

Arsenic uptake in plants can occur through aqueous transfer at the roots or from absorption of colloids or dissolved species through leaves. The cycling of As in the near-surface environment thus occurs not only through solution transfer but also through the decay of As-bearing vegetation and recycling of this material (Fig. 4).

Table 4. Biogeochemistry of arsenic in vegetation from mine sites.

Region	Range As (mg kg^{-1})	Refs.
Ashanti, Ghana	18 – 4,800	[1]
South West England	3 – 2,700	[2]
Kilimafeza, Tanzania	0.5 – 26.3	[3]
Flin Flon, Manitoba, Canada	400	[4]
Nova Scotia, Canada	36 – 738	[4] [5]

References: [1] Bowell (1991); [2] Abrahams and Thornton (1987); [3] Bowell et al. (1995); [4] Brooks et al. (1981); [5] Wong et al. (1999)

SUMMARY

Arsenic is common in the near-surface environment but concentrations in water, solids, and biota are highly variable. The distribution of As in the environment is dependent on source, mineralogy, speciation, biological interactions, and geochemical controls. This paper

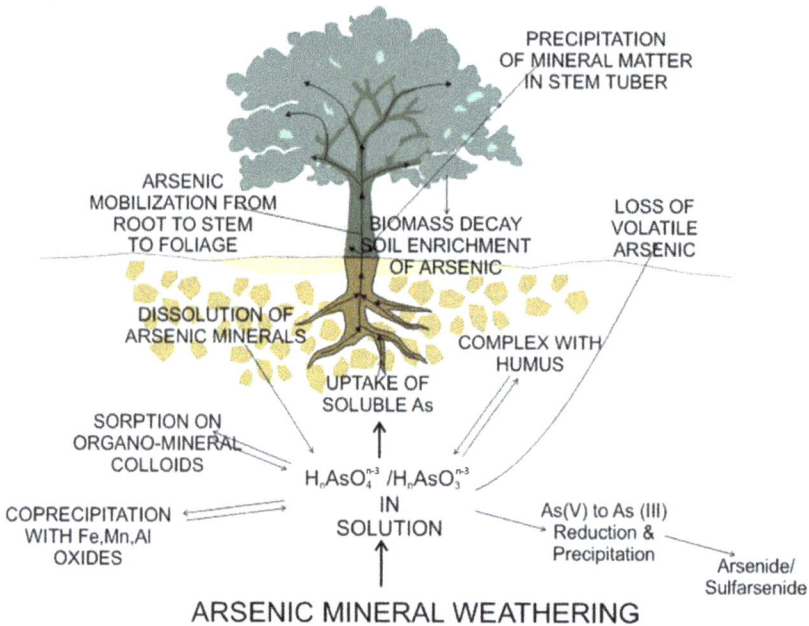

Figure 4. Conceptual model of arsenic cycling in the near-surface environment (after similar schemes by Stopinski 1976).

has attempted to characterize the distribution and cycling of As in the environment and to introduce the main biogeochemical controls on its speciation and mobilization.

Arsenic-rich environments are typically associated with chalcophile mineral deposits or geothermal activity. However, As-rich sediments and soils can also occur due to sedimentary and hydrogeological cycling of As-rich materials. The biological transformation and cycling of As can lead to oxidation or reduction of species that mobilize or attenuate As. Methylation and demethylation may also occur and this may promote transfer of As in the food chain, affecting ecological toxicity.

Predictive calculations of arsenic cycling are limited by the accuracy and precision of thermodynamic data for some minerals and aqueous species (Nordstrom et al. 2014, this volume). Data for global fluxes are limited for some natural processes, such as volcanism and soil degassing. Pedosphere and hydrosphere studies have tended to focus on areas of known "enriched" arsenic anomalies but regional mapping studies (such as Reimann et al. 2009) have greatly improved estimates for baseline or background estimates of As. The natural variation in As concentrations is large (2-3 orders of magnitude) and typically displays little variation in a vertical soil profile. Anthropogenic contamination is highly localized and provides notable hot spots.

In understanding the dispersion of As in the environment, geological materials provide a basic framework for characterizing As concentrations in ecosystems. Large variations can be observed on all spatial scales influenced by a variety of natural processes including non-geological influences such as climate and vegetation. These natural variations must be documented and the processes causing them understood before reliable estimates of the anthropogenic impact on the natural environment can be made.

ACKNOWLEDGMENTS

The authors wish to thank Stephen Day and Michael Parsons for reviews that greatly improved the manuscript.

REFERENCES

Abernathy CO, Calderon RL, Chappell WR (eds) (1997) Arsenic: Exposure and Health Effects II. Chapman & Hall, London, 429 p

Abrahams PW, Thornton I (1987) Distribution and extent of land contaminated by arsenic and associated metals in the mining regions of southwest England. Trans Inst Min Metall 96:B1-B8

Alpers CN, Hunerlach MP, Gallanthine SK, Taylor HE, Rye RO, Kester CL, Marvin-DiPasquale MP, Bowell RJ, Perkins WT, Humphreys RD (2002) Environmental legacy of the California gold rush: 2. Acid mine drainage from abandoned placer-gold mines. Geol Soc Am, Abstracts with Programs, Paper 92-1

Alpers CN, Myers PA, Millsap D, Regnier TB (2014) Arsenic associated with historical gold mining in the Sierra Nevada foothills: case study and field trip guide for Empire Mine State Historic Park, California. Rev Mineral Geochem 79:553-587

Amend JP, Saltikov C, Lu G-S, Hernandez J (2014) Microbial arsenic metabolism and reaction energetics. Rev Mineral Geochem 79:391-433

Anwar J (2000) Arsenic Poisoning in Bangladesh. Palash Media and Publisher, Dhaka, 336 p

Aphuja S (2008) Arsenic Contamination of Groundwater. John Wiley & Sons, Hoboken, New Jersey, 387 p

Bhattacharjee H, Mukhopadhyay R, Thiyagarajan S, Rosen BP (2008) Aquaglyceroporins: ancient channels for metalloids. J Biol 7:33-39

Bhattacharya P, Mukherjee AB, Bundschuh J, Zevenhoven R, Loeppert RH (eds) (2007) Arsenic in Soil and Groundwater Environment: Biogeochemical Interactions, Health Effects and Remediation. Trace Metals and Other Contaminants in the Environment, 9. Elsevier, New York, 653 p

Bianchelli T (ed) (2003) Arsenic Removal from Drinking Water. Nova Science Publishers, Hauppauge, NY, 150 p

Bowell RJ (1991) The Mobility of Gold in Tropical Rain Forest Soils. PhD thesis, Univ of Southampton, UK

Bowell RJ (1994) Arsenic sorption by iron oxyhydroxides and oxides. Appl Geochem 9:279-286

Bowell RJ (2014) Hydrogeochemistry of the Tsumeb deposit: implications for arsenate mineral stability. Rev Mineral Geochem 79:589-627

Bowell RJ, Craw D (2014) The management of arsenic in the mining industry. Rev Mineral Geochem 79:507-532

Bowell RJ, Warren A, Minjera HA, Kimaro N (1995) Environmental impact of former gold mining on the Orangi river, Serengeti N.P., Tanzania. Biogeochem 28:131-160.

Bowell RJ, Rees SB, Barnes A, Prestia A, Warrender R, Dey BM (2013) Geochemical assessment of arsenic toxicity in mine sites along the proposed Mineral Tamway Project, Camborne, Cornwall. Geochem Explor Environ Anal 13:145-158

Boyle RW, Jonasson IR (1973) The geochemistry of As and its use as an indicator element in geochemical prospecting. J Geochem Explor 2:251-296

Brooks RR, Holzbecher J, Ryan DE (1981) Horsetails (*Equisetum*) as indirect indicators of gold mineralization, J Geochem Explor 16:21-26

Bundschuh J, Bhattacharya P, Chandrasekharam D (eds) (2005) Natural Arsenic in Groundwater: Occurrence, Remediation and Management. A.A. Balkema, Leiden, The Netherlands, 339 p

Bundschuh J, Armienta MA, Birkle P, Bhattacharya P, Matschullat J, Mukherjee AB (eds) (2009) Natural Arsenic in Groundwaters of Latin America. CRC Press, Boca Raton, Florida, 742 p

Caley ER, Richards JFC (1956) Theophrastus, On Stones. Introduction, Greek Text, English Translation and Commentary. Columbus, Ohio: Ohio State University, 238 p

Campbell KM, Nordstrom DK (2014) Arsenic speciation and sorption in natural environments. Rev Mineral Geochem 79:185-216

Chakraborti D, Rahman MM, Das B, Murrill M, Dey S, Mukherjee SC, Dhar RK, Biswas BK, Chowdhury UK, Roy S, Sorif S, Selim M, Rahman M, Quamruzzaman Q (2010) Status of groundwater arsenic contamination in Bangladesh: A 14-year study report. Water Res 44:5789-5802

Chappell WR, Abernathy CO, Cothern CR (eds) (1994) Arsenic: Exposure and Health Effects I. Science and Technology Letters, Norwood, UK, 321 p

Chappell WR, Abernathy CO, Calderon RL (eds) (1999) Arsenic Exposure and Health Effects III. Elsevier, New York, 416 p

Chappell WR, Abernathy CO, Calderon RL (eds) (2001) Arsenic Exposure and Health Effects IV. Elsevier Science, Oxford, UK, 492 p

Chappell WR, Abernathy CO, Calderon RL, Thomas DJ (eds) (2003) Arsenic Exposure and Health Effects V. Elsevier, New York, 533 p

Chappells H, Campbell N, Drageb J, Fernandezc CV, Parkera L, Dummer TJB (2014) Understanding the translation of scientific knowledge about arsenic risk exposure among private well water users in Nova Scotia. Sci Total Environ (in press), doi: 10.1016/j.scitotenv.2013.12.108

Chen CJ, Chiou HY (eds) (2011) Health Hazards of Environmental Arsenic Poisoning: From Epidemic to Pandemic. World Scientific, Singapore, 247 p

Clifford DA, Karori S, Ghurye G, Samanta G (2004) Field speciation method for arsenic inorganic species. AWWA Research Foundation, US Environmental Protection Agency, Association of California Water Agencies, Denver, CO

Cohen DR, Bowell RJ (2014) Exploration Geochemistry. *In*: Treatise on Geochemistry (2nd ed), Scott SD (ed) Elsevier, Oxford, 13(24):624-649

Craw D, Bowell RJ (2014) The characterization of arsenic in mine waste. Rev Mineral Geochem 79:473-506

Cullen WR (2008) Is Arsenic an Aphrodisiac? The Sociochemistry of an Element. The Royal Society, Cambridge, UK, 412 p

Cullen WR, Reimer KJ (1989) Arsenic speciation in the environment. Chem Rev 89:713-764

Drahota P, Filippi M (2009) Secondary arsenic minerals in the environment: A review. Environ Int 35:1243-1255

Drahota P, Pačes, T, Pertold Z, Mihaljevič M, Skřivan P (2006) Weathering and erosion fluxes of arsenic in watershed mass budgets. Sci Total Environ 372:306-316

Drahota P, Rohovec J, Filippi M, Mihaljevič M, Rychlovskŷ P, Červenŷ V, Pertold Z (2009) Mineralogical and geochemical controls of arsenic speciation and mobility under different redox conditions at the Mokrosko-West gold deposit, Czech Republic. Sci Total Environ 407:3372-3384

Drahota P, Nováková B, Matoušsek T, Mihaljevič M, Rohovec J (2013) Diel variation of arsenic, molybdenum and antimony in a stream draining a natural As geochemical anomaly. Appl Geochem 31:84-93

Ellis AJ, Mahon WAJ (1977) Chemistry and Geothermal Systems, Academic Press, NY

Finkelman RB (2004) Potential health impacts of burning coal beds and waste banks. Int J Coal Geol 59:19-24

Finkelman RB, Belkin HE, Zheng B (1999) Health impacts of domestic coal use in China. Proc Nat Acad Sci USA 96:3427-3431

Finkelman RB, Orem W, Castranova V, Tatu CA, Belkin HE, Zheng B, Lerch HE, Maharaj SV, Bates AL (2002) Health impacts of coal and coal use: possible solutions. J Coal Geol 50:425-44

Francesconi KA, Edmonds JS (1997) Arsenic and marine organisms. Adv Inorg Chem 44: 147–185

Francesconi KA, Kuehnelt D (2004) Determination of arsenic species: a critical review of methods and applications, 2000-2003. Analyst 129:373-395

Frankenberger WT Jr (ed) (2002) Environmental Chemistry of Arsenic. Marcel Dekker, New York, 391 p

Gammons CH, Grant TM, Nimick DA, Parker SR, DeGrandpre MD (2007) Diel changes in water chemistry in an arsenic-rich stream and treatment-pond system. Sci Tot Environ 384:433-451

Garelick H, Jones H (2009) Arsenic Pollution and Remediation: An International Perspective. Rev Environ Contam Toxicol 197. Springer, New York, 194 p

Hem JD (1985) Study and Interpretation of the Chemical Characteristics of Natural Water. US Geological Survey Water Supply Paper 2254, 263 p

Henke KR (2009) Arsenic: Environmental Chemistry, Health Threats and Waste Treatment. Wiley, Chichester, UK

Hu Z, Gao S (2008) Upper crustal abundances of trace elements: A revision and update. Chem Geol 253:205-221

IMA (2014) The new IMA list of minerals. A work in progress. Updated: March 2014. *http://pubsites.uws.edu. au/ima-cnmnc/IMA_Master_List_(2014-03).pdf* (accessed June 2014)

Jamieson HE (2014) The legacy of arsenic contamination from mining and processing refractory gold ore at Giant Mine, Yellowknife, Northwest Territories, Canada. Rev Mineral Geochem 79:533-551

Jean J-S, Bundschuh J, Bhattacharya P (eds) (2010) Arsenic in Geosphere and Human Diseases: Arsenic 2010: Proceedings of the Third International Congress on Arsenic in the Environment. CRC Press, Boca Raton, FL

Kempton JH, Lindberg RD, Runnels DD (1990) Numerical modeling of platinum Eh measurements by using heterogeneous electron-transfer kinetics. *In*: Chemical Modeling of Aqueous Systems, II. Melchior DC, Basset RL (eds) Am Chem Soc Symp Ser 419:339-349

Kuhn A, Sigg L (1993) Arsenic cycling in eutrophic Lake Greifen, Switzerland: Influence of seasonal redox processes. Limnol Oceanog 38:1052-1059

Le XC (2001) Speciation of Arsenic in Water and Biological Matrices. American Water Works Association (AWWA) and AWWA Research Foundation, Denver, 145 p

Le XC, Weinfeld M (2004) Cellular Responses to Arsenic: DNA Damage and Defense Mechanisms. AWWA Research Foundation, Denver, 147 p

Lewis EA, Hansen LD, Baca EJ, Temer DJ (1976) Effects of alkyl chain length on the thermodynamics of proton ionization from arsonic and arsinic acids. J Chem Soc Perkin II, 125-128

Lloyd JR, Oremland RS (2006) Microbial transformations of arsenic in the lake environment: From soda lakes to aquifers. Elements 2:85-90

Ma LQ, Komar KM, Tu C, Zhang W, Kennelley ED (2001) A fern that hyperaccumulates arsenic. Nature 409:579

Maeda S (1994) Biotransformation of arsenic in the freshwater environment. *In:* Arsenic in the Environment, Part I: Cycling and characterization. Nriagu JD (ed) John Wiley & Sons, New York, p 155-188

Majzlan J, Plášil J, Škoda R, Gescher J, Kögler F, Rusznyak A, Küsel K, Neu TR, Mangold S, Rothe J (2014) Arsenical acid mine drainage with colossal arsenic concentration: mineralogy, geochemistry, microbiology. Environ Sci Technol (*submitted*)

Matschullat J (2000) Arsenic in the geosphere – a review. Sci Total Environ 249:297-312

Matschullat J (2011) The global arsenic cycle revisited. *In:* Arsenic: Natural and Anthropogenic. (Arsenic in the Environment, Vol 4). Deschamps E, Matschullat J (eds) CRC Press, Balkema, p 43-26

McGuigen CF, Hamula CIA, Huang S, Gabos S, Le XC (2010) A review on arsenic concentrations in Canadian drinking water. Environ Rev 18:291-307

Meharg AA (2005) Venomous Earth: How arsenic caused the world's worst mass poisoning. Macmillan, New York, 192 p

Meliker JR (2007) Lifetime Exposure to Arsenic in Drinking Water. VDM Verlag, Saarbrucken, Germany, 310 p

Meng YL (2004) As(III) and Sb(III) uptake by G1pF and efflux by ArsB in Escherichia coli. J Biol Chem 18:18334-18341

MFG (2009) Data transmittal and evaluation report for Historic Mine and Mill Sites, 2008 Work Plan, Empire Mine State Historic Park, October 2009. MFG, Inc, Fort Collins, Colo. Prepared for Newmont USA, Ltd, *http://www.envirostor.dtsc.ca.gov/public/profile_report.asp?global_id=29100003* (accessed July 2014)

Morin G, Calas G (2006) Arsenic in soils, mine tailings and former industrial sites. Elements 2:97-102

Mukhopadhyay R, Rosen BP, Phung LT, Silver S (2002) Microbial arsenic: from geocycles to genes and enzymes. FEMS Microbiol Rev 26:311-325

Murphy T, Guo J (2003) Aquatic Arsenic Toxicity and Treatment. Leiden Backhuys Publishers, Kerkwerve, The Netherlands, 165 p

Naidu R, Smith E, Owens G, Bhattacharya P, Nadebaum P (eds) (2006) Managing Arsenic in the Environment: From Soil to Human Health. CSIRO Publishing, Collingwood, Australia

NRC (1977) Arsenic: Medical and Biologic Effects of Environmental Pollutants. National Research Council. National Academy Press, Washington D.C., 332 p

NRC (1999) Arsenic in Drinking Water. National Research Council. National Academy Press, Washington, D.C. 310 p

NRC (2001) Arsenic in Drinking Water: 2001 Update. National Research Council. National Academy Press, Washington, D.C. 225 p

Ng JC, Noller BN, Naidu R, Bundschuh J, Bhattacharya P (eds) (2012) Understanding the Geological and Medical Interface of Arsenic: As 2012, Proceedings of the 4th International Congress on Arsenic in the Environment, 22-27 July 2012, Cairns, Australia. CRC Press, Boca Raton, Florida, 616 p

Nordic Ministers Council (1999) Arsenic in Impregnated Wood: A Compilation of Data from Nordic Reviews. Nordisk Råd, 94 p

Nordstrom DK (2000) An overview of arsenic mass-poisoning in Bangladesh and West Bengal, India. *In:* Minor Elements 2000 – Processing and environmental aspects of As, Sb, Se, Te, and Bi. Young C (ed) Proceedings, Society for Mining, Metallurgy and Exploration Meeting, Salt Lake City, 21-30

Nordstrom DK, Alpers CN (1999) Negative pH, efflorescent mineralogy, and consequences for environmental restoration at the Iron Mountain Superfund site, California. Proc Natl Acad Sci USA 96:3455-3462

Nordstrom DK, Archer D (2003) Arsenic Thermodynamic data and environmental geochemistry. *In:* Arsenic in Ground Water: Geochemistry and Occurrence. Welch AH, Stollenwerk KG (eds) Kluwer Academic Publishers, Boston 1-27

Nordstrom DK, Majzlan J, Königsberger E (2014) Thermodynamic properties for arsenic minerals and aqueous species. Rev Mineral Geochem 79:217-255

Nriagu JO (ed) (1994a) Arsenic in the Environment. Part I: Cycling and Characterization. Wiley-Interscience, New York, 430 p

Nriagu JO (ed) (1994b) Arsenic in the Environment. Part II: Human Health and Ecosystem Effects. Wiley-Interscience, New York, 293 p

Neumann T, Scholz F, Kramar U, Ostermaier M, Rausch N, Berner Z (2013) Arsenic in framboidal pyrite from recent sediments of a shallow water lagoon of the Baltic Sea. Sedimentology, 60: 1389-1404

O'Day PA, Vlassopoulos D, Meng X, Benning LG (eds) (2005) Advances in arsenic research: Integration of experimental and observational studies and implications for mitigation. Am Chem Soc Symp Ser 915. Am Chem Soc, Washington, D.C., 433 p

Oremland RS, Dowdle PR, Hoeft S, Sharp JO, Schaefer JK, Miller LG, Switzer BJ, Smith RL, Bloom NS, Wallschlaeger D (2000) Bacterial dissimilatory reduction of arsenate and sulfate in meromictic Mono Lake, California. Geochim Cosmochim Acta 64:3073-3084

Parker JN, Parker PM (2004) Arsenic Poisoning: A 3-in-1 Medical Reference. ICON Group International, San Diego, CA, 72 p

Pettine M, Camusso M, Martinotti W (1992) Dissolved and particulate transport of arsenic and chromium in the Po River, Italy. Sci Tot Environ 119:253-280

Ravenscroft P, Brammer H, Richards K (2009) Arsenic Pollution. Wiley-Blackwell, Chichester, UK, 588 p

Reich M, Kesler S, Utsunomiya S, Palenik S, Chryssoulis SL (2005) Solubility of gold in arsenian pyrite. Geochim Cosmochim Acta 69:2781–2796

Reimann C, Matschullat J, Birke M, Salminen R (2009) Arsenic distribution in the environment: The effects of scale. Appl Geochem 24:1147-1167

Riveros PA, Dutrizac JE, Spencer P (2001) Arsenic disposal practices in the metallurgical industry. Can Metallurg Quart 40:395-420

Roberts FI (1982) Trace element chemistry of pyrite: A useful guide to the occurrence of sulfide base metal mineralization. J Geochem Explor 17:49-62

Samanta G, Clifford DA (2006) Preservation of Arsenic Species. AWWA Research Foundation, Denver, CO, 120 p

Santini JM, Ward SA (eds) (2012) The metabolism of arsenite. *In:* Arsenic in the Environment 5. CRC Press, Boca Raton, Florida, 189 p

Savage KS, Bird DK, Ashley RP (2000) Legacy of the California Gold Rush: Environmental geochemistry of arsenic in the southern Mother Lode Gold District. Int Geol Rev 42:385-415

Simon G, Huang H, Penner-Hahn JE, Kesler, SE, Kao IS (2013) Oxidation state of gold and arsenic in gold-bearing arsenian pyrite. Am Mineral 84: 1071–1079

Smedley PL, Kinniburgh DG (2002) A review of the source, behaviour and distribution of arsenic in natural waters. Appl Geochem 17:517-568

Sorlini S, Collivignarellli C (2011) Arsenic in Water from Human Consumption. Lambert Academic Publishing, 144 p

Srivastava M, Ma LQ, Santos JAG (2006) Three new arsenic hyperaccumulating ferns. Sci Total Environ 364:24-31

Stopinski O (1976) Arsenic. U.S. Environmental Protection Agency. Environmental Health Effects Research Series. EPA-600/1-76-036 North Carolina. 488 p

Száková J, Havlík J, Valterová B, Thlustoš P, Goessler W (2010) The contents of risk elements, arsenic speciation, and possible interactions of elements and betalains in beetroot (*Beta vulgaris*, L.) growing in contaminated soil. Central Eur J Biol 5:692-701

Theodore T, Kotlyar BB, Singer DA, Berger VI, Abbott EW, Foster AL (2003) Applied geochemistry, geology and mineralogy of the northernmost Carlin trend, Nevada. Econ Geol 98:287-316

Thompson TB, Teal L, Meeuwig RO (eds) (2002) Gold deposits of the Carlin trend. Nevada Bureau of Mines and Geology, Bulletin 111, 204 p

Tourtelot HA (1964) Minor-element composition and organic carbon content of marine and nonmarine shales of Late Cretaceous age in the western interior of the United States. Geochim Cosmochim Acta 28:1579-1604

van Lis R, Nitschke W, Duval S, Schoepp-Cothenet B (2013) Arsenics as bioenergetic substrates. Biochim Biophys Acta Bioenerg 1827:176-188

Webster JG, Nordstrom DK (2003) Geothermal Arsenic. *In:* Arsenic in Ground Water: Geochemistry and Occurrence. Welch AH, Stollenwerk KG (eds) Kluwer Academic Publishers, Boston, p 101-126

Welch AH, Lico MS, Hughes JL (1988) Arsenic in ground water of the western United States. Ground Water 26:333-347

Welch AH, Stollenwerk KG (eds) (2003) Arsenic in Ground Water: Geochemistry and Occurrence. Kluwer Academic Publishers, Boston, 475 p

Wharton JC (2010) The Arsenic Century: How Victorian Britain was poisoned at home, work and play. Oxford University Press, Oxford, UK, 412 p

Wilkin RT, Barnes HL (1997) Formation processes of framboidal pyrite. Geochim Cosmochim Acta 61:323-339

Wong HKT, Gauthier A, Nriagu JO (1999) Dispersion and toxicity of metals from abandoned gold mine tailings at Goldenville, Nova Scotia, Canada. Sci Total Environ 228:35-47

Yang HC, Fu HL, Lin YF, Rosen BP (2012) Pathways of arsenic uptake and efflux. *In:* Metal Transporters. Lutsenko S, Argüello JM (eds) Academic Press, Inc.; Current Topics in Membranes 69:325-358

Zhao FJ, McGrath SP, Meharg AA (2010) Arsenic as a food chain contaminant: mechanisms of plant uptake and metabolism and mitigation strategies. Ann Rev Plant Biol 61:535-59

Zhu YG, Yoshinaga M, Zhao FJ, Rosen BP (2014) Earth abides arsenic biotransformations. Ann Rev Earth Planet Sci 42:443-467

Reviews in Mineralogy & Geochemistry
Vol. 79 pp. 17-184, 2014
Copyright © Mineralogical Society of America

Parageneses and Crystal Chemistry
of Arsenic Minerals

Juraj Majzlan

Institute of Geosciences
Friedrich-Schiller-Universität
Jena, Germany

Juraj.Majzlan@uni-jena.de

Petr Drahota

Institute of Geochemistry, Mineralogy and Mineral Resources
Charles University
Prague, Czech Republic

petr.drahota@natur.cuni.cz

Michal Filippi

Institute of Geology
Academy of Sciences of the Czech Republic
Prague, Czech Republic

filippi@gli.cas.cz

INTRODUCTION

The labyrinthine world of arsenic minerals has piqued the curiosity of many researchers in mineralogy, geochemistry, chemistry, and environmental sciences. Arsenic was known to the ancient civilizations; there are written Greek, Roman, and Chinese reports about minerals and substances of this element (Emsley 2001). The discovery of elemental arsenic is attributed to Albertus Magnus (1193-1280) who prepared it by reduction of As_2O_3. The common public association of arsenic and poison is the heritage of a long history of eliminating unwanted and unloved ones with compounds of this element. Mary Ann Cotton (1832-1873) was charged with murder of her mother, three husbands, a lover, eight of her own children, and seven stepchildren, all of them with an arsenic-based de-worming compound (Emsley 2005). Kořínek (1675) gave a vivid and frightening account on how a natural ferric sulfo-arsenate (bukovskýite) was used to poison the German armies of Albrecht Habsburg who invaded Bohemia in 1304. An arsenic derivative called lewisite (2-chlorovinyl-dichloroarsine) was used in the World War I (Emsley 2001). On the other hand, brightly-colored arsenic compounds were used in all imaginable products well into the 20th century. Arsenic whetted the appetite of many children as green arsenical chemicals were used as cake decorations and coatings of sugar sweets (Emsley 2005). The death of Napoleon Bonaparte has been regarded for a long time as a consequence of ingested or inhaled arsenical compouds (e.g., Aldersey-Williams 2011), however there are alternative interpretations (Lugli et al. 2011). Accidental mass arsenic poisoning occurred in Manchester in 1900 when many men drank beer contaminated with arsenic. The arsenic was tracked back to pyrite which was used to produce sulfuric acid which was employed in the manufacture of glucose for this batch of beer (Emsley 2005). Despite its toxicity, arsenic finds a few uses in alloys with lead in ammunition, in special types of glasses, in car batteries, and as a wood preservative. Arsenic is used as a dopant to silicon to optimize its properties as a semiconductor or as a major component of semiconducting phases such as GaAs, $Al_xGa_{1-x}As$, InAs, and others.

Arsenic is a metalloid, atomic number 33 of the group 15 in the periodic table. Its ground-state electronic configuration is $[Ar]3d^{10}4s^24p^3$ and it has a moderate Pauling electronegativity of 2.2. It can exist in the $-III$, $-I$, 0, III, or V oxidation states. Native arsenic is a grey, brittle crystalline solid, with hardness of 3.5 on the Mohs scale, and sublimation temperature of 890 K under atmospheric pressure. Estimated average concentrations of As are 1.8 mg kg^{-1} (Mason and Moore 1982) and 5.7 mg kg^{-1} in the upper crust (Hu and Gao 2008). The anthropogenic fluxes of arsenic out of the crust of ~30 thousand metric tons per year rival the natural fluxes of arsenic release by volcanoes (17,000 t/y) and submarine volcanism (~5,000 t/y) (Matschullat 2000). The fluxes due to weathering were not estimated owing to lack of data. The weathering fluxes, as one could expect, should roughly balance the sedimentation (46,000 t/y) and subduction (38,000 t/y) fluxes into the crust (the numerical values reported by Matschullat 2000).

This chapter has a twofold purpose. First, typical assemblages of arsenic minerals are discussed for the settings in which these minerals are found. Second, the crystal chemistry of arsenic minerals is described. In both parts, we focus on the minerals in which arsenic is the central cation of oxyanions, that is, arsenates or arsenites. Sulfides and sulfosalts of arsenic are also described but not in much detail. The reason for this focus is that sulfides and sulfosalts of arsenic are generally the primary reactants that produce the environmental contamination, but not the secondary products which store and release As in the environment. In terms of crystal chemistry, sulfide minerals have been recently reviewed (Makovicky 2006) and we feel that we can add little to that review.

The chapter is concluded by an index of As minerals which should aid in locating a particular mineral in the sea of mineralogical, geochemical, and crystal chemical wisdom amassed by many scientists. A full list of arsenic minerals and their chemical formulae can be found in, i.e., Appendix 1 of this chapter. Likewise, Appendix 2 lists minerals mentioned in this chapter whose nominal composition does not include As but frequently contain traces to minor amounts of As. Hence, chemical formulae are usually not given in the text, striving to avoid unnecessary repetition at the expense of having to refer the reader to the appendices.

PARAGENESES OF MINERALS OF ARSENIC

More than 560 minerals contain As as an essential structural constituent (IMA 2014). Arsenic minerals display a remarkable structural and chemical diversity. Approximately 58% of all known arsenic minerals are arsenates, approximately 24% are sulfides and sulfosalts, nearly 8% are oxides and arsenites, and the rest are arsenides, the native element, and metal alloys (Strunz and Nickel 2001). This chemical diversity, especially at the Earth's surface, results from the different chemical conditions under which arsenic minerals are formed. Arsenic minerals are therefore good indicators of geochemical environments, which are closely related to geochemical element cycles.

Arsenic does not readily substitute into the structures of the major rock-forming minerals, including silicates and carbonates. The concentration of arsenic in the rock is rather controlled by the presence of other minerals, such as sulfides, oxides, or phosphates, which may include arsenic as a trace to major component in their structure (Smedley and Kinniburgh 2002; Garelick et al. 2008). The oxidation and dissolution of these arsenic minerals delivers arsenic to hydrothermal and meteoric geochemical fluids. Under mildly reducing conditions, arsenic can migrate as arsenous acid (H_3AsO_3) or thioarsenite complexes many kilometers from its source in altered rocks or buried sediments, until changes in physico-chemical properties or solution chemistry lead to precipitation of arsenic sulfides and other classes of arsenic minerals (Anthony et al. 2000; Goldhaber et al. 2001; Strunz and Nickel 2001; Smedley and Kinniburgh 2002; Cleverley et al. 2003; Webster and Nordstrom 2003; Fendorf et al. 2010). Where conditions are sufficiently oxidizing to stabilize pentavalent arsenic and its complexes in the solution, the anion and the complexes may be transported in the aqueous phase or precipitate a variety of arsenate minerals (Anthony et al. 2000; Strunz and Nickel 2001; Drahota and Filippi 2009). In addition to natural sources, human activities have greatly contributed to local arsenic contamination, e.g., through activities associated with mining, ore processing, waste disposal, coal burning, indiscriminate use of certain pesticides and herbicides, as well as the manufacture of As-containing chemicals (Azcue and Nriagu 1994; Garelick et al. 2008; Reimann et al. 2009). The difficulty in providing a robust assessment of arsenic mobility at the Earth's surface comes from the complex interplay among the various forms of As and geochemical fluids, as well as the inseparable role of biotic processes in surface natural systems.

Precipitation of arsenic minerals in the crust or near the Earth's surface can occur in a wide variety of environments, resulting in an impressive diversity of arsenic minerals. These may represent (i) magmatic and metamorphic systems, often with distinct traces of hydrothermal action, (ii) hydrothermal systems, (iii) oxidation zones of ore deposits and mineralized rocks, (iv) hot springs and hot gaseous sublimates, (v) coal basins, (vi) soils and fluvial systems, (vii) mine wastes and tailings, and (viii) former industrial sites. Understanding the conditions of arsenic mineral formation in these environments is an important part of understanding the geochemical behavior of arsenic.

The interest in the structures and parageneses of arsenate and arsenite minerals has arisen historically and continues to be driven by their roles as alteration products of arsenides, sulfarsenides, and arsenic sulfides under oxidizing conditions (Morin and Calas 2006; Lattanzi et al. 2008; Drahota and Filippi 2009; Walker et al. 2009). Many arsenate and arsenite compounds have also been important substances in agriculture and chemical industry (Azcue and Nriagu 1994; Folkes et al. 2001; Hingston et al. 2001; Embrick et al. 2005) and may exert control over groundwater concentration of arsenic in contaminated soils and sediments (Johnston and Singer 2007; Fitzmaurice et al. 2009; Burnol and Charlet 2010). The following sections review variations in the origin and occurrence of the more common arsenic oxides, arsenites, and arsenates in many different environments. We focus mainly on arsenic minerals where arsenic is the central cation in oxyanions and largely disregard arsenides and arsenic sulfides which have been thoroughly reviewed by many authors (Schaufelberger 1994; Nordstrom and Archer 2003; Makovicky 2006; O'Day 2006; Vaughan 2006 and authors therein). We deliberately avoid the discussion of the oxidation zones of ore deposits in this chapter; this topic is handled by other chapters in this volume (e.g., Craw and Bowell 2014, this volume).

Definitions

Paragenesis and assemblage refer to the regular occurrence of minerals that are related to common physico-chemical formation conditions and implies that equilibrium existed among those phases at the time of formation. The term association is used for a usually wider group of minerals that occur in one place but did not necessarily originate from a common process.

For the natural settings, we conventionally attribute the primary arsenic minerals to those that have not been altered chemically or structurally since their deposition and crystallization from primary fluids; either melts, supercritical fluids, liquids, or gases. In general, these primary fluids originate during magma emplacement and crystallization, magmatic outgassing, prograde metamorphism, and hydrothermal processes. Secondary minerals result from the *in situ* decomposition of the primary minerals or from the precipitation from an aqueous solution which carries the decomposition products of the primary minerals. The aqueous solutions which transport the solutes are commonly typified by their low temperatures (< 50 °C) although fluids with higher temperatures (50-200 °C, commonly at water-saturation pressure), for example in epithermal system, may also alter primary minerals. These definitions apply to natural assemblages and there is usually no problem in assigning the minerals to the primary or secondary group.

The identification of phases from the field settings as primary and secondary may encounter difficulties once man-made materials and wastes are considered. In a strict sense, many of these phases are not minerals because a mineral is "a naturally occurring solid with a highly ordered atomic arrangement and a definite (but not fixed) chemical composition, usually formed by inorganic processes" (Klein 2002). They fulfill almost all conditions for being classified as a mineral but they are not "naturally occurring," in a sense that their formation would not take place without a prior human intervention (e.g., by mining). Yet, these phases are found at many polluted sites and broadly referred to as minerals in the scientific literature. We include such phases in our descriptions of mineral assemblages because they may provide important clues to mineralogical and geochemical cycling of arsenic under natural conditions.

Another complication arises when the terms primary and secondary are applied to man-made materials and waste and their decomposition products. We too use this broad description here, being fully aware that it may be confusing. However, from the literature available, it is hardly ever possible to discern if an arsenate, described from a polluted site, was produced naturally as a secondary mineral, during processing of the ores, or by weathering of the waste. The terms tertiary or quaternary may be applied in the latter cases but we do not feel that they lift the confusion. The man-made materials and wastes are then considered to be the primary source of arsenic and other elements. For example, gossans developed on massive-sulfide deposits may be mined for gold and contain abundant goethite, scorodite, or beudantite. These three minerals, if stored in a tailings pond, would then be considered primary in a technological sense although they are clearly secondary in the mineralogical sense (Jambor and Blowes 1998; Jambor 2003). Similarly, the metallurgical wastes (slags, roasting products, and flue dust) or remnants of arsenic-bearing chemicals are primary in the technological sense.

Where it is possible to do so, mineralogical terminology is used to distinguish primary and secondary phases for the sake the clarity. Where not otherwise possible, the reader is warned that the definition of primary and secondary phases deviates from the usual mineralogical conventions.

Magmatic-hydrothermal arsenic minerals

Arsenates are rare as magmatic minerals. The apparent lack of arsenates as minerals of magmatic rocks is most likely owing to both the tendency of the magmatic fluids to be reducing, at least with respect to arsenic and the fractionation of arsenic into late magmatic, aqueous fluids (e.g., Pokrovski et al. 2002; Mustard et al. 2006). Another reason for the scarcity of magmatic arsenates may be simply the relatively low abundance of arsenic in the crust relative to other tetrahedrally coordinated cations, such as Si^{4+} or Al^{3+}.

In the following text, arsenates directly related to magmatic rocks and their origin are described. As outlined below, arsenates appear in the late-magmatic or high-temperature hydrothermal stages of the evolution of the magmatic systems. In all the cases discussed here, the link between magmatic rocks, the fluids expelled from them, and the arsenates is clear and unambiguous. Such systems are usually reduced and if arsenic is present, it will be precipitated as arsenopyrite or rarer sulfides of arsenic and other cations.

Arsenates of rare-earth elements have been reported from several late- and postmagmatic systems, such as Permian rhyolite bodies from Slovakia (Ondrejka et al. 2007) or Variscan leucogranites from Germany (Förster et al. 2011). In the Slovak rhyolites, primary monazite-(Ce) [(Ce,La,Nd,Th)PO_4] and xenotime-(Y) (YPO_4) were transformed into secondary monazite-(Ce)

– gasparite-(Ce) and xenotime-(Y) – chernovite-(Y) solid solutions. Miscibility over a substantial portion of the solid solution between the phosphate and arsenate end-members was observed (Fig. 1). The minerals associate with Fe-Ti oxides and zircon (Fig. 2). In addition to the As-P substitution, As^{5+} is also replaced by Nb^{5+}, Si^{4+}, or S^{6+}. Beside REE and Y, these minerals may contain substantial amounts of Th. The granites from Zinnwald, Germany (Förster et al. 2011) contain accessory arsenoflorencite, chernovite, solid solution between hydrated xenotime and chernovite, and As-containing silicates thorite, coffinite, and zircon.

Although these minerals are not strictly magmatic, that is, they did not crystallize from the silicate-rich melt, they originated in processes which directly followed the crystallization from melt. Such processes that commonly alter magmatic rocks and related deposits (e.g., greisen deposits) are considered to be closely related to the magmatic activity itself. The arsenates are then formed by alteration of the primary magmatic phosphates, such as monazite and xenotime. Other minerals, for example thorite, may also be enriched in As during these processes. The common As-bearing mineral in these rocks is arsenopyrite, with As in the reduced state (as As^- in the covalently bonded AsS groups). The presence of arsenates, with the (SO_4) group partially substituting for the (AsO_4) group, suggests that the postmagmatic fluids in these rare cases were oxidizing and arsenopyrite was therefore unstable (Förster et al. 2011). An alternative explanation was proposed by Breiter et al. (2009). These authors observed that the initial stages of post-magmatic alteration of the granites caused the precipitation of arsenopyrite and pyrite. Later oxidizing fluids dissolved these minerals, oxidized arsenic, and altered the magmatic accessory phosphates and silicates.

Pentavalent arsenic (As^{5+}) can be stored in silicates with a charge-balanced berlinite-like substitution $As^{5+} + Al^{3+} = 2Si^{4+}$. The extent of the substitution can be large when As^{5+} is available in large amounts. Some natural specimens of filatovite (Filatov et al. 2004) reach the end-member composition $(Na,K)Al_2Si(P,As)O_8$ (Fig. 3). Kotelnikov et al. (2011) succeeded in synthesizing the pure Al-As end-member as well as the intermediate Si-Al-As feldspars. Thus, it is likely that a

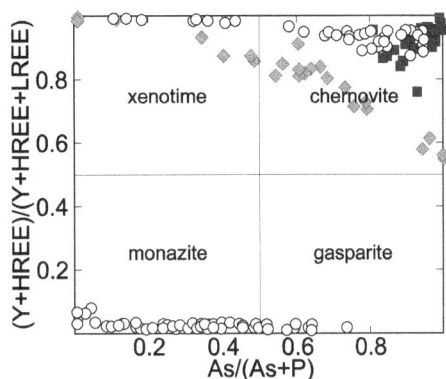

Figure 1. Compositional ranges of natural REE-Y phosphates and arsenates from magmatic systems. Squares: unaltered chernovite, diamonds: hydrated chernovite-xenotime; triangles and circles: post-magmatic monazite-gasparite and xenotime-chernovite solid solutions, respectively. Redrawn from Ondrejka et al. (2007) and Förster et al. (2011).

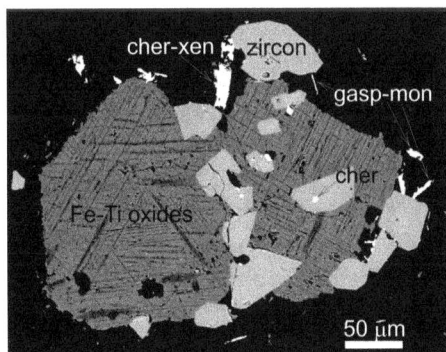

Figure 2. Back-scattered electron (BSE) image of Fe-Ti oxides (magnetite, ilmenite, rutile), zircon, and magmatic REE-Y phosphates-arsenates. Abbreviations: cher: chernovite-Y; cher-xen: chernovite-Y – xenotime-Y solid solution; gasp-mon: gasparite-Ce – monazite-Ce solid solution. This assemblage was found in a rhyolite body in Tisovec-Rejkovo (Slovakia, see Ondrejka et al. 2007). Figure courtesy of M. Ondrejka.

significant portion of the trace amount of As^{5+} in silicate rocks is bound in feldspars. Alternatively, As^{5+} could be also stored in accessory iron oxides. We are not aware of any detailed study on arsenic speciation in normal (i.e., not As-enriched) rocks.

Accounts of arsenates (e.g., scorodite) in magmatic rocks are scattered in collector literature, including the internet. Little genetic information is usually provided, though sometimes clues about the secondary origin of the arsenates are provided. The primary mineral is usually arsenopyrite, and less commonly is löllingite. There are reports, however, of primary pegmatitic arsenates, such as yanomamite and In-bearing scorodite from Mangabeira, Brazil (Moura et al. 2007). Botelho et al. (1994) proposed that these arsenates did replace early arsenopyrite but the replacement was coeval with the main cassiterite (SnO_2) deposition event rather than the usual supergene weathering.

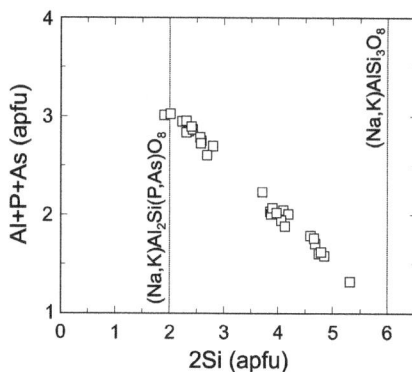

Figure 3. Extent of the $Al^{3+} + (P,As)^{5+} = 2Si^{4+}$ substitution in natural filatovite specimens from Tobalchik volcano, Russia. Redrawn from Kotelnikov et al. (2011).

Metamorphic-hydrothermal arsenic minerals

A number of localities with metamorphic-hydrothermal As-rich or As-enriched assemblages have been described. A strict division between metamorphic and hydrothermal origin is very difficult. The ore bodies described below are metamorphic in their origin but they were invariably affected by fluid flow, most likely commencing in large scale after the peak of the metamorphic event or events. Many minerals are therefore hydrothermal, although their existence is intimately related to the metamorphic events and processes.

The most striking feature of all these occurrences is the geochemical association of metamorphic As minerals with manganese. Given the number of such occurrences around the world and in various geotectonic settings, this union cannot be a coincidence. It must be clarified, however, that the metamorphic arsenates and arsenites commonly contain cations other than Mn. The association is thus foremost of geochemical nature; to explain this association, the authors usually invoke adsorption of As onto Mn or Fe oxides as the most likely enrichment mechanism.

The two prime examples of such metamorphosed Mn-rich ore bodies with As anomalies are Långban, Sweden and Franklin-Sterling Hill, New Jersey, USA. Both are briefly described below and the As minerals found there are listed in Tables 1 and 2. Unfortunately, the quest after rare minerals overshadowed systematic geological, geochemical, and metallogenetic work at these deposits and a rigorous explanation of the origin of these deposits is still required.

The Franklin and Sterling Hill deposits are located 80 km NW from New York City. They are thought to represent strongly metamorphosed, originally Precambrian stratiform deposits of sedimentary origin (Frondel and Baum 1974). The regional sillimanite-grade metamorphism caused recrystallization and mineral transformation. Although Franklin and Sterling Hill were famous as Zn deposits, it is often overlooked that the ores were predominantly Fe-dominated (Dunn 1995). The ores are therefore described as Fe^{3+}-Mn-Zn-Si-O units, separated from silicate assemblages with abundant Ca and lesser Al, Mn, Mg, Fe^{3+}, K, and Na. Only about 25 minerals were assigned to the action of the metamorphic processes *sensu stricto*; the other

Table 1. Arsenic minerals described from Franklin and Sterling Hill deposits (after Dunn 1995). For the chemical composition of the minerals, see Appendix 1.

Arsenides, sulfarsenides, sulfides (incl. sulfosalts):
   arsenopyrite, baumhauerite, domeykite, gersdorffite, löllingite, nickeline, pararammelsbergite, rammelsbergite, realgar, safflorite, seligmannite, skutterudite, tennantite

Arsenites:
   magnussonite, synadelphite

Arsenate-arsenite-silicates:
   kraisslite, mcgovernite, nelenite, schallerite

Arsenates:
   adamite, adelite, akrochordite, allactite, annabergite, arseniosiderite, austinite, bariopharmacosiderite, brandtite, chlorophoenicite, clinoclase, conichalcite, duftite, erythrite, euchroite, eveite, flinkite, fluckite, guérinite, haidingerite, hedyphane, holdenite, jarosewichite, johnbaumite, köttigite, kolicite, legrandite, liroconite, magnesium-chlorophoenicite, manganberzeliite, manganohörnesite, metalodèvite, metazeunerite, mimetite, ogdensburgite, ojuelaite, parabrandtite, parasymplesite, pharmacolite, pharmacosiderite, picropharmacolite, retzian-(La), retzian-(Nd), sarkinite, scorodite, sterlinghillite, tilasite, turneaureite, uranospinite, villyaellenite, wallkilldellite, wendwilsonite, yukonite

---

ones, counting more than 200 species, postdate the metamorphism. According to Frondel and Baum (1974), the only primary arsenic minerals of the ores were löllingite, arsenopyrite, and the calcium arsenate svabite. Dunn (1995) proposed that löllingite may have been the source of As for many of the secondary As minerals. The major primary minerals of the ores were olivine solid solution (including tephroite, Mn_2SiO_4), franklinite-magnetite-jacobsite solid solution, calcite ($CaCO_3$), willemite (Zn_2SiO_4), zincite (ZnO), and a few others (see Frondel and Baum 1974, their Table 2). The surrounding calc-silicate rocks contain As^{3+}-bearing phyllosilicates (ferroschallerite) and the Ca-Pb arsenates hedyphane and svabite as the primary As carriers. The major phases of the calc-silicate rocks are andradite [$Ca_3Fe_2(SiO_4)_3$], feldspar, hendricksite [$K(Zn,Mn)_3(Si,Al)_4O_{10}(OH)_2$], and calcite.

In general, minerals of As are rare at Franklin and Sterling Hill. Almost all of the arsenites and arsenates are postmetamorphic, hydrothermal minerals, found in veins and fissures. Enigmatic is the relative paucity of Zn arsenates, even though the ores are rich in Zn. The dominant cation of many arsenite and arsenate minerals is Mn; many Mn arsenates are extremely rare and have been described from a few grains, never to be found again. Some of the Mg and Ca arsenates described from Franklin and Sterling Hill are post-mining, crystallizing at the dumps.

Table 2. Arsenic minerals described from Långban (after Holtstam and Langhof 1999). For the chemical composition of the minerals, see Appendix 1.

Sulfides and sulfarsenides:
   algodonite, cobaltite, domeykite, koutekite, löllingite, orpiment, rammelsbergite, realgar, tennantite

Arsenites:
   armangite, ecdemite, finnemanite, freedite, heliophyllite, magnussonite, manganarsite, paulmooreite, rouseite, stenhuggarite, synadelphite, trigonite

Arsenite-arsenates:
   hematolite

Arsenates:
   adelite, akrochordite, allactite, arseniopleite, arsenoclasite, bergslagite, berzeliite, brandtite, caryinite, eveite, flinkite, gabrielsonite, hedyphane, hörnesite, manganberzeliite, mimetite, parwelite, sahlinite, sarkinite, svabite, tilasite, turneaureite

The Långban deposit (Holtstam and Langhof 1999) is located 210 km WNW from Stockholm and is also hosted by Precambrian rocks, older than those in Franklin (1800 Ma for Långban, 1000 Ma for Franklin). Ores similar to those found in Långban occur in the wide region at Sjögruvan, Pajsberg-Harstigen, Jakobsberg, and Nordmark. The host rocks are also marbles but more dolomitic than in Franklin (Lundström 1999). The minerals present in Långban were divided into the primary minerals and their recrystallization products, ore and skarn minerals formed at the peak of metamorphism, cavity minerals, and fissure minerals (Magnusson 1930 *in* Sandström and Holtstam 1999). The principal primary ore minerals in Långban were hematite (Fe_2O_3), braunite ($Mn_7O_8SiO_4$), and hausmannite (Mn_3O_4). This assemblage and the spatial separation of Fe and Mn are in a stark contrast to the ores in Franklin where Fe and Mn are united in franklinite-magnetite-jacobsite solid sultions. The chief constituents of the skarns are pyroxenes, pyroxenoids, and manganoan varieties of olivine, garnets, phlogopite, and richterite. These skarns are a home of frequent As minerals berzeliite, hedyphane, svabite, and tilasite. The greatest mineralogical diversity, just as in Franklin, is found among the minerals of veins and fissures. While the skarn As minerals are usually anhydrous arsenates, the vein arsenates contain hydrogen and are represented by allactite, sarkinite, and many others. Reducing conditions are documented by minerals with Fe^{2+} or As^{3+}, such as gabrielsonite, magnussonite, or ecdemite. The formation conditions of these minerals are largely unknown; fluid inclusion studies are missing and the the available geothermobarometers are usually not applicable, owing to the unusual overall geochemistry of the rocks and mineral assemblages.

A number of small occurrences of Mn ores with As minerals is well documented from the Alps in Switzerland, Italy, and Austria, with a few examples described below. Some of them are lenses of only a few meters in size, therefore the term "ore deposit" certainly does not apply to them. They are always located in Mesozoic sediments of the Tethys ocean, are thought to represent syngenetic exhalative Fe-Mn accumulations, and were metamorphosed during the Alpine orogeny (Abrecht 1990; Brugger and Meisser 2006). Metamorphic grade varied from very low to eclogitic facies. The lenses of Fe-Mn rich rocks are embedded in Triassic or Jurassic platform carbonates or Jurassic deep-sea sediments. Upon metamorphism, they developed complex assemblages with minor phases whose principal cations are As, Sb, V, REE, and Be (Brugger and Gieré 1999). At the Pipjigletscher occurrence in the Swiss Alps, an unusually strongly reduced assemblage developed, with manganosite (MnO) and As^{3+}-As^{5+} minerals, e.g., turtmannite (Brugger et al. 2001). Other occurrences with As^{3+} minerals are known, e.g., asbecasite, cafarsite, fetiasite, or even As^{3+}-enriched titanite (Fig. 4, Krzemnicki and Gieré 1996). Otherwise, the minerals present are arsenates, e.g., manganlotharmeyerite (Brugger et al. 2002), fianelite (Brugger and Berlepsch 1996), or nabiasite (Brugger et al. 1999). Brugger and Meisser (2006) also offered a comparison of the Alpine Fe, Mn, and Fe-Mn ores in terms of their genesis and metamorphic grade (see their Table 7). In the Ködnitz valley in Austria, the metamorphosed manganese cherts contain a prograde Mn-silicate assemblage with schallerite as the As carrier, and a retrograde assemblage with sarkinite and tiragalloite (Abrecht 1990).

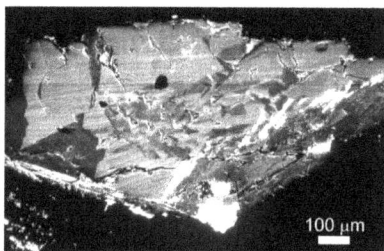

Figure 4. BSE image of As-REE-titanite with a complex intergrowth pattern (Krzemnicki 1997). The figure courtesy of M. Krzemnicki.

Brugger and Meisser (2006) argued that the chemical composition of the rocks (at least at the Pipji occurrence) reflects the pre-metamorphic state, where the syngenetic exhalative sediments adsorbed copious quantities of As and other elements from the hydrothermal fluids. Cabella et al. (1999) reported that the abundance of arsenates reduces sharply as a function of proximity of the Fe-Mn ores, essentially implying the same conclusion as Brugger and Meisser (2006). The same genetic conclusion was presented for Mn-rich metamorphic rocks (greenschist facies) and the Mn-As minerals in the Hoskins manganese mine, New South Wales, Australia (Ashley 1989). Metamorphosed Mn-rich rocks in the Iberian Massif, Spain, acquired their As from hydrothermal input (Jiménez-Millán and Velilla 1994) and the main As carrier is caryinite.

Young or modern analogues of such ores are known; oxidic ores of Mn and Fe precipitate in the active Matupi Harbor hydrothermal system (southwest Pacific) (Ferguson and Lambert 1972; Bonatti 1975). Not only do these ores contain Fe and Mn, but the two elements are efficiently separated along a redox gradient, accounting for widely variable Fe/Mn ratios in the ores. This variability of the Fe/Mn ratios is a typical feature of Mn ores with arsenate or arsenite minerals. High adsorption capacity of As was also described from such settings, at least for the Fe-rich materials (e.g., Rancourt et al. 2001). Young hydrothermal Mn deposits were drilled in the sediments near Galapagos (Lalou et al. 1983), with mineralogy dominated by todorokite $[(Na,Ca,K)_2(Mn^{4+},Mn^{3+})_6O_{12}{\cdot}nH_2O]$, nontronite $[Na_{0.3}Fe_2Si_3AlO_{10}(OH)_2{\cdot}4H_2O]$, and barite $(BaSO_4)$, and some As enrichment reported. Zantop (1981) described a significant As enrichment from Mn ores at the Jalisco deposit in Mexico, As being most strongly correlated with Ba and Mg. The ores were deposited from hydrothermal solutions in a small lacustrine basin during a quiescent period during the Tertiary with relatively little volcanic activity. Such ores, at least in terms of their chemical composition, may be analogues of the metamorphosed Mn ores in the occurrences described above.

On the other hand, Perseil and Latouche (1989) observed botryoidal microstructures in the ores at Falotta in the Swiss Alps and likened these structures to those observed in manganese nodules. These ores were metamorphosed only in the zeolite facies and the preservation of the primary sedimentary textures could be expected. Brugger and Meisser (2006), however, assigned the ores at Falotta also to the group of syngenetic-exhalative deposits. Reinecke et al. (1985) proposed that the Mn-rich metamorphic rocks (blueschist facies) from Andros, Greece were originally Mn-nodule-bearing, siliceous clayey sediments. The enrichment in a suite of elements, including As, was explained in this case by mobilization during diagenesis or early metamorphism from the sedimentary column. The metamorphic As carrier in these rocks is ardennite.

The locality Nežilovo (Macedonia) is located in Precambrian, polymetamorphosed rocks (Bermanec et al. 1993). A strong Ba anomaly was detected in quartz-cymrite $(BaAl_2Si_2(O,OH)_8{\cdot}H_2O)$ schists and dolomitic marbles; additionally, elevated contents of Zn, Pb, Cu, Mn, Ti, and As were measured. Tilasite was documented as the As-containing phase in these rocks (Bermanec 1994).

Haruna et al. (2002) described the Nagasawa deposit (Japan) situated in Mesozoic metacherts. The deposit was strongly overprinted by a contact metamorphic event associated with a granitic body. The authors argue that minor element characteristics are typical for ferromanganese nodules. The minor elements, among them also As, were scavenged by the nodules from seawater, pore waters in the sediments, or hydrothermal fluids, redistributed into chlorite and pyrite during diagenesis, and remobilized during the metamorphic event. The contact-metamorphic fluids, however, were reducing and deposited the minor elements in a form of native elements, sulfides, or tellurides.

In summary, two basic models are presented for the metamorphic deposits with As enrichment. One model is the exhalative origin of Fe-Mn ores, with enrichment of As and other elements from the hydrothermal fluids. The other model is the sedimentary deposition of Mn-rich sediments, mostly in the form of manganese nodules, and the enrichment of As from sea water, pore fluids during diagenesis, or hydrothermal input. Both models are feasible. The similarities of the rocks described here indicate that the different processes may converge to similar mineralogy under similar bulk geochemistry and *P-T* conditions.

Arsenic minerals in hot springs and fumarolic gases

Arsenic is a common companion of geothermal fluids, whether aqueous or gaseous. Webster and Nordstrom (2003) compiled numerous data sources and showed that the As content in aqueous geothermal fluids ranges from negligible to 50 mg kg^{-1} (see their Table 1). The solids deposited from these springs may contain prodigious amounts of As, up to almost 10% (McKenzie et al. 2001). Probably the most common crystalline precipitate is orpiment, in addition to amorphous As_2S_3, realgar, and other As sulfides. Under oxidizing conditions, Fe and As are found in their highest oxidation states and usually precipitate into X-ray amorphous hydrous ferric oxide with copious quantities of adsorbed arsenate. Crystalline arsenates scorodite, annabergite, pharmacosiderite, pharmacolite, and beudantite have been also described from hot springs (Onishi 1969 *in* Pfeifer et al. 2004) or paleosprings (Denning 1943). Chemistry and speciation of As in hot springs and volcanic emission gases have been also reviewed by Henke (2009). The arsenic mineralogy of the hot springs is controlled by their redox and pH state, ion concentrations, and water/precipitate ratio.

Volcanic gases around the world contain variable amounts of As with up to 30 μg g^{-1} (Signorelli 1997). An extensive analysis of several volcanic gaseous discharges around the world showed no correlation of As content with gas temperature, in disagreement with previous studies on individual volcanic systems (Signorelli 1997 and references therein). Even within a single volcanic system, the gaseous As concentration can vary considerably (Signorelli et al. 1998). Even though As concentrations in the gases have been measured several times, there are far fewer investigations on the mineralogy of the associated volcanic systems.

Stoiber and Rose (1969) reported high (up to 2% As) As concentrations in the fumarolic precipitates at the Santiaguito Volcano, Guatemala. High As content was documented from the hottest (~700 °C) to the coolest (< 250 °C) zone. Unfortunately, the As minerals were not identified but only assumed to be arsenates. The dominant minerals in the hottest zone were identified as hematite and thenardite and it is here suggested that arsenic could be stored in angelellite, a mineral macroscopically very similar to hematite. Previous observations at this volcano gave accounts of orpiment crystallizing together with native Se, Te, and S from the fumarolic gases.

There are sparse reports of other arsenates from fumaroles. Angelellite was described as crystalline and globular incrustations in andesite at Cerro Pululus (Argentina) (Ramdohr et al. 1959). There, it is believed to have been deposited from the vapor phase. Another mineral found in a paleo-fumarole is nickenichite, described from cavities in scoria at Nickenicher Sattel (Germany) together with cerussite ($PbCO_3$), chrysocolla [$(Cu,Al)_2H_2Si_2O_5(OH)_4 \cdot nH_2O$], duhamelite [$Cu_4Pb_2Bi(VO_4)_4(OH)_3 \cdot 8H_2O$], malachite [$Cu_2(CO_3)(OH)_2$], and vanadinite [$Pb_5(VO_4)_3Cl$] (Auernhammer et al. 1993).

Active fumaroles in the Russian Far East yielded a number of rare and new minerals, among them the arsenates alarsite, filatovite, bradaczekite, lammerite, coparsite, johillerite (Filatov et al. 1984, 2001; Semenova et al. 1994; Glavatskikh and Bykova 1998; Vergasova et al. 1999, 2004), all of them from the Tolbachik volcano (Kamchatka, Russia). Zelenski and Bortnikova (2005) collected discharging volcanic gases from the Mutnovsky volcano (Kamchatka, Russia) in silica tubes and allowed them to deposit the solid phases. They identified a series of phases as a function of the gas temperature. The As-containing phases included sulfosalts and sulfosalt-halides deposited from gas with temperature of 300-450 °C. At lower temperatures (< 300 °C), the deposits contained amorphous As_2S_3 and As-S-I compounds.

Arsenic minerals in coal

The average As content for bituminous coals and lignites is 9.0±0.8 and 7.4±1.4 mg kg^{-1}, respectively (Yudovich and Ketris 2005). Coals strongly enriched in As above the average level are widespread and include, for example, bituminous coals and lignites in Eastern Germany,

Czech Republic, and southeastern China (Bouška and Pešek 1999; Yudovich and Ketris 2005; Kang et al. 2011). In the Neogene coal seams of the Anlong county in Guizhou Province (SW China), arsenic has been found at levels as high as 35,000 mg kg^{-1} (Ding et al. 2001). Yudovich and Ketris (2005) distinguished four genetic types of As accumulation in coal: (i) epigenetic hydrothermal As enrichment (*Chinese type*), (ii) epigenetic As enrichment from ground waters draining As-bearing tufa host rocks (*Dakota type*); (iii) syngenetic As enrichment from As-bearing waters entering coal-forming peat bogs from sulfide deposit aureoles (*Bulgarian type*), and (iv) syngenetic As enrichment in coal-forming peat bogs from volcanogenic exhalations, brines, and ash (*Turkish type*).

Beside the important association of As with organic matter, especially in low-rank coals, Fe disulfides (pyrite and to a lesser extent marcasite) have been considered to be the main carriers of As in coal (e.g., Kolker and Finkelman 1998; Kolker et al. 2000; Ruppert et al. 2005; Yudovich and Ketris 2005; Liu et al. 2007; Rieder et al. 2007; Zielinski et al. 2007; Huggins et al. 2009; Kolker 2012). The concentration of As in Fe disulfides can significantly vary on the scale of individual grains (Fig. 5), with the highest concentrations in *Chinese type* coals. Maximum concentrations for As in the FeS$_2$ phases in coal are approximately 5 wt% (Kolker and Finkelman 1998; Ding et al. 2001; Ruppert et al. 2005). Huggins and Huffman (1996) and Rieder et al. (2007) indicated that As substitutes for S in the structure of pyrite. Blanchard et al. (2007) proposed that substitution for disulfide occurs as molecular ions, such as (As$_2$)$^{2-}$ or (AsS)$^{2-}$. Other minor sulfides may occur locally, such as arsenopyrite, realgar, orpiment, and enargite (Palmer et al. 1993; Huffman et al. 1994; Ding et al. 2001; Brownfield et al. 2005; Masalehdani et al. 2007). Beside the arsenic sulfides, native arsenic and (AsO$_4$) substitution for the (PO$_4$) group in the crandallite-group minerals (arsenate members of the alunite supergroup) was observed in the Franklin No. 12 coal bed, Washington, USA (Brownfield et al. 2005). Coals in the Anlong County, Gouizhou province, China, contain as much as 3.5 wt% of As and are mineralogically exceptional. In addition to the arsenic sulfides, these coals contain small grains and veinlets of As-bearing sulfates, clays, and phosphates (Ding et al. 2001).

Huffman et al. (1994), Huggins et al. (2002), and Kolker and Huggins (2007) demonstrated that arsenian pyrite in fresh coal, in which the As oxidation state is As(−I), undergoes spontaneous oxidation in contact with atmosphere, with conversion of the As originally present in pyrite or other primary minerals to the arsenate form. Arsenic appears to be retained preferentially in

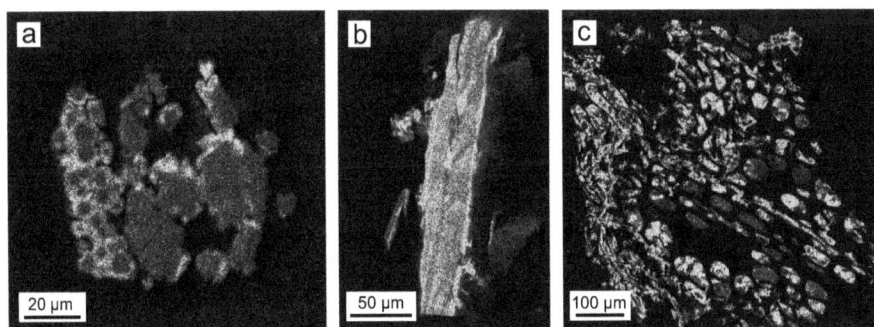

Figure 5. Grayscale wavelength-dispersive electron microprobe arsenic maps of Fe disulfides. (a) A map of pyrite from Black Warrior Basin, Alabama, USA, showing As-rich overgrowths on clustered framboids (?) (left-hand side of the image) and As-rich domains within subhedral/euhedral pyrite (right), with 0.09 to 0.54 wt% As (based on 4 analyses). (b) A map of cleat pyrite from Artema Mine, Donets Basin, Ukraine, showing variable As concentrations parallel to the cleat propagation, with 0.40 to 2.17 wt% As (based on 5 analyses). (c) A map of pyrite from Glubokaya Mine, Donets Basin, Ukraine, showing patchy As enrichment and deformation, with 0.13 to 2.23 wt% As (based on 8 analyses). All figures from Kolker (2012).

the newly formed jarosite (Kolker and Huggins 2007), in which arsenate substitutes for sulfate. These results are consistent with other mineralogical studies of natural wastes showing that As-bearing jarosite is an important constituent of mine wastes derived from arsenian pyrite (Foster et al. 1998; Savage et al. 2000; Asta et al. 2009).

Arsenic minerals as products of coal combustion

The exothermic oxidation of sulfides and organic matter in coal leads to a significant temperature increase and may trigger its spontaneous ignition. These processes act naturally in coal beds, underground workings, open pit faces, waste rock dumps, and slag heaps (Sinha and Singh 2007; Sokol and Volkova 2007). Burning of coal can result in the formation of secondary minerals which form from a liquid or gas phase within or along gas vents and ground fissures. Mechanisms of formation have been recently reviewed by Stracher (2007) and include isochemical processes (gas condensation to a liquid or solid phase, possible subsequent precipitation from the liquid) and more complex mass-transfer processes (e.g., reactions between gas and/or liquid phase and substrate). These processes are commonly considered to be analogous to the exhalation-condensation processes in which minerals form in solfataric or fumarolic environments (see section *Arsenic Minerals in Hot Springs and Fumarolic Gases*).

Several hundreds of different chemical compounds have been described from burning coal waste piles and other coal-burning settings. Focusing only on the secondary As phases, the number of the occurrences is significantly reduced to several localities where the coal is sufficiently rich in As (Table 3). As an exception to the general scarcity of secondary arsenic phases as products of coal burning, realgar and orpiment are relatively abundant and occur at numerous sites associated with coal combustion (e.g., Lapham et al. 1980; Witzke 1996; Sejkora et al. 1997; Žáček 1998; Masalehdani et al. 2007). The chemical composition of the simple As sulfides at the coal burning sites is shown in Figure 6. The richest assemblage of As minerals has been documented in the burning waste pile of the Kateřina coal mine in Radvanice, Czech Re-

Table 3. Arsenic phases described from burning coal waste sites.

Mineral or phase	Localization and reference
native arsenic	Kladno, Czech Republic (Žáček 1998); St. Etiene, France (Masalehdani et al. 2007)
realgar	eastern Pennsylvania, USA (Lapham et al. 1980); Radvanice, Czech Republic (Sejkora et al. 1997); Kladno, Czech Republic (Žáček 1998); St. Etiene, France (Masalehdani et al. 2007)
alacranite	Oelsnitz, Saxony, Germany (Witzke 1996); Radvanice, Czech Republic (Sejkora et al. 1997)
pararealgar	St. Etiene, France (Masalehdani et al. 2007)
orpiment	Burnside, Pennsylvania, USA (Lapham et al. 1980; Dunn et al. 1986a); Radvanice, Czech Republic (Dubanský et al. 1991; Žáček and Ondruš 1997); Oelsnitz, Saxony, Germany (Witzke 1996); St. Etiene, France (Masalehdani et al. 2007)
laphamite	Burnside, Pennsylvania, USA (Dunn et al. 1986a; Bindi et al. 2008)
As-S alloy	Radvanice, Czech Republic (Sejkora et al. 1997; Žáček and Ondruš 1997)
Ge-Se-As-Sb-S alloy	Radvanice, Czech Republic (Dubanský et al. 1991; Žáček and Ondruš 1997)
Sb-Bi-As-S alloy	St. Etiene, France (Masalehdani et al. 2007)
arsenolite	Burnside, Pennsylvania, USA (Dunn et al. 1986a); Radvanice, Czech Republic (Žáček and Ondruš 1997); St. Etiene, France (Masalehdani et al. 2007)

public (Dubanský et al. 1991; Sejkora et al. 1997; Žáček and Ondruš 1997). The site was remediated and the fires extinguished in 2005 (Klink and Martinec 2013). In total, six secondary arsenic phases have been recognized at a depth of 10-30 cm below the surface, including four arsenic minerals (alacranite, arsenolite, orpiment, and realgar) and two As-bearing alloys (amorphous As-S alloy and Ge-Se-As-Sb-S alloy) (Fig. 7). The formation of these phases has been attributed to a variety of processes; deposition process resulted in the formation of euhedral realgar crystals, whereas different rates of gas cooling (condensation) were attributed to the formation of alacranite or As-S alloy (Sejkora et al. 1997).

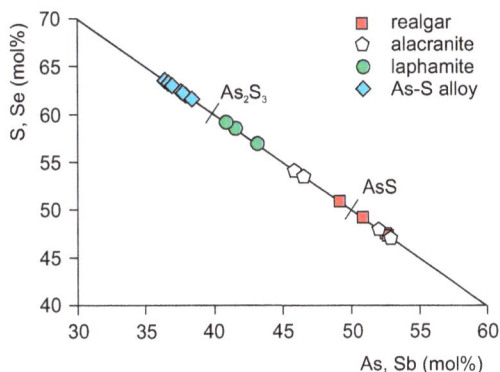

Figure 6. Compositional range of arsenic-bearing phases from the As-S system in coal burning sites. Compositional data from Dunn et al. (1986a), Sejkora (1997), and Bindi et al. (2008).

The observed relationships between various phases provided information about the sequence of formation of the arsenic compounds. For example, partial melting and transformation of alacranite into amorphous As-S alloy can occur in response to local temperature increase (Sejkora et al. 1997). Another unique coal-fire-gas arsenic mineral assemblage has been described by Lapham et al. (1980) from several burning waste piles in eastern Pennsylvania. They identified a similar arsenic phase assemblage as at Radvanice (realgar, orpiment, and arsenolite). Later, an

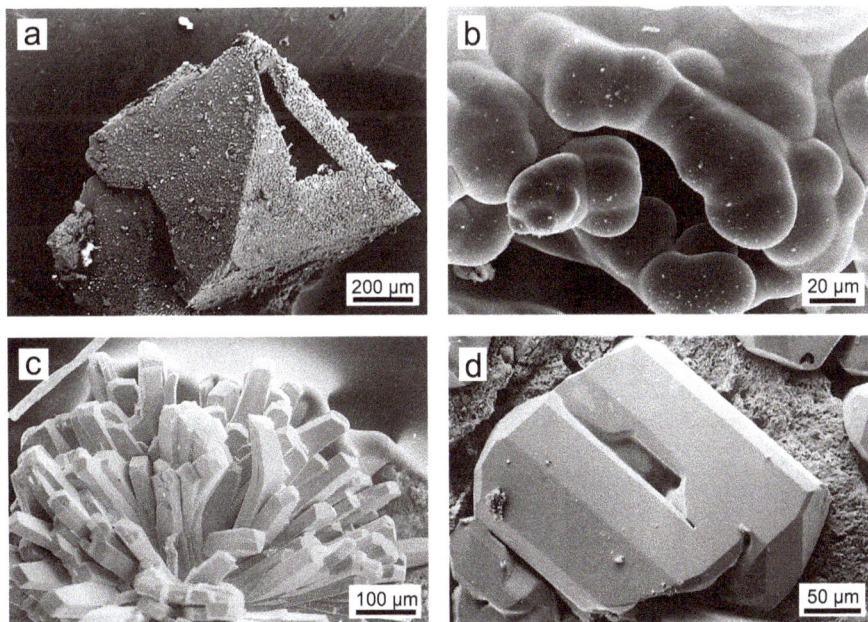

Figure 7. Secondary electron (SE) images of arsenolite (a), As-S alloy (b), orpiment (c), and realgar (d) from burned waste pile of the Kateřina coal mine in Radvanice, Czech Republic. All figures courtesy of J. Sejkora.

investigation of the burning anthracite dumps at Burnside (eastern Pennsylvania) has resulted in the description of new mineral laphamite (Dunn et al. 1986a). Condensation from the gas phase was also the main mechanism for the formation of realgar, pararealgar, orpiment, arsenolite, native arsenic, and a Sb-Bi-As-S compound of uncertain nature in a burning culm bank in eastern central France (Masalehdani et al. 2007). These phases were associated with salammoniac (NH_4Cl), mascagnite [$(NH_4)_2SO_4$], native sulfur, and a fluorine-sulfate phase around the rims of the gas vents and inside the fissures at a depth of 0-40 cm below surface. Precipitation temperatures from the cooling gas were estimated at 94.5-572.7 °C.

During coal combustion at power plants, most As in coal volatilizes and therefore escapes to the gas and aerosol phase, with only a minor portion remaining in bottom ash. After coming in contact with the ambient air, volatile As can condense to form arsenic trioxides (arsenolite, claudetite) or orpiment and realgar. However, most of the escaping As is captured by fly ash. Speciation of As in the fly ash suggests significant combustion-induced transformation from the reducing As forms in the coal (e.g., as those present in the FeS_2 minerals) to the dominant As(V) forms in the fly ash (Huffman et al. 1994; Goodarzi and Huggins 2005; Shah et al. 2007; Luo et al. 2011; Bolanz et al. 2012). The partitioning of As between the vapor and solid phase is determined by the interaction of As vapor with fly ash compounds in the post combustion zone of the power station. Illite [$(K,H_3O)Al_2(Si,Al)_4O_{10}(OH)_2$] or calcite, if present in the fly ash, may interact with volatile As_2O_3 generated in the combustion zone and form stable refractory arsenates such as potassium arsenate, calcium arsenate (Eqn. 1), and calcium pyroarsenate (Eqn. 2) (Huffman et al. 1994; Goodarzi and Huggins 2005; Yudovich and Ketric 2005; Zielinski et al. 2007; Luo et al. 2011).

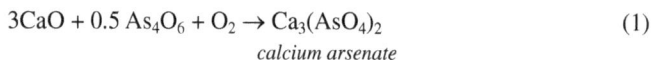

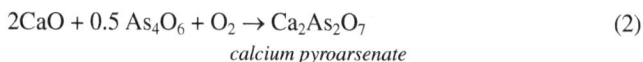

$$3CaO + 0.5\ As_4O_6 + O_2 \rightarrow Ca_3(AsO_4)_2 \qquad (1)$$
$$\textit{calcium arsenate}$$

$$2CaO + 0.5\ As_4O_6 + O_2 \rightarrow Ca_2As_2O_7 \qquad (2)$$
$$\textit{calcium pyroarsenate}$$

Arsenic can also enter crystalline phases and glasses of the fly ash (Huffman et al. 1994; Goodarzi and Huggins 2005; Bolanz et al. 2012). Several studies on fly ash produced in the gasification processes have shown that the mode of occurrence of As in such materials is quite different from those seen in the conventional coal combustion fly ashes. The proportion of As in the gas phase could be controlled by the levels of Ni in the feed fuel, available to form nickeline in this gasification fly ash. A limited number of studies revealed that As occurred in the gasification fly ash mainly as Ni and Fe sulfarsenides and Ni arsenates (Font et al. 2005, 2010).

Arsenic minerals in soil and fluvial systems

Soil is the unconsolidated mineral or organic material on the immediate surface of the terrestrial continents that lies at the interface between two important reservoirs in the global biogeochemical cycle of As, the lithosphere and the atmosphere. Furthermore, the soil interacts with the main transporting medium and another important natural reservoir, the hydrosphere (Matschullat 2000). Soil acts as a natural biogeochemical barrier that may scavenge As from percolating water and thus strongly accumulate the element. Accordingly, soils contain highly variable As concentrations. The global average concentration of As in uncontaminated soils is 5 to 10 mg kg^{-1}, with variations of more than an order of magnitude, depending on the kind of soil considered (Smith et al. 1998; Mandal and Suzuki 2002; Smedley and Kinniburgh 2002; Reimann et al. 2009). The arsenic concentration in the contaminated soils can reach up to several tens of thousands of mg kg^{-1} of As (Smith et al. 1998 and references therein; Garelick et al. 2008 and references therein; Bowell et al. 2014, this volume).

Under the usual oxidizing conditions in most soils, As is sorbed to or co-precipitated with iron, aluminum, and manganese oxides and hydroxides, clay minerals, and organic matter (Sadiq

1997; Stollenwerk 2003; Wang and Mulligan 2006a). It may also be directly substituted into the crystal structures of the minerals (Savage et al. 2005). The precipitation of secondary arsenic minerals may occur if supersaturation with respect to these phases is reached. This condition is usually achieved when (i) the amount of As exceeds the capacity of the available surface ligand-bonding sites, (ii) the concentration of dissolved arsenate or arsenite and metal cations exceeds the solubility product of a secondary arsenic mineral and (iii) mineral precipitation rates are fast compared with rate of groundwater flow (O'Day et al. 2004). The supersaturation of the entire volume of soil solution with respect to secondary arsenic minerals is very rare, even in the highly contaminated environments, because the sorption capacity of most soil types is usually very high and controls the dissolved As concentration at low levels far from saturation (Bowell 1994; De Brouwere et al. 2004; Yang et al. 2005; Wang and Mulligan 2006a; Drahota et al. 2012). Despite the undersaturation in the total soil solution, the formation of secondary arsenic minerals may occur in the vicinity of mineral surfaces owing to local supersaturation in the immediate vicinity of a surface with a dense population of adsorbed As. Recently, experiments with goethite, Zn, and As(V) have shown that functional groups on goethite can act as nucleation sites for adamite and koritnigite (Gräfe et al. 2004; Gräfe and Sparks 2005). Similar behavior has been observed with Cu and As(V) on kaolinite [$Al_2Si_2O_5(OH)_4$], jarosite, and goethite promoting the precipitation of euchroite- and clinoclase-like phases, respectively (Gräfe et al. 2008a). Tournassat et al. (2002) observed the formation of a krautite-like phase at the birnessite surface following the oxidation of arsenite by birnessite [$(Na,Ca,K)_x(Mn^{4+},Mn^{3+})_2O_4 \cdot nH_2O$]. However, more concrete work is required to explain how precipitates form from seemingly undersaturated solution conditions and whether they also occur on other mineral surfaces. Supersaturation and precipitation of secondary arsenic minerals in soil and fluvial environments has been most commonly found on the surfaces of the primary arsenic sulfides or arsenides at the bottom of the soil profiles (when derived from bedrock) or in the top horizons (when derived from superficial deposition). As a result, arsenic precipitates are often seen as reaction rims on arsenic sulfides and arsenides rather than isolated particles. After the depletion of reactants in the As-rich micro-environments, e.g., due to complete oxidation of arsenic sulfide or exhaustion of the superficial As supply accompanied by continuous pedogenesis, the transient secondary arsenic minerals usually become unstable and dissolve or transform to more stable compounds (e.g., transformation of ferric and lead arsenates or arsenolite to As-bearing hydrous ferric oxides) (Morin et al. 2002, Cancès et al. 2005, 2008; Arai et al. 2006; Drahota et al. 2009; Haffert and Craw 2009). These facts point to a conclusion that the initial secondary arsenic minerals could be unstable under the prevailing long-term conditions. Alternatively, their persistence in soils is related to the combination of the slow kinetics of their dissolution and persistently supersaturated micro-environments that enable the precipitation of the initial secondary arsenic minerals.

Although the secondary arsenic minerals in soil are relatively rare with respect to other As-rich environments such as mine wastes and mine tailings (see section *Arsenic Minerals in Mine Wastes*), the main factor of the data paucity for soil environments is the use of routine determinative mineralogical and geochemical techniques that are usually not able to identify scarce arsenic minerals in bulk soil samples. The use of these techniques means that As speciation cannot be determined due to relatively low concentrations and the wide variety of chemical forms in a given soil, including many surface complexes. In favorable cases, the studies describing and even quantifying As speciation in such heterogeneous media as soils have employed some type of preconcentration of arsenic minerals-rich fractions and/or utilization of more sophisticated microanalysis and/or spectroscopic techniques on these fractions or bulk samples (Filippi et al. 2004; Cancès et al. 2005; Frau et al. 2005; Arai et al. 2006; Gräfe et al. 2008b; Gómez-Parrales et al. 2011). However, *in situ* determination of the arsenic mineralogy in soils still remains a largely unaddressed challenge.

Natural geochemical anomalies from parental rock weathering. Arsenic is sometimes concentrated in soils as a result of natural weathering of As-mineralized parental bedrocks.

Such occurrences are examples of geogenic, regional and long-term environmental contamination. The authigenic formation of arsenic minerals usually commences at the primary arsenic mineral-water interface in the slightly altered parental rock or in the saprolite and soil. The primary arsenic minerals are most commonly arsenides, sulfarsenides, and arsenic sulfides, and the extent of their oxidation is controlled primarily by the access of descending oxygenated groundwater, which could be enhanced by suitable geologic characteristics, higher topographic relief, and relatively dry climate (where the water table tends to be relatively deep). The environmentally important physical parameters that maintain the supersaturation with respect to the secondary arsenic minerals in the slightly altered rock and deep soil horizons are those that control groundwater flow into these environments. The ability of a rock to transmit groundwater is termed its hydraulic conductivity and is a function of its porosity and permeability. Published hydraulic conductivity values for given rock types can vary over many orders of magnitude (Plumlee 1999) but some generalizations can be made for different rock types. Igneous and metamorphic rocks tend to have low porosity and permeability and groundwater flow occurs mainly along fractures and joints within the rocks; they can therefore easily transmit oxygenated groundwater over great distances along a few major flow pathways. In the unfractured parts of mineralized rock, however, water flow would be very slow and access of oxygen hindered. Slow supply of oxygen and water over geologic time may be suitable for secondary arsenic mineral growth. On the other hand, clastic sedimentary rocks such as conglomerates and sandstone have a primary porosity and permeability that is a function of the size and interconnection of the pore spaces between their grains. The greatest porosities and permeabilities occur in rocks such as gravels that have the coarsest grain size, greatest sorting and lowest interstitial cement content, so they can transmit groundwater to and from primary arsenic minerals in great volume and over great distances, thus avoiding supersaturation.

Chemical composition of authigenic As-bearing minerals is primarily related to the prevailing chemical system in the saprolite and the soil (Drahota and Filippi 2009). The parent material for saprolite and soils is generally dominated (in mass) by eight major elements (O, Si, Al, Fe, Ca, Mg, Na, K). Oxygen is an essential constituent of all oxidized arsenic compounds; among the major cations in the common chemical systems in saprolite and soil, Fe(III) and Ca are the most frequent major structural constituents of secondary arsenic minerals, depending on the bulk chemistry of parental rocks. In addition, elevated concentration of As in the parental mineralized rocks are usually accompanied by a suite of other elements, such as Ag, Co, Cu, Ni, Pb, Zn (Cox and Singer 1986; Plumlee 1999). These transition metal ions (Ag, Co, Cu, Ni, Zn) and large divalent cations, such as Pb, readily bond to the arsenate tetrahedra to form a variety of secondary arsenate minerals (Anthony et al. 2000). Consequently, diverse secondary arsenic mineral associations have been found in the saprolite and soil overlying mineralized geochemical anomalies or particular mineral deposits (Table 4).

The first example includes an association of secondary arsenic minerals in the chemical system K-Ba±Ca-Fe(III)-As(V)-H$_2$O. The major components of the system correspond to those in weathering solutions of alumosilicate rocks under oxidizing conditions and temperate climate with the addition of As (Brantley 2003). Field examples include oxidation zones of sulfide-poor deposits whose ores are dominated by primary As minerals and which are commonly developed in metamorphic terrains. Secondary arsenic minerals in this chemical system developed in the saprolite and soil overlying the granodiorite or metamorphic rocks with the low-sulfide quartz veins at the Mokrsko gold district, central Czech Republic (Filippi et al. 2004, 2007; Drahota et al. 2009) and Saint-Yrieix-la-Perche gold district, SW France (Bossy et al. 2010) or with Li, Be, Nb, Ta, Sn and W mineralization at the Echassières geochemical anomaly, France (Morin et al. 2002). All these sites are characterized by temperate climate (average temperatures 8-12 °C, rainfall 550-1300 mm) that lead to relatively slow rates of chemical weathering (White and Blum 1995). Arsenic is primarily derived from the natural alteration of arsenopyrite or löllingite

Table 4. Secondary arsenic mineral associations in the saprolite and soil related to mineralized geochemical anomalies or particular mineral deposits.

Chemical system	Secondary arsenic minerals	Mineralization/ primary arsenic minerals	Soil type; pH	Climatic regime (avg. rainfall in mm yr⁻¹)	Locality	Reference
K-Ba-Ca-Fe(III)-As(V)-H$_2$O	pharmacosiderite, arseniosiderite, HFO, goethite, hematite, scorodite	low-sulfide quartz veins in granodiorite/arseno-pyrite	vadose zone of saprolite and cambisol; pH 6.0-7.5	temperate (555)	Mokrsko, Psí hory gold district, central Czech Rep.	Filippi et al. (2004, 2007), Drahota et al. (2009)
Ba-Fe(III)-As(V)-H$_2$O	bariopharmacosiderite, ferrihydrite, goethite, hematite	low-sulfide quartz veins in granodiorite / arsenopyrite, arsenian pyrite	vadose zone of alocrisol and arenosol; pH 4.0-4.7	temperate (1300)	Saint-Yrieix-la-Perche gold district, SW France	Bossy et al. (2010)
K-Ba-Fe(III)-As(V)-H$_2$O	pharmacosiderite, bariopharmacosiderite, HFO	Li, Be, Nb, Ta, Sn and W mineralization in micaschist /arsenopyrite, löllingite	vadose zone of soil and saprolite	temperate	Echassières, Allier, France	Morin et al. (2002)
Fe(III)-As(V)-S(VI)- H$_2$O	Fe oxides, scorodite, bukovskýite, kaňkite, arsenolite	massive sulfides in granodiorite/arsenopyrite	wet vadose zone of deep laterite saprolite	tropical (1250)	Ashanti, Ghana, western Africa	Bowell (1992, 1994)
Pb-Ba-Fe(III)-As(V)-H$_2$O	goethite, lepidocrocite, scorodite, angelellite (?), schultenite, dussertite	granodiorites with Cu-Pb-Sb and Sn-W/ arsenopyrite	vadose zone of leptosol, cambisol, and luvisol; pH 4.6-5.5		province of Córdoba "Valle de los Pedroches", Spain	Gómez-Parrales et al. (2011)
Pb-Fe(III)-As(V)-H$_2$O	goethite, hematite, beudantite-corkite	Iberian type of massive sulfides	vadose zone of HFO-rich saprolite "gossan"		Rio Tinto and Tharsis ore districts, Spain	Viñals et al. (1995), Capitán et al. (2003), Nieto et al. (2003)

in saprolite and four secondary arsenic minerals were described from these sites: pharmaco-siderite, bariopharmacosiderite, arseniosiderite, and scorodite. At all these localities, As also associates with ferrihydrite, goethite, and hematite (up to 15 wt% As associated with goethite). The mineralogical evolution of As-bearing minerals in the weathering profiles starts in the deep saprolite where primary arsenic minerals alter *in situ* to scorodite or poorly crystalline As-rich hydrous ferric oxides (HFO) (Carlson et al. 1992; Nesbitt et al. 1995; Cruz et al. 1997) (Eqns. 3 and 4, respectively).

$$\text{FeAsS} + 3\text{H}_2\text{O} + 3.5\text{O}_2 \rightarrow \text{FeAsO}_4 \cdot 2\text{H}_2\text{O} + \text{SO}_4^{2-} + 2\text{H}^+ \tag{3}$$
$$\quad\quad \textit{arsenopyrite} \quad\quad\quad\quad\quad\quad \textit{scorodite}$$

$$\text{FeAsS} + 3\text{H}_2\text{O} + 3.5\text{O}_2 \rightarrow \text{FeOOH} + \text{H}_2\text{AsO}_4^- + \text{SO}_4^{2-} + 3\text{H}^+ \tag{4}$$
$$\quad\quad \textit{arsenopyrite} \quad\quad\quad\quad\quad\quad \textit{HFO}$$

Minerals of the pharmacosiderite supergroup probably form upon aging, remobilization, and recrystallization of As-rich HFO under near-neutral conditions (Fig. 7), under which the HFO is less soluble than scorodite (Haffert et al. 2010) (Eqn. 5).

$$4\text{FeOOH} + 3\text{H}_2\text{AsO}_4^- + 0.5\text{Ba}^{2+} + \text{H}_2\text{O} + 2\text{H}^+ \rightleftharpoons \text{Ba}_{0.5}\text{Fe}_4(\text{OH})_4(\text{AsO}_4)_3 \cdot 5\text{H}_2\text{O} \tag{5}$$
$$\quad\quad \textit{HFO} \quad\quad\quad\quad\quad\quad\quad\quad\quad\quad\quad\quad\quad \textit{bariopharmacosiderite}$$

At all studied sites, quantitative mineralogical analyses have shown that the proportion of As hosted by the pharmacosiderite minerals decreases systematically from the saprolite to topsoil; in the upper soil horizons, the system evolves toward As-rich Fe oxides, probably via solid-state reaction (Morin et al. 2002; Drahota et al. 2009; Bossy et al. 2010). These observations imply that pharmacosiderite forms in the deep saprolite where constantly high dissolved As and Fe concentrations are maintained owing to the slow groundwater flow. The increasing Fe/As molar ratio and/or decreasing Fe and As concentrations during pedogenesis destabilizes pharmaco-siderite and drives its transformation into As-rich Fe oxides (reversed Eqn. 5).

Although the As-rich Fe oxides are the end-products in this chemical system at all localities, arseniosiderite has previously formed at the Mokrsko soil (Drahota et al. 2009). It is always younger than pharmacosiderite and scorodite which are evidently replaced by massive aggregates and veins of younger arseniosiderite (Fig. 8; Eqn. 6). Similar findings have been described by Paktunc et al. (2004) in the Ketza River mine tailings.

$$4\text{Ba}_{0.5}\text{Fe}_4(\text{AsO}_4)_3(\text{OH})_5 \cdot 5\text{H}_2\text{O} + 9\text{Ca}^{2+} \rightarrow$$
$$\quad \textit{bariopharmacosiderite} \quad\quad\quad 3\text{Ca}_3\text{Fe}_4(\text{OH})_6(\text{AsO}_4)_4 \cdot 3\text{H}_2\text{O} + 4\text{Fe}(\text{OH})_3^0 + 2\text{Ba}^{2+} + 14\text{H}^+$$
$$\quad\quad\quad\quad\quad\quad\quad\quad\quad\quad\quad \textit{arseniosiderite} \tag{6}$$

At the Mokrsko site, Ca in the weathering solution is derived from the dissolution of carbonate minerals during pedogenesis. Carbonates buffer the pH of the solutions at about 6-8 and thus the formation of arseniosiderite occurs at near-neutral to slightly alkaline condition. As pedogenesis proceeds, the ratio of arseniosiderite to pharmacosiderite in the soil profiles increases from the saprolite upward, however, arseniosiderite disappears in the topsoil horizons (Drahota et al. 2009). In the topsoils, arseniosiderite most likely transforms to As-bearing Fe oxides while buffering acidity (Eqn. 7).

$$\text{Ca}_3\text{Fe}_4(\text{OH})_6(\text{AsO}_4)_4 \cdot 3\text{H}_2\text{O} + 2\text{H}^+ \rightarrow 4\text{FeOOH(s)} + 3\text{Ca}^{2+} + 4\text{H}_2\text{AsO}_4^- + \text{H}_2\text{O} \tag{7}$$
$$\quad\quad \textit{arseniosiderite} \quad\quad\quad\quad\quad\quad\quad \textit{goethite}$$

As shown in Fig. 9, the chemical composition of HFO and arsenate minerals in the Mokrsko saprolite is highly variable. Composition of the As-HFO projects along the Fe-As join up to ~30 wt% As_2O_5 but is clearly separated from the composition of arseniosiderite and pharmaco-siderite (Fig. 9). In summary, the evolution of secondary arsenic minerals in this system can be described by the sequence (arsenopyrite, löllingite) → (scorodite, As-rich HFO) → (pharmaco-siderite) → (arseniosiderite) → (As-bearing Fe oxide) (Fig. 10).

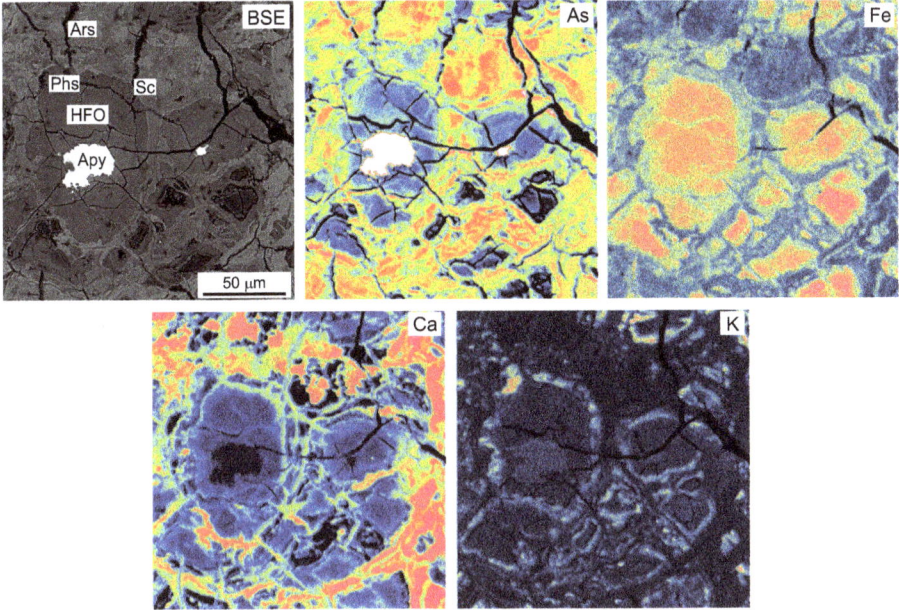

Figure 8. BSE image and the distribution maps of As, Fe, Ca, and K in secondary arsenic minerals in soils overlying the Mokrsko geochemical anomaly. The hydrous ferric oxide (HFO; dark grey) pseudomorphoses after arsenopyrite crystals with enclosed arsenopyrite relics (Apy; light) are surrounded by pharmacosiderite (Phs; dark grey). Scorodite (Sc; middle grey) forms cement between the pharmacosiderite rims. The youngest arseniosiderite (Ars; pale grey) fills the cracks between the minerals. The mineral identification is based on electron microprobe and Raman microspectroscopy data.

The oxidation of arsenopyrite in the gold-bearing sulfide lodes at Ashanti (Ghana, western Africa) led to precipitation of Fe oxide, HFA (amorphous hydrous ferric arsenate), bukovskýite, kaňkite, scorodite, and arsenolite (Bowell 1992) within the chemical system Fe(III)-As(III,V)-S(VI)-H_2O. The minerals formed at the bottom of the thick saprolite through lateritic weathering of the arsenopyrite-rich bedrock. In the tropical environment of Ashanti, high annual rainfall (1250 mm), alternating wet and dry seasons, and high average annual temperatures (27 °C) lead to rapid rates of chemical weathering. Certainly, the combination of high chemical weathering rate of sulfide-rich bedrock and relatively low groundwater flow in the deep saprolite re-

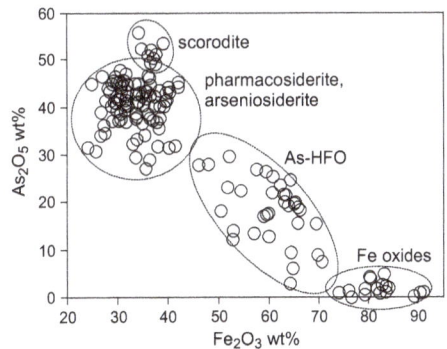

Figure 9. Variations of As_2O_5 as a function of Fe_2O_3 in secondary arsenic-bearing minerals from Mokrsko saprolite. Redrawn from Filippi et al. (2007) and Drahota et al. (2009).

sulted in supersaturation with respect to the secondary arsenic minerals that are more common for extremely arsenopyrite-rich mine wastes (Drahota and Filippi 2009; Haffert et al. 2010). In the uppermost part of the mature saprolite and in the soil profiles that are of the ferrasol type, these secondary arsenic minerals alter further to As-goethite and As-hematite.

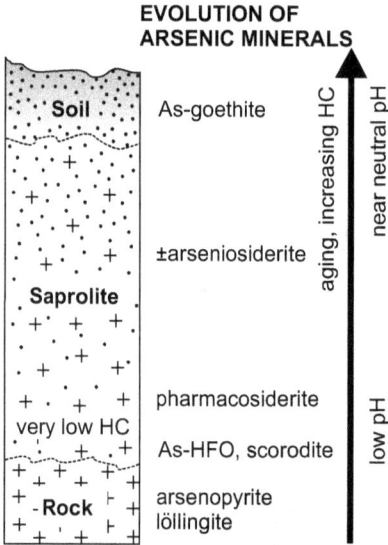

**EVOLUTION OF
ARSENIC MINERALS**

Soil — As-goethite

±arseniosiderite

Saprolite

pharmacosiderite

very low HC

As-HFO, scorodite

Rock — arsenopyrite
löllingite

aging, increasing HC

near neutral pH

low pH

Figure 10. A schematic mineralogical model of As carriers in the chemical system K-Ba±Ca-Fe(III)-As(V)-H$_2$O of naturally mineralized saprolites and soils in the Mokrsko district (Czech Republic), Saint-Yrieix-la-Perche and Echassières (France). HC: hydraulic conductivity.

Weathering of complex sulfide ores with arsenic and lead can result in precipitation of secondary arsenic minerals in the chemical system Pb±Ba-Fe(III)-As(V)-H$_2$O. An example of a mineral assemblage in this system is the acidic soils in northern Córdoba, Spain (Gómez-Parrales et al. 2011). Ferric arsenate (scorodite and probably angelellite), lead arsenate (schultenite), barium arsenate (dussertite) and As-bearing Fe oxides (goethite and lepidocrocite) were identified in the upper 40 cm of soil profiles developed above different granitic rock types. The precursors of the secondary minerals in the soil are different ore mineralizations that are abundant in the bedrock north of Córdoba (e.g., argentiferous Pb mineral assemblage with abundant barite). Arsenopyrite is probably the main As carrier in the gangue (Gómez-Parrales et al. 2011).

The massive sulfide deposits at the Rio Tinto and Tharsis ore districts have been exposed at the surface at least since the Miocene (Nieto et al. 2003). The deposits have undergone extensive oxidation processes and developed a thick saprolite cemented by Fe oxides known as a gossan. The most abundant mineral phases in the gossans are As- and Pb-bearing goethite (up to 2.26 wt% of As), hematite (up to 1.19 wt% of As), and beudantite (a mineral from the alunite-jarosite supergroup) (Fig. 11a) (Viñals et al. 1995; Capitán et al. 2003; Nieto et al. 2003). Beudantite is present throughout the whole gossan/saprolite profile, although it is concentrated especially at the bottom of the profile (up to 40% by volume), where massive beudantite rests directly on the top of unoxidized massive sulfides (Nieto et al. 2003). In the upper part of the profile, beudantite is often associated with As-rich corkite (Fig. 11b).

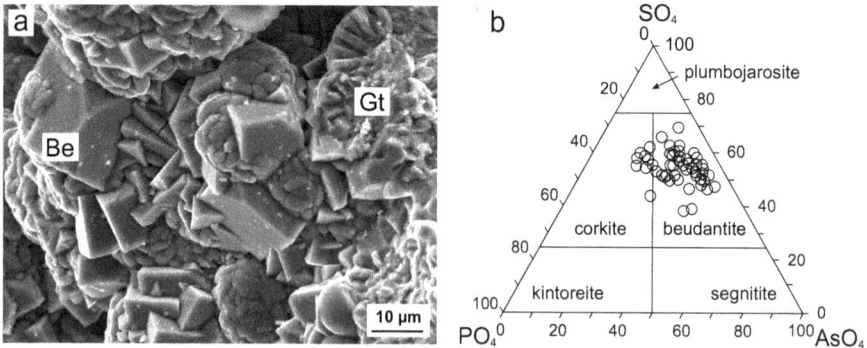

Figure 11. (a) SE image of euhedral beudantite (Be) and botryoidal goethite (Gt) from the central part of gossan in the Filón Sur deposit, Tharsis ore district, Spain. (b) Compositional ranges of beudantite-corkite solid solution from the Filón Sur deposit. All figures courtesy of J. M. Nieto. [Used by permission of Maney publishing, from Nieto et al. (2003) *Appl Earth Sci Trans IMM B*, Vol. 112, Figs. 2,3, p. B295]

Anthropogenic contamination. Although As occurs naturally in soils, human activities have greatly increased As contamination in the environment through release associated with mining and smelting of ores, waste disposal, coal burning, manufacture and use of chromate copper arsenate (CCA) wood preservatives, As-bearing agricultural products as well as the manufacture of other As-bearing chemicals. The major As enrichment in soils is attributable to mining, beneficiation and smelting activities that have been estimated at approximately 70% of the total anthropogenic input of As to soil (Nriagu and Pacyna 1988; Matschullat 2000). Other sources in close proximity to contaminated sites include agriculture (pesticides and manure) and the wood preserving industry (Smith et al. 1998; Mandal and Suzuki 2002; Smith et al. 2003; Wang and Mulligan 2006b; Garelick et al. 2008; Garcia-Sanchez et al. 2010; Camacho et al. 2011). These contamination sources generate local positive anomalies, with up to thousands mg kg^{-1} As, which may persist as secondary As sources over decades or centuries. The remaining anthropogenic source of arsenic to the soils is mostly atmospheric, through emissions related to fossil-fuel combustion, metallurgical industry and the pulp and paper industry. These sources together constitute ~20% of the total anthropogenic input of As to soil (Nriagu and Pacyna 1988; Matschullat 2000). Atmospheric inputs play a role particularly in the long-distance transport and global redistribution of As.

In contrast to the natural geochemical anomalies with various distribution of As along the soil profile, the anthropogenic arsenic contamination is often reflected by high concentration of As in the topsoil horizon as a consequence of superficial deposition. The humus and mineralized surface soil layers have high sorption capacity and are usually oxic. They suppress the percolation of the As and accumulate this element strongly. The formation of secondary arsenic minerals in these soil layers has usually been observed at sites of major contamination (more than hundreds mg kg^{-1}) derived from mining operations (Table 5) and pesticide application (Table 6). With respect to mining environments, the formation of secondary arsenic minerals in soils is most commonly attributed to the alteration of primary arsenic sulfides or arsenides that have been previously transported to soil or fluvial sediments from mine wastes and mine tailings by water or wind erosion (Hudson-Edwards et al. 1999; Williams 2001; Black et al. 2004; Frau and Ardau 2004; Cancès et al. 2008; Jamieson 2014, this volume). In addition to mobilization of sulfide or arsenide minerals, other As phases, where present, are likely to supply As to the contaminated soils (Corriveau et al. 2011a). The most common example is probably represented by arsenolite release through roasting of As-bearing ores (Ashley and Lottermoser 1999; Folkes et al. 2001; Impellitteri 2005, Greif et al. 2008; Haffert and Craw 2008, 2009). The upper soil horizons may also contain As phases inherited from the contaminated industrial sites, such as schultenite, manufactured for inorganic pesticides (Folkes et al. 2001; Cancès et al. 2005; Arai et al. 2006). All these source minerals and phases in the soil are unstable and commonly alter to more stable secondary arsenic phases, such as ferric or calcium arsenates (Foster et al. 1997; Impellitteri 2005; Cancès et al. 2008; Niazi et al. 2011). The alteration of arsenic sulfides and precipitation of secondary arsenic minerals is faster in the upper soil horizons than in the saprolite of the natural geochemical anomalies. The most feasible explanation is the greater humidity and, particularly, the more oxic conditions in the upper horizons. For example, the most intense alteration of arsenopyrite measured as the amount of ferric arsenate (e.g., scorodite) formed has been found in the upper soil layers that were rich in organic matter (Fig. 12) (Mihaljevič et al. 2010). However, a detailed analysis of As speciation throughout soil profiles revealed that most of the As was frequently present as arsenate scavenged by HFO. This trapping process appeared to take place after the breakdown of the primary source minerals and phases such as arsenopyrite, arsenolite, schultenite (Folkes et al. 2001; Frau and Ardau 2004; Cancès et al. 2005; Arai et al. 2006, Haffert and Craw 2009).

Soils and fluvial sediments impacted by mining and processing operations. Arsenic is often an integral component of lead-, zinc-, copper-, and gold-bearing ores and consequently may contaminate soils and fluvial sediments from a variety of sources associated with mining,

Table 5. Secondary arsenic mineral associations in soil and fluvial sediments impacted by mining and processing operations.

Arsenic source	Secondary arsenic minerals	Sample characterization	Locality	References
dam failure of pyrite-containing sludge	sainfeldite, scorodite, HFO, jarosite	alluvium at the depth of 0-40 cm; pH 2.7-9.5	Vado del Quema, 31 km downstream of the ruptured dam at Aznalcóllar, Spain	Hudson-Edwards et al. (2005); Álvarez-Ayuso et al. (2008)
arsenopyrite in tailings discharge	scorodite, Ca-Fe arsenate (arseniosiderite), K-Fe arsenate (pharmacosiderite), HFO, jarosite	fluvial sediments; pH 7.4-8.0	Baccu Locci stream, Villaputzu, Italy	Frau and Ardau (2003, 2004); Frau et al. (2005)
superficial arsenopyrite contamination	arseniosiderite, HFO	clay loam layer of inceptisol soil at the depth of 0-15 cm; pH 6-7	Auzon, Haute-Loire, France	Cancès et al. (2008)
beudantite particles from irrigation water	Fe oxide, beudantite	vadose zone of clay loam and Albaqualfs; pH 5.1-7.1	Guandu Plain, northern Taiwan	Chiang et al. (2010)
arsenolite from calciner	Fe₂(AsO₄)₃, scorodite (?), HFA (?), Pb arsenate (ludlockite ?)	clay loam at the depth of 0-30 cm; pH 4.8-7.5	Camborne-Redruth-St. Day orefield, Cornwall, UK	Camm et al. (2003, 2004); Arčon et al. (2005); Van Elteren et al (2006)
arsenic from smelter	HFO, As-bearing phosphates	topsoil at the depth of 0-2 cm; pH 6.0-7.4	Anaconda copper smelter in Montana, USA	Davis et al. (1996)
arsenic from smelter	scorodite, arsenolite, goethite	soil/0-90 cm	Asarco lead smelter in East Helena, Montana, USA	Impellitteri (2005)
arsenolite from smelter	scorodite, HFA, HFO	sandy and silt loam in the B horizon	Mole River mine, New South Wales, Australia	Ashley and Lottermoser (1999)

Table 6. Secondary arsenic mineral associations in soils
contaminated by arsenic-bearing chemicals.

Arsenic-bearing chemical	Secondary arsenic minerals	Soil depth/pH	Localization	Reference
arsenical pesticides	goethite, ferrihydrite, scorodite	0-10 cm/pH < 7	New South Wales, Australia	Niazi et al. (2011)
arsenical pesticides	schultenite, alumopharmacosiderite, HFO	inceptisol soil/pH 4.5-6.1	Auzon, Haute- -Loire, France	Cancès et al. (2005, 2008)
arsenical pesticides	schultenite, Ca arsenate coprecipitates, HFO	loamy sand at a depth of 2.75-4 m/pH 7.1	northeastern USA	Arai et al. (2006)
copper chromate arsenate	Cu arsenate precipitates, scorodite, adamite, ojuelaite, goethite and gibbsite	loamy, siliceous, hyperthermic Grossarenic Paleudults at a depth of 0-20 cm/ pH 7.0-7.5	Gainesville, Florida, USA	Gräfe et al. (2008b)

processing, and metallurgical operations. There are many ways that As enters soils and fluvial sediments within the mining-affected environments. The release of As into the soils and fluvial sediments generally results from (i) acid mine drainage (Williams 2001; Morin et al. 2003), (ii) the direct release of waste slurries containing solute and particulate-associated As (Pirrie et al. 2003; Frau and Ardau 2004; Blackwood and Edinger 2007), (iii) dumping of mine and processing waste that is subsequently leached or dispersed downstream or downhill (Juillot et al. 1999; Cancès et al. 2008; Haffert and Craw 2008; Corriveau et al. 2011a; Grosbois et al. 2011; Drahota et al. 2012), (iv) mine tailings dam failures (Hudson-Edwards et al. 1999; Jurkovič et al. 2011; Mayes et al. 2011) and (v) release of As-bearing fumes into the atmosphere (Davis et al. 1996; Ashley and Lottermoser 1999; Camm et al. 2003, 2004; Impellitteri 2005; Vaněk et al. 2008). Despite the variety of potential As sources stated above, *in situ* formation of secondary arsenic minerals and phases has been mainly documented in the contaminated soils and fluvial

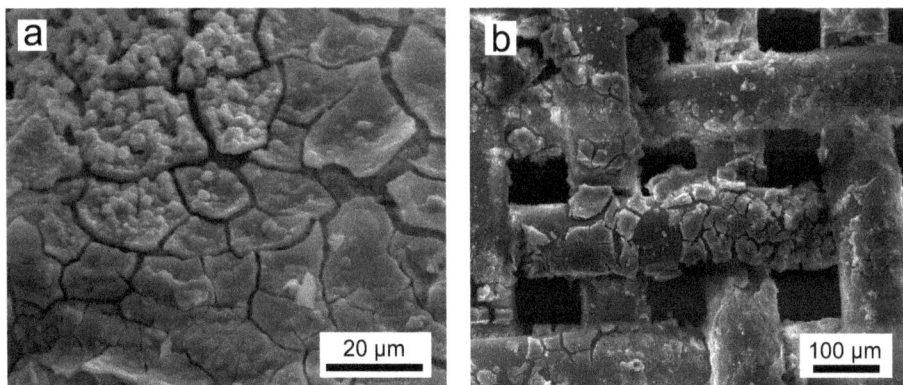

Figure 12. SE images of ferric arsenates developed on the surface of arsenopyrite after one-year *in situ* weathering experiment in organic-rich soil. (a) Botryoidal coatings and spherical aggregates of ferric arsenates on the surface of an arsenopyrite grain. (b) Coatings of ferric arsenate on the fibers of the experimental bag. [Used by permission of Elsevier Limited, from Mihaljevič et al. (2010) *Sci Total Environ*, Vol. 408, Figs. 3b,d, p. 1291]

sediments affected by deposition of As-rich particles released into fluvial systems or dispersed to topsoil layers (Morin and Calas 2006; Drahota and Filippi 2009). The mineralogy of As-rich particles released into these environments usually corresponds to the primary minerals in ores such as arsenopyrite and arsenian pyrite (Hudson-Edwards et al. 1999; Pirrie et al. 2003; Black et al. 2004; Frau and Ardau 2004; Blackwood and Edinger 2007; Cancès et al. 2008). Other studies documented the release of other mineral groups, such as As oxides (mainly arsenolite) and arsenates (scorodite and other ferric-arsenate and calcium-arsenate precipitates) into the soil environment (Davis et al. 1996; Ashley and Lottermoser 1999; Camm et al. 2003; Impellitteri 2005; Haffert and Craw 2008; Corriveau et al. 2011a) and fluvial systems (Hudson-Edwards et al. 1999; Grosbois et al. 2011). For example, the arsenic minerals present in the windblown and vehicle-raised dust (0.15-16 μm) from unvegetated As-rich abandoned mine tailings from Nova Scotia, Canada, were similar to those observed in the near-surface tailings. These arsenic minerals were HFA (~75%), As-HFO (~12%), scorodite (~7%), and As-bearing chlorites (Corriveau et al. 2011a). In another recent study, Grosbois et al. (2011) used a combination of mineralogical methods to identify the As-bearing phases in suspended particulate matter from the Isle River which drains a former gold district in France. The most common As-bearing phases here were clays and HFO associated with Fe-rich clay aggregates, either as micro- to nano-sized particles or as coatings. HFO, Mn hydroxides and rare arsenate minerals had higher As concentration than the clays. The study has also shown how the relative proportions of the As carriers in the suspended particulate fraction depends on the hydrological factors, such as high- and low-flow conditions. After the deposition of the As-bearing phases (i.e., As phases which could be considered primary in the technological sense), chemical remobilization and/or diagenesis often results in the release of As from the host phases and/or the formation of secondary As-bearing phases in the fluvial sediments and soils that may be chemically more stable within the weathering environment than the original minerals (Table 5).

Studies of overbank floodplain sediments (Hudson-Edwards et al. 1999, 2005; Frau and Ardau 2004; Oyarzun et al. 2004; Frau et al. 2005; Álvarez-Ayuso et al. 2008) showed that only a small proportion of particle-bound As remained in the form of the original sulfide or arsenide ore minerals. The majority of As alteration products were present as secondary As-bearing Fe oxides or less frequent secondary arsenate minerals. For example, in the overbank sediments of the Baccu Locci stream watershed (Sardinia, Italy), the primary arsenopyrite grains were largely converted to scorodite (Fig. 13a) and poorly identified arsenical phases containing Ca and Fe (arseniosiderite?) (Fig. 13b) or K and Fe (pharmacosiderite?) (Fig. 13c) (Frau and Ardau 2003, 2004; Frau et al. 2005). These minerals are associated with As-bearing HFO and As-bearing plumbojarosite. Primary arsenopyrite is rare in the sediments and occurs generally as unaltered micro-crystals enclosed within quartz or arsenate mineral grains (Fig. 13b). The authors argue that scorodite and plumbojarosite are unstable under slightly alkaline conditions of the sediment (pH 7.4-8.0) and decompose to As-bearing HFO which tend to be finally converted into goethite and hematite with time.

Failures of tailings dams can release very large quantities of arsenical fluids and As-rich particles to fluvial systems. The discharge of these tailings often greatly exceeds the sediment-transport capacity of a given stream and results in considerable channel and floodplain aggradation of fine-grained material (Macklin et al. 2006). This material can be leached and dispersed into the underlying floodplain sediments. These processes may consequently promote the precipitation of secondary arsenic minerals in the contaminated soils and fluvial sediments. One of the largest mine-related discharges of As to the fluvial environment in the recent history occurred on April 25, 1998 as a result of the breaching of the Aznalcóllar tailings dam, 45 km west of Seville, Spain. An area with a width of 300-400 m on each side of the Agrio and Guadiamar rivers and length of 50 km downstream of the dam was covered with 2-40 cm thick layer of pyritic sludge (4300-5000 mg kg^{-1} of As). In addition to As-bearing pyrite, the sludge contained other arsenic minerals such as arsenopyrite and scorodite (Hudson-Edwards et al. 1999; Ál-

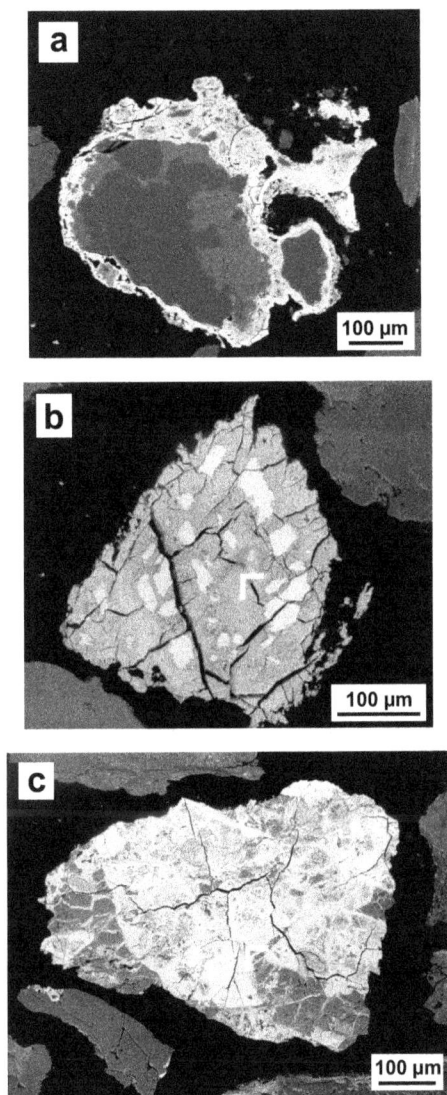

Figure 13. BSE images of the secondary arsenic compounds from the stream-bed sediments at the Baccu Locci river, Sardinia, Italy. (a) Scorodite and As-bearing HFO coatings (brighter rim) on quartz and a K-Al-Si phase (darker core). (b) Arsenopyrite microcrystals (brighter phase) embedded in a matrix consisting mainly of Ca-Fe-As phase (arseniosiderite ?). (c) A complex grain consisting mainly of a K-Fe-As phase (pharmacosiderite ?) and a K-Al-Si phase (darker zones). All figures courtesy of F. Frau. [Used by permission of The Mineralogical Society, from Frau and Ardau (2004) *Mineral Mag*, Vol. 68, Figs. 3a,b,d, p. 21]

varez-Ayuso et al. 2008). The mineralogy of the soil underneath the sludge was affected by sulfide weathering and penetration of metal- and metalloid-rich acid solution (pH 2.3-3.5) as well as by dispersion of arsenic sulfides (Hudson-Edwards et al. 2005). Thus, traces of As-bearing pyrite and arsenopyrite were detected in the uppermost soil layers where newly-formed As-bearing HFO and rare secondary arsenic minerals such as sainfeldite and, probably, beudantite precipitated following the dissolution of the primary sulfides (Hudson-Edwards et al. 2005; Álvarez-Ayuso et al. 2008). In other fluvial systems, however, rapid burial and persistent reducing conditions resulted in the preservation of the arsenic sulfides. Such cases were described in the surface sediments of the Fal Estuary, Cornwall, UK (Pirrie et al. 2003), in the active stream beds in the Shag River and Deepdell Creek, both New Zealand (Black et al. 2004), and in the sediments in the Buyat and Tokok Rivers, North Sulawesi, Indonesia (Blackwood and Edinger 2007). Thus, the formation of secondary arsenic minerals has not been documented in those locations.

Both the alteration of original arsenic minerals and precipitation of secondary arsenic minerals in the oxidizing upper soil layers are relatively rapid processes. Many mineralogical studies that investigated the solid-state speciation of As in the soils affected by erosion and weathering of the mine waste dumps and tailings with primary arsenic sulfides and secondary arsenic minerals have found As-bearing HFO and Fe oxides as the only As-bearing phases (Filippi et al. 2004; Patinha et al. 2004; Haffert and Craw 2009; Walker et al. 2009; Meunier et al. 2010; Drahota et al. 2012). There are, however some, rather limited, attributions of arsenic minerals in soils. For example, Cancès et al. (2008) identified secondary arseniosiderite, formed by the oxidation of arsenopyrite, in the near-neutral topsoil horizons near an industrial waste site at the Auzon, France. Another example is provided by slightly acidic paddy soils in the Guandu Plain, northern

Taiwan (Chiang et al. 2010). The As and Pb contamination in soil (up to 290 mg kg^{-1} and 1300 mg kg^{-1}, respectively) was due to 50-100 years of irrigation by water from the creek of the Beitou Hot Spring with high concentrations of As, Pb, and SO_4^{2-} (4-6 mg L^{-1}, 0.3-0.4 mg L^{-1}, 2-4 g L^{-1}, respectively). Beudantite has been identified in the clay fraction of the soil. This finding supports the idea that some amount of beudantite has been introduced into the soil from the fine-particle fraction in the creek-water suspension in which this phase was detected (Chiang et al. 2010).

More common examples of secondary arsenic phases in soils are related to distribution of particulate As in the form of arsenic trioxide. This compound is easily volatilized during the smelting of lead, zinc, copper, gold, and concentrates and fractionates into the flue dust (Davis et al. 1996; Morales et al. 2010). Up until the early part of the 19th century, As-bearing fumes escaped freely to the atmosphere. When the early uses for As were devised, the metalloid was recovered, usually in the form of arsenic oxide by roasting the ore, sublimation, and condensation in collection chambers built specifically for this purpose. As a result of the release of arsenic trioxide from both the processing waste from previous mining activity and from calcinations of As-bearing ores, As soil contamination is widespread in many areas in England, USA, and Australia (Ashley and Lottermoser 1999; Potts et al. 2002; Camm et al. 2004; Impellitteri 2005). The major As-bearing constituents of flue dust, arsenolite and claudetite, are not stable in soils (Yue and Donahoe 2009). They alter rapidly to secondary, more thermodynamically stable assemblages. The most common alteration products of the arsenic trioxide precursors in the soil environment are As-bearing Fe oxides and hydrated ferric arsenates that correspond mainly to scorodite or HFA (Davis et al. 1996; Ashley and Lottermoser 1999; Camm et al. 2003; Haffert and Craw 2008). Davis et al. (1996) carried out a detailed investigation of As solid-state speciation in circumneutral (pH 6.0-7.4) Anaconda soil, Montana, USA, affected by historical smelting of copper sulfide ore. They showed that the relative contributions of As sources follow the order: smelting-derived (As oxides, ferric arsenate, and slag) > soil alteration phases (As-bearing phosphates and Fe-Mn oxides) > residual ore concentrate (e.g., enargite). Chemistry of the ferric arsenate phases was similar to that of scorodite. The ferric arsenates were then replaced by As-bearing phosphates and As-bearing ferric oxides in the soil environment. Similarly, the arsenolite precursor was mostly transformed to As-bearing Fe oxides and scorodite in the highly contaminated soils from the Asarco lead smelter in East Helena, Montana, USA (Impellitteri 2005) and Mole River mine, New South Wales, Australia (Ashley and Lottermoser 1999). Strongly contaminated fluvial sediments at the Mole River mine (As > 1 wt%) contained smelting-derived arsenolite and HFA (Ashley and Lottermoser 1999). Historical processing and refinement of As in purpose-built calciner in Cornwall, UK, has resulted in a significant soil contamination (up to 4000 mg kg^{-1} of As, Camm et al. 2004). Arsenic-bearing phases in the Cornwall soils, identified using a SEM-EDS, have been dominated by arsenic oxides, hydrated ferric arsenates, lead arsenates, and As-bearing HFO (Camm et al. 2003). The most common As phases were ferric arsenates which occurred as rims of and filled microfractures in large (>500 µm) coke/fly ash and silicate particles and cemented pore space between mineral grains (Camm et al. 2003). Arčon et al. (2005) used X-ray absorption spectroscopy to retrieve arsenic molecular information from the ferric arsenate phases in the Cornwall soils. Their results suggested that the secondary arsenic phases have been formed by (co)precipitation of arsenate leading to X-ray amorphous or poorly crystalline HFA. This phase is probably related to a crystalline Fe(III) tris(arsenato)-arsenite, $Fe_2(As^{3+}(As^{5+}O_4)_3)$, identified and characterized in a recent study of Van Elteren et al. (2006) but not described as a mineral. In addition, arsenopyrite was present as a minor phase and there were discrete particles chemically corresponding to ludlockite (Camm et al. 2003). A phase chemically similar to ludlockite was also found in the soil in vicinity of a lead smelter in Příbram, Czech Republic (Vaněk et al. 2008).

Soils contaminated by arsenic-bearing chemicals. Until the 1970's, approximately 80% of the As produced was used to manufacture pesticides (Matschullat 2000). Due to the poisonous nature of these compounds, their application has gradually diminished with the introduction of dichlorodiphenyltrichlorethane (DDT) and 2,4-dichlorophenoxyacetic acid after World War II (IPCS 1979). Arsenical pesticides faded from the scene in the 1980's, and were officially banned in the US on August 1st, 1988. But the outdoor use of arsenical herbicides persisted until today (e.g., Cai et al. 2002). The extensive use of calcium arsenate [$Ca_3(AsO_4)_2$], lead hydrogen arsenate ($PbHAsO_4$), sodium arsenite ($NaAsO_2$), and many other arsenic compounds as pesticides/insecticides in orchards, cattle and sheep dips has supplied As to soils in many countries, such as Australia, Canada, France, Great Britain, New Zealand, and USA (Peryea 1998; Smith et al. 1998; Wang and Mulligan 2006b; Yang and Donahoe 2007; Zevenhoven et al. 2007; Garelick et al. 2008). Although most of this contamination is of a diffuse nature, there are incidences of point-source contamination, largely at the former pesticide plants and storage sites. For example, sodium arsenite was applied to control southward migration of cattle tick across New South Wales in Australia from the early 1900's to 1955. Over 1500 dips were constructed along the eastern coast of Australia with many of these being located in the northern New South Wales. The repeated application of the arsenite pesticide led to concentrations of up to 3000 mg kg^{-1} of As in soils (Smith et al. 1998). Recently, Niazi et al. (2011) investigated the solid-state speciation of As in the contaminated surface soils (0-10 cm, pH less than 7) sampled from 18 historical cattle dip sites using XANES spectroscopy. Arsenic associates here mostly with ferrihydrite and goethite but the newly formed scorodite has accounted for up to 22% of the total As in 10 of the 18 soil samples studied. Similarly, application of an As_2O_3-based herbicide has contaminated soils in southeastern USA (near the coast of Gulf of Mexico) with up to 900 mg kg^{-1} of As (Yang and Donahoe 2007). Most of As was observed in association with Fe and Al oxides, however, some newly-formed As-rich particles detected in Ca-rich sandy loam (pH 7.5) were tentatively identified as phaunouxite (Yang and Donahoe 2007). In another example, Voigt et al. (1996) and Foster et al. (1997) detected hörnesite in the neutral, As-rich soil (pH 7.5, up to 2200 mg kg^{-1} of As) contaminated by smelter waste. Arsenic-bearing waste was stockpiled at this site (Palo Alto, California, USA) for up to 69 years for the manufacture of pesticides. In the coastal Maine watershed (New England, USA), Ayuso and Foley (2008) and Foley and Ayuso (2008) attributed the calcium, lead, and sodium arsenates identified in acidic, As-rich soils contaminated by As (up to 3960 mg kg^{-1}) to be remnants of the previously applied pesticides.

The most extensively used inorganic arsenical insecticide was lead hydrogen arsenate (Pb-HAsO$_4$), also called acid lead arsenate. This compound corresponds to the mineral schultenite and replaced early on more soluble and more phytotoxic inorganic pesticides, such as Paris green [$Cu(CH_3COO)_2\cdot3Cu(AsO_2)_2$], and became the principal arsenical pesticide used in France and many other countries (Peryea 1998). Several recent studies have investigated the solid-state speciation of As in the highly contaminated soil at the former arsenical pesticide plants that produced the acid lead arsenate pesticide. The first example includes As-rich soil near a former arsenical pesticide plant at Auzon, Haute-Loire, France (Cancès et al. 2005, 2008). The industrial activities of the plant started at the beginning of the 20th century with roasting of arsenic sulfides to produce arsenolite. Subsequently, arsenical pesticides (lead, copper, and aluminum arsenates) were manufactured at the site until 1949 when the factory was closed (Cancès et al. 2005). The As concentration is very high at the top of soil profiles (up to 8780 mg kg^{-1}) and decreases rapidly downward to a few hundreds of milligram per kilogram of As (Cancès et al. 2005, 2008). The thin white and blue layers found at a depth of 10 and 15 cm were identified as schultenite and alumopharmacosiderite, respectively. Schultenite represents the end-product of pesticide production at the Auzon site (Cancès et al. 2005), whereas alumopharmacosiderite is likely a transformation product of an insecticide called *arsalumine* (Cancès et al. 2008), a mixture of sodium arsenate and aluminum sulfate. These arsenical pesticides together with arsenic sulfides (realgar, orpiment, and arsenopyrite) used initially for manufacturing pesticides were likely the sources of As at this site (Cancès et al. 2005, 2008). The results of selective chemical extractions and XAS analyses confirmed that arsenate is mainly (at least 80 wt%) associated with X-ray amorphous As-HFO in the slightly acidic to neutral soil profiles. The proportion of schultenite in the topsoil horizon does not account for more than 10 wt% of total As (Cancès et al. 2005). Almost similar situation has been detected in the vicinity of a former arsenical pesticide plant in the northeastern USA. During its 96-year history of operation, this plant manufactured inorganic arsenical pesticides such as $PbHAsO_4$, organic pesticides, and many other chemicals (Arai et al. 2006) that led to As concentration of up to 254 mg kg^{-1} in the soil profiles. Multiscale spectroscopic techniques showed that decades of contamination and weathering resulted in an alteration of the original As contaminant, lead hydrogen arsenate. The neutral oxic soil contained predominantly arsenate adsorbed to HFO (~71%), followed by the residual schultenite (~29%). In the neutral semi-reduced soil, schultenite was absent. Instead of within schultenite, 46% of the total As was present as arsenate-calcium coprecipitates (most likely weilite), followed by arsenic adsorbed on HFO (~28%) and amorphous As_2S_3 (~25%) (Arai et al. 2006).

Another potential contributing source of As-bearing chemical in soil is the use of chromate copper arsenate (CCA) as an inorganic water-borne wood preservative. Despite the current gradual reduction of CCA-treated wood applications since 2003, this compound was used extensively for more than 60 years in North America and Europe (Zevenhoven et al. 2007; Garelick et al. 2008). Many sites were used by the timber industry to treat wood or to dispose of the waste; they are invariably contaminated with As and metals (Garelick et al. 2008 and references therein). Gräfe et al. (2008b) have determined the solid-state speciation of As in the CCA-contaminated upper soil horizon (0-20 cm) containing up to 880 mg kg^{-1} As, 1300 mg kg^{-1} Cu, and 700 mg kg^{-1} Zn from a former timber/lumber treatment site near the University of Florida, Gainesville, USA. The As and metal cations (Fe, Cu, Zn, etc.) in the circumneutral soil accumulated in distinct and isolated domains (not larger than 50 μm) that have been identified as prevailing Cu arsenate (63-75%) and to a lesser extent as Zn arsenate and Fe(III) arsenate precipitates corresponding to adamite, ojuelaite, and scorodite, respectively. The remaining As occurred as arsenate complexes adsorbed on the surfaces of goethite and gibbsite.

Arsenic minerals in mine wastes

This section provides a brief overview of the secondary arsenic minerals and phases known from environments contaminated by mining, ore processing, and metallurgical operations. The waste generated by these processes is collectively referred to as mine waste and defined as "solid, liquid, or gaseous by-products of mining, mineral processing, and metallurgical extraction. They are unwanted, have no current economic value and accumulate at mine sites" (Lottermoser 2003). Several chapters in this volume also cover general or specific topics regarding As in mine wastes (Alpers et al. 2014; Basta and Juhasz 2014; Bowell 2014; Campbell and Nordstrom 2014; Craw and Bowell 2014; Foster and Kim 2014; Jamieson 2014). This section provides the general context of arsenic mineral parageneses in different mine waste settings at and near the Earth's surface and some basic information on the occurrence of secondary arsenic minerals or phases in these materials.

Mine wastes are well known as substantial local sources of As on the Earth's surface. At tens of thousands of sites of active or abandoned mines around the globe, waste is stored in piles, tailings impoundments, and slag heaps and contains up to several wt% of As (e.g., Pichler et al. 2000; Roussel et al. 2000; Cappuyns et al. 2002; Filippi 2004; Ettler et al. 2009c; Walker et al. 2009; DeSisto et al. 2011). Nriagu and Pacyna (1988) estimated that the As input from solid mine wastes into the soil is approximately 16 % of the whole anthropogenic arsenic input. This represents nearly 5 thousands metric tons of As per year.

Mine wastes are commonly classified according to their physico-chemical properties and according to their source (Lottermoser 2003). In the following text, we follow this classification and focus on those types of mine wastes which contain secondary arsenic minerals or phases. Mine wastes can be divided into three groups: i) mining and processing wastes, ii) metallurgical wastes, and iii) mine drainage. Despite the variety of arsenate and arsenite minerals documented in literature (see section *Crystal Chemistry of Arsenic*), only a limited number of them have been repeatedly identified in the mine wastes (Riveros et al. 2001; Drahota and Filippi 2009). These arsenic minerals or phases must therefore be considered as the "environmentally important" ones. They are widespread in the wastes and often control As mobility within and around the waste reservoirs. The apparent lack of diversity among the most abundant and most common secondary arsenic minerals in mine wastes is most likely due to small variability among the major primary minerals (both those which contain As and those which do not contain As) and the similar physico-chemical conditions within the waste (Drahota and Filippi 2009).

Mining and processing wastes. Mining is the first step in the exploitation of a mineral or metal commodity and represents the extraction of the material from the uppermost portions of the crust. Mining generates a large volume of waste that either does not contain the target ore minerals or in which the concentration of the target commodity is subeconomic. Solid mining wastes consist of overburden and waste rocks but also can include low-grade ore stockpiles and process tailings. Waste rocks excavated from coal mining are referred to as spoils. Solid mining wastes are usually stored in piles around and near mine sites and consist of highly heterogeneous material ranging from clay-size particles to boulder-size fragments. Grain size exerts important control on both geochemical and physical processes in mining waste dumps as the finer fractions with larger surface area are more reactive than the coarser ones. When reactive minerals such as arsenopyrite are encapsulated inside the waste fragments, oxygen, water, and other reactants must first be transported through the large particle to the reactive mineral by diffusion. This so-called mass-transfer limitation is responsible for slower reaction rates of larger particles in mining wastes, because the time required for diffusion of reactants into the particles and products out of the particles must be added to the intrinsic reaction time. Grain size is also a determinant of the behavior of unsaturated flow that dominates in the mining waste dumps. In fact, the groundwater level in most mining waste dumps is below the base of the piles. Generally, under relatively dry conditions, water occupies and preferentially flows in the smallest pores (Smith and Beckie 2003) which occur in the volume of the waste with the finest grain size. Larger pores are filled with air and if the moisture-vapor transport is significant, evaporation of pore water causes formation of secondary precipitates on the pore walls in the form of coatings, crusts and eventually cement. These precipitates will accumulate until dissolved by the relatively diluted water from the next infiltration event. If the rate at which the secondary precipitates are produced is fast relative to the frequency of the infiltration events, then a significant amount of secondary minerals or phases will accumulate in the larger pores, and *vice versa*. This information emphasizes the role of climate on the formation of secondary arsenic minerals and the potential leaching of As from the mining wastes. The initial secondary precipitates tend to be poorly crystalline and/or metastable with respect to the later phases and the transformation between the initial and final phases may occur in one or several steps. An example of such processes is the transformation of HFA or other metastable ferric arsenates (kaňkite and parascorodite) and sulfo-arsenates (bukovskýite) to scorodite (Paktunc et al. 2008; Filippi et al. 2009; Haffert et al. 2010; Majzlan et al. 2012a,b).

Coal spoils consist especially of fine-grained sedimentary rocks and possess different physical properties than waste rock from metal-mining operations. The sediments are usually mechanically less stable and contain a larger amount of clay the fraction. Both of these factors affect the hydrological as well as geochemical regimes in the spoils. In fact, secondary arsenic mineralogy of these waste forms is restricted mainly to adsorption processes on HFO and clay minerals, and substitution in jarosite (see section *Arsenic Minerals in Coal*).

After extraction from the ground, the raw material is processed and split into the valuable fraction (ore) and the fraction with no or limited value (gangue). In the physical processes of beneficiation or mineral processing, the ore minerals are collected as ore concentrate. Mineral processing of hard-rock ores thus involves size reduction (crushing, grinding, milling) and separation of the ore mineral(s) from gangue (gravity, magnetic, electrical, and optical sorting; flotation techniques). These treatment operations generate processing wastes generally referred to as tailings. In nonferrous metal mining, the ore concentrate may represent only a small fraction (usually less than 5%) of the extracted material, and thus the vast remainder of the mined mineralized rock will be converted to tailings. Thus, tailings and waste rock represent the most voluminous waste forms at metal-mine sites (Lottermoser 2003). In the past, tailings were discharged directly to rivers or wetlands or were deposited close to the mills. Currently, these fine-grained sediment-water slurries are mostly pumped into different types of tailings ponds or may be stored in underground workings (backfill) and mined-out open pits. Tailings are also currently disposed into rivers, lakes, and marine environments. In those cases, tailings may increase turbidity and contaminate the lacustrine, riverine, or marine sediments with metals and metalloids. The occasional failure of tailings dams causes contamination of downstream environments from direct discharge and indirect fluvial erosion of tailings, and the dispersal of fine-grained material into waterways. The arsenic mineralogy of fluvial sediments and overbank floodplain sediments contaminated by processing wastes has been discussed previously in section *Arsenic Minerals in Soil and Fluvial Systems*. The hydrogeological characteristics of the tailings ponds are controlled by the tailings grain size and deposition history. Tailings are fine-grained, ranging between <25 μm to 1 mm (Robertson 1994). The grain size depends on the liberation characteristics of the ore and gangue minerals and the applied crushing and grinding process. Differential settling results in the sand-size fraction depositing near the high-energy environment adjacent to the discharge point, with the clay-size fraction accumulating in the stagnant part of the impoundment. The hydraulic conductivity of tailings may thus vary between 10^{-6} and 10^{-2} cm s^{-1} (Robertson 1994). During deposition, much of the impoundment is covered and saturated with processing water. After the deposition of the tailings is complete and the discharge of processing water ceases, the groundwater level usually descends below the surface of the tailings, and from that time, water balance of the impoundment is controlled by precipitation, evapotranspiration, surface runoff, and seepage of water. As the tailings oxidize, secondary minerals and phases accumulate on grain surfaces as coatings or within pore spaces among the tailings particles. One of the specific features of tailings may be the development of surface or subsurface layers of secondary precipitates with a substantial lateral extent. Such layers may dry out and become cemented with additional precipitates and are then referred to as hardpans. Another specific feature may include seasonal surface precipitates that are commonly referred to as efflorescence.

Despite certain different physical and depositional characteristics of mining wastes and processing wastes (tailings) that are mentioned in the above paragraphs, biogeochemical processes involved in the formation of secondary arsenic minerals are similar in both environments. Therefore, we summarize the information for both waste forms together. Primary arsenic minerals are usually unwanted components of the mined raw material and thus end up in mining waste dumps or in tailings. The concentration of As in the dumps and tailings commonly attains thousands of mg kg^{-1} and in specific cases may exceed 10 wt% (e.g., Ashley and Lottermoser 1999; Roussel et al. 2000; Ashley et al. 2004; Ahn et al. 2005; Walker et al. 2009; Haffert and Craw 2010; DeSisto et al. 2011). Less frequently, the primary arsenic minerals themselves have been the targeted fraction and ore concentrates with extremely high As content (up to several tens of wt%) may be found as stockpiles at the mining or processing sites (Filippi 2004; Mains and Craw 2005; Salzsauler et al. 2005; Drahota et al. 2012). The most common groups of primary arsenic minerals in mining wastes and tailings are undoubtedly sulfarsenides and sulfides (e.g., arsenopyrite and arsenian pyrite), and less commonly arsenides (e.g., löllingite). Beside the dominating sulfides, sulfarsenides, and arsenides, primary arsenic minerals in mining and processing wastes may also include supergene arsenic minerals. In this case, these minerals are considered primary in a technological sense (see above for definitions). The distinction of the supergene pre-mining primary minerals and the post-mining secondary minerals is a challenging task because some arsenic minerals, particularly soluble arsenates and HFO, may have formed during both stages.

Sulfides, arsenides, and sulfarsenides are unstable under surface oxidizing conditions and the oxidative dissolution of these minerals results in the release of acidity, dissolved As, Fe, SO_4^{2-}, and other metals and metalloids. Most common culprits for the generation of the acid are iron sulfides (pyrite, FeS_2, and pyrrhotite, $Fe_{1-x}S$); other minerals play a minor or negligible role in these processes. Mining and processing wastes may also contain primary non-sulfide minerals and secondary minerals formed by *in situ* weathering. The former group of minerals usually consists of silicates and carbonates which have an important role in neutralizing acid drainage (Jambor 2003). Secondary minerals comprise a large group of mine-waste minerals that have been thoroughly reviewed by Jambor et al. (2000), Jambor (2003), and Lottermoser (2003). Supersaturation with respect to a secondary mineral is initially most commonly achieved in the vicinity of other mineral surfaces, where the solution chemistry is altered by solute-surface interactions. As a result, arsenic precipitates are often seen as reaction rims on the parent As minerals (e.g., arsenopyrite, arsenian pyrite) rather than isolated particles. Examples of reaction rims on arsenopyrite can, for example, include scorodite, HFA, and As-HFO (e.g., Petrunic et al. 2006; Majzlan et al. 2007; Courtin-Nomade et al. 2010; Corriveau et al. 2011b). Secondary arsenic minerals and phases also occur as cement and masses within the waste and as cemented layers or hardpans at or near the surface of the waste. Supersaturation followed by precipitation of arsenic minerals in such cases usually takes place in response to one these following processes (modified after Lottermoser 2003):

- Oxidation of the dissolved anions ($H_xAs^{3+}O_3^{x-3}$) and/or cations (Fe^{2+})

- Hydrolysis of the dissolved cations (e.g., Fe^{3+}, Al^{3+})

- Oxidation of arsenides and sulfarsenides in humid air

- Reaction of As-bearing waste solutions with solutions with significantly different pH

- Concentration of the solutes due to evaporation

The mechanism of As-bearing cement formation can vary from one site to another, but most generally involves prolonged transport of large volumes of dissolved constituents and their precipitation at the surface or at a particular depth of mine waste. Arsenic-bearing cemented layers and masses usually occur at the transition between oxidized and reduced zones, which often occurs between saturated and unsaturated zones (e.g., Craw et al. 2002; Gieré et al. 2003; Alakangas and Öhlander 2006; Graupner et al. 2007; DeSisto et al. 2011). At these tran-

sitions, concentrations of aqueous As(V) and Fe(III) increase *via* oxidation of primary minerals and/or reduced aqueous species so that supersaturation is reached. Arsenic minerals and phases documented from the cemented layers and masses correspond mainly to amorphous and poorly crystalline HFA (Fig. 14), As-HFO, As-jarosite, and scorodite (McGregor and Blowes 2002; Courtin-Nomade et al. 2003; Gieré et al. 2003; Salzsauler et al. 2005; Sidenko et al. 2005; Alakangas and Öhlander 2006; Lottermoser and Ashley 2006; Graupner et al. 2007; Filippi et al. 2009; DeSisto et al. 2011). Hardpans were observed in the oxic zones of capillary fringes where agglomeration of particles is driven by capillary transport. Precipitation may also occur in the oxida-

Figure 14. Amorphous hydrous ferric arsenate (HFA) fills the intergranular pores and cements the fragments of a historic mining waste dump, Giftkies arsenic mine (Czech Republic). HFA is locally accompanied by scorodite, kaňkite, and hydrous ferric oxide (HFO).

tion zone at a depth where pore waters react with carbonates, resulting in increased pH, which allows precipitation of HFO (Blowes et al. 1991). Indeed, many case studies supposed that As-HFO is the most common arsenic-bearing mineral in the hardpans (McGregor and Blowes 2002; Courtin-Nomade et al. 2003; Salzsauler et al. 2005; Alakangas and Öhlander 2006; Lottermoser and Ashley 2006). Rapid formation of the hardpans as physical and chemical barriers causes encapsulating of the acid-producing or -buffering minerals (in the case of As mostly arsenopyrite and arsenian pyrite) and thus strongly slows down the weathering and AMD generation (Lottermoser 2003).

Secondary arsenic mineralogy in mining and processing wastes may vary among the mine sites (Table 7) as it often reflects the dominant components of the chemical system in the specific waste (Fig. 15). As noted above, the most abundant and most common secondary arsenates display limited diversity and can be therefore used as markers for specific systems. We decided to distinguish three principal chemical systems that repeatedly occur in mining wastes and tailings and correspond, in our opinion, to three common end-members in the composition of primary ore mineralogy. Such division has undoubtedly its limitations because of the great variability of the mine wastes that rarely, if ever, precisely fit one of these three chemical systems. We feel, nevertheless, that many actual chemical systems in mining wastes and tailings can be considered as a combination of these end-members. Thus, this empirical division provides a useful starting point for describing the secondary arsenic mineral assemblages in such environments. These three chemical systems are: 1. Fe(III)-As(V)-S(VI)-H_2O, 2. Ca-Fe(III)-As(V)-H_2O, and 3. (Co, Cu, Ni, Pb, Zn)-Fe(III)-As(V)-S(VI)-H_2O and are described below.

System Fe(III)-As(V)-S(VI)-H_2O. Arsenopyrite and/or arsenian pyrite are the dominant sources of As in most mining and processing wastes. Their oxidative dissolution in carbonate- and other reactive minerals-poor wastes produces low pH with a concomitant release of As, Fe^{2+}, and SO_4^{2-}. In such environments, these components dominate the weathering solution and are substantial constituents of the newly formed secondary arsenic minerals and phases in the chemical system Fe(III)-As(V)-S(VI)-H_2O. Based on a literature review, this chemical system is undoubtedly the most common one in the mining and processing wastes. Secondary arsenic minerals in this chemical system include mainly As-HFO, Fe oxides, jarosite, scorodite,

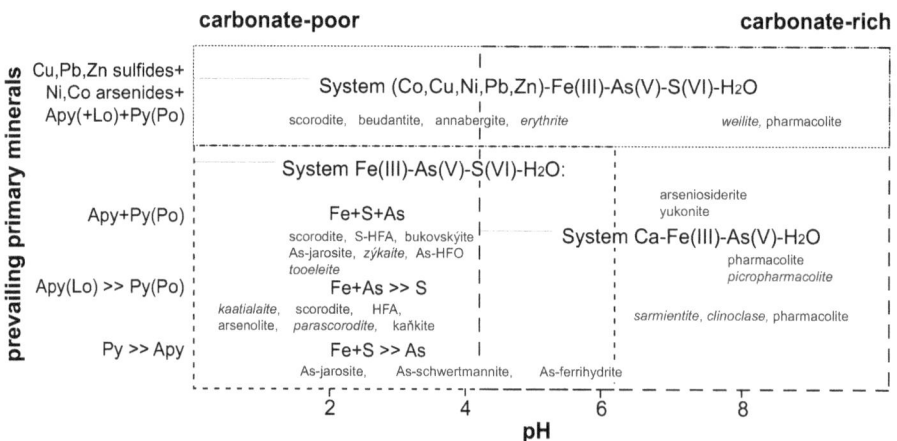

Figure 15. A schematic empirical model showing the most important chemical systems and the distribution of abundant and rare secondary arsenic minerals in mining and processing wastes as a function of the primary ore mineralogy and pH. Note that borders of the chemical systems are only approximate. Italics denote minor or rare minerals. Abbreviations: Py – pyrite, Po – pyrrhotite, Apy – arsenopyrite, Lo – löllingite.

Table 7. Chemical systems and their secondary arsenic mineral assemblages in mining and processing wastes. Minerals listed in italics are minor or rare.

Secondary arsenic minerals	Precursor minerals	Characterization	Localities and reference
Fe(III)-As(V)-S(VI)-H₂O chemical system			
HFO, jarosite	pyrite, arsenopyrite	historic tailings; pH 2.7-3.8	abandoned Jumna mill (closed 1989) processing the tin ore from Irvine bank deposits, Australia (Lottermoser and Ashley 2006)
HFO	pyrite, arsenopyrite	historic tailings; pH 5.9-7.7	abandoned antimony Pezinok deposit (closed 1992), Slovakia (Majzlan et al. 2007)
scorodite, arsenolite, *kaatialaite*	löllingite, arsenopyrite	löllingite-arsenopyrite concentrate; pH~1	abandoned tin deposit (closed 1957) at Přebuz, Czech Republic (Filippi 2004)
HFO, HFA, jarosite	pyrite, arsenopyrite	historic mining waste dump; pH 3.0	abandoned Seobo tungsten mine (closed 1970), Korea (Lee et al. 2005)
scorodite, kaňkite, jarosite, HFO	pyrite, arsenopyrite	historic mining dump	abandoned Suzukura mine (closed approx. 1964), Japan (Kato et al. 1984)
scorodite, HFA, kaňkite, HFO, *beudantite*	arsenopyrite, *chalcopyrite, galena*	historic mining waste; pH 3.3-5.5	historic Giftkies arsenic mine (closed in 1770's), Jáchymov ore district, Czech Republic (Filippi et al. 2009, 2014; Drahota et al. 2012)
HFA, jarosite, HFO	pyrite, arsenopyrite	historic tailings; pH 2.9-4.5	historic Muenzbachtal tailings impoundment (closed 1968), Freiberg polymetallic sulfide district, Germany (Graupner et al. 2007)
scorodite, jarosite, HFA	arsenopyrite, pyrrhotite, pyrite	arsenopyrite residue stockpile; pH 6.7-8.6	abandoned Noc-Areme Gold Mine (closed 1959) in Snow Lake, Manitoba, Canada (Flemming et al. 2005; Salzsauler et al. 2005)
scorodite, HFA, kaňkite, jarosite-beudantite, HFO	pyrite, arsenopyrite, sphalerite, pyrrhotite	historic mining waste dump; pH 1.9-3.4	abandoned Dlouhá Ves polymetallic deposit (closed 1960), Czech Republic (Kocourková et al. 2011)
scorodite, HFA, HFO, pharmacosiderite	arsenopyrite	historic mining waste dumps; pH 3.5 (Barruecopardo), 7.8 (Terrubias)	abandoned tungsten Barruecopardo and Terrubias mines (closed 1983), Spain (Murciego et al. 2011)
jarosite, HFO, HFA, *scorodite, arsenowaylandite*	arsenopyrite, pyrite	historic tailings; pH~3	abandoned Enguialès tungsten deposit (closed 1979), France (Courtin-Nomade et al. 2002, 2003)

Table 7. *continued*

Secondary arsenic minerals	Precursor minerals	Characterization	Localization and reference
jarosite, HFO, goethite, *zýkaite*, *bukovskýite*, *sarmientite*, *tooeleite*	arsenopyrite, pyrite, pyrrhotite	recent processing waste from the bacterial oxidation process	São Bento gold deposit, Brazil (Márquez et al. 2006)
scorodite, tooeleite, jarosite	arsenopyrite, pyrite	arsenopyrite- and pyrite-rich mining waste dump	abandoned gold and arsenic U.S. Mine at Gold Hill, Tooele County, Utah, USA (Cesbron and Williams 1992)
HFA, kaňkite, scorodite, *zýkaite*, HFO	arsenopyrite, pyrite	eroded historic tailings; pH 3-6	abandoned Bullendale historic gold mine (closed 1902), New Zealand (Haffert and Craw 2010)
scorodite, HFO, *bukovskýite*	arsenopyrite, pyrite	pyrite-rich arsenopyrite concentrate; pH 2.2-3	abandoned Golden Point historic battery (closed 1940), New Zealand (Mains and Craw 2005)
jarosite, HFA, scorodite, kaňkite, *bukovskýite*, HFO, *zýkaite*, *parascorodite*	arsenopyrite, pyrite, pyrrhotite	highly weathered, heterogeneous mining waste dump; pH (2.4-3.9)	Medieval silver mines at Kaňk near Kutná Hora (15th–16th century), Czech Republic (Novák et al. 1967; Čech et al. 1976, 1978; Ondruš et al. 1999; Majzlan et al. 2012b)
HFA, jarosite-beudantite, As incorporation into copiapite and rhomboclase	pyrite, arsenopyrite	high-sulfide processing waste dump; pH 1.7	abandoned Berikurul gold mine (closed 1991), western Siberia, Russia (Gieré et al. 2003; Sidenko et al. 2005)
scorodite, HFA, HFO, pharmacosiderite, yukonite, kaňkite, *jarosite, tooeleite, Fe oxides, realgar*	arsenopyrite, pyrite	historic tailings; pH 2.4-7.1	historic Montague, Whiteburn, Brookfield, Oldham, Goldenville, Caribou, Lower Seal Harbour gold districts (closed 1940), Nova Scotia, Canada (Walker et al. 2009; Meunier et al. 2010; Corriveau et al. 2011b; DeSisto et al. 2011)
Ca-Fe(III)-As(V)-H₂O chemical system			
HFO, arseniosiderite	arsenopyrite	historic tailings; pH 7.0-9.4	low-sulfide gold Lava Cap deposit (closed 1943), California, USA (Foster et al. 2011)
arseniosiderite, yukonite, HFO, Ca arsenate	pyrite	historic tailings	abandoned Ruth gold-silver Mine (closed 1942), California, USA (Rytuba et al. 2011)

Table 7. *continued*

Secondary arsenic minerals	Precursor minerals	Characterization	Localization and reference
HFO, scorodite, pharmacosiderite, arseniosiderite, yukonite	arsenopyrite, *pyrite, pyrrhotite*	historic mine tailings; pH 6.5-7.0	abandoned Ketza River gold mine (closed 1990), Canada (Soprovich 2000; Paktunc et al. 2003, 2004)
scorodite, HFO, arsenolite, pharmacolite, clinoclase	arsenopyrite	historic mining dump; pH 4.1-4.8	abandoned Mole River arsenic mine (closed 1938), New South Wales, Australia (Ashley and Lottermoser 1999)

(Co, Cu, Ni, Pb, Zn)-Fe(III)-As(V)-S(VI)-H$_2$O chemical system

Secondary arsenic minerals	Precursor minerals	Characterization	Localization and reference
erythrite, scorodite, *annabergite* (erythrite, pharmacolite, weilite, brassite occurred only in the concentrate or furnace residue near the tower on Nipissing Hill)	nickeline, cobaltite, safflorite, löllingite, etc.	historic mine tailings (Nipissing mill site, Cart Lake, Crosswise Lake, Peterson Lake; Mill Creek, Farr Creek and the Bucke Park campground); pH 6.4-9.6	abandoned silver, cobalt, nickel deposits in the Cobalt area (closed 1989), Ontario, Canada (Percival et al. 2004)
cabrerite-annabergite, scorodite, HFA, HFO	nickeline, gersdorffite	processing uranium ores in the JEB mill; pH 2-8	uranium ore from the Athabasca basin of northern Saskatchewan, Canada (Langmuir et al. 1999; Mahoney et al. 2007)
beudantite, jarosite, HFO	pyrite, sphalerite, galena, *pyrrhotite, chalcopyrite*	mining waste; pH~2.5	abandoned Santa Lucia Pb-Zn mine (closed 1998), Cuba (Romero et al. 2010)
scorodite, beudantite	arsenopyrite, galena, berthierite	historic sulfide-rich tailings; pH ~ 3.2	processing of gold ore at La Petite Faye tailings (closed 1960s), France (Roussel et al. 2000; Courtin-Nomade et al. 2002)
HFO, beudantite	pyrite, sphalerite, galena	historic mine tailings; pH 2.9-4.0	abandoned Pb and Zn flotation tailings at "El Fraile" impoundments (closed 1973), Taxco, Mexico (Romero et al. 2007)
HFO, HFA, Bi-Pb-As-O phase, Fe-Zn-As-O phase	arsenopyrite, löllingite, sphalerite, wolframite, molybdenite, galena, bismuth	historic mine tailings; pH 5.3-10.7	abandoned Mount Pleasant W mine (closed 1985), New Brunswick, Canada (Petrunic et al. 2005, 2006, 2009)

HFA and less commonly arsenolite, bukovskýite, kaatialaite, kaňkite, parascorodite, sarmientite, tooeleite, and zýkaite. The direct identification of secondary arsenic minerals and phases as well as the observation of the surrounding environmental settings suggest that the mineralogical composition of secondary arsenic assemblages is controlled by several variables. The most important variable is the relative proportion of the primary sulfides, sulfarsenides, and arsenides in the waste material (Fig. 15). Three different waste types and their weathering products can be distinguished within the chemical system Fe(III)-As(V)-S(VI)-H$_2$O:

- *Fe+S >> As* — It is probably the most widespread scenario for As-bearing mining and processing wastes. These waste forms contain abundant pyrite and/or pyrite is the dominant primary As carrier. Note that these two conditions are not necessarily equal, as some coal spoils may be rich in pyrite but arsenic is supplied by minor amounts of sulfides such as orpiment or realgar. If Fe and S greatly exceed As, the dominant secondary As carriers are As-HFO, As-bearing Fe oxides, and As-jarosite. Such a mineralogical association develops from weathering solutions with high molar Fe/As ratio (> approx. 3) that is basically controlled by relatively lower proportion of primary As precursors in the waste with respect to pyrite or pyrrhotite (e.g., arsenian pyrite or complex ore mineralization with negligible amounts of arsenides or sulfarsenides). Representative environments for such secondary mineral assemblage are typically coal spoils (Murad and Rojík 2005; Kolker and Huggins 2007). The other examples include mine tailings where As was adsorbed or coprecipitated with HFO (e.g., McGregor and Blowes 2002; Petrunic and Al 2005; Lottermoser and Ashley 2006; Flores and Rubio 2010) and jarosite (e.g., Slowey et al. 2007; Flores and Rubio 2010).

- *Fe+As>>S* — Oxidative weathering of waste forms where arsenopyrite and/ or löllingite are the dominant ore minerals results in acidic, high-As weathering solutions with relatively low dissolved sulfate and low molar Fe/As ratio (< approx. 3). In these systems, Fe and As are abundant but the concentration of sulfur is low. Such conditions enable formation of ferric arsenates such as scorodite, HFA, kaňkite, and rarely kaatialaite. An unusual example of oxidation products of extremely acidic (pH ~ 1) and As-rich processing waste was provided by Filippi (2004). A stockpile of finely milled, almost pure arsenopyrite-löllingite concentrate has been oxidizing for ~20 years at the Přebuz abandoned mine, Czech Republic. The alteration products developed during that time precipitate in the succession: arsenolite → scorodite and native sulfur → very rare kaatialaite. In the less altered parts of the pile, arsenopyrite and löllingite grains are cemented by scorodite in association with minor arsenolite in the matrix (Fig. 16a). Kaatialaite was found on the surface of the pile together with powdery arsenolite (Fig. 16b). Highly altered parts of the pile do not include arsenolite and arsenopyrite in the matrix. The cavities after dissolved arsenopyrite grains are filled with native sulfur (Fig. 16c). No other study of secondary phases in naturally-altered extremely As-rich concentrates (Flemming et al. 2005; Salzsauer et al. 2005; Walker et al. 2009; Haffert and Craw 2010) has documented kaatialaite. This fact points at a necessity of very specific conditions needed for its formation (see the section *Arsenic Minerals in Underground Spaces*), which may include enormous amounts of dissolved As and pH close to 1 (Drahota and Filippi 2009). According to Zhu and Merkel (2001), scorodite can alter into kaatialaite under such extremely acidic conditions (Eqn. 8):

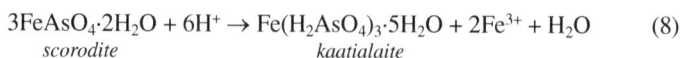

$$3FeAsO_4 \cdot 2H_2O + 6H^+ \rightarrow Fe(H_2AsO_4)_3 \cdot 5H_2O + 2Fe^{3+} + H_2O \qquad (8)$$
$$\underset{\textit{scorodite}}{} \qquad \underset{\textit{kaatialaite}}{}$$

Most of the similar studies identified scorodite, HFA, and As-HFO as the major secondary arsenic minerals in the tailings and mining wastes, and As-jarosite, kaňkite as the less widespread phases (Čech et al. 1976; Kato et al. 1984; Craw et al. 2000; Courtin-Nomade et al. 2002; Courtin-Nomade et al. 2003; Gieré et al. 2003; Ashley

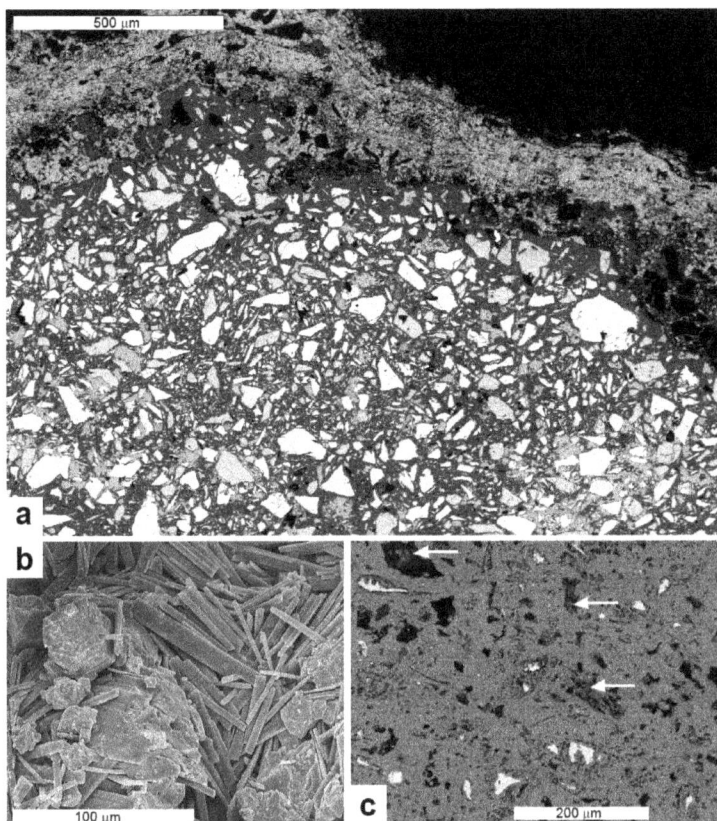

Figure 16. SEM images of naturally weathered arsenopyrite-löllingite concentrate from the processing plant at the Přebuz tin mine, Czech Republic. (a) BSE image of a slightly weathered concentrate. The white grains are arsenopyrite and löllingite (undistinguishable), the light grey surficial crusts and grains consist of arsenolite, the dark gray matrix is scorodite. (b) SE image of acicular crystals of kaatialaite with corroded crystals of arsenolite on the surface of the stockpile. (c) BSE image of highly weathered concentrate. The white grains are relics of arsenopyrite, the dark gray matrix consist of scorodite, some of black cavities are partially filled by native sulfur (marked by white arrows). [Used by permission of Elsevier Limited, from Filippi (2004) *Sci Total Environ*, Vol. 322, Fig. 5, pp. 277].

et al. 2004; Filippi et al. 2004; Ahn et al. 2005; Flemming et al. 2005; Lee et al. 2005; Salzsauer et al. 2005; Graupner et al. 2007; Filippi et al. 2009; Walker et al. 2009; Haffert and Craw 2010; Corriveau et al. 2011b; DeSisto et al. 2011; Kocourková et al. 2011; Parviainen et al. 2012). The early alteration products of arsenopyrite under acidic pH conditions most commonly correspond to an amorphous ferric arsenate (Filippi et al. 2009; Courtin-Nomade et al. 2010). This amorphous material (called pitticite, AFA or HFA and considered to be an amorphous equivalent of scorodite) has no well-defined stoichiometry (Dunn 1982), widely varying contents of major components, and substantial amounts of S, Si, Al, and P (e.g., Dunn 1982; Gieré et al. 2003; Salzsauler et al. 2005; Graupner et al. 2007; Walker et al. 2009). Maturation of this material under persistently acidic conditions leads to the formation of more crystalline scorodite (Paktunc et al. 2008).

Kaňkite has been described more often from mining wastes (Čech et al. 1976; Kato et al. 1984; Hyršl and Kaden 1992; Witzke and Hocker 1993; Pažout 2004; Filippi et al. 2009; Kocourková et al. 2011) than from tailings (Walker et al. 2009; Haffert and Craw 2010; Parviainen et al. 2012). It usually encrusts both the HFA and scorodite, forming botryoidal crusts consisting of lamellar crystals in the micro-scale. The deposition settings of kaňkite suggest that it forms preferentially in the larger pores of the waste where the moisture-vapor transport dominates after the infiltration events. Kaňkite is metastable with respect to scorodite and water or water vapor (Majzlan et al. 2012a) and it appears to persist for only a limited time. Perhaps the explanation is that an increase in pH and/or or Fe/As molar ratio in this system would cause ferric arsenates (scorodite and kaňkite) to transform to As-HFO and/or As-jarosite, respectively (Gieré et al. 2003; Courtin-Nomade et al. 2003; DeSisto et al. 2011). In most cases, As-jarosite and As-HFO represent the youngest minerals in the arsenic mineral associations.

- *High concentration of Fe, As, and S* — Waste with high contents of arsenopyrite and simultaneously copious pyrite or pyrrhotite may develop acidic weathering solutions which attain great concentrations of Fe and As (more than tens of mg L^{-1}), and, above all, enormous concentration of dissolved sulfate (usually tens of g L^{-1}). Characteristic compounds of the secondary arsenic mineral assemblage formed from these solutions include sulfo-arsenates (bukovskýite, zýkaite, and very rarely sarmientite), As-bearing sulfates, and S-bearing HFA. The sulfo-arsenate group of minerals is not widespread and their occurrence is restricted to a few sites. Bukovskýite and zýkaite were for the first time described from the Medieval mining wastes at Kaňk near Kutná Hora (Novák et al. 1967; Čech et al. 1978) where they occur with ferric arsenates (scorodite, kaňkite, HFA, parascorodite) and less commonly with sulfates (e.g., gypsum, jarosite) (Ondruš et al. 1999; Majzlan et al. 2012b). The nodules of bukovskýite crystallize here from Fe-, As-, and S-rich Si-Al gel-like matrix which occurs strictly in the clayey parts of the waste dump (Fig. 17), and is chemically close to the stoichiometry of bukovskýite (Majzlan et al. 2012b). The authors suppose that bukovskýite forms in acidic microenvironments rich in Fe^{3+}, SO_4^{2-}, and AsO_4^{3-}, high concentrations being maintained for a long time by the elements released from the associated clay minerals. However, the mechanisms which control the formation, chemical composition, and crystallization of the gels are unknown. In contrast, zýkaite precipitates strictly in open spaces of the rocky portions in the dumps, often in intimate association with kaňkite. Bukovskýite and zýkaite were later identified in other mining wastes (Hyršl and Kaden 1992; Witzke and Hocker 1993) and processing wastes (Gieré et al. 2003; Mains and Craw 2005; Márquez et al. 2006; Haffert and Craw 2010). Due to the fact that bukovkýite and zýkaite have never been found in an intimate association in the environments considered here, we suggest that they form in response to different mechanisms. In addition, precipitation of sarmientite and the sulfo-arsenite tooeleite was confirmed during bacterial oxidation of the arsenopyrite-bearing processing waste (Márquez et al. 2006). Tooeleite was also found in the mining waste dump of the U.S. Mine, Utah, where it encrusts both jarosite and scorodite (Cesbron and Williams 1992), and in the historic tailings from Montague gold district (Walker et al. 2009).

A different secondary arsenic mineral assemblage may also arise in this chemical system and was described from the processing waste at the Berikul abandoned gold mine, Russia (Gieré et al. 2003; Sidenko et al. 2005). The major secondary As carriers in the sulfide-rich processing waste (40-45 wt% of sulfides) were minerals of jarosite-beudantite solid solution and HFA with 5.4-5.8 wt% S (Fig. 18). Sulfur-bearing HFA accumulated in the hardpan located ~1 m below the surface, while the mineral proportion of jarosite increased upwards to reach approx. 19 wt% just above the hardpan and 50 wt% in the surface zone (Sidenko et al. 2005). Gieré et al. (2003) also found that sulfate minerals incorporated As (average 0.27, 0.87 and 0.64 wt% in copiapite, rhomboclase, and dietrichite, respectively). Secondary minerals precipitated

Figure 17. (a) Medieval mining waste dump at Kaňk, Czech Republic, rich in bukovskýite, scorodite, and parascorodite; (b) Detail of the exposed mining waste dump with scorodite-rich rock fragments *in situ*; (c) Detail of the clayey part of the mining waste dump with bright nodules of bukovskýite *in situ*.

in the sequence: gypsum → jarosite-beudantite → HFA. We can only speculate that the reason of the absence of crystalline sulfo-arsenate minerals is the low average annual temperature (−1.5 °C) that apparently suppresses evaporation processes in the waste dump.

System Ca-Fe(III)-As(V)-H$_2$O. If a waste rock contains a larger amount of calcite, dolomite, or ankerite to buffer the acidity released from sulfide weathering, the waste and associated aqueous phase will be circumneutral and Ca is another dominant component. The most important sink for As under circumneutral conditions is sorption to HFO or Fe oxides (e.g., goethite). The literature is replete with examples; a few of the many recent pertinent studies are those by Moldovan et al. (2003), Morin and Calas (2006), Majzlan et al. (2007), and references therein. Under low Fe/As ratios, dissolved As may combine with Ca^{2+} to form Ca arsenates. Arseniosiderite and yukonite are the most common representative minerals of this system (Paktunc et al. 2003, 2004; Walker et al. 2009; Foster et al. 2011; Rytuba et al. 2011). X-ray absorption spectroscopy suggested that arseniosiderite is a significant reservoir of oxidized As in submerged and dry tailings at the Lava Cap gold Mine Superfund Site (Foster et al. 2011) and Ruth gold-silver mine (Rytuba et al. 2011), both in California. Foster et al.

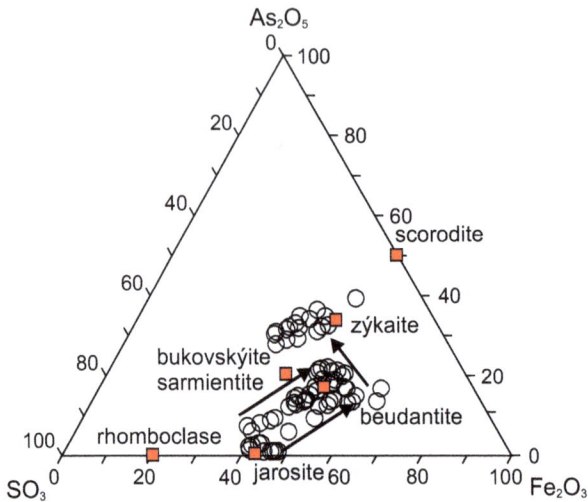

Figure 18. A triangular plot (in mol%) of the theoretical composition of minerals (solid squares) and the analyzed compositions of jarosite-beudantite solid solutions and amorphous ferric arsenates (HFA) from the high-sulfide Berikul tailings, Russia. The arrows show the sequence of phase precipitation. Redrawn from Gieré et al. (2003). [Used by permission of Elsevier Limited, from Gieré et al. (2003) *Appl Geochem*, Vol. 18, Fig. 5, p. 1354]

(2011) suggested that the pore waters in the Lost Lake tailings and water in the tailings pond, both adjacent to Lava Cap Mine, are saturated with respect to arseniosiderite. These waters are alkaline (pH 7.0-9.4), with Fe/As molar ratios between 0.01 and 13 (average 6.7) and with the dissolved Ca concentrations up to 65.5 mg L^{-1}. These conditions are similar to those observed in the pore water of arseniosiderite-containing regolith at the Mokrsko As-geochemical anomaly (Drahota et al. 2009). Arseniosiderite has been also proposed to form nanoparticles during co-precipitation of Ca and As along with HFO in the ore from the Ketza River gold mine, based on spectroscopic and microprobe data (Paktunc et al. 2004). Interestingly enough, a replacement of scorodite or pharmacosiderite by arseniosiderite and yukonite was observed in the same place (Paktunc et al. 2003, 2004), suggesting a similar sequence of secondary arsenic minerals to that observed in the Mokrsko As-geochemical anomaly (Fig. 10). Arseniosiderite is commonly associated with yukonite (Paktunc et al. 2003, 2004; Rytuba et al. 2011). Yukonite occurs most often as intensive fractured, gel-like aggregates often coprecipitated with HFO and other phases. Yukonite was also found in the tailings of a gold mining operation in Nova Scotia, Canada by Walker et al. (2009), Meunier et al. (2010), and Corriveau et al. (2011b). Yukonite occurred here as a rims on arsenopyrite or as irregular deposits on Fe-rich carbonates. Yukonite was also reported as a coprecipitate with HFO, HFA, and pharmacosiderite (Walker et al. 2009).

The waste environments of the Ca-rich chemical system discussed above were almost free of gypsum; in the absence of the SO_4^{2-} ions, Ca-Fe arsenates are expected to form (Riveros et al. 2001). Conversely, in relatively concentrated sulfate solutions, arseniosiderite and yukonite are not the principal As carriers; this observation has been made by many authors (Swash and Monhemius 1995; Mahoney et al. 2007; Pantuzzo and Ciminelli 2010). Although the presence of Ca-Fe arsenate minerals in mine wastes is generally common, Ca arsenates are seldomly reported from mining and processing wastes. Pharmacolite, however, was documented in the oxidized mining waste dump at the Mole River arsenic mine (Ashley and Lottermoser 1999). Another example of pharmacolite was reported from a pile of ore concentrate or furnace residue in the Nipissing Hill, Cobalt area, Ontario (Percival et al. 2004). Pharmacolite in the white crusts within this pile was associated with gypsum, weilite, and brassite. The presence of Ca arsenates in mine wastes was more frequently indicated indirectly, for example by selective extraction (Pichler et al. 2000) and thermodynamic modeling (Donahue and Hendry 2003; Shaw et al. 2011).

System (Co, Cu, Ni, Pb, Zn)-Fe(III)-As(V)-S(VI)-H₂O. Sulfides and sulfosalts of transition metal ions (e.g., Co, Cu, Ni, Zn) and large divalent cations (e.g., Pb) commonly accompany primary arsenic minerals in the waste rock. These elements readily bond to arsenate tetrahedra to form countless arsenate minerals, hence, the complex system (Co, Cu, Ni, Pb, Zn)-Fe(III)-As(V)-S(VI)-H₂O. Despite the great number of arsenate minerals that can originate in this system (see the section *Crystal Chemistry of Arsenic*), these compounds usually play a negligible role in sequestering As in mining wastes and tailings. If present, they occur as thin rims or powdery aggregates and crusts on the primary sulfides (Horák 2000; Petrunic et al. 2006, 2009) rather than large masses and cements within the waste. The explanation for the relatively low abundance of these compounds is the combined constraints of the relatively low dissolved concentrations of Co, Cu, Ni, Pb, and Zn in the waste and relatively high solubility of the majority of arsenate minerals which contain these metals as the main components (Drahota and Filippi 2009; Nordstrom et al. 2014, this volume). Since these arsenate minerals are usually of low environmental significance, their evidence in mine wastes is commonly restricted to studies focused only on the identification of minerals, with little or no information about their formation and stability. The number of relevant studies concerning secondary arsenates from this system is thus low. The most important secondary arsenic mineral from this system, repeatedly documented in mine wastes as an important sink of As, is beudantite (Roussel et al. 2000; Romero et al. 2007, 2010; Kocourková et al. 2011). For example, in La Petite Faye tailings in France, beudantite is suggested to be the most important secondary sink of As along with scorodite (Roussel et al. 2000; Courtin-Nomade et al. 2002). Substantial trapping of As by the jarosite-beudantite solid solution was documented in the tailings impoundments in western Siberia (Russia), Mexico, and Cuba (Gieré et al. 2003; Romero et al. 2007, 2010, respectively). Beside the jarosite-beudantite solid solutions, the precipitation of erythrite or annabergite could also exert a significant control over the dissolved concentration of As in some processing wastes (Jambor and Dutrizac 1995). Annabergite has been identified in association with scorodite and HFO in uranium mine tailings (Langmuir et al. 1999; Mahoney et al. 2007) and it has also been associated with the prevailing erythrite in the abandoned tailings after the historic silver production at Cobalt, Canada (Percival et al. 2004).

Detailed mineralogical studies, especially transmission electron microscopy, reveal variable As mineralogy down at nanoscale dimensions. Petrunic et al. (2006) analyzed the As minerals in the oxidation zone of W-mine tailings. They found that oxidation of arsenopyrite led to the formation of a Fe-As-O phase, probably corresponding to HFA (Fig. 19a). As the oxidation of arsenopyrite and other sulfides proceeded, unidentified crystalline Bi-Pb-As-O phase (Fig. 19b,c,d) and wulfenite (PbMoO₄) precipitated in the voids within the coatings, suggesting that these compounds precipitated from the local pore waters. In another study, Petrunic et al. (2009) found an amorphous Fe-Zn-As-O phase that dominated the secondary coatings on the sphalerite (ZnS) and tennantite grains. These results show that many specific arsenic phases may be restricted to microenvironments and thus require nanoscale characterization.

Metallurgical wastes. Non-ferrous metallurgy produces waste forms that can contain high levels of As bound in a variety of phases (Riveros et al. 2001). The ore processing technologies leading to the separation of desired metals/metalloids include hydrometallurgy, pyrometallurgy, and electrometallurgy.

Hydrometallurgy (mainly used for Al, Au, U and to some extent Cu, Ni, P, Zn) involves the use of solvents to dissolve and separate the element of interest (Lottermoser 2003). For example, gold is commonly retrieved from the ores by leaching with alkaline cyanide solutions,

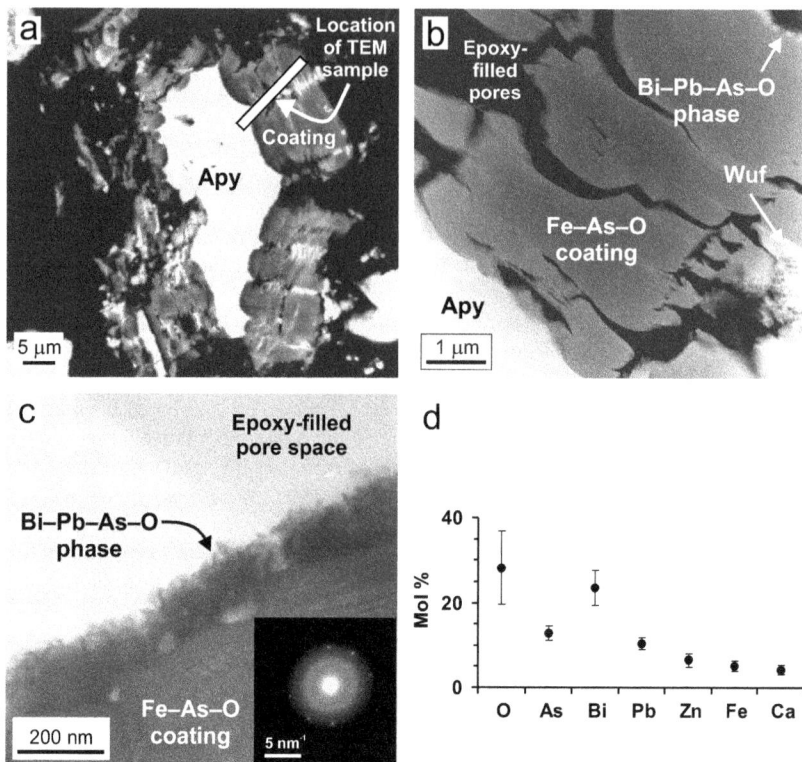

Figure 19. (a) BSE image of an oxidized arsenopyrite (Apy) grain from W mine tailings (Mount Pleasant tungsten mine, Canada). The rectangle represents the location where a thin sample for TEM (transmission electron microscope) analysis was prepared using a focused-ion beam instrument. (b) High-angle annular dark-field image showing the location of a Bi-Pb-As-O phase and wulfenite (Wuf) which were identified in the secondary coating. (c) TEM image of the Bi-Pb-As-O phase at the outer margin of the secondary coating where the Fe-As-O coating has ruptured. The inset is a selected-area electron diffraction pattern demonstrating the crystalline nature of the phase. (d) Average values of STEM-EDS spot analyses ($n = 11$) of the Bi-Pb-As-O phase, using a nominal probe size of 5 nm. [Used by permission of Elsevier Limited, from Petrunic et al. (2006) *Appl Geochem*, Vol. 21, Figs. 2b,f,11, pp. 1262, 1270]

whereas the process chemicals for uranium extraction include acidic or alkaline solvents and oxidants (Lottermoser 2003). As a result of the extraction processes used to recover gold or uranium, significant concentrations of As (up to thousands mg L^{-1}) usually dissolve in the raffinate (Mahoney et al. 2007). After neutralization, the slurry of residual minerals (either insoluble, such as silicates, or minerals which dissolve too slowly, e.g., arsenides) and newly precipitated phases (arsenates and As-bearing HFO and Fe oxides) is pumped into the tailings facilities for long-term disposal.

Pyrometallurgy (used for the extraction of Cu, Fe, Ni, Pb, Sn, Zn) and electrometallurgy (Al, Zn) are based on the breakdown of the structure of an ore mineral by heat and electricity, respectively (Lottermoser 2003). During pyrometallurgical operations, such as roasting, smelting, and converting, most of the As is volatilized as arsenic trioxide (Eqn. 9) and As(III) emerges from the furnaces either as flue dust (as As_2O_3) or vapor (Davis et al. 1996; Ettler et al. 2005b; Morales et al. 2010). Thus, a significant amount of arsenic trioxide may enter the atmosphere and only some of the arsenic trioxide remains trapped by precipitation along the flue gas cleaning system or as atmospheric fallout (Smith 1977; Ashley and Lottermoser 1999;

Haffert and Craw 2008; Power et al. 2009):

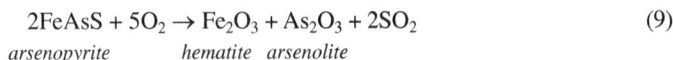

$$2FeAsS + 5O_2 \rightarrow Fe_2O_3 + As_2O_3 + 2SO_2 \tag{9}$$
$$\text{arsenopyrite} \qquad \text{hematite arsenolite}$$

Modern smelting plants collect the arsenical compounds in electrostatic precipitators or wet gas scrubbers (Riveros et al. 2001) and, if relatively pure, a limited amount of arsenic trioxide-bearing dust was sold for the manufacture of pesticides and wood preservatives. These uses are, however, no longer profitable.

Other types of waste forms resulting from metallurgy are slags (wastes resulting from solidification of a silicate melt) and mattes (wastes resulting from solidification of a sulfide melt) from the recovery of base metals by fusion in smelting furnaces. These two types of waste, in particular slags, are produced in large amounts. Arsenic occurs primarily in the sulfide, arsenide, and metallic and glass phases (Kucha et al. 1996; Ettler et al. 2001, 2009a,c; Ettler and Johan 2003; Lottermoser 2002; Puziewicz et al. 2007; Vítková et al. 2010). Slags are generally considered to be chemically inert, because As and other potential contaminants are encapsulated in relatively low-solubility compounds such as silicates, oxides, and glass. Recent studies have shown, however, that As could be extensively released from slag waste to soil and waters by natural weathering processes or incorporated into the structure of newly formed secondary phases (Lottermoser 2002, 2005; Ettler et al. 2003a, 2009c; Sáez et al. 2003; Bril et al. 2008).

Roasting products. High levels of As in mineral concentrates have a detrimental effect on smelting of ores. Traditionally, before the ores were sent to smelters to extract and refine the metals of interest (e.g., Ag, Au, Cu, Zn), the arsenical impurities were often removed. The method of choice was usually heating in roasting ovens and allowing the volatile As to enter the atmosphere or condense in specially constructed collection chambers. The temperature of roasting was empirically kept at approximately 543-593 °C: a temperature range where the most efficient sublimation of As occurs (Klinck et al. 2005). With the increasing evidence of lethal properties of As on nearby ecosystems coupled with growing demand for As, steps were taken to collect the As, either for disposal or for use. Because the sublimation temperature of As_2O_3 is 193 °C, the flue gas was initially passed through a zigzag cooling chamber (so-called "lambreth flues") where arsenic trioxide condensed. This As-rich soot was then scraped off the walls and either stockpiled, sold, or processed on site. An amazing use of such soot was described from Styrian Alps (Austria) where the peasants were sprinkling the arsenic trioxide just like salt on bread and were called the "arsenic-eaters" (Emsley 2005). A description of older cooling chambers (active until early 1950's) from southwest England, designed to trap the condensed arsenic oxides on their walls, is briefly given by Klinck et al. (2005) and Power et al. (2009). Some of these historical arsenic calciner buildings coated with secondary mineral efflorescence are still preserved in southwest England (Klinck et al. 2005; Power et al. 2009) and Australia (Ashley and Lottermoser 1999). Power et al. (2009) studied the efflorescence mineralogy from the masonry of eleven calciner sites throughout Cornwall and Devon in England. They found very diverse range of efflorescent minerals (Fig. 20) including sulfates, carbonates, arsenates, and oxides that have formed from the reaction between the furnace condensates and the bricks, mortar, and limestone (addition of limestone into the flue helped reduce the SO_2 emissions). By far the most abundant and widespread efflorescent mineral is gypsum that contains from below detection (nominally 0.06 wt% As) up to ~7 wt% As (Fig. 21), although two analyses give As concentrations as high as 10.7 and 22.4 wt% As. The ability of arsenate to substitute for sulfate in the gypsum structure conforms to the experimental results of Fernández-Martínez et al. (2008) who found that substitution is favored under alkaline condition, when the $HAsO_4^{2-}$ species predominates. Alternating concentric layers of gypsum and calcium arsenates (Fig. 22a) have been attributed to variations of dry and wet seasons. During increasingly dry weather, gypsum precipitated first, followed by pharmacolite, then haidingerite, and finally weilite (Eqns. 10 and 11). Other abundant arsenic minerals include scorodite and arsenolite (Fig. 22b). Bukovskýite

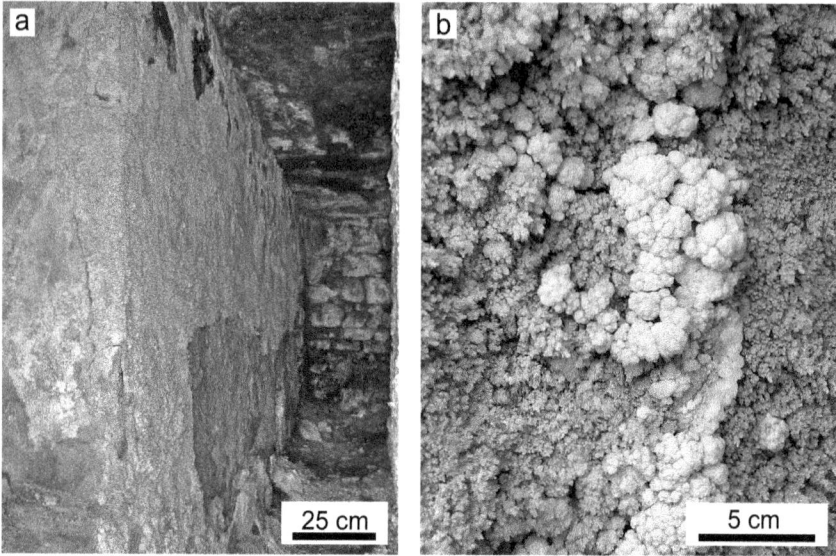

Figure 20. Arsenic-rich efflorescence in a historic calciner building from southeast England. (a) White calcium arsenates (pharmacolite, haidingerite, and weilite) coat the internal walls of the lambreth plasterwork. (b) White colloform masses of pharmacolite grow on grey translucent gypsum aggregates on the external brickwork wall of the calciner. [Used by permission of The Mineralogical Society, from Power et al. (2009) *Mineral Mag*, Vol. 73, Figs. 4c,d, p. 31]

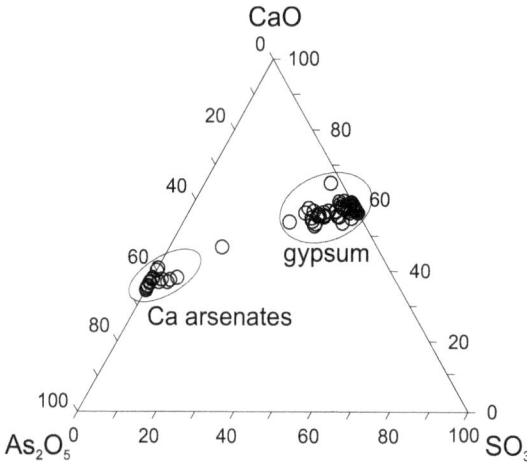

Figure 21. A triangular plot of the chemical compositions of the As-bearing gypsum (up to ~ 7 wt% of As) and calcium arsenates (up to ~ 4 wt% of S) from efflorescence on the walls of a historic arsenical calciner building in SW England. Redrawn from Power et al. (2009). [Used by permission of The Mineralogical Society, from Power et al. (2009) *Mineral Mag*, Vol. 73, Fig. 8, p. 39]

has been found at one site as pale-yellow crusts that coat white efflorescent growths of scorodite and gypsum. Power et al. (2009) also detected a wide variety of unidentified As-bearing minerals, including aluminum-arsenic-sulfate phase, As-rich fluorine-bearing phase (Fig. 22c), and As-bearing potassium aluminum phase. Secondary mineral efflorescence on the brickwork of

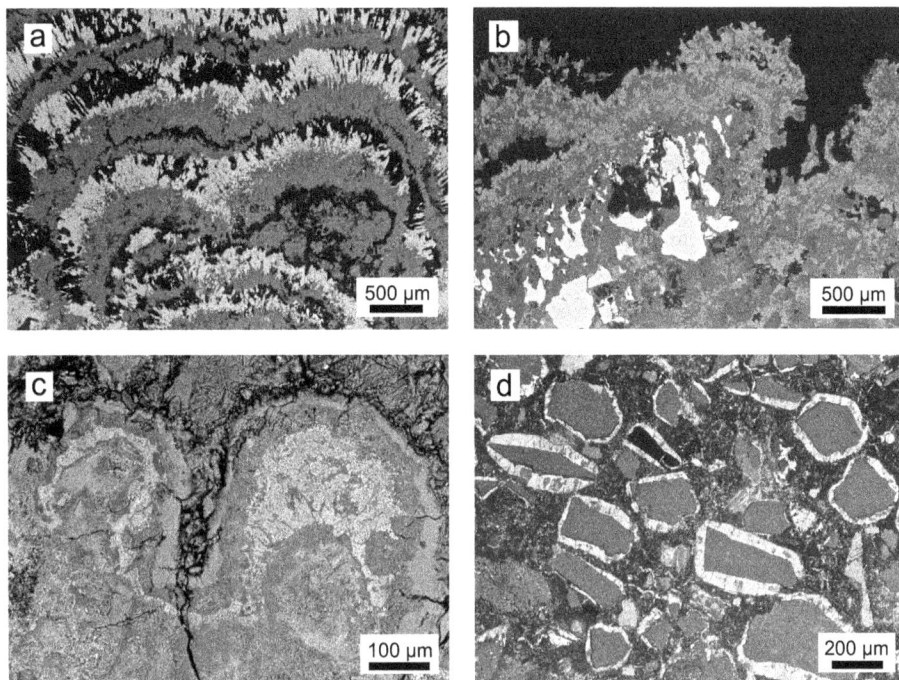

Figure 22. BSE images of efflorescence from a historic calciner building from southeast England. (a) Alternating growth of calcium arsenate (white) and gypsum (grey). (b) Arsenolite grains (white) enclosed in calcium arsenate (pale grey) and gypsum (mid-grey). (c) Concentric zones of As-rich (white) and relatively As-poor (dark grey) Ca-Na-K-Al-F phase. (d) Calcium arsenate rims (white) surrounding quartz (dark grey) from plasterwork. [Used by permission of The Mineralogical Society, from Power et al. (2009) *Mineral Mag*, Vol. 73, Figs. 5c,e,f,h, p. 34]

the furnace and condensing flue at the Mole River mine, northern New South Wales, Australia, is composed of gypsum, pharmacolite, and krautite (Ashley and Lottermoser 1999). The material collected within the flue is composed of both arsenic trioxide polymorphs (arsenolite and minor claudetite), along with gypsum and pharmacolite.

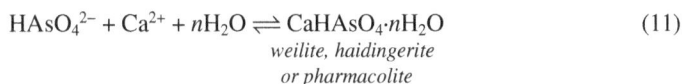

$$H_2AsO_4^- + Ca^{2+} + nH_2O \rightleftharpoons CaHAsO_4 \cdot nH_2O + H^+ \tag{10}$$

$$HAsO_4^{2-} + Ca^{2+} + nH_2O \rightleftharpoons CaHAsO_4 \cdot nH_2O \tag{11}$$
$$\textit{weilite, haidingerite}$$
$$\textit{or pharmacolite}$$

During current smelting operations, vaporized As is precipitated as either arsenic trioxide or arsenic-metal oxides, and is therefore concentrated within the flue dust. Roasting and smelting of pyrite and arsenopyrite also produces As-bearing Fe oxides (hematite and maghemite) that may contain up to 18.6 wt% of As (Paktunc et al. 2006). Furthermore, Walker et al. (2005) and Fawcett and Jamieson (2011) found that As associated with roaster-derived Fe oxides is hosted in both oxidation states of III and V, thus pointing at the complex nature of the solid-gas phase reactions occurring in the roaster. Most of these raw materials originate from primary cobalt smelting, although some also come from primary and secondary lead smelting, secondary aluminum operations, and other smelters (EPA 1998).

Smelting flue dust. Arsenic trioxides are typical As-bearing species in the flue gas of the smelting furnaces. For example, the mineralogical composition of the copper smelter flue dusts containing up to 19 wt% of As indicated the presence of oxidized (arsenolite, claudetite, complex metal-arsenic oxides containing Cu, Fe and Pb in varying ratios) and sulfidic (enargite) forms of arsenic (Davis et al. 1996; Shih and Lin 2003; Morales et al. 2010). Another example of the As-rich complex copper smelting residue (19.5 wt% As) in flue dust sampled from the waste stockpile in Japan indicated the occurrence of arsenolite, scorodite, and schultenite, along with chalcopyrite ($CuFeS_2$), gypsum, and quartz (Shibayama et al. 2010). However, scorodite and schultenite represent probably secondary compounds formed from arsenolite alteration in the stockpile. Arsenic-bearing flue dusts from smelting and refining operations are usually recovered by different strategies (Leist et al. 2000; Sullivan et al. 2010). Residues from copper smelting and refining processes can be stabilized/solidified by the addition of lime or cement, which results in the formation of calcium hydrogen arsenite ($CaHAsO_3$) (Dutré and Vandecasteele 1998; Vandecasteele et al. 2002), $NaCaAsO_4 \cdot 7.5H_2O$ (Akhter et al. 1997) or poorly soluble calcium arsenate [$Ca_3(AsO_4)_2$] under alkaline conditions (Vandecasteele et al. 2002). The flue dust from secondary lead smelting is usually recycled for additional metallurgical recovery of lead and other valuable elements. Ettler et al. (2005b) studied the mineralogy in a secondary lead smelter which processes old scrap (mainly used car batteries; e.g., Orisakwe et al. 2002). The flue gases are allowed to condense, are collected and sintered in a rotary furnace (300-500 °C). The product contains 4090 mg kg^{-1} of As and Ettler et al. (2005b) identified traces of arsenolite in such recycled material. Ettler et al. (2010) suggested that the solid speciation of As in such materials is in general controlled by arsenolite. Recently, mimetite was reported as the major As-controlling phase in weathered air-pollution control (APC) residue from secondary lead smelting (Ettler et al. 2010, 2012). Mimetite is thought to originate during the interaction of APC with water or after exposure in soils.

Complete removal of As from any of the dust-collecting chambers was difficult to achieve and thus arsenic trioxide dust was commonly left lying around historical arsenic processing sites (Mains and Craw 2005; Haffert and Craw 2008, 2009). Prohibition Mill and Snowy River Battery sites in the vicinity of Waiuta, New Zealand, processed gold ore (crushing, flotation, cyanidation, and roasting) at different times from the same major quartz vein rich in arsenopyrite with only minor pyrite (Haffert and Craw 2008, 2009). Arsenic concentrations in processing residues at the Prohibition Mill site and the Snowy River Battery site were very high, reaching up to 40 and 26 wt% of As, respectively (Haffert and Craw 2008). The dominant mineral was arsenolite, with some minor hematite. Arsenolite had partially dissolved in the wet climate under slightly reducing conditions and maintained the dissolved As concentrations at high levels (up to 52 mg L^{-1} at places, Haffert and Craw 2008). Scorodite cements the waste material and encapsulates altered arsenolite crystals. Oxidation of the arsenite oxyanions, which are released during arsenolite dissolution, caused acidification of the substrate down to pH 3 (Haffert and Craw 2008). Once all the available arsenolite had dissolved, the solubility-controlling phase was scorodite. A different situation has been encountered in the Golden Point Battery, a small abandoned processing plant in the Otago mining area, New Zealand (Mains and Craw 2005). A sulfide concentrate of pyrite and arsenopyrite was inefficiently roasted and the waste was deposited off the site. The oxidation of pyrite relics caused the chemical remobilization and acidification of the tailings (pH 2.2-3) that led to cementation of the outermost volume of the tailings by newly formed ferrihydrite and scorodite. Bukovskýite occurred only locally as cement and veinlets in the interior of the pile (Mains and Craw 2005).

Slags and mattes. Arsenic-bearing slags and mattes mostly result from smelting of non-ferrous ores and ore concentrates. The pyrometallurgy of copper and lead, for instance, is designed to remove the Fe and Si from the gangue minerals by separating them as iron-silicate slags which float on the top of the melted charge in the smelting furnace. Slags almost always contain a mixture of phases including various metal oxides (equivalents of spinel-group minerals, maghemite, and hematite), silicates (equivalents of olivine-group minerals), sulfide/metallic inclusions or droplets, and glass. Compared to mining waste and mine tailings, well-smelted slags usually do not contain primary ore and gangue minerals (except for relics of quartz or rare sulfides) and are composed of phases which solidified directly from the furnace melt during cooling. Most of the slags derived from non-ferrous metallurgy are rich in various metals such as Co, Cu, Ni, Pb, and Zn, and also some metalloids such as As and Sb (e.g., Ettler et al. 2001, 2009a,c; Parsons et al. 2001; Cappuyns et al. 2002; Sáez et al. 2003; Piatak et al. 2004; Puziewicz et al. 2007; Navarro et al. 2008; Piatak and Seal 2010; Vítková et al. 2010). Arsenic concentrations in non-ferrous smelting slags attain usually tens to hundreds of mg kg^{-1}. Nevertheless, very high As concentrations were documented in some localities such as in slags and mattes produced between 1963-1970 from copper and lead smelting in the Tsumeb smelter, Namibia (up to 7.6 wt% of As, Ettler et al. 2009c; Bowell 2014, this volume) and in slags produced between 1897-1970 from an arsenic refinery at the industrial site Bocholt-Reppel, Belgium (up to 16.4 wt% of As, Cappuyns et al. 2002). The most commonly observed primary arsenic phases are arsenides and intermetallic compounds belonging to the Fe-As (e.g., löllingite), Cu-As (e.g., koutekite, α-domeykite), and Ni-As (e.g., nickeline) systems (Table 8). Sulfide mattes contain up to 0.15 wt% As with similar phases to those present in slags (Ettler et al. 2009b) (Table 8) and the low As concentration in the mattes and slags in comparison to the composition of the primary ores most likely reflects the tendency of arsenic to fractionate into the gaseous phase during heat treatment. The substitutions of Co, Pb, Sb, and other elements in the structures of the above-mentioned arsenides were commonly observed. Conversely, substitution of As has been documented in the structure of pyrite (< 2.5 wt%), pyrrhotite (< 0.14 wt%), breithauptite (< 6 wt%), cuprostibite (< 2.3 wt%), iron metal (< 1.0 wt%), lead metal (< 4.8 wt%) and antimony (< 3.8 wt%) (Kucha et al. 1996; Ettler et al. 2003a, 2009a; Navarro et al. 2008; Sivry et al. 2010). Arsenides, arsenic sulfides, and intermetallic compounds usually form irregular or rounded matte inclusions trapped within the silicate matrix. The skeletal crystals and symplectite textures of metal- and metalloid-bearing phases in slags and mattes indicate rapid cooling of the melt. In addition, rapid cooling of the silicate slag melts enriched in metals and metalloids (e.g., granulation under a water stream) generally leads to the formation of silicate glass with unusually high levels of many elements including As (e.g., up to 2.99 wt% of As_2O_3 in slags from Namibia; Ettler et al. 2009c).

Many studies on weathering and leaching behavior of smelting slags (e.g., Ettler et al. 2005a, 2009c; Ganne et al. 2006; Navarro et al. 2008; Rosado et al. 2008) have shown that the potential release of arsenic and metals to their surroundings depends not only on the quantities of pollutants present in the dumps, but also on the stability of the phases holding these pollutants. Sulfides, arsenides, and glasses react more rapidly than silicates and oxide phases under conditions prevailing in the dumps (Ettler et al. 2001; Parsons et al. 2001). The extent of weathering and release of arsenic and other contaminants is thus related to the porosity and texture of the waste form and also to the degree to which the reactive contaminant-bearing glass, sulfides, and arsenides are exposed or encapsulated. Further important factors which determine the rate and degree of slag alteration and secondary arsenic minerals formation are physico-chemical properties of the weathering solutions (e.g., pH, Eh, CO_2 partial pressure) and residence time. For example, the presence of various associations of secondary phases within the slag waste indicates a wide range of pH, oxidizing conditions, sulfate activity, and other chemical factors in the weathering solutions (e.g., Lottermoser 2005; Bril et al. 2008; Ettler et al. 2009c). At some smelting sites, climatic conditions are of prime importance for the weathering and release of metals and metalloids. Seasonal variations between dry and wet periods can lead to the dissolution of the primary As-bearing phases and formation of secondary efflorescent minerals which are almost always highly soluble (Lottermoser 2005; Bril et al. 2008; Ettler et al. 2009c). As the secondary weathering products commonly include soluble As-bearing sulfates (Lottermoser 2005) and arsenates (Sáez et al. 2003; Bril et al. 2008; Ettler et al. 2009c), significant amounts of As can be flushed out from the slag during rainy seasons. Furthermore, precipitation of As-bearing phases due to evaporation emphasizes the

Table 8. Primary arsenic-bearing phases from smelting slags and mattes.

Mineral or phase	Source and locality	Reference
Fe_2As	copper and cobalt smelting (1931-2009) at Tsumeb, Namibia	Ettler et al. (2009c)
Fe_2As	lead smelting (1850-1970s) at Lhota near Příbram, Czech Rep.	Ettler and Johan (2003), Ettler et al. (2009b)
FeAs	copper and cobalt smelting (1931-2009) at Tsumeb, Namibia	Ettler et al. (2009c)
FeAs	lead and silver smelting (1880-1935) at Clayton, Idaho, USA	Piatak et al. (2004)
FeAs	copper, lead, and silver smelting (13th -14th century) at Massa Marittima, Tuscany, Italy	Manasse and Mellini (2002)
FeAs	zinc and lead smelting (from Medieval ages until the beginning of 20th century) at Plombières, Belgium	Kucha et al. (1996)
löllingite	lead smelting (1850-1970s) at Lhota near Příbram, Czech Rep.	Ettler and Johan (2003), Ettler et al. (2001, 2009b)
$(Fe, Co)_2As$	copper and cobalt smelting (1931-2009) at Nkana, Copperbelt province, Zambia	Vítková et al. (2010)
α-Fe with As, Sb metal with As	lead smelting (1850-1970s) at Lhota near Příbram, Czech Rep.	Ettler et al. (2009b)
α-domeykite, Cu_3As	copper and cobalt smelting (1931-2009) at Tsumeb, Namibia	Ettler et al. (2009c)
α-domeykite, Cu_3As	lead and silver smelting (1880-1935) at Clayton, Idaho, USA	Piatak et al. (2004)
koutekite, Cu_5As_2	lead smelting (1850-1970s) at Lhota near Příbram, Czech Rep.	Ettler and Johan (2003), Ettler et al. (2001, 2009b)
nickeline, NiAs	lead smelting (1850-1970s) at Lhota near Příbram, Czech Rep.	Ettler and Johan (2003), Ettler et al. (2009b)
$PbSnAs_2$ $Pb_3Sn(As,Sb)_4$ $Pb_2Sn_3(As,Sb)_2$	lead smelting (1850-1970s) at Lhota near Příbram, Czech Rep.	Ettler and Johan (2003), Ettler et al. (2009b)
Fe-Co-As-Cu-Ni alloys	copper and cobalt smelting (1931-2009) at Nkana, Copperbelt province, Zambia	Vítková et al. (2010)
As-Sn-Pb alloy	copper, lead, and silver smelting (between the end of 19th century and 1940's) at Queensland, Australia	Lottermoser (2002)
Fe-As-Cu alloy	copper, lead, and silver smelting (between the end of 19th century and 1940's) at Queensland, Australia	Lottermoser (2002)
arsenian pyrite (up to 2.5 wt% of As)	zinc smelting (1842-1987) at Viviez near Aveyron, France	Sivry et al. (2010)

importance of hot and dry periods. The influence of these factors on the behavior of potentially hazardous elements has been recently reviewed by Kierczak et al. (2010). When toxic elements are released from the primary waste material, they can be retained by secondary phases or become available to seepage water. Secondary As-bearing phases occur as efflorescence on slag surfaces, cement precipitates at the seepage points at the base of the dump, fillings within slag pores, and coatings in fractures. Secondary As-bearing phases identified in slags are mostly represented by base-metal arsenates and As-rich HFO and Fe oxides (Table 9).

Oxidizing conditions allow precipitation of HFO and Fe oxides, the most common As-bearing secondary phases observed both at the field sites (Ettler et al. 2001, 2003a; Parsons et al. 2001) and during leaching laboratory experiments (Ettler et al. 2003b, 2005a). Newly formed HFO adsorbs As strongly (Fig. 23) and this phenomenon seems to be the most common mechanism that controls As mobility in the slag wastes (Kucha et al. 1996; Ettler et al. 2001;

Table 9. Secondary As-bearing alteration products from smelting slags.

Mineral or phase	Source and locality	Reference
lead-chloride arsenite, $Pb_5(As^{3+}O_3)Cl_7$	silver smelting (500-200 B.C.E.) at Lavrion, Punta Zeza area, Greece	Siidra et al. (2011)
scorodite	copper, lead, and silver smelting (between the end of 19th century and 1940's) at Queensland, Australia	Lottermoser (2002)
scorodite	nickel and cobalt smelting (1897-1970) at Bocholt-Reppel, Flanders, Belgium	Cappuyns et al. (2002)
lammerite	lead and copper smelting (1907-1948) at Tsumeb, Namibia	Ettler et al. (2009c)
olivenite	lead and copper smelting (1907-1948) at Tsumeb, Namibia	Ettler et al. (2009c)
conichalcite	copper smelting (2873-2274 B.C.E.) at Cabezo Juré near Huelva, Spain	Sáez et al. (2003)
lavendulan	lead and copper smelting (1907-1948) at Tsumeb, Namibia	Ettler et al. (2009c)
bayldonite	lead and copper smelting (1907-1948) at Tsumeb, Namibia)	Ettler et al. (2009c)
mimetite	zinc smelting (1858-1974) at Świętochłowice, Upper Silesia, Poland	Bril et al. (2008)
hedyphane	zinc smelting (1858-1974) at Świętochłowice, Upper Silesia, Poland	Bril et al. (2008)
calcium-lead arsenate, $(Pb,Ca,Fe)_3(AsO_4)_2 \cdot H_2O$	lead and copper smelting (1907-1948) at Tsumeb, Namibia	Ettler et al. (2009c)
legrandite	zinc smelting (1858-1974) at Świętochłowice, Upper Silesia, Poland	Bril et al. (2008)
ojuelaite	zinc smelting (1858-1974) at Świętochłowice, Upper Silesia, Poland)	Bril et al. (2008)
iron-zinc arsenate, $(Zn,Fe)_x(AsO_4)_3(OH)_y$	zinc smelting (1858-1974) at Świętochłowice, Upper Silesia, Poland)	Bril et al. (2008)
HFO (up to 4.6 wt% of As)	lead smelting at Lhota near Příbram, Czech Rep.	Ettler et al. (2003a)

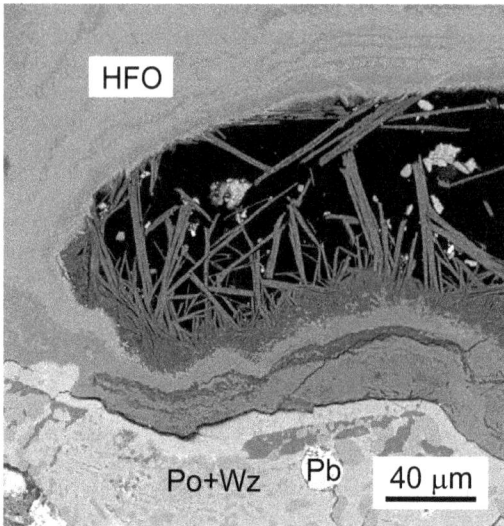

Figure 23. A BSE image of secondary HFO with up to 4.63 wt% of As developed in the cavity during the alteration of a matte from lead smelting near Příbram, Czech Republic. Abbreviations: Pb: native lead; Po: pyrrhotite, Wz: wurtzite. [Used by permission of Elsevier SAS, from Ettler et al. (2003a) *C R Geosci*, Fig. 2b, p. 1017]

2003a,b, 2005a; Parsons et al. 2001; Lottermoser 2002). The coprecipitation and/or substitution of As within the crystal structures of phases was also documented in the sulfate efflorescence in the slag waste at the historical Río Tinto smelter site, southwest Spain (Lottermoser 2005). Mineral efflorescence occurs mainly at seepage points at the base of the slag dump and includes calcium and magnesium sulfates (blödite: $Na_2Mg(SO_4)_2 \cdot 4H_2O$, gypsum, epsomite: $MgSO_4 \cdot 7H_2O$, hexahydrite: $MgSO_4 \cdot 6H_2O$) and mixed ferrous/ferric hydrated sulfates (copiapite, römerite: $Fe^{2+}Fe^{3+}_2(SO_4)_4 \cdot 14H_2O$). The highest concentration of As (up to 2790 mg kg^{-1}, mean: 1280 mg kg^{-1}) has been found in the copiapite-rich mixtures, whereas calcium- and magnesium-rich sulfates contain less than 100 mg kg^{-1} of As (mean: 25 mg kg^{-1}) (Lottermoser 2005). Preferential arsenic storage in copiapite group minerals is consistent with As distribution in other localities such as Iron Mountain, California (Jamieson et al. 2005) and Peña del Hierro, southeast Spain (Romero et al. 2006).

Under oxidizing conditions, formation of various secondary arsenate and rare arsenite phases has been documented in several historical waste dumps. Lottermoser (2002) identified scorodite in association with As-bearing Fe oxides, metal sulfates, carbonates, and chlorides in the slag dumps in north Queensland, Australia. Scorodite was found in the altered slag buried in the soils at the industrial site of Reppel-Bocholdt, northern Belgium (Cappuyns et al. 2002).

Inefficient pyrometallurgical processes and metal recovery technologies, prevalent until the early 20th century, produced smelting slags with elevated metal contents (particularly of the metals Cu, Pb and Zn where several wt% has been observed) (e.g., Ettler et al. 2001, 2009c; Parsons et al. 2001; Cappuyns et al. 2002; Lottermoser 2002; Sáez et al. 2003; Piatak et al. 2004; Puziewicz et al. 2007; Piatak and Seal 2010). Subsequently, chemical alteration of these slags has often resulted in formation of copper, lead, and zinc arsenates or arsenites (Table 9). Abundant associations of such secondary alteration products have been reported from the slag dumps in Świętochłowice, southern Poland (Bril et al. 2008). Microcrystalline aggregates and crystals of mimetite, ojeulaite, hedyphane, legrandite, and unknown iron-zinc arsenate in association with secondary sulfates, carbonates, oxides, and silicates were found as millimeter-thick coatings at the surfaces of the slag fragments. Mimetite is the most common arsenate, often associated with barite in the spinel-, melilite-, and willemite-bearing slags, locally associated with iron-zinc arsenate. However, precise characterization of some secondary alteration products, e.g., ojeulaite, hedyphane, and legrandite is often complicated due to their small size (Bril et al. 2008). At the Cabezo Juré site near Huelva, Spain, ancient metallurgical activity produced copper-rich slags. Supergene alteration of the slag led to formation of conichalcite and other secondary phases including copper carbonates and iron-manganese hydroxides (Sáez et al. 2003). Other examples include As-, Cu-, and Pb-rich historical slags from Tsumeb copper smelting in Namibia (Ettler et al. 2009c). The oldest Tsumeb slag (1907-1948) contains calcium-lead arsenate enclosed in the slag matrix, chemically corresponding to tsumcorite and implicates significant and deep alteration of the slags on the dumps. These slags are also commonly coated by green copper and copper-lead arsenates corresponding to lammerite and bayldonite, respectively (Fig. 24a). Other Cu-bearing secondary precipitates were documented for the most Cu-rich slag type (1963-1970). Green olivenite aggregates cover altered parts of the slag surface rich in copper arsenides and sulfides (Fig. 24b), whereas blue crusts of lavendulan are commonly associated with gypsum (Fig. 24c). Ettler et al. (2009c) suggested that the variety of secondary minerals indicates differences in the chemical microenvironments on the slag surfaces. For example, lammerite precipitates from acidic conditions (pH < 3), whereas olivenite and bayldonite are generally formed under slightly acidic to neutral conditions (Ettler et al. 2009c). The historic sites in the Lavrion mining district in Greece are famous for the presence of more than 100 secondary phases (Gelaude et al. 1996) that have formed during the last 2500 years by the alteration of metal-rich slags. Siidra et al. (2011) described two lead chloride

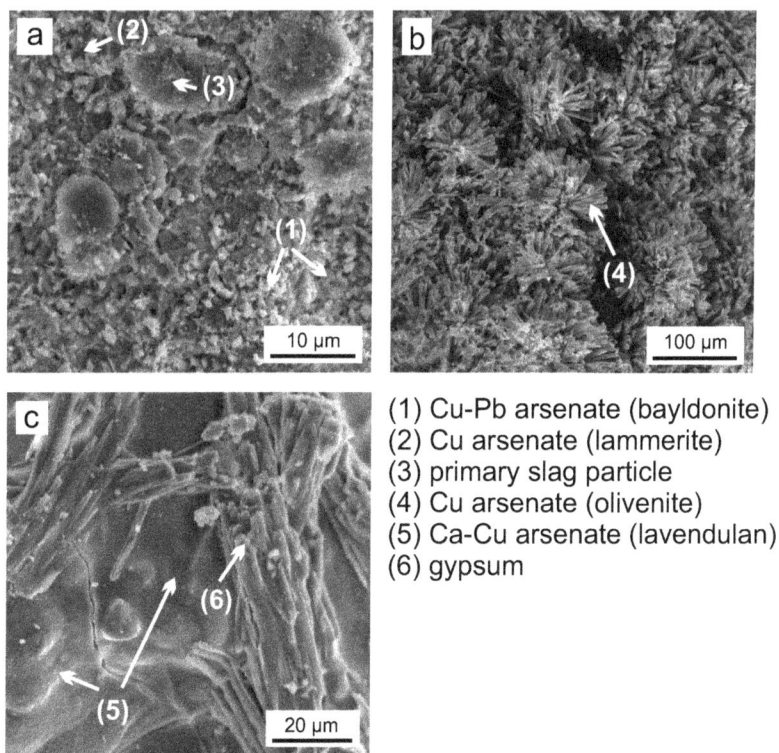

(1) Cu-Pb arsenate (bayldonite)
(2) Cu arsenate (lammerite)
(3) primary slag particle
(4) Cu arsenate (olivenite)
(5) Ca-Cu arsenate (lavendulan)
(6) gypsum

Figure 24. SE images of secondary precipitates developed on surface of slags from copper smelting in Tsumeb, Namibia. The phase identifications are based on EDS and XRD results. Modified after Ettler et al. (2009c). [Used by permission of Elsevier Limited, from Ettler et al. (2009c) *Appl Geochem*, Vol. 24, Figs. 4a,c,d, p. 10]

arsenites from the shoreline at the Punta Zeza area in this area. These phases formed by reaction of seawater with Pb-rich glassy material in the slag.

Leached residues. Cyanide leaching is currently the standard process to extract gold from low-grade precious metal ores. As a result, exceptionally large quantities of cyanide-bearing waste are produced from this highly efficient hydrometallurgical method. Currently, the hydrometallurgical cyanidation wastes are deposited in the form of heap leach residues and tailings. Gold is often locked as inclusions in sulfides or is substituted in the structure in sulfides, especially in arsenopyrite. These refractory ores require pretreatment before cyanide leaching (Riveros et al. 2001). Apart from roasting operations thoroughly discussed above, new processes have been developed for the treatment of refractory ores, including pressure oxidation, biooxidation, whole ore roasting, ultrafine grinding, nitric acid oxidation, and fine milling combined with low-pressure oxidation (Riveros et al. 2001). Because most of the refractory gold ores are arsenical, autoclave systems have a distinct environmental advantage in that, unlike roasting, sulfur and arsenic cannot simply escape. Acid pressure oxidation employs temperatures above the melting point of sulfur (119 °C), generally in the range of 200 to 250 °C and high oxygen overpressures up to 2000 kPa (Papangelakis and Demopoulos 1990). In pressure oxidation and biooxidation, arsenic is transferred mostly to ferric arsenates and sulfo-arsenates (Riveros et al. 2001). The mineralogical transformations with respect to arsenic during these processes have been described by Márquez et al. (2006). They identified arsenical jarosite and subordinate sarmientite, zýkaite, bukovskýite, and tooeleite in the arsenopyrite-rich flotation ore concentrate from São Bento gold deposit (Quadrilátero Ferrífero, Brasil) that was treated by a combination of bacterial and pressure oxidation.

Sulfidic ores are leached with high-pH solution (pH ~ 10.3) which introduces a significant neutralizing factor. The leached ore may contain carbonate minerals and therefore, the piles and tailings may be strongly alkaline once ore leaching has ceased (e.g. the Ketza River mine tailings in south-central Yukon in Canada; Paktunc et al. 2003, 2004). At the Ketza River mine, arsenical Fe oxides were the dominant species in the oxidized gold ore and process tailings with lesser quantities of scorodite, HFA, arseniosiderite, probably yukonite, and traces of pharmacosiderite and arsenical jarosite. Close association of Fe oxides with variable content of As (from ~ 2 to 22 wt%) and ferric arsenates in successive bands suggested precipitation and coprecipitation in close intervals. Scorodite was replaced by arseniosiderite and yukonite and precipitation of arseniosiderite appeared to postdate yukonite as evidenced by the occurrence of arseniosiderite around yukonite spheroids. The As levels in Fe oxides varies as function of calcium content and As/Ca molar ratio agrees roughly with those typical for arseniosiderite and yukonite. Paktunc et al. (2004) determined that the As atoms in their samples have Ca neighbors at a distance of 4.14-4.17 Å. They interpreted this result in terms of the precipitation of arseniosiderite-like nanoclusters in their samples rather than simple adsorption of Ca onto Fe oxides. There were no apparent differences in the arsenic mineralogy of the gold ore and tailings samples after cyanidation. Nevertheless, sustained alkaline conditions during the cyanidation process promoted significant As desorption and Paktunc et al. (2004) suggested that ferric arsenates, calcium-ferric arsenates, and As-rich Fe oxides will not be stable in the tailings impoundment. Most of these findings (i.e., evolution and precipitation sequence of secondary arsenic minerals and the distribution of Ca and As content in Fe oxides) are consistent with observations from the natural chemical system K-Ba±Ca-Fe(III)-As(V)-H$_2$O in saprolite and soil environments (see the section *Arsenic Minerals in Soil and Fluvial Systems*).

In conventional uranium extraction, the ore is first crushed or powdered and then leached by acid or alkaline solvents. The leaching operations are applied to heap leach piles, or more commonly, under controlled conditions in a hydrometallurgical plant. Several mineralogical studies have showed that HFO are the stable solid phases in the neutral to alkaline As-rich uranium mine tailings that control the solubility of As (Moldovan et al. 2003; Moldovan and Hendry 2005). The laboratory-scale studies that involved lime neutralization of highly acidic hydrometallurgical solutions (pH ≤ 1.5) containing As, Ni, and Fe showed the formation of HFA (precipitated especially in the pH range 2-4) with smaller amounts of annabergite/cabrerite (near pH ~ 5-6) and As-HFO (Langmuir et al. 1999; Mahoney et al. 2007).

Acid mine drainage. Many As minerals form by weathering of mine waste material or as efflorescence on mine walls, but only a few of these have been related to acid mine drainage (AMD). AMD is generated by oxidation of pyrite, marcasite, pyrrhotite and occasionally other sulfides (Bigham and Nordstrom 2000; Jambor 2003; Nordstrom 2011). The acidity generated in this process may cause leaching and dissolution of other minerals. Progressive neutralization of acid waters and microbial oxidation of Fe(II) frequently leads to precipitation of iron sulfates and HFO that can trap As by adsorption or coprecipitation (Williams 2001). However, when As/Fe ratios are high enough in AMD, other arsenic minerals can precipitate (Juillot et al. 1999; Morin et al. 2003; Triantafyllidis and Skarpelis 2006).

In the previous paragraphs, case studies on acidic waste forms were mentioned. In many instances, ores or metallurgical products may contain the AMD-generating sulfides and pH may be depressed locally. Hence, some overlap of this and previous sections is inevitable. In the following paragraphs, we describe a few case studies of As minerals from AMD systems. Here, we restrict our attention to those minerals which precipitated directly from AMD waters which were physically decoupled from their sources, that is, AMD waters which are the discharges or outflows from pit walls, mining wastes, or bedrock.

Reigous creek near Gard (France) drains tailings with arsenian pyrite of the Carnoulès lead-zinc mine. Acidic waters (pH 2.7-3.4) contain high loads of dissolved Fe (up to 2230 mg L^{-1}), As (up to 262 mg L^{-1}) and SO$_4^{2-}$ (up to 6700 mg L^{-1}). Over the first 40 meters from the mine discharge, biologically-induced precipitation of nanocrystalline tooeleite and/or amorphous Fe(III)-As(III) hydroxosulfates was observed in the bed of the creek (Casiot et al. 2003; Morin et al. 2003). Further downstream, these phases are replaced by schwertmannite and ferrihydrite (Egal et al. 2010). Field and laboratory studies conducted with isolated bacterial strains from the Carnoulès AMD suggest that the As(III)- and As(V)-bearing phases form via distinct mechanisms. Oxidation of Fe(II) by *Acidithiobacillus ferrooxidans*, which is unable to oxidize As(III), leads to the formation of the amorphous Fe(III)-As(III) hydroxosulfates, tooeleite, or As(III)-bearing schwertmannite (Casiot et al. 2003; Duquesne et al. 2003; Morin et al. 2003; Egal et al. 2010). In contrast, oxidation of both Fe(II) and As(III), for example by *Thiomonas* strains, leads to the formation of amorphous Fe(III)-As(V) hydroxosulfates (Morin et al. 2003; Egal et al. 2010). Triantafyllidis and Skarpelis (2006) found scorodite and bukovskýite in association with jarosite and other sulfate minerals in the yellowish unconsolidated mineral precipitate that has formed on the floor of the acid pit lake (pH ~3) of the Kirki deposit in northeastern Greece.

Another example is provided by an old industrial site located on limestone near Marseille in southern France (Juillot et al. 1999). Crusts, films, or particles of calcium and calcium-magnesium secondary arsenates have formed as a result of the interaction of the runoff waters with As-bearing sulfidic waste and the limestones of the bedrock. Runoff waters are initially strongly acidified (down to pH 2.1) and enriched in SO$_4^{2-}$ (up to 10.5 g L^{-1}), Fe, and As (up to 236 mg L^{-1}) as a result of oxidation of pyrite, arsenopyrite (Eqns. 3 and 4) and dissolution of arsenolite (Eqn. 12).

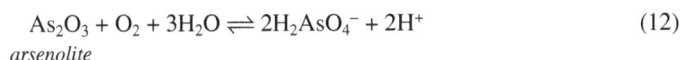

$$As_2O_3 + O_2 + 3H_2O \rightleftharpoons 2H_2AsO_4^- + 2H^+ \qquad (12)$$
arsenolite

The subsequent interaction with the underlying limestone causes pH increase to circumneutral values and also raises the aqueous Ca (up to 546 mg L^{-1}) and Mg (up to 165 mg L^{-1}) concentrations. As the pH rises, the following sequence of secondary arsenic minerals has been observed at this site and also in the laboratory: weilite, haidingerite, and pharmacolite (Fig. 25)

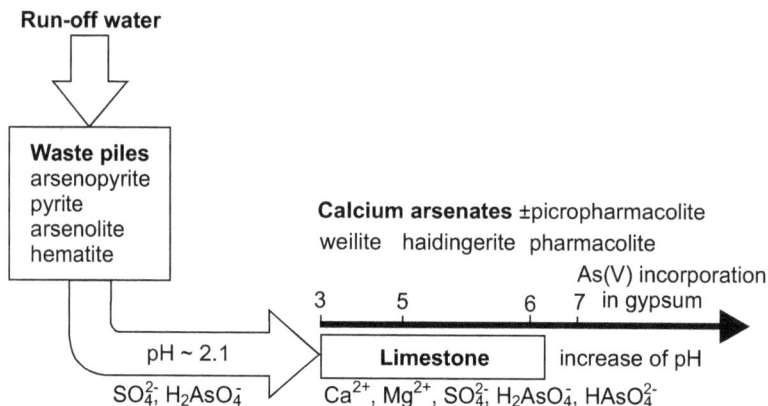

Figure 25. A schematic mineralogical and geochemical model of As at an industrial site near Marseille in southern France. Modified after Juillot et al. (1999).

(Pierrot 1964; Juillot et al. 1999). Picropharmacolite was also reported as a minor phase in one solid sample. Depending on the pH and As(V) speciation ($H_2AsO_4^-$ or $HAsO_4^{2-}$; pK_{a2}=6.99, Nordstrom and Archer 2003), precipitation of calcium arsenates could occur through equations 10 and 11 (Juillot et al. 1999). In all solid samples, gypsum was always present in association with the calcium arsenates. Electron microprobe analyses of mineral precipitates suggested the substitution of SO_4^{2-} ions by AsO_4^{3-} is favored under the neutral to alkaline conditions (Fernández-Martínez et al. 2008; Fig. 25). Such conditions occurred here after prolonged interaction of AMD with limestone (Juillot et al. 1999).

Arsenic minerals in underground spaces

Arsenic mineralogy in underground mines. Exposed mineralized wall rock and rock piles in underground adits, drifts, levels, and shafts are often characterized by a multitude of secondary minerals, including those with arsenic. The range of secondary arsenic minerals in underground spaces is comparable to that found in mining and processing wastes (see the section *Arsenic Minerals in Mine Wastes*) because their formation is basically controlled by the same primary precursors. There are some important features, however, which do differentiate the formation and stability of secondary arsenic minerals in the underground mines from those in mining and processing wastes.

First, the underground mine spaces offer a long-term stable environment with high relative humidity in the range of 90-100% and a constant temperature (Palmer 2007). The air and water temperature in mines may be significantly higher (up to or more than 50 °C), either in deep mines owing to the terrestrial geothermal gradient (e.g., Wanger et al. 2006) or in mines where exothermic oxidation reactions generate a large amount of heat (e.g., Nordstrom and Alpers 1999). The stable climate conditions (air humidity, temperature, and moisture) and additional physico-chemical factors (see below) provide a specific environment that allows for the formation and preservation of rare, metastable, or water-soluble secondary As phases (Figs. 26, 27).

The mineral zýkaite is a prominent example of such a metastable phase. This hydrated sulfo-arsenate occurs only rarely at the surface but is a relatively frequent weathering product of arsenopyrite in the moist underground mine environment, if the carbonate content in the rocks is low and concentrations of SO_4, As, and Fe are high. Zýkaite is known from piles of mineralized rock fragments in several abandoned mines of variable age (Kratochvíl 1960; Řídkošil and Slavíček 1985; Witzke and Hocker 1993; Ondruš et al. 1997a,b; Siuda 2004; Filippi et al. 2014). The associated minerals are always other hydrated Fe(III) (sulfo-)arsenates kaňkite,

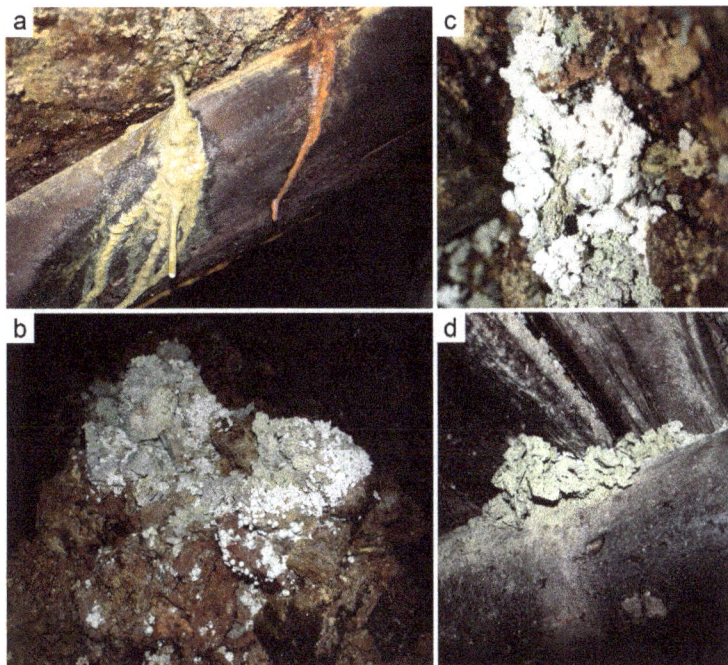

Figure 26. Examples of As-bearing weathering products from the underground mine environment, all sites are located in Czech Republic: (a) stalactites and curtains of amorphous hydrous ferric arsenate (HFA) (approx. 10 cm long; left) and hydrous ferric oxides (HFO) (right) on the timbering in the Giftkies mine, Jáchymov ore district; (b) crusts of kaňkite and scorodite accompanied by white globular aggregates of zýkaite, coatings of HFO and powdery jarosite aggregates. Approximately 1 m wide pile of rock fragments mineralized by weathering arsenopyrite on the floor of an abandoned adit in the Mikulov mine near Teplice; (c) identical secondary mineral association as in (b) (approx. 10 cm long detail of the zýkaite globular aggregates on kaňkite, scorodite, and HFA masses, surrounded by HFO coatings and powdery jarosite aggregates) from the Giftkies mine, Jáchymov ore district; (d) kaňkite crusts (approx. 15 cm in longer dimension) precipitating on the timbering situated beneath the pile of weathered arsenopyrite-rich ore, the Giftkies mine, Jáchymov ore district.

scorodite, HFA, and less commonly bukovskýite. If pyrite, pyrrhotite, or other Fe sulfides are abundant, these minerals are accompanied by jarosite and usually also gypsum because many rocks contain sufficient amounts of Ca. Zýkaite typically forms white to greenish, gooey, soft, and plastic nodules up to several centimeters in diameter which become brittle after drying (Fig. 26b,c). The formation and persistence of zýkaite in underground mines is dependent on the high and stable humidity, a condition which is rarely fulfilled in most mining dumps.

A sharp geochemical boundary between the mine atmosphere and the adit walls enables massive oxidation of the sulfides and sulfarsenides and the precipitation of newly formed arsenic minerals from aqueous solutions (Table 10). These solutions could be derived from (i) condensation of water on the walls, or (ii) waters percolating through the host rocks. The condensed water may evolve into extremely acidic solutions with simple chemical composition, influenced only by the primary mineralogy at a particular site. In contrast, the waters percolating through the mineralized rock (mostly along fractures) can have complex chemical composition dictated by all reactive minerals which came into the contact with this water.

A unique paragenesis consisting of As(III) oxides and hydrated arsenates was found in the Svornost mine of the hydrothermal vein deposit Jáchymov, Czech Republic. Historically, the

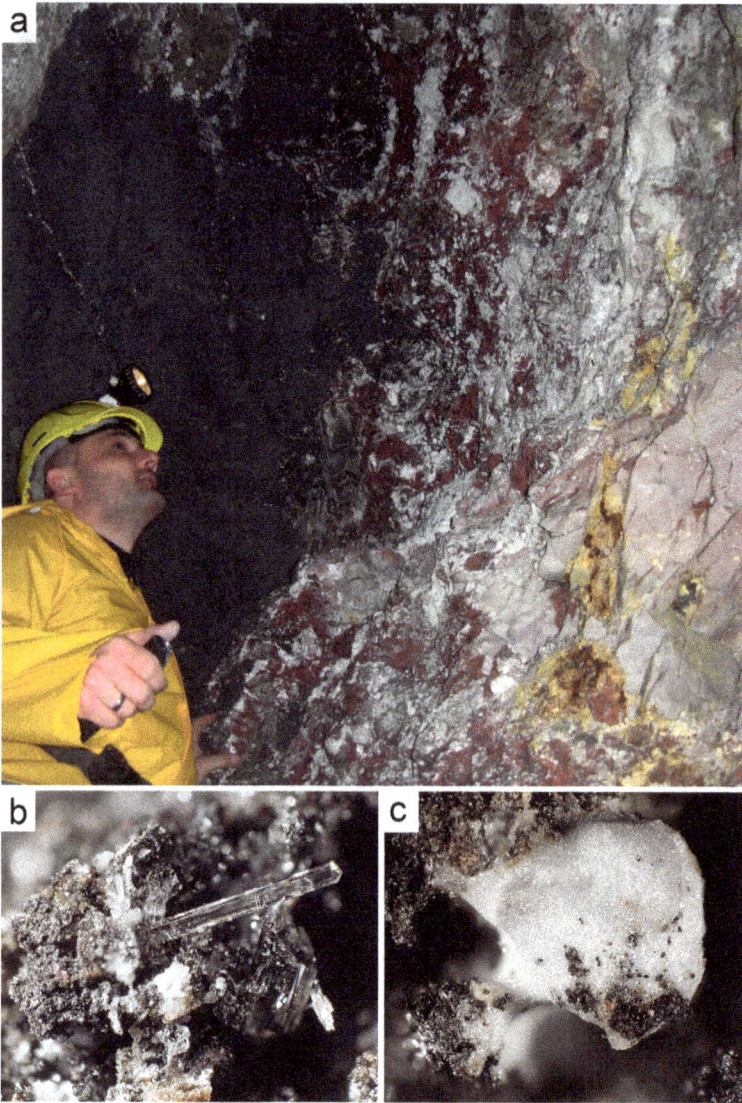

Figure 27. (a) A lens of native As (black) enclosed in the Geschieber quartz vein on the 10th floor of the Svornost mine, Jáchymov ore district, Czech Republic. Contact of the lens with mineralogically very simple gangue represents an interesting environment with abundance of concentrated arsenic acid and extremely high concentrations of arsenic. (b) Up to approx. 2 mm large claudetite crystals on weathered native As; (c) approx. 2 mm large aggregate of very fine kaatialaite crystals on native As. Photographs (b) and (c) by P. Škácha.

ores were rich in and targeted for Ag, As, Co, Ni, Bi, and U. A chemically simple weathering solution formed due to the condensation of air moisture on a ~2 m wide and ~5 m long lens of native As enclosed in the Geschieber quartz vein (Ondruš et al. 1997a,b) (Fig. 27). The absence of carbonates and arsenides and the unusual scarcity of sulfides (marcasite and pyrite) in this lens caused the production of concentrated arsenic acid, similarly to the production of sulfuric acid in the usual acid mine drainage derived from pyrite or pyrrhotite ores. The extremely acidic waters with colossal As concentrations (Majzlan et al. 2014) induce precipitation of colorful blooms of arsenolite, claudetite, kaatialaite, and locally also scorodite accompanied by rösslerite and brassite. This paragenesis originates from the condensing water underground and may be locally chemically more diverse, reflecting local variations of the primary mineralogy. Ondruš et al. (2002) described major scorodite, arsenolite, kaňkite, annabergite, and köttigite, along with rare parascorodite and vajdakite, from fractures in proximity of this lens. They report that the aqueous solutions there were mostly concentrated sulfuric acid with a high As concentration.

In an underground mine environment, the secondary As minerals usually occur as crusts and aggregates overgrowing the decomposed primary ore and the host rocks. In the open spaces of underground mines, ferric arsenates and HFO precipitate from As-rich aqueous solutions as speleothem-like formations or cement the rock fragments. Unique curly speleothem crusts composed of green kaňkite, curtains and straw stalactites consisting of greyish-green to brownish HFA, accompanied by rusty brown HFO stalactites are found on the timbering in the Giftkies mine, Czech Republic (Filippi et al. 2014, Fig. 26a, d). Dark to reddish brown HFA stalagmites with waxy luster up to 3 cm in height were described from the Wilhelm Mine, Stara Góra deposit, Poland (Siuda 2004). Foshag and Clinton (1927) described rich botryoidal aggregates of HFA at adit bottom and walls or rarely coatings on melanterite ($FeSO_4 \cdot 7H_2O$) stalactites inside the White Caps Mine in Nevada, USA. All these unusual HFA occurrences in speleothem-like formations localized at some distance from the arsenic source clearly demonstrate short-distance As transport (certainly in units of cm to m) *via* the mine waters (Siuda et al. 2004; Filippi et al. 2014). Similarly as in the caves, the formation of speleothems in mines is caused by the slow, continual and focused supply of mineralized solutions. Incremental growth of the speleothems is caused by the permanent evaporation of the mineralized solutions leading to temporary supersaturation.

Although there are no firm data to support this claim, we believe that the amorphous Fe^{3+}-As^{5+} phases are the most common As-bearing secondary products in the underground mine environment. One of these compounds is the HFA, mentioned above. Another one is As-HFO which precipitates, for example, from waters with high molar Fe/As ratios in the abandoned mines of the Barewood district (New Zealand) in places where arsenopyrite and abundant pyrite oxidize simultaneously (Craw et al. 2000).

Pharmacosiderite of two different origins has been documented from abandoned adits at the Barewood mining area under the perennially damp conditions. The first type precipitated on and near the mineralized quartz veins over the last century since the mines were abandoned (Craw et al. 2000). This pharmacosiderite was observed to precipitate as thin encrustations from shallow groundwater (1-14 mg K L^{-1}; <10 mg SO_4 L^{-1}) pervading the host schists (Craw and Nelson 2000). The second type formed by crystallization of As-HFO for more than 50 years in mine waters with pH 7-8, ~23 mg As L^{-1}, <10 mg K L^{-1}, and molar Fe/As ratio >1 (Haffert et al. 2010).

Arsenates of Ca and Ca-Mg (or Ca-Mn) form in those mines where the aqueous solutions are significantly affected by carbonate-rich ore veins or host rocks. With their characteristic acicular morphology and the radial aggregates ("fuzzy balls"), these minerals include pharmacolite, haidingerite, picropharmacolite, weilite, and brassite (Foshag and Clinton 1927; Pierrot 1964; Řídkošil and Slavíček 1985). In samples from the hydrothermal (Ag-Bi-Co-Ni-As-U) Jáchymov ore deposit (Czech Republic), Ondruš et al. (1997a,b,c) determined the succession and environmental conditions of formation of these minerals. The earliest mineral in the Ca-Mg arsenate paragenesis is picropharmacolite, found in places with rich water supply. Picropharmacolite often occurs in mixture with an unnamed Ca-Mg-AsO_4-H_2O phase, an early weathering product of the arsenide ore. Pharmacolite follows in the sequence, sometimes mixed with weilite. Therefore, weilite is assumed to form by dehydration of pharmacolite. Rösslerite

Table 10. Selected examples of some rich secondary arsenic mineral parageneses and associated minerals identified on samples from the underground mine spaces. (CZ = Czech Republic).

Parageneses/major associated minerals	Major precursor minerals, settings	Deposit type, locality, and reference
Parageneses of oxides and arsenates of the Fe(III)-As(III,V)-SO₄-H₂O system		
arsenolite, claudetite, kaatialaite ±scorodite/ geminite, sulfur	weathering of native arsenic, ±pyrite, ±marcasite; pH < 1, very high As concentrations, very simple element composition (absence of carbonates and arsenides)	hydrothermal Ag, As, Co, Ni, Bi + U deposit, Geschieber vein in the Svornost mine at the Jáchymov ore district, CZ (Ondruš et al. 1997a,b; Majzlan et al. 2014)
Parageneses of (sulfo-)arsenates of the Fe(III)-As(III,V)-SO₄-H₂O system		
scorodite, kaňkite, HFA, ±zýkaite/ pharmacolite, gypsum	weathering of simple primary mineralization with arsenopyrite, ±löllingite, ±pyrite, low pH (3.7–3.9 for the Stara Góra mine)	base metals + Sn-W, Brand-Erbisdorf, Freiberg, and Schwarzenberg ore districts, Germany (Witzke and Hocker 1993); Jáchymov, CZ (Ondruš et al. 1997a,b); Pb, Zn, Cu deposit, Stara Góra mine, Poland (Siuda 2004); As deposit, Giftkies mine, Jáchymov, CZ (Filippi et al. 2014)
Parageneses of the Ca, Ca-Mg arsenates		
picropharmacolite, pharmacolite, weilite, rösslerite, brassite, haidingerite, sainfeldite/ fluckite, villyaellenite, talmessite, annabergite, köttigite, parasymplesite, erythrite, hörnesite, and other minerals	weathering of arsenopyrite, ±pyrite or native arsenic, ±arsenides of Ni, Co, Fe or tennantite in veins with abundant carbonates	Jáchymov, CZ (Ondruš et al. 1997a,b)
Examples of mixed paragenesis of arsenates of Pb, Cu, Zn, Ni, Co, and other elements		
mottramite-duftite, bayldonite, zeunerite, mimetite, hidalgoite, mixite	weathering of arsenides of Ni, Co, Fe, native arsenic, galena, uraninite	Jáchymov, CZ (Ondruš et al. 1997a,b)
mimetite, beudantite–segnitite, pharmacosiderite, scorodite, cesàrolite, carminite	weathering of galena, arsenopyrite, cassiterite, wolframite in an abandoned adit 10 m below the surface, in a steep, tectonically crushed and altered zone	Sn-W + base metals, Stříbrná adit, Krupka, CZ (Sejkora et al. 2009)

Table 10. *continued*

Parageneses/major associated minerals	Major precursor minerals, settings	Deposit type, locality, and reference
lemanskiite/lammerite, olivenite, mansfieldite, senarmontite, a mineral of the crandallite group, and other minerals	weathering of sulfides, mostly enargite, in a quartz vein	Abundancia gold mine, El Guanaco mining district, Antofagasta Province, Chile (Ondruš et al. 2006)
attikaite, arsenocrandalite, arsenogoyazite, conichalcite, olivenite, philipsbornite, azurite, malachite, carminite, beudantite, goethite, quartz, and allophane	supergene zone of sulfide-quartz vein	base metal deposit, Christina Mine, Lavrion District, Attiki Prefecture, Greece (Chukanov et al. 2007a)
rruffite, mansfieldite, alumopharmacosiderite, conichalcite, metazeunerite, barahonaite-(Al), and barite	weathering of pyrite, stibnite, native arsenic, quartz, barite, and ankerite–siderite in volcaniclastic rocks	Cu-As Maria Catalina mine, Tierra Amarilla, Chile (Yang et al. 2011)
zdeněkite/anglesite, an unknown mineral of the tsumcorite group, geminite, and olivenite	weathering of tennantite and covellite on approx. 100-year-old adit wall damped by oxidizing low-temperature solutions	a "red-bed" type Cu-Pb deposit, Cap Garonne mine, Var, France (Chiappero et al. 1995)
vajdakite, scorodite and arsenolite/parascorodite, kaňkite, annabergite, köttigite	weathering of nickelskutterudite, löllingite, and native As, pyrite and marcasite in an concentrated sulfuric acid-rich environment and a high concentration of As_2O_3	Geschieber vein in the Svornost mine, Jáchymov, CZ (Ondruš et al. 2002)
segnitite, beudantite, carminite, mimetite, bayldonite, agardite-(Y) / goethite, coronadite, alunite-jarosite minerals	weathering of galena and sphalerite, minor chalcopyrite, and arsenopyrite-löllingite in a quartz-rich gangue	Broken Hill, New South Wales, Australia (Birch et al. 1992)
veselovskýite / strashimirite, amorphous Cu arsenate	not stated	Geister vein in the Rovnost mine, Jáchymov, CZ (Sejkora et al. 2010a)
ondrušite / lindackerite, geminite, lavendulan, slavkovite, strashimirite, olivenite, picropharmacolite and köttigite	weathering of tennantite and chalcopyrite disseminated in a quartz gangue, strongly acidic conditions	Geister vein in the Rovnost mine and Geschieber vein in the Svornost mine, Jáchymov, CZ (Sejkora et al. 2011)

follows after pharmacolite, however, it is stable only in specific and restricted depth in the mine, loses water, and converts to brassite. According to Ondruš et al. (1997a), the small amounts of sulfuric acid in the solutions stabilize rösslerite. Haidingerite, sainfeldite, fluckite, and (Ca-)villyaellenite are minor minerals identified in this paragenesis. Talmessite, annabergite, köttigite, hörnesite, and some other arsenates are rare associated minerals.

Arsenates of Pb, Cu, Zn, Ni, Co, and other metals represent the most variable group of As weathering products in the mines where waters percolate through the multicomponent primary mineralizations. Most of these arsenates are found in mines which exploited hydrothermal poly-metallic sulfides (usually Pb-Zn-Cu+As), skarn or pneumatolytic mineralizations (Sn-W-Mo), five-element-type (Ag-Bi-Co-Ni-As) or uranium deposits. The primary mineralogy is often very variable (native elements, oxides, sulfides, sulfosalts), both in terms of the chemical composition and spatial distribution of the minerals. Hence, the parageneses of the secondary minerals may dramatically change over short distances and reflect the local physico-chemical conditions.

Ondruš et al. (1997a,b,c) studied secondary minerals inside the Svornost mine, Jáchymov (Czech Republic). They identified a wide range of arsenic secondary minerals and divided them into several parageneses. Copper-rich arsenates of the beudantite–segnitite series accompanied by other arsenates have been found in the abandoned adits of the Sn-W-Mo deposit Horní Krupka (Czech Republic) (Sejkora et al. 2009). In fact, a great deal of secondary As metal arsenates have been identified from samples collected in the underground mines (see Table 10). Detailed mineralogical and geochemical studies of such assemblages from underground spaces are largely missing, most likely owing to safety and security concerns linked to working in abandoned or active mines. We have located a number of studies with limited scope, mostly only identification of minerals, with little or no information about their formation, detailed relationship to the primary mineralization, geochemical parameters of the aqueous solutions, or the physical parameters of the underground microclimate. If the collection of such samples is possible, a thorough scientific study of these assemblages, not a mere enumeration of the mineral species, should be carried out and encouraged.

Arsenic mineralogy in caves. By definition, a cave mineral is "a secondary deposit precipitated inside a human-sized natural cavity," where the secondary deposit refers to a mineral derived from a primary mineral existing in the bedrock or cave sediment through various physico-chemical reactions (Hill and Forti 1997). Although the vast majority of the chemical deposits precipitated in caves consist of calcium carbonate, cave environments accommodate a wide range of authigenic minerals from all chemical classes (Onac and Forti 2011b). Although only a limited number of caves have been investigated in a detail with respect to their mineralogy, 15 cave arsenic minerals have already been recognized and described (Table 11). Their formation is caused by the interaction of aqueous fluids and the wall rocks or cave sediments with primary arsenic minerals. The fluids may be the percolating waters that flow or seep into the caves or hydrothermal, sometimes sulfide-rich solutions. Although fluids are definitely the key vehicles of As and other chemical components into and out of the cave environment, the cave itself is a fundamental factor in the formation of different minerals. The influence of other factors, such as cave temperature, relative air humidity, fluid pH, and redox conditions, on the generation of cave minerals including secondary arsenic minerals has been recently reviewed by Onac and Forti (2011a). Most of the arsenic cave minerals probably result from the low-temperature oxidation of primary arsenic sulfides in the wall rock or in the overlying strata. Secondary arsenic minerals in caves include arseniosiderite, beudantite, conichalcite, mimetite, olivenite, talmessite and yukonite (Dietrich 1960; Smol'yaninova and Senderova 1963; Smol'yaninova 1970; Feraud et al. 1976; Forti and Urbani 1995; Hill and Forti 1997; De Waele and Forti 2005; Garavelli et al. 2009).

Some processes are restricted to more peculiar cave conditions (e.g., higher temperatures, unusual fluid chemistry). Orpiment covers cave walls as crumbling crusts associated with

Table 11. Secondary arsenic minerals from caves (modified after Onac and Forti 2011b).

Mineral	Locality	Reference
realgar	Chauvai, Kyrgyzstan	Hill and Forti (1997)
orpiment	Aghia Paraskevi caves, Greece	Lazarides et al. (2011)
arsenolite	Corkscrew Cave, Arizona, USA	Onac et al. (2007)
claudetite	Corkscrew Cave, Arizona, USA	Onac et al. (2007)
arseniosiderite	Tyuya-Muyun Cave, Kyrgysztan	Smol'yaninova and Senderova (1963)
beudantite	Island Ford Cave, USA	Dietrich (1960)
hedyphane	cave in Santa Barbara mine, Sardinia, Italy	De Waele and Forti (2005)
hörnesite	Corkscrew Cave, Arizona, USA	Onac et al. (2007)
manganberzeliite	Cave Alfredo Jahn, Venezuela	Forti and Urbani (1995)
mimetite	Bisbee mine cave, USA	Dietrich (1960)
olivenite	cavities in the Tintic district, Utah, USA	Dietrich (1960)
pharmacolite	Corkscrew Cave, Arizona, USA	Onac et al. (2007)
strashimirite	Dupkata na Mare Cave, Bulgaria	Minceva-Stefanova (1968)
talmessite	Corkscrew Cave, Arizona, USA	Feraud et al. (1976); Onac et al. (2007)
yukonite	Grotta della Monaca, Italy	Garavelli et al. (2009)

pickeringite ($MgAl_2(SO_4)_4 \cdot 22H_2O$), tamarugite ($NaAl(SO_4)_2 \cdot 6H_2O$), and gypsum in the active sulfuric acid caves of Aghia Paraskevi on the Kassandra peninsula, northern Greece (Lazarides et al. 2011). Orpiment accumulated here from vapors above the surface of a thermal water cave pool that has a temperature of 39.2 °C, pH of 6.38, and As concentration of 3.4 mg L^{-1} (i.e., sub-aerial conditions at relatively low temperatures through an evaporation-condensation process and fluid cooling). Realgar contemporaneous with late, low-temperature hydrothermal karstification was described from the Chauvai Hg-Sb ore deposit in Kyrgyzstan (Hill and Forti 1997). Hydrothermal leaching of barite and carbonates with lead-zinc and silver sulfide ores in several mining districts in central Peru has produced open caves with collapse breccias (e.g., in the Cerro de Pasco Mine) (Lacy and Hosmer 1956). Sulfides were deposited in the residual box works in the final stage of the hydrothermal activity, locally accompanied by complex fine-grained botryoidal and stalactitic forms of closer non-specified Pb-As-containing mineralization.

Strashimirite described from the Dupkata na Mare Cave (exposed in the Zapachitsa mine in Bulgaria) is considered to be a hydrothermal weathering product of As-bearing copper ores under slightly alkaline conditions (Minceva-Stefanova 1968). Manganberzeliite of an unclear origin was found mixed with amorphous Fe and Mn oxides by Forti and Urbani (1995) inside the Cave Alfredo Jahn in Venezuela. Arsenic oxides arsenolite and claudetite in association with hörnesite, pharmacolite, and talmessite were described by Onac et al. (2007) to form rare complex speleothems in the Corkscrew Cave in Arizona, USA. The source of this unusual mineralization is a mineralized breccia-pipe body with Co, Mo, Ni, Pb, U, and Zn ores with primary arsenic minerals enargite, tennantite, and arsenopyrite. The formation of secondary arsenic minerals is related to a multistage process comprising leaching from the mineralized breccia-pipes, transport by neutral to alkaline, oxidized low-temperature waters, and precipitation at a redox interface inside the cave.

CRYSTAL CHEMISTRY OF ARSENIC MINERALS

Only a few elements have such a rich crystal chemistry as arsenic. The common oxidation states of As are –I, III, and V. Within the group 15 of the periodic table, the immediate neighbors of arsenic, phosphorus and antimony, share some crystal chemical and geochemical features but there are also striking differences. Some parallels can be drawn to the element sulfur. The differences in the proportions of the reduced and oxidized forms of arsenic, phosphorus, antimony, and sulfur in nature (Fig. 28) are remarkable. For arsenic, arsenates clearly dominate, but arsenides, sulfides, sulfosalts, as well as arsenite minerals are not negligible. Phosphorus, in many ways similar in its crystal chemistry to arsenic, is in nature largely restricted to phosphates.

Reduced arsenic is abundant in arsenides and sulfarsenides. Not so many minerals belong to this group but its most common member arsenopyrite (FeAsS) is a hallmark of many mesothermal ore deposits and almost invariably present together with gold ores in these deposits. It is this association which is the cause of many environmental problems related to arsenic release. Extensive substitution between S, As, and to some extent Sb, has been reported from minerals in this group (e.g., Hem and Makovicky 2004). The analogous antimonides and sulfantimonides (e.g., gudmundite, FeSbS) are rarer than the arsenides and sulfarsenides but probably only because the natural abundance of Sb is lower than that of As. Phosphides, on the other hand, are extremely rare (Fig. 28) and found mostly in meteorites (e.g., Zolensky et al. 2008).

Trivalent arsenic has two principal modes of occurrence, either coordinated by reduced sulfur or by oxygen. Coordinated by reduced sulfur, As^{3+} is found in many sulfides and sulfosalts. Naturally, sulfur plays a major role in these compounds but a role very different from that of As. The stereochemistry of these structures is dictated by the necessity to position the lone electron pair on As^{3+}, similar to the lone electron pairs on Sb^{3+} or Bi^{3+}. The simple molecular structure of realgar, As_4S_4, has no analog with Sb or Bi. Likewise, the structure of orpiment, As_2S_3 is completely different from that of stibnite, Sb_2S_3, and bismuthinite, Bi_2S_3; stibnite and bismuthinite are isostructural. In sulfosalts, As, Sb, and Bi are housed in larger coordination polyhedra defined by sulfur atoms; these polyhedra are commonly capped by additional sulfur atoms at larger distances.

Trivalent arsenic can be also coordinated by three oxygen atoms to form rare arsenites. Similarly, sulfite minerals are very rare. Most of the arsenic minerals are arsenates (Fig. 28), with nominally pentavalent As coordinated with four oxygen atoms. This valence state and coordination is also typical for abundant sulfates and phosphates. In contrast, Sb^{5+} is almost always coordinated octahedrally and its crystal chemistry deviates much from that of As^{5+}.

The general crystal-chemical features of these groups of minerals are described separately below. Crystal chemistry and crystallography of sulfides and related minerals (collectively called chalcogenides, including arse-

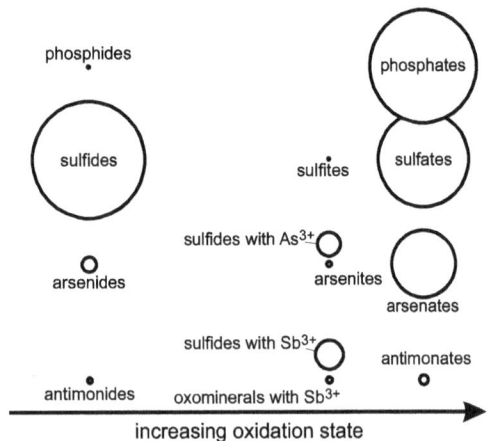

Figure 28. Relative proportions of minerals of arsenic, phosphorus, antimony, and sulfur. Size of the circles shows the number of mineral species in each group. Note that some minerals may be counted in two groups, for example sartorite, $PbAs_2S_4$, would be counted as a sulfide and as a sulfide with As^{3+}.

nides, sulfarsenides, and sulfosalts) was extensively described in a recent excellent summary by Makovicky (2006) and there is no need for another exhaustive review here. Only the structure of very few common chalcogenide minerals with As will be discussed here.

For the structures of arsenites and arsenates, we follow the bond-valence approach developed and applied to sulfates (Hawthorne et al. 2000) and phosphates (Huminicki and Hawthorne 2002) by F.C. Hawthorne and his co-workers. The stereochemistry of the As-ϕ polyhedra ($\phi = O_2^-$, OH⁻) are of interest for the establishment of links between crystal structure and chemical-physical properties of phases, links between crystal structure and thermodynamics (stability) of phases, and for prediction of unknown and validation of known structures. The details of the approach will be further discussed below.

Hierarchical organization of crystal structures

A mineral is defined as a naturally occurring solid with defined (not necessarily precisely fixed) chemical composition and crystal structure, usually of inorganic origin (Klein 2002). It is the interplay of chemical composition and crystal structure which defines the thermodynamic, physical, and optical properties of each given mineral species. There is an intimate relationship between the two factors (chemical composition and crystal structure) as the changes in chemical composition may induce a phase transition or appearance of a completely new structure; on the other hand, a given crystal structure limits the nature and range of possible substitutions, therefore dictating the possible ranges of chemical composition.

Pauling's rule of parsimony describes structures as sets of a small, limited number of available sites on which, in case of a need, substitution may take place. These sites, surrounded by their neighbors, define the coordination polyhedra. The higher bond-valence polyhedra, in turn, may and will further condense into homo- or heteropolyhedral clusters which constitute the fundamental building blocks (FBB) (Hawthorne 1983). The FBB is repeated, often polymerized, by translational symmetry operators to form the structural unit. The structural unit is then a polyhedral array, usually with negative charge which is balanced by interstitial species, these being large, low bond-valence cations (Hawthorne 1985a). One-polyhedron clusters may remain unconnected to the rest of the structure. Polyhedra may only condense into finite clusters, infinite chains, infinite sheets, and infinite frameworks. These are the five possible modes of cluster polymerization.

Crystal chemistry of arsenates

Selected geometric features of the Asϕ_4 groups are summarized and statistically evaluated below. Not all minerals discussed in the text below were included in this evaluation for a very prosaic reason. Because the crystal structures were solved mostly on natural samples, common and ample amount of P^{5+} on the As^{5+} site invalidates many well-refined structures for such purposes. Occasionally, other cations such as V^{5+} or Sb^{5+} may have intervened. When reviewing the literature, great care was taken to filter out those structures for which such substitution was documented by chemical analyses.

Variations in As-ϕ and <As-ϕ> distances. An examination of the scatter of As-ϕ ($\phi = O^{2-}$, OH⁻) distances gives insights in the stereochemical behavior of the structures and establishes the range of variations. The distribution of the individual As-ϕ distances is shown in Figs. 29a,b. For the As-O bonds, the average length is 1.685 Å. The average length for the As-OH bonds is 1.727 Å. Note that these As-OH distances exclude all cases when H disorder was detected or assumed, or the hydrogen bond was symmetric.

The variations in <As-ϕ> (where <> denotes the arithmetic mean) distances are shown in Fig. 29c. The mean value of all considered <As-ϕ> distances is 1.687 Å, marginally larger than the mean of the As-O distances. This value (1.687 Å), however, includes mean values for AsO$_4$, AsO$_3$OH and AsO$_2$(OH)$_2$ tetrahedra. Only one <As-ϕ> value slightly below 1.66 Å was

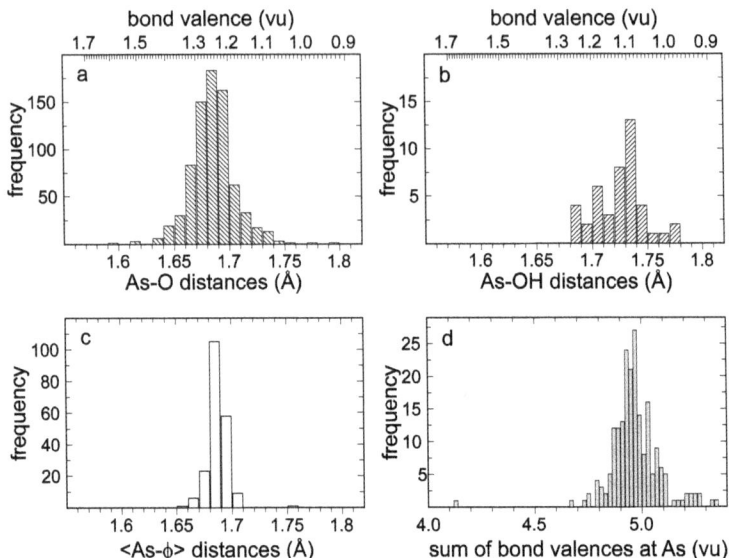

Figure 29. (a) As-O, (b) As-OH, and (c) <As-φ> distances in the structures of arsenate minerals, and (d) the sum of the bond valences received by the central As cation. The upper edge of the graphs for the As-O and As-OH distances shows bond valence of the As-O bonds, calculated as $exp((1.767 - d)/0.37)$, where d is the As-O bond length in Å. The parameters were taken from Brown and Altermatt (1985).

found (1.659 Å in picropharmacolite). The single anomalously long value of 1.756 Å comes from the structure of kuznetsovite, $Hg_3Cl(AsO_4)$, an arsenate structure with metal clusters and an intervening Cl^- anion. The sum of the bond valences at the central cation scatters around 5 vu (Fig. 29d).

Polyhedral distortion in minerals with Asϕ_4 groups. Polyhedral distortion can be summarized as the displacement of the cation from the centroid of a polyhedron and the deviation of the polyhedron from sphericity (Balić-Žunić and Makovicky 1996).

Sphericity of a polyhedron is related to the standard deviation of the distance between the centroid and the ligands. Fig. 30 shows that the displacement of the As^{5+} from the centroid is most commonly ~0.03 Å. Displacements larger than 0.05 Å are rare but are documented (Fig. 30). The sphericity of the Asϕ_4 tetrahedra is always nearly 1 (> 0.9999) which means that the tetrahedra can be almost perfectly inscribed into a sphere. Balić-Žunić and Makovicky (1996) calculated much lower sphericity values for cations with stereochemically active s^2 lone electron pairs (e.g., Tl^+, As^{3+}, Sb^{3+}, Bi^{3+}); because As^{5+} does not belong to this group of cations, almost perfect sphericity was expected.

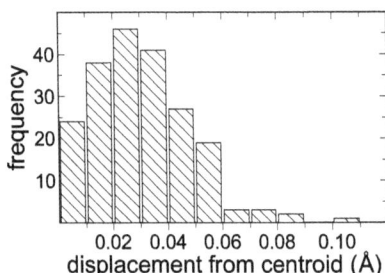

Figure 30. Displacement of the As cation from the centroid of the tetrahedra. The coordinates of the centroid were calculated as shown by Balić-Žunić and Makovicky (1996).

Very significant displacement was reported from sabelliite where the pentavalent cations As^{5+} and Sb^{5+} are located near the base of the tetrahedron. Consequently, Olmi et al. (1995) described the coordination as trigonal-pyramidal, not tetrahedral. Because of the considerable Sb^{5+}-As^{5+} substitution, this mineral is not represented in Fig. 30 and in the statistical analysis.

Distortions in tetrahedral oxyanions have been used as an empirical predictive tool in crystal chemistry (Baur 1974; Griffen and Ribbe 1979). These authors defined the following distortion parameters, where TO are the distances between the central tetrahedral cation (T) and oxygen ligand; OO are the distances between the oxygen ligands; OTO is the tetrahedral O-T-O angle (Fig. 31):

$$DI(TO) = \frac{\overset{4}{\Sigma}|TO - <TO>|}{4 <TO>}$$

$$DI(OO) = \frac{\overset{6}{\Sigma}|OO - <OO>|}{6 <OO>}$$

$$DI(OTO) = \frac{\overset{6}{\Sigma}|OTO - <OTO>|}{6 <OTO>}$$

$$BLDP = \frac{1}{<TO>}\left[\frac{\overset{4}{\Sigma}(TO - <TO>)^2}{N-1}\right]^{1/2} \quad N = 4$$

$$ELDP = \frac{1}{<OO>}\left[\frac{\overset{6}{\Sigma}(OO - <OO>)^2}{N-1}\right]^{1/2} \quad N = 6$$

where DI stands for distortion index, BLDP for bond-length distortion parameter and ELDP for edge-length distortion parameter. All of these parameters (DI(TO), DI(OO), DI(OTO), BLDP, ELDP) are some measure of the deviation of the observed quantity (TO, OO, OTO) from the mean value. While investigating the phosphate tetrahedra, Baur (1974) found average values of

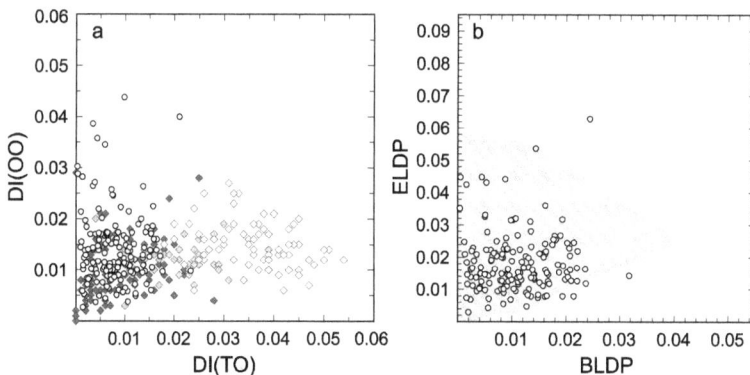

Figure 31. a) Tetrahedral distortion parameters DI(TO) and DI(OO), as defined by Baur (1974) (see text). The circles are calculated from arsenate structures in this review with R < 7%. Diamonds represent the data for phosphates, taken from Baur (1974, his Table 15). Open diamonds: organic phosphates, ring phosphates, polyphosphates; light grey diamonds: diphosphates; dark grey diamonds: orthophosphates, acid orthophosphates. b) Tetrahedral distortion parameters BLDP (bond-length distortion parameter) and ELDP (edge-length distortion parameter), as defined by Griffen and Ribbe (1979) (see text). The circles are calculated from arsenate structures in this review with R < 7%. The shaded area is the field populated by the structures considered by Griffen and Ribbe (1979).

DI(TO) and DI(OO) of 0.021 and 0.012, respectively. These indices mean that the distortion in P-O distances is much larger than the distortion of the O-O distances, and hence "phosphate group can be viewed, to a first approximation, as a rigid regular arrangement of O atoms, with the P atoms displaced from their centroid" (Baur 1974). The average DI(TO) and DI(OO) values for arsenates considered here are 0.008 and 0.015, respectively. For the arsenates, therefore, the situation appears to be reversed because the O-O distances are distorted more than the As-O distances. The arsenate tetrahedra, therefore, could be viewed as bodies where the central As^{5+} cation lies near the centroid of the tetrahedron, and the oxygen atoms of the tetrahedron are displaced to accommodate the external strain in the structures. There is, however, not such a great difference between phosphates and arsenates. Baur (1974) considered several subsamples of phosphates, that is, orthophosphates, acid orthophosphates, organic phosphates, ring phosphates, and polyphosphates. A closer look at his data reveals that the large values of DI(TO) can be tracked to the organic phosphates, ring phosphates, and polyphosphates (Fig. 31a). The orthophosphates and acid orthophosphates, most similar to the inorganic natural arsenates reviewed here, have the average DI(TO) and DI(OO) values of 0.010 and 0.010. The (PO_4) tetrahedra in these phosphates thus behave similarly as the (AsO_4) tetrahedra in the arsenate minerals.

The values of BLDP and ELDP, as defined by Griffen and Ribbe (1979) and specified above, are plotted in Fig. 31b. The shaded field in Fig. 31b is the field which was populated by data points in the analysis of Griffen and Ribbe (1979) and our points fall into this field, with a few exceptions. Griffen and Ribbe (1979) concluded that As^{5+} is a tetrahedral cation with a relatively large spread of both BLDP and ELDP, similar to Si^{4+}. The degree of distortion is related to the structural relationship of the tetrahedra to other coordination polyhedra. The reasons for the distortions may be tracked more easily for oxyanions with small BLDP values (e.g., LiO_4, GaO_4) or small ELDP values (e.g., PO_4, SO_4). For oxyanions with larger BLDP and ELDP values (e.g., AsO_4, SiO_4), the influence of the coordination polyhedra on the distortion may be recognized but is not as clear as for the other oxyanions. It is interesting to note that As^{5+} fits snugly into an AsO_4 tetrahedron, that is, As^{5+} is very close or at the "no-rattle" limit for tetrahedrally coordinated cations (see Griffen and Ribbe 1979, their Fig. 7b).

The average coordination number for the oxygen ligands of the AsO_4 tetrahedra is 3.24 and the average electronegativity of the neighboring cations is 1.84. We calculated these values for all structures considered in this work with R < 7%. Inspection of the correlation between the calculated variables revealed only a weak negative correlation between <CN> and <χ> (r = –0.37). In line with the discussion in the previous paragraph, there is a weak correlation between BLDP and <χ> (r = +0.40) and ELDP and <CN> (r = +0.41). All other correlations are insignificant. The pair-distribution function for the oxygens which are ligands of the As^{5+} cations is shown in Fig. 32. The first sharp peak corresponds to the $O-As^{5+}$ neighbours, the second broader peak to other O-cation pairs. The first coordination sphere of these oxygen ligands terminates at distances of around 3 Å.

A structural hierarchy for arsenate minerals. Despite the dazzling chemical complexity of arsenate minerals, the variations in the coordination numbers of cations are meager. The principal coordination polyhedra are tetrahedra, octahedra or similar

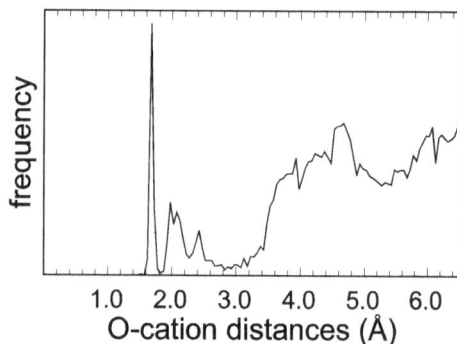

Figure 32. Pair-distribution function of the O-cation pairs of oxygen atoms which are the ligands of As^{3+} in the structures of arsenite minerals.

polyhedra, and large-cation polyhedra (Hawthorne 1983). Hence, we can divide arsenates into three unequally populated classes:

1. Structures with polymerized tetrahedra,

2. Structures with polymerized tetrahedra and octahedra or similar polyhedra (the latter two collectively denoted as $M\phi_N$, see below),

3. Structures with polymerized tetrahedra and large-cation polyhedra.

Class (1) may contain octahedra or large-cation polyhedra, as long as the tetrahedra in the structure are polymerized. Similarly, class (2) can contain large-cation polyhedra and, of course, contains tetrahedra, as long tetrahedra and octahedra or similar polyhedra are polymerized. Class (3) should not contain octahedra; only tetrahedra and large-cation polyhedra occur and they polymerize. The $M\phi_N$ polyhedra in class (2) comprise octahedra and the loosely defined "similar polyhedra." These are coordination polyhedra other than octahedra for cations with six-fold coordination, e.g., trigonal prisms, and coordination polyhedra for cations with five-fold coordination or four-fold coordination other than tetrahedral, for example square planar. The latter two are common in some arsenates of copper. Therefore, the coordination number N in this case is usually six but could be also four, five, and in exceptional cases described in this section, also higher than six.

During the compilation of the data for this chapter, a question arose as to what extent of distortion does an "octahedron" qualify to be still be called an octahedron. In many original papers, the authors label the coordination polyhedra of cations with a strong tendency to distort its local environment differently. This issue is of relevance especially to cations with a strong Jahn-Teller distortion such as Cu^{2+} where axially distorted octahedra are sometimes labeled as square bipyramids. Here, we understand octahedra as geometric bodies composed of eight triangular faces, four of which meet at each vertex. Such bodies do not have to, and hardly ever do, attain the point symmetry O_h. Octahedra with a large deviation from the point symmetry O_h can be referred in the text as "distorted" to emphasize the magnitude of the deviation.

The definitions of coordination polyhedra become blurry for cations with a stereochemically active lone electron pair, especially Pb^{2+}. The coordination of Pb is variable, with usually large coordination numbers. On the other hand, more or less distorted octahedra are also known. In the case of a pronounced one-sidedness of the Pb^{2+} coordination polyhedra, the coordination number may be as low as 4 for this large cation. We considered such structures case by case, inspecting all polyhedra of Pb^{2+} and then assigning the structure under consideration to a specific class.

The division between the three classes is not as clear-cut as it may seem. In some cases, we see a gradual increase in the coordination number while the basic topological features of the structures remain untouched. A superb example from class (2) is krautite and related minerals with heteropolyhedral layers. Krautite itself, $Mn(H_2O)(AsO_3OH)$, contains layers in which $Mn\phi_6$ octahedra polymerize with AsO_4 tetrahedra. The same topology, however, is seen in haidingerite, $Ca(H_2O)(AsO_3OH)$, except that the $Mn\phi_6$ octahedra from krautite are replaced by $Ca\phi_7$ polyhedra. In pharmacolite, $Ca(H_2O)_2(AsO_3OH)$, the crude features of the layers are preserved, but with $Ca\phi_8$ polyhedra. In such cases, all these minerals are considered to belong to class (2); their division between class (2) and (3) would only obscure their structural relationships.

In rare cases, the large monovalent or divalent cations are also found in six-fold, although not necessarily octahedral coordination. An example is lavendulan, $NaCaCu_5(H_2O)_5Cl(AsO_4)_4$, and related minerals. The layers in this structure are a result of polymerization of $Cu\phi_5$ and AsO_4 polyhedra. Because of the low bond valence of the Na-O bonds within the $Na\phi_6$ polyhedra, these are not considered to be a part of the fundamental building block. Instead, they are considered to be interstitial species, as in other structures.

In line with the structural classification of sulfates (Hawthorne et al. 2000) and phosphates (Huminicki and Hawthorne 2002), the structures are arranged within each class according to the increasing connectivity of the polyhedra in the structural unit. The possibilities are

1. unconnected polyhedra,

2. finite clusters of polyhedra,

3. infinite chains or rods of polyhedra,

4. infinite sheets or layers of polyhedra,

5. infinite frameworks of polyhedra.

For the minerals in class (2), we assigned the minerals to the groups a) through e) and further sorted them in the text according to the increasing linkage of the polyhedra. Polyhedra which share a corner are denoted by a simple dash (M-M, M-T); edge sharing is indicated by a double dash (M=M). Rare face sharing is shown by a triple dash (M≡M). The polyhedral linkage is specified at the beginning of each mineral description in class (2).

Most structure descriptions are accompanied by a figure that shows a parallel projection of a polyhedral model of the structure. Owing to constraints which limit the size of this chapter, more than one figure per structure is exceptional. Some very complicated structures are not depicted at all. The projections were chosen to document the connectivity and position of the As-containing units, not necessarily the overall structural features. In all projections, arsenate tetrahedra are hatched. Other tetrahedral units have a different fill. Octahedra, similar polyhedra, and large-cation polyhedra are usually shaded, sometimes hatched. In the projections of the minerals from classes (1) and (2), the large-cation polyhedra, if present and if their depiction seemed to be suitable, are usually shown by a ball-and-stick model.

Polymerization of (Asϕ_4) and other (Tϕ_4) tetrahedra. Pauling's second rule defines an approximate value of bond valence as a ratio of the magnitude of formal valence of the central ion and its coordination number. Furthermore, the sum of bond valences from all bonds an ion is receiving should be equal to the formal valence of that ion. Hence, in an AsO_4 group, the bond valence of each As-O bond is 5/4 and, obviously, the sum of bond valences on the central As is exactly 5. In such a hypothetical isolated AsO_4 group, the sum of bond valences on each O atom is 5/4 = 1.25, far away from the magnitude of formal charge of 2. Therefore, each O atom should receive another 0.75 valence units (vu) from another cation(s) to satisfy its bonding requirements.

Just like for phosphates (see Huminicki and Hawthorne 2002), the immediate consequence is that polymerization of (Asϕ_4) units in the structures of arsenates is unlikely. Yet, there are two rare minerals which defy this prediction, petewilliamsite, $(Ni,Co)_{30}(As_2O_7)_{15}$ and theoparacelsite, $Cu_3(OH)_2As_2O_7$. In addition, polymerization of the AsO_4 and VO_4 units were described from fianelite, $Mn_2(H_2O)_2[V(V,As)O_7]$. On the other hand, polymerization with tetrahedral units with cations of lower formal valence (+2, +3) is possible and also realized, although not too commonly (Table 12).

Table 12. Minerals with polymerized Tϕ_4 groups.

Polymerized Tϕ_4 groups	Minerals
AsO_4-AsO_4	petewilliamsite, theoparacelsite
AsO_4-VO_4	fianelite
AsO_4-AlO_4	urusovite, filatovite, alarsite
AsO_4-Beϕ_4	bearsite, bergslagite
AsO_4-Znϕ_4	holdenite, dugganite, philipsburgite, kolicite, warikahnite, sabelliite, chlorophoenicite and related minerals

Similarly as for phosphates, hydrated pyroarsenates are known, for example $Mn_2As_2O_7·2H_2O$ (Stock et al. 2001). The existence of these phases suggests that absence of hydrogen or water is not a necessary prerequisite for the AsO_4-AsO_4 polymerization, even though it may be a limiting factor.

To our knowledge, there are no examples among the known minerals with AsO_4-SiO_4 polymerization. In the minerals where these units occur together (holdenite, kolicite), they are bridged by $Zn\phi_4$ polyhedra. In filatovite, $K[(Al,Zn)_2(As,Si)_2O_8]$, extensive As^{5+}-Si^{4+} substitution was observed, and the tetrahedral cations observe the aluminum-avoidance rule of Loewenstein (1954).

Polymerization of (Asϕ_4) and other (Mϕ_N) polyhedra. Here, the analysis of phosphate structures (Huminicki and Hawthorne 2002) applies with no exceptions and will be briefly summarized. The As^{5+}-O bond, just as the P^{5+}-O bond, has a formal valence of 1.25 vu. Each oxygen requires therefore additional 0.75 vu to satisfy its bonding requirements. Since oxygen in oxysalt structures is usually coordinated by 3-4 cations, the average bond valence of the bonds outside of the AsO_4 tetrahedron should be 0.75/3 = 0.25 vu or 0.75/2 = 0.38 vu. The most likely candidates for the formation of these bonds are divalent (Mg, Fe^{2+}, Mn^{2+}) and trivalent (Al, Fe^{3+}) cations in six-fold coordination, or monovalent (Na, K) and divalent (Ca) cations in seven- to nine-fold coordination. Additional 0.1-0.3 vu may be supplied by hydrogen bonds in structures with OH or H_2O groups.

Similarity to other groups of minerals. Many structures of arsenates, especially the complicated ones, are unique and not seen in other groups of minerals. On the other hand, there are many structures which have counterparts, close or distant, elsewhere in the kingdom of minerals. A brief comparison to common mineral structures may perhaps convey the ways in which arsenates could be allied to them.

Without doubt, the closest group to arsenates are the phosphates. Many arsenates have their phosphate counterparts. A complete P^{5+}-As^{5+} miscibility is in some cases known, in other ones assumed. Groups of minerals traditionally considered as predominantly phosphates, such as apatites, have a number of arsenate members.

A substitution or miscibility between P^{5+}-S^{6+} is limited but possible. Some flexible sulfate structures, especially those in the alunite-jarosite supergroup, have a number of arsenate members. On the other hand, in some structures, a very strict division of P^{5+} and S^{6+} over the available tetrahedral sites is observed (e.g., bukovskýite, leogangite).

Despite the common belief that As^{5+} and Sb^{5+} are intimate companions because they are both metalloids of the group V in the periodic table, very few minerals are found with As^{5+}-Sb^{5+} substitution. These are sabelliite and perhaps also theisite.

Although the capability of AsO_4 tetrahedra to polymerize is curbed by the bonding requirements of the potential bridging oxygen, many major groups of rock-forming silicates have their representatives among the arsenates. The structure of xanthiosite, $Ni_3(AsO_4)_2$, is a cation-deficient olivine arrangement where one fourth of the available octahedral sites are vacant, resulting in the change of the stoichiometry from A_2BO_4 (as in Mg_2SiO_4) to $Ni_{1.5}(AsO_4)$ or $Ni_3(AsO_4)_2$. Filatovite, $K[(Al,Zn)_2(As,Si)_2O_8]$, is isostructural with the feldspar mineral celsian, $Ba[Al_2Si_2O_8]$. The structure of alarsite, $AlAsO_4$, is analogous to that of quartz with the exception of Al^{3+}- As^{5+} ordering over the tetrahedral sites. There are no natural chain arsenates known but the synthetic pyroxene $LiAsO_3$ has been described (Hilmer 1956). The topology of the tetrahedral sheet in philipsburgite, $Cu_5(OH)_5(H_2O)[Zn(OH)(As,P)O_4]$, and bergslagite, $CaBe(OH)AsO_4$, is similar to the topology seen in the sheet silicate apophyllite, $KCa_4(Si_4O_{10})_2F·8H_2O$. Finally, the minerals berzeliite, $(NaCa_2)Mg_2(AsO_4)_3$ and manganberzeliite, $(NaCa_2)Mn_2(AsO_4)_3$, adopt the well-known garnet structure.

Structures with polymerized (Tφ₄) groups.

Structures with finite clusters of tetrahedra. The structure of **petewilliamsite**, $(Ni,Co)_{30}(As_2O_7)_{15}$, (Roberts et al. 2004) is closely related to those of the synthetic compounds $Ni_2As_2O_7$ and $Co_2As_2O_7$ (Buckley et al. 1990). The large unit cell of petewilliamsite is a supercell of the synthetic arsenates. There are sixteen independent M (M = Ni, Co, Cu) sites and fifteen independent As sites in the structure. The M coordination polyhedra are tetrahedra, square bipyramids, and octahedra, and combine to form complicated (010) sheets. The pyroarsenate (As_2O_7) units attach to the sheets above and below vacant sites within the sheets (Fig. 33a) and amalgamate the structural fragments into a three-dimensional network.

A large number of atoms in the asymmetric unit and weak diffraction data precluded a refinement with the accuracy usually encountered in the structures of minerals. The reported As-O distances vary between 1.54-2.02 Å. The As-O distances in synthetic pyroarsenates do not scatter so much, for example, are 1.70 Å for the terminal As-O bonds and 1.76 Å for the bridging As-O bonds in $Mn_2As_2O_7 \cdot 2H_2O$ (Stock et al. 2001), or 1.68 Å for the terminal As-O bonds and 1.65 Å for the bridging As-O bonds in $Ni_2As_2O_7$ (Buckley et al. 1990).

The structure of **fianelite**, $Mn_2(H_2O)_2[V(V,As)O_7]$, comprises pairs of tetrahedra and octa-hedral chains (Brugger and Berlepsch 1996) (Fig. 33b). The dominant tetrahedral cation is V^{5+}. As^{5+} substitutes for V^{5+} only at one tetrahedral position and the chemical analyses of Brugger and Berlepsch (1996) indicate that As^{5+} can exceed V^{5+} at this site. The $V(V,As)O_7$ pairs cross-link zigzag octahedral chains which house Mn^{2+}. These chains extend in the $[\bar{1}01]$ direction and contain two independent Mn^{2+} sites. The structure contains continuous [001] channels.

The seemingly simple structure of **theoparacelsite**, $Cu_3(OH)_2As_2O_7$, with a small unit cell contains one Cu and one As position (Sarp and Černý 2001). Cu is found in a distorted octahedral coordination. The octahedra share edges to form [010] chains and the adjacent chains attach

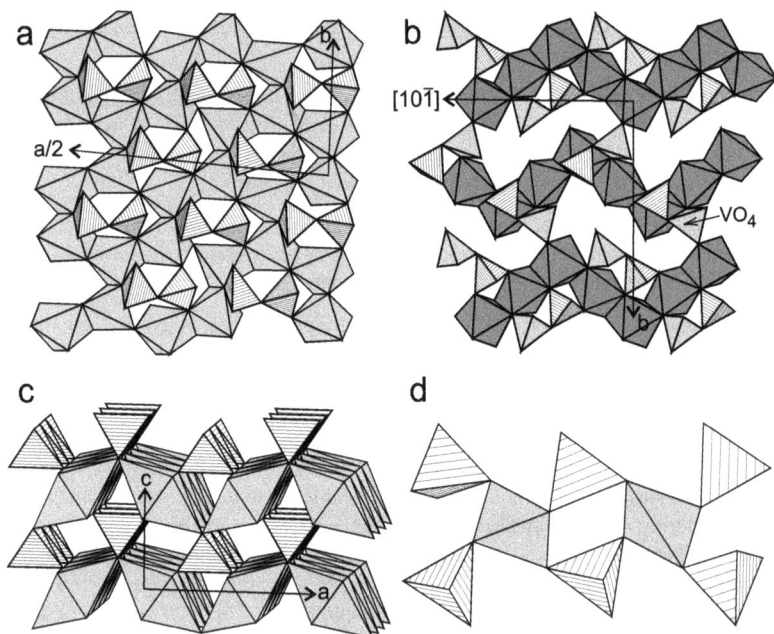

Figure 33. The structures of arsenates with tetrahedral clusters: (a) petewilliamsite, (b) fianelite, (c) theoparacelsite, (d) warikahnite.

via corners into (001) sheets. The space between the sheets is occupied by arsenate tetrahedra (Fig. 33c). Sarp and Černý (2001) described the arrangement of the tetrahedra as [010] chains. The As occupancy in these chains is only 2/3; Sarp and Černý (2001) found no indication of ordering of the vacancies within the chains. Therefore, it is possible that the "chains" are disordered As_2O_7 dimers, perhaps even longer oligomers of arsenate tetrahedra. Sarp and Černý (2001) neither located the hydrogen atom nor proposed its position in the structure; it is likely that the hydrogen is spatially correlated with the vacancies in the arsenate "chains."

The simple chemical formula of **warikahnite**, $Zn_3(H_2O)_2(AsO_4)_2$, does not express properly the complexity of its structure. There are 6 Zn positions whose coordination varies from octahedral, distorted octahedral, trigonal bipyramidal to tetrahedral (Riffel et al. 1980). If only the connectivity of the tetrahedra is considered, the ZnO_4 and AsO_4 groups define clusters of the composition $[Zn_2As_6O_{24}]$ (Fig. 33d). These clusters are interconnected by the $Zn\phi_5$ and $Zn\phi_6$ polyhedra which share corners and edges among themselves.

Structures with infinite chains of tetrahedra. The structure of **bearsite**, $Be_2(OH)$ $(H_2O)_2(AsO_4)\cdot2H_2O$, consists of [100] tetrahedral chains of the composition $[Be_2(OH)$ $(H_2O)_2(AsO_4)]$ (Fig. 34a) (Harrison et al. 1993). Both Be^{2+} and As^{5+} are tetrahedrally coordinated as $BeO_2(OH)(H_2O)$ and AsO_4, respectively. The arsenate tetrahedra attach with all four O atoms to four neighboring $Be\phi_4$ tetrahedra. Each $Be\phi_4$ tetrahedron shares two corners with adjacent arsenate tetrahedra. The OH^- group bridges two $Be\phi_4$ tetrahedra and the H_2O group of the $Be\phi_4$ tetrahedra points into the interchain space. There are also two H_2O molecules located in this space. Harrison et al. (1993) described the chains as a structural unit made of alternating three- and four-membered tetrahedral rings.

The crystal structure of **holdenite**, $Mn_6Zn_3(OH)_8(AsO_4)_2(SiO_4)$, is based on cubic close-packing of oxygen anions with 3/10 of the available octahedral interstitial sites and 3/20 of the available tetrahedral interstitial sites filled by cations (Moore and Araki 1977a). Zinc, As, and Si are coordinated tetrahedrally. The $Zn1(OH)_4$ tetrahedron does not connect to any other tetrahedral units in the structure. The As and Zn2 tetrahedra share a corner to form $AsZn\phi_7$ dimers which are further linked by the SiO_4 tetrahedra into ladder-like rods with a general [001] direction (Fig. 34b). The rods do not attach to each other but coalesce into (010) slabs. The tetrahedral rods and slabs are interconnected by $Mn\phi_6$ octahedra. The octahedra share edges and define short branching chain fragments.

The peculiar structure of **chlorophoenicite**, $Mn_3Zn_2(OH)_6[As(O,OH)_6]$, consists of thick (100) slabs. The slabs comprise infinite [010] three-octahedra wide strips of $Mn\phi_6$ octahedra.

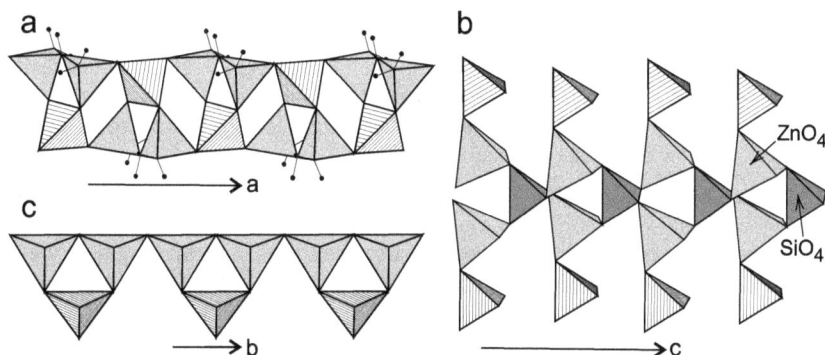

Figure 34. The structures of arsenates with infinite chains or rods of tetrahedra: (a) bearsite, (b) holdenite, (c) chlorophoenicite.

These strips are structural fragments of the pyrochroite ($Mn(OH)_2$) (or brucite) structure, an arrangement similar to that seen in akrochordite and related structures. The strips are bridged by [010] chains of $Zn\phi_4$ tetrahedra and arsenate tetrahedra (Fig. 34c). The local environment of As^{5+} in chlorophoenicite is rather strange. If the As^{5+} positions were fully occupied, this mineral would belong to the group of chain arsenates. According to Moore (1968b), $As\phi_4$ tetrahedra and O_4H_4 tetrahedral units are disordered within these chains; however, only an average structure could have been determined. The formula proposed by Moore (1968b) suggests that As and H substitute for each other. The substitution does operate in this structure but is heterotopic and, therefore, we have adopted a different formula from that proposed by Moore (1968b). Hence, the tetrahedral chains are constituted by the $Zn\phi_4$ tetrahedra, with the $As\phi_4$ tetrahedra attaching to these chains.

Little is known about **magnesium-chlorophoenicite**, $Mg_3Zn_2(OH)_6[As(O,OH)_6]$. Dunn (1981a) listed some data about the chemical composition of the mineral and Bayliss and Warne (1987) gave powder X-ray diffraction pattern.

The structure of **jarosewichite**, $Mn^{3+}Mn_3{}^{2+}(AsO_4)(OH)_6$, is unknown. Dunn et al. (1982) speculated, based on its stoichiometry and refined lattice parameters that the mineral is structurally close to chlorophoenicite. They also provided an incomplete description of yet another phase which could be structurally related to chlorophoenicite. This phase contains predominantly Mn, Zn, and As and Dunn et al. (1982) proposed that it could be "oxidized and hydrated chlorophoenicite."

Structures with infinite sheets of tetrahedra. The structure of **dugganite**, $Pb_3Te^{6+}(Zn_3As_2O_{14})$, is built by tetrahedrally coordinated As^{5+} and Zn^{2+}, octahedrally coordinated Te^{6+}, and Pb^{2+} in a regular eight-fold coordination (Lam et al. 1998). The tetrahedra share edges and coalesce into (0001) sheets (Fig. 35a). Each (AsO_4) tetrahedron shares three edges with neighboring (ZnO_4) tetrahedra. Each (ZnO_4) tetrahedron shares only two edges with the arsenate tetrahedral neighbors. The $Te^{6+}O_6$ octahedra attach to the tetrahedral sheets via sharing three edges with three (ZnO_4) tetrahedra in the sheet above and three edges with three (ZnO_4) tetrahedra below the octahedron. The (ZnO_4) and (TeO_6) polyhedra define $[Zn_3TeO_{12}]$ chains with the same topology as the chains in kaatialaite. The remaining space between the tetrahedral sheets serves to accommodate Pb^{2+}. Despite the ideal formula of $Pb_3Zn_3Te^{6+}As_2O_{14}$ given by Lam et al. (1998), their crystals had the As/P ratio near 1, with As being only slightly more abundant than P.

The structure of **philipsburgite**, $Cu_5(OH)_5(H_2O)[Zn(OH)(As,P)O_4]$, is assumed to be isostructural with kipushite (Peacor et al. 1985) although a structural analysis of philipsburgite was never carried out. The structure of kipushite, $Cu_5(OH)_5(H_2O)[Zn(OH)PO_4]$, is built by (001) sheets in a sequence [T-O1-T'-O2] (Piret et al. 1985). The T sheets in kipushite contain ZnO_3OH and PO_4 tetrahedra and have the overall composition $[Zn(OH)PO_4]$. These tetrahedra share corners and form four- and eight-membered rings with a sheet topology similar to that in the mineral apophyllite, $KCa_4Si_8O_{20}F\cdot8H_2O$. The O1 and O2 sheets are both brucite-like octahedral sheets with 1/6 of the available octahedral positions vacant. The $Cu\phi_6$ octahedra are distorted. The T' are sheets comprised of isolated PO_4 tetrahedra which attach to the O1 and O2 sheets above or below the vacant octahedral sites.

The structure of **bergslagite**, $CaBe(OH)AsO_4$, is built by tetrahedral berylloarsenate sheets (Fig. 35b) and $Ca\phi_8$ polyhedra sandwiched between the sheets (Hansen et al. 1984). The sheets have the same topology as the zincoarsenate sheets in philipsburgite or the sheets in datolite, $CaB(OH)SiO_4$, or herderite, $CaBeFPO_4$ (Hansen et al. 1984).

The crystal structure of **kolicite**, $Mn_7(OH)_4[As_2Zn_4Si_2O_{16}(OH)_4]$, is based on cubic close-packing of oxygen anions, closely related to the structures of holdenite and gerstmannite, $(Mn,Mg)Mg(OH)_2[ZnSiO_4]$ (Moore and Araki 1977a). Overall, 7/24 of the available octahedral interstices and 1/6 of the available tetrahedral interstices are filled with cations.

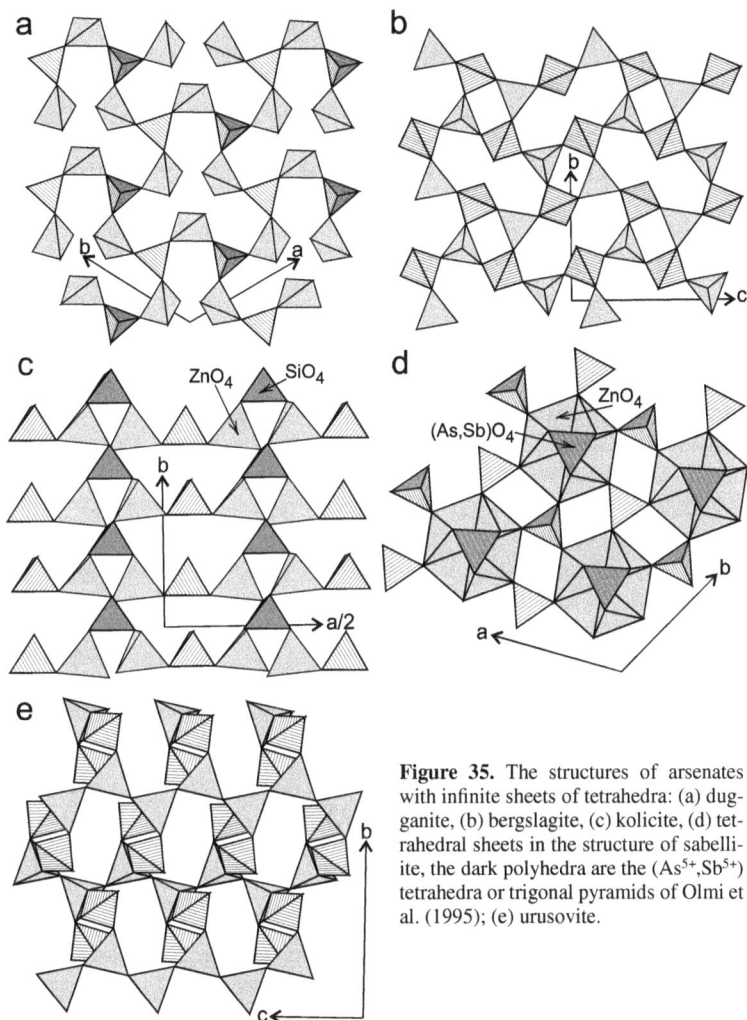

Figure 35. The structures of arsenates with infinite sheets of tetrahedra: (a) dugganite, (b) bergslagite, (c) kolicite, (d) tetrahedral sheets in the structure of sabelliite, the dark polyhedra are the (As^{5+}, Sb^{5+}) tetrahedra or trigonal pyramids of Olmi et al. (1995); (e) urusovite.

The tetrahedral cations As, Zn, and Si define (001) sheets (Fig. 35c). Within the sheets, As and Zn combine into Zn-As-Zn trimers; the trimers are amalgamated by the SiO_4 tetrahedra. The overall chemical composition of the sheets is $As_2Zn_4Si_2O_{16}(OH)_4$. The $Mn\phi_6$ octahedra share edges to merge into seven-membered clusters, thus filling the large voids between the units built by tetrahedra.

The structure of **sabelliite**, $(Cu,Zn)_2Zn(OH)_3[(As,Sb)O_4]$, contains a few awkward features. Its principal feature, seen also in the structures of many other minerals, are (001) brucite-like sheets occupied predominantly by Cu^{2+} and Zn^{2+} (Olmi et al. 1995). The two cations were assigned to the octahedral sites within these sheets based on geometric arguments, that is, the degree of distortion of the octahedra. The Cu^{2+} ions occupy the distorted octahedra in a 4+2 coordination, whereas the Zn^{2+} ions are located in the regular octahedra. Owing to the similar scattering power of these two elements, no refinement of the Cu and Zn fractions at these sites was possible.

A peculiar feature of the interlayer between the brucite-like sheets are clusters of eight edge-sharing tetrahedra. Six of the tetrahedra house Zn^{2+} and two should be occupied by (As^{5+}, Sb^{5+}). These clusters are linked by arsenate tetrahedra into (001) sheets (Fig. 35d). The (As^{5+}, Sb^{5+}) polyhedra are described as trigonal pyramids by Olmi et al. (1995) because the cation is located near the base of the polyhedron, not near its geometric center. The face sharing between the (As^{5+}, Sb^{5+}) polyhedron and the $Zn\phi_6$ octahedron in the (001) sheets is awkward, with a (As^{5+}, Sb^{5+})-Zn^{2+} distance of 1.717 Å. Olmi et al. (1995) rationalized this situation by vacancies at the (As^{5+}, Sb^{5+}) and Zn^{2+} sites and additional H atoms in the structure which could be suspected in the electron density maps. The exact structure, however, remains unclear.

The crystal structure of **urusovite**, $CuAlO(AsO_4)$, consists of layers parallel to (100) and CuO_5 polyhedra between these layers (Fig. 35e) (Krivovichev et al. 2000). Within the layers, Al^{3+} and As^{5+} are both tetrahedrally coordinated and ordered over the available sites. The layers have the composition $[AlAsO_5]$ and can be described as [001] chains of Al^{3+} tetrahedra bridged by As^{5+} tetrahedra. In other words, an arsenate tetrahedron does not share any of its corners with another arsenate tetrahedron. The CuO_5 polyhedra, sandwiched between the tetrahedral layers, are tetragonal pyramids and knit the structure into a three-dimensional network.

Structures with infinite frameworks of tetrahedra. The tetrahedral framework in the structure of **filatovite**, $K[(Al,Zn)_2(As,Si)_2O_8]$, has a feldspar-like topology (Fig. 36a) (Filatov et al. 2004). Filatovite is isostructural with the feldspar mineral celsian, $Ba[Al_2Si_2O_8]$. The cations Al^{3+}, Zn^{2+}, As^{5+}, and Si^{4+} are ordered over four crystallographic tetrahedral sites in the crankshaft chains in the framework. In the specimen studied by Filatov et al. (2004), two tetrahedral sites are populated by Al^{3+} and a small amount (~4%) of Zn^{2+}. Another two sites house mostly As^{5+} (55%) and Si^{4+} (45%). The K^+ cations are located on a single crystallographic site and are coordinated by nine O atoms. The authors observed that the distribution of Al^{3+} and As^{5+}/Si^{4+} over the tetrahedral sites obeys the aluminum-avoidance rule of Loewenstein (1954).

In the structure of **alarsite**, $AlAsO_4$, both Al^{3+} and As^{5+} are tetrahedrally coordinated (Fig. 36b). The structure is analogous to that of quartz (Machatschki 1936) with the exception that ordering of Al^{3+} and As^{5+} over the tetrahedral sites leads to doubling of the lattice parameter c. High-pressure behavior, which is similar to that of SiO_2, was studied by Sowa (1991).

Structures with polymerized $(As\phi_4)$ and $(M\phi_N)$ polyhedra.

Structures with unconnected $(As\phi_4)$ tetrahedra. This small group of minerals consists only of a few members. Three of them contain metallic clusters whose essential element is mercury. Hence, in contrast to sulfates or phosphates for which common minerals with isolated tetrahe-

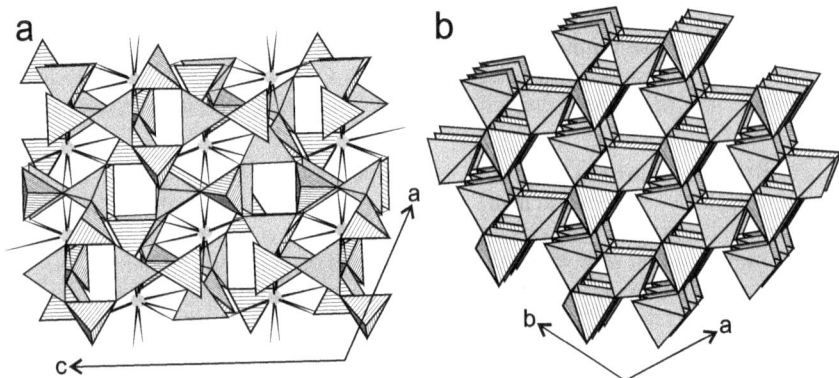

Figure 36. The structures of arsenates with infinite frameworks of tetrahedra: (a) filatovite, (b) alarsite.

dral units are known (e.g., epsomite and struvite, respectively), there are only a few of these arsenate minerals and those that are known are only rarely observed in the natural environment.

Linkage T. The unusual structure of **chursinite**, $(Hg_2)_3(AsO_4)_2$, is built by $(Hg_2)^{2+}$ pairs and AsO_4 tetrahedra (Fig. 37a) (Kamenar and Kaitner 1973). There are two crystallographically independent $(Hg_2)^{2+}$ pairs with a Hg-Hg distance of 2.535 Å. Each Hg atom is coordinated by the neighboring Hg atom and three oxygens from the arsenate tetrahedra. In contrast to other known structures of Hg^+, the O-Hg-Hg-O moiety deviates significantly from linearity (Kamenar and Kaitner 1973). Each oxygen atom of the AsO_4 tetrahedra binds to 2 Hg atoms with the exception of O4 which is bound to three Hg atoms, causing the unusually long As-O4 distance of 1.777 Å.

Linkage T. The structure of **kuznetsovite**, $Hg_3Cl(AsO_4)$, is built by $(Hg_3)^{4+}$ clusters, arsenate tetrahedra and Cl^- anions (Fig. 37b) (Romanenko et al. 1999). The $(Hg_3)^{4+}$ clusters are equilateral triangles; each Hg atom is located near an oxygen (2.17, 2.28, 2.60 Å) and chlorine (2.84 Å) atom. Therefore, each Hg could be considered as coordinated by other 2 Hg atoms, 3 O and 1 Cl atom, this coordination polyhedron being rather irregular. A much more regular polyhedron is observed when the entire $(Hg_3)^{4+}$ group is considered as a single cation. The polyhedron can be described as two interpenetrated tetrahedra, with the faces of the dominant tetrahedron being capped by the Cl^- anions. The distances from the center of gravity of the $[Hg_3]^{4+}$ cluster to O or Cl atoms vary between 3.18 and 3.83 Å. We also note an unusually large As-O distance ranging from 1.74 to 1.79 Å.

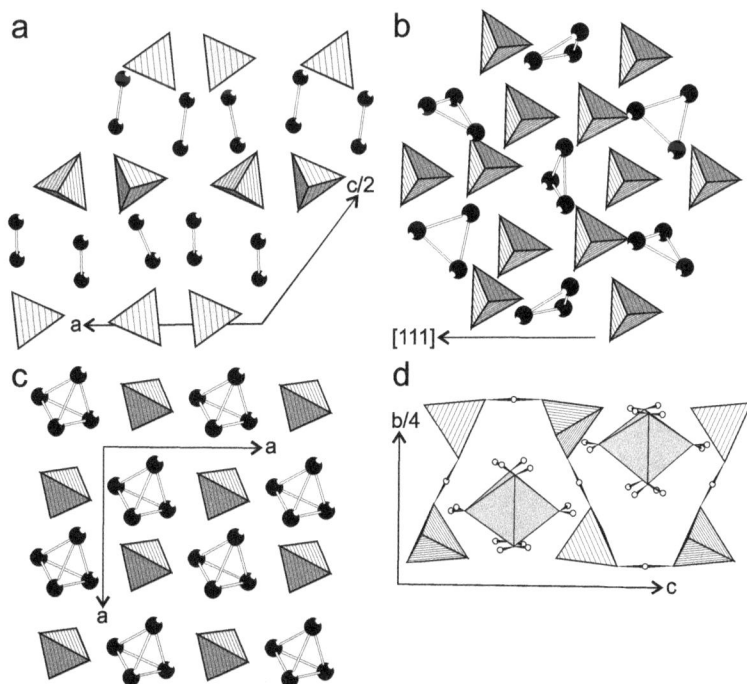

Figure 37. The structures of arsenates with unconnected (Asϕ_4) groups: (a) chursinite, with arsenate tetrahedra and the $(Hg_2)^{2+}$ pairs; (b) kuznetsovite, with arsenate tetrahedra and the $(Hg_3)^{4+}$ clusters; (c) tillmannsite, with arsenate tetrahedra and the tetrahedral Ag_3Hg clusters; (δ) rösslerite, with arsenate tetrahedra connected by the symmetric H bonds.

Linkage T. The structure of **tillmannsite**, $(Ag_3Hg)(V,As)O_4$, contains isolated $(V,As)O_4$ tetrahedra interspersed with tetrahedral clusters of average composition Ag_3Hg (Fig. 37c) (Sarp et al. 2003). The four metal atoms are randomly distributed over the four available sites.

Linkage M, T. The structure of **rösslerite**, $Mg(H_2O)_6(AsO_3OH)\cdot H_2O$, contains $Mg(H_2O)_6$ octahedra and arsenate tetrahedra (Ferraris and Franchini-Angela 1973). Both types of polyhedra are connected with the rest of the structure exclusively via hydrogen bonds (Fig. 37d). The structure can be described as an alternate sequence of two types of units (slabs) stacked along the b axis. One slab consists of the $Mg(H_2O)_6$ octahedra and an additional H_2O molecule. The second slab (Fig. 37d) includes the $Mg(H_2O)_6$ octahedra and arsenate tetrahedra. The octahedral units in both slabs facilitate hydrogen bonding within and between the slabs. A fascinating feature of the structure, however, is the role of hydrogen which bonds to the arsenate groups. There are two H atoms which protonate the arsenate groups and both form symmetrical hydrogen bonds, with the O-O distance in the O-H-O groups of 2.49 and 2.54 Å. When these symmetrical hydrogen bonds are taken into account, the arsenate tetrahedra can be described as building crankshaft chains with a general [001] direction.

Structures with finite clusters of tetrahedra and $M\phi_N$ polyhedra. Another sparsely populated group among the arsenate minerals is the one with heteropolyhedral clusters. Among these minerals, there are both clusters and isolated (AsO_4) tetrahedra in picropharmacolite.

Linkage T, M-T. The structure of **picropharmacolite**, $Ca_4Mg(H_2O)_7(AsO_3OH)_2(AsO_4)_2\cdot 4H_2O$, consists of (100) hydrogen-bonded, corrugated slabs (Catti et al. 1981). In these slabs, the $MgO_2(H_2O)_4$ octahedra are flanked by two arsenate tetrahedra to form $[Mg(H_2O)_4(AsO_3OH)(AsO_4)]$ clusters (Fig. 38a). Additional structural elements within the slabs are (AsO_3OH) and (AsO_4) tetrahedra. The clusters and isolated tetrahedra are linked by four independent $Ca\phi_{6-7}$ polyhedra. The $Ca\phi_6$ polyhe-

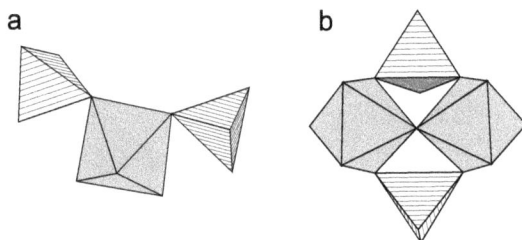

Figure 38. The structures of arsenates with finite clusters of tetrahedra and $M\phi_N$ polyhedra: (a) picropharmacolite, (b) esperanzaite.

dron deviates significantly from a regular octahedron. If the longer Ca-O distances (of 3.14-3.48 Å) are considered, the coordination of Ca is higher. There is only one direct hydrogen bond between the polyhedra between neighboring slabs, the rest of the hydrogen bonding is mediated by four crystallographically independent H_2O molecules located between the slabs. Picrophar-macolite is chemically and structurally related to guérinite and ferrarisite (Catti et al. 1981).

Linkage M-M, M-T. **Esperanzaite**, $NaCa_2Al_2F_4(OH)(H_2O)_2(AsO_4)_2$, is one of a few arsenate minerals with heteropolyhedral clusters. The clusters (Fig. 38b) consist of four polyhedra—two arsenate tetrahedra and two $AlF_2O_2(OH)(H_2O)$ octahedra (Foord et al. 1999). The two Al octahedra are bridged via an $(OH)^-$ group and the arsenate tetrahedra attach to the octahedra via corner sharing of the O^{2-} ligands. The overall composition of a cluster is $[Al_2F_4(OH)(H_2O)_2(AsO_4)_2]$. The clusters are connected via $Na\phi_5$ polyhedra into [010] chains and these chains are knitted via the $Ca\phi_8$ polyhedra into a relatively dense structure. Foord et al. (1999) noted that the Na positions are not fully occupied but the structural reasons for the Na deficiency are not understood.

Structures with infinite chains of tetrahedra and $M\phi_N$ polyhedra. Another meagerly inhabited group is the group with heteropolyhedral arsenate chains. The chains are cross-linked either by hydrogen bonds or by large polyhedra of divalent Ca or Pb.

Linkage M-T. The structure of **kaatialaite**, $Fe[AsO_2(OH)_2]_3 \cdot 5H_2O$, consists of infinite chains which extend in the direction of the c axis (Fig. 39a) (Boudjada and Guitel 1981). The composition of these chains is $Fe[AsO_2(OH)_2]_3$; they comprise $Fe^{3+}O_6$ octahedra and (H_2AsO_4) tetrahedra. The chains are interconnected only by an intricate network of hydrogen bonds mediated by five independent H_2O molecules. The topology of these chains is very similar, but not identical, to the chains in the structure of aluminocoquimbite (Demartin et al. 2010).

Linkage M-T. **Brassite**, $Mg(H_2O)_4(AsO_3OH)$, contains [010] heteropolyhedral chains (Fig. 39b). In these chains, the $Mg(H_2O)_4O_2$ octahedra alternate with AsO_3OH tetrahedra. The chains are cross-linked by a network of hydrogen bonds (Protas and Gindt 1976).

Linkage M-T. The structures of all arsenate **minerals with kröhnkite-type chains** (Table 13) are based on an octahedral-tetrahedral chain of the composition $[M(AsO_4)_2(H_2O)_2]$ (Fig. 39c,d) (Catti et al. 1977; Fleck et al. 2002). A similar chain can be found in the structurally related sulfate mineral kröhnkite $[Na_2Cu(H_2O)_2(SO_4)_2]$ or the phosphate mineral collinsite $[Ca_2(Mg,Fe)(H_2O)_2(PO_4)_2]$. In these chains, the $MO_4(H_2O)_2$ octahedra alternate regularly with a pair of (AsO_4) tetrahedra (Fig. 39d). The chains are linked by hydrogen bonds and by $Ca\phi_{7-8}$ polyhedra.

This group of structurally close minerals can be subdivided to monoclinic and triclinic phases (Table 13). Fleck et al. (2002) classified many natural and synthetic phases in this group

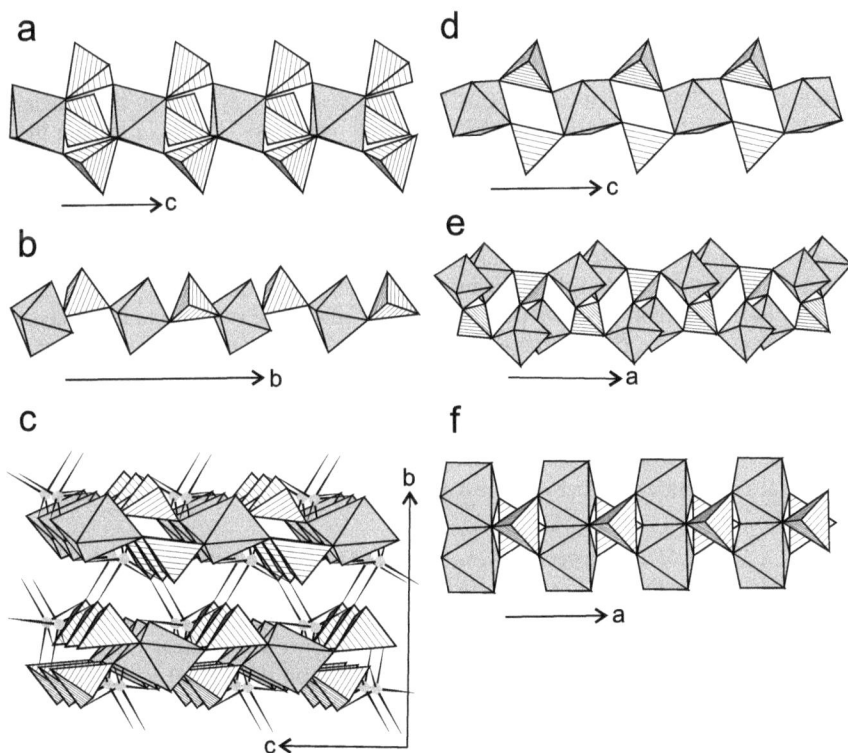

Figure 39. The structures of arsenates with infinite chains of tetrahedra and $M\phi_N$ polyhedra: (a) kaatialaite, (b) brassite, (c,d) brandtite and talmessite as members of the structures with kröhnkite-type chains; (c) shows the entire structure of brandtite with chains (polyhedral model) and the Ca cations (ball-and-stick models) between the chains, (d) shows the chains in talmessite; (e) bukovskýite, (f) burgessite.

Table 13. Arsenate minerals with the kröhnkite-type chains (after Fleck et al. 2002, with additional data from Keller et al. 2004, Kolitsch and Fleck 2006, Chukanov et al. 2009, Yang et al. 2011).

Monoclinic ($P2_1/c$)	Triclinic ($P\overline{1}$)
roselite $Ca_2(Co,Mg)(H_2O)_2(AsO_4)_2$	β-roselite $Ca_2Co(H_2O)_2(AsO_4)_2$
brandtite $Ca_2Mn(H_2O)_2(AsO_4)_2$	parabrandtite $Ca_2Mn(H_2O)_2(AsO_4)_2$
zincroselite $Ca_2Zn(H_2O)_2(AsO_4)_2$	gaitite $Ca_2Zn(H_2O)_2(AsO_4)_2$
wendwilsonite $Ca_2Mg(H_2O)_2(AsO_4)_2$	talmessite $Ca_2Mg(H_2O)_2(AsO_4)_2$
rruffite $Ca_2Cu(H_2O)_2(AsO_4)_2$	nickeltalmessite $Ca_2Ni(H_2O)_2(AsO_4)_2$

and divided them into six subgroups, some of them populated only by synthetic phases. Keller et al. (2004) considered the $[M(AsO_4)_2(H_2O)_2]$ chains in combination with the $Ca\phi_{7-8}$ polyhedra as rods whose packing determines the monoclinic or triclinic symmetry of the phases in this group.

There are several general observations valid for the minerals in this group (Keller et al. 2004). The $MO_4(H_2O)_2$ octahedra are usually distorted; in the triclinic phases, they are compressed; in the monoclinic phases, they are elongated. The coordination of Ca^{2+} is eight-fold in the triclinic phases but only seven-fold (or 7+1-fold) in the monoclinic phases. The bond valence calculations are usually in relatively poor agreement with the nominal charges of the ions, as already discussed by Hawthorne and Ferguson (1977).

Keller et al. (2004) noted that another hydrogen position may exist in the structure of gaitite. If so, the hydrogen atoms could be disordered which would perhaps explain the discrepancies in the bond valence calculations, not only for the gaitite structure but also for the structures of other phases in this group.

Linkage M-M, M-T. The dominant feature of the structure of **bukovskýite**, $[Fe_2(H_2O)_6(OH)(AsO_4)](SO_4)\cdot3H_2O$, are [100] chains made of Fe^{3+} octahedra and arsenate tetrahedra (Fig. 39e) (Majzlan et al. 2012b). Within these chains, we find dimers of $FeO_2(OH)(H_2O)_3$ octahedra bridged by the OH groups. Arsenate tetrahedra attach with all their ligands to the dimers, thus creating a chain with the overall composition $Fe_2(AsO_4)(H_2O)_6(OH)$. Sulfate tetrahedra and three crystallographically independent H_2O molecules are located in the space between these chains and are bonded to them by an intricate network of hydrogen bonds. The structure of bukovskýite shows a very strict distribution of As and S over the available tetrahedral sites; essentially no substitution between the two tetrahedral cations is observed, although these are known to substitute for each other in limited extent in some structures (e.g., in jarosite; Savage et al. 2005).

Linkage M=M, M-T. **Burgessite**, $Co_2(H_2O)_4[AsO_3(OH)]_2\cdot H_2O$, contains chains of Co octahedra and As tetrahedra (Fig. 39f) (Cooper and Hawthorne 2009). Two Co octahedra condense into an edge-sharing dimer $Co_2O_6(H_2O)_4$. The adjacent dimers are connected by two arsenate tetrahedra. Each tetrahedron shares two O atoms with a dimer, one O atom with the neighboring dimer, and the last ligand of the As^{5+} cation is an (OH) group. The overall composition of the chains is $[Co_2(H_2O)_4][AsO_3(OH)]_2$. There is an additional H_2O molecule, not bonded directly to any cation, in the structure. This H_2O molecule site is disordered off the center symmetry and half-occupied. Cooper and Hawthorne (2009) discussed the relationship between burgessite and the sheet structure of **erythrite**, $Co_3(H_2O)_8(AsO_4)_2$. In the latter structure, topologically similar but not identical chains connect into sheets.

Linkage M=M, M-T. The structural feature shared by the minerals of the **arsenbrackebuschite group** are the octahedral chains decorated by tetrahedra. The octahedra share edges and house smaller cations Fe^{3+}, Zn^{2+}, Cu^{2+}; the tetrahedra contain As^{5+}, Cr^{5+}, or S^{6+}. The chains are linked by large Pb^{2+} polyhedra.

The structure of **arsenbrackebuschite**, $Pb_2(Fe^{3+},Zn)(OH,H_2O)(AsO_4)_2$, is built by [010] chains of (Fe^{3+},Zn) $O_4(OH_2)_2$ octahedra. Four ligands bridge the (Fe^{3+},Zn) cations to the adjacent arsenate tetrahedra (Fig. 40a), giving the chains a resemblance of four-membered pinwheels when viewed along the screw twofold axis. The chains are linked by two distinct Pb polyhedra; Pb1 was reported in a six-fold coordination (Hofmeister and Tillmanns 1978). The coordination of Pb2 is more difficult to ascertain as the Pb-O distances scatter relatively uniformly over a range from 2.50 to 3.14 Å. Hofmeister and Tillmanns (1978) proposed that Pb2 is coordinated by at least eight oxygen atoms.

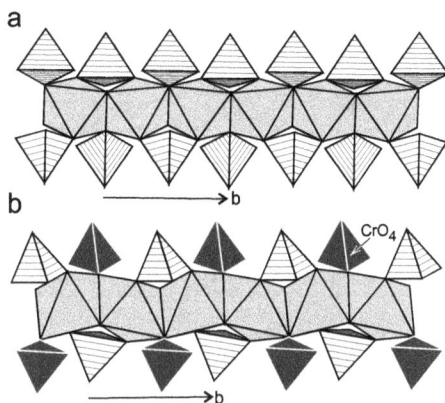

Figure 40. The structures of arsenates with infinite chains of tetrahedra and $M\phi_N$ polyhedra: (a,b) arsenbrackebuschite and fornacite as members of the arsenbrackebuschite group. Note the tetrahedral cation ordering (arsenate tetrahedra are hatched, chromate tetrahedra shaded dark grey) in (b).

As^{5+} and S^{6+} are disordered over the tetrahedral sites in **arsentsumebite**, $Pb_2Cu(OH)[(As,S)O_4]_2$. The distorted Cu octahedra are axially compressed, not elongated, as usually typical for the Jahn-Teller distorted Cu^{2+} octahedra. Such geometry is known only from a few Cu compounds (see Burns and Hawthorne 1996; Zubkova et al. 2002). Two independent Pb atoms are coordinated by 10 and 11 ligands.

In contrast to arsentsumebite, the tetrahedral cations As^{5+} and Cr^{6+} are ordered in **fornacite**, $Pb_2Cu(OH)(CrO_4)(AsO_4)$ (Fig. 40b) (Cocco et al. 1967). The Cu^{2+} octahedra are axially elongated. Lead is 9-fold coordinated and the large, irregular polyhedra are the only structural elements which interconnect the chains and knit the structure into a dense three-dimensional network. The phosphate analogue of fornacite is the mineral vauquelinite.

Based on the powder X-ray diffraction data, Clark et al. (1997b) proposed that **feinglosite**, $Pb_2(Zn,Fe)(H_2O)[(As,S)O_4]$, is monoclinic and structurally related to arsenbrackebuschite.

Structures with infinite sheets of tetrahedra and $M\phi_N$ polyhedra.

Linkage M-M, M-T (assumed). **Maghrebite**, $MgAl_2[(OH)(AsO_4)]_2 \cdot 8H_2O$, was reported to be isostructural with laueite (Meisser and Brugger 2006), with no further details. Structural features of the laueite group were described by Huminicki and Hawthorne (2002). All minerals in this group contain heteropolyhedral sheets.

Linkage M=M, M-T. **Akrochordite**, $(Mn,Mg)_5(OH)_4(H_2O)_4(AsO_4)_2$, and **guanacoite**, $Cu_2Mg_2(Mg_{0.5}Cu_{0.5})(OH)_4(H_2O)_4(AsO_4)_2$, are isostructural. They are an appealing example of combination of structural fragments into larger units. The basic building blocks of the structures are infinite octahedral ribbons which extend in the *a* direction (Moore et al. 1989; Witzke et al. 2006). The ribbons are alternatively two- and three-octahedra wide and their topology is identical to the octahedral ribbons found in the structures of amphiboles. The ribbons are cross-linked by arsenate octahedra into thick (010) slabs (Fig. 41a). Within the slabs, the ribbons are imbricated such that their long axes are always [100] but the short axes alternate between [011]

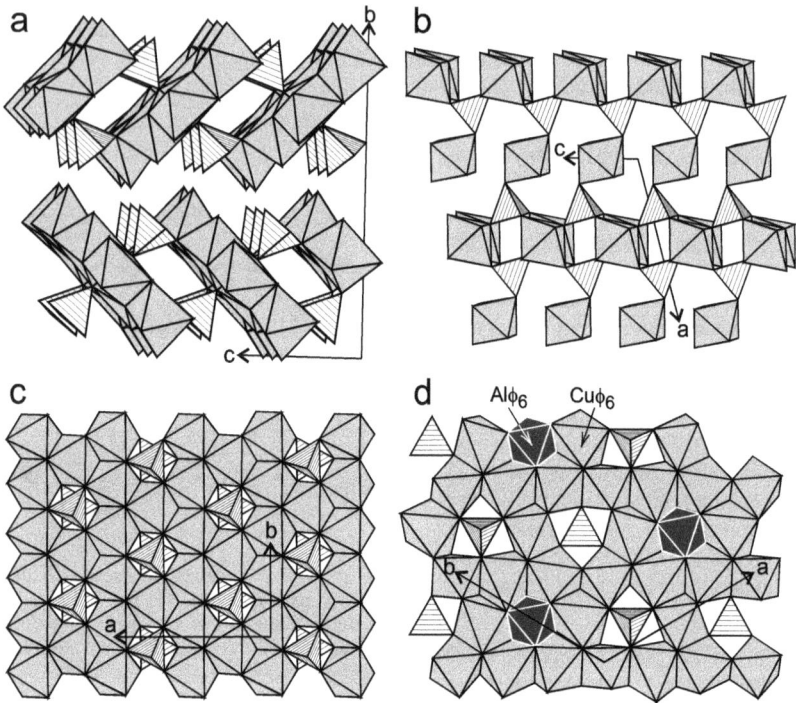

Figure 41. The structures of arsenates with infinite sheets of tetrahedra and Mϕ_N polyhedra: (a) side view of two adjacent layers in akrochordite, (b) annabergite as a member of the vivianite group (top view), (c) bayldonite (top view), (d) chalcophyllite (top view).

and [01$\bar{1}$] between adjacent slabs. The slabs are bonded to each other only via hydrogen bonds; hydrogen positions have been determined for the structure of guanacoite (Witzke et al. 2006).

Linkage M=M, M-T. The arsenate members of the **vivianite group** are listed in Table 14. The structure of these minerals consists of $MO_2(OH_2)_4$ octahedra and $M_2O_6(OH_2)_4$ double octahedral groups (Hill 1979; Schmetzer et al. 1982; Wildner et al. 1996; Capitelli et al. 2007). These octahedra and clusters of octahedra are linked by arsenate tetrahedra to form relatively open sheets stacked along [010] (Fig. 41b). There are only hydrogen bonds between the sheets.

There is some confusion about the names symplesite, parasymplesite, and ferrisymplesite in the literature. **Symplesite** is reported as triclinic, **parasymplesite** is reported as monoclinic

Table 14. Arsenate members of the vivianite group (for references see text).

annabergite	$Ni_3(H_2O)_8(AsO_4)_2$
erythrite	$Co_3(H_2O)_8(AsO_4)_2$
köttigite	$Zn_3(H_2O)_8(AsO_4)_2$
hörnesite	$Mg_3(H_2O)_8(AsO_4)_2$
manganohörnesite	$(Mn,Mg)_3(H_2O)_8(AsO_4)_2$
symplesite or parasymplesite	$Fe_3(H_2O)_8(AsO_4)_2$
metaköttigite	$(Zn,Fe^{3+})(Zn,Fe^{3+},Fe^{2+})_2(H_2O,OH)_8(AsO_4)_2$

and isostructural with vivianite. Unfortunately, the structure reported by Mori and Ito (1950) is monoclinic and named symplesite. **Ferrisymplesite** is an X-ray amorphous ferric arsenate. It was named for its supposed relationship to symplesite.

Linkage M=M, M-T. The structure of **bayldonite**, $Cu_3Pb(OH)_2(AsO_4)_2$, is built by brucite-like octahedral sheets which comprise three independent Cu^{2+} positions (Fig. 41c) (Ghose and Wan 1979). Each Cu^{2+} is coordinated by a Jahn-Teller distorted octahedron. A fourth of the available octahedral positions are vacant and the arsenate tetrahedra attach *via* three oxygen atoms to the sheets above and below the vacant positions. The fourth oxygen of each arsenate tetrahedron points in the space between the sheets which is populated by Pb^{2+} cations. Pb^{2+} is coordinated by eight ligands that define a square antiprism. Ghose and Wan (1979) refined the position of the hydrogen atom but this atom appears to be disordered between O4 and O5 with O-H distances of 1.3 and 1.4 Å, respectively. Ghose and Wan (1979) also speculated that for compositions richer in Zn, an ordered derivative of the bayldonite structure may be possible, with the composition $Cu_2ZnPb(OH)_2(AsO_4)_2$ and Zn located on the most regular Cu2 site.

Linkage M=M, M-T. The principal feature of the structure of **chalcophyllite**, $Cu_9Al(OH)_{12}$ $(H_2O)_6(SO_4)_{1.5}(AsO_4)_2 \cdot 12H_2O$, are octahedral (0001) sheets (Fig. 41d) (Sabelli 1980). These brucite-like sheets are constructed from distorted $Cu\phi_6$ and regular $Al\phi_6$ octahedra. Three quarters of the available octahedral positions are occupied by Cu^{2+}, 1/12 by Al^{3+}, and 1/6 are vacant. Arsenate tetrahedra attach to the octahedral sheets above and below the vacant positions. The space between the octahedral-tetrahedral layers is occupied by H_2O molecules and SO_4 tetrahedra. Hence, the layers are held together by an intricate network of hydrogen bonds. Furthermore, the occupancy of the SO_4 units was found to be only ~3/4. The missing SO_4 groups may be replaced by two H_2O molecules; the exact nature of charge compensation for the missing SO_4 groups is unknown.

Linkage M=M, M-T, with exceptions: M=M, M=T, M-T (haidingerite, pharmacolite); *M=M, M-M, M-T* (pushcharovskite). Minerals of the **krautite group** (Table 15, Fig. 42) are not isostructural but closely related in terms of their structures. All of these minerals have layer structures with layers being linked by hydrogen bonds. The coordination of the divalent cations, on the other hand, varies among these phases while the principal features of the sheets remain the same. For this reason, we include into this group structures whose heteropolyhedral sheets contain $M\phi_{6-8}$ polyhedra and arsenate tetrahedra.

The structure of **krautite** is built by (010) layers (Fig. 42a) (Catti and Franchini-Angela 1979). The layers are composed of chains of $Mn\phi_6$ octahedra which run in the $[10\bar{1}]$ direction. The chains are interconnected by the $As\phi_4$ tetrahedra. Each of the four independent arsenate tetrahedra attaches to the octahedral chains via 3 oxygen atoms; the fourth oxygen is protonated and participates in hydrogen bonding between the layers. Adjacent layers are held together only by hydrogen bonds, either from the As-OH or $Mn-OH_2$ groups, resulting in the excellent (010) cleavage of krautite. The layers are stacked along and related by the 2_1 axis.

Table 15. Minerals of the krautite group (for references see text).

krautite	$Mn(H_2O)(AsO_3OH)$
haidingerite	$Ca(H_2O)(AsO_3OH)$
pharmacolite	$Ca(H_2O)_2(AsO_3OH)$
fluckite	$CaMn(H_2O)_2(AsO_3OH)_2$
koritnigite	$Zn(H_2O)(AsO_3OH)$
cobaltkoritnigite	$(Co,Zn)(H_2O)(AsO_3OH)$
geminite	$Cu(H_2O)(AsO_3OH)$
pushcharovskite	$Cu(H_2O)(AsO_3OH) \cdot 0.5H_2O$
related structures	
yvonite	$Cu(H_2O)(AsO_3OH) \cdot H_2O$
schultenite	$Pb(AsO_3OH)$

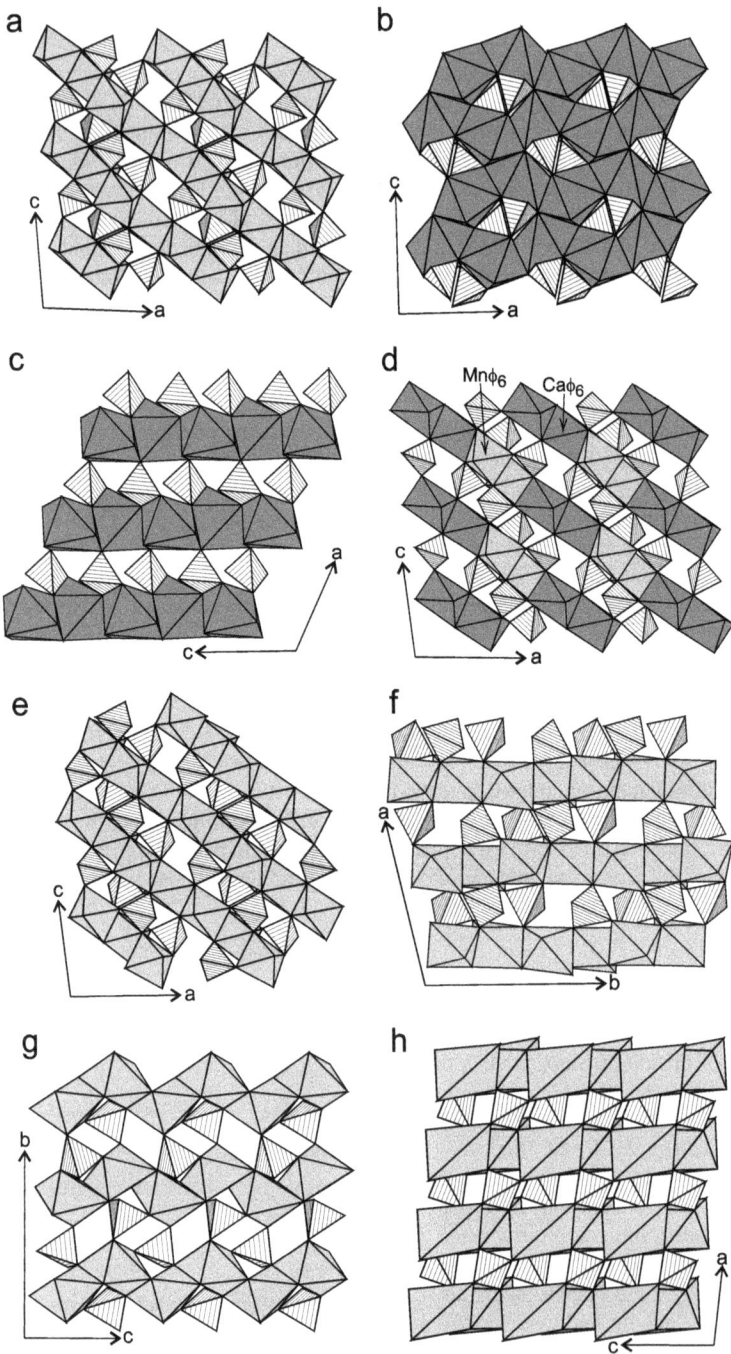

Figure 42. Top view of the sheets in the structures of the members of krautite group (arsenates with infinite sheets of tetrahedra and Mφ$_N$ polyhedra): (a) krautite, (b) haidingerite, (c) pharmacolite, (d) fluckite, (e) koritnigite, (f) geminite, (g) yvonite, (h) schultenite. Note the increase of the coordination number of the M cations from krautite to haidingerite and pharmacolite. The M cations are ordered in fluckite.

The layered structure of **haidingerite** (Ferraris et al. 1972a) is related to that of krautite. The principal difference among the two structures is the coordination of the divalent cations in the (010) layers. Distorted octahedral coordination of Mn^{2+} in krautite is replaced by irregular seven-fold coordination of Ca^{2+} in the structure of haidingerite, with Ca-O distances between 2.34 and 2.74 Å (Fig. 42b). The Ca polyhedra condense into [100] chains which attach to each other to form two-dimensional framework. The arsenate tetrahedra join the octahedral layers by corner- and edge-sharing via 3 oxygens. The fourth oxygen is protonated, points into the interlayer space, and participates along with the H_2O ligands of the Ca^{2+} cations in the hydrogen bonding between the layers. The layers are stacked along the *b* axis and related by a glide plane.

Pharmacolite has a layered structure composed of arsenate tetrahedra and large, irregular polyhedra which house Ca^{2+} (Ferraris 1969). Calcium is coordinated by 7 oxygens; Ferraris (1969) argued that an eighth oxygen, at a distance of 2.962 Å from the Ca^{2+} ion, also belongs to its coordination sphere. The chains of large Ca polyhedra extend in the [001] direction and are held together by arsenate tetrahedra which share both edges and corners with the Ca polyhedra (Fig. 42c). As in the case of krautite and haidingerite, the layers interact only via hydrogen bonds (Ferraris et al. 1971). Pharmacolite is isostructural with brushite, $Ca(PO_3OH)\cdot2H_2O$.

The structure of **fluckite** is closely related to that of krautite (Catti et al. 1980a). In fluckite, the octahedral chains parallel to [10$\bar{1}$] host Ca and Mn in an ordered fashion Ca-Ca-Mn-Mn (Fig. 42d). The Ca^{2+} coordination polyhedra are slightly larger than those with Mn^{2+}. The chains are interlinked by arsenate tetrahedra into sheets parallel with (010). The only significant difference between fluckite and krautite is the stacking of the sheets. In krautite, the sheets are related by the 2_1 screw axis, whereas in fluckite, the adjacent sheets are related only by the **b** vector, that is, stacked onto each other.

The pseudomonoclinic, triclinic structure of **koritnigite** is closely related to the structure of krautite (Keller et al. 1980). Zn^{2+} octahedra form chains along [10$\bar{1}$] and are interlinked by arsenate tetrahedra into sheets (Fig. 42e). The structure of **cobaltkoritnigite** (Zettler et al. 1979) is very similar to that of koritnigite. There are differences, however, in the hydrogen positions and hydrogen bonding schemes, between the two compounds (Keller et al. 1980).

Sheet structure of **geminite** (Fig. 42f) (Cooper and Hawthorne 1995a) similar to that of krautite. The significant difference between the structure of geminite and those of the minerals in this group is the strong Jahn-Teller distortion of the Cu^{2+} octahedra (Prensipe et al. 1996).

Sheet structure of **pushcharovskite** (Pushcharovsky et al. 2000) has a topology similar to that seen in geminite. The Cu chains are built by distorted octahedra and square pyramids. In contrast to geminite, the chains are also bridged by additional $Cu\phi_5$ or $Cu\phi_6$ polyhedra. The large unit cell contains 15 As and 18 Cu positions. H_2O molecules, and a small amount of K^+ also occurs in the interlayer portion of the structure. Owing to the large number of atoms in the unit cell and disorder on many interlayer positions, the exact formula of the mineral is uncertain.

Linkage M=M, M-T. The structure of **yvonite**, $Cu(H_2O)(AsO_3OH)\cdot H_2O$, is composed of (100) sheets which are made of [001] zigzag chains of strongly distorted $Cu\phi_6$ octahedra (Fig. 42g) (Sarp and Černý 1998). The cations in the chains fill only half of the available octahedral positions within the sheets. The vacant sites between the chains are populated by two independent (AsO_3OH) tetrahedra which link the chains into the sheets. There are additional H_2O molecules in the interlayer portion of the structure. They participate in the hydrogen bonding between the sheets but are not bonded directly to any cation. Sarp and Černý (1998) drew attention to the similarity of the sheets in the structures of yvonite, geminite, and fluckite.

Linkage M=M, M-T. The crystal structure of **schultenite**, $Pb(AsO_3OH)$, is built by (010) layers reminiscent of the layers in the minerals of the krautite group (Fig. 42h). Within the layers, there are infinite [001] chains composed of irregular Pb polyhedra (Effenberger and

Pertlik 1986). Lead is found in 6-fold coordination and the polyhedra can be described as strongly distorted trigonal antiprisms. The chains are linked by arsenate tetrahedra into the (010) layers which are bonded only via a system of hydrogen bonds. Hydrogen atoms are disordered over two positions at room temperature which is close to the temperature of ferro- to paraelectric transition in this phase (Wilson et al. 1991).

Linkage M=M, M-T, with exception *M, M=M, M-T* (richelsdorfite). The **lavendulan group** (Table 16) contains several structurally related but not necessarily isostructural compounds, all of them complex arsenates of copper and other cations. These structures are characterized by heteropolyhedral sheets (Giester et al. 2007) with Cu^{2+} cations in five-fold coordination, that is, there are no $Cu\phi_6$ units in these structures. The sheets are bonded to each other via coordination polyhedra around mono- or divalent cations and additional hydrogen bonds. The $Na\phi_6$ polyhedra are not considered to be a constituent of the heteropolyhedral sheets because of the low bond valence of the Na-O bonds, comparable to a bond valence attributable to a hydrogen bond.

Table 16. Minerals of the lavendulan group (after Giester et al. 2007).

heteropolyhedral sheets, *linkage M=M, M-T*	
lavendulan	$NaCaCu_5(H_2O)_5Cl(AsO_4)_4$
zdeněkite	$NaPbCu_5(H_2O)_5Cl(AsO_4)_4$
lemanskiite	$NaCaCu_5(AsO_4)_4Cl\cdot5H_2O$
shubnikovite	$Ca_2Cu_8(AsO_4)_6Cl(OH)\cdot7H_2O$
mahnertite	$(Na,Ca)Cu_{2.75}(H_2O)_{3.63}Cl_{0.62}(AsO_4)_2$
heteropolyhedral sheets, *linkage M, M=M, M-T*	
richelsdorfite	$Ca_2Cu_5Sb^{5+}Cl(OH)_6(H_2O)_6(AsO_4)_4$
lavendulan-like heteropolyhedral frameworks	
andyrobertsite	$KCdCu_5(AsO_4)_4[As(OH)_2O_2]\cdot2H_2O$
calcio-andyrobertsite	$KCaCu_5(AsO_4)_4[As(OH)_2O_2]\cdot2H_2O$

In **lavendulan** itself, the heteropolyhedral sheets comprise clusters $[Cu_4O_{12}Cl]$ of four edge-sharing tetragonal pyramids (Fig. 43a). Each $[Cu_4O_{12}Cl]$ cluster connects to eight neighboring arsenate tetrahedra (Giester et al. 2007). Another CuO_5 tetragonal pyramid attaches to the sheets via corner-sharing with the arsenate tetrahedra. There are trigonal-prismatic $Na\phi_6$ and irregular $Ca\phi_7$ polyhedra in the space between the sheets.

Zdeněkite (Zubkova et al. 2003) is isostructural with lavendulan. **Lemanskiite** is a polymorph of lavendulan and was reported to be tetragonal (Ondruš et al. 2006). Given that lavendulan and zdeněkite are strongly pseudo-tetragonal, lemanskiite may also be monoclinic. **Mahnertite** is tetragonal (Pushcharovsky et al. 2004), with sheet topology very similar to that in lavendulan. **Shubnikovite** is an inadequately described mineral, likely related to lavendulan (Giester et al. 2007).

There are additional $Sb\phi_6$ octahedra in the space between the sheets in the mineral **richelsdorfite** (Fig. 43b) (Süsse and Tillmann 1987). Several minerals related to richelsdorfite, with minor variation in the chemical composition, were reported (see Giester et al. 2007). The details on the structural variations of these phases are unknown.

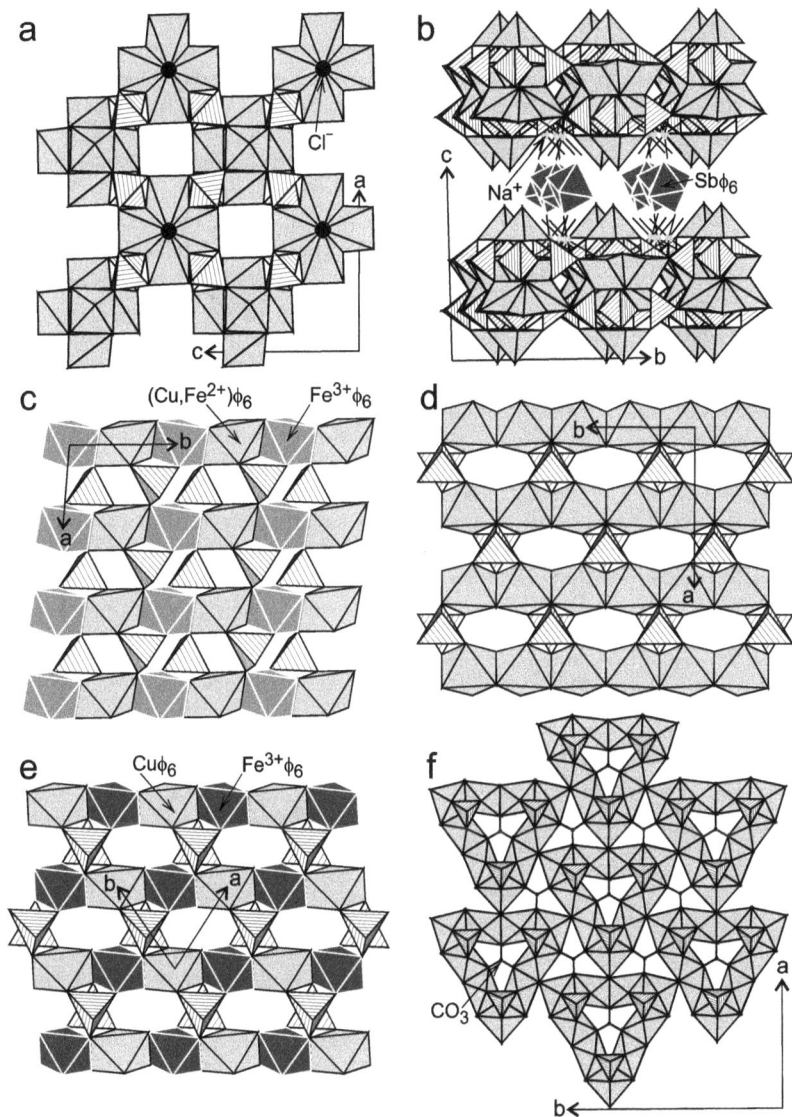

Figure 43. The structures of the arsenates with infinite sheets of tetrahedra and $M\phi_N$ polyhedra: (a) lavendulan, top view of the sheet. (b) richelsdorfite, side view of the sheets, the $Sb\phi_6$ octahedra and Na polyhedra between the sheets. (c) medenbachite (top view), (d,e) tsumcorite and gartrellite as members of the tsumcorite group. Note the cation ordering in (e) gartrellite. (f) sailaufite as a member of the mitridatite group.

The minerals **andyrobertsite** and **calcio-andyrobertsite** are closely related to lavendulan. However, the sheets in these minerals are directly connected to each other, therefore forming a heteropolyhedral network.

Linkage M=M, M-T. The minerals **medenbachite**, $Bi_2Fe(Cu,Fe)(O,OH)_2(OH)_2(AsO_4)_2$, **neustädtelite**, $Bi_2Fe_2^{3+}O_2(OH)_2(AsO_4)_2$, and **cobaltneustädtelite**, $Bi_2Fe^{3+}Co^{2+}O(OH)_3(AsO_4)_2$ are isostructural (Fig. 43c). Their crystal structures consist of [010] chains of edge-sharing octahedra (Krause et al. 1996, 2002b). In medenbachite, two types of octahedra alternate regularly in the chains, a regular octahedron, assumed to be occupied only by Fe, and a strongly distorted octahedron, assumed to house Cu^{2+} and a small amount of Fe. Some Fe/Co ordering within the octahedral chains in cobaltneustädtelite was noted based on the variations of the R values. Based on the refined Fe-O distances, most of the iron should be trivalent in medenbachite. The presence of ferrous iron was suspected (Krause et al. 1996) but later refuted (Krause et al. 2002b). If ferric iron substitutes for Cu^{2+} in medenbachite, the O/OH ratio must change accordingly, as indicated in the chemical formula of the mineral. This substitution scheme was confirmed by Krause et al. (2002b) for neustädtelite and cobaltneustädtelite where it can be written as $Fe^{3+}O^{2-} = Co^{2+}(OH)^-$. The octahedral chains are cross-linked by arsenate tetrahedra into (001) sheets. The space between the sheets is occupied by large $Bi\phi_7$ polyhedra in which Bi atoms are statistically distributed over two positions. Partial ordering of the Bi atoms in medenbachite and cobaltneustädtelite doubles the volume of the unit cell.

Linkage M=M, M-T. The phases in the **tsumcorite group** (Table 17) share a general formula $AB_2(OH,OH_2)_2(TO_4)_2$, where A = Pb, Ca, Na, Bi; B = Fe^{3+}, Mn^{3+}, Cu, Zn, Co, Ni, Al; and T = P, As, V, S (Krause et al. 1998). Most of the minerals are monoclinic (space group $C2/m$), a few are triclinic (P-1). The structures are built by chains of edge-sharing $B\phi_6$ octahedra. Arsenate tetrahedra (for T = As, of course) attach to the chains and connect them into (001) sheets of $[B(OH,OH_2)AsO_4]$ composition (Fig. 43d,e). The sheets are bonded together only via large $A\phi_8$ polyhedra and hydrogen bonds. The A cation is located at the symmetry center which means that the usual one-sidedness in the $Pb\phi_8$ or $Bi^{3+}\phi_8$ polyhedra (owing to the lone $6s^2$ electron pair) is not observed.

Both A and B sites are able to accommodate di- and trivalent cations. The variation in the M^{2+}/M^{3+} ratio on either the A or the B site is coupled to the OH-H_2O substitution, rarely to AsO_4-SO_4 substitution (Krause et al. 1998). The OH/H_2O ratio in the tsumcorite-group minerals tends to be close to 1.0 (Krause et al. 2002a), most likely because a particular, probably stable hydrogen-bond network can be established. A few minerals deviate from this ratio but in these cases, the ratio is either very small or very large. Therefore, although the substitution on the cation sites may be almost unlimited, the extent of the solid solutions may be limited by the possible departures of the OH/H_2O ratio from ~1.

Ordering within the octahedral chains was described in **gartrellite** and **zincgartrellite** (Fig. 43e) (Effenberger et al. 2000) and assumed in **lukrahnite** (Krause et al. 2001b). Significant numbers of vacancies on the B site were reported for **manganlotharmeyerite** by Brugger et al. (2002). They proposed the substitution $[Mn^{3+} + (AsO_4)^{3-} + (OH)^-] = [\square + AsO_2(OH)_2^- + H_2O]$. Kampf et al. (1984) suggested the substitution $[(AsO_4)^{3-} + H_2O] = [(AsO_3OH)^{2-} + (OH)^-]$ for **lotharmeyerite**. The protonation state of the arsenate group remains unclear.

Nomenclature of the minerals of the lotharmeyerite subgroup was proposed by Brugger et al. (2002). The relationship between gartrellite, zincgartrellite, and **helmutwinklerite** was discussed by Effenberger et al. (2000).

Linkage M=M, M-M, M-T. The shared structural feature of the minerals of the **mitrid-atite group** are pseudotrigonal sheets with nonamers of octahedra (Fig. 43f) (see Wildner et al. 2003). Arsenate minerals in this group are **arseniosiderite**, $Ca_3Fe_4(OH)_6(AsO_4)_4 \cdot 3H_2O$, and **sailaufite**, $(Ca,Na,)_2Mn_3O_2(H_2O)_2(CO_3)(AsO_4)_2 \cdot H_2O$,. The affiliation of arseniosiderite is only assumed, albeit on the basis of strong arguments (Moore and Ito 1974), because the structure of this mineral was never directly solved. Other members of the group are the phosphates mitrid-atite, robertsite, and pararobertsite (Moore and Araki 1977b; Wildner et al. 2003).

As indicated above, the sheets in these structures are composed of triangular nonamers of edge-sharing octahedra of a general composition $[M_9\phi_{36}]$ (Fig. 43f). The nonamers attach to each other via corner sharing. In mitridatite as a model structure for arseniosiderite, the octa-

Table 17. Minerals of the tsumcorite group (Tillmanns and Gebert 1973; Pring et al. 1989; Kharisun et al. 1997; Krause et al. 1998, 1999, 2001a,b, 2002a,b; Brugger et al. 2000, 2002; Effenberger et al. 2000).

tsumcorite	$Pb(Zn,Fe^{3+},Fe^{2+})_2(H_2O,OH)_2(AsO_4)_2$
mawbyite	$PbFe_2(OH)_2(AsO_4)_2$
helmutwinklerite	$PbZn_2(H_2O)_2(AsO_4)_2$
thometzekite	$Pb(Cu,Zn)_2(H_2O)_2(AsO_4)_2$
gartrellite	$PbCuFe^{3+}(OH)(H_2O)(AsO_4)_2$
zincgartrellite	$Pb(Zn,Cu)(Zn,Fe^{3+})(H_2O,OH)_2(AsO_4)_2$
rappoldite	$Pb(Co,Ni)_2(H_2O)_2(AsO_4)_2$
cobalttsumcorite	$Pb(Co,Fe^{3+})_2(H_2O,OH)_2(AsO_4)_2$
lotharmeyerite	$Ca(Zn,Mn^{3+})_2(H_2O,OH)_2(AsO_4)_2$
cobaltlotharmeyerite	$Ca(Co,Fe^{3+})_2(H_2O,OH)_2(AsO_4)_2$
nickellotharmeyerite	$Ca(Ni,Fe^{3+})_2(H_2O,OH)_2(AsO_4)_2$
ferrilotharmeyerite	$Ca(Fe^{3+},Zn)(OH,H_2O)_2(AsO_4)_2$
manganlotharmeyerite	$Ca(Mn^{3+},Mg)_2(OH,H_2O)_2\{AsO_4,[AsO_2(OH)_2]\}_2$
cabalzarite	$Ca(Mg,Al,Fe)_2(OH,H_2O)_2(AsO_4)_2$
lukrahnite	$CaCuFe^{3+}(OH)(H_2O)(AsO_4)_2$
schneebergite	$BiCo_2(OH)(H_2O)(AsO_4)_2$
nickelschneebergite	$BiNi_2(OH)(H_2O)(AsO_4)_2$

hedra house Fe^{3+} and are relatively regular. In sailaufite, the octahedra house Mn^{3+} and show a strong Jahn-Teller distortion. Arsenate tetrahedra attach to the octahedral sheets. In sailaufite, the central gap in the nonamers is occupied by the carbonate ions.

The space between the octahedral-tetrahedral layers is filled by $Ca\phi_7$ or $(Ca,Na,)\phi_7$ polyhedra and two crystallographically distinct H_2O molecules, not directly bonded to any cation.

Linkage M=M, M-M, M-T. The structures of both polytypes (1*M* and 2*M*) of **tyrolite**, $Ca_2Cu_9(OH)_8(CO_3)(H_2O)_{11}(AsO_4)_4 \cdot xH_2O$, $x = 0$-1, contain five independent Cu positions, two As positions, and one Ca position. Cu is mostly six-fold coordinated, with the usual strong Jahn-Teller distortion of the coordination polyhedra. Some of the ligands, however, show only partial occupancy, so that the coordination of Cu may be locally five-fold – square pyramidal. The Ca^{2+} cation resides in a $CaO_5(H_2O)_2$ polyhedron.

The structure of tyrolite is based on complex slabs (Fig. 44) with a thickness of 2.6 nm, hence called "nanolayers" by Krivovichev et al. (2006). These authors elucidated the topology of the slabs and divided them into

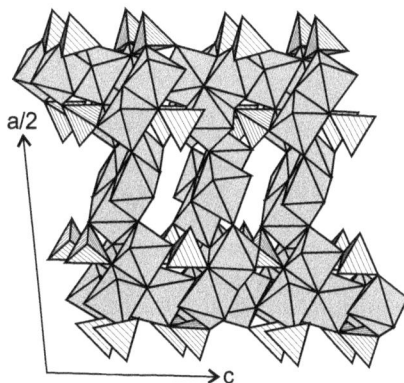

Figure 44. The structure of tyrolite, a side view of a nanolayer.

sublayers A and B. Sublayer B is the central part of the slabs (or nanolayers) and consists of octahedral chains extending in the [010] direction. These chains function essentially as pillars between the neighboring sublayers A. The sublayers A are built by octahedral trimers which are interconnected via corner sharing with arsenate tetrahedra. The sublayers A are capped by an additional sublayer made of the $Ca\phi_7$ polyhedra. The overall schematic sequence of the nanolayers is Ca-A-B-A-Ca. The space between the nanolayers is filled by H_2O molecules and $(CO_3)^{2-}$ anions. The structures of tyrolite-$1M$ and tyrolite-$2M$ differ only by the stacking sequence of the nanolayers, as expected for polytypes.

Structures with infinite frameworks of tetrahedra and $M\phi_N$ polyhedra. This group is the largest one among the arsenates, reflecting the tendency of the arsenate oxyanions to polymerize with other polyhedral units in the structures. The framework structures range from very dense frameworks, such as in nabiasite, xanthiosite, or grischunite, to minerals with essentially layered structure but with layers interconnected with sparse additional structural fragments, such as in leogangite or rollandite. There are structures with large channels, such as those of mixite or pharmacosiderite. In addition to the usual linkages $M=M$, $M-M$, $M-T$, there are also examples of edge sharing between tetrahedra and octahedra (linkage $M=T$) in ferrarisite, arsenoclasite, lammerite, and parwelite, and face sharing of octahedra (linkage $M\equiv M$) in allactite and wilhelmkleinite.

Linkage M-T. **Scorodite**, $Fe(H_2O)_2AsO_4$, **mansfieldite**, $Al(H_2O)_2AsO_4$, (Harrison 2000; Zoppi and Pratesi 2009) and **yanomamite**, $In(H_2O)_2AsO_4$, (Tang et al. 2002) are isostructural. The crystal structure of scorodite, one of the most common secondary minerals of As^{5+}, is a relatively dense framework of $FeO_4(OH_2)_2$ octahedra and AsO_4 tetrahedra (Fig. 45a) (Kitahama et al. 1975). Four O ligands of the Fe^{3+} cations and all four O ligands of the As^{5+} cations participate in the polymerization of the polyhedra into the framework structure. The Fe^{3+} octahedra share vertices only with arsenate tetrahedra, i.e., no ligands bridge two Fe^{3+} cations. Full polymerization of the structure is precluded by the presence of H_2O as ligands of Fe^{3+}. Earlier suggestions that the scorodite structure may contain the H_3O^+ moieties were refuted by Hawthorne (1976a). Other isostructural arsenates or phosphates have been reviewed by Botelho et al. (1994).

Linkage M-T. The crystal structure of **parascorodite**, $Fe(H_2O)_2AsO_4$, belongs to the framework structures but can be described in terms of infinite columns (or chains) in the [001] direction, interconnected by additional polyhedral units (Fig. 45b) (Perchiazzi et al. 2004). The columns have the composition $[Fe_2(AsO_4)_6]$ and their axes coincide with the threefold symmetry axes. The FeO_6 octahedra share all their vertices with arsenate tetrahedra and the columns have topology similar to that found in kaatialaite $[Fe(H_2AsO_4)_3 \cdot 5H_2O]$. The columns are connected by $FeO_3(OH_2)_3$ octahedra, thus creating a framework structure.

Linkage M-T. **Berzeliite**, $(NaCa_2)Mg_2(AsO_4)_3$, and **manganberzeliite**, $(NaCa_2)$ $Mn_2(AsO_4)_3$, adopt the well-known garnet structure (Hawthorne 1976b), with Na and Ca occupying the large dodecahedral sites, Mg (and a minor amount of Mn) the octahedral sites, and As^{5+} the tetrahedral sites (Fig. 45c). In a garnet structure, all cations are located at special positions with no adjustable coordinates. Therefore, the variation in cation size must be accommodated by varying the only general position, that of the single oxygen atom. In terms of the polyhedral arrangement, such variation corresponds to the rotation of the polyhedra. The angle α, which is a measure of tetrahedral rotations in garnets, is larger in berzeliite than in silicate garnets. Hawthorne (1976b) assigned the large value of α to the presence of a pentavalent cation (As^{5+}) at the tetrahedral site as opposed to Si^{4+} in the silicate garnets.

Linkage M-M, M-T. **Tilasite**, $CaMgFAsO_4$, **maxwellite**, $NaFeFAsO_4$, and **durangite**, $NaAlFAsO_4$ are isostructural with titanite, $CaTiOSiO_4$, or kieserite, $Mg(H_2O)SO_4$ (Fig. 45d). The earliest descriptions of this structure are due to Strunz (1937) and Kokkoros (1938).

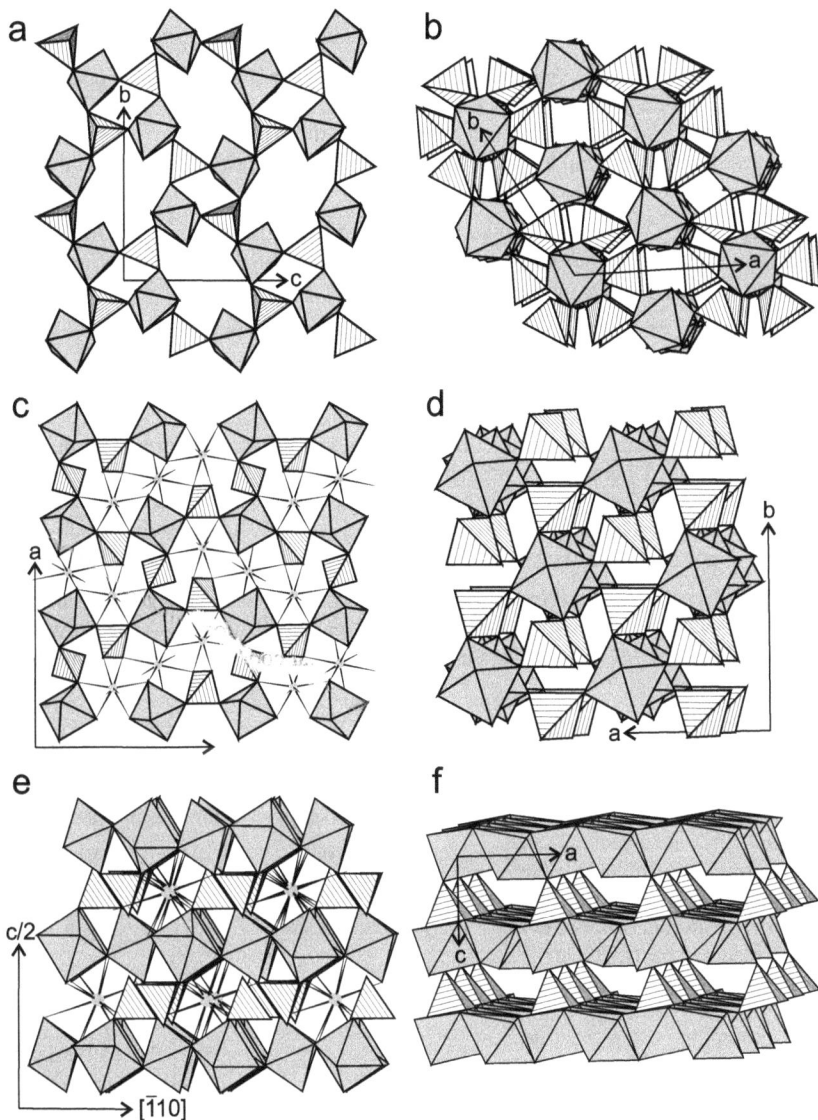

Figure 45. The structures of the arsenates with infinite frameworks of tetrahedra and Mϕ$_N$ polyhedra: (a) scorodite, (b) parascorodite, (c) berzeliite, (d) durangite, (e) beudantite, (f) gilmarite.

The structure of tilasite is similar or identical to that of titanite, CaTiSiO$_5$, although the details of the structure were a matter of a two-paper debate (Bladh et al. 1972; Bermanec 1994). Strunz (1937) proposed that tilasite and titanite are isostructural, although titanite crystallizes in a centrosymmetric space group *C2/c* and tilasite was shown to be piezoelectric (Smith and Prior 1911). Based on the fact that tilasite was shown to be piezoelectric, Bladh et al. (1972) refined the structure in a space group *Cc*, with the structural features being very close to those in titanite. The main argument for the non-centrosymmetric space group was a slight offset of the Ca atoms from the two-fold axis. Bermanec (1994) inspected another sample of tilasite and concluded that the mineral is indeed centrosymmetric, stating that "piezoelectric and pyroelectric effects are not infrequent in strongly pseudo-centrosymmetric minerals."

The structure of tilasite is built by edge-sharing octahedral chains which house the Mg^{2+} cations. The orientation of the chains is [001] in the $C2/c$ setting or [101] in the Cc setting. The edges shared between the MgO_4F_2 octahedra are the F^- atoms. The chains are cross-linked by arsenate tetrahedra, resulting in a three-dimensional skeleton. Large voids within the network are occupied by CaO_6F groups.

For maxwellite, Cooper and Hawthorne (1995b) discuss the possibility of local clustering of the $CaMg(OH)(AsO_4)$ and $Na(Fe,Al)F(AsO_4)$ compositional units, that is, the preferred coordination polyhedra $CaO_6(OH)$ and NaO_6F.

Linkage M-M, M-T. The phases of the **alunite supergroup** with a general formula $AB_3(TO_4)_2(OH)_6$ possess a structure common for many sulfates (alunite-jarosite group) and phosphates (Jambor 1999). The structure can be described as (0001) octahedral sheets linked by tetrahedra (Fig. 45e). The B sites are the octahedral ones, the T sites the tetrahedral ones. Within the sheets, octahedra share corners and the metal ions lie on the nodes of a triangular *kagomé* lattice. The sheets are decorated by tetrahedra which may house S^{6+}, P^{5+}, or As^{5+}. Large cavities between the sheets are the A sites and contain the mono- or divalent cations. Many arsenate or sulfo-arsenate minerals belong to this group (Table 18) and their basic features conform well to the structure of alunite. Some deviations from this basic structure are described below.

Among the A-site cations, Pb^{2+} shows a specific behavior owing to its lone electron pair. In **beudantite**, Pb occupies six sites around the "ideal" A-cation position (Szymański 1988). Other cations, for example Ba^{2+}, are found in a much more regular coordination environment (e.g., in **dussertite**, Kolitsch et al. 1999).

There is extensive substitution at the octahedral B sites. The most common octahedral cations in this group are Fe^{3+} and Al^{3+}. Other rarer cations, such as Ga^{3+}, Sb^{5+}, are also mentioned in the literature (see Table 18). Ge^{4+} was reported from an unnamed phase from Tsumeb, Namibia (Jambor et al. 1996) and assigned to the B site. The charge-compensating mechanisms for the substitutions of trivalent cations for those with higher valence are usually rebuffed with speculations about H atoms vacancies.

Substitution of As^{5+} by P^{5+} or S^{6+} at the T site is very common. In some cases, no ordering of As/S/P was observed. Ordering on the tetrahedral positions was reported in **gallobeudantite** (Jambor et al. 1996) with the associated symmetry reduction to $R3m$. The charge-compensating mechanism for the As^{5+}-S^{6+} exchange is not precisely known.

A novel structure based on the alunite parent structure is that of

Table 18. Arsenate and sulfo-arsenate minerals of the alunite supergroup (after Bayliss et al. 2010).

arsenogorceixite	$BaAl_3[AsO_{3.5}(OH)_{0.5}]_2(OH)_6$
arsenocrandallite	$CaAl_3[AsO_{3.5}(OH)_{0.5}]_2(OH)_6$
arsenoflorencite-(Ce)	$CeAl_3(AsO_4)_2(OH)_6$
arsenoflorencite-(La)	$LaAl_3(AsO_4)_2(OH)_6$
philipsbornite	$PbAl_3[AsO_{3.5}(OH)_{0.5}]_2(OH)_6$
arsenogoyazite	$SrAl_3[AsO_{3.5}(OH)_{0.5}]_2(OH)_6$
dussertite	$BaFe_3[AsO_{3.5}(OH)_{0.5}]_2(OH)_6$
graulichite-(Ce)	$CeFe_3(AsO_4)_2(OH)_6$
segnitite	$PbFe_3[AsO_{3.5}(OH)_{0.5}]_2(OH)_6$
arsenoflorencite-(Nd)*	$NdAl_3(AsO_4)_2(OH)_6$
arsenowaylandite*	$BiAl_3(AsO_4)_2(OH)_6$
weilerite	$BaAl_3(As_{0.5}S_{0.5}O_4)_2(OH)_6$
hidalgoite	$PbAl_3(As_{0.5}S_{0.5}O_4)_2(OH)_6$
kemmlitzite	$SrAl_3(As_{0.5}S_{0.5}O_4)_2(OH)_6$
beudantite	$PbFe_3(As_{0.5}S_{0.5}O_4)_2(OH)_6$
gallobeudantite	$PbGa_3(As_{0.5}S_{0.5}O_4)(SO_4)(OH)_6$
kolitschite	$Pb[Zn_{0.5},M_{0.5}]Fe_3(AsO_4)_2(OH)_6$

*questionable as minerals

kolitschite (Grey et al. 2008). Zinc is a major component here although zinc was long considered to be incompatible with the alunite-type structure. The zinc atoms are ordered in trigonal bipyramidal sites in half of the available six-membered rings of the layers of octahedra. Ordered displacements of the Pb atoms occur in response to the Zn–M ordering.

Linkage M=M, M-T. The structure of **gilmarite**, $Cu_3(OH)_3(AsO_4)$, is formed by [010] chains in which distorted octahedra $Cu\phi_6$ and square pyramids $Cu\phi_5$ share edges (Sarp and Černý 1999). These chains are linked by $Cu\phi_5$ square pyramids into (001) sheets. The sheets are connected by arsenate tetrahedra, resulting in an open framework structure (Fig. 45f).

The structure of **arhbarite**, $Cu_2Mg(OH)_3(AsO_4)$, can be derived from that of gilmarite (Krause et al. 2003). Mg occupies the octahedral position, Cu is found in five-fold coordination. Because of the lack of suitable single crystals, no complete structural model is available for arhbarite.

Linkage M=M, M-T. The minerals **chenevixite**, $Cu_2Fe_2(OH)_4(H_2O)(AsO_4)_2$, and **luetheite**, $Cu_2Al_2(OH)_4(H_2O)(AsO_4)_2$, are isostructural although a full structural solution and refinement was carried out only for chenevixite (Burns et al. 2000b). Cu^{2+}, Fe^{3+}, and Al^{3+} are sixfold coordinated. Cu^{2+} resides in Jahn-Teller distorted octahedra which share edges and form [100] chains. The $(Al,Fe^{3+})\phi_6$ octahedra are attached to opposing sides of the chain and the chains are cross-linked by AsO_4 tetrahedra. Owing to twinning, the structure was solved in a non-standard space group $B2_1$ and we were unable to retrieve the structure from the data given in the paper by Burns et al. (2000b).

Linkage M=M, M-T. The open framework structure of **rollandite**, $Cu_3(H_2O)_2(AsO_4)_2·2H_2O$, is assembled from (010) sheets decorated by arsenate tetrahedra and cross-linked by [100] chains (Fig. 46a) (Sarp and Černý 2000). The sheets are modulated and made of distorted $Cu\phi_6$ octahedra which share edges. The topology of the sheets is similar to the topology of the brucite sheets with additional vacancies on the cation sites. Arsenate tetrahedra attach to the sheets above and below the vacant sites via two oxygen atoms. The third oxygen atom of each arsenate tetrahedron is shared with [100] chains of $Cu\phi_5$ square pyramids which share edges. The fourth oxygen atom of each tetrahedron points into [100] cavities between the $Cu\phi_5$ chains which are populated by H_2O molecules.

Linkage M=M, M-T. The backbone of the framework structure of **euchroite**, $Cu_2(OH)(H_2O)_3(AsO_4)$, are chains of edge-sharing $Cu1\phi_6$ octahedra (Fig. 46b) (Eby and Hawthorne 1989). These octahedra show a pronounced Jahn-Teller distortion but are flanked by even more Jahn-Teller distorted $Cu2\phi_6$ octahedra. The two crystallographically unique octahedra are linked by arsenate tetrahedra into an open tetrahedral-octahedral framework. Eby and Hawthorne (1989) note two significant distortions in this structure, namely the commensurate modulation along [010] and the strong distortion of the Cu2 octahedra which leads to a small shift of adjacent chains along [001].

Linkage M=M, M-T. One of the most complex structures among natural arsenate minerals, **bouazzerite**, $Bi_6(Mg,Co)_{11}Fe_{14}O_{12}(OH)_4(H_2O)_{52}(AsO_4)_{18}·34H_2O$, can be described as a zeolite-like framework structure (Brugger et al. 2007). The key structural constituent are $[Bi_3Fe_7O_6(OH)_2(AsO_4)_9]$ clusters. The geometric center of the cluster is a FeO_6 trigonal prism. All six edges of the basal planes of the trigonal prism are shared with edges of six FeO_6 octahedra. These heptamers of FeO_6 polyhedra are decorated by nine AsO_4 tetrahedra and three large, irregular $Bi\phi_{6-8}$ polyhedra. The irregular, one-sided nature of the Bi polyhedra is a result of the stereochemical activity of the $6s^2$ lone electron pair on Bi^{2+}. These clusters are linked together by $Mg(H_2O)_6$ octahedra. The channels in the structure of bouazzerite are ~4.8 Å wide and house a number of crystallographically distinct H_2O molecules.

Linkage M=M, M-T. The crystal structure of **cornwallite**, $Cu_5(OH)_4(AsO_4)_2$, is similar to that of its polymorph cornubite (Artl and Armbruster 1999). There are sheets of distorted $Cu\phi_6$

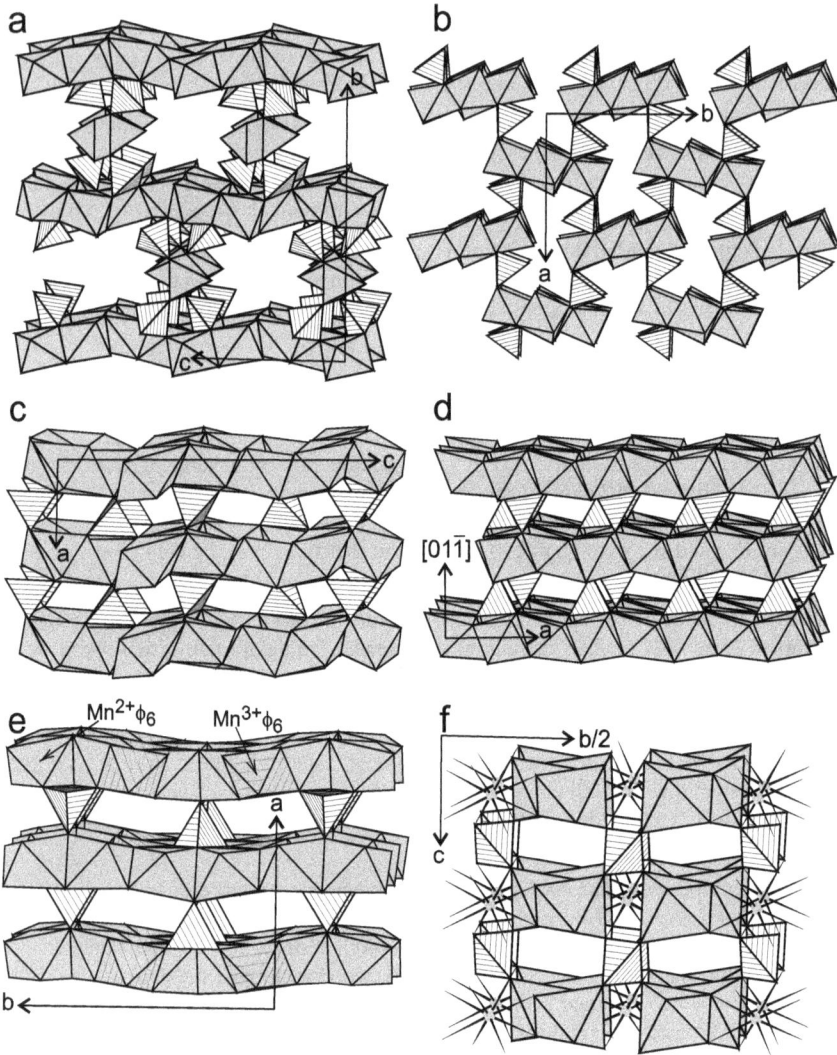

Figure 46. The structures of the arsenates with infinite frameworks of tetrahedra and $M\phi_N$ polyhedra: (a) rollandite, (b) euchroite, (c) cornwallite, (d) cornubite, (e) flinkite, (f) retzian.

octahedra in the structure of cornwallite (Fig. 46c). The sheets are cross-linked by arsenate tetrahedra. The tetrahedra share two corners with one sheet and another two with an adjacent sheet. This arrangement leads to a stronger distortion within the octahedral sheets than in cornubite. In cornubite, the tetrahedra share one corner with one sheet and three corners with the adjacent sheet.

Linkage M=M, M-T. The prominent feature of the structure of **cornubite**, $Cu_5(OH)_4(AsO_4)_2$, are the sheets parallel to $(0\bar{1}1)$ which are connected into a three-dimensional framework (Fig. 46d). Within the sheets, three crystallographically independent copper cations are coordinated in a Jahn-Teller distorted octahedral arrangement. In the sheets, five of the six available octahedral positions are occupied by Cu^{2+} and the sixth one is vacant. The arsenate tetrahedra attach by three oxygen atoms to the vacant positions from both sides of the sheets; each of the three oxygen atoms bridges one As^{5+} with two Cu^{2+} cations. The fourth oxygen bridges the As^{5+} cation with three Cu^{2+} cations. The arsenate tetrahedra link the octahedral sheets into a framework structure. Tillmanns et al. (1985) provided a detailed discussion of related natural and synthetic compounds.

Linkage M=M, M-T. The principal feature of the structure of **flinkite**, $Mn_2^{2+}Mn^{3+}(OH)_4(AsO_4)$, are the (100) octahedral sheets cross-linked by arsenate tetrahedra (Fig. 46e) (Moore 1967b). The sheets are constructed from [010] chains of $Mn^{2+}\phi_6$ octahedra; the chains are linked by $Mn^{3+}\phi_6$ octahedra. The Jahn-Teller distortion of the $Mn^{3+}\phi_6$ octahedron is fairly strong; the $Mn^{2+}\phi_6$ octahedron is relatively regular (Kolitsch 2001). The (100) sheets are topologically similar to sheets in the structures of pyrochroite, $Mn(OH)_2$, or brucite, $Mg(OH)_2$, with the exception that some octahedral positions in the sheets of flinkite are vacant to allow the attachment of the arsenate tetrahedra. In addition, the sheets in flinkite are slightly modulated.

Linkage M=M, M-T. **Retzian**, $Mn_2REE(OH)_4(AsO_4)$, contain sheets of $Mn\phi_6$ and $REE\phi_8$ polyhedra, cross-linked by the arsenate tetrahedra (Fig. 46f) (Moore 1967b). The backbone of the sheets are [100] zigzag octahedral chains which house the Mn^{2+} cations. The octahedra share edges and meld into a sheet via additional edge sharing with the $REE\phi_8$ square antiprism. The adjacent $REE\phi_8$ square antiprism do not attach to other; instead, small square-shaped cavities are left between the antiprisms. Arsenate tetrahedra attach to sheets above and below the square-shaped cavities. The basic structural features of retzian are reminiscent of the structure of flinkite, $Mn_2^{2+}Mn^{3+}(OH)_4(AsO_4)$, although marked differences exist, especially in the coordination of REE^{3+} in retzian versus Mn^{3+} in flinkite and the connectivity of polyhedra within the sheets (Moore 1967b).

Linkage M=M, M-T. **Mapimite**, $Zn_2Fe_3O_{12}(OH)_4(H_2O)_4(AsO_4)_3 \cdot 6H_2O$, is an example of an open heteropolyhedral framework (Fig. 47a). The structure is built by (001) octahedral-tetrahedral slabs which are interconnected by arsenate tetrahedra and a network of hydrogen bonds (Ginderow and Cesbron 1981). The slabs are constructed by octahedral clusters with composition $[Zn_2Fe_3O_{12}(OH)_4(H_2O)_4]$. Within these compact clusters, the octahedra share only edges. The clusters are knitted together via corner sharing with arsenate tetrahedra into the (001) slabs. Additional arsenate tetrahedra are located in the space between the slabs and provide the bonding between them, in addition to the hydrogen bonds. The large cavities between slabs are populated by H_2O molecules.

Linkage M=M, M-T. The overall cubic symmetry and the basic features of the structure of the **pharmacosiderite group** (Table 19) were established in the early studies of Zemann (1947) and Buerger et al. (1967). Buerger et al. (1967) probably investigated a sample of hydronium-dominated pharmacosiderite as no alkalis were found, either by chemical or by X-ray diffraction analysis. In spite of the reported cubic symmetry, Zemann (1948) and Buerger et al. (1967) gave accounts of parts of the investigated pharmacosiderite crystals being optically anisotropic.

The structure of **pharmacosiderite** can be described as an open heteropolyhedral framework (Fig. 47b). The key structural fragments in pharmacosiderite are clusters of $[Fe_4(OH)_4O_{12}]$ (or $[Al_4(OH)_4O_{12}]$ in **alumopharmacosiderite**). They consist of edge-sharing octahedra and each cluster is centered on a lattice node of a primitive cubic lattice.

Table 19. Minerals of the pharmacosiderite group (Zemann 1948; Buerger et al. 1967; Peacor and Dunn 1985; Walenta 1994).

pharmacosiderite	$KFe_4(OH)_4(AsO_4)_3 \cdot 5H_2O$
natropharmacosiderite	$NaFe_4(OH)_4(AsO_4)_3 \cdot 5H_2O$
bariopharmacosiderite	$Ba_{0.5}Fe_4(OH)_4(AsO_4)_3 \cdot 5H_2O$
alumopharmacosiderite	$KAl_4(OH)_4(AsO_4)_3 \cdot 5H_2O$

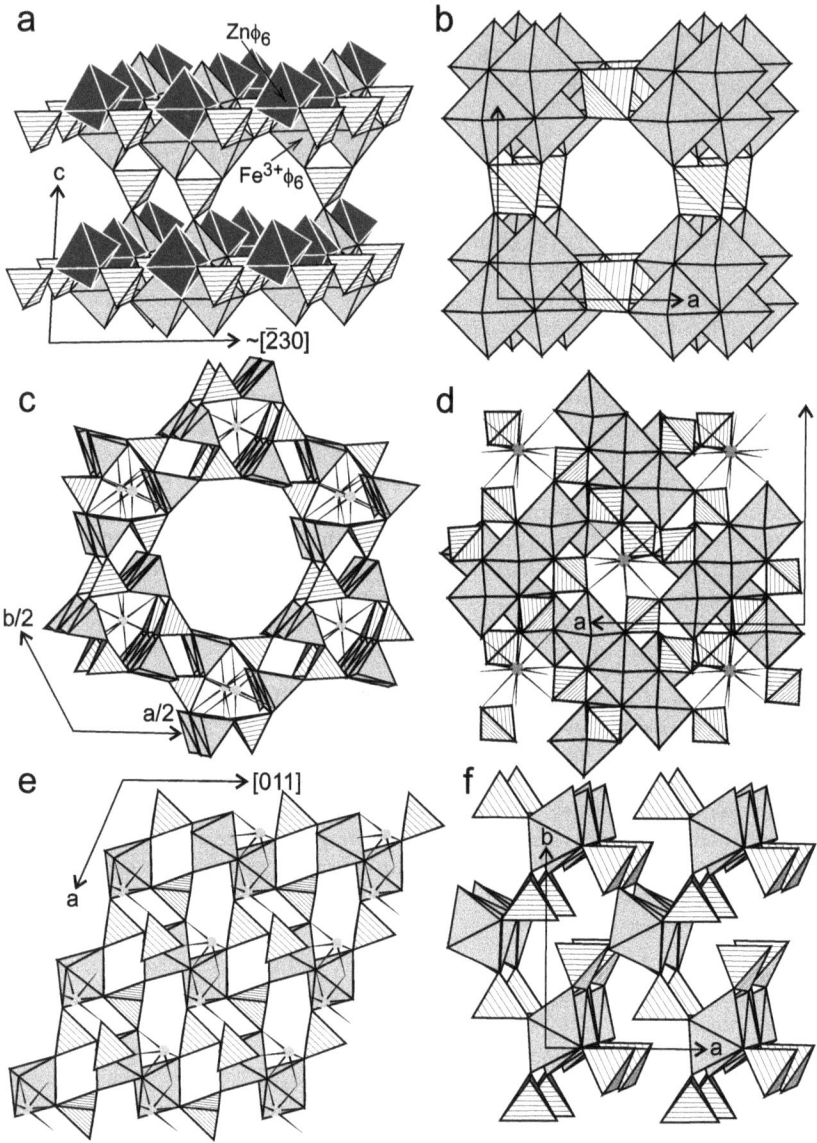

Figure 47. The structures of the arsenates with infinite frameworks of tetrahedra and $M\phi_N$ polyhedra: (a) mapimite, (b) pharmacosiderite, (c) mixite, (d) nabiasite, (e) paganoite, (f) adelite.

The clusters are bridged by arsenate tetrahedra. This arrangement results in a three-dimensional network of channels, about 4.4 Å in diameter. The water content of pharmacosiderite is uncertain, perhaps variable.

Linkage M=M, M-T. The minerals in the **mixite group** (Table 20) have a general formula $Cu_6A(OH)_6(TO_4)_3 \cdot 3H_2O$, where A = Ca, Pb, Bi, Y, rare-earth elements, and T = As, P (Aruga and Nakai 1985). The group is mostly populated by arsenates, a phosphate member is the mineral petersite (Peacor and Dunn 1982).

The honeycomb-like, zeolitic framework (Fig. 47c) is constructed by assembling [001] chains of $Cu\phi_5$ polyhedra and cross-linking them by the arsenate tetrahedra and the large $A\phi_9$ polyhedra. The $Cu\phi_5$ polyhedra are square pyramids; Aruga and Nakai (1985) proposed that Cu is six-fold coordinated, with the longest Cu-O distance of 3.14 Å. This distal ligand, however, has very large thermal parameters, hinting that it may be mobile. The heteropolyhedral walls enclose tubular, one-dimensional channels which extend along the *c* axis in the structure of mixite and related minerals.

Table 20. Minerals of the mixite group (Wise 1978; Aruga and Nakai 1985; Mereiter and Preisinger 1986; Sejkora et al. 1999; Walenta and Theye 2004).

agardite (agardite-(Nd), agardite-(Y), agardite-(La), agardite-(Ce))	$Cu_6REE(OH)_6(AsO_4)_3 \cdot 3H_2O$
plumboagardite	$Cu_6Pb(OH)_6(AsO_3OH)(AsO_4)_2 \cdot 3H_2O$
mixite	$Cu_6Bi(OH)_6(AsO_4)_3 \cdot 3H_2O$
goudeyite	$Cu_6Al(OH)_6(AsO_4)_3 \cdot 3H_2O$
zálesíte	$Cu_6Ca(OH)_6(AsO_3OH)(AsO_4)_2 \cdot 3H_2O$

In mixite, the stereochemical inactivity of the lone electron pair on Bi^{3+} preserves the hexagonal symmetry of the structure and the $Bi\phi_9$ coordination polyhedron (tricapped trigonal prism) (Miletich et al. 1997).

The reversible loss and gain of H_2O in the channels has been studied, from a crystallographic point of view, by Miletich et al. (1997).

Linkage M=M, M-T. The dense structure of **nabiasite**, $BaMn_9(OH)_2[(V,As)O_4]_6$, is based on a nearly cubic close-packed arrangement of the oxygen anions (Fig. 47d) (Brugger et al. 1999). The close-packed planes are parallel to (111) of the cubic unit cell. The stacking of the layers is ABB' where B and B' are related by inversion. Mn^{2+} is coordinated octahedrally and there are two independent MnO_6 octahedra and an $MnO_5(OH)$ octahedron. The $Mn\phi_6$ octahedra define a framework whose cavities are populated by $Ba[(V,As)O_6]$ groups. Ba^{2+} is found in 12-fold coordination and the six pairs of its ligands bridge Ba^{2+} and the six neighboring vanadate-arsenate tetrahedra. The remaining two oxygen atoms of the tetrahedra, i.e., those which are not bonded to the Ba^{2+}, connect the tetrahedra to the octahedral framework.

Linkage M=M, M-T. The structure of **paganoite**, $NiBiOAsO_4$, contains dimers of distorted Ni^{2+} octahedra of the composition Ni_2O_{10}; the two octahedra share an edge (Fig. 47e) (Roberts et al. 2001). All corners of the dimer, except for the two which define the shared edge, bridge the dimer to the adjacent arsenate tetrahedra, thus forming an open framework. The large voids of the framework house the Bi^{3+} cations in a five-fold coordination. This coordination polyhedron could be described as a square pyramid but the Bi^{3+} cation is located outside of the pyramid, below its base. This location is the result of the presence of a lone s^2 pair on the Bi^{3+} cation.

Linkage M=M, M-T. The minerals of the **adelite group** share the general formula $AB(OH)AsO_4$, where A = Ca and B = Co, Cu, Zn, M, Ni *or* A = Pb and B = Cu, Zn, Fe (Table 21) (e.g., Clark et al. 1997b). The basic features of the structure were established by Qurashi and Barnes (1963). The structure of these minerals is built by chains of edge-sharing octahedra (Fig. 47f). Because of the possibility of exchanging the vectors **a**, **b**, and **c** in the space group $P2_12_12_1$, different authors report different orientations of the chains. The octahedra house the B cations and are defined by four O atoms and two OH groups. The size and degree of distortion vary as a function of the nature of the B cation. The size variations were recently reviewed by Sakai et al. (2009); the distortion is greatest, as expected, for Cu^{2+}. Adjacent octahedral chains are cross-linked by arsenate tetrahedra, thus creating an octahedral-tetrahedral framework. The

Table 21. Minerals of the adelite group (Qurashi and Barnes 1963; Moore 1967a; Cesbron et al. 1987; Giuseppetti and Tadini 1988; Clark et al. 1997a; Kharisun et al. 1998; Witzke et al. 2000; Effenberger et al. 2002; Keller et al. 2003; Yang et al. 2007; Henderson et al. 2008; Sakai et al. 2009).

adelite	$CaMg(OH)AsO_4$
conichalcite	$CaCu(OH)AsO_4$
austinite	$CaZn(OH)AsO_4$
cobaltaustinite	$CaCo(OH)AsO_4$
nickelaustinite	$CaNi(OH)AsO_4$
gottlobite	$CaMg(OH)(V,As)O_4$
arsendescloizite	$PbZn(OH)AsO_4$
duftite	$PbCu(OH)AsO_4$
gabrielsonite	$PbFe(OH)AsO_4$

voids in the framework are occupied by the large *A* cations (either Ca or Pb) in eight-fold coordination of a distorted square antiprism. Extensive solid solutions among the end-members have been reported (see the references in Sakai et al. 2009). There is a phase in this group, β-**duftite**, (Pb,Ca)$Cu(OH)AsO_4$, which appears to be a simple chemical intermediate between **duftite** and **conichalcite**, but is an incommensurately modulated superstructure (Kharisun et al. 1998).

The membership of **gabrielsonite** to the adelite group is somewhat unclear. Moore (1967a) reported that his gabrielsonite crystals "yielded excellent [data]," yet no structural model is given. He assigned the crystals to the space group $P2_1ma$, noting that the symmetry is nearly *Pnma*. Quarashi and Barnes (1963) refuted *Pnma* as the space group for the adelite-group phases and found that the correct space group is $P2_12_12_1$, based on a few weak reflections. Since no work has appeared on gabrielsonite since 1967, it is questionable whether the mineral does belong to the adelite group or not.

Linkage M=M, M-T. The crystal structure of **lindackerite** and related minerals (Table 22) is based on sheets and additional structural constituents (Hybler et al. 2003). The topology of the sheets is similar to the sheets of the minerals of the krautite group (Fig. 48a, compare with Fig. 42). Within the sheets in lindackerite, there are infinite [110] chains made of Jahn-Teller distorted $Cu\phi_5$ and $Cu\phi_6$ polyhedra. The $Cu\phi_5$ polyhedra are distorted square pyramids with the composition CuO_5. The $Cu\phi_6$ polyhedra are distorted octahedra with the composition $CuO_4(OH)(H_2O)$, where the oxygen of the OH group is being shared with the neighboring arsenate tetrahedron. The chains are connected by arsenate tetrahedra (AsO_4) and (AsO_3OH) to form the (001) sheets. The space between the sheets is occupied by $MO_2(H_2O)_4$ octahedra

Table 22. Members and related phases of the lindackerite group (for references see text).

lindackerite	$[Cu(H_2O)_4][Cu_4(H_2O)_4(AsO_3OH)_2(AsO_4)_2]\cdot H_2O$
geigerite	$[Mn(H_2O)_4][Mn_4(H_2O)_4(AsO_3OH)_2(AsO_4)_2]\cdot 2H_2O$
chudobaite	$[(Mg,Zn)H_2O)_4][(Mg,Zn)_4(H_2O)_4(AsO_3OH)_2(AsO_4)_2]\cdot 2H_2O$
klajite	$[Mn(H_2O)_4][Cu_4(H_2O)_4(AsO_3OH)_2(AsO_4)_2]\cdot H_2O$
pradetite	$[Co(H_2O)_4][Cu_4(H_2O)_4(AsO_3OH)_2(AsO_4)_2]\cdot H_2O$
veselovskýite	$[Zn(H_2O)_4][Cu_4(H_2O)_4(AsO_3OH)_2(AsO_4)_2]\cdot H_2O$
ondrušite	$[Ca(H_2O)_4][Cu_4(H_2O)_4(AsO_3OH)_2(AsO_4)_2]\cdot H_2O$
Related phases	
ferrarisite	$[Ca(H_2O)_4][Ca_4(H_2O)_4(AsO_3OH)_2(AsO_4)_2]\cdot H_2O$
slavkovite	$[Cu(H_2O)_4][Cu_{12}(H_2O)_{14}(AsO_4)_6(AsO_3OH)_4]\cdot 7H_2O$

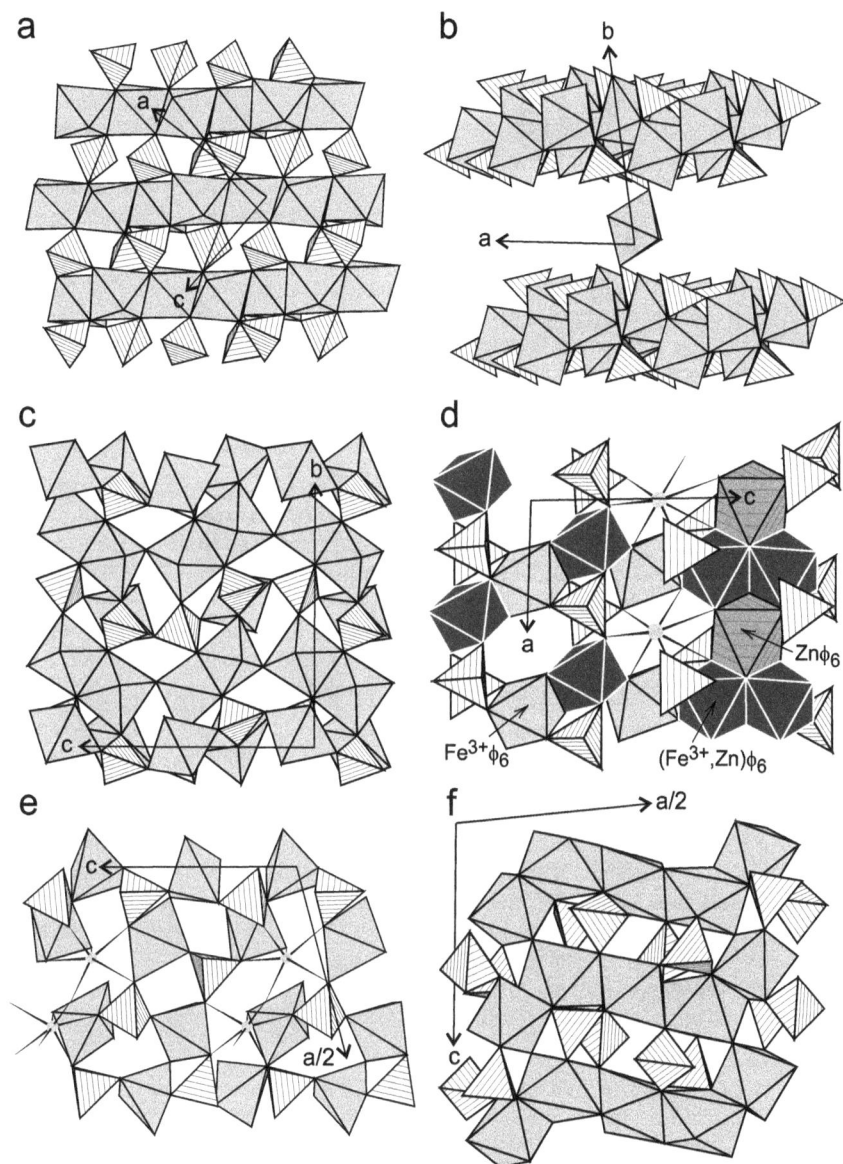

Figure 48. The structures of the arsenates with infinite frameworks of tetrahedra and $M\phi_N$ polyhedra: (a,b) geigerite as a member of the lindackerite group (a) top view of the sheets in geigerite, note the topology of the sheets is the same as in the krautite group and (b) side view of the sheets in geigerite, showing the octahedra in the intersheet space which connect the sheets into an open framework. (c) sarkinite, (d) jamesite, (e) prosperite, (f) villyaellenite, showing three octahedral pentamers cross-linked by arsenate tetrahedra.

(Fig. 48b) and H_2O molecules. In lindackerite, there are two equally likely orientations of the M octahedra. The occupancy of one of the interlayer H_2O molecules is only ~0.4 and, therefore, lindackerite has only 9 H_2O molecules per formula unit, one fewer in comparison with the related minerals geigerite and chudobaite.

In **geigerite**, all Mn^{2+} cations are found in a near-regular octahedral coordination (Graeser et al. 1989). The [101] chains comprise $MnO_4(OH)(H_2O)$ and $MnO_5(H_2O)$ octahedra and are linked by arsenate tetrahedra (AsO_4) and (AsO_3OH) into (010) sheets (Fig. 48a). The space between the sheets is populated by $MnO_2(H_2O)_4$ octahedra (Fig. 48b) and one crystallographically independent H_2O molecule. Only a single, fully ordered position was reported for the Mn octahedron between the sheets. The structural formulae of lindackerite and geigerite are slightly different (with respect to the assignment of the H_2O molecules to different structural units) because of different coordination of Cu^{2+} and Mn^{2+} in lindackerite and geigerite, respectively, and different occupancy of the free H_2O molecules.

Chudobaite is very likely isostructural with geigerite (Dorner and Weber 1976; Graeser et al. 1989). Some discrepancy remains because in the original description of the mineral, Strunz (1960) reported substantial amounts of Na, K, and Ca in chudobaite. The presence or absence of these elements in this mineral was discussed by Graeser et al. (1989).

Pradetite (Burke et al. 2007), **veselovskýite** (Sejkora et al. 2010a), **klajite** (Szakall et al. 2011), and **ondrušite** (Sejkora et al. 2011) are isostructural with lindackerite. Their formulae are listed in Table 22.

Linkage M=M, M-M, M=T, M-T. The structure of **ferrarisite** is built by (001) heteropolyhedral layers joined by additional structural units. Its stoichiometry strikingly resembles that of the minerals of the lindackerite group. Structurally, the sheets in ferrarisite bear similarity to the sheets in haidingerite; in this mineral, for reasons elucidated above, we decided to consider the $Ca\phi_7$ polyhedra as "similar polyhedra (to octahedra)". The same reasoning, namely structural similarity to the octahedral-tetrahedral frameworks of the lindackerite group, led us to place ferrarisite in this group of arsenate minerals.

The sheets in ferrarisite are constructed by zigzag chains of $Ca\phi_7$ polyhedra with a general [110] direction. The basic motif of the chains is a branching edge-sharing cluster of four $Ca\phi_7$ polyhedra. Within these clusters, two polyhedra attach in the direction of the chain, hence, the chain is one polyhedron wide. The other two polyhedra cause the chain to branch, that is, the chain is two polyhedra wide. The neighboring chains connect to each other via edge sharing of the $Ca\phi_7$ polyhedra. This arrangement creates larger and smaller cavities within the sheets. The arsenate tetrahedra attach to the sheets above and below the larger and smaller cavities. The heteropolyhedral layers are joined by an interstitial Ca-centered polyhedron, in this case a nearly regular octahedron. The interlayer space is also populated by a H_2O molecule, disordered over two centrosymmetric positions. Consequently, Catti et al. (1980b) assumed that the positions of other H atoms in H_2O molecules are also disordered.

Linkage M=M, M-M, M-T. There are seven $Cu\phi_{5-6}$ crystallographically distinct polyhedra in the structure of **slavkovite** (Sejkora et al. 2010b). The sheets have topology similar to those in the lindackerite group; there are additional polyhedra attaching to the $Cu\phi_{5-6}$ chains within the sheets. This feature is similar to the sheets in pushcharovskite.

Linkage M, M=M, M-M, M-T. The structure of **betpakdalite**, $\{Mg(H_2O)_6\}$ $Ca_2(H_2O)_{13}[Fe_3Mo_8O_{28}(OH)(AsO_4)_2]\cdot4H_2O$, was initially solved by Schmetzer et al. (1984) but later critically re-investigated by Moore (1992) and Cooper and Hawthorne (1999a). Cooper and Hawthorne (1999a) identified and listed several unsatisfactory aspects of the earlier model of Schmetzer et al. (1984), although the basic structural features were retained. Moore (1992) discussed the structure in terms of anion close packing, vacancies on the anion sites, and filling of the interstices in the close-packed arrays with cations. The structure can be described as constructed by (001) sheets with large spacing (~ 11 Å). The space between the sheets is filled by a complex array of coordination polyhedra. The sheets are assembled from large, 12-membered rings of alternating $Fe^{3+}\phi_6$ octahedra and AsO_4 tetrahedra. The centers of these rings are occupied by $Mg(H_2O)_6$ octahedra, connected to the rest of the structure only via hydrogen bonds of its H_2O ligands. The space between the sheets is pillared by Mo^{6+} octahedra. The octahedra condense into tight $[Mo_4\phi_{16}]$ tetramers in which all octahedra share edges. In each tetramer, three octahedra (three central Mo^{6+} cations) lie in the (001) plane and the fourth one is located above or below the three. Two tetramers share a single corner, a disordered O^{2-} anion. In addition, one of the ligands of the tetramers is occupied by $O_{0.5}(OH)_{0.5}$, depending on the occupancy of a nearby H_2O molecule. Pairs of the tetramers pillar the sheets and the voids between the tetramers are filled by large irregular $Ca(H_2O)_{7-8}$ polyhedra. Only half of the Ca sites are occupied; Cooper and Hawthorne (1999a) discussed in length and detail the ordering scheme and coordination numbers of the Ca cations in the structure of betpakdalite.

Linkage M=M, M-M, M-T. **Sarkinite**, $Mn_2(OH)(AsO_4)$, has a complicated and dense three-dimensional framework structure (Fig. 48c) (Dal Negro et al. 1974). There are eight independent Mn positions and four independent As positions in the asymmetric unit. Four Mn^{2+} cations are coordinated octahedrally, the other four in a trigonal bipyramidal coordination. The Mn^{2+} polyhedra compose dense slabs stacked along [001], made of interwoven two-dimensional frameworks of the octahedra and the trigonal bipyramids. It is possible to describe the structure in terms of chains of the Mn^{2+} polyhedra although the three-dimensional network character clearly prevails. The arsenate tetrahedra link the Mn^{2+} polyhedra further. Sarkinite is isostructural with wagnerite, $Mg_2(PO_4)F$. Sarkinite is not isostructural with paradamite $[Zn_2(OH)(AsO_4)]$, although **eveite** $[Mn_2(AsO_4)(OH)]$ and **adamite** $[Zn_2(AsO_4)(OH)]$ are isostructural.

Linkage M=M, M-M, M-T. The structure of **jamesite**, $Pb_2ZnFe^{3+}_2(Fe^{3+}_{2.8}Zn_{1.2})$ $(OH)_8[(OH)_{1.2}O_{0.8}](AsO_4)_4$, can be described in terms of two chain types (Fig. 48d) (Cooper and Hawthorne 1999b). Within these chains, Zn^{2+} and Fe^{3+} are coordinated octahedrally. One chain is an [100] octahedral band, one- and two-octahedra wide, with edge-sharing among the octahedra. The other chain is built by short linear trimers of edge-sharing octahedra mutually connected by arsenate tetrahedra. These two chain types, designated as A and B by Cooper and Hawthorne (1999b), alternate regularly along the [010] direction to form complicated (001) slabs. The slabs are sewn together by two types of polyhedra. One of them is another octahedron, occupied by Fe^{3+}, and sharing corners with octahedra in both A and B chains. The space between the slabs is sufficient to accommodate Pb^{2+} on a split site in a seven-fold coordination.

Linkage M=M, M-M, M-T. The crystal structure of **prosperite**, $Ca_2Zn_4(H_2O)(AsO_4)_4$, is built by five distinct polyhedra, namely two ZnO_5 trigonal bipyramids, two AsO_4 tetrahedra, and a large $Ca\phi_9$ polyhedron (Fig. 48e) (Keller et al. 1982). One of the trigonal bipyramids ($Zn1O_5$) is strongly distorted, with 4 shorter and 1 long Zn-O distance (Zn-O 2.567 Å), the $Zn2O_5$ bipyramid is relatively regular. The bond valence calculation confirm that the polyhedron around Zn1 should be described as a distorted trigonal bipyramid and not as a tetrahedron. Two $Zn2O_5$ bipyramids share an edge and furthermore corners with two adjacent $Zn1O_5$ bipyramids, thus creating a cluster $[Zn_4O_{16}]$. These clusters are interconnected by the arsenate tetrahedra into a framework whose cavities house the large $Ca\phi_9$ polyhedra.

Linkage M=M, M-M, M-T. **Villyaellenite** is isostructural with **sainfeldite, nyholmite, miguelromeroite** (Table 23), and the phosphate mineral hureaulite $[Mn_5(PO_4)(HPO_4)_2 \cdot 4H_2O]$

Table 23. Minerals isostructural with villyaellenite (for references see text).

villyaellenite	$(Mn,Ca)Mn_2Ca_2(H_2O)_4(AsO_4)_2(AsO_3OH)_2$
sainfeldite	$Ca_5(H_2O)_4(AsO_4)_2(AsO_3OH)_2$
nyholmite	$Cd_3Zn_2(H_2O)_4(AsO_4)_2(AsO_3OH)_2$
miguelromeroite	$Mn_5(H_2O)_4(AsO_4)_2(AsO_3OH)_2$

(Ferraris and Abbona 1972; Stock et al. 2002; Elliott et al. 2009; Kampf 2009). The structure is made of infinite bands extending in the direction of the c axis (Ferraris and Abbona 1972). Within the bands, five distorted octahedra condense into M2-M3-M1-M3-M2 pentamers (Fig. 49f) which are interconnected by corner sharing between the octahedra. The pentamers form the backbone of the bands and the adjacent bands are further connected via edge-sharing between the octahedra. The loose octahedral framework has the composition $M_5\phi_{18}$, where $\phi = 4OH_2 + 14O$. The presence of H_2O as ligands of Ca at the M2 position precludes condensation of the bands into sheets. The arsenate tetrahedra knit the pentamers together and interconnect also the adjacent bands. One of the arsenate tetrahedra shares all its corners with the Mn octahedra, the other one, the AsO_3OH octahedron, shares only three corners. The bands are arranged into a three-dimensional framework structure. Studies on natural crystals hint at the partitioning of Ca and Mn over the three octahedral positions (Kampf and Ross 1988), indicating that a complete solid solution between villyaellenite and sainfeldite does not exist (Stock et al. 2002). This supposition was confirmed by a recent work of Kampf (2009) who redefined villyaellenite as an ordered intermediate phase between sainfeldite and miguelromeroite.

Linkage M=M, M-M, M-T. The crystal structure of **legrandite**, $Zn_4(OH)_2(H_2O)_2(AsO_4)_2$, is a complicated network built by four distinct Zn coordination polyhedra and arsenate tetrahedra (Fig. 49a) (McLean et al. 1971). The zinc atoms are four- to six-coordinated and their coordination polyhedra aggregate into larger structural units. These units may be described in several ways, depending on whether a single Zn-O distance of 2.99 Å between two units is seen as a chemical bond or not. The long Zn-O distance is macroscopically manifested by the cleavage of legrandite. McLean et al. (1971) described the structure in terms of kinked chains of Zn polyhedra which extend in the direction of the a axis. If the 2.99 Zn-O distance is considered as a bond, then these chains are continuous, otherwise, they are broken. An alternative description of the structure is that of thick undulating slabs parallel to (100); there are additional Zn polyhedra attached to the slabs and these polyhedra extend into the space between the slabs in an interlocking manner. Using either description, arsenate tetrahedra attach with all their apices to the Zn polyhedra and complete the three-dimensional network of the legrandite structure. Hydrogen bonding in legrandite was discussed in a detail by Hawthorne et al. (2013).

Linkage M=M, M-M, M-T. The crystal structure of **leogangite**, $Cu_{10}(OH)_6(H_2O)_4(AsO_4)_4$ $(SO_4)\cdot4H_2O$, (Lengauer et al. 2004) could be viewed as a loose framework structure or as a layered structure with sparse structural elements which link the layers (Fig. 49b). The layers or thick slabs are stacked along the crystallographic a axis. The slabs are constructed by CuO_5 square pyramids in a complicated arrangement. One of the Cu^{2+} polyhedra could also be described as a distorted trigonal bipyramid. The arsenate tetrahedra are positioned within the slabs and further link the Cu^{2+} polyhedra. The slabs are connected not only by the hydrogen bonds mediated by H_2O molecules in the space between them, but also by sulfate tetrahedra. Because of these sulfate tetrahedra, the structure could be regarded as a loose heteropolyhedral framework.

Linkage M=M, M-M, M-T. The minerals **adamite**, $Zn_2(OH)(AsO_4)$, **olivenite**, $Cu_2(OH)$ (AsO_4), libethenite, $Cu_2(OH)(PO_4)$, **eveite**, $Mn_2(OH)(AsO_4)$ are isostructural with andalusite, $Al_2O(SiO_4)$. A continuous series between adamite and a synthetic Co analogue has been reported (Keller 1971).

The structure of adamite is built by infinite chains of edge-sharing Zn octahedra which extend along the c axis (Fig. 49c). Each arsenate tetrahedron shares two corners with two Zn octahedra in one chain, another corner with an octahedron in an adjacent chain, and the fourth corner with a dimer of Zn atoms in trigonal-bipyramidal (five-fold) coordination. These dimers with composition $Zn_2O_6(OH)_2$ are located in the tunnels defined by the octahedral chains. Both Hill (1976) and Hawthorne (1976c) located the H atom in the structure of adamite based on electrostatic bond strength considerations. Hill (1976) included the position in the final list

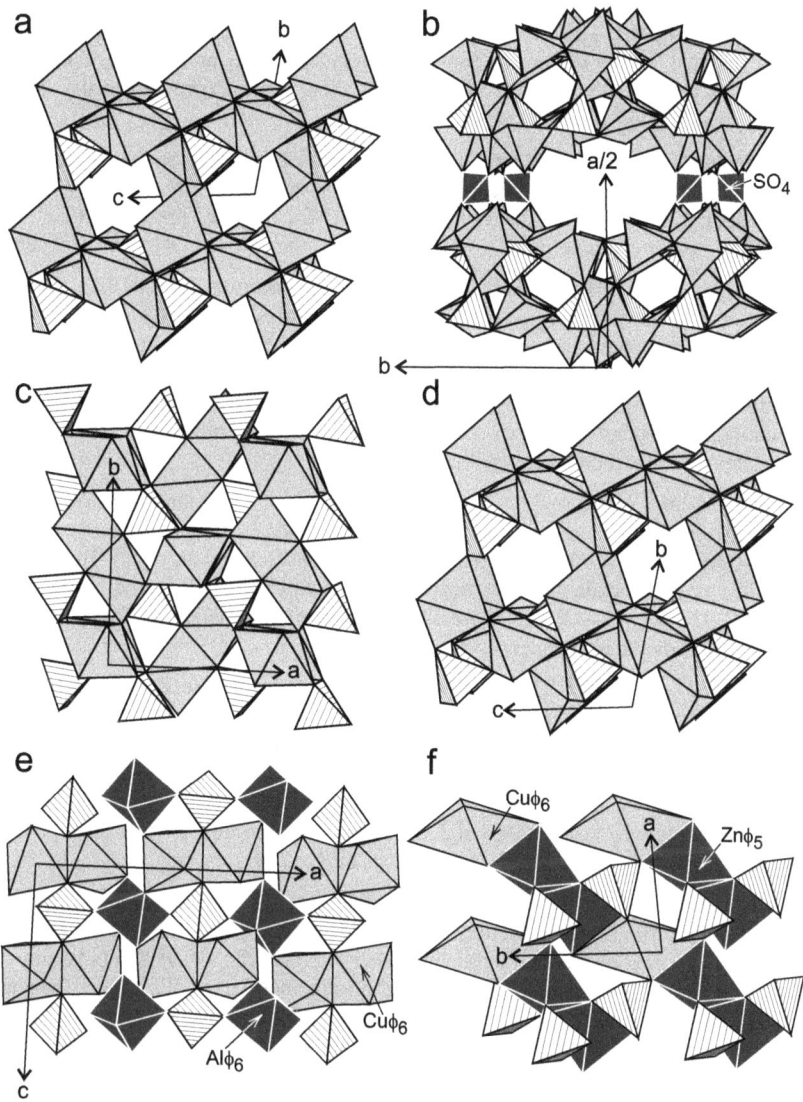

Figure 49. The structures of the arsenates with infinite frameworks of tetrahedra and Mϕ_N polyhedra: (a) legrandite, (b) leogangite, (c) adamite, (d) paradamite, (e) liroconite, (f) stranskiite.

of coordinates but did not refine it. The structure of eveite was reported by Moore and Smyth (1968).

The structure of olivenite was studied early by Toman (1977) and later in a detail by powder X-ray diffraction and subsequent Rietveld refinement (Burns and Hawthorne 1995). The results of the refinement indicate a slight, but statistically significant deviation from the orthorhombic symmetry. The resulting monoclinic symmetry is ascribed to the Jahn-Teller distortion of the Cu^{2+} octahedra, although the authors entertain the possibility that even adamite may be monoclinic.

Zincolivenite, $CuZn(OH)(AsO_4)$, is an ordered derivative of the olivenite structure (Chukanov et al. 2007b). The Cu^{2+} cations populate the infinite octahedral chains, the Zn^{2+} cations are located in the dimers of trigonal bipyramids. The preference of Cu^{2+} for the octahedral sites is caused by the fact that these octahedra can accommodate the Jahn-Teller distortion typical for Cu^{2+}.

The structure of **auriacusite**, $Fe^{3+}CuO(AsO_4)$, is closely related to that of zincolivenite. The infinite octahedral chains are occupied by Cu^{2+} cations, just as in zincolivenite (Mills et al. 2010). The trigonal bipyramidal sites house Fe^{3+} and Cu^{2+} and the presence of Fe^{3+} (instead of Cu^{2+} or Zn^{2+}, as for example in zincolivenite) requires the coupled $OH^- = O^{2-}$ substitution.

Linkage M=M, M-M, M-T. The crystal structure of **paradamite**, $Zn_2(OH)(AsO_4)$, is similar to the structure of its polymorph adamite in that both structures are built by chains of polyhedra interconnected by arsenate tetrahedra and dimers of trigonal-bipyramidal polyhedra (Fig. 49d) (Kato and Miura 1977). While the chains comprise $Zn\phi_6$ and $Zn\phi_5$ polyhedra in adamite, the chains in the structure of paradamite consist of edge-sharing trigonal bipyramids which house the Zn^{2+} cations and extend along the *a* axis (Hawthorne 1979). Paradamite is isostructural with the phosphate mineral tarbuttite, $Zn_2(OH)(PO_4)$.

Linkage M=M, M-M, M-T. A principal motif of the structure of **liroconite**, $Cu_2Al(OH)_4(H_2O)_4(AsO_4)$, are [100] octahedral-tetrahedral chains with a repeat unit $[Al_2(AsO_4)_2(OH)_4]$ (Fig. 49e) (Kolesova and Fesenko 1968; Burns et al. 1991). The Al^{3+} cations are found in a regular octahedral coordination. The chains are cross-linked by edge-sharing dimers $[Cu_2O_2(OH)_4(H_2O)_4]$. Within these dimers, the Cu^{2+} cations are six-fold coordinated in the usual, strongly distorted manner.

Linkage M=M, M-M, M-T. The framework structure of **stranskiite**, $Zn_2Cu(AsO_4)_2$, consists of dimers of Zn^{2+} cations in distorted trigonal-bipyramidal coordination, mutually linked by strongly Jahn-Teller distorted Cu^{2+} polyhedra and arsenate tetrahedra (Fig. 49f) (Keller et al. 1979). The two trigonal bipyramids occupied by Zn^{2+} share an edge. Cu is coordinated by 4+2 oxygen atoms; on the basis of bond-valence calculations, Keller et al. (1979) argued that the distant two oxygens, even at a distance of 3.134 Å from the central Cu^{2+} ion, must be included in the coordination shell of this ion.

Linkage M=M, M-M, M-T. The principal feature of the structure of **angelellite**, $Fe_4O_3(AsO_4)_2$, are [001] staggered chains of edge-sharing octahedra which house Fe^{3+} (Moore and Araki 1978a). The chains condense by corner sharing into (010) sheets which can be described as fragments of a cubic close-packed arrangement (Fig. 50a). One oxygen atom within the chains links to four Fe atoms, three O atoms to 2 Fe atoms, and two O atoms to 1 Fe atom. The sheets are connected into a three-dimensional network by arsenate tetrahedra.

Interestingly, the (010) sheets in angelellite are topologically identical to the sheets found in the structure of dumortierite, $Si_3B(Al_7O_{18})$. Another interesting property of angelellite is its ability to form epitaxial intergrowths with hematite (Weber 1959; Moore and Araki 1978a).

Linkage M=M, M-M, M-T. There are four different coordination polyhedra for the divalent ions (Cu, Zn, Cd) and arsenate tetrahedra in the structure of **keyite**, $Cu_3Zn_4Cd_2(H_2O)_2(AsO_4)_6$ (Cooper and Hawthorne 1996a). Cu^{2+} is found either in a square-planar or distorted octahedral coordination. Zn^{2+} is coordinated by an almost regular octahedral arrangement of oxygen atoms. The coordination of Cd^{2+} is also sixfold but the polyhedron is rather irregular, "probably closest to a trigonal prism" (Cooper and Hawthorne 1996a). The high degree of connectivity of the polyhedra allows for several alternative descriptions of the structure (Fig. 50b). Here we describe the structure as built by two types of sheets, alternating in the [100] direction.

One sheet comprises (Zn_2O_{10}) dimers cross-linked by arsenate tetrahedra. The second sheet is more complex. This sheet contains [001] corner-sharing $Cu1\phi_6$ octahedral chains in which

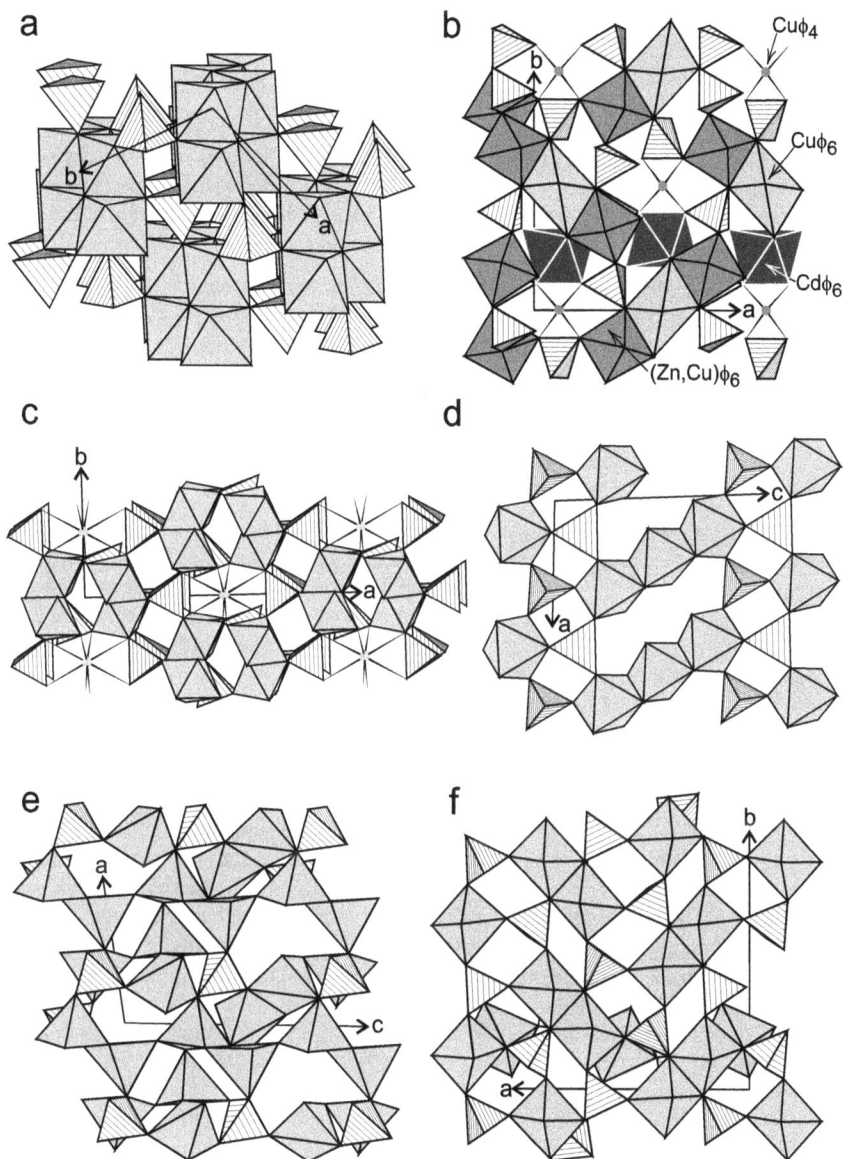

Figure 50. The structures of the arsenates with infinite frameworks of tetrahedra and $M\phi_N$ polyhedra: (a) angelellite, (b) keyite, (c) sewardite, (d) xanthiosite, a slab from the structure showing similarity with the structure of olivine, (e) clinoclase, (f) grischunite.

only a half of the Cu sites are occupied. The remaining vacancies appear not to be ordered as this would lead to the doubling of the lattice parameter c. Since two ligands of Cu1 are H_2O molecules, the presence of vacancies at these sites should lead to re-orientation of the H_2O molecules. The Cu1ϕ_6 chain attach to the neighboring [CdAsO$_8$] chains via corner sharing. The square-planar Cu2ϕ_4 units are sandwiched between the adjacent Cdϕ_6 polyhedra, thus forming a slab with a chain sequence Cu1ϕ_6 – [CdAsO$_8$] – Cu2ϕ_4 – [CdAsO$_8$] – Cu1ϕ_6.

Linkage M=M, M-M, M-T. The structures of **carminite**, $PbFe_2(OH)_2(AsO_4)_2$, and **sewardite**, $CaFe_2(OH)_2(AsO_4)_2$, consist of octahedral chains which extend in the c direction (Fig. 50c) (Finney 1963; Olmi and Sabelli 1995; Roberts et al. 2002). In these chains, the $Fe\phi_6$ octahedra are connected alternately by sharing an edge or a corner, essentially segmenting the chains into dimers. There are two crystallographically distinct arsenate tetrahedra in these structures. One of them, $As2O_4$, links the neighboring chains by sharing all its corners with the ligands of Fe^{3+}. The $As1O_4$ tetrahedron also attaches to the octahedral chains but one of its corners is shared only with the large (Ca^{2+}, Pb^{2+}) cations. The large cations reside in the cavities of the octahedral-tetrahedral framework and are eight-fold coordinated.

Linkage M=M, M-M, M-T. The dense three-dimensional framework of the **xanthiosite**, $Ni_3(AsO_4)_2$, structure is built by NiO_6 octahedra and AsO_4 tetrahedra (Fig. 50d) (Barbier and Frampton 1991). The structure is based on a mixed hexagonal and cubic close packing of oxygen atoms with a sequence hc. The NiO_6 octahedra are distorted and assembled by edge sharing into a framework. The arsenate tetrahedra share all their corners with adjacent Ni octahedra. Barbier and Frampton (1991) described the structure in terms of a cation-deficient olivine arrangement where one fourth of the available octahedral sites are vacant, resulting in the change of the stoichiometry from A_2BO_4 (as in Mg_2SiO_4) to $Ni_{1.5}(AsO_4)$ or $Ni_3(AsO_4)_2$.

Linkage M=M, M-M, M-T. Three independent Cu^{2+} polyhedra and arsenate tetrahedra condense into a complicated three dimensional framework in **clinoclase**, $Cu_3(OH)_3(AsO_4)$. Whereas Ghose et al. (1965) assigned five-fold coordination to all three Cu^{2+} cations, Eby and Hawthorne (1990) argued that Cu1 and Cu3 are coordinated by a very distorted octahedral arrangement, the longest Cu^{2+}-ϕ bond measured 2.871 and 2.995 Å, respectively. The coordination polyhedron of Cu2 is a square pyramid. The structure can be described as (100) slabs which consist of edge-sharing $Cu2\phi_5$ and $Cu1\phi_6$ polyhedra (Fig. 50e). These slabs are linked by edge-sharing with $Cu3\phi_6$ octahedra and corner-sharing with the arsenate tetrahedra. Because of the high degree of polymerization, other descriptions of the structure can be found (Ghose et al. 1965).

Linkage M=M, M-M, M-T. The dense framework structure of **grischunite**, $NaCa_2Mn_4$ $(Mn^{2+}_{0.5}Fe^{3+}_{0.5})_2(H_2O)_2(AsO_4)_6$, is based on an octahedral framework around the Mn^{2+} and Fe^{3+} cations (Fig. 50f) (Bianchi et al. 1987). There are three octahedral sites in this structure, two occupied by Mn^{2+} and the third one by Mn^{2+} and Fe^{3+}. Two Mn^{2+} octahedra share an edge to form a $[Mn_2O_{10}]$ cluster. These clusters connect via corners with the neighboring clusters into zigzag chains with a general [001] direction. The chains are stacked in the [100] direction but they are not connected into a sheet. Instead, the chains are cross-linked by the $(Mn^{2+},Fe^{3+})O_6$ octahedra into the octahedral framework. Additional cross-linking is provided by the arsenate tetrahedra and the Na cations in an irregular six-fold coordination. Ca^{2+} is coordinated by eight ligands. Bianchi et al. (1987) expect a certain degree of Ca^{2+}, Na^+, and Mn^{2+} substitution in the structure and the presence of vacancies, mostly at the Na^+ site. Ordering of Mn^{2+} and Fe^{3+} on one octahedral position is possible but could not have been proven by the X-ray diffraction experiments.

Linkage M=M, M-M, M-T. The minerals in the **arthurite group** (Table 24) have a general formula $AFe^{3+}_2(OH)_2(H_2O)_4(AsO_4)_2$, where A = Cu, Co, Zn, Fe^{2+}. After the isostructural phos-

Table 24. Minerals of the arthurite group (Keller and Hess 1978; Hughes et al. 1996; Kampf 2005; Kolitsch et al. 2010).

arthurite	$CuFe^{3+}_2(OH)_2(H_2O)_4(AsO_4)_2$
cobaltarthurite	$CoFe^{3+}_2(OH)_2(H_2O)_4(AsO_4)_2$
ojuelaite	$ZnFe^{3+}_2(OH)_2(H_2O)_4(AsO_4)_2$
bendadaite	$Fe^{2+}Fe^{3+}_2(OH)_2(H_2O)_4(AsO_4)_2$

phate whitmoreite, this group may be referred to in literature also as the whitmoreite group. A possible Mn^{2+}-Mn^{3+} member of the arthurite group has been mentioned (see Kolitsch et al. 2010).

The crystal structure of the minerals of the arthurite group is built by corrugated (100) octa-hedral sheets (Fig. 51a). In these sheets, two edge-sharing octahedra form a dimer $[Fe_2O_6(OH)_4]$. The dimers link via the (OH) groups to form the sheets which are decorated from above and

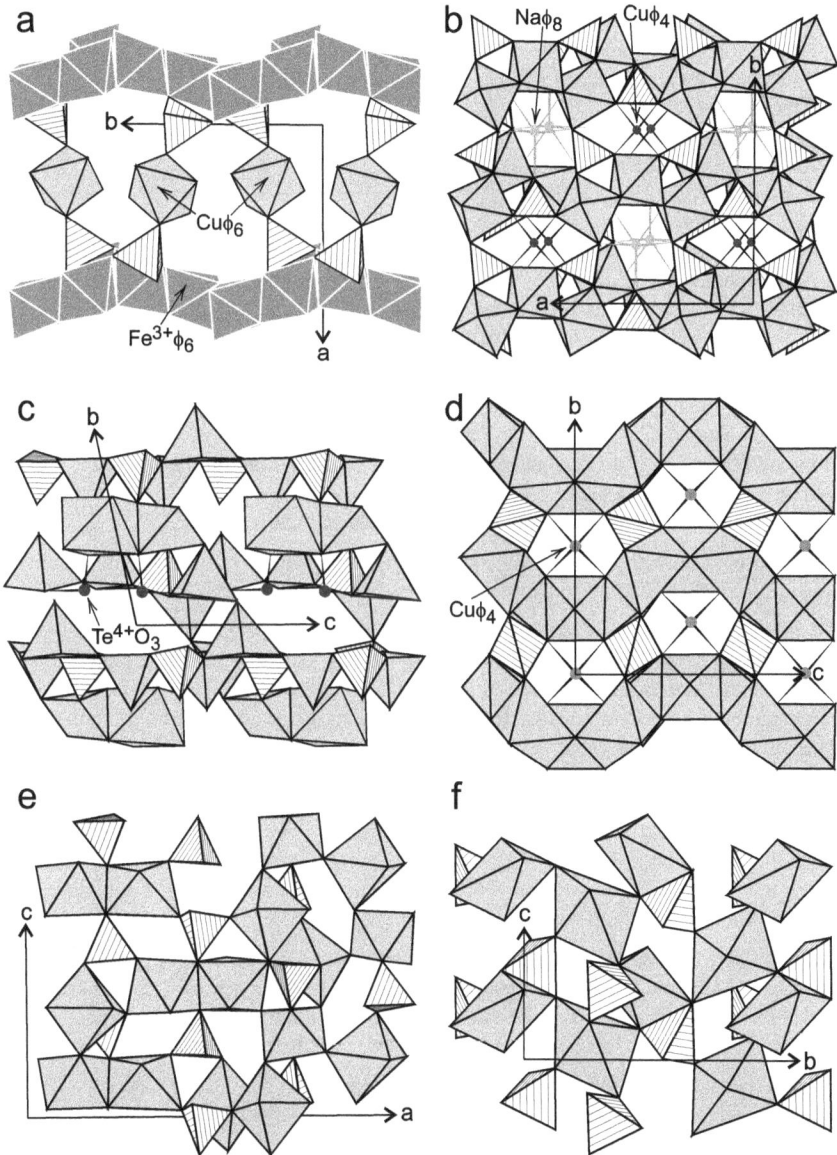

Figure 51. The structures of the arsenates with infinite frameworks of tetrahedra and $M\phi_N$ polyhedra: (a) arthurite, (b) johillerite as a member of the alluaudite group, (c) juabite, (d) coparsite, (e) arsenoclasite, (f) lammerite.

below by arsenate tetrahedra. The sheets are bridged by $AO_2(H_2O)_4$ octahedra. A possibility of the incomplete occupancy of the A site (i.e., vacancies) was discussed at length by Kolitsch et al. (2010).

Linkage M=M, M-M, M-T. The arsenate members of the **alluaudite** $[CaNaMn^{2+}Fe^{2+}_2(PO_4)_3]$ **group** (Table 25) share common structural features and a general formula $X1X2M1M2_2(AsO_4)_3$. The structures are formed by thick (010) slabs which meld by corner-sharing among the polyhedra in the slabs into a three-dimensional network (Fig. 51b) (e.g., Tait and Hawthorne 2003). The slabs comprise MO_6 octahedra and arsenate tetrahedra.

Table 25. Arsenate members of the alluaudite group (Keller and Hess 1988; Ercit 1993; Auerhammer et al. 1993; Filatov et al. 2001; Tait and Hawthorne 2003).

arseniopleite	$CaNaMn^{2+}Mn^{2+}_2(AsO_4)_3$
caryinite	$CaNaCaMn^{2+}_2(AsO_4)_3$
nickenichite	$Na_{0.8}Ca_{0.4}Cu_{0.4}(Mg,Fe,Al)_3(AsO_4)_3$
bradaczekite	$NaCuCu_3(AsO_4)_3$
o'danielite	$NaH_2(Zn,Mg)_3(AsO_4)_3$
johillerite	$NaCu(Mg,Zn)_3(AsO_4)_3$

Three MO_6 octahedra share edges and condense into trimers which join each other to form infinite [101] chains. Arsenate tetrahedra interconnect the chains. The large cations X1 and X2, predominantly Ca and Na, are housed in channels between the slabs and coordinated to seven or eight oxygen atoms. Other cations, for example Cu or H, may also be found in the channels.

The crystal structure of **arseniopleite** is able to accommodate a relatively wide range of cations. Among the two crystallographically independent octahedral sites, one was found to be occupied only by Mn^{2+}, but the second one accepts Mn^{2+}, Mg, Fe^{2+} or perhaps Fe^{3+} (Tait and Hawthorne 2003). Two independent sites with high coordination numbers house Ca, Na, Ba, Pb^{2+}. In arseniopleite, the channel I (as described in Auerhammer et al. 1993) is occupied by Na, Ba, and Pb. Channel II is populated by Ca and a minor amount of Na.

Caryinite is isostructural with alluaudite and arseniopleite (Ercit 1993). The Ca site in the channels was split into two partially occupied sites.

In **nickenichite**, the two octahedral positions contain mostly Mg with a minor amount of Fe and traces of Al and Mn. One of these positions shows a greater preference for Mg, the other one accommodates more Fe and Mn. The nature and occupancy of the cation sites in the channels is more complicated than in the related phases. Channel I contains Na ions. Channel II has Ca, Cu, and presumably also a small amount of H atoms. The Ca atoms occupy the same positions as Ca atoms in the related structures (for example, arseniopleite). The Cu atoms, on the other hand, occupy a new position. Because the distance between neighboring Ca and Cu positions is only 1.662 Å, it is impossible that two such neighboring positions would be occupied. Auerhammer et al. (1993) considered several ordering schemes of Ca, Cu, and vacancies in channels in the structure of nickenichite and concluded that the actual structure is an intermediate between the possible ways of ordering.

The structure of **bradaczekite** has three independent Cu^{2+} positions. Two Cu^{2+} positions are six-fold coordinated in strongly distorted octahedra (Krivovichev et al. 2001) and build up the alluaudite-like framework. One of the Cu^{2+} cations is coordinated by 4 oxygen atoms in a square planar coordination and resides in the channels which are otherwise occupied by Ca.

One of the arsenate tetrahedra shows an unusual spread of As-O distances of 2×1.665 and 2×1.736 Å. These variations are caused by edge-sharing between this tetrahedron and a Cu square in the channels.

Square-planar coordination of Cu^{2+} in the channels is also observed in **johillerite** (Keller and Hess 1988). For both bradaczekite and johillerite, the $Cu\phi_4$ units are considered to be interstitial species because they replace the large Ca-centered polyhedra.

In o'danielite, one of the channels is populated by H atoms (Keller and Hess 1988). An interesting feature of this structure is extremely short hydrogen bonds. It is possible that the hydrogen is shared between two O atoms instead of an OH group and an acceptor O.

Linkage: M=M, M-M, M-T. The heteropolyhedral structure of **juabite**, $CaCu_{10}(OH)_2(H_2O)_4$ $(Te^{4+}O_3)_4(AsO_4)_4$, can be described as a framework structure (Fig. 51c) although the dominant feature of the structure are (010) layers interconnected by polyhedra located in the interlayer space (Burns et al. 2000a). The (010) layers consist of two symmetrically identical sheets and [001] chains sandwiched between the two sheets. The chains consist of $Cu\phi_5$ and $Ca\phi_6$ polyhedra. The two symmetrically equivalent sheets have a complicated architecture. The sheet is an assembly of large eight-membered rings of three $Cu\phi_5$ polyhedra, an AsO_4 tetrahedron, three $Cu\phi_5$ polyhedra, and another AsO_4 tetrahedron. All $Cu\phi_5$ polyhedra in this structure are square pyramids. The central portion of each ring is populated by two crystallographically distinct Te-centered polyhedra. Tellurium in this structure was shown to be tetravalent and forms three short bonds to O^{2-} anions. The Te^{4+} and three O^{2-} ions define a trigonal pyramid. Each of the two Te^{4+} cations places its lone electron pairs in the interlayer and forms additional weaker bonds to more distal O^{2-} anions. The interlayer space is occupied by $Cu\phi_5$ square pyramids which link the layers but are not connected to each other.

Linkage M=M, M-M, M-T. The crystal structure of **coparsite**, $Cu_4O_2Cl[(As,V)O_4]$, is built by three independent Cu polyhedra and arsenate tetrahedra (Fig. 51d) (Starova et al. 1998). The $Cu2O_4Cl_2$ and $Cu3O_5Cl$ polyhedra can be described as strongly distorted octahedra. Starova et al. (1998) reported only five-fold coordination for Cu3. However, if assuming six-fold coordination, the most distant oxygen would be located 2.849 Å from the central Cu3 cation, which is not unusual for copper arsenates. The $Cu2O_4Cl_2$ and $Cu3O_5Cl$ polyhedra constitute strongly modulated layers, in general parallel to (010). The space between the layers is occupied by arsenate-vanadate tetrahedra and Cu1 in planar square coordination. Although the As/(As+V) ratio is close to 0.5, no As/V ordering was detected. Starova et al. (1998) described the structure of coparsite in terms of chains of oxygen-centered, edge-sharing tetrahedra OCu_4.

Linkage M=M, M-M, M-T. A prominent feature of the structure of **braithwaiteite**, $NaCu_5(Sb^{5+}Ti^{4+})O_2(H_2O)_8(AsO_4)_4(AsO_3OH)_2$, are heteropolyhedral (001) sheets composed of two different chain types (Hawthorne et al. 2008). One chain has composition $[(SbTi)(AsO_4)_4O_2]$, the (Sb,Ti) octahedra share corners and are decorated by the (AsO_4) groups. The other chain with composition $[Cu_2(AsO_3OH)_2O_4]$ consists of edge-sharing $Cu\phi_6$ octahedra, decorated by additional (AsO_4) groups. These two chains alternate regularly and define the (001) sheets. The sheets are cross-linked by Cu and Na octahedra and a network of hydrogen bonds. Hawthorne et al. (2008) also provided a detailed analysis of $[M(TO_4)_2\phi_4]$ chain types in the structures of sulfates, phosphates, arsenates, and silicates.

Linkage M=M, M-M, M=T, M-T. The complicated structure of **arsenoclasite**, $Mn_5(OH)_4(AsO_4)_2$, has five Mn positions and two As positions (Fig. 51e) and its solution was "a miserable problem" (Moore and Molin-Case 1971). Moore and Molin-Case (1971) postulated that the coordination numbers of the Mn^{2+} cations are 4, 5, and 6 and described the structure as "constructed of a double hexagonal close-packed array of oxygen atoms with a stacking sequence ($\cdots ch \cdots$)". In a study of the isostructural phosphates $Co_5(OH)_4(PO_4)_2$ and $Mn_5(OH)_4(PO_4)_2$, the coordination of all five crystallographically independent divalent cations was described as octahedral (Ruszala et al. 1977). Ruszala et al. (1977) also re-analyzed the structure reported by Moore and Molin-Case (1971) and concluded that all Mn^{2+} coordination polyhedra in arsenoclasite are octahedra.

The structure of arsenoclasite is built by two types of chains. The first chain comprises the $Mn2\phi_6$, $Mn3\phi_6$, and $Mn4\phi_6$ octahedra and is alternately one and two octahedra wide. The $Mn1\phi_6$ and $Mn5\phi_6$ octahedra and AsO_4 tetrahedra share edges and corners in the second, more complicated chain. Both chains extend in the direction of the *b* axis and are cross-linked by corner sharing between neighboring polyhedra.

Linkage M=M, M-M, M=T, M-T. The framework structure of **lammerite**, $Cu_3(AsO_4)_2$, is built by Jahn-Teller distorted CuO_6 octahedra and arsenate tetrahedra (Fig. 51f) (Hawthorne 1986). There are two Cu sites in the structure; the coordination polyhedra of the Cu2 site (after Hawthorne 1986) condense into [100] zigzag chains. These chains are cross-linked by the Cu1 polyhedra and arsenate tetrahedra. Hawthorne (1986) recognized the distorted cubic close packing of the oxygen atoms in the structure in which Cu and As fill the octahedral and tetrahedral cavities, respectively. The strong Jahn-Teller distortion of the Cu octahedra causes strong deviation of the oxygen arrays from planarity; instead, these arrays are modulated.

Linkage M=M, M-M, M=T, M-T. The complicated and dense framework structure of **parwelite**, $Mn_5SbO_4(SiO_4)(AsO_4)$ (Fig. 52a), can be described in several ways. Perhaps the simplest description encompasses two types of sheet alternating along [001]. One sheet is built by MnO_{4-8} polyhedra, including a highly-distorted tetrahedron, distorted trigonal bipyramids, octahedra, and a distorted cube. The other sheet includes zigzag [100] chains of distorted MnO_6 octahedra, more regular SbO_6 octahedra, and silicate tetrahedra. These chains are decorated and cross-linked by arsenate tetrahedra. Moore and Araki (1977c) described this complicated structure as an anion-deficient fluorite derivative. We note that the original space group A*a* reported by Moore and Araki (1977c) was corrected to A2/*a* by Marsh and Schomaker (1979).

Linkage M≡M, M=M, M-M, M-T. The structure of **allactite**, $Mn_7(OH)_8(AsO_4)_2$, is composed of sheets stacked along the *a* axis (Fig. 52b) (Moore 1968a). The sheets are made of two-octahedra wide bands which are cut out from the structure of pyrochroite, $Mn(OH)_2$,

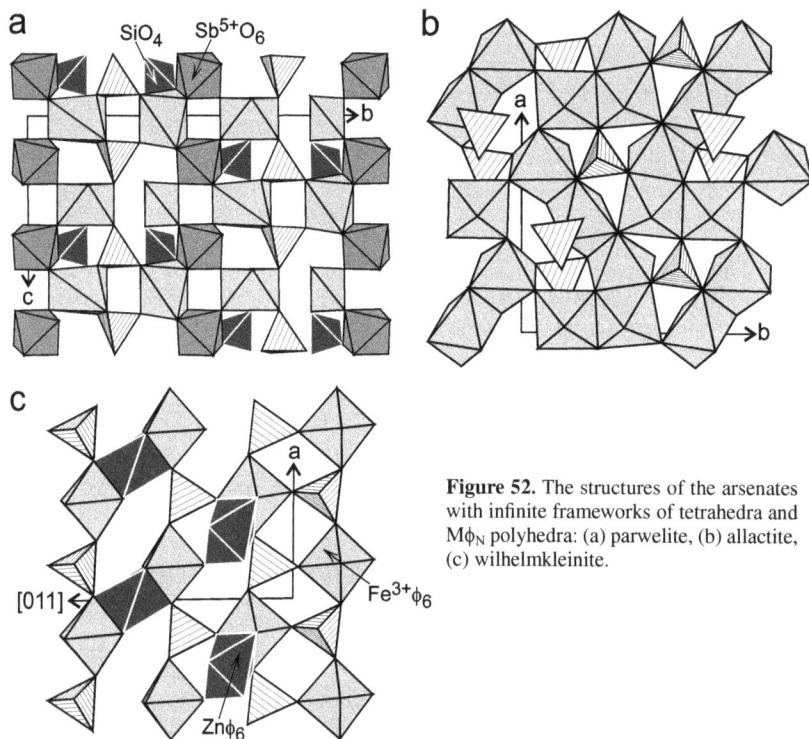

Figure 52. The structures of the arsenates with infinite frameworks of tetrahedra and $M\phi_N$ polyhedra: (a) parwelite, (b) allactite, (c) wilhelmkleinite.

and the bands are interconnected by corner-sharing octahedra. There are [010] chains of edge-sharing $Mn2\phi_6$ octahedra and AsO_4 tetrahedra between these sheets. The $Mn2\phi_6$ octahedra share corners, edges, and faces with the neighboring $Mn\phi_6$ octahedra and corners with the AsO_4 tetrahedra. Moore (1968a) noted that it is remarkable that this complicated arrangement is found in the structure of the most common basic manganese arsenate from Långban where many Mn arsenates occur.

Linkage M≡M, M-M, M-T. The structure of **wilhelmkleinite**, $ZnFe_2(OH)_2(AsO_4)_2$, is characterized by chains of corner-sharing octahedra which house the Fe^{3+} (Fig. 52c) (Adiwidjaja et al. 2000). The chains extend along the *b* axis and the shared corner is an OH group. Adjacent chains are connected by arsenate tetrahedra and $Zn\phi_6$ octahedra. The $Zn\phi_6$ octahedra share faces with the adjacent $Fe\phi_6$ octahedra; the structure could be alternatively described as made of trimers of one $Zn\phi_6$ and two adjacent $Fe\phi_6$ octahedra with the composition $[ZnFe_2(OH)_4O_8]$ (Adiwidjaja et al. 2000). These trimers are interconnected via the arsenate tetrahedra. Using either description, the structure is a relatively dense network of octahedra and tetrahedra.

Structures with polymerized ($T\phi_4$) groups and large-cation polyhedra.

Structures with infinite sheets of tetrahedra and large-cation polyhedra. The structures of **phaunouxite**, $Ca_3(H_2O)_8(AsO_4)_2 \cdot 3H_2O$, and **rauenthalite**, $Ca_3(H_2O)_8(AsO_4)_2 \cdot 2H_2O$, are closely related and their stability is likely a finely balanced function of temperature and air humidity (Catti and Ivaldi 1983). Both structures contain (100) slabs composed of $Ca\phi_{7-8}$ polyhedra and arsenate tetrahedra (Fig. 53a,b). The space between the slabs is occupied by H_2O molecules not bonded to the cations.

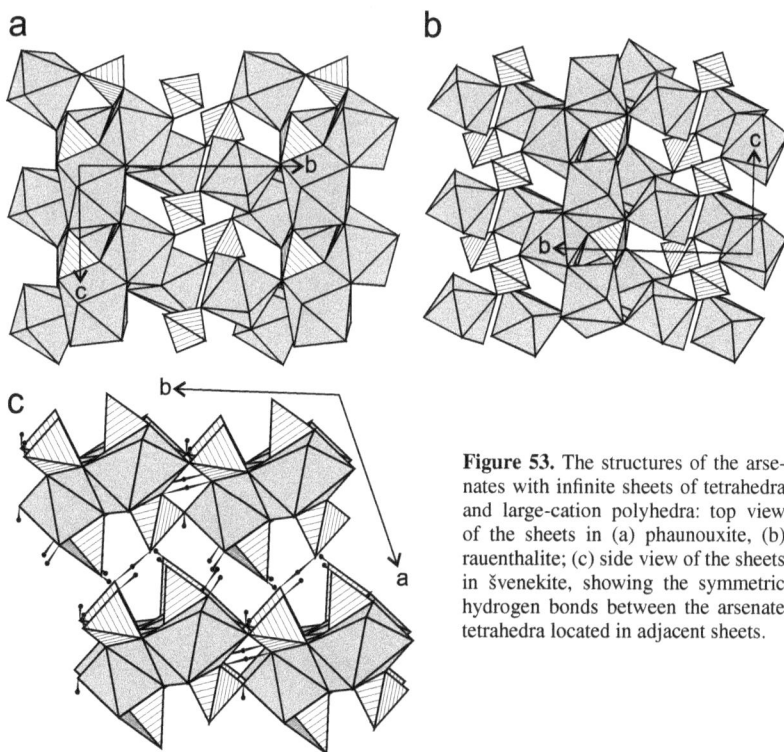

Figure 53. The structures of the arsenates with infinite sheets of tetrahedra and large-cation polyhedra: top view of the sheets in (a) phaunouxite, (b) rauenthalite; (c) side view of the sheets in švenekite, showing the symmetric hydrogen bonds between the arsenate tetrahedra located in adjacent sheets.

Within the (100) slabs, the Ca^{2+} polyhedra and $As2O_4$ tetrahedra condense into branching stripes parallel to the *c* axis. The neighboring stripes are bridged by the $As1O_4$ tetrahedra. Hydrogen bonding scheme is thoroughly discussed by Catti and Ivaldi (1983).

The basic features of the structure of **švenekite**, $Ca[AsO_2(OH)_2]$, are relatively simple (Fig. 53c). The $Ca\phi_8$ polyhedra and arsenate tetrahedra condense into (100) layers (Ferraris et al. 1972b). The backbone of these layers are zigzag chains of the large $Ca\phi_8$ polyhedra. The layers, however, are not flat, but a series of steps, each step being made of one $Ca\phi_8$ zigzag chain and the accompanying $As\phi_4$ tetrahedra.

A specific feature of this structure is the role and position of hydrogen atoms (Ferraris et al. 1972b). All H atoms bind to the arsenate tetrahedra. Three of the five distinct H atoms constitute the usual OH groups. The remaining two H atoms are located at an inversion center; each of these two H atoms forms two equally long bonds to two O neighbors related by the inversion center (Ferraris et al. 1972b).

Structures with infinite frameworks of tetrahedra and large-cation polyhedra. The intricate structure of **guérinite**, $Ca_5(H_2O)_{8.2}(AsO_3OH)_2(AsO_4)_2 \cdot 0.8H_2O$, is built by six independent $Ca\phi_{7-8}$ polyhedra interwoven with (AsO_3OH) and (AsO_4) tetrahedra (Fig. 54a) (Catti and Ferraris 1974). A further complication is the disorder and occupancies of the cavities in the structure.

The $Ca\phi_{7-8}$ polyhedra share corners, edges, and faces to form $(\overline{1}01)$ slabs. Arsenate tetrahedra join the slabs. There are five independent As positions in the structure. Of those, As2 and As3 form (AsO_4) tetrahedra and As5 an (AsO_3OH) tetrahedron. As4 is also coordinated

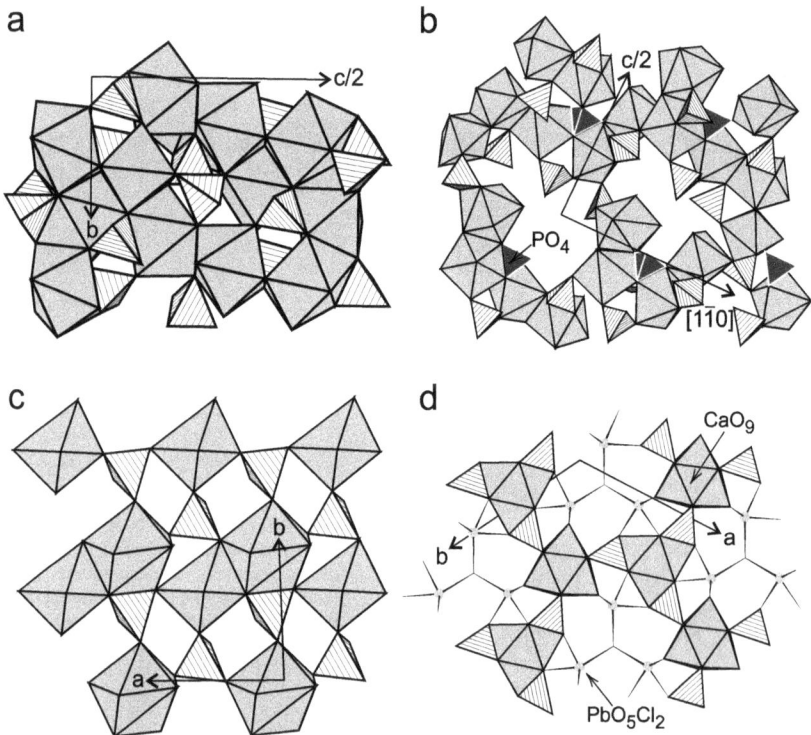

Figure 54. The structures of the arsenates with infinite frameworks of tetrahedra and large-cation polyhedra: (a) guérinite, (b) machatschkiite, (c) weilite, (d) hedyphane.

by 3 O and 1 OH ligands but the position of the H atom is statistically distributed among two oxygens. A half of the $As1\phi_4$ tetrahedra are (AsO_4) units, the other half (AsO_3OH) units. Since the H positions were not located in the structure refinement, the assignment of the H atoms was done according to bond-valence calculations (Catti and Ferraris 1974).

The slabs are bonded by hydrogen bonds and disordered Ca^{2+} cations. The hydrogen bonds are donated and accepted by H_2O molecules and OH^- groups within the slabs and an isolated H_2O molecule between the slabs. The disordered Ca^{2+} cations populate cavities between the slabs (two sites with occupancies of 0.17 and 0.08). Within the cavities, there are only two Ca-O distances below 3.0 Å but space is available for five additional H_2O molecules. Catti and Ferraris (1974) hypothesized that the Ca^{2+} cations, if present in the cavities, carry the additional five H_2O molecules to achieve a coordination number of seven.

The structure of **machatschkiite**, $Ca_6(AsO_4)(AsO_3OH)_3(PO_4)(H_2O)_{15}$, can be essentially viewed as a framework of $Ca\phi_{7-8}$ polyhedra, with the cavities in this framework occupied by isolated (AsO_3OH), (AsO_4), and (PO_4) tetrahedra and isolated H_2O molecules (Fig. 54b) (Effenberger et al. 1982). Three $Ca\phi_7$ and three $Ca\phi_8$ polyhedra connect *via* corners and edges to form a cluster $[Ca_6O_{19}(OH)_3(H_2O)_{15}]$. The six Ca^{2+} cations in these clusters lie essentially in the (001) plane. The center of the clusters coincide with the intersection of the threefold axes and the (001) planes. The clusters are then stacked in the c direction such that the centers of neighboring clusters are offset by a $(a/3,-b/3,c/6)$ vector. Effenberger et al. (1982) also noted that three O atoms of the As2O$_4$ group are acceptors of three hydrogen bonds from the adjacent As1O$_3$OH group, an unusual feature for the arsenate structures.

The structure of **weilite**, $CaHAsO_4$, contains Ca in seven- and eight-fold coordination. The $Ca\phi_7$ polyhedra can be described as distorted pentagonal bipyramids. The $Ca\phi_7$ and $Ca\phi_8$ polyhedra form $[1\overline{1}0]$ chains which attach to each other by corner sharing between the adjacent $Ca\phi_7$ and $Ca\phi_8$ units. In addition, the chains are cross-linked by arsenate tetrahedra to form thick slabs stacked along the c axis. The slabs are tightly connected by corner sharing between arsenate tetrahedra and the Ca polyhedra into a dense three-dimensional framework (Fig. 54c).

The location and distribution of hydrogen atoms in the structure of weilite is unknown. Identical arrangement of the heavy atoms in weilite and monetite, $CaHPO_4$, and similarity between the infrared spectra of the two minerals (Ferraris and Chiari 1970) suggest that the hydrogen atom positions are similar in these two phases. One of the hydrogen positions has been assigned to O1, an oxygen atom which bridges As1 and Ca1. The other oxygen atoms in the As1 tetrahedron bridge As1 and two Ca1 atoms. Another hydrogen atom (H2) should form a symmetrical bridge between O7 and O15. The H3 atom is supposed to be located in a more complicated geometry, either being statistically distributed between O6 and O14, or relating O6-O14 and O8-O16.

In discussion of the possible space group, either $P1$ or $P\overline{1}$, Ferraris and Chiari (1970) concluded that the centrosymmetry applies to heavy atoms and may be violated only by the hydrogen atoms.

The generic formula of the minerals in the **apatite supergroup** is $M1_4M2_6(TO_4)_6X_2$ (Pasero et al. 2010). The M1 and M2 cations are usually nine- and seven-fold coordinated, respectively. The seven-fold coordination of the M2 site is observed especially when the M2 cation is Ca^{2+}; with other cations, for example Pb^{2+}, or variable anions at the X site, the coordination may be eight- or nine-fold and more irregular than in the case of Ca^{2+}. The T site is always tetrahedral and occupied by P, As, V, S, or Si (Fig. 54d). The X site can accommodate F^-, Cl^-, $(OH)^-$, or O^{2-}. For the purposes of this review, only the phases where As^{5+} predominates on the T site are of interest and these phases are listed in Table 26.

The M1 and M2 sites may host the same cation (e.g., in **mimetite**) or two different cations (e.g., in **morelandite**, Table 26). Little is known about ordering of these cations in the arsenate

Table 26. Arsenate minerals of the apatite supergroup (after Pasero et al. 2010). The names with the suffices -*M* represent the monoclinic polymorphs and these names should replace those given here according to the suggestions of Pasero et al. (2010). As the nomenclature of the apatite supergroup changes almost habitually (three times during 2000-2010), we have retained the "old" names as they may be handy after the next revision.

svabite	$Ca_5(AsO_4)_3F$
turneaureite	$Ca_5(AsO_4)_3Cl$
johnbaumite	$Ca_5(AsO_4)_3(OH)$
fermorite (johnbaumite-*M*)	$Ca_5(AsO_4)_3(OH)$
mimetite	$Pb_5(AsO_4)_3Cl$
clinomimetite (mimetite-*M*)	$Pb_5(AsO_4)_3Cl$
hedyphane	$Ca_2Pb_3(AsO_4)_3Cl$
hydroxylhedyphane	$Ca_2Pb_3(AsO_4)_3(OH)$
morelandite	$Ca_2Ba_3(AsO_4)_3Cl$

members of the apatite supergroup as the single-crystal studies have concentrated on mimetite (e.g., Dai et al. 1991). **Hedyphane** is the only apatite-type arsenate with different M1 and M2 for which a good structural model exists (Rouse et al. 1984).

The hexagonal phases with the apatite supergroup are considered to be the "parent structures," the structures with different symmetry being derived from them (Pasero et al. 2010).

Atelestite, $Bi_2O(OH)AsO_4$, contains a mesh of two systems of chains of $Bi\phi_8$ polyhedra (Fig. 55a). The $Bi1\phi_8$ polyhedra form [100] chains. The chains do not attach to each other and the adjacent chains are offset by *b*/2. The [100] chains are entangled with [010] chains of $Bi2\phi_8$ polyhedra. They also do not attach to each other, hence, the [100] and [010] chains build together a three-dimensional mesh whose cavities are occupied by arsenate tetrahedra.

The crystal structure of **preisingerite**, $Bi_3O(OH)(AsO_4)_2$, is assembled from three distinct Bi polyhedra and arsenate tetrahedra (Fig. 55b) (Bedlivy and Mereiter 1982a). The Bi^{3+} ions are nine-fold coordinated, with four more proximal and five more distal ligands. The one-sided coordination in all three Bi polyhedra is dictated by the presence of the lone electron pair on this ion. In the structure of preisingerite, six Bi polyhedra form $Bi_6O_2(OH)_2$ groups and these groups are connected by arsenate tetrahedra into a dense framework.

Rooseveltite, $BiAsO_4$, has a monazite-type structure with arsenate tetrahedra and Bi^{3+} in irregular eight-fold coordination (Fig. 55c) (Bedlivy and Mereiter 1982b). In the structure of monazite, $CePO_4$, and related phases, the trivalent ions are usually found in a nine-fold coordination. The eight-fold coordination of Bi^{3+} in rooseveltite results from the presence of the lone electron pair on Bi^{3+}. The lone electron pair points presumably in the direction of O2 and the Bi-O2 distance of 3.24 Å is too long to be considered as a bond (Bedlivy and Mereiter 1982b).

The Bi^{3+} polyhedra share corners and form [010] chains which are interconnected by arsenate tetrahedra into (100) layers. Similar chains and layers can be found in monazite *s.s.* but the large polyhedra share edges in this structure. Adjacent layers in the structure of rooseveltite (and monazite) condense into a dense framework by edge-sharing between the BiO_8 polyhedra.

Gasparite-Ce, $CeAsO_4$, is another mineral isostructural with monazite and rooseveltite (Brahimi et al. 2002; Kolitsch et al. 2004). **Chernovite-Y**, $YAsO_4$, or **chernovite-Lu**, $LuAsO_4$,

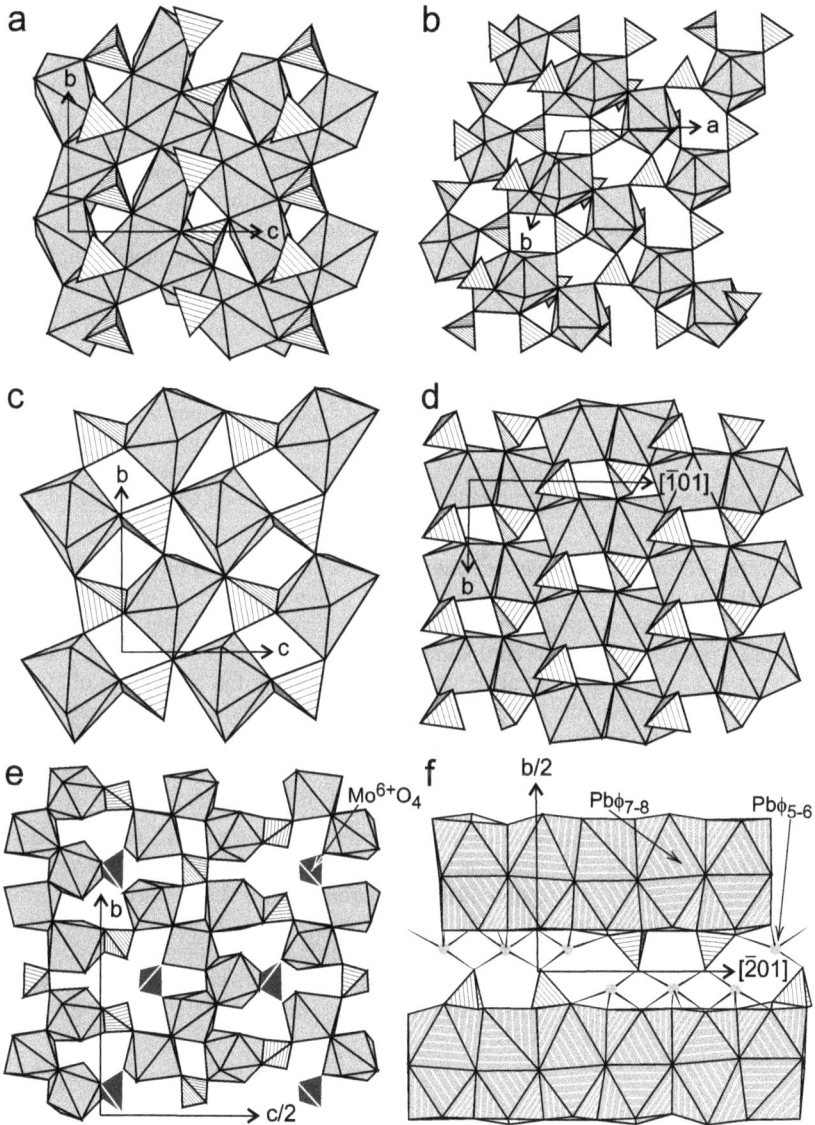

Figure 55. The structures of the arsenates with infinite frameworks of tetrahedra and large-cation polyhedra: (a) atelestite, (b) preisingerite, (c) rooseveltite, (d) tetrarooseveltite, (e) schlegelite, (f) sahlinite.

are isostructural with xenotime, $REEPO_4$, or zircon, $ZrSiO_4$ (Lohmueller et al. 1973). Kolitsch and Holtstam (2004) reviewed the monazite- and zircon-type phase fields, solid solution ranges and phase transitions of $REEXO_4$ (X = P, As, V) compounds, and discuss implications for the natural occurrences of such compounds.

Tetrarooseveltite, $BiAsO_4$, is isostructural with scheelite, $CaWO_4$ (Mooney 1948). Bi^{3+} is eight-fold coordinated and the large polyhedra share edges to create chains parallel to the *a* axis (Fig. 55d). The tunnels between the chains are populated by As^{5+} cations in the usual tetrahedral coordination.

The crystal structure of **schlegelite**, $Bi_7O_4(MoO_4)_2(AsO_4)_3$, is constructed by Bi-, Mo-, and As-centered polyhedra (Fig. 55e) (Krause et al. 2006). The coordination of As is tetrahedral. Bi is eight-fold coordinated in large, one-sided polyhedra. The coordination of Mo, however, can be a matter of debate. Each of the two independent Mo atoms forms four short bonds with its neighboring O atoms, defining two distorted tetrahedra. Bond valence calculations suggest a contribution from two more distal O atoms for each of the Mo atoms and hence an octahedral coordination of Mo. Krause et al. (2006) remarked that "bonds lengths and bond valences indicate a predominantly tetrahedral coordination with marked interactions to two further ligands in both cases". If tetrahedral coordination is assumed for both As and Mo, the structure can be described as built by isolated arsenate and molybdate tetrahedra, linked by the large BiO_8 polyhedra into a framework.

There are 7 independent Pb^{2+} positions in **sahlinite**, $Pb_{14}O_9Cl_4(AsO_4)_2$, with coordination numbers varying from 5 to 8 (Fig. 55f) (Bonaccorsi and Pasero 2003). All polyhedra are irregular owing to the presence of the lone electron pair on Pb^{2+}. The Pb polyhedra share corners and edges with arsenate tetrahedra and form thick (010) slabs of the composition $[Pb_{14}O_9(AsO_4)_2]$. The structural fragment which connects the slabs are the Cl^- ions arranged almost perfectly in a (010) plane between the slabs. The $[Pb_{14}O_9(AsO_4)_2]$ slabs are structurally related to litharge, PbO, and can be derived from this structure with substitution of Pb with As, and introduction and ordering of O vacancies.

Structures with octahedrally coordinated As^{5+}. Octahedral coordination of As^{5+} has been recorded only for one mineral, aerugite (see below). Synthetic phases with octahedral As^{5+} are known but are also not abundant. An example is $LiAsO_3$ (Driss and Jouini 1989), with the structure derived from that of corundum. The As-O distance is aerugite is 1.835 Å, the sum of bond valences on the central As cation is 4.99 (using the parameters of Brown and Altermatt 1985). In $LiAsO_3$, the As-O bond lengths are 1.816 and 1.848 Å. In both structures, the octahedra have relative high symmetry dictated by the overall symmetry of the structure. In aerugite, the As^{5+} cation is displaced 0.033 Å from the centroid of the octahedron and the angles within the octahedron (O-As-O) are very close to the ideal 90° (89.54, 90.46°).

The only structure of a mineral with octahedrally coordinated As^{5+} is that of **aerugite**, $Ni_{8.5}O_2(AsO_4)_2(AsO_6)$ (Fig. 56). This structure based on a cubic closed-packed array of oxygen atoms stacked along the [001] direction (Fleet and Barbier 1989). The tetrahedral and octahedral vacancies are filled by Ni^{2+} and As^{5+}. The structure comprises three-layer units built by NiO_6 octahedra and AsO_4 tetrahedra and a rock-salt-like single layer composed of NiO_6 and AsO_6 octahedra. The electrostatic balance requires that the NiO_6octahedra within the single layer units are only partially occupied. Structurally related arsenates, germanates, and vanadates were discussed by Fleet and Barbier (1989).

[111]

Figure 56. The structure of aerugite, a mineral with tetrahedrally and octahedrally coordinated As^{5+}. All As^{5+} polyhedra (i.e., both tetrahedra and octahedra) are hatched, the Ni octahedra are shaded.

Crystal chemistry of arsenites and arsenites-arsenates

Arsenites and arsenite-arsenates are presented here together. Both of these groups contain only a few minerals. The arsenite-arsenates can be viewed as arsenites with a "special" tetrahedrally coordinated cation As^{5+}. Among the arsenite-arsenates, only two (radovanite and synadelphite) do not belong to the complex sheet structures, described below separately.

Selected geometric features of the $As^{3+}\phi_3$ groups are summarized and statistically evaluated below. Data were screened for inconsistencies and aberrant coordination polyhedra. Fourfold coordination of As^{3+} with oxygen was proposed for the structure of schallerite (Kato and Watanabe 1991) but is perhaps only an artifact of the experimental difficulties rather than a representation of the actual atomic environment in this mineral. Essentially no substitutions for As^{3+} were reported, most likely owing to the coordination specific to cations with stereochemically active lone electron pairs. Similar homovalent cations (Sb^{3+}, Bi^{3+}) could take the place of As^{3+} but are probably too large for a broad miscibility with As^{3+}. An example of such behavior is the mineral stenhuggarite, $CaFeSbAs_2O_7$, where As^{3+} and Sb^{3+} occur together but partition strictly to separate structural sites. Some substitution between As^{3+} and Sb^{3+} is possible, though, for example in hemloite, $(Ti,V,Fe,Al)_{12}(As,Sb)_2O_{23}(OH)$.

Variations in As-ϕ and <As-ϕ> distances and ϕ-As-ϕ angles. The distribution of the As-ϕ ($\phi = O^{2-}$, OH^-) distances is shown in Fig. 57a. The average As^{3+}-O bond length is 1.782 Å. The shortest As-O distances (all less than 1.7 Å) and several of the longest ones were observed in gebhardite. The longest individual As-O distance (1.914 Å) was reported in georgiadèsite. There are only four minerals with (AsO_2OH) groups: radovanite, trigonite, armangite, and schallerite, with As-OH distances of 1.808, 1.885, and 1.819 Å for the first three minerals, respectively. The variations in <As-ϕ> (where <> denotes the arithmetic mean) distances are shown in Fig. 57b. The mean value of all considered <As-ϕ> distances is 1.782 Å. The O-As-O angles scatter relatively widely (Fig. 57c) with a mean value of 97.09°.

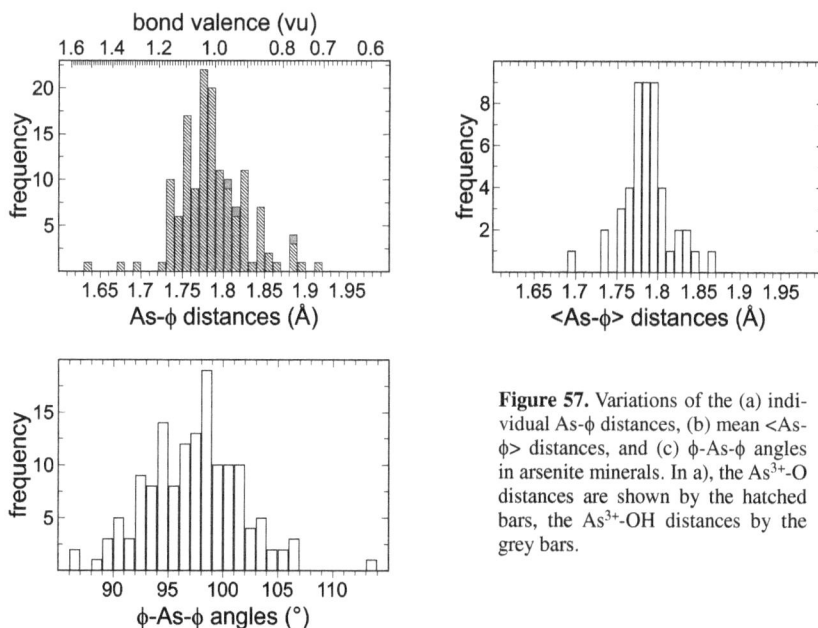

Figure 57. Variations of the (a) individual As-ϕ distances, (b) mean <As-ϕ> distances, and (c) ϕ-As-ϕ angles in arsenite minerals. In a), the As^{3+}-O distances are shown by the hatched bars, the As^{3+}-OH distances by the grey bars.

Geometry of the Asϕ_3 pyramids and coordination of the oxygen ligands. As for the arsenates, the distortion of the coordination polyhedra in the arsenites can be quantified as

$$BLDP = \frac{1}{<TO>}\left[\frac{\overset{3}{\Sigma}(TO-<TO>)^2}{N-1}\right]^{1/2} \quad N=3$$

$$ELDP = \frac{1}{<OO>}\left[\frac{\overset{3}{\Sigma}(OO-<OO>)^2}{N-1}\right]^{1/2} \quad N=3$$

$$ADP = \frac{1}{<OTO>}\left[\frac{\overset{3}{\Sigma}(OTO-<OTO>)^2}{N-1}\right]^{1/2} \quad N=3$$

In addition, the coordination number of each oxygen ligand (CN), the average coordination number for a Asϕ_3 pyramid (<CN>), and the average cation electronegativity (<χ>) can be calculated. A similar analysis for the arsenites has been undertaken by Hawthorne (1985b). The bond-length and edge-length distortions are quantitatively similar to the values calculated for the arsenates (Fig. 58). The average coordination number for the oxygen ligands is 3.20; the average electronegativity of the neighboring cations is 1.84. These values are remarkably similar to the ones determined for arsenates (see above). Inspection of the correlation between the calculated variables revealed only the expected negative correlation between <CN> and <χ>. There is a weaker negative correlation between <CN> and <TO> ($r = -0.54$); other correlations are insignificant. The pair-distribution function for the oxygens that are ligands of the As^{3+} cations is shown in Fig. 59. The first sharp peak corresponds to the O-As^{3+} neighbors, the second broader peak to other O-cation pairs. The first coordination sphere of these oxygen ligands terminates at distances of around 3 Å.

A structural hierarchy for arsenite minerals. As for the arsenate minerals, we examine the polymerization of the Asϕ_3 units in the structures of arsenites and arsenite-arsenates. Pauling's bond valence for each bond in an As$^{3+}\phi_3$ is 1. If the ligand ϕ is an oxygen anion with a nominal charge of -2, its bonding requirement would be perfectly matched by bridging two As^{3+} cations. Indeed, polymerization of the Asϕ_3 arsenite units is observed in a much greater proportion among the arsenite minerals than polymerization of the tetrahedral Asϕ_4 units among the arsenates.

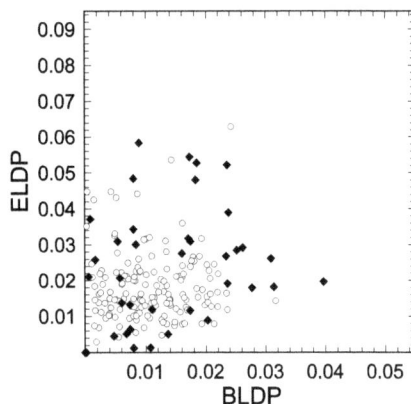

Figure 58. Edge-length and bond-length distortion parameters (ELDP and BLDP) for the Asϕ_3 trigonal pyramids in the structures of arsenite minerals (black diamonds). For comparison, the ELDP and BLDP parameters for the arsenate minerals are plotted as open circles (see also Fig. 31). For definitions of ELDP and BLDP, see text. Note that the parameters are defined slightly differently for arsenates and arsenites.

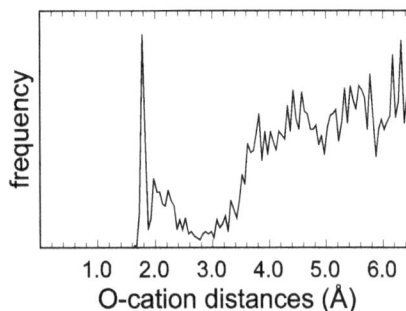

Figure 59. Pair-distribution function of the O-cation pairs of oxygen atoms which are the ligands of As^{3+} in the structures of arsenite minerals.

If the polymerization of the $As\phi_3$ units does not occur, the oxygen ligands will establish additional bonds to the neighboring cations. In this case, each O atom should receive another valence unit (vu) from another cation(s).

In none of the structures considered here, polymerization of $As^{3+}O_3$ and $As^{5+}O_4$ is seen. Polymerization between the $As^{3+}O_3$ polyhedra and tetrahedra is known, examples being the structures of leiteite and reinerite. In both of these structures, $As^{3+}O_3$ trigonal pyramids polymerize with ZnO_4 tetrahedra. Polymerization of $As^{3+}O_3$ trigonal pyramids and SiO_4 tetrahedra could be expected, as the Pauling's bond valence of both As-O and Si-O bonds is 1, the sum being then 2. Arsenite-silicate minerals are rare. Pairs of AsO_3 trigonal pyramids and SiO_4 tetrahedra, as well as AsO_3 pyramids and BeO_4 tetrahedra are known from asbecasite. AsO_3-SiO_4 pairs exist perhaps in mcgovernite, kraisslite, or turtmannite, although the exceeding complexity of these structures may conceal the existence of such pairs.

Polymerization with other coordination polyhedra with low coordination numbers can also be encountered. Examples are polymerization with the one-sided PbO_4 polyhedra in paulmooreite, one-sided $Sb^{3+}O_4$ polyhedra in stenhuggarite, or the square-planar FeO_4 units in cafarsite.

Most commonly, the oxygen anions of the $As^{3+}O_3$ groups bond to octahedrally coordinated cations, especially Mn^{2+}, Fe^{2+}, Fe^{3+}, Cu^{2+}, amongst others. Structures with oxygen ligands of As^{3+} which bond to large cations (e.g., Pb^{2+}) are also quite common.

For the arsenite minerals, we use a similar classification scheme as for the arsenates. We consider

1. structures with polymerized $As\phi_3$ trigonal pyramids,

2. structures with polymerized $As\phi_3$ trigonal pyramids and $M\phi_{4-6}$ polyhedra,

3. structures with polymerized $As\phi_3$ trigonal pyramids and large-cation polyhedra.

Within each class, the structures should be arranged according to the increasing connectivity of the polyhedra in the structural unit. The possibilities are

•   unconnected polyhedra,

•   finite clusters of polyhedra,

•   infinite chains or rods of polyhedra,

•   infinite sheets or layers of polyhedra,

•   infinite frameworks of polyhedra.

Among the arsenites, however, we do not find all these types of connectivities. The absence of some is perhaps caused by the small number of the known arsenite structures and perhaps by the specific requirements of the $As\phi_3$ units to accommodate their lone electron pairs in the structure. The specific features for the classes (1), (2), and (3) are briefly discussed in the introduction to each class. Here we only note that there are some families of minerals which have unusual structural features and these are described in small separate sections. These are phyllosilicates with As^{3+}, arsenites with metal clusters, and complex sheet arsenite-arsenate structures.

Structures with polymerized ($As\phi_3$) trigonal pyramids. Most of these structures contain an (As_2O_5) dimer. The only exceptions are stenhuggarite with the four-membered (As_4O_8) rings and ludlockite with (As_5O_{11}) pentamers of arsenite groups. Cooper and Hawthorne (1996b) pointed out that the arsenite groups condense into dimers, tetramers, and pentamers. There is then a gap in the size of the oligomers until infinite chains appear in the minerals trippkeite and leiteite. Cooper and Hawthorne (1996b) noted that this behavior is very similar to that observed for silicates.

Structures with finite clusters of trigonal pyramids. The structure of **schneiderhöhnite**, $Fe^{2+}Fe^{3+}_3As^{3+}_5O_{13}$, is a union of Fe octahedra and (AsO_3) trigonal pyramids in a dense heteropolyhedral framework (Fig. 60a) (Hawthorne 1985b). There are five independent As^{3+} positions; two of the trigonal pyramid share a corner to form a (As_2O_5) dimer, the other ones do not polymerize with each other. The Fe polyhedra condense into complicated sheets. Within the sheets, the Fe^{3+} octahedra share edges and form zigzag chains. These chains are bridged into sheets with additional dimers of Fe^{3+} octahedra and the sheets are decorated by Fe^{2+} octahedra.

The massive framework of **gebhardite**, $Pb_8OCl_6(As_2O_5)_2$, is built by eight distinct $Pb\phi_{6-8}$ polyhedra (Fig. 60b) (Klaska and Gebert 1982). The one-sided geometry, characteristic for Pb^{2+}, is found in many of them. The Pb polyhedra form corrugated slabs, roughly parallel to (100) and the slabs join each other to define the framework. The [010] tunnels in this framework

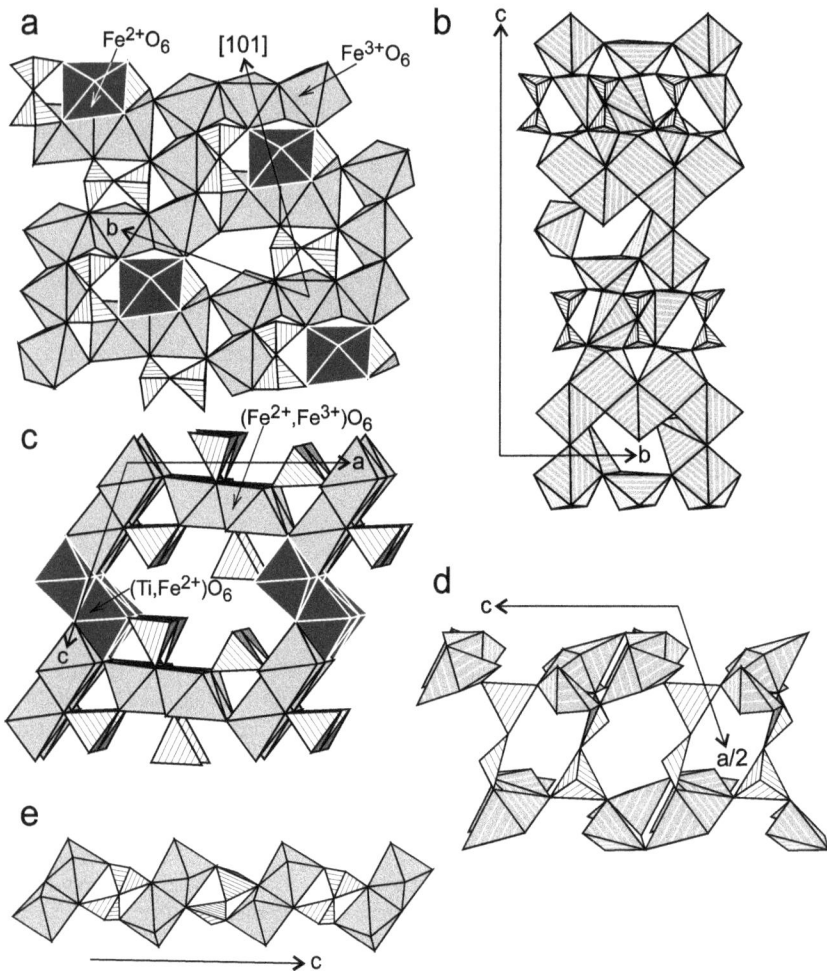

Figure 60. The structures of the arsenites with dimers of $As\phi_3$ trigonal pyramids: (a) schneiderhöhnite, (b) gebhardite, (c) fetiasite; the As^{3+} polyhedra appear as tetrahedra because some of the As^{3+} sites are vacant; the true coordination of As^{3+} is trigonal pyramidal, as in all other arsenites; (d) paulmooreite; all polyhedra were constructed with anions (O^{2-}) and cations (As^{3+}, Pb^{2+}) as their vertices, (e) vajdakite, the chain with alternating dimers of octahedra and $As\phi_3$ pyramids.

house (As_2O_5) dimers. Klaska and Gebert (1982) discussed the symmetry of the slabs, possible superstructures and polytypism of this mineral.

Fetiasite, $(Fe,Ti)_3O_2(As_2O_5)$, is an octahedral framework structure (Fig. 60c) (Graeser et al. 1994). Octahedrally coordinated Fe^{2+}, Fe^{3+}, and Ti share edges and corners and the framework enclose tunnels along [010]. Each tunnel is large enough to house four rows of As^{3+} cations, each row being parallel to the axis of the tunnel. Each As^{3+} is thought to be coordinated by three oxygen atoms; two of them belong to the octahedral walls of the tunnels, the third is a disordered O position inside the tunnels. The occupancy of the O atoms inside the tunnels is 0.5 (Graeser et al. 1994). Such arrangement leads necessarily to the presence of dimers (As_2O_5), with a perfect one-dimensional order in each row. There is a lack of correlation, however, between the adjacent tunnels or perhaps rows, so that no superstructure based on the orientation of the arsenite dimers was reported.

The structure of **paulmooreite**, $Pb_2(As_2O_5)$, is a result of an enticing puzzle to arrange the lone electron pairs of Pb^{2+} and As^{3+} in a periodic array within the available space (Fig. 60d) (Araki et al. 1980). Both Pb^{2+} and As^{3+} polyhedra are one-sided. Araki et al. (1980) reported four-fold coordination of Pb^{2+}. These polyhedra are distorted tetragonal pyramids, with four apices being the oxygen ligands, the fifth one the Pb^{2+} cation. The Pb polyhedra coalesce into [010] chains *via* corner and edge sharing. The As^{3+} cations are found in the usual trigonal-pyramidal coordination. Two trigonal pyramids share a corner to form an (As_2O_5) dimer. These dimers link the Pb polyhedral chains into a framework.

The structure of **vajdakite**, $(MoO_2)_2(H_2O)_2As_2O_5 \cdot H_2O$, is built by heteropolyhedral chains with [001] direction (Fig. 60e) (Ondruš et al. 2002). One component of the chains are $[Mo_2O_8(H_2O)_2]$ dimers of edge-sharing Mo octahedra. The other component are (As_2O_5) dimers of two trigonal pyramids with As^{3+}. These two components alternate regularly along the direction of the chains. The chains are linked by hydrogen bonds into loose (010) sheets. The space between the sheets is occupied by additional H_2O molecules.

The structure of **stenhuggarite**, $CaFeSbAs_2O_7$, is a rare example of a cyclo-arsenite (Coda et al. 1977). The backbone of the structure are bulky, zigzag chains of CaO_8 polyhedra which extend in the [001] direction. Sb^{3+} is coordinated by four oxygen atoms which define a corrugated base of a pyramid, Sb^{3+} cation being the apex of that pyramid. The lone electron pair on Sb^{3+} is the cause of the one-sided geometry of the coordination polyhedron. The SbO_4 polyhedra attach with an edge to the CaO_8 chains. These chains are further interconnected by $Fe^{3+}O_5$ trigonal bipyramids and the four-membered arsenite rings (Fig. 61a).

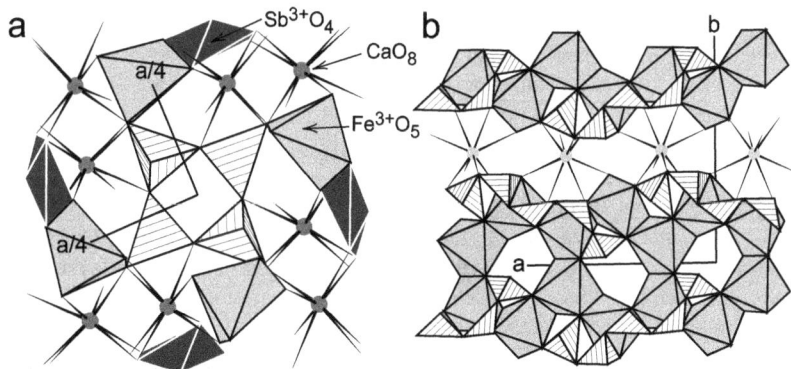

Figure 61. The structures of the arsenites with finite clusters of $As\phi_3$ trigonal pyramids: (a) stenhuggarite, (b) ludlockite.

Majzlan, Drahota, Filippi

Ludlockite, $PbFe_4As_{10}O_{22}$, has a structure based on polyhedral sheets with pentamers of arsenite groups attached to these sheets (Fig. 61b) (Cooper and Hawthorne 1996b). The sheets can be derived from dioctahedral sheets in gibbsite; instead of the continuous octahedral sheets in gibbsite, the sheets in ludlockite are built by strips of Fe^{3+} octahedra, joined via large $Pb\phi_8$ polyhedra. Cooper and Hawthorne (1996b) discuss in detail the omissions and substitutions necessary to derive the ludlockite sheets. Pentamers of the arsenite groups, with an overall composition (As_5O_{11}) attach to the octahedral portions of the sheets. Cooper and Hawthorne (1996b) described the relationship between the pentamers and octahedral strips as the one of "geometrical spiders consuming their prey." There are two unique but topologically very similar pentamers and the structure has a strong pseudomonoclinic symmetry. The interactions between the sheets are facilitated by long As-O bonds, resulting in perfect cleavage of the mineral; there are no atoms located in the space between the sheets.

Structures with infinite chains of trigonal pyramids. The building blocks of the structure of **leiteite**, $ZnAs_2O_4$, are Zn^{2+} tetrahedral sheets and As^{3+} pyramidal chains (Fig. 62a) (Ghose et al. 1987). The (001) tetrahedral sheets consist of regular ZnO_4 polyhedra. The chains of the (AsO_3) groups attach to the tetrahedral sheets and extend in the *a* direction. The sheets and the chains thus define (001) layers with an interlayer spacing of 8.8 Å. The layers are held together by long (2.84-3.34 Å) and weak As-O bonds, thus accounting for the excellent cleavage of the mineral, reminiscent of the cleavage of micas (Ghose et al. 1987).

Trippkeite, $CuAs_2O_4$, has a simple but elegant structure based on octahedral chains and chains of AsO_3 trigonal pyramids (Fig. 62b) (Pertlik 1975). The octahedral chains house Cu^{2+} in the usual Jahn-Teller distorted coordination and edge sharing between the octahedra. All oxygen atoms of the octahedral chains are shared with adjacent trigonal-pyramidal AsO_3 groups. These groups meld into chains which extend in the same direction as the octahedral chains. The two types of chains define a framework structure with large channels along the four-fold axis; the channels are most likely populated by the lone electron pairs of the As^{3+} cations.

Structures with polymerized $(As\phi_3)$ and $(M\phi_{4-6})$ polyhedra. This group comprises structures with $(As\phi_3)$ monomers which connect to other polyhedra in the structures of arsenite minerals. In two instances (trigonite and schallerite), the $(As\phi_3)$ unit is a (AsO_2OH) group, otherwise the usual (AsO_3) group is found. In almost all cases, the connectivity is established by the $(M\phi_{4-6})$ groups. This means, for example, that the $(M\phi_{4-6})$ groups define a framework and the $(As\phi_3)$ monomers take up the available space within the framework. Only in trigonite, cafarsite, and zimbabweite, $(As\phi_3)$ groups actually participate in the construction of the framework by cross-linking structural units of the $(M\phi_{4-6})$ groups. Most commonly, the arsenite groups line, occupy, nest, and fill cavities, tunnels, voids and holes within frameworks of the $(M\phi_{4-6})$ polyhedra.

Loosely attached to this group are several other arsenite minerals whose structural features render their assignment to small separate sets. These are arsenites with metal clusters, phyllosilicates with As^{3+}, and complex arsenite-arsenate-silicates. They are described below in separate sections.

Structures with finite clusters of trigonal pyramids and $M\phi_{4-6}$ polyhedra. The structural constituents in **nealite**, $Pb_4Fe(H_2O)_2Cl_4(AsO_3)_2$, are the (AsO_3) trigonal pyramids, $Pb\phi_{8-9}$ polyhedra, and $Fe[O_2Cl_2(H_2O)_2]$ octahedra (Pertlik and Schnorrer 1993). The large $Pb\phi_{8-9}$ polyhedra condense into corrugated layers, in general parallel with (100). Clusters of one octahedron and two (AsO_3) pyramids (Fig. 63a) are squeezed between the layers and share edges and corners with the $Pb\phi_{8-9}$ polyhedra. The massive framework of Pb polyhedra constricts narrow [001] tunnels where the arsenite groups are found.

Structures with infinite sheets of trigonal pyramids and $M\phi_{4-6}$ polyhedra. The structure of **tooeleite**, $Fe_6(OH)_4(H_2O)_4(AsO_3)_4(SO_4)$, consists of (001) heteropolyhedral sheets (Fig. 63b) (Morin et al. 2007). The sheets are built by $Fe\phi_6$ octahedra and are decorated by AsO_3 trigonal pyramidal groups. Two Fe2ϕ_6 octahedra share an edge to form a Fe$_2$O$_6$(OH)$_2$(OH$_2$)$_2$ cluster. The H$_2$O ligands point in the space between the sheets. The (OH) ligands facilitate the edge sharing to two adjacent octahedral monomer units (the Fe1O$_6$ octahedra) in the direction of the *a* axis. Four O ligands are shared (edge shared) with the Fe1O$_6$ octahedra in the direction of the *b* axis. Hence, each dimer is connected *via* edge-sharing with four neighboring Fe octahedra which, in turn, share edges with the other dimers, thus creating the octahedral sheet. The (AsO$_3$) group only decorate the sheets, i.e., the octahedra alone knit the sheets. The space between the sheets

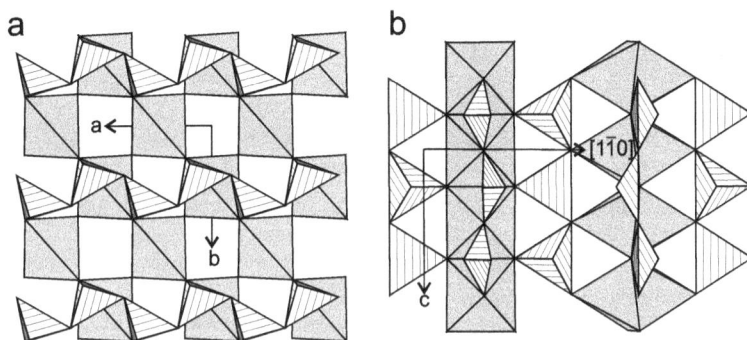

Figure 62. The structures of the arsenites with infinite chains or rods of Asϕ_3 trigonal pyramids: (a) leiteite, (b) trippkeite.

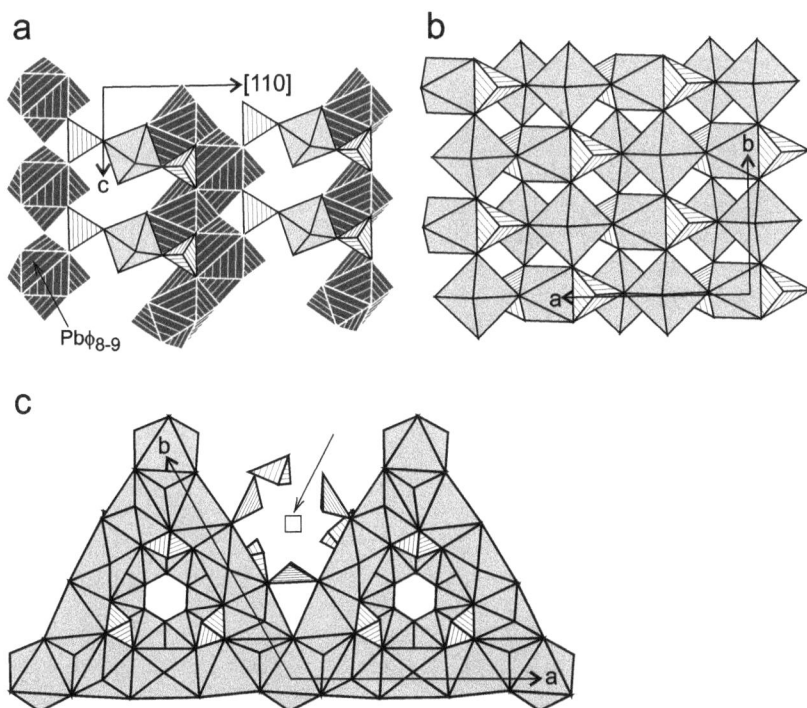

Figure 63. The structures of the arsenites with finite clusters, infinite sheets, or infinite frameworks of Asϕ_3 and Mϕ_{4-6} polyhedra: (a) nealite, (b) tooeleite, (c) armangite. The square highlighted by an arrow is the empty position inside a cavity into which six Asϕ_3 groups position their lone electron pairs.

is populated by disordered SO_4 groups. The nature of the disorder (static *versus* dynamic) is not known.

Structures with infinite frameworks of trigonal pyramids and Mϕ_{4-6} polyhedra. **Armangite**, $Mn_{26}[As_6(OH)_4O_{14}](As_6O_{18})_2(CO_3)$, has a structure that can be derived from the simple fluorite structure (Fig. 63c) (Moore and Araki 1979a). The unit cell of armangite contains 12 fluorite unit cells and has an overall composition $X_{48}\phi_{96}$, where X are cations or cation vacancies and ϕ are anions or anion vacancies. Five independent Mn^{2+} sites are six-fold coordinated. For three of them, the coordination polyhedron is a distorted octahedron, for the fourth one it is a distorted trigonal prism. The fifth Mn position is also coordinated octahedrally, with one of the ligands being disordered over two positions. The $Mn\phi_6$ polyhedra share edges and corners and constrct an elaborate framework. There are two cavities or nests in this framework per unit cell, centered at (1/3,2/3,2/3) and (2/3,1/3,1/3). These nests are lined by six (AsO_3) groups which point their lone electron pairs into the geometric center of the nests. The nests are interconnected by a small hexagonal opening centered at (0,0,0) where the rotationally disordered (CO_3) groups reside.

Synadelphite, $Mn_9(OH)_9(H_2O)_2(AsO_3)(AsO_4)_2$, one of Nature's inorganic architectural masterpieces (Moore 1970), has a structure based on edge-sharing octahedral clusters, arsenate tetrahedra and AsO_3 trigonal pyramids. The underlying motif of the structure is a nonamer of Mn octahedra with overall composition $Mn_9\phi_{37}$. This nonamer was described as a "wedge" (see Fig. 1 in Moore 1970) and has a point symmetry *m*. The octahedra share edges and build a massive octahedral framework. AsO_3 pyramids are positioned in the apices of the wedges, the arsenate tetrahedra flank the wedges. The unit cell is constructed from the octahedral wedges with AsO_4 and AsO_3 groups; within the unit cell, the wedges are related by inversion. Moore (1970) also commented on earlier accounts of this mineral having abundant Pb and discussed anomalous optical properties of this mineral.

The structure of **radovanite**, $Cu_2Fe^{3+}(As^{5+}O_4)(As^{3+}O_2OH)_2\cdot H_2O$, is assembled from (100) sheets and [010] chains (Sarp and Guenée 2002). The repeat unit of the sheets is $[Cu_2As^{3+}_2O_6(OH)_3]$. The sheets are constructed of $[Cu\phi_8]$ dimers of edge-sharing square pyramids which are interconnected by the AsO_3 trigonal pyramids. Each AsO_3 group shares an edge with one dimer and a corner with the neighboring dimer. The sheets are interconnected by [010] chains with a repeat unit $[FeAs^{5+}O_8]$. In these chains, Fe^{3+} is coordinated octahedrally, As^{5+} tetrahedrally. The octahedra and tetrahedra share corners and alternate regularly along the chain. This arrangement forms [100] tunnels, approximately 5.8 Å wide (in the *b* direction). The tunnels are narrower in the *c* direction because they are cluttered by the arsenate tetrahedra. These tunnels house the lone electron pairs of the As^{3+} cations and a H_2O molecule.

The structure of **reinerite**, $Zn_3(AsO_3)_2$, is based on a framework of ZnO_4 tetrahedra (Fig. 64a) (Ghose et al. 1977). The framework is constructed from two types of ZnO_4 tetrahedral units. One of the units is a dimer made of edge-sharing tetrahedra. This unusual connectivity results in a significant distortion of the tetrahedra in the dimer. The other units are infinite tetrahedral chains, extending along the *a* axis. The arsenite groups only beautify the framework. Having their usual trigonal-pyramidal geometry, the arsenite groups line the [001] channels in the framework, with their lone electron pairs pointed into the channels.

The complicated, high-symmetry structure of **cafarsite**, $Ca_{5.9}Mn_{1.7}Fe_{3.0}Ti_{3.0}(AsO_3)_{12}\cdot4-5H_2O$, consists of several types of polyhedral clusters, knitted together into a heteropolyhedral framework (Edenharter et al. 1977). Ti, Mn, and part of Fe (position Fe2) are coordinated octahedrally. Similarly, one of the Ca positions (Ca1) is housed in a large, distorted octahedron. The Ca2 and Fe2 octahedra meld into large spherical clusters with the composition $Ca_4Fe_6\phi_{36}$, in which the Ca and Fe octahedra alternate and share edges. Two Ti octahedra attach *via* edge sharing to the opposite edges of the Fe1 polyhedron. Fe1 is found here in a four-fold, almost square-planar coordination. These clusters have the composition $FeTi_2\phi_{12}$. The $Mn\phi_6$ octahedra

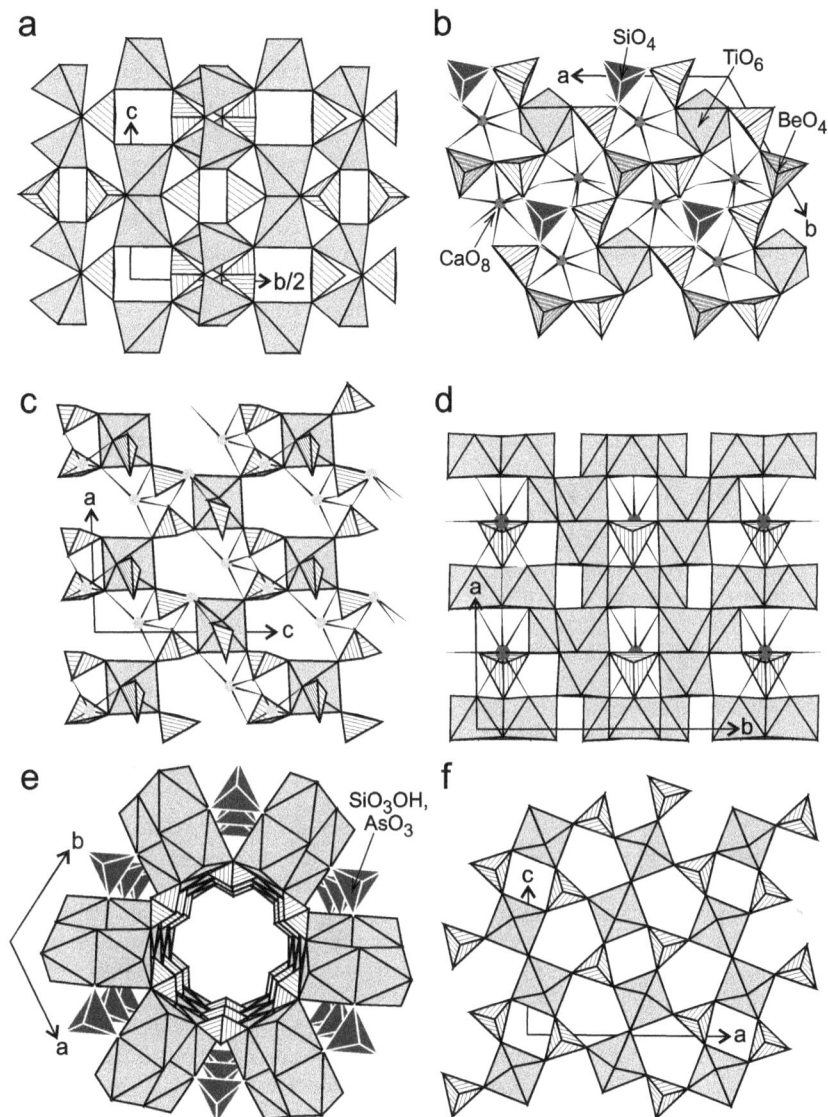

Figure 64. The structures of the arsenites with infinite frameworks of Asϕ_3 and Mϕ_{4-6} polyhedra: (a) reinerite, (b) asbecasite, (c) trigonite, (d) tomichite as a member of the derbylite group, (e) ekatite, a perspective view into the tunnels lined by Asϕ_3 groups, (f) zimbabweite.

do not connect to the octahedral clusters in the structure. These three types of octahedral units are cross-linked by the (AsO$_3$) groups into a framework. The AsO$_3$ groups do not polymerize with each other. The remaining cavities in the structure are filled by large Ca2ϕ_8 polyhedra. Both Fe and Mn are assumed to be present in the oxidation states +2 and +3; the substitution mechanisms which could maintain the electrostatic balance are unknown.

The structure of **asbecasite**, Ca$_3$Ti(As$_6$Be$_2$Si$_2$O$_{20}$), is a complex heteropolyhedral framework (Fig. 64b) (Sacerdoti et al. 1993). Six arsenite trigonal pyramids attach with their corners to a TiO_6 octahedron, creating a $TiAs_6O_{18}$ cluster. Both Be^{2+} and Si^{4+} are coordinated tetrahedrally and share a corner in a dimer $BeSiO_7$. The bases of the tetrahedral dimer opposite to the oxygen shared by the two tetrahedra link the $TiAs_6O_{18}$ clusters into (001) slabs. The axes of the dimers align with the c axis, i.e., perpendicular to the slabs. The $BeSiO_7$ dimers bridge the slabs. Large spaces with the slabs are filled with CaO_8 polyhedra.

The structure of **trigonite**, $Pb_3Mn(AsO_3)_2(AsO_2OH)$, is based on an open heteropolyhedral framework of MnO_6 and $As\phi_3$ trigonal pyramidal groups (Fig. 64c) (Pertlik 1978). The distorted MnO_6 polyhedra share all six ligands with $As\phi_3$ groups. The structure could be described in terms of (010) heteropolyhedral sheets which consist of the MnO_6 and $As\phi_3$ units and are cross-linked by additional $As\phi_3$ groups. The Pb cations reside in irregular $Pb\phi_5$ and $Pb\phi_7$ polyhedra in the large cavities of the framework. The presence of Mn^{2+} and an OH group, as opposed to Mn^{3+} and an O^{2-} anion, was deciphered based on considerations of bond distances and distortions of coordination polyhedra (Pertlik 1978).

Minerals of the **derbylite group** (Table 27) have a complex octahedral framework architecture. Derbylite $[SbTi_3(V,Fe)_4O_{13}(OH)]$, **tomichite** (Fig. 64d), **barian tomichite**, and **graeserite** are closely structurally related. The structure of hemloite is more complex, with tomichite-like structural fragments (Harris et al. 1989).

The structure of the derbylite-group members is based on a close packing of the oxygen anions. With the exception of hemloite, fourteen of the available 30 octahedral sites in the unit cell are filled by Ti^{4+}, V^{3+}, Fe^{3+}, or Fe^{2+}, with no ordering of the cations over the sites. The octahedra condense into two types of layers which alternate regularly along the a axis. One layer consists of V_3O_5-like chains that are one- and two-octahedra wide. The chains within one layer are not connected to each other. The other layer is two octahedra thick; the octahedral dimers build double octahedral α-PbO_2-like chains which are also not connected within a single layer. This arrangement creates tunnels with large cubo-octahedral cavities. In tomichite, each tunnel is populated by arsenite trigonal pyramidal groups with all lone electron pairs pointing in one direction. In the structure of barian tomichite (Grey et al. 1987), the tunnels host not only the arsenite groups but also Ba in twelve-fold, cubo-octahedral coordination. The Ba^{2+} and As^{3+} cations are ordered in the tunnels. In barian tomichite, one cubo-octahedral cavity can host two (AsO_3) groups with their lone pairs pointing towards each other. These two groups are referred to as (As_2) in the formula of barian tomichite. Furthermore, barian tomichite can contain isolated derbylite-like domains; derbylite itself contains Sb^{3+} instead of As^{3+} (Moore and Araki 1976). In graeserite, there are Pb^{2+} and As^{3+} cations in the cubo-octahedral cavities. In contrast to barian tomichite, Pb^{2+} in graeserite is displaced from the center of the cavity and only five-fold coordinated. The lowering of coordination number is caused by the preference of Pb^{2+} for a one-sided coordination environment. Berlepsch and Armbruster (1998) detected additional, meagerly populated sites for Fe and As in the structure of graeserite. Berlepsch and Armbruster (1998) also gave a detailed structural comparison of the derbylite-group minerals.

Table 27. Arsenite minerals of the derbylite group.

tomichite	$AsTi_3(V,Fe)_4O_{13}(OH)$
barian tomichite	$Ba_{0.5}(As_2)_{0.5}Ti_2(V,Fe)_5O_{13}(OH)$
graeserite	$Pb_{0.14}As(Fe,Ti)_7O_{12+x}(OH)_{2-x}$
hemloite	$(Ti,V,Fe,Al)_{12}(As,Sb)_2O_{23}(OH)$

Ekatite, $(Fe^{3+},Fe^{2+},Zn)_{12}(OH)_6(AsO_3)_6(AsO_3,SiO_3OH)_2$, has an impressive architecture of octahedral chains or rods and tunnels (Fig. 64e) (Keller 2001). Fe^{3+}, Fe^{2+}, and Zn are randomly distributed over a single octahedral site. Two adjacent octahedra share an edge to form a dimer and the dimers meld into chains in the direction of the c axis. These chains enclose two types of channels. The larger hexagonal tunnels are lined with trigonal pyramidal (AsO_3) groups whose lone electron pairs point into the channels. The narrow trigonal channels host the (AsO_3)

and (SiO_3OH) groups which replace each other. Ordering of As^{3+} and Si^{4+} was not detected, the positions are occupied by ~2/3 As^{3+} and ~1/3 Si^{4+}. Keller (2001) pointed out the structural relationship of ekatite to ellenbergerite, a high-pressure Mg-Al phase, and other similar phases.

Zimbabweite, $Na(Pb,Na,K)_2As_4(Ta,Nb,Ti)_4O_{18}$, belongs to the group of compounds with corner-linked octahedral sheets, so-called bronzes (Fig. 64f) (Duesler et al. 1988). In zimbabweite, the (100) sheets are formed by $(Ta,Nb,Ti)O_6$ octahedra, connected exclusively *via* corner sharing. There are four- and eight-membered octahedral rings within the sheets. The central portions of the eight-membered rings host Na. The octahedral sheets are cross-linked by AsO_3 groups with large voids between the sheets. The voids above and below the peripheral portions of the eight-membered rings are filled with the lone electron pairs of the As^{3+} cations. The voids above and below the four-membered rings are the sites of (Pb,Na,K) in a distorted trigonal-prismatic coordination.

Arsenite structures with metal clusters. This small but most peculiar group of arsenite minerals contains structures with metal clusters, that is, with bonds between As^{3+} and another metal cation. The existence of all these clusters is rationalized by the 18-electron rule because the cations within these clusters possess 18 electrons altogether. The bonds are facilitated by the lone electron pairs of As^{3+} which all point toward the central cation (Fig. 65). Note that a similar arrangement of the arsenite groups is also found in armangite but in this case, the center is vacant (Fig. 63c).

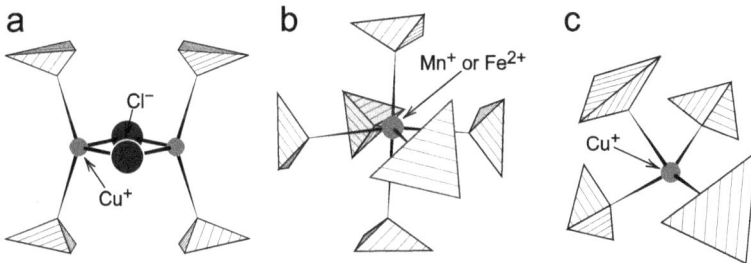

Figure 65. The structures of the arsenites with metal clusters: (a) a $[(AsO_3)_2(Cu^+Cl)_2(AsO_3)_2]$ cluster in freedite; (b) an octahedral cluster, either $[Mn^+(AsO_3)_6]$ in magnussonite or $[Fe^{2+}(AsO_3)_6]$ cluster in nanlingite; (c) a $[Cu^+(AsO_3)_4]$ cluster in dixenite.

Freedite, $Pb_8Cu(AsO_3)_2O_3Cl_5$, has a structure built by a massive framework of five independent $Pb\phi_{7-8}$ polyhedra. The ligands are four Cl atoms in one hemisphere and three to four O atoms in the other hemisphere. Small, not interconnected cavities are squeezed in the framework and populated by $[(AsO_3)_2(CuCl)_2(AsO_3)_2]$ clusters (Fig. 65a). The arsenite groups are situated at the top and bottom of the cavities and point their lone electron pairs into the cavity center. The center is occupied by two Cu^+ cations which share two Cl^- ligands. Therefore, each Cu^+ is coordinated by two Cl^- and two As^{3+} ions. The short Cu-As distance of 2.32 Å is unusual for two cations but known from other related structures (see Pertlik 1987). Pertlik (1987) discussed also the nature of the Cu-As bond as covalent, as opposed to the proposed metallic Cu-As bond in dixenite (Araki and Moore 1981).

The structure of **magnussonite**, $Mn^{2+}_{18}(As^{3+}_6Mn^+O_{18})_2Cl_2$, is a complex derivative of the simple fluorite structure (Moore and Araki 1979b). The unit cell of magnussonite contains 64 fluorite unit cells with a general formula $X_{32}O_{36}\square_{28}$, where $\square$ are ordered anion vacancies. The $Mn^{2+}O_{4-8}$ coordination polyhedra, with minor substitutions of other cations for Mn^{2+}, define a complex framework. There are MnO_8 cubes, MnO_6 trigonal prisms, and MnO_6 octahedra, with a minor Mg^{2+} substitution for Mn^{2+}. The square planar MnO_4 units incorporate minor amounts of Cu^{2+}. The most peculiar feature of the structure are the $[Mn^+(AsO_3)_6]$ clusters (Fig. 65b), located in large cavities centered at (0,0,0). The walls of these cavities are lined with six AsO_3 groups and the lone electron pairs of all of them point into the center of the cavity. The refinements showed conclusively that there is an Mn atom located very near (0.30 Å) the cavity center. Hence, the Mn atom in a formal monovalent state is octahedrally coordinated by six As^{3+} atoms. The Mn-As distance in magnussonite is similar to that in MnAs (NiAs structural type). The Cl^-/OH^- anions are positioned in cavities centered at (1/8, 1/8, 1/8), displaced 0.46 Å from the cavity center with partial occupancies.

The structure of **nanlingite**, $Na(Ca_5Li)Mg_{12}(AsO_3)_2[Fe(AsO_3)_6]F_{14}$, has several levels of complexity (Yang et al. 2010). In terms of connectivity of octahedra and the AsO_3 groups, the structure consists of elaborate octahedral slabs, in general parallel to (001). The MgF_2O_4 octahedra coalesce into six-membered rings which form the base of large pyramidal clusters, each counting 12 $Mg\phi_6$ octahedra; the axis of these pyramidal clusters coincides with the crystallographic c axis. The octahedral slabs are interconnected by the AsO_3 groups into a framework, with abundant spaces between the slabs. These spaces are occupied by large NaF_8 and CaF_4O_4 polyhedra, commonly seen in other structures of arsenite or arsenate minerals. The peculiar feature of this structure, however, are $[Fe^{2+}(AsO_3)_6]$ clusters (Fig. 65b). These clusters consist of six (AsO_3) groups which point their lone electron pairs to the geometric center of the cluster. The center, however, is not void, as in armangite, but filled with an Fe atom. Yang et al. (2010) argued that the central atom is an Fe^{2+} cation in low-spin state, with Fe-As distances of 2.40 Å, similar to the Fe-As distance in löllingite, $FeAs_2$.

Dixenite, $CuMn_{14}Fe(OH)_6(AsO_3)_5(SiO_4)_2(AsO_4)$, is an arsenite-silicate-arsenate mineral with metallic $[Cu^+(AsO_3)_4]$ clusters (Araki and Moore 1981). Structurally, the mineral is close to hematolite. In dixenite, octahedrally coordinated Mn^{2+} and Fe^{3+} join the tetrahedral cations Si^{4+}, As^{5+}, and Mn^{2+}, and the trigonal $As\phi_3$ pyramids in five distinct layers. Three of the layers are the same as in hematolite. One layer, however, is completely unlike the layers in hematolite or any other complex sheet arsenite-arsenate structures. This layer contains the $[Cu^+(AsO_3)_4]$ clusters (Fig. 65c). All lone electron pairs of the As^{3+} cations point into the cavity toward the Cu^+ cation. Araki and Moore (1981) proposed that dixenite represents a transition from ionic oxides to sulfides and sulfosalts. Another unusual feature of the structure is the broad miscibility of As^{5+} and Si^{4+} at the tetrahedral sites.

Phyllosilicates with As^{3+}. Phyllosilicates with lone-electron pair cations Sb^{3+} and Bi^{3+} are chapmanite and bismutoferrite, respectively. They are closely related to the dioctahedral T-O phyllosilicates of the kaolinite group (Ballirano et al. 1998), with Sb^{3+} and Bi^{3+} fixed in their interlayer and the lone electron pairs pointing across the interlayer into the center of the six-membered ring of tetrahedra in the adjacent tetrahedral sheet. To our knowledge, such a mineral with As^{3+} has not been described, although it could be an elegant explanation of enrichment of some clayey sediments with As. There are, however, two Mn-rich phyllosilicates with essential As^{3+}, described below.

Schallerite, $Mn_{16}Si_{12}O_{30}(OH)_{17}(As_3O_6)$, and **nelenite**, $(Mn,Fe)_{16}Si_{12}O_{30}(OH)_{17}(As_3O_6)$, belong to the group of T-O phyllosilicates. The T-O layers contain the usual trioctahedral sheet with Mn^{2+} or Fe^{2+} in the octahedral positions. The tetrahedral sheet differs from that found in most common phyllosilicates. In common phyllosilicates such as kaolinite, all tetrahedra within a single sheet point with their apical oxygen in one direction and each tetrahedron is shared by three adjacent tetrahedral six-membered rings. In schallerite, the tetrahedral sheets contain small four-membered and large twelve-membered rings. Half of the tetrahedra in one sheet point in one direction (up), the other half down. The arsenite groups reside within the large twelve-membered rings. The As^{3+} positions are only partially occupied. One of the positions (As1) should be coordinated by an OH group in an octahedral sheet, an O atom in a tetrahedral sheet, and two additional O atoms, hence four-fold coordinated, unusual for an As^{3+}. The As2 position has the usual three-fold coordination. Kato and Watanabe (1991) proposed that those OH groups which are bonded to As^{3+} deprotonate and convert to an O^{2-} anion. Peacor et al. (1986) described additional phases which may belong to this group but did not provide their crystal structures.

Complex (silicate-) arsenite-arsenate sheet structures. There are several structures of arsenite-arsenates, both with and without silicate, with a complex sequence of sheets. The cations in these structures are predominantly Mn^{2+} with Mg, Al, and Fe^{3+} (Table 28). In each of these structures, there are packets of sheets which are repeated, with varying cation occupancy, to populate the entire unit cell.

The number of sheets in the unit cell of each mineral is listed in Table 28. The formulae in this table also indicate the coordination of cations. Mn^{2+}, Mg, Al, and Fe^{3+} are predominantly octahedrally coordinated, although Mn^{2+} and Zn can be tetrahedrally coordinated in arakiite and Zn is found is square-pyramidal coordination in mcgovernite. As^{5+} and Si^{4+} are in the usual tetrahedral coordination and As^{3+} shows the trigonal-pyramidal coordination geometry. Turtmannite contains V^{5+} in addition to other major cations.

The sequences in these structures can be described in terms of close-packing of the oxygen anions, resulting usually in the overall hexagonal or trigonal symmetry of the structures. The exception is the monoclinic structure of arakiite where a non-close packed sheet causes lowering of the symmetry.

The structures of **hematolite** (Moore and Araki 1978b) and **arakiite** (Cooper and Hawthorne 1999c) are closely related. The packets in both structures contain 5 sheets, out of which 4 are topologically identical, although not the same in terms of cation ordering. The structure of **kraisslite** is unknown but the mineral certainly belongs to this group. The gargantuan structure of **mcgovernite** (Wuensch 1968) was believed to be "right around the corner" when the structure of hematolite was reported (Moore and Araki 1978b). Yet, the structure remains unpublished although the solution was presented on a poster of an impressive height at a Goldschmidt conference (Cooper and Hawthorne 2001). The structure of **turtmannite** was, however, already described (Brugger et al. 2001). Both of these minerals (mcgovernite and turtmannite) have a lattice parameters c of 204 Å, making this dimension of the cell the largest among known inorganic substances. An outstanding question, apart from the crystal structures, is how does nature manage to repeat these sequences over such an incredible distance. Here, it is interesting to note that the structure of turtmannite was reported to be well ordered, providing excellent diffraction data (within reason with such large lattice parameters), meaning that there is no substantial disorder over large distances in these crystals (see Brugger et al. 2001).

Structures with (AsO₃) trigonal pyramids and large-cation polyhedra. The structure of **finnemanite**, $Pb_5(AsO_3)_3Cl$, is closely related to the apatite structure (Effenberger and Pertlik 1979), described above for the more common arsenate members.

The structure of **georgiadèsite**, $Pb_4Cl_4(OH)(AsO_3)$, is constructed by four distinct Pb polyhedra and the AsO_3 group (Pasero and Vacchiano 2000). One of the Pb polyhedra is a distorted $Pb\phi_6$ octahedron, the ligands being one O and five Cl atoms. Two polyhedra are strongly one-sided, with Pb atoms coordinated by five ligands, with one or two additional distant ligands. The fourth Pb polyhedron is a bicapped trigonal prism $Pb\phi_8$. The structure can be described as a sequence of layers of $Pb\phi_6$ octahedra and of $Pb\phi_8$ trigonal prisms. The structural cavities are occupied by Pb polyhedra with the five-fold coordination and the AsO_3 groups. The arsenite groups attach to the adjacent Pb polyhedra but do not polymerize with each other.

Table 28. Arsenite-arsenate-silicates with complex sheet structures. N is the number of sheets in an unit cell. Manganese is always divalent, iron always trivalent. Oxidation state of arsenic is given only when As substitutes for another cation. The square brackets give the coordination number. For references, see text.

Mineral	N						
arakiite	10	[4](Zn,Mn)	[6](Fe,Al)$_2$	(AsO$_3$)	(AsO$_4$)$_2$		(OH)$_{23}$
hematolite	15	[6]Mn^{2+}	[6](Al,Fe)$_2$	(AsO$_3$)	(AsO$_4$)$_2$		(OH)$_{23}$
kraisslite*	18	[3](Mn$_{22}$Mg$_{1.9}$Zn$_{3.2}$Fe$_{0.7}$)		(AsO$_3$)$_2$	(AsO$_4$)$_{2.7}$	(SiO$_4$)$_{6.4}$	(OH)$_{18}$
mcgovernite	84	[6](Mn,Mg,Fe,Al)$_{42}$	[4](Zn,Mn)$_3$	[([3]As^{3+},[5]Zn)O$_3$]$_2$	(AsO$_4$)$_4$	[(Si,As^{5+})O$_4$]$_8$	(OH)$_{42}$
turtmannite**	84	[6](Mn,Mg)$_{42}$	[4]Mn$_3$	(AsO$_3$)$_2$	[(V,As^{5+})O$_4$]$_6$	(SiO$_4$)$_6$	(OH)$_{42}$

* structure is unknown; only an empirical formula is reported here.
** one of the end-member compositions listed by Brugger et al. (2001); the coefficients from Brugger et al. (2001) were doubled for an easier comparison with mcgovernite; this end-member is the second most abundant but is listed here because only this one contains both As^{3+} and As^{5+}

Uranyl arsenates and arsenate-arsenites

Arsenates and arsenate-arsenites of hexavalent uranium are listed in Table 29. Their structural classification is based on the coordination of the uranyl [(UO$_2$)$^{2+}$] ion in tetragonal, pentagonal, and hexagonal bipyramids, and the connectivity of those bipyramids among each other and with other polyhedra (Burns 1999). Those interested in the structural classification of these minerals should consult Burns (1999) or Krivovichev and Plášil (2013).

Arsenates and arsenites with unknown structure

In the following text, arsenites and arsenates with unknown structure are listed and briefly discussed.

Attikaite – Ca$_3$Cu$_2$Al$_2$(AsO$_4$)$_4$(OH)$_4$·2H$_2$O. The mineral should be orthorhombic, with possible space groups *Pban*, *Pbam*, or *Pba*2 (Chukanov et al. 2007a). Viñals et al. (2008) considered this mineral to be structurally close to barahonaite-Al.

Bulachite – Al$_2$(AsO$_4$)(OH)$_3$·3H$_2$O. Indexing of the powder X-ray diffraction data gave an orthorhombic cell (Frau and Da Pelo 2001).

Camgasite – CaMg(AsO$_4$)(OH)·5H$_2$O. The space group *P*2$_1$/*m* was reported as probable (Walenta and Dunn 1989).

Céruléite – Cu$_2$Al$_7$(OH)$_{13}$(AsO$_4$)$_4$·11.5H$_2$O. Indexing of powder X-ray diffraction data gave a triclinic cell (Schmetzer et al. 1976).

Coralloite – Mn^{2+}Mn$^{3+}_2$(AsO$_4$)$_2$(OH)$_2$·4H$_2$O. Reported triclinic and related to arthurite and whitmoreite (Callegari et al. 2010).

Ecdemite – Pb$_6$As$_2$O$_7$Cl$_4$.

Heliophyllite – Pb$_6$As$_2$O$_7$Cl$_4$. Structures of both minerals are unknown. Ecdemite was reported to be tetragonal, heliophyllite orthorhombic, pseudotetragonal (Anthony et al. 2000). It is possible that these two species are actually one mineral.

Gerdtremmelite – (Zn,Fe)(Al,-Fe)$_2$(OH)$_5$(AsO$_4$). Found in one specimen from

Table 29. Arsenates and arsenate-arsenites of hexavalent uranium.

Arsenates

abernathyite	$K(UO_2)_2(AsO_4)\cdot 6H_2O$
arsenovanmeersscheite	$U^{4+}(OH)_4[(UO_2)_3(AsO_4)_2(OH)_2]\cdot 4H_2O$
arsenuranospathite	$AlH(UO_2)_4(AsO_4)_4\cdot 40H_2O$
arsenuranylite	$Ca(UO_2)_4(AsO_4)_2(OH)_4\cdot 6H_2O$
asselbornite	$(Pb,Ba)(UO_2)_6(BiO)_4(AsO_4)_2(OH)_{12}\cdot 3H_2O$
chistyakovite	$Al(UO_2)_2(AsO_4)_2(F,OH)\cdot 6.5H_2O$
hallimondite	$Pb_2(UO_2)(AsO_4)_2\cdot nH_2O$
heinrichite	$Ba(UO_2)_2(AsO_4)_2\cdot 10H_2O$
hügelite	$Pb_2(UO_2)_3(AsO_4)_2O_2\cdot 5H_2O$
kahlerite	$Fe^{2+}(UO_2)_2(AsO_4)_2\cdot 12H_2O$
kamitugaite	$PbAl(UO_2)_5[(P,As)O_4]_2(OH)_9\cdot 9.5H_2O$
metanováčekite	$Mg(UO_2)_2(AsO_4)_2\cdot 4\text{-}8H_2O$
metauranospinite	$Ca(UO_2)_2(AsO_4)_2\cdot 8H_2O$
metakahlerite	$Fe^{2+}(UO_2)_2(AsO_4)_2\cdot 8H_2O$
metakirchheimerite	$Co(UO_2)_2(AsO_4)_2\cdot 8H_2O$
metazeunerite	$Cu(UO_2)_2(AsO_4)_2\cdot 8H_2O$
metarauchite	$Ni(UO_2)_2(AsO_4)_2\cdot 8H_2O$
metalodèvite	$Zn(UO_2)_2(AsO_4)_2\cdot 10H_2O$
metaheinrichite	$Ba(UO_2)_2(AsO_4)_2\cdot 8H_2O$
nováčekite	$Mg(UO_2)_2(AsO_4)_2\cdot 12H_2O$
orthowalpurgite	$Bi_4O_4(UO_2)(AsO_4)_2\cdot 2H_2O$
sodium-uranospinite	$(Na_2,Ca)(UO_2)_2(AsO_4)_2\cdot 5H_2O$
trögerite	$(H_3O)(UO_2)(AsO_4)\cdot 3H_2O$
uramarsite	$(NH_4,H_3O)_2(UO_2)_2[(As,P)O_4]_2\cdot 6H_2O$
uranospinite	$Ca(UO_2)_2(AsO_4)_2\cdot 10H_2O$
zeunerite	$Cu(UO_2)_2(AsO_4)_2\cdot 12H_2O$
walpurgite	$Bi_4O_4(UO_2)(AsO_4)_2\cdot 2H_2O$

Arsenites-arsenates

seelite	$Mg[(UO_2)(AsO_3)_{0.7}(AsO_4)_{0.3}]_2\cdot 7H_2O$

Tsumeb (Schmetzer and Medenbach 1985). Powder X-ray diffraction data indexed with a triclinic cell.

Irhtemite – $Ca_4MgH_2(AsO_4)_4$. Irhtemite was reported to be monoclinic (Pierrot and Schubnel 1972). The mineral can be synthesized by dehydration of picropharmacolite. It is therefore possible to prepare specimens of this rare mineral for structural determination.

Juanitaite – $(Cu,Ca,Fe)_{10}Bi(AsO_4)_4(OH)_{11}\cdot 2H_2O$. Reported to be tetragonal, space group $P4_2/nnm$ (Kampf et al. 2000).

Kolfanite – $Ca_2Fe_3O_2(AsO_4)_3 \cdot 2H_2O$. This mineral is considered to be monoclinic, structurally perhaps close to arseniosiderite (Voloshin et al. 1982).

Lazarenkoite – $(Ca,Fe^{2+})Fe^{3+}As_3^{3+}O_7 \cdot 3H_2O$. Reported to be orthorhombic. Yakhontova et al. (1983) speculated about the structure based on refractometric data.

Liskeardite – $(Al,Fe)_3(AsO_4)(OH)_6 \cdot 5H_2O$. Exact chemical composition and structure uncertain. Perhaps orthorhombic.

Manganarsite – $Mn_3As_2O_4(OH)_4$. Wicks et al. (1986) provided some evidence that manganarsite is structurally closely related to manganpyrosmalite, $(Mn,Fe^{2+})_8Si_6O_{15}(OH,Cl)_{10}$.

Mcnearite – $NaCa_5H_4(AsO_4)_5 \cdot 4H_2O$. Reported to be triclinic (Sarp et al. 1981).

Ogdensburgite – $Ca_2Fe^{3+}_4(Zn,Mn)^{2+}(AsO_4)_4(OH)_6 \cdot 6H_2O$. The structure of ogdensburgite is unknown despite the efforts of Kampf and Dunn (1987). They were able to narrow the space group search to the possibilities $Bmmm$, $B222$, $B2mm$, $Bm2m$, or $Bmm2$. They also speculated that the structure could be based on the sheets of Fe^{3+} octahedra, based on the unit-cell dimensions.

Parnauite – $Cu_9(AsO_4)_2(SO_4)(OH)_{10} \cdot 7H_2O$. The structure is unknown. The original description by Wise (1978) proposed a hexagonal cell. Lai et al. (1997) reported an orthorhombic cell.

Pitticite – amorphous hydrated ferric arsenate with poorly defined chemical composition.

Rouseite – $Pb_2Mn(AsO_3)_2 \cdot 2H_2O$. Reported to be triclinic, structure is unknown (Dunn et al. 1986b).

Sarmientite – $Fe_2(AsO_4)(SO_4)(OH) \cdot 5H_2O$. Structure unknown.

Barahonaite-(Al) – $(Ca,Cu,Na,Al)_{12}Al_2(AsO_4)_8(OH)_x \cdot nH_2O$.

Barahonaite-(Fe) – $(Ca,Cu,Na,Fe^{3+},Al)_{12}Fe^{3+}_2(AsO_4)_8(OH)_x \cdot nH_2O$.

Smolianinovite – $(Co,Ni,Mg,Ca)_3(Fe^{3+},Al)_2(AsO_4)_4 \cdot 11H_2O$.

Fahleite – $CaZn_5Fe^{3+}_2(AsO_4)_6 \cdot 14H_2O$. Smolianinovite is a poorly crystalline arsenate. Barahonaite-(Al) and -(Fe) were described by Viñals et al. (2008). No crystals suitable for a single-crystal structure determination were found. The authors assume that these minerals are related to smolianinovite. Fahleite may be the Zn analogue of smolianinovite (Medenbach et al. 1988).

Sterlinghillite – $Mn_3(AsO_4)_2 \cdot 4H_2O$. At the type locality, only a minute amount of the mineral was found, with no single crystal suitable for structure determination (Dunn 1981b). Matsubara et al. (2000) reported the second occurrence of this mineral, with a somewhat different composition, especially with respect to the water content.

Strashimirite – $Cu_4(AsO_4)_2(OH)_2 \cdot 2.5H_2O$. The mineral was reported to be monoclinic, space group $P2/m$.

Theisite – $Cu_5Zn_5(OH)_{14}[(As,Sb)O_4]_2$. Trigonal symmetry suspected (Bonazzi and Olmi 1989).

Vladimirite – $Ca_5H_2(AsO_4)_4 \cdot 5H_2O$. The structure of vladimirite appears to be unknown. The crystallographic databases point to work of Catti and Ivaldi (1981) who solved the structure of the triclinic polymorph of vladimirite, not vladimirite itself. Vladimirite is reported to be monoclinic.

Walentaite – $Ca_4Fe^{3+}_{12}H_4(AsO_4)_{10}(PO_4)_6 \cdot 28H_2O$. The mineral is supposed to be orthorhombic, with a centered cell (Dunn et al. 1984).

Wallkilldellite – $Ca_4Mn_6As_4O_{16}(OH)_8 \cdot 18H_2O$. The mineral was reported to be hexagonal (Dunn and Peacor 1983). It occurs in milligram quantities on the type specimen.

Yukonite – $Ca_2Fe_3(AsO_4)_3(OH)_4 \cdot 4H_2O$. Yukonite is a nanocrystalline hydrated arsenate of calcium and iron. Neither the structure nor the chemical composition are well constrained. The best crystallographic study appears to be that of Nishikawa et al. (2006) who used selected-area electron diffraction and high-resolution transmission electron microscopy to study natural yukonite samples. They found that the electron diffraction patterns can be indexed mostly as orthorhombic, less commonly as hexagonal. Based on X-ray absorption near-edge structure (XANES) spectroscopy, Gomez et al. (2010) stated that the local environment of Ca, Fe, and As in yukonite is similar to that found in arseniosiderite.

Zykaite – $Fe_4^{3+}(AsO_4)_3(SO_4)(OH) \cdot 15H_2O$. Reported to be orthorhombic.

Arsenides, sulfarsenides, and sulfides with arsenic, including sulfosalts

The most common mineral among arsenides and sulfarsenides is undoubtedly **arsenopyrite**, FeAsS. Although there are no firm data to support this claim, the most abundant chalcogenide mineral with arsenic could be **pyrite**, nominally FeS_2. Kolker and Nordstrom (2001) reported a wide range of As contents in both sedimentary and hydrothermal pyrite, ranging from trace values to almost 10 wt%. Arsenic in pyrite is reduced, found in the pyrite structure and in the marcasite-like lamellae (Dódony et al. 1996). The only report of oxidized arsenic (as As^{3+}) in the pyrite structure was published by Deditius et al. (2008).

The structure of pyrite can be most easily derived by a substitution of Fe^{2+} for Na^+ and the coupled substitution of $(S_2)^{2-}$ dumbbells for Cl^- in the structure of halite. The structure of marcasite, a dimorph of FeS_2, can be derived from that of pyrite. In both pyrite and marcasite, cations are octahedrally coordinated by sulfur atoms from the $(S_2)^{2-}$ dumbbells. A number of arsenides and sulfarsenides adopt these structures; for some of them, both ordered and disordered varieties have been described (e.g., CoAsS derived from pyrite, Makovicky 2006). The disulfides and diarsenides differ clearly by the S-S and As-As distances in the covalently bonded pairs (dumbbells) (Fig. 66). The structure of arsenopyrite is a superstructure of marcasite (Fig. 67a).

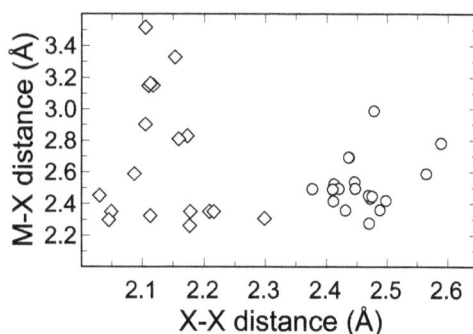

Figure 66. Variations of the X-X and M-X distances in disulfides (diamonds) and diarsenides (circles) with a general formula MX_2 (X is either As *or* S, i.e., no sulfarsenides are shown).

The most common sulfides of arsenic are realgar and orpiment. The structure of **realgar**, As_4S_4, is a member of the group of molecular structures with basket-like As_mS_n molecules (Fig. 67b). The basket-like molecules in realgar are eight-membered and the interactions between the molecules are mediated by weak van-der-Waals bonds. The lone electron pairs of the As^{3+} ions point always outwards from the As_mS_n molecules and these structures were therefore classified as inverted lone electron pair micelles (Makovicky 2006).

Orpiment, As_2S_3, has a layered structure (Fig. 67c), with the lone electron pairs pointing into the interlayer space. The layers are built by AsS_5 pyramids with three shorter (2.2-2.3 Å) and two much longer (3.2-3.5 Å) As-S bonds.

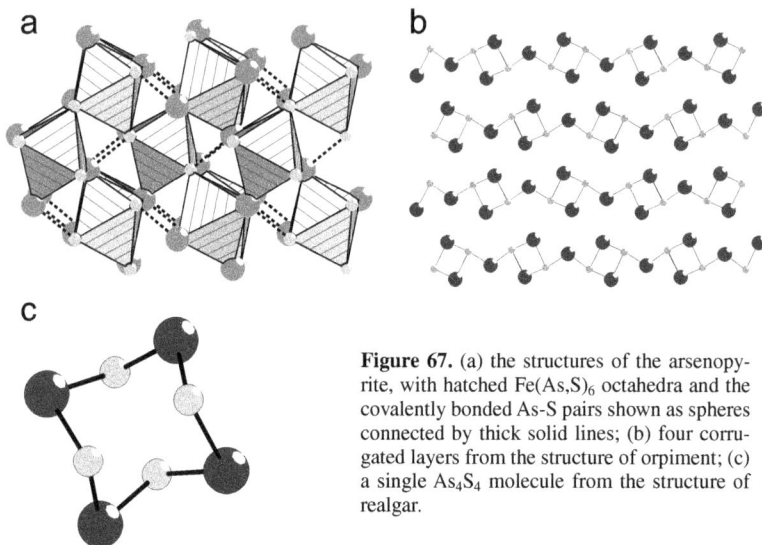

Figure 67. (a) the structures of the arsenopyrite, with hatched Fe(As,S)₆ octahedra and the covalently bonded As-S pairs shown as spheres connected by thick solid lines; (b) four corrugated layers from the structure of orpiment; (c) a single As₄S₄ molecule from the structure of realgar.

Sulfosalts are complex sulfides in which the metalloid cations As^{3+}, Sb^{3+}, and Bi^{3+} combine with different cations. The different cation is in most cases Pb^{2+}, less commonly Cu, Ag, Fe, Tl, Mn, Sn, and others (Makovicky 1997, 2006). The specific feature of the metalloid cations in the structures of sulfosalts is the stereochemical activity of their lone electron pairs.

A classification based on specific polyhedral types and the polymerization of these polyhedra, such as that for natural silicates or arsenates, is bound to fail for sulfosalts. The reason is a) variable coordination of As^{3+}, Sb^{3+}, and Bi^{3+} and b) variable polymerization of the As^{3+}, Sb^{3+}, and Bi^{3+} polyhedra, even in structures with very similar chemical composition and properties. Therefore, Makovicky (1989, 1997) defined modules which constitute the structures of many sulfosalts, both natural and synthetic. These modules can be viewed as fragments of simpler, carefully selected structures, so-called archetypes. For sulfosalts, such archetype structures are those of PbS and SnS (Fig. 68a).

The fragments—blocks, rods, layers—with the archetype-like structure are assembled together via interfaces of variable nature and chemical composition. For sulfosalts with As^{3+}, the SnS archetype is commonly adopted. With the heavier metalloid cations, the PbS archetype is preferred.

The structures which share the same fragment type and interface type can be grouped in accretional or variational homologous series. For the As sulfosalts, an example is the accretional homologous series of sartorite, PbAs₂S₄ (Fig. 68b). The assembly of the fragments can also proceed by keeping one type of fragment but varying the nature of the other type among related structures. Such structures are then grouped in plesiotypic or merotypic families (Makovicky 1997). An example is the hutchinsonite, TlPbAs₅S₉, family (Fig. 68c). An interested reader is referred to the work of Makovicky (1989, 1997, 2006) for further details and examples.

ACKNOWLEDGMENTS

We thank Rob Bowell for the invitation to write this chapter and Vojtěch Ettler, Heather Jamieson, Chris Brough and an anonymous reviewer for the reviews that improved the chapter. The contribution of J.M. was supported by the project INFLUINS (grant Nr. 03IS2091A) by

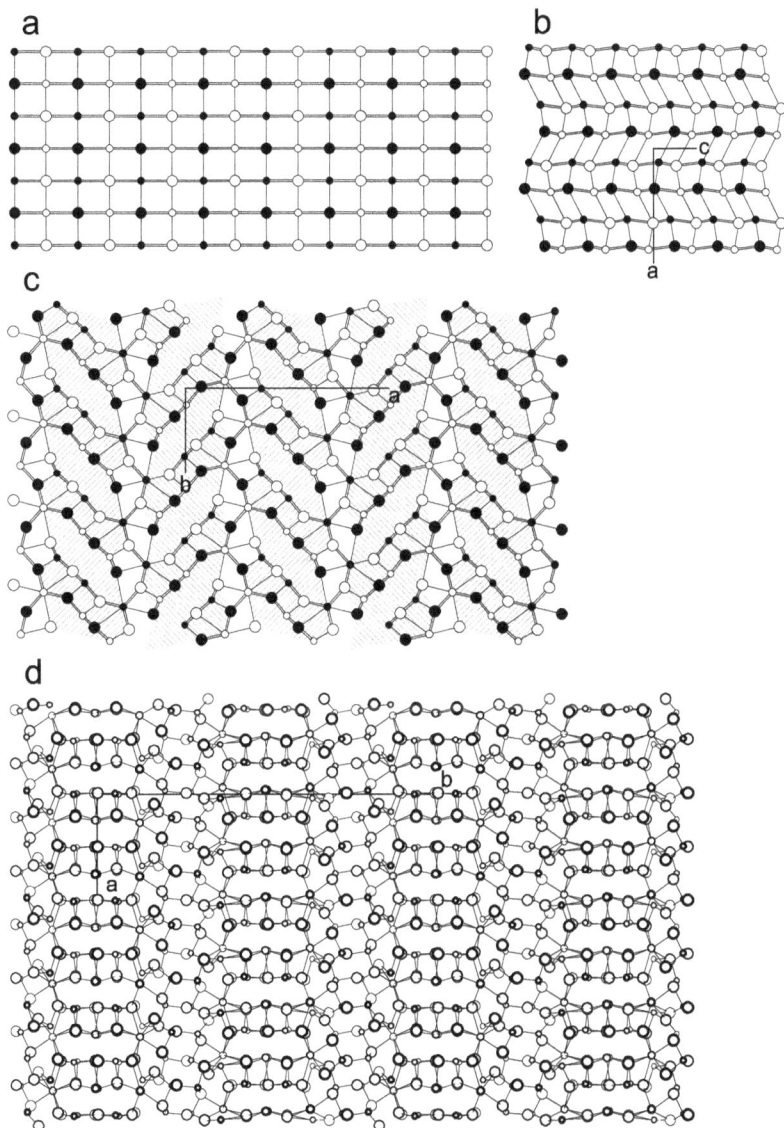

Figure 68. The two archetype crystal structures for sulfosalts (see Makovicky 1997) (a) PbS, (b) SnS (del Bucchia et al. 1981). Large circles represent sulfur anions, small circles cations. The shading (black, white) shows atoms at two different levels above the projection plane. (c) structure of sartorite, $PbAs_2S_4$ (Iitaka and Nowacki 1961). Symbols have the same meaning as in (a) and (b). (d) structure of hutchinsonite, (Tl, $Pb)_2As_5S_9$ (Takeuchi et al. 1965). Large circles represent sulfur anions, small circles cations. Thickness of the outlines of the balls (atoms) and the sticks (bonds) indicates the height above the projection plane.

the German Ministry of Education and Research (BMBF) within the program *Spitzenforschung und Innovation in den Neuen Ländern*. Some research addressed here was supported by Czech Science Foundation project (GAČR 210/10/P096) and Czech Ministry of Education, Youth and Sports (MSM 0021620855) to P.D. and by Institution Research Programs Nr. AVOZ30130516 and Nr. RVO67985831 to M.F.

REFERENCES

Abrecht J (1990) An As-rich manganiferous mineral assemblage from the Ködnitz Valley (Eastern Alps, Austria): Geology, mineralogy, genetic considerations, and implications for metamorphic Mn deposits. Neues Jahrb Mineral Monatsh 363-375

Adiwidjaja G, Friese K, Klaska K-H, Moore PB, Schlüter J (2000) The crystal structure of the new mineral wilhelmkleinite – $ZnFe^{3+}_2(OH)_2(AsO_4)_2$. Z Kristallogr 215:96-101

Ahn JS, Park YS, Kim JY, Kim KW (2005) Mineralogical and geochemical characterization of arsenic in an abandoned mine tailings of Korea. Environ Geochem Health 27:147-157

Akhter H, Cartledge FK, Roy A, Tittlebaum ME (1997) Solidification/stabilization of arsenic salts: effects of long cure time. J Hazard Mater 52:247-264

Alakangas L, Öhlander B (2006) Formation and composition of cemented layers in low-sulphide mine tailings, Laver, northern Sweden. Environ Geol 50:809-819

Aldersey-Williams H (2011) Periodic Tales: The curious lives of the elements. Penguin Books, London 428 p

Alpers CN, Myers PA, Millsap D, Regnier TB (2014) Arsenic associated with historical gold mining in the Sierra Nevada foothills: case study and field trip guide for Empire Mine State Historic Park, California. Rev Mineral Geochem 79:553-587

Álvarez-Ayuso E, García-Sánchez A, Querol X, Moyano A (2008) Trace element mobility in soils seven years after the Aznalcóllar mine spill. Chemosphere 73:1240-1246

Anthony JW, Bideaux RA, Bladh KW, Nichols MC (2000) Handbook of Mineralogy. Mineralogical Society of America

Arai Y, Lanzirotti A, Sutton SR, Newville M, Dyer J, Sparks DL (2006) Spatial and temporal variability of arsenic solid-state speciation in historically lead arsenate contaminated soils. Environ Sci Technol 40:673-679

Araki T, Moore PB (1981) Dixenite, $Cu^{1+}Mn^{2+}_{14}Fe^{3+}(OH)_6(As^{3+}O_3)_5(Si^{4+}O_4)_2$ ($As^{5+}O_4$): metallic [$As^{3+}_4Cu^{1+}$] clusters in an oxide matrix. Am Mineral 66:1263-1273

Araki T, Moore PB, Brunton GD (1980) The crystal structure of paulmooreite, $Pb_2[As_2O_5]$: dimeric arsenite groups. Am Mineral 65:340-345

Arčon I, Van Elteren JT, Glass HJ, Kodre A, Šlejkovec Z (2005) EXAFS and XANES study of arsenic in contaminated soil. X-Ray Spectrom 34:435-438

Artl T, Armbruster T (1999) Single-crystal X-ray structure refinement of cornwallite, $Cu_5(AsO_4)_2(OH)_4$: a comparison with its polymorph cornubite and the PO_4-analogue pseudomalachite. Neues Jahrb Mineral Monatsh X 468-480

Aruga A, Nakai I (1985) Structure of Ca-rich agardite, $(Ca_{0.40}Y_{0.31}Fe_{0.09}Ce_{0.06}La_{0.04}Nd_{0.01})Cu_{6.19}[(AsO_4)_{2.42}(HAsO_4)_{0.49}](OH)_{6.38} \cdot 3H_2O$. Acta Crystallogr C 41:161-163

Ashley PM (1989) Geochemistry and mineralogy of tephroite-bearing rocks from the Hoskins manganese mine, New South Wales, Australia. Neues Jb Miner Abh 161:85-111

Ashley PM, Lottermoser BG (1999) Arsenic contamination at the Mole River mine, northern New South Wales. Aust J Earth Sci 46:861-874

Ashley PM, Lottermoser BG, Collins AJ, Grant CD (2004) Environmental geochemistry of the derelict Webbs Consols mine, New South Wales, Australia. Environ Geol 46:591-604

Asta MP, Cama J, Martínez M, Giménez J (2009) Arsenic removal by goethite and jarosite in acidic conditions and its environmental implications. J Hazard Mater 171:965-972

Auernhammer M, Effenberger H, Hentschel G, Reinecke Th, Tillmanns E (1993) Nickenichite, a new arsenate from Eifel, Germany. Miner Petrol 48:153-166

Ayuso RA, Foley NK (2008) Anthropogenic and natural lead isotopes in Fe-hydroxides and Fe-sulphates in a watershed associated with arsenic-enriched groundwater, Maine, USA. Geochem Explor Environ Anal 8:77-89

Azcue JM, Nriagu JO (1994) Arsenic: Historical perspectives. *In:* Arsenic in the Environment. Part I. Cycling and Characterization. Nriagu JO (ed) Wiley, New York, p 1-17

Balić-Žunić T, Makovicky E (1996) Determination of the centroid or 'the best centre' of a coordination polyhedron. Acta Crystallogr B52:78-81

Ballirano P, Maras A, Marchetti F, Merlino S, Perchiazzi N (1998) Rietveld refinement of chapmanite $SbFe_2Si_2O_8OH$, a TO dioctahedral kaolinite-like mineral. Powder Diffr 13:44-49

Barbier J, Frampton C (1991) Structures of orthorhombic and monoclinic $Ni_3(AsO_4)_2$. Acta Crystallogr B 47:457-462

Basta NT, Juhasz A (2014) Using in vivo bioavailability and/or in vitro gastrointestinal bioaccessibility testing to adjust human exposure to arsenic from soil ingestion. Rev Mineral Geochem 79:451-472

Baur WH (1974) The geometry of polyhedral distortions. Predictive relationships for the phosphate group. Acta Crystallogr B 30:1195-1215

Bayliss P, Kolitsch U, Nickel EH, Pring A (2010) Alunite supergroup: recommended nomenclature. Mineral Mag 74:919-927

Bayliss P, Warne SStJ (1987) Powder x-ray diffraction data of magnesium-chlorophoenicite. Powder Diffr 2:225-6

Bedlivy D, Mereiter K (1982a) Preisingerite, $Bi_3O(OH)(AsO_4)_2$, a new species from San Juan Province, Argentina: Its description and crystal structure. Am Mineral 67:833-840

Bedlivy D, Mereiter K (1982b) Structure of α-BiAsO$_4$ (rooseveltite). Acta Crystallogr B 38:1559-1561

Berlepsch P, Armbruster T (1998) The crystal structure of Pb^{2+}-bearing graeserite, $Pb_{0.14}(Fe,Ti)_7AsO_{12+x}(OH)_{2-x}$, a mineral of the derbylite group. Schweiz Miner Petrog 78:1-9

Bermanec V (1994) Centro-symmetric tilasite from Nežilovo, Macedonia: A crystal structure refinement. Neues Jahrb Mineral Monatsh 289-294

Bermanec V, Balen D, Šćavničar S, Tibljaš D (1993) Zn-rich magnetoplumbite from Nežilovo, Macedonia. Eur J Mineral 5:957-960

Bianchi R, Pilati T, Mannucci G (1987) Crystal structure of grischunite. Am Mineral 72:1225-1229

Bigham JM, Nordstrom DK (2000) Iron and aluminum hydroxysulfates from acid sulfate waters. Rev Mineral Geochem 40:351-404

Bindi L, Bonazzi P, Spry PG (2008) Effects of sulfur-for-selenium substitution on the structure of laphamite, $As_2(Se,S)_3$. Can Mineral 46:269-274

Birch WD, Pring A, Gatehouse BM (1992) Segnitite, $PbFe_3H(AsO_4)_2(OH)_6$, a new mineral in the lusungite group from Broken Hill, New South Wales, Australia. Am Mineral77:656-659

Black A, Craw D, Youngson JH, Karubaba J (2004) Natural recovery rates of a river system impacted by mine tailing discharge: Shag River, East Otago, New Zealand. J Geochem Explor 84:21-34

Blackwood GM, Edinger EN (2007) Mineralogy and trace element relative solubility patterns of shallow marine sediments affected by submarine tailings disposal and artisanal gold mining, Buyat-Ratototok district, North Sulawesi, Indonesia. Environ Geol 52:803-818

Bladh KW, Corbett RK, McLean WJ, Laughon RB (1972) The crystal structure of tilasite. Am Mineral 57:1880-1884

Blanchard M, Alfredsson M, Brodholt J, Wright K, Catlow CRA (2007) Arsenic incorporation into FeS_2 pyrite and its influence on dissolution: A DFT study. Geochim Cosmochim Acta 71:624-630

Blowes DW, Reardon EJ, Jambor JL, Cherry JA (1991) The formation and potential importance of cemented layers in inactive sulfide mine tailings. Geochim Cosmochim Acta 55:965-978

Bolanz RM, Majzlan J, Jurkovič Ľ, Göttlicher J (2012) Mineralogy, geochemistry, and arsenic speciation in coal combustion waste from Nováky, Slovakia. Fuel 94:125-136

Bonaccorsi E, Pasero M (2003) Crystal structure refinement of sahlinite, $Pb_{14}(AsO_4)_2O_9Cl_4$. Mineral Mag 67:15-21

Bonatti E (1975) Metallogenesis at oceanic spreading centers. Annu Rev Earth Planet Sci 3:401-431

Bonazzi P, Olmi F (1989) Theisite from Forno (Alpi Apuane) and from Sa Duchessa (Sardinia), Italy. Neues Jahrb Mineral Monatsh 241-244

Bossy A, Grosbois C, Beauchemin S, Courtin-Nomade A, Hendershot W, Bril H (2010) Alteration of As-bearing phases on a high grade arsenic-geochemical anomaly (French Massif Central). Appl Geochem 25:1889-1901

Botelho N, Roger G, d'Yvoire F, Moëlo Y, Volfinger M (1994) Yanomamite, $InAsO_4 \cdot 2H_2O$, a new indium mineral from topaz-bearing greisen in the Goiás Tin Province, Brazil. Eur J Mineral 6:245-254

Boudjada A, Guitel JC (1981) Structure cristalline d'un orthoarséniate acide de fer(III) pentahydraté: $Fe(H_2AsO_4)_3 \cdot 5H_2O$. Acta Crystallogr B 37:1402-1405

Bouška V, Pešek J (1999) Quality parameters of lignite of the North Bohemian Basin in the Czech Republic in comparison with the world average lignite. Int J Coal Geol 40:211-235

Bowell RJ (1992) Supergene gold mineralogy at Ashanti, Ghana: Implications for the supergene behaviour of gold. Miner Mag 56:545-560

Bowell RJ (1994) Sulphide oxidation and arsenic speciation in tropical soils. Environ Geochem Health 16:84

Bowell RJ (2014) Hydrogeochemistry of the Tsumeb deposit: implications for arsenate mineral stability. Rev Mineral Geochem 79:589-627

Brahimi A, Mohamed Mongi F, Amor H (2002) Cerium arsenate, $CeAsO_4$. Acta Crystallogr E 58:i98-i99

Brantley SL (2003) Reaction kinetics of primary rock-forming minerals under ambient conditions. *In:* Treatise on Geochemistry. Volume 5. Surface and Ground Water, Weathering, and Soils. Holland HD, Turekian KK (eds) Elsevier, p 73-117

Breiter K, Čopjaková R, Škoda R (2009) The involvement of F, CO_2, and As in the alteration of Zr-Th-REE-bearing accessory minerals in the Hora Svaté Kateřiny A-type granite, Czech Republic. Can Mineral 47:1375-1398

Bril H, Zainoun K, Puziewicz J, Courtin-Nomade A, Vanaecker M, Bollinger JC (2008) Secondary phases from the alteration of a pile of zinc-smelting slag as indicators of environmental conditions: an example from Świętochłowice, Upper Silesia, Poland. Can Mineral 46:1235-1248

Brown ID, Altermatt D (1985) Bond-valence parameters obtained from a systematic analysis of the Inorganic Crystal Structure Database. Acta Crystallogr B 41:244-247

Brownfield ME, Affolter RH, Cathcart JD, Johnson SY, Brownfield IK, Rice CA (2005) Geologic setting and characterization of coals and the modes of occurrence of selected elements from the Franklin coal zone, Puget Group, John Henry No. 1 mine, King County, Washington, USA. Int J Coal Geol 63:247-275

Brugger J, Berlepsch P (1996) Description and crystal structure of fianelite, $Mn_2V(V,As)O_7\cdot2H_2O$, a new mineral from Fianel, Val Ferrera, Graubünden, Switzerland. Am Mineral 81:1270-1276

Brugger J, Gieré R (1999) As, Sb, Be and Ce enrichment in minerals from a metamorphosed Fe-Mn deposit, Val Ferrera, eastern Swiss Alps. Can Mineral 37:37-52

Brugger J, Meisser N (2006) Manganese-rich assemblages in the Barrhorn unit, Turtmanntal, central Alps, Switzerland. Can Mineral 44:229-248

Brugger J, Bonin M, Schenk KJ, Miesser N, Berlepsch P, Ragu A (1999) Description and crystal structure of nabiasite, $BaMn_9[(V,As)O_4]_6(OH)_2$, a new mineral from the Central Pyrénées (France). Eur J Mineral 11:879-890

Brugger J, Meisser N, Schenk K, Berlepsch P, Bonin M, Armbruster T, Nyfeler D, Schmidt S (2000) Description and crystal structure of cabalzarite $Ca(Mg,Al,Fe)_2(AsO_4)_2(H_2O,OH)_2$, a new mineral of the tsumcorite group. Am Mineral 85:1307-1314

Brugger J, Armbruster T, Meisser N, Hejny C, Grobety B (2001) Description and crystal structure of turtmannite, a new mineral with a 68 Å period related to mcgovernite. Am Mineral 86:1494-1505

Brugger J, Krivovichev SV, Kolitsch U, Meisser N, Andrut M, Ansermet S, Burns PC (2002) Description and crystal structure of manganlotharmeyerite, $Ca(Mn^{3+},Mg)_2\{AsO_4,[AsO_2(OH)_2]\}_2(OH,H_2O)_2$, from the Starlera Mn deposit, Swiss Alps, and a redefinition of lotharmeyerite. Can Mineral 40:1597-1608

Brugger J, Meisser N, Krivovichev S, Armbruster T, Favreau G (2007) Mineralogy and crystal structure of bouazzerite from Bou Azzer, Anti-Atlas, Morocco: Bi-As-Fe nanoclusters containing Fe^{3+} in trigonal prismatic coordination. Am Mineral 92:1630-1639

Buckley AM, Bramwell ST, Day P (1990) Structural properties of transition metal pyroarsenates $M_2As_2O_7$ (M = Co,Mn,Ni). J Solid State Chem 86:1-15

Buerger MJ, Dollase WA, Garaycochea-Wittke I (1967) The structure and composition of the mineral pharmacosiderite. Z Kristallogr 125:92-108

Burke EAJ, Sejkora J, Sarp H, Chiappero P-J (2007) Revalidation of pradetite as a mineral. Arch Sci 60:51-54

Burnol A, Charlet L (2010) Fe(II)-Fe(III)-bearing phases as a mineralogical control on the heterogeneity of arsenic in southeast Asian groundwater. Environ Sci Technol 44:7541-7547

Burns PC (1999) The crystal chemistry of uranium. Rev Mineral 38:23-90

Burns PC, Hawthorne FC (1995) Rietveld refinement of the crystal structure of olivenite: a twinned monoclinic structure. Can Mineral 33:885-888

Burns PC, Hawthorne FC (1996) Static and dynamic Jahn-Teller effects in Cu^{2+} oxysalt minerals. Can Mineral 34:1089-1105

Burns PC, Eby RK, Hawthorne FC (1991) Refinement of the structure of liroconite, a heteropolyhedral framework oxysalt mineral. Acta Crystallogr C 47:916-919

Burns PC, Clark CM, Gault RA (2000a) Juabite, $CaCu_{10}(Te^{4+}O_3)_4(AsO_4)_4(OH)_2$ $(H_2O)_4$: Crystal structure and revision of the chemical formula. Can Mineral 38:809-816

Burns PC, Smith JV, Steele IM (2000b) Arizona porphyry copper/hydrothermal deposits I. The structure of chenevixite and luetheite. Mineral Mag 64:25-30

Cabella R, Lucchetti G, Marescotti P (1999) Occurrence of LREE- and Y-arsenates from a Fe-Mn deposit, Ligurian Briançonnais domain, Maritime Alps, Italy. Can Mineral 37:961-972

Cai Y, Cabrera JC, Georgiadis M, Jayachandran K (2002) Assessment of arsenic mobility in the soils of some golf courses in South Florida. Sci Total Environ 291:123-134

Callegari AM, Boiocchi M, Ciriotti ME, Balestra C (2010) Coralloite, IMA 2010-012. CNMNC Newsletter, June 2010, page 579. Mineral Mag 74:577-579

Camacho LM, Gutiérrez M, Alarcón-Herrera MT, Villalba ML, Deng S (2011) Occurrence and treatment of arsenic in groundwater and soil in northern Mexico and southwestern USA. Chemosphere 83:211-225

Camm GS, Butcher AR, Pirrie D, Hughes PK, Glass HJ (2003) secondary mineral phases associated with a historic arsenic calciner identified using automated scanning electron microscopy; a pilot study from Cornwall, UK. Miner Eng 16:1269-1277

Camm GS, Glass HJ, Bryce DW, Butcher AR (2004) Characterisation of a mining-related arsenic-contaminated site, Cornwall, UK. J Geochem Explor 82:1-15

Campbell KM, Nordstrom DK (2014) Arsenic speciation and sorption in natural environments. Rev Mineral Geochem 79:185-216

Cancès B, Juillot F, Morin G, Laperche V, Alvarez L, Proux O, Hazemann J-L, Brown GE Jr, Calas G (2005) XAS Evidence of As(V) association with iron oxyhydroxides in a contaminated soil at a former arsenical pesticide processing plant. Environ Sci Technol 39:9398–9405

Cancès B, Juillot F, Morin G, Laperche V, Polya D, Vaughan DJ, Hazemann J-L, Proux O, Brown GE Jr, Calas G (2008) Changes in arsenic speciation through a contaminated soil profile: A XAS based study. Sci Total Environ 397:178-186

Capitán MÁ, Nieto JM, Sáez R, Almodóvar GR (2003) Characterización textural y mineralogical de Filón Sur (Tharsis, Huelva). Bol Soc Esp Miner 26:45-58. (in Spanish)

Capitelli F, Elaatmani M, Lalaoui MD, Piniella JF (2007) Crystal structure of a vivianite-type mineral: Mg-rich erythrite, $(Co_{2.16}Ni_{0.24}Mg_{0.60})(AsO_4)_2 \cdot 8H_2O$. Z Kristallogr 222:676-679

Cappuyns V, Van Herreweghe S, Swennen R, Ottenburgs R, Deckers J (2002) Arsenic pollution at the industrial site of Reppel-Bocholt (north Belgium). Sci Total Environ 295:217-240

Carlson L, Lindstrom EB, Hallberg KB, Tuovinen OH (1992) Solid-phase products of bacterial oxidation of arsenical pyrite. Appl Environ Microbiol 58:1046-1049

Casiot C, Morin G, Juillot F, Bruneel O, Personné JC, Leblanc M, Duquesne K, Bonnefoy V, Leblanc V, Duquesne K, Bonnefoy V, Elbaz-Poulichot F (2003) Bacterial immobilization and oxidation of arsenic in acid mine drainage (Carnoulès creek, France). Water Res 37:2929-2936

Catti M, Ferraris G (1974) Crystal structure of $Ca_5(HAsO_4)_2(AsO_4)_2 \cdot 9H_2O$ (guérinite). Acta Crystallogr B 30:1789-1794

Catti M, Franchini-Angela M (1979) Krautite, $Mn(H_2O)(AsO_3OH)$: crystal structure, hydrogen bonding and relations with haidingerite and pharmacolite. Am Mineral 64:1248-1254

Catti M, Ivaldi G (1981) Mechanism of the reaction $Ca_5H_2(AsO_4)_4 \cdot 9H_2O$ (ferrarisite) → $Ca_5H_2(AsO_4)_4 \cdot 5H_2O$ (dimorph of vladimirite), and structure of the latter phase. Z Kristallogr 157:119-130

Catti M, Ivaldi G (1983) On the topotactic dehydration $Ca_3(AsO_4)_2 \cdot 11H_2O$ (phaunouxite) → $Ca_3(AsO_4)_2 \cdot 10H_2O$ (rauenthalite) and the structures of both minerals. Acta Crystallogr B 39:4-10

Catti M, Ferraris G, Ivaldi G (1977) Hydrogen bonding in the crystalline state. Structure of talmessite, $Ca_2(Mg,Co)(AsO_4)_2 \cdot 2H_2O$, and crystal chemistry of related minerals. Bull Soc Fr Mineral Cr 100:230-236

Catti M, Chiari G, Ferraris G (1980a) Fluckite, $CaMn(HAsO_4)_2(H_2O)_2$, a structure related by pseudo-polytypism to krautite $MnHAsO_4(H_2O)$. Bull Mineral 103:129-134

Catti M, Chiari G, Ferraris G (1980b) The structure of ferrarisite, $Ca_5(HAsO_4)_2(AsO_4)_2 \cdot 9H_2O$ disorder, hydrogen bonding, and polymorphism with guerinite. Bull Mineral 103:541-546

Catti M, Ferraris G, Ivaldi G (1981) The crystal structure of picropharmacolite, $Ca_4Mg(HAsO_4)_2(AsO_4)_2 \cdot 11H_2O$. Am Mineral 66:385-391

Čech F, Jansa J, Novák F (1976) Kaňkite, $FeAsO_4 \cdot 3.5H_2O$, a new mineral. Neues Jahrb Mineral Monatsh 9:426-436

Čech F, Jansa J, Novák F (1978) Zýkaite, $Fe_4^{3+}(AsO_4)_3(SO_4)(OH) \cdot 15H_2O$, a new mineral. Neues Jahrb Mineral Monatsh 3:134-144

Cesbron FP, Ginderow D, Giraud R, Pelisson P, Pillard F (1987) La nickelaustinite $Ca(Ni,Zn)(AsO_4)(OH)$: Nouvelle espèce minérale du district cobalto-nickelifère de Bou-Azzer, Maroc. Can Mineral 25:401-407

Cesbron FP, Williams SA (1992) Tooeleite, a new mineral from the U.S. Mine, Tooele County, Utah. Mineral Mag 56:71-73

Chiang KY, Lin KC, Lin SC, Chang TK, Wang MK (2010) Arsenic and lead (beudantite) contamination of agricultural rice soils in the Guandu Plain of northern Taiwan. J Hazard Mater 181:1066-1071

Chiappero PJ, Sarp H (1995) Zdeněkite, $NaPbCu_5(AsO_4)_4Cl \cdot 5H_2O$, a new mineral from the Cap Garonne Mine, Var France. Eur J Mineral 7:553-557

Chukanov NV, Pekov IV, Zadov AE (2007a) Attikaite, $Ca_3Cu_2Al_2(AsO_4)_4(OH)_4 \cdot 2H_2O$, a new mineral species. Geol of Ore Dep 49:720-726

Chukanov NV, Puscharovsky DYu, Zubkova NV, Pekov IV, Pasero M, Merlino S, Möckel S, Rabadanov MKh, Belakovskiy DI (2007b) Zincolivenite $CuZn(AsO_4)(OH)$: A new adamite-group mineral with ordered distribution of Cu and Zn. Dokl Earth Sci 415A:841-845

Chukanov NV, Mukhanova AA, Moeckel Sh, Belakovskii DI, Levitskaya LA (2009) Nickeltalmessite, $Ca_2Ni(AsO_4)_2 \cdot 2H_2O$ – a new fairfieldite-group mineral from Bou Azzer, Morocco. Zapiski Ross Mineral Obsh 138:32-39

Clark LA, Pluth JJ, Steele I, Smith JV, Sutton SR (1997a) Crystal structure of austinite, $CaZn(AsO_4)OH$. Mineral Mag 61:677-683

Clark AM, Criddle AJ, Roberts AC, Bonardi M, Moffatt EA (1997b) Feinglosite, a new mineral related to brackebuschite, from Tsumeb, Namibia. Mineral Mag 61:285-289

Cleverley JS, Benning LG, Mountain RW (2003) Reaction path modeling in the As-S system: a case study for geothermal As transport. Appl Geochem 18:1325-1345

Cocco G, Fanfani L, Zanazzi PF (1967) The crystal structure of fornacite. Z Kristallogr 124:385-397

Coda A, Dal Negro A, Sabelli C, Tazzoli V (1977) The crystal structure of stenhuggarite. Acta Crystallogr B 33:1807-1811

Cooper MA, Hawthorne FC (1995a) The crystal structure of geminite, $Cu^{2+}(AsO_3OH)(H_2O)$, a heteropolyhedral sheet structure. Can Mineral 33:1111-1118

Cooper MA, Hawthorne FC (1995b) The crystal structure of maxwellite. Neues Jahrb Mineral Monatsh 97-104

Cooper MA, Hawthorne FC (1996a) The crystal structure of keyite, $Cu_3^{2+}(Zn,Cu^{2+})_4Cd_2(AsO_4)_6(H_2O)_2$, an oxysalt mineral with essential cadmium. Can Mineral 34:623-630

Cooper MA, Hawthorne FC (1996b) The crystal structure of ludlockite, $PbFe^{3+}_4As^{3+}_{10}O_{22}$, the mineral with pentameric arsenite groups and orange hair. Can Mineral 34:79-89

Cooper MA, Hawthorne FC (1999a) The crystal structure of betpakdalite, and a new chemical formula: $\{Mg(H_2O)_6\}$ $Ca_2(H_2O)_{13}$ $[Mo^{6+}_8As^{5+}_2Fe^{3+}_3O_{36}(OH)]$ $(H_2O)_4$. Can Mineral 37:61-66

Cooper MA, Hawthorne FC (1999b) Local Pb^{2+}- disorder in the crystal structure of jamesite, $Pb_2ZnFe^{3+}_2(Fe^{3+}_{2.8}Zn_{1.2})(AsO_4)_4(OH)_8[(OH)_{1.2}O_{0.8}]$, and revision of the chemical formula. Can Mineral 37:53-60

Cooper MA, Hawthorne FC (1999c) The effect of differences in coordination on ordering of polyvalent cations in close-packed structures: The crystal structure of arakiite and comparison with hematolite. Can Mineral 37:1471-1482

Cooper MA, Hawthorne FC (2001) The biggest mineral: The crystal structure of mcgovernite. Eleventh Annual V.M. Goldschmidt Conference, May 20-24. Abstract # 3446

Cooper MA, Hawthorne FC (2009) The crystal structure of burgessite, $Co_2(H_2O)_4[AsO_3(OH)]_2$ (H_2O), and its relation to erythrite. Can Mineral 47:165-172

Corriveau MC, Jamieson HE, Parsons MB, Campbell JL, Lanzirotti A (2011a) Direct characterization of airborne particles associated with arsenic-rich mine tailings: Particle size, mineralogy and texture. Appl Geochem 26:1639-1648

Corriveau MC, Jamieson HE, Parsons MB, Hall GEM (2011b) Mineralogical characterization of arsenic in gold mine tailings from three sites in Nova Scotia. Geochem Explor Environ Anal 11:179-192

Courtin-Nomade A, Neel C, Bril H, Davranche M (2002) Trapping and mobilization of arsenic and lead in former mine tailings – Environmental condition effect. B Soc Geol Fr 173:479-485

Courtin-Nomade A, Bril H, Neel C, Lenain JF (2003) Arsenic in iron cements developed within tailings of a former metalliferous mine-Enguialès, Aveyron, France. Appl Geochem 18:295-408

Courtin-Nomade A, Bril H, Bény JM, Kunz M, Tamura N (2010) Sulfide oxidation observed using micro-Raman spectroscopy and micro-X-ray diffraction: The importance of water/rock ratios and pH conditions. Am Mineral 95:582-591

Cox DP, Singer DA (1986) Mineral Deposit Models. U.S. Geological Survey Bulletin 1963, 379 p

Craw D, Bowell RJ (2014) The characterization of arsenic in mine waste. Rev Mineral Geochem 79:473-506

Craw D, Chappell D, Reay A, Walls D (2000) Mobilisation and attenuation of arsenic around gold mines, east Otago, New Zealand. New Zeal J Geol Geophys 43:373-383

Craw D, Koons PO, Chappell DA (2002) Arsenic distribution during formation and capping of an oxidized sulphidic mine soil, Macraes mine, New Zealand. J Geochem Explor 76:13-29

Craw D, Nelson M (2000) Geochemical signatures of discharge waters, Macraes mine flotation tailings, east Otago, New Zeal. J Mar Freshwat Res 34:597-613

Cruz R, Lázaro L, Rodríguez JM, Monroy M, González L (1997) Surface characterization of arsenopyrite in acidic medium by triangular scan voltammetry on carbon paste electrode. Hydrometallurgy 46:303-319

Dai Y, Hughes J, Moore PB (1991) The crystal structures of mimetite and clinomimetite, $Pb_5(AsO_4)_3Cl$. Can Mineral 29:369-376

Dal Negro A, Guiseppetti G, Martin Pozas JM (1974) The crystal structure of sarkinite, $Mn_2AsO_4(OH)$. Mineral Petrol 21:246-260

Davis A, Ruby MV, Bloom M, Schoof R, Freeman G, Bergstrom PD (1996) Mineralogic constraints on the bioavailability of arsenic in smelter-impacted soils. Environ Sci Technol 30:392-399

De Brouwere K, Smolders E, Merckx R (2004) Soil properties affecting solid-liquid distribution of As(V) in soils. Eur J Soils Sci 55:165-173

De Waele J, Forti P (2005) Mineralogy of mine caves in Sardinia. Proc 14th Int Con Speleo, Kalamos, Greece: 306-311

Deditius AP, Utsunomiya S, Renock D, Ewing RC, Ramana CV, Becker U, Kesler SE (2008) A proposed new type of arsenian pyrite: Composition, nanostructure and geological significance. Geochim Cosmochim Acta 72:2919-2933

del Bucchia S, Jumas JC, Maurin M (1981) Contribution a l'etude de composes sulfures d'etain (II): Affinement de la structure de SnS. Acta Crystallogr B 37:1903-1905

Demartin F, Castellano C, Gramaccioli CM, Campostrini I (2010) Aluminum-for-iron substitution, hydrogen bonding, and a novel structure-type in coquimbite-like minerals. Can Mineral 48:323-333

Denning RM (1943) Aluminum-bearing scorodite from Hobart Butte, Oregon. Am Mineral 28:55-57

DeSisto SL, Jamieson HE, Parsons MB (2011) Influence of hardpan layers on arsenic mobility in historical gold mine tailings. Appl Geochem 26:2004-2018

Dietrich RV (1960) Virginia mineral localities. Virginia Polytechnic Institute, Engineering Station Series, Bulletin 138

Ding Z, Zheng B, Long J, Belkin HE, Finkelman RB, Chen C, Zhou D, Zhou Y (2001) Geological and geochemical characteristics of high arsenic coals from enedemic arsenosis areas in southwestern Guizhou Province, China. Appl Geochem 16:1353-1360

Dódony I, Pósfai M, Buseck PR (1996) Structural relationship between pyrite and marcasite. Am Mineral 81:119-125

Donahue R, Hendry MJ (2003) Geochemistry of arsenic in uranium mine mill tailings, Saskatchewan, Canada. Appl Geochem 18:1733-1750

Dorner R, Weber K (1976) Die Kristallstruktur von Chudobait, $(Mg,Zn)_5H_2[AsO_4]_4 \cdot 10H_2O$. Naturwissenschaften 63:243

Drahota P, Filippi M (2009) Secondary arsenic minerals in the environment: A review. Environ Int 35:1243-1255

Drahota P, Rohovec J, Filippi M, Mihaljevič M, Rychlovský P, Červený V, Pertold Z (2009) Mineralogical and geochemical controls on arsenic speciation and mobility under different redox conditions in soil, sediment and water at the Mokrsko-West gold deposit, Czech Republic. Sci Total Environ 407:3372-3384

Drahota P, Filippi M, Ettler V, Rohovec J, Mihaljevič M, Šebek O (2012) Natural attenuation of arsenic in soils near a highly contaminated medieval mine waste dump. Sci Total Environ 414:546-555

Driss A, Jouini T (1989) Structure cristalline d'une nouvelle variete polymorphique de $LiAsO_3$. J Solid State Chem 78:126-129

Dubanský A, Langrová A, Dvořáček P, Čejka UJ, Kouřimský J (1991) Minerals of the kaustic metamorphosis from the Kateřina mine. Geol Průzk 6:172-173 (in Czech)

Duesler EN, Chakoumakos BC, Foord EE (1988) Zimbabweite, $Na(Pb,Na,K)_2As_4$ $(Ta,Nb,Ti)_4O_{18}$, an arsenite-tantalate with a novel corner-linked octahedral sheet. Am Mineral 73:1186-1190

Dunn PJ (1981a) Magnesium-chlorophoenicite redefined and new data on chloropheonicite. Can Mineral 19:333-336

Dunn PJ (1981b) Sterlinghillite, a new hydrated manganese arsenate mineral from Ogdensburg, New Jersey. Am Mineral 66:182-184

Dunn PJ (1982) New data for pitticite and a second occurrence of yukonite at Sterling Hill, New Jersey. Mineral Mag 46:261-264

Dunn PJ (1995) Franklin and Sterling Hill, New Jersey: The world's most magnificent mineral deposits. The Franklin-Ogdensburg Mineralogical Society

Dunn PJ, Peacor DR (1983) Kittatinnyite and wallkilldellite, silicate/arsenate analogues containing calcium and manganese, from Franklin and Sterling Hill, New Jersey. Am Mineral 68:1029-1032

Dunn PJ, Peacor DR, Leavens PB, Simmons WB (1982) Jarosewichite and a related phase: Basic manganese arsenates of the chlorophoenicite group from Franklin, New Jersey. Am Mineral 67:1043-1047

Dunn PJ, Peacor DR, Roberts WL, Campbell TJ, Ramik RA (1984) Walentaite, a new calcium iron arsenate phosphate from the White Elephant Mine, Pringle, South Dakota. Neues Jahrb Mineral Monatsh 169-174

Dunn PJ, Peacor DR, Criddle AJ, Finkelman RB (1986a) Laphamite, an arsenic selenide analogue of orpiment, from burning anthracite deposits in Pennsylvania. Mineral Mag 50:279-282

Dunn PJ, Peacor DR, Sturman BD, Wicks FJ (1986b) Rouseite, a new lead manganese arsenite from Langban, Sweden. Am Mineral 71:1034-1036

Duquesne K, Lebrun S, Casiot C, Bruneel O, Personné JC, Leblanc M, Elbaz-Poulichet F, Morin G, Bonnefoy V (2003) Immobilization of arsenite and ferric iron by *Acidithiobacillus ferrooxidans* and its relevance to acid mine drainage. Appl Environ Microbiol 69:6165-6173

Dutré C, Vandecasteele C (1998) Immobilization mechanisms of arsenic in waste solidified using cement and lime. Environ Sci Technol 32:2782-2787

Eby RK, Hawthorne FC (1989) Euchroite, a heteropolyhedral framework structure. Acta Crystallogr C 45:1479-1482

Eby RK, Hawthorne FC (1990) Clinoclase and the geometry of [5]-coordinate Cu^{2+} in minerals. Acta Crystallogr C 46:2291-2294

Edenharter A, Nowacki W, Weibel M (1977) Zur Struktur und Zusammensetzung von Cafarsit. Cafarsit ein As(III)-Oxid, kein Arsenat. Schweiz Mineral Petrogr 57:1-17

Effenberger H, Pertlik F (1979) Die Kristallstruktur des Finnemanits, $Pb_5Cl(AsO_3)_3$, mit einem Vergleich zum Strukturtyp des Chlorapatits, $Ca_5Cl(PO_4)_3$. Mineral Petrol 26:95-107

Effenberger H, Mereiter K, Pimminger M, Zemann J (1982) Machatschkiite: Crystal structure and revision of the chemical formula. Mineral Petrol 30:145-155

Effenberger H, Pertlik F (1986) Schultenit, $PbHAsO_4$, und $PbHPO_4$: Syntheses und Kristallstrukturen nebst einer Diskussion zur Symmetrie. Mineral Petrol 35:157-166

Effenberger H, Krause W, Bernhardt H-J, Martin M (2000) On the symmetry of sumcorite group minerals based on the new species rappoldite and zincgartrellite. Mineral Mag 64:1109-1126

Effenberger H, Krause W, Bernhardt H-J (2002) Structural investigation of adelite and cobaltaustinite, two members of the adelite-descloizite group. Ninth International Symposium on Experimental Mineralogy, Petrology and Geochemistry, 30

Egal M, Casiot C, Morin G, Elbaz-Poulichet F, Cordier MA, Bruneel O (2010) An updated insight into the natural attenuation of As concentrations in Reigous Creek (southern France). Appl Geochem 25:1949-1957

Elliott P, Turner P, Jensen P, Kolitsch U, Pring A (2009) Description and crystal structure of nyholmite, a new mineral related to hureaulite, from Broken Hill, New South Wales, Australia. Mineral Mag 73:723-735

Embrick LL, Porter KM, Pendergrass A, Butcher DJ (2005) Characterization of lead and arsenic contamination at Barber Orchard, Haywood County, NC. Microchem J 81:117-121

Emsley J (2001) Nature's Building Blocks. An A-Z Guide to the Elements. Oxford University Press

Emsley J (2005) The Elements of Murder. A History of Poison. Oxford University Press

EPA (1998) Location and estimating air emissions from sources of arsenic and arsenic compounds. EPA-454/R-98-013, US Environmental Protection Agency

Ercit TS (1993) Caryinite revisited. Mineral Mag 57:721-727

Ettler V, Johan Z (2003) Mineralogy of metallic phases in sulphide mattes from primary lead smelting. C R Geosci 335:1005-1012

Ettler V, Legendre O, Bodénan F, Touray JC (2001) Primary phases and natural weathering of old lead-zinc pyrometallurgical slag from Příbram, Czech Republic. Can Mineral 39:873-888

Ettler V, Johan Z, Hradil D (2003a) Natural alteration products of sulphide mattes from primary lead smelting. C R Geosci 335:1013-1020

Ettler V, Piantone P, Touray JC (2003b) Mineralogical control on inorganic contaminant mobility in leachate from lead-zinc metallurgical slag: experimental approach and long-term assessment. Mineral Mag 67:1269-1283

Ettler V, Jehlička J, Mašek V, Hruška J (2005a) The leaching behaviour of lead metallurgical slag in high-molecular-weight (HMW) organic solutions. Mineral Mag 69:737-747

Ettler V, Johan Z, Baronnet A, Jankovský F, Gilles C, Mihaljevič M, Šebek O, Strnad L, Bezdička P (2005b) Mineralogy of air-pollution-control residues from a secondary lead smelter: environmental implications. Environ Sci Technol 39:9309-9316

Ettler V, Červinka R, Johan Z (2009a) Mineralogy of medieval slags from lead and silver smelting (Bohutín, Příbram district, Czech Republic): Towards estimation of historical smelting conditions. Archaeometry 51:987-1007

Ettler V, Johan Z, Bezdička P, Drábek M, Šebek O (2009b) Crystallization sequences in matte and speiss from primary lead metallurgy. Eur J Mineral 21:837-854

Ettler V, Johan Z, Křibek B, Šebek O, Mihaljevič M (2009c) Mineralogy and environmental stability of slags from the Tsumeb smelter, Namibia. Appl Geochem 24:1-15

Ettler V, Mihaljevič M, Šebek O (2010) Antimony and arsenic leaching from secondary lead smelting air-pollution-control residues. Waste Manage Res 28:587-595

Ettler V, Mihaljevič M, Šebek O, Valigurová R, Klementová M (2012) Differences in antimony and arsenic from lead smelter fly ash in soils. Chem Erde/Geochem 72:15-22

Fawcett SE, Jamieson HE (2011) The distinction between ore processing and post-depositional transformation on the speciation of arsenic and antimony in mine waste and sediment. Chem Geol 283:109-118

Fendorf S, Michael HA, Geen A (2010) Spatial and temporal variations of groundwater arsenic in South and Southeast Asia. Science 328:1123-1127

Feraud J, Pillard F, Vernet J (1976) La talmesite $Ca_2Mg(AsO_4)_2 \cdot 2H_2O$ du karst antéalbien a barytine de Luceram (Alpes-Martitimes). Bull Soc Franç Minéral Cristallogr 99:331-333

Ferguson J, Lambert IB (1972) Volcanic exhalations and metal enrichments at Matupi Harbor, New Britain, T.P.N.G. Econ Geol 67:25-37

Fernández-Martínez A, Cuello GJ, Johnson MR, Bardelli F, Román-Ross G, Charlet L, Turrillas X (2008) Arsenate incorporation in gypsum probed by neutron, x-ray scattering and density functional theory modeling. J Phys Chem A 112:5159-5166

Ferraris G (1969) The crystal structure of pharmacolite, $CaH(AsO_4) \cdot 2H_2O$. Acta Crystallogr B 25:1544-1550

Ferraris G, Abbona F (1972) The crystal structure of $Ca_5(HAsO_4)_2(AsO_4)_2(H_2O)_4$ (sainfeldite). B Soc Fr Mineral Cr 95:33-41

Ferraris G, Chiari G (1970) The crystal structure of $CaHAsO_4$ (weilite). Acta Crystallogr B 26:403-410

Ferraris G, Franchini-Angela M (1973) Hydrogen bonding in the crystalline state. Crystal structure of $MgHAsO_4 \cdot 7H_2O$, roesslerite. Acta Crystallogr B 28:286-292

Ferraris G, Jones DW, Yerkess J (1971) Determination of hydrogen atom positions in the crystal structure of pharmacolite, $CaHAsO_4 \cdot 2H_2O$ by neutron diffraction. Acta Crystallogr B 27:349-354

Ferraris G, Jones DW, Yerkess J (1972a) A neutron and X-ray refinement of the crystal structure of $CaHAsO_4 \cdot H_2O$ (haidingerite). Acta Crystallogr B 28:209-214

Ferraris G, Jones DW, Yerkess J (1972b) A neutron diffraction study of the crystal structure of calcium bis(hydrogen arsenate), $Ca(H_2AsO_4)_2$. Acta Crystallogr B 28:2430-2437

Filatov SK, Gaidamako IM, Glavatskikh SF, Starova GL, Sorokin ND (1984) Exhalation lammerite, $Cu_3[(As,P)O_4]_2$ (Kamchatka). Dokl Akad Nauk SSSR 279:197-200

Filatov SK, Vergasova LP, Gorskaya MG, Krivovichev SV, Burns PC, Ananiev VV (2001) Bradaczekite, $NaCu_4(AsO_4)_3$, a new mineral species from the Tolbachik volcano, Kamchatka Peninsula, Russia. Can Mineral 39:1115-1119

Filatov SK, Krivovichev SV, Burns PC, Vergasova LP (2004) Crystal structure of filatovite, K[(Al,Zn)$_2$(As,Si)$_2$O$_8$], the first arsenate of the feldspar group. Eur J Mineral 16:537-543

Filippi M (2004) Oxidation of the arsenic-rich concentrate at the Přebuz abandoned mine (Erzgebirge Mts, CZ): mineralogical evolution. Sci Total Environ 322:271-282

Filippi M, Goliáš V, Pertold Z (2004) Arsenic in contaminated soils and anthropogenic deposits at the Mokrsko, Roudný, and Kašperské Hory gold deposits, Bohemian Massif CZ. Environ Geol 45:716–730

Filippi M, Doušová B, Machovič V (2007) Mineralogical speciation of arsenic in soils above the Mokrsko-west gold deposit, Czech Republic. Geoderma 139:154-170

Filippi M, Machovič V, Drahota P, Böhmová V (2009) Raman microspectroscopy as a valuable additional method to X-ray diffraction and electron microscope-microprobe analysis in the study of iron arsenates in environmental samples. Appl Spectr 63(6):621-626

Filippi M, Drahota P, Machovič V, Böhmová V, Mihaljevič M (2014) Arsenic mineralogy and mobility in the arsenic-rich historical mine waste dump. (in preparation)

Finney JJ (1963) The crystal structure of carminite. Am Mineral 48:1-13

Fitzmaurice AG, Bilgin AA, O`Day PA, Illera V, Burris D, Reisinger HJ, Hering JG (2009) Geochemical and hydrologic controls on the mobilization of arsenic derived from herbicide application. Appl Geochem 24:2152-2162

Fleck M, Kolitsch U, Hertweck B (2002) Natural and synthetic compounds with kröhnkite-type chains: Review and classification. Z Kristallogr 217:435-443

Fleet ME, Barbier J (1989) Structure of aerugite (Ni$_{8.5}$As$_3$O$_{16}$) and interrelated arsenate and germanate structural series. Acta Crystallogr B 45:201-205

Flemming RL, Salzsauler KA, Sherriff BL, Sidenko NV (2005) Identification of scorodite in fine-grained, high-sulfide, arsenopyrite mine-waste using micro X-ray diffraction (μXRD). Can Mineral 43:1243-1254

Flores AN, Rubio LMD (2010) Arsenic and metal mobility from Au mine tailings in Rodalquilar (Almería, SE Spain). Environ Earth Sci 60:121-138

Foley NK, Ayuso RA (2008) Mineral sources and transport pathways for arsenic release in a coastal watershed, USA. Geochem Explor Environ Anal 8:59-75

Folkes DJ, Helgen SO, Litle RA (2001) Impacts of historic arsenical pesticide use in residential soils in Denver, Colorado. *In:* Arsenic Exposure and Health Effects. Chappel WR, Abernathy CO, Calderon RL (eds) Elsevier Science, p 97-113

Font O, Querol X, Huggins FE, Chimenos JM, Fernández AI, Burgos S, Peña FG (2005) Speciation of major and selected trace elements in IGCC fly ash. Fuel 84:1364-1371

Font O, Querol X, Izquierdo M, Alvarez E, Moreno N, Diez S, Álvarez-Rodríguez R, Clemente-Jul C, Coca P, Garcia-Peña F (2010) Partitioning of elements in a entrained flow IGCC plant: Influence of selected operational conditions. Fuel 89:3250-3261

Foord EE, Hughes JM, Cureton F, Maxwell CH, Flaster AU, Sommer AJ, Hlava PF (1999) Esperanzaite, NaCa$_2$Al$_2$(As^{5+}O$_4$)$_2$F$_4$(OH)·2H$_2$O, a new mineral species from the La Esperanza Mine, Mexico: Descriptive mineralogy and atomic arrangement. Can Mineral 37:67-72

Förster H-J, Ondrejka M, Uher P (2011) Mineralogical responses to subsolidus alteration of granitic rocks by oxidizing As-bearing fluids: REE arsenates and As-rich silicates from the Zinnwald granite, eastern Erzgebirge, Germany. Can Mineral 49:913-930

Forti P, Urbani PF (1995) I nuovi minerali di grotta scoperti nella "Cueva Alfredo Jahn", Venezuela. Atti Cong Naz Spel 1994:155-159

Foshag WF, Clinton HG (1927) An occurrence of pitticite in Nevada. Am Mineral 12:290-292

Foster AL, Brown GE Jr, Parks GA, Tingle TN, Voigt DE, Brantley SL (1997) XAFS determination of As(V) associated with Fe(III): oxyhydroxides in weathered mine tailings and contaminated soil from California, U.S.A. J Phys IV 7:815-816

Foster AL, Brown GE Jr, Tingle TN, Parks GA (1998) Quantitative arsenic speciation in mine tailings using X-ray absorption spectroscopy. Am Mineral 83:553-568

Foster AL, Ashley RP, Rytuba JJ (2011) Arsenic species in weathering mine tailings and biogenic solids at the Lava Cap Mine Superfund Site, Nevada City, CA. Geochem Trans 12:1

Foster AL, Kim CS (2014) Arsenic speciation in solids using X-ray absorption spectroscopy. Rev Mineral Geochem 79:257-369

Frau F, Ardau C (2003) Geochemical controls on arsenic distribution in the Baccu Locci stream catchment (Sardinia, Italy) affected by past mining. Appl Geochem 18:1373-1386

Frau F, Ardau C (2004) Mineralogical controls on arsenic mobility in the Baccu Locci stream catchment Sardinia, Italy affected by past mining. Mineral Mag 68:15–30

Frau F, Da Pelo S (2001) Bulachite, a rare aluminum arsenate from Sardinia, Italy: The second world occurrence. Neues Jahrb Mineral Monatsh 18-26

Frau F, Rossi A, Ardau C, Biddau R, Da Pelo S, Atzei D, Licheri C, Cannas C, Capitani G (2005) Determination of arsenic speciation in complex environmental samples by the combined use of TEM and XPS. Microchim Acta 151:189-201

Frondel C, Baum J (1974) Structure and mineralogy of the Franklin zinc-iron-manganese deposit, New Jersey. Econ Geol 69:157-180

Ganne P, Cappuyns V, Vervoort A, Buvé L, Swennen R (2006) Leachability of heavy metals and arsenic from slags of metal extraction industry at Angleur (eastern Belgium). Sci Total Environ 356:69-85

Garavelli A, Pinto D, Vurro F, Mellini M, Viti C, Balic-Zunic T, Della Ventura G (2009) Yukonite from the Grotta della Monaca Cave, Sant'Agata di Esaro, Italy: characterization and comparison with cotype material from the Daulton Mine, Yukon, Canada. Can Mineral 47:39-51

Garcia-Sanchez A, Alonso-Rojo P, Santos-Frances F (2010) Distribution and mobility of arsenic in soils of a mining area (Western Spain). Sci Total Environ 408:4194-4201

Garelick H, Jones H, Dybowska A, Valsami-Jones E (2008) Arsenic pollution sources. *In:* Reviews of Environmental Contamination. Volume 197. Whitacre DM (ed) Springer, p 17-59

Gelaude P, Kalmthout P, Rewitzer C (1996) Lavrion, the Minerals in the Historic Slags. Janssen Print, Nijmegen, The Netherlands

Ghose S, Fehlmann M, Sundaralingam M (1965) The crystal structure of clinoclase, $Cu_3AsO_4(OH)_3$. Acta Crystallogr 18:777-787

Ghose S, Boving P, LaChapelle WA, Wan C (1977) Reinerite, $Zn_3(AsO_3)_2$: an arsenite with a novel type of Zn-tetrahedral double chain. Am Mineral 62:1129-1134

Ghose S, Gupta PKS, Schlemper EO (1987) Leiteite, $ZnAs_2O_4$: A novel type of tetrahedral layer structure with arsenite chains. Am Mineral 72:629-632

Ghose S, Wan C (1979) Structural chemistry of copper and zinc minerals. VI. Bayldonite, $(Cu,Zn)_3Pb(AsO_4)_2(OH)_2$: A complex layer structure. Acta Crystallogr B 35:819-823

Gieré R, Sidenko NV, Lazareva EV (2003) The role of secondary minerals in controlling the migration of arsenic and metals from high-sulfide wastes (Berikul gold mine, Siberia). Appl Geochem 18:1347-1359

Giester G, Kolitsch U, Leverett P, Turner P, Williams PA (2007) The crystal structures of lavendulan, sampleite, and a new polymorph of sampleite. Eur J Mineral 19:75-93

Ginderow D, Cesbron F (1981) Structure de la Mapimite, $Zn_2Fe_3(AsO_4)_3(OH)_4 \cdot 10H_2O$. Acta Crystallogr B 37:1040-1043

Giuseppetti G, Tadini C (1988) The crystal structure of austinite, $CaZn(AsO_4)(OH)$, from Kamareza, Laurion (Greece). Neues Jahrb Mineral Monatsh 159-166

Glavatskikh SF, Bykova EYu (1998) First finding of exhalative johillerite (Kamchatka). Dokl Akad Nauk+ 361:795-798

Goldhaber MB, Lee RC, Hatch JR, Pashin JC, Treworgy J (2001) Role of large scale fluid-flow in subsurface arsenic enrichment. *In:* Arsenic in Groundwaters: Geochemistry and Occurrences. Welch A, Stollenwerk K (eds) Kluwer Academic Publishers, Boston, p 127-164

Gomez MA, Becze L, Blyth RIR, Cutler JN, Demopoulos GP (2010) Molecular and structural investigation of yukonite (synthetic & natural) and its relation to arseniosiderite. Geochim Cosmochim Ac 74:5835-5851

Gómez-Parrales I, Bellinfante N, Tejada M (2011) Study of mineralogical speciation of arsenic in soils using X ray microfluorescence and scanning electronic microscopy. Talanta 84:853-858

Goodarzi F, Huggins FE (2005) Speciation of arsenic in feed coals and their ash byproducts from Canadian power plants burning sub-bituminous and bituminous coals. Energy Fuel 19:905-915

Graeser S, Schwander H, Bianchi R, Pilati T, Gramaccioli CM (1989) Geigerite, the Mn analogue of chudobaite: Its description and crystal structure. Am Mineral 74:676-684

Graeser S, Schwander H, Demartin F, Gramaccioli CM, Pilati T, Reusser E (1994) Fetiasite, $(Fe^{2+}, Fe^{3+}, Ti)_3O_2[As_2O_5]$, a new arsenite mineral: Its description and structure determination. Am Mineral 79:996-1002

Gräfe M, Nachtegaal M, Sparks DL (2004) Formation of metal-arsenate precipitates at the goethite-water interface. Environ Sci Technol 38:6561-6570

Gräfe M, Sparks DL (2005) Kinetics of zinc and arsenate co-sorption at the goethite-water interface. Geochim Cosmochim Acta 69:4573-4595

Gräfe M, Beattie DA, Smith E, Skinner WM, Singh B (2008a) Copper and arsenate co-sorption at the mineral-water interfaces of goethite and jarosite. J Colloid Interface Sci 322:399-413

Gräfe M, Tappero RV, Marcus MA, Sparks DL (2008b) Arsenic speciation in multiple metal environments II. Micro-spectroscopic investigation of a CCA contaminated soil. J Colloid Interface Sci 321:1-20

Graupner T, Kassahun A, Rammlmair D, Meima JA, Kock D, Furche M, Fiege A, Schippers A, Melcher F (2007) Formation of sequences of cemented layers and hardpans within sulfide-bearing mine tailings (mine district Freiberg, Germany). Appl Geochem 22:2486-2508

Greif A, Klemm W, Klemm K (2008) Influence of arsenic from anthropogenic loaded soils on the mine water quality in the tin district Ehrenfriedersdorf, Erzgebirge (Germany). Eng Life Sci 8:631-640

Grey IE, Madsen IC, Harris DC (1987) Barian tomichite, $Ba_{0.5}(As_2)_{0.5}Ti_2(V,Fe)_5O_{13}$ (OH), its crystal structure and relationship to derbylite and tomichite. Am Mineral 72:201-208

Grey IE, Mumme WG, Bordet P, Mills SJ (2008) A new crystal-chemical variation of the alunite-type structure in monoclinic $PbZn_{0.5}Fe_3(AsO_4)_2(OH)_6$. Can Mineral 46:1355-1364

Griffen DT, Ribbe PH (1979) Distortions in the tetrahedral oxyanions of crystalline substances. Neues Jahrb Mineral Abh 137:54-73

Grosbois C, Courtin-Nomade A, Robin E, Bril H, Tamura N, Schäfer, Blanc G (2011) Fate of arsenic-bearing phases during the suspended transport in a gold mining district (Isle river Basin, France). Sci Total Environ 409:4986-4999

Haffert L, Craw D (2008) Mineralogical controls on environmental mobility of arsenic from historic mine processing residues, New Zealand. Appl Geochem 23:1467-1483

Haffert L, Craw D (2009) Field quantification and characterization of extreme arsenic concentrations at a historic mine processing site, Waiuta, New Zealand. New Zeal J Geol Geop 52:261-272

Haffert L, Craw D (2010) Geochemical processes influencing arsenic mobility at Bullendale historic gold mine, Otago, New Zealand. New Zeal J Geol Geop 53:129-142

Haffert L, Craw D, Pope J (2010) Climatic and compositional controls on secondary arsenic mineral formation in high-arsenic mine wastes, South Island, New Zealand. New Zeal J Geol Geop 53:91-101

Hansen S, Faelth L, Johnson O (1984) Bergslagite, a mineral with tetrahedral berylloarsenate sheet anions. Z Kristallogr 166:73-80

Harris DC, Hoskins BF, Grey IE, Criddle AJ, Stanley CJ (1989) Hemloite (As, Sb)$_2$(Ti,V,Fe,Al)$_{12}$O$_{23}$OH: A new mineral from the Hemlo gold deposit, Hemlo, Ontario, and its crystal structure. Can Mineral 27:427-440

Harrison WTA (2000) Synthetic mansfieldite, AlAsO$_4$·2H$_2$O. Acta Crystallogr C 56:e421

Harrison WTA, Nenoff TM, Gier TE, Stucky GD (1993) Tetrahedral-atom 3-ring groupings in 1-dimensional inorganic chains: Be$_2$AsO$_4$OH·4H$_2$O and Na$_2$ZnPO$_4$OH ·7H$_2$O. Inorg Chem 32:2437-2441

Haruna M, Satoh H, Banno Y, Kono M, Bunno M (2002) Mineralogical and oxygen isotopic constraints on the origin of the contact-metamorphosed bedded manganese deposit at Nagasawa, Japan. Can Mineral 40:1069-1089

Hawthorne FC (1976a) The hydrogen positions in scorodite. Acta Crystallogr B 32:2891-2892

Hawthorne FC (1976b) Refinement of the crystal structure of berzeliite. Acta Crystallogr B 32:1581-1583

Hawthorne FC (1976c) A refinement of the crystal structure of adamite. Can Mineral 14:143-148

Hawthorne FC (1979) Paradamite. Acta Crystallogr B 35:720-722

Hawthorne FC (1983) Enumeration of polyhedral clusters. Acta Crystallogr A 39:724-736

Hawthorne FC (1985a) Towards a structural classification of minerals: the ^{vi}M^{iv}T$_2$O$_n$ minerals. Am Mineral 70:455-473

Hawthorne FC (1985b) Schneiderhöhnite, Fe^{2+}Fe$^{3+}_3$As$^{3+}_5$O$_{13}$, a densely packed arsenite structure. Can Mineral 23:675-679

Hawthorne FC (1986) Lammerite, Cu$_3$(AsO$_4$)$_2$, a modulated close-packed structure. Am Mineral 71:206-209

Hawthorne FC, Ferguson RB (1977) The crystal structure of roselite. Can Mineral 15:36-42

Hawthorne FC, Krivovichev SV, Burns PC (2000) The crystal chemistry of sulfate minerals. Rev Mineral Geochem 40:1-112

Hawthorne FC, Cooper MA, Paar WH (2008) The crystal structure of braithwaiteite. J Coord Chem 61:15-29

Hawthorne FC, Abdu YA, Tait KT (2013) Hydrogen bonding in the crystal structure of legrandite: Zn$_2$(AsO$_4$) (OH)(H$_2$O). Can Mineral 51:233-241

Hem SR, Makovicky E (2004) The system Fe-Co-Ni-As-S. II. Phase relations in the (Fe,Co,Ni)As$_{1.5}$S$_{0.5}$ section at 650 ° and 500 °C. Can Mineral 42:63-86

Henderson RR, Yang H, Downs RT, Jenkins RA (2008) Redetermination of conichalcite, CaCu(AsO$_4$)(OH). Acta Crystallogr E 64:i53-i54

Henke KR (ed) (2009) Arsenic. Environmental Chemistry, Health Threats and Waste Treatment. Wiley

Hill CA, Forti P (1997) Cave Minerals of the World. National Speleological Society, Huntsville

Hill RJ (1976) The crystal structure and infrared properties of adamite. Am Mineral 61:979-986

Hill RJ (1979) The crystal structure of köttigite. Am Mineral 64:376-382

Hilmer W (1956) Die Kristallstruktur von Lithiumpolyarsenat (LiAsO$_3$)$_x$. Acta Crystallogr 9:87-88

Hingston JA, Collins CD, Murphy RJ, Lester JN (2001) Leaching of chromate copper arsenate wood preservatives: a review. Environ Pollut 111:53-66

Hofmeister W, Tillmanns E (1978) Strukturelle untersuchungen an arsenbrackebuschit. Mineral Petrol 25:153-163

Holtstam D, Langhof J (eds) (1999) Långban. The Mines, Their Minerals, Geology and Explorers. Christian Weise Verlag

Horák V (2000) Mineral collecting and collectors at the deposit Jáchymov in Krušné Hory Mts. since the 16th century until today. Minerál 8:236-254 (in Czech)

Hudson-Edwards KA, Schell C, Macklin MG (1999) Mineralogy and geochemistry of alluvium contaminated by metal mining in the Rio Tinto area, southwest Spain. Appl Geochem 14:1015-1030

Hudson-Edwards KA, Jamieson HE, Charnock JM, Macklin MG (2005) Arsenic speciation in waters and sediment of ephemeral floodplain pools, Ríos Agrio-Guadiamar, Aznalcóllar, Spain. Chem Geol 219:175-192

Huffman GP, Huggins FE, Shah N, Zhao J (1994) Speciation of arsenic and chromium in coal and combustion ash by XAFS spectroscopy. Fuel Process Technol 39:47-62

Huggins FE, Huffman GP (1996) Modes of occurrence of trace elements in coal from XAFS spectroscopy. Int J Coal Geol 32:31-53

Huggins FE, Huffman GP, Kolker A, Mroczkowski SJU, Palmer CA, Finkelman RB (2002) Combined application of XAFS spectroscopy and sequential leaching for determination of arsenic speciation in coal. Energy Fuel 16:1167-1172

Huggins FE, Seidu LBA, Shah N, Huffman GP, Honaker RQ, Kyger JR, Higgins BL, Robertson JD, Pal S, Seehra MS (2009) Elemental modes of occurrence in an Illinois #6 coal and fractions prepared by physical separation techniques at a coal preparation plant. Int J Coal Geol 78:65-76

Hughes JM, Bloodaxe ES, Kobel KD, Drexler JW (1996) The atomic arrangement of ojuelaite, $ZnFe^{3+}_2(AsO_4)_2(OH)_2 \cdot 4H_2O$. Mineral Mag 60:519-521

Huminicki DMC, Hawthorne FC (2002) The crystal chemistry of the phosphate minerals. Rev Mineral Geochem 48:123-254

Hu Z, Gao S (2008) Upper crustal abundances of trace elements: A revision and update. Chem Geol 253:205-221

Hybler J, Ondruš P, Císařová I, Petříček V, Veselovský F (2003) Crystal structure of lindackerite, (Cu,Co,Ni)$Cu_4(AsO_4)_2(AsO_3OH)_2 \cdot 9H_2O$ from Jáchymov, Czech Republic. Eur J Mineral 15:1035-1042

Hyršl J, Kaden M (1992) Eine paragenese von eisen-arsenaten von Kaňk bei Kutná Hora in Böhmen und Munzig bei Meißen in Sachsen. Aufschluss 43:95-102

Iitaka Y, Nowacki W (1961) A refinement of the pseudo crystal structure of scleroclase $PbAs_2S_4$. Acta Crystallogr 14:1291-1292

IMA (2014) The new IMA list of minerals-A work in progress. Updated: March 2014. *http://pubsites.uws.edu.au/ima-cnmnc/IMA_Master_List_(2014-03).pdf*

Impellitteri CA (2005) Effects of pH and phosphate on metal distribution with emphasis on As speciation and mobilization in soils from a lead smelting site. Sci Total Environ 345:175-190

IPCS (1979) DDT and its derivatives. World Health Organization, International Programme on Chemical Safety (Environmental Health Criteria 9), Geneva

Jambor JL (1999) Nomenclature of the alunite supergroup. Can Mineral 37:1323-1341

Jambor JL, Dutrizac JE (1995) Solid solutions in the annabergite-erythrite-hörnesite synthetic system. Can Mineral 33:1063-1071

Jambor JL (2003) Mine-waste mineralogy and mineralogical perspectives of acid-base accounting. *In:* Environmental Aspects of Mine Wastes. Short Cource Series, Vol 31. Jambor JL, Blowes DW, Ritchie AIM (eds) Mineralogical Association of Canada, p 117-145

Jambor JL, Blowes DW (1998) Theory and applications of mineralogy in environmental studies of sulfide-bearing mine wastes. *In:* Modern Approaches to Ore and Environmental Mineralogy. Short Cource Series, Vol 27. Cabri LJ, Vaughan DJ (eds) Mineralogical Association of Canada, p 367-401

Jambor JL, Owens, DR, Grice, JD, Feinglos MN (1996) Gallobeudantite, $PbGa_3[(AsO_4),(SO_4)]_2(OH)_6$, a new mineral species from Tsumeb, Namibia, and associated new gallium analogues of the alunite-jarosite family. Can Mineral 34:1305-1315

Jambor JL, Blowes DW, Ptacek CJ (2000) Mineralogy of mine wastes and strategies for remediation. *In:* Environmental Mineralogy. EMU Notes in Mineralogy, Vol 2. Vaughan DJ, Wogelius RA (eds) European Mineralogical Union, p 197-252

Jamieson HE, Robinson C, Alpers CN, McCleskey RB, Nordstrom DK, Peterson RC (2005) Major and trace element composition of copiapite-group minerals and coexisting water from Richmond mine, Iron Mountain, California. Chem Geol 215:387-405

Jamieson HE (2014) The legacy of arsenic contamination from mining and processing refractory gold ore at Giant Mine, Yellowknife, Northwest Territories, Canada. Rev Mineral Geochem 79:533-551

Jiménez-Millán J, Velilla N (1994) Mineralogy and geochemistry of reduced manganese carbonate-silicate rocks from the Aracena area (Iberian Massif, SW Spain). Neues Jahrb Mineral Abh 166:193-209

Johnston RB, Singer PV (2007) Solubility of symplesite (ferrous arsenate): implications for reduced groundwaters and other environments. Soil Sci Soc Am J 71:101-107

Juillot F, Ildefonse Ph, Morin G, Calas G, Kersabiec AM, Benedetti M (1999) Remobilization of arsenic from buried wastes at an industrial site: mineralogical and geochemical control. Appl Geochem 14:1031-1048

Jurkovič Ľ, Hiller E, Veselská V, Peťková K (2011) Arsenic concentrations in soils impacted by dam failure of coal-ash pond in Zemianske Kostolany, Slovakia. Bull Environ Contam Toxicol 86:433-437

Kamenar B, Kaitner B (1973) The crystal structure of mercury(I) orthoarsenate. Acta Crystallogr B 29:1666-1669

Kampf AR (2005) The crystal structure of cobaltarthurite from the Bou Azzer district, Morocco: The location of hydrogen atoms in the arthurite structure-type. Can Mineral 43:1387-1391

Kampf AR (2009) Miguelromeroite, the Mn analogue of sainfeldite, and redefintion of villyaellenite as an ordered intermediate in the sainfeldite-miguelromeroite series. Am Mineral 94:1535-1540

Kampf AR, Dunn PJ (1987) Ogdensburgite from Mapimi and new data for the species. Am Mineral 72:409-412

Kampf AR, Ross CR II (1988) End-member villyaellenite from Mapimi, Durango, Mexico: Descriptive mineralogy, crystal structure, and implications for the ordering of Mn and Ca in the type villyaellenite. Am Mineral 73:1172-1178

Kampf AR, Shigley JE, Rossman GR (1984) New data on lotharmeyerite. Mineral Record 15:223-226

Kampf AR, Wise WS, Rossman GR (2000) Juanitaite: A new mineral from Gold Hill Utah. Mineral Record 31:301-305

Kang Y, Liu G, Chou CL, Wong MH, Zheng L, Ding R (2011) Arsenic in Chinese coals: Distribution, modes of occurrence, and environmental effects. Sci Total Environ 412-412:1-13

Kato A, Matsubara S, Nagashima K, Nakai I, Shimizu M (1984) Kaňkite from the Suzukura mine, Enzan city, Yamanashi Prefecture, Japan. Mineral J 12:6-14

Kato T, Miura Y (1977) The crystal structure of adamite and paradamite. Mineral J (Japan) 8:320-328

Kato T, Watanabe I (1991) The crystal structures of schallerite and friedelite. Yamaguchi University, College of Arts Bulletin 26:51-63

Keller P (1971) Darstellung und Eigenschaften von $Co_2[OH|AsO_4]$. Neues Jahrb Mineral Monatsh 560-564

Keller P (2001) Ekatite, $(Fe^{3+},Fe^{2+},Zn)_{12}(OH)_6[AsO_3]_6[AsO_3,HOSiO_3]_2$, a new mineral from Tsumeb, Namibia, and its crystal structure. Eur J Mineral 13:769-777

Keller P, Hess H (1978) Die Kristallstruktur von Arthurit, $CuFe^{3+}_2[(H_2O)_4|(OH)_2|(AsO_4)_2]$. Neues Jahrb Mineral Abh 133:291-302

Keller P, Hess H (1988) Die Kristallstrukturen von O'Danielit, $Na(Zn,Mg)_3H_2(AsO_4)_3$, und Johillerit, $Na(Mg,Zn)_3Cu(AsO_4)_3$. Neues Jahrb Mineral Monatsh 395-404

Keller P, Hess H, Dunn PJ (1979) Die Ladungsbilanz für eine verfeinerte Kristallstruktur von Stranskiit, $Zn_2Cu(AsO_4)_2$. Mineral Petrol 26:167-174

Keller P, Hess H, Riffel H (1980) Die Kristallstruktur von Koritnigit, $Zn[H_2O|HOAsO_3]$. Neues Jahrb Mineral Abh 138:316-332

Keller P, Riffel H, Hess H (1982) Die Kristallstruktur von Prosperit, $Ca_2Zn_4[H_2O|(AsO_4)_4]$. Z Kristallogr 158:33-42

Keller P, Lissner F, Schleid T (2003) The crystal structure of arsendescloizite, $PbZn(OH)[AsO_4]$, from Tsumeb (Namibia). Neues Jahrb Mineral Monatsh 374-384 }

Keller P, Lissner F, Schleid T (2004) The crystal structures of zincroselite and gaitite: Two natural polymorphs of $Ca_2Zn[AsO_4]_2 \cdot 2H_2O$ from Tsumeb, Namibia. Eur J Mineral 16:353-359

Kharisun, Taylor MR, Bevan DJM, Rae AD, Pring A (1997) The crystal structure of mawbyite, $PbFe_2(AsO_4)_2(OH)_2$. Mineral Mag 61:685-691

Kharisun, Taylor MR, Bevan DJM (1998) The crystal chemistry of duftite, $PbCuAsO_4(OH)$ and the β-duftite problem. Mineral Mag 62:121-130

Kierczak J, Bril H, Neel C, Puziewicz J (2010) Pyrometallurgical slags in Upper and Lower Silesia (Poland): from environmental risks to use of slag-based products – a review. Arch Environ Prot 36:111-126

Klika Z, Martinec P (2013) Czech Republic coal fires and waste piles. In: Coal and Peat Fires: A Global Perspective, Stracher G, Prakash A, Sokol EV (eds), Elsevier, p 80-114

Kitahama K, Kiriyama R, Baba Y (1975) Refinement of the crystal structure of scorodite. Acta Crystallogr B 31:322-324

Klaska R, Gebert W (1982) Polytypie und struktur von gebhardit - $Pb_8OCl_6(As_2O_5)_2$. Z Kristallogr Kristallgeom Kristallphys Kristallchem 159:75-76

Klein C (2002) The Manual of Mineral Science. 22nd edition. Wiley

Klika Z, Martinec P (2013) Czech Republic coal fires and waste piles. In: Coal and Peat Fires: A Global Perspective, Stracher G, Prakash A, Sokol EV (eds), Elsevier, p 80-114

Klinck BA, Palumbo P, Cave M, Wragg J (2005) Arsenic dispersal and bioaccessibility in mine contaminated soil: a case study from an abandoned arsenic mine in Devon, UK. British Geological Survey Research Report, RR/04/03

Kocourková E, Sracek O, Houzar S, Cempírek J, Losos Z, Filip J, Hršelová P (2011) Geochemical and mineralogical control on the mobility of arsenic in a waste rock pile at Dlouhá Ves, Czech Republic. J Geochem Explor 110:61-73

Kokkoros P (1938) Ueber die Struktur des Durangit $NaAlF(AsO_4)$. Z Kristallogr Kristallgeom Kristallphys Kristallchem 99:38-49

Kolesova RV, Fesenko EG (1968) Determination of the crystal structure of liroconite, $Cu_2Al(AsO_4)(OH)_4(H_2O)_4$. Kristallografiya 13:396-402

Kolitsch U (2001) Redetermination of the mixed-valence manganese arsenate flinkite, $Mn^{II}_2Mn^{III}(OH)_4(AsO)_4$. Acta Crystallogr E 57:i115-i118

Kolitsch U, Fleck M (2006) Third update on compounds with kröhnkite-type chains: the crystal structure of wendwilsonite $[Ca_2Mg(AsO_4)_2 \cdot 2H_2O]$ and the new triclinic structure types of synthetic $AgSc(CrO_4)_2 \cdot 2H_2O$ and $M_2Cu(Cr_2O_7)_2 \cdot 2H_2O$ (M = Rb, Cs). Eur J Mineral 18:471-482

Kolitsch U, Holtstam D (2004) Crystal chemistry of REEXO$_4$ compounds (X = P, As, V). II. Review of REEXO$_4$ compounds and their stability fields. Eur J Mineral 16:117-126

Kolitsch U, Slade PG, Tiekink ERT, Pring A (1999) The structure of antimonian dussertite and the role of antimony in oxysalt minerals. Mineral Mag 63:17-26

Kolitsch U, Holtstam D, Gatedal K (2004) Crystal chemistry of REEXO$_4$ compounds (X = P, As, V). I. Paragenesis and crystal structure of phosphatian gasparite-(Ce) from the Kesebol Mn-Fe-Cu deposit, Västra Götaland, Sweden. Eur J Mineral 16:111-116

Kolitsch U, Atencio D, Chukanov NV, Zubkova NV, Menezes Filho LAD, Coutinho JMV, Birch WD, Schlüter J, Pohl D, Kampf AR, Steele IM, Favreau G, Nasdala L, Möckel S, Giester G, Pushcharovsky DYu (2010) Bendadaite, a new iron arsenate mineral of the arthurite group. Mineral Mag 74:469-486

Kolker A (2012) Minor element distribution in iron disulfides in coal: A geochemical review. Int J Coal Geol 94:32-43

Kolker A, Finkelman RB (1998) Potentially hazardous elements in coal: Modes of occurrence and summary of concentration data for coal components. Int J Coal Prep Util 19:133-157

Kolker A, Huggins FE (2007) Progressive oxidation of pyrite in five bituminous coal samples. An As XANES and Fe-57 Mossbauer spectroscopic study. Appl Geochem 22:778-787

Kolker A, Nordstrom DK (2001) Occurrence and micro-distribution of arsenic in pyrite. Unpublished USGS report

Kolker A, Huggins FE, Palmer CA, Shah N, Crowley SS, Huffman GP, Finkelman RB (2000) Mode of occurrence of arsenic in four US coals. Fuel Process Technol 63:167-178

Kořínek J (1675) The Old Memories of Kutná Hora. (in Czech)

Kotelnikov AR, Ananiev VV, Kovalsky AM, Suk NI (2011) Synthesis of phosphorus- and arsenic-bearing framework silicates similar to feldspar. Vestnik Otdelenia Nauk o Zemle RAN 3, doi: 10.2205/2011NZ000177

Kratochvíl (1960) Topographic mineralogy of Czech Republic III. ČSAV, Praha. (in Czech)

Krause W, Bernhardt H-J, Gebert W, Graetsch H, Belendorf K, Petitjean K (1996) Medenbachite, Bi$_2$Fe(Cu,Fe)(O,OH)$_2$(OH)$_2$(AsO$_4$)$_2$. Am Mineral 81:505-512

Krause W, Belendorff K, Bernhardt H-J, McCammon C, Effenberger H, Mikenda W (1998) Crystal chemistry of the tsumcorite-group minerals. New data on ferrilotharmeyerite, tsumcorite, thometzekite, mounanaite, helmutwinklerite, and a redefinition of gartrellite. Eur J Mineral 10:179-206

Krause W, Effenberger H, Bernhardt H-J, Martin M (1999) Cobaltlotharmeyerite, Ca(Co,Fe,Ni)$_2$(AsO$_4$)$_2$(OH,H$_2$O)$_2$, a new mineral from Schneeberg, Germany. Neues Jahrb Mineral Monatsh 505-517

Krause W, Bernhardt H-J, Effenberger H, Martin M (2001a) Cobalttsumcorite and nickellotharmeyerite, two new minerals from Schneeberg, Germany: Description and crystal structure. Neues Jahrb Mineral Monatsh 558-576

Krause W, Blass G, Bernhardt H-J, Effenberger H (2001b) Lukrahnite, CaCuFe^{3+}(AsO$_4$)$_2$[(H$_2$O)(OH)], the calcium analogue of gartrellite. Neues Jahrb Mineral Monatsh 481-492

Krause W, Bernhardt H-J, Effenberger H, Witzke T (2002a) Schneebergite and nickelschneebergite from Schneeberg, Saxony, Germany: The first Bi-bearing members of the tsumcorite group. Eur J Mineral 14:115-126

Krause W, Bernhardt H-J, McCammon C, Effenberger H (2002b) Neustädtelite and cobaltneustädtelite, the Fe^{3+}- and Co^{2+}-analogues of medenbachite. Am Mineral 87:726-738

Krause W, Bernhardt H-J, Effenberger H, Kolitsch U, Lengauer Ch (2003) Redefinition of arhbarite, Cu$_2$Mg(AsO$_4$)(OH)$_3$. Mineral Mag 67:1099-1107

Krause W, Bernhardt H-J, Effenberger H (2006) Schlegelite, Bi$_7$O$_4$(MoO$_4$)$_2$(AsO$_4$)$_3$, a new mineral from Schneeberg, Saxony, Germany. Eur J Mineral 18:803-811

Krivovichev SV, Molchanov AV, Filatov SK (2000) Crystal structure of urusovite Cu[AsAsO$_5$]: A new type of tetrahedral aluminoarsenate polyanion. Crystallography Reports (translated from Kristallografiya) 45:793-797

Krivovichev SV, Filatov SK, Burns PC (2001) The Jahn-Teller distortion of copper coordination polyhedra in the alluaudite structural type. Crystal structure of bradaczekite, NaCu$_4$(AsO$_4$)$_3$. Zapiski Vserossiiskogo Mineralogicheskogo Obshchestva 130:1-8 (in Russian)

Krivovichev SV, Chernyshov DYu, Döbelin N, Armbruster T, Kahlenberg V, Kaindl R, Ferraris G, Tessadri R, Kaltenhauser G (2006) Crystal chemistry and polytypism of tyrolite. Am Mineral 91:1378-1384

Krivovichev SV, Plášil J (2013) Mineralogy and crystallographuy of uranium. *In*: Uranium: Cradle to Grave. Burns PC, Sigmon GE (eds) Mineralogical Association of Canada Short Course Series 43:15-120

Krzemnicki M (1997) Mineralogical investigations on hydrothermal As- and REE-bearing minerals within the gneisses of the Monte Leone nappe (Binntal region, Switzerland). Unpublished PhD thesis, University of Basel

Krzemnicki M, Gieré R (1996) As-REE-bearing titanite from the Monte Leone nappe (Binntal, Switzerland). Schweiz Miner Petrog 76:117-118

Kucha H, Martens A, Ottenburgs R, De Vos W, Viaene W (1996) Primary minerals of Zn-Pb mining and metallurgical dumps and their environmental behavior at Plombières, Belgium. Environ Geol 27:1-15

Lacy WC, Hosmer HL (1956) Hydrothermal leaching in central Peru. Econ Geol 51:69-79

Lai L, Li Y, Shi N (1997) Discovery and study of parnauite. Yanshi Kuangwuxue Zazhi 16:50-55

Lalou C, Brichet E, Jehanno C, Perez-Leclaire H (1983) Hydrothermal manganese oxide deposits from Galapagos mounds, DSDP Leg 70, hole 509B, and "Alvin" dives 729 and 721. Earth Planet Sci Lett 63:63-75

Lam AE, Groat LA, Ercit TS (1998) The crystal structure of dugganite, $Pb_3Zn_3Te^{6+}As_2O_{14}$. Can Mineral 36:823-830

Langmuir D, Mahoney J, MacDonald A, Rowson J (1999) Predicting arsenic concentrations in the porewaters of buried uranium mill tailing. Geochim Cosmochim Acta 63:3379-3394

Lapham MD, Barnes JH, Downey WF Jr, Finkelman RB (1980) Mineralogy associated with burning anthracite deposits of eastern Pennsylvania. Miner Resour Rep 78, Harrisburg, p. 82

Lattanzi P, DaPelo S, Musu E, Atzei D, Elsener B, Fantauzzi M, Rossi A (2008) Enargite oxidation: A review. Earth-Sci Rev 86:62-88

Lazaridis G, Melos V, Papadopoulou L (2011) The first cave occurrence of orpiment (As_2S_3) from the sulfuric acid caves of Aghia Paraskevi (Kassandra Peninsula, N. Greece). Int J Speleol 40:133-139

Lee P, Kang MJ, Choi SH, Touray JC (2005) Sulfide oxidation and the natural attenuation of arsenic and trace metals in the waste rocks of the abandoned Seobo tungsten mine, Korea. Appl Geochem 20:1687-1703

Leist M, Casey RJ, Caridi D (2000) The management of arsenic wastes: problems and prospects. J Hazard Mater 7:125-138

Lengauer CL, Giester G, Kirchner E (2004) Leogangite, $Cu_{10}(AsO_4)_4(SO_4)(OH)_6 \cdot 8H_2O$, a new mineral from the Loegang mining district, Salzburg province, Austria. Miner Petrol 81:187-201

Liu GJ, Zheng LG, Zhang Y, Qi CC, Yiwei CW, Peng ZC (2007) Distribution and mode of occurrence of As, Hg and Se and sulfur in coal Seam 3 of the Shanxi Formation, Yanzhou Coalfield, China. Int J Coal Geol 71:371-385

Loewenstein W (1954) The distribution of aluminium in the tetrahedra of silicates and aluminates. Am Mineral 39:92-96

Lohmueller G, Schmidt G, Deppisch B, Gramlich V, Scheringer C (1973) Kristallstrukturen von yttrium vanadat, lutetium phosphat und lutetium arsenat. Acta Crystallogr B 29:141-142

Lottermoser BG (2002) Mobilization of heavy metals from historical smelting slag dumps, north Queensland, Australia. Mineral Mag 66:475-490

Lottermoser BG (2003) Mine Wastes. Characterization, Treatment, Environmental Impacts. 2nd ed. Springer

Lottermoser BG (2005) Evaporative mineral precipitates from a historical smelting slag dump, Río Tinto, Spain. Neues Jahrb Mineral Abh 181:183-190

Lottermoser BG, Ashley PM (2006) Mobility and retention of trace elements in hardpan-cemented cassiterite tailings, north Queensland, Australia. Environ Geol 50:835-846

Lugli A, Clemenza M, Corso PE, di Costanzo J, Dirnhofer R, Fiorini E, Herborg C, Hindmarsh JT, Orvini E, Piazzoli A, Previtali E, Santagostino A, Sonnenberg A, Genta RM (2011) The medical mystery of Napoleon Bonaparte: An interdisciplinary exposé. Adv Anat Pathol 18:152-158

Lundström I (1999) General geology of the Bergslagen ore region. *In:* Långban. The Mines, Their Minerals, Geology and Explorers. Holtstam D, Langhof J (ed) Christian Weise Verlag, p 19-27

Luo Y, Giammar DE, Huhmann BL, Catalano JG (2011) Speciation of selenium, arsenic, and zinc in class C fly ash. Energy Fuel 25:2980-2987

Machatschki F (1936) Die Kristallstruktur von Tiefquarz SiO_2 und Aluminiumorthoarsenat $AlAsO_4$. Z Kristallogr 94:222-230

Macklin MG, Brewer PA, Hudson-Edwards KA, Bird G, Coulthard TJ, Dennis I, Lechler PJ, Miller JR, Turner JN (2006) A geomorphological approach to the management of rivers contaminated by metal mining. Geomorphology 79:423-447

Mahoney J, Slaughter M, Langmuir D, Rowson J (2007) Control of As and Ni releases from a uranium mill tailings neutralization circuit: Solution chemistry, mineralogy and geochemical modeling of laboratory study results. Appl Geochem 22:2758-2776

Mains D, Craw D (2005) Compositions and mineralogy of historic gold processing residues, east Otago, New Zealand. New Zeal J Geol Geop 48:641-647

Majzlan J, Lalinská B, Chovan M, Jurkovič L, Milovská S, Göttlicher J (2007) The formation, structure, and ageing of As-rich hydrous ferric oxide at the abandoned Sb deposit Pezinok (Slovakia). Geochim Cosmochim Acta 71:4206-4220

Majzlan J, Drahota P, Filippi M, Grevel KD, Kahl WA, Plášil J, Boerio-Goates J, Woodfield BF (2012a) Thermodynamic properties of scorodite and parascorodite, ($FeAsO_4 \cdot 2H_2O$), kaňkite ($FeAsO_4 \cdot 3.5H_2O$), and $FeAsO_4$. Hydrometallurgy 117-118:47-56

Majzlan J, Lazic B, Armbruster T, Johnson MB, White MA, Fisher RA, Plášil J, Loun J, Škoda R, Novák M (2012b) Crystal structure, thermodynamic properties, and paragenesis of bukovskýite, $Fe_2(AsO_4)(SO_4)(OH) \cdot 9H_2O$). J Mineral Petrol Sci 107:133-148

Majzlan J, Plášil J, Škoda R, Gescher J, Kögler F, Rusznyak A, Küsel K, Neu TR, Mangold S, Rothe J (2014) Arsenical acid mine drainage with colossal arsenic concentration: mineralogy, geochemistry, microbiology. Environ Sci Technol (*submitted*)

Makovicky E (1989) Modular classification of sulphosalts - current status. Definition and application of homologous series. Neues Jahrb Mineral Abh 160:269-297

Makovicky E (1997) Modular crystal chemistry of sulphosalts and other complex sulphides. EMU Notes in Mineralogy 1:237-271

Makovicky E (2006) Crystal structures of sulfides and other chalcogenides. Rev Mineral Geochem 61:7-125

Manasse A, Mellini M (2002) Chemical and textural characterization of medieval slags from the Massa Marritima smelting sites. J Cult Heritage 3:187-198

Mandal BK, Suzuki KT (2002) Arsenic round the world: a review. Talanta 58:201-235

Márquez M, Gaspar J, Bessler KE, Magela G (2006) Process mineralogy of bacterial oxidized gold ore in São Bento Mine (Brasil). Hydrometallurgy 83:114-123

Marsh RE, Schomaker V (1979) Some incorrect space groups in Inorg Chem, Volume 16. Inorg Chem 18:2331-2336

Masalehdani MNN, Paquette Y, Bouchardon JL, Guy B, Stracher GB, Chalier J (2007) Vapor deposition of arsenic-bearing minerals originating from a burning culm bank: Saint-Etienne, the Loire region, France. Geol Soc Am, Abstr Programs 39:298

Mason B, Moore CB (1982) Principles of Geochemistry. John Wiley and Sons

Matschullat J (2000) Arsenic in the geosphere – a review. Sci Total Environ 249:297-312

Matsubara S, Miyawaki R, Mouri T, Kitamine M (2000) Sterlinghillite, a rare manganese arsenate, from the Gozaisho mine, Fukushima Prefecture, Japan. Bull. National Sci. Museum, Tokyo, Ser. C, 26:1-7

Mayes WM, Jarvis AP, Burke IT, Walton M, Feigl V, Klebercz O, Gruiz K (2011) Dispersal and attenuation of trace contaminants downstream of the Ajka Bauxite Residue (Red Mud) Depository Failure, Hungary. Environ Sci Technol 45:5147-5155

McGregor RG, Blowes DW (2002) The physical, chemical and mineralogical properties of three cemented layers within sulfide-bearing mine tailings. J Geochem Explor 76:195-207

McKenzie EJ, Brown KL, Cady SL, Campbell KA (2001) Trace metal chemistry and silicification of microorganisms in geothermal sinter, Taupo Volcanic Zone, New Zealand. Geothermics 30:483-502

McLean WJ, Anthony JW, Finney JJ, Laughon RB (1971) The crystal structure of legrandite. Am Mineral 56:1147-1154

Medenbach O, Schmetzer K, Abraham K (1988) Fahleite from Tsumeb/Namibia, a new mineral belonging to the smolianinovite group. Neues Jahrb Mineral Monatsh 167-171

Meisser N, Brugger J (2006) Bouazzerit und Maghrebit, zwei neue Arsenatmineralien aus dem Revier Bou Azzer, Marokko. Lapis 31:69-71

Mereiter K, Preisinger A (1986) Kristallstrukturdaten der wismutminerale atelestit, mixit und pucherit. Österreische Akademie der Wissenschaften, Mathematich-Naturwissenschaftliche Klasse, Sitzungsberichte 123:79-81

Meunier L, Walker SR, Wragg J, Parsons MB, Koch I, Jamieson HE, Reimer KJ (2010) Effects of soil composition and mineralogy on the bioaccessibility of arsenic from tailings and soil in gold mine districts of Nova Scotia. Environ Sci Technol 44:2667-2674

Mihaljevič M, Ettler V, Šebek M, Drahota P, Strnad L, Procházka R, Zeman J, Šráček O (2010) Alteration of arsenopyrite in soil under different vegetation covers. Sci Total Environ 408:1286-1294

Miletich R, Zemann J, Nowak M (1997) Reversible hydration in synthetic mixite, $BiCu_6(OH)_6(AsO_4)_3 \cdot nH_2O$ ($n \leq 3$): Hydration kinetics and crystal chemistry. Phys Chem Miner 24:411-422

Mills SJ, Kampf AR, Poirier G, Raudsepp M, Steele IM (2010) Auriacusite, $Fe^{3+} Cu^{2+}AsO_4O$, the first M^{3+} member of the olivenite group, from the Black Pine mine, Montana, USA. Miner Petrol 99:113-120

Minceva-Stefanova J (1968) Strashimirite – novyj vodnyj arsenat medi. Zap Vses Mineral Ob-Va 97:470-477 (in Russian)

Moldovan BJ, Hendry MJ (2005) Characterizing and quantifying controls on arsenic solubility over a pH range of 1-11 in a uranium mill-scale experiment. Environ Sci Technol 39:4913-4920

Moldovan BJ, Jiang DT, Hendry J (2003) Mineralogical characterization of arsenic in uranium mine tailings precipitated from iron-rich hydrometallurgical solutions. Environ Sci Technol 37:873-879

Mooney RCL (1948) Crystal structure of tetragonal bismuth arsenate, $BiAsO_4$. Acta Crystallogr 1:163-165

Moore PB (1967a) Gabrielsonite, $PbFe(AsO_4)(OH)$, a new member of the descloizite-pyrobelonite group, from Långban. Arkiv Mineral Geol 4:401-407

Moore PB (1967b) Crystal chemistry of the basic manganese arsenate minerals 1. The crystal structures of flinkite, $Mn_2^{2+}Mn^{3+}(OH)_4(AsO_4)$ and retzian, $Mn_2^{2+}Y^{3+}(OH)_4(AsO_4)$. Am Mineral 52:1603-1613

Moore PB (1968a) Crystal chemistry of the basic manganese arsenate minerals: II. The crystal structure of allactite. Am Mineral 53:733-741

Moore PB (1968b) The crystal structure of chlorophoenicite. Am Mineral 53:1110-1119

Moore PB (1970) Crystal chemistry of the basic manganese arsenates: IV. Mixed arsenic valences in the crystal structure of synadelphite. Am Mineral 55:2023-2037

Moore PB (1992) Betpakdalite unmasked, and a comment on bond valences. Aust J Chem 45:1335-1354

Moore PB, Araki T (1976) Derbylite, $Fe^{3+}_4Ti^{4+}_3Sb^{3+}O_{13}(OH)$, a novel close-packed oxide structure. Neues Jahrb Mineral Abh 126:292-303

Moore PB, Araki T (1977a) Holdenite, a novel cubic close-packed structure. Am Mineral 62:513-521

Moore PB, Araki T (1977b) Mitridatite, $Ca_6(H_2O)_6[Fe^{III}_9O_6(PO_4)_9]\cdot3H_2O$. A noteworthy octahedral sheet structure. Inorg Chem 16:1096-1106

Moore PB, Araki T (1977c) Parwelite, $Mn^{II}_{10}Sb^V_2As^V_2Si_2O_{24}$, a complex anion-deficient fluorite derivative structure. Inorg Chem 16:1839-1847

Moore PB, Araki T (1978a) Hematolite: A complex dense-packed sheet structure. Am Mineral 63:150-159

Moore PB, Araki T (1978b) Angelellite, $Fe_4^{3+}O_3(As^{5+}O_4)_2$: a novel cubic close-packed oxide structure. Neues Jahrb Mineral Abh 132:91-100

Moore PB, Araki T (1979a) Armangite, $Mn^{2+}_{26}[As^{3+}_6(OH)_4O_{14}][As^{3+}_6O_{18}]_2[CO_3]$, a fluorite derivative structure. Am Mineral 64:748-757

Moore PB, Araki T (1979b) Magnussonite, manganese arsenite, a fluorite derivative structure. Am Mineral 64:390-401

Moore PB, Gupta PKS, Schlemper EO (1989) Akrochordite, $(Mn,Mg)_5(OH)_4(H_2O)_4(AsO_4)_2$: A sheet structure with amphibole walls. Am Mineral 74:256-262

Moore PB, Ito J (1974) Isotypy of robertsite, mitridatite, and arseniosiderite. Am Mineral 59:48-59

Moore PB, Molin-Case J (1971) Crystal chemistry of the basic manganese arsenates: V. Mixed manganese coordination in the atomic arrangement of arsenoclasite. Am Mineral 56:1539-1552

Moore PB, Smyth JR (1968) Crystal chemistry of the basic manganese arsenates: III. The crystal structure of eveite, $Mn_2(OH)(AsO_4)$. Am Mineral 53:1841-1845

Morales A, Cruells M, Roca A, Bergó R (2010) Treatment of copper flash smelter flue dusts for copper and zinc extraction and arsenic stabilization. Hydrometallurgy 105:148-154

Mori H, Ito T (1950) The structure of vivianite and symplesite. Acta Crystallogr 3:1-6

Morin G, Calas G (2006) Arsenic in soils, mine tailings, and former industrial sites. Elements 2:97-101

Morin G, Lecocq D, Juillot F, Calas G, Ildefonse Ph, Beline S, Briois V, Dillmann Ph, Chevallier Ch, Gauthier Ch, Sole A, Petit P-E, Borensztajn S (2002) EXAFS evidence of sorbed arsenic(V) and pharmacosiderite in a soil overlying the Echassières geochemical anomaly, Allier, France. B Soc Geol Fr 173:281–291

Morin G, Juillot F, Casiot C, Bruneel O, Personné JC, Elbaz-Poulichet F, Leblanc M, Ildefonse Ph, Calas G (2003) Bacterial formation of tooleite and mixed arsenic(III) or arsenic(V)-iron(III) gels in the Carnoulès Acid Mine Drainage, France. A XANES, XRD, and SEM Study. Environ Sci Technol 37:1705-1712

Morin G, Rousse G, Elkaim E (2007) Crystal structure of tooeleite, $Fe_6(AsO_3)_4SO_4(OH)_4\cdot4H_2O$, a new iron arsenite oxyhydroxysulfate mineral relevant to acid mine drainage. Am Mineral 92:193-197

Moura MA, Botelho NF, de Mendonça FC (2007) The indium-rich sulfides and rare arsenates of the Sn-In-mineralized Mangabeira A-type granite, central Brazil. Can Mineral 45:485-496

Murad E, Rojík P (2005) Iron mineralogy of mine-drainage precipitates as environmental indicators: review of current concepts and a case study from the Sokolov Basin, Czech Republic. Clay Miner 40:427-440

Murciego A, Álvarez-Ayuso E, Pellitero E, Rodríguez MA, García-Sánchez A, Tamayo A, Rubio J, Rubin J (2011) Study of arsenopyrite weathering products in mine wastes from abandoned tungsten and tin exploitations. J Hazard Mater 186:590-601

Mustard R, Ulrich T, Kamenetsky VS, Mernagh T (2006) Gold and metal enrichment in natural granitic melts during fractional crystallization. Geology 34:85-88

Navarro A, Cardellach E, Mendoza JL, Corbella M, Domènech LM (2008) Metal mobilization from base-metal smelting slag dumps in Sierra Almagrera (Almería, Spain). Appl Geochem 23:895-913

Nesbitt HW, Muir LJ, Pratt AR (1995) Oxidation of arsenopyrite by air and air-saturated, distilled water and implications for mechanisms of oxidation. Geochim Cosmochim Acta 59:1773-1786

Niazi NK, Singh B, Shah P (2011) Arsenic speciation and phytoavailability in contaminated soils using a sequential extraction procedure and XANES spectroscopy. Environ Sci Technol 45:7135-7142

Nieto JM, Capitán MÁ, Sáez R, Almodóvar GR (2003) Beudantite: a natural sink for as and Pb in sulphide oxidation processes. Appl Earth Sci Trans IMM B 112:293-296

Nishikawa O, Okrugin V, Belkova N, Saji I, Shiraki K, Tazaki K (2006) Crystal symmetry and chemical composition of yukonite: TEM study of specimens collected from Nalychevskie hot spring, Kamchatka, Russia and from Venus Mine, Yukon Territory, Canada. Mineral Mag 70:73-81

Nordstrom DK (2011) Mine waters: acidic to circumneutral. Elements 7:383-398

Nordstrom DK, Alpers CN (1999) Negative pH, efflorescent mineralogy, and consequences for environmental restoration at the Iron Mountain Superfund site, California. Proc Natl Acad Sci USA 96:3455-3462

Nordstrom DK, Archer DG (2003) Arsenic thermodynamic data and environmental geochemistry. *In:* Arsenic in Groundwaters: Geochemistry and Occurrences. Welch A, Stollenwerk K (eds) Kluwer Academic Publishers, Boston, p 1-25

Nordstrom DK, Majzlan J, Königsberger E (2014) Thermodynamic properties for arsenic minerals and aqueous species. Rev Mineral Geochem 79:217-255

Novák F, Povondra P, Vtělenský J (1967) Bukovskýite, $Fe_2^{3+}(AsO_4)(SO_4)(OH)\cdot7H_2O$, from Kaňk, near Kutná Hora – a new mineral. Acta Univ Carol Geol 4:297-325

Nriagu JO, Pacyna JM (1988) Quantitative assessment of worldwide contamination of air, water and soils by trace metals. Nature 333:134-139

O'Day PA (2006) Chemistry and mineralogy of arsenic. Elements 2:77-83

O'Day PA, Vlassopoulos D, Root R, Rivera N (2004) The influence of sulfur and iron on dissolved arsenic concentrations in the shallow subsurface under changing redox conditions. Proc Natl Acad Sci USA 101:13703-3708

Olmi F, Sabelli C (1995) Carminite from three localities of Sardinia (Italy). Crystal structure refinements. Neues Jahrb Mineral Monatsh 553-562

Olmi F, Sabelli C, Trosti-Ferroni R (1995) The crystal structure of sabelliite. Eur J Mineral 7:1331-1337

Onac BP, Forti P (2011a) Minerogenetic mechanisms occurring in the cave environment: an overview. Int J Speleol 40:79-98

Onac BP, Forti P (2011b) State of the art and challenges in cave minerals studies. Studia UBB Geologia 56:33-42

Onac BP, Hess JW, White WB (2007) The relationship between the mineral composition of speleothems and mineralization of breccia pipes: Evidence from Corkscrew Cave, Arizona, USA. Can Mineral 45:1177-1188

Ondrejka M, Uher P, Pršek J, Ozdín D (2007) Arsenian monazite-(Ce) and xenotime-(Y), REE arsenates and carbonates from the Tisovec-Rejkovo rhyolite, Western Carpathians, Slovakia: Composition and substitutions in the (REE,Y)XO$_4$ system (X = P, As, Si, Nb, S). Lithos 95:116-129

Ondruš P, Skála R, Císařová I, Veselovský F, Frýda J, Čejka J (2002) Description and crystal structure of vajdakite, [(Mo^{6+}O$_2$)$_2$(H$_2$O)$_2$As$^{3+}_2$O$_5$]·H$_2$O – A new mineral from Jáchymov, Czech Republic. Am Mineral 87:983-990

Ondruš P, Skála R, Viti C, Veselovský F, Novák F, Jansa J (1999) Parascorodite, FeAsO$_4$·2H$_2$O–a new mineral from Kaňk near Kutná Hora, Czech Republic. Am Miner 84:1439-1444

Ondruš P, Veselovský F, Hloušek J (1997a) A review of mineral associations and paragenetic groups of secondary minerals of the Jáchymov (Joachimsthal) ore district. J Czech Geol Soc 42 (4):109-114

Ondruš P, Veselovský F, Hloušek J, Skála R, Vavřín I, Frýda J, Čejka J, Gabašová A (1997b) Secondary minerals of the Jáchymov (Joachimsthal) ore district. J Czech Geol Soc 42(4):3-76

Ondruš P, Veselovský F, Skála R, Císařová I, Hloušek J, Frýda J, Vavřín I, Čejka J, Gabašová (1997c) New naturally occurring phases of secondary origin from Jáchymov (Joachimsthal). J Czech Geol Soc 42(4):77-108

Ondruš P, Veselovský F, Skála R, Sejkora J, Pažout R, Frýda J, Gabašová A, Vajdak J (2006) Lemanskiite, NaCaCu$_5$(AsO$_4$)$_4$Cl·5H$_2$O, a new mineral species from the Abundancia Mine, Chile, Can Mineral 44:523-531

Orisakwe OE, Asomugha R, Afonne OJ, Anisi CN, Obi E, Dioka CE (2004) Impact of effluents from a car battery manufacturing plant in Nigeria on water, soil, and food qualities. Arch Environ Health 59:31-36

Oyarzún R, Lillo J, Higueras P, Oyarzún J, Maturana H (2004) Strong arsenic enrichment in sediments from the Elqui watershed, Northern Chile: industrial (gold mining at El Indio-Tambo district) vs. geologic processes. J Geochem Explor 84:53-64

Paktunc D, Dutrizac J, Gertsman V (2008) Synthesis and phase transformations involving scorodite, ferric arsenate and arsenical ferrihydrite: Implications for arsenic mobility. Geochim Cosmochim Acta 72:2649-2672

Paktunc D, Foster A, Heald S, Laflamme G (2004) Speciation and characterization of arsenic in gold ores and cyanidation tailings using X-ray absorption spectroscopy. Geochim Cosmochim Acta 68:969-983

Paktunc D, Foster A, Laflamme G (2003) Speciation and characterization of arsenic in Ketza River Mine tailings using X-ray absorption spectroscopy. Environ Sci Technol 37:2067-2074

Paktunc D, Kingston D, Pratt A, McMullen J (2006) Distribution of gold in pyrite and in products of its transformation resulting from roasting of refractory gold ore. Can Mineral 44:213-227

Palmer AN (2007) Cave Geology. Allen Press, Lawrence Kansas, U.S.A

Palmer CA, Krasnow MR, Finkelman RB, D'Angelo WM (1993) An evaluation of leaching to determine modes of occurrence of selected toxic elements in coal. Int J Coal Geol 12:135-141

Pantuzzo FL, Ciminelli VST (2010) Arsenic association and stability in long-term disposed arsenic residues. Water Res 44:5631-5640

Papangelakis VG, Demopoulos GP (1990) Acid pressure oxidation of arsenopyrite: part I, reaction chemistry. Can Metall Quart 29:1-11

Parsons MB, Bird DK, Einaudi MT, Alpers CN (2001) Geochemical and mineralogical controls on trace element release from the Penn Mine base-metal slag dump, California. Appl Geochem 16:1567-1593

Parviainen A, Lindsay MBJ, Perez-Lopez R, Gibson BD, Ptacek CJ, Blowes DW, Loukola-Ruskeeniemi K (2012) Arsenic attenuation in tailings at a former Cu-W-As mine, SW Finland. Appl Geochem 27:2289-2299

Pasero M, Kampf AR, Ferraris C, Pekov IV, Rakovan J, White TJ (2010) Nomenclature of the apatite supergroup minerals. Eur J Mineral 22:163-179

Pasero M, Vacchiano D (2000) Crystal structure and revision of the chemical formula of georgiadesite, Pb$_4$(AsO$_3$)Cl$_4$(OH). Mineral Mag 64:879-884

Patinha C, Silva EF, Fonseca EC (2004) Mobilisation of arsenic at the Talhadas old mining area-Central Portugal. J Geochem Explor 84:167-180

Pažout R (2004) New finds of secondary minerals in the Kutná Hora ore district: Valentinite and brochantite from the Gruntecko-Hloušecké zone, kaňkite from Turkaňk zone. Bull Mineral-Petrol Odd Nár Muz Praha 12:155-158 (in Czech)

Peacor DR, Dunn PJ (1982) Petersite, a REE and phosphate analog of mixite. Am Mineral 67:1039-1042

Peacor DR, Dunn PJ, Ramik RA, Sturman BD, Zeihen LG (1985) Philipsburgite, a new copper zinc arsenate hydrate related to kipushite, from Montana. Can Mineral 23:255-258

Peacor DR, Dunn PJ, Simmons WB, Wicks FJ (1986) Arsenites related to layer silicates: Manganarsite, the arsenite analogue of manganpyrosmalite, and unnamed analogues of friedelite and schallerite from Långban, Sweden. Am Mineral 71:1517-1521

Peacor DR, Dunn PJ (1985) Sodium-pharmacosiderite, a new analog of pharmacosiderite from Australia and new occurrences of barium-pharmacosiderite. Mineral Record 16:121-124

Perchiazzi N, Ondruš P, Skála R (2004) Ab initio X-ray powder structure determination of parascorodite, Fe(H$_2$O)$_2$AsO$_4$. Eur J Mineral 16:1003-1007

Percival JB, Kwong YTJ, Dumaresq CG, Michel FA (2004) Transport and attenuation of arsenic, cobalt and nickel in an alkaline environment (Cobalt, Ontario). Geological Survey of Canada, Open File 1680, p 1-30

Perseil EA, Latouche L (1989) Découverte de microstructures de nodules polymétalliques dans les minéralisations manganésifères métamorphiques de Fallota et de Parsettens (Grisons-Suisse). Mineral Deposita 24:111-116

Pertlik F (1975) Verfeinerung der kristallstruktur von synthetischem trippkeit, CuAs$_2$O$_4$. Mineral Petrol 22:211-217

Pertlik F (1978) The crystal structure of trigonite, Pb$_3$Mn(AsO$_3$)$_2$(AsO$_2$OH). Mineral Petrol 25:95-105

Pertlik F (1987) The structure of freedite, Pb$_8$Cu(AsO$_3$)$_2$O$_3$Cl$_5$. Miner Petrol 36:85-92

Pertlik F, Schnorrer G (1993) A re-appraisal of the chemical formula of nealite, Pb$_4$Fe(AsO$_3$)$_2$Cl$_4$·2H$_2$O, on the basis of a crystal structure determination. Miner Petrol 48:193-200

Peryea FJ (1998) Historical use of lead arsenate insecticides, resulting soil contamination and implications for soil remediation. 16th World Congress of Soil Science, Montpellier, France

Petrunic BM, Al TA (2005) Mineral/water interactions in tailings from a tungsten mine, Mount Pleasant, New Brunswick. Geochim Cosmochim Acta 69:2469-2483

Petrunic BM, Al TA, Weaver L (2006) A transmission electron microscopy analysis of secondary minerals formed in tungsten-mine tailings with an emphasis on arsenopyrite oxidation. Appl Geochem 21:1259-1273

Petrunic BM, Al TA, Weaver L, Hall D (2009) Identification and characterization of secondary minerals formed in tungsten mine tailings using transmission electron microscopy. Appl Geochem 24:2222-2233

Pfeifer H-R, Gueye-Girardet A, Reymond D, Schlegel C, Temgoua E, Hesterberg DL, Chou JW (2004) Dispersion of natural arsenic in the Malcantone watershed, southern Switzerland: field evidence for repeated sorption-desorption and oxidation-reduction processes. Geoderma 122:205-234

Piatak NM, Seal II RR (2010) Mineralogy and the release of trace elements from slag from the Hegeler Zinc smelter, Illinois (USA). Appl Geochem 25:302-320

Piatak NM, Seal II RR, Hammarstrom JM (2004) Mineralogical and geochemical controls on the release of trace elements from slag produced by base- and precious-metal smelting at abandoned mine sites. Appl Geochem 19:1039-1064

Pichler T, Hendry MJ, Hall GEM (2000) The mineralogy of arsenic in uranium mine tailings at the Rabbit Lake In-pit facility, northern Saskatchewan, Canada. Environ Geol 40:495-506

Pierrot R (1964) Contribution à la minéralogie des arséniates calciques et calcomagnésiens naturels. Bull Soc Fr Mineral Crist 87:169-211. (in French)

Pierrot R, Schubnel HJ (1972) Irhtemite, a new hydrated calcium magnesium arsenate. Bull Soc Fr Mineral Cristallogr 95:365-70

Piret P, Deliens M, Piret-Meunier J (1985) Occurrence and crystal structure of kipushite, a new copper-zinc phosphate from Kipushi, Zaire. Can Mineral 23:35-42

Pirrie D, Power MR, Rollinson G, Camm GS, Hughes SH, Butcher AR, Hughes P (2003) The spatial distribution and source of arsenic, copper, tin and zinc within the surface sediments of the Fal Estuary, Cornwall, UK. Sedimentology 50:579-595

Plumlee GS (1999) The environmental geology of mineral deposits. *In:* The Environmental Geochemistry of Mineral Deposits Part A: Processes, Techniques and Health Issues. Plumlee GS, Logsdon MJ (eds) Society of Economic Geologists, Littleton, CO. Rev Econ Geol 6B 71-116

Pokrovski GS, Zakirov IV, Roux J, Testemale D, Hazemann J-L, Bychkov AYu, Golikova GV (2002) Experimental study of arsenic speciation in vapor phase to 500 °C: implications for As transport and fractionation in low-density crustal fluids and volcanic gases. Geochim Cosmochim Acta 66:3453-3480

Potts PJ, Ramsey MH, Carlisle J (2002) Portable X-ray fluorescence in the characterisation of arsenic contamination associated with industrial buildings at a heritage arsenic works site near Redruth, Cornwall, UK. J Environ Monitor 4:1017-1024

Power MR, Pirrie D, Camm GS, Andersen JCØ (2009) The mineralogy of efflorescence on As calciner building in SW England. Mineral Mag 73:27-42

Prensipe M, Pushcharovskii DYu, Sarp H, Ferraris G (1996) Comparative crystal chemistry of geminite Cu[AsO₃OH]H₂O and minerals related to it. Vestnik Moskovskogo Universiteta, Geologiya 4:66-74. (in Russian)

Pring A, McBriar EM, Birch WD (1989) Mawbyite, a new arsenate of lead and iron related to tsumcorite and carminite, from Broken Hill, New South Wales. Am Mineral 74:1377-1381

Protas J, Gindt R (1976) Structure cristalline de la brassite, MgHAsO₄·4H₂O, produit de déshydratation de la roesslérite. Acta Crystallogr B 32:1460-1466

Pushcharovsky DYu, Teat SJ, Zaitsev VN, Zubkova NV, Sarp H (2000) Crystal structure of pushcharovskite. Eur J Mineral 12:95-104

Pushcharovsky DYu, Zubkova NV, Teat SJ, MacLean EJ, Sarp H (2004) Crystal structure of mahnertite. Eur J Mineral 16:687-692

Puziewicz J, Zainoun K, Bril H (2007) Primary phases in pyrometallurgical slags from a zinc-smelting waste dump, Upper Silesia, Poland. Can Mineral 45:1189-1200

Qurashi MM, Barnes WH (1963) The structures of the minerals of the descloizite and adelite groups: IV – Descloizite and conichalcite (Part 2) the structure of conichalcite. Can Mineral 7:561-577

Ramdohr P, Ahlfeld F, Berndt F (1959) Angelellite, a natural triclinic iron arsenate, 2Fe₂O₃.As₂O₅. Neues Jahrb Mineral Monatsh 145-51

Rancourt DG, Fortin D, Pichler T, Thibault PJ, Lamarche G, Morris RV, Mercier PHJ (2001) Mineralogy of a natural As-rich hydrous ferric oxide coprecipitate formed by mixing of hydrothermal fluid and seawater: implications regarding surface complexation and color banding in ferrihydrite deposits. Am Mineral 86:834-851

Reimann C, Matschullat J, Salminen R (2009) Arsenic distribution in the environment: The effects of scale. Appl Geochem 24:1147-1167

Reinecke T, Okrusch M, Richter P (1985) Geochemistry of ferromanganoan metasediments from the island of Andros, Cycladic blueschist belt, Greece. Chem Geol 53:249-278

Řídkošil T, Slavíček P (1985) Mineralization in the Renner adit near Mikulov in the Krušné Hory Mts. Čas Mineral Geol 30(4):433 (in Czech)

Rieder M, Crelling JC, Šustai O, Drábek M, Weiss Z, Klementová M (2007) Arsenic in iron disulfides in a brown coal from the North Bohemian Basin, Czech Republic. Int J Coal Geol 71:115-121

Riffel H, Keller P, Hess H (1980) Die kristallstruktur von warikahnit. Mineral Petrol 27:187-199

Riveros PA, Dutrizac JE, Spencer P (2001) Arsenic disposals practices in the metallurgical industry. Can Metall Quart 40:395-420

Roberts AC, Burns PC, Gault RA, Criddle AJ, Feinglos MN (2004) Petewilliamsite, (Ni,Co)₃₀(As₂O₇)₁₅, a new mineral from Johanngeorgenstadt, Saxony, Germany: Description and crystal structure. Mineral Mag 68:231-240

Roberts AC, Burns PC, Gault RA, Criddle AJ, Feinglos MN, Stirling JAR (2001) Paganoite, NiBi³⁺As⁵⁺O₅, a new mineral from Johanngeorgenstadt, Saxony, Germany: Description and crystal structure. Eur J Mineral 13:167-175

Roberts AC, Cooper MA, Hawthorne FC, Criddle AJ, Stirling JAR (2002) Sewardite, CaFe³⁺₂(AsO₄)₂(OH)₂, the Ca-analogue of carminite, from Tsumeb, Namibia: Description and crystal structure. Can Mineral 40:1191-1198

Robertson WD (1994) The physical hydrogeology of mill-tailings impoundments. *In:* The Environmental Geochemistry of Sulfide Mine-Wastes. Short Course Series, Vol 22. Jambor JL, Blowes DW (eds) Mineralogical Association of Canada, p 1-17

Romanenko GV, Pervukhina NV, Borisov SV, Magarill SA, Vasiliev VI (1999) Crystal structure of kuznetsovite Hg₃(AsO₄)Cl. J Structural Chem 40:270-275

Romero A, Gonzáles I, Galán E (2006) The role of the efflorescent sulfates in the storage of trace elements in stream waters polluted by acid mine-drainage: the case of Peña del Hierro, southwest Spain. Can Mineral 44:1431-1446

Romero FM, Armienta MA, González-Hernández G (2007) Solid-phase control on the mobility of potentially toxic elements in an abandoned lead/zinc mine tailings impound, Taxco, Mexico. Appl Geochem 22:109-127

Romero FM, Prol-Ledesma RM, Canet C, Alvares LN, Perez-Vazquez R (2010) Acid drainage at the inactive Santa Lucia mine, western Cuba: Natural attenuation of arsenic, barium and lead, and geochemical behavior of rare earth elements. Appl Geochem 25:716-727

Rosado L, Morais C, Candeias AE, Pinto AP, Guimarães F, Mirão J (2008) Weathering of S. Domingos (Iberian Pyritic Belt) abandoned mine slags. Mineral Mag 72:489-494

Rouse RC, Dunn PJ, Peacor DR (1984) Hedyphane from Franklin, New Jersey and Långban, Sweden: Cation ordering in an arsenate apatite. Am Mineral 69:920-927

Roussel C, Néel C, Bril H (2000) Minerals controlling arsenic and lead solubility in an abandoned gold mine tailings. Sci Total Environ 263:209-219

Ruppert LF, Eble JC, Eble CF (2005) Arsenic-bearing pyrite and marcasite in the Fire Clay coal bed, Middle Pennsylvanian Breathitt Formation, eastern Kentucky. Int J Coal Geol 63:27-35

Ruszala FA, Anderson JB, Kostiner E (1977) Crystal structures of two isomorphs of arsenoclasite: $Co_5(PO_4)_2(OH)_4$ and $Mn_5(PO_4)_2(OH)_4$. Inorg Chem 16:2417-2422

Rytuba JJ, Kim CS, Goldstein DN (2011) Review of Samples of Tailings, Soils, and Stream Sediments Adjacent to and Downstream from the Ruth Mine, Inyo County, California. US Geological Survey Open-File Report 2011-1105, p 1-38

Sabelli C (1980) The crystal structure of chalcophyllite. Z Kristallogr 151:129-140

Sacerdoti M, Parodi GC, Mottana A, Maras A, della Ventura G (1993) Asbecasite: Crystal structure refinement and crystal chemistry. Mineral Mag 57:315-322

Sadiq M (1997) Arsenic chemistry in soils: an overview of thermodynamic predictions and field observations. Water Air Soil Pollut 93:117-136

Sáez R, Nocete F, Nieto JM, Capitán MÁ, Rovira S (2003) The extractive metallurgy of copper from Cabezo Juré, Huelva, Spain: Chemical and mineralogical study of slags dated to the third millennium B.C. Can Mineral 41:627-638

Sakai S, Yoshiasa A, Sugiyama K, Miyawaki R (2009) Crystal structure and chemistry of conichalcite, $CaCu(AsO_4)(OH)$. J Mineral Petrol Sci 104:125-131

Salzsauler KA, Sidenko NV, Sherriff BL (2005) Arsenic mobility in alteration products of sulfide-rich, arsenopyrite-bearing mine wastes, Snow Lake, Manitoba, Canada. Appl Geochem 20:2303-2314

Sandström F, Holtstam D (1999) Geology of the Långban deposit. *In:* Långban. The Mines, Their Minerals, Geology and Explorers. Holtstam D, Langhof J (eds) Christian Weise Verlag, 29-41

Sarp H, Guenée L (2002) Radovanite, $Cu_2Fe^{3+}(AsO_4)(As^{3+}O_2OH)\cdot2H_2O$, a new mineral: its description and crystal structure. Archives des Sciences 55:47-55

Sarp H, Černý R (1998) Description and crystal structure of yvonite, $Cu(AsO_3OH)\cdot2H_2O$. Am Mineral 83:383-389

Sarp H, Černý R (1999) Gilmarite, $Cu_3(AsO_4)(OH)_3$, a new mineral: Its description and crystal structure. Eur J Mineral 11:549-555

Sarp H, Černý R (2000) Rollandite, $Cu_3(AsO_4)_2\cdot4H_2O$, a new mineral: Its description and crystal structure. Eur J Mineral 12:1045-1050

Sarp H, Černý R (2001) Theoparacelsite, $Cu_3(OH)_2As_2O_7$, a new mineral: its description and crystal structure. Arch Sciences 54:7-14

Sarp H, Deferne J, Liebich BW (1981) Mcnearite, $NaCa_5H_4(AsO_4)_5\cdot4H_2O$, a new hydrous arsenate of calcium and sodium. Schweiz Mineral Petrogr Mitt 61:1-6

Sarp H, Pushcharovsky DYu, MacLean EJ, Teat SJ, Zubkova NV (2003) Tillmannsite, $(Ag_3Hg)(V,As)O_4$, a new mineral: Its description and crystal structure. Eur J Mineral 15:177-180

Savage KS, Bird DK, O'Day PA (2005) Arsenic speciation in synthetic jarosite. Chem Geol 215:473-498

Savage KS, Tingle TN, O'Day PA, Waychunas GA, Bird DK (2000) Arsenic speciation in pyrite and secondary weathering phases, Mother Lode Gold District, Tuolumne County, California. Appl Geochem 15:1219-1244

Schaufelberger FA (1994) Arsenic minerals formed at low temperatures. *In:* Arsenic in the Environment, Part I: Cycling and characterization. Nriagu JO (ed) Wiley, New York, p 403-415

Schmetzer K, Amthauer G, Stähle V, Medenbach O (1982) Metaköttigite, $(Zn,Fe^{3+})(Zn,Fe^{3+},Fe^{2+})_2(AsO_4)_2\cdot8(H_2O,OH)$, a new mineral from Mapimi, Mexico. Neues Jahrb Mineral Monatsh 506-518

Schmetzer K, Berdesinski W, Bank H, Krouzek E (1976) New investigations of ceruleite. Neues Jahrb Mineral Monatsh 418-425

Schmetzer K, Medenbach O (1985) Gerdtremmelite, $(Zn,Fe)(Al,Fe)_2[(AsO_4)|(OH)_5]$, a new mineral from Tsumeb, Namibia. Neues Jahrb Mineral Monatsh 1-6

Schmetzer K, Nuber B, Tremmel G (1984) Betpakdalit aus Tsumeb, Namibia: Mineralogie, Kristallchemie und Struktur. Neues Jahrb Mineral Monatsh 393-403

Sejkora J (1997) Mineral phases of the As-S system. Bull min-petr odd NM v Praze 4-5:106-112 (in Czech)

Sejkora J, Šrein V, Litochleb J (1997) Alacranite, arsensulphur and realgare from burning spoil of the Kateřina mine in Randvanice near Trutnov. Bull min-petr odd NM v Praze 4-5:194-200. (in Czech)

Sejkora J, Řídkošil T, Šrein V (1999) Zálesíite, a new mineral of the mixite group, from Zálesí, Rychlebské hory Mts, Czech Republic. Neues Jahrb Mineral Abh 175:105-124

Sejkora J, Škovíra J, Čejka J, Plášil J (2009) Cu-rich members of the beudantite–segnitite series from the Krupka ore district, the Krušné hory Mountains, Czech Republic. J Geosci 54:355-371

Sejkora J, Ondruš P, Novák M (2010a) Veselovskyite, triclinic (Zn,Cu,Co)Cu$_4$ (AsO$_4$)$_2$(AsO$_3$OH)$_2$·9H$_2$O, a Zn-dominant analogue of lindackerite. Neues Jahrb Mineral Abh 187:83-90

Sejkora J, Plášil J, Ondruš P, Veselovský F, Císařová I, Hloušek J (2010b) Slavkovite, Cu$_{13}$(AsO$_4$)$_6$(AsO$_3$OH)$_4$·23H$_2$O, a new mineral species from Horní Slavkov and Jáchymov, Czech Republic: Description and crystal-structure determination. Can Mineral 48:1157-1170

Sejkora J, Plášil J, Veselovský F, Císařová I, Hloušek J (2011) Ondrušite, CaCu$_4$(AsO$_4$)$_2$(AsO$_3$OH)$_2$·10H$_2$O, a new mineral species from the Jáchymov ore district, Czech Republic: Description and crystal-structure determination Can Mineral 49(3): 885-897

Semenova TF, Vergasova LP, Filatov SK, Ananev VV (1994) Alarsite AlAsO$_4$ is a new mineral from volcanic sublimates. Dokl Akad Nauk 338:501-505

Shah P, Strezov V, Stevanov C, Nelson PF (2007) speciation of arsenic and selenium in coal combustion products. Energy Fuel 21:506-512

Shaw SA, Hendry MJ, Essilfie-Dughan J, Kotzer T, Wallschläger D (2011) Distribution, characterization, and geochemical controls of elements of concern in uranium mine tailings, Key Lake, Saskatchewan, Canada. Appl Geochem 26:2044-2056

Shibayama A, Takasaki Y, William T, Yamatodani A, Higuchi Y, Sunagawa S, Ono E (2010) Treatment of smelting residue for arsenic removal and recovery of copper using pyro-hydrometallurgical process. J Hazmat Mater 181:1016-1023

Shih CJ, Lin CF (2003) Arsenic contaminated site at an abandoned copper smelter plant: waste characterization and solidification/stabilization treatment. Chemosphere 53:691-703

Sidenko NV, Lazareva EV, Bortnikova SB, Kireev AD, Sherriff BL (2005) Geochemical and mineralogical zoning of high-sulfide mine-waste at the Berikul mine-site, Kemerovo region, Russia. Can Mineral 43:1141-1156

Signorelli S (1997) Arsenic in volcanic gases. Environ Geol 32:239-244

Signorelli S, Buccianti A, Martini M, Piccardi G (1998) Arsenic in fumarolic gases of Vulcano (Aeolian Islands, Italy) from 1978 to 1993: Geochemical evidence from multivariate analysis. Geochem J 32:367-382

Siidra OI, Krivovichev SV, Chukanov NV, Pekov IV, Magganas A, Katerinopoulos A, Voudouris P (2011) The crystal structure of Pb$_5$(As^{3+}O$_3$)Cl$_7$ from the historic slags of Lavrion, Greece – a novel Pb(II) chloride arsenite. Mineral Mag 75:337-345

Sinha A, Singh VK (2007) Spontaneous Coal Sem fires: a Global Phenomenon. ERSEC Ecological Book Series-4:42-66

Siuda R (2004) Iron arsenates from the Stara Góra deposit at Radzimowice in Kaczawskie Mountains, Poland – a preliminary report. Polsk Tower Mineral – Prac Spec Mineral Soc Poland 24:345-347

Sivry Y, Munoz M, Sappin-Didier V, Riotte J, Denaix L, Parseval P, Destrigneville C, Dupré B (2010) Multimetallic contamination from Zn-ore smelter: solid speciation and potential mobility in riverine floodbank soils of the upper Lot River (SW France). Eur J Mineral 22:679-691

Slowey AJ, Johnson SB, Newville M, Brown Jr GE (2007) Speciation and colloid transport of arsenic from mine tailings. Appl Geochem 22:1884-1898

Smedley PL, Kinniburgh DG (2002) A review of the source behavior and distribution of arsenic in natural waters. Appl Geochem 17:517-568

Smith E, Naidu R, Alston AM (1998) Arsenic in the Soil Environment: A Review. Adv Agron 64:149-195

Smith E, Smith J, Smith L, Biswas T, Cowwell R, Naidu R (2003) Arsenic in Australian environment: An overview. J Environ Sci Heal A 38:223-229

Smith GFH, Prior GT (1911) On fermorite, a new arsenate and phosphate of lime and strontia and tilasite, from the manganese-ore deposits of India. Mineral Mag 16:84-96

Smith L, Beckie R (2003) Hydrologic and geochemical transport processes in mine waste rock. *In:* Environmental Aspects of Mine Wastes. Short Course Series, Vol. 31. Jambor JL, Blowes DW, Ritchie AIM (eds) Mineralogical Association of Canada, Ottawa, p 51-72

Smith RA (1977) Dispersion of particulate arsenic waste from the chimney of Giant Yellowknife Gold Mines Ltd. Sci Total Environ 7:227-233

Smoľyaninova NN (1970) Nekotorye dannye po mineralogii i gezisu mestorozdenia Tyuya-Muyun. Izd nauk SSSR, Nauka, Moskva (in Russian)

Smoľyaninova NN, Senderova VM (1963) Arseniosiderit iz Juznoj Kirghizii. Trans Min Mus Akad Nauk SSSR 14:250-258 (in Russian)

Sokol EV, Volkova NI (2007) Combustion metamorphic events resulting from natural coal fires. *In:* Geology of Coal Fires: Case Studies from Around the World. Stracher GB (ed) Rev Eng Geol 18, p. 97-115

Soprovich EA (2000) Arsenic release from oxide tailings containing scorodite, Fe-Ca arsenates, and As-containing goethites. *In:* Tailings and Mine Waste '00. Balkema, Rotterdam, p 277-287

Sowa H (1991) The crystal structure of AlAsO$_4$ at high pressure. Z Kristallogr 194:291-304

Starova GL, Krivovichev SV, Filatov SK (1998) Crystal chemistry of inorganic compounds based on chains of oxocentered tetrahedra. II. Crystal structure of Cu$_4$O$_2$[(As,V)O$_4$]Cl. Z Kristallogr 213:650-653

Stock N, Stucky GD, Cheetham AK (2001) Synthesis and characterization of the manganese pyroarsenate Mn$_2$As$_2$O$_7$·2H$_2$O. Z Naturforsch 56b:359-363

Stock N, Stucky GD, Cheetham AK (2002) Synthesis and characterization of the synthetic minerals villyaellenite and sarkinite, Mn$_5$(AsO$_4$)(HAsO$_4$)$_2$·4H$_2$O and Mn$_2$(AsO$_4$)(OH). Z Anorg Allg Chem 628:357-362

Stoiber RE, Rose WI (1969) Recent volcanic and fumarolic activity at Santiaguito volcano, Guatemala. Bull Volcanol 33:475-502

Stollenwerk KG (2003) Geochemical processes controlling transport of arsenic in groundwater: a review of adsorption. *In:* Arsenic in Groundwaters: Geochemistry and Occurrences. Welch A, Stollenwerk K (eds) Kluwer Academic Publishers, Boston, p 67-100

Stracher GB (2007) The origin of gas-vent minerals: Isochemical and mass-transfer processes. *In:* Geology of Coal Fires: Case Studies from Around the World. Stracher GB (ed) Reviews Rev Eng Geol 18, p. 91-96

Strunz H (1937) Titanit und tilasit. Z Kristallogr 96:7-14

Strunz H (1960) Chudobait, ein neues mineral von Tsumeb. Neues Jahrb Mineral Monatsh 1-7

Strunz H, Nickel EH (2001) Strunz Mineralogical Tables. Chemical-Structural Mineral Classification System. 9th ed. Schweizerbart`sche Verlagsbuchhandlung, Stuttgart

Sullivan C, Tyrer M, Cheeseman CR, Graham NJD (2010) Disposal of water treatment wastes containing arsenic – A review. Sci Total Environ 408:1770-1778

Süsse P, Tillmann B (1987) The crystal structure of the new mineral richelsdorfite, Ca$_2$Cu$_5$Sb(Cl/(OH)$_6$/(AsO$_4$)$_4$)·6H$_2$O. Z Kristallogr 179:323-334

Swash PM, Monhemius AJ (1995) Synthesis, characterization and solubility testing of solids in the Ca-Fe-AsO$_4$ system. *In:* Sudbury '95 – Mining and the Environment. Hynes TP, Blanchette MC (eds) CANMET, Canada, p 17-28

Szakáll S, Fehér B, Bigi S, Mádai F (2011) Klajite from Recsk (Hungary), the first Mn-Cu arsenate mineral. Eur J Mineral 23:829-836

Szymański JT (1988) The crystal structure of beudantite, Pb(Fe,Al)$_3$[(As,S)O$_4$]$_2$(OH)$_6$. Can Mineral 26:923-932

Tait KT, Hawthorne FC (2003) Refinement of the crystal structure of arseniopleite: Confirmation of its status as a valid species. Can Mineral 41:71-77

Takeuchi Y, Ghose S, Nowacki W (1965) The crystal structure of hutchinsonite, (Tl, Pb)$_2$As$_5$S$_9$. Z Kristallogr Krist 121:321-348

Tang X, Gentiletti MJ, Lachgar A (2002) Synthesis and crystal structure of indium arsenate and phosphate dihydrates with variscite and metavariscite structure types. J Chem Crystallogr 31:45-50

Tillmanns E, Gebert W (1973) The crystal structure of tsumcorite, a new mineral from the Tsumeb mine, S.W. Africa. Acta Crystallogr B 29:2789-2794

Tillmanns E, Hofmeister W, Petitjean K (1985) Cornubite, Cu$_5$(AsO$_4$)(OH)$_4$, first occurrence of single crystals, mineralogical description and crystal structure. Bull Geol Soc Finland 57:119-127

Toman K (1977) The symmetry and crystal structure of olivenite. Acta Crystallogr B 33:2628-2631

Tournassat C, Charlet L, Bosbach D, Manceau A (2002) Arsenic(III) oxidation by birnessite and precipitation of manganese(II) arsenate. Environ Sci Technol 36:493-500

Triantafyllidis S, Skarpelis N (2006) Mineral formation in an acid pit lake from a high-sulfidation ore deposit: Kirki, NE Greece. J Geochem Explor 88:68-71

Van Elteren JT, Šlejkovec Z, Arčon I, Glass HJ (2006) An interdisciplinary physical-chemical approach for characterization of arsenic in a calciner residue dump in Cornwall (UK). Environ Pollut 139:477488

Vandecasteele C, Dutré V, Geysen D, Wauters G (2002) Solidification/stabilisation of arsenic bearing fly ash from the metallurgical industry. Immobilisation mechanisms of arsenic. Waste Manage 22:143-146

Vaněk A, Ettler V, Grygar T, Borůvka L, Šebek O, Drábek O (2008) Combined chemical and mineralogical evidence for heavy metal binding in mining- and smelting-affected alluvial soils. Pedosphere 18:464-478

Vaughan DJ (ed) (2006) Sulfide Mineralogy and Geochemistry. Reviews in Mineralogy & Geochemistry, Volume 61. Mineralogical Society of America

Vergasova LP, Starova GL, Krivovichev SV, Filatov SK, Ananiev VV (1999) Coparsite, Cu$_4$O$_2$[As,V)O$_4$]Cl, a new mineral species from the Tolbachik volcano, Kamchatka Peninsula, Russia. Can Mineral 37:911-914

Vergasova LP, Krivovichev SV, Britvin SN, Burns PC, Ananiev VV (2004) Filatovite, K[(Al,Zn)$_2$(As,Si)$_2$O$_8$], a new mineral species from the Tolbachik volcano, Kamchatka peninsula, Russia. Eur J Mineral 16:533-536

Viñals J, Jambor JL, Raudsepp M, Roberts AC, Grice JD, Kokinos M, Wise WS (2008) Barahonaite-(Al) and barahonaite-(Fe), new Ca-Cu arsenate mineral species from Murcia province, southeastern Spain, and Gold Hill, Utah. Can Mineral 46:205-217

Viñals J, Roca A, Cruells M, Núñez C (1995) Characterization and cyanidation of Rio Tinto gossan ores. Can Metall Quart 34:115-122

Vítková M, Ettler V, Křibek B, Šebek O, Mihaljevič M (2010) primary and secondary phases in copper-cobalt smelting slags from the Copperbelt Province, Zambia. Mineral Mag 74:581-600

Voigt DE, Brantley SL, Hennet RJC (1996) Chemical fixation of arsenic in contaminated soils. Appl Geochem 11:633-643

Voloshin AV, Menshikov YuP, Polezhaeva LI, Lentsi AA (1982) Kolfanite, a new mineral from granite pegmatite, Kola Peninsula. Mineralogicheskiy Zhurnal 4:90-95 (in Russian)

Walenta K (1994) Über den Barium-Pharmakosiderit. Aufschluss 45:73

Walenta K, Dunn PJ (1989) Camgasit, ein neues calcium-magnesiumarsenatmineral der zusammensetzung CaMg(AsO₄)(OH)·5H₂O von Wittichen im mittleren Schwarzwald. Aufschluss 40:369-372

Walenta K, Theye T (2004) Agardit-(Ce) von der Grube Clara im mittleren Schwarzwald. Aufschluss 55:17

Walker SR, Jamieson HE, Lanzirotti A, Andrade CF, Hall GEM (2005) The speciation of arsenic in iron oxides in mine wastes from the Giant gold mine, N.W.T.: Application of synchrotron micro-XRD and micro XANES at the grain scale. Can Mineral 43:1205-1224

Walker SR, Parsons MB, Jamieson HE, Lanzirotti A (2009) Arsenic mineralogy of near-surface tailings and soils: Influences on arsenic mobility and bioaccessibility in the Nova Scotia gold mining districts. Can Mineral 47:533-556

Wang S, Mulligan CN (2006a) Natural attenuation processes for remediation of arsenic contaminated soils and groundwater. J Hazard Mater 138:459-470

Wang S, Mulligan CN (2006b) Occurrence of arsenic contamination in Canada: sources, behavior and distribution. Sci Total Environ 366:701-721

Wanger G, Southam G, Onstott TC (2006) Structural and chemical characterization of a natural fracture surface from 2.8 kilometers below land surface: Biofilms in the deep subsurface. Geomicrob J 23(6):443-452

Weber K (1959) Eine kristallographische Untersuchung des Angelellits, 2Fe₂O₃·As₂O₅. Neues Jahrb Mineral Monatsh 152-158

Webster JG, Nordstrom KD (2003) Geothermal arsenic. *In:* Arsenic in Ground Water. Welch AH, Stollenwerk KG (eds) Kluwer Academic Publishers, p 101-125

White AF, Blum AE (1995) Effects of climate on chemical-weathering in watersheds. Geochim Cosmochim Acta 59: 1729-1747

Wicks FJ, Simmons WB, Dunn PJ, Peacor DR (1986) Arsenites related to layer silicates: Manganarsite, the arsenite analogue of manganpyrosmalite, and unnamed analogues of friedelite and schallerite from Långban, Sweden. Am Mineral 71:1517-1521

Wildner M, Giester G, Lengauer CL, McCammon CA (1996) Structure and crystal chemistry of vivianite-type compounds: Crystal structure of erythrite and annabergite with a Mössbauer study of erythrite. Eur J Mineral 8:187-192

Wildner M, Tillmanns E, Andrut M, Lorenz J (2003) Sailaufite, (Ca,Na,)₂Mn₃O₂(AsO₄)₂(CO₃)·3H₂O, a new mineral from Hartkoppe hill, Ober-Sailauf (Spessart Mountains, Germany), and its relationship to mitridatite-group minerals and pararobertsite. Eur J Mineral 15:555-564

Williams M (2001) Arsenic in mine waters: an international study. Environ Geol 40:267-278

Wilson CC, Cox PJ, Stewart NS (1991) Structure and disorder in schultenite, lead hydrogen arsenate. J Crystallogr Spectrosc Res 21:589-593

Wise W (1978) Parnauite and goudeyite, two new copper arsenate minerals from the Majuba Hill Mine, Pershing County, Nevada. Am Mineral 63:704-708

Witzke T (1996) Die Minerale der brennenden Halde der Steinkohlengrube „Deutschlandschacht" in Oelsnitz bei Zwickau. Aufschluss 47:41-48 (in German)

Witzke T, Hocker M (1993) Neue vorkommen von bukovskyit, zykait und kankit. Lapis 18:49-50. (in German)

Witzke T, Steins M, Doering T, Kolitsch U (2000) Gottlobite, CaMg(VO₄,AsO₄)(OH), a new mineral from Friedrichroda, Thuringia, Germany. Neues Jahrb Mineral Monatsh 444-454

Witzke T, Kolitsch U, Krause W, Wiechowski A, Medenbach O, Kampf AR, Steele IM, Favreau G (2006) Guanacoite, Cu₂Mg₂(Mg₀.₅Cu₀.₅)(OH)₄(H₂O)₄(AsO₄)₂, a new arsenate mineral species from the El Guanaco Mine, near Taltal, Chile: Description and crystal structure. Eur J Mineral 18:813-821

Wuensch BJ (1968) Comparison of the crystallography of dixenite, mcgovernite and hematolite. Z Kristallogr 127:309-318

Yakhontova LK, Poroshina IA, Plyunina II (1983) Structural model of lazarenkoite according to refractometric analysis data. Probl Kristallokhim Genezisa Miner 145-148 (in Russian)

Yang L, Donahoe RJ (2007) The form, distribution and mobility of arsenic in soils contaminated by arsenic trioxide, at sites in southeast USA. Appl Geochem 22:320-341

Yang JK, Barnett MO, Zhuang JL, Fendorf SE, Jardine PM (2005) Adsorption, oxidation, and bioaccessibility of As(III) in soils. Environ Sci Technol 39:7102-7110

Yang H, Costin G, Keogh J, Lu R, Downs RT (2007) Cobaltaustinite, CaCo(AsO₄)(OH). Acta Crystallogr E 63:i53-i55

Yang Z, Giester G, Ding K, Tillmanns E (2010) Crystal structure of nanlingite – the first mineral with a [Fe(AsO₃)₆] configuration. Eur J Mineral 23:63-71

Yang HX, Jenkins RA, Downs R.T, Evans SH, Tait KT (2011) Rruffite, $Ca_2Cu(AsO_4)_2 \cdot 2H_2O$, a new member of the roselite group, from Tierra Amarilla, Chile. Can Mineral 49:877-884

Yudovich YaE, Ketris MP (2005) Arsenic in coal: a review. Int J Coal Geol 61:141-196

Yue Z, Donahoe RJ (2009) Experimental simulation of soil contamination by arsenolite. Appl Geochem 24:650-656

Žáček V (1998) Recent secondary mineralization at the dumps and mines in the Kladno district. Bull Min-Petr Odd NM v Praze 6:161-175 (in Czech)

Žáček V, Ondruš P (1997) Mineralogy of recently formed sublimates from Kateřina colliery in Radvanice, eastern Bohemia, Czech Republic. Věst Čes Geol Úst 72:289-302

Zantop H (1981) Trace elements in volcanogenic manganese oxides and iron oxides: The San Francisco manganese deposit, Jalisco, Mexico. Econ Geol 76:545-555

Zelenski M, Bortnikova S (2005) Sublimate speciation at Mutnovsky volcano, Kamchatka. Eur J Mineral 17:107-118

Zemann J (1947) Über die struktur des pharmakosiderits. Experientia 3:452

Zemann J (1948) Formel und strukturtyp des pharmakosiderits. Mineral Petrol 1:1-13

Zettler F, Riffel H, Hess H, Keller P (1979) Cobalthydrogenarsenate-monohydrat. Darstellung und kristallstruktur. Z Anorg Allg Chem 454:134-144

Zevenhoven R, Mukherjee AB, Bhattacharya P (2007) Arsenic flows in the environment of the European Union: a synoptic review. *In:* Trace Metals and other Contaminants in the Environment. Vol 9, Bhattacharya P, Mukherjee AB, Bundschuh J, Zevenhoven R, Loeppert RH (eds) Elsevier B.V, p 527-547

Zhu Y, Merkel BJ (2001) The dissolution and stability of scorodite, $FeAsO_4 \cdot 2H_2O$ evaluation and simulation with PHREEQC2. Wiss Mitt Inst Geol TU Bergakedemie Freiberg 18:1-2

Zielinski RA, Foster AL, Meeker GP, Brownfield IK (2007) Mode of occurrence of arsenic in feed coal and its derivative fly ash, Black Warrior Basin, Alabama. Fuel 86:560-572

Zolensky M, Gounelle M, Mikouchi T, Ohsumi K, Le L, Hagiya K, Tachikawa O (2008) Andreyivanovite: A second new phosphide from the Kaidun meteorite. Am Mineral 93:1295-1299

Zoppi M, Pratesi G (2009) Rietveld refinement of a natural cobaltian mansfieldite from synchrotron data. Acta Crystallogr E 65:i6-i7

Zubkova NV, Puscharovsky DYu, Giester G, Tillmanns E, Pekov IV, Kleimenov DA (2002) The crystal structure of arsentsumebite, $Pb_2Cu[(As,S)O_4]_2(OH)$. Miner Petrol 75:79-88

Zubkova NV, Pushcharovsky DYu, Sarp H, Teat SJ, MacLean EJ (2003) Crystal structure of zdenekite $NaPbCu_5(AsO_4)_4Cl \cdot 5H_2O$. Crystallogr Rep 48:939-943

APPENDIX 1

Index and list of arsenates, arsenites, and a few common As sulfides and arsenides. Note that we deliberately omitted any information about the oxidation state of the elements because in many cases, the oxidation state is not clear, not definitely proven, or may be variable.

Name	Formula
abernathyite	$K(UO_2)(AsO_4) \cdot 3H_2O$
adamite	$Zn_2(OH)(AsO_4)$
adelite	$CaMg(OH)AsO_4$
aerugite	$Ni_{8.5}O_2(AsO_4)_2(AsO_6)$
agardite	$Cu_6REE(OH)_6(AsO_4)_3 \cdot 3H_2O$
akrochordite	$(Mn,Mg)_5(OH)_4(H_2O)_4(AsO_4)_2$
alacranite	AsS
alarsite	$AlAsO_4$
algodonite	$(Cu_{1-x}As_x)$
allactite	$Mn_7(OH)_8(AsO_4)_2$
alumopharmacosiderite	$KAl_4(OH)_4(AsO_4)_3 \cdot 5H_2O$
andyrobertsite	$KCdCu_5(AsO_4)_4(H_2AsO_4) \cdot 2H_2O$
angelellite	$Fe_4O_3(AsO_4)_2$
annabergite	$Ni_3(H_2O)_8(AsO_4)_2$
arakiite	$(Zn,Mn)(Mn,Mg)_{12}(Fe,Al)_2(AsO_3)(AsO_4)_2(OH)_{23}$
ardennite	$Mn_4(Al,Mg)_6(OH)_6(Si_3O_{10})(SiO_4)_2(AsO_4,VO_4)$
arhbarite	$Cu_2Mg(OH)_3(AsO_4)$
armangite	$Mn_{26}[As_6(OH)_4O_{14}](As_6O_{18})_2(CO_3)$
arsenbrackebuschite	$Pb_2(Fe,Zn)(OH,H_2O)(AsO_4)_2$
arsendescloizite	$PbZn(OH)AsO_4$
arsenic, native	As
arseniopleite	$CaNaMnMn_2(AsO_4)_3$
arseniosiderite	$Ca_3Fe_4(OH)_6(AsO_4)_4 \cdot 3H_2O$
arsenoclasite	$Mn_5(OH)_4(AsO_4)_2$
arsenocrandallite	$CaAl_3[AsO_{3.5}(OH)_{0.5}]_2(OH)_6$
arsenoflorencite-(Ce)	$CeAl_3(AsO_4)_2(OH)_6$
arsenoflorencite-(La)	$LaAl_3(AsO_4)_2(OH)_6$
arsenogorceixite	$BaAl_3[AsO_{3.5}(OH)_{0.5}]_2(OH)_6$
arsenogoyazite	$SrAl_3[AsO_{3.5}(OH)_{0.5}]_2(OH)_6$
arsenolite	As_2O_3
arsenopyrite	$FeAsS$
arsenovanmeerssscheite	$U(OH)_4[(UO_2)_3(AsO_4)_2(OH)_2] \cdot 4H_2O$

Name	Formula
arsentsumebite	$Pb_2Cu(OH)[(As,S)O_4]_2$
arsenuranospathite	$AlH(UO_2)_4(AsO_4)_4 \cdot 40H_2O$
arsenuranylite	$Ca(UO_2)_4(AsO_4)_2(OH)_4 \cdot 6H_2O$
arthurite	$CuFe_2(OH)_2(H_2O)_4(AsO_4)_2$
asbecasite	$Ca_3Ti(As_6Be_2Si_2O_{20})$
asselbornite	$(Pb,Ba)(UO_2)_6(BiO)_4(AsO_4)_2(OH)_{12} \cdot 3H_2O$
atelestite	$Bi_2O(OH)AsO_4$
attikaite	$Ca_3Cu_2Al_2(AsO_4)_4(OH)_4 \cdot 2H_2O$
auriacusite	$FeCuO(AsO_4)$
austinite	$CaZn(OH)AsO_4$
barahonaite-(Al)	$(Ca,Cu,Na,Al)_{12}Al_2(AsO_4)_8(OH)_x \cdot nH_2O$
barahonaite-(Fe)	$(Ca,Cu,Na,Fe,Al)_{12}Fe_2(AsO_4)_8(OH)_x \cdot nH_2O$
barian tomichite	$Ba_{0.5}(As_2)_{0.5}Ti_2(V,Fe)_5O_{13}(OH)$
bariopharmacosiderite	$Ba_{0.5}Fe_4(OH)_4(AsO_4)_3 \cdot 5H_2O$
baumhauerite	$Pb_3As_4S_9$
bayldonite	$Cu_3Pb(OH)_2(AsO_4)_2$
bearsite	$Be_2(OH)(H_2O)_2(AsO_4) \cdot 2H_2O$
bendadaite	$FeFe_2(OH)_2(H_2O)_4(AsO_4)_2$
bergslagite	$CaBe(OH)AsO_4$
berzeliite	$(NaCa_2)Mg_2(AsO_4)_3$
beta-roselite	$Ca_2(Co,Mg)(H_2O)_2(AsO_4)_2$
betpakdalite	$\{Mg(H_2O)_6\}Ca_2(H_2O)_{13}[Fe_3Mo_8O_{28}(OH)(AsO_4)_2] \cdot 4H_2O$
beudantite	$Pb(Fe,Al)_3[(As,S)O_4]_2(OH)_6$
bouazzerite	$Bi_6(Mg,Co)_{11}Fe_{14}O_{12}(OH)_4(H_2O)_{52}(AsO_4)_{18} \cdot 34H_2O$
bradaczekite	$NaCuCu_3(AsO_4)_3$
braithwaiteite	$NaCu_5(SbTi)O_2(H_2O)_8(AsO_4)_4(AsO_3OH)_2$
brandtite	$Ca_2Mn(H_2O)_2(AsO_4)_2$
brassite	$Mg(H_2O)_4(AsO_3OH)$
bukovskýite	$[Fe_2(H_2O)_6(OH)(AsO_4)](SO_4) \cdot 3H_2O$
bulachite	$Al_2(AsO_4)(OH)_3 \cdot 3H_2O$
burgessite	$Co_2(H_2O)_4[AsO_3(OH)]_2 \cdot H_2O$
cabalzarite	$Ca(Mg,Al,Fe)_2(OH,H_2O)_2(AsO_4)_2$
cafarsite	$Ca_{5.9}Mn_{1.7}Fe_{3.0}Ti_{3.0}(AsO_3)_{12} \cdot 4\text{-}5H_2O$
calcio-andyrobertsite	$KCaCu_5(AsO_4)_4(H_2AsO_4) \cdot 2H_2O$
camgasite	$CaMg(AsO_4)(OH) \cdot 5H_2O$
carminite	$PbFe_2(OH)_2(AsO_4)_2$

Name	Formula
caryinite	$CaNaCaMn_2(AsO_4)_3$
céruléite	$Cu_2Al_7(OH)_{13}(AsO_4)_4 \cdot 11.5H_2O$
chalcophyllite	$Cu_9Al(OH)_{12}(H_2O)_6(SO_4)_{1.5}(AsO_4)_2 \cdot 12H_2O$
chenevixite	$Cu_2Fe_2(OH)_4(H_2O)(AsO_4)_2$
chernovite-Lu	$LuAsO_4$
chernovite-Y	$YAsO_4$
chistyakovite	$Al(UO_2)_2(AsO_4)_2(F,OH) \cdot 6.5H_2O$
chlorophoenicite	$Mn_3Zn_2(OH)_6[As(O,OH)_6]$
chudobaite	$[(Mg,Zn)H_2O)_4][(Mg,Zn)_4(H_2O)_4(AsO_3OH)_2 \\ (AsO_4)_2] \cdot 2H_2O$
chursinite	$(Hg_2)_3(AsO_4)_2$
claudetite	As_2O_3
clinoclase	$Cu_3(OH)_3(AsO_4)$
clinomimetite	$Pb_5(AsO_4)_3Cl$
cobaltarthurite	$CoFe_2(OH)_2(H_2O)_4(AsO_4)_2$
cobaltaustinite	$CaCo(OH)AsO_4$
cobaltite	$CoAsS$
cobaltkoritnigite	$(Co,Zn)(H_2O)(AsO_3OH)$
cobaltlotharmeyerite	$Ca(Co,Fe)_2(H_2O,OH)_2(AsO_4)_2$
cobaltneustädtelite	$Bi_2FeCoO(OH)_3(AsO_4)_2$
cobalttsumcorite	$Pb(Co,Fe)_2(H_2O,OH)_2(AsO_4)_2$
conichalcite	$CaCu(OH)AsO_4$
coparsite	$Cu_4O_2Cl[(As,V)O_4]$
coralloite	$MnMn_2(AsO_4)_2(OH)_2 \cdot 4H_2O$
cornubite	$Cu_5(OH)_4(AsO_4)_2$
cornwallite	$Cu_5(OH)_4(AsO_4)_2$
dixenite	$CuMn_{14}Fe(OH)_6(AsO_3)_5(SiO_4)_2(AsO_4)$
domeykite	Cu_3As
duftite	$PbCu(OH)AsO_4$
dugganite	$Pb_3Te(Zn_3As_2O_{14})$
durangite	$NaAlFAsO_4$
dussertite	$BaFe_3[AsO_{3.5}(OH)_{0.5}]_2(OH)_6$
ecdemite	$Pb_6As_2O_7Cl_4$
ekatite	$(Fe,Zn)_{12}(OH)_6(AsO_3)_6(AsO_3,SiO_3OH)_2$
enargite	Cu_3AsS_4
erythrite	$Co_3(H_2O)_8(AsO_4)_2$
esperanzaite	$NaCa_2Al_2F_4(OH)(H_2O)_2(AsO_4)_2$

Name	Formula
euchroite	$Cu_2(OH)(H_2O)_3(AsO_4)$
eveite	$Mn_2(OH)(AsO_4)$
fahleite	$CaZn_5Fe_2(AsO_4)_6 \cdot 14H_2O$
feinglosite	$Pb_2(Zn,Fe)(H_2O)[(As,S)O_4]$
fermorite	$Ca_5(AsO_4)_3(OH)$
ferrarisite	$Ca(H_2O)_4Ca_4(H_2O)_4(AsO_3OH)_2(AsO_4)_2 \cdot H_2O$
ferrilotharmeyerite	$Ca(Fe,Zn)(OH,H_2O)_2(AsO_4)_2$
ferrisymplesite	poorly crystalline ferric arsenate
fetiasite	$(Fe,Ti)_3O_2(As_2O_5)$
fianelite	$Mn_2(H_2O)_2[V(V,As)O_7]$
filatovite	$K[(Al,Zn)_2(As,Si)_2O_8]$
finnemanite	$Pb_5(AsO_3)_3Cl$
flinkite	$Mn_2Mn(OH)_4(AsO_4)$
fluckite	$CaMn(H_2O)_2(AsO_3OH)_2$
fornacite	$Pb_2Cu(OH)(CrO_4)(AsO_4)$
freedite	$Pb_8Cu(AsO_3)_2O_3Cl_5$
gabrielsonite	$PbFe(OH)AsO_4$
gaitite	$Ca_2Zn(H_2O)_2(AsO_4)_2$
gallobeudantite	$PbGa_3(As_{0.5}S_{0.5}O_4)(SO_4)(OH)_6$
gartrellite	$PbCuFe(OH)(H_2O)(AsO_4)_2$
gasparite-Ce	$CeAsO_4$
gebhardite	$Pb_8OCl_6(As_2O_5)_2$
geigerite	$[Mn(H_2O)_4][Mn_4(H_2O)_4(AsO_3OH)_2(AsO_4)_2] \cdot 2H_2O$
geminite	$Cu(H_2O)(AsO_3OH)$
georgiadesite	$Pb_4Cl_4(OH)(AsO_3)$
gerdtremmelite	$(Zn,Fe)(Al,Fe)_2(OH)_5(AsO_4)$
gersdorffite	$NiAsS$
gilmarite	$Cu_3(OH)_3(AsO_4)$
gottlobite	$CaMg(OH)(V,As)O_4$
goudeyite	$Cu_6Al(OH)_6(AsO_4)_3 \cdot 3H_2O$
graeserite	$Pb_{0.14}As(Fe,Ti)_7O_{12+x}(OH)_{2-x}$
graulichite-(Ce)	$CeFe_3(AsO_4)_2(OH)_6$
grischunite	$NaCa_2Mn_4(Mn_{0.5}Fe_{0.5})_2(H_2O)_2(AsO_4)_6$
guanacoite	$Cu_2Mg_2(Mg_{0.5}Cu_{0.5})(OH)_4(H_2O)_4(AsO_4)_2$
guérinite	$Ca_5(H_2O)_{8.2}(AsO_3OH)_2(AsO_4)_2 \cdot 0.8H_2O$
haidingerite	$Ca(H_2O)(AsO_3OH)$
hallimondite	$Pb_2(UO_2)(AsO_4)_2 \cdot nH_2O$

Name	Formula
hedyphane	$Ca_2Pb_3(AsO_4)_3Cl$
heinrichite	$Ba(UO_2)_2(AsO_4)_2 \cdot 10H_2O$
heliophyllite	$Pb_6As_2O_7Cl_4$
helmutwinklerite	$PbZn_2(H_2O)_2(AsO_4)_2$
hematolite	$Mn(Mn,Mg)_{12}(Al,Fe)_2(AsO_3)(AsO_4)_2(OH)_{23}$
hemloite	$(Ti,V,Fe,Al)_{12}(As,Sb)_2O_{23}(OH)$
HFA	hydrous ferric arsenate, composition uncertain
HFO	hydrous ferric oxide, composition uncertain
hidalgoite	$PbAl_3(As_{0.5}S_{0.5}O_4)_2(OH)_6$
holdenite	$Mn_6Zn_3(OH)_8(AsO_4)_2(SiO_4)$
hörnesite	$Mg_3(H_2O)_8(AsO_4)_2$
hügelite	$Pb_2(UO_2)_3(AsO_4)_2O_2 \cdot 5H_2O$
hutchinsonite	$TlPbAs_5S_9$
hydroxylhedyphane	$Ca_2Pb_3(AsO_4)_3(OH)$
irhtemite	$Ca_4MgH_2(AsO_4)_4$
jamesite	$Pb_2ZnFe_2(Fe_{2.8}Zn_{1.2})(OH)_8[(OH)_{1.2}O_{0.8}](AsO_4)_4$
jarosewichite	$MnMn_3(AsO_4)(OH)_6$
johillerite	$NaCu(Mg,Zn)_3(AsO_4)_3$
johnbaumite	$Ca_5(AsO_4)_3(OH)$
juabite	$CaCu_{10}(OH)_2(H_2O)_4(TeO_3)_4(AsO_4)_4$
juanitaite	$(Cu,Ca,Fe)_{10}Bi(AsO_4)_4(OH)_{11} \cdot 2H_2O$
kaatialaite	$Fe[AsO_2(OH)_2]_3 \cdot 5H_2O$
kahlerite	$Fe(UO_2)_2(AsO_4)_2 \cdot 12H_2O$
kamitugaite	$PbAl(UO_2)_5[(P,As)O_4]_2(OH)_9 \cdot 9.5H_2O$
kemmlitzite	$SrAl_3(As_{0.5}S_{0.5}O_4)_2(OH)_6$
kaňkite	$Fe(H_2O)_{3.5}AsO_4$
keyite	$Cu_3Zn_4Cd_2(H_2O)_2(AsO_4)_6$
klajite	$[Mn(H_2O)_4][Cu_4(H_2O)_4(AsO_3OH)_2(AsO_4)_2] \cdot H_2O$
köttigite	$Zn_3(H_2O)_8(AsO_4)_2$
kolfanite	$Ca_2Fe_3O_2(AsO_4)_3 \cdot 2H_2O$
kolicite	$Mn_7(OH)_4[As_2Zn_4Si_2O_{16}(OH)_4]$
kolitschite	$Pb[Zn_{0.5,\ 0.5}]Fe_3(AsO_4)_2(OH)_6$
koritnigite	$Zn(H_2O)(AsO_3OH)$
koutekite	Cu_5As_2
kraisslite	$Mn_{22}Mg_{1.9}Zn_{3.2}Fe_{0.7}(AsO_3)_2(AsO_4)_{2.7}(SiO_4)_{6.4}(OH)_{18}$
krautite	$Mn(H_2O)(AsO_3OH)$
kuznetsovite	$Hg_3Cl(AsO_4)$

Name	Formula
lammerite	$Cu_3(AsO_4)_2$
laphamite	$As_2(Se,S)_3$
lavendulan	$NaCaCu_5(H_2O)_5Cl(AsO_4)_4$
lazarenkoite	$(Ca,Fe)FeAs_3O_7 \cdot 3H_2O$
legrandite	$Zn_4(OH)_2(H_2O)_2(AsO_4)_2$
leiteite	$ZnAs_2O_4$
lemanskiite	$NaCaCu_5(AsO_4)_4Cl \cdot 5H_2O$
leogangite	$Cu_{10}(OH)_6(H_2O)_4(AsO_4)_4(SO_4) \cdot 4H_2O$
lindackerite	$[Cu(H_2O)_4][Cu_4(H_2O)_4(AsO_3OH)_2(AsO_4)_2] \cdot H_2O$
liroconite	$Cu_2Al(OH)_4(H_2O)_4(AsO_4)$
liskeardite	$(Al,Fe)_3(AsO_4)(OH)_6 \cdot 5H_2O$
löllingite	$FeAs_2$
lotharmeyerite	$Ca(Zn,Mn)_2(H_2O,OH)_2(AsO_4)_2$
ludlockite	$PbFe_4As_{10}O_{22}$
luetheite	$Cu_2Al_2(OH)_4(H_2O)(AsO_4)_2$
lukrahnite	$CaCuFe(OH)(H_2O)(AsO_4)_2$
machatschkiite	$Ca_6(AsO_4)(AsO_3OH)_3(PO_4)(H_2O)_{15}$
maghrebite	$MgAl_2[(OH)(AsO_4)]_2 \cdot 8H_2O$
magnesium-chlorophoenicite	$Mg_3Zn_2(OH)_6[As(O,OH)_6]$
magnussonite	$Mn_{18}(As_6MnO_{18})_2Cl_2$
mahnertite	$(Na,Ca)Cu_{2.75}(H_2O)_{3.63}Cl_{0.62}(AsO_4)_2$
manganarsite	$Mn_3As_2O_4(OH)_4$
manganberzeliite	$(NaCa_2)Mn_2(AsO_4)_3$
manganohörnesite	$(Mn,Mg)_3(H_2O)_8(AsO_4)_2$
manganlotharmeyerite	$Ca(Mn^{3+},\square,Mg)_2(OH,H_2O)_2\{AsO_4,[AsO_2(OH)_2]\}_2$
mansfieldite	$Al(H_2O)_2AsO_4$
mapimite	$Zn_2Fe_3O_{12}(OH)_4(H_2O)_4(AsO_4)_3 \cdot 6H_2O$
mawbyite	$PbFe_2(OH)_2(AsO_4)_2$
maxwellite	$NaFeFAsO_4$
mcgovernite	$(Zn,Mn)_3(Mn,Mg,Fe,Al)_{42}$ $[(As,Zn)O_3]_2(AsO_4)_4[(Si,As)O_4]_8(OH)_{42}$
mcnearite	$NaCa_5H_4(AsO_4)_5 \cdot 4H_2O$
medenbachite	$Bi_2Fe(Cu,Fe)(O,OH)_2(OH)_2(AsO_4)_2$
metaheinrichite	$Ba(UO_2)_2(AsO_4)_2 \cdot 8H_2O$
metakahlerite	$Fe(UO_2)_2(AsO_4)_2 \cdot 8H_2O$
metakirchheimerite	$Co(UO_2)_2(AsO_4)_2 \cdot 8H_2O$
metaköttigite	$(Zn,Fe)(Zn,Fe)_2(H_2O,OH)_8(AsO_4)_2$

Name	Formula
metalodèvite	$Zn(UO_2)_2(AsO_4)_2 \cdot 10H_2O$
metanováčekite	$Mg(UO_2)_2(AsO_4)_2 \cdot 4\text{-}8H_2O$
metarauchite	$Ni(UO_2)_2(AsO_4)_2 \cdot 8H_2O$
metauranospinite	$Ca(UO_2)_2(AsO_4)_2 \cdot 8H_2O$
metazeunerite	$Cu(UO_2)_2(AsO_4)_2 \cdot 8H_2O$
miguelromeroite	$Mn_5(H_2O)_4(AsO_4)_2(AsO_3OH)_2$
mimetite	$Pb_5(AsO_4)_3Cl$
mixite	$Cu_6Bi(OH)_6(AsO_4)_3 \cdot 3H_2O$
morelandite	$Ca_2Ba_3(AsO_4)_3Cl$
nabiasite	$BaMn_9(OH)_2[(V,As)O_4]_6$
nanlingite	$Na(Ca_5Li)Mg_{12}(AsO_3)_2[Fe(AsO_3)_6]F_{14}$
natropharmacosiderite	$NaFe_4(OH)_4(AsO_4)_3 \cdot 5H_2O$
nealite	$Pb_4Fe(H_2O)_2Cl_4(AsO_3)_2$
nelenite	$(Mn,Fe)_{16}Si_{12}O_{30}(OH)_{17}(As_3O_6)$
neustädtelite	$Bi_2Fe_2O_2(OH)_2(AsO_4)_2$
nickelaustinite	$CaNi(OH)AsO_4$
nickeline	$NiAs$
nickellotharmeyerite	$Ca(Ni,Fe)_2(H_2O,OH)_2(AsO_4)_2$
nickelschneebergite	$BiNi_2(OH)(H_2O)(AsO_4)_2$
nickeltalmessite	$Ca_2Ni(H_2O)_2(AsO_4)_2$
nickenichite	$Na_{0.8}Ca_{0.4}Cu_{0.4}(Mg,Fe,Al)_3(AsO_4)_3$
nováčekite	$Mg(UO_2)_2(AsO_4)_2 \cdot 12H_2O$
nyholmite	$Cd_3Zn_2(H_2O)_4(AsO_4)_2(AsO_3OH)_2$
o'danielite	$NaH_2(Zn,Mg)_3(AsO_4)_3$
ogdensburgite	$Ca_2Fe_4(Zn,Mn)(AsO_4)_4(OH)_6 \cdot 6H_2O$
ojuelaite	$ZnFe_2(OH)_2(H_2O)_4(AsO_4)_2$
olivenite	$Cu_2(OH)(AsO_4)$
ondrušite	$[Ca(H_2O)_4][Cu_4(H_2O)_4(AsO_3OH)_2(AsO_4)_2] \cdot H_2O$
orpiment	As_2S_3
orthowalpurgite	$Bi_4O_4(UO_2)(AsO_4)_2 \cdot 2H_2O$
paganoite	$NiBiOAsO_4$
parabrandtite	$Ca_2Mn(H_2O)_2(AsO_4)_2$
paradamite	$Zn_2(OH)(AsO_4)$
pararammelsbergite	$NiAs_2$
pararealgar	AsS
parascorodite	$Fe(H_2O)_2AsO_4$
parasymplesite	$Fe_3(H_2O)_8(AsO_4)_2$

Name	Formula
parnauite	$Cu_9(AsO_4)_2(SO_4)(OH)_{10} \cdot 7H_2O$
parwelite	$Mn_5SbO_4(SiO_4)(AsO_4)$
paulmooreite	$Pb_2(As_2O_5)$
petewilliamsite	$(Ni,Co)_{30}(As_2O_7)_{15}$
pharmacolite	$Ca(H_2O)_2(AsO_3OH)$
pharmacosiderite	$KFe_4(OH)_4(AsO_4)_3 \cdot 5H_2O$
phaunouxite	$Ca_3(H_2O)_8(AsO_4)_2 \cdot 3H_2O$
philipsbornite	$PbAl_3[AsO_{3.5}(OH)_{0.5}]_2(OH)_6$
philipsburgite	$Cu_5(OH)_5(H_2O)[Zn(OH)(As,P)O_4]$
picropharmacolite	$Ca_4Mg(H_2O)_7(AsO_3OH)_2(AsO_4)_2 \cdot 4H_2O$
pitticite	amorphous hydrated ferric arsenate with poorly defined chemical composition
plumboagardite	$Cu_6Pb(OH)_6(AsO_3OH)(AsO_4)_2 \cdot 3H_2O$
pradetite	$[Co(H_2O)_4][Cu_4(H_2O)_2(AsO_3OH)_2(AsO_4)_2] \cdot H_2O$
preisingerite	$Bi_3O(OH)(AsO_4)_2$
prosperite	$Ca_2Zn_4(H_2O)(AsO_4)_4$
pushcharovskite	$Cu(H_2O)(AsO_3OH) \cdot 0.5H_2O$
radovanite	$Cu_2Fe(AsO_4)(AsO_2OH)_2 \cdot H_2O$
rammelsbergite	$NiAs_2$
rappoldite	$Pb(Co,Ni)_2(H_2O)_2(AsO_4)_2$
rauenthalite	$Ca_3(H_2O)_8(AsO_4)_2 \cdot 2H_2O$
realgar	AsS
reinerite	$Zn_3(AsO_3)_2$
retzian	$Mn_2REE(OH)_4(AsO_4)$
richelsdorfite	$Ca_2Cu_5SbCl(OH)_6(H_2O)_6(AsO_4)_4$
rollandite	$Cu_3(H_2O)_2(AsO_4)_2 \cdot 2H_2O$
rooseveltite	$BiAsO_4$
roselite	$Ca_2(Mg,Co)(H_2O)_2(AsO_4)_2$
rösslerite	$Mg(H_2O)_6(AsO_3OH) \cdot H_2O$
rouseite	$Pb_2Mn(AsO_3)_2 \cdot 2H_2O$
rruffite	$Ca_2Cu(H_2O)_2(AsO_4)_2$
sabelliite	$(Cu,Zn)_2Zn(OH)_3[(As,Sb)O_4]$
sahlinite	$Pb_{14}O_9Cl_4(AsO_4)_2$
safflorite	$(Co,Fe)As_2$
sailaufite	$(Ca,Na,)_2Mn_3O_2(H_2O)_2(CO_3)(AsO_4)_2 \cdot H_2O$
sainfieldite	$Ca_5(H_2O)_4(AsO_4)_2(AsO_3OH)_2$
sarkinite	$Mn_2(OH)(AsO_4)$

Name	Formula
sarmientite	$Fe_2(AsO_4)(SO_4)(OH)\cdot5H_2O$
sartorite	$PbAs_2S_4$
schallerite	$Mn_{16}Si_{12}O_{30}(OH)_{17}(As_3O_6)$
schlegelite	$Bi_7O_4(MoO_4)_2(AsO_4)_3$
schneebergite	$BiCo_2(OH)(H_2O)(AsO_4)_2$
schneiderhöhnite	$FeFe_3As_5O_{13}$
schultenite	$Pb(AsO_3OH)$
scorodite	$Fe(H_2O)_2AsO_4$
skutterudite	$(Co,Fe,Ni)As_{2-3}$
seelite	$Mg[(UO_2)(AsO_3)_{0.7}(AsO_4)_{0.3}]_2\cdot7H_2O$
segnitite	$PbFe_3[AsO_{3.5}(OH)_{0.5}]_2(OH)_6$
seligmannite	$PbCuAsS_3$
sewardite	$CaFe_2(OH)_2(AsO_4)_2$
shubnikovite	$Ca_2Cu_8(AsO_4)_6Cl(OH)\cdot7H_2O$
slavkovite	$Cu_{13}(H_2O)_{16}(AsO_4)_6(AsO_3OH)_4\cdot7H_2O$
smolianinovite	$(Co,Ni,Mg,Ca)_3(Fe,Al)_2(AsO_4)_4\cdot11H_2O$
sodium-uranospinite	$(Na_2,Ca)(UO_2)_2(AsO_4)_2\cdot5H_2O$
stenhuggarite	$CaFeSbAs_2O_7$
sterlinghillite	$Mn_3(AsO_4)_2\cdot4H_2O$
stranskiite	$Zn_2Cu(AsO_4)_2$
strashimirite	$Cu_4(AsO_4)_2(OH)_2\cdot2.5H_2O$
svabite	$Ca_5(AsO_4)_3F$
švenekite	$Ca[AsO_2(OH)_2]$
symplesite	$Fe_3(H_2O)_8(AsO_4)_2$
synadelphite	$Mn_9(OH)_9(H_2O)_2(AsO_3)(AsO_4)_3$
talmessite	$Ca_2Mg(H_2O)_2(AsO_4)_2$
tennantite	$Cu_6[Cu_4(Fe,Zn)_2]As_4S_{13}$
tetrarooseveltite	$BiAsO_4$
theisite	$Cu_5Zn_5(OH)_{14}[(As,Sb)O_4]_2$
theoparacelsite	$Cu_3(OH)_2As_2O_7$
thometzekite	$Pb(Cu,Zn)_2(H_2O)_2(AsO_4)_2$
tilasite	$CaMgFAsO_4$
tillmannsite	$(Ag_3Hg)(V,As)O_4$
tiragalloite	$Mn_4(HAsSi_3O_{13})$
tomichite	$AsTi_3(V,Fe)_4O_{13}(OH)$
tooeleite	$Fe_6(OH)_4(H_2O)_4(AsO_3)_4(SO_4)$
trigonite	$Pb_3Mn(AsO_3)_2(AsO_2OH)$

Name	Formula
trippkeite	$CuAs_2O_4$
trögerite	$(H_3O)(UO_2)(AsO_4)\cdot 3H_2O$
tsumcorite	$Pb(Zn,Fe)_2(OH, H_2O)_2(AsO_4)_2$
turneaureite	$Ca_5(AsO_4)_3Cl$
turtmannite	$(Mn,Mg)_{22.5}Mg_{3-3x}[(V,As)O_4]_3(SiO_4)_3$ $(AsO_3)_xO_{5-5x}(OH)_{20+x}$
tyrolite-1M	$Ca_2Cu_9(OH)_8(CO_3)(H_2O)_{11}(AsO_4)_4\cdot xH_2O$
tyrolite-2M	$Ca_2Cu_9(OH)_8(CO_3)(H_2O)_{11}(AsO_4)_4\cdot xH_2O$
uramarsite	$(NH_4,H_3O)_2(UO_2)_2[(As,P)O_4]_2\cdot 6H_2O$
uranospinite	$Ca(UO_2)_2(AsO_4)_2\cdot 10H_2O$
urusovite	$CuAlO(AsO_4)$
vajdakite	$(MoO_2)_2(H_2O)_2As_2O_5\cdot H_2O$
veselovskýite	$[Zn(H_2O)_4][Cu_4(H_2O)_4(AsO_3OH)_2(AsO_4)_2]\cdot H_2O$
villyaellenite	$(Mn,Ca)Mn_2Ca_2(H_2O)_4(AsO_4)_2(AsO_3OH)_2$
vladimirite	$Ca_5H_2(AsO_4)_4\cdot 5H_2O$
walentaite	$Ca_4Fe_{12}H_4(AsO_4)_{10}(PO_4)_6\cdot 28H_2O$
wallkilldellite	$Ca_4Mn_6As_4O_{16}(OH)_8\cdot 18H_2O$
walpurgite	$Bi_4O_4(UO_2)(AsO_4)_2\cdot 2H_2O$
warikahnite	$Zn_3(H_2O)_2(AsO_4)_2$
weilerite	$BaAl_3(As_{0.5}S_{0.5}O_4)_2(OH)_6$
weilite	$CaHAsO_4$
wendwilsonite	$Ca_2Mg(H_2O)_2(AsO_4)_2$
wilhelmkleinite	$ZnFe_2(OH)_2(AsO_4)_2$
xanthiosite	$Ni_3(AsO_4)_2$
yanomamite	$In(H_2O)_2AsO_4$
yukonite	$Ca_2Fe_3(AsO_4)_3(OH)_4\cdot 4H_2O$
yvonite	$Cu(H_2O)(AsO_3OH)\cdot H_2O$
zálesíite	$Cu_6Ca(OH)_6(AsO_3OH)(AsO_4)_2\cdot 3H_2O$
zděněkite	$NaPbCu_5(H_2O)_5Cl(AsO_4)_4$
zeunerite	$Cu(UO_2)_2(AsO_4)_2\cdot 12H_2O$
zimbabweite	$Na(Pb,Na,K)_2As_4(Ta,Nb,Ti)_4O_{18}$
zincgartrellite	$Pb(Zn,Cu)(Zn,Fe)(OH,H_2O)_2(AsO_4)_2$
zincolivenite	$CuZn(OH)(AsO_4)$
zincroselite	$Ca_2Zn(H_2O)_2(AsO_4)_2$
zýkaite	$Fe_4(AsO_4)_3(SO_4)(OH)\cdot 15H_2O$

APPENDIX 2

Index and list of As-bearing minerals which do not contain arsenic as in their nominal chemical composition but may include trace to major amount of this element, either in their crystal structures or adsorbed onto the surfaces. Note that we deliberately omitted any information about the oxidation state of the elements in the formulae as in many cases, the oxidation state is not clear, not definitely proven, or may be variable.

Mineral	Formula
Sulfides, antimonides	
breithauptite	NiSb
marcasite	FeS_2
pyrite	FeS_2
pyrrhotite	$Fe_{1-x}S$
Oxides	
ferrihydrite	$5Fe_2O_3 \cdot 9H_2O$
goethite	$\alpha\text{-FeO(OH)}$
gibbsite	$\gamma\text{-Al(OH)}_3$
hematite	$\alpha\text{-Fe}_2O_3$
lepidocrocite	$\gamma\text{-FeO(OH)}$
maghemite	$\gamma\text{-Fe}_2O_3$
Mn oxides	oxyhydroxides of Mn^{4+}, Mn^{3+} and additional cations
Sulfates	
copiapite	$(Fe,Mg)Fe_4(SO_4)_6(OH)_2 \cdot 20H_2O$
dietrichite	$(Zn,Fe,Mn)Al_2(SO_4)_4 \cdot 22H_2O$
jarosite	$KFe_3(SO_4)_2(OH)_6$
rhomboclase	$HFe(SO_4)_2 \cdot 4H_2O$
schwertmannite	$Fe_8O_8(OH)_6SO_4$
Phosphates	
corkite	$Pb(Fe,Al)_3[(P,S)O_4]_2(OH)_6$
Silicates	
chlorites	complex sheet silicates of Mg,Fe,Al
coffinite	$USiO_4$
crandallite	$CaAl_3(PO_4)_2(OH_5) \cdot H_2O$
thorite	$ThSiO_4$
titanite	$CaTi(SiO_4)\,O$
zircon	$ZrSiO_4$

Reviews in Mineralogy & Geochemistry
Vol. 79 pp. 185-216, 2014
Copyright © Mineralogical Society of America

Arsenic Speciation and Sorption in Natural Environments

Kate M. Campbell and D. Kirk Nordstrom

U.S. Geological Survey
National Research Program
Boulder, Colorado 80303, U.S.A.

kcampbell@usgs.gov　　　*dkn@usgs.gov*

INTRODUCTION

Aqueous arsenic speciation, or the chemical forms in which arsenic exists in water, is a challenging, interesting, and complicated aspect of environmental arsenic geochemistry. Arsenic has the ability to form a wide range of chemical bonds with carbon, oxygen, hydrogen, and sulfur, resulting in a large variety of compounds that exhibit a host of chemical and biochemical properties. Besides the intriguing chemical diversity, arsenic also has the rare capacity to capture our imaginations in a way that few elements can duplicate: it invokes images of foul play that range from sinister to comedic (e.g., "inheritance powder" and arsenic-spiked elderberry wine). However, the emergence of serious large-scale human health problems from chronic arsenic exposure in drinking water has placed a high priority on understanding environmental arsenic mobility, toxicity, and bioavailability, and chemical speciation is key to these important questions. Ultimately, the purpose of arsenic speciation research is to predict future occurrences, mitigate contamination, and provide successful management of water resources.

Chemical speciation is fundamental to understanding mobility and toxicity. Speciation affects arsenic solubility and solid-phase associations, and thus the mobility, of arsenic in natural waters. It is also critical to designing treatment strategies, understanding human exposure routes, and even developing medical applications (e.g., as a treatment for acute promyelocytic leukemia; Antman 2001). As single- and multi-celled organisms are exposed to various forms of arsenic, they often alter its speciation to either utilize the arsenic for energy or to mitigate the detrimental effects of intracellular arsenic (detoxification). Some organisms can accumulate arsenic in cell material, which can be a concern if it accumulates in a human food product such as rice or seafood, but could be a potential remediation solution in hyper-accumulating plants (Ma et al. 2001).

It is important to quantify speciation in addition to total amount of arsenic because arsenic speciation determines the toxicity and physiological effects of consuming arsenic-contaminated food and water. Geochemists often rely on computational geochemical speciation codes to calculate various forms of arsenic species based on available major solute and trace element composition. The applicability of these calculations depends heavily upon reliable thermodynamic data and careful updating of the thermodynamic constants as new species are identified and measured (Nordstrom 2000; Nordstrom and Archer 2003); an updated discussion of arsenic stability constants is presented by Nordstrom et al. (2014, this volume). Geochemical modeling often takes an equilibrium-based approach, but the kinetics of geochemical arsenic transformations can be an important factor in the observed speciation, resulting in numerous co-existing arsenic species. For example, it is not uncommon to measure oxidized and reduced arsenic oxidation states simultaneously in water samples, yet arsenic would be predicted to exist as only one oxidation state based on equilibrium calculations.

The purpose of this review is to provide an overview of aqueous arsenic speciation in natural waters, adsorption of arsenic on minerals, and redox reactions of arsenic. Additionally, we present speciation calculations for a set of reported water compositions (acid mine drainage, geothermal water, and groundwater) to illustrate the behavior of arsenic speciation over a wide range of natural conditions and to identify critical knowledge gaps.

AQUEOUS INORGANIC ARSENIC SPECIES

Complexation

Arsenic is stable in five oxidation states (−III, −I, 0, III, V), although it is not commonly found in its elemental form outside of hypogene ore deposits and there are no known stable As(0) aqueous species. Arsenic may also occur in the −I oxidation state in sulfarseniide minerals such as arsenopyrite and arsenian pyrite. Arsenite (As(III)) and arsenate (As(V)) tend to form oxyanionic and thioanionic species in aqueous solution. Arsenate behaves similarly to its chemical analog, phosphate (Tamaki and Frankenberger 1992), and arsenite behaves somewhat similarly to aqueous boron species. However, while reduced forms of phosphorous are not abundant in the environment, reduced arsenic, primarily as As(III), is often observed. Arsine (As(−III), Table 1, compound 3) only forms under extremely reducing conditions or as the result of biological activity, and tends to partition into the gas phase.

Oxyanions of As(III) and As(V) are stable forms of inorganic arsenic across a wide range of environmentally relevant aqueous conditions. The hydrolysis of H_3AsO_4 and H_3AsO_3 (Table 1, compounds 1, 2) has been extensively studied and the hydrolysis constants are known with a high degree of confidence (Nordstrom and Archer 2003). Arsenite exists as a neutral species ($pK_{a1} = 9.2$) over the majority of naturally occurring water conditions, whereas arsenate ($pK_{a1} = 2.3$, $pK_{a2} = 7.0$) tends to form univalent or divalent anions. Thus, the behavior and mobility of arsenic is fundamentally controlled by its oxidation state and its degree of hydrolysis.

Arsenic complexation with major cations (Na, Ca, K, Mg, Table 1, compound 4) and trace metals (Al, Cd, Co, Fe, Mn, Ni, Pb, Sr, Zn) commonly found in natural waters has recently been measured or estimated (Marini and Accornero 2007; Nordstrom et al. 2014, this volume). These aqueous complexes had been hypothesized to exist based on phosphate analogs, but direct measurement or thermodynamic estimation of the arsenic species allows for inclusion of these species into geochemical models. Both As(III) and As(V) can form metal-arsenic complexes, but the metal-As(V) complexes should be more common in an environmental context than the metal-As(III) complexes because arsenate is more ionized than arsenite. In fact, Na-, Mg-, or Ca-As(V) complexes can be the dominant aqueous species under some geochemical conditions, as demonstrated in the speciation calculations of several natural waters, presented in Figure 1 and later in this chapter (Tables 3 and 4). Iron-arsenic complexes may be important under low pH, high-iron conditions, such as acid rock drainage (Fig. 1).

Arsenic speciation in sulfidic environments may be dominated by thioarsenic species (Table 1, compounds 5, 6), in which the oxygen atoms bonded to arsenic are replaced by sulfur atoms. These species are important in environments with abundant reduced sulfur, such as organic-rich surface- and ground waters, landfill leachate, hydrocarbon plumes, reduction-based engineered groundwater remediation, lacustrine and marine sediments, and geothermal waters. Thioarsenic species have been observed in geothermal waters in Yellowstone National Park (Wyoming, USA), with high abundance (up to 83% of total arsenic) in high temperature, high sulfide, and high pH waters (Planer-Friedrich et al. 2007). In Champagne Pool (a geothermal spring in Waiotapu, New Zealand), di- and tri-thioarsenates composed up to 69% of total arsenic (Ullrich et al. 2013). Speciation showed a strong diurnal trend, and di- and tri-

Table 1. Inorganic aqueous arsenic species.

#	Name	Chemical & Structural Formula
1	arsenic acid, hydrolytic anions	H_3AsO_4, $H_2AsO_4^-$, $HAsO_4^{2-}$, AsO_4^{3-}
2	arsenous acid (arsenious acid), hydrolytic anions	H_3AsO_3, $H_2AsO_3^-$, $HAsO_3^{2-}$, AsO_3^{3-}
3	arsine	AsH_3
4	divalent metal arsenates	$Me^{II}HAsO_4^{\circ}$, $Me^{II}H_2AsO_4^+$ Me = Na, Ca, K, Mg, Al, Mn, Cu, Co, Cd, Fe, Ni, Pb, Sr, Zn
5	thioarsenic acid, hydrolytic anions; thioarsenates	H_3AsS_4, $H_2AsS_4^-$, $HAsS_4^{2-}$, AsS_4^{3-}; $AsO_{4-x}S_x^{3-}$
6	thioarsenous acid, hydrolytic anions; thioarsenites	H_3AsS_3, $H_2AsS_3^-$, $HAsS_3^{2-}$, AsS_3^{3-}; $AsO_{3-x}S_x^{3-}$

thioarsenates were dominant at night when excess sulfide was present. Increasing amounts of monothioarsenate were present during the day due to phototrophic sulfur oxidizing bacteria producing elemental sulfur that reacted with arsenite to form monothioarsenate (Ullrich et al. 2013). Thioarsenic species have also been measured in Mono Lake (California, USA), a hyper-saline, alkaline lake with high arsenic concentrations and elevated sulfide at depth. When the water contained high sulfide, thioarsenic species were the dominant form of arsenic in the lake water, and the amount of sulfhydryl substitution (mono-, di-, and tri-thioarsenic) increased with increasing sulfide concentrations (Hollibaugh et al. 2005). During organic carbon amendment for treatment of a metal-contaminated aquifer (Rifle, Colorado, USA), thioarsenic species were measured during sulfate reduction, with more than 50% of the total arsenic occurring as mono-, di-, and tri-thioarsenic species (Stucker et al. 2013). The complexity of thioarsenic species may even extend to trace metals. For example, Clarke and Helz (2000) describe the possibility of a ternary copper-thioarsenic aqueous complex that may increase mobility of copper under sulfidic conditions. However, characterization of ternary metal-thioarsenic species is very limited, and more information is needed to understand their behavior and stability.

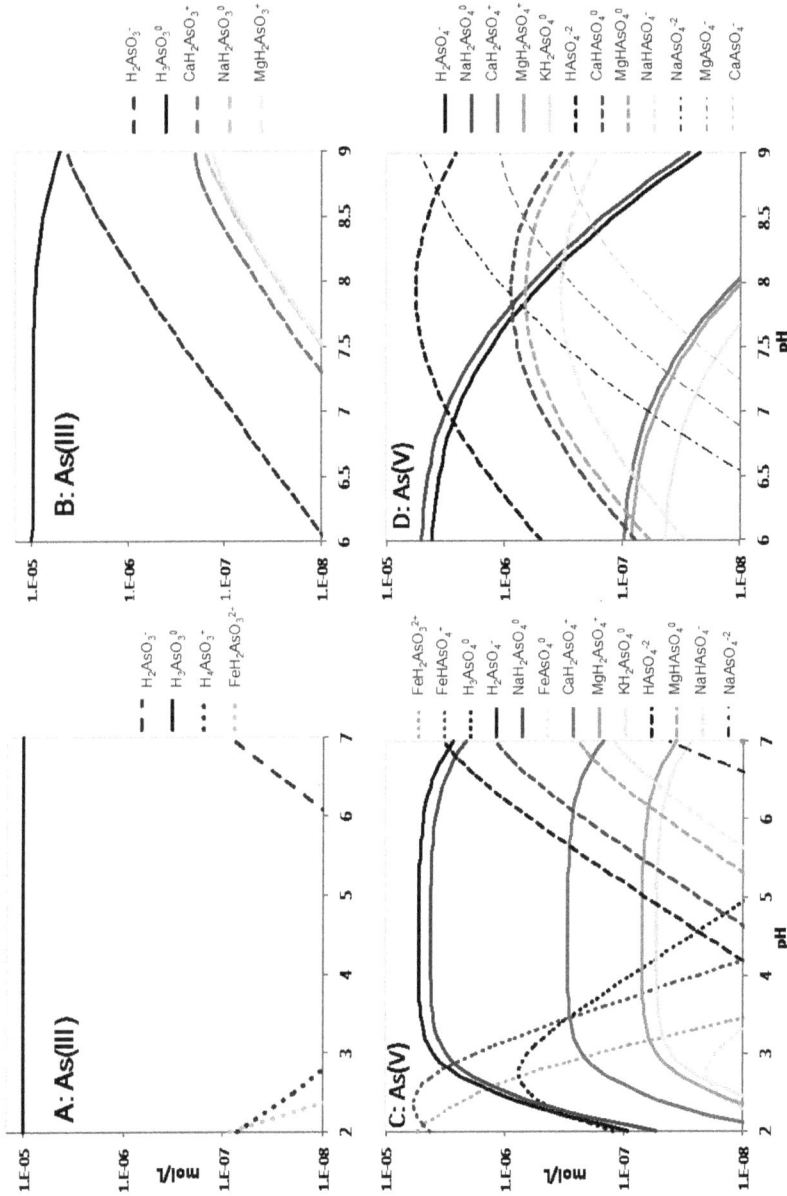

Figure 1. Abundance of arsenite (A, B) and arsenate (C, D) species in a generic high sulfate, acid-to-neutral water in panels A and C (10 mmol L^{-1} As, 15 mmol L^{-1} sulfate, 7 mmol L^{-1} Ca, 1 mmol L^{-1} Mg, 0.2 mmol L^{-1} K, 20 mmol L^{-1} Na, 90 mmol L^{-1} Fe or Fe in equilibrium with ferrihydrite, 400 ppm CO$_2$) and an alkaline sodium-type water in panels B and D (10 mmol L^{-1} As, 2 mmol L^{-1} sulfate, 2 mmol L^{-1} Ca, 1 mmol L^{-1} Mg, 0.2 mmol L^{-1} K, 30 mmol L^{-1} Na, 400 ppm CO$_2$). Speciation was calculated with the PHRE-EQC geochemical code (Parkhurst and Appelo 2013) with the WATEQ4F database modified to include constants for arsenic–metal complexes (Nordstrom et al. 2014, this volume).

Measurement of thioarsenic species is extremely difficult, as they are susceptible to alteration during sampling, preservation, and analysis. Proper preservation requires anoxic flash freezing of the water sample with minimal headspace after addition of a stabilizing EDTA solution (Suess et al. 2011), and speciation determination must be performed with specialized analytical capabilities (Wilkin et al. 2003; Bostick et al. 2005; Stauder et al. 2005; Wallschläger and Stadey 2007; Suess et al. 2011). As a result, there has been considerable debate about the identification of thioarsenites versus thioarsenates in natural waters. It is clear that thioarsenic species are environmentally relevant and further research is needed to understand their formation, transformation, and reactivity under various conditions.

Polymerization

Some evidence for the existence of polymeric inorganic As(III) has been reported in the literature (Garrett et al. 1940), but the experiments were conducted at relatively high pH compared to conditions in most natural waters, and are probably not directly applicable to environmental systems. There is also evidence for synthetic As_n polymers as ligands to a variety of metal ions, such iron, cobalt, and tungsten, which are chemical analogs to phosphorous polymers (Scherer 1999). Although the chemistry of these compounds is quite interesting, they do not appear to be particularly stable in air, and to our knowledge have not yet been adapted for a commercial use that would distribute them in the environment. Therefore, their environmental relevance is limited.

In environments where thioarsenic species are present, a trimer species $(As_3S_4(SH)_2^-)$ is possible in systems saturated with $As_2S_{3(am)}$ (Spycher and Reed 1989; Webster 1990; Helz et al. 1995). Trimeric thioarsenite species were measured by Raman spectroscopy in mildly acidic solutions (pH 4) of As(V) reacted with excess aqueous sulfide, but the species completely dissociated into arsenite and polysulfides/elemental sulfur after several days (Rochette et al. 2000). Trimeric thioarsenites were also observed with X-ray absorption spectroscopy in a laboratory study when sulfide concentrations were at least three times higher than arsenic concentrations. Trimeric species $(As_3S_{3+x}(SH)_{3-x}^{-x})$ composed up to 40% of the total arsenic (Bostick et al. 2005). However, the trimeric species appeared to hydrolyze into thioarsenite during chromatographic separation, suggesting that trimers are unstable and difficult to quantify. The relevance and stability of polymeric thioarsenic species in the environment has not been determined.

ORGANIC ARSENIC COMPOUNDS AND INTERACTIONS WITH ORGANIC MATTER

Organic arsenical compounds

A wide variety of methylated arsenic compounds can be found naturally as the byproducts of bacterial, fungal, or eukaryotic enzymatic processes (Ehrlich 2002). It is generally thought that many organisms methylate arsenic as a detoxification strategy (Tamaki and Frankenberger 1992; Oremland and Stolz 2003), although recent data suggests that some methylated As(III) species can be as toxic or more toxic than inorganic arsenic in humans (Petrick et al. 2000). In eukaryotes, some plants, and many bacteria, production of monomethylarsonic acid (MMAs^V), monomethylarsonous acid (MMAs^{III}), dimethylarsinic acid (DMAs^V), and dimethylarsenous acid (DMAs^{III}) is common, while methanogenic archaea and some fungi can produce volatile di- and trimethylarsine (Table 2, compounds 1-5) (Tamaki and Frankenberger 1992; Oremland and Stolz 2003). Trimethylarsine oxide (TMAO) and tetramethylarsonium ion (TETRA; Table 2, compounds 6-7) have more methyl groups than MMAs^V, and are common detoxification products that occur in the same metabolic series. Reduced arsenic in methylated species (MMAs^{III} and DMAs^{III}) is unstable under oxidizing conditions and is readily oxidized (Gong et al. 2001). Methylation may be pH dependent and is favored at slightly acidic pH conditions (Tamaki and Frankenberger 1992). Methylated arsenic species tend to be persistent in the environment and are generally not metabolized or bioaccumulated by higher eukaryotes. However, many bacteria are capable of demethylation, in which organic arsenicals are converted to inorganic arsenic and carbon dioxide (Rosen et al. 2012). Other types of naturally occurring organic arsenic compounds include arsenobetaine, arsenocholine, trimethylarsoniopropionate (TMAP), and arsenic substituted sugars, which are commonly found in shellfish, marine animals, and plants (Table 2, compounds 8-10, 18) (Cullen and Reimer 1989; Oremland and Stolz 2003).

Historically, there have been many uses for manufactured organic arsenic compounds. Organoarsenical compounds were also used for military purposes, including the warfare agents diphenylchlorarsine, diphenylcyanoarsine, and phenyldicholoroarsine, as well as cacodylic acid (DMAs^V) as a defoliation agent (Bissen and Frimmel 2003a). Organic arsenical compounds such as arsanilic acid, Nitrarsone, phenylarsinic acid, and Roxarsone have been used in feedlots (poultry and swine) to control parasites (Table 2, compounds 11-14), although the FDA has announced a voluntary withdrawal of some of these compounds from the market. Salvarsan, carbarsone, and tryparsamide (Table 2, compounds 15-17) were revolutionary in chemotherapeutic medicine and were developed for treatment of spirochete and protozoan infections. One of the most common anthropogenic sources of arsenic to the environment is the agricultural use of herbicides, fungicides, and insecticides such as monosodium methylarsonate (MSMA), disodium methylarsonate (DSMA), dimethylarsinic acid (cacodylic acid), and calcium methylarsonate (Bissen and Frimmel 2003a; EPA 2013). The herbicides MSMA and DMSA have been widely applied to cotton fields. In a study of agricultural water and soil near cotton fields where these herbicides had been applied via crop-dusting, elevated concentrations of methyarsenic species were found in drainage ditches and irrigation wells, though adsorption onto soils and dilution maintained low concentrations of total arsenic in surrounding groundwater (Bednar et al. 2002).

Arsenic complexation and reaction with natural organic matter

Natural organic matter (NOM) is generally derived from decaying plant, animal, and microbial material, resulting in a large diversity of molecular weights, functional groups, and reactivity. It is a mixture of phenols, fatty acids, carbohydrates, sugars, and amino acids, and contains a wide variety of active functional groups. Molecules of NOM can be truly dissolved, colloidal, bound to mineral surfaces in soils and sediments, or found as a solid organic phase (e.g., peat). The chemical reactivity of NOM is complex, and varies with its molecular weight, elemental composition, functional group composition, and aromaticity. NOM may serve as a ligand for metal and metalloid binding (complexation), as a growth substrate for microbes, as an electron shuttle (transports electrons from an electron donor to acceptor), and as a ternary complexing agent on mineral surfaces. The wide range in composition and reactivity of NOM make it extremely important in speciation and redox cycling of arsenic in the environment. One of the most important roles of NOM is its ability to be a substrate for microbial activity, resulting in reducing conditions that greatly affect arsenic speciation and mobility.

Solid-phase association of arsenic and NOM is complex. It can promote arsenic mobilization by competitive desorption from mineral surfaces and it can accumulate arsenic through direct complexation on solid organic phases. NOM will bind to solid iron minerals through functional groups such as carboxylic acids, phenols, and amines or as outer sphere complexes (Kaiser et al. 1997; Grafe et al. 2002; Saada et al. 2003; Buschmann et al. 2006). Adsorption of NOM on iron (hydr)oxides tends to decrease arsenic adsorption, although the extent of this process depends on the functional groups and size of the carbon compound. Organic matter, especially humic and fulvic acids, can inhibit arsenate and arsenite adsorption to Fe(III) oxide surfaces because of steric and/or electrostatic effects (Xu et al. 1991; Grafe et al. 2001, 2002; Redman et al. 2002; Ko et al. 2004; Bauer and Blodau 2009).

Table 2. Aqueous organic arsenicals.

#	Name	Chemical/Structural Formula
1	methylated arsines	CH_3AsH_2, $(CH_3)_2AsH$, $(CH_3)_3As$
2	monomethylarsonous acid [MMAs(III)]	$CH_3As(OH)_2$
3	dimethylarsenous acid [DMAs(III)]	$(CH_3)_2AsOH$
4	monomethylarsonic acid [MMAs(V)]	$CH_3AsO(OH)_2$
5	dimethylarsinic acid [DMAs(V), cacodylic acid]	$(CH_3)_2AsO(OH)$
6	trimethylarsine oxide (TMAO)	$(CH_3)_3AsO$
7	tetramethylarsonium ion (TETRA)	$(CH_3)_4As^+$
8	arsenobetaine (AB, trimethylarsonioacetate)	$(CH_3)_3As^+CH_2COO^-$

Table 2 continued. Aqueous organic arsenicals.

#	Name	Chemical/Structural Formula
9	arsenocholine (AC)	$(CH_3)_3As^+C_2H_4OH$
10	trimethylarsonio-propionate (TMAP)	$(CH_3)_3As^+C_2H_4CO_2H$
11	(4-aminophenyl)-arsonic acid (arsanilic acid, *p*-aminobenzene-arsonic acid)	$NH_2C_6H_4AsO(OH)_2$
12	4-nitrophenylarsonic acid (*p*-nitrophenylarsonic acid, Nitarsone)	$NO_2(OH)C_6H_3As(OH)_2$
13	3-nitro-4-hydroxy-phenylarsinic acid	$NO_2(OH)C_6H_3As(OH)_2$
14	3-nitro-4-hydroxy-phenylarsonic acid (Roxarsone)	$NO_2(OH)C_6H_3AsO(OH)_2$

Table 2 continued. Aqueous organic arsenicals.

#	Name	Chemical/Structural Formula
15	4,4-arsenobis (2-aminophenol) dihydrochloride (arsphenamine, salvarsan)	$NH_3HCl(OH)C_6H_3As_2NH_2HCl(OH)$
16	[4-[aminocarbonyl-amino]phenyl] arsonic acid (carbarsone, *N*-carbamoylarsanilic acid)	$NH_2CONHC_6H_4AsO(OH)_2$
17	[4-[2-amino-2-oxoethyl) amino]-phenyl] arsonic acid (tryparsamide)	$NH_2COCH_2NHC_6H_4AsO(OH)_2$
18	dimethyl arsenosugars (dimethylarsenoyl-ribosides)	$(CH_3)_2AsOCH_2[CHO(CHOH)_2CH]R$ (R = organic functional group)

Natural organic matter indirectly affects arsenic by binding metal cations as aqueous complexes and stabilization of metal colloids, especially iron (Ritter et al. 2006). Organic matter can stabilize ferric iron in solution through direct complexation (oligomeric Fe(III) binding) or through stabilization of ferric iron polymers and colloids (prevention of aggregation and precipitation) (Bauer and Blodau 2009). Arsenic has been found to associate with dissolved organic carbon primarily through metal bridging with complexed iron (Ko et al. 2004; Lin et al. 2004; Warwick et al. 2005; Ritter et al. 2006; Mikutta and Kretzschmar 2011; Hoffmann et al. 2013). When arsenate and arsenite complexation with six NOM samples was tested, four formed arsenic-NOM complexes, and the extent of complexation was positively correlated to iron content of the NOM (Redman et al. 2002). Spectroscopic evidence indicates that As(V) and As(III) bind to oligomeric Fe(III) or small Fe(III) clusters of low nuclearity within the NOM through ternary complex formation (Mikutta and Kretzschmar 2011; Hoffmann et al. 2013). As a result, the effect of pH is generally similar to arsenic adsorption to Fe oxides, although more data are required to understand the combined effects of pH, Fe and NOM functional group composition on arsenic binding. Arsenate tends to form stronger complexes than As(III) on a variety of NOM substrates (Thanabalasingam and Pickering 1986a; Warwick et al. 2005; Buschmann et al. 2006), although this trend depends on the composition of the NOM (Ko et al. 2004). Warwick et al. (2005) also showed that some organic arsenic species bind less strongly than inorganic arsenic species. Other metal cations such as Al and Mn can also form NOM-metal complexes and potentially As-metal-OM ternary complexes. However, Al does not appear to form ternary complexes to the same extent as iron (Buschmann et al. 2006). There are also proposed mechanisms for direct NOM-As binding, such as via amino groups (Thanabalasingam and Pickering 1986a) and phenolic and carboxylate groups (Buschmann et al. 2006). The extent of these direct complexation reactions compared to the ternary complexes most likely depends on the functional group composition of the NOM.

Stabilization of Fe oxide colloids and particles by a NOM coating ultimately increases the amount of arsenic associated with the particles, making arsenic more mobile (Bauer and Blodau 2009). The importance of arsenic-associated NOM-stabilized Fe(III) colloids in the environment was demonstrated in a study of a German stream that is fed by deep, oxic groundwater during base flow conditions, but has significant inputs of surface water that flows through organic-rich soil layers during storm events (Neubauer et al. 2013). During storm events, the concentrations of iron, NOM, and arsenic were higher in the stream. The high dissolved iron and arsenic concentrations were attributed to NOM-stabilized Fe(III) colloids.

Numerous wetland and peat environments show accumulation of arsenic that cannot be explained by arsenic mineral precipitation alone (e.g., Gonzalez et al. 2006), indicating the importance of direct NOM-As interactions in these environments. In a spectrographic study of As(III) associations with peat from an arsenic-enriched wetland, Langner et al. (2011) found that As(III) was bound to organic sulfur groups, especially in reduced NOM. This observation also suggests that the oxidation state of the arsenic and the functional groups in the NOM may be the key to understanding the ultimate binding of arsenic to NOM in a particular environment.

Many functional groups in NOM can participate in redox reactions by acting as an electron donor and/or acceptor. For example, humic acids can act as an electron acceptor for microbial growth (e.g., Lovley et al. 1996), and quinone functional groups can accept and release electrons, allowing NOM to serve as an electron shuttle (e.g., Newman and Kolter 2000). This reactivity opens the possibility of NOM participating in a variety of redox reactions with arsenic, and there is evidence that both the oxidation of As(III) and reduction of As(V) can be carried out by NOM. Bauer and Blodau (2006) observed that As(V) was reduced by peat NOM, but that wetland-derived NOM oxidized As(III). In a study by Redman et al. (2002), out of six NOM samples, one sample reduced As(V) while the other five samples oxidized As(III), and the metal content of the NOM showed no significant correlation with the ability to transfer electrons. Redman et al. (2002) speculate that the NOM molecules were the redox active agents. This hypothesis suggests that the redox activity of the NOM is dependent on the redox state of the functional groups. Mineral-associated NOM can also participate in redox reactions, and both solid phase As(V) reduction (Redman et al. 2002) and As(III) oxidation (Ko et al. 2004) have been enhanced by NOM adsorbed onto hematite. In general, the molecular mechanisms involved in arsenic-NOM and arsenic-NOM-mineral redox reactions are not well understood, primarily because of the wide diversity of functional groups and their correspondingly varied ability to participate in electron transfer.

Natural organic matter is therefore an important consideration in understanding arsenic speciation and mobility in a wide variety of environments, from wetland, lacustrine, and riparian areas to groundwater aquifers. The pathways for NOM to affect arsenic fate and transport can be generally grouped into four categories: (1) mobilization by aqueous complexation and competitive adsorption, (2) oxidation and reduction of arsenic in solution and on mineral surfaces, (3) sequestration by association with solid phase NOM, and (4) indirect effects such as serving as an electron donor for microbial respiration. Dissolved and surface associated NOM tend to mobilize arsenic into solution directly through complexation and sorption inhibition. Redox cycling by NOM can have varied effects on the ultimate mobility of arsenic depending on the available surfaces, the type of NOM, and the direction of electron transfer. Although it is difficult to generalize the effect of redox cycling, it can have significant impacts on other reactions such as sorption, binding, and bioavailability of arsenic. Accumulation in reduced solid-phase organic carbon-rich areas such as wetlands and in localized lenses of NOM in groundwater aquifers may be an important source of arsenic to the environment if the physical or redox state of those areas should change naturally or due to hydrologic modifications (e.g., cessation of arsenic-rich mine discharge to a downstream wetland). This effect was observed in a study of arsenic mobilization into a stream near a peat soil that experienced seasonal wetting/drying cycles (Rothwell et al. 2009). In permanently wetted peats, the arsenic remained immobile, but in peats that were periodically dried and oxidized, significant amounts of arsenic were mobilized into the surrounding water. One of the most environmentally relevant examples of indirect effects of NOM is mineralization of NOM as the substrate for microbial iron reduction and reductive dissolution of host phases for arsenic. This mechanism is a primary process for arsenic release in a variety of situations, including groundwater in West Bengal (e.g., McArthur et al. 2004), lacustrine sediments (e.g., Campbell et al. 2008a), and in groundwater undergoing engineered bioremediation by organic carbon amendment (Stucker et al. 2013). The presence of organic carbon stimulates microbial activity, which in turn drives the local redox chemistry to more reducing conditions, until iron (hydr)oxides reductively dissolve, potentially releasing adsorbed arsenic to the water column, porewater, or aquifer.

SURFACE COMPLEXATION AND COMPETITIVE SORPTION

An important process controlling arsenic speciation is sorption, in which arsenic forms or loses chemical bonds with functional groups on a mineral surface. In many soil environments, adsorption or desorption is the primary process controlling arsenic transport (Smith et al. 1998). Sorption reactions are subject to laws of mass action, similar to aqueous equilibrium reactions, and the equilibrium constant is a product of a term corresponding to the chemical free energy of binding and a coulombic term, which describes the effect of electrostatic surface charge. Surface complexes are sensitive to pH, similarly to aqueous species. Surfaces also exhibit acid-base reactions via protonation-deprotonation of surface sites. Adsorption can be categorized as inner-sphere (covalent ligand-exchange reaction) and outer-sphere (electrostatic interaction). Outer-sphere complexes are often affected strongly by ionic strength because of the effect on coulombic attraction to the surface. Arsenic forms inner-sphere complexes, often with surface hydroxyl groups on metal oxides (Dzombak and Morel 1990), supported by the insensitivity of the surface species to ionic strength (Pierce and Moore 1982; Hsia et al. 1994; Jain et al. 1999; Goldberg and Johnston 2001) and by more detailed spectroscopic data. However, Catalano et al. (2008) found evidence for outer-sphere As(V) complexes together with inner-sphere complexes, which may have implications for arsenic sorption kinetics.

Sorption onto various mineral and organic phases has important consequences for the fate and transport of arsenic in a wide variety of natural environments. Adsorption is commonly thought of as a process that sequesters arsenic onto a solid phase, but it does not always immobilize arsenic in the environment: as previously noted, arsenic adsorbed onto Fe(III) colloidal particles, especially when stabilized by NOM, can potentially make arsenic more mobile in an aquatic environment (Bauer and Blodau 2009; Neubauer et al. 2013). It also has application to a variety of engineered systems, including permeable reactive barriers for groundwater remediation, natural attenuation, and drinking water treatment. A discussion of the numerous available technologies utilizing arsenic adsorption is available in Bissen and Frimmel (2003b), Wang and Mulligan (2006), and Mohan and Pittman (2007). These technologies are applied from household-level drinking water treatment systems to community-level treatment facilities and large scale water treatment and distribution systems.

The effect of pH is a key factor in understanding arsenic sorption behavior. Adsorption isotherms (adsorbed vs. dissolved arsenic) and adsorption edges (adsorbed arsenic vs. pH) on iron oxides (ferrihydrite, goethite, hematite) have shown that arsenate adsorption decreases with increasing pH, especially above pH 7-8, with a maximum adsorption at low pH (~pH 3), whereas arsenite has a broad maximum adsorption around pH 7, with decreasing sorption at low and high pH (Fig. 2; Pierce and Moore 1982; Raven et al. 1998; Goldberg 2002; Dixit and Hering 2003). Arsenite is often considered to be more mobile than arsenate, but this generalization is not always true as the relative affinity depends on multiple factors including pH and surface coverage (Manning and Goldberg 1997b; Dixit and Hering 2003). In the case of iron

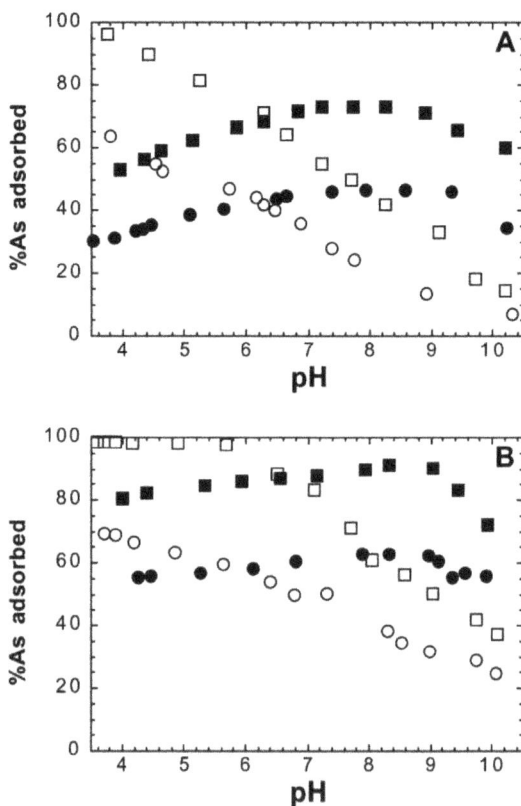

Figure 2. Comparison of As(V) (open symbols) and As(III) (closed symbols) adsorption edges onto (A) ferrihydrite and (B) goethite for total arsenic concentrations of 100 μM (circles) and 50 μM (squares). Reprinted with permission from Dixit and Hering (2003). Copyright 2003 American Chemical Society.

oxides, the relative affinity for adsorption depends strongly on pH as well as the presence of other adsorbed ions (e.g., phosphate, silicate, carbonate, NOM).

Adsorption of organic arsenic compounds (primarily mono- and di-methyl arsenic) has been studied on iron oxides (goethite, hematite, lepidocrocite, ferrihydrite, magnetite) (Ghosh and Yuan 1987; Bowell 1994; Lafferty and Leoppert 2005; Gimenez et al. 2007). On iron oxides, adsorption affinity decreases in the order: $H_2As^VO_4^- > DMAs^V > MMAs^V > H_3As^{III}O_3$ below pH 7 and $HAs^VO_4^{2-} > H_3As^{III}O_3 > MMAs^V = DMAs^V$ above pH 7 (Bowell 1994; Lafferty and Leoppert 2005). Methylated As(III) compounds ($DMAs^{III}$ and $MMAs^{III}$) were not significantly adsorbed to ferrihydrite or goethite at any pH, even though arsenite was strongly adsorbed (Lafferty and Leoppert 2005). Organic and inorganic As(V) compounds were desorbed effectively by the adsorption of phosphate and, to a lesser extent, sulfate. Increased methylation (mono-, di-, and tri-methyl species) resulted in decreased adsorption and increased release of arsenic from the surface in the presence of competing ions.

Characterization and chemistry of arsenic surface species

The molecular structure of arsenic surface species affects surface charge. Whether arsenic binds as mononuclear or binuclear surface species affects surface charge, because arsenic exists as a weak acid and binuclear complexes have fewer acidic protons due to an additional bond with the mineral surface. There have been extensive studies of inorganic arsenic binding to iron oxides because of the ubiquity and importance in field-scale arsenic fate and transport. Adsorption on aluminum oxides, iron oxyhydroxysulfates, and clay minerals has been studied to a lesser extent. Numerous studies have found a positive correlation between arsenic adsorbed and ammonium-oxalate or citrate-dithionate extractable iron and aluminum concentrations and clay content of soils, demonstrating the importance of iron oxides, aluminum oxides, and clays in arsenic mobility in soil environments (e.g., Livesey and Huang 1981; Manning and Goldberg 1997b).

Of the three possible configurations (mononuclear monodentate, mononuclear bidentate (edge-shared), or binuclear bidentate (corner-shared)), X-ray absorption spectroscopy (XAS) studies have shown that As(V) is adsorbed on iron oxides primarily as a binuclear bidentate complex, with an average As-Fe bond length between 3.26 and 3.30 Å (Waychunas et al. 1993; Sherman and Randall 2003). The predominant species for As(III) is also a binuclear bidenate species with an average As-Fe bond length of 3.3 Å with a less important mononuclear bidentate complex (Manning et al. 1998; Ona-Nguema et al. 2005). Binuclear bonding of both As(V) and As(III) has been supported by Fourier transform infrared spectroscopy (FTIR) studies (Sun and Doner 1996). The predominant surface species of mono- and tetra-thioarsenate adsorbed onto ferrihydrite and goethite was a monodentate, inner-sphere complex, and adsorption of the thioarsenic was lower compared to arsenite and arsenate (Couture et al. 2013).

There are often differences in the amount of sorbed arsenic depending upon whether the arsenic was adsorbed to an existing mineral surface or co-precipitated with the mineral. Arsenic adsorption onto a mineral surface (iron, aluminum oxides) is initially fast (~2 h), then followed by a period of slower uptake (>100 h) (Pierce and Moore 1982; Ghosh and Yuan 1987; Fuller et al. 1993; Raven et al. 1998; Dou et al. 2013). The kinetics of adsorption have been explained by arsenic diffusion into an aggregate, with sorption sites on the surface accounting for fast sorption and sites located in the interior of the aggregate requiring longer diffusion times (Fuller et al. 1993). When arsenate was co-precipitated with ferrihydrite, the uptake of arsenate was significantly higher initially than arsenate adsorbed after the mineral had precipitated. However, arsenic was slowly released as aging and aggregation occurred and reached steady state concentrations similar to arsenate adsorbed onto presynthesized ferrihydrite (Fuller et al. 1993). Similar steady-state arsenic concentrations between the two conditions suggest that the difference between adsorption and co-precipitation is primarily uptake kinetics, and that the controlling chemistry is similar.

Arsenic adsorption behavior on other mineral phases has been less extensively studied than for iron oxides, but it is still paramount to understanding arsenic sorption under natural conditions. Arsenic adsorption onto aluminum oxides differs from iron oxides primarily in that As(III) exhibits primarily weak, outer-sphere adsorption from pH 3 to 11 (Ghosh and Yuan 1987; Goldberg 2002). Arsenate forms binuclear bidentate inner-sphere complexes on the surface of aluminum oxides, which may be marginally stronger than bonds formed on iron oxides, and the adsorption edge extends to higher pH values (pH 9) (Goldberg 2002; Kappen and Webb 2013). In general, clays exhibit a maximum arsenate adsorption at low pH, with a sharp decrease between pH 3 to 7, depending upon the mineral (Manning and Goldberg 1997a; Goldberg 2002). Arsenite exhibits strong adsorption on clays up to pH ~ 9, such that arsenite adsorption may be greater than arsenate adsorption between pH 7 and 9 (Manning and Goldberg 1997a). The extent of adsorption varied between clay phases (e.g., halloysite, chlorite, illite, kaolinite, and montmorillonite), with halloysite and chlorite exhibiting the highest affinity (Lin and Puls 2000). X-ray absorption spectroscopy analysis of arsenate adsorbed on kaolinite showed arsenate coordinated to aluminum octahedral sites via bidentate binuclear binding (~3.1 Å) (Arai 2010). There is also evidence for arsenite adsorption onto mackinawite (amorphous FeS) with subsequent transformation to an orpiment-like phase (Burton et al. 2013). Sorption on ferrous iron sulfide phases is harder to identify because of the tendency for arsenic to form a solid solution. In addition, ferrous sulfide phases tend to reduce adsorbed thioarsenic species to arsenite in laboratory experiments (Couture et al. 2013). Adsorption on manganese oxides has been extensively studied, and because of their redox activity, is described in more detail below.

In low pH, acid-sulfate systems, iron oxyhydroxysulfates, such as jarosite ($KFe_3(OH)_6(SO_4)_2$) and schwertmannite (ideally $Fe_8O_8(OH)_6SO_4$), become important phases for arsenic adsorption. Schwertmannite has a high adsorptive capacity for As(III) and As(V), partly because of its amorphous nature and high surface area, similar to ferrihydrite (Fukushi et al. 2003a, 2004; Burton et al. 2009). The mechanism of arsenic adsorption onto schwertmannite is through exchange with sulfate, both adsorbed to the surface as a monodentate complex and incorporated into interior tunnel structures (Fukushi et al. 2004). Arsenate has higher sorption affinity at low pH than As(III) due to surface charge effects (Burton et al. 2009). Arsenic surface species on jarosite also exist as inner sphere complexes, exchanged for surface-bound sulfate. Jarosite has a lower sorption capacity than schwertmannite, but adsorbs more arsenic than goethite at low pH (Asta et al. 2009).

Effect of competitively adsorbing ions

The presence of other ions that adsorb to mineral surfaces can substantially change the adsorption behavior of arsenic through electrostatic, steric, or competitive mechanisms. Because arsenic surface species are primarily inner-sphere complexes, competitive effects are not due to ionic strength, but rather a physiochemical interaction with arsenic and mineral surface sites. The main components of natural waters that are known to affect arsenic adsorption are pH, phosphate, sulfate, silicate, inorganic carbon, organic carbon, calcium, and magnesium. Phosphate has the largest effect, due to its similar molecular structure and charge to the arsenate ion and its surface complexation chemistry (inner-sphere complexation at the same type of surface sites). It effectively competes for surface sites with both arsenite and arsenate, inhibiting adsorption of arsenate slightly more than arsenite (Manning and Goldberg 1996; Liu et al. 2001; Dixit and Hering 2003; Campbell et al. 2008b). Arsenate adsorbs marginally more strongly to iron oxides than phosphate, and arsenate sorption is slightly less pH dependent than phosphate (Jain and Loeppert 2000; Hongshao and Stanforth 2001; Liu et al. 2001; Violante and Pigna 2002; Antelo et al. 2005). Kinetic effects are also important in competitive sorption of phosphate and arsenic, as the sorbate added first will adsorb to a greater extent than the competing sorbate (Hongshao and Stanforth 2001; Liu et al. 2001).

Other oxyanions also compete with arsenite and arsenate for surface adsorption due to similar bonding affinities and surface charge effects. In a study of arsenate adsorption onto goethite and gibbsite, molybdate decreased arsenate adsorption below pH 6, whereas phosphate decreased arsenate adsorption in the entire pH range studied (pH 2-11) (Manning and Goldberg 1996). Tungstate, selenate, and chromate could also potentially compete with arsenic for surface sites. However, molybdenum, selenium, chromium, and tungsten tend to occur at much lower concentrations in the environment, and are less likely to be an important driver of arsenic adsorption, especially compared to phosphate. It has been observed in several field studies that Mo and W often co-occur with arsenic because of the similar chemical behavior of these oxyanions (Campbell 2007; Mohajerin et al. 2014).

Dissolved inorganic carbon (DIC, dissolved CO_2 as H_2CO_3 and its acid dissociation products) is often elevated in natural waters due to carbonate mineral dissolution and microbial respiration. Although DIC is a much weaker inhibitor of arsenic adsorption than phosphate, it is an important competitive adsorbent because of high concentrations found in groundwater or in high pH settings. Dissolved inorganic carbon forms only a very weak aqueous complex with As(III), and sorption inhibition is due to surface effects (Neuberger and Helz 2005). Inorganic carbon adsorption onto iron oxide surfaces has been shown to be a combination of inner and outer sphere complexes, and the competitive effects on arsenic adsorption are likely due to electrostatic effects of specifically adsorbed bicarbonate species (van Geen et al. 1994; Su and Suarez 1997; Villalobos and Leckie 2000, 2001; Bargar et al. 2005). The pH affects inorganic carbon adsorption, with a maximum at circumneutral pH (e.g., pH 6 on goethite; van Geen et al. 1994), partly because bicarbonate (HCO_3^-) has a greater affinity for surfaces than carbonic acid (H_2CO_3) or carbonate (CO_3^{2-}). Ternary complexation of sodium with adsorbed inorganic carbon on the surface of iron oxides has been observed to affect surface electrostatics and may impact arsenic adsorption (Villalobos and Leckie 2001).

At atmospheric CO_2 concentrations, the effect of DIC on arsenic adsorption is minimal, but at concentrations typical of groundwater aquifers (1-10% CO_2), arsenate and arsenite adsorption was suppressed (Fuller et al. 1993; Meng et al. 2000; Arai et al. 2004; Radu et al. 2005). In studies with arsenite pre-equilibrated onto goethite-coated sand, high concentrations of arsenite were desorbed in the presence of 22 mmol L^{-1} DIC (10% CO_2), and DIC had a larger effect on arsenite than arsenate (Radu et al. 2005), even though arsenate was predicted to be more affected based on surface charge effects (Appelo et al. 2002). Bicarbonate has been demonstrated to mobilize arsenic from natural sediments from the Bengal Delta (Anawar et al. 2004).

Sulfate has been shown to adsorb to iron oxides, but appears to occupy different surface sites than arsenic (Jain and Loeppert 2000). Similarly to DIC, sulfate has a much weaker effect than phosphate, but can be important in environments with high concentrations of sulfate, such as marine or brackish waters and acid rock drainage; in alkaline water, sulfate tends to be limited by precipitation of gypsum in groundwater. Sulfate has a very slight effect on arsenate adsorption on iron oxides (e.g., goethite, ferrihydrite, hematite) but can decrease arsenite adsorption below pH 7.5 (Wilkie and Hering 1996; Jain and Loeppert 2000; Meng et al. 2000). In low pH systems with high iron sulfate concentrations (e.g., acid rock drainage), the primary iron minerals are jarosite, goethite, and schwertmannite. Because arsenic adsorbs to iron oxyhydroxysulfates by exchange of surface and structural sulfate, elevated concentrations of sulfate can compete with arsenic on iron oxides and is an important consideration in acid-sulfate conditions (Asta et al. 2009). At low pH, this effect is likely to be more pronounced for arsenate than arsenite because of the shape of adsorption envelopes on schwertmannite (Burton et al. 2009). At slightly acidic pH (pH 5), arsenate adsorption onto schwertmannite exhibited similar inhibition as arsenate adsorption to iron oxides in the presence of phosphate, DIC, and silicate in several simulated groundwaters, and no effect was observed for chloride, sulfate, carbonate, and nitrate (Dou et al. 2013).

Silicate adsorption is more complex than most other competing ions because of its ability to form polymers on mineral surfaces at high concentrations (Swedlund and Webster 1999; Holm 2002; Swedlund et al. 2009; Gao et al. 2013). Silicate forms inner sphere complexes on ferrihydrite with a broad maximum adsorption between pH 8 and 10, and transitions from an adsorbed monomer to a polymer as Si:Fe ratios increase (Swedlund et al. 2009; Gao et al. 2013). Adsorbed silicate generally decreases arsenic adsorption at high Si:Fe ratios and with greater effects on arsenite than arsenate, although at low concentrations the effect of silicate is less pronounced or negligible (Swedlund and Webster 1999; Meng et al. 2000; Waltham and Eick 2002). At high concentrations, silicate forms surface polymers that decrease arsenic adsorption due to both steric and electrostatic effects. It is also possible to form colloidal Si-Fe(III) polymers at high silicate concentrations. If arsenic adsorbs to these colloids, it may be more mobile, but the relevance of this process in environmental systems could be limited to very unique circumstances (Meng et al. 2000; Davis et al. 2001).

Cation adsorption has also been shown to affect arsenic surface binding. In a study by Wilkie and Hering (1996), addition of Ca^{2+} and arsenate to a slurry of ferrihydrite increased arsenate adsorption at circumneutral pH, primarily through electrostatic effects. A similar effect was observed with Ca^{2+} and Mg^{2+} for arsenate and phosphate adsorption onto goethite, but there was no effect of cations on arsenite adsorption. Geochemical modeling of the system supports an electrostatic mechanism (Stachowicz et al. 2008). Meng et al. (2000) found that addition of Ca^{2+} or Mg^{2+} counteracted the competitive effect of silicate on arsenic adsorption on iron oxides, supporting the electrostatic mechanism of both cations and silicate. However, arsenate adsorption onto kaolinite was decreased in the presence of Ca^{2+}, most likely due to competitive ion exchange for the same edge sites (Arai 2010). Further research is needed to better understand cation-arsenic-mineral surface interactions. Cation effects could be important in interpreting arsenic adsorption in groundwater systems, where cation concentrations are often elevated and arsenic occurs primarily as an adsorbed species.

Effect of mineral transformations on arsenic speciation and vice-versa

Mineral transformations can occur as metastable phases reorganize to thermodynamically stable phases or during variations in redox conditions. Generally, as a less crystalline phase transforms into a more crystalline phase, surface area decreases, resulting in fewer surface sites for arsenic adsorption and potential mobilization into solution. The relationship between mineral transformation, arsenic mobility, and speciation is complex, but we can identify two general categories: direct effects of arsenic on mineral transformation and the consequences of mineral transformation on arsenic adsorption.

Many adsorbed ions, such as phosphate, carbonate, silicate, calcium, nickel, and cobalt, can poison, slow, or alter the products of the transformation of ferrihydrite, especially in the presence of Fe^{2+} (Zachara et al. 2002; Hansel et al. 2003, 2004, 2005). A possible mechanism for this effect is structural blockage of surface sites by adsorbed ions, preventing dissolution and recrystallization of a new phase. Adsorbed arsenic has the same effect, thus stabilizing less-crystalline phases with high-adsorptive capacity. This stabilization has been demonstrated for schwertmannite, which is metastable with respect to jarosite and goethite. In laboratory experiments, the nucleation of other phases was almost completely inhibited by adsorbed As(III) on schwertmannite, which is important because the transition to jarosite or goethite tends to release arsenic into solution (Asta et al. 2009; Johnston et al. 2011). Arsenate has a similar effect on synthetic schwertmannite (Waychunas et al. 1995; Fukushi et al. 2003b; Regenspurg and Peiffer 2005) and natural schwertmannite (Fukushi et al. 2003a). In addition, the dissolution of arsenic-containing jarosites resulted in accumulation of adsorbed arsenic on the mineral surface, inhibiting further dissolution of the jarosite as well as reducing the crystallinity of the transformation products such as goethite and maghemite (Kendall et al. 2013). The presence of other adsorbed ions can have a similar stabilizing effect, thus preserving or slowing the transformation of a high-surface-area mineral phase. However, the ultimate speciation of arsenic (solution vs. adsorbed) depends on the mineralogy, oxidation state, and concentrations of these competing ions for sorption sites. In a study of arsenic adsorbed onto schwertmannite in the presence of variable amounts of sulfate under reducing conditions, high concentrations of sulfate stabilized schwertmannite, resulting in a persistent As(V)-schwertmannite phase in spite of active sulfate-reducing conditions (Burton et al. 2013).

Transformation of minerals can occur spontaneously or can be driven by changing redox conditions. For example, reductive dissolution of ferric iron phases is a common process that has significant implications for arsenic. Adsorbed arsenic is stable as long as the host phase has sufficient adsorptive capacity, which can change as bulk mineral is lost to reductive dissolution or as more crystalline phases with fewer surface sites are formed. Ferrous iron in combination with co-sorbed ions control mineral transformations, and is often driven by microbial activity and the presence of NOM. In addition, kinetics are important in controlling which phases form (e.g., Egal et al. 2009).

Adsorbed arsenic may retard or inhibit mineral transformation and it may have secondary effects on rates and extent of microbial processes. Adsorption of arsenate onto iron and aluminum oxides substantially slowed the kinetics of arsenate reduction by *Shewanella putrefaciens*, a known arsenate- and iron-reducing bacterium (Huang et al. 2011). Similarly, in a flooding study using natural soils, adsorptive capacity controlled the rates of desorption and microbial reduction of As(V) (McGeehan and Naylor 1994). In addition, the presence of arsenic can change the aggregation properties of minerals, which in turn may affect the bioavailability of the mineral. The presence of high concentrations of As(III) adsorbed on ferrihydrite increased the rate of microbial iron reduction, most likely due to particle aggregation effects from changes in surface charge (Campbell et al. 2006). Although the importance of this effect has not been demonstrated with natural sediments or in a field setting, it illustrates the interrelationship between arsenic adsorption, microbial redox processes, and mineral transformation.

ARSENIC OXIDATION-REDUCTION REACTIONS AND SPECIATION EFFECTS

Biogenic redox reactions of arsenic

Microorganisms capable of metabolizing arsenic are found in a wide range of conditions, including extremes in pH, salt content, and metal concentrations. They exhibit a wide phylogenetic diversity of As(III)-oxidizing and As(V)-reducing bacteria and archaea (Oremland and Stolz 2003). The metabolic pathways utilized for As(III) oxidation and As(V) reduction processes are varied and often complex. More detail on the specific pathways relating to arsenic redox transformations and methylation can be found in Amend et al. (2014, this volume). Generally, the effect of microbial redox cycling is to accelerate the kinetics of reactions, which can considerably affect the fate and speciation of arsenic in the environment. Arsenic redox behavior in many environmental systems is usually controlled by a combination of biotic and abiotic processes, as we have demonstrated in numerous examples throughout this review.

Abiotic redox reactions of arsenic

Aqueous redox reactions. Some dissolved gases are possible reactants in arsenic redox chemistry, but are generally only kinetically reactive under extreme pH conditions. Oxidation of arsenite by dissolved oxygen is thermodynamically favorable, but the abiotic reaction kinetics are slow (Cherry et al. 1979; Kuehnelt et al 1997; Manning and Goldberg 1997a; Hug and Leupin 2003). Similarly, arsenate reduction by dissolved sulfide is favorable, but only observed under acidic conditions (pH < 4); under circumneutral pH conditions, it is kinetically limited (Cherry et al. 1979; Rochette et al. 2000).

Abiotic oxidation of arsenite by nitrate appears to occur to a limited extent (Oremland et al. 2002). However, in many lakes and aquifers, nitrate is indirectly a key component in arsenic redox chemistry through its bioavailability for numerous microbial pathways. Nitrate can be present at relatively high concentrations in some natural waters. Microbially mediated anaerobic arsenite oxidation coupled to nitrate reduction has been observed in an organism isolated from Mono Lake (California, USA), a high pH (>9), saline lake (Oremland et al. 2002; Oremland and Stolz 2003). Microbially mediated nitrate-dependent arsenite oxidation has also been observed in an aquifer in Bangladesh (Harvey et al. 2002), in a freshwater lake (Senn and Hemond 2002, 2004), and during denitrification in anaerobic sludges (Sun et al. 2008), indicating that this process is widespread. Nitrate can also be coupled to anaerobic microbial Fe(II) oxidation at circumneutral pH, producing fresh ferric oxides which serve as a strong adsorbent for dissolved arsenic, as demonstrated in an aquifer at Cape Cod (Massachusetts, USA) (Hohn et al. 2006) and in a eutrophic lake (Senn and Hemond 2004). Photolysis of nitrate or nitrite has also been demonstrated to oxidize arsenite under adequate solar irradiation (Kim et al. 2014).

Oxidation of arsenite by oxygen in the presence of dissolved Fe(II) is potentially an important pathway over a range of conditions. At circumneutral pH, dissolved Fe(II) facilitated partial arsenite oxidation in the presence of dissolved oxygen through a theoretical reactive Fe(IV) intermediate species or carbonate radicals (Hug and Leupin 2003). This pathway may explain co-existing As(III) and As(V) species in circumneutral pH waters where oxygen is being introduced into Fe(II)-containing waters. Under acidic conditions, photochemical reduction of dissolved Fe(III) can react to oxidize dissolved arsenite, via hydroxyl radical and an As(IV) intermediate (Woods 1966; Emett and Khoe 2001; Kocar and Inskeep 2003; Leuz et al. 2006). The most active wavelengths of light for this reaction are in the near UV range. The simplified reaction scheme is presented in Equations (1)-(6), and is based on experiments of Kocar et al. (2003), Woods (1966), and Emett and Khoe (2001). By using scavenging agents (e.g., 2-propanol), the hydroxyl radical mechanism was confirmed for pH < 4; it is possible that other unknown intermediates are active at pH > 4 (Hug and Leupin 2003; Kocar and Inskeep 2003; Leuz et al. 2006). However, these reactions are mainly relevant at pH < 4, because of the low solubility of dissolved Fe(III) at higher pH values. The rate of oxidation is controlled by the rate of photon absorption, the rate of reaction of the unstable As(IV) intermediate, and the other compounds in solution that can scavenge free radicals and As(IV) species (Eqns. 1-6). It can also be dependent upon the rate of acidophilic microbial Fe(II) oxidation (Campbell et al. 2012). Even in the presence of NOM, As(III) oxidation was still observed (Kocar and Inskeep 2003). Photochemical Fe(III) reduction coupled to As(III) oxidation may be an important pathway under certain circumstances, especially in acid rock drainage in areas that receive ample sunlight.

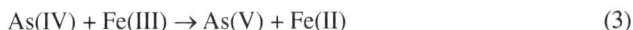

$$Fe(III) + h\nu \rightarrow Fe(II) + \cdot OH \tag{1}$$

$$As(III) + \cdot OH \rightarrow As(IV) + OH^- \tag{2}$$

$$As(IV) + Fe(III) \rightarrow As(V) + Fe(II) \tag{3}$$

$$As(IV) + O_2 \rightarrow As(V) + \cdot HO_2 \tag{4}$$

$$As(IV) + Fe(II) \rightarrow As(III) + Fe(III) \tag{5}$$

$$\cdot HO_2 + Fe(III) \rightarrow O_2 + Fe(II) \tag{6}$$

Thioarsenic species are highly reactive and tend to degrade into arsenic oxyanions. Because of the complex redox chemistry of sulfur, the degradation pathways may involve redox transitions of arsenic. Microbial growth has also been observed to be promoted by the presence of thioarsenic species (Fisher et al. 2008), and the observed transformation of thioarsenic in the environment most likely involves a combination of biotic and abiotic pathways at temperatures conducive to life (Planer-Friedrich et al. 2007; Hartig and Planer-Friedrich 2012).

Solid-phase redox reactions. Solid-phase electron transfer is another important pathway for redox transformation of adsorbed arsenic. Oxidation of adsorbed As(III) has been observed for a variety of minerals and natural soils, as described in detail below. Arsenite oxidation is also important in water treatment processes, in which more efficient removal of As(V) has been observed on a variety of applicable sorbents such as Fe-, Al-, and Mn-oxides and resins (Bissen and Frimmel 2003b; Wang and Mulligan 2006; Mohan and Pittman 2007).

One of the most environmentally relevant set of minerals that can oxidize adsorbed As(III) are manganese oxides (e.g., poorly-crystalline birnessite (δ-MnO$_2$), pyrolusite (β-MnO$_2$), hausmannite (Mn$_3$O$_4$), and manganite (MnOOH)). In the absence of microbes, abiotic As(III) oxidation occurs by sorption onto the mineral surface followed by electron transfer; the resulting As(V) may stay adsorbed onto the surface, or be released into solution depending on the solution conditions and surface site availability. The reaction is clearly between As(III) and Mn(III or IV) because of the lack of dependence on pO$_2$ and the increase of Mn(II) species (Scott and Morgan 1995; Chiu and Hering 2000). As manganese is reduced, new surface sites are created for Mn(II) and/or As(V) (Manning et al. 2002). The reaction decreases with increasing pH (Thanabalasingam and Pickering 1986b). Phosphate decreases the rate of As(III) oxidation by competing for surface sites (Chiu and Hering 2000). Various manganese oxides have different surface-area-normalized reactivities, most likely related to point of zero charge (PZC) of the particular mineral (Feng et al. 2006). In a set of laboratory experiments comparing biotic and abiotic As(III) oxidation, biotic oxidation appeared to be faster than oxidation by manganese oxides (Katsoyiannis et al. 2004), but the relative rates in a field setting may depend on the specific mineralogy and the microbial community. The importance of manganese in As(III) oxidation in natural soils with varied mineralogy has been demonstrated with samples taken from the U.S. Geological Survey (USGS) research site at Cape Cod (Massachusetts, USA). Arsenite oxidation was observed both in field- and laboratory-scale experiments, and the rate of oxidation was directly related to the amount of total manganese content of the solid samples (Amirbahman et al. 2006). The key manganese phase was micrometer to submicrometer aggregates of Mn-bearing goethite coatings on silicate mineral grains (Singer et al. 2013), showing that the importance of manganese-driven As(III) oxidation may extend beyond the occurrence of primary manganese mineral phases.

Heterogeneous As(III) photo-induced oxidation has been observed in the presence of ferric oxides and titanium dioxide, and explored as a useful reaction for water treatment technologies. Arsenite adsorbed on ferrihydrite and goethite has been studied in batch reactors with incident light with a 600 nm maximum wavelength (Bhandari et al. 2011, 2012). Partial oxidation of As(III) was measured with co-incident production of Fe(II). In the case of ferrihydrite, less than 10% of the ferrihydrite was reduced because of a self-terminating reaction with adsorbed As(V). Goethite exhibited a higher oxidation rate than ferrihydrite under similar reaction conditions, with adsorbed Fe(II) facilitating electron transfer to the adsorbed As(III). There has been extensive research on As(III) oxidation when adsorbed onto commercially available TiO$_2$ in the presence of UV light because of its application to water treatment. Titanium dioxide has two common polymorphs, anatase and rutile. Anatase is typically more photoreactive than rutile, but many commercial products are a mix of the two phases (e.g., the mixed-phase photocatalyst, Degussa P25). The details of the mechanisms, interferences, and performance of TiO$_2$-based As(III) oxidation for water treatment are presented in a review by Guan et al. (2012). Naturally occurring kaolin containing uniquely high concentrations of titanium have also been shown to oxidize As(III) (Foster et al. 1998).

Carbonate green rust (Fe$^{II}_4$Fe$^{III}_2$(OH)$_{12}$·[CO$_3$]), a phase that is stable only under strictly anoxic conditions, can oxidize As(III) in laboratory experiments, albeit incompletely (Su and

Wilkin 2005). Although green rust has the capacity to reduce other metals, As(V) reduction by green rust has not been observed (Randall et al. 2001; Ruby et al. 2006). The importance of green rust (carbonate or sulfate forms) in redox cycling of arsenic in soil or sedimentary settings is unclear. A variety of clay minerals can also oxidize adsorbed As(III). In studies of As(III) adsorbed to halloysite, kaolinite, illite, montmorillonite, and chlorite, all clay minerals showed complete or nearly complete oxidation of As(III) over several months, even in the absence of oxygen (Manning and Goldberg 1997a; Lin and Puls 2000). The mechanism of electron transfer in clays has been debated, but may proceed via multiple pathways depending on the clay structure and composition. Several studies suggest that the presence of trace amounts of manganese or other impurities in the clay samples could be responsible for As(III) oxidation (Manning and Goldberg 1997a; Lin and Puls 2000). Foster et al. (1998) identified an anatase-like phase in a Ti-rich clay sample to be the primary oxidant of adsorbed As(III) in the presence of light and oxygen, with a smaller contribution by trace amounts of manganese.

ARSENIC SPECIATION CALCULATIONS FOR SEVERAL NATURAL WATER COMPOSITIONS

Aqueous arsenic speciation can be extremely complex, with a wide range of possible concentrations and forms depending upon geological, chemical, hydrological, and microbial conditions (Smedley and Kinniburgh 2002). We selected a set of eight natural water compositions that encompassed a range of pH conditions, arsenic concentrations and oxidation states, and water types (arsenic-containing groundwater, acid mine drainage, and geothermal water). The calculations included two acidic samples: an acid mine discharge from the Leviathan Mine (California, USA, pH 1.8; Ball and Nordstrom 1989), and an acidic geothermal hot spring from Yellowstone National Park (Wyoming, USA; "Lifeboat" Spring in Norris Geyser basin, pH 3.23; Ball et al. 2010). Alkaline geothermal water (pH 9-9.5) from a Yellowstone hot spring and its drainage was also selected (Mound Spring and 55 m downstream from Mound Spring; Ball et al. 2006); CO_2 degassing, increasing pH, and cooling of the water occurred as the geothermal water flowed down the drainage. Groundwater samples included average water chemistry from high-arsenic, neutral-pH, shallow (<30 m) tube-wells in southern Bangladesh (Lakshmipur, pH 7.17) and central Bangladesh (Faridpur, pH 6.94; British Geological Survey 2001), and neutral-to-alkaline groundwater (pH 7.1-9.3) in the Carson Desert (Nevada, USA). The Carson Desert groundwater was influenced by recharge of evaporated water from an arid, alkaline region (Welch and Lico 1998). Details of the sampling and analytical methods are described in the specific references.

For speciation calculations, the data were tabulated (Table 3) and input to PHREEQC, a geochemical code distributed by the USGS (Parkhurst and Appelo 2013), with the WATEQ4F database modified to include metal-arsenic aqueous complexes as reviewed in Nordstrom et al. (2014, this volume). There are no reliable thioarsenic stability constants available, and therefore, these species were not included in the speciation calculations, even though they are likely to be important in geothermal water. Future quantification of these constants is necessary to complete representative speciation calculations for these types of natural waters. The Carson Desert groundwater samples had detectable Fe(III) concentrations after 0.1 mm filtration even though the water had neutral to basic pH values (Table 3), suggesting that Fe(III) colloids may be present. The authors report humic-acid-like carbon present in the groundwater samples, which may have helped to stabilize the Fe(III) colloids. Presence of iron colloids has significant implications for the interpretation of the arsenic data. However, because the colloids were not identified or characterized, surface complexation of arsenic onto the Fe colloids was not included in the speciation calculations, and iron was included in the model as total dissolved iron (i.e., sum of measured Fe(II) and Fe(III)). Even though this information gap potentially affects the speciation calculation, more importantly, it points to the importance of understanding NOM-Fe-As colloidal interactions in field situations. The calculations presented here are illustrative of the types of speciation expected in several types of natural waters.

Arsenic speciation results are presented in Table 4, and show the dominant (>0.01%) species for As(III) and As(V) in each water sample. Generally, the most important species were arsenate, arsenite, and various major metal-arsenic complexes (Ca, Mg, Na, K, Fe), while most trace-metal complexation (e.g., Co, Cu, Ni, Sr) was insignificant for both arsenic and the respective trace elements. The As(III) speciation was predominantly arsenite (H_3AsO_3, $H_2AsO_3^-$) in all water types, and metal-arsenite complexes comprised <2% of the total As(III) in each sample, even in the acid mine drainage. Speciation of As(V) was more complex, with a wide range of metal-arsenate complexes forming depending on the water composition. Arsenate was present (>10% of total As(V)) in all samples. Sodium-arsenate complexes were important in nearly all of the samples, with pH as the primary driver of formation; at low and high pH, the sodium-arsenate complexes were formed at higher proportions than at neutral pH. At low pH (acidic hot spring), the NaH_2AsO_4 complex composed 37% of the total As(V), whereas $NaAsO_4^{2-}$ was the dominant species in high pH waters with 70-75% of the total As(V) (Mound Spring, drainage, and DR16B). The pH appears to be the most important control in the formation of the sodium-arsenate species, as demonstrated by differences in speciation in waters from the same region: DR15B (pH 7) and DR16B (pH 9). The neutral-pH groundwaters contain relatively high concentrations of calcium and magnesium, which is reflected in the increased importance of the $MgH_2AsO_4^+$ and $CaH_2AsO_4^+$ species in these samples. These waters also have a higher diversity of complexes present, occurring at lower overall abundance, including K, Fe, Mn, and additional Ca and Na complexes.

These calculations demonstrate how aqueous speciation changes as a function of water chemistry and the importance of metal-arsenic complexation, especially for As(V) species under a variety of conditions. They also highlight several key areas where a better understanding of biogeochemical arsenic processes would improve the relevance and applicability of geochemical modeling. Specifically, thioarsenic species in high-sulfide environments are definitely important and improved constants and reaction schemes for these species in natural waters would greatly increase accuracy of speciation calculations. Cation-mineral-arsenic interactions could affect arsenic adsorption behavior, but the details of these interactions could be expanded to improve prediction of arsenic adsorption on soils and sediments. In addition, NOM-As and NOM-colloid-As interactions can be complex but are key to understanding arsenic mobility in certain environments. It is critical to measure and report this fraction when measuring NOM-rich water, and this experimental information will greatly improve speciation conceptual and numerical models.

ACKNOWLEDGMENTS

The authors thank Dr. JoAnn Holloway, Dr. Jennifer Webster, Dr. Michael Hay, Dr. Rob Bowell and A.S.C. for their extremely helpful comments and editing of this work. Chemical structures were drawn using the software programs ChemDraw and Avogadro. This work was supported by the USGS National Research Program. Any use of trade, product, or firm names is for descriptive purposes only and does not imply endorsement by the U.S. Government.

Table 3. Water chemistry data used in geochemical speciation calculations, including acid mine drainage from the Leviathan Mine, California, USA (Ball and Nordstrom 1989), average groundwater composition from shallow tube wells in Bangladesh (British Geological Survey 2001), geothermal hot springs in Yellowstone National Park, Wyoming, USA (Ball et al. 2006, 2010), and groundwater from the Carson Desert area of Nevada, USA (Welch and Lico 1998).

Location	Acid mine drainage, Leviathan Mine, CA	As-contaminated groundwater in Bangladesh		Hot Springs in Yellowstone National Park, WY			Carson Desert, NV, Groundwater	
Sample	Tunnel 5 drainage*	Lakshmipur region, southern Bangladesh	Faridpur region, central Bangladesh	Acid spring in Norris Geyser Basin ("Lifeboat" spring)**	Mound Spring***	Mound Spring drainage −55 m downstream	DR 15B	DR 16B
Field parameters								
T (°C)	12	25.4	26.2	60.1	94	24.5	16	16.5
pH	1.8	7.17	6.94	3.23	8.98	9.5	7.1	9.3
well depth (m)	—	10	39	—	—	—	9	20
Chemical constituents (mmol L^{-1})								
Al	16.2	—	—	0.09	0.014	0.012	—	0.035
As(T)	0.47	1.4×10^{-3}	5.2×10^{-4}	0.019	0.012	0.016	4.3×10^{-4}	4.7×10^{-3}
As(III)	—	6.9×10^{-4}	1.2×10^{-4}	3.5×10^{-4}	0.011	2.8×10^{-3}	4.7×10^{-5}	4.7×10^{-3}
As(V)	—	6.8×10^{-4}	4.0×10^{-4}	0.018	9.3×10^{-5}	1.3×10^{-2}	3.8×10^{-4}	0.030
Ba	5.5×10^{-5}	1.5×10^{-4}	1.2×10^{-3}	3.4×10^{-4}	—	—	—	—
B	9.3×10^{-3}	0.014	2.8×10^{-3}	0.79	0.29	0.33	—	—
Br	—	—	—	0.040	0.010	8.8×10^{-4}	—	—
Cd	2.5×10^{-3}	—	—	—	—	—	—	—
Ca	3.27	1.22	2.52	0.13	8.2×10^{-3}	7.6×10^{-3}	2.74	0.030

Cl	0.24	1.92	0.26	17.8	6.79	0.031	0.96	0.96
Co	0.087	—	—	—	—	—	—	—
Cu	0.084	—	—	1.1×10^{-5}	—	—	—	—
F	0.18	—	—	0.28	1.49	14.4	—	—
Fe(T)	28.1	0.031	0.10	0.022	—	—	1.5×10^{-3}	7.5×10^{-4}
Fe(II)	25.8	—	—	0.017	—	—	6.1×10^{-4}	7.4×10^{-4}
Fe(III)	2.33	—	—	5.4×10^{-3}	—	—	9.3×10^{-4}	1.1×10^{-5}
bicarbonate	—	7.64	8.79	—	4.82	5.34	10.5	17.9
I	—	2.4×10^{-4}	7.6×10^{-5}	—	—	—	—	—
K	0.35	0.26	0.012	1.50	0.25	0.28	0.11	0.28
Li	0.014	0.014	—	0.67	0.24	0.27	—	—
Mg	2.21	1.60	1.28	5.1×10^{-3}	—	—	0.91	0.033
Mn	0.19	0.010	8.7×10^{-3}	4.7×10^{-4}	—	—	0.027	1.1×10^{-3}
Mo	2.5×10^{-4}	—	—	—	—	—	—	—
Na	0.84	3.00	0.81	16.3	12.2	14.0	6.96	21.7
Ni	0.20	—	—	—	—	—	—	—
P	3.6×10^{-4}	0.032	0.048	—	—	—	3.6×10^{-3}	0.16
Pb	3.88	—	—	—	—	—	—	—
Si	0.032	0.50	0.57	7.53	5.52	5.42	0.62	0.48
Sr	78.5	3.3×10^{-3}	4.3×10^{-3}	1.6×10^{-4}	9.1×10^{-6}	8.0×10^{-6}	—	—
sulfate	0.021	—	—	1.04	0.14	0.19	2.60	1.98
Zn	—	3.4×10^{-4}	1.4×10^{-4}	1.2×10^{-4}	—	1.5×10^{-5}	—	—
U	—	—	—	—	—	—	6.7×10^{-5}	5.5×10^{-4}

*=82WA118
**=08WA125
***=05WA111 and 05WA117
— value of 0 used in model calculations

Table 4. Results of arsenic speciation calculations with the PHREEQC geochemical code using chemistry data from Table 3 and the WATEQ4F database modified to include additional arsenic aqueous complexes reported in Nordstrom et al. (2014, this volume). Redox potential was calculated based on the measured As(III)-As(V) redox couple, if reported, or on the Fe(II)-Fe(III) couple if arsenic oxidation state was not measured. The * denotes the sum of all other calculated As(III) or As(V) species, each individually at an abundance <0.01%, and "—" denotes the calculated abundance of the particular species was <0.01% and was therefore not reported.

Location	Acid mine drainage, Leviathan Mine, CA	As-contaminated groundwater in Bangladesh		Acid spring in Norris Geyser Basin ("Lifeboat" spring)	Hot Springs in Yellowstone National Park, WY		Carson Desert, NV, Groundwater	
Sample	Tunnel 5 drainage	Lakshmipur region, southern Bangladesh	Faridpur region, central Bangladesh		Mound Spring	Mound Spring drainage - 55m downstream	DR 15B	DR 16B
pH	1.8	7.17	6.94	3.23	8.98	9.5	7.1	9.3
pe (calculated)	0.61	−0.01	0.03	0.3687	−0.4787	−0.2611	0.0205	−0.2374
calculated charge balance error	2.1%	1.4%	7.6%	−0.8%	4.0%	−7.7%	−3.3%	−8.5%
As(III)								
As(III) (mmol L^{-1})	$4.67{\times}10^{-1}$	$6.93{\times}10^{-4}$	$1.20{\times}10^{-4}$	$3.47{\times}10^{-4}$	$1.05{\times}10^{-2}$	$2.84{\times}10^{-3}$	$4.67{\times}10^{-5}$	$4.73{\times}10^{-3}$
As(III) (mg L^{-1})	35003	52	9	26	786	213	4	355
As(III)/As(T)	100%	50.5%	23.1%	2%	99%	18%	11%	14%
As(III) species:								
$H_2AsO_3^-$	—	1.16%	0.68%	—	84.93%	70.81%	1.00%	61.37%
$H_3AsO_3^0$	98.71%	98.70%	99.17%	99.94%	13.63%	27.79%	98.87%	36.89%
$H_4AsO_3^+$	1.05%	—	—	0.03%	—	—	—	—
$AlH_2AsO_3^{+2}$	0.23%	—	—	—	—	—	—	—
$CaH_2AsO_3^+$	—	0.06%	0.07%	—	—	—	0.09%	0.03%
$MgH_2AsO_3^+$	—	0.09%	0.04%	—	—	—	0.03%	0.04%
NaH_2AsO_3	—	—	—	—	1.41%	1.36%	0.01%	1.66%
other As(III) species*	0.01%	0.00%	0.04%	0.02%	0.03%	0.04%	0.00%	0.00%

As(V)

	$<1\times10^{-7}$	6.80×10^{-4}	4.00×10^{-4}	1.83×10^{-2}	9.33×10^{-5}	1.26×10^{-2}	3.83×10^{-4}	3.02×10^{-2}
As(V) (mmol L^{-1})	$<1\times10^{-7}$	6.80×10^{-4}	4.00×10^{-4}	1.83×10^{-2}	9.33×10^{-5}	1.26×10^{-2}	3.83×10^{-4}	3.02×10^{-2}
As(V) (mg L^{-1})	<1	51	30	1374	7	947	29	2265
As(V)/As(T)	0%	49.5%	76.9%	98%	1%	82%	89%	86%
As(V) species:								
$H_3AsO_4^0$	—	—	—	6.98%	—	—	—	—
$H_2AsO_4^-$	—	28.82%	38.65%	51.33%	0.18%	0.07%	30.52%	0.11%
$HAsO_4^{-2}$	—	40.91%	32.35%	0.01%	23.08%	23.85%	39.84%	25.24%
AsO_4^{-3}	—	—	—	—	0.38%	0.32%	—	0.25%
$CaH_2AsO_4^+$	—	0.55%	1.50%	0.09%	—	—	1.03%	—
$CaHAsO_4^0$	—	7.06%	11.44%	—	—	0.06%	10.99%	0.03%
$CaAsO_4^-$	—	0.15%	0.31%	—	—	—	0.12%	0.05%
$FeAsO_4^0$	—	—	0.30%	—	—	—	0.02%	—
$FeAsO_4^-$	—	—	0.79%	—	—	—	0.01%	—
$FeH_2AsO_4^+$	—	0.57%	2.03%	—	—	—	0.01%	—
$FeHAsO_4$	—	—	—	—	—	—	0.01%	—
$KAsO_4^{-2}$	—	—	—	—	1.11%	1.06%	—	0.69%
$KHAsO_4^-$	—	0.46%	0.29%	—	—	—	0.01%	0.01%
$KH_2AsO_4^0$	—	0.80%	0.29%	4.36%	—	—	0.19%	—
$MgAsO_4^-$	—	1.26%	1.34%	—	—	—	0.27%	0.50%
$MgH_2AsO_4^+$	—	14.19%	8.87%	0.01%	—	—	0.59%	—
$MgHAsO_4^0$	—	0.21%	—	—	—	—	5.46%	0.06%
$MnHAsO_4^0$	—	—	—	—	—	—	0.40%	—
$MnAsO_4^-$	—	—	—	—	—	—	0.01%	—
$NaH_2AsO_4^0$	—	4.17%	1.51%	37.22%	0.10%	—	9.65%	0.10%
$NaHAsO_4^-$	—	0.32%	—	—	0.68%	0.82%	0.65%	1.15%
$NaAsO_4^{-2}$	—	—	—	—	74.36%	73.71%	0.24%	71.79%
other As(V) species *	—	0.53%	0.35%	0.00%	0.11%	0.11%	0.00%	0.01%

REFERENCES

Amend JP, Saltikov C, Lu G-S, Hernandez J (2014) Microbial arsenic metabolism and reaction energetics. Rev Mineral Geochem 79:391-433

Amirbahman A, Kent D, Curtis G, Davis J (2006) Kinetics of sorption and abiotic oxidation of arsenic(III) by aquifer materials. Geochim Cosmochim Acta 70:533-547

Anawar HM, Akai J, Sakugawa H (2004) Mobilization of arsenic from subsurface sediments by effect of bicarbonate ions in groundwater. Chemophere 54:753-762

Antelo J, Avena M, Fiol S, Lopez R, Arce F (2005) Effects of pH and ionic strength on the adsorption of phosphate and arsenate at the goethite-water interface. J Colloid Interface Sci 285:476-486

Antman KH (2001) Introduction: The history of arsenic trioxide in cancer therapy. Oncologist 6:1-2

Appelo CAJ, Van der Weiden MJJ, Tournassat C, Charlet L (2002) Surface complexation of ferrous iron and carbonate on ferrihydrite and the mobilization of arsenic. Environ Sci Technol 36:3096-3103

Arai Y (2010) Effects of dissolved calcium on arsenate sorption at the kaolinite-water interface. Soil Sci 175:207-213

Arai Y, Sparks DL, Davis JA (2004) Effects of dissolved carbonate on arsenate adsorption and surface speciation at the hematite-water interface. Environ Sci Technol 38:817-824

Asta MP, Cama J, Martinez M, Gimenez J (2009) Arsenic removal by goethite and jarosite in acidic conditions and its environmental implications. J Hazard Mater 171:965-972

Ball JW, Nordstrom DK (1989) Final revised analyses of major and trace elements from acid mine waters in the Leviathan mine drainage basin, California and Nevada - October 1981 to October 1982. U.S. Geological Survey Water Resources Investigation Report 89-4138: 46

Ball JW, McCleskey RB, Nordstrom DK, Holloway JM (2006) Water-chemistry data for selected springs, geysers and streams in Yellowstone National Park, Wyoming, 2003-2005. U.S. Geological Survey Open File Report 2006-1339: 137

Ball JW, McCleskey RB, Nordstrom DK (2010) Water-chemistry data for selected springs, geysers, and streams in Yellowstone National Park, Wyoming, 2006-2008. U.S. Geological Survey Open File Report 2010-1192: 109

Bargar JR, Kubicki JD, Reitmeyer R, Davis JA (2005) ATR-FTIR spectroscopic characterization of coexisiting carbonate surface complexes on hematite. Geochim Cosmochim Acta 69:1527-1542

Bauer M, Blodau C (2006) Mobilization of arsenic by dissolved organic matter from iron oxides, soils, and sediments. Sci Total Environ 354:179-190

Bauer M, Blodau C (2009) Arsenic distribution in the dissolved, colloidal and particulate size fraction of experimental solutions rich in dissolved organic matter and ferric iron. Geochim Cosmochim Acta 73:529-542

Bednar AJ, Garbarino JR, Ranville JF, Wildeman TR (2002) Presence of organoarsenicals used in cotton production in agricultural water and soil of the southern United States. J Agric Food Chem 50:7340-7344

Bhandari N, Reeder RJ, Stongin DR (2011) Photoinduced oxidation of arsenite to arsenate on ferrihydrite. Environ Sci Technol 45:2783-2789

Bhandari N, Reeder RJ, Stongin DR (2012) Photoinduced oxidation of arsenite to arsenate in the presence of goethite. Environ Sci Technol 46:8044-8051

Bissen M, Frimmel FH (2003a) Arsenic - a review. Part I: Occurrence, toxicity, speciation, mobility. Acta Hydrochem Hydrobio 31:9-18

Bissen M, Frimmel FH (2003b) Arsenic - a review. Part II: Oxidation of arsenic and its removal in water treatment. Acta Hydrochem Hydrobio 31:97-107

Bostick BC, Fendorf S, Brown GE (2005) In situ analysis of thioarsenite complexes in neutral to alkaline arsenic sulphide solutions. Mineral Mag 69:781-795

Bowell RJ (1994) Sorption of arsenic by iron oxides and oxyhydroxides in soils. Appl Geochem 9:279-268

British Geological Survey (2001) Groundwater Studies for Arsenic Contamination in Bangladesh. Phase 1: Rapid Investigation Phase. Volume S4: Hydrogeochemistry of the Special Study Areas, 32 p

Burton ED, Bush RT, Johnston SG, Watling KM, Hocking RK, Sullivan LA, Parker GK (2009) Sorption of arsenic(V) and arsenic(III) to schwertmannite. Environ Sci Technol 43:9202-9207

Burton ED, Johnston SG, Kraal P, Bush RT, Claff S (2013) Sulfate availability drives divergent evolution of arsenic speciation during microbially mediated reductive transformation of schwertmannite. Environ Sci Technol 47:221-2229

Buschmann J, Kappeler A, Lindauer U, Kistler D, Berg M, Sigg L (2006) Arsenite and arsenate binding to dissolved humic acids: Influence of pH, type of humic acid, and aluminum. Environ Sci Technol 40:6015-6020

Campbell KM (2007) Biogeochemical mechanisms of arsenic mobilization in Haiwee Reservoir sediments. PhD Dissertation, California Institute of Technology, Pasadena

Campbell KM, Malasarn D, Saltikov CW, Newman DK, Hering JG (2006) Simultaneous microbial reduction of iron(III) and arsenic(V) in suspensions of hydrous ferric oxide. Environ Sci Technol 40:5950-5955

Campbell KM, Root R, O'Day PA, Hering JG (2008a) A gel probe equilibrium sampler for measuring arsenic porewater profiles and sorption gradients in sediments: II. Field application to Haiwee Reservoir sediment. Environ Sci Technol 42:504-510

Campbell KM, Root R, O'Day PA, Hering JG (2008b) A gel probe equilibrium sampler for measureing arsenic porewater profiles and sorption gradients in sediments: I. Laboratory development. Environ Sci Technol 42:497-503

Campbell KM, Hay MB, Nordstrom DK (2012) Kinetic modeling of microbial Fe(II) oxidation, Fe(III) hydrolysis, and As(III) oxidation in acid waters. *In*: Understanding the Geological and Medical Interface of Arsenic – As 2012: Proceedings of the 4th International Conference on Arsenic in the Environment, 22-27 July 2012, Cairns, Australia. Ng JC, Noller BN, Naidu R, Bundschuh J, Bhattacharya P (ed), CRC Press, London, p 461-462

Catalano JG, Park C, Fenter P, Zhang Z (2008) Simultaneous inner- and outer-sphere arsenate adsorption on corundum and hematite. Geochim Cosmochim Acta 72:1986-2004

Cherry JA, Shaikh AU, Tallman DE, Nicholson RV (1979) Arsenic species as an indicator of redox conditions in groundwater. J Hydrol 43:373-392

Chiu VQ, Hering JG (2000) Arsenic adsorption and oxidation at manganite surfaces. 1. Method for simultaneous determination of adsorbed and dissolved arsenic species. Environ Sci Technol 34:2029-2034

Clarke MB, Helz GR (2000) Metal-thiometalate transport of biologically acive trace elements in sulfidic environments: 1. Experimental evidence for copper thioarsenite complexing. Environ Sci Technol 34:1477-1482

Couture RM, Rose J, Kumar N, Mitchell K, Wallschläger D, van Cappellen P (2013) Sorption of arsenite, arsenate, and thioarsenates to iron oxides and iron sulfides: A kinetic and spectroscopic investigation. Environ Sci Technol 47:5652-5659

Cullen WR, Reimer KJ (1989) Arsenic speciation in the environment. Chem Rev 89:713-764

Davis CC, Knocke WR, Edwards M (2001) Implications of aqueous silica sorption to iron hydroxide: mobilization of iron colloids and interference with sorption arsenate and humic substances. Environ Sci Technol 35:3158-3162

Dixit S, Hering JG (2003) Comparison of arsenic(V) and arsenic(III) sorption onto iron oxide minerals: Implications for arsenic mobility. Environ Sci Technol 37:4182-4189

Dou X, Mohan D, Pittman CU (2013) Arsenate adsorption on three types of granular schwertmannite. Water Res 47:2938-2948

Dzombak DA, Morel FMM (1990) Surface Complexation Modeling: Hydrous Ferric Oxide. John Wiley & Sons, New York

Egal M, Casiot C, Morin G, Parmentier M, Bruneel O, Lebrun S, Elbaz-Poulichet F (2009) Kinetic control on the formation of tooeleite, schwertmannite and jarosite by *Acidithiobacillus ferrooxidans* strains in an As(III)-rich acid mine water. Chem Geol 265:432-441

Ehrlich HL (2002) Geomicrobiology. Marcel Dekker Inc., New York

Emett MT, Khoe GH (2001) Photochemical oxidation of arsenic by oxygen and iron in acidic solutions. Water Res 35:649-656

EPA (2013) Organic Arsenicals. Pesticides: Reregistration. U.S. Environmental Protection Agency. *http://www.epa.gov/pesticides/reregistration/organic_arsenicals_fs.html* (accessed July 2014)

Feng X-H, Zu Y-Q, Tan W-F, Liu F (2006) Arsenite oxidation by three types of manganese oxides. J Environ Sci 18:292-298

Fisher JC, Wallschläger D, Planer-Friedrich B, Hollibaugh JT (2008) A new role for sulfur in arsenic cycling. Environ Sci Technol 42:81-85

Foster AL, Brown GE, Parks GA (1998) X-ray absorption fine structure spectroscopy study of photocatalyzed, heterogeneous As(III) oxidation on kaolin and anatase. Environ Sci Technol 32:1444-1452

Fukushi K, Sasaki M, Sato T, Yanase N, Amano H, Ikeda H (2003a) A natural attenuation of arsenic in drainage from an abandoned arsenic mine dump. Appl Geochem 18:1267-1278

Fukushi K, Sato T, Yanase N (2003b) Solid-solution reactions in As(V) sorption by schwertmannite. Environ Sci Technol 37:3581-3586

Fukushi K, Sato T, Yanase N, Minato J, Yamada H (2004) Arsenate sorption on schwertmannite. Am Mineral 89:1728-1734

Fuller CC, Davis JA, Waychunas GA (1993) Surface chemistry of ferrihydrite: Part 2. Kinetics of arsenate adsorption and coprecipitation. Geochim Cosmochim Acta 57:2271-2282

Gao X, Root RA, Farrell J, Ela W, Chorover J (2013) Effect of silicic acid on arsenate and arsenite retention mechanisms on 6-l ferrihydrite: A spectroscopic and batch adsorption approach. Appl Geochem 38:110-120

Garrett AB, Holmes O, Laube A (1940) The solubility of arsenious oxide in dilute solutions of hydrochloric acid and sodium hydroxide: The character of the ions of trivalent arsenic and evidence for polymerization of arsenious acid. J Am Chem Soc 62:2024-2028

Ghosh MM, Yuan JR (1987) Adsorption of inorganic arsenic and organoarsenicals on hydrous oxides. Environ Prog 6:150-157

Gimenez J, Martinez M, Pablo Jd, Rovira M, Duro L (2007) Arsenic sorption onto natural hematite, magnetite, and goethite. J Hazard Mater 141:575-580

Goldberg S (2002) Competitive adsorption of arsenate and arsenite on oxides and clay minerals. Soil Sci Soc Am J 66:413-421

Goldberg S, Johnston CT (2001) Mechanisms of arsenic adsorption on amorphous oxides evaluated using macroscopic meaurements, vibrational spectroscopy, and surface complexation modeling. J Colloid Interface Sci 234:204-216

Gong Z, Lu X, Cullen W, Le X (2001) Unstable trivalent metabolites, monomethylarsonous acid and dimethylarinous acid. J Anal At Spectrom 16:1409-1413

Gonzalez ZI, Krachler M, Cheburkin AK, Shotyk W (2006) Spatial distribution of natural enrichments of arsenic, selenium, and uranium in a minerotrophic peatland, Gola di Lago, Canton Ticino, Switzerland. Environ Sci Technol 40:6568-6574

Grafe M, Eick MJ, Grossl PR (2001) Adsorption of arsenate (V) and arsenite (III) on goethite in the presence and absence of dissolved organic carbon. Soil Sci Soc Am J 65:1680-1687

Grafe M, Eick MJ, Grossl PR, Saunders AM (2002) Adsorption of arsenate and arsenite of ferrihydrite in the presence and absence of dissolved organic carbon. J Environ Qual 31:1115-1123

Guan X, Du J, Meng X, Sun Y, Sun B, Hu Q (2012) Application of titanium dioxide in arsenic removal from water: A review. J Hazard Mater 215-216:1-16

Hansel CM, Benner SG, Ness J, Dohnalkova A, Kukkadapu RK, Fendorf S (2003) Secondary mineralization pathways induced by dissimilatory iron reduction of ferrihydrite under advective flow. Geochim Cosmochim Acta 67:2977-2992

Hansel CM, Benner SG, Nica P, Fendorf S (2004) Structural constraints of ferric (hydr)oxides on dissimilatory iron reduction and the fate of Fe(II). Geochim Cosmochim Acta 68:3217-3229

Hansel CM, Benner SG, Fendorf S (2005) Competing Fe(II)-induced mineralization pathways of ferrihydrite. Environ Sci Technol 39:7147-7153

Hartig C, Planer-Friedrich B (2012) Thioarsenate transformation by filamentous micobial mats thriving in an alkaline, sulfidic hot spring. Environ Sci Technol 46:4348-4356

Harvey CF, Schwartz CH, Budruzzaman ABM, Keon-Blute N, Yu W, Ali MA, Jay J, Beckie R, Ashfaque KN, Islam S, Hemmond HF, Ahmed MF (2002) Arsenic mobility and groundwater extraction in Bangladesh. Science 298:1602-1606

Helz GR, Tossell JA, Charnock JM, Pattrick RAD, Vaughan DJ, Garner CD (1995) Oligomerization in As(III) sulfide solutions: Theoretical constraints and spectroscopic evidence. Geochim Cosmochim Acta 59:4591-4604

Hoffmann M, Mikutta C, Kretzschmar R (2013) Arsenite binding to natural organic matter: Specroscopic evidence for ligand exchange and ternary complex formation. Environ Sci Technol 47:12165-12173

Hohn R, Isenbeck-Schroter M, Kent DB, Davis JA, Jakobsen R, Jann S, Niedan V, Scholz C, Stadler S, Tretner A (2006) Tracer test with As(V) under variable redox conditions controlling arsenic transport in the presence of elevated ferrous iron concentrations. J Contam Hydrol 88:36-54

Hollibaugh JT, Carini S, Gurleyuk H, Jellison R, Joye S, LeCleir G, Meile C, Vasquez L, Wallschläger D (2005) Arsenic speciation in Mono Lake, California: Response to seasonal stratification and anoxia. Geochim Cosmochim Acta 69:1925-1937

Holm TR (2002) Effects of CO_3^{2-}/bicarbonate, Si, and PO_4^{3-} on arsenic sorption to HFO. J Am Water Works Assoc 94:174-181

Hongshao Z, Stanforth R (2001) Competitive adsorption of phosphate and arsenate on goethite. Environ Sci Technol 35:4753-4757

Hsia T-H, Lo S-L, Lin C-F, Lee D-Y (1994) Characterization of arsenate adsorption on hydrous iron oxide using chemical and physical methods. Colloids Surf A 85:1-7

Huang JH, Voegelin A, Pombo SA, Lazzaro A, Zeyer J, Kretzschmar R (2011) Influence of arsenate adsorption to ferrihydrite, goethite, and boehmite on the kinetics of arsenate reduction by *Shewanella putrefaciens* strain CN-32. Environ Sci Technol 45:7701-7709

Hug SJ, Leupin O (2003) Iron-catalyzed oxidation of arsenic(III) by oxygen and by hydrogen peroxide: pH-dependent formation of oxidants in the fenton reaction. Environ Sci Technol 37:2734-2742

Jain A, Raven KP, Loeppert RH (1999) Arsenite and arsenate adsorption on ferrihydrite: Surface charge reduction and net OH- release stoichiometry. Environ Sci Technol 33:1179-1184

Jain A, Loeppert RH (2000) Effect of competing anions on the adsorption of arsenate and arsenite by ferrihydrite. J Environ Qual 29:1422-1430

Johnston SG, Keene AF, Burton ED, Bush RT, Sullivan LA (2011) Iron and arsenic cycling in intertidal surface sediments during wetland remediation. Environ Sci Technol 45:2179-2185

Kaiser K, Guggenberger G, Haumaier L, Zech W (1997) Dissolved organic matter sorption on subsoils and minerals studied by ^{13}C-NMR and drift spectroscopy. Eur J Soil Sci 48:301-310

Kappen P, Webb J (2013) An EXAFS study of arsenic bonding on amorphous aluminum hydroxide. Appl Geochem 31:79-83

Katsoyiannis IA, Zouboulis AI, Jekel M (2004) Kinetics of bacterial As(III) oxidation and subsequent As(V) removal by sorption onto biogenic manganese oxides during groundwater treatment. Ind Eng Chem Res 43:486-493

Kendall MR, Madden AS, Madden MEE, Hu Q (2013) Effects of arsenic incorporation on jarosite dissolution rates and reaction products. Geochim Cosmochim Acta 112:192-207

Kim DH, Lee J, Ryu J, Kim K, Choi, W (2014) Arsenite oxidation initated by the UV photolysis of nitrite and nitrate. Environ Sci Technol 48:4030-4037

Ko I, Kim J-Y, Kim K-W (2004) Arsenic speciation and sorption kinetics in the As-hematite-humic acid system. Colloids Surf A 234:43-50

Kocar BD, Inskeep WP (2003) Photochemical oxidation of As(III) in ferrioxalate solutions. Environ Sci Technol 37:1581-1588

Kuehnelt D, Goessler W, Irgolic KJ (1997) The oxidation of arsenite in aqueous solutions. *In:* Arsenic: Exposure and Health Effects. Abernathy CO, Calderon RL, Chappell WR (eds) Chapman and Hall, London, p 45-54

Lafferty BJ, Leoppert RH (2005) Methyl arsenic adsorption and desorption behavior on iron oxides. Environ Sci Technol 39:2120-2127

Langner P, Mikutta C, Kretzschmar R (2011) Arsenic sequestration by organic sulphur in peat. Nature Geosci 5:66-73

Leuz A, Hug SJ, Wehrli B, Johnson C (2006) Iron-mediated oxidation of antimony(III) by oxygen and hydrogen peroxide compared to arsenic(III) oxidation. Environ Sci Technol 40:2565-2571

Lin H-T, Wang MC, Li G-C (2004) Complexation of arsenate with humic substance in water extract of compost. Chemophere 56:1105-1112

Lin Z, Puls RW (2000) Adsorption, desorption, and oxidation of arsenic affected by clay minerals and aging process. Environ Geol 39:753-759

Liu F, Cristofaro AD, Violante A (2001) Effect of pH, phosphate and oxalate on the adsorption/desorption of arsenate on/from geothite. Soil Sci 166:197-208

Livesey NT, Hyang PM (1981) Adsorption of arsenate by soils and its relation to selected chemical properties and anions. Soil Sci 131:88-94

Lovley DR, Coates JD, Blunt-Harris EL, Phillips EJP, Woodward JC (1996) Humic substances as electron acceptors for microbial respiration. Nature 382:445-448

Ma LQ, Komar KM, Tu C, Zhang W, Cai Y, Kennelley ED (2001) A fern that hyperaccumulates arsenic. Nature 409:579

Manning BA, Goldberg S (1996) Modeling competitive adsorption of arsenate with phosphate and molybdate on oxide minerals. Soil Sci Soc Am J 60:121-131

Manning BA, Goldberg S (1997a) Adsorption and stability of arsenic(III) at the clay mineral-water interface. Environ Sci Technol 31:2005-2011

Manning BA, Goldberg S (1997b) Arsenic(III) and Arsenic(V) adsorption on three California soils. Soil Sci 162:886-895

Manning BA, Fendorf SE, Goldberg S (1998) Surface structures and stability of arsenic(III) on geothite: Spectroscopic evidence for inner-sphere complexes. Environ Sci Technol 32:2383-2388

Manning BA, Fendorf S, Bostick BC, Suarez DL (2002) Arsenic (III) oxidation and arsenic (V) adsorption reactions on synthetic birnessite. Environ Sci Technol 36:976-981

Marini L, Accornero M (2007) Prediction of the thermodynamic properties of metal-arsenate and metal-arsenite aqueous complexes to high temperatures and pressures and some geological consequences. Environ Geol 52:1343-1363

McArthur JM, Banerjee DM, Hudson-Edwards KA, Mishra R, Purohit R, Ravenscroft P, Cronin A, Howarth RJ, Chatterjee A, Talukder T, Lowry D, Houghton S, Chadha DK (2004) Natural organic matter in sedimentary basins and its relation to arsenic in anoxic ground water: The example of West Bengal and its worldwide implications. Appl Geochem 19:1255-1293

McGeehan SL, Naylor DV (1994) Sorption and redox transformation of arsenite and arsenate in two flooded soils. Soil Sci Soc Am J 58:337-342

Meng X, Bang S, Korfiatis GP (2000) Effects of silicate, sulfate, and carbonate on arsenic removal by ferric chloride. Water Res 34:1255-1261

Mikutta C, Kretzschmar R (2011) Spectroscopic evidence for ternary complex formation between arsenate and ferric iron complexes of humic substances. Environ Sci Technol 45:9550-9557

Mohajerin TJ, Neal AW, Telfeyan K, Sasihharan SM, Ford S, Yang N, Chevis DA, Grimm DA, Datta S, White CD, Johannesson HH (2014) Geochemistry of tungsten and arsenic in aquifer systems: A comparative study of groundwaters from West Bengal, India, and Nevada, USA. Water, Air, Soil, Poll 225:1792

Mohan D, Pittman CU (2007) Arsenic removal from water/wastewater using adsorbents - a critcal review. J Hazard Mater 142:1-53

Neubauer E, Kammer Fvd, Knorr K, Peiffer S, Reichert M, Hofmann T (2013) Colloid-associated export of arsenic in stream water during stormflow events. Chem Geol 352:81-91

Neuberger CS, Helz GR (2005) Arsenic(III) carbonate complexing. Appl Geochem 20:1218-1225

Newman DJ, Kolter R (2000) A role for excreted quinones in extracellular electron transfer. Nature 405:94-97

Nordstrom DK (2000) Thermodynamic properties of environmental arsenic species: Limitations and needs. *In*: Minor elements 2000: Processing and environmental aspects of As, Sb, Se, Te, and Bi. Young C (ed) Society for Mining, Metallurgy, and Exploration, Littleton, CO, p 325-331

Nordstrom DK, Archer DG (2003) Arsenic thermodynamic data and environmental geochemistry. *In*: Arsenic in groundwater: Geochemistry and occurance. Welch AH, Stollenwerk KG (eds) Kluwer Academic Press, Norwell, p 1-25

Nordstrom DK, Majzlan J, Königsberger E (2014) Thermodynamic properties for arsenic minerals and aqueous species. Rev Mineral Geochem 79:217-255

Ona-Nguema G, Morin G, Juillot F, Calas G, Brown G Jr (2005) EXAFS analysis of arsenite adsorption onto two-line ferrihydrite, hematite, goethite, and lepidocrocite. Environ Sci Technol 39:9147-9155

Oremland R, Hoeft S, Santini J, Bano N, Hollibaugh R, Hollibaugh J (2002) Anaerobic oxidation of arsenite in mono lake water and by a facultative, arsenite-oxidixing chemoautotroph, strain MLHE-1. Appl Environ Microbiol 68:4795-4802

Oremland RS, Stolz JF (2003) The ecology of arsenic. Science 300:393-944

Parkhurst DL, Appelo CAJ (2013) Description of input and examples for PHREEQC version 3—a computer program for speciation, batch-reaction, one-dimensional transport, and inverse geochemical calculations. *In*: U.S. Geological Survey Techniques and Methods, Book 6, chap. A43, Denver, CO, p 497

Petrick JS, Ayala-Fierro F, Cullen W, Carter D, Aposhian H (2000) Monomethylarsonous acid (MMA) is more toxic than arsenite in chang human hepatocytes. Toxicol Appl Pharmacol 163:203-207

Pierce ML, Moore CB (1982) Adsorption of arsenite and arsenate on amorphous iron hydroxide. Water Res 16:1247-1253

Planer-Friedrich B, London J, McCleskey RB, Nordstrom DK, Wallschläger D (2007) Thioarsenates in geothermal waters of Yellowstone National Park: Determination, preservation, and geochemical importance. Environ Sci Technol 41:5245-5251

Radu T, Subacz JL, Phillippi JM, Barnett MO (2005) Effects of dissolved carbonate on arsenic adsorption and mobility. Environ Sci Technol 39:7875-7882

Randall SR, Sherman DM, Ragnarsdottir KV (2001) Sorption of As(V) on green rust (Fe$_4$(II)Fe$_2$(III) (OH)$_{12}$SO$_4$·3H$_2$O) and lepidocrocite (γ-FeOOH): Surface complexes from EXAFS spectroscopy. Geochim Cosmochim Acta 65:1015-1023

Raven KP, Jain A, Leoppert RH (1998) Arsenite and arsenate adsorption on ferrihydrite: Kinetics, equilibrium, and adsorption envelopes. Environ Sci Technol 32:344-349

Redman AD, Macalady DL, Ahmann D (2002) Natural organic matter affects arsenic speciation and sorption onto hematite. Environ Sci Technol 36:2889-2896

Regenspurg S, Peiffer S (2005) Arsenate and chromate incorporation in schwertmannite. Appl Geochem 20:1226-1239

Ritter K, Aiken GR, Ranville JF, Bauer M, Macalady DL (2006) Evidence for the aquatic binding of arsenate by natural organic matter-suspended Fe(III). Environ Sci Technol 40:5380-5387

Rochette EA, Bostick BC, Li G, Fendorf S (2000) Kinetics of arsenate reduction by dissolved sulfide. Environ Sci Technol 34:4714-4720

Rosen B, Marapakala K, Abdul Salam AA, Packianathan C, Yoshinaga M (2012) Pathways of arsenic biotransformations: The arsenic methylation cycle. *In*: Understanding the geological and medical interface of arsenic. Ng JC, Noller BN, Naidu R, Bundschuh J, Bhattacharya P (ed) CRC Press, London, p 185-188

Rothwell JJ, Taylor KG, Ander EL, Evans MG, Daniels SM, Allott TEH (2009) Arsenic retention and release in ombrotrophic peatlands. Sci Total Environ 407:1405-1417

Ruby C, Upadhyay C, Gehin A, Ona-Nguema G, Genin J-MR (2006) In situ redox flexibility of Fe(II)-(III) oxyhydroxycarbonate green rust and fougerite. Environ Sci Technol 40:4696-4702

Saada A, Breeze D, Crouzet C, Cornu S, Baranger P (2003) Adsorption of arsenic(V) on kaolinite and on kaolinite-humic acid complexes: Role of humic acid nitrogen groups. Chemophere 51:757-763

Scherer OJ (1999) P$_n$ and As$_n$ ligands: A novel chapter in the chemistry of phosphorous and arsenic. Acc Chem Res 32:751-762

Scott MJ, Morgan JJ (1995) Reactions at oxide surfaces. 1. Oxidation of As(III) by synthetic birnessite. Environ Sci Technol 29:1898-1905

Senn DB, Hemond HF (2002) Nitrate controls on iron and arsenic in an urban lake. Science 296:2373-2376

Senn DB, Hemond HF (2004) Particulate arsenic and iron during anoxia in a eutrophic, urban lake. Environ Toxicol Chem 23:1610-1616

Sherman DM, Randall SR (2003) Suface complexation of arsenic(V) to iron(III) (hydr)oxides: Structural mechanism from ab initio molecular geometries and EXAFS spectroscopy. Geochim Cosmochim Acta 67:4223-4230

Singer DM, Fox PM, Guo H, Marcus MA, Davis JA (2013) Sorption and redox reactions of As(III) and As(V) within secondary mineral coatings on aquifer sediment grains. Environ Sci Technol 47:11569-11576

Smedley PL, Kinniburgh DG (2002) A review of the source, behaviour and distribution of arsenic in natural waters. Appl Geochem 17:517-568

Smith E, Naidu R, Alson AM (1998) Arsenic in the soil environment: A review. Adv Agron 64:149-195

Spycher, NF, Reed, MH (1989) As(III) and Sb(III) sulfide complexes: an evaluation of stoichiometry and stability from existing experimental data. Geochim Cosmochim Acta 53:2185-2194

Stachowicz M, Hiemstra T, Hiemsdijk WH (2008) Multi-competitive interaction of As(III) and As(V) oxyanions with Ca^{2+}, Mg^{2+}, PO_4^{3-}, and CO_3^{2-} ions on goethite. J Colloid Interface Sci 320:400-414

Stauder S, Raue B, Sacher F (2005) Thioarsenates in sulfidic waters. Environ Sci Technol 39:5933-5939

Stucker VK, Williams KH, Robbins MJ, Ranville JF (2013) Arsenic geochemistry in a biostimulated aquifer: An aqeous speciation story. Environ Toxicol Chem 32:1216-1223

Su C, Suarez DL (1997) In situ infrared speciation of adsorbted carbonate on aluminum and iron oxides. Clays Clay Miner 45:814-825

Su C, Wilkin RT (2005) Arsenate and arsenite sorption on and arsenite oxidation by iron(II,III) hydroxycarbonate green rust. *In*: Advances in arsenic research: Integration of experimental and observational studies and implications for mitigation. O'Day PA, Vlassopoulos D, Meng X, Benning LG (eds) Americal Chemical Society, Washington, DC, p 25-40

Suess E, Wallschläger D, Planer-Friedrich B (2011) Stabilization of thioarsenates in iron-rich waters. Chemosphere 83:1524-1531

Sun W, Sierra R, Field J (2008) Anoxic oxidation of arsenite linked to denitrification in sludges and sediments. Water Res 42:4569-4577

Sun X, Doner HE (1996) An investigation of arsenate and arsenite bonding structures on goethite by FTIR. Soil Sci 161:865-872

Swedlund PJ, Webster JG (1999) Adsorption and polymerization of silicic acid on ferrihydrite and its effect on arsenic adsorption. Water Res 33:3413-3422

Swedlund PJ, Miskelly GM, McQuillan AJ (2009) An attenutated total reflectance IR study of silicic acid adsorbed onto a ferric oxyhydroxide surface. Geochim Cosmochim Acta 73:4199-4214

Tamaki S, Frankenberger WT (1992) Environmental biogeochemistry of arsenic. Rev Envrion Contam Toxicol 124:79-110

Thanabalasingam P, Pickering WF (1986a) Arsenic sorption by humic acids. Environ Pollut (Series B) 12:233-246

Thanabalasingam P, Pickering WF (1986b) Effect of pH on interaction between As(III) or As(V) and manganese(IV) oxide. Water Air Soil Pollut 29:205-216

Ullrich MK, Pope JG, Seward TM, Wilson N, Planer-Friedrich B (2013) Sulfur redox chemistry governs diurnal antimony and arsenic cycles at Champagne Pool, Waiotapu, New Zealand. J Volcanol Geotherm Res 262:164-177

van Geen A, Robertson AP, Leckie JO (1994) Complexation of carbonate species at the goethite surface: Implications for adsorption of metal ions in natural waters. Geochim Cosmochim Acta 58:2073-2086

Villalobos M, Leckie JO (2000) Carbonate adsorption on goethite under closed and open CO_2 conditions. Geochim Cosmochim Acta 64:3787-3802

Villalobos M, Leckie JO (2001) Surface complexation modeling and FTIR study of carbonate adsorption to goethite. J Colloid Interface Sci 235:15-32

Violante A, Pigna M (2002) Competitive sorption of arsenate and phosphate on different clay minerals and soils. Soil Sci Soc Am J 66:1788-1796

Wallschläger D, Stadey CJ (2007) Determination of (oxy)thioarsenate in sulfidic waters. Anal Chem 79:3873-3880

Waltham CA, Eick MJ (2002) Kinetics of arsenic adsorption on goethite in the presence of sorbed silicic acid. Soil Sci Soc Am J 66:818-825

Wang S, Mulligan CN (2006) Effect of natural organic matter on arsenic release from soils and sediments into groundwater. Environ Geochem Health 28:197-214

Warwick P, Inam E, Evans N (2005) Arsenic's interaction with humic acid. Environ Chem 2:119-124

Waychunas GA, Rea BA, Fuller CC, Davis JA (1993) Surface chemistry of ferrihydrite: Part 1. EXAFS studies of the geometry of coprecipitated and adsorbed arsenate. Geochim Cosmochim Acta 57:2251-2269

Waychunas GA, Davis JA, Fuller CC (1995) Geometry of sorbed arsenate on ferrihydrite and crystalline FeOOH: Re-evaluation of EXAFS results and topological factors in predicting sorbate geometry, and evidence for monodentate complexes. Geochim Cosmochim Acta 59:3655-3661

Webster JG (1990) The solubility of As_2S_3 and speciation of As in the dilute and sulphide-bearing fluids at 25 and 90 °C. Geochim Cosmochim Acta 54:1009-1017

Welch AH, Lico MS (1998) Factors controlling As and U in shallow groundwater, southern Carson Desert, Nevada. Appl Geochem 13:521-539

Wilkie J, Hering JG (1996) Adsorption of arsenic onto hydrous ferric oxide: Effects of adsorbate/adsorbent ratios and co-occurring solutes. Colloids Surf A 107:97-110

Wilkin RT, Wallschläger D, Ford RG (2003) Speciation of arsenic in sulfidic waters. Geochem Trans 4:1-7

Woods R (1966) Arsenic(IV) as an intermediate in the photochemical oxidation of ferrous sulfate in the presence of arsenic acid. J Phys Chem 70:1446-1452

Xu H, Allard B, Grimvall A (1991) Effects of acidification and natural organic materials on the mobility of arsenic in the environment. Water Air Soil Pollut 57-58:269-278

Zachara JM, Kukkadupu RK, Fredrickson JK, Gorby YA, Smith SC (2002) Biomineralization of poorly crystalline Fe(III) oxides by dissimilatory metal reducing bacteria (DMRB). Geomicrobiol J 19:179-207

Reviews in Mineralogy & Geochemistry
Vol. 79 pp. 217-255, 2014
Copyright © Mineralogical Society of America

4

Thermodynamic Properties for Arsenic Minerals and Aqueous Species

D. Kirk Nordstrom

U.S. Geological Survey
3215 Marine St.
Boulder, Colorado 80303, U.S.A.

dkn@usgs.gov

Juraj Majzlan

Friedrich-Schiller-Universität Jena
Chemisch-Geowissenschaftliche Fakultät
Institut für Geowissenschaften
Carl-Zeiss-Promenade 10
07745 Jena, Germany

Juraj.Majzlan@uni-jena.de

Erich Königsberger

Chemical and Metallurgical Engineering and Chemistry
Murdoch University
Murdoch, WA 6150, Australia

E.Koenigsberger@murdoch.edu.au

INTRODUCTION

Quantitative geochemical calculations are not possible without thermodynamic databases and considerable advances in the quantity and quality of these databases have been made since the early days of Lewis and Randall (1923), Latimer (1952), and Rossini et al. (1952). Oelkers et al. (2009) wrote, *"The creation of thermodynamic databases may be one of the greatest advances in the field of geochemistry of the last century."* Thermodynamic data have been used for basic research needs and for a countless variety of applications in hazardous waste management and policy making (Zhu and Anderson 2002; Nordstrom and Archer 2003; Bethke 2008; Oelkers and Schott 2009). The challenge today is to evaluate thermodynamic data for internal consistency, to reach a better consensus of the most reliable properties, to determine the degree of certainty needed for geochemical modeling, and to agree on priorities for further measurements and evaluations.

Recent attention has been directed to arsenic (As) thermodynamic data, partly because of the worldwide recognition of arsenic poisoning in more than 70 countries (Nordstrom 2002; Ravenscroft et al. 2009) and the need to interpret As mobility more quantitatively in groundwater and surface-water systems. Unfortunately, not as many useful thermodynamic measurements have been made on reactions involving As compared to other major solutes and trace elements. Grenthe et al. (1992), when reviewing As data for the Organization for Economic Cooperation and Development/Nuclear Energy Agency (OECD/NEA) thermodynamic database on U, stated *"Although needed, a complete reanalysis of the chemical thermodynamic data for arsenic species is not within the scope of the current review."* Although a complete reanalysis is not feasible at this time, this chapter reviews As thermodynamic data with a focus on internal consistency and the quality of the original measurements, updates with new data, points out some areas of discrepancies, makes recommendations for the resolution of some properties, and suggests avenues for further investigations.

METHODOLOGICAL APPROACH TO
INTERNALLY CONSISTENT DATA

Wagman et al. (1982) described thermodynamic data as an over-determined network because of the replicate measurements of the same system and because some reactions or properties were measured that were already possible to calculate from available data. In a strictly theoretical sense, the thermodynamic network is often over-determined, but in another sense it is under-determined because uncertainties can be large and inconsistencies persist that may require more careful measurements or further evaluation. For example, when calorimetric or electrochemical measurements are compared to a solubility product constant for the same species in a solubility reaction using thermodynamic relationships, the difference can be substantially greater than the stated experimental errors. Furthermore, species and reactions of interest to the aqueous geochemistry of many trace elements and some major components still have not been measured.

Wagman et al. (1982) pointed out that typically two methods have been used to evaluate thermodynamic data, the sequential method and the simultaneous fit method. The National Bureau of Standards (NBS) tables were done with a combination of both techniques although primarily with the sequential method. There is a "Standard Order of Arrangement of the Elements," an ordered sequence through the periodic table, that was followed in that project, and some subsets of specific elements (not arsenic) were simultaneously fit. The Committee on Data for Science and Technology (CODATA) book on Ca compounds (Garvin et al. 1987) simultaneously fit most measurements of the major compounds and species of Ca to provide what is probably the most consistent set available for Ca data. Several iterations of simultaneous fits were necessary to screen outliers and to find the best weighting for reliable data, a very important aspect of evaluating data. Some combination of both sequential and simultaneous fitting is usually required to obtain the most reliable data. Internal consistency has been defined by Wagman et al. (1982) and reiterated by Nordstrom and Munoz (1994); one of the criteria is an appropriate choice of starting point in the network. The starting point should be a reaction or compound property that is the most reliable for the chosen network set (presumed to be a subset of a larger network) but evaluating the most reliable original measurements can be challenging and laborious. If a highly reliable property is found, that property would have the highest weight in the fitting procedure and could be considered an anchor or cornerstone for the network. Defining the most reliable measurements is challenging. The evaluator has to consider the accuracy and precision of the analytical techniques, whether enough constituents were determined, whether the thermodynamic measurement method was adequate to define the system under study, how many replicates or independent investigations were done, and any sources of error or uncertainty that may compromise or bias the result.

In many cases there may be more than one anchor; for example, the entropies of elements used to calculate the standard state entropy of formation from the elements for a compound are needed to calculate almost any other property. Elemental entropies need to be periodically updated or at least rechecked as well. Fortunately, the heat capacity and entropy for elemental arsenic has been re-evaluated recently (Nordstrom and Archer 2003). In this paper we evaluate the solubility of both arsenolite (As_2O_3, cubic) and scorodite ($FeAsO_4 \cdot 2H_2O$) as points of comparison to establish an anchor with the thermodynamic properties of the arsenate (AsO_4^{3-}) aqueous species.

FINDING AN ANCHOR FOR STANDARD STATE
PROPERTIES AT 298.15 K, 1 BAR

For the system As-H_2O-S-Fe there are several evaluations to be considered and some new data that should be included as confirmation for a standard state anchor that would include both minerals and aqueous species. Pokrovski et al. (1996) and Perfetti et al. (2008) have applied the HKF equations of state (Helgeson-Kirkham-Flowers; Helgeson et al. 1981) to the evaluation of arsenolite, claudetite (As_2O_3, monoclinic), orpiment (As_2S_3), realgar (AsS), elemental As, and the aqueous species that form during aqueous solubility equilibrium to 350 °C and 300 bars. The solubility of arsenolite or claudetite would be a desirable anchor because these represent one of the most direct pathways to the aqueous species. Nordstrom and Archer (2003) considered the same system (also without Fe). Another brief evaluation is that of Grenthe et al. (1992) which drew substantially from the NBS tables (Wagman et al. 1982). Another past evaluation is that of Gaskova et al. (2001) at the French Geological Survey (BRGM). The most recent work is that on scorodite solubility and the recognized consistency among the $FeAsO_4$-H_2O properties (Majzlan et al. 2012a). We have evaluated these systems by first considering the most reliable data and then comparing additional sources of data where differences exist. In our analysis, we have avoided compilations and used original measurements for the most reliable data wherever possible.

Standard state thermodynamic properties of arsenolite, As_2O_3 (cubic)

The thermodynamic properties for solid arsenolite seem to be in good agreement. Although there are limited data, there are some replicate measurements and most of the measurements are of high quality. Two investigations have measured the electrochemical potential for the reaction,

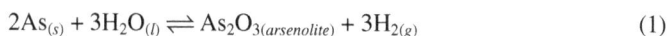

$$2As_{(s)} + 3H_2O_{(l)} \rightleftharpoons As_2O_{3(arsenolite)} + 3H_{2(g)} \tag{1}$$

The results were $\Delta G_r^\circ = 135.52 \pm 0.060$ kJ mol^{-1} (Schuhmann 1924) and $\Delta G_r^\circ = 135.149 \pm 0.065$ kJ mol^{-1} (Kirschning and Plieth 1955). The very close agreement between these two studies suggests that we can safely average them to get $\Delta G_r^\circ = 135.33$ kJ mol^{-1}. By the error propagation method (e.g., Bevington 1969), the error would be ± 0.044 kJ mol^{-1} but for better accuracy the error is calculated by half the difference of the two values (± 0.185 kJ mol^{-1}). This approach to estimating error is used in this report whenever there are only 2 or 3 property measurements because standard deviations in such cases are not meaningful. Using the $\Delta G_f^\circ{}_{(H_2O,l)}$ from CODATA (Cox et al. 1989) we calculate $\Delta G_f^\circ{}_{As_2O_3} = -576.09$ kJ mol^{-1} with an estimated error of ± 0.185 kJ mol^{-1}. This is the most direct and most reliable path for obtaining the Gibbs free energy of arsenolite. Heat capacity measurements were obtained by Chang and Bestul (1971) and the S° at 298.15 K was 107.41 ± 0.064 J K^{-1} mol^{-1}. From these data the ΔH_f° can be calculated using $S^\circ{}_{(As,s)} = 35.63$ J K^{-1} mol^{-1} from Nordstrom and Archer (2003), which is nearly identical to that in Hultgren et al. (1973), and the entropy for oxygen gas from CODATA. Earlier heat capacity measurements of arsenic trioxide by Anderson (1930) gave $S^\circ = 107.11 \pm 1.67$ J K^{-1} mol^{-1} which is nearly identical to the value from Chang and Bestul (1971) but with a larger error. Gorgoraki and Tarasov (1965) reported 116.69 ± 0.84 J K^{-1} mol^{-1} which is inconsistent with other measurements. Hence, we recommend $\Delta H_f^\circ = -657.061 \pm 0.185$ kJ mol^{-1} which compares very favorably with heat measurements and evaluations from several investigators (Table 1).

Table 2 summarizes reference state (25 °C, 1 bar, not necessarily the most stable form) thermodynamic values for arsenolite. These values are also in good agreement with the compilations of Wagman et al. (1982) and Robie and Hemingway (1995) because the same original measurements were used. We see little need for improvement in these numbers.

Table 1. Comparison of standard state enthalpies
of formation for arsenolite.

$\Delta H_f°$ (kJ mol^{-1})	Reference
−654.38	Bertholet (1897)
−647.26	Thomsen (1882-85)
−644.04	Schuhmann (1924)
−654.80	Anderson (1930) and Schuhmann (1924)
−647.26	de Passille (1936)
−656.70	Rossini et al. (1952)
−656.05 ± 2.76	Beezer et al. (1965) corrected[*]
−669.44 ± 12.55	Beezer and Mortimer (1966)
−658.4	Minor et al. (1971)
−656.97 ± 8.0	Grenthe et al. (1992); Wagman et al. (1982)
−656.60	Pokrovski et al. (1996)
−657.27	Nordstrom and Archer (2003)
−657.061 ± 0.185	This study

[*]corrected to the appropriate number of moles as discussed in Archer and Nordstrom (2002)

Table 2. Summary and comparison of standard state thermodynamic properties for arsenolite.

$\Delta G_f°$ (kJ mol^{-1})	$\Delta H_f°$ (kJ mol^{-1})	$S°$ (J K^{-1} mol^{-1})	$C_p°$ (J K^{-1} mol^{-1})	Reference
−575.96 ± 0.29	—	—	—	Schuhmann (1924), corrected[*]
−576.225 ± 8.015	−656.97 ± 8.00	107.1 ± 1.2	95.645 ± 0.4	Grenthe et al. (1992)
−576.215	−656.97	107.1	95.645	Wagman et al. (1982)
−576.0 ± 1.9	−657.0 ± 1.7	107.4 ± 0.1	—	Robie and Hemingway (1995)
−576.09 ± 0.185	−657.061 ± 0.185	107.41 ± 0.06	96.88 ± 0.06	This study

[*]Schuhmann incorrectly calculated the standard electrode potential from his value for the solubility and the measured electrode potential.

Arsenolite solubility

Pokrovski et al. (1996; Table 1) measured and compiled data on the solubility of arsenolite and claudetite for 0-250 °C in terms of log m_{As}. We have compiled the arsenolite values for temperatures at or near 25 °C along with other data in terms of molal solubilities (not logarithmic) in Table 3. An atomic weight of 74.92 for As and 15.9994 for O were used where needed (Weiser and Coplen 2011). These data suggest there is good agreement for most of the studies except those for Wood (1908), Stranski et al. (1958), Aja et al. (1992), and Pokrovski et al. (1996). Stranski et al. (1958) and Schulman and Schumb (1943) made their solubility measurements in 1 N HCl which would affect their values compared to those in pure water. Unfortunately this comparison questions the reliability of the evaluation by Pokrovski et al. (1996) and Perfetti et al. (2008) because their listed value for arsenolite solubility is considerably discrepant from most of the other values. Solubility equilibrium for arsenolite, it has been noted, is quite slow to achieve and this observation may explain why the outlier values are consistently low. The values interpolated linearly from Schreinemakers and deBaat (1917, cited by Linke and Seidell 1958), Anderson and Story (1923), Schuhmann (1924),

Table 3. Solubility data for arsenolite in pure water
(as As in molality) at temperatures of 20-30 °C.

Solubility (mol As/kg$_{H_2O}$)	T (°C)	Reference
0.157, 0.151	ambient	Wood (1908)
0.234	30	Schreinemakers and deBaat (1917)
0.183	20	Schreinemakers and deBaat (1917)
0.209	25	Schreinemakers and deBaat (1917, by interpolation)
0.207	25	Anderson and Story (1923)
0.2067 ± 0.003	25	Zawidski (1903)
0.183	20	de Carli (1932)
0.2070 ± 0.002	25	Garrett et al. (1940)
0.176 ± 0.001	25	Schulman and Schumb (1943) in 1 N HCl
0.209	25	Tourky and Mousa (1949)
0.2080	25	Jozefowicz et al. (1950)
0.138	21	Stranski et al. (1958) as cited in Pokrovski et al. (1996)
0.148	22	Pokrovski et al. (1996)
0.207 ± 0.002	25	This study

Garrett et al. (1940), Tourky and Mousa (1949), Jozefowicz et al. (1950), are in very good agreement, are all at 25 °C, and suggest the solubility should be between 0.206 and 0.209 m. Because of the excellent agreement among Zawidski (1903, cited by Linke and Seidell 1958), Anderson and Story (1923), and Garrett et al. (1940), we take the solubility in pure water to be 0.207 ± 0.002 m. Baes and Mesmer (1976) in their evaluation of hydrolysis constants adopted a solubility of 0.206 m at 25 °C. This value (0.207 m) converts to a log K_{sp} = log m = −0.684 ± 0.004 and $\Delta G_r°$ = 3.905 ± 0.023 kJ mol^{-1} (per mole of As), if the activity coefficient is assumed to be unity (a reasonable assumption for an uncharged aqueous species, $H_3AsO_3°$). With the properties of arsenolite well characterized, this solubility value allows us to calculate the thermodynamic properties of the aqueous species $H_3AsO_3°$.

Aqueous arsenite and arsenate species

From the K_{sp} and the Gibbs free energy of formation of arsenolite, the Gibbs free energy of arsenous acid can be derived assuming the equilibrium solubility reaction,

$$As_2O_{3(cr)} + 3H_2O_{(l)} \rightleftharpoons H_3AsO_{3(aq)}° \qquad (2)$$

which results in −639.85 ± 0.5 kJ mol^{-1} using the Gibbs free energy of water from CODATA. The next step is to examine measurements made for the oxidation of arsenous acid to arsenic acid and to obtain the Gibbs free energy of arsenic acid. An alternate path might consider the solubility and Gibbs free energy of As_2O_5 but the solubility is quite high (~260 g As L^{-1}; Linke and Seidell 1958), few data are available, and a thermodynamic analysis would not be feasible. Also, this compound has not been found in the environment, to the best of our knowledge.

Foerster and Pressprich (1927) measured the equilibrium potential of an electrochemical cell containing arsenous and arsenic acids, hydrochloric acid, and potassium iodide at 18 °C. The reference electrode was mercury/mercuric sulfate in different concentrations of sulfuric acid. The reference voltage was with 2 N H_2SO_4 and compared well with a value calculated from Harned and Hamer (1935). After making activity coefficient corrections and using entropies from Wagman et al. (1982), the equilibrium potential was calculated for 298.15 K (0.573 V) and the consequent Gibbs free energy of reaction is -110.59 ± 0.46 kJ mol^{-1} for the reaction

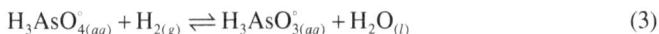

$$H_3AsO_{4(aq)}^{\circ} + H_{2(g)} \rightleftharpoons H_3AsO_{3(aq)}^{\circ} + H_2O_{(l)} \tag{3}$$

More recently, Pesavento (1989) measured the electrochemical potential for the same reaction at 25 °C and utilized iodide to catalyze the reaction rate and extrapolated the data. She varied the background electrolyte, sodium and hydrogen perchlorate solutions to extrapolate to infinite dilution. Her results gave identical results to the corrected data of Foerster and Pressprich (1927). These data are the most accurate and the most direct path to obtain the Gibbs free energy of arsenic acid.

Using the previously derived Gibbs free energy value for arsenous acid and the Gibbs free energy of liquid water from CODATA, the Gibbs free energy of arsenic acid is -766.42 ± 0.35 kJ mol^{-1}. The assigned error is based on that from Pesavento (1989).

Further confirmation of this value comes from the study by Washburn and Strachan (1913) in which they measured the equilibrium constant for the reaction

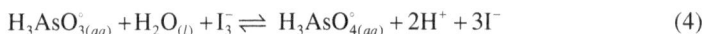

$$H_3AsO_{3(aq)}^{\circ} + H_2O_{(l)} + I_3^{-} \rightleftharpoons H_3AsO_{4(aq)}^{\circ} + 2H^{+} + 3I^{-} \tag{4}$$

The equilibrium constant is $K = 0.0554$ and $\Delta G_r^{\circ} = 7.17$ kJ mol^{-1}. Using Gibbs free energy data on aqueous iodide ion from CODATA and the triiodide ion from Wagman et al. (1982), the Gibbs free energy of arsenic acid is -766.05 kJ mol^{-1}. Randall (1930) pointed out a calculation error in the Washburn and Strachan (1913) paper and revised the K to 0.057 or $\Delta G_r^{\circ} = 7.10$ kJ mol^{-1}. This correction leads to $\Delta G_f^{\circ}{}_{H_3AsO_4} = -766.12$ kJ mol^{-1}. Both of these experiments were carefully done using reliable reversible reactions and the results are in excellent agreement. However, we note that there is a 1.6 kJ mol^{-1} difference in the enthalpy of the iodide ion given by Wagman et al. (1982) compared to that in CODATA and we have no confirmation of the free energy for the triiodide ion. Consequently, we have chosen the more direct and precise path using the data of Foerster and Pressprich (1927) and Pesavento (1989). Another study on the arsenite oxidation by triiodide equilibrium was done by Roebuck (1905) and evaluated and corrected by Liebhafsky (1931) which produced $\Delta G_f^{\circ}{}_{H_3AsO_4} = -769.02$ kJ mol^{-1} but this value was obtained without activity coefficient corrections and at 0 °C.

Obtaining reliable enthalpy and entropy values for arsenous and arsenic acids is a little more challenging. Anderson and Story (1923) measured the solubility of arsenolite as a function of temperature between 0 and 98.5 °C from which the ΔS_r and ΔH_r can be calculated. However, because these are derivative functions, their uncertainty increases. A careful analysis of their data indicate that a linear fit of log $m_{As_2O_3}$ vs $1/T$ gives an improved correlation coefficient ($r^2 = 0.999945$) if data points at 0, 15, and 39.8 °C are removed. The result is $\Delta H_r^{\circ} = 34.661$ kJ mol^{-1} with an estimated error of ± 1.0 kJ mol^{-1}. Thomsen (1882-1885) measured the heat of solution of As$_2$O$_3$ at 17.5 °C and after correcting for the temperature difference, the ΔH_r° is 31.48 ± 2.4 kJ mol^{-1} assuming a calorimetric error of 2%. These two studies, the only ones that give a direct measure of enthalpy for arsenolite solubility, are surprisingly close considering the uncertainty inherent in each study. Washburn and Strachan (1913) also calculate an enthalpy for the oxidation reaction with triiodide but they used the same value from Thomsen (1882-1885) for the heat of arsenous acid dissolution. These measurements are not only the best available for the temperature dependence of arsenolite solubility but they are consistent with the 25 °C value established previously. Combining the Anderson and Story (1923) enthalpy value derived from the temperature dependence of arsenolite solubility with the enthalpy of arsenolite dissolution results in $\Delta H_f^{\circ}{}_{H_3AsO_3} = -740.00$ kJ mol^{-1} and $S^{\circ}{}_{H_3AsO_3} = 203.7$ J K^{-1} mol^{-1} (Table 4).

Table 4. Standard state thermodynamic properties for arsenous and arsenic acids.

Species	ΔG_f° (kJ mol⁻¹)	ΔH_f° (kJ mol⁻¹)	S° (J K⁻¹ mol⁻¹)	Reference
$H_3AsO_{3(aq)}$	−639.90	−741.82	196.65	Rossini et al. (1952)
$H_3AsO_{3(aq)}$	−639.80	−742.2	195.0	Wagman et al. (1982)
$H_3AsO_{3(aq)}$	−640.03	−742.36	195.83	Nordstrom and Archer (2003)
$H_3AsO_{3(aq)}$	−639.8 ± 0.5	−737.3 ± 2.0	212.4 ± 10.0	Perfetti et al. (2008)
$H_3AsO_{3(aq)}$	−639.85 ± 0.5	−740.00 ± 1.0	203.7 ± 0.5	This study
$H_3AsO_{4(aq)}$	−769.02	−898.723	206.27	Rossini et al. (1952)
$H_3AsO_{4(aq)}$	−766.0	−902.5	184	Wagman et al. (1982)
$H_3AsO_{4(aq)}$	−766.75	−903.45	183.07	Nordstrom and Archer (2003)
$H_3AsO_{4(aq)}$	−766.3 ± 3.0	−898.6 ± 3.0	198.3 ± 15.0	Perfetti et al. (2008)
$H_3AsO_{4(aq)}$	−766.42 ± 0.35	−901.76 ± 0.325	188.0 ± 2.0	This study

Enthalpy measurements have been made of the arsenite-arsenate redox reaction from which the enthalpies and entropies of arsenous and arsenic acids can be obtained. Bjellerup et al. (1957) measured the heat of arsenous acid oxidation by elemental bromine and obtained −238.15 ± 0.15 kJ mol⁻¹ for the reaction

$$2H_3AsO_{3(aq)}^\circ + 2H_2O_{(l)} + 2Br_{2(l)} \rightleftharpoons 2H_3AsO_{4(aq)}^\circ + 4HBr_{(aq)} \tag{5}$$

Sunner and Thorén (1964) measured the same reaction but using chlorine as the oxidizing agent instead of bromine. Using CODATA enthalpy values for bromide and chloride aqueous ions and the previously derived enthalpy for H_3AsO_3, Bjellerup et al. (1957) give $\Delta H_f^\circ{}_{H_3AsO_4}$ = −901.435 ± 0.64 kJ mol⁻¹ and Sunner and Thorén (1964) give $\Delta H_f^\circ{}_{H_3AsO_4}$ = −902.084 ± 0.50 kJ mol⁻¹. These values are nearly within the experimentally reported errors and we averaged the results. From the enthalpy and Gibbs free energy, we calculated the entropy (Table 2).

Standard state thermodynamic properties of claudetite, As₂O₃ (monoclinic)

Two important pieces of information help to constrain the thermodynamic properties of claudetite. First, sufficient evidence exists to demonstrate that claudetite (monoclinic) is more stable than arsenolite (cubic) at 25 °C (e.g., Schulman and Schumb 1943). Second, the careful heat capacity measurements and literature critique of Chang and Bestul (1971) provide an accurate and precise value for the entropy difference between claudetite and arsenolite of 5.92 J K⁻¹ mol⁻¹. Hence, the free energy of claudetite must be more negative than that of arsenolite and the entropy of claudetite must be 5.92 J K⁻¹ mol⁻¹ greater than that of arsenolite for the standard state temperature of 25 °C.

Kirschning and Plieth (1955) determined the difference in Gibbs free energy between claudetite and arsenolite by electrochemical measurements of As-As₂O₃ electrodes using a standard hydrogen electrode as the reference. They found the free energy difference to be −0.761 kJ mol⁻¹. When combined with the entropy difference, an enthalpy of transition is calculated to be 0.998 kJ mol⁻¹. This enthalpy is lower than any measured enthalpy for this reaction and makes the free energy suspect of being biased slightly on the low side. The difference in solubility measured by Schulman and Schumb (1943) at 25 °C results in a free energy change of −0.510 kJ mol⁻¹. Using this value and the entropy difference produces an enthalpy change of 1.255 kJ mol⁻¹ which is at odds with their own enthalpy values of 1.715 kJ mol⁻¹ and 2.211 kJ mol⁻¹ by solubility and calorimetry, respectively. Chang and Bestul (1971) indicated an enthalpy change of 1.25 kJ mol⁻¹, based on their measurements and unpublished data from the NBS. Another estimate for this enthalpy change can be made using the arsenolite and claudetite solubility measurements of Stranski et al. (1958) for 23-99 °C, resulting in a value of 1.36 kJ mol^{-1}.

For the Gibbs free energy change of the conversion of arsenolite to claudetite to have the correct negative sign, the enthalpy of this transition must be less than 1.765 kJ mol^{-1} if the results of Chang and Bestul (1971) are accepted for the heat capacity and entropy. Therefore, the enthalpy data of Minor et al. (1971), of Schulman and Schumb (1943) by calorimetry, and of Karutz and Stranski (1957) from vapor pressure measurements are biased high because they are all greater than this. Using simultaneous regression, Nordstrom and Archer (2003) derived an enthalpy change of 1.597 kJ mol^{-1} and a Gibbs free energy change of -0.190 kJ mol^{-1}. In this chapter, a network built on highest quality data while maintaining the basic thermodynamic relations has been developed without simultaneous regression. To accept external regressed data could introduce an inconsistency. For thermodynamic properties of claudetite, which are quite close to those of arsenolite (hence it is one source of difficulty in making measurements), a mean value optimized between the enthalpy of transition from Chang and Bestul (1971) and from Stranski et al. (1958) was selected (1.305 kJ mol^{-1}). This choice leads to a Gibbs free energy of transition of -0.454 kJ mol^{-1} which is within 0.265 kJ mol^{-1} of the regressed value of Nordstrom and Archer (2003) and within the errors of the measurements. The consequent properties of claudetite are shown in Table 5.

Hydrolysis constants for arsenous and arsenic acids

Arsenous acid deprotonates at a somewhat high pH according to the reaction

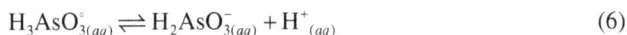

$$H_3AsO^{\circ}_{3(aq)} \rightleftharpoons H_2AsO^{-}_{3(aq)} + H^{+}_{(aq)} \tag{6}$$

and the equilibrium constant is designated as pK_1 ($-$log K_1). Literature references compiled in Table 6 reflect fair to good agreement. Excluding those studies with variable or high ionic strength ($I \geq 0.1$ m) and large precision error (≥ 0.1 pK), the average pK_1 of the remaining 12 values is 9.23 ± 0.12 (2 standard deviations) which is considered excellent agreement. Alternatively, the more recent measurements of Yamazaki et al. (1993), Raposo et al. (2003), and Zakaznova-Herzog et al. (2006) are in even better agreement with each other and with the careful measurements of Britton and Jackson (1934). When these 4 values are averaged, the result is p$K_1 = 9.24 \pm 0.02$ and $\Delta G_r^{\circ} = 52.749 \pm 0.115$ kJ mol^{-1}.

Higher dissociation constants for arsenous acid (pK_2, pK_3) have not been adequately measured to derive thermodynamic quantities and they occur at rather high pH values (pK_2 =13 to 14, Sillén and Martell 1964; Perrin 1982). Consequently, solubility product constants of arsenite minerals cannot be reliably derived from calorimetric data. Solubility measurements are necessary.

Hydrolysis of arsenic acid occurs in 3 stages

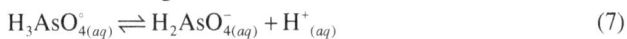

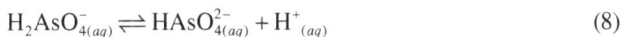

$$H_3AsO^{\circ}_{4(aq)} \rightleftharpoons H_2AsO^{-}_{4(aq)} + H^{+}_{(aq)} \tag{7}$$

$$H_2AsO^{-}_{4(aq)} \rightleftharpoons HAsO^{2-}_{4(aq)} + H^{+}_{(aq)} \tag{8}$$

$$HAsO^{2-}_{4(aq)} \rightleftharpoons AsO^{3-}_{4(aq)} + H^{+}_{(aq)} \tag{9}$$

for which the successive negative logarithms of the hydrolysis constants are designated as pK_1, pK_2, and pK_3. Literature references compiled in Table 7 reflect fair to good agreement. The pK_1 ranges from 2.2 to 2.36 if the spurious value of Chukhlantsev (1959), and the high ionic strength value of Khoe and Robins (1988), and the value of Omarova et al. (1980) based on dissociation enthalpies are omitted. The average of the remaining pK_1 values is 2.28 ± 0.05. However, the quality of the studies by Britton and Jackson (1934), Agafanova and Agafanov (1953), Tossidis (1976), and Raposo et al. (2002) is superior to the others and their pK_1 average is 2.25 ± 0.04 (2 standard deviations).

Table 5. Summary of standard state thermodynamic properties for claudetite (Chang and Bestul 1971; This study).

ΔG_f° (kJ mol^{-1})	ΔH_f° (kJ mol^{-1})	S° (J K^{-1} mol^{-1})	C_p° (J K^{-1} mol^{-1})
-576.54 ± 0.3	-655.876	113.33	96.98

Table 6. First hydrolysis constants of arsenous acid; T = temperature and I = ionic strength.

Comments	pK$_1$	Reference
T = ambient, I = variable	9.20, 8.58	Wood (1908)
T = ambient	9.22	Kolthoff (1920)
T = 25 °C, I = 0	9.22	Hughes (1928)
T = 25 °C, I = variable	9.05	Cernatescu and Mayer (1932)
T = ambient, I = variable	9.4 ± 0.3	Goldfinger and von Schweinitz (1932)
T = 18 °C, I = variable	9.26, 9.23	Britton and Jackson (1934)
T = 25 °C, I = 0	9.08	Ishikawa and Aoki (1940)
T = ambient, 1 m NaCl	8.85	Carpeni and Souchay (1945)
T = 25 °C, I = variable	8.13	Tourky and Mousa (1949)
T = 25 °C, I = 0	9.294	Antikainen and Rossi (1959)
T = 25 & 45 °C, I = 0	9.18	Antikainen and Tevanen (1961)
T = 20 °C, I = variable but dilute	9.38	Hargreaves and Stevinson (1965)
T = 25 °C, I = 0	9.279	Salomaa et al. (1969)
T = 22 °C adjusted to 25 °C, I = 0	9.15	Ivakin et al. (1976)
T = ?, I = 0	9.15 ± 0.01	Vesala and Saloma (1977)
T = 20-40 °C, I = 0.1 m, ΔH_r°, S°	9.45	Nishida and Kimura (1989)
T = 25-65 °C. I = 0-3 m, ΔH_r°	9.243 ± 0.003	Yamazaki et al. (1993)
T = 25 °C, I = 0	9.22 ± 0.01	Raposo et al. (2003)
T = 25-300 °C, I = ~0, ΔH_r°, S°	9.25 ± 0.05	Zakaznova-Herzog et al. (2006)
T = 25 °C, I = 0	9.17	Nordstrom and Archer (2003)
T = 25 °C, I = 0	9.24 ± 0.02	This study

The pK$_2$ value is relatively consistent and omitting Chukhlantsev's (1959) value, the mean of the remaining values is 6.958 ± 0.113. Calculating a mean based on the most reliable measurements of Britton and Jackson (1934), Agafanova and Agafanov (1953), Mader (1958), Tossidis (1976), Wauchope (1976), Gresser et al. (1986), and Raposo et al. (2002) changes the result only slightly to pK$_2$ = 6.98 ± 0.11.

We could only find 6 reported values for the pK$_3$ and the two most reliable values are Britton and Jackson (1934) and Raposo et al. (2002). These two give an average of 11.70 ± 0.12 where the error covers the range of the two results. Alternatively, and usually preferably, one or the other study should be considered more reliable instead of using a mean value. The study by Raposo et al. (2002) made careful considerations of ionic strength effects on activity coefficients and utilized mean activity coefficient data on NaH$_2$AsO$_4$ solutions that were not available to Britton and Jackson (1934). Bearing in mind these aspects, we recommend the results from Raposo et al. (2002) as being slightly superior and we assume an error of about 5 times that reported by them so that pK$_3$ = 11.58 ± 0.05.

Table 7. Hydrolysis constants of arsenic acid at or near 25 °C where T = temperature, I = ionic strength, adj→25 °C = temperature adjusted to 25 °C, adj→0 = adjusted to infinite dilution.

Comments	pK_1	pK_2	pK_3	Reference
$T = 25$ °C	2.35	—	—	Walden (1888)
$T = 25$ °C	2.30	—	—	Luther (1907)
$T = 25$ °C, $I = 0$	2.36 ± 0.1	—	—	Washburn and Strachan (1913)
$T = 25$ °C, $I = 0$	—	7.08	—	Hughes (1928)
$T = 18$ °C, T adj→25 °C, I adj→0	2.28	6.89	11.81	Britton and Jackson (1934)
$T = 25$ °C, $I = 0$	2.223	6.979	—	Agafanova and Agafanov (1953)
$T = 25$ °C, I adj→0	—	7.00	—	Mader (1958)
	2.66	5.89	11.5	Chukhlantsev (1959)
$T = 25$ °C, $I = 0$	2.19	6.94	—	Flis et al. (1959)
$T = 25$ °C, $I = 1$ M NaClO$_4$	—	6.90 ± 0.05	—	Beech and Lincoln (1971)
$T = 25$ °C, $I = 0$	2.301	—	—	Salomaa et al. (1964)
Measured dissociation enthalpies	2.25	6.77	11.53	Sellers et al. (1964)
Literature evaluation	2.21	6.93	11.51	Khodakovsky et al. (1968)
$T = 20$ °C, $I < 0.1$	2.22	—	—	Hodgkin and Kruus (1973)
$T = 25$ °C, $I = 3$ M NaClO$_4$	2.31 ± 0.1	—	—	Secco et al. (1970)
$T = 25$ °C, $I = 0.1$ M KCl	2.26	6.76	11.29	Tossidis (1976)
$T = 25$ °C, $I = 0$	—	7.089 ± 0.01	—	Wauchope (1976)
Dissociation enthalpies	2.49	—	11.51	Omarova et al. (1980)
$T = 25$ °C, I adj→0	—	7.07	—	Gresser et al. (1986)
$T = 25$ °C, $I = 3$ m NaNO$_3$	2.128 ± 0.002	—	—	Khoe and Robins 1988
$T = 25$ °C, $I = 0$	2.25 ± 0.01	7.06 ± 0.01	11.58 ± 0.01	Raposo et al. (2002)
Literature evaluation	2.31	7.05	11.9	Goldberg et al. (2002)
Literature evaluation	2.26 ± 0.078	6.99 ± 0.1	11.80 ± 0.1	Nordstrom and Archer (2003)
$T = 25$ °C, $I = 0$	2.25 ± 0.04	6.98 ± 0.11	11.58 ± 0.05	This study

The relationship between the 3 hydrolysis constants of arsenic acid expressed as the pK (Table 7) is remarkably uniform, i.e., the difference is close to −4.7 between each pair of constants. This seems to follow a general rule (Butler 1964) that was formulated by Kossiakoff and Harker (1938) and refined by Ricci (1948) for oxy-acids where all OH groups are bonded to the same atom.

From these data, values for the Gibbs free energy can be calculated for each dissociated species (Table 8). Deriving enthalpy and entropy values for these aqueous species is more problematic. There are only two studies in which the change in the dissociation constants with temperature were measured (Agafanova and Agafanov 1953; Flis et al. 1959) and one of these (Flis et al. 1959) showed a very inconsistent trend with temperature. There are measured heats of ionization for arsenic acid (Sellers et al. 1964; Omarova et al. 1980) but the comparison is still not consistent. Sellers et al. (1964) reported an ionization enthalpy for K_1 of −7.07 kJ mol^{-1} compared to −11.44 kJ mol^{-1} from a linear temperature-dependent fit of Agafanova and Agafanov's data (limited to 15-35 °C) and Omarova et al. (1980) reported −18.5 kJ mol^{-1}. Nordstrom and Archer (2003; based on Archer and Nordstrom 2002) pointed out that the heat capacity of the first deprotonation reaction calculated from Agafanova and Agafanov's equilibrium constant values, −250 J K^{-1} mol^{-1}, is not quite in agreement with the heat capacity

Table 8. Gibbs free energies, enthalpies, and
entropies of dissociated species of arsenic acid.

Species	ΔG_f° (kJ mol^{-1})	ΔH_f° (kJ mol^{-1})	S° (J K^{-1} mol^{-1})
$H_3AsO_4^\circ{}_{(aq)}$	-766.420 ± 0.35	-901.76 ± 0.325	188.0 ± 2.0
$H_2AsO_4^-{}_{(aq)}$	-753.575 ± 0.35	-913.20 ± 4.30	106.55
$HAsO_4^{2-}{}_{(aq)}$	-713.727 ± 0.35	-910.34	-17.52
$AsO_4^{3-}{}_{(aq)}$	-647.618 ± 1.5	-892.14	-178.20

changes for similar deprotonation reactions, approximately -130 J K^{-1} mol^{-1} (Smith et al. 1986). They averaged the results from Agafanova and Agafanov (1953) with those from Sellers et al. (1964) along with the heat capacity value of -130 J K^{-1} mol^{-1}. The problem with that approach is that it tends to compromise the reliability of the pK_1 at 25 °C. If the 25 °C value is fixed at the average discussed earlier then both the Sellers et al. (1964) data and the Agafanova and Agafanov (1953) data do not agree with the result over the temperature range of measurements. Also, the Britton and Jackson (1934) pK_1 value obtained at 18 °C agrees better with the temperature-dependence of Agafanova and Agafanov (1953) than that indicated by the enthalpy of Sellers et al (1964). Furthermore, the enthalpy data of Sellers et al. (1964) was based on uncorrected ionic strengths of 0.1-0.3 M and the degree of hydrolysis was not precisely known. The enthalpy extracted from Agafanova and Agafanov (1953) is quite close to the average of the calorimetric heats measured by Sellers et al. (1964) and Omarova et al. (1980). Consequently, we have chosen the results from Agafanova and Agafanov (1953) as the best available for the enthalpy and we have calculated the ΔH_f° for $H_2AsO_4^-{}_{(aq)}$ based on that (Table 8). From the Gibbs free energy and enthalpy ($\Delta G_r^\circ = \Delta H_r^\circ - T\Delta S_r$) we calculated the entropy of $H_2AsO_4^-{}_{(aq)}$.

Maintaining consistency with the previous arguments, we have followed the same procedure for finding the ΔH_f° for $HAsO_4^{2-}{}_{(aq)}$ (2.86 kJ mol^{-1}) based on the temperature dependence of the pK_2 for arsenic acid from Agafanova and Agafanov (1953). We used only the most linear points nearest 25 °C (15-30 °C). We calculated the entropy in the same manner as before.

For the third dissociation of arsenic acid there are no data for the temperature dependence of the equilibrium constant. There are only the dissociation heats of Sellers et al. (1964), 18.2 kJ mol^{-1}, and of Omarova et al. (1980), 74.23 kJ mol^{-1}. To resolve this large discrepancy we have estimated ΔH_f° values by assuming the ΔH_r° follows a similar trend as the ΔG_r° trend discussed by Kossiakoff and Harker (1938) and Ricci (1948) for oxyacids and modified it with the trend found in the results from this paper. Specifically, we assume that the difference in enthalpy between the enthalpy for pK_2 and the enthalpy for pK_3 is the same as the enthalpy difference between pK_1 and pK_2. That calculation gives an enthalpy of 17.16 kJ mol^{-1} for the reaction embodied in pK_3, surprisingly close to that measured by Sellers et al. (1964). Doing the same comparison for Gibbs free energy we find that there is less than 1 kJ mol^{-1} difference between the Gibbs free energy difference of pK_2-pK_1 and that for pK_3-pK_2 which supports this approach. Consequently, we have accepted the enthalpy measurement of Sellers et al. (1964) as the most reliable for this reaction and completed the calculations accordingly (Table 8).

The thermodynamic properties of arsenolite and the aqueous species of arsenite and arsenate are considered fundamental to the solubility of many other arsenic minerals and have been critically reviewed in this section. The pathways of this evaluation are outlined in the network shown in Figure 1 with abbreviated primary literature references shown using a similar convention to Sillén and Martell (1964).

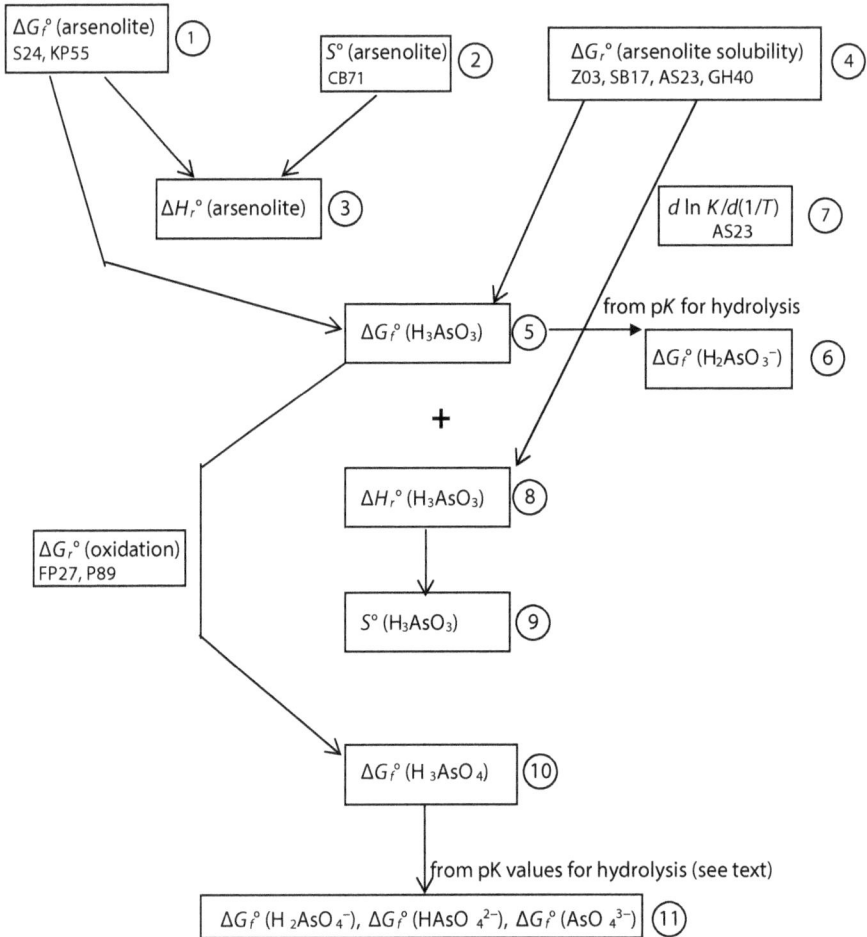

Figure 1. Thermodynamic pathways for deriving properties of arsenolite and arsenite and arsenate aqueous ions. References are abbreviated as S24 = Schuhmann (1924); KP55 = Kirschning and Plieth (1955); CB71 = Chang and Bestul (1971); Z03 = Zawidski (1903); SB17 = Schreinemakers and de Baat (1917); AS23 = Anderson and Story (1923); GH40 = Garrett et al. (1940); FP27 = Foerster and Pressprich (1927); and P89 = Pesavento (1989).

SOLUBILITY EQUILIBRIA IN THE Fe(III)-As(V)-H₂O SYSTEM

Including the evaluation of heterogeneous (solubility) equilibria in the thermodynamic analysis introduces an additional degree of complexity. The system Fe(III)-As(V)-H₂O, which has been investigated in many studies because of its role in the sequestration of As in natural and industrial processes, provides a pertinent example. It has been shown that the solubility of sparingly soluble Fe(III)-arsenate phases strongly depends on the method of preparation, aging and equilibration times, degree of crystallinity, particle size, and possibly surface defects. All of these factors have an impact on the measured solubility, leading to variations in the analytically determined concentration by several orders of magnitude. It transpires that well-crystallized scorodite, $FeAsO_4 \cdot 2H_2O_{(cr)}$, has the lowest solubility and is the most stable solid phase in this system, whereas fine-grained scorodite specimens, amorphous ferric arsenates, or metastable phases such as kaňkite, $FeAsO_4 \cdot 3.5H_2O_{(cr)}$, have higher solubility. In strongly acidic solutions, scorodite dissolves congruently (with respect to the original Fe/As ratio). At higher pH, scorodite can become metastable with respect to various Fe(III) (oxy-)hydroxides. This means that, due to the precipitation of these phases, incongruent dissolution of scorodite will occur at pH > 2 to 4, depending on the stability (solubility) of the ferric (oxy-)hydroxides. With rising pH, increasing amounts of these phases will precipitate, resulting in a progressive release of arsenate into the solution. The solubility phenomena involving scorodite and other (metastable) arsenate phases in the Fe(III)-As(V)-H_2O system have been discussed, among others, by Tozawa et al. (1978), Dove and Rimstidt (1985), Robins (1987), Krause and Ettel (1988), Robins (1990), Nishimura and Robins (1996), Welham et al. (2000), Zhu and Merkel (2001), Langmuir et al. (2006), Bluteau and Demopoulos (2007), and Majzlan et al. (2012a).

The thermodynamic properties of scorodite have traditionally been evaluated from solubility data, in analogy to numerous other systems involving equilibria between solid and aqueous electrolytes. One of the first comprehensive sets of standard Gibbs free energy data for solid phases and aqueous species in the Fe(III)-As(V)-H_2O system was reported by Robins (1990). This database also includes stability constants for ferric arsenate complexes, for which independent information is available from potentiometric titrations (Khoe and Robins 1988) and spectrophotometric measurements (Raposo et al. 2006). However, the speciation model selected by Robins (1990) was based on his solubility data. The formation constants for ferric arsenate complexes so obtained agree within ±0.6 log units with values obtained by other methods. The corresponding standard Gibbs free energy values (Table 9) were made consistent with the auxiliary data used in this work and given in previous sections or in Table 9 (Parker and Khodakovsky 1995), for $Fe_{(cr)}$ and $Fe^{3+}_{(aq)}$; Nordstrom and Archer (2003), for $As_{(cr)}$; Cox et al. (1989), for $H_2O_{(l)}$, $O_{2(g)}$, and $H_{2(g)}$). Standard Gibbs free energies for ferrihydrite and the Fe(III)-hydroxo complexes were derived from the equilibrium constants reported by Stefansson (2007).

Given the large scatter of scorodite solubility data indicated above, reported values for standard-state Gibbs free energy of formation differ within tens of kJ mol^{-1}. Only very recently, Majzlan et al. (2012a) have reported calorimetrically determined standard-state enthalpy of formation, standard-state entropy and standard-state heat capacity for various Fe(III)-arsenate phases including kaňkite and carefully prepared, well-crystallized scorodite. These data have been incorporated into the present model to assess thermodynamic consistency with solubility data. The calculations were performed using the thermochemical software ChemSage (Eriksson and Hack 1990), which employs a Gibbs free energy minimizer in conjunction with various

Table 9. Standard state Gibbs free energy, enthalpy, and entropy of solid phases and aqueous species in the Fe(III)-As(V)-H_2O system at 25 °C.

Species	ΔG_f° (kJ mol^{-1})	ΔH_f° (kJ mol^{-1})	S° (J K^{-1} mol^{-1})	Ref.
$Fe_{(cr)}$	0	0	27.319 ± 0.002	[1]
$Fe^{3+}_{(aq)}$	−16.28 ± 1.1	−49.0 ± 1.5	−278.4 ± 7.7	[1]
$FeOOH_{(cr,\ goethite)}$	−488.51 ± 1.7	−559.3 ± 1.7	60.4 ± 10.6	[1]
$FeAsO_{4(aq)}$	−772.88 ± 1.2	—	—	[2]
$FeHAsO_4^+{}_{(aq)}$	−787.55 ± 1.2	—	—	[2]
$FeH_2AsO_4^{2+}{}_{(aq)}$	−792.97 ± 1.2	—	—	[2]
$FeAsO_4 \cdot 2H_2O_{(cr,\ scorodite)}$	−1284.8 ± 2.9	−1508.9 ± 2.9	−188.0 ± 2.1	[3]
$FeAsO_4 \cdot 3.5H_2O_{(cr,\ kaňkite)}$	−1629.6 ± 2.9	−1940.2 ± 2.8	−247.6 ± 2.8	[3]

Reference: [1] Parker and Khodakovsky (1995); [2] Robins (1990); [3] Majzlan et al. (2012a).

excess Gibbs free energy models to compute thermodynamic equilibrium compositions of multicomponent, multiphase systems from Gibbs free energy data for all species present in the system. Because the solutions considered in this study are rather dilute, the in-built Davies equation was employed to calculate the activity coefficients of the aqueous species.

Results and discussion

In Figure 2, two simulations allowing either ferrihydrite or goethite precipitation are compared with experimental data of Krause and Ettel (1988) and Majzlan et al. (2012a). In the region of congruent dissolution (pH < 2), the calculated values provide a lower limit to measured solubilities, apparently corresponding to 'bulk' scorodite. As discussed by Langmuir et al. (2006), solubility data obtained from less well-crystallized scorodite samples and amorphous ferric arsenates are significantly higher. However, there might also be possible problems with the data shown in Figure 2. There are large discrepancies between the two data sets at 2 < pH < 3 and, as stressed by Majzlan et al. (2012a), $[Fe]_T$ is actually larger than $[As]_T$ at pH < 2. However, Nishimura and Robins (1996) have described the precipitation of $FeH_3(AsO_4)_2 \cdot 10H_2O_{(cr)}$ at pH < 1.5, which might explain the deviation from the $[Fe]_T/[As]_T = 1$ ratio expected for congruent scorodite dissolution.

The total As concentrations in the region of incongruent dissolution (pH > 2) are compatible with precipitation of ferrihydrite rather than goethite (Fig. 2). The lowest Fe concentration predicted by the model is determined solely by the stability constant of the $FeAsO_{4(aq)}$ complex and is independent of the nature of the precipitating Fe(III) (oxy-)hydroxide. The $Fe(OH)_{3(aq)}$ and $Fe(OH)_4^-{}_{(aq)}$ complexes are insignificant in the pH range considered; thus the present model does not predict the observed increase in $[Fe]_T$ at pH > 6. To explain this behavior,

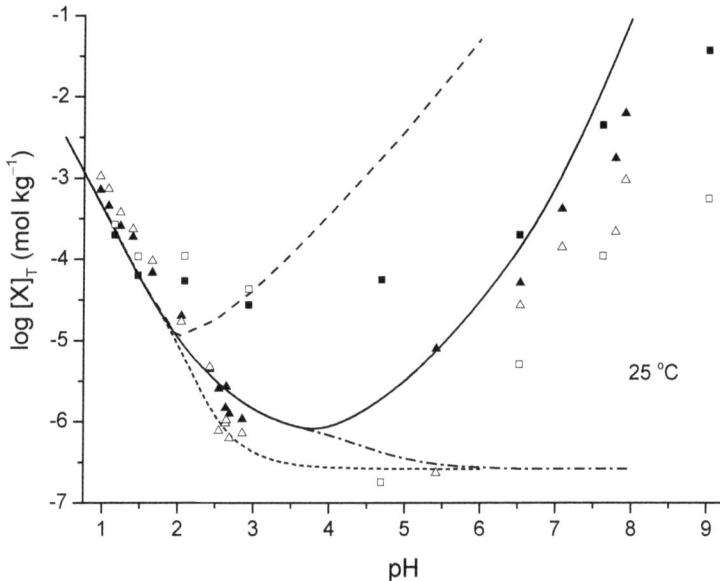

Figure 2. Calculated equilibria of $FeAsO_4 \cdot 2H_2O_{(cr, scorodite)}$ with aqueous solutions, allowing precipitation of either ferrihydrite (solid lines) or goethite (dashed lines), are compared with analytically determined, total Fe (open symbols) and As (filled symbols) concentrations reported by Krause and Ettel (1988, triangles) and Majzlan et al. (2012a, squares). Although Nishimura and Robins' (1996) solubility data were also obtained using well-crystallized scorodite, their results cannot be plotted in this representation because their measurements were performed in an excess of $H_3AsO_{4(aq)}$.

Robins (1990) proposed a $Fe(AsO_4)_2^{3-}{}_{(aq)}$ complex. However, test calculations using a range of $\Delta G_f°$ values close to the datum reported by Robins (1990) revealed a very steep increase in $[Fe]_T$ within a very narrow pH range, which is in contrast to the observed data. Langmuir et al. (2006) selected data for significantly less stable ferrihydrate and ferric arsenate phases to approximately model the increase in both $[Fe]_T$ and $[As]_T$ by extending the region of congruent dissolution to higher pH.

A note on solubility constants

Whenever metal salts of basic anions are concerned, the common definition of the solubility product involves the activity of the completely deprotonated anion. However, in the pH ranges employed for solubility measurements, the concentrations of these deprotonated species are often negligible. For similar reasons, the corresponding deprotonation constants may carry large errors, which propagate into the solubility product values. In any case, it is necessary to ensure consistency between solubility and deprotonation constants when solubility calculations are performed. While the most prominent and problematic systems in this regard concern sulfides, similar but less pronounced problems occur in the case of scorodite.

The usual definition of the standard solubility product, $K_{sp}°$, of scorodite corresponds to the reaction

$$FeAsO_4 \cdot 2H_2O_{(cr,\ scorodite)} \rightleftharpoons Fe^{3+}_{(aq)} + AsO_4^{3-}{}_{(aq)} + 2H_2O_{(l)} \tag{10}$$

Another scorodite solubility constant, $*K_{s0}°$, can be defined via the predominant arsenic acid species,

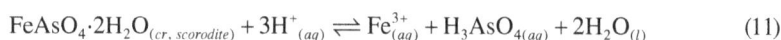

$$FeAsO_4 \cdot 2H_2O_{(cr,\ scorodite)} + 3H^+{}_{(aq)} \rightleftharpoons Fe^{3+}_{(aq)} + H_3AsO_{4(aq)} + 2H_2O_{(l)} \tag{11}$$

Then, the solubility product, $K_{sp}°$, and the solubility constant, $*K_{s0}°$, are related by the formation constant of arsenic acid from its ions, $\beta_3°$, corresponding to Equation (12)

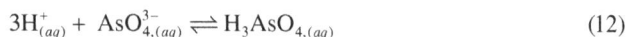

$$3H^+_{(aq)} + AsO_4^{3-}{}_{(aq)} \rightleftharpoons H_3AsO_{4(aq)} \tag{12}$$

Because

$$\log \beta_3° = pK_1° + pK_2° + pK_3° \tag{13}$$

where $pK_i° = -\log K_i°$ and the $K_i°$ are the standard hydrolysis constants of arsenic acid, it follows that

$$\log *K_{s0}° = \log K_{sp}° + \log \beta_3° \tag{14}$$

Langmuir et al. (2006) demonstrated the effect of different speciation models on the numerical value of the solubility product. Reevaluating the experimental data sets of Krause and Ettel (1988) and Nishimura and Robins (1996), Langmuir et al. (2006) calculated two different values of the solubility product of scorodite, $\log K_{sp}° = -25.83$ and $\log K_{sp}° = -26.12$, when they used Baes and Mesmer's (1976) and Nordstrom and Archer's (2003) hydrolysis constants of arsenic acid respectively. In terms of $\log *K_{s0}°$, these two evaluations agree within 0.10 log units (Table 10). Similarly, Majzlan et al. (2012a) used Nordstrom and Archer's (2003) standard-state Gibbs free energy for the arsenate ion and derived $\log K_{sp}° = -25.83$, noting excellent agreement with the first value of Langmuir et al. (2006). Combining the thermodynamic data of Majzlan et al. (2012a) for scorodite and of the present work for the aqueous arsenate species, the result is $\log K_{sp}° = -25.68$. These two values correspond to $\log *K_{s0}° = -4.74$ and -4.86, respectively. The last value is recommended as best estimate for $\log *K_{s0}°{}_{(scorodite,\ cr)}$ because it is based on independently measured properties of crystalline scorodite (Majzlan et al. 2012a) and the critically evaluated speciation model for arsenic acid proposed in this review. Although its uncertainty of ± 0.52 is derived from the experimental

Table 10. Solubility products, log $K_{sp}°$ (Eqn. 10), and solubility constants, log $*K_{s0}°$ (Eqn. 11), of $FeAsO_4·2H_2O_{(cr, scorodite)}$, hydrolysis constants, $pK_n°$, of $H_3AsO_{4(aq)}$, corresponding to the reactions $H_{4-n}AsO_4^{1-n}{}_{(aq)} \rightleftharpoons H^+{}_{(aq)} + H_{3-n}AsO_4^{-n}{}_{(aq)}$, and formation constant of $H_3AsO_{4(aq)}$, log $β_3°$ (Eqn. 12), as reported in various studies for 25 °C.

Reference	$pK_1°$	$pK_2°$	$pK_3°$	log $β_3°$	log $K_{sp}°$	log $*K_{s0}°$
This work [a]	2.25	6.98	11.58	20.81	−25.68 ± 0.52	−4.86 ± 0.20
Majzlan et al. (2012a) [b]	2.30	6.99	11.80	21.09	−25.83 ± 0.52	−4.74 ± 0.52
Bluteau and Demopoulos (2007) [c]	2.24	6.96	11.50	20.70	−25.4 ± 0.5	−4.7 ± 0.5
Langmuir et al. (2006) [c]	2.24	6.96	11.50	20.70	−25.83 ± 0.07	−5.13 ± 0.07
Langmuir et al. (2006) [b]	2.30	6.99	11.80	21.09	−26.12 ± 0.06	−5.03 ± 0.06
Robins (1990) [d]	2.25	6.76	11.59	20.60	−24.60 [e]	−4.00 [e]
Krause and Ettel (1988) [f]	2.24	6.86	11.49	20.59	−24.41 ± 0.15 [g]	−3.82 ± 0.15 [g]

[a] Thermodynamic data from Tables 8 and 9; [b] $pK_n°$ from Nordstrom and Archer (2003); [c] $pK_n°$ from Baes and Mesmer (1976); [d] $pK_n°$ agree with values calculated from thermodynamic data of Wagman et al. (1982); [e] activity coefficient model not specified; [f] $pK_n°$ from Dove and Rimstidt (1985), incorrectly attributed by them to Wagman et al. (1982); [g] activity coefficients assumed to be 1.

uncertainties of the underlying calorimetric measurements, the present review of solubility constants for well-crystallized scorodite (Table 10) suggests a more precise value:

$$\log *K_{s0}°{}_{(scorodite, cr)} = -4.86 ± 0.20$$

It should be noted that the rather small uncertainties reported by Langmuir et al. (2006) (Table 10) are consistent with the scatter of their evaluated log $K_{sp}°$ values; however, they hardly reflect the propagation of errors from the speciation models employed in the evaluations. For homogeneous reactions, representations of equilibrium constants in terms of predominant species are also advantageous. Discrepancies similar to those discussed for solubility constants were observed by Langmuir et al. (2006) regarding the stability constants of ferric arsenate (and other) complexes. Originally, formation constants were defined by Robins (1990) via reactions analogous to Equation (11), e.g.,

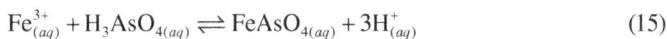

$$Fe^{3+}_{(aq)} + H_3AsO_{4(aq)} \rightleftharpoons FeAsO_{4(aq)} + 3H^+_{(aq)} \tag{15}$$

The $\Delta G_f°$ values reported by Robins result in log $K°(15) = -1.71$. Langmuir et al. (2006) apparently used reactions analogous to Equation (10), e.g.,

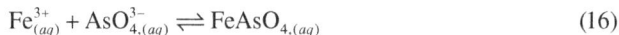

$$Fe^{3+}_{(aq)} + AsO^{3-}_{4,(aq)} \rightleftharpoons FeAsO_{4,(aq)} \tag{16}$$

The $\Delta G_f°$ values reported by Robins result in log $K°(16) = 18.89$. Obviously, log $K°(16) =$ log $K°_{(15)}$ + log $β_3°$, which gives log $β_3° = 20.60$. Because the same value for log $β_3°$ can also be calculated from Robins' (1990) $\Delta G_f°$ values, the three equilibrium constants are internally consistent (Table 10). However, Langmuir et al. (2006) apparently then combined Robins' value of log $K°_{(16)} = 18.89$ with log $β_3°$ values of 20.70 (Baes and Mesmer 1976) and 21.09 (Nordstrom and Archer 2003), which results in log $K°_{(15)}$ values of −1.81 and −2.20 respectively. These numbers agree within 0.01 log units with the values given in their Table 3 (Langmuir et al. 2006). This example demonstrates that thermodynamic inconsistencies inevitably arise when data sets from different sources are combined. Preferably, Robins' (1990) original value of log $K°_{(15)} = -1.71$ should have been retained.

Finally, it should also be noted that earlier evaluations of scorodite solubility product values should be treated with caution. For example, activity coefficient models for the aqueous species were either not specified (Robins 1990) or activity coefficients were neglected altogether

(Krause and Ettel 1988). In a qualitative attempt to relate solubility products to particle size, Robins (1990) concluded that he had investigated the coarsest scorodite specimens because his log $K_{sp}°$ value was the lowest yet reported (Table 10). However, the thermodynamically more rigorous evaluation of Langmuir et al. (2006) suggested that increasing values of log $K_{sp}°$, in the order log $K_{sp}°$ (Krause and Ettel 1988) ≈ log $K_{sp}°$ (Nishimura and Robins 1996) < log $K_{sp}°$ (Robins 1990) < log $K_{sp}°$ (Tozawa et al. 1978), correlate with a decreasing degree of crystallinity or decreasing particle size, ranging from coarse crystals to amorphous material. In any case, a systematic experimental study about grain size effects on solubility, similar to the work of Schindler et al. (1965), would be desirable for scorodite.

AQUEOUS METAL ARSENATE COMPLEXES

There are very few actual measurements of aqueous stability constants for arsenate and arsenite complexes. Consequently, researchers have estimated them using correlation techniques. In the few instances where comparisons of estimations with reduced data from measurements are possible, the results are usually comparable, at least for the range of conditions likely encountered in natural waters. Table 11 is a compilation of divalent metal arsenate stability constants. Table 12 follows with the stability constants for trivalent metal arsenates and Table 13 follows with monovalent metal arsenates. Finally, Table 14 compiles metal arsenite stability constants.

Whiting (1992) used the Fuoss model of ion pairing (Fuoss 1958) that considers ion charge and size (the Born equation; Born 1920), the Brown-Silva electroneutrality principle (BSEP; Brown et al. 1985), and correlation techniques by chemical analogy such as those outlined by Langmuir (1979). Marini and Accornero (2007) carried these estimation techniques further, partly by including boric acid as a chemical analog to arsenous acid and by including guidelines of the revised HKF model for aqueous complexes (Shock and Helgeson 1988). They included temperature-dependent stability constants for 0-300 °C. Estimates by Nordstrom (2000) simply assumed the same stability constants for iron-arsenate complexes as those reported for iron-phosphate complexes and should not be considered further. Data reported by Lee and Nriagu (2007) disagree with other estimates and the authors admitted to several potential sources of error in their measurements as well as the omission of some data that clearly had not reached equilibrium; these results should not be considered further. Bothe and Brown (1999) adjusted the experimental results of Mironov et al. (1995) from 40 °C to 23 °C.

These stability constants should be used to refine solubility measurements of the corresponding precipitating phases but have rarely ever been so used. Further work should evaluate the effect of including these ion pairs in deriving thermodynamic K_{sp} values and in applications to water-mineral reactions.

SUMMARY OF THERMODYNAMIC DATA FOR
ARSENATE MINERALS AND RELATED PHASES

This summary contains tables and a brief discussion of thermodynamic data for arsenate minerals which may be present in natural systems. The summary is by no means exhaustive; it is not the goal of this overview to include every value ever measured. We focused on the data presented in original papers and relevant to natural systems. In our opinion, further research is warranted to yield thermodynamic data that is reliable enough to be used with confidence in calculations. Additional data for Gibbs free energy of formation of various arsenic compounds are found in a long list in Appendix 3 of Henke (2009). Most of these data, however, were taken from other compilations. Some of these compilations, unfortunately, contain no references to

Table 11. Logarithms of association constants for divalent metal arsenate complexes.

Reaction	log K (25 °C)	Reference
$Ca^{2+} + H_2AsO_4^- \rightleftharpoons CaH_2AsO_4^+$	1.06	Whiting (1992)
$Ca^{2+} + H_2AsO_4^- \rightleftharpoons CaH_2AsO_4^+$	1.30	Bothe and Brown (1999); Mironov et al. (1995)
$Ca^{2+} + H_2AsO_4^- \rightleftharpoons CaH_2AsO_4^+$	1.4	Nordstrom (2000)
$Ca^{2+} + H_2AsO_4^- \rightleftharpoons CaH_2AsO_4^+$	1.495	Marini and Accornero (2007)
$Ca^{2+} + HAsO_4^{2-} \rightleftharpoons CaHAsO_4^\circ$	2.69	Whiting (1992)
$Ca^{2+} + HAsO_4^{2-} \rightleftharpoons CaHAsO_4^\circ$	2.74	Nordstrom (2000)
$Ca^{2+} + HAsO_4^{2-} \rightleftharpoons CaHAsO_4^\circ$	2.66	Bothe and Brown (1999); Mironov et al. (1995)
$Ca^{2+} + HAsO_4^{2-} \rightleftharpoons CaHAsO_4^\circ$	2.51	Marini and Accornero (2007) adj.[1]
$Ca^{2+} + AsO_4^{3-} \rightleftharpoons CaAsO_4^-$	6.22	Whiting (1992)
$Ca^{2+} + AsO_4^{3-} \rightleftharpoons CaAsO_4^-$	4.36	Bothe and Brown (1999); Mironov et al. (1995)
$Ca^{2+} + AsO_4^{3-} \rightleftharpoons CaAsO_4^-$	5.94	Marini and Accornero (2007) adj.[1]
$Mg^{2+} + H_2AsO_4^- \rightleftharpoons MgH_2AsO_4^+$	1.52	Whiting (1992)
$Mg^{2+} + H_2AsO_4^- \rightleftharpoons MgH_2AsO_4^+$	1.9	Nordstrom (2000)
$Mg^{2+} + H_2AsO_4^- \rightleftharpoons MgH_2AsO_4^+$	1.76	Marini and Accornero (2007)
$Mg^{2+} + HAsO_4^{2-} \rightleftharpoons MgHAsO_4^\circ$	2.86	Whiting (1992)
$Mg^{2+} + HAsO_4^{2-} \rightleftharpoons MgHAsO_4^\circ$	2.91	Nordstrom (2000)
$Mg^{2+} + HAsO_4^{2-} \rightleftharpoons MgHAsO_4^\circ$	2.68	Marini and Accornero (2007) adj.[1]
$Mg^{2+} + AsO_4^{3-} \rightleftharpoons MgAsO_4^-$	6.34	Whiting (1992)
$Mg^{2+} + AsO_4^{3-} \rightleftharpoons MgAsO_4^-$	6.07	Marini and Accornero (2007) corr.[1]
$Sr^{2+} + H_2AsO_4^- \rightleftharpoons SrH_2AsO_4^+$	0.825	Marini and Accornero (2007)
$Sr^{2+} + H_2AsO_4^- \rightleftharpoons SrH_2AsO_4^+$	1.72	Lee and Nriagu (2007)
$Sr^{2+} + HAsO_4^{2-} \rightleftharpoons SrHAsO_4^\circ$	1.83	Marini and Accornero (2007) adj.[1]
$Sr^{2+} + HAsO_4^{2-} \rightleftharpoons SrHAsO_4^\circ$	0.777	Lee and Nriagu (2007)
$Sr^{2+} + AsO_4^{3-} \rightleftharpoons SrAsO_4^-$	4.98	Marini and Accornero (2007) adj.[1]
$Fe^{2+} + H_2AsO_4^- \rightleftharpoons FeH_2AsO_4^+$	2.68	Whiting (1992)
$Fe^{2+} + H_2AsO_4^- \rightleftharpoons FeH_2AsO_4^+$	2.0	Nordstrom (2000)
$Fe^{2+} + H_2AsO_4^- \rightleftharpoons FeH_2AsO_4^+$	2.795	Marini and Accornero (2007)
$Fe^{2+} + HAsO_4^{2-} \rightleftharpoons FeHAsO_4^\circ$	3.54	Whiting (1992)
$Fe^{2+} + HAsO_4^{2-} \rightleftharpoons FeHAsO_4^\circ$	3.6	Nordstrom (2000)
$Fe^{2+} + HAsO_4^{2-} \rightleftharpoons FeHAsO_4^\circ$	3.37	Marini and Accornero (2007) adj.[1]
$Fe^{2+} + AsO_4^{3-} \rightleftharpoons FeAsO_4^-$	7.06	Whiting (1992)
$Fe^{2+} + AsO_4^{3-} \rightleftharpoons FeAsO_4^-$	7.41	Marini and Accornero (2007) adj.[1]

the original data sources and it is therefore impossible to decide whether such data are of any relevance for modeling of natural systems or not. More Gibbs free energy data were compiled by Wagman et al (1982) and by Drahota and Filippi (2009), who cite original sources.

Most of the available data come from solubility studies. The problems encountered in this review were improper characterization of the solid before and/or after the experiment, missing proof of congruent dissolution, omission of or incorrect activity coefficients, and neglect of ion pairing. Those aspiring to perform solubility experiments should be forewarned that such studies are not easy and many pitfalls await. The scatter of the data attests the truth of this statement.

Table 11. continued.

Reaction	log K (25 °C)	Reference
$Mn^{2+} + H_2AsO_4^- \rightleftharpoons MnH_2AsO_4^+$	1.01	Marini and Accornero (2007)
$Mn^{2+} + HAsO_4^{2-} \rightleftharpoons MnHAsO_4^{\circ}$	3.74	Whiting (1992)
$Mn^{2+} + HAsO_4^{2-} \rightleftharpoons MnHAsO_4^{\circ}$	2.92	Marini and Accornero (2007) adj.[1]
$Mn^{2+} + AsO_4^{3-} \rightleftharpoons MnAsO_4^-$	6.13	Whiting (1992)
$Mn^{2+} + AsO_4^{3-} \rightleftharpoons MnAsO_4^-$	6.23	Marini and Accornero (2007) adj.[1]
$Zn^{2+} + H_2AsO_4^- \rightleftharpoons ZnH_2AsO_4^+$	0.526	Marini and Accornero (2007)
$Zn^{2+} + H_2AsO_4^- \rightleftharpoons ZnH_2AsO_4^+$	−1.62	Lee and Nriagu (2007)
$Zn^{2+} + HAsO_4^{2-} \rightleftharpoons ZnHAsO_4^{\circ}$	3.21	Whiting (1992)
$Zn^{2+} + HAsO_4^{2-} \rightleftharpoons ZnHAsO_4^{\circ}$	3.28	Lee and Nriagu (2007)
$Zn^{2+} + HAsO_4^{2-} \rightleftharpoons ZnHAsO_4^{\circ}$	3.03	Marini and Accornero (2007) adj.[1]
$Zn^{2+} + AsO_4^{3-} \rightleftharpoons ZnAsO_4^-$	7.40	Marini and Accornero (2007) adj.
$Cd^{2+} + HAsO_4^{2-} \rightleftharpoons CdHAsO_4^{\circ}$	3.71	Whiting (1992)
$Cu^{2+} + H_2AsO_4^- \rightleftharpoons CuH_2AsO_4^+$	1.86	Marini and Accornero (2007)
$Cu^{2+} + HAsO_4^{2-} \rightleftharpoons CuHAsO_4^{\circ}$	3.68	Whiting (1992)
$Cu^{2+} + HAsO_4^{2-} \rightleftharpoons CuHAsO_4^{\circ}$	3.83	Marini and Accornero (2007) adj.[1]
$Cu^{2+} + AsO_4^{3-} \rightleftharpoons CuAsO_4^-$	9.33	Marini and Accornero (2007) adj.[1]
$Pb^{2+} + H_2AsO_4^- \rightleftharpoons PbH_2AsO_4^+$	1.53	Whiting (1992)
$Pb^{2+} + H_2AsO_4^- \rightleftharpoons PbH_2AsO_4^+$	1.60	Marini and Accornero (2007)
$Pb^{2+} + HAsO_4^{2-} \rightleftharpoons PbHAsO_4^{\circ}$	3.04	Whiting (1992)
$Pb^{2+} + HAsO_4^{2-} \rightleftharpoons PbHAsO_4^{\circ}$	2.87	Marini and Accornero (2007) adj.[1]
$Pb^{2+} + AsO_4^{3-} \rightleftharpoons PbAsO_4^-$	6.89	Marini and Accornero (2007) adj.[1]
$Ni^{2+} + H_2AsO_4^- \rightleftharpoons NiH_2AsO_4^+$	1.64	Marini and Accornero (2007)
$Ni^{2+} + HAsO_4^{2-} \rightleftharpoons NiHAsO_4^{\circ}$	2.90	Whiting (1992)
$Ni^{2+} + HAsO_4^{2-} \rightleftharpoons NiHAsO_4^{\circ}$	2.71	Marini and Accornero (2007) adj.[1]
$Ni^{2+} + AsO_4^{3-} \rightleftharpoons NiAsO_4^-$	7.84	Marini and Accornero (2007) adj.[1]
$Co^{2+} + H_2AsO_4^- \rightleftharpoons CoH_2AsO_4^+$	0.28	Marini and Accornero (2007)
$Co^{2+} + H_2AsO_4^- \rightleftharpoons CoH_2AsO_4^+$	−0.79	Lee and Nriagu (2007)
$Co^{2+} + HAsO_4^{2-} \rightleftharpoons CoHAsO_4^{\circ}$	3.00	Whiting (1992)
$Co^{2+} + HAsO_4^{2-} \rightleftharpoons CoHAsO_4^{\circ}$	1.50	Lee and Nriagu (2007)
$Co^{2+} + HAsO_4^{2-} \rightleftharpoons CoHAsO_4^{\circ}$	2.93	Marini and Accornero (2007) adj.[1]
$Co^{2+} + AsO_4^{3-} \rightleftharpoons CoAsO_4^-$	6.96	Marini and Accornero (2007) adj.[1]

[1] Values from Marini and Accornero (2007) were adjusted to be consistent with the hydrolysis constants of arsenic acid in this study.

There are markedly few calorimetric studies on arsenates. The derivation of data for the Gibbs free energy of formation of arsenates by a route other than solubility studies is desirable; comparison of such studies may reveal problems with respect to the accuracy of the data. However, the calculation of a solubility product constant from free energy values based on calorimetric data usually involves the difference of large numbers, e.g., the free energy of the solid phase in a solubility reaction is often a much larger value than the free energy of the reaction itself such that the error in the free energy of the solid phase is comparable to the free energy of the reaction. A large uncertainty is associated with this approach. Ultimately,

Table 12. Logarithm of association constants for trivalent metal arsenate complexes.

Reaction	log K (25 °C)	Reference
$Fe^{3+} + H_2AsO_4^- \rightleftharpoons FeH_2AsO_4^{2+}$	4.06	Whiting (1992)
$Fe^{3+} + H_2AsO_4^- \rightleftharpoons FeH_2AsO_4^{2+}$	5.43	Nordstrom (2000)
$Fe^{3+} + H_2AsO_4^- \rightleftharpoons FeH_2AsO_4^{2+}$	0.63	Lee and Nriagu (2007)
$Fe^{3+} + H_2AsO_4^- \rightleftharpoons FeH_2AsO_4^{2+}$	4.265	Marini and Accornero (2007)
$Fe^{3+} + HAsO_4^{2-} \rightleftharpoons FeHAsO_4^+$	9.76	Whiting (1992)
$Fe^{3+} + HAsO_4^{2-} \rightleftharpoons FeHAsO_4^+$	9.92	Nordstrom (2000)
$Fe^{3+} + HAsO_4^{2-} \rightleftharpoons FeHAsO_4^+$	4.88	Lee and Nriagu (2007)
$Fe^{3+} + HAsO_4^{2-} \rightleftharpoons FeHAsO_4^+$	9.95	Marini and Accornero (2007) adj.[1]
$Fe^{3+} + HAsO_4^{2-} \rightleftharpoons FeHAsO_4^+$	9.21	Raposo et al. (2006)
$Fe^{3+} + AsO_4^{3-} \rightleftharpoons FeAsO_4^0$	18.85	Whiting (1992)
$Fe^{3+} + AsO_4^{3-} \rightleftharpoons FeAsO_4^0$	13.97	Marini and Accornero (2007) adj.[1]
$Al^{3+} + H_2AsO_4^- \rightleftharpoons AlH_2AsO_4^{2+}$	3.07	Whiting (1992)
$Al^{3+} + H_2AsO_4^- \rightleftharpoons AlH_2AsO_4^{2+}$	3.20	Marini and Accornero (2007)
$Al^{3+} + HAsO_4^{2-} \rightleftharpoons AlHAsO_4^+$	7.29	Whiting (1992)
$Al^{3+} + HAsO_4^{2-} \rightleftharpoons AlHAsO_4^+$	7.17	Marini and Accornero (2007) adj.[1]
$Al^{3+} + AsO_4^{3-} \rightleftharpoons AlAsO_4^0$	14.1	Whiting (1992)
$Al^{3+} + AsO_4^{3-} \rightleftharpoons AlAsO_4^0$	11.19	Marini and Accornero (2007) adj.[1]

[1]Values from Marini and Accornero (2007) were adjusted to be consistent with the hydrolysis constants of arsenic acid in this study.

Table 13. Logarithm of association constants for monovalent metal arsenate complexes (Marini and Accornero 2007)[1].

Reaction	log K (25 °C)
$Na^+ + H_2AsO_4^- \rightleftharpoons NaH_2AsO_4^0$	1.78
$K^+ + H_2AsO_4^- \rightleftharpoons KH_2AsO_4^0$	1.89
$Na^+ + HAsO_4^{2-} \rightleftharpoons NaHAsO_4^-$	0.69
$K^+ + HAsO_4^{2-} \rightleftharpoons KHAsO_4^-$	0.55
$Na^+ + AsO_4^{3-} \rightleftharpoons NaAsO_4^{2-}$	4.71
$K^+ + AsO_4^{3-} \rightleftharpoons KAsO_4^{2-}$	4.57

[1]Values from Marini and Accornero (2007) were adjusted to be consistent with the hydrolysis constants of arsenic acid in this study.

Table 14. Logarithm of association constants for metal arsenite complexes (Marini and Accornero 2007).

Reaction	log K (25 °C)
$Na^+ + H_2AsO_3^- \rightleftharpoons NaH_2AsO_3^0$	0.25
$Ca^{2+} + H_2AsO_3^- \rightleftharpoons CaH_2AsO_3^+$	1.81
$Mg^{2+} + H_2AsO_3^- \rightleftharpoons MgH_2AsO_3^+$	1.88
$Sr^{2+} + H_2AsO_3^- \rightleftharpoons SrH_2AsO_3^+$	0.37
$Ba^{2+} + H_2AsO_3^- \rightleftharpoons BaH_2AsO_3^+$	1.43
$Cu^{2+} + H_2AsO_3^- \rightleftharpoons CuH_2AsO_3^+$	7.11
$Pb^{2+} + H_2AsO_3^- \rightleftharpoons PbH_2AsO_3^+$	5.20
$Fe^{3+} + H_2AsO_3^- \rightleftharpoons FeH_2AsO_3^{2+}$	7.28
$Al^{3+} + H_2AsO_3^- \rightleftharpoons AlH_2AsO_3^{2+}$	7.82

thermodynamic principles require that the values must agree regardless of the type of measurement, once the errors are minimized.

A peculiar shortcoming of a number of papers is that the data were derived for compounds which may not exist or have never been described in nature. Examples are the "phases" $AlAsO_4$ or $FeAsO_4$ and curiously, studies predict such phases to form and store arsenic in nature (e.g., Sadiq 1997). Hence, indiscriminate use of some solubility products presented in the literature could lead to predictions that have no practical application and may not even relate to observed phenomena.

Ca arsenates

A number of solubility products were reported for arsenates of calcium (Table 15) although it is not clear whether all arsenates with such stoichiometry indeed do exist. Different studies report the presence of various Ca arsenates in their precipitates. These discrepancies may be due to the variable water content of these phases, especially of $Ca_3(AsO_4)_2 \cdot xH_2O$, for which $2.25 \leq x \leq 4.25$ have been reported (Table 15). Several studies inform that traces of impurities (e.g., Mg, CO_2) may considerably change the products of the syntheses but a systematic investigation is lacking. This concerns particularly $Ca_5(AsO_4)_3OH$ (johnbaumite, a member of the apatite group), which was found to precipitate only when Mg^{2+} was absent (Bothe and Brown 1999). Bothe and Brown (2002) compared some of the thermodynamic data with the solubility results from several sources and presented phase diagrams for the system $CaO-As_2O_5-H_2O$.

We have incorporated Bothe and Brown's (1999) solubility products for various calcium-arsenate phases into the present arsenate speciation model, which also includes the (average) values for the calcium (hydrogen) arsenate complexes given in Table 11. While most of Bothe and Brown's (1999) values (Table 15) were retained, the log $K_{sp}°$ values for $Ca_5(AsO_4)_3OH$ and $Ca_4(AsO_4)_2(OH)_2 \cdot 4H_2O$ were changed to more negative values (−39.05 and −30.05 respectively) to better represent the available solubility data. Since this adjustment of log $K_{sp}°$ was required only for the two phases that are stable at the highest pH range, it might be due to inconsistencies resulting from different arsenate speciation models used by Bothe and Brown (1999) and in this study (see the discussion in the section on scorodite solubility). Figures 3 and 4 present two different representations of solubility equilibria in the ternary system $CaO-As_2O_5-H_2O$, which were calculated from this model and compared with experimental data of Nishimura and Robins (1998) and Bothe and Brown (1999). Data from the more recent study of Zhu et al. (2006) cannot be depicted in these diagrams because they refer to quaternary systems (due to the addition of either KOH or HNO_3 to adjust the pH). It seems that the log $[As]_T$ vs. pH diagram (Fig. 3) is more sensitive in revealing discrepancies between calculated values and experimental data. Not surprisingly, there is good agreement between Bothe and Brown's (1999) model and their data, whereas Nishimura and Robins' (1998) measurements significantly depart from the calculated values.

While there is a broadly consistent pattern of phases precipitating in certain pH ranges, some significant differences in the precipitation sequences were reported by various workers. Figures 3 and 4 show that these uncertainties concern (i) the formation of $Ca_5(AsO_4)_3OH$ (johnbaumite), which was observed in the long-term measurements (4 years) of Bothe and Brown (1999) but not at equilibration times of 1 month (Nishimura and Robins 1998), (ii) the relative stabilities of $Ca_3(AsO_4)_2 \cdot 4.25H_2O$, ferrarisite and guerinite (both $Ca_5H_2(AsO_4)_4 \cdot 9H_2O$) around neutral pH and (iii) the relative stabilities of various $Ca_3(AsO_4)_2 \cdot xH_2O$ hydrates. Although Bothe and Brown (1999) reported thermodynamic data for $Ca_3(AsO_4)_2 \cdot 3.666H_2O$ and guerinite (both of which were found experimentally), these phases never become stable in the model calculations. To resolve these discrepancies, careful solubility studies and a subsequent refinement of the thermodynamic model are required.

In addition, a complete data set that may explain the transformation relationships and sequences between pharmacolite, haidingerite, and weilite, is still missing. Although the studies reviewed above indicate that only haidingerite precipitates from aqueous solution, this set of minerals appears to be common at sites where arsenate-enriched solutions come in contact with limestones and these minerals should be therefore targeted in future thermodynamic studies. When combined with thermodynamic data for carbonate species and phases (Königsberger et al. 1999), the present model predicts that $Ca_3(AsO_4)_2 \cdot 4.25H_2O$ + calcite are at equilibrium with atmospheric $CO_2(g)$ and an aqueous phase of pH = 8.0 and $[As]_T = 0.004$ mol kg^{-1}.

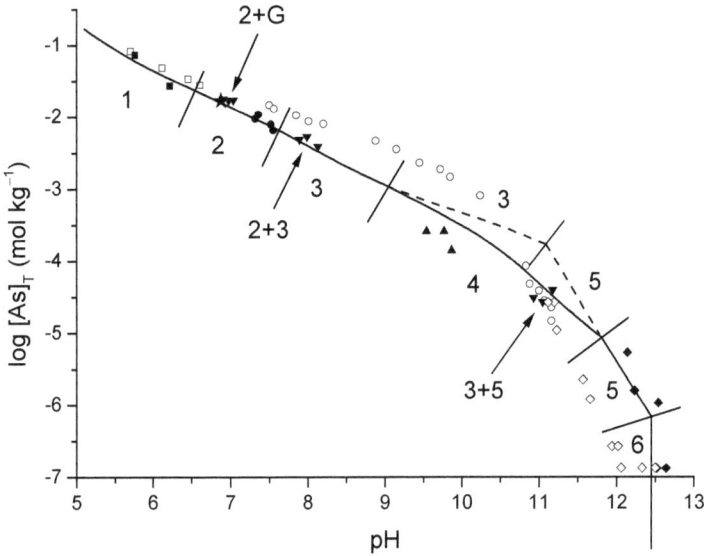

Figure 3. Total [As] as a function of pH at equilibrium with various calcium arsenate phases in the ternary CaO-As$_2$O$_5$-H$_2$O system: 1, CaHAsO$_4$·H$_2$O (haidingerite); 2, Ca$_5$H$_2$(AsO$_4$)$_4$·9H$_2$O (ferrarisite); 3, Ca$_3$(AsO$_4$)$_2$·4.25H$_2$O; 4, Ca$_5$(AsO$_4$)$_3$OH (johnbaumite); 5, Ca$_4$(AsO$_4$)$_2$(OH)$_2$·4H$_2$O; 6; Ca(OH)$_2$; G, Ca$_5$H$_2$(AsO$_4$)$_4$·9H$_2$O (guerinite). Short lines indicate three-phase equilibria. Experimental values: dots, Nishimura and Robins (1998); inverted triangles, Bothe and Brown (1999). Two- and three-phase equilibria are indicated by the numbers representing the solid phases. Dashed lines depict metastable equilibria when the formation of johnbaumite is suppressed.

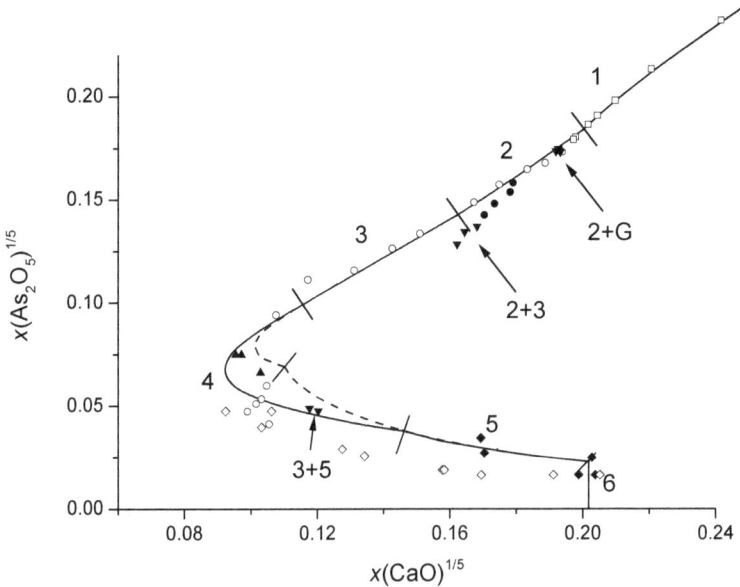

Figure 4 Partial phase diagram (solutus curves only) of the ternary CaO-As$_2$O$_5$-H$_2$O system in the representation using (mole fractions)$^{1/5}$ as suggested by Bothe and Brown (2002). Phase numbers, dashed lines, and symbols of experimental data as in Figure 3.

Ba arsenates

Solubility products for $Ba_3(AsO_4)_2$ and $BaHAsO_4 \cdot H_2O$ were reported by various researchers (Table 15). In a recent study, Zhu et al. (2005) have carefully characterized the solid phases before and after equilibration with the aqueous phase. Their values are in reasonable agreement with solubility data of Essington (1988) and Davis (2000), whereas older solubility product values for $Ba_3(AsO_4)_2$, especially by Chukhlantsev (1956b), are 25 orders of magnitude more negative. Based on Chukhlantsev's (1956b) data, Essington (1988) derived a solubility product value for $BaHAsO_4 \cdot H_2O$ that is 19 orders of magnitude more negative than that of Zhu et al. (2005) (Table 15).

Cu arsenates

Data for four copper arsenates have been reported by Magalhães et al. (1988) (Table 15). They used natural samples which are hardly to be expected to be chemically pure. Additional data were recently reported by Nelson et al. (2011). They found that their samples were more soluble than the same phases investigated by Magalhães et al. (1988), most likely because Nelson et al. (2011) used synthetic precipitates with relatively poor crystallinity. Verification of the data and additional work are definitely needed here.

Fe arsenates

The thermodynamic data for scorodite are summarized above and appear to converge to one set of values. Data for other phases such as parascorodite, kaňkite, bukovskýite are given in Table 15. Because of its sheer predominance over other minerals in nature, the data for scorodite suffice for modeling of most natural systems with abundant Fe(III) and As(V).

Substances in the system $Fe_2O_3-As_2O_5-H_2O$ are common at hazardous waste sites such as mining and mineral processing facilities. These substances, however, are somewhat different from those in other systems in that a poorly crystalline compound, hydrous ferric oxide with substantial amounts of associated As(V), occurs very commonly. The structure of this compound is somewhat uncertain and probably variable, the chemical composition in terms of the Fe/As ratio and H_2O content is certainly highly variable (e.g., Paktunc et al. 2008). Complex mixtures of fine-grained poorly crystalline phases and hydrous ferric oxides that contain sorbed As(V) are a considerable challenge to characterize mineralogically, let alone thermodynamically.

A number of studies investigated a mysterious compound $FeAsO_4$, prepared by mixing of Fe(III) and As(V) chemicals in an aqueous solution and subsequent precipitation. The X-ray diffraction pattern of this compound published by Lee and Nriagu (2007) resembles an XRD pattern of ferrihydrite, thus confirming poor crystallinity of the material and suggesting variable chemical composition. It is implausible that such material does not contain any water and the neglect of H_2O content in this material over 50 years of its investigation is puzzling. Majzlan (2011) used calorimetric experiments on X-ray amorphous As(V)-bearing hydrous ferric oxide to determine its solubility. Langmuir et al. (2006) evaluated critically all earlier data sets on X-ray amorphous ferric arsenates with the Fe/As ratio close to 1 and proposed the best value for the solubility of such compound.

There is also a significant confusion about the solubility of the ferrous arsenate symplesite. The reported K_{sp} values vary by 8 orders of magnitude (Khoe et al. 1991; Johnston and Singer 2007; Shan 2008).

Mg arsenates

A few data available for the Mg arsenates are given in Table 15. There are thermodynamic data for a number of "normal" Mg arsenates (i.e., those with the stoichiometry $Mg_3(AsO_4)_2 \cdot xH_2O$)

although, similar to the Ca arsenates, it is not clear if they all really exist. The stoichiometry for rösslerite was previously given as $Mg_3(AsO_4)_2 \cdot 7H_2O$ but has been shown by crystallographic studies to be $MgHAsO_4 \cdot 7H_2O$. Such change in stoichiometry will change the expression for the stability constant and subsequently may alter the interpretation of the solubility studies; the system should be revisited and the data for brassite and rösslerite re-measured.

Pb arsenates

An extensive discussion of the thermodynamic data for $Pb_3(AsO_4)_2$ was provided by Magalhães and Silva (2003) because this phase appears to be rather soluble, which may have environmental implications for the mobility of Pb and As(V). They noted that the value of Chukhlantsev (1956a) is used widely to predict the behavior of lead and arsenate in the environment, although he provided only a very crude characterization of the material used for solubility study. Most of the arsenate solubility data reported by Chukhlantsev (1956a) have been shown to be poor estimates (Magalhães et al. 1988; Flis et al. 2011). Later studies concluded that the precipitate used by Chukhlantsev (1956a) was probably a mixture of phases. Unfortunately, some recent studies (e.g., Lee and Nriagu 2007) also used the procedures described by Chukhlantsev (1956a) to synthesize arsenate phases for thermodynamic measurements.

Several polymorphs of $Pb_3(AsO_4)_2$ are known (von Hodenberg 1974; Viswanathan and Miehe 1978) but to the best of our knowledge, no such phases have been reported as naturally occurring. Therefore, modeling of Pb-As(V) solubility with the phase $Pb_3(AsO_4)_2$ is a low priority.

At many sites polluted with Pb and As(V), schultenite is the mineral which carries both elements (see Majzlan et al. 2014, this volume). A solubility product for this phase was reported by Magalhães et al. (1988). Independently of the quality of the available thermodynamic data, the modeling of the mobility of Pb and As(V) (in the absence of Cl) should be done with respect to schultenite, not the elusive phase $Pb_3(AsO_4)_2$.

If arsenate were to immobilize with lead in the presence of chloride, mimetite would be the most likely phase. The thermodynamics of mimetite and the mimetite-pyromorphite solid solution were investigated in the solubility studies of Bajda (2010) and Flis et al. (2011), respectively.

Zn arsenates

Data for three zinc arsenates are available, all determined by solubility studies (Table 15). From the data and from natural observations, it seems that adamite is the most stable zinc arsenate. Further work is needed in this system, especially to work out the conditions of stability and transformation between adamite, hydrozincite ($Zn_5(CO_3)_2(OH)_6$), smithsonite ($ZnCO_3$), and hemimorphite ($Zn_4Si_2O_7(OH)_2 \cdot H_2O$).

Arsenates with multiple cations

The data for arsenates with multiple cations and one sulfate-arsenate (beudantite) are summarized at the bottom of Table 15. The use of natural samples, seen in several studies, is very problematic as no chemical analyses, proof of homogeneity, or lack of microscopic impurities were provided.

Table 15. Solubilities of metal arsenates reported in literature, primarily but not exclusively minerals.

Phase (mineral) and its dissolution reaction	log K_{sp}	Reference
Ca arsenates		
$Ca_5(AsO_4)_3OH$ (johnbaumite) $\rightleftharpoons 5Ca^{2+} + 3AsO_4^{3-} + OH^-$	−38.04	Bothe and Brown (1999)
$Ca_5(AsO_4)_3OH$ (johnbaumite) $\rightleftharpoons 5Ca^{2+} + 3AsO_4^{3-} + OH^-$	−40.12	Zhu et al. (2006)
$Ca_5(AsO_4)_3F$ (svabite) $\rightleftharpoons 5Ca^{2+}\ 3AsO_4^{3-} + F^-$	−39.21	Zhu et al. (2011)
$CaHAsO_4 \cdot 2H_2O$ (pharmocolite) $+ H^+ \rightleftharpoons Ca^{2+} + H_2AsO_4^- + 2H_2O$	−4.68	Rodríguez–Blanco (2007)
$CaHAsO_4 \cdot H_2O$ (haidingerite) $\rightleftharpoons Ca^{2+} + HAsO_4^{2-} + H_2O$	−4.79	Bothe and Brown (1999)
$CaHAsO_4 \cdot H_2O$ (haidingerite) $+ H^+ \rightleftharpoons Ca^{2+} + H_2AsO_4^- + H_2O$	+3.23 ±0.07	Nishimura and Robins (1998)
$CaHAsO_4$ (weilite) $+ H^+ \rightleftharpoons Ca^{2+} + H_2AsO_4^-$	+2.36	Mahapatra et al. (1986)
$Ca_5H_2(AsO_4)_4 \cdot 9H_2O$ (ferrarisite) $\rightleftharpoons 5Ca^{2+} + 2AsO_4^{3-} + 2HAsO_4^{2-} + 9H_2O$	−31.49	Bothe and Brown (1999)
$Ca_5H_2(AsO_4)_4 \cdot 9H_2O$ (guerinite) $\rightleftharpoons 5Ca^{2+} + 2AsO_4^{3-} + 2HAsO_4^{2-} + 9H_2O$	−30.69	Bothe and Brown (1999)
$Ca_3(AsO_4)_2 \cdot 2.25H_2O \rightleftharpoons 3Ca^{2+} + 2AsO_4^{3-} + 2.25H_2O$	−21.40	Zhu et al. (2006)
$Ca_3(AsO_4)_2 \cdot 3H_2O \rightleftharpoons 3Ca^{2+} + 2AsO_4^{3-} + 3H_2O$	−21.14	Zhu et al. (2006)
$Ca_3(AsO_4)_2 \cdot 3.67H_2O \rightleftharpoons 3Ca^{2+} + 2AsO_4^{3-} + 3.67H_2O$	−21.00	Bothe and Brown (1999)
$Ca_3(AsO_4)_2 \cdot 4H_2O + 2H^+ \rightleftharpoons 3Ca^{2+} + 2HAsO_4^{2-} + 4H_2O$	+5.58 ±0.40	Nishimura and Robins (1998)
$Ca_3(AsO_4)_2 \cdot 4.25H_2O \rightleftharpoons 3Ca^{2+} + 2AsO_4^{3-} + 4.25H_2O$	−21.00	Bothe and Brown (1999)
$Ca_4(AsO_4)_2(OH)_2 \cdot 4H_2O \rightleftharpoons 4Ca^{2+} + 2AsO_4^{3-} + 2OH^- + 4H_2O$	−29.20	Bothe and Brown (1999)
$Ca_4(AsO_4)_2(OH)_2 \cdot 4H_2O \rightleftharpoons 4Ca^{2+} + 2AsO_4^{3-} + 2OH^- + 4H_2O$	−27.49	Zhu et al. (2006)
$Ca_2(AsO_4)(OH) \cdot 2H_2O + H^+ \rightleftharpoons 2Ca^{2+} + AsO_4^{3-} + 3H_2O$	+1.29 ±0.68	Nishimura and Robins (1998)
Cu arsenates		
$Cu_2AsO_4(OH)$ (olivenite) $+ 3H^+ \rightleftharpoons 2Cu^{2+} + H_2AsO_4^- + H_2O$	+2.38 ± 0.23	Magalhães et al. (1988)
$Cu_5(AsO_4)_2(OH)_4$ (cornubite) $+ 8H^+ \rightleftharpoons 5Cu^{2+} + 2H_2AsO_4^- + 4H_2O$	+12.40 ± 0.58	Magalhães et al. (1988)
$Cu_3AsO_4(OH)_3$ (clinoclase) $+ 5H^+ \rightleftharpoons 3Cu^{2+} + H_2AsO_4^- + 3H_2O$	+10.10 ± 0.38	Magalhães et al. (1988)

Table 15. *continued.*

Phase (mineral) and its dissolution reaction	log K_{sp}	Reference
$Cu_2AsO_4(OH) + 2H^+ \rightleftharpoons 2Cu^{2+} + HAsO_4^{2-} + H_2O$	-2.91 ± 0.03	Nelson et al. (2011)
$Cu_2AsO_4(OH) \cdot 3H_2O$ (euchroite) $+ 3H^+ \rightleftharpoons 2Cu^{2+} + H_2AsO_4^- + 4H_2O$	$+3.28 \pm 0.17$	Magalhães et al. (1988)
$Cu_3(AsO_4)_2$ (lammerite) $\rightleftharpoons 3Cu^{2+} + 2AsO_4^{3-}$	-35.17	Wagman et al. (1982)
$Cu_3(AsO_4)_2$ (lammerite) $+ 2H^+ \rightleftharpoons 3Cu^{2+} + 2HAsO_4^{2-}$	-13.52 ± 0.08	Nelson et al. (2011)
$Cu_3(AsO_4)_2 \cdot 4H_2O$ (rollandite) $\rightleftharpoons 3Cu^{2+} + 2AsO_4^{3-} + 4H_2O$	-38.88	Charykova et al. (2010)
Fe arsenates		
$FeAsO_4$ (crystalline) $\rightleftharpoons Fe^{3+} + AsO_4^{3-}$	-21.66	Majzlan et al. (2012a)
$FeAsO_4 \cdot xH_2O$ (amorphous) $\rightleftharpoons Fe^{3+} + AsO_4^{3-} + xH_2O$	-23.3 ± 0.3	Langmuir et al. (2006)
$FeAsO_4 \cdot 3.5H_2O$ (kaňkite) $\rightleftharpoons Fe^{3+} + AsO_4^{3-} + 3.5H_2O$	-23.92 ± 0.51	Majzlan et al. (2012a)
$FeAsO_4 \cdot 2H_2O$ (parascorodite) $\rightleftharpoons Fe^{3+} + AsO_4^{3-} + 2H_2O$	-25.44	Majzlan et al. (2012a)
$fFe_2O_3 \cdot xAs_2O_5 \cdot hH_2O$ (As–hydrous ferric oxide) $+ (6f-6x)H^+ \rightleftharpoons 2fFe^{3+} + 2xAsO_4^{3-} + (3f+h-3x)H_2O$, where $f = 0.2090$, $x = 0.04536$, $h = 0.7456$	-1.41	Majzlan (2011)*
$Fe_2(AsO_4)(SO_4)(OH) \cdot 9H_2O$ (bukovskýite) $+ H^+ \rightleftharpoons 2Fe^{3+} + SO_4^{2-} + AsO_4^{3-} + 10H_2O$	-30.63	Majzlan et al. (2012b)
$Fe_3(AsO_4)_2 \cdot 8H_2O$ (symplesite) $+ 2H^+ \rightleftharpoons 3Fe^{2+} + 2HAsO_4^{2-} + 8H_2O$	-33.25 ± 0.46	Johnston and Singer (2007)
Mg arsenates		
$MgHAsO_4 \cdot 4H_2O$ (brassite) $\rightleftharpoons Mg^{2+} + H^+ + AsO_4^{3-} + 4H_2O$	-5.66 ± 0.07	Raposo et al. (2004)
$Mg_3(AsO_4)_2 \cdot 8H_2O \rightleftharpoons 3Mg^{2+} + 2AsO_4^{3-} + 8H_2O$	-22.32 ± 0.09	Raposo et al. (2004)
$Mg_3(AsO_4)_2 \cdot 10H_2O \rightleftharpoons 3Mg^{2+} + 2AsO_4^{3-} + 10H_2O$	$+3.00$	Nishimura et al. (1988)
$MgHAsO_4 \cdot 7H_2O$ ("rösslerite") $+ H^+ \rightleftharpoons Mg^{2+} + H_2AsO_4^- + 7H_2O$	$+3.60$	Nishimura et al. (1988)
Ba arsenates		
$Ba_3(AsO_4)_2 \rightleftharpoons 3Ba^{2+} + 2AsO_4^{3-}$	-23.53	Zhu et al. (2005)
$Ba_3(AsO_4)_2 \rightleftharpoons 3Ba^{2+} + 2AsO_4^{3-}$	-21.62	Essington (1988)

Table 15. *continued.*

Phase (mineral) and its dissolution reaction	log K_{sp}	Reference
$Ba_3(AsO_4)_2 \rightleftharpoons 3Ba^{2+} + 2AsO_4^{3-}$	−21.57	Davis (2000)
$Ba_3(AsO_4)_2 \rightleftharpoons 3Ba^{2+} + 2AsO_4^{3-}$	−50.11	Chukhlantsev (1956b)
$Ba_3(AsO_4)_2 \rightleftharpoons 3Ba^{2+} + 2AsO_4^{3-}$	−16.58	Robins (1985)
$BaHAsO_4 \cdot H_2O \rightleftharpoons Ba^{2+} + HAsO_4^{2-} + H_2O$	−5.60	Zhu et al. (2005)
$BaHAsO_4 \cdot H_2O \rightleftharpoons Ba^{2+} + HAsO_4^{2-} + H_2O$	−5.51	Davis (2000)
$BaHAsO_4 \cdot H_2O \rightleftharpoons Ba^{2+} + HAsO_4^{2-} + H_2O$	−4.70	Robins (1985)
$BaHAsO_4 \cdot H_2O \rightleftharpoons Ba^{2+} + HAsO_4^{2-} + H_2O$	−0.8	Orellana et al. (2000)
$BaHAsO_4 \cdot H_2O \rightleftharpoons Ba^{2+} + HAsO_4^{2-} + H_2O$	−24.64	Essington (1988), based on Chukhlantsev (1956b)
Pb arsenates		
$Pb_5(AsO_4)_3Cl$ (mimetite) $\rightleftharpoons 5Pb^{2+} + 3AsO_4^{3-} + Cl^-$	−76.35 ± 1.01	Bajda (2010), Flis et al. (2011)
$Pb_5(AsO_4)_3Cl$ (mimetite) $\rightleftharpoons 5Pb^{2+} + 3AsO_4^{3-} + Cl^-$	K_{sp} not given	Huang et al. (2014)
$Pb_5(AsO_4)_3Cl$ (mimetite) $+ 6H^+ \rightleftharpoons 5Pb^{2+} + 3H_2AsO_4^- + Cl^-$	−28.24	Inegbenebor et al. (1989)
$Pb_5(AsO_4)_3Cl$ (mimetite) $+ 6H^+ \rightleftharpoons 5Pb^{2+} + 3H_2AsO_4^- + Cl^-$	−17.95	Comba et al. (1988)
$Pb_5(PO_4)_3Cl$ (pyromorphite) $\rightleftharpoons 5Pb^{2+} + 3PO_4^{3-} + Cl^-$	−79 ± 1.01	Flis et al. (2011)
$Pb_5(AsO_4)_x(PO_4)_{3-3x}Cl$ (mimetite–pyromorphite solid solution) $\rightleftharpoons 5Pb^{2+} + xAsO_4^{3-} + (3-3x)PO_4^{3-} + Cl^-$	−76 to −79	Flis et al. (2011)
$Pb_5(AsO_4)_3(OH)$) (hydroxymimetite)$\rightleftharpoons 5Pb^{2+} + 3AsO_4^{3-} + OH^-$	−76.14	Lee and Nriagu (2007)
$PbHAsO_4$ (schultenite) $+ H^+ \rightleftharpoons Pb^{2+} + H_2AsO_4^-$	−5.41 ± 0.02	Magalhães et al. (1988)
$PbHAsO_4 + H^+ \rightleftharpoons Pb^{2+} + H_2AsO_4^-$	−5.08	Lee and Nriagu (2007)
$PbHAsO_4 \cdot 2H_2O \rightleftharpoons Pb^{2+} + HAsO_4^{2-} + 2H_2O$	−10.70	Liu et al. (2009)
$Pb_5(AsO_4)_3OH \cdot H_2O \rightleftharpoons 5Pb^{2+} + 3AsO_4^{3-} + OH^- + H_2O$	−81.75	Liu et al. (2009)
$Pb_3(AsO_4)_2 \rightleftharpoons 3Pb^{2+} + 2AsO_4^{3-}$	−33.83	Liu et al. (2009)

Table 15. *continued.*

Phase (mineral) and its dissolution reaction	log K_{sp}	Reference
$Pb_3(AsO_4)_2 \rightleftharpoons 3Pb^{2+} + 2AsO_4^{3-}$	−35.39	Chukhlantsev (1956a)
$Pb_8As_2O_{13} + 16H^+ \rightleftharpoons 8Pb^{2+} + 2H_3AsO_4 + 5H_2O$	−174.24	Liu et al. (2009)
Zn arsenates		
$Zn_2AsO_4(OH)$ (adamite) $+ 3H^+ \rightleftharpoons 2Zn^{2+} + H_2AsO_4^- + H_2O$	+5.71 ± 0.22	Magalhães et al. (1988)
$Zn_2AsO_4(OH){\cdot}H_2O$ (legrandite) $+ 3H^+ \rightleftharpoons 2Zn^{2+} + H_2AsO_4^- + 2H_2O$	+5.97 ± 0.40	Magalhães et al. (1988)
$Zn_3(AsO_4)_2{\cdot}8H_2O$ (köttigite) $\rightleftharpoons 3Zn^{2+} + 2AsO_4^{3-} + 8H_2O$	−32.40	Lee and Nriagu (2007)
Al arsenates		
$AlAsO_4{\cdot}3.5H_2O$ (amorphous phase) $\rightleftharpoons Al^{3+} + AsO_4^{3-} + 3.5H_2O$	−18.06	Pantuzzo et al. (2014)
arsenates with multiple cations		
$PbCu_3(AsO_4)_2(OH)_2$ (bayldonite) $+ 6H^+ \rightleftharpoons Pb^{2+} + 3Cu^{2+} + 2H_2AsO_4^- + 2H_2O$	+0.03 ± 0.05	Magalhães et al. (1988)
$CaZnAsO_4(OH)$ (austinite) $+ 3H^+ \rightleftharpoons Ca^{2+} + Zn^{2+} + H_2AsO_4^- + H_2O$	+6.88 ± 0.21	Magalhães et al. (1988)
$CaCuAsO_4(OH)$ (conichalcite) $+ 3H^+ \rightleftharpoons Ca^{2+} + Cu^{2+} + H_2AsO_4^- + H_2O$	+1.29 ± 0.33	Magalhães et al. (1988)
$PbCuAsO_4(OH)$ (duftite) $+ 3H^+ \rightleftharpoons Pb^{2+} + Cu^{2+} + H_2AsO_4^- + H_2O$	−1.97 ± 0.05	Magalhães et al. (1988)
$Cu_2Al_7(AsO_4)_4(OH)_{13}{\cdot}12H_2O$ (ceruleite) $\rightleftharpoons$ $2Cu^{2+} + 7Al^{3+} + 4AsO_4^{3-} + 13OH^- + 12H_2O$	−154.32	Lee and Nriagu (2007)
$Pb_2Cu(AsO_4)(CrO_4)(OH)$ (formacite) $\rightleftharpoons 2Pb^{2+} + Cu^{2+} + AsO_4^{3-} + CrO_4^{2-} + OH^-$	−44.66	Lee and Nriagu (2007)
$(H_3O)_{0.68}Pb_{0.32}Fe_{2.86}(SO_4)_{1.69}(AsO_4)_{0.31}[(OH)_{5.59}{\cdot}(H_2O)_{0.41}]$ (beudantite) $+ 4.91H^+ \rightleftharpoons$ $0.32Pb^{2+} + 2.86Fe^{3+} + 1.69SO_4^{2-} + 0.31AsO_4^{3-} + 6.68H_2O$	−13.94 ± 1.89	Forray et al. (2014)

* several other compositions of As–hydrous ferric oxide were investigated in this work

ARSENIDES AND SULFIDES

Thermodynamic data for arsenide and arsenic-sulfide minerals have been challenging to obtain by either calorimetry or solubility measurements. Inconsistencies, often arising from experimental difficulties, still hinder attempts at evaluation. In this section, data are presented for the main minerals in Table 16; readers are encouraged to read the original papers to reach their own conclusions on data reliability.

Solubility studies (Mironova et al. 1984; Webster 1990; Eary 1992) are inherently difficult because the solution speciation is not well characterized. Undoubtedly, there should be thioarsenates and/or thioarsenites in solution (Suess and Planer-Friedrich 2012) in equilibrium with arsenic-sulfide minerals, but reliable stability constants are not available. Combustion calorimetry measurements (Johnson et al. 1980; Bryndzia and Kleppa 1988) are difficult because of incompletely and poorly characterized products.

GASEOUS SPECIES

Thermodynamic properties for the two gases, arsine (AsH_3) and tetraarsenic hexaoxide (As_4O_6), have been measured and calculated from quantum mechanical principles. Table 17 tabulates reported values for standard state thermodynamic properties of these gases. Properties of arsine and As_4O_6 gas were reported by Wagman et al. (1982) without references or method of calculation. Mayer et al. (1997) performed *ab initio* molecular orbital calculations (G2 level) for arsine with the GAUSSIAN 94 program. Behrens and Rosenblatt (1972) measured vapor pressures of $As_4O_{6(g)}$ in equilibrium with arsenolite using the Knudsen effusion method with a vacuum microbalance. The reported values by Behrens and Rosenblatt (1972) for $As_4O_{6(g)}$ were based on precursor data to, and the same as, the data of Wagman et al. (1982) for arsenolite. Recalculating those results using our evaluated enthalpy of formation for arsenolite makes no difference in the reported property. We recommend the standard state properties reported by Wagman et al. (1982) for arsine for better consistency and completeness and those reported by Behrens and Rosenblatt (1972) for $As_4O_{6(g)}$ because of the sensitive nature of the technique, the directness of the measurement (only auxiliary data needed is that of arsenolite), the temperature range of the measurements (94-156 °C), and a careful evaluation using both second-law and third-law methods. It would appear that Wagman et al. (1982) missed Behrens and Rosenblatt's study in their compilation.

CONCLUDING REMARKS

A critical evaluation of original literature sources on standard-state thermodynamic properties of selected arsenic species has led to improved values for the entropy, enthalpy and Gibbs free energy of arsenolite, claudetite, arsenic acid and its hydrolysis species, arsenous acid and its hydrolysis species, scorodite and its solubility, and the solubility of selected metal arsenate compounds. The approach taken was to (1) select high quality data based on type of measurement and agreement between different investigators and different techniques, (2) evaluate original sources, and (3) create a thermodynamic network with the most direct paths possible and with minimal auxiliary data. This approach found that thermodynamic properties of arsenolite provided the best anchor to develop a network leading to arsenate aqueous ions that are essential to the evaluation of solubilities for metal arsenates. Very few measurements of metal arsenate and metal arsenite complexes are reported from the literature but by using a consistent chemical model and the same equation of state, several metal arsenate and arsenite complexes have been estimated. Numerous solubilities of metal arsenates have been reported and are compiled but many of these are of variable quality and often in need of better measurements and further evaluation. Scorodite solubility is fairly well characterized with

Table 16. Standard state thermodynamic properties of arsenides and sulfides.

	ΔH_f° (kJ mol^{-1})	S° (J K^{-1} mol^{-1})	C_p (J K^{-1} mol^{-1})	ΔG_f° (kJ mol^{-1})
As$_2$S$_3$, orpiment				
Barton (1969)	−96.2	163.2	—	−96.2
Bryndzia and Kleppa (1988)	−83.0	—	—	—
Johnson et al. (1980)	−91.6 ± 4.8	164.7 ± 4.2	—	−90.7 ± 5.0
O'Hare (1993)	−88.7	—	—	—
Nordstrom and Archer (2003)	−85.8	163.78	163	−84.91
Romanovsky and Tarasov (1960); Tarasov and Zhdanov (1970)	—	163.18	116.2	—
Britzke et al. (1933)	−145.6	—	—	—
Johnson et al. (1980)	−91.6	—	—	−90.7
Mironova et al (1984)	−86.5	—	—	−99.7
Spycher and Reed (1989)	−110.9	—	—	−101.9
Webster (1990)	−139.3	—	—	−96.3
Pankratz et al. (1987)	−91.6	163.18	116.2	−90.3
As$_2$S$_3$, amorphous orpiment				
Eary (1992)	−88.1	—	—	—
Nordstrom and Archer (2003)	−66.9	200	—	−76.8
AsS, realgar (α)				
Weller and Kelley (1964)	—	61.9	—	—
Barton (1969)	−36.4	63.6	—	−35.6
Johnson et al. (1980)	−34.5 ± 0.8	61.9	—	−32.9 ± 1.7
Nordstrom and Archer (2003)	−31.3	62.9	—	−31.8
AsS, realgar (β)				
Weller and Kelley (1964)	—	63.5	—	—
Johnson et al. (1980)	−33.6 ± 0.8	63.5	47.02	−32.45 ± 1.7
Pankrantz et al. (1987)	−33.7	63.5	47.0	−32.4
Robie and Hemingway (1995)	−30.9	63.5	47.02	−29.6
Nordstrom and Archer (2003)	−30.9	63.5	47	−31.0
FeAsS, arsenopyrite				
Barton (1969)	−105.44	108.4	—	−109.62
Pokrovski et al. (2002)	—	68.5 ± 0.9	68.44	−141.6 ± 6.0
FeAs$_2$, loellingite				
Barton (1969)	−43.51	127.2	—	−52.3
Pashinkin et al. (1991)	—	80.06	70.83	—

consistent results at ambient temperatures. Some of the calcium arsenate solubilities are well characterized and show agreement between investigations but inconsistencies still exist and some phases are poorly characterized or not measured at all.

Thermodynamic data are necessary for the construction of pe-pH (or Eh-pH) diagrams that are useful for outlining the equilibrium stability or predominance fields for both minerals

Table 17. Standard state thermodynamic properties of the gaseous species, arsine and tetraarsenic hexaoxide.

Species	ΔG_f° (kJ mol^{-1})	ΔH_f° (kJ mol^{-1})	S° (J K^{-1} mol^{-1})	C_p° (J K^{-1} mol^{-1})	Ref.
AsH$_{3(g)}$	—	68.4	—	—	[1]
AsH$_{3(g)}$	68.93	66.44	222.78	38.07	[2]
As$_4$O$_{6(g)}$	-1092.3 ± 3.8	-1196.2 ± 3.8	409.24 ± 10.46	173.6 ± 4.2	[3]
As$_4$O$_{6(g)}$	-1097.8	-1209.2	381	—	[2]

References: [1] Mayer et al. (1997); [2] Wagman et al. (1982); [3] Behrens and Rosenblatt (1972)

and aqueous species. One example of an aqueous species predominance field mostly consistent with the thermodynamic data in this paper for arsenic follows in Figure 5 (Nordstrom and Archer 2003). The upper and lower limits delineate the equilibrium stability field for water. The four predominance fields for aqueous arsenate species occur in the upper part of the diagram separated by vertical lines at the pK values for each hydrolysis equilibrium (Table 7). The lower part of the diagram portrays the reduced species of arsenite, H$_3$AsO$_3$ and H$_2$AsO$_4^-$, with a vertical boundary reflecting the pK for arsenite hydrolysis. The second pK for arsenite is not shown because it is an uncertain number somewhere between 13 and 14 and barely on the diagram (Sillen and Martell 1964, 1971).

An example of a pe-pH diagram outlining solid phase stability fields follows in Figure 6 (Nordstrom and Archer 2003). This diagram was developed for a total dissolved sulfur concentration of 10^{-4} m and no iron present. With iron present there would be a stability field for scorodite in the upper left-hand corner of the diagram with a maximum pH of around 2.5. At higher

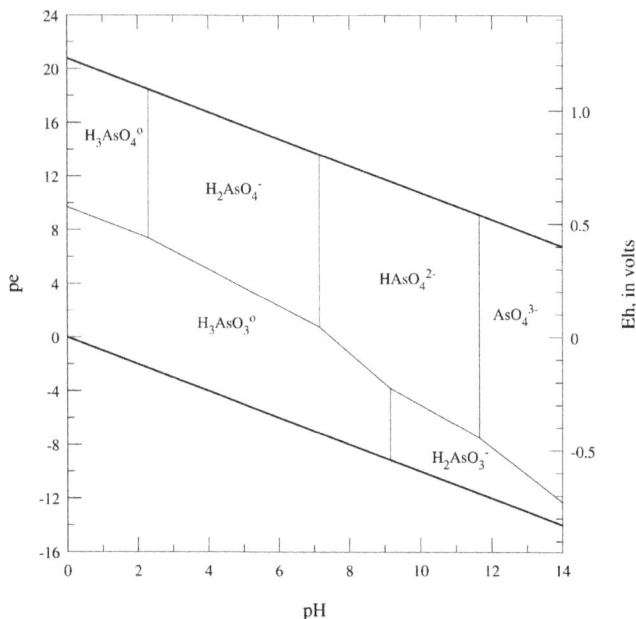

Figure 5. pe-pH diagram for predominant aqueous species of arsenic at equilibrium and 25 °C and 1 atmosphere pressure (Nordstrom and Archer 2003).

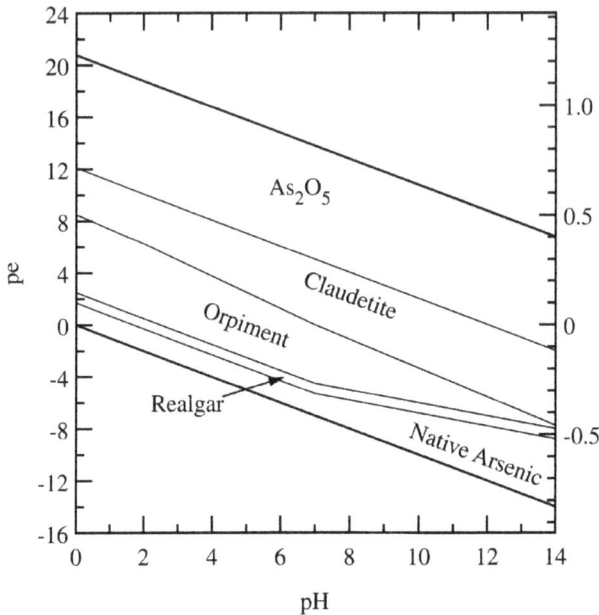

Figure 6. pe-pH diagram for equilibrium mineral stabilities in the As-O-S-H$_2$O system at 25 °C and 1 atmosphere pressure (modified from Nordstrom and Archer 2003). Total dissolved sulfur = 10^{-4} m.

pH values, minerals such as goethite and schwertmannite would be stable. Note that orpiment is stable under more oxidizing conditions than realgar, which is determined by a reaction such as

$$2AsS + HS^- \rightleftharpoons As_2S_3 + 2e^- + H^+ \tag{17}$$

The pe-pH diagram is a useful heuristic tool that is one of many applications of thermodynamics to the understanding of mineral reactions. However, these diagrams are developed for a constant temperature, constant activity (not concentration) of components, void of most complexes, and usually limited to just a few components. If too many components are added then it becomes impossible to show anything meaningful in the diagram. For these reasons it makes little sense to plot water analyses on these diagrams. Water analyses are given in terms of total dissolved concentrations not activities, at variable temperatures, variable compositions, and variable amounts of complexing. Therefore, speciation codes that can take temperature dependence and simultaneous complexing of multicomponent solutions into account are better to interpret water analyses and water-mineral interactions.

Some speciation codes provide the possibility of plotting pe-pH diagrams but users must be aware that if the database in the code holds inconsistent thermodynamic data, these diagrams can be incorrect. During the preparation of this volume arsenic pe-pH diagrams were prepared from The Geochemist's Workbench® which produced unlikely results such as realgar more oxidizing that orpiment and scorodite stable from pH values of 4.5 to 14. Such results can happen if the thermodynamic data is internally inconsistent and inconsistent with equilibrium stoichiometry. Another possibility is that important phases, such as native arsenic or goethite or ferrihydrite, might be left out and that can greatly affect the appearance of stable phases. Databases can have serious errors and those who interpret water-mineral interactions with geochemical codes must be aware of possible sources of errors from thermodynamic data for their particular system under study. Some of the more consistent and reliable data for mineral reactions have been compiled in Nordstrom et al. (1990) and in Appendix D of Nordstrom and Munoz (1994). For other reactions not in these reports, searches have to be done in the literature and among the major compilations of data such as Sillén and Martell (1964, 1971), Wagman et al. (1982), Robie and Hemingway (1995), Garvin et al. (1987), Chase (1998), and Cox et al. (1989).

Future improvements needed in the arsenic system include measurements of metal arsenate and arsenite stability constants, better measurements on metal arsenate mineral solubilities, especially their temperature dependence, and field tests of these equilibria when stability constant data are more robust. As better thermodynamic data becomes available they should be updated in databases of geochemical codes and tested to be sure that they are computing correctly.

ACKNOWLEDGMENTS

The senior author is grateful for the support of the National Research Program of the U.S. Geological Survey. The authors appreciate the careful reviews by Drs. D. Archer, Q. Guo, and F.L. Forray and the careful editing by Dr. C.N. Alpers and K. Lucey. Any use of trade, product, or firm names is for descriptive purposes only and does not imply endorsement by the U.S. Geological Survey. The contribution of J.M. was supported by the project INFLUINS (grant Nr. 03IS2091A) by the German Ministry of Education and Research (BMBF) within the program *Spitzenforschung und Innovation in den Neuen Ländern*. The contribution of E. K. was supported by the Australian Research Council (Linkage project LP130100991) and Rio Tinto Technology and Innovation.

REFERENCES

Agafanova AL, Agafanov IL (1953) Temperature dependence of the dissociation constants of electrolytes. II. First and second dissociation constants of arsenic acid. Zh Fiz Khim 27:1137-1144
Aja SU, Wood SA, Williams-Jones AE (1992) On estimating the thermodynamic properties of silicate minerals. Eur J Mineral 4:1251-1263
Anderson CT (1930) The heat capacities of arsenic, arsenic trioxide and arsenic pentoxide at low temperatures. J Am Chem Soc 52:2296-2300
Anderson E, Story LG (1923) Studies on certain physical properties of arsenic trioxide in water solutions. J Am Chem Soc 45:1102-1105
Antikainen PJ, Rossi VMK (1959) Chelation of arsenious acid with polyols. Suom Kemistil 32B:185-189
Antikainen PJ, Tevanen K (1961) The effect of temperature on the ionization of arsenious acid. Suom Kemistil 34B:3-4
Archer D, Nordstrom DK (2002) Thermodynamic properties of some arsenic compounds of import to groundwater and other applications. J Chem Eng Data (in proofs) (unpublished, censored by NIST)
Baes CF, Mesmer RM (1976) The Hydrolysis of Cations. Wiley-Interscience, New York
Bajda T (2010) Solubility of mimetite $Pb_5(AsO_4)_3Cl$. Environ Chem 7:268-278
Barton PB (1969) Thermochemical study of the system Fe-As-S. Geochim Cosmochim Acta 33:841-857
Beech TA, Lincoln SF (1971) Arsenato complexes of cobalt(III). Aust J Chem 24:1065-1070
Beezer AE, Mortimer CT (1966) Heats of formation and bond energies. XIV. Heat of combustion of arsenic. J Chem Soc:330-332
Beezer AE, Mortimer CT, Tyler, EG (1965) Heats of formation and bond energies. Part XIII. Arsenic, tribromide, arsenious and arsenic oxides, and aqueous solutions of sodium arsenite and sodium arsenate. J Chem Soc 1965:4471-4478
Behrens RG, Rosenblatt GM (1972) Vapor pressure and thermodynamics of octahedral arsenic trioxide (arsenolite). J Chem Thermodynamics 4:175-190
Bertholet PEM (1897) Thermochemie. Gauthier-Villars, Paris
Bethke CM (2008) Geochemical and Biogeochemical Reaction Modeling. 2nd ed., Cambridge University Press, Cambridge, UK
Bevington PR (1969) Data Reduction and Error Analysis for the Physical Sciences. McGraw-Hill, New York

Bjellerup L, Sunner S, Wadsö I (1957) The heat of oxidation of arsenious oxide to arsenic oxide in aqueous solutions. Acta Chem Scand 11:1761-1765

Bluteau M-C, Demopoulos GP (2007) The incongruent dissolution of scorodite: Solubility, kinetics and mechanism. Hydrometall 87:163-177

Born M (1920) Volumen und Hydratationswärme der Ionen. (Volumes and hydration heats of ions) Z Physik 1:45-48

Bothe JV Jr, Brown PW (1999) The stabilities of calcium arsenates at 23 ± 1 °C. J Hazard Mater B69:192-197

Bothe JV Jr, Brown PW (2002) CaO-As$_2$O$_5$-H$_2$O system at 23 ± 1 °C. J Am Ceram Soc 85:221-224

Britton HT, Jackson P (1934) Physiochemical studies of complex formation involving weak acids. X. Complex formation between tartaric acid and (a) arsenic acid, (b) arsenious acid, (c) antimonous hydroxide, in acid and alkaline solutions. The dissociation constants of arsenious and arsenic acids. J Chem Soc 1934:1048-1055

Britzke EV, Kapustinsky AF, Tschenzowa-Moskau LG (1933) Affinity of metals to sulfur. III. Heats of combustion and formation of arsenic sulfides and of the compounds As$_2$O$_3$·As$_2$O$_5$ and As$_2$O$_3$·SO$_3$. Z Anorg Allg Chem 213:58-64

Brown PL, Sylva RN, Ellis J (1985) An equation for predicting the formation constants of hydroxy-metal complexes. J Chem Soc, Dalton Trans 1985:723-730

Bryndzia LT, Kleppa OJ (1988) Standard molar enthalpies of formation of realgar (α-AsS) and orpiment (As$_2$S$_3$) by high temperature direct-synthesis calorimetry. J Chem Thermodyn 20:755-764

Butler JN (1964) Ionic Equilibrium: A Mathermatical Approach. Addison-Wesley, Reading, Massachussetts

Carpeni G, Souchay P (1945) The experimental laws of the variation of pH with dilution. J Chim Phys Phys-Chim Biol 42:149-167

Cernatescu R, Mayer A (1932) Alkali arsenites. Z Physik Chem 160:305-326

Chang SS, Bestul AB (1971) Heat capacities of cubic, monoclinic, and vitreous arsenious oxide from 5 to 360 K. J Chem Phys 55:933-936

Charykova MV, Krivovichev VG, Depmeir W (2010) Thermodynamics of arsenates, selenites, and sulfates in the oxidation zone of sulfide ores: I. Thermodynamic constants at ambient conditions. Geol Ore Dep 52:689-700

Chase MW Jr (1998) NIST-JANAF Thermochemical Tables. Am Chem Soc Am Inst Phys, Woodbury, NY

Chukhlantsev VG (1956a) The solubility products of a number of arsenates. J Anal Chem (USSR) 11:565-571

Chukhlantsev VG (1956b) Solubility products of arsenates. J Inorg Chem (USSR) 1:1975-1982

Chukhlantsev VG (1959) Determination of the dissociation constant of arsenic acid. Zh Fiz Khim 33:3-7

Comba P, Dahnke DR, Twidwell LG (1988) Removal of arsenic from process and wastewater solutions. *In*: Arsenic Metallurgy: Fundamentals and Applications. Reddy GR, Queneau PB (eds) TMS-AIME Symp Proc, Warrendale, Pennsylvania, USA

Cox JD, Wagman DD, Medvedev VA (1989) CODATA Key Values for Thermodynamics. Hemisphere Publ Corp, New York

Davis J (2000) Stability of metal-arsenic solids in drinking water systems. Pract Period Hazard Toxic Radioact Waste Manag 4:31-35

de Carli F (1932) The solubility of sodium gluconate in the presence of sodium phosphate and arsenious acid. Atti Accad Lincei 15:579-583

de Passille A (1936) Ammonium salts of arsenic, phosphoric and antimonic acids and the direct determination of the heats of oxidation of arsenic. Ann Chim 5:83-146

Dove PM, Rimstidt JD (1985) The solubility and stability of scorodite, FeAsO$_4$·2H$_2$O. Am Mineral 70:838-844

Drahota P, Filippi M (2009) Secondary arsenic minerals in the environment: A review. Environ Int 35:1243-1255

Eary LE (1992) The solubility of amorphous As$_2$S$_3$ from 25 to 90 °C. Geochim Cosmochim Acta 56:2267-2280

Eriksson G, Hack K (1990) ChemSage - A computer program for the calculation of complex chemical equilibria. Metall Mater Trans B 21:1013-1023

Essington ME (1988) Solubility of barium arsenate. Soil Sci Soc Am J 52:1566-1570

Flis IE, Mishchenko KP, Tumanova TA (1959) The dissociation of arsenic acid. J Inorg Chem 4:120-125

Flis J, Manecki M, Bajda T (2011) Solubility of pyromorphite Pb$_5$(PO$_4$)$_3$Cl - mimetite Pb$_5$(AsO$_4$)$_3$Cl solid solution series. Geochim Cosmochim Acta 75:1858-1868

Foerster F, Pressprich H (1927) Über das elektromotorische Verhalten von Arsensäure-Arsenigsäurelösungen. (On the electromotive behavior or the arsenic acid – arsenous acid solutions). Z Elektrochem 33:176-181

Forray FL, Smith AML, Navrotsky A, Wright K, Hudson-Edwards KA, Dubbin WE (2014) Synthesis, characterization and thermochemistry of synthetic Pb-As, Pb-Cu and Pb-Zn jarosites. Geochim Cosmochim Acta 127:107-119

Fuoss RM (1958) Ionic association. III. The equilibrium between ion pairs and free ions. J Am Chem Soc 80:5059-5061

Garrett AB, Holmes O, Laube A (1940) The solubility of arsenious oxide in dilute solutions of hydrochloric acid and sodium hydroxide. The character of the ions of trivalent arsenic. Evidence for polymerization of arsenious acid. J Am Chem Soc 62:2024-2028

Garvin D, Parker VB, White HJ Jr (eds) (1987) CODATA Thermodynamic Tables. Selections for Some Compounds of Calcium and Related Mixtures: A Prototype Set of Tables. Hemisphere Publ Corp, New York

Gaskova O, Azaroual M, Piantone P, Lassin A (2001) Arsenic behaviour in subsurface hydrogeochemical systems - a critical review of thermodynamic data for arsenic minerals and aqueous species – a compilation of arsenic surface complexation reactions. BRGM Report 51356-FR

Goldberg RN, Kishore N, Lennen RM (2002) Thermodynamic quantities for the ionization reactions of buffers. J Phys Chem Ref Data 31:231-370

Goldfinger P, von Schweinitz HD (1932) Autoxidation. VII. Absorption spectrum and dissociation constants of arsenious acid. Z Physik Chem B19:219-227

Gorgoraki EA, Tarasov VV (1965) Low temperature heat capacity and certain thermochemical data on the sequioxides of As and Sb. Te Mosk Khim Teckhnol Inst 49:11-15

Grenthe I, Fuger J, Konings RJM, Lemire RJ, Muller AB, Nguyen-Trung C, Wanner H (1992) Chemical Thermodynamics of Uranium. North-Holland, Amsterdam

Gresser MJ, Tracey AS, Parkinson KM (1986) Vanadium(V) oxyanions: The interactions of vanadate with pyrophosphate, phosphate, and arsenate. J Am Chem Soc 108:6229-6234

Hargreaves MK, Stevinson EA (1965) The apparent dissociation constants of various weak acids in mixed aqueous solvents. J Chem Soc:4582-4583

Harned HS, Hamer WJ (1935) The thermodynamics of aqueous sulfuric acid solutions from electromotive force measurements. J Am Chem Soc 57:27-33

Helgeson HC, Kirkham DH, Flowers GC (1981) Theoretical prediction of the thermodynamic behavior of aqueous electrolytes at high pressures and temperatures: IV. Calculation of activity coefficients, osmotic coefficients and relative partial molal properties to 600 °C and 5 kb. Am J Sci 281:1249-1516

Henke K (2009) Arsenic: Environmental Chemistry, Health Threats, and Waste Treatment. Wiley, Chichester, UK

Hodgkin DC, Kruus P (1973) Proton transfer kinetics in medium-strong inorganic acids. Can J Chem 51:2297-2305

Huang YH, Zhu ZQ, Zhang ZL, Zhu YN, Tan LL, Dai LQ, Wei CC (2014) Characterization, dissolution, and solubility of mimetite [$Pb_5(AsO_4)_3Cl$] at 25 °C. Appl Mech Mat 448-453:15-18

Hughes WS (1928) On Haber's glass cell. J Chem Soc:491-506

Hultgren RR, Desai PD, Hawkins DT, Gleiser M, Kelley KK, Wagman DD (1973) Selected Values of the Thermodynamic Properties of the Elements. Am Soc Metals, Metals Park, Ohio

Inegbenebor A, Thomas JH, Williams PA (1989) The chemical stability of mimetite and distribution coefficients for pyromorphite-mimetite solid solutions. Mineral Mag 53:363-371

Ishikawa F, Aoki I (1940) Dissociation constants of arsenious acid and hydroxylamine. Bull Inst Phys Chem Res (Tokyo) 19:136-141

Ivakin AA, Vorob'eva SV, Gertman EM, Voronova EM (1976) Acid-base and self-association in arsenous acid solutions. Russ J Inorg Chem 21:237-240

Johnson GK, Paptheodorou GN, Johnson CE (1980) The enthalpies of formation and high-temperature thermodynamic function of As_4S_4 and As_2S_3. J Chem Thermodyn A-167:545-557

Johnston RB, Singer PC (2007) Solubility of symplesite (ferrous arsenate): Implications for reduced groundwaters and other geochemical environments. Soil Sci Soc Am 71:101-107

Jozefowicz E, Witekowa S, Zubranska W (1950) Solubility of arsenious oxide in aqueous electrolyte solutions. Rozniki Chem 24:64-76

Karutz I, Stranski IN (1957) Über die Verdampfung von Arsenolith und Claudetit. (On the stability of arsenolite and claudetite) Z Anorg Allg Chem 292:330–342

Khodakovsky IL, Ryzhenko BN, Naumov GB (1968) Thermodynamics of aqueous electrolyte solutions at elevated temperatures (Temperature dependence of the heat capacities of ions in aqueous solution). Geokhimiya 12:1486-1503

Khoe GH, Huang JC, Robins RG (1991) Precipitation chemistry of the aqueous ferrous-arsenate system, EPD Congr 91 Proc Symp TMS Ann Mtg, 103-115

Khoe GH, Robins RG (1988) The complexation of iron(III) with sulphate, phosphate, or arsenate ion in sodium nitrate medium at 25°C. J Chem Soc Dalton Trans 1988:2015-2021

Kirschning HJ, Plieth K (1955) Electrochemical determination of the transition point between the cubic and monoclinic modifications of arsenic trioxide. Z Anorg Allg Chem 280:346-352

Kolthoff IM (1920) The significance of dissociation constants in identifying acids and detecting impurities. Pharm Weekblad 57:87-796

Königsberger E, Königsberger L-C, Gamsjäger H (1999) Low-temperature thermodynamic model for the system Na_2CO_3 - $MgCO_3$ - $CaCO_3$ - H_2O. Geochim Cosmochim Acta 63:3105-3119

Kossiakoff A, Harker D (1938) The calculation of the ionization constants of inorganic oxygen acids from their structures. J Am Chem Soc 60:2047-2055

Krause E, Ettel VA (1988) Solubility and stability of scorodite $FeAsO_4 \cdot 2H_2O$: New data and further discussion. Am Mineral 73:850-854

Langmuir D (1979) Techniques of estimating thermodynamic properties for some aqueous complexes of geochemical interest. *In*: Chemical Modeling in Aqueous Systems: Speciation, Sorption, Solubility, and Kinetics. Jenne, EA (ed) Am Chem Soc Symp Ser 93:353-387

Langmuir D, Mahoney J, Rowson J (2006) Solubility products of amorphous ferric arsenate and crystalline scorodite ($FeAsO_4 \cdot 2H_2O$) and their application to arsenic behavior in buried mine tailings. Geochim Cosmochim Acta 70:2942-2956

Latimer WM (1952) Oxidation Potentials. Prentice-Hall, New York

Lee JS, Nriagu JO (2007) Stability constants for metal arsenates. Environ Chem 4:123-133

Lewis GN, Randall M (1923) Thermodynamics. McGraw-Hill, New York

Liebhafsky HA (1931) The reaction between arsenious acid and iodine. J Phys Chem 35:1648-1654

Liu H-L, Zhu Y-N, Yu H-X (2009) Solubility and stability of lead arsenates at 25 °C. J Environ Sci Health A: Tox Hazard Subst Environ Eng 44:1465-1475

Linke WF, Seidell A (1958) Solubilities of Inorganic and Metal-organic Compounds. Vol. I. Am Chem Soc, Washington, DC

Luther R (1907) The dissociation of sulphuric acid and arsenic acid. Z Elektrochim 13:294-297

Mader PM (1958) Kinetics of the hydrogen peroxide-sulfite reaction in alkaline solution. J Am Chem Soc 80:2634-2639

Magalhães MCF, de Jesus JDP, Williams PA (1988) The chemistry of formation of some secondary arsenate minerals of Cu(II), Zn(II) and Pb(II). Mineral Mag 52:679-690

Magalhães MCG, Silva MCM (2003) Stability of lead(II) arsenates. Monatsh Chem 134:735-743

Mahapatra PP, Mahapatra LM, Mishra B (1986) Solubility of calcium hydrogen arsenate in aqueous medium. Ind J Chem A 25:647-649

Majzlan J (2011) Thermodynamic stabilization of hydrous ferric oxide by adsorption of phosphate and arsenate. Environ Sci Technol 45:4726-4732

Majzlan J, Drahota P, Filippi M, Grevel K-D, Kahl W-A, Plasil J, Boerio-Goates J, Woodfield BF (2012a) Thermodynamic properties of scorodite and parascorodite ($FeAsO_4 \cdot 2H_2O$), kaňkite ($FeAsO_4 \cdot 3.5H_2O$), and $FeAsO_4$. Hydrometall 117-118:47-56

Majzlan J, Lazic B, Armbruster T, Johnson MB, White MA, Fisher RA, Plasil J, Skoda R, Novak M (2012b) Crystal structure, thermodynamic properties, and paragenesis of bukovskyite, $Fe_2(AsO_4)(SO_4)(OH) \cdot 9H_2O$. J Mineral Petrol Sci (Japan) 107:133-148

Majzlan J, Drahota P, Filippi M (2014) Parageneses and crystal chemistry of arsenic minerals. Rev Mineral Geochem 79:17-184

Marini L, Accornero M (2007) Prediction of the thermodynamic properties of metal-arsenate and metal-arsenite aqueous complexes to high temperatures and pressures and some geological consequences. Environ Geol 52:1343-1363

Mayer PM, Gal J-F, Radom L (1997) The heats of formation, gas-phase acidities, and related thermochemical properties of the third-row hydrides AsH_3, GeH_3, SeH_2 and HBr from G2 ab initio calculations. Int J Mass Spec Ion Processes 167/168:689-696

Minor JI, Gilliland AA, Wagman DD (1971) The enthalpies of hydrolysis of arsenious oxide and arsenic trichloride in 1 N sodium hydroxide: The enthalpy of arsenic trichloride. Natl Bur Stand Report 10-860

Mironov VE, Kiselev VP, Egizaryan MB, Golovnev NN, Pashkov GL (1995) Ion association in aqueous solutions of calcium arsenate. Russ J Inorg Chem 40:1690

Mironova GD, Zotov AV, Gul'ko NV (1984) Determination of the solubility of orpiment in acid solutions at 25-150 °C. Geochem Int 21:53-59

Nelson HF, Shchukarev A, Sjöberg S, Lövgren L (2011) Composition and solubility of precipitated copper(II) arsenates. Appl Geochem 26:696-704

Nishida S, Kimura M (1989) Kinetic studies of the oxidation reaction of arsenic(III) to arsenic(V) by peroxydisulfate ion in aqueous alkaline media. J Chem Soc Dalton Trans 1989(2):357-360

Nishimura M, Ito C, Tozawa K (1988) Stabilities and solubilities of metal arsenites and arsenates in water and effect of sulfate and carbonate ions on their solubility. *In*: Arsenic Metallurgy Fundamentals and Applications. Reddy RG, Queneau RB (ed) Metallurgical Society, p 77-98

Nishimura T, Robins RG (1996) Crystalline phases in the system Fe(III)-As(V)-H_2O at 25 °C. *In*: Iron Control and Disposal. Dutrizac JE, Harris GB (eds) Canadian Institute of Mining, Metallurgy and Petroleum, p 521-534

Nishimura T, Robins RG (1998) A re-evaluation of the solubility and stability regions of calcium arsenites and arsenates in aqueous solution at 25°C. Mineral Process Extr Metall Rev 18:283-308

Nordstrom DK (2000) Thermodynamic properties of environmental arsenic species: Limitations and needs. *In*: Minor Elements 2000. Young C (ed) Society for Mining, Metallurgy, and Exploration, Littleton, CO, p 325-331

Nordstrom DK (2002) Worldwide occurrences of arsenic in groundwater. Science 296:2143-2145

Nordstrom DK, Archer D (2003) Arsenic thermodynamic data and environmental geochemistry. *In*: Arsenic in Ground Water. Welch AH, Stollenwerk KG (eds) Kluwer Academic Publishers, Amsterdam, p 1-25

Nordstrom DK, Munoz JL (1994) Geochemical Thermodynamics. Blackwell Scientific Publications, Boston

Nordstrom DK, Plummer LN, Langmuir D, Busenberg E, May HM, Jones BF, Parkhurst DL (1990) Revised chemical equilibrium data for major water-mineral reactions and their limitations. *In*: Chemical Modeling of Aqueous Systems II. Melchior DC and Bassett RL (eds.) Am Chem Series 416, Washington, DC, p 398–413

O'Hare PAG (1993) Calorimetric measurements of the specific energies of reaction of arsenic and of selenium with fluorine. Standard molar enthalpies of formation at the temperature 298.15 K of AsF_5, SeF_6, As_2Se_3, and As_2S_3. Thermodynamic properties of AsF_5 and SeF_6 in the ideal-gas state. Critical assessment of $\Delta H_f^\circ{}_m$ (AsF_3, l) and the dissociation enthalpies of As-F bonds. J Chem Thermodyn 25:391-402

Oelkers EH, Schott J (eds) (2009) Thermodynamics and Kinetics of Water-Rock Interaction. Reviews in Mineralogy and Geochemistry, Volume 70. Mineralogical Society of America, Chantilly VA

Oelkers EH, Benezeth P, Pokrovski GS (2009) Thermodynamic databases for water-rock interaction. *In*: Thermodynamics and Kinetics of Water-Rock Interaction. Oelkers EH, Schott J (eds) Rev Mineral Geochem 70:1-46

Omarova FM, Sharipov MS, Shaikhina LS (1980) Possibility for the calorimetric determination of the dissociation constant of multibase acids, for example arsenic acid. Deposited document VINITI 5358-80

Orellana F, Ahumada E, Suarez C, Cote G, Lizama H (2000) Thermodynamics studies of parameters involved in the formation of arsenic(V) precipitates with barium(II). Bol Soc Chil Quim 45:415-422

Paktunc D, Dutrizac J, Gertsman V (2008) Synthesis and phase transformations involving scorodite, ferric arsenate and arsenical ferrihydrite: Implications for arsenic mobility. Geochim Cosmochim Acta 72:2649-2672

Pankratz LB, Mah AD, Watson SW (1987) Thermodynamic properties of sulfides. US Bur Mines Bull 689

Pantuzzo FL, Santos LRG, Ciminellie VST (2014) Solubility-product constant of an amorphous aluminum-arsenate phase ($AlAsO_4 \cdot 3.5H_2O$) at 25°C. Hydrometall 144-145:63-68

Parker VB, Khodakovsky IL (1995) Thermodynamic properties of the aqueous ions (2+ and 3+) of iron and the key compounds of iron. J Phys Chem Ref Data 24:1699-1745

Pashinkin AS, Muratova VA, Moiseyev NV, Bazhenov JV (1991) Heat capacity and thermodynamic functions of iron diarsenide in the temperature range 5 K to 300 K. J Chem Thermodyn 23:827-830

Perfetti E, Pokrovski G, Ballerat-Busserolles K, Majer V, Gibert F (2008) Densities and heat capacities of aqueous arsenious and arsenic acid solutions to 350 °C and 300 bar, and revised thermodynamic properties of $As(OH)_3{}^\circ{}_{(aq)}$, $AsO(OH)_3{}^\circ{}_{(aq)}$ and iron sulfarsenide minerals. Geochim Cosmochim Acta 72:713-731

Perrin DD (1982) Ionisation Constants of Inorganic Acids and Based in Aqueous Solution. Pergamon Press, Oxford, UK

Pesavento M (1989) Potentiometric determination of the standard potential of the As(V)/As(III) couple. Talanta 36:1059-1063

Pokrovski G, Gout R, Schott J, Zotov A, Harrichoury JC (1996) Thermodynamic properties and stoichiometry of As(III) hydroxide complexes at hydrothermal conditions. Geochim Cosmochim Acta 60:737-749

Pokrovski G, Zakirov IV, Roux J, Testemale D, Hazeman J-L, Bychkov A Yu, Golokova GV (2002) Experimental study of arsenic speciation in vapor phase to 500 °C: Implications for As transport and fractionation in low-density crustal fluids and volcanic gases. Geochim Cosmochim Acta 66:3453–3480

Randall M (1930) Free energy of chemical substances, activity coefficients, partial molal quantities, and related constants. *In*: International Critical Tables of Numerical Data, Physics, Chemistry, and Technology. Washburn EW, West CJ, Dorsey NE (eds) McGraw-Hill, New York, p 224-313

Raposo JC, Sanz J, Zuloaga O, Olazabal MA, Madariaga JM (2002) The thermodynamic model of inorganic arsenic species in aqueous solutions: Potentiometric study of the hydrolitic equilibrium of arsenic acid. Talanta 57:849-857

Raposo JC, Sanz J, Zuloaga O, Olazabal MA, Madariaga JM (2003) Thermodynamic model of inorganic arsenic species in aqueous solutions. Potentiometric study of the hydrolytic equilibrium of arsenious acid. J Solution Chem 32:253-264

Raposo JC, Zuloaga O, Olazabal MA, Madariaga JM (2004) Study of the precipitation equilibria of arsenate anion with calcium and magnesium in sodium perchlorate at 25 °C. Appl Geochem 19:855-862

Raposo JC, Olazabal MA, Madariaga JM (2006) Complexation and precipitation of arsenate and iron species in sodium perchlorate solutions at 25 °C. J Solution Chem 35:79-94

Ravenscroft P, Brammer H, Richards K (2009) Arsenic Pollution. Wiley-Blackwell, Chichester, UK

Ricci JE (1948) The aqueous ionization constants of inorganic oxygen acids. J Am Chem Soc 70:109-113

Robie RA, Hemingway B (1995) Thermodynamic properties of minerals and related substances at 298.15 K and 1 bar (10^5 pascals) pressure and higher temperatures. US Geol Survey Bull 2131

Robins RG (1985) The solubility of barium arsenate: Sherritt's barium arsenate process. Metall Trans 16B:404-406

Robins RG (1987) Solubility and stability of scorodite, $FeAsO_4 \cdot 2H_2O$: discussion. Am Mineral 72:842-844

Robins RG (1990) The stability and solubility of ferric arsenate. Proc Extr Proc Div, Miner Metals Mat Soc p 93-104

Rodriguez-Blanco (2007) Oriented overgrowth of pharmacolite ($CaHAsO_4 \cdot 2H_2O$) on gypsum ($CaSO_4 \cdot 2H_2O$). Crystal Growth Design 7:2756-2763

Roebuck JR (1905) The rate of reaction between arsenious acid and iodine in acid solution; the rate of the reverse reaction; and the equilibrium between them. J Phys Chem 9:727-763

Romanovsky VA, Tarasov VV (1960) Heat capacity of the trisulfides of arsenic, antimony, and bismuth in connection with their structural and physical-chemical properties. Sov Phys, Solid State 2:1170-1175

Rossini FD, Wagman DD, Evans WH, Levine S, Jaffe I (1952) Selected values of chemical thermodynamic properties. Natl Bur Stand US Circ 500

Sadiq M (1997) Arsenic chemistry in soils: An overview of thermodynamic predictions and field observations. Water Air Soil Poll 93:117-136

Salomaa P, Hakala R, Vesala S, Aalto T (1969) Solvent deuterium isotope effects of acid-base reactions. III. Relative acidity constants of inorganic oxyacids in light and heavy water, kinetic applications. Acta Chem Scand 23:2116-2126

Salomaa P, Schaleger LL, Long FA (1964) Solvent deuterium isotope effects on acid-base equilibria. J Am Chem Soc 86:1-7

Schindler P, Althaus H, Hofer F, Minder W (1965) Solubility product of metal oxides and hydroxides. 10. Solubility product of zinc oxide, copper hydroxide and copper oxide in relation to grain size and molar surface area. A contribution to the thermodynamics of the solid-liquid interface. Helv Chim Acta 48:1204-1215 {German}

Schreinemakers FAH, de Baat WC (1917) Compounds of arsenious oxide and salts. I. Chem Weekblad 14:141-146

Schuhmann R (1924) Free energy and heat content of arsenic trioxide and the reduction potential of arsenic. J Am Chem Soc 46:1444-1449

Schulman JH, Schumb WC (1943) The polymorphism of arsenious oxide. J Am Chem Soc 65:878-883

Secco F, Indelli A, Bonora PL (1970) Kinetic study of the reaction of arsenic acid with iodide ions. Inorg Chem 9:337-342

Sellers PS, Sunner S, Wadso I (1964) Heats of ionization of arsenious and arsenic acids. Acta Chem Scand 18:202-206

Shan J (2008) Stabilization of arsenic in iron-rich residuals by crystallization to a stable phase of arsenic mineral. PhD Dissertation, University of Arizona, Tucson, Arizona

Shock EL, Helgeson HC (1988) Calculation of the thermodynamic and transport properties of aqueous species at high pressures and temperatures: Correlation algorithms for ionic species and equation of state predictions to 5 kb and 1000°C. Geochim Cosmochim Acta 52:2009-2036

Sillén LG, Martell AE (1964) Stability Constants of Metal-Ion Complexes. Spec Pub No 17, The Chemical Society, London

Sillén LG, Martell AE (1971) Stability Constants of Metal-Ion Complexes. Supplement No 1, Spec Pub No 25, The Chemical Society, London

Smith RW, Popp CJ, Norman DI (1986) The dissociation of oxy-acids at elevated temperatures. Geochim Cosmochim Acta 50:137-142

Spycher NF, Reed MH (1989) As(III) and Sb(III) sulfide complexes: An evaluation of stoichiometry and stability from existing experimental data. Geochim Cosmochim Acta 53:2185-2194

Stefansson A (2007) Iron(III) hydrolysis and solubility at 25 °C. Environ Sci Technol 41:6117-6123

Stranski IN, Plieth K, Zoll I (1958) Über die Auflösung, die Löslichkeit und die Umwandlung der beiden Arsenik-Modifikationen in Wasser und wässrigen Lösungen (On the dissolution, solubility and transitions of both arsenic modifications arsenolite and claudetite in water and aqueous solutions). Z Elektrochem 62:366-372

Suess E, Planer-Friedrich B (2012) Thioarsenate formation upon dissolution of orpiment and arsenopyrite. Chemosph 89:1390-1398

Sunner S, Thorén SA (1964) The heat of oxidation of aqueous arsenious oxide with gaseous chlorine. Acta Chem Scand 18:1528-1532

Tarasov VV, Zhdanov VM (1970) Low-temperature heat capacity of vitreous arenic trisulphide. Zh Fiz Khim 44:2384-2387

Thomsen JJ (1882-1885) Thermochemische Untersuchungen (Thermochemical Investigations). Barth, Leipzig

Tossidis I (1976) About the dissociation of phosphoric and arsenic acids in water-organic mixed solutions. Inorg Nucl Chem Letters 12:609-615

Tourky AR, Mousa AA (1949) Studies on some metal electrodes. VIII. The bearing of the properties of arsenious oxide on the behavior of the arsenic electrode. J Chem Soc:1305-1308

Tozawa K, Umetsu Y, Nishimura T (1978) Hydrometallurgical recovery or removal of arsenic from copper smelter by-product. 107th AIME Meeting Am Inst Min Eng, Denver, CO

Vesala A, Saloma E (1977) Determination of basicity constants by potentiometric titration. Finn Chem Lett 160-163

Viswanathan K, Miehe G (1978) The crystal structure of low temperature $Pb_3(AsO_4)_2$. Z Kristallogr 148:275-280

von Hodenberg R (1974) Zur den Modifikationen im System $Pb_3(PO_4)_2$-$Pb_3(VO_4)_2$-$Pb_3(AsO_4)_2$ (On modifications in the system $Pb_3(PO_4)_2$-$Pb_3(VO_4)_2$-$Pb_3(AsO_4)_2$). Ber Deut Keram Gesell 51:64-68

Wagman DD, Evans WH, Parker VB, Schuman PH, Halow I, Bailey SM, Churney KL, Nuttall RL (1982) The NBS tables of chemical thermodynamic properties: selected values for inorganic and C_1 and C_2 organic substances in SI units. J Phys Chem Ref Data 11:1-392

Walden P (1888) Über die Bestimmung der Molekulargrössen von Salzen aus der elektrischen Leitfähigkeit ihrer wässerigen Lösungen (On the determination of the size of molecules of salts from the conductivity of their aqueous solutions). Z Physik Chemie 2:49-77

Washburn EW, Strachan EK (1913) The laws of "concentrated" solutions. V. Part I: The equilibrium between arsenious acid and iodine in aqueous solution; Part II: A general law for chemical equilibrium in solutions containing ions; Part III: The energetics of the reaction between arsenious acid and iodine. J Am Chem Soc 35:681-714

Wauchope D (1976) Acid dissociation constants of arsenic acid, methylarsonic acid (MAA), dimethylarsinic acid (cacodylic acid), and *N*-(phosphonomethyl)glycine (glyphosate). J Agric Food Chem 24:717-721

Webster JG (1990) The solubility of As_2S_3 and speciation of As in dilute and sulphide-bearing fluids at 25 and 90 °C. Geochim Cosmochim Acta 54:1009-1017

Weiser ME, Coplen TB (2011) Atomic weights of the elements 2009 (IUPAC Technical Report). Pure Appl Chem 83:359-396

Welham NJ, Malatt KA, Vukcevic S (2000) The stability of iron phases presently used for disposal from metallurgical systems – a review. Miner Eng 13:911-931

Weller WW, Kelley KK (1964) Low-temperature heat capacities and entropies at 298.15 K of sulfides of arsenic, germanium, and nickel. US Bur Mines Rept Invest 6511

Whiting KS (1992) The thermodynamics and geochemistry of arsenic, with application ot subsurface waters at the Sharon Steel superfund site at Midvale, Utah. PhD Dissertaion, Colorado School of Mines, Golden, Colorado

Wood JK (1908) Amphoteric metallic hydroxides. J Chem Soc 93:411-423

Yamazaki H, Sperline RP, Freiser H (1993) Spectrophotometric determination of the dissociation constant (pK_a) of arsenous acid. Anal Chim Acta 284:379-384

Zakaznova-Herzog VP, Seward TM, Suleimenov OM (2006) Arsenous acid ionisation in aqueous solutions from 25 to 300°C. Geochim Cosmochim Acta 70:1928-1938

Zawidski J (1903) Zur Kenntnis der arsenigen Säure (On understanding arsenous acid). Ber Deut Chem Gesell 36:1427-1436

Zhu C, Anderson GM (2002) Environmental Applications of Geochemical Modelling, Cambridge University Press, Cambridge, UK

Zhu Y, Zhang X, Xie Q, Chen Y, Wang D, Liang Y, Lu J (2005) Solubility and stability of barium arsenate and barium hydrogen arsenate at 25 °C. J Haz Mat A120:37-44

Zhu YN, Zhang XH, Xie QL, Wang DQ, Cheng GW (2006) Solubility and stability of calcium arsenates at 25 °C. Water Air Soil Poll 169:221–238

Zhu Y, Zhang X, Zeng H, Liu H, He N, Qian M (2011) Characterization, dissolution, and solubility of synthetic svabite [$Ca_5(AsO_4)_3F$] at 25-45 °C. Environ Chem Let 9:339-345

Zhu Y, Merkel BJ (2001) The dissolution and solubility of scorodite, $FeAsO_4 \cdot 2H_2O$. Evaluation and simulation with PHREEQC2. Wiss Mitt Inst Geol TU Bergakedemie Freiberg, Germany 18:1-12

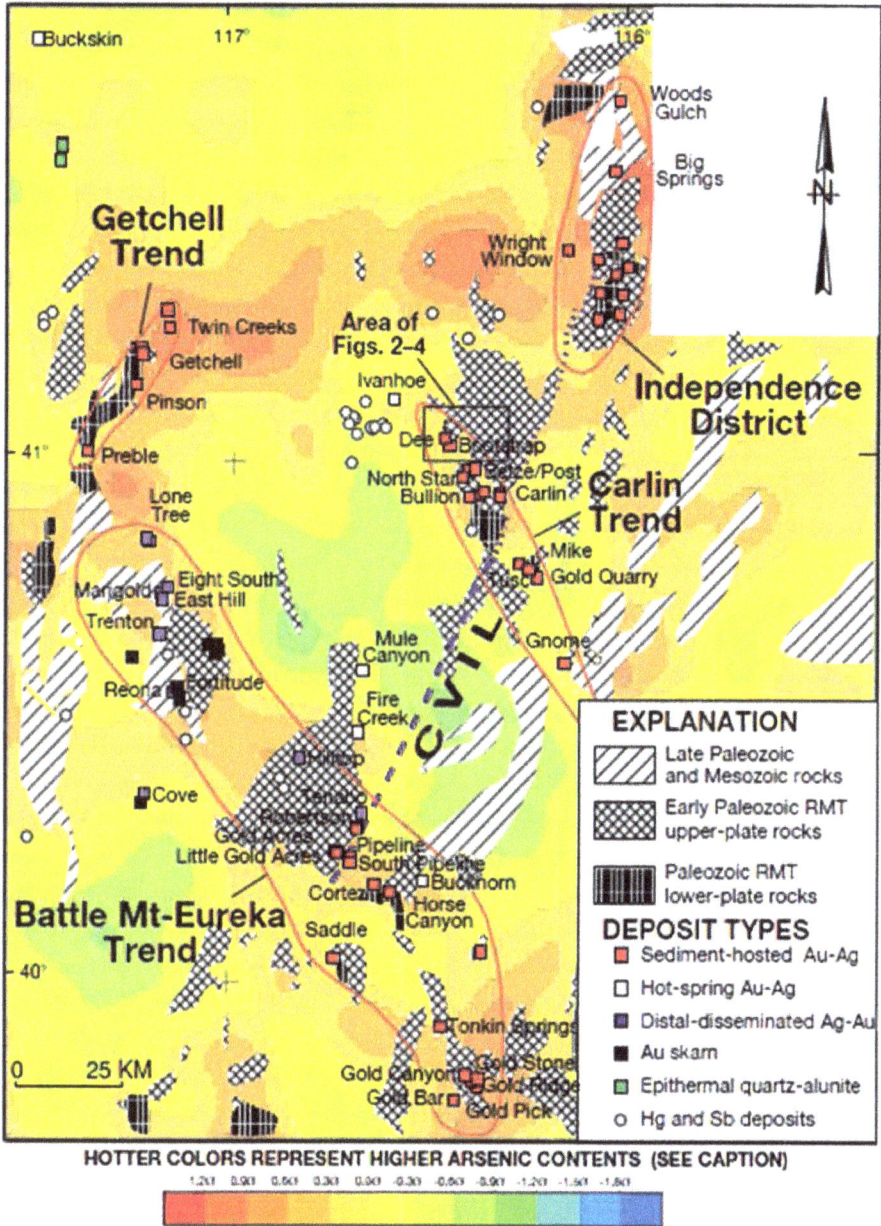

Plate 1. (*see Fig. 1 and associated text in Chapter 1*) Index map showing relationship of alluvium As concentrations in northernmost Carlin trend to bedrock, Crescent Valley-Independence Lineament (CVIL), and major mineral deposits in north-central Nevada (Theodore et al. 2003). [Reproduced with permission of the Society of Economic Geologists from Theodore et al. (2003), *Econ Geol*, Vol. 98, Fig. 1, p. 288.]

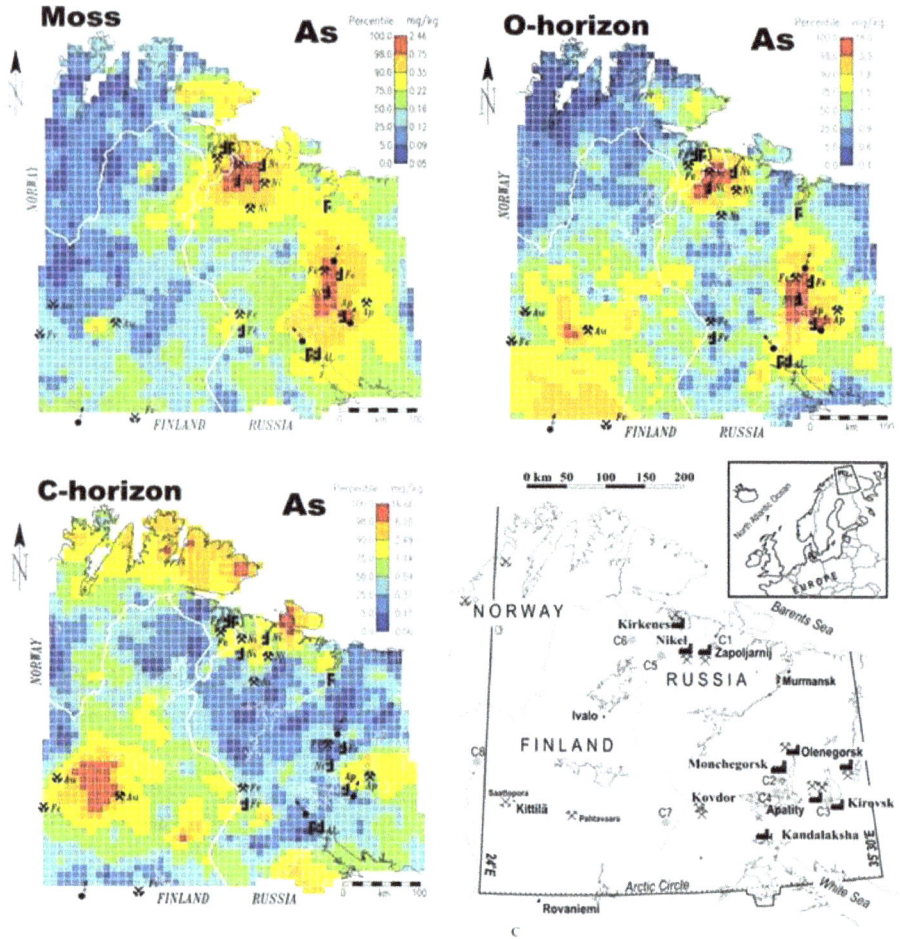

Plate 2. (*see Fig. 2 and associated text in Chapter 1*) Regional distribution of As in moss and O- and C-horizon soil samples in the Kola Project area. [Reproduced with permission of Elsevier from Reimann et al. (2009) *Appl Geochem*, Vol. 24, Fig. 9, p. 1158.]

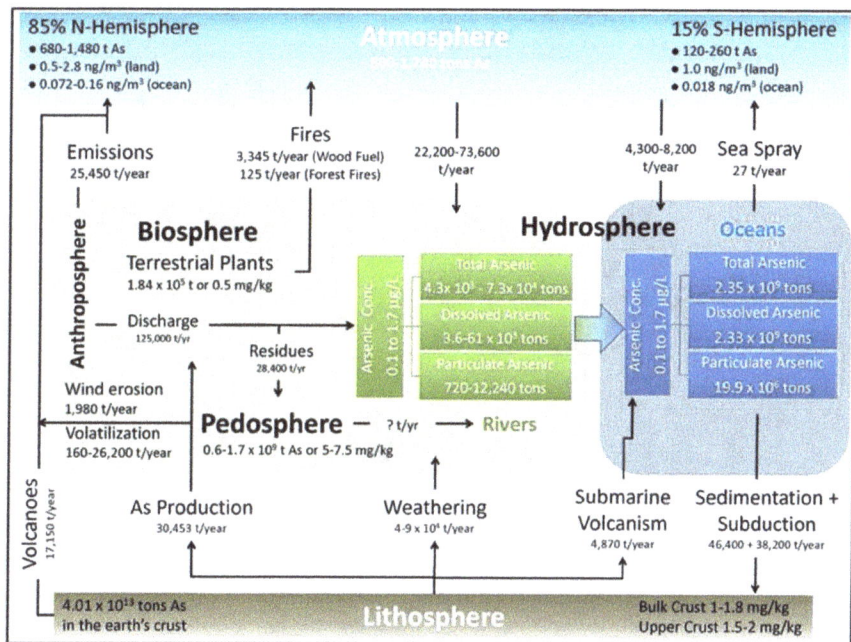

Plate 3. (*see Fig. 3 and associated text in Chapter 1*) Global arsenic cycle. Redrawn and simplified from Matschullat (2000), Zhu et al (2014).

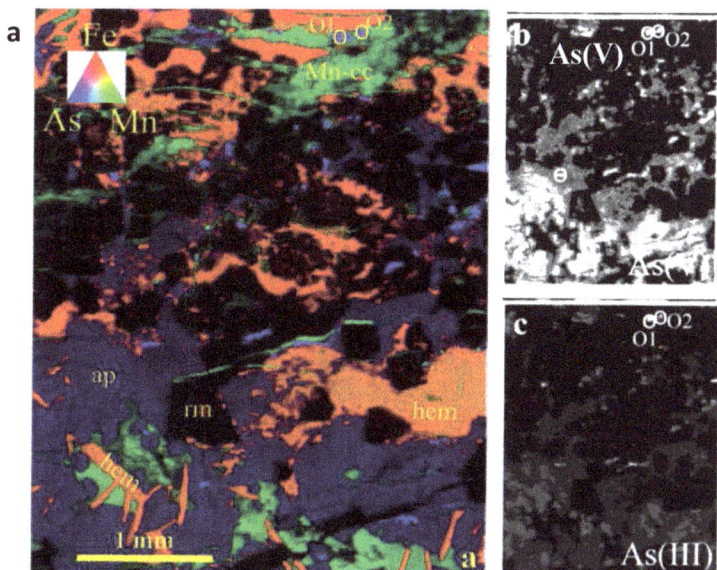

Plate 4. (*see Fig. 15 and associated text in Chapter 5*) Micro-XRF mapping of Fe-Mn ore sample from Val Ferrera sedimentary exhalative deposit, Switzerland. (a) 2.1 million pixel RGB image (3.3×4 mm^2) shows As speciation as substitution in apatite (ap = apatite, Mn-cc = Mn-calcite, hem = hematite, rm=romeite). Grayscale oxidation state maps of As^{5+}(b) and As^{3+}(c) reveal small amounts of As^{3+} in this highly oxidized system. From Etschmann et al. (2010).

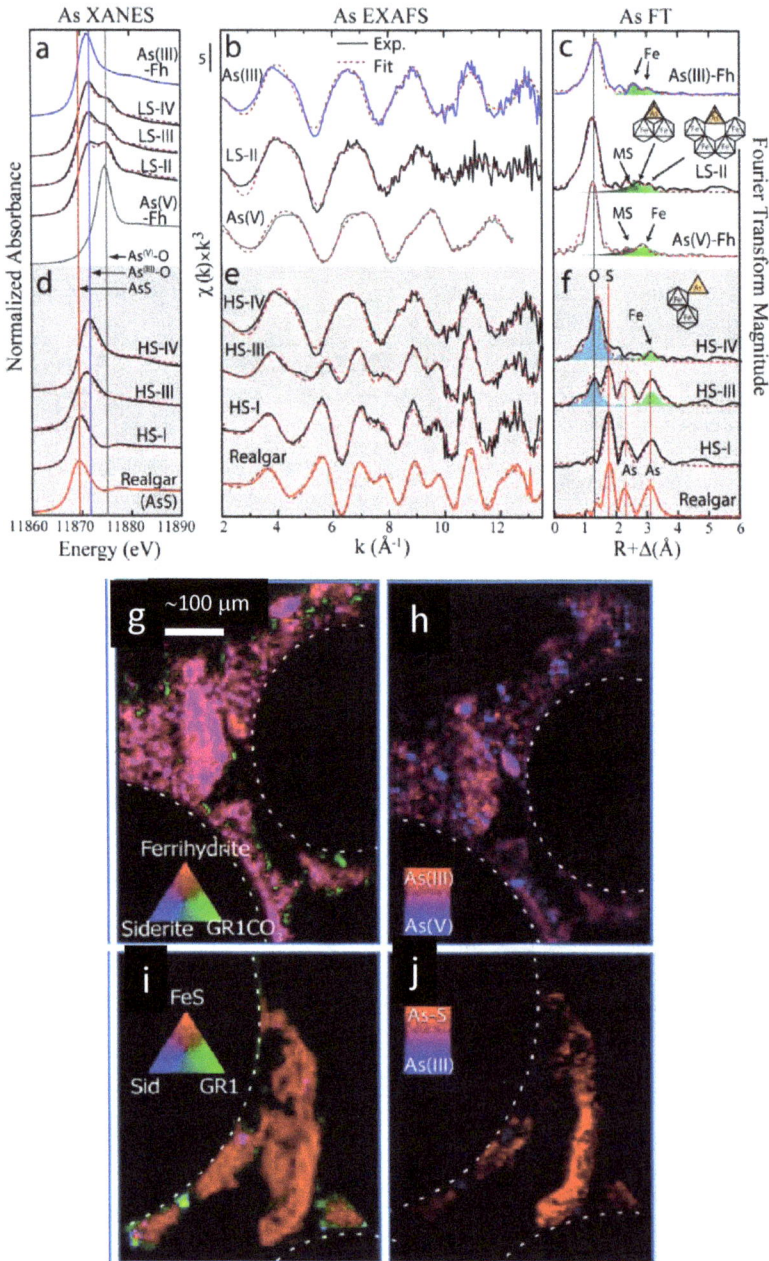

Plate 5. (*see Fig. 14 and associated text in Chapter 5*) (a-f): Results of bulk As-XAS analysis of low-SO$_4$ (LS) and high-SO$_4$ (HS) columns containing As^{5+}ferrihydrite and inoculated with a natural, anoxic Fe-reducing microbial consortium (realgar phase forms in HS experiments as shown in panel f); (g-j): qualitative µ-XRF speciation maps of Fe (g, i) and As (h, j) in sediment from LS column (g, h) and HS column (i, j). Dotted circles outline glass beads. [Used by permission of American Chemical Society, from Root et al. (2013) *Environmental Science and Technology*, Vol 47, Fig. 1, p. 12994 and Fig. 3, p. 12996]

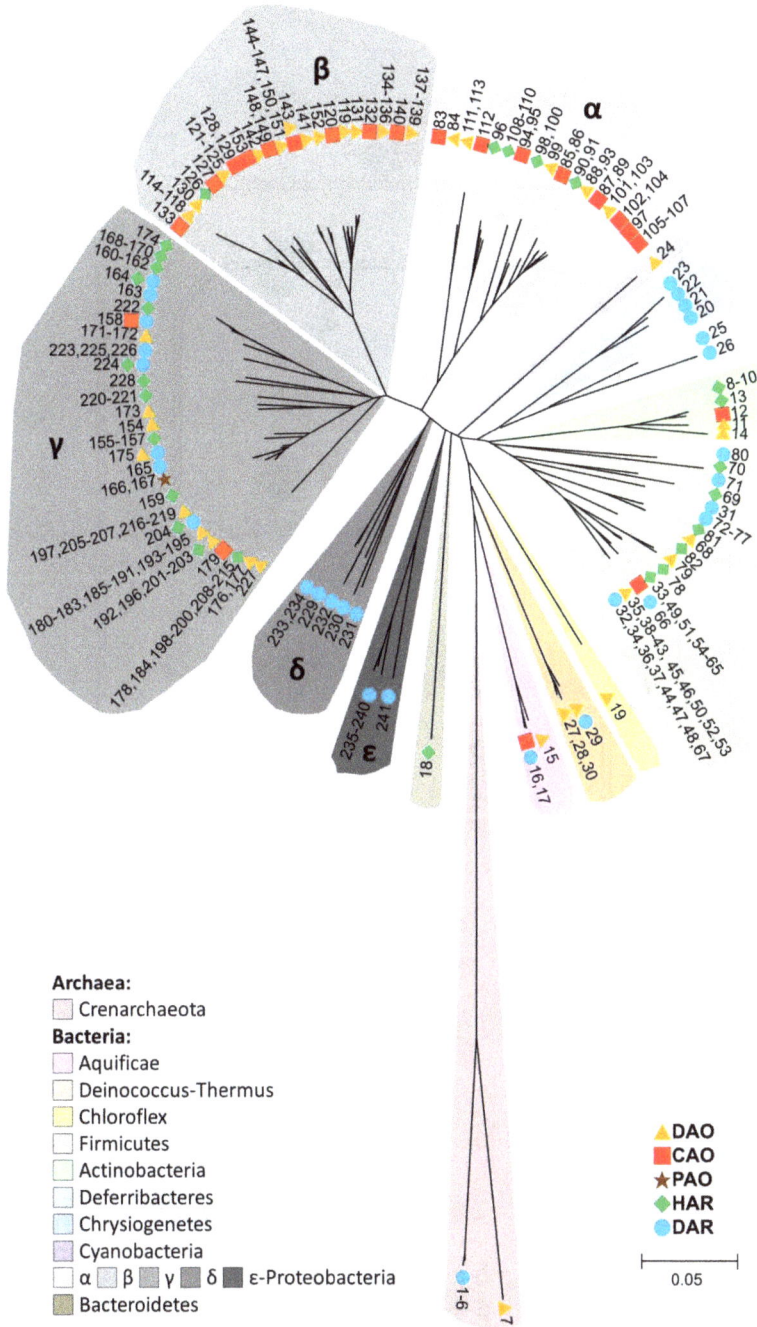

Plate 6. (*see Fig. 1 and associated text in Chapter 7*) Phylogenetic tree based on 16S rRNA gene sequences of cultivable Archaea and Bacteria that are known to transform of arsenic. Detoxifying arsenite oxidizers (DAOs) are indicated by gold triangles, chemolithotrophic arsenite oxidizers (CAOs) by red squares, phototrophic arsenite oxidizers (PAO) by brown stars, heterotrophic arsenate reducers (HARs) by green diamonds, and dissimilatory arsenate reducers (DARPs) by blue circles. The scale bar = substitution of 5% divergence.

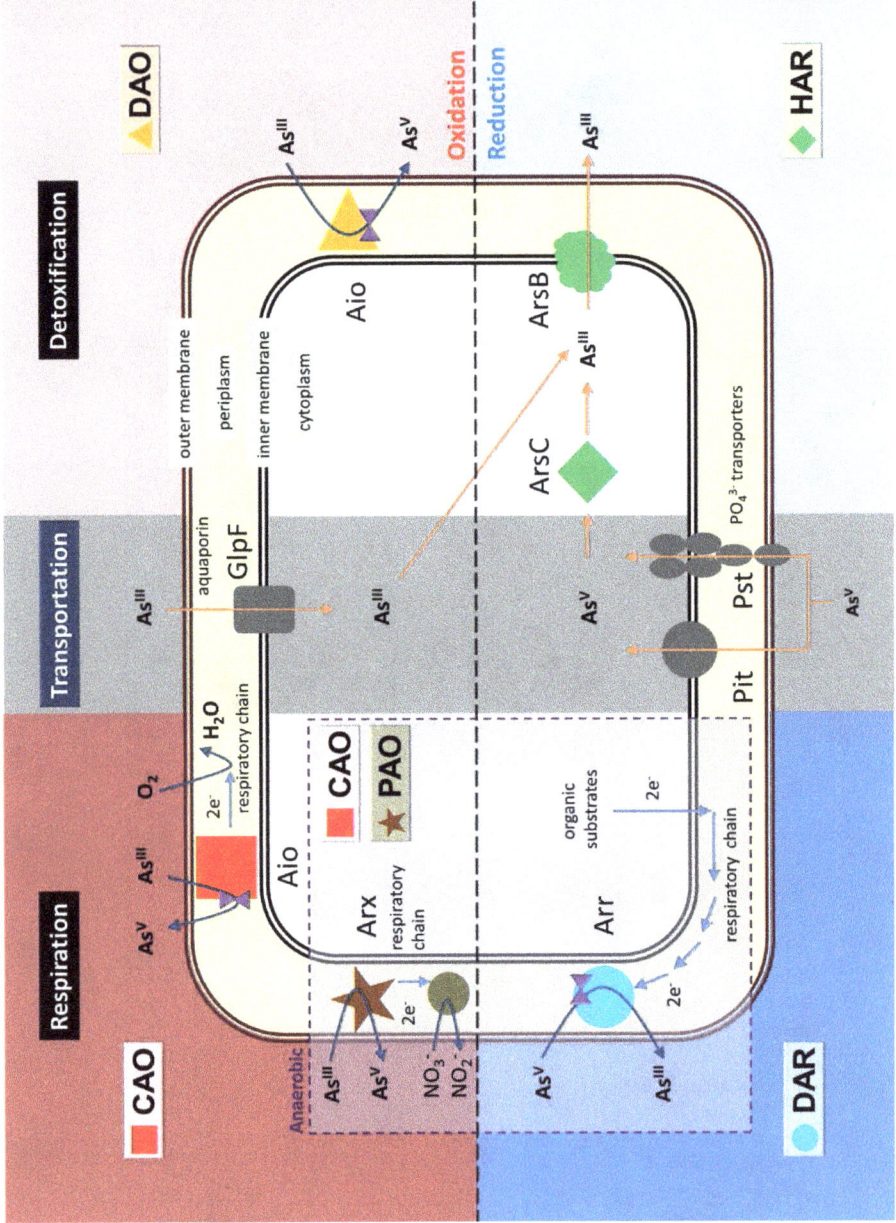

Plate 7. (*see Fig. 2 and associated text in Chapter 7*) Schematic of inorganic arsenic respiration (*left*), transportation (*center*), and detoxification (*right*) in a composite cell of CAOs, DAOs, HARs, DARs, and PAOs, with oxidative processes in red tones (*top*) and reductive processes in blue tones (*bottom*). Modified from Paez-Espino et al. (2009).

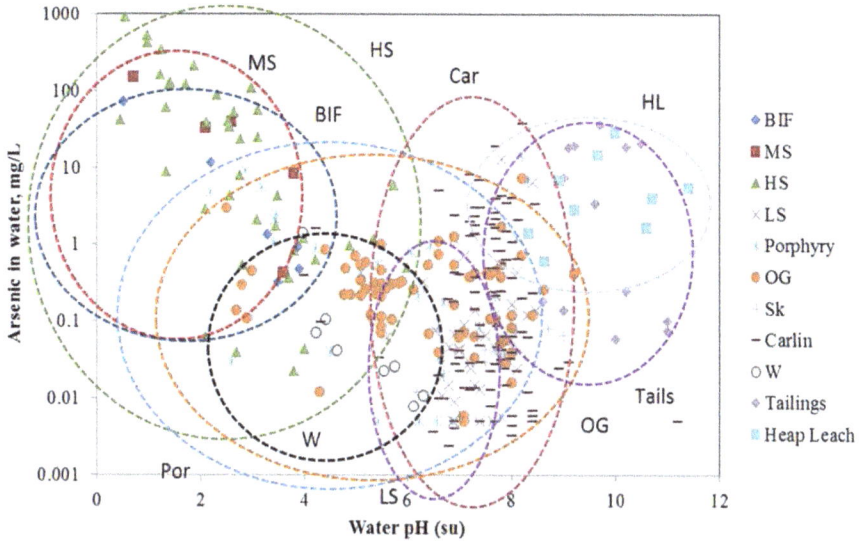

Plate 8. (*see Fig. 5 and associated text in Chapter 10*) Arsenic concentration versus pH in gold mine waters from published studies. Key: BIF – banded Fe formation; MS – massive sulfide; W – Witwatersrand; Carlin - Carlin type; HS – high-sulfidation epithermal; LS – low-sulfidation epithermal; POR – gold-rich porphyries; Sk – skarn-hosted gold deposits; OG – orogenic or shear zone hosted gold deposits; TCN – cyanide tailings or heap leach facilities (data sourced from; Bowell et al. 1994; Azcue and Nriagu 1995; Grimes et al. 1995; Schwartz 1995; Leblanc et al. 1996; Smedley et al. 1996; Odor et al. 1998; Bennett and Tempel 2000; Bowell et al. 2000, 2003; Savage et al. 2000a,b; Stichbury et al. 2000; Welch et al. 2000; Bowell 2001, 2002; Williams 2001; Lazareva et al. 2002; Wong et al. 2002; Loredo et al. 2003; Craw et al. 2004; Oyarzun et al. 2004; Bowell and Parshley 2005; Haffert and Craw 2008b; Tutu et al. 2008; Bowell et al. 2009; Cidu et al. 2009; Coetzee et al. 2010; Warrender et al. 2012; Drahota et al. 2013).

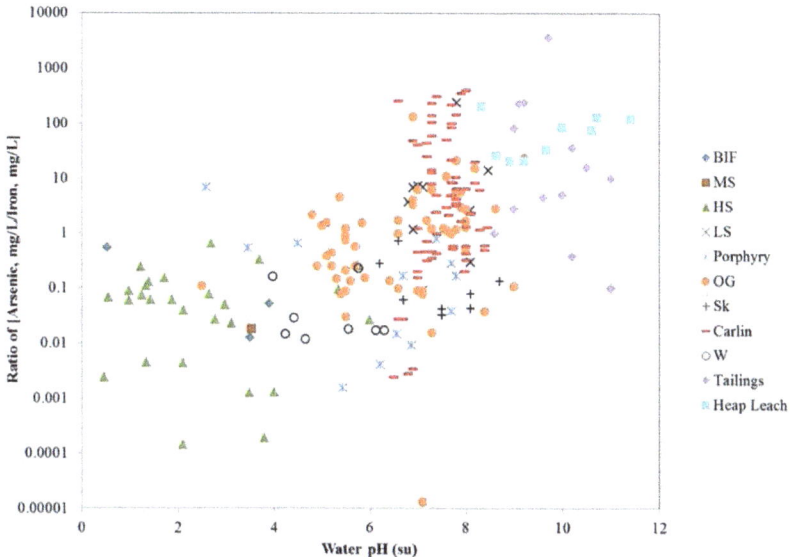

Plate 9. (*see Fig. 6 and associated text in Chapter 10*) Correlation of pH to As/Fe ratio in mine waters. Same data sources and key as Plate 8.

Plate 10. *(see Fig. 2 and associated text in Chapter 12)* Air photo of the Giant Mine property ca. 2000. Data show total As concentration in outcrop soil samples from Bromstad (2011). The industrial remediation guideline for Giant mine NWT is 340 mg kg^{-1}.

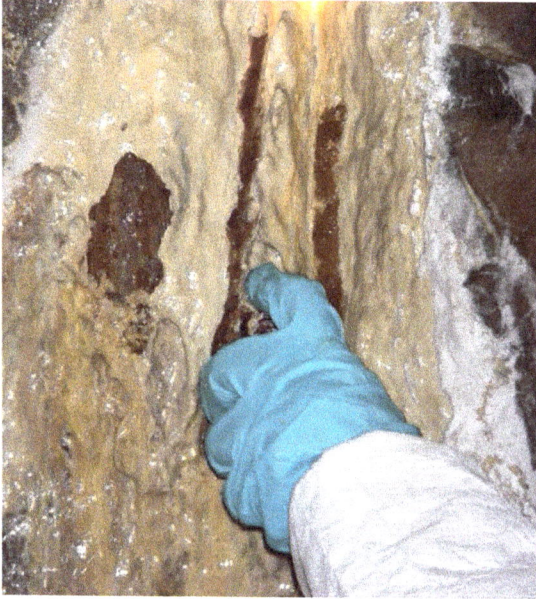

Plate 11. (*see Fig. 5 and associated text in Chapter 12*) Biofilm developed on underground stope wall where As-rich water drips from an drillhole. Mineral precipitate is yukonite $Ca_7Fe_{12}(AsO_4)_{10}(OH)_{20} \cdot 15H_2O$.

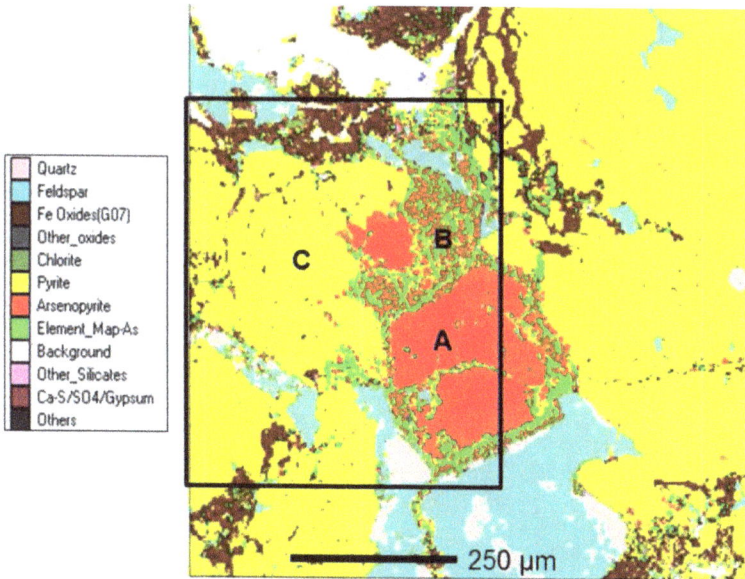

Plate 12. (*see Fig. 15 and associated text in Chapter 13*) QEMSCAN® image of a polished thin section of waste-rock from the Prescott Shaft area, Empire Mine SHP. (A) arsenopyrite; (B) hydrous ferric arsenate (HFA); (C) arsenian pyrite and hydrous ferric oxide (HFO). From Burlak (2012).

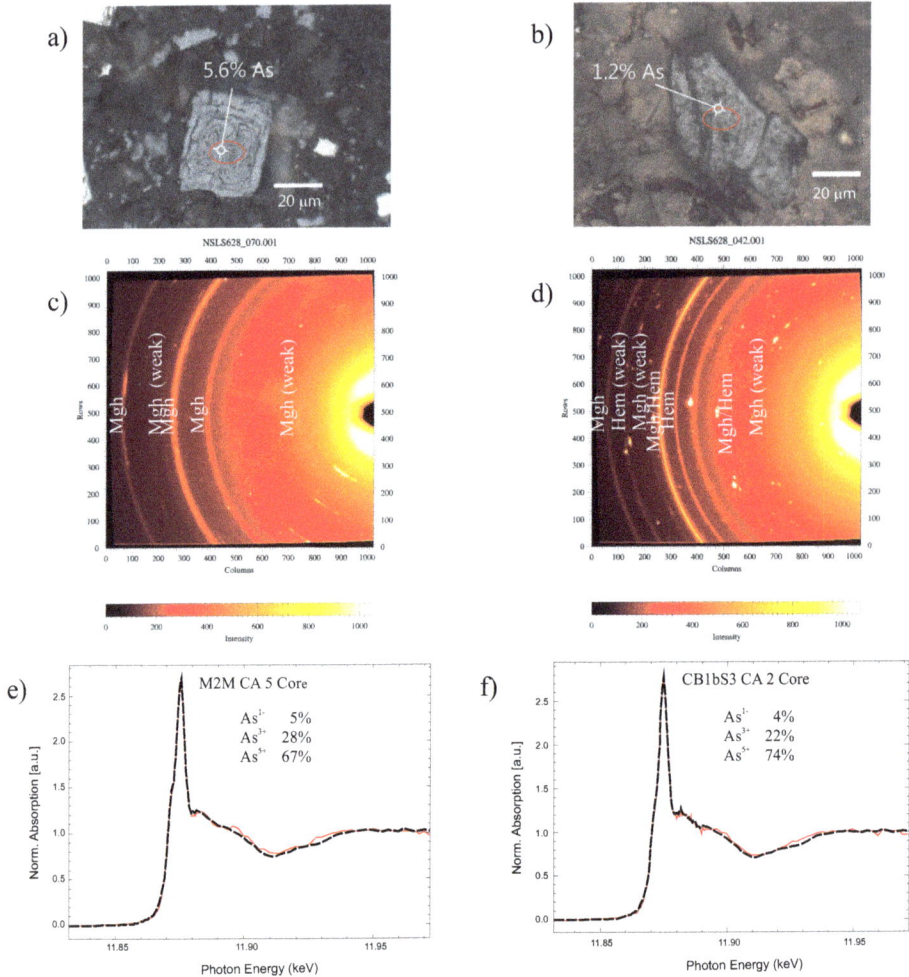

Plate 13. (*see Fig. 6 and associated text in Chapter 12*) Selected analyses of two grains of roaster iron oxides. a) Reflected-light photomicrograph of square concentric roaster iron oxide from calcine sample (M2M). Total As established by EPMA (designated in white). b) Transmitted- and reflected light photomicrograph of target grain from shoreline tailings sample (CB1bS3). Total As by EPMA as indicated. c) micro-XRD image of target indicated by ellipse in (a). Pattern corresponds to maghemite. Three arcs in lower right-hand corner are chlorite reflections. d) Micro-XRD image of target indicated by ellipse in (b). Pattern is a mixture of maghemite and hematite. e) Micro-XANES analysis of target in (a). Sample spectrum is lighter undashed line, dashed line is best-fit linear combination for result shown. f) Micro-XANES analysis of target in (b). [Reprinted with permission of the Mineralogical Association of Canada, from Walker et al. (2005) *Can Mineral*, Vol. 43, Fig. 8. p. 1218.]

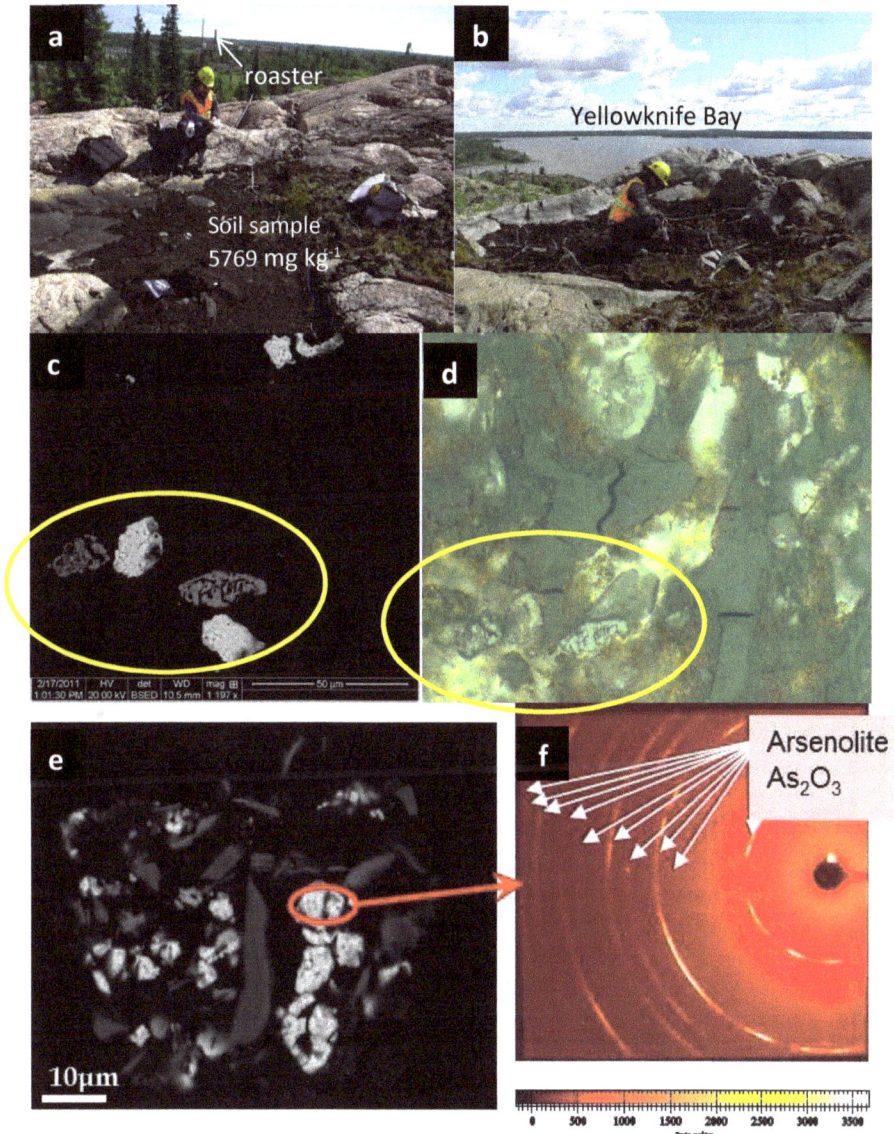

Plate 14. (*see Fig. 7 and associated text in Chapter 12*) Roaster-derived As-bearing particles in soils near Giant mine. (a) and (b) show typical soil pockets on large outcrops, the sites that contained the highest concentration of roaster-derived soil. (c) back-scattered electron image of a thin section of soil showing light (As_2O_3) and dark (roaster-generated Fe oxide) particles. (d) petrographic image of the same area as (c), combined transmitted and reflected light. (e) As_2O_3 particles in soil (f) microXRD of one of these particles indicating the pattern corresponding to arsenolite. From Wrye (2008) and Bromstad (2011).

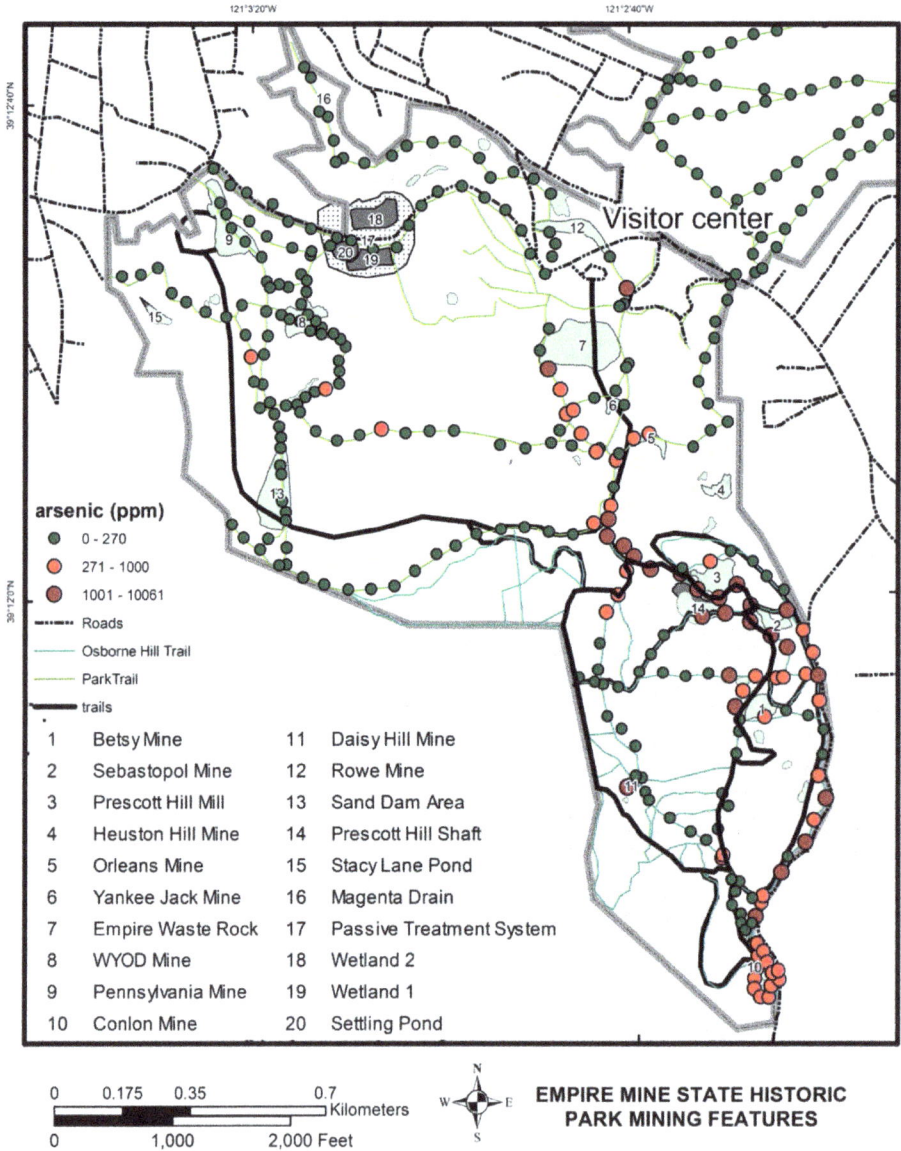

Plate 15. (*see Fig. 9 and associated text in Chapter 13*) Map of Empire Mine State Historic Park showing mining features. Waste-rock piles shown in light grey (green). Data for arsenic concentration on trails prior to remediation determined by field X-ray fluorescence. Modified from MFG (2008b).

Mineral	As:Fe (For Sample 10: Prescott Shaft Area)	Masks (Fe vs As in Point Counts)	Fe ■ 30μm	As ■ 30μm	
Arseno-pyrite	0.84 – 0.94	Slope 0.30-0.32			A
HFA	0.30 – 0.79	Slope 0.34-0.38			B
Arsenian Pyrite + HFO	0.01 – 0.21	Slope 0.39-1.40			C

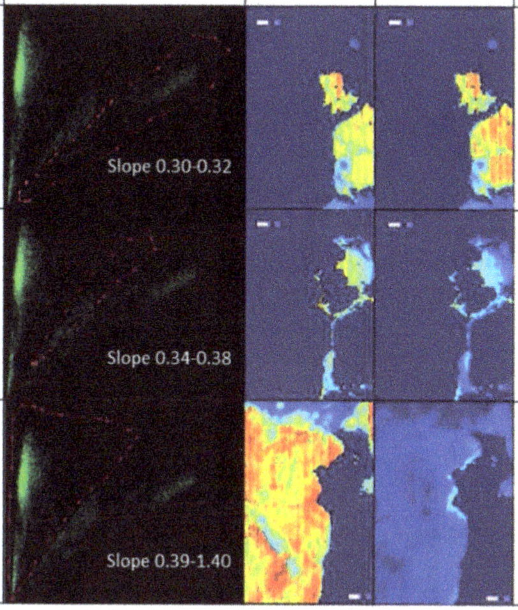

Plate 16. (*see Fig. 16 and associated text in Chapter 13*) µXAS images of a thin section of waste rock from Prescott Shaft (same view as QEMSCAN® image in Fig. 16). Linear trends are non-quantitative point counts for As vs. Fe collected from Beamline 2-3 at the Stanford Synchrotron Radiation Laboratory. Slopes are for the upper and lower bound of each encircled linear trend. The slopes for pyrite and HFO could not be completely separated, so they are combined. From Burlak (2012).

Plate 17. *(see Fig. 4 and associated text in Chapter 14)* Examples of arsenites from Tsumeb. *(left)* Schnei-derhöhnite with minor ludlockite. Scale bar = 5 mm. *(right)* Reinerite, probably the finest ever collected. ~5 cm in height. Crystal Classics specimen, photograph by Ed Loye. Reproduced with permission of Crystal Classics.

Plate 18. *(see Fig. 11 and associated text in Chapter 14)* *(left)* Aragonite associated with seepage zone in Tsumeb mine. These precipitates *(right)* are associated with bayldonite, such as the one shown.

Plate 19. (*see Fig. 18 and associated text in Chapter 14*) Examples of secondary arsenate mineral assemblages from Tsumeb. (a) Bayldonite and arsentsumebite on azurite, Scale bar = 1 cm. (b) Olivenite on cuprian adamite, Scale bar = 2.5 cm. (c) Balydonite psuedomorphs of mimetite. Scale bar = 1.5 cm. (d) Keyite, Level Scale bar = 0.3 mm. (e) Olivenite with gartrellite. Scale bar = 1 cm. (f) Conichalcite, molbdofornacite with dioptase. Scale bar = 1 cm. (g) Legrandite on matrix. Scale bar = 1 mm. (h) Mimetite. Scale bar = 1 cm.

Plate 20. (*see Fig. 24 and associated text in Chapter 14*) Azurite and olivenite from Tsumeb. Scale bar is 1 cm.

Plate 21. (*see Fig. 26 and associated text in Chapter 14*) (a) Conichalcite on cerussite, malachite and calcite. Scale bar = 10 mm. (b) Conichalcite with malachite as inclusions in calcite. Scale bar = 8 mm.

Plate 22. (*see Fig. 28 and associated text in Chapter 14*) Scorodite from Tsumeb. Scale bar = 1 cm.

Reviews in Mineralogy & Geochemistry
Vol. 79 pp. 257-369, 2014
Copyright © Mineralogical Society of America

Arsenic Speciation in Solids Using X-ray Absorption Spectroscopy

Andrea L. Foster

U.S. Geological Survey
Menlo Park, California 94025, U.S.A.

afoster@usgs.gov

Christopher S. Kim

Chapman University,
Orange, California 92866, U.S.A.

cskim@chapman.edu

INTRODUCTION

Synchrotron-based X-ray absorption spectroscopy (XAS) is an *in situ*, minimally-destructive, element-specific, molecular-scale structural probe that has been employed to study the chemical forms (species) of arsenic (As) in solid and aqueous phases (including rocks, soils, sediment, synthetic compounds, and numerous types of biota including humans) for more than 20 years. Although several excellent reviews of As geochemistry and As speciation in the environment have been published previously (including recent contributions in this volume), the explosion of As-XAS studies over the past decade (especially studies employing microfocused X-ray beams) warrants this new review of the literature and of data analysis methods.

This review has two main sections. The first is a presentation of methods for sample preparation and for the collection, processing and analysis of As-XAS spectra. Since several more comprehensive reviews of the X-ray absorption theory and data collection methodology exist, this section is brief and focused specifically on As. The second section is a critical review of the As-XAS literature, arranged by sample type and accompanied by summary tables (collected as appendices at the end of the chapter).

One of the most important aims of this review is to clarify the different types of analysis that are performed on As-XAS spectra, and to describe the benefits, drawbacks, and limitations of each. Arsenic XAS spectra are analyzed to obtain one or more of the following types of information (in increasing order of sophistication):

1. Identification of oxidation state (+ quantification of relative abundance, if multiple oxidation states are present)

2. Selection of the appropriate number and type of model compound As-XAS spectra to use in speciation fits (see 3);

3. Quantification of the relative abundance of species in a multi-species sample;

4. Identification of the molecular-scale (~ 4-7 Å) attributes of the predominant (one or two) As species.

The type of As-XAS spectral analysis ultimately employed and the utility of each type depends on several factors, including data collection mode, spectral quality, the type of

1529-6466/14/0079-0005$15.00

sample, quality of reference spectral libraries, information from ancillary data, and the study objectives.

X-ray absorption spectroscopy (XAS): the "gold standard" for determination of arsenic species in solid phases

The mobility, bioavailability, and toxicity of As in the environment is strongly dependent on its chemical form, which varies widely (Fig. 1 and Table 1). Identification and measurement of the chemical forms of As in all phases of matter is therefore critical to advance our understanding of its environmental cycling, to model its behavior under specific conditions, and to assess and mitigate its harmful effects on humans and biota. Most chemists use the term "species" instead of "chemical form" to refer collectively to the oxidation state and coordination chemistry/geometry of an element, a convention we will also follow (for more information on this topic, see Reeder et al. 2006). We will also use the term "speciation" to describe the distribution of As species in a sample and/or the process of identifying/quantifying the relative abundance of species.

Methods of measuring gaseous and dissolved inorganic and organic As species have proliferated and improved over the past 20 years, with different techniques available for samples based on expected As concentration range or matrix composition (Cullen and Reimer 1989; Melamed 2005; Wolf et al. 2011; Planer-Friedrich and Thomas 2014). Dissolved species previously only predicted from theory or produced synthetically are still being identified in natural systems using wet chemical techniques (e.g., thioarsenates, Suess et al. 2011; "Arsenicin A," Mancini et al. 2006; Fig. 1]. Over the same time period, synchrotron-based X-ray absorption spectroscopy (XAS) has emerged as the "gold standard" for determining As speciation in solid phases (and for elemental speciation in solids generally; Table 1). X-ray absorption spectroscopy is an element-specific molecular structure probe that can provide species information on As present in bulk concentrations of mg kg^{-1} to wt% (see review articles in Table 1). As-XAS can be conducted *in situ*; that is, on samples that are hydrated, anaerobic, living, or manifest any of several other characteristics that would preclude *in situ* analysis by other techniques. X-ray absorption spectroscopy has often been described as non-destructive, although this is not universally true; beam-induced conversion of As species in biological and/or wet samples has been observed at ambient temperature by multiple investigators (e.g., Smith et al. 2005; Miot et al. 2008; Paktunc 2008a). These transformations can be minimized by collecting As-XAS data at cryogenic temperatures (Calvin 2013).

The information provided by As-XAS often complements or extends that obtained by conventional techniques for the analysis of solids [including chemical digestion/analysis, X-ray diffraction (XRD) transmission electron microscopy (TEM), electron probe microanalysis (EPMA), X-ray photoelectron spectroscopy (XPS), proton-induced X-ray emission (PIXE), (micro) IR spectroscopy (IR), and (micro) Raman spectroscopy]. These techniques have been successfully used to identify As species in a variety of solids, but can be inadequate for some types of samples due to factors including high detection limits (e.g., IR, Raman), unsuitable sample preparation requirements (e.g., vacuum for XPS and EPMA; specially-thinned or powdered and dried samples for XRD, TEM, and EPMA).

The use of XAS spectra to understand As speciation in solids from the natural or modeled environment has grown dramatically in the past 10 years (Fig. 2). Many of the recent studies employ micron-sized X-ray beams to collect XAS spectra on the micron scale (μ-XAS). These spectra are processed and interpreted in a manner identical to XAS spectra collected with beams in the millimeter dimension (called "bulk" XAS to distinguish from the "micro" form; see Table 2). Micro-X-ray fluorescence (μ-XRF) mapping and micro-X-ray diffraction (μ-XRD) are two additional techniques performed with micron-sized beams that have provided key information about As speciation in solid phases. Detailed explanation of these techniques is

Inorganic Arsenic

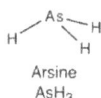

Arsine
AsH₃

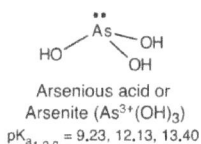

Arsenious acid or
Arsenite (As³⁺(OH)₃)
pK$_{a_{1,2,3}}$ = 9.23, 12.13, 13.40

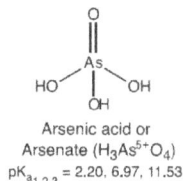

Arsenic acid or
Arsenate (H₃As⁵⁺O₄)
pK$_{a_{1,2,3}}$ = 2.20, 6.97, 11.53

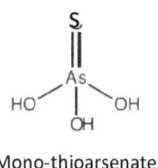

Mono-thioarsenate

Methylated Arsenic Compounds

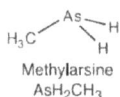

Methylarsine
AsH₂CH₃

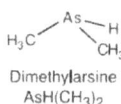

Dimethylarsine
AsH(CH₃)₂

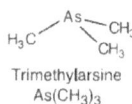

Trimethylarsine
As(CH₃)₃

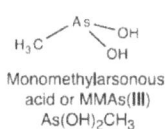

Monomethylarsonous
acid or MMAs(III)
As(OH)₂CH₃

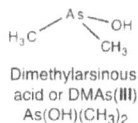

Dimethylarsinous
acid or DMAs(III)
As(OH)(CH₃)₂

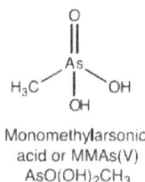

Monomethylarsonic
acid or MMAs(V)
AsO(OH)₂CH₃

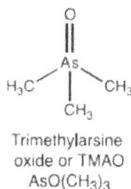

Trimethylarsine
oxide or TMAO
AsO(CH₃)₃

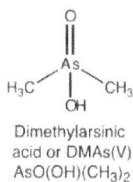

Dimethylarsinic
acid or DMAs(V)
AsO(OH)(CH₃)₂

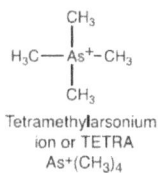

Tetramethylarsonium
ion or TETRA
As⁺(CH₃)₄

Organoarsenic Compounds

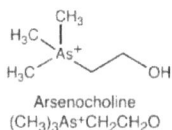

Arsenocholine
(CH₃)₃As⁺CH₂CH₂O

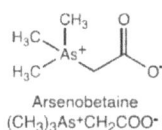

Arsenobetaine
(CH₃)₃As⁺CH₂COO⁻

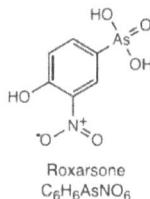

Roxarsone
C₆H₆AsNO₆

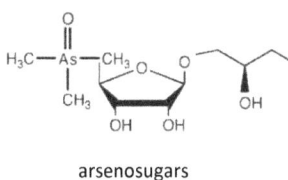

arsenosugars

R = OH
R = OP(O)(O·)OCH₂CH(OH)CH₂OH
R = (SO₃)²⁻
R = (OSO₃)¹⁻

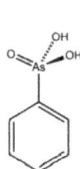

Phenylarsonic
acid

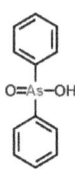

Diphenylarsinic
acid

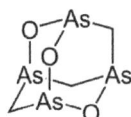

Arsenicin A

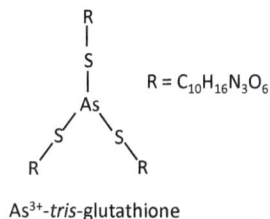

As³⁺-*tris*-glutathione

R = C₁₀H₁₆N₃O₆

Figure 1. Molecular representations of major inorganic and organic As species.
Modified from Pickering et al. (2000), O'Day (2006), and Wikimedia Commons (2014).

Table 1. Summary of review articles pertaining to synchrotron-based studies
of arsenic in environmentally-relevant solid phases

Subject	References
Arsenic toxicity, health effects, and epidemiology	Morton and Dunnette 1994; Yamauchi and Fowler 1994; Reeder et al. 2006; Hopenhayn 2006; Liu and Cai 2007; Mitchell 2014 (many more available)
Arsenic aqueous speciation	Cullen and Reimer 1989; Nordstrom and Archer 2003; O'Day 2006; Campbell and Nordstrom 2014, this volume; Leybourne et al. 2014, this volume; Majzlan et al. 2014, this volume
As mineralogy	Essington 1988; Sadiq 1997; O'Day 2006; Drahota and Filippi 2009; Majzlan et al. 2014, this volume
Reviews of geomicrobiology pertaining to As transformations in the environment	Lovley 2000; Lloyd and Oremland 2006; Fendorf and Kocar 2009; Benzerara et al. 2011; Amend and Saltikov 2014
Applications of synchrotron techniques to molecular environmental science	Brown et al. 1999; Brown and Sturchio 2002; Brown and Calas 2012
XAS theory/XAS data collection/XAS data analysis	Koningsberger and Prins 1988; Sayers 1988; Osanna and Jacobsen 2000; Kelly et al. 2008; Bunker 2010; Calvin 2013; Henderson et al. 2014; Newville 2014
Software: XAS data processing + analysis	WinXAS (Ressler 1998); EXAFSPAK (George 2000); Athena/Artemis/Hephaestus (Ravel and Newville 2005); Sixpack (Webb 2005) Also Viper, EXCURVE, EDA, MAX, and several others at *http://xafs.org/Software*
Software: *Ab initio* modeling of XANES and EXAFS spectra	FEFF9 (Rehr et al. 2010) and several other programs under "theory" header at *http://xafs.org/Software*
μ-XAS/μ-XRF/μ-XRD data collection/data analysis (some topical)	Manceau et al. 2002; Sutton et al. 2002; Hornberger et al. 2008; Dillon 2012; Sarret et al. 2013
Software: μ-XRF/μ-XRD data processing and analysis	μ-XRF: SMAK (Webb 2006) + many other beamline-specific programs available, check with beamline scientist μ-XRD: Fit2D (Hammersley 2004); Area Diffraction Machine (Lande and Webb 2007); Datasqueeze (Heiney 2002)
Reviews of XAS-focused studies of As species in model systems and/or environmental solids	Foster 2003; Morin and Calas 2006; Fendorf et al. 2007; Liu and Cai 2007; Fendorf et al. 2010; Jamieson 2011
Reviews of XAS-focused studies of As speciation in biota	Smith et al. 2005 (biology); George and Pickering 2007 (biology and chemistry); George et al. 2009 (seafood); Dillon 2012 (medicinal chemistry); Sarret et al. 2013 (plants); Salt et al. 2002 (plants)

Search Terms: "Arsenic" and "X-ray absorption spectroscopy"

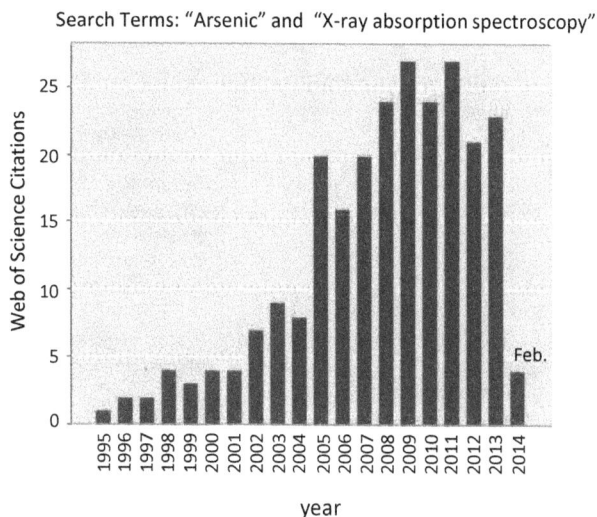

Figure 2. Publications by year, using terms "arsenic" and
"X-ray absorption spectroscopy" in Web of Science as of February, 2014.

outside the scope of this review; for more information, the reader is referred to the review articles listed in Table 1 and to the many specific investigations cited in the text and listed in Appendices 1-9.

As we will demonstrate in the sections to follow, the As-XAS literature reveals a fairly high degree of consistency in the number and types of solid-phase species that predominate under generally similar physical and chemical conditions. This suggests that there is good potential for transferability of information learned from site-specific studies to similar locations, and for modeling of As transport and transformation in the environment. However, there are also limits on the level of speciation information obtained from analysis of As-XAS spectra; these limits are specific to the chemistry of the system, type of XAS data being analyzed, and the type of analysis being performed. The topical format of this review provides a means to discuss some of these system-specific limits in context.

PREPARING, COLLECTING, AND PROCESSING BULK XAS DATA AND μ-XAS, XRF, AND XRD DATA

Characterizing samples prior to beamtime

In light of the high value and short duration of beamtime, extensive sample characterization by standard techniques prior to beamtime is strongly recommended. At a minimum, the total As concentration for samples should be known, since the spectral data quality can be highly dependent on this parameter. Information on mineralogy or chemical composition at the appropriate scale (bulk or micro depending on experiment) can guide the selection of model compounds or the construction of highly probable structural models. In general, the better the pre-characterization information, the better designed and executed the As-XAS experiment will be. Sediment separations on the basis of size, density, or magnetism can enrich As concentrations, improving the quality of data obtained and/or the ease of interpreting it; in order to optimize beamtime, these fractions should be pre-characterized as much as possible. For biota, XAS data quality often benefits from concentration/segregation as well: roots,

Table 2. Comparison of bulk and microbeam XAS/XRF experiments.

Parameter	Bulk	Microbeam
Sample	Usually ground to < 10 micron Diluted if As > 1wt%	Sediment, rocks, soils, tailings, dust, etc.: grains on tape or fixed in epoxy and polished (thin sectioning preferred) Biological: Intact or sliced frozen, freeze dried, or living
Detector	< ~ 1000 ppm As: solid-state detector > ~1000 ppm: Ar-, Kr-, or Xe-filled ionization chamber or photodiode	Energy dispersive: single or multi-element solid-state detector
Calibration	Simultaneous or sequential possible	Sequential (before or after sample)
XANES	Yes; bulk As ~10 ppm[*]	Yes
EXAFS	Yes: total As 100 ppm[*]	Usually no[*]
Beam size	Usually rectangular, defined by slits; 1-2 mm vertical, 5-15 mm horizontal	Variable: typically 2-100 μm; adjustable in real time on some beamlines
Spatial Resolution	mm-scale (i.e., poor)	μm-scale; beam size and scan step size dependent
Temperature/ Redox control	< 77 K cryostat (holds one or more samples in a cassette) or Peltier-cooled stage Move sample after each scan	cryojet (blows ultracold N_2 at sample) and Peltier-cooled stage options available at some beamlines Move sample after each scan
XRF data	1 multichannel analyzer (MCA) spectrum; can be saved, retrieved, or analyzed (similar to SEM EDAX)	N MCA spectra, where N = # of map pixels; elemental data in each channel can be retrieved and analyzed in detail (bivariate scatter plots, PCA analysis)
Origin of XAS spectrum	Multiple species contribute to spectrum in weighted sum	Spectrum often derived from 1 species (sometimes 2 due to mixtures or beam-induced redox)
Species Quantification (relative abundance)	Operational detection limit of species in a mixture: 5-10% (XANES and EXAFS, but system-dependent); quantification error estimated at 2-30% (usually ±10% is reported)	Accuracy relative to bulk depends on # of particles interrogated. Better for ID of low-abundance, As-rich phases. Oxidation state mapping possible: N oxidation states; $N+1$ maps. Total As map @ 11 880; see Table 3 for other map energies
Detection limit (As concentration in bulk sample)	1-10 mg kg^{-1} (XANES) 100 mg kg^{-1} (EXAFS) (operational)	< ~ 1 mg kg^{-1} (bulk), depends on time and success in locating pre-selected targets or high-As spots from mapping

[*]Time-dependent parameter; can improve if more data are collected

stems, leaves, xylem (plants), as well as various tissues (biota) or subcellular fractions (cells) can be analyzed separately to gain maximum information about As speciation. One of the great benefits of XAS spectroscopy is that small amounts of dilute samples can be analyzed that are not amenable to quantitative chemical analysis due to volume requirements. The typical mass of sample needed for bulk XAS analysis is usually only a few hundred milligrams; even less sample is required for μ-XAS. The sample(s) to be selected for analysis by XAS should be representative of a larger suite of samples, and should represent key end-members of the environment or process under study, since beamtime is very limited. Given the recommendation to collect at least 3 consecutive scans for evaluation of As-EXAFS spectra (Calvin 2013), and the roughly 45 min collection time for a typical As-EXAFS scan (see Table 2 and discussion below), a minimum of 2.25 h would be spent for each sample. In a 24-h period, this would ideally result in the collection of data from 10 samples. Furthermore, the time required for data collection is greater for samples containing trace As levels (e.g., 500-1000 mg kg^{-1} As), which can require 8 or more individual scans (6 h) to achieve acceptable signal-to-noise levels in the averaged spectrum (the time required will also vary due to beamline-specific factors).

Preparing samples for bulk As-XAS data collection

Considerably less sample preparation is required for bulk (millimeter-sized beam) XAS measurements than for many common analytical methods. In fact, it is often possible to prepare and analyze samples under conditions comparable to those of the natural environment (e.g., as damp pastes, solutions, etc.). Oxygen-sensitive samples can also be collected or prepared in a way as to preserve their redox state (e.g., in an anaerobic chamber), then analyzed under inert atmosphere with continual gas purging and/or in a frozen state.

Grinding inorganic samples and models to a fine powder (~ 10 micron, if hand-grinding) is a common preparation step for bulk XAS data collection that minimizes unwanted spectral artifacts arising from sample inhomogeneities and particle scattering (Kelly et al. 2008; Newville 2014). Inorganic samples sensitive to oxidation or heat-induced transformations can be ground under acetone to preclude these effects. Biological samples are often ground under liquid N_2 to minimize transformations (Pickering et al. 2000).

Samples with As concentrations < 1 wt% can typically be analyzed directly after loading into a sample holder. Typical holders for bulk XAS of dilute samples are made of Teflon or polycarbonate with a window of about 30 mm wide by 5 mm tall (dimensions can vary substantially). Holder thickness for dilute samples is usually 1-3 mm; thicker holders should be avoided to minimize self-absorbance artifacts, which dampen the XAS signal (Kelly et al. 2008).

Concentrated samples (> 1 wt% As) require thinning after grinding to minimize self-absorbance. Several methods are available to achieve a thinner sample. If the composition of the sample is known or can be estimated, direct absorption length calculations can be made specific to the dimensions of the sample holder to be used, and the sample can be diluted and well mixed with an inert, X-ray transparent material (e.g., boron nitride; Bargar 2004; Ravel and Newville 2005; Kelly et al. 2008). If the sample composition is not known, non-quantitative methods of sample thinning can be used, such as spreading sample powders on tape (and analyzing multiple overlapping tape layers), or "painting" samples on tape by mixing powder with an inert binder. These methods can achieve comparable results to dilution based on absorption length calculations, and have the advantage of being easier to prepare.

The XAS spectra of aqueous As species can be substantially different from their solid-phase analogs, but these data can be challenging to collect due to beam-induced alteration. As a result, data for aqueous standards (and samples such as plant extracts) are often collected at temperatures at or below 77 K (liquid nitrogen freezing point) to avoid such artifacts. Glycerol can also be added to aqueous model compounds to prevent solute segregation that can alter the appearance of XAS spectra collected at cryogenic temperature (Pickering et al. 2000).

Preparation methods for microscale X-ray studies are more variable, sample-specific, and depend on the type of information sought. Grinding is avoided given that the investigator wishes to characterize the spatial distribution and association of As in a heterogeneous matrix at the microscale. The simplest preparation method for sediment/soil involves adhering particles onto the adhesive side of X-ray-transparent tape and presenting this surface to the beam, but this method has several undesirable characteristics. If samples are predominantly inorganic, it is preferable to analyze them as grains that have been embedded in epoxy, sectioned, and polished (commercial services are available). A low-curing temperature, low-metal epoxy such as EPOTEK 301™ or equivalent should be used. Standard petrographic "thin sections" are 30 μm thick but other thicknesses can be requested based on sample grain-size considerations. High purity fused quartz (not glass) slides should be used for As μ-XAS analysis whenever possible. If conventional glass is used, the slides should be checked for As content first. Some types of borosilicate glass contain appreciable As that can interfere with μ-XAS/μ-XRF measurements at the As K-edge. Alternatively, the sections can be removed from the glass after preparation, providing that specific (cyanoacrylate-based) adhesives were used in their preparation (Walker et al. 2009).

Analyzing As species in biota presents special challenges that have been overcome in different ways over time. Historically, when bulk XAS was the only technique, biota samples were usually homogenized prior to data collection at low temperature. Analyzing subsamples such as sub-cellular fractions, xylem sap, and root tissue was one way of obtaining a degree of spatially-localized information. Many biological structures are on the micron scale and therefore well-suited to analysis by μ-XAS/μ-XRF techniques. Cryo-ultramicrotomy and other methods are now being employed to produce small, thin sections of macrobiota, and collection of low-temperature μ-XAS/μ-XRF data is also available at some facilities to aid in the preservation of *in situ* As speciation. Sample preparation of biological specimens is nearly eliminated by tomographic X-ray methods, which allow 3D chemical imaging of As speciation (e.g., oxidation state mapping and tomography) in whole tissues (Seyfferth et al. 2010).

Embedding paraffin, waxes and other mounting media are usually transparent to the X-ray beam at the As K-edge, but the trace element composition should be checked, as contaminants can interfere with μ-XRF, μ-XAS, and μ-XRD measurements. Thicker sections may also be analyzed successfully, as well as intact organisms (in one instance, a moth larva immobilized in a capillary tube; Andrahennadi and Pickering 2008). For biological samples, the low degree of X-ray attenuation (comparable to that of cellulose) at the As K-edge (11,867 electron volts) means that full penetration of even mm-thick biological samples is likely; the information collected from the experiment will also be from the entire volume of the sample penetrated. In some cases, this can be beneficial, as when searching for small As-rich particles dispersed in lung tissue (Foster et al. 2009a); in other cases it can be detrimental to the desired analysis (e.g., when two As phases are present at different depths in the analyzed volume, and fluorescence from both overlap in the recorded signal).

XAS spectra for model compounds (standards, see Table 3) collected in bulk mode can be used to analyze μ-XAS spectra, but it is also useful to have a set of model compounds in a holder suitable for μ-XAS. This set can be used to test the validity of oxidation state or species maps, or to check calibration. A set of elemental standards is also useful for semi-quantitative analysis of elemental abundance data from μ-XRF mapping, and can be prepared from commercially available XRF standards deposited on Mylar film. However, note that extraction of quantitative elemental abundance from XRF data requires correction for matrix effects.

Micro-XRD data can be collected in reflectance or transmission mode; the former is compatible with samples mounted on glass slides, whereas the latter either requires that the sample be without backing or that the scattered glass background be subtracted post-process. Some cyanoacrylate adhesive formulations are amenable to removal with acetone (Walker et

Table 3. Important parameters of As K-edge XAS experiments.

Parameter	Details
Monochromator crystal set	Higher order cuts = better resolution (311 > 220 > 111) at expense of flux; but intrinsic resolution at 3rd generation sources surpasses these differences; detune to lower harmonics or use rejection mirror
Crystal set cut phi = 0, phi = 90	Check glitch curves for specific crystals, if available; avoid glitches >1% if possible; otherwise detuning can help lower glitch intensity
Energy calibrants/ typical model compounds and K-edge absorption energies (calibrated on As^0)	As^{-1}: arsenopyrite or pyrite: ~ 11 867 eV (oxidizes) As^0 = 11 867.0 eV *by definition* (oxidizes) As^{3+}-S (inorg): As$_2$S$_3$: ~ 11 869 eV As^{3+}-O (inorg): NaAsO$_2$, As$_2$O$_3$: ~ 11 871 eV FeAs^{5+}O$_4$·2H$_2$O (scorodite) ~ 11 875 eV As^{5+} sorbed to ferrihydrite: ~ 11 875 eV NaHAs^{5+}O$_4$·7H$_2$O ~ 11 875 eV Au^0 L$_{III}$ edge = 11 919.0 eV *by definition*
Mono hutch slits (bulk only)	1-1.5 mm vertical aperture
Sample hutch (table) slits (bulk only)	0.75-1 mm vertical aperture; horizontal as dictated by sample/beam dimensions and detector counts
Step size (typical values)	10 eV step in pre-edge region 0.2 eV step in XANES, ~5 eV step in EXAFS
Count time	Pre-edge: 1 s XANES: usually 1 s (+ if low concentration) EXAFS: 1-30 s (15 max is standard), cubic weighting applied as function of *k*
Z-1 filter (bulk only)	Ge 3 or 6 μm thickness
Fe fluorescence blocker	1-6 layers of household Al foil (or Teflon)

al. 2009). Model compounds for μ-XRD can be prepared on tape as described above or loaded in capillary tubes for analysis; a strongly diffracting standard reference material (e.g., Al$_2$O$_3$, CeO$_2$, LaB$_6$) is required to calibrate the diffraction setup. Collection of μ-XRD patterns from standard reference materials similar to phases in the unknown sample may be useful to test analytical procedures, but due to crystallinity and grain size differences of phases in sample and reference materials, μ-XRD patterns from the two may not be directly comparable.

Collecting XAS spectra, μ-XRF maps, and μ-XRD patterns

The following review of data collection procedures and overview of the X-ray absorption phenomenon is brief and As-specific; the reader is referred to the reviews listed in Table 1 for detailed summaries of the physics behind the X-ray absorption process, the mathematical formalism used to understand it quantitatively, schematics of various XAS beamline setups, and data collection considerations for bulk and microbeam experiments.

A range of possible experimental configurations to conduct bulk and micro-beam X-ray absorption spectroscopy measurements exists; the specific configuration is highly dependent on the X-ray facility, availability of instrumentation, etc. and is custom-built to a considerable extent. Generalized schematic configurations for bulk XAS and microbeam XAS/XRF/XRD data collection are shown in Figures 3a and 3b, respectively. In both cases, X-rays generated from the synchrotron pass through a diffracting double crystal monochromator to obtain the desired X-ray beam energy. The monochromator crystal set choice should be made to minimize the occurrence of glitches (energy-dependent fluctuations in beam intensity) in the spectral region to be analyzed. The first gas-filled ionization chamber (I_0) records the incident beam intensity. In the bulk configuration, the beam size is defined by collimating slits prior to entering I_0 (Fig. 3a; Table 2); in the microbeam setup, the beam is focused using any of several types of optics, which are mounted either upstream or downstream of I_0 (Fig. 3b). Next, the

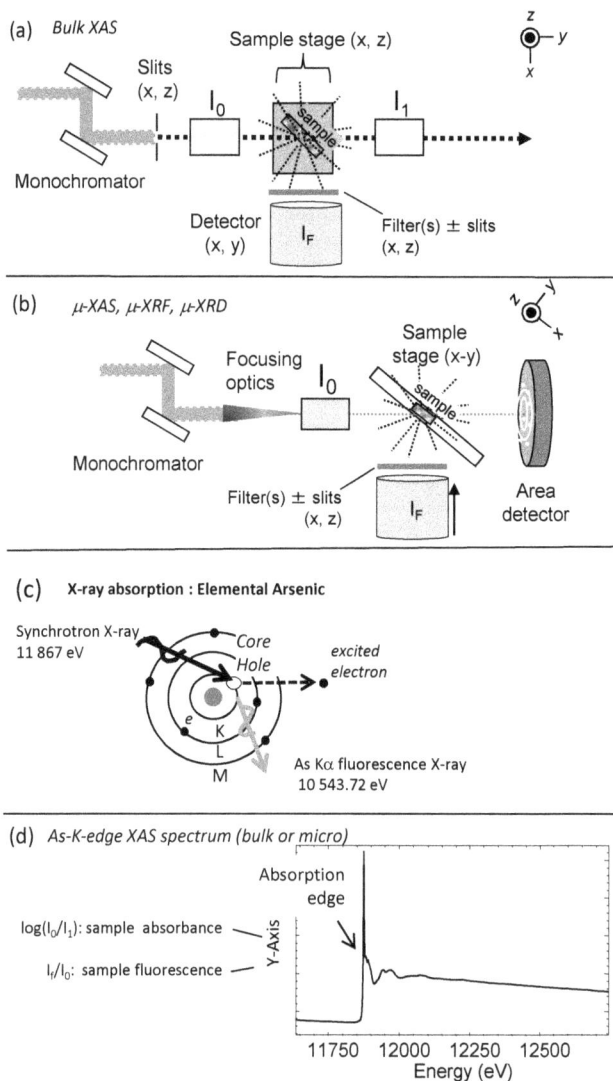

Figure 3. Standard beamline configurations for (a) bulk XAS and (b) micro XAS/XRF/XRD; (c) schematic of X-ray absorption As K-edge; (d) raw As-XAS spectrum.

beam strikes the sample, which is oriented at a 45° angle to the beam for bulk and μ-XAS measurements collected in fluorescence mode (details below; Fig. 3a,b) or perpendicular to the beam for bulk XAS measurements collected in transmission mode.

At the atomic scale, illumination of the sample by the X-ray beam in either configuration at or above a quantized "binding energy" results in absorption by a core-level electron, which leads to its excitation and ejection from the atom (Fig. 3c). The energy required to eject electrons from the K shell of metallic As is 11,867 eV by definition (Williams 2001; Ravel and Newville 2005). It should be noted that all elements whose characteristic "binding energies" are at or lower than 11,867 eV will also be undergoing X-ray absorption events, the significance of which is discussed below.

Ejection of the excited core-level electron from the As atom leads to a series of electronic and scattering events, including the emission of a "characteristic" fluorescence X-ray with specific energy matched to that of the absorption process as the empty core level shell is filled by a higher energy electron (10,543.72 eV for As K_{a1}; Fig. 3c) this signal is collected by a fluorescence detector. Several detector types are available, from simple gas-filled and photodiode models to solid-state, multi-element, wavelength- or energy-dispersive types (Fig. 3a,b). In the bulk XAS measurement, the post-sample beam intensity (attenuated by absorption) is measured by a second gas-filled ionization chamber (I_1; Fig. 3a).

The other elements in the sample undergoing X-ray absorption will also emit characteristic X-rays and absorb some of the incident beam intensity. For a bulk XAS or μ-XAS measurement, only the As characteristic X-ray fluorescence is desired; accordingly, fluorescence from other elements and direct beam scattering off the sample is minimized by placing a Z-1 filter, collimating (Soller) slits, and/or layers of Al foil between the sample and the detector (Fig. 3a,b; Table 3; Kelly et al. 2008). An energy-discriminating detector featuring multichannel analyzers (MCAs) will further allow the separation of As fluorescence from that of other elements. In the case of μ-XRF mapping, the entire dispersive fluorescence spectrum with characteristic "lines" from several elements in addition to As is desired and collected by the detector as the sample is moved in x-y space (Fig. 3b). The fluorescence intensity from multiple elements can then be viewed in real time by windowing on characteristic lines (Webb 2006).

In a typical XAS measurement, the monochromator is stepped through the binding energy of the absorber of interest to record the XAS spectrum. For As K-edge EXAFS spectroscopy, an energy range spanning ~200 eV below the K-edge to ~1000 eV above the absorption edge covers a range of ~16 inverse angstroms (Fig. 3d). The sample transmission XAS spectrum is conventionally represented by log (I_0/I_1) and its fluorescence spectrum by I_f/I_0. The energy step size chosen changes with XAS spectral region to capture the shape of spectral features, and count times also vary in each region to optimize signal-to-noise (Table 3); for further details on these topics, the reader is referred to the review articles referenced above. Multiple repeated scans are usually collected and averaged together to produce a XAS spectrum with adequate quality for data analysis (ideally < 1% noise; Kelly et al. 2008).

Micro-XRF, μ-XAS, and μ-XRD data collection is typically sequential: first an "exploratory" μ-XRF map is collected from a region of interest (ROI); after analysis of the map (in real time, at the beamline), either higher resolution maps are collected within that area, or points of interest (POI) are selected for collection of μ-XANES spectra. Each μ-XRD pattern is typically collected immediately following the acquisition of a μ-XANES spectrum from the same spot. Micro-XRF maps collected at closely-spaced energies are used to construct oxidation-state or speciation maps, in which the relative abundance of As oxidation states or species can be visualized on a per-pixel basis. Beam size, pixel size, and dwell time are all important considerations when setting up μ-XRF maps, and the choice is dependent on the scale of sample features and the As concentration at the micron scale. These parameters are best determined upon consultation with the beamline scientist prior to beginning the experiment.

Micro X-ray diffraction in transmission mode is schematically illustrated in Figure 3b. The X-ray beam is usually raised to a high energy such as 17 keV for μ-XRD data collection in order to provide higher Q range and reduce sample absorption; thus the measurement is sequential to μ-XAS data collection rather than simultaneous. The sample is illuminated by the X-ray beam as described above for X-ray absorption, and an area detector downstream of the sample records the diffracted X-rays through the sample. The resultant 2D diffraction pattern is usually composed of full or partial rings and/or spots representing the continuum between random crystallite orientation and single crystal diffraction, respectively, and whose spacing relates to the arrangement of atoms in ordered materials. This pattern can be integrated to produce a 1D pattern suitable for mineral phase identification by standard methods of powder XRD analysis (Manceau et al. 2002; Dillon 2012).

Analysis of μ-XRD patterns is complicated by several factors. If mineral particles are similar or larger than the X-ray beam, a spot pattern will be obtained due to diffraction from individual crystal faces rather than the preferred ring-pattern indicative random crystallite orientation. Nanoscale phase mixing is the rule rather than the exception in μ-XRD patterns from environmental samples, and results in μ-XRD patterns in which rings are derived from multiple species. Identification of crystalline phases in these mixed patterns is somewhat more complex, but completely analogous to the analysis of bulk, multicomponent XRD patterns using peak search-fit procedures available in most commercial XRD software packages. Finally, highly disordered phases will provide diffuse scattering and are not easily identified with peak matching, but this knowledge can itself be a very important piece of information. When combined with element ratios obtained by microanalysis (synchrotron XRF or electron microprobe, for example), amorphous patterns can be used to distinguish compositionally-similar phases such as scorodite ($FeAsO_4 \cdot 2H_2O$) and hydrous ferric arsenate (HFA; Walker et al. 2009). Further details and examples of μ-XRD pattern analysis are given in the references in Table 3 and in research papers where the technique has been successfully used to aid in As speciation (e.g., Walker et al. 2005, 2009; Corriveau et al. 2011a; Jamieson et al. 2011).

Processing XAS spectra

Figure 4 illustrates the standard processing steps for As K-edge XAS spectra, using the XAS spectrum of scorodite ($FeAsO_4 \cdot 2H_2O$) as an example. If a solid-state, multi-element detector was used to collect the As K_α fluorescence, the spectra should first be corrected for detector deadtime. Post-collection spectral re-calibration is required if the monochromator energy was not well calibrated or was unstable during the experiment. If multiple scans have been collected for averaging, calibration can take place prior or subsequent to this procedure. However, if beam instability has been a problem during the experimental period, it is recommended that individual spectra be re-calibrated prior to averaging. Pre- and post-edge backgrounds are then subtracted and spectra are normalized (at the edge jump, E_0; Fig. 4a). The X-ray absorption near edge structure" (XANES) spectrum is generally considered to include the region from ~50 eV below the absorption edge to ~100-300 eV above it (Fig. 4b,c)]. The shape of As-XANES spectra arises from both scattering between As and neighboring atoms and electronic transitions within the As atom. These two processes are convolved in a way that precludes quantitative analysis of XANES spectra, so they are often used as "spectral fingerprints" of As species, a topic we will discuss further in the section on data analysis. Micro-XAS data collection is often limited to XANES, extended XANES, or "mu" spectra (Fig. 4c) due to time or other considerations (Manceau et al. 2002; Dillon 2012).

The extended X-ray absorption fine structure (EXAFS) region of the X-ray absorption spectrum refers to modulations in the absorption spectrum observed at energies well above the absorption edge (> ~50 eV; Fig. 4c). The EXAFS are extracted from the mu spectrum by choosing an energy value to define the start of the EXAFS region (E_0; two possibilities are listed in Fig. 4c), and then subtracting a "spline" function whose curvature is defined

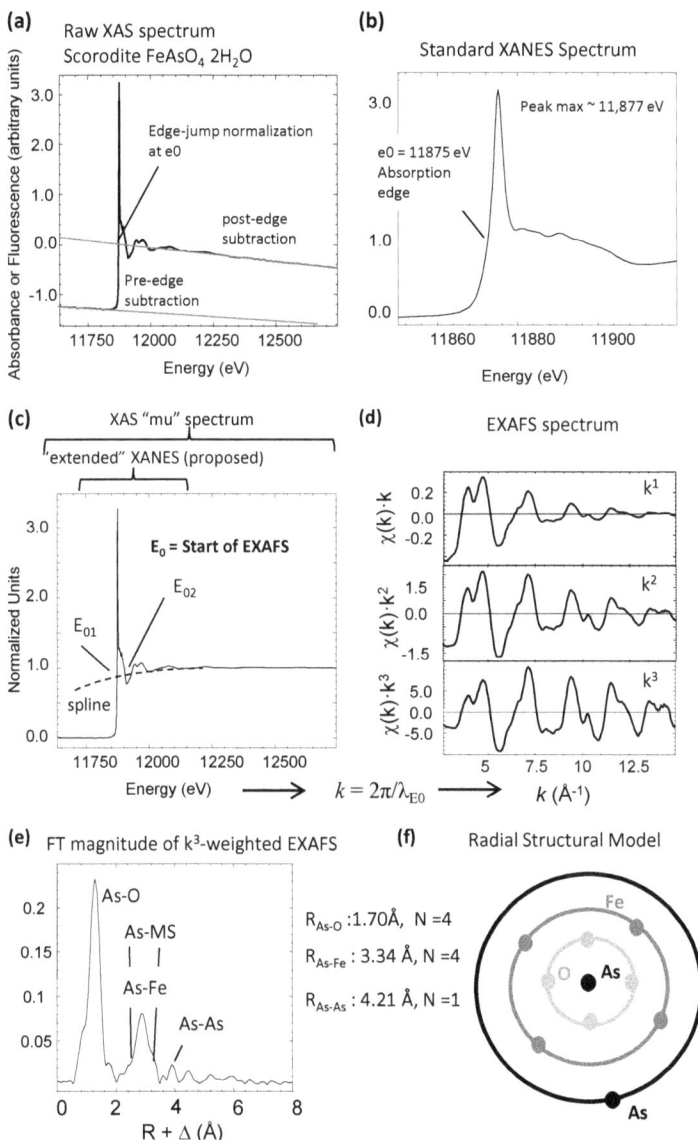

Figure 4. XAS data processing illustrated for scorodite (FeAsO$_4$·2H$_2$O): (a) pre-and post-background subtraction and normalization; (b) the standard XANES spectrum; (c) EXAFS extraction; (d) EXAFS spectrum as a function of k-weighting; (e) Fourier transform magnitude (distances not corrected for phase shift); (f) model of the radial structure of scorodite, derived from the reported crystal structure.

by the number of polynomial segments it contains. The units of the spectrum are converted from energy (in electron volts) to momentum (k, in units of Å^{-1}) space using the relation $k = 2\pi/\lambda_{E_0}$, where λ_{E_0} is the wavelength of the electron at the designated onset of the EXAFS region. This procedure normalizes EXAFS spectra to a "per atom" basis, which removes concentration effects (Sayers and Bunker 1988). For this reason, the quantification of As species using EXAFS spectra is in terms of relative (rather than absolute) abundance. EXAFS spectra are often viewed in k^3-weighted form to enhance the high-k end of the spectrum, which contains more information about more distant atoms (compare weightings in Fig. 4d).

The EXAFS spectrum is a sum of sinusoidal frequencies, and as such is a type of data amenable to Fourier analysis (Sayers et al. 1971; Sayers and Bunker 1988). The Fourier transformation (FT) spectrum has magnitude, real, and imaginary parts, though typically only the magnitude is presented. Fourier transformation of an EXAFS spectrum visually separates the absorber-backscatterer pair frequencies in *inverse* k space, which has units of angstroms (Å; Fig. 4e). When evaluating FT magnitude spectra in the literature, it is important to note that, unless phase corrections are applied, the peak positions are ~0.5 Å lower than true distances (the Δ symbol in Fig. 4e denotes that phase correction has not been applied). Phase corrections are usually not applied to FT data presented in publications because they result in unwanted distortions in FT peak shapes (George 2000).

Although commonly called a radial distribution function (RDF), the FT magnitude is not a true RDF because the position, shape, and height of the peaks are affected by additional physical properties (Sayers and Bunker 1988; Calvin 2013). However, inspection of the FT magnitude can give a reasonable picture of radial coordination around As. The uncorrected FT magnitude of the scorodite EXAFS spectrum contains 2 large peaks at ~1.3 Å and ~2.9 Å radial distance (R) that are well above noise "ripples", denoting two strong frequency signals in the data. Inspection of the scorodite crystal structure shows these peaks to represent the closest O and Fe atoms to As (Fig. 4f; Hawthorne 1976). Although both shells contain 4 atoms, the intensity of the closer O shell is much greater, demonstrating the sensitivity of the FT magnitude to the distance from the absorbing atom. In addition to these variables, FT peak intensity also depends on atomic number (N) and the thermal/static disorder (represented in simplest form by the Debye-Waller parameter, σ^2). The scorodite crystal structure report indicates that a small peak that is only slightly above the background is actually derived from a single As atom at 4.21 Å. The FT peak shows additional "sidelobes" on the Fe peak that do not correspond to atomic positions from the structure report; these peaks are due to multiple scattering (MS) events between the ejected electron and neighboring atoms. Although less important to the overall EXAFS than single scattering, MS events are nevertheless very important contributions to the As EXAFS spectra of several crystalline solid-phase species, and will be discussed further in later sections.

Fourier filtering (FF) is a technique used to separate the sinusoidal frequencies represented by FT peaks so that they can be fit separately (not shown). Fourier-filtered data can be used to compare the quality of fits based on different structural models and to perform preliminary analysis of overlapping peaks. Analysis of FF spectra is typically used as an intermediate step in the analysis of complex spectra, after which a final fit is made to the raw data.

The data processing techniques described above are all included in the common XAS data analysis programs (see references in Table 1). Most data analysis programs are freeware, but a few analysis packages have been bundled into commercial software. Detailed discussion of the procedures and pitfalls of data processing (also called data reduction) are given in several recent reports and in many of the guides for commonly-used XAS analysis programs (see references in Table 1).

Processing μ-XRF maps and μ-XRD patterns

Each μ-XRF map collected requires a set of processing steps to be applied so that different maps can be compared quantitatively or qualitatively. Multiple (usually beamline-specific) computer programs are available to perform these operations, and both homemade and commercial packages are utilized. We are most familiar with MicroToolkit, a versatile program which accepts data from several different beamlines (Webb 2013), and the unnamed software package developed by M. Marcus for use on μ-XAS beamline 10.3.2 at the ALS (*http://xraysweb.lbl.gov/uxas/Index.htm*). The data analysis procedures described below are largely derived from options available in either or both of those programs.

Per-pixel, element-specific fluorescence counts in the μ-XRF map are deadtime corrected, then divided by I_0 to account for any variability in incident beam flux. Additional data processing steps may be required to improve or expand data interpretation, and include (1) adjusting thresholds or other display parameters in false color maps to highlight desired features, (2) re-binning multi-channel analyzer (MCA) data to analyze signals that were not initially windowed, (3) importing energy-specific channels for principal component analysis (PCA) or quantitative speciation analysis, and (4) applying smoothing functions to channels suffering from low counts. When processing maps of the same area at multiple energies, registration (or alignment) of maps may be required (this is beamline-dependent). Semi-quantification of elemental concentrations using XRF standards can be performed at this point, as discussed previously. Correlations between elements can be viewed in 2-color or 3-color (RGB) maps, bivariate plots, or by performing PCA on the element channels (more on PCA below).

Microscale XRD patterns are very quick to collect (milliseconds to few minutes), and, in systems where crystalline phases are present, they can provide conclusive identification of As mineral residence (Walker et al. 2005, 2009; Corriveau et al. 2011a,b). The data (2D area detector ring and/or spot patterns) are corrected for camera distortions and calibrated using the diffraction pattern from a calibration standard as previously discussed. After fitting values for these energy-dependent parameters, they are fixed. At this point, identification of phases can be made directly based on comparison of the calibrated *d*-spacings of the ring pattern. If the 2D patterns are somewhat complicated, they can be radially integrated and analyzed by standard XRD powder pattern methods. Several computer programs are available to perform these procedures on 1D or 2D patterns; perhaps the best known of these is Fit2D (Hammersley 2004). Analysis of μ-XRD patterns typically occurs offline (after data collection).

XAS DATA ANALYSIS

Figure 5 presents a flowchart of data analysis methods based on spectral type, complexity of speciation, and the information derived. All of the methods can be applied to bulk XAS data, while μXAS analysis is usually restricted to methods (1), (2), and (4) as enumerated in the introduction. The following paragraphs briefly discuss the advantages and drawbacks of each method; additional details and considerations are discussed in the data analysis review articles collected in Table 1.

Identifying As oxidation states

Loss of one or more valence shell electron (those that participate in bonding with other atoms), increases the binding energy of the remaining core-level electrons, and is observed in As K-edge XANES spectra as a shift to higher absorption edge energy with increasing formal oxidation state (e.g., Fig. 6a). For example, ~8 more eV is required to eject a K-shell electron from As^{5+} than from native As (Fig. 6b; Lowers et al. 2007). Significant differences in bond covalence can also produce binding energy shifts within a given formal oxidation state; the best example is the 1-3 eV increase in binding energy for As^{3+} bound to oxygen (in $NaAsO_2$ $4H_2O$) relative to As^{3+} bound to sulfur in As_2S_3 (Fig. 6b). In general, reduced As in sulfide and arsenide minerals have greater covalent character and edge positions at or below that of elemental As (Fig. 6 a,b).

Qualitative oxidation state analysis for As can be applied to samples of low spectral quality, since the energy position of the absorption edge can often be determined even low-quality spectra. Calibrated spectra may be visually compared, overlain, or actually fit using oxidation state (rather than full speciation) models (e.g., sodium salts or oxides of As^{3+} and As^{5+}, that are

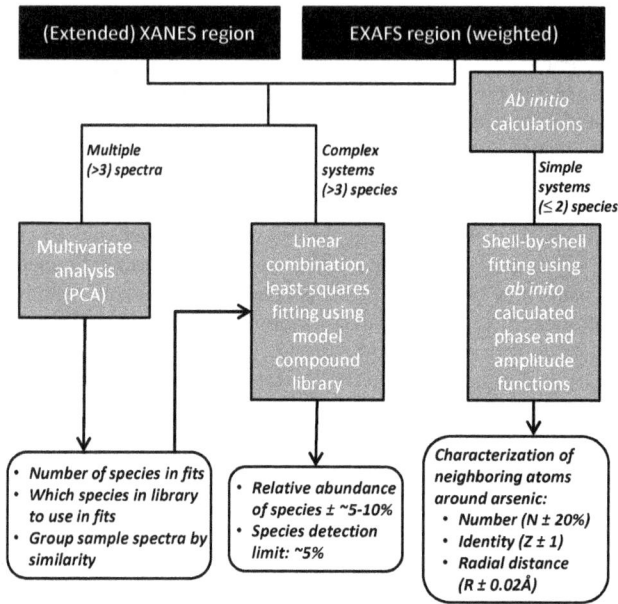

Figure 5. Flowchart of data analysis options for XAS spectra.

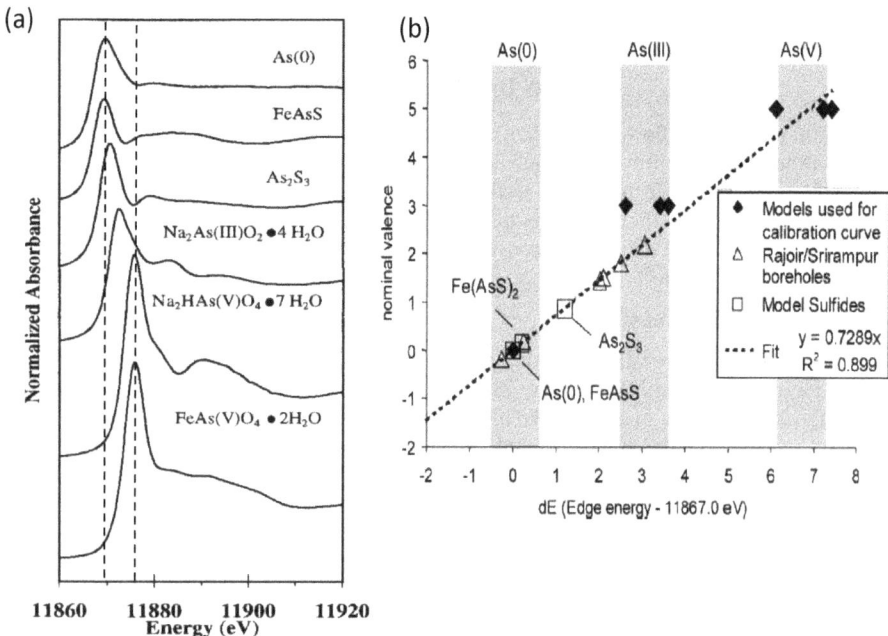

Figure 6. (a) stack plot of representative As-XANES spectra spanning a range of oxidation states; (b) regression of a plot of edge shift vs. nominal valence can be used to determine nominal valence in unknown samples (energy shift is defined relative to native As⁰ edge position at 11867 eV. [(b) Used by permission of Elsevier, from Lowers et al. (2007) *Geochimica et Cosmochimica Acta*, Vol 71, Fig. 6, p. 2706]

not likely to be present in most environmental samples, can still be used for simple oxidation state quantification).

Determining the relative abundance of As oxidation states or As species

Linear combination, least-squares fitting (LC-LSF) provides quantitative determination of species relative abundance in samples containing multiple As species (Fig. 5). LS-LSF results for a set of As K-edge spectra may differ considerably based on the type of spectrum being fit (i.e., XANES vs. EXAFS, k^3-weighted EXAFS vs. unweighted EXAFS; Fig. 4d) due to different spectral sensitivity to As speciation differences. Due to the greater degree of structural detail in the EXAFS region, and due to the high similarity of As-XANES spectra of some important classes of As species (Fig. 7a), LC-LSF on EXAFS spectra is the preferred analysis.

The bulk XAS spectrum represents a sum of all the As species present, weighted by their relative abundances. If good representative (model) spectra of the individual species are available, then the unknown spectrum can be reconstructed from the appropriate weighted sum of the models. The quality of the speciation information returned from LC-LSF fits is dependent on the quality and breadth of the model compound library used in fitting and on the type of sample under investigation. For example, if analyzing soils or sediments, relevant crystalline and amorphous As model phases as well as As sorbed to representative minerals should be included (Fig. 7b,c). Alternatively, if biota samples are to be analyzed, a collection of relevant inorganic, methylated, and organoAs species should be included (e.g., Fig. 1). If redox variations are anticipated in the set of samples to be examined it is also important that the model spectra library include representatives from the entire range of possible natural As oxidation states from −1 to +5 (and for As^{3+} models, representatives of both As^{3+}-O and As^{3+}-S binding). A review of the literature in the area of interest usually gives a good idea of which model compounds are most important for inclusion into a library. All spectra should be carefully (and similarly) calibrated and, if XANES are being analyzed, should have the same resolution for proper analysis.

One or more model spectra can be fit simultaneously to the unknown spectrum using LC-LSF routines. Fit quality is judged by visual inspection and by goodness-of-fit parameters such as the R-factor, χ^2, or reduced-χ^2 value (Ressler 1998; Ravel and Newville 2005; Webb 2005; Kelly et al. 2008). x-axis (energy) shifts should also be small (<1 eV) for fitted parameters (model spectra) and for samples. Complicating the LC-LSF process is the fact that the number and identity of appropriate model spectra is usually not known *a priori*, and there is a risk of overfitting with this type of analysis. If several ($N > 3$) sample spectra are available, principal component analysis (PCA) with target transformation (TT) can be used to constrain the number and identity of model compounds to use in fits (Osanna and Jacobsen 2000; Ressler et al. 2000; Beauchemin et al. 2002; Malinowski 2002); PCA will be discussed further in a later section. If multiple related spectra are not available for PCA, semi-automated methods are available in most programs that can iteratively test the fits produced by: (1) all possible combinations (number and type) of model spectra or (2) a user-specified maximum number of models, trying all possible type combinations (Webb 2005; Ravel and Newville 2005). The numerical fit quality indicators are tabulated for inspection after the fitting completes. The "cycle fit" approach is a hybrid of automated and user-directed fitting which sequentially and systematically adds fit components as long as the fit quality with each additional component improves by a designated proportion, e.g., 10% or more (Kim et al. 2013). Another test of the significance is the "F" test, which is bundled in some data analysis programs (Ressler 1998; Ravel and Newville 2005; Downward et al. 2006).

Experimental spectra collected from physical mixtures of As compounds have been analyzed by several investigators to assess error and detection limits. It should be noted that physical mixtures contain an additional amount of error arising from weighing that is rarely

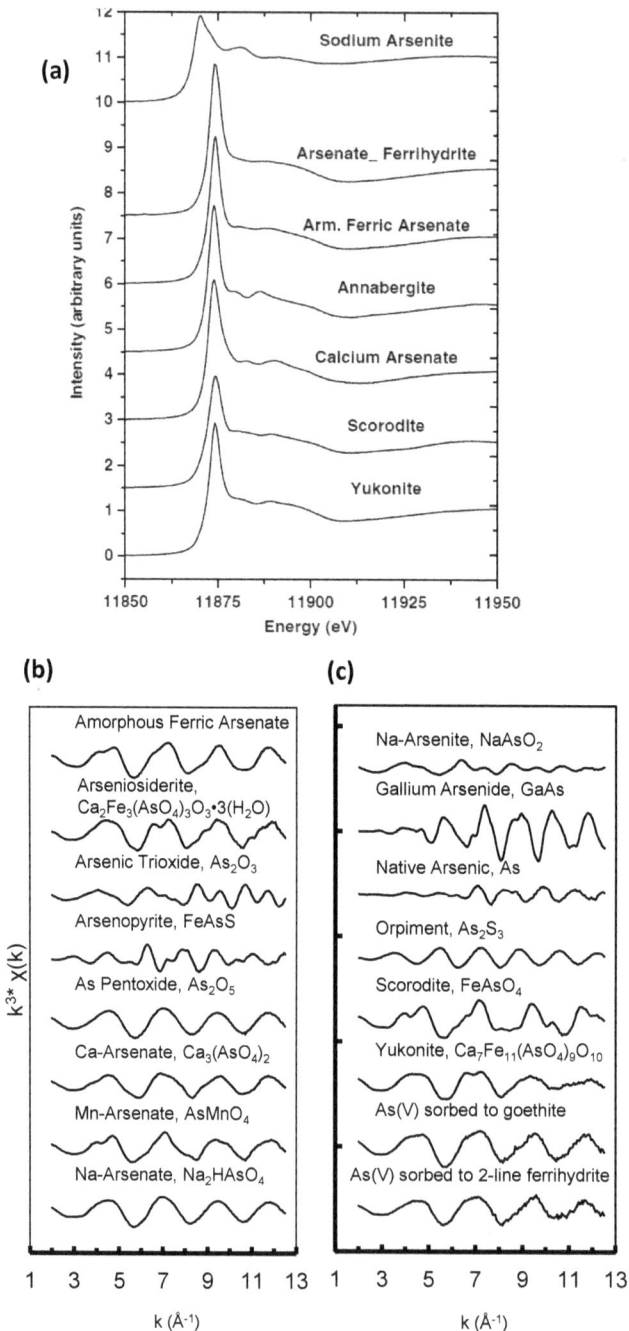

Figure 7. Stack plots of normalized As-XANES spectra (a) and As-EXAFS spectra (b), (c) of several important model As species. XANES spectra of As^{5+}/Fe^{3+} -bearing phases are very similar, but the EXAFS spectra are substantially different. [(a) Used by permission of Elsevier, from Essilfie-Dughan et al. (2013) *Geochimica et Cosmochimica Acta* vol 96, Fig. 9e, p. 345; (b) and (c) Used by permission of American Chemical Society, from Kim et al. (2013) *Environmental Science and Technology*, Vol 47, Fig. 5, p. 8168]

separated from the true quantification error and may lead to fit error underestimation. Arsenic-specific LC-LSF quantification errors range widely depending on the species composition, with greater errors evident in mixtures containing species of a single oxidation state. Morin et al. (2003) reported an error of ±2% for fits to XANES spectra of mixtures of As^{3+} and As^{5+} coprecipitates with ferrihydrite, but reported errors as large as ±30% for fits containing mixtures of As^{3+} compounds. Several investigators have reported errors in the range of ±5-10% for XANES fits to bimodal model mixtures of different As species (Foster et al. 1998a; Takahashi et al. 2003; Campbell et al. 2007; Parsons et al. 2009), and the ±10% value is commonly reported.

Another issue potentially complicating the results of LC-LSF fits to As-XANES spectra is the effect on quantification arising from the peak height differences between the XANES spectra of As^{3+} and As^{5+} species (see Na-arsenite and Na-arsenate spectra in Fig. 6a). While some studies have reported good agreement between fitted and known values (Takahashi et al. 2003), other studies suggest that large errors can be obtained in fits to certain combinations of XANES-spectra (O'Day et al. 2004; Rowland et al. 2005; Campbell et al. 2008; Choi et al. 2009). Campbell et al. (2008) found that fit error varied with composition: when As^{3+}-O was the dominant species, fit errors were below 5%, but when As^{5+}-O became predominant, fit errors increased, up to a maximum of 15%. This error is primarily due to inaccurate fitting of the closely-spaced As^{3+} and As^{5+} peaks in the raw XANES spectrum. Quantitative corrections can be applied to XANES fits to circumvent this problem (O'Day et al. 2004; Campbell et al. 2008). Fitting the 1st or 2nd derivative of the XANES spectrum is another approach that better separates the As^{3+} and As^{5+} XANES peak intensities and leads to better quantification (Foster et al. 1998a; Pickering et al. 2000).

To our knowledge, there have not been similar published studies quantifying errors in fits to EXAFS spectra of physical model compound mixtures. The presence of sulfide-associated As can be clearly distinguished from oxygen-bound As in EXAFS spectra, but oxygen-bound As^{5+} and As^{3+} may be difficult to distinguish by EXAFS alone and is best accomplished using XANES spectra (e.g., Parsons et al. 2009).

Mapping arsenic oxidation states using μXRF

Mapping As oxidation states is achieved by taking advantage of the energy-dependent differences in the fluorescence intensity of As species representing the main oxidation states (see spectra in Figs. 6a and 7a). For a map collected at a given energy, only a subset of the As model XANES spectra will show strong fluorescence. For example, at a map energy of 11,867 eV, As^0 and As^{-1} phases (e.g., As-sulfide/arsenide minerals) will provide the majority of the As fluorescence to the As MCA channel. In As oxidation state mapping, several ($N+1$, where N = number of oxidation states/species) complete XRF maps are collected at a series of closely-spaced energies. A "total" fluorescence map is also collected at ~10-50 eV higher energy, where signal is approximately equivalent from all oxidation states. In the case of As, 5 maps of the same area are required to span all possible oxidation states (Table 3; Root et al. 2013). A matrix of fluorescence intensity at each mapped energy is constructed from the processed XANES spectra of the key model compounds, and this matrix is subsequently used to construct a single map with separate channels for each As oxidation state, allowing a per-pixel view of As speciation in the map. The contribution of up to 3 oxidation states in a mapped area can be viewed in tri-color RGB plots, or PCA analysis can be applied (discussed in the section on multivariate analysis below).

We make a distinction between mapping oxidation states and mapping speciation, because obviously more than one species can be present in a single oxidation state. For As^{3+}, a distinction can be made between bonding to O and bonding to S atoms due to the substantial shift in edge positions (Fig. 6; Table 3). Detailed As speciation mapping (e.g., mapping several

As^{5+} species) remains to be explored in future studies, but may ultimately be limited by the similarity of many As^{5+} XANES spectra (see Fig. 7a). These limitations do not exist for all elements, however: up to 7 different Fe^{2+} and Fe^{3+} species have been quantified in μ-XRF maps (Mayhew and Templeton 2011). The success of speciation mapping depends on sufficient variability in XANES spectral shapes of the model compounds.

Characterizing the structure of solid-phase arsenic species at the molecular-scale

The typical information obtained in a full characterization of a solid phase As species is the identity (Z), number (N), radial distance (R), and information on the magnitude of the root-mean-square thermal/positional disorder (σ^2) of the atoms surrounding As which, from the perspective of XAS, is positioned at the center of an atomic cluster. A good characterization is obtained by reproducing the features of the experimental spectrum (using non-linear, least-squares optimization procedures) using a set of phase and amplitude functions representing scattering between As and the neighboring atoms (see review articles in Table 1). Phase and amplitude functions are generated from theoretical calculations and require information from one or more radial structural models as input (e.g., Fig. 4f). In practice, this means that analysis of EXAFS spectra using theoretical functions is usually restricted to systems in which (1) As concentrations are high enough to collect quality EXAFS spectra and (2) As speciation in the sample is dominated by one to two species with distinctive bonding environments. When the number of species is greater, or when species are too similar, the abundance of non-unique fits to the EXAFS spectrum often confounds this type of analysis.

Several programs are available to generate the theoretical phase and amplitude functions required to fit EXAFS spectra. These programs model core level spectroscopic processes (not limited to EXAFS) from theory (*ab initio*). The program FEFF, in its many versions, is the best-known of these in the US, with some version of it bundled within most data analysis packages (Rehr et al. 2010). Other theoretical packages are available, however (see the "theory" programs listed at *http://xafs.org/Software*). Given the positional and compositional data for an atomic cluster (in our case, with As at the center), *ab initio* EXAFS programs calculate all possible scattering paths between As and neighboring atoms and determine which paths contribute the most to the resulting (theoretical) spectrum. *Ab initio* calculations of XANES spectra are typically performed separately from EXAFS, due to the additional processes (e.g., electronic transitions) that contribute to the shape of the XANES (Bianconi 1988; Durham 1988; Henderson et al. 2014).

Data from these models show that in the EXAFS region of the XAS spectrum, the strongest scattering arises from single interactions between an excited electron and a neighboring atom ("single scattering"). However, in some cases, strong scattering may arise from multiple scattering (MS) of the electron off a single atomic neighbor or off several neighbors sequentially. Multiple scattering manifests in the EXAFS spectrum as one or more individual sinusoidal frequencies and appears in the Fourier transform magnitude as "virtual peaks" which can overlap with other peaks of interest (Fig. 4e). Although MS path amplitudes are usually weak and can often be ignored, the importance of specific types of MS can only be assessed by *ab initio* modeling. Such modeling indicates that MS arising within the As oxyanions and between As polyhedral and second-neighbor atoms (Fig. 8) can contribute significantly to the EXAFS spectra of these species when they are in condensed form (Foster et al. 1998b; Manceau et al. 2007; Voegelin et al. 2007; Wang et al. 2011). The importance of MS in specific systems will be discussed further below.

Once a representative set of phase and amplitude functions is derived from *ab initio* calculations, fitting of an unknown EXAFS spectrum can begin. The EXAFS equation that is minimized in the fit routine is not presented here; the user is referred to the various forms presented in past reviews (Sayers and Bunker 1988; Kelly et al. 2008; Newville 2014) and to data analysis packages for specifics on algorithms (George 2000; Ravel and Newville 2005).

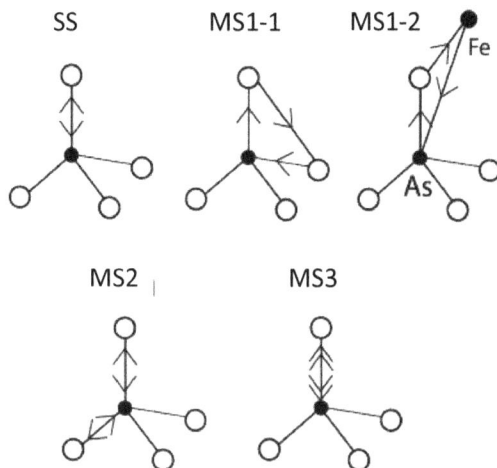

Figure 8. Schematic representation of single-scattering (SS) path As^{5+}-O and the 3- and 4-legged MS paths arising from multiple scattering (MS) in the regular tetrahedron of AsO_4 (MS1-1, MS2, MS3). (Several authors have suggested that MS involving As, O, and neighboring metal atoms (e.g., MS1-2) can also important contributor to scattered intensity. Modified from Manceau et al. (2007).

The phase and amplitude functions (implicitly specific for atomic neighbor Z at a resolution of ±1) are parameterized with values for N, R and σ^2. Two more "global" variables are also important in fits: the $S0_2$ parameter (a global variable) accounts for element-specific "many body" effects that impact the amplitude of EXAFS oscillations, and ΔE_0 corrects x-axis offset between the theoretical functions and the experimental data. ΔE_0 is not technically a global parameter, but since its value is dominated by the nearest neighbor atoms (e.g., the oxygen atoms of As oxyanions), in practice ΔE_0 is refined for the 1st path only, and the ΔE_0 values of more distant paths are fixed at the first path's value (Sayers and Bunker1988; Calvin 2013).

Prior to and in the process of fitting, the user also defines which variables are fixed or allowed to vary during iterative fits, and defines relationships between variables (such as linking values of related parameters). Care must be taken that the number of floating variables in fits does not exceed the number of independent data points (N_{idp}) in the EXAFS spectrum to be fitted (Sayers and Bunker 1988; Kelly et al. 2008). N_{idp} can be maximized by maximizing the k-range of the EXAFS spectrum. It has been noted that for typical EXAFS spectra, a maximum of about 10 independent parameters can be refined (Calvin 2013). Care also must be taken in the co-refinement of highly correlated variables. Strong correlations exist among the variables N, σ^2, and $S0_2$, all of which affect the amplitude of EXAFS oscillations. The variables R and E_0 are also strongly correlated. Some programs provide the correlation matrix for inspection, and some also provide warning flags for correlations greater than 0.8 (George 2000; Ravel and Newville 2005). In practice, variables are often linked or fixed in order to minimize N_{idp} and to avoid artifacts from highly correlated variables.

Fitting EXAFS spectra is an iterative process. First, it is recommended to fit a spectrum of a material with known structure: since R and N are known for each path and can be fixed, values for $S0_2$, ΔE_0, and σ^2 (additional path-specific variables along with N and R) can be determined with minimal correlation to other parameters. Most researchers use a single value of $S0_2$ for all EXAFS fits to a given element (e.g., all fits to As K-edge EXAFS spectra); this value is usually determined after fitting several model compounds of known structure (Calvin 2013), and is usually between 0.8 and 0.9 for most elements including As.

The first shell of nearest neighboring atoms is fit first (raw EXAFS, Fourier-filtered EXAFS, or the FT imaginary part can be fit), with "initial values" parameters for the other variables based on previous fits to known structures. $S0_2$ was discussed above; typical refined values for ΔE_0 are usually between 10 and -10 eV, so 0 is a good starting guess. Typical refined values for σ^2_{As-O} (first-shell oxygen in the As oxyanion) range from 0.001-0.003 Å^2; however, note that temperature strongly influences σ^2 so that values determined from fits to spectra collected at cryogenic temperature are often substantially lower than fits to the same data at room temperature (especially for crystalline materials). Initial guesses for R and N are made from the literature; N is usually held constant in initial refinement so that a good value for σ^2 can be determined.

After the first shell is refined satisfactorily, one or more shells at greater distances are added sequentially, with the fit re-optimized after each addition. The ΔE_0 value for second and higher shells is linked (set to equal) that of the first shell. Typically no more refinements are made to ΔE_0 until the final fit (discussed below). Whole-number guesses for N are appropriate, and are made based on the type of species hypothesized to be present; for example, adsorbed complexes typically have 1 or 2 second-shell neighbors. Initial values for σ^2 are taken from fits to models compounds or from the literature. Typical refined values for second-shell σ^2_{As-M} (M = metal atom in adsorption complex) range from 0.005-0.008 Å^2. The dampening effects of disorder increase with distance from the central As atom, so σ^2 should grow larger with successive shells; however, values above $\sim$0.1 Å^2 force abnormally large N values to compensate, and are therefore regarded with suspicion (recall that N and σ^2 are highly correlated variables). Fits are often very sensitive to σ^2 values, and for fits with multiple neighboring atoms beyond the first shell, σ^2 values are usually fixed or linked to prevent the fit from minimizing on unreasonable (e.g., negative) values. After refining fits to Fourier filtered data, or after sequential addition of paths to the raw data fit, the final fit should be made to raw EXAFS data with as many of the parameters floated (and/or linked) as possible (Calvin 2013).

When fitting As-EXAFS spectra of the As oxyanions, paths due to multiple scattering (MS) within the As tetrahedron (Fig. 8) should be included because they overlap with single scattering (SS) paths from neighboring atoms in the 3-4 Å region. Voegelin et al. (2007) demonstrated that these MS paths can be included in the fit without adding a single additional parameter. N_{MS} is known from geometrical considerations, and in this case, is a fixed number for each path type (MS1, MS2, etc.; see Fig. 8). Usually, one representative from the different MS path groups is chosen to represent the rest; its R_{MS} and σ^2_{MS} values are defined in terms of the corresponding parameters of the single-shell path using geometric considerations (for R_{MS}) and correlations between individual path leg lengths (for σ^2_{MS}).

All data analysis programs report errors of and correlations among the floated variables. Typical "average" reported errors are $N \pm 20\%$, and $R \pm 0.02$ Å (it is not common to report "average" errors for other values). Most programs also give variable-specific error estimates that should be reported in fits. These errors are likely to underestimate the true values, but rigorous error analysis is difficult for XAS spectra as has been discussed in several reviews (George 2000; Ravel and Newville 2005; Kelly et al. 2008; Newville 2014). All of the common data analysis/fitting software packages will allow basic fitting to EXAFS spectra as described above, but they also vary widely in their architecture, flexibility, and functionality. For example, most programs allow constraints that can be placed on floated parameters (such as non-negativity). Table 1 provides a website link where descriptions and downloads of current programs can be found.

Several suggestions have been made regarding best-practice procedures for estimating and reporting *ab initio* fit errors (and uncertainties) as well as testing the statistical significance of different possible fit results (Teo 1986; Sayers and Bunker 1988; IXS Standards and Criteria Committee 2000; Ravel and Newville 2005; Grafe et al. 2008b; Calvin 2013; Newville 2014).

In addition to the F-test mentioned previously for LC-LSF fits (Downward et al. 2006), for *ab initio* fits some have used a criterion that added shells must improve the goodness of fit parameter $\chi^2_{reduced}$ by at least 20%; others have used equivalence of the fit residual magnitude to that of the noise level of the EXAFS as a stopping point for the inclusion of additional shells to *ab initio* fits (Couture et al. 2013).

Analyzing bulk and microbeam datasets using multivariate techniques

The application of multivariate data exploration/analysis techniques to XAS and XRF data sets collected at synchrotrons has increased in recent years. Principal component analysis is by far the most popular technique employed to analyze bulk XAS data sets as well as μ-XAS and μ-XRF map data sets, and modules to perform the analysis are included in several popular data analysis packages (Ressler 1998; Ravel and Newville 2005; Webb 2005, 2006). Details of the mathematics of PCA (and other multivariate techniques applied in spectroscopy) are given in Malinowski (2002), and a mathematical formalism of PCA specifically applied to XAS spectra was presented in Beauchemin et al. (2002). The application of PCA to μ-XRF map datasets is completely analogous to its use in exploration of other types of spatially-resolved spectroscopic datasets (Osanna and Jacobsen 2000; Lerotic et al. 2004).

Given a set of processed experimental XAS spectra (XANES or EXAFS, bulk or microbeam), PCA provides several pieces of information without requiring model compound spectra or theoretical calculations based on model structures. First and most important of these is the number of *significant* (i.e., principal) components (a subset of the total number of components, which is always equal to the number of experimental spectra in the set). Although these components are mathematical constructs with no physical significance, they are created from the experimental spectra, and therefore can be used to reconstruct them precisely. Further, the number of *significant* components is assumed to be equivalent to the number of *unique* spectra present in the dataset, and therefore to the number of unique chemical species present. Determining the cutoff between significant and insignificant components is typically very easy in model systems but can be less obvious in datasets that contain significant noise and/or spectral outliers (Brown et al. 2010), or in natural systems (Foster et al. 2011). Differences in calibration and normalization can also introduce artifacts into the PCA that can skew the results significantly. Techniques that help to locate the cutoff between significant and insignificant components in an intuitive manner include: (1) visual examination of the components, (2) spectral reconstruction (determining the minimum # of components required to reconstruct all spectra), and (3) plots of variance accounted for by the sum of components (i.e., "scree" plots). There is also the option of inspecting the IND (indicator) parameter, an empirical function developed for this purpose that is calculated by some PCA programs (Malinowski 2002; Ressler 1998; Webb 2005). Minimization of IND occurs at the last significant component (Malinowski 2002).

The next most common application of PCA to the analysis of XAS spectra is to aid in the identification of appropriate model compounds to use in linear combination, least-squares fits to the experimental spectra. Through the process of target transformation (TT), each potential model spectrum is tested for suitability by the degree to which it can be reconstructed using the principal components determined from a PCA run (Malinowski 2002). The quality of the model compound reconstruction can be checked visually, or by using the empirical SPOIL parameter which has values classified as follows: <1.5 = excellent; 1.5-3.0 = good; 3.0-4.5 = fair; 4.5-6.0 = poor; >6.0 = not acceptable (Malinowski 2002). Complicating the analysis of TT results is the fact that models of somewhat similar spectra can have similar reconstruction quality and SPOIL values, making it difficult to select among them, and requiring testing by LC-LSF to determine which provides the best fits to the actual data. In some cases, models will be so similar that LC-LSF fits using these models are essentially equivalent, and selecting among them is not possible; in this case one model spectrum should be chosen to represent the group of similar spectra, and the results should be stated so as to reflect this grouping (e.g., As^{5+} sorbed to gibbsite and As^{5+} coprecipitated with $Al(OH)_3$ might be grouped together under the title "As^{5+} sorbed to Al oxyhydroxides").

Applied to the analysis of μ-XRF map data (either of elements, As oxidation states, or As species), PCA can aid in the location of chemically-distinct points of interest for subsequent collection of μ-XAS spectra. Furthermore, since most modern beamlines regularly collect complete fluorescence spectra per map pixel, PCA can be used to identify the primary chemical components in the sample and define the number of different chemical groupings in the system (often these represent chemically-distinct minerals, particles, or regions of the sample). Since real-time μ-XRF map analysis is required for the selection of POI for μ-XANES data collection during beamtime periods, the speed and simplicity of PCA analysis is a great asset. Specific examples of μ-XRF PCA applications will be presented in later sections.

Use of PCA results to construct variance plots that visually display the relationships among individual samples (bulk XAS) or analyzed volumes within a sample (microXAS) is common in ecological studies but rare in XAS studies (Ressler et al. 2000). Foster et al. (2011) used a PCA-derived variance plot to separate As-EXAFS spectra from a mining-impacted area into groups based on spectral similarity alone. After subsequent LC-LSF using a library of model compound spectra, the PCA-derived spectral groups could be explained on the basis of differences in As speciation (Fig. 9a,b).

XAS STUDIES OF ARSENIC SPECIATION IN SOLIDS

Arsenic mineralogy

Hundreds of As minerals have been identified (see references in Table 1 and chapter 2 in this volume), but of these only a few are commonly found in the environment (Welch et al. 1988). By definition, As minerals are crystalline solids with defined stoichiometry; therefore they can be studied by a variety of electron- and X-ray based techniques that will not be covered here. However, high-As poorly-crystalline and X-ray amorphous phases (not true minerals) can also be extremely important in some systems. Below, we summarize the means by which As minerals are identified by XAS at the bulk and micron scale. In later sections, we will mention specific uses of μ-XRD for identification of As minerals but will not discuss the techniques; the reader is instead referred to reviews of the topic (Table 1) and recent applications (Walker et al. 2005, 2009; Jamieson et al. 2011).

Bulk and microscale XAS (and μ-XRF/XRD) have been particularly valuable for studies of As mineralogy in at least four ways: (1) identification of As minerals that are present at levels below detection by conventional XRD; (2) documentation of mineral alteration at the micron scale to better understand the processes behind As mobility and sequestration; (3) characterization of X-ray amorphous or poorly crystalline phases at the molecular scale to identify crystalline analogs and gain insight into (re)crystallization processes/phase transformations; and (4) testing and validation of XAS theoretical fit procedures and evaluation of the importance of multiple scattering (MS) contributions to As-EXAFS spectra. Topics 1 and 2 will be covered below; topics 3 and 4 will be discussed in the case studies described in later sections.

Metal arsenides, arsenic sulfides, and native As. Arsenopyrite (FeAsS), orpiment (As_2S_3), and realgar (AsS or As_4S_4) are the members of this group most commonly identified in environmental samples, typically forming at hydrothermal or magmatic temperatures (>150 °C), and often found in association with other sulfide minerals (O'Day 2006). The molecular-scale coordination of As in this group is diverse enough that EXAFS spectral fingerprinting can usually be used to distinguish them from one another and also from As-substituted iron

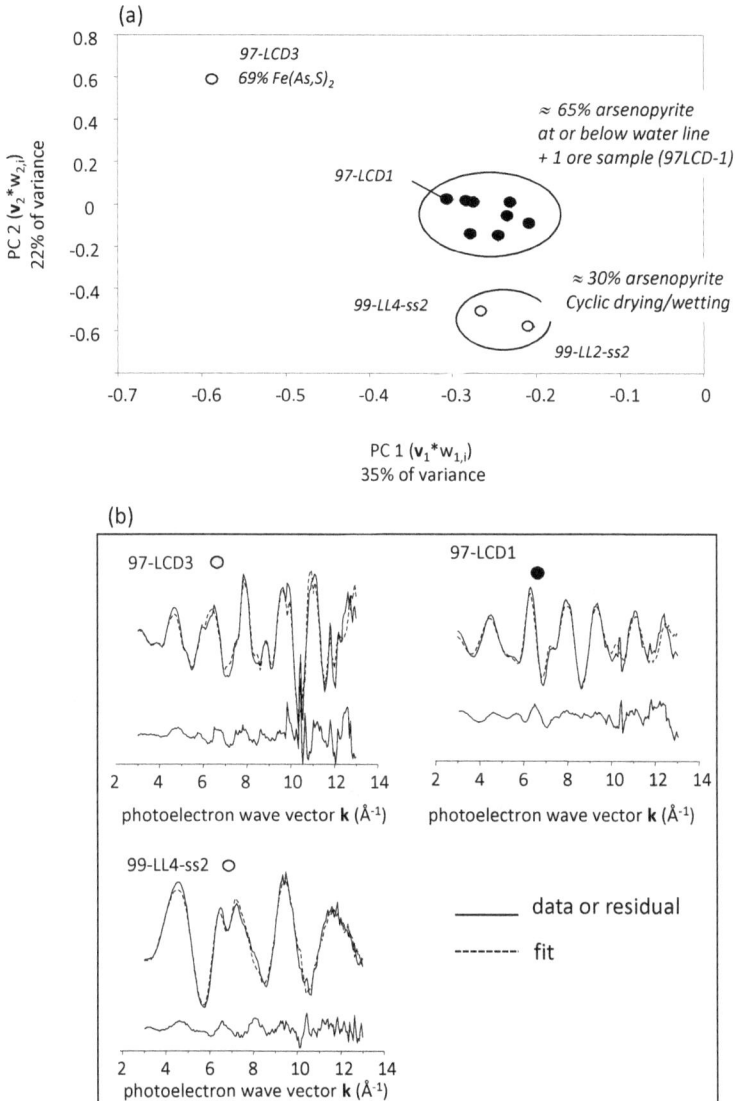

Figure 9. (a) PCA-generated spectral similarity plot of k^3-weighted EXAFS spectra of samples from a mine-impacted area; (b) species determined from LC-LSF can be used to explain the observed PCA groupings. [Used by permission of Chemistry Central Ltd, from Foster et al. (2011) *Geochemical Transactions*, Vol 12, Fig. 3, p. 10 and Fig. 5, p. 12]

sulfides (pyrite, marcasite; Fig. 10a and 10b; Savage et al. 2000b). Within the metal arsenides, formal oxidation state assignments are generally not helpful since bonding is strongly covalent (O'Day 2006). For example, the absorption edge position (typically measured at the edge inflection point) of the orpiment XANES spectrum occurs at higher energy than As^0 or FeAsS but at lower energy than sodium metaarsenite $[NaAs^{3+}O_2)_x]$, which has the same formal oxidation state (Fig. 6a,b). *Ab initio* calculations of the EXAFS spectra of As^0, FeAsS, and orpiment (ambient temperature) indicate that multiple scattering (MS) paths are not required

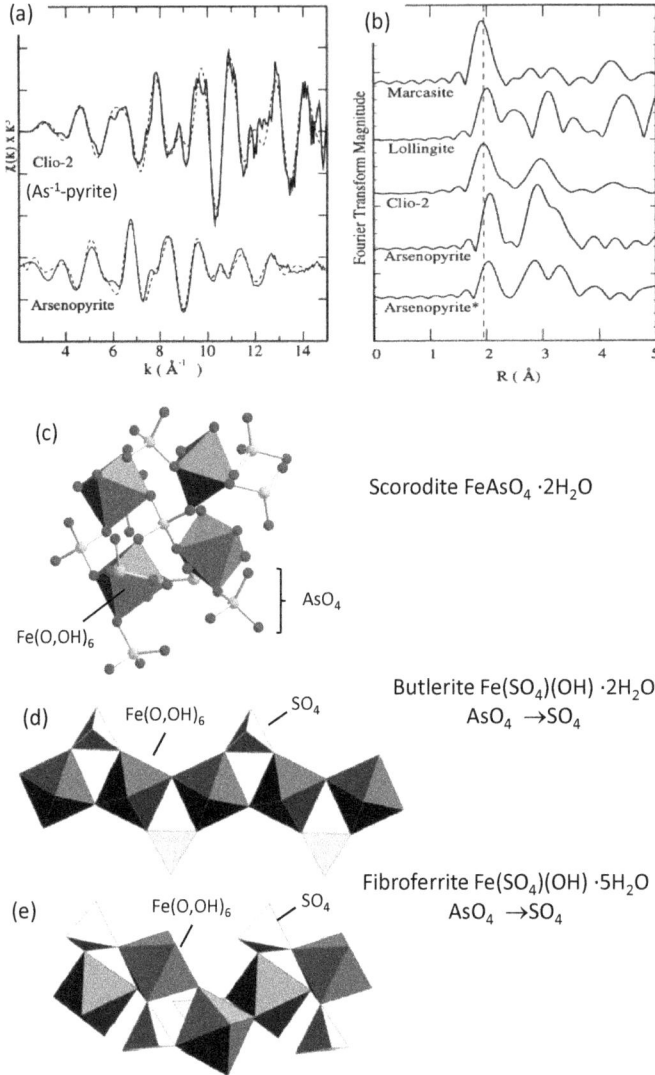

Figure 10. (a) k^3-weighted As-EXAFS spectra (solid lines) and least-squares fits (dotted lines) of Clio-2 sample (As^{-1} pyrite) and arsenopyrite; (b) FT magnitude of As EXAFS spectra representing several sulfide coordination environments [including those in (a)]; (c) short-range structure of As in scorodite includes 4 Fe atoms at 3.34 Å; (d, e) proposed structures for amorphous ferric arsenate based on sulfate minerals [Figures (a, b): Used by permission of Elsevier, from Savage et al. (2000b) *Applied Geochemistry*, Vol 15, Fig. 4, p. 1229. Figure (c): Used by permission of Elsevier, from Savage et al. (2005) *Chemical Geology*, Vol 215, Fig. 11, p. 493. Figure (d, e): Used by permission of Elsevier, from Paktunc et al. (2008) *Geochimica et Cosmochimica Acta*, Vol 72, Fig. 19, p. 2667]

for the reproduction of the major features of their experimental EXAFS spectra (Foster et al. 1998b); this finding may be generally true for other arsenides and As sulfides, but has not to our knowledge been tested. Realgar contains As-As bonds and orpiment does not, so they can be clearly distinguished by As-EXAFS (Helz et al. 1995). Furthermore, amorphous orpiment can be distinguished from its crystalline isomorph because its FT magnitude lacks peaks beyond the closest S neighbors (Helz et al. 1995). Several specific examples of the identification of arsenide, As sulfide, and sulfosalt minerals by As-XAS and/or μ-XRD in environmental samples will be provided later in the chapter.

Arsenite minerals. Several alkali-, alkaline earth, and metal arsenite minerals have been described (see references in Table 1), but outside of unique mineral deposits (e.g., Tsumeb mine; Bowell 2014, this volume), they are typically found at sites impacted by mining or industrial contamination.

Arsenolite (As_2O_3) is probably the most common arsenite mineral in the environment, and is primarily of anthropogenic origin. Although it can form as a weathering product of As sulfides, and has also been found in hot spring or volcanic fumarole deposits, its presence in common soils is overwhelmingly due to the production, storage, and application of arsenical pesticides/herbicides, as well as from the practice of roasting and/or smelting As-bearing ore to enhance metal recovery or for As_2O_3 manufacture (Morin and Calas 2006; O'Day 2006; Jamieson 2014, this volume). The high symmetry of this phase gives rise to distinctive characteristic XANES and EXAFS spectra (Fig. 7b). *Ab initio* analysis of the MS features in As_2O_3 has not been performed to our knowledge, but we would expect MS to be significant in this phase due to its high symmetry. $As^{3+}O_3$ groups polymerize in sodium metaarsenite $(NaAsO_2)_x$, which brings the two linked As atoms closer together than in most oxyanion minerals and has a strong effect on the shape of the EXAFS spectrum. The ability of $As^{3+}O_3$ to form polymerized solid-phase species distinguishes it from AsO_4 tetrahedra and affects the types of complexes it forms on mineral surfaces (Morin et al. 2009; Wang et al. 2010).

The ferric arsenite-sulfate mineral tooleite [$Fe^{3+}_6(As^{3+}O_3)_4(SO_4)(OH)_4 \cdot 4H_2O$] has been identified in cemented layers of precipitated As^{3+}-rich Fe^{3+} oxyhydroxides formed at a French site impacted by Pb-Zn-Sb mining (Morin et al. 2003; identified by XAS) and at a Canadian site impacted by gold mining (DeSisto et al. 2011; identified by μ-XRD). Tooleite was considered to be a ferric arsenate phase before analysis of its XANES and EXAFS spectra. Other, as-yet unidentified, arsenite phases may be present in some environments, as suggested by EXAFS data (Morin et al. 2003).

Arsenate minerals. The arsenate minerals are a highly diverse group that generally forms isomorphous analogs with phosphate minerals and are the most common As minerals found in the environment (O'Day 2006). They are often identified as products of weathering and oxidation of As-rich deposits, but their industrial production and use was widespread in the past. The solubility of arsenate minerals is quite variable across the group (Robins 1981; Essington 1988); for this reason, identification of specific phases in environmental samples is key to the geochemical modeling of As fate and transport in systems where they occur. A variety of arsenate minerals including Fe^{3+}-, Ca-, Ca-Fe-, Mg, Pb-, and Zn forms have been identified in environmental samples using μ-XAS/μ-XRD/μ-XRF. Because of their particular importance in many natural systems, the Fe and Ca/Ca-Fe arsenates are discussed further below. Other types of metal arsenates (including Mg-, Cu-, and Zn-arsenates) have been studied by XAS (primarily at sites of industrial contamination) and are discussed in the topical sections to follow.

Ferric arsenates and ferric arsenate-sulfate minerals are important in mining-impacted environments (Camm et al. 2003; Jamieson et al. 2008; Kim et al. 2013) and in some processes designed to lower dissolved As concentrations in mine-waste effluents (Paktunc et al. 2008; Paktunc and Bruggeman 2010; Paktunc et al. 2013). Several polymorphs and hydration states of ferric arsenate exist, and their variable solubility has serious implications for the mobility of As, especially under various mine waste treatment scenarios (Langmuir et al. 2006; Paktunc et al. 2008). As-XANES spectra of the ferric arsenate polymorphs (including amorphous variants) are too similar to distinguish among these phases when present in multi-phase mixtures (Fig. 7a), but the EXAFS spectra of the same phases are often significantly different (Fig. 7b,c).

Under synthesis conditions at ambient temperature and pressure, and with Fe/As ratio between 1 and 4, a material known either as hydrous ferric arsenate (HFA) or amorphous ferric arsenate (AFA) forms, with the approximate formula $FeAsO_4 \cdot 4\text{-}7H_2O$ (Paktunc et al. 2008). HFA can also form on the surface of ferrihydrite under acidic conditions (Mikutta et al. 2013a). HFA is a precursor to scorodite ($FeAsO_4 \cdot 2H_2O$), which can form by aging HFA or directly in acidic solutions with Fe/As ≤ 1 (Paktunc et al. 2008). Prior to As-XAS analysis, HFA was presumed to be an amorphous equivalent of scorodite based on 2 broad reflections in XRD patterns. Currently, there are 2 models for the structure of HFA/AFA, both of which are based on theory-based EXAFS fits at both the As- and Fe K-edges and supported by additional diffraction and/or scattering techniques. The "framework" model hypothesizes that HFA/AFA is composed of nanocrystalline scorodite, in which Fe coordination will be reduced from the ideal 4 neighbors in crystalline scorodite (Fig. 10c; Mikutta et al. 2013a,b). The alternative "chain" model proposes a completely different structural motif for HFA/AFA based on known ferric sulfate minerals in which AsO_4 is bound through apical oxygens of corner-sharing $Fe(O,OH)_6$ octahedra linked in a chain-like structure (Fig. 10d,e; Paktunc et al. 2008; Paktunc and Manceau 2013). The current debate in the literature hinges on the relative importance of multiple scattering paths between As, O, and neighboring Fe atoms (Mikutta et al. 2013b; Paktunc and Manceau 2013).

In a study of the potential solid-phase products of pressure oxidation of sulfide-rich refractory gold ore, Paktunc et al. (2013) identified the formation of ferric orthoarsenate $FeAsO_4 \cdot 0.75H_2O$, As-substituted basic ferric sulfate (BFS), and ferric arsenate sulfate (FAS) minerals as potential reaction products. The EXAFS-derived local structure of BFS and FAS was consistent with the above-described "chain" model for HFA/AFA, with SO_4 and AsO_4 occupying the same structural site.

Ca-arsenate minerals are present in some types of coal fly ash (Huggins et al. 2007; Zielinski et al. 2007; Catalano et al. 2012) and have been documented forming in surface soils at several industrial sites (Arai et al. 2006; Yang and Donahoe 2007; Cancès et al. 2008). Several polymorphs also exist in this group, including one that is isostructural with apatite, $Ca_5(PO_4)_3(F,Cl,OH)$. The XANES spectrum of Ca-arsenate orthoarsenate, $Ca_3(AsO_4)_2$, is easily distinguished from the XANES spectra of ferric arsenates (Fig. 7a) and from spectra of As^{5+}sorbed on various phases, including calcite. As a result, several studies have simply used the XANES spectrum for identification of Ca-arsenate in heterogeneous samples. Conversely, the XANES spectra of Ca-Fe arsenates (e.g., arseniosiderite, yukonite) are very similar to that of scorodite, though they can be discriminated on the basis of EXAFS spectral shape (Fig. 7a-c). A recent study that included EXAFS analysis presents strong evidence that the two phases known as arseniosiderite and yukonite may actually be a single phase exhibiting different habits (Gomez et al. 2010). Obtaining pure end-members appears to be an impediment to the resolution of the controversy regarding the structures of arseniosiderite and yukonite.

A distinctive splitting can be observed in the first EXAFS oscillation of many arsenate minerals that can be used to aid in their identification or quantification in multi-species samples: it is very strong in crystalline Fe^{3+} and Mn^{4+} arsenates, is weaker in amorphous analogs of those phases (such as HFA/AFA) and in crystalline Ca-and Mg-arsenates, and is not observed in Na-arsenate (Fig. 7b,c). *Ab initio* calculations indicate that the splitting is a combination of MS paths and neighboring atoms at ≥4 Å radial distance (Foster et al. 1998b, 2003).

Arsenic oxyanion sorption on mineral surfaces and soils

Sorption is an important means by which As is scavenged from the aqueous phase and sequestered in the solid phase, thus reducing its bioavailability and potential toxicity. The extent, pH behavior, and reversibility of As sorption are dependent on sorption mechanism. Figure 11 schematically represents 4 of 5 important classes of As sorption: outer-sphere adsorption (OS), inner-sphere (IS) adsorption, surface (co-) precipitation, and isomorphic substitution; the 5th

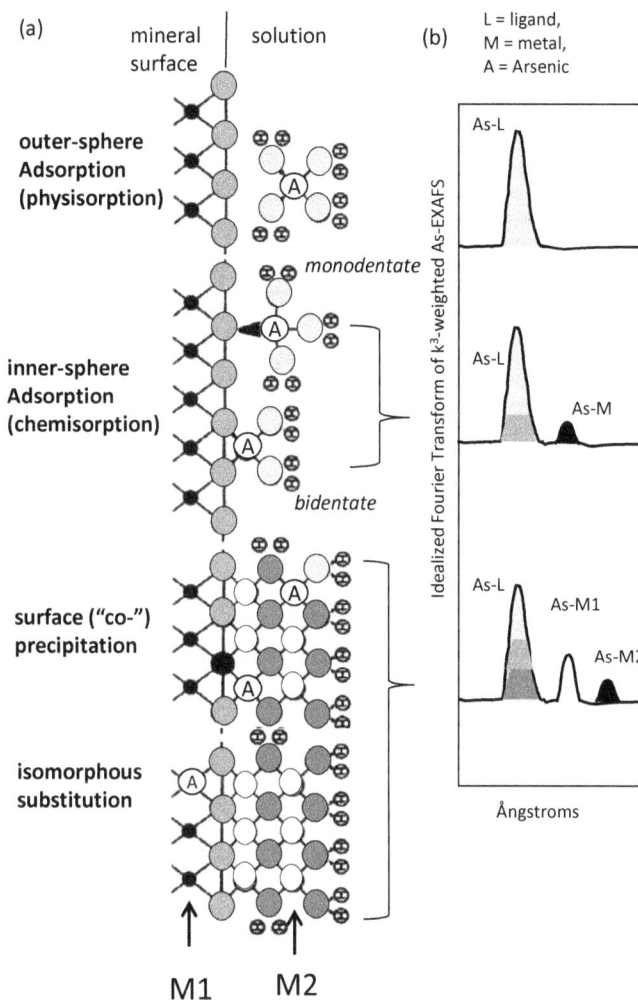

Figure 11. (a) Schematic of the 3 basic modes of As sorption on mineral surfaces and schematic of isomorphic substitution; (b) schematic illustration of FT magnitude associated with each type of sorption complex. With increasing number and/or ordering of atomic neighbors, more peaks in the FT can be observed. Modified from Brown et al. (1995).

type, occlusion, is not shown. The element specificity of XAS, its sensitivity to short range structure, and mg kg^{-1}-level detection limit makes it a particularly effective tool for examining sorption complexes on mineral surfaces (Brown et al. 1999; Brown and Sturchio 2002; Brown and Calas 2012; Wogelius 2013), and has been extensively applied to model system studies of As sorption on various mineral substrates, as summarized in Appendix 1.

Adsorption is defined as an accumulation of sorbate at the interface between an aqueous solution and a solid adsorbent phase without the development of a three-dimensional molecular arrangement (Sposito 1984). Outer-sphere adsorption (also called non-specific adsorption or physisorption) is characterized by the retention of one (or more) hydration shell(s) of water molecules between the adsorbate and the surface (Fig. 11a); attraction is through a combination of hydrogen bonding and electrostatic forces, and is very sensitive to the charge on the adsorbate and on the surface, which are both dependent on solution conditions. The EXAFS spectrum of an OS As oxyanion sorption complex [e.g., the arsenate anion (AsO_4^{3-})] will contain one frequency derived from the nearest neighbor ligands (in this case, 4 oxygen atoms), and its corresponding FT magnitude will display a single peak (Fig. 11b). Occlusion (non-specific As retention in pore structures), anion exchange in mineral interlayers (e.g., As retention in layered double hydroxide minerals), and substitution of As in fully-hydrated sites in mineral structures [e.g., AsO_4 for SO_4 in ettringite $Ca_6Al_2(SO_4)_3(OH)_{12}·26H_2O$] can also produce EXAFS spectra consistent with outer-sphere adsorption (Myneni et al. 1997; Wu et al. 2013).

Inner-sphere adsorption (also called specific adsorption or chemisorption) occurs through the loss of hydration waters and the formation of one or more direct chemical bonds with the mineral surface (Fig. 11a). The FT magnitude of an isolated IS As oxyanion surface complex will contain one or more peaks beyond the nearest neighbor peak; the distance(s) are often characteristic of the binding geometry (Fig. 11b). Often multiple IS complexes exist in a single sample, in which case 1, 2, or even 3 (usually overlapping) peaks in the FT can be seen at distances characteristic geometry of the sorption complexes. EXAFS-derived adsorption complex geometries for As adsorbed to metal hydroxide mineral surfaces are shown in Figure 12 and include monodentate corner sharing (^1V, one bond formed at a vertex site), bidentate corner-sharing (^2C, two bonds at a corner site), bidentate, edge-sharing (^2E), and tridentate complexes (^3C, three bonds).

Co-precipitation occurs when As is present in solution during formation of other minerals, or when As is present at high concentrations near mineral surfaces (Fig. 11a). Arsenic can substitute randomly or non-specifically in the structure of the forming mineral (occlusion), or can form a new mineral, the structure of which is sometimes epitaxially-controlled (e.g., Bolanz et al. 2013). Isomorphous substitution of As in specific crystallographic sites of non-As minerals is also common and has been documented in metal sulfides, sulfates, and other minerals (Fig. 11a; Savage et al. 2000b; Paktunc and Dutrizac 2003; Savage et al. 2005; Paktunc et al. 2013). Arsenic concentrations can range from trace to weight percent in these two types of sorption complexes, depending on the solution conditions. The FTs of co-precipitate EXAFS spectra often contain multiple peaks due to multiple shells of neighboring atoms in the precipitate structure (Fig. 11b). The EXAFS-derived As-Fe interatomic distances of co-precipitated As^{5+} in ferrihydrite are too similar to the As-Fe distances of As^{5+}sorbed to ferrihydrite to allow distinction between the two (Appendix 1). However, isomorphic substitution usually produces distinct features in the EXAFS spectrum and distances in the FT magnitude that can be used to distinguish it from other types of sorption.

General conclusions regarding As oxyanion sorption. The large number of model-system studies in this area have almost exclusively employed curve fitting with theoretical standards as the primary means of data analysis. Review of these studies (collected in Appendix 1) leads to several general conclusions regarding the behavior of As oxyanions during sorption on mineral surfaces. First, both $As^{5+}O_4$ and $As^{3+}O_3$ predominantly form IS complexes on mineral surfaces, rather than OS complexes. Second, the two oxyanions maintain their respective coordination geometries in all 5 of the sorption modes listed above, as indicated by the relatively invariant parameters of coordination number (N) and radial distance (R) obtained in EXAFS fits to the first shell oxygen atoms. Specifically, typical R_{As-O} values for $As^{5+}O_4$ sorption complexes range from 1.68-1.72 ± 0.02 Å, and for $As^{3+}O_3$ sorption complexes range from 1.74-1.79 ± 0.02 Å. The observed consistency in values may in fact be due to the nearly universal practice of fitting the O of the oxyanion species as one average shell, rather than considering them separately, or creating subgroups. Sherman and Randall (2003) presented theoretical calculations and EXAFS fits suggesting that upon sorption, O coordination is distorted so as to give short and long distances (in the case of AsO_4, the short distances occur at 1.62 and 1.67 Å and the 2 long distances are equivalent at 1.71 Å).

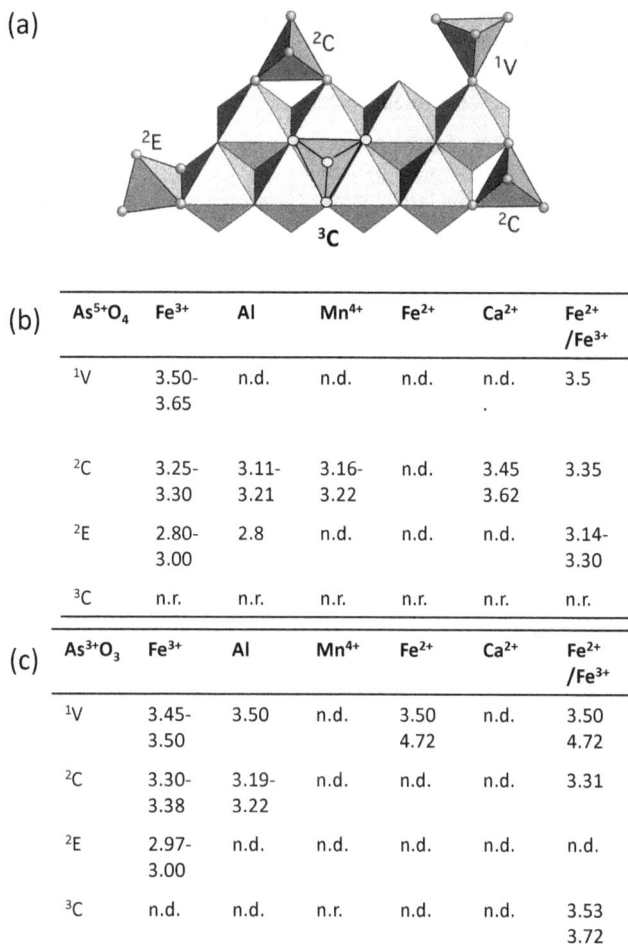

Figure 12. (a) Polyhedral representations of $As^{5+}O_4$ inner-sphere sorption (chemisorption) geometries on a model metal hydroxide (geometries are analogous for $As^{3+}O_3$ complexes). The shorthand notation assigned to various bonding configurations is used in much of the sorption literature, and throughout this text. (b) and (c) Compilation of EXAFS-determined interatomic distances (R) between AsO_4 (b) and AsO_3 (c) and various metal atoms (coordinated to oxygen/hydroxyl) resulting from the bonding configurations in (a). [Used by permission of Elsevier, from Sherman and Randall (2003) *Geochimica et Cosmochimica Acta*, Vol 67, Fig. 1, p. 4223]

(b)

$As^{5+}O_4$	Fe^{3+}	Al	Mn^{4+}	Fe^{2+}	Ca^{2+}	Fe^{2+}/Fe^{3+}
^1V	3.50-3.65	n.d.	n.d.	n.d.	n.d.	3.5
^2C	3.25-3.30	3.11-3.21	3.16-3.22	n.d.	3.45-3.62	3.35
^2E	2.80-3.00	2.8	n.d.	n.d.	n.d.	3.14-3.30
^3C	n.r.	n.r.	n.r.	n.r.	n.r.	n.r.

(c)

$As^{3+}O_3$	Fe^{3+}	Al	Mn^{4+}	Fe^{2+}	Ca^{2+}	Fe^{2+}/Fe^{3+}
^1V	3.45-3.50	3.50	n.d.	3.50-4.72	n.d.	3.50-4.72
^2C	3.30-3.38	3.19-3.22	n.d.	n.d.	n.d.	3.31
^2E	2.97-3.00	n.d.	n.d.	n.d.	n.d.	n.d.
^3C	n.d.	n.d.	n.r.	n.d.	n.d.	3.53-3.72

The third conclusion to be drawn from the wealth of existing studies is that As^{5+} and As^{3+} oxyanions usually adopt the same dominant coordination geometry on trivalent and tetravalent metal hydroxide phases (Fig. 12; Appendix 1). Despite the abundance of possible sorption complexes that could occur on metal hydroxides based on geometric considerations, only four have actually been determined by EXAFS analysis, and of those, the ^2C (bidentate, corner-sharing) complex (Fig. 12) is predominant for both oxyanions over a broad range of solution conditions. The reported differences in interatomic distances between the As^{5+} or As^{3+} atom and the nearest neighboring metal (Me) atoms (Fe^{3+}, Al^{3+}, Mn^{4+}) are only related to the different average As-O and Me-O distances in the coordination polyhedra; the sorption complex is identical (Fig. 12; Appendix 1). Waychunas et al. (1995) suggested that the consistency of the values arises from the need for adsorbing As oxyanions to align themselves in anion close-packed mode with surface O/OH groups on ferric oxyhydroxides; this alignment forces a $0°$ tilt angle of the oxyanion with respect to the surface. This theory was presented for As adsorption on Fe^{3+} hydroxides, but might be generally true for all or most metal hydroxides.

A fourth conclusion is that adsorbate concentration (also known as surface loading) exerts a greater influence on the predominant surface complexes formed by the oxyanions than the metal hydroxide structure. Particularly for $As^{3+}O_3$, formation and adsorption of dimers has been observed at high surface loading.

A final conclusion from model system studies of As sorption is that EXAFS data should be collected as far in energy as reasonably possible to aid accurate fitting; for example, collecting data out to $k = 16$ Å^{-1} is recommended when possible. This provides the data range required to best characterize As oxyanion sorption complexes, given the need to select among potential models in which the atomic number (Z) of metal atom neighbors is invariant (Manceau 1995). The paragraphs below provide additional details regarding As sorption on various classes of representative minerals.

Aluminum oxides and hydroxides. As-XAS has been used to examine As^{3+} and/or As^{5+} uptake on γ-Al_2O_3, gibbsite ($Al(OH)_3$), and amorphous Al hydroxide as well as As^{3+} and As^{5+} co-precipitation in amorphous Al hydroxide (Appendix 1). In most of these systems, the ^2C (binuclear, bidentate) complex was dominant, but "minor" (unquantified) amounts of ^2E complexes have also been identified (Duarte et al. 2012). Arai et al. (2001) determined that ~33% of As^{3+} sorbed on γ-Al_2O_3 at a pH 8 was OS by fitting XANES spectra using aqueous As^{3+} as the OS model and a pH 5 sorption sample as the IS model (Appendix 1). Resonant surface X-ray scattering experiments (another synchrotron-based technique) have demonstrated that OS complexes are also present in addition to ^2C IS complexes for As^{5+} adsorbed on γ-Al_2O_3 (Catalano et al. 2008). Linkages among the $Al(O,OH)_6$ octahedra of the various $Al(OH)_3$ polymorphs constrain the possible As-Al interatomic distances of sorption complexes (Fig. 12), with theoretical interatomic distances as low as 3.11 Å for As^{5+} on bayerite and as high as 3.22 Å on gibbsite possible for the ^2C complex (Arai 2010). The main conclusions to be gained from these studies are that although both IS and OS species exist on Al-oxide and hydroxide surfaces, the IS species (^2C) is predominant.

Aluminosilicate clay minerals and micas. The ^2C complex also appears to be the dominant IS sorption mode in As^{5+}-kaolinite and As^{3+}-kaolinite systems, based on R_{As-Al} values that are identical (within error and the large range of possible values) to the R_{As-Al} values determined for the Al oxides/hydroxides (above, Appendix 1). However, based on FT magnitude peak heights, fitted coordination numbers (N) and macroscopic data, investigators concluded that OS complexes probably constitute a greater fraction of sorbed As^{3+} and/or As^{5+} on kaolinite than on the Al oxides/hydroxides (Foster 1999, 2003; Arai 2010). Outer sphere adsorption appears to be the dominant mode for As^{3+} and As^{5+} adsorption on montmorillonite, based on the lack of FT peaks attributable to As-Al scattering and the similarity of EXAFS spectra to those of the RT aqueous species (Foster 1999, 2003).

Biotite is an aluminosilicate mica mineral with structural ferrous iron (see formula in Appendix 1). It is a major component of the sediment in the Bengal delta, so Chakraborty et al. (2011) investigated As^{3+} and As^{5+} sorption to this mineral. They found no evidence for heterogeneous redox reactions, and found ^2E complexes dominating over ^2C complexes in a 5:1 ratio in experiments with As^{3+} (Appendix 1). They explained the lack of As^{5+} ^1V sorption complexes by steric constraints.

Iron oxides, hydroxides, and hydroxy-sulfates. The common naturally-occurring Fe^{3+} oxides/hydroxides/hydroxysulfates include hematite (α-Fe_2O_3), ferrihydrite ($Fe_2O_3 \cdot 0.5H_2O$), goethite (α-$FeOOH$), akaganeite (β-$FeOOH$), lepidocrocite (γ-$FeOOH$), and schwertmannite

$[Fe_{16}O_{16}(OH)_{12}(SO_4)_2]$. Common mixed-valence Fe oxides/hydroxides include magnetite (Fe_3O_4), maghemite $(\gamma\text{-}Fe_2O_3)$, and the green rusts $[Fe^{2+}_{(1-x)}Fe^{3+}_x(OH)_2]^{x+}(CO_3,Cl,SO_4)^{x-}]$. Interconversion among the phases of this group is well known, and controlled by environmental redox conditions, pH, temperature, and the presence/absence of specific counter-ions. Several recent XAS studies have focused on defining the molecular geometry of As sorption complexes on these phases due to their ubiquity in the environment and their known strong affinity for As species, (Appendix 1).

Most EXAFS studies of As^{3+} and As^{5+} adsorption on Fe^{3+} hydroxide phases (including Fe^{3+} hydroxysulfates such as schwertmannite) indicate that the ^2C-type complex is predominant on all phases in this subgroup (Fig. 12; Appendix 1). Within a single oxyanion (e.g., AsO_3), the theoretical $R_{As\text{-}Fe}$ distances for the ^2C complex vary according to different Fe octahedral linkages in the structures; these findings have largely been supported by experimental results. For example, the $R_{As\text{-}Fe}$ distance of the ^2C complex for As^{5+} sorption on goethite and lepidocrocite is shifted about ~0.05 Å higher than the same complex on ferrihydrite due to geometric constraints (Waychunas et al. 1993, 1995). A compilation of the results of previous studies indicates that for ^2C complexes, the $R_{As\text{-}Fe}$ distances can be as much as ~0.08 Å longer in $As^{3+}O_3$ systems than in $As^{5+}O_4$ systems, a fact which has been attributed to the longer $R_{As\text{-}O}$ distance in $As^{3+}O_3$ (Fig. 12b,c). However, further inspection of the results in Figure 12 indicates that there is substantial overlap among the fit-derived $R_{As\text{-}Fe}$ distances indicative of ^2E and ^1V complexes for the two oxyanions. This may be due to significant shortening of an As-O bond in the $As^{3+}O_3$ trigonal pyramid to accommodate the same bonding configurations as $As^{5+}O_4$, but has not been tested, to our knowledge. Individual fitting of each oxygen atom would be required, as was performed in Sherman and Randall (2003).

Some controversy exists in the literature regarding the relative importance of ^2E and ^1V complexes for $As^{5+}O_4$ sorbed on Fe^{3+} hydroxides as determined by EXAFS fit results. Leaving aside potential measurable differences in adsorbed As speciation that could be derived from experimental variability, there appear to be at least three aspects of spectral analysis that could also play a role in the discrepancies among fit results: (1) analysis of data with a k-range of < 15 Å^{-1} misses key features in the EXAFS spectrum needed to properly identify the ^2E complex (Manceau 1995); (2) inclusion (or lack of) oxyanion MS paths in fits affects relative contribution of the ^2E complex (Sherman and Randall 2003); and (3) choice of the low-k end of the EXAFS data range over which the FT magnitude is calculated affects the relative contribution of MS paths in the critical 3-4 Å range of the FT (Sherman and Randall 2003). The choice of E_0, the point at which $k = 0$, may also influence aspect (3) listed above. This issue would be aided if reporting of the relative abundance of ^2C, ^2E and ^1V complexes (or at least the relative importance of their phase/amplitude functions to the overall constructed EXAFS spectrum) became standard procedure for this type of fitting. A normalized percent contribution of each scattering path to the overall fit is part of the default output of some data analysis programs (e.g., EXAFSPAK; see Table 1).

Both ^2C and ^1V adsorption complexes were observed in an EXAFS study of phenyl- and diphenyl-substituted organoarsenate species adsorbed on ferrihydrite (Tanaka et al. 2014). The organic portion of the species is oriented away from the surface in such a way so as to allow standard As bonding geometries (Fig. 13a,b), and there is good agreement between refined $R_{As\text{-}Fe}$ values of the inorganic and organic ^2C complexes as well as with the results of density functional calculations; however, the refined ^1V $R_{As\text{-}Fe}$ value of 3.44-3.46 Å for both species is ~0.1 Å less than that determined in the inorganic system analog (Fig. 12) and by density functional calculations (Tanaka et al. 2014).

The ^1V adsorption complex has been proposed to dominate adsorption of monothioarsenate, $(MTAs^{5+})$ and tetrathioarsenates $(TTAs^{5+})$ on several Fe^{3+} hydroxide polymorphs (Couture et al. 2013). The S atoms in the anion tetrahedron coordinating As^{5+} produce $R_{As\text{-}S}$ distances of

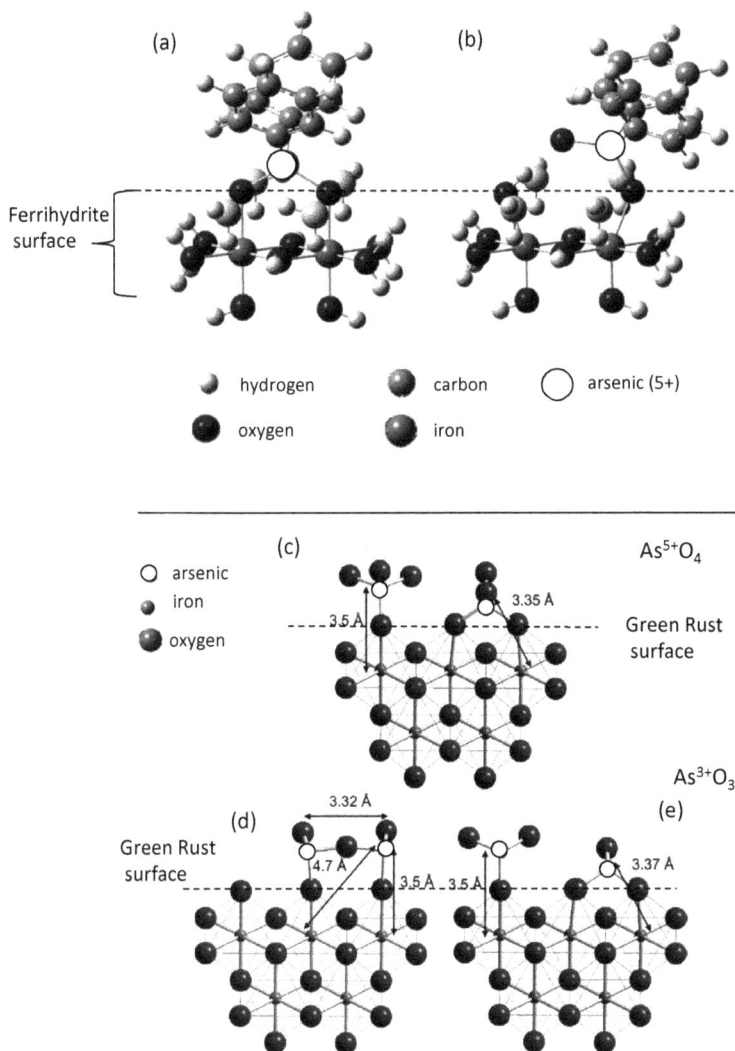

Figure 13. Optimized geometries of adsorbed diphenyl arsenate on ferrihydrite as calculated from density functional theory: (a) ^{2}C complex and (b) ^{1}V complex. Proposed complexes for As oxyanion adsorption on green rusts include (c) As^{5+} ^{1}V and ^{2}C complexes, (d) As^{3+} polymerized species, and (e) As^{3+} ^{1}V and ^{2}C complexes). [(a, b): Used by permission of Elsevier, modified from Tanaka et al. (2014) *Journal of Colloid and Interface Science*, Vol 415, Fig. 2, p. 15; (c, d, e) Used by permission of American Chemical Society, from Wang et al. (2010) *Environmental Science and Technology*, Vol 44, Fig. 3, p. 113]

2.10-2.15 Å in both mono- and tetrathioarsenate species, and an additional R_{As-S} distance of 2.18 Å in $TTAs^{5+}$. Disproportionation of greater than 50% of added $TTAs^{5+}$ to $MTAs^{5+}$ was observed during adsorption to Fe^{3+} hydroxides. The refined R_{As-Fe} distances for the proposed ^{1}V complex on Fe^{3+} hydroxides (3.30-3.32 Å) is more consistent with a ^{2}C complex for pure $As^{5+}O_4$ oxyanion sorption, and although the refined coordination number for the complex was approximately 1.0, the large error associated with this value precludes meaningful distinction based on those values. Additional studies may be needed to confirm the geometry of adsorbed thioarsenates on iron hydroxide phases and on the broader class of model phases representing reactive surfaces in soils.

EXAFS studies suggest that ^2E and ^1V complex geometries are more common and consistently observed for $As^{3+}O_3$ adsorption on Fe^{3+} hydroxide phases than for AsO_4 adsorption (however, the ^2C complex is still a substantial component of adsorbed complexes in all systems). In addition, some systematic behavior has been observed: As^{3+} forms ^2E complexes on ferrihydrite and hematite, but not on goethite and lepidocrocite; on the latter two phases, the ^1V geometry is favored (Ona-Nguema et al. 2005; Morin et al. 2008). The same factors as listed above that potentially confound identification of As^{5+} sorption complexes by EXAFS are probably also important for interpretation of As^{3+}EXAFS spectra of the same systems.

Arsenite oxidation during sorption to Fe^{3+} hydroxide polymorphs has been observed in some studies. Initially attributed to Mn impurities (Sun and Doner 1998), it was later shown that Fenton reactions (involving molecular O_2 and adsorbed Fe^{2+}) can oxidize $As^{3+}O_3$ even on Mn-free substrates (Ona-Nguema et al. 2010). Photocatalyzed oxidation of As^{3+} on goethite in the presence of O_2 has also been observed (Zhao et al. 2011; Bhandari et al. 2012). We can conclude that, in systems containing Fe^{2+}, rigorous exclusion of O_2 is required to obtain As^{3+} EXAFS spectra free from oxidation.

EXAFS studies of As oxyanion sorption on mixed-valence Fe oxides and hydroxides indicate the formation of similar surface complex geometries as on Fe^{3+} oxides/hydroxides, but different R_{As-Fe} distances due to the longer Fe^{2+}-O bonds. Similar to hematite, the highly condensed oxide magnetite has fewer surface sites available for bonding AsO_4 and therefore a greater relative abundance of OS to ^2C complexes in comparison to green rust, which has a more open structure that also allows AsO_4 to form ^1V complexes (Wang et al. 2008; Ona-Nguema et al. 2010). When coprecipitated with magnetite, $As^{3+} O_3$ forms a tridentate ^3C species over structural vacancies (Fig. 12; Wang et al. 2008). When adsorbed to green rusts and coprecipitated with $Fe^{2+}(OH)_2$ phases, As^{3+} forms polymerized species (dimers and/or oligomers; Ona-Nguema et al. 2009; Wang et al. 2010). Each $As^{3+} O_3$ group bonds in the ^1V geometry, resulting in three characteristic distances: $R_{As-As} = 3.32$ Å, $R_{As-Fe1} = 3.50$, and $R_{As-Fe2} = 4.7$Å (Appendix 1; Fig. 13d). As^{5+} sorption to maghemite, an oxidation product of magnetite, suggests that ^2C complexes are predominant, but As^{3+} sorption to the same material has been interpreted as a mix of ^2C and ^1V complexes (Appendix 1). Oxygen must be strictly excluded from these experiments; As^{3+} is rapidly oxidized by magnetite in the presence of O_2 (Ona-Nguema et al. 2010).

Manganese oxides. Arsenite is oxidized to As^{5+} on synthetic and natural Mn-oxides (Scott and Morgan 1995; Manning et al. 2002, 2003; Deschamps et al. 2003). Arsenate adopts the ^2C complex geometry on birnessite and vernadite surfaces, with $R_{As-Mn} = 3.16$-3.22 Å, (Manning et al. 2002; Foster et al. 2003). The relatively large range in R_{As-Mn} may be due to the same issues discussed previously As^{5+} complexes on Fe^{3+} oxyhydroxides, or may have another source, such as structural differences among Mn oxide phases used in the studies. Neither Foster et al. (2003b) nor Manning et al. (2002) included MS in their fits to adsorption samples, so this is also a possible reason for the shorter bond distances relative to sorption on Fe^{3+} hydroxides.

Sulfides. The limited data for As oxyanion sorption to iron sulfide minerals (Appendix 1) indicates that at the two oxyanions adsorb primarily as OS complexes, but approach close enough to the surface in a complex sufficiently ordered so that EXAFS signals from non-bonded S and Fe neighbors at ~3.1 Å and 3.4-3.5 Å from As can be observed (Farquhar et al. 2002). The finding of predominantly OS sorption complexes agrees with the results of solution studies (Wolthers et al. 2005), but the more distant fitted shells suggest a degree of ordering rarely observed in outer-sphere complexes.

Metal sulfide minerals have a greater metallic character than metal oxides/hydroxides and therefore more readily transfer electrons to adsorbed species. At high initial As concentrations on mackinawite and pyrite, and during coprecipitation experiments with mackinawite, both oxyanions showed evidence of partial reduction and formation of a surface-associated species with coordination similar to As_2S_3 (Farquhar et al. 2002). Other sorption studies with $As^{3+}O_3$ on pyrite and troilite show its rapid reduction and formation of a FeAsS-like precipitate (Bostick and Fendorf 2003).

Thioarsenates adsorbed to FeS and FeS_2 with less decomposition than noted during their adsorption on Fe^{3+} hydroxides (Couture et al. 2013). Decomposed thioarsenate is reduced to As^{3+} on Fe sulfides, however, and substitutes for S in the structures (Appendix 1). Adsorbed thioarsenates were proposed to adopt an inner-sphere, ^1V adsorption geometry, based on comparison of fitted R_{As-Fe} distances with those of $As^{5+}O_4$ on a variety of minerals.

Arsenic sorption on other sulfide minerals exhibits similar behavior: on both sphalerite (ZnS) and galena (PbS), $As^{3+}O_3$ exchanges O for S with surface groups, and forms multinuclear As-sulfide surface complexes with proposed stoichiometry $As_3S_3(SH)_3$ (Bostick et al. 2003). These complexes are stable on sphalerite but photoredox-sensitive on galena.

Sulfates, carbonates, and phosphates. EXAFS studies have shown that As oxyanions sorb to secondary sulfate, carbonate, and phosphate minerals via adsorption, isomorphous substitution, and occlusion reactions. In the realm of adsorption, the type of sorption configurations and resultant interatomic distances will be dictated by the mineral structure, which can be quite distinct from the octahedral linkage-dominated structures of metal hydroxides. Adsorption of AsO_4 on jarosite, $KFe_2(SO_4)_2(OH)_6$, produces an As-Fe distance consistent with the ^2C-complex on Fe^{3+} oxyhydroxides (Appendix 1), but AsO_4 adsorption on calcite results in two non-equivalent Ca shells (Appendix 1; Fig. 12; Alexandratos et al. 2007). In the calcite system, geometric considerations were not enough to distinguish between the possibility of 2 distinct ^1V complexes or 1 distorted ^2C complex (Alexandratos et al. 2007). Redox and/or surface precipitation can also occur during As sorption to sulfates and phosphates when specific counter ions are present: a phase similar to euchroite, $Cu_2(AsO_4)(OH)\cdot3H_2O$, has been documented by XAS during As sorption to jarosite in the presence of Cu^{2+} (Grafe et al. 2008a), and both redox and precipitation of As-substituted Fe^{3+} hydroxy phosphate phases were described via XAS analysis of the As^{3+}-Fe^{2+}-hydroxyapatite system (Sahai et al. 2007).

If As is present during the precipitation of sulfate, carbonate, or phosphate minerals, sorption typically occurs via isomorphous substitution of As in the oxyanion structural position. Structural distortions must take place to accommodate the larger AsO_4 tetrahedron; when the mineral can accommodate no further strain due to distortion, new phases form (e.g., scorodite during jarosite synthesis at high initial As; Savage et al. 2005) Isomorphous substitution often produces a characteristic set of interatomic distances that can be used to distinguish this mode of sorption from adsorption or coprecipitation to mineral surfaces. In the case of jarosite, the coordination of AsO_4 in the ^2C adsorbed complex geometry is too similar to the coordination of AsO_4 substituted for structural SO_4 to be distinctive on the basis of R_{As-Fe} distances alone, but the number of neighboring Fe atoms is only 2 in the adsorbed complex versus 4 in the sulfate position (Appendix 1; Paktunc and Dutrizac 2003; Savage et al. 2005; Paktunc et al. 2013). The unusual case of ettringite, in which isomorphous substitution resembles outer sphere adsorption, was discussed previously.

Natural soils. Analysis of As-XANES spectra has been used to monitor changes in As redox state upon adsorption of oxyanion species to naturally-occurring, low-As soils, and during simulation of the redox cycling that occurs in some soils (Reynolds et al. 1999; Appendix 1). Under aerated conditions, oxidation of As^{3+} and sorption of both As^{3+} and As^{5+} (primarily the latter) were noted (Deschamps et al. 2003; Manning 2005). Subtle difference in the XANES spectra of As^{5+} sorbed to Al- and Fe-hydroxides exist and were utilized in fits indicating that between 50 -100% of adsorbed As was associated with Al hydroxides (the rest with Fe hydroxides). However given the limitations of XANES spectra to discriminate many As^{5+} species, without EXAFS analysis it is difficult to be certain about these assignments. An EXAFS study of As^{5+} sorption on low-As, red soils from PR China largely supports XANES evidence: ^2C complexes were predominant and scattering from both Fe hydroxide and Al hydroxide/aluminosilicate was indicated (MS not considered; Luo et al. 2006).

Effects of microbial activity on sorbed arsenic species

Microbial oxidation and microbial reduction are processes central to As cycling in the environment, as noted in recent reviews (see Table 1). As-XAS analysis has proven useful for studies of natural and constructed systems dominated by microbial activity, as described below and summarized in Appendix 2. Over time, model system EXAFS studies of the effect of microbes on As speciation have increased in complexity, reflecting advances in understanding of the often coupled geochemical and redox cycling of As and Fe (Kocar et al. 2005; Fendorf et al. 2010; Masue-Slowey et al. 2013), as well as increased understanding of the influence of microorganisms on As speciation in the environment (see references in Table 1).

A major conclusion to be drawn from the many studies in this area is that microbial community composition (or at least the dominant microbial physiology) is a major determinant of solid phase As speciation.

Microbial Fe and S oxidation. Biominerals produced by microbial sulfur and iron oxidation (most commonly, oxidation of sulfide minerals and/or Fe^{2+}_{aq}) are arguably the most important sorbents of dissolved As under oxidizing environmental conditions. The minerals produced (usually the Fe^{2+}/Fe^{3+} and Fe^{3+}oxides/hydroxides listed previously) are typically poorly crystalline and are micron- to nanometer scale particles (Benzerara et al. 2011). Several As-EXAFS studies have concluded that the molecular-scale As speciation in microbially-produced iron hydroxide and hydroxysulfate phases is not appreciably different from abiotically-produced analogs, with the ^2C complex commonly observed for As^{3+} or As^{5+} adsorbed or co-precipitated with biogenic Fe^{3+} oxyhydroxides (Foster and Ashley 2002; Morin et al. 2003; Maillot et al. 2013). One study attributed the greater relative abundance of ^2E complexes relative to ^2C in experiments to the blockage of ^2C sorption sites by microbial exopolymers during the buildup of unnaturally high microbe densities in batch culture experiments (Hohmann et al. 2011).

Abiotic model system studies suggest that pH, sulfate concentration, and the dissolved As/Fe^{3+} ratio dictate the solid-phase species formed by microbial iron and/or sulfide oxidation. The pH- and SO_4-dependent sequence of iron mineral precipitation is (from acidic to basic): jarosite [$KFe_2(SO_4)_2(OH)_6$; high SO_4], schwertmannite [$Fe_8O_8(OH)_6SO_4$; high SO_4], ferrihydrite [$Fe_5(OH)_8·4H_2O$; low SO_4], and goethite [α-FeOOH; low SO_4; Maillot et al. 2013]. One study used EXAFS results to conclude that below a solid-phase As/Fe^{3+} ratio of ~0.25, As partitions into the solid phase via sorption (adsorption/coprecipitation). Formation of the aforementioned phases are inhibited where As/Fe^{3+} ratio exceeds ~0.25; instead, amorphous or crystalline ferric arsenates can form, the specific type being controlled by the As/Fe ratio as described previously (Maillot et al. 2013 and references therein; Paktunc et al. 2008; Drahota and Filippi 2009).

When As^{3+} is the dominant solution species the resulting solid phase speciation can be quite different, but is still dependent on As/Fe ratios. As^{3+}/Fe ratios > ~0.6 are required to inhibit the precipitation of schwertmannite in model systems designed to simulate the acidic As^{3+} and Fe^{3+}-rich waters at the Carnoulès Pb-Zn-Sb mine acid drainage in France; this value is three times higher than for the As^{5+} equivalent (Maillot et al. 2013). Tooleite, the rare ferric arsenite mineral discussed previously, forms at Carnoulès at As^{3+}/Fe ratios > ~0.6 (Morin et al. 2003).

Microbial respiration of sulfur and/or metal(loid)s. Several studies have employed synchrotron techniques (usually bulk As-XAS, but also As oxidation state mapping by μ-XRF) to explore the role that metal-reducing bacteria [especially Fe-reducing bacteria (FRB), sulfate-reducing bacteria, (SRB) and As-reducing bacteria (ARB)] play in transformation of solid phase As, particularly the adsorbed oxyanion species discussed in the section above (Table 1; Appendix 2). These microbial groups have the potential to increase As mobility by enhancing dissolution of As-rich phases, particularly those containing SO_4, Fe^{3+}, and Mn^{4+}, and/or reducing As^{5+} to As^{3+}. However, as we will discuss below, some of the newly-formed phases produced from reductive dissolution can actually re-sequester As. Some of the studies summarized in Appendix 2 were conducted in batch, and others in flow-through columns; all included microbes with known FRB, SRB, and/or ARB function and an As-rich substrate. Due to its ubiquity in the environment, As^{3+} or As^{5+} adsorbed or coprecipitated to ferrihydrite has been the main substrate studied, but systems with As^{5+} adsorbed to lepidocrocite and As^{5+} substituted in jarosite have also been examined (Ona-Nguema et al. 2009 and Johnston et al. 2012, respectively). Mixed-mineral systems, such as As^{5+} adsorbed to ferrihydrite mixed with (initially) As-free MnO_2 have also been investigated (Ying et al. 2013).

Taken together, these studies lead to three main conclusions. The first is that FRB activity leads to the sequestration of As released by reductive dissolution of Fe^{3+} phases (Kocar et al. 2006, 2010; Burnol et al. 2007; Ona-Nguema et al. 2009; Johnston et al. 2012; Root et al. 2013; Ying et al. 2013). Analysis of batch systems by XAS suggests that several potential solid phase products of microbial iron reduction can be reservoirs for the released As (partially or totally reduced to As^{3+} in most systems), including Fe^{2+}-hydroxide (Ona-Nguema et al. 2009), Fe^{2+}/Fe^{3+} oxide (magnetite; Kocar et al. 2006; Cismasu et al. 2008), green rust (Ona-Nguema et al. 2009; Root et al. 2013), siderite (Root et al. 2013), and vivianite ($Fe^{2+}_3(PO_4)_2 \cdot 8H_2O$; (Burnol et al. 2007).

The second conclusion is that sulfidization (via SRB or abiotic means) of Fe^{3+}-bearing minerals increases As mobility due to formation of amorphous FeS (Kocar et al. 2010; Johnston et al. 2012; Burton et al. 2013), and results in mainly weak, OS sorption complexes with dissolved $As^{3+} O_3$ and $As^{5+} O_4$ (Appendix 1). Flow-through column experiments additionally show that mobilized As is displaced along with dissolved Fe^{2+} from the zone of SO_4 reduction, but is immobilized again as a coprecipitate with magnetite and newly-formed Fe^{3+} hydroxide in the suboxic-to-oxic environment outside this zone (Kocar et al. 2010; Burton et al. 2013). These experiments additionally suggest that released As has low affinity for neo-formed green rust phases under high-As, sulfidogenic conditions, in contrast to results described above for FRB conditions.

The third conclusion is that ARB activity in the absence of FRB activity increases As mobility substantially because no new phases are formed for sorption (Kocar et al. 2006). The ARB used in Kocar et al. (2006) was limited physiologically to As reduction; although such bacteria are abundant (and can even dominate) some extreme environments (Lloyd and Oremland 2006), most natural reducing environments will contain an anaerobic microbial community composed of FeRB, SRB, ARB, fermenting bacteria, and other groups that could influence As speciation as described in the previous paragraphs. In addition, several microorganisms are capable of changing their physiology depending on environmental conditions, and this could affect endpoint As speciation. For example, bacteria capable of both AsO_4 and SO_4 reduction are known, and in culture they precipitate amorphous As-sulfide (Lloyd and Oremland 2006). Therefore, while these experiments provide great insight into the interplay between microbial physiology and mineral transformation/As sequestration, the activity of mixed microbial communities also requires study to evaluate the relative importance of the various microbial groups to the observed As speciation.

Such a study was recently conducted using an enrichment culture of an anaerobic, metal-reducing microbial consortia, and yielded somewhat different results for fate As released by the reductive dissolution of As^{5+}-ferrihydrite (Root et al. 2013). Under low SO_4 conditions (0.064 mM), As^{5+} is reduced to As^{3+} and partitioned onto solids in the order: residual ferrihydrite > siderite > green rust (Fig. 14a-c,g,h). This result coincides with the results of previous field observations (Haiwee Reservoir studies, Appendix 2). Under the high SO_4 conditions (2.1 mM), As^{5+} is reduced to As^{3+}, then coprecipitates with sulfide, forming a solid with realgar-like local structure (Fig. 14d-f); this finding varies substantially from model system results using pure cultures of SRB or ARB, but is supported by XAS observations of similar phases in natural systems (O'Day et al. 2004; Root et al. 2009). Arsenic and Fe speciation mapping at the micron scale shows that nearly As-free amorphous FeS can precipitate in close association with the realgar-like phase in this system (Fig. 14g,h). This result does match the pure-system studies in the low sorption affinity of As species for FeS surfaces.

The oxidation of As^{3+} by Mn^{4+} oxide (MnO_2), which can proceed in the absence of oxygen (Scott and Morgan 1995), has been suggested as a mechanism to explain As^{3+} oxidation at Fe^{3+} hydroxide surfaces under reducing conditions in laboratory (Sun and Doner 1998; Sun et al. 1999) and natural systems (Foster et al. 2000b; Singer et al. 2013). A set of model experiments employing As-XAS and/or μ-XRF have examined the interplay between one or more of the following factors: presence of model FRB/ARB, oxidation of As^{3+} by MnO_2, and diffusion-limited or preferential flowpath-directed transport of reaction products (Masue-Slowey et al. 2011, 2013; Ying et al. 2011, 2012, 2013). Taken together, these experiments demonstrate that As^{3+} oxidation by MnO_2 can drive the formation of As^{5+}-Fe^{3+}oxyhydroxides under oxic conditions, but under anoxic conditions and high microbial activity, formation of $MnCO_3$ passivates MnO_2 surfaces from further oxidative transformation of As (Ying et al. 2011, 2013). Experiments with natural Mn-goethite-coated aquifer sands indicated that reactive surface area limited the extent of As^3 oxidation (Singer et al. 2013).

XAS studies of rocks and related soils

X-ray absorption spectroscopy studies of As speciation in rocks and rock-forming fluids have been conducted to test theories about the transfer of fluid-soluble elements in subduction zones, to understand aspects of orogenesis, to identify potential water-rock interactions in bedrock aquifers, geothermal areas, and karst systems, to evaluate As dispersion from marine and terrestrial active hydrothermal systems, and to understand chemical and mineralogical relationships between soil and its parent bedrock. These studies are briefly reviewed below, and summarized in Appendix 3.

Hydrothermal arsenian pyrite. Pyrite is probably the single most important As-bearing mineral in rocks formed from low-temperature hydrothermal fluids. Although it contains less As than other sulfides (e.g., arsenopyrite, FeAsS), it is far more abundant, and often contains several weight percent As (Savage et al. 2000b; Deditius et al. 2008; Paktunc 2008b). Hydrothermal arsenian pyrite is of great interest for two reasons: (1) it hosts gold in several types of ore deposits, and (2) in a variety of rock types, its chemical breakdown is a source of As to the environment.

Relatively new data indicates that there are two types of As speciation in hydrothermal pyrite (As-pyrite): As^{-1}-pyrite is the most common form, and the only one studied by XAS, to our knowledge. All sedimentary pyrite studied by XAS (to our knowledge) is As^{-1}-pyrite (Appendix 3). The second form, As^{3+}-pyrite, is rarer; it appears to form under more oxidizing conditions than typical for pyrite formation (Deditius et al. 2008). Although we mentioned previously that nominal valence concept is of little relevance in sulfides, it is useful here for the following reasons: electron microprobe analysis of As^{-1}-pyrite reveals a negative correlation of As with S suggesting substitution for S and the generalized formula $Fe(As,S)_2$, whereas

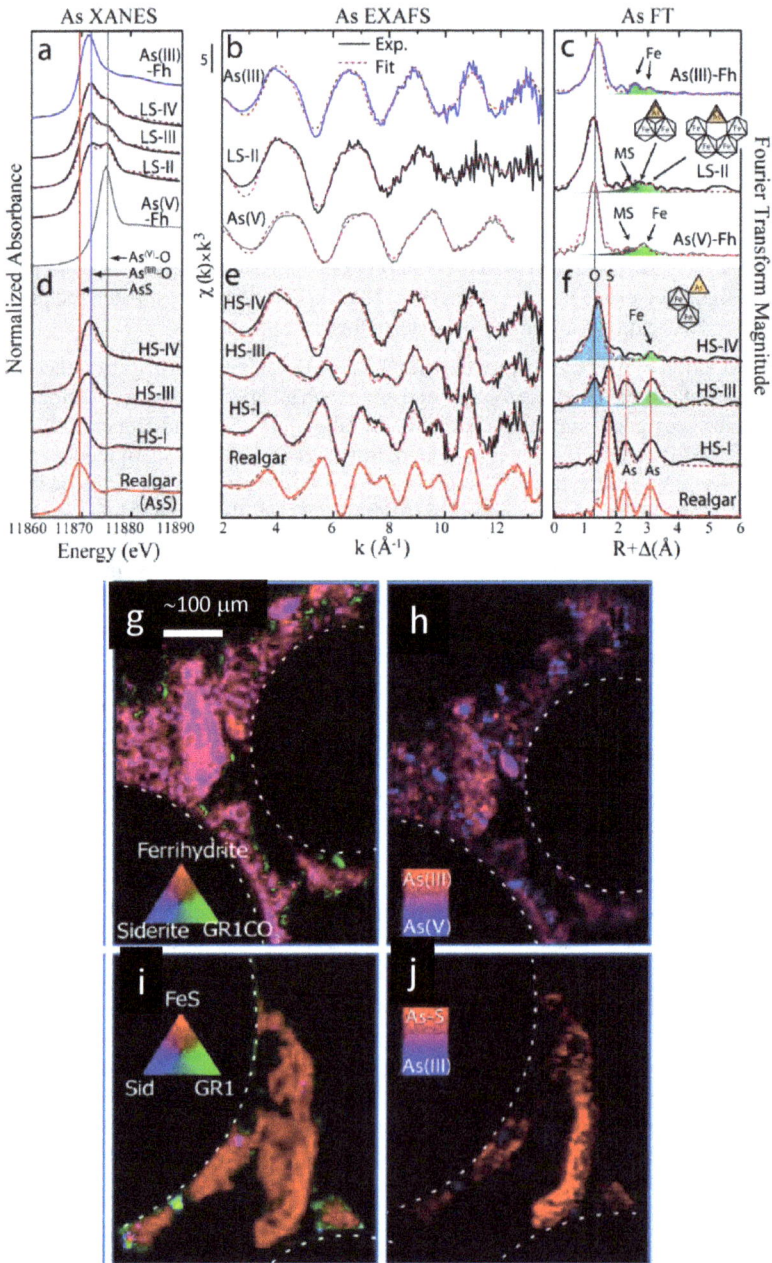

Figure 14. (*for color see Plate 5*) (a-f): Results of bulk As-XAS analysis of low-SO$_4$ (LS) and high SO$_4$ (HS) columns containing As^{5+}ferrihydrite and inoculated with a natural, anoxic Fe-reducing microbial consortium (realgar phase forms in HS experiments as shown in panel f); (g-j): qualitative μ-XRF speciation maps of Fe (g, i) and As (h, j) in sediment from LS column (g, h) and HS column (i, j). Dotted circles outline glass beads. This figure is also reproduced in the color figures insert. [Used by permission of American Chemical Society, from Root et al. (2013) *Environmental Science and Technology*, Vol 47, Fig. 1, p. 12994 and Fig. 3, p. 12996]

the same analyses of As^{3+}-pyrites reveal negative correlation between As and Fe, suggesting substitution for Fe and the generalized formula $(Fe,As)S_2$ (Deditius et al. 2008). As^{-1}-pyrite is distinct from FeAsS (arsenopyrite), which as a different crystal structure and in which As resides in distinct crystallographic positions from sulfur.

Savage et al. (2000b) used XAS to study the local structure of As in a sample of hydrothermal As^{-1}-pyrite (~1.2 wt%) from the Clio mine tailings in the southern Mother Lode Gold District (California, USA). Visual comparison of bulk As^{-1}-pyrite to arsenopyrite (FeAsS), marcasite (FeS), and loellingite ($FeAs_2$) EXAFS spectra ruled out these phases as predominant hosts for As in the pyrite (Fig. 10a,b). Curve fitting the As^{-1}-pyrite EXAFS spectrum gave a coordination environment similar to that expected for isomorphous substitution of As for S in the pyrite structure, but with important variations: (1) the interatomic distances between As and neighboring atoms in As-pyrite were ~ 0.1 Å longer than interatomic distances between S and neighboring atoms in As-free pyrite, and (2) fits were greatly improved with the addition of neighboring As atoms, which would not be expected if As was randomly substituted in the structure at 1.2 wt% (Savage et al. 2000b). Ab initio calculations of As-substituted pyrite were used to conclude that: (1) As atoms cluster together in As^{-1}-pyrite, with ~30% substitution of As for the nearest-neighbor S atoms, and 43% substitution of As for more distant S atoms, and (2) As substitution for S at the observed level expands the pyrite unit cell by about 2.6% over As-free pyrite (Savage et al. 2000b).

A subsequent, non-XAS study demonstrated that pyrite grains from Carlin-type and epithermal Au deposits contain nanoparticles of several other minerals and phases including realgar, orpiment, other Fe-bearing sulfides, and complex metallic phases containing Au, As, Ag, and other elements (Deditius et al. 2011). The As clustering observed by Savage et al. (2000b) does not appear to be an artifact of the presence of these phases, because the pyrite used showed no structural defects on the nanoscale by transmission electron microscopy. It was also from a different type of Au deposit (low-sulfide, quartz-hosted Au).

As-pyrite is more reactive than pure pyrite, with higher oxidation rates and different electrochemical behavior (Lehner et al. 2007; Lehner and Savage 2008). The oxidized As would presumably be located on the surfaces that were exposed to air or aerated solutions. A μ-XAS study on a wider variety of hydrothermal As^{-1}-pyrite (massive, semi-massive, inclusion-rich, and framboidal; 6-8 wt% As) confirmed this: As^{5+} was detected in most samples (Paktunc 2008b). This result reveals an important distinction between bulk- and micro micro-XAS: a small amount of As oxidation on the pyrite surface might be entirely undetected by bulk XAS due to the low relative abundance of the As^{5+} species. At the micron scale, however, the analyzed area is smaller, and therefore the relative contribution of surface As^{5+} species to the total analyzed volume is greater. In our experience, a brief polish with 800 grit or higher paper of a small-pebble-sized grain immediately prior to microanalysis was sufficient to remove the surface As^{5+} oxidation layer from As-pyrite.

Fluid inclusions. Analysis of micron-scale fluid inclusions found within minerals from ore deposits can provide important information about the transport and precipitation of metal(loid) s at elevated temperature and pressure in the Earth's crust. James-Smith et al. (2010) studied As-rich fluid inclusions from 3 types of Au ore deposits *in situ* at temperatures ranging from ambient to 200 °C. Reduced As (as As^{-1} or As^0) and As^{3+} were the initial species in all samples as determined by As-XANES comparison with model compound spectra. However, beam-induced oxidation of reduced As species was observed in all inclusions except the one with fluid As concentrations much greater than 1000 mg kg^{-1} (ppm). Water radiolysis generated by the X-ray beam and elevated temperature were the culprits: even the As^{3+}-rich inclusion was oxidized to As^{5+} at 200 °C (James-Smith et al. 2010). The fitted As coordination environment in the most concentrated inclusion (ambient temperature), contained 3 O at 1.76 Å, consistent with As^{3+}. The typical low-temperature (<77 K) data collection procedure for preventing beam-induced sample transformation could not be performed in this technically-difficult experiment, and may preclude analysis of redox-sensitive elements at elevated temperature in fluid inclusions by conventional synchrotron-based XAS, but might be possible with ultrafast experiments.

Terrestrial hydrothermal systems. Terrestrial and marine near-surface hydrothermal systems often contain elevated As in water and rocks (Wilkie and Hering 1998; Breier et al. 2012; Alsina et al. 2013; Bundschuh and Pichler 2013; Price et al. 2013a). Arsenic-XAS studies of these systems have mainly been conducted at sites where some threat to human or environmental health is of concern (Pichler et al. 1999; Kocar et al. 2004; Bundschuh and Pichler 2013; Price et al. 2013b).

Alsina et al. (2013) used As-μXAS to identify As species in sinter mineralization (rocks formed by hot fluids) at El Tatio geothermal field, Chile. Linear-combination/Least-squares fitting of μ-XANES spectra indicated a speciation of 63% As^{5+}-HFO and 37% of a reduced As phase present as ~10 μm micronodules in the sample (μ-XRF); its μ-XRD pattern was equivalent to loellingite (FeAs$_2$). Formation of loellingite was taken to indicate the presence of micro-anoxic zones, where reducing conditions could develop, similar to the findings of a previous study in which loellingite micronodules around bacterial cells were identified by TEM (Tazaki et al. 2003). Loellingite was not identified by a previous bulk XAS study of material from the same location, only As^{5+}-HFO (Alsina et al. 2008). The difference in As speciation observed might be derived from the samples used, which were not identical, but is also very likely due to the different level of speciation detail provided by bulk vs. microscale synchrotron techniques (Table 2).

Geothermal areas can also contain elevated As in rocks and water (Ilgen et al. 2011). The highly alkaline geothermal waters of the Chalkidiki peninsula in *N.* Greece contain up to 3,760 mg L^{-1} As, up to ~25% of which precipitates out with travertine (limestone) forming from saturated solution (up to 913 mg kg^{-1} As; Winkel et al. 2013). Bulk and μ- As XAS as well as μ-XRF analyses were used to show that As^{5+} is the predominant species, and that it substitutes in the calcite structure rather than forming a Ca-arsenate phase or partitioning into Fe-rich phases.

Marine hydrothermal systems. Arsenic is one of the few elements whose concentration is greater in seawater than in river water, largely due to submarine hydrothermal inputs (Breuer and Pichler 2013). Several shallow-sea hydrothermal systems have been described that discharge high-As fluids, presenting a potential hazard to local ecosystems (Bundschuh and Pichler 2013; Ruiz-Chancho et al. 2013).

Synchrotron-based μ-XRF/μ-XAS/μ-XRD has been used to study several modern or ancient marine hydrothermal systems, including a deep-sea hydrothermal plume near the East Pacific Rise (Breier et al. 2012) and a chemically-unusual, ancient marine hydrothermal system (Val Ferrera, in the Swiss Alps; Etschmann et al. 2010). Arsenic is associated with both Fe^{3+} oxyhydroxides and Fe sulfide plume particulates at the East Pacific Rise, and these phases as well as As$_2$O$_3$, and orpiment-like phases have been documented in particulates from ultra-high As marine hydrothermal systems (Bundschuh and Pichler 2013). Arsenic speciation mapping by μ-XRF of material from the Val Ferrera sedimentary-exhalative deposit reveals that fluorapatite, is the primary host for As^{3+}, whereas hematite and the unusual mineral kemmlitzite [(Sr,Ce)Al$_3$(AsO$_4$)(SO$_4$)(OH)$_6$] are the primary hosts of As^{5+} (Etschmann et al. 2010; Fig. 15a-c).

Metamorphic rocks. Metamorphic rocks are produced by thermal and/or pressure-induced mineralogical and structural alteration of other rocks. Hattori et al. (2005) studied As speciation in serpentinite, a low-temperature metamorphic rock formed during subduction of mafic rocks in the presence of water. The samples studied contained 6-275 ppm As, which was primarily hosted within magnetite and antigorite—ideal formula: (Mg,Fe^{2+})$_3$(Si$_2$O$_5$(OH)$_4$—as

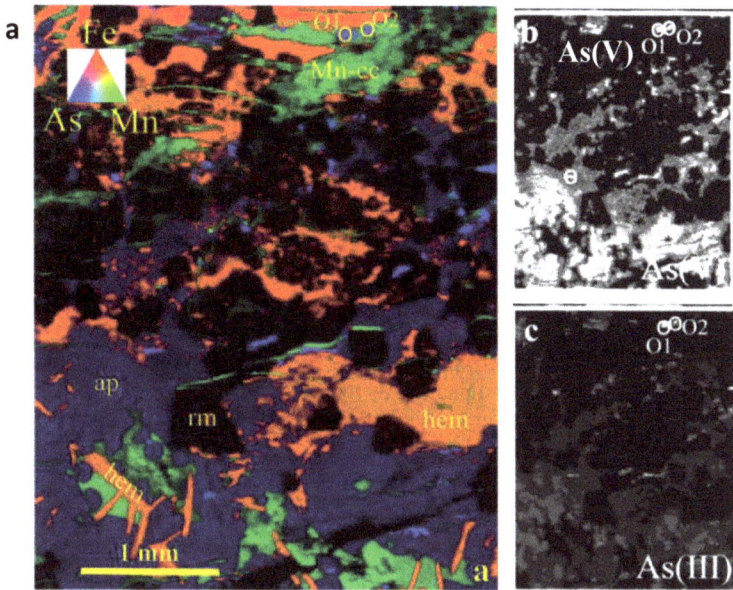

Figure 15. (*for color see Plate 4*) Micro-XRF mapping of Fe-Mn ore sample from Val Ferrera sedimentary exhalative deposit, Switzerland. (a) 2.1 million pixel RGB image (3.3 × 4 mm^2) shows As speciation as substitution in apatite (ap = apatite, Mn-cc = Mn-calcite, hem = hematite, rm=romeite). Grayscale oxidation state maps of As^{5+}(b) and As^{3+}(c) reveal small amounts of As^{3+} in this highly oxidized system. This figure is also reproduced in the color figures insert. [Used by permission of American Mineralogist, from Etschmann et al. (2010) *American Mineralogist*, Vol 95, Fig. 3, p. 887]

indicated by electron microprobe analyses. Pentavalent As was the primary species in three of the six samples studied, but minor amounts of As^{3+} and/or As^{-1} were present in some samples. The authors explained the predominance of As^{5+} by invoking oxidizing subduction solutions and/or contribution of oxygen by gypsum in subducted evaporite beds.

Sedimentary rocks. Sedimentary rocks are formed by the deposition of material at the earth's surface (including underwater) and its subsequent lithification. Types of deposited material that can be lithified include mineral and rock particles, organic remains of plants and animals, and newly-formed precipitates forming from saturated solutions.

Shale is a common type of sedimentary rock that forms from lithification of small mineral particles; black shales are a subset of the type characterized by higher amounts of unoxidized carbon (Allaby and Allaby 1990). Substantial incorporation of As and other trace elements in black shale is well-known, and is typically concentrated within pyrite grains (Paikaray 2012). As-XAS analysis of the products of long-term (1-3 yr) humidity cell tests designed to mimic natural shale weathering showed > 88% conversion of As^{-1} species to As^{5+} at the endpoint of the experiment; no evidence for intermediate As^{3+} was observed (Yu et al. 2014). In arid climates, natural weathering and soil development on shale produces acid-sulfate soils containing As-enriched soil nodules containing hydrous ferric oxide and jarosite: a μ-XRF study of these nodules from an acid-sulfate soil in *N*. California, USA demonstrated that As^{5+} is partitioned strongly in the hydrous ferric oxide, but not into jarosite (Strawn et al. 2002).

Oil shales of the western US have typical As concentrations in the 10-200 ppm range; at these levels, As can be detrimental to catalysts used in the refining process (Cramer et al. 1988). One of the earliest As-XAS studies was focused on determining As speciation in several oil shales. Variable As speciation was determined from XANES fits: As^{-1} (~60%) and As^{5+}(~40%) in Green River (US) oil shale; ~100% As^{-1} in Irati oil shale (Brazil); and ~100% As^{5+} in Montana Phosphoria (Cramer et al. 1988). Spent shale contained relatively more As^{5+} than its fresh equivalent. The XANES and EXAFS spectral analysis was not diagnostic for the reduced species, but SEM-EDS work identified skutterudite [(Co,Fe,Ni)As_3] and safflorite [(Co,Fe)As_2] in the oil shales. The variability in As speciation derived from XANES analysis may indicate oxidation of reduced As species in oil shales during storage rather than inherent speciation differences.

Sedimentary rocks can also form by direct chemical precipitation from supersaturated waters. Frierdich and Catalano (2012) investigated the speciation of As and other trace elements in Fe and Mn precipitates forming in a temperate, carbonate-karst system in SW Illinois, USA. Microbeam-XRF mapping indicated that As partitioned into Fe-rich and Mn-rich precipitate layers, unlike the other trace elements examined (Ni, Cu, Zn, Co, and Se), which strongly associated with either the Fe or Mn phase. Fits to bulk As-EXAFS data indicated As^{5+} as the sole oxidation state, and the derived structural parameters were consistent with the formation of ^{2}C-type complexes on both Fe and Mn precipitates.

Coal is a type of biological sedimentary rock formed primarily from vegetative matter subjected to heat, pressure, and reacting fluids during diagenesis (Allaby and Allaby 1990). Pyrite is a common inorganic phase found in coal, and often contains As at the weight percent level on a micro-scale (Kolker and Huggins 2007), leading to typical bulk concentrations of As in coal of a few 10s of ppm. The late Permian, high As (3.2 wt%) coals of Guizhou Province, PR China, are an exception to the generally low As concentrations in coal (Zhao et al. 1998). Oxidative weathering of pyrite in coal, coal tailings, and mineralized waste rock has created well-publicized problems of acid mine drainage in the eastern U.S. and elsewhere in the world.   Arsenic speciation differs in the two main types of coal. Bituminous (high thermal maturity) coal is common in the Eastern US and is dominated by As^{-1} pyrite (Huggins et al. 1993, 1996; Huffman et al. 1994; Kolker et al. 2000; Kolker and Huggins 2007; Zielinski et al. 2007). Lignite and sub-bituminous coal (low thermal maturity) are common in the Western US and contain predominantly As^{3+} which is presumed to be in an organic form, but whose structure has not been studied in detail (Huggins et al. 1996; Zhao et al. 1998; Kolker et al. 2000). A study of a high-As lignite coal from Slovakia identified realgar and orpiment as As^{3+} hosts by electron microprobe (Veselska et al. 2013).

Substantial oxidation of As in both coal types has been noted by XAS in samples stored under ambient conditions (Huggins et al. 1993; Huffman et al. 1994; Kolker et al. 2000; Zielinski et al. 2007). Weathering tests on several US bituminous coals analyzed by As-XANES indicate that As oxidation rate in coal is more influenced by environmental conditions (periodic wetting, humidity, presence/absence of O_2) than by the average As concentration in pyrite from each sample (Kolker and Huggins 2007). As^{-1} pyrite is the dominant species in cleaned coal as well as coal tailings fractions of bituminous coal (Huggins et al. 2009).

Arsenic species in ore, mine wastes, and mining-impacted soil

Arsenic is enriched in many types of base and precious metal ores (including antimony, copper, gold, iron, lead, silver, tungsten, uranium, and zinc) through the same geochemical processes that caused enrichment and emplacement of the ore elements (Guilbert and Park 1985). However, mining and processing can leave As enriched in a variety of solid wastes, including ore, tailings, sulfide concentrates, roaster/calciner/smelter wastes, and waste rock. Subsequently, these wastes can serve as a long-term source of As contamination to the surrounding environment, even when responsibly managed (Paktunc and Bruggeman 2010). Given the widespread occurrence of As in the huge volumes of rock that have been moved, crushed, and exposed to a range of extraction methods at historic and modern mines (Lottermoser 2010; Nordstrom 2011), the challenge of identifying the solid phase As species in mine waste in order to effectively mitigate As dispersion into the wider environment is very likely to remain with humanity well into the future (Hudson-Edwards et al. 2011).

Extensive literature on mine waste issues exists; the 2nd and 7th volumes of *Elements* magazine (April 2006 and Dec 2011) are good general-audience introductions to this significant topic, whereas the recent book by Lottermoser (2010) explores mine waste issues in more detail. Several chapters in this volume also treat general or specific topics concerning As in mine wastes (Alpers et al. 2014; Basta and Juhasz 2014; Bowell 2014; Campbell and Nordstrom 2014; Craw and Bowell 2014; Majzlan et al. 2014; Jamieson 2014). Given the large literature on the various environmental impacts of different types of mines and mining/ processing techniques, and the numerous site-specific XAS studies that have been conducted on As in various mining-impacted environments (Appendix 4), we have attempted below to provide a summary of the key contributions that As-XAS (and μ-XRF/μ-XRD) studies have made to the body of knowledge on As speciation in mine wastes. We also attempt to highlight areas where additional work is needed to address outstanding questions.

Due to the chemical heterogeneity and multiple As species typically present in mine wastes, LC-LSF fitting has been the primary method used to quantify As speciation from bulk As-EXAFS spectra of these materials. The complementary application of μ-XAS/μ-XRF/μ-XRD has helped to corroborate bulk XAS analysis, resolve potential ambiguities in bulk analyses, and provide additional details about low-abundance species, chemically-similar species, and spatial relationships among species. The relative insensitivity of As-XANES to structural differences among species containing both As^{5+} and Fe^{3+} (discussed previously; Fig. 7a) is a serious impediment to As^{5+}-speciation based solely on bulk-XANES or μ-XANES spectra, but adding complimentary analyses can overcome this limitation. For example, combining μ-XRD with μ-XANES analysis of the same spot allowed identification of As^{5+}-goethite, As^{5+}-lepidocrocite, and hydrous ferric arsenate (Walker et al. 2009). Also chemically-similar species can be identified in μ-XRF maps by discrete Fe:As ratios observed in bivariate plots of pixel fluorescence data (often hundreds or thousands of points); the species represented by these ratios can then be related to those obtained from analysis of corresponding bulk EXAFS spectra (Fig. 16a,b) and to phases identified by Fe:As ratio differences using electron microprobe (Paktunc et al. 2003, 2004).

Appendix 4 provides a summary of As-XAS studies of ore, mine wastes, and mining-impacted soils/sediments. Consistent speciation trends can be identified across these studies, although there are also variations based on processing history/waste types, geology, climate, ore type, and other variables. These trends are described in more detail in the following paragraphs.

Sulfides and arsenides (specifically As^{-1}-pyrite and arsenopyrite) are the most common As hosts in in ore and waste rock from many types of mines (Hudson-Edwards et al. 2005; Burlak et al. 2010; Corriveau et al. 2011b; Kossoff et al. 2012; Fig. 17). Many historic mine sites contain abandoned ore and/or waste rock piles that have been weathering in place for decades to centuries (Arcon et al. 2005; Hudson-Edwards et al. 2011).

Ore processing increases reactive surface area and/or transforms refractory species to enhance extraction and recovery of the mineral commodity (Lottermoser 2010). These processes have the cumulative effect of oxidizing the material and in many cases dissolving or transforming the primary As-bearing ore minerals, with corresponding changes in dominant As oxidation state and speciation in the spent ore. Pentavalent As is usually the dominant oxidation state observed in ore processing wastes, but As^{3+} is also commonly observed in waste types that were subjected to high heat (e.g., calcines, roaster dusts, and smelters; Beaulieu and Savage 2005; Hudson-Edwards et al. 2005; Walker et al. 2005). Small amounts of relic

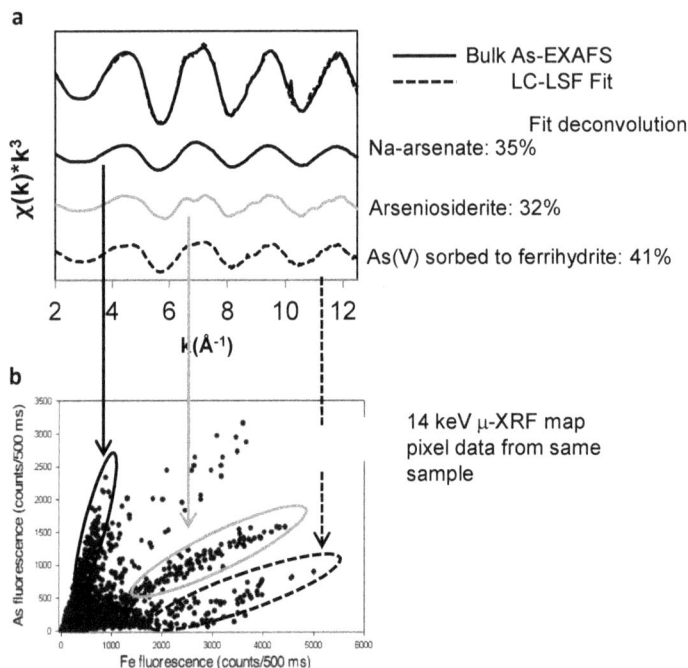

Figure 16. (a) Bulk As K-edge EXAFS spectrum of Descarga mine tailings (Mojave Desert, CA) < 20 micron size fraction (solid line) and LC-LSF Fit (dotted line). Fit deconvolution, showing three main species (two of which contain both As^{5+} and Fe^{3+}; see text for formulae); (b) bivariate plot of per-pixel fluorescence recorded in As and Fe channels during μ-XRF mapping of the same sample, showing 4 near-constant As/Fe ratios that corresponded to 3 of the species used in the bulk LC-LSF fits. [Used by permission of American Chemical Society, from Kim et al. (2013) *Environmental Science and Technology*, Vol 47, Fig. 5, p. 8168]

primary phases are also commonly observed in processed ore due to incomplete oxidation (Hudson-Edwards et al. 2005; Walker et al. 2005). During high-temperature roasting, As^{-1}-pyrite and arsenopyrite are transformed to As_2O_3 as well as As^{5+} and As^{3+} species sorbed to maghemite and hematite (Walker et al. 2005; Paktunc et al. 2006; Fawcett and Jamieson 2011). It should be emphasized that the As-solid phase species formed from high temperature and/or high pressure ore processing treatment are distinct from those formed by chemical weathering of mine wastes under ambient conditions.

Solid mine wastes are dispersed primarily via wind or water transport (Hudson-Edwards et al. 2011; Plumlee and Morman 2011). Several catastrophic dispersals of large volumes of mine wastes from tailings impoundments have been reported, and As-XAS has been used as part of the overall impact assessment at several of these sites (Hudson-Edwards et al. 2005; Foster et al. 2011; Burke et al. 2012). As-XAS has also been used to study the efficacy of dispersion minimization techniques such as subaqueous or underground storage of tailings (Paktunc et al. 2003, 2004).

During chemical weathering of mine wastes under ambient conditions, arsenate typically becomes the dominant solid-phase species (except in reducing, aquatic environments). Increases in the relative abundance of the following species groups have been noted: (1) As sorbed to Fe and/or Al hydroxides or aluminosilicate clays; (2) As substituted in secondary sulfates; and (3) poorly crystalline secondary arsenate phases (Appendix 4). The relative abundance of these species depends on site-specific factors including the total As, mineralogy,

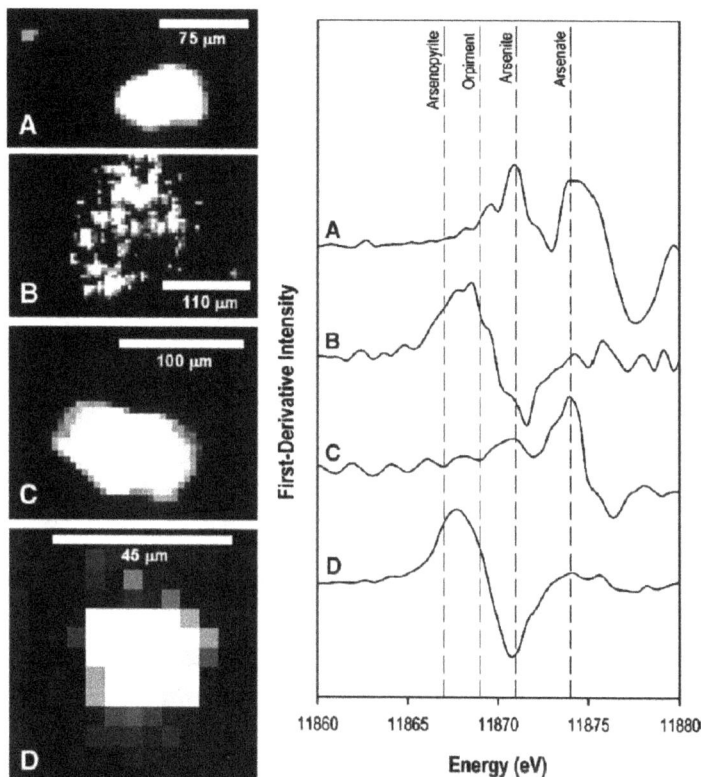

Figure 17. Micro-XRF maps (left, a-d) and corresponding 1ˢᵗ derivatives of μ-XANES spectra (right, a-d) of individual As-bearing grains in aquifer sediment from central Bangladesh, documenting the occurrence of reduced As species in large and small grains. [Used by permission of Elsevier, from Polizzotto et al. (2006) *Chemical Geology*, Vol 228, Fig. 1, p. 101]

hydrologic regime, and geochemistry. Also, the rates of many chemical weathering reactions increase with temperature and precipitation (Maher and Chamberlain 2014), therefore the local climate may exert an important control on the proportions of As phases observed in soils and sediments containing dispersed mine waste materials. For example, the highly oxidizing conditions in desert environments lead to near-complete oxidation of the original As-bearing sulfide mineral hosts (Kim et al. 2013).

Mine pit lakes form by water table recovery at sites of former open pit mining. Pit lake chemistry can vary widely, and is influenced by factors including wall rock mineralogy, water depth, water sources, and climate. At the Harvard Pit, a high As (>1000 μg L⁻¹), actively-infilling alkaline pit lake in the Mother Lode Gold District, Calif., weathering of As⁻¹-pyrite on pit walls produces As⁵⁺-bearing goethite, jarosite, and copiapite (Fe²⁺/Fe³⁺ sulfate hydrate), as determined by As-XANES analysis (Savage et al. 2009). A mass-balance model further indicated that wash-in and subsequent dissolution of these secondary, As-bearing phases into the pit lake was a significant source of As to the system.

XAS studies suggest that the relative redox condition of water and waste dictates whether major changes in As speciation will occur when they are in contact. If redox conditions are the same for both, and mineral solubility is low, As speciation is much more likely to remain stable for longer periods. Little oxidation of As-bearing sulfidic tailings is observed after decades of storage if they are maintained in a sub-aqueous, sub- to anoxic condition, in keeping with previous studies documenting long term arsenopyrite preservation in subaqueous sediment (Craw et al. 2003). However, substantial and rapid oxidation is observed in sulfide-bearing tailings that undergo cyclic wetting/drying (Savage et al. 2000a; Foster et al. 2011; Fig. 9a,b). If tailings are located in an area with variable redox conditions, e.g., a wetland ecosystem, As can undergo seasonal fluctuations in oxidation state, reducing to As^{3+} or even As^{-1} when anoxic conditions prevail, but rapidly reoxidizing back to As^{5+} once conditions become more oxidizing (Beauchemin and Kwong 2006). In contrast to the behavior described above for sulfide-bearing tailings, roaster wastes (in which sulfides have been nearly completely oxidized at high temperature to maghemite and hematite) appear to contain very stable forms of As^{3+} and As^{5+}. The XAS-determined ratio of these species in roaster waste from the Giant mine (Canada) has remained near-constant over decades of storage under subaerial conditions (Walker et al. 2005).

Porewater in tailings retention structures has reacted with the waste for a considerable period, and is usually elevated in As; if the porewater is chemically reduced, As^{3+} is usually the dominant species in solution. As-XAS has been used to study As- and Fe-rich porewaters seeping from the base of tailings dams in California and France (Appendix 4). In California, the near-neutral porewater contains both As^{3+} and Fe^{2+} and supports an Fe-oxidizing microbial community dominated by sheath-forming members of the group *Leptothrix* (Foster and Ashley 2002; Foster et al. 2011). Arsenate (and less abundant As^{3+}) is the species sorbed to the newly-formed Fe^{3+} hydroxides formed on bacterial sheaths and in solution, indicating that oxidation is taking place prior or subsequent to adsorption. In France, acidic porewater seeping from a tailings dam holding Pb-Zn-Sb wastes also contains As^{3+} and Fe^{2+}, but the low pH water supports an acidophilic Fe-oxidizing microbial community dominated by *Acidothiobacillus* species. The temperate climate also leads to the precipitation of predominantly As^{3+}-Fe^{3+} oxyhydroxide in the winter and predominantly As^{5+}-Fe^{3+} oxyhydroxide in the summer (Morin et al. 2003).

Arsenic species in floodplain, aquatic, and organic-rich sediments

As-XAS is particularly useful for studying changes in As speciation that accompany dynamic, often seasonal, redox conditions that can exist in aquatic sediments. As-XAS is also useful for the study of As speciation in organic-rich matrices, which have traditionally presented a challenge for chemical analysis. Appendix 5 provides an overview of As-XAS studies of floodplain/river, pond/lake/reservoir, and organic-rich sediments (wetland and peat).

A prevailing characteristic of floodplain sediments is cyclic oxidation/reduction, with oxidation fronts extending far into sediments during dry periods. Although differing in the specifics, the general conclusions with respect to As speciation in floodplain sediments are similar for these studies. First, the predominant oxidized As species in the solid phase is As^{5+}-Fe^{3+} oxyhydroxide, (^2C sorption complex; Fig. 12a). Since sorbed species are predominant, inclusion of MS paths arising within the As^{5+} tetrahedron in theoretical fits is critical for distinguishing As^{5+} sorption on the various metal hydroxide and aluminosilicate phases common in river/floodplain sediments (Voegelin et al. 2007; Adra et al. 2013). Analysis of sediment size fractions has also revealed important speciation details: in one study of flood plain sediments impacted by arsenopyrite mining, the silt and larger size fractions contained appreciable arsenopyrite, but the clay-sized fraction only contained As^{5+}-Fe^{3+} hydroxide (Mandaliev et al. 2014).

Analyses of As (and Fe) XANES spectra of floodplain sediments from several areas in Bangladesh demonstrate that As^{5+} reduction to As^{3+} (and Fe^{3+} reduction to Fe^{2+}) occurs within a few meters of the ground surface (Foster et al. 2000b; Breit et al. 2004; Polizzotto et al. 2005, 2006). In east-central Bangladesh, this reduction profile coexists with pronounced accumulation of As^{5+}-Fe^{3+} hydroxide within one or more subhorizontal layers in the oxic zone of the floodplain sediments (usually <1.5 m depth; Breit et al. 2004; Foster et al. 2005). Work in eastern Bangladesh has also demonstrated that during the summer months, groundwater discharge to rivers leads to the oxidative precipitation of As^{5+}-Fe^{3+} hydroxide that can produce bulk As sediment concentrations locally as high as 1000 mg kg^{-1} in near-surface sediments (Datta et al. 2009). The implication of these studies is that a large portion of the As^{5+}-Fe^{3+} oxyhydroxide in near-surface sediment in Bangladesh is recycled in the near surface, rather than being a product of transport and deposition. A μ-XRF/μ-XANES study of delta/wetland sediment from Kandal province, Cambodia, arrived at similar conclusions with respect to rapid reduction of As^{5+} in shallow sediment. However, in this highly organic-rich sediment, "hot spots" of sulfide-As^{-1} were also identified (Kocar et al. 2008; Polizzotto et al. 2008).

One of the predominant themes of As-XAS studies of freshwater aquatic sediment (river, lake/reservoir, subaqueous tailings ponds) is the transformation of solid phase As^{5+} (predominantly As^{5+}-Fe^{3+} hydroxide) to more reduced forms (Appendix 5). The depth at which these transformations occur and the reduced, solid-phase As species produced depend on site-specific geochemical conditions and the chemical gradients prevailing on both sides of the sediment-water interface. As-XAS studies document reduction of As^{5+} to As^{3+} at relatively shallow depths (< ~10 cm) in aquatic, Fe^{3+}-rich sediment (Kneebone et al. 2002; Root et al. 2007; Moriarty et al. 2014). This reduction leads to an increase in dissolved porewater As concentrations in many systems (e.g., Campbell and Hering 2008), but also can lead to a re-partitioning of As on newly-formed solid phases (e.g., green rusts, Fe sulfides, As sulfides), as discussed previously in the section on microbial transformations of solid-phase As species.

Shallow sub-oxic reservoir sediment (<4-10 cm depth) with low organic matter, low SO$_4$, and high inputs of As^{5+}-Fe^{3+} hydroxide from the water column contains As^{3+}-Fe^{3+} hydroxide as the primary phase (Kneebone et al. 2002; Root et al. 2007), indicating the preferential reduction of As^{5+} over Fe^{3+} in this zone (Campbell et al. 2006; Root et al. 2013). At greater depths As^{3+} sorption to green rust phases (containing Fe^{2+} and Fe^{3+}) indicates the onset of Fe reduction (Root et al. 2007).

Arsenic-XAS studies of high SO$_4$ aquatic sediments present indicate that As is predominantly incorporated into solid phases containing reduced sulfur (La Force et al. 2000; Beauchemin and Kwong 2006; Wilkin and Ford 2006). The specific species formed are dictated by prevailing Fe-As-S ratios (Root et al. 2013; Langner et al. 2013) or site-specific abundances of other elements (Beauchemin and Kwong 2006). Wetlands and peatland sediments represent an extreme of the aquatic sediment category in which the S is abundant in both oxidized and reduced forms (the latter associated with fresh organic matter). The following reduced As species have been documented in high-S aquatic, wetland, and peatland sediments: (1) orpiment-like species; (2) realgar, α-As_4S_4; (3) safflorite, (Co,Fe)As_2; (4) greigite, Fe_3S_4; (5) FeAsS + As^{-1} pyrite, and (6) As sorbed to sulfide-rich particulate organic matter (see Appendix 5 for citations and details). Studies of freshwater peatland sediments from Gola di Lago, Switzerland suggest a transition from realgar-like phases in near-surface sediment, to organo-As (complexed with S) at greater depths (Langner et al. 2012a). Microbeam XAS/XRF was also key to identifying low-abundance FeAsS + Fe(As,S)$_2$ species in peat (Langner et al. 2013). Slow oxidation of peatlands by natural processes leads to oxidation and redistribution of As and Fe with accumulation of As^{5+}-Fe^{3+} hydroxide in the uppermost sediment horizon (Bauer et al. 2008).

Estuarine and marine sediment also fall under the category of high SO$_4$-aquatic sediments; As-XAS studies indicated that under sub-oxic to anoxic conditions, As undergoes similar transformations to sulfide-associated species as described above. An orpiment-like phase was identified in surficial marine sediment (contaminated by Zn smelter waste) incubated under artificial anaerobic conditions for 60 days (Xu et al. 2011), whereas As^{-1} pyrite dominated the As speciation in the pyrite-rich (> 7% modal abundance) sediments of the Achterwasser lagoon (NE Germany).

Arsenic species in aquifer sediments

Naturally-occurring, high-As groundwater is a common occurrence worldwide (Smedley et al. 2001; Mandal and Suzuki 2002; Smedley 2003; Nriagu et al. 2007; Barringer and Reilly 2013). The processes resulting in high dissolved As concentrations vary as a function of As mineralogy and aquifer chemical/microbiological characteristics (Smedley 2003). As-XAS studies of aquifer sediment (Appendix 6) have generally taken one of two approaches, or used them in combination: (1) analysis of aquifer material "as is" to understand spatial/temporal dynamics of As solid-phase speciation, and (2) analysis of aquifer material after laboratory incubations designed to test mechanisms of As retention or release (e.g., Stollenwerk et al. 2007; Singer et al. 2013).

Reducing aquifers. Anoxic aquifer systems differing in SO_4 concentration can produce very different As speciation in sediments. Aquifers low in SO_4 and high in Fe produce dissolved As^{3+} that can repartition to existing minerals or new phases forming in the aquifer. Shallow (<150 m) aquifer sediments of northwestern, central, and eastern Bangladesh and in West Bengal, India fit this category. Dowling et al (2002) reported that mica minerals constitute as much as 15% mineral modal abundance in some sediment horizons, and that mica concentrates from Bangladesh aquifers are ~4× more enriched in As over the bulk sediment. The association of As^{3+} with secondary Fe phases (Fe-carbonate, Fe-phosphate) formed on Fe-rich mica in gray, Holocene-age aquifer sediment from eastern Bangladesh has been documented by μ-XAS/μ-XRF (Foster et al. 2000b, 2005, 2010); others have also noted the association of As and Fe as discrete coatings on mica grains by u-XRF (Hudson-Edwards et al. 2004).

Pentavalent As has also observed by As-XANES in rust-colored, Fe^{3+} oxyhydroxide-coated aquifer sands of Pleistocene-age in eastern Bangladesh. Since these sediments underlie the gray, Holocene-age sediments and are in contact with highly reducing groundwater, Mn^{4+} in oxide form or substituted in Fe^{3+} oxyhydroxides was suggested by Foster et al. (2000b) as the mechanism for maintaining As^{5+} in the Pleistocene-age sands, in accordance with several XAS-studies described earlier. Another study employing bulk As- and Mn-XANES provides additional evidence that Mn^{4+} oxides in subsurface aquifer sediment can oxidize As^{3+} as it is formed, converting it back to As^{5+} that adsorbs to mineral surfaces (Choi et al. 2009).

Other researchers working in central Bangladesh have documented the existence of primary and secondary sulfide minerals in Bangladesh aquifer sediment and used mass balance approaches to suggest that a combination of relic and neo-formed sulfides can contain up to 60% of the solid-phase (Polizzotto et al. 2005, 2006; Fig. 17a-d). The observed differences in speciation in central and eastern Bangladesh may reflect different source materials, since river channels have altered course multiple times in this tectonically-active region. Additional studies are needed to reconcile differences in As speciation observed in shallow aquifer sediment from central and eastern parts of Bangladesh.

Anoxic aquifers with high SO_4 can become sulfidic; this is the case in southern Bangladesh, where the shallow aquifer system is brackish due to mixing with seawater. However, the deep aquifer (>150 m) contains potable, low As water. A bulk and microbeam As-XAS study of shallow and deep aquifer sediment from southern Bangladesh (450 m maximum depth) showed the overwhelming predominance of As^{-1}-pyrite in aquifer sediments; Comparison of micro and bulk XANES spectra from the same sediment horizon allowed the definitive identification of As^{-1} pyrite as a major host for As, but also showed that As^{3+} and As^{5+} were also present in additional species which were not conclusively identified (Lowers et al. 2007). Isotope and microprobe evidence indicated the active formation of framboidal pyrite by bacterial processes in the aquifer.

Reducing aquifers high in DOC can support different types of As species, which may influence human health. Several regions of Taiwan are plagued with high As groundwater, but the distinct peripheral vascular disease illness ("blackfoot disease") occurs only in a population living in the southeast region of the country where, highly reducing, As-rich groundwater also contains high levels of DOC (Jean and Lee 2006; Liu et al. 2009). To our knowledge, a distinct organoAs species has not been determined by XAS or other method in samples from this region.

The Hetao basin in Inner Mongolia, PR China, also contains DOC-rich groundwater in an aquifer that also is growing increasingly saline and anoxic (Guo et al. 2011a). Micro-XRF data suggest that As is at least partially mobilized through the formation of colloidal species containing reduced S. Micro-XRF was used to compare As concentration in 4 colloidal size fractions examined (100, 30, 10, 5 kDa): As was highest in the 10-5 kDa fraction and was positively correlated with S, whereas Fe was highest in the > 100 kDa fraction (Guo et al. 2011b). Water chemistry was used to infer that As^{5+} was the predominant species in the organic colloids, but was not directly verified with XAS.

Redox-stratified aquifers. Coastal aquifer systems can have characteristics of oxidizing and reducing aquifers, as XAS studies reveal. A stratified aquifer system adjacent to the San Francisco Bay characterized by brackish, sulfate-rich zones overlying zones of freshwater contained three subsurface zones, each with different solid phase As speciation (O'Day et al. 2004; Root et al. 2009). Reduced S-rich sediment in the shallow aquifer contained As^{3+}-S in either orpiment-like or realgar-like structure with variable As^{3+}-O or As^{5+}-O. An intermediate-depth zone with redox-transitional sediment contained As^{3+}-O or As^{5+}-O but no sulfide-As. The lower, oxidized, freshwater aquifer contained only As^{5+}-O. These results were used to help constrain a model whereby As is effectively retained by precipitation of sulfides in the shallow aquifer and sorption on oxyhydroxides in the deeper aquifer, but has the potential for release in the intermediate aquifer (O'Day et al. 2004; Root et al. 2009). A model was also proposed to explain the occurrence of both realgar and orpiment like As-sulfides in the system: realgar is suggested to be the thermodynamically stable form to form in the presence of pyrite and As^{3+}, and realgar is proposed to form under more reducing conditions than orpiment (Root et al. 2009).

The Pearl River Delta (PRD) region, PR China, is another high-As coastal aquifer system characterized by a reducing aquifer in the shallow subsurface and a more oxidizing, basal aquifer at depth (about 30 m). As-XANES was used to produce depth profiles of As speciation at two locations in the PRD where bulk As ranged from 1-40 mg kg^{-1} (Wang et al. 2013). Reduced As^{-1} was common in sediment (20-40% relative abundance) and it was most abundant in silty clay or silt units, least abundant in the basal sandy aquifer, and absent in the shallowest (oxidized) samples (3 and 4 m depth). As^{3+} and As^{5+} were ubiquitous in aquifer sediment from all depths.

XAS studies have also shown some dependence of As speciation with particle size in aquifer sediment. Sand-sized sediment has been is found to contain more As^{5+} than silt- or clay-sized sediment, which contains more As^{3+} (Rowland et al. 2005; Wang et al. 2013).

Oxidizing aquifers. Arsenic sorption to natural coatings on sand grains from an aquifer in Cape Cod, MA, USA was studied using μ-XRF, μ-XAS, and transmission electron microscopy (Singer et al. 2013). Aquifer sediment was reacted with As^{3+} or As^{5+} in laboratory batch experiments, and coatings on individual grains after 0.5, 2, and 24 h were examined by spectroscopic techniques. Micro-XRF maps demonstrated that zones of Mn-rich goethite coatings on sand grains provided sites of preferential As sorption. Moreover, μ-XANES spectra revealed partial oxidation of As^{3+} in experiments that was attributed to Mn^{3+} present in the goethite. The authors suggested that oxidation was partial due to limited surface area exposed on the coating particles and the finite amount of Mn substitution in goethite relative to the abundance of As^{3+} used in the experiment (Singer et al. 2013).

Arsenic species in wastes from industry, agriculture, and the built environment

XAS studies in this subject area are many and varied in scope. They are related to both historic and modern uses of As in agriculture, industry, and consumer products, and include laboratory and field-based studies. Particulates (dust and aerosols) form urban landscapes and home interiors have also been examined by As-XAS. The results of XAS studies in this broad area are summarized in Appendix 7 and highlighted below.

Arsenical pesticides/herbicides. The organic arsenical pesticides/herbicides include monosodium methanearsonate (MSMA), disodium methanearsonate (DSMA), calcium acid methanearsonate (CAMA), and cacodylic acid and its sodium salt (some structures shown in Fig. 1). They have been used on cotton, citrus, and grape crops (the latter two during non-fruiting condition only), in forestry, on ornamental lawns and turf, and on railroad rights-of-way, drainage ditch banks, fence rows, and storage yards (USEPA 2006). Although the organoarsenicals are less toxic than inorganic As in toxicology tests (USEPA 2006), the possibility of their subsequent decomposition to inorganic As is a major concern. In 2006, the US Environmental Protection Agency concluded that all 4 organoarsenicals should be phased out of use by 2012; by 2009, this was complete for all but MSMA (USEPA 2013c). The phase-out of MSMA was halted pending the results of a National Academy of Sciences/National Research Council review of the data on cancer risk from exposure to inorganic As, which is ongoing at the time of this writing (USEPA 2013c).

Inorganic-As compounds were utilized in the past for many of the uses for which organoarsenic compounds are now employed: in fact, low-solubility As solids were favored for herbicides due to their long effectiveness period relative to liquid organoarsenicals (Smith et al. 2006). Inorganic arsenicals were also widely used in animal husbandry: for example, use of inorganic arsenical compounds to control fleas, ticks, and lice in cattle and sheep populations was mandatory in New Zealand for almost 100 years prior to 1980, when it was replaced by other chemicals. There was similar use of arsenical compounds in other countries with large herds of grazing livestock, including Australia and the USA (Gaw and McBride 2010).

Most of the As-XAS studies of arsenical pesticides/herbicides have been conducted at sites where inorganic As compounds were manufactured or stored. In most of these studies, the primary issue of concern is to assess the speciation and fate of the fraction of As that has become labile due to partial or complete dissolution of the primary As compound. In many cases, XAS has been used to show that in addition to adsorbed species, $As^{5+}O_4$ has combined with elements that reach relatively high dissolved levels in oxidizing soil pore water solutions (e.g., Ca, Mg, and Al) to form metal arsenate precipitates.

Formation of Ca-arsenate or Ca-Fe arsenates have been documented at sites where inorganic metal arsenates were manufactured or stored: one is a site in the northeastern US where schultenite ($PbHAsO_4$) was stockpiled for manufacture of As-based compounds (Arai et al. 2006), and the second is a site in France where Pb- and Al-arsenates were manufactured (Cancès et al. 2008). Arsenate sorbed to Fe^{3+}-hydroxide is the dominant species in surface or oxidizing soils from both sites, but at the French site, up to 35% arseniosiderite was found in the topsoil (0-5 cm depth; not detected in deeper sediment). Direct comparison of LC-LCF fit quality with and without arseniosiderite as a component was used to demonstrate the presence of this phase (Fig. 18). In contrast, at the US site, As^{5+} associated with one or more unidentified Ca minerals [potential phases: calcite, gypsum, and weilite ($CaHAsO_4$)] was an important component only in "semi-reduced" sediments.

Stockpiling As_2O_3 leads to the formation of secondary arsenate minerals and/or As sorbed complexes on soil minerals. The rare mineral hoernesite ($Mg_3(AsO_4)_2 \cdot 8H_2O$) was produced in surface soils from reactions between As_2O_3 and estuarine water of the San Francisco Bay (Voigt et al. 1996; Foster 1999). On the southeastern US coastline, application of As_2O_3 as

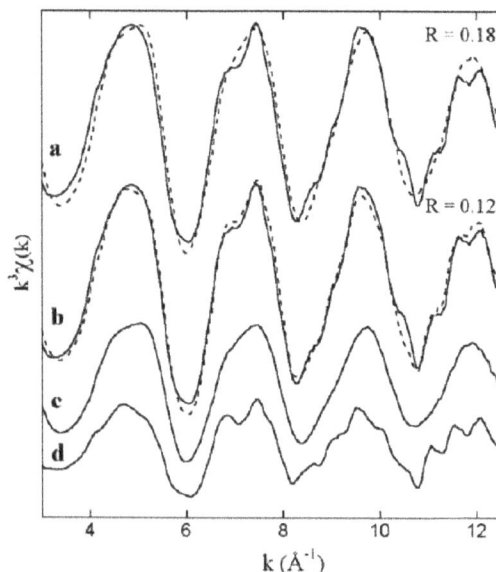

Figure 18. Comparison of 1- and 2- component LC-LSF fits to bulk, k^3-weighted As-EXAFS spectra of a topsoil sample from a site of inorganic arsenical compound manufacture. Fit **a** uses As^{5+}coprecipitated with Fe^{3+}-hydroxide (spectrum labeled **c**), whereas fit (**b**) additionally contains a component of arseniosiderite (spectrum labeled **d**).The residual (*R*) value shows that fit **b** is ~33% better than fit **a**. [Used by permission of Elsevier, from Cancès et al. (2008) *Science of the Total Environment*, Vol 397, Fig. 7, p. 186]

an herbicide to soils than 50 years ago has resulted in the formation of one or more labile As^{5+} forms sorbed to amorphous and crystalline Fe/Al hydroxides as well as rare particles of a Ca-arsenate phase similar to phaunouxite [$Ca_3(AsO_4)_2 \cdot 11H_2O$), determined from μ-XRD and μ-XAS; Yang and Donahoe 2007]. No trace of the original As_2O_3 was found in soils from 10-60 cm depth below ground surface, even though these soils contain As concentrations as high as 900 mg kg^{-1}. Application of As_2O_3 at Tyndall Air Force Base, Florida, USA produced comparable results, with soils containing up to 280 mg kg^{-1} As in a labile form and no evidence for the parent As_2O_3 material by As-XAS. The predominant As species derived from theoretical fits to EXAFS spectra was As^{5+} sorbed to Al-hydroxide surface coatings or sorbed onto aluminosilicate minerals (Fitzmaurice et al. 2009).

In contrast to these observations, As_2O_3 inadvertently deposited over the landscape near the Giant Mine roaster (Canada) has remained stable for more than 75 years with no evidence of conversion to other phases (Bromstad and Jamieson 2012; Jamieson 2014, this volume). Climate is one big difference that might explain the differences in As_2O_3 stability observed at US sites (sub-tropical) vs. Canadian site (artic/sub-artic). Other potential reasons include the formulation (especially particle size) of the materials; in the US studies, the As_2O_3 was the commercial product used for its intended purpose, whereas the material deposited around the Giant Mine was largely derived from roaster stack emissions.

The studies described above show that applied inorganic arsenicals can undergo complete transformation in the environment in some conditions, but can be well-preserved in others; the reasons for this starkly different behavior need further investigation. In cases where inorganic arsenicals applied to soils do dissolved over time, secondary arsenate minerals can form where conditions supply enough counterions; Ca- and Mg-arsenates have been documented in samples, though not always as the primary species. However, they may be more common and important than has been appreciated in previous studies of As_2O_3 herbicide application. Labile As^{5+} sorbed to soil minerals is a key species to quantify in these sediments, because it is more susceptible to release by small changes in pore water pH and/or composition.

Arsenical drugs and feed additives. The structurally-related organoarsenic compounds nitarsone, roxarsone, arsanilic acid, and carbasone have been used as drugs in veterinary medicine in the US (some are shown in Fig. 1). Roxarsone, which is banned for veterinary use in the European Union, was one of the three organoarsenic drugs voluntarily withdrawn from the US market by its manufacturer in 2013 (only nitarsone is currently available; USFDA 2013). The withdrawal was prompted by a study showing higher accumulation of As in liver and breast tissue of conventional US poultry over certified organic US poultry, which prohibits the use of roxarsone in its products (Nachman et al. 2013).

Roxarsone was used extensively in the US prior to the recent ruling; many studies assessing the impacts of its use have focused on the Delmarva Peninsula, a poultry-producing coastal region shared by the states of Delaware, Maryland, and Virginia. In this region, As-rich poultry fecal matter (AKA "poultry litter"; up to 50 mg kg^{-1} As) is routinely applied to agricultural soils as a means of disposal and soil fertilization. Both As^{3+} and As^{5+} species have been identified (1:1 ratio) in poultry litter associated with abundant needle-shaped microscopic particles also containing Ca, Cu, Fe, S, Cl, and Zn as determined by μ-XRF (Arai et al. 2003). The As^{3+} species was similar to orpiment based on white line position, but not conclusively identified from μ-XANES spectra. Roxarsone was used to fit the As^{5+} component with good results, suggesting that at a substantial fraction of the drug is eliminated by poultry in an unchanged form.

Covey et al. (2010) used μ-XRF/μ-XAS to study roxarsone transformation by earthworms in laboratory soil mesocosms. Based on fingerprinting comparison of As XANES spectra, they concluded that roxarsone had been transformed to methylated arsenicals on soil particles after 30 days.

Copper chromium arsenate-treated wood. Residential use of wood treated with copper-chromium-arsenate (CCA) preservative solutions has recently been phased out in the US, but given its ~25 year service life, it remains a potential source of As, Cr, and Cu to the environment (Katz and Salem 2005). Furthermore, use of CCA-treated wood is still allowed for electric utility poles and agricultural fence posts. An estimated 6-10 million m^3 of CCA-treated timber waste is disposed of each year in construction and debris (C&D) landfills; current projections forecast disposal at this rate to continue through 2030 (Jambeck et al. 2008). Approximately one-half of the states in the US do not require impermeable liners on C&D landfills, raising concerns about offsite migration of Cu, Cr, and As. There is also concern about element migration from sites where CCA solutions were applied, because often these solutions leached into soils, causing high concentrations of Cu, Cr, and As (e.g., Hopp et al. 2008).

There are several types of CCA treatments, each varying in the proportion of Cu, Cr, and As introduced (Schwer and McNear 2011). Sludge rich in particulates also forms during the pressurized reaction of CCA solutions with treated wood (Nico et al. 2006). Theoretical fits to As-K-edge EXAFS (and Cr K-edge EXAFS) spectra have demonstrated that the sludge particulates are composed of a $CrAsO_4 \cdot nH_2O$-type phase with a characteristic R_{As-Cr} distance of 2.97 Å (Bull et al. 2000; Nico et al. 2006). CCA-treated wood, however, contains a strongly-bound polymerized species whose fundamental repeating unit contains AsO_4 bound to 2 edge-sharing $Cr(OH)_6$ octahedra in a ^2C-type geometry with a characteristic R_{As-Cr} distance of 3.25 Å (Fig. 19a; Nico et al. 2004, 2006). Binding to the wood cellulose occurs through apical oxygen/hydroxyl atoms of the Cr octahedron, and linkages among the Cr octahedra form the polymer network. This species is very stable and has been found in fresh wood, aged wood, and in wood particulates (Fig. 19b). Copper appears to have a completely different speciation unrelated to Cr or As in CCA-treated wood (Bull et al. 2000; Nico et al. 2006).

a

b

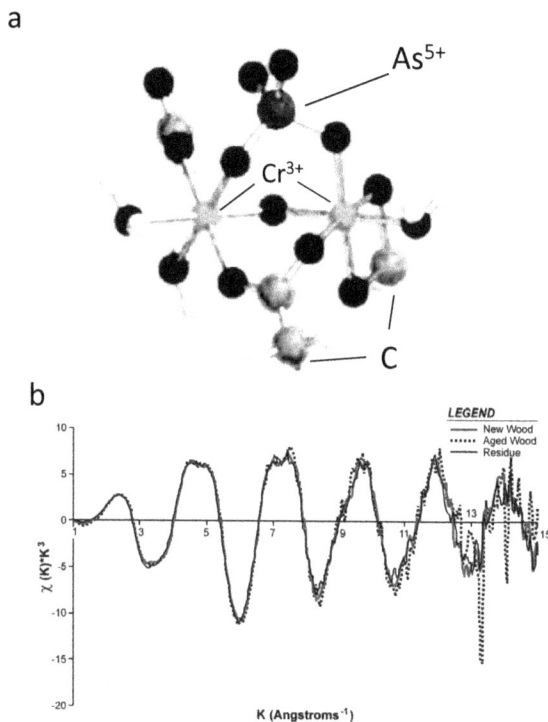

Figure 19. (a) Predominant As speciation in CCA-treated wood based on theoretical fits to As-EXAFS spectra. Arsenate binds to polymerized Cr octahedra in the ^{2}C geometry (Cr octahedra, in turn, bind to wood cellulose); (b) EXAFS spectra of aged CCA wood, new CCA wood, and particulates scraped from CCA wood residue show equivalent speciation. [Used by permission of American Chemical Society, from Nico et al. (2004) *Environmental Science and Technology*, Vol 38, Fig. 2, p. 5256 and Fig. 4, p. 5258]

The extent of human exposure to inorganic As via contact with CCA-treated wood has been a subject of study using As-XAS. As speciation in residues from CCA-treated wood collected on human subjects' hands or by wire brush (standard residue collection technique) was examined before and after reaction with human sweat (Nico et al. 2006). Hand-collected and brush-collected residue had identical speciation (both identical to the Cr-As cluster described above). Sweat extracted only about 12% of the total As present in wood samples, but the extracted As was 100% inorganic As^{5+}, and Fe rather than Cr was found in extraction solutions. Based on these findings, the authors concluded that the leached As^{5+} was probably derived from desorption of As^{5+} from Fe^{3+} oxyhydroxide particle surfaces rather than CCA-treated wood fragments. The origin of these particles was uncertain, but the authors speculated that simple dust deposition on the CCA-treated wood might have been the source.

Due to the higher levels of CCA sludge and waste liquids, the speciation and mobility of As (and Cr) in soils from industrial sites where CCA was applied to wood is likely to differ from sites where CCA-treated wood was placed in service and reacted with soil over time. Grafe et al. (2008c) studied As speciation from 2 depth horizons at a site in Florida (USA) that was formerly used for CCA wood preserving. A combination of bulk and μ-XRF/XAS with PCA and LC-LSF analysis was used to identify several solid-phase As species, including ferric and zinc arsenate hydrates (scorodite, adamite, ojuelaite), Cu-arsenates, and As^{5+} sorbed to goethite and gibbsite.

Hopp et al. (2008) studied As speciation in soils from a wood preserving site in Germany. This site differed from that studied by Grafe et al. (2008c) in several ways, including contamination by tar creosote in addition to CCA, and the fact that soluble Fe^{2+} had been applied to soils as a remediation strategy prior to the XAS study. In the upper 65 cm of the soil, As^{5+} was still associated with Cr in wood fragments. In lower soil horizons, As^{5+} was primary associated with amorphous allophane (an aluminosilicate).

Schwer and McNear (2011) measured As dispersion from CCA-treated fenceposts in KY, where the horse husbandry industry has constructed hundreds of miles of fencing across large areas. Speciation was dominated by As^{5+} sorbed to Al-hydroxides (represented by amorphous $AlAsO_4$ and As^{5+}-gibbsite) and As^{5+} sorbed to Fe^{3+} hydroxides, but reversed with pH: in acidic soils (pH ≤ 4.5), bonding of As^{5+} to Al-hydroxides was favored, whereas at higher pH (up to 7), bonding to Fe^{3+}-hydroxides was dominant.

Carbon-based fuel combustion residuals. Industrial wastes remaining after burning coal for electricity include fly ash, bottom ash, boiler slag, and flue gas desulfurized gypsum (FGD gypsum), and are collectively called coal combustion residuals (CCRs; USEPA 2013a). CCRs are one of the largest waste streams generated in the United States (USEPA 2013a); since coal combustion is a primary source of electricity generation worldwide, this is likely to be true in other countries as well. In the US, CCRs are typically disposed of in liquid form at large surface impoundments or in solid form at landfills (USEPA 2013b). In a 2009 survey, 15 of 42 US states did not have minimum liner requirements for CCR landfills, and 24 of 36 states did not have liner requirements for surface impoundments (USEPA 2013a,b). A 2008 coal ash spill from a surface impoundment in Tennessee flooded 300 acres of land, damaged homes and property, filled large areas of rivers, and resulted in fish kills (Dewan 2008). This event has prompted the US Environmental Protection Agency to propose the first national rules regulating the disposal of CCRs (USEPA 2013b).

Coal fly ash has been the main form of CCR studied by As-XAS. It is the predominant material in surface impoundments, yet examination of its trace-element speciation is difficult by other techniques because the elements of concern (As, Cr, Cd, Hg, Se) are typically present at concentrations below detection by many conventional solid analysis techniques (0.1-100 ppm) and often the fly ash matrix is composed of small crystalline, and amorphous particles that are difficult to characterize.

Several XAS investigations of coal fly ash have shown that As speciation varies with parent coal type. "Class F" fly ash is derived from bituminous coals, produces acidic waters, and primarily contains As^{5+} associated with one or more Fe^{3+}-bearing phases (Zielinski et al. 2007; Catalano et al. 2012; Fig. 20a,b). Between 10 and 15% As^{3+} has also been observed by As-XANES in high-sulfur bituminous coal fly ashes (Huggins et al. 2007; Catalano et al. 2012). The reported details of As^{5+} speciation in F type fly ash differ based on the type of As-EXAFS spectral analysis performed. Using LC-LSF fits to bulk EXAFS spectra, Zielinski et al. (2007) reported than an F-type fly ash contained roughly equivalent amounts of As^{5+} substituted in jarosite and As^{5+} sorbed on goethite and ferrihydrite. However, anhydrous $FeAsO_4$ or angelellite [$Fe_4O_3 \cdot (AsO_4)_2$] was proposed as the primary As host in F-type fly ash on the basis of theoretical fits (Catalano et al. 2012). Examined in detail, the As coordination determined in theoretical fits to the F-type fly ash (~2 Fe neighbors at 3.26 Å; Catalano et al. 2012) is not matched by the coordination environment of jarosite, anhydrous $FeAsO_4$,or angelellite, because these latter phases contain approximately twice the number of neighboring Fe atoms. Detailed studies of fly ash mineralogy also did not identify any of these phases (Valentim et al. 2009; Catalano et al. 2012). Given the relatively low concentration of As in coal fly ash (100s of ppm), and the previously mentioned reaction products of pyrite oxidation at high T and/or high P, sorption of As^{5+} on one of the Fe oxides identified in coal fly ash (hematite, maghemite) might be the most probable coordination environment for As^{5+} in

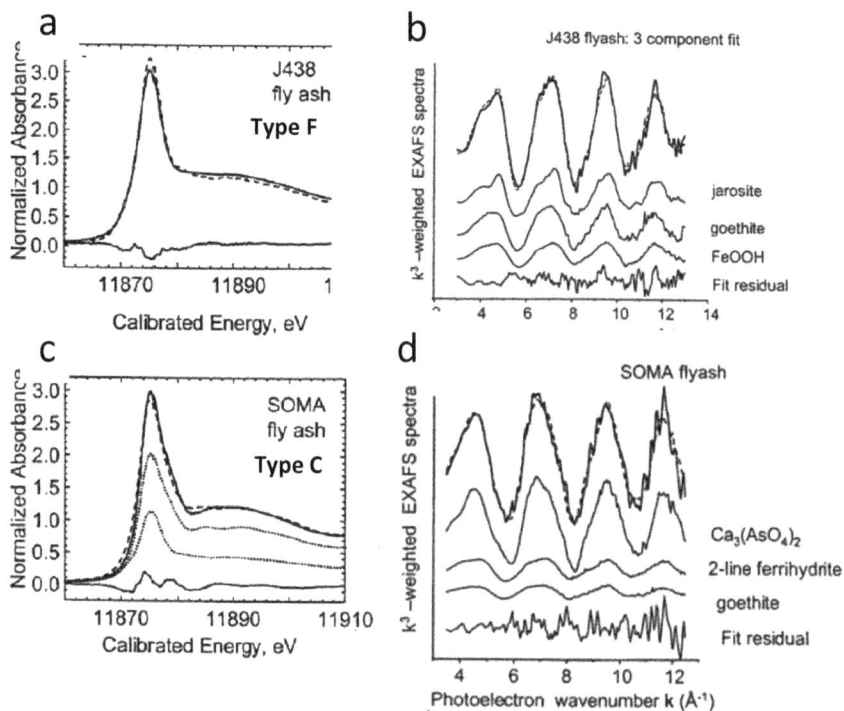

Figure 20. LC-LSF fits to As-XANES spectra (a, c) and As-EXAFS (b, d) of F-type coal fly ash ("J438", panels a and b) and C-type fly ash ("SOMA", panels c and d). The number of species required to fit XANES spectra is fewer than required for fits to EXAFS. [Used by permission of Elsevier, from Zielinski et al. (2007) *Fuel*, Vol 86, Fig. 7, p. 567]

F-type fly ash. However, the splitting in the first oscillation of the EXAFS spectra of type F fly ash suggests an As coordination environment of higher symmetry than typically observed in sorption complexes (Fig. 20b). The recent discovery that As^{5+} forms epitaxially-grown angelellite during coprecipitation with hematite in aqueous solution may provide a mechanism for angelellite formation in F-type fly ash (Bolanz et al. 2013; Appendix 1).

"Class C" fly ash is derived from sub-bituminous/lignite coal, produces alkaline water, and primarily contains As^{5+} in a phase with similar structure to synthetic Ca-pyroarsenate (Zielinski et al. 2007; Luo et al. 2011; Catalano et al. 2012; Fig. 20c,d). Microscale XANES mapping of C-type fly ash from Slovakia indicated that although As^{5+} is dominant, As^{3+} is present in rare vesiculated and/or Ca-rich particles (Veselska et al. 2013). These results highlight the complementary results provided by bulk and microscale XAS information, and suggest that both be employed wherever possible to characterize heterogeneous natural materials.

CCRs have been generally treated as non-hazardous waste (with lower standards for environmental protection), because standard leach tests recover levels of As and other potentially-toxic trace elements that are below regulatory thresholds. However, the leach tests do not attempt to mimic the environmental conditions experienced by fly ash during aging in lagoons, after burial, or during human ingestion/inhalation. Detailed As-EXAFS analysis of C- and F-type fly ashes subjected to leaching by simulated rainfall under aerobic conditions showed significant leaching of As, Se, Cr, Cu, Zn, and Ba from F-type ashes during the initial weeks, but no further change in As speciation over a period of ~60 weeks (Catalano et al. 2012).

These results were similar to an earlier aging study of fly ashes (Zielinski et al. 2007). Another study saw no changes in As-XANES speciation between freshly lagooned C-type coal fly ash and material from the same plant that had been buried under an agricultural field for ~45 years (Veselska et al. 2013).

Fuel oil generally refers to the heavier fractions obtained from petroleum distillation. It is typically used for industrial heat production or in diesel engines (Huffman et al. 2000). XANES analysis was used to examine As speciation in particulates derived from burning residual fuel oil in an industrial boiler (Huffman et al. 2000). Arsenate was the sole species identified (and was postulated to be sequestered in one or more of the several crystalline metal sulfates identified by XRD and XANES analysis of other metal K-edges.

The built environment. The speciation of As in the "built environment" (a term including but not limited to urban areas, cars, and buildings) is an area of active study since it has been widely appreciated that the man-made environment can contain high levels of As and other potentially toxic trace elements (Plumlee and Ziegler 1999).

Smog is a part of the built environment that presents a clear risk to human health on several fronts. Increasing levels of airborne particulate matter in the urban environment is a worldwide problem and is exacerbated by population, climate, and geographic factors. Godelitsas et al. (2011) studied the distribution of several heavy elements and the speciation of As in airborne particulates collected in Athens, Greece. Arsenic was always associated with Pb and not with other elements in both the larger (10-2.5 μm) and smaller (<2.5 μm) size fractions studied. Fingerprint comparison of As-μXANES spectra to those of $As^{3+}_2O_3$ and $As^{5+}_2O_5$ models indicates the presence of both oxidation states. The authors postulated diesel exhaust, residential central heating, and industrial activity (coal burning power plants) as likely particulate sources (Godelitsas et al. 2011).

Indoor dusts have long been recognized as sources of potentially toxic metals, but it is not widely appreciated that in urban environments indoor dusts can have higher metal concentrations than exterior dust and outdoor soil, or that that metal bioaccessibility may be higher in indoor dusts (MacLean et al. 2010 and references therein). Walker et al. (2011) used μ-XRF/μ-XAS/μ-XRD to characterize metal speciation in dusts from 3 rooms of a residential home in Ottawa, Canada. Only the living room dust contained As-rich particles. A strong correlation of As and Cr by μ-XRF, identification of wood cellulose by μ-XRD pattern suggested that CCA-treated wood particles originating from an unknown source outside the house was the origin of the As in living-room dust. The overall conclusion was that rooms can have different sources of metal-rich dusts, with the living rooms generally having more dust sourced from outside the home, and bedrooms having more metal-rich dusts associated with painting and other renovations.

Arsenic is present in tobacco and has been measured at levels as high as 10s of ppm in filtered cigarette smoke (Liu et al. 2012). As many as six inorganic and organic As species have been identified in water-soluble extracts of tobacco and smoke using chemical methods (Liu et al. 2012). However, obtaining and preparing a concentrated sample of smoke suitable for XAS analysis and minimizing redox reactions in collected smoke has been a challenge. Liu et al. (2012) developed a simple cold-trapping method for collecting and preserving smoke and airborne particulates from burning cigarettes for subsequent XAS analysis. The tobacco contained 94% As^{5+}, whereas the cigarette ash contained 100% As^{3+}. Cryo-trapped smoke condensate contained 10% more As^{3+} than room-temperature smoke, which had roughly equivalent proportions of As^{5+} and As^{3+} (Liu et al. 2012). Chemical analysis of water soluble tobacco indicated the presence of several organoAs species in addition to inorganic As^{5+} and As^{3+}.

Arsenic treatment wastes. Due to the 2001 lowering of the As drinking water standard from 50 to 10 μg L^{-1}, many US municipal water systems will need to add As removal treatment to their potable water supply systems (MacPhee et al. 2003). Arsenic removal technologies can produce a variety of liquid and solid waste types including sludges, brine streams, backwash slurries, and spent adsorbent media (USEPA 2002). Water treatment residuals (WTRs) are solid waste products generated during drinking water treatment processes. They are heterogeneous mixtures of inorganic and organic solids generally formed by coagulation and flocculation after addition of a chemical agent to stimulate precipitation (Makris et al. 2007). They are classified into 3 types based on the chemical agent used to scavenge the As: Al-WTRs (Al-salts), Ca-WTRs (Ca-salts), and Fe-WTRs (Fe-salts; Makris et al. 2009).

Fe-WTRs utilize precipitation of hydrous ferric oxides to affect As removal. Arsenite and As^{5+} adsorbed strongly to Fe-WTR from a municipal sewage treatment plant synthesized by $Fe_2(SO_4)_3$ enhanced coagulation, even in the presence of high phosphate concentrations typical in sewage treatment (Makris et al. 2007). The solid phase species formed were invariant with time (10 min vs. 48 h), and the As-EXAFS theoretical curve fits were consistent with sorption of both oxyanion species in the ^2E geometry. These results are highly unusual because the ^2E geometry has never been fit as the overwhelmingly dominant species for As^{5+} coprecipitated with Fe^{3+} hydroxide in model system studies. The same group also studied Al-WTRs but in this case no MS shells were added, and 3 different As-Al shells were fit to the data consistent with As^{3+} and As^{5+} oxyanions in a mixture of all three "standard" coordination geometries (^2E, ^2C, and ^1V) (Makris et al. 2009).

Spent adsorbents are another type of solid waste generated from As water treatment. In this case, preconditioned water is reacted with fine particulate material (usually packed in columns). A variety of heterogeneous reactions are exploited to remove As (and other dissolved species) from solution, including adsorption, coprecipitation, oxidation/reduction, and ion exchange. Several compositions of adsorbents have been investigated (Appendix 7).

Release of species from WTRs and spent adsorbents after landfill disposal or during storage in aqueous surface impoundments is a concern, particularly due to the microbially-generated reducing conditions which can predominate in these environments. The effect of two year anaerobic, abiotic incubations on the fate of As^{5+} in a variety of spent adsorbents was investigated by As-EXAFS (Liu et al. 2008). Greater than 75% of As^{5+} was reduced to As^{3+} in hydrous ferric oxide (HFO) spent media, in TiO_2 spent media and in "modified" activated alumina media (HFO added). However, only 35% of As^{5+} initially sorbed to "unmodified" activated alumina was reduced, suggesting that resistance of the substrate to reductive dissolution might preserve As^{5+} species as well. In the anaerobic reactors, As^{3+} sorbed to HFO in the standard ^2C bidentate geometry (MS paths not considered). These findings contrast significantly with those of Root et al. (2013), which were described in the section on biogenic minerals and microbial transformations. That study differed in its inclusion of an active microbial consortium that reduced the Fe^{3+} to Fe^{2+} more completely in the spent adsorbent media, producing a variety of new Fe^{2+} phases with less affinity for As (Root et al. 2013).

Natural siderite ($FeCO_3$) is under consideration as a low-cost sorbent for As^{3+} removal in Inner Mongolia, PR China, where abundant deposits are found (Zhao and Guo 2014). Under anoxic conditions siderite persists as the sole mineral during As^{3+} removal, but under oxic conditions siderite undergoes partial transformation to goethite, and As^{3+} is also partially oxidized. However, partially-oxidized siderite was a better sorbent for As^{3+} and As^{5+} than unoxidized siderite. As-EXAFS fits were consistent with the formation of ^1V and ^2C sorption complexes for both oxyanions on both sorbents in oxic experiments. In the absence of oxygen, the fits indicated predominantly ^2C complex formation (Zhao and Guo 2014).

Nanoparticulate zero-valent iron (nZVI) is a highly effective sorbent of As oxyanions, but the details of As speciation on the material are not completely understood. In aqueous solution, the surface of nZVI particles is continuously corroded, producing a layer of ferrihydrite that was initially thought to be the site of As adsorption. A time-dependent As-EXAFS study following the reaction of As^{3+} with nZVI *in situ* over a period of 22 h demonstrated a sequence of reactions: first, As^{3+} sorbs to the outer oxide coating, then diffuses through the coating where it is reduced and accumulates at the nZVI-ferrihydrite interface (Yan et al. 2012). Based the edge position of the As-nZVI XANES spectrum (which was lower in energy than As^0, and fits to As-EXAFS spectrum, which showed no evidence for neighboring As or S atoms, formation of an As-Fe intermetallic phase was postulated (Yan et al. 2012).

Layered double hydroxide phases (LDHs) exhibit high anion sorption capacity due to interlayer anion exchange. Paikaray et al. (2013) investigated the structure of As^{5+}, Se^{6+}, and Mo^{6+} oxyanion sorption complexes on several hydrotalcite-like LDHs (in the system Mg-Fe-Al-SO_4-CO_3) using As-EXAFS. Second-neighbor scattering was evident from the magnitude FT for As^{5+}, but not for Se^{6+} or Mo^{6+}. Fits to As-EXAFS yielded structural parameters consistent with formation of ^2E type complexes between As^{5+} and Fe^{3+} octahedra. A separate examination of As and NOM removal by Mg-Al LDHs found no evidence for second-neighbor scattering features beyond the first-neighbor tetrahedral oxygen atoms (CO_3^{2-} and Cl^- varieties; Wu et al. 2013). Apparently, on LDHs lacking Fe, As^{5+} sorbs by the same interlayer, anion-exchange mechanism postulated for Se^{6+} and Mo^{6+} oxyanions.

Several studies have used As-XAS for direct characterization of solid phases generated from field-scale treatment of As-contaminated water or soil. Root et al. (2007) studied the Fe-rich products of As removal by coagulation/flocculation at Haiwee Reservoir, as previously described (also see Appendix 5). Paktunc (2008a or b?) studied As speciation in an experimental passive anaerobic treatment system designed to encourage sulfate reduction as the main biological process removing As and other metal(loid)s. Based on geochemical considerations, sequestration of As in newly-formed sulfide minerals would be a reasonable hypothesis for As speciation in the anaerobic cells. However, analysis of As-XANES revealed that organoarsenic species dominated in the sediment (primarily composed of biosolids and sand). The dominant As^{3+} species was most similar to mono- and di-methylarsenic variants of 2,3-dimercapto-1-propane sulfonic acid (DMPS). Dimethyl arsenic acid ($DMAs^{5+}$) was proposed as the dominant $organoAs^{5+}$ species. The results point to potentially important differences in As speciation between natural and constructed biotic treatment systems.

Foster et al. (1995) used As-XAS to identify As species resulting from a 2-step "fixation" procedure using ferric chloride and Portland cement that was effective in the immobilization of highly soluble forms of arsenic in soil from an industrial site (Voigt et al. 1996). Prior to fixation, As speciation in soil was dominated by hoernesite, the Mg-arsenate mineral discussed previously. After the procedure, the EXAFS spectra and FTs of reacted samples showed features indicative of a coordination environment with Al, Ca, or Mg atoms (difficult to distinguish in EXAFS theoretical fits) instead of Fe, which was expected react with As to form a ferric arsenate phase. Although the theoretical EXAFS analysis was limited (due to weak scattering amplitude of Al, Ca, and/or Mg atoms) XANES spectra of reacted soil were best fit by the spectrum of As^{5+} coprecipitated in ettringite, suggesting that fixed soil contains As^{5+} substituting in structural sites in ettringite or other Ca-rich hydrous cementitious phases where it is fully coordinated by water molecules, as previously discussed (Foster 1999).

Arsenic species in bioaccessibility and bioavailability test materials

Several attempts have been made to relate As speciation derived from XAS to As solubility (through sequential extractions), As bioaccessibility [through a physiologically-based extraction test (PBET) that yields *in vitro* bioaccessibility (IVBA)], or As bioavailability [through *in vivo* animal exposure experiments, which are quantified as absolute or relative bioavailability (ABA, RBA)]. Previously defined by operational or empirical results, such studies are now incorporating XAS methodologies (specifically LS-LCF of XANES and

EXAFS spectra) to correlate As release with the removal of specific As phases and/or oxidation states, thus providing insight into how As speciation plays a key role in the potential toxicity of As. However, a number of physical and chemical sample characteristics can complicate the ability to directly correlate As speciation with As mobility, including the sample's geochemistry and organic content, particle size distribution, and the location of As-bearing phases (e.g., at particle surfaces vs. inclusions within larger sample grains). Incorporation of μ-XAS and μ-XRF methods allows the additional dimension of investigating spatial relationships at the microscale of As on or within particles, which can help address this latter concern; however, "encapsulation" or "nugget" effects, exacerbated in the coarse particle size fraction, can hinder the potential of As speciation to predict As mobility. Despite these complicating factors, some broad conclusions regarding the impact of As speciation on solubility, bioaccessibility, and bioavailability can still be drawn from the relatively limited number of systems studied.

Appendix 8 summarizes the major findings of published studies that combine XAS with solubility, bioaccessibility, and bioavailability assays. In several cases, specific As species have been identified and qualitatively associated with either low, intermediate, or high arsenic mobility. For example, gold mine tailings and soils featuring dominant proportions of mineral species including arsenopyrite and scorodite exhibit low IVBAs (<1% As release) when subjected to a PBET designed to mimic the gastric and intestinal systems (Meunier et al. 2010; Plumlee and Mormon 2011; Basta and Juhasz 2014, this volume). The low values result from the sparingly soluble nature and slow kinetic release of As from these phases under physiological conditions. Other studies show a moderate bioaccessibility (1-10% As release) associated with As-bearing Fe-(hydr)oxides and amorphous ferric arsenate (AFA) (Beak et al. 2006; Meunier et al. 2010). The highest levels of As bioaccessibility (>10% As release) in natural samples were associated with the identification of one or more of the following phases: Ca-Fe arsenate (likely amorphous or poorly-crystalline), As-bearing schwertmannite, As-bearing jarosite, and As-bearing clay minerals (Meunier et al. 2010). A μXRF analysis of individual sediment grains demonstrates that As thought to be associated with mafic silicate or phyllosilicate mineral phases (e.g., biotite, chlorite) exhibits much lower solubility during a phosphate leach extraction step than As associated with Fe-oxyhydroxide grain coatings, which dissolve in this step and release As into solution (Eiche et al. 2010). Organic matter addition (as peat moss) to arsenic-contaminated soils followed by a two-year aging period increased As bioaccessibility, but only for those samples initially containing >25% organic matter and exhibiting >30% As bioaccessibility to begin with (Meunier et al. 2011a).

Attempts to find significant correlative relationships between the relative abundance of As species determined from fits to bulk EXAFS spectra and RBA data from mine-impacted soils have met with little success in both mouse (Bradham et al. 2011) and juvenile pig (Foster et al. 2014) animal models. However, significant correlative relationships between As speciation by bulk EXAFS and IVBA and/or other measured sample properties (physical, chemical, or mineralogical) have been found in some studies. Koch et al. (2011) determined that bioaccessible arsenic in a range of traditional Indian medicines was correlated with the sum of As^{5+}-O and As^{3+}-O determined by fits to As-XANES spectra; in this case, the oxygen-bound As species were oxidation products of the dominant As_2S_3 material used in the medicinal concoctions (Fig. 21). However, As oxidation state alone is not necessarily sufficient to predict bioaccessibility: Meunier et al. (2010) identified As^{5+} in samples exhibiting both the lowest and highest bioaccessibility. Bradham et al. (2011) found no strong correlations between As species by EXAFS and several physiochemical parameters in mine-impacted soils, but Foster et al. (2014), who also studied mine-impacted soils, found significant ($p < 0.05$) Pearson (R^2) correlations between the amount of As released during IVBA and the EXAFS-derived relative abundance of: As^{5+} sorbed to gibbsite, As^{5+} substituted in jarosite (positive correlations) and arsenopyrite (negative correlation).

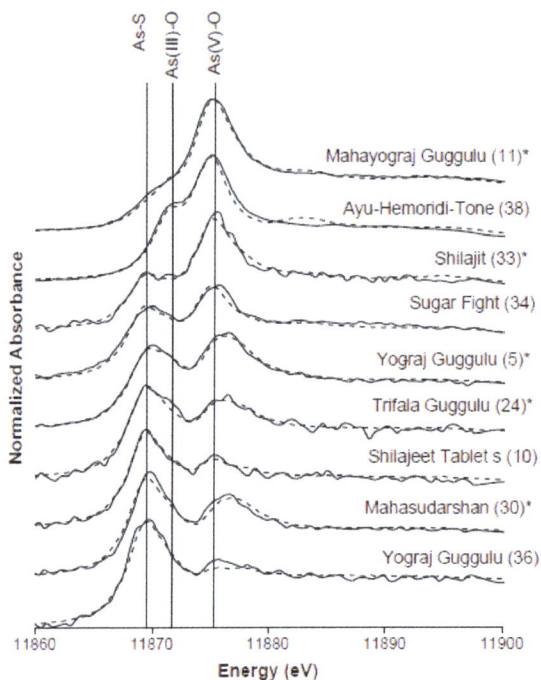

Figure 21. As-XANES spectra (solid lines) of several traditional Ayurvedic medicines, with LC-LSF fits (dotted lines) used to quantify oxidation states. [Used by permission of Elsevier, from Koch et al. (2011) *Science of the Total Environment*, Vol 409, Fig. 1, p. 4548]

Arsenic species in biota

In situ (and, in some studies, *in vivo*) As-XAS studies of biota have helped support investigations of (1) the physiology of biota (plants, microorganisms) adapted to extreme As concentrations; (2) the pathways and mechanisms of As toxicology and As resistance at the microbial, cellular, and tissue levels; (3) trophic transfer of As in aquatic and terrestrial food webs; (4) bioremediation of As contamination; (5) bioavailability of As in foodstuffs; and (6) the use of biotic tissues as records of time-dependent or spatially-dependent As exposure (Appendix 9). Simple methylated As species (mono- through tetra-methylated forms) are known to be produced from inorganic As by biochemical pathways, and their presence in biota is generally assumed to represent metabolism of bioavailable As (Smith et al. 2005). The methylated forms can be ingested by higher organisms and are thought to be the precursors for more complex organoarsenicals such as arsenobetaine and arsenosugars found in higher trophic level biota (Smith et al. 2005).

The low internal concentration of As typically present in biota from the environment has presented a considerable challenge in some experiments and often limits analysis and interpretation to As-XANES or extended XANES. Further complicating matters is the high level of similarity among the As-XANES spectra of organoAs species (Smith et al. 2005). As explained in George and Pickering (2007), As-XANES spectra of organoAs compounds are sensitive to differences in the coordination of nearest-neighbor C groups), but insensitive to the different aliphatic substituents on the C backbone. In other words, aliphatic and aromatic organoAs species can be distinguished (e.g., arsenobetaine and roxarsone), but compounds differing only in the type of aliphatic substituents cannot (e.g., arsenobetaine and arsenocholine;

Fig. 1). There are additionally some cases of overlap between the As-XANES edge position of organoAs^{3+} and organoAs^{5+} species (e.g., trimethylarsine oxide (R$_3$As^{5+}-O, R = CH$_3$) and arsenobetaine (R$_4$As^{3+}; George et al. 2009; Fig. 22a). Similarly, As-XAS (XANES and EXAFS) of As^{3+} bound to a variety of low molecular weight thiols cannot be distinguished (Pickering et al. 2000; Castillo-Michel et al. 2012). The ability to distinguish among these species in biota might reveal key features of the metabolic pathway used by the organism for detoxification and/or sequestration of arsenic.

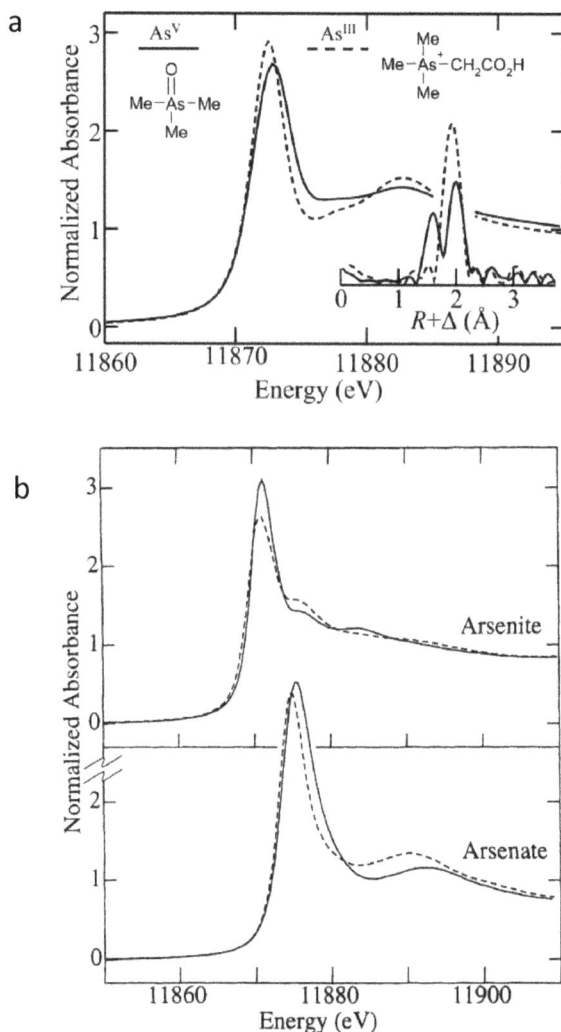

Figure 22. (a) Comparison of XANES spectrum of trimethylarsine oxide (R$_3$AsO, solid line) and arseno-betaine, (R$_4$As$^+$, dotted line). Although the edge positions of XANES spectra overlap, the FTs of EXAFS spectra indicated differences that could be diagnostic. (b) Spectra of As^{3+} and As^{5+} oxyanions as a function of pH, showing differences (5.5 = solid, 9 = dashed; spectra collected at cryogenic temperature). [(a): Used by permission of Wiley VCH, from George et al. (2009) *Molecular Nutrition and Food Research*, Vol 53, Fig. 3, p. 554], (b): Used by permission of American Association of Plant Physiologists, from Pickering et al. (2000) *Plant Physiology*, Vol 122, Fig. 1, p. 1173]

Another important consideration in the analysis of As-XANES spectra collected from biota is the pH-dependent change in the shape of aqueous As^{3+}, As^{5+} XANES spectra (Pickering et al. 2000; Fig. 22b). This shape change is most pronounced in samples collected at cryogenic temperature, a common collection condition for biotic samples.

Prokaryotes and simple eukaryotes. Inorganic As species (As^{5+}-O, As^{3+}-O, As^{3+}-S) have been the primary intracellular forms of As identified in As-XAS studies of Prokaryotes and Eukaryotic microbes (Appendix 9). Prokaryotes isolated from high As environments contain predominantly oxygen-bound As^{5+} or As^{3+} (Patel et al. 2007; Wolfe-Simon et al. 2010). As^{5+}-O was the primary species identified in native phytoplankton from mining-impacted lakes Caumette et al. (2011), but As^{3+}-S was the dominant species in *Euglena gracilis* cultures grown in the presence of As^{5+} or As^{3+} (Miot et al. 2008, 2009). Inorganic As^{3+}-S is widely assumed to represent a detoxification product resulting from As^{3+} binding to low molecular weight thiols (glutathione, phytochelatins, or metallothioneins are commonly proposed in biotic systems depending on the specifics of the system; Miot et al. 2009).

Arsenic speciation in zooplankton collected from mining-impacted lakes followed the trend As^{5+}-O > As^{3+}-S > organoAs species (~10%), but zooplankton from pristine lakes had a higher relative abundance of organoAs species Caumette et al. (2011). A separate study of field-collected *Daphnia pulex*, a zooplankton commonly used in toxicity testing, indicated the predominance of sulfate arsenosugar and arsenobetaine (by HPLC-ICP-MS), and therefore the ingestion of primarily biotic As rather than inorganic As by these invertebrates (Caumette et al. 2012b).

Additional details of intracellular As speciation in single-celled organisms has been obtained in some systems by analysis of subcellular fractions (bulk XAS) or application of μ-XRF mapping. For example, inorganic As^{5+} was not detected in bulk XAS spectra of whole *Euglena gracilis* cells, but was present in a subcellular fraction associated with positively-charged nuclear membrane material, where it was presumed to be substituting for PO_4 (Miot et al. 2009; Fig. 23). Also, μ-XRF mapping of field-collected *Daphnia pulex* revealed As to be 10-fold more concentrated in the gut than in surrounding tissues (Caumette et al. 2012b).

Seaweed is a multicellular algae, several species of which are consumed by humans. As^{5+}-O, As^{3+}-O, and a third species that might represent an arsenosugar or a tetra-alkyl-arsonium species were identified in fresh algae collected along the Pacific coast as well as dried, commercial seaweed (George et al. 2009). Fresh and dried red algae had more As^{5+} than other seaweed species. These findings are relevant to the assessment of the hazards associated with seaweed consumption, since trivalent arsenosugars are known to be more toxic than pentavalent forms (George et al. 2009).

Fungi are uni- and multicellular Eukaryotes that have long been known to form volatile methylated As species. Inorganic As^{3+}-O was the dominant species (~50-60%) observed in spores and mycelia of 3 fungal species (Su et al. 2012). The remaining As was predominantly organic, but differed among the species: two contained ~30% dimethyl arsenic acid (DMA) and minor monomethyl arsonic acid (MMA) as additional species, whereas the 3rd contained ~40% MMA, no DMA and a small amount of As^{5+}-O (Su et al. 2012). As speciation in edible white button mushrooms (*Agaricus bisporus*) was studied by Smith et al. (2007). The fruiting bodies (caps) primarily contain As^{3+}-O, DMA, and arsenobetaine (AB), with minor amounts of As^{5+}-O and trimethylarsine oxide (TMAO). These results helped to demonstrate that AB is produced biochemically by this species rather than being taken up from the growth medium; they also supported the hypothesis that AB plays a role in nutrient translocation and maintenance of turgor pressure in the mushroom cap (Smith et al. 2007).

Plants: overview. As-XAS studies of plants have primarily focused on their importance as foodstuffs and on phytoremediation studies. Although the information derived from As-XAS

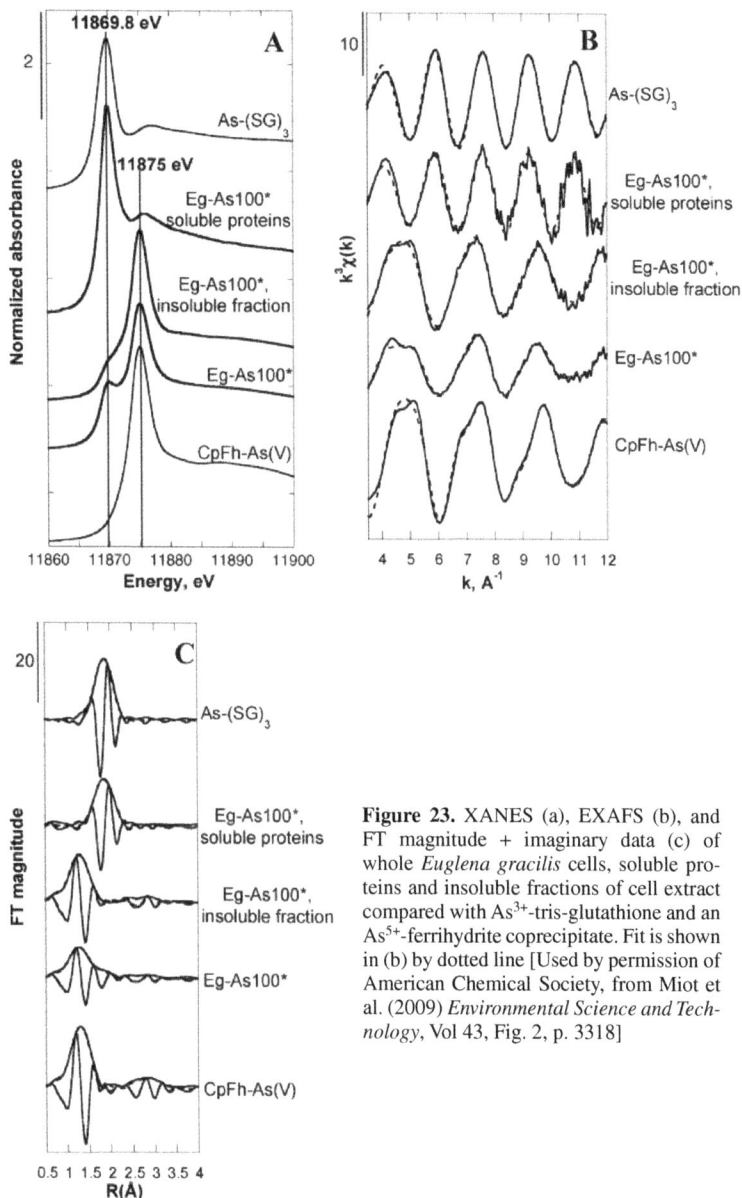

Figure 23. XANES (a), EXAFS (b), and FT magnitude + imaginary data (c) of whole *Euglena gracilis* cells, soluble proteins and insoluble fractions of cell extract compared with As^{3+}-tris-glutathione and an As^{5+}-ferrihydrite coprecipitate. Fit is shown in (b) by dotted line [Used by permission of American Chemical Society, from Miot et al. (2009) *Environmental Science and Technology*, Vol 43, Fig. 2, p. 3318]

studies is limited, as described above, chemical extraction methods from plants (particularly leaves) are acknowledged to be incomplete (Smith et al. 2008a), making *in situ* information derived from As-XAS analysis even more valuable.

The potential for sample preparation artifacts in several published studies of As speciation in plants exists. Many studies were conducted on bulk, ground tissues that were freeze-dried prior to data collection at room temperature. Although this approach limits effects due to photo-induced water hydrolysis, the effects of freeze-drying plant tissue on As speciation have not been examined, to our knowledge. The effect of sample grinding has been investigated, however, and has been shown to affect As speciation. Smith et al. (2008a) found that wet grinding of fresh tissue (under liquid N_2) and dry grinding after drying tissue at 70 °C changed As speciation in radish leaves, but had little effect on speciation in radish stems and roots.

Arsenic-accumulating plants. Several arsenic-accumulating ferns (most famously, the hyperaccumulating Brake fern, *Pteris vittata*), have been studied by As-XAS (Appendix 9). Inorganic As^{3+}-O is the predominant species in ground fresh *P. vittata* leaves under low-As (1000 mg kg^{-1}) growth conditions; under high As growth conditions (10,000 mg kg^{-1}) the leaves also contain As^{3+}-S (EXAFS fits consistent with As^{3+}-*tris*-glutathione; Webb et al. 2003). Dried ground *P. vittata* leaves contained only As^{5+}-O, suggesting that the plant uses an active mechanism to maintain As in the trivalent state in living tissue (Webb et al. 2003). Studies on ground root tissue have differed with respect to the As speciation determined: As^{3+}-O was dominant in *P. vittata* roots (Huang et al. 2004), indicating complete reduction of added As^{5+}, but only partial reduction was observed in roots of *P. cretica* (Kashiwabara et al. 2010) and *P. calomelanos var. austroamericana*; (Kachenko et al. 2010). Complete reduction and formation of As^{3+}-S was observed in rhizoids of *Athyrium yokoscense*, (a fern, but not a hyperaccumulator species; Kashiwabara et al. 2010).

Additional details of As speciation in *P. vittata* were provided by a µ-XRF/µ-XAS study of live, 1 month old gametophytes and 4 month old sporophytes (Pickering et al. 2006). The overwhelming majority of leaf-associated As is isolated in vacuoles; other important cellular components (cell wall, rhizoid) did not accumulate As. Mapping As^{5+}-O, As^{3+}-O, and As^{3+}-S in the sporophyte, As^{5+}-O was found only in vascular tissue and was rimmed by a layer of As^{3+}-S (Fig. 24) These results led to a hypothesis that *P. vittatta* synthesizes phytochelatins only in response to As^{5+} and localizes the phytochelatin-bound As (now reduced to As^{3+}-S) in tubular sheaths around the vascular system. The results of this study support a theory that the lack of (dominant) metal-thiol coordination observed in hyperaccumulator plants generally is an evolutionary characteristic (Pickering et al. 2006).

Brassica juncea (Indian mustard) has also been evaluated for potential use in phytoremediation of As contamination. Bulk As-EXAFS analysis of roots, xylem sap exudates, and leaves revealed a very different distribution and speciation of As in *B. juncea* than observed in *P. vittata* and other hyperaccumulating fern species (Pickering et al. 2000). First, the majority of the

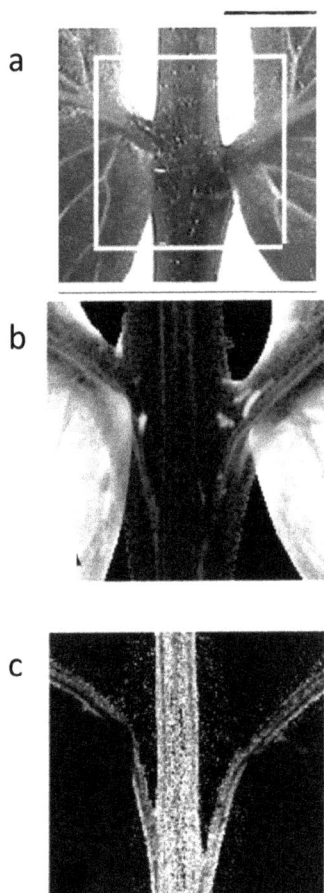

Figure 24. (a) Grayscale optical image of *P. Vittata* stem and leaves analyzed by µ-XRF oxidation state mapping; (b) log[As^{3+}] concentration, showing accumulation in leaf and specific areas of transport channels in stem; (c) map of As^{5+} distribution, showing localization in transport channels in stem and anticorrelation with As^{3+}. Optical image of plant in (a) is analogous, but not identical to, the X-ray images in (b) and (c). [Used by permission of American Chemical Society, from Pickering et al. (2006) *Environmental Science and Technology*, Vol 20, Fig. 3, p. 5012]

As accumulated by *B. juncea* is stored in the roots, not in the above-ground biomass as in *Pteris* species. Second, the primary species in both root and leaf is an As^{3+}-S species (local structure consistent with As^{3+}-*tris*-glutathione; Pickering et al. 2000).

Several desert plant species can accumulate arsenic in their biomass to concentrations exceeding that of the growth medium and therefore have been considered for phytoremediation in arid climates (Castillo-Michel et al. 2011). A series of papers using bulk and μ-XANES reveal similar arsenic speciation in Mesquite (*Prosopis* spp.; Aldrich et al. 2007, Castillo-Michel et al. 2012), desert willow (*Chilopsis linearis*; Castillo-Michel et al. 2009), tumbleweed (*Salsola kali*; De la Rosa et al. 2006), and *Parkinsonia florida* (Castillo-Michel et al. 2011). Although differing somewhat in the amount of above ground transport and site of As accumulation, As^{3+}-S is the predominant species in all these desert plants (as As^{3+}-*tris* glutathione).

Edible plants. There is current interest in As accumulation and speciation in edible plants, given recent findings of elevated As concentrations some foods, including in organic brown rice syrup (used as a sugar substitute in some foods; Jackson et al. 2012) and arsenic-rich lipids in fish oil (identified by non-spectroscopic methods; Taleshi et al. 2008).

Edible plants exposed to dissolved As^{5+} in growth medium appear to convert it rapidly to As^{3+}-S forms for detoxification, but have different strategies for localization. Cherry Belle radishes (*Raphanus sativus*) grown in mine waste or in liquid culture preferentially accumulate As in leaves, but the majority of As in pea plants (*Pisum sativum*) and rice (*Oryza sativa*) is accumulated in the roots (Castillo-Michel et al. 2007; Smith et al. 2008a, 2009). In radishes, As^{5+}-O was only detected in the outer red rind of roots, and in the xylem of stem vascular channels (Smith et al. 2008a). Arsenic is bound to Fe^{3+} plaques on mature rice roots, but the plaques are not believed to offer much protection to the rice plant from As uptake generally, because young rice roots have little to no Fe^{3+} plaque (Seyfferth et al. 2010, 2011). Within root cross sections, As^{5+}-O, As^{3+}-O, DMA, and As^{3+}-S have been identified (Seyffert et al. (2011)).

A μ-XRF/μ-XAS investigation of rice grains showed that As speciation and localization is heterogeneous (Seyfferth et al. 2011). The starchy interior of the grain (endosperm) contained no detectable As, but As was imaged in the germ and outer bran layers. The germ layer contained As^{5+}-O and As^{3+}-O, whereas the bran layer contained only As^{5+}-O. The bran layer is removed for the production of "white" rice, but retained in brown rice, and therefore is the likely source of the observed As enrichment in brown rice and its derivative products.

Marine invertebrates. A variety of organic and inorganic As species have been detected in marine invertebrates (Price et al. 2013a), but where examined, the body burden of inorganic As appears to be more sensitive to prevailing environmental conditions than does the body burden of organoAs. The concentration of inorganic As species in blue mussels (*Mytilus edulis*) and sea snails (*Littorina littorea*) was positively correlated with environmental As concentrations, but the levels of organoAs species (AB, arsenosugars, $MMAs^{3+,5+}$ and $DMAs^{3+,5+}$ were not (Whaley-Martin et al. 2012, 2013). Investigations coupling μ-XRF and μ-XAS revealed differences in the localization of inorganic and organoAs species in mussels: As^{3+}-S (the predominant inorganic species) was concentrated in the digestive gland, whereas AB (the dominant organoAs species) was found in all tissue types. Its ubiquity suggests that it serves a physiological function in the organism (AB is commonly thought to be an osmolytic regulator; Whaley-Martin et al. 2012).

Similar trends were noted in sea snails from a environment contaminated by mining: bulk As-XAS analysis indicated inorganic As^{3+}-S and As^{3+}-O constituted ~75% of the As body burden of the gastropods (Whaley-Martin et al. 2013). There may be differences among species or environments, however: gastropods from a shallow-water marine hydrothermal system contain 2 organoAs species whose abundance did correlate positively with inorganic As exposure (Ruiz-Chancho et al. 2013). Variability in the occurrence and abundance of

As^{3+}-S species in gastropods is uncertain, but might also be due to partial oxidation of reduced species during sample processing (grinding) as indicated for plants.

Freshwater and terrestrial invertebrates. Microbeam XAS/XRF studies have been particularly useful for the study of As speciation in terrestrial invertebrates. Inorganic As (predominantly As^{3+}-S) was the predominant species in : (1) freshwater midges (*Chironomus riparius*; Mogren et al. 2013); (2) bertha armyworm moths (*Mamestra configurata;* Andrahennadi and Pickering 2008); (3) lab-grown and field-collected earthworms (*Lumbricus rubellus*; Covey et al. 2010; Button et al. 2011); and (4) eight other orders of invertebrates collected from the environment (including grasshoppers, mosquitos, ants, slugs, and spiders; Moriarty et al. 2009). Non-XAS methods (including polyclonal antibodies) have been used in earthworms to identify the ambiguous As^{3+}-S species as a metallothionein (MT)-like species (Langdon et al. 2005).

There is some indication in invertebrates of changes in (1) sensitivity to As and (2) dominant As speciation with lifestage. Midge larvae contained dominantly As^{5+}-O, but adults contained dominantly As^{3+}-S (Mogren et al. 2013). Furthermore, excretion of As during growth was indicated by lower concentrations of As in midge pupae. High concentrations of As affected larval survival in armyworm moths, but did not affect development of mosquito larvae (Andrahennadi and Pickering 2008; Mogren et al. 2013).

µ-XRF/XAS indicates that localization of As in invertebrate tissues is variable and may be most influenced by diet or As exposure route. As^{3+}-S was localized in the midgut of midge larvae, but was limited to the thorax in adults (Mogren et al. 2013). Armyworm larvae also concentrated As^{3+}-S in the midgut, but metamorphosis did not appreciably alter As speciation (Andrahennadi and Pickering 2008). Further, As storage in granules or lysosomes and/ or sorption to the peritrophic membrane (a protective chitinous sheet lining the midgut) were suggested as possible locations within the armyworm midgut for As (Andrahennadi and Pickering 2008). Arsenic is also concentrated in the gut of earthworms, as indicated by polyclonal antibody imaging (Langdon et al. 2005) and µ-XRF mapping (Button et al. 2011). Antibody analysis further localized As to the peri-intestinal region and intestinal wall cells of earthworms, suggesting that As induces metallothionein synthesis as a detoxification mechanism (Langdon et al. 2005).

Spectroscopic methods have also been used to test traditional chemical extraction techniques for As speciation in invertebrates. One study showed that methanol/water extracts traditionally used to extract As from invertebrate tissues can transform As^{3+}-S species to As^{3+}-O species (Moriarty et al. 2009).

Vertebrates exclusive of humans. Bulk As-XAS of dogfish (a marine fish) indicates a species similar to arsenobetaine (George et al. 2009); this result is consistent with chemical analytical methods that indicate the predominance of arsenobetaine in marine and freshwater fish (Slejkovec et al. 2004). Arsenic speciation in terrestrial vertebrates has focused on its sequestration in keratin, a highly stable carbon compound containing reduced S groups that is known to sequester As and can serve as a record of exposure to the element (Smith et al. 2008b). Extraction of As from keratin-rich tissues without its transformation has been difficult due to the extreme stability of the material (Smith et al. 2008b). Keratin in rodent fur (vole, deer mouse, squirrel), bird feathers (jay, tree sparrow, and dark-eyed junco), and tortoise scute collected from specimens in areas with elevated As have been examined by bulk and microbeam synchrotron techniques. Given the abundance of thiol groups in keratin, the dominance of an As^{3+}-S species would be expected, but XAS evidence to date does not fully support this model. As^{3+}-S was highly variable in fur and feathers (5-58%) and the microXANES spectrum of As in tortoise scute did not match that of orpiment, a model for As-S bonding (Foster et al. 2009a). Exogenous As was difficult to distinguish from endogenous As in feathers, but was relatively easy to distinguish in cross sections of tortoise scute due to morphology differences (exogenous As was in particulate form; Foster et al. 2009a). Direct reduction of aqueous As^{5+} by solid keratin from several species has also been observed (Smith et al. 2008b).

Humans. Most metabolized inorganic As is methylated for detoxification purposes and excreted in urine (Yamauchi and Fowler 1994), but some As is retained in the human body. Human hair and nails contain keratin, and, as with other vertebrates, these tissues retain a longer record of exposure to As than does urine. Three studies on human keratin (two hair and one toenail) came to similar conclusions with respect to As speciation (Appendix 9). The predominant species in human keratin was identified as As^{3+}, possibly with mixed sulfur and methyl coordination (Pearce et al. 2010). Hair and nail was a good record of exposure in individuals exposed to elevated As levels in water (Gault et al. 2008) or soil/airborne particles (Pearce et al. 2010), even in long-deceased, mummified individuals (Kakoulli et al. 2014; Fig. 25). Localization of As^{3+} in human nail was irregular, unlike the case for tortoise scute (Pearce

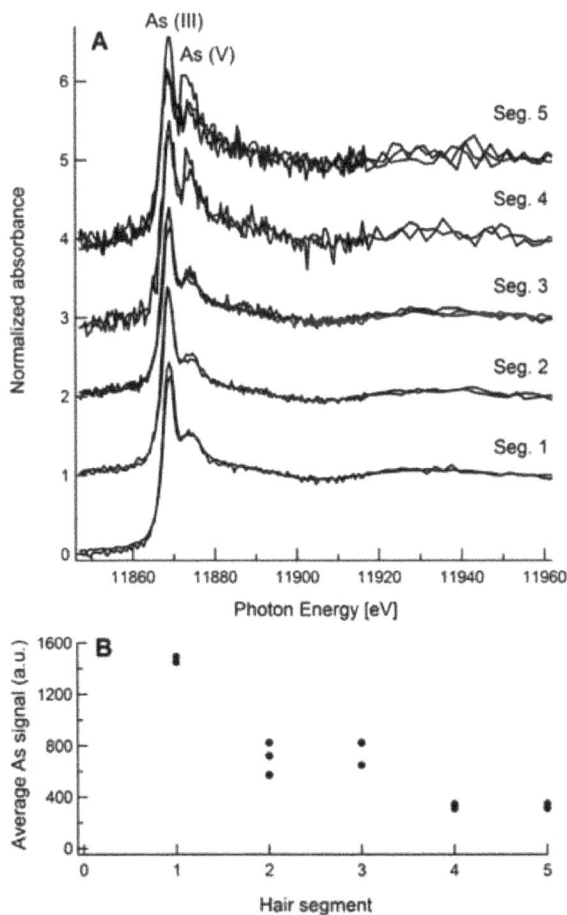

Figure 25. (A) As K-edge μ-XANES spectra collected from 5 locations (segments) along a strand of human hair from a Pre-Columbian period mummy. (B) An overall decrease in spectral quality and increase in As^{5+} content from younger to older hair (0 – 5 on hair segment) indicates lower total As and As^{3+} oxidation in older hair. [Used by permission of American Chemical Society, from Kakoulli et al. (2014) *Analytical Chemistry*, Vol 86, Fig. 7, p. 524]

et al. 2010). Aqueous As^{5+} can be reduced by human nail keratin (Pearce et al. 2010), but appears to re-oxidize in aging human hair (Kakoulli et al. 2014; Fig. 25).

Although its mode of action is unknown at this time, As_2O_3 is used to treat leukemia and has been proposed for treatment of ovarian and other cancers (Bacquart et al. 2010). Model cancer-prone human cell lines, especially HepG2 cells, have been a focus of XAS studies attempting to elucidate the mechanisms behind this activity (Del Razo et al. 2001; Munro et al. 2008; Bacquart et al. 2010). Cells exposed to inorganic As^{3+}-O as As_2O_3 or dissolved forms in cultures accumulated inorganic As^{3+}-O in the cytosol, nucleus, and mitochondrial network. Nuclear structures contained As^{5+} in one of 9 cells examined, indicating an unknown oxidative process; furthermore, the As^{5+} was localized in a narrow region hypothesized to correspond to micronuclei (known to be formed as a consequence of As exposure in this cell line; Bacquart et al. 2010).

CONCLUSIONS AND RECOMMENDATIONS FOR FUTURE RESEARCH DIRECTIONS IN AS XAS SPECTROSOCPY

In large part, progress in As-XAS research has generally followed developments in synchrotron science. Starting with a few reports in the late 1980s related mainly to energy research, the field began to expand rapidly in the mid- to late-1990s when bulk XAS studies of metal adsorption to mineral surfaces became a major focus. While studies of As sorption to mineral surfaces have remained a key area of study, in the late 1990s the application of As-XAS to natural materials became a major focus of synchrotron studies and may currently be the major application. Since the mid-2000s, the increased availability of µ-XAS (and µ-XRF/µ-XRD) has opened up entirely new areas of research in microscale As speciation, particularly in biota. Another, less heralded trend is the combination of XAS studies of multiple metal(loid)s (e.g., coupling As and Fe EXAFS). Below we summarize the conclusions derived from As-XAS investigations by general topic, and provide some recommendations for future research based on our own experiences and our review of the literature.

Model systems

Our understanding of the molecular-scale structures of As species sorbed to mineral surfaces has been greatly enhanced by the many As-XAS studies in this area. There is now a clear understanding of the dominant sorption complexes formed on the most relevant mineral phases in soils and aquatic sediments (particularly ferrous and ferric phases with which As is strongly associated in many systems). Although the bidentate bridging ^2C complex is the dominant adsorbed species for As^{5+} and As^{3+} on many surfaces, questions remain about the importance of other complexes that have been identified by EXAFS fitting analysis. Some of the variability in As speciation results (for the same species and mineral, of course) probably lies in the variable ways in which multiple scattering (MS) paths are handled in fits, when they are included at all. Evaluation of the importance of MS paths and consistent treatment of N_{MS}, R_{MS}, and σ^2_{MS} (discussed in the section on data analysis) should become required practice for theoretical fits (model systems and natural samples) that include atoms beyond the first shell of nearest neighbors.

Assessment of the importance of As^{5+}/As^{3+} binding to Al-phases in the environment (mica minerals, Al-oxides/hydroxides, poorly-crystalline aluminosilicate precursors phases, and double layer hydroxides) is an area that merits greater research, since formation of some of these phases has been suggested by field studies and leach tests (e.g., Landrot et al. 2012; Beaulieu and Savage 2005). This remains challenging because Al has less electron density than Fe or Mn and therefore does not scatter ejected photoelectrons as effectively (a similar problem exists for As in coordination with Ca and Mg; Myneni 2002).

Additional As-XAS model system studies are required to define the sorption characteristics of As³⁺ to an equivalent level of detail as As⁵⁺. Recent studies on mixed-valent Fe minerals show that As³⁺ can have distinctly different sorption behavior from As⁵⁺ in some geochemical environments. Finally, additional studies that utilize XAS in concert with other techniques to evaluate the effect of As adsorption complex geometry on sorption complex lability are also needed.

Environmental samples

Sediments and soils. The multiple studies of As speciation in sediments and soils from the environment point to a high degree of transferability of knowledge gained in one site-specific study to other, similar sites. The importance of ferric hydroxides in As⁵⁺ sorption is observed across many different natural systems, for example. However, there are still many knowledge gaps that could be filled by additional As-XAS studies. Although this list is not comprehensive, we see a need for better understanding of: (1) the controls on As³⁺ oxidation by MnO_2 in aquifer sediments, (2) the speciation of As in mildly to strongly reducing sediments across a range of Fe, S, and organic carbon concentrations, (3) the transformation of organoarsenical herbicides and pesticides in the sediments, (4) links between climate and As speciation in mine wastes and mining-impacted sediments, and (5) the role of particulate and colloidal organic matter in sorption and transport of As in aquifers and organic-rich sediment.

Wastes from industry, agriculture, and the built environment. The problem of management, disposal or beneficial use of solid wastes originating from these sources ongoing and speciation studies will be needed to drive the science forward. The speciation of As in the built environment is an area that we believe will be of growing importance in coming years as increased frequency and/or severity of climate-related events such as fire and flooding disturb the delicate balance that humans have with the environment they construct around them. The effect of fire on As speciation in consumer products in the home and products used in construction (e.g., drywall, concrete, CCA-treated wood) is one example. Another is the fate of (GaAs) semiconductor chips in cellular phones. Although each chip is small, cell phones are abundant, product life is short, and recycling is underdeveloped, leading to disposal of many cell phones as municipal waste and ultimately landfill or incineration (Uryu et al. 2003; Dagan et al. 2007; Lim and Schoenung 2010). Another example is the speciation and fate of As in waste automobiles. One study estimated the As concentration in shredded automobile waste to be 25 mg kg⁻¹; perhaps not much, but the volume of material is quite large when scaled to the mass of disposed vehicles each year (Zevenhoven et al. 2007).

Biota. Despite the aforementioned limitations of As-XANES for precise identification of organoAs compounds, the information provided from micron-scale As-XAS/XRF studies of biota has provided unique details of intracellular As-S chemistry that have advanced our understanding of As detoxification, accumulation, and toxicity. That being said, there are also some areas that need further refinement and development. For example, the ability to discriminate among organoAs species might be improved if data were collected and analyzed inclusive of the first oscillation of the EXAFS spectrum (extended XANES); this is also generally true for many systems that have been studied by As-XANES analysis.

Understanding the trophic transfer of As in food webs and the relative importance of ingested vs. internally-synthesized organoAs compounds for organismal body burden is an active area of research (Caumette et al. 2012a). As previously noted, the extraction efficiency of reduced sulfur in low-molecular weight thiols and/or proteins is low by standard chemical methods. In addition, elution of unidentified species from HPLC columns during chemical analysis has also been described in several reports. In the case of non-eluting species, the column material itself could be analyzed by As-XAS for identification, and unknown eluting species, could be concentrated for As-XAS data collection. Additional XAS studies to aid development of phytoremediation or biosorption techniques will certainly be required in the future, as will studies of organisms with novel mechanisms for managing high tissue arsenic levels.

Continued validation of extraction and bioaccessibility methods. Past attempts to relate As speciation by bulk XAS with sequential extraction steps, *in vitro* bioaccessibility, and *in vivo* bioavailability have not produced the desired significant positive correlative relationships between one or more XAS-derived species and *in vivo* or *in vitro* test results. Additional studies analyzing samples pre- and post-extraction and observing the relative changes in proportions of different As species could help confirm the preferential removal of the more soluble phases and the persistence (or secondary formation) of more insoluble phases. However, extending XAS studies to IVBA and RBA experiments with natural samples remains challenging because most soils/sediments contain a preponderance of low solubility As species, resulting in relatively low-% As released in bioaccessibility extraction experiments, and increasing the difficulty of detecting differences between pre- and post-extracted samples.

The ability of μ-XRF/μ-XAS to detect low-abundance phases is a clear benefit to these studies (e.g., Eiche et al. 2010) and rapid μ-XRF imaging with improved methods of As quantification might allow a better assessment of As speciation and documentation of As disappearance from specific phases across a statistically significant number of particles. Rapid μ-XRF imaging would also facilitate studies of the relationship between speciation and particle size, which can vary in some systems (Meunier et al. 2011a).

General recommendations

Refinements in theoretical modeling of As XANES and EXAFS spectra. Theoretical (*ab initio*) modeling of As-XAS spectra is an area needing further research for several reasons. First, programs available for *ab initio* calculations have continued to improve, and are now much better at simulating the features that dominate As-XANES spectra of oxyanion species. Second, effective resolution of As-XANES spectra has increased at 3rd generation sources with the result that minor features are now apparent in XANES spectra that were not before (i.e., spectra have become less "similar" and therefore old assumptions about the limitations of XANES analysis may not always be true). Reliance on analysis of As-XANES spectra for As speciation in solids will continue due to limitations of μ-XAS beamlines and due to the constraints imposed by limited beamtime and low As concentrations (for both μ-XAS and bulk XAS). Extracting as much information as possible from the XANES spectrum is therefore imperative. Finally, as mentioned earlier, the 1st oscillation of the As-EXAFS spectrum contains information about the positional "order" of As^{5+} and As^{3+} species that aid in the identification of species with otherwise similar XANES spectra (e.g., crystalline scorodite and amorphous ferric arsenate). The origin of this feature may be better understood by further *ab initio* modeling studies.

Quantifying error and uncertainty in LC-LSF fits. Linear combination, least-squares analysis has been the primary means of determining the relative abundance of As species in environmental samples. We see the need for improvement in systematically exploring the limitations, uncertainties, and errors of the LC-LSF technique, especially because the results from As-XANES and As-EXAFS spectra of the same samples can differ considerably as previously described. Analysis of physical mixtures of As model compounds is the standard method of assessing error, but it includes weighing errors that cannot be readily isolated from the fit error, and therefore is a less-than-ideal method. We propose that analysis of *in silico* mixtures of XAS spectra of "pure" As model compounds can be used in concert with or in lieu of physical mixtures to: (1) determine the "detection limits" of one species of As in a mixture of one or more others; (2) determine the effects of experimental noise on fit error, and (3) generate error analyses that are system-specific.

Need for a shared As XAS spectral database. Our review of the As-XAS literature reinforces our opinion that a shared database of As-XAS spectra would be extremely beneficial to the worldwide community of researchers engaged in these studies. Several XAS spectral databases already exist (see *http://xafs.org/Databases*), but they are not constructed with the aim of providing information on resolution, collection conditions, and sample preparation/purity that researchers would need to use unknown spectra with some confidence. Although it is true that certain systems require the analysis of model compounds under conditions identical to samples, it is also true that this is not universally the case. Differences in spectral resolution are much more important for XANES spectra than for EXAFS spectra, for example, and high-resolution XANES spectra can be transformed to lower resolution through smoothing or other techniques. The beamtime required to amass deep and diverse spectral databases is prohibitive for most researchers, and furthermore it is redundant for researchers to continually produce the As-XAS spectra of commonly-occurring As species. We envision a database somewhat akin to the National Center for Biotechnology Information (NCBI) database of DNA sequences where data could be up- and downloaded by users, and searched by spectral similarity. We also propose that each database entry should contain several types of associated metadata (e.g., chemistry, XRD patterns, and reference to original publication, similar to the RRUFF Raman spectroscopy database at *http://rruff.info/*). In order to add incentive to upload spectra, it should become standard practice to refer to the spectra used in publications by their database record number. Such a database would not necessarily have to be As-specific, but given the large amount of work on As worldwide, it might be a good place to start.

REFERENCES

Adra A, Morin G, Ona-Nguema G, Menguy N, Maillot F, Casiot C, Bruneel O, Lebrun S, Juillot F, Brest J (2013) Arsenic scavenging by aluminum-substituted ferrihydrites in a circumneutral pH river impacted by acid mine drainage. Environ Sci Technol 47:12784-12792

Aldrich MV, Peralta-Videa JR, Parsons JG, Gardea-Torresdey JL (2007) Examination of arsenic(III) and (V) uptake by the desert plant species mesquite (*Prosopis* spp.) using X-ray absorption spectroscopy. Sci Total Environ 379:249-255

Alexandratos VG, Elzinga EJ, Reeder RJ (2007) Arsenate uptake by calcite: macroscopic and spectroscopic characterization of adsorption and incorporation mechanisms. Geochim Cosmochim Acta 71:4172-4187

Allaby A, Allaby M (eds) (1990) The Concise Oxford Dictionary of Earth Sciences. Oxford University Press, New York

Alpers CN, Burlak TL, Foster AL, Basta NT, Mitchell VL (2012) Arsenic and old gold mines: mineralogy, speciation, and bioaccessibility. VM Goldschmidt Conference, Montreal

Alpers CN, Myers PA, Millsap D, Regnier TB (2014) Arsenic associated with historical gold mining in the Sierra Nevada foothills: case study and field trip guide for Empire Mine State Historic Park, California. Rev Mineral Geochem 79:553-587

Alsina MA, Saratovsky I, Gaillard J-F, Pastén PA (2008) Arsenic speciation in solid phases of geothermal fields. *In:* Adsorption of Metals by Geomedia II: Variables, Mechanisms, and Model Applications. Barnett MO, Kent DB (eds) Elsevier, Amsterdam, p 417-440

Alsina MA, Zanella L, Hoel C, Pizarro GE, Gaillard JF, Pasten PA (2013) Arsenic speciation in sinter mineralization from a hydrothermal channel of El Tatio geothermal field, Chile. J Hydrol. In press, corrected proof available online 25 April 2013, doi: 10.1016/j.jhydrol.2013.04.012

Amend JP, Saltikov C, Lu G-S, Hernandez J (2014) Microbial arsenic metabolism and reaction energetics. Rev Mineral Geochem 79:391-433

Andrahennadi R, Pickering IJ (2008) Arsenic accumulation, biotransformation and localisation in bertha armyworm moths. Envrion Chem 5:413-419

Arai Y (2010) Effects of dissolved calcium on arsenate sorption at the kaolinite-water interface. Soil Science 175:207-213

Arai Y, Elzinga EJ, Sparks DL (2001) X-ray absorption spectroscopic investigation of arsenite and arsenate adsorption at the aluminum oxide-water interface. J Colloid Interface Sci 235:80-88

Arai Y, Lanzirotti A, Sutton S, Davis JA, Sparks DL (2003) Arsenic speciation and reactivity in poultry litter. Environ Sci Technol 37:4083-4090

Arai Y, Lanzirotti A, Sutton SR, Newville M, Dyer J, Sparks DL (2006) Spatial and temporal variability of arsenic solid-state speciation in historically lead arsenate contaminated soils. Environ Sci Technol 40:673-679

Arcon L, van Elteren JT, Glass HJ, Kodre A, Slejkovec Z (2005) EXAFS and XANES study of arsenic in contaminated soil. X-Ray Spectrom 34:435-438

Asaoka S, Takahashi Y, Araki Y, Tanimizu M (2012) Comparison of antimony and arsenic behavior in an Ichinokawa River water-sediment system. Chem Geol 334:1-8

Bacquart T, Deves G, Ortega R (2010) Direct speciation analysis of arsenic in sub-cellular compartments using micro-X-ray absorption spectroscopy. Environ Res 110:413-416

Bardelli F, Benvenuti M, Costagliola P, Di Benedetto F, Lattanzi P, Meneghini C, Romanelli M, Valenzano L (2011) Arsenic uptake by natural calcite: An XAS study. Geochim Cosmochim Acta 75:3011-3023

Bargar J (2004) MEIS: Guide to XAFS Measurements at SSRL. 2014. *https://www-ssrl.slac.stanford.edu/mes/xafs/index.html* (accessed July 2014)

Barringer JL, Reilly PA (2013) Arsenic in groundwater: A summary of sources and the biogeochemical and hydrologic factors affecting arsenic occurrence and mobility. *In:* Current Perspectives in Contaminant Hydrology and Water Resources Sustainability. Bradley PM (ed). InTEch, Rijeka, Croatia, p 83-116

Basta NT, Juhasz A (2014) Using in vivo bioavailability and/or in vitro gastrointestinal bioaccessibility testing to adjust human exposure to arsenic from soil ingestion. Rev Mineral Geochem 79:451-472

Bauer M, Fulda B, Blodau C (2008) Groundwater derived arsenic in high carbonate wetland soils: Sources, sinks, and mobility. Sci Total Environ 401:109-120

Beak DG, Basta NT, Scheckel KG, Traina SJ (2006) Bioaccessibility of arsenic(V) bound to ferrihydrite using a simulated gastrointestinal system. Environ Sci Technol 40:1364-1370

Beauchemin S, Hesterberg D, Beauchemin M (2002) Principal component analysis approach for modeling sulfur K-XANES spectra of humic acids. Soil Soc Am J 66:83-91

Beauchemin S, Kwong YTJ (2006) Impact of redox conditions on arsenic mobilization from tailings in a wetland with neutral drainage. Environ Sci Technol 40:A6297-6303

Beaulieu BT, Savage KS (2005) Arsenate adsorption structures on aluminum oxide and phyllosilicate mineral surfaces in smelter-impacted soils. Environ Sci Technol 39:3571-3579

Benzerara K, Miot J, Morin G, Ona-Nguema G, Skouri-Panet F, Ferard C (2011) Significance, mechanisms and environmental implications of microbial biomineralization. Comptes Rendus Geosci 343:160-167

Bhandari N, Reeder RJ, Strongin DR (2012) Photoinduced oxidation of arsenite to arsenate in the presence of goethite. Environ Sci Technol 46:8044-8051

Bianconi A (1988) XANES spectroscopy. *In:* X-ray absorption: principles, applications, techniques of EXAFS, SEXAFS, and XANES. Vol 92. Koningsberger DC, Prins R, (eds). John Wiley and Sons, New York, p 573-662

Bolanz RM, Wierzbicka-Wieczorek M, Caplovicova M, Uhlik P, Goettlicher J, Steininger R, Majzlan J (2013) Structural Incorporation of As^{5+} into hematite. Environ Sci Technol 47:9140-9147

Bostick BC, Fendorf S (2003) Arsenite sorption on troilite (FeS) and pyrite (FeS2). Geochim Cosmochim Acta 67:909-921

Bostick BC, Fendorf S, Manning BA (2003) Arsenite adsorption on galena (PbS) and sphalerite (ZnS). Geochim Cosmochim Acta 67:895-907

Bowell RJ (2014) Hydrogeochemistry of the Tsumeb deposit: implications for arsenate mineral stability. Rev Mineral Geochem 79:589-627

Bradham KD, Scheckel KG, Nelson CM, et al. (2011) Relative bioavailability and bioaccessibility and speciation of arsenic in contaminated soils. Environ Health Perspect 119:1629-1634

Breier JA, Toner BM, Fakra SC, Marcus MA, White SN, Thurnherr AM, German CR (2012) Sulfur, sulfides, oxides and organic matter aggregated in submarine hydrothermal plumes at 9° 50' N East Pacific Rise. Geochim Cosmochim Acta 88:216-236

Breit GN, Foster AL, Perkins RB, Yount JC, Whitney JW, Welch AH, Islam MN (2000) Arsenic cycling in eastern Bangladesh: the role of phyllosilicates. Geol Soc Am Abstr Prog 32:A192

Breit GN, Foster AL, Perkins RB, Yount JC, King T, Welch AH, Whitney WJ, Uddin MN, Muneemn AA, Alam MM (2004) As-rich ferric oxyhydroxide enrichments in the shallow subsurface of Bangladesh. *In:* Proceedings of the Eleventh International Symposium on Water-Rock Interaction. Wanty RB, Seal R (eds). A.A. Balkema, New York, p 1457-1461

Breuer C, Pichler T (2013) Arsenic in marine hydrothermal fluids. Chem Geol 348:2-14

Bromstad MJ, Jamieson HE (2012) Giant Mine, Yellowknife, Canada: Arsenite waste as the legacy of gold mining and processing. *In:* The Metabolism of Arsenite. Santini JA, Ward SA (eds) CRC Press ISSN. p 25-41

Brown GE, Calas G (2012) Mineral-aqueous solution interfaces and their impact on the environment. Geochem Perspect 1:483-732

Brown GE, Parks GA, O'Day PA (1995) Sorption at mineral-water interfaces: macroscopic and microscopic perspectives. *In:* Mineral Surfaces. Vaughan DJ, Pattrick RAD (eds). Chapman and Hall, London, p 129-183

Brown GE, Foster AL, Ostergren JD (1999) Mineral surfaces and bioavailability of heavy metals: A molecular-scale perspective. Proc Natl Acad Sci USA 96:3388-3395

Brown GE, Sturchio NC (2002) An overview of synchrotron radiation applications to low temperature geochemistry and environmental science. Rev Mineral Geochem 49:1-115

Brown A, Foster AL, Alpers CN, Hansel C, Lentini C, Kim CS (2010) Factors affecting principal component analysis (PCA) if X-ray absorption fine structure spectral datasets of arsenic and iron compounds. Geol Soc Am Abstr Prog 42:615

Bull DC, Harland PW, Vallance C, Foran GJ (2000) XAFS study of chromated copper arsenate timber preservative in wood. J Wood Sci 46:248-252

Bundschuh J, Pichler T (2013) Arsenic in marine hydrothermal systems: Source, fate and environmental implications. Chem Geol 348:1-1

Bunker G (2010) Introduction to XAFS. Cambridge University Press, New York

Burke IT, Mayes WM, Peacock CL, Brown AP, Jarvis AP, Gruiz K (2012) Speciation of arsenic, chromium, and vanadium in red mud samples from the Ajka spill site, Hungary. Environ Sci Technol 46:3085-3092

Burke IT, Peacock CL, Lockwood CL, Stewart DI, Mortimer RJG, Ward MB, Renforth P, Gruiz K, Mayes WM (2013) Behavior of aluminum, arsenic, and vanadium during the neutralization of red mud leachate by HCl, gypsum, or seawater. Environ Sci Technol 47:6527-6535

Burlak TL (2012) The mineralogical fate of arenic during weathering of sulfides in gold-quartz veins: a microbeam analytical study. MS Thesis, California State University, Sacramento

Burlak TL, Alpers CN, Foster AL, Brown A, Hammersley LC, Petersen E (2010) Tracking the mineralogical fate of arsenic in weathered sulfides from the Empire Mine gold-quartz vein deposit using microbeam analytical techniques. 2010 Fall AGU Meeing Abstr V51C-2220

Burnol A, Garrido F, Baranger P, Joulian C, Dictor M-C, Bodenan F, Morin G, Charlet L (2007) Decoupling of arsenic and iron release from ferrihydrite suspension under reducing conditions: a biogeochemical model. Geochem Trans 8: doi:10.1186/1467-4866-1188-1112

Burton ED, Johnston SG, Planer-Friedrich B (2013) Coupling of arsenic mobility to sulfur transformations during microbial sulfate reduction in the presence and absence of humic acid. Chem Geol 343:12-24

Busbee MW, Kocar BD, Benner SG (2009) Irrigation produces elevated arsenic in the underlying groundwater of a semi-arid basin in Southwestern Idaho. Appl Geochem 24:843-859

Button M, Moriarty MM, Watts MJ, Zhang J, Koch I, Reimer KJ (2011) Arsenic speciation in field-collected and laboratory-exposed earthworms *Lumbricus terrestris*. Chemosphere 85:1277-1283

Calvin S (2013) XAFS for Everyone. CRC Press, Boca Raton

Camm GS, Butcher AR, Pirrie D, Hughes PK, Glass HJ (2003) Secondary mineral phases associated with a historic arsenic calciner identified using automated scanning electron microscopy; a pilot study from Cornwall, UK. Miner Eng 16:1269-1277

Campbell KM, Hering JG (2008) Biogeochemical mechanisms of arsenic mobilization and sequestration. *In:* Arsenic Contamination of Groundwater: Mechanism, Analysis, and Remediation. Ahuja S (ed) John Wiley and Sons, New York, p 95-121

Campbell KM, Malasarn D, Saltikov CW, Newman DK, Hering JG (2006) Simultaneous microbial reduction of iron(III) and arsenic(V) in suspensions of hydrous ferric oxide. Environ Sci Technol 40:5950-5955

Campbell KM, Nordstrom DK (2014) Arsenic speciation and sorption in natural environments. Rev Mineral Geochem 79:185-216

Campbell KM, Root R, O'Day PA, Hering JG (2008) A gel probe equilibrium sampler for measuring arsenic porewater profiles and sorption gradients in sediments: I. Laboratory development. Environ Sci Technol 42:497-503

Cancès B, Juillot F, Morin G, Laperche V, Polya D, Vaughan DJ, Hazemann JL, Proux O, Brown GE Jr, Calas G (2008) Changes in arsenic speciation through a contaminated soil profile: a XAS based study. Sci Total Environ 397:178-189

Castillo-Michel H, Parsons JG, Peralta-Videa JR, Martinez-Martinez A, Dokken KM, Gardea-Torresdey JL (2007) Use of X-ray absorption spectroscopy and biochemical techniques to characterize arsenic uptake and reduction in pea (*Pisum sativum*) plants. Plant Physiol Biochem 45:457-463

Castillo-Michel HA, Zuverza-Mena N, Parsons JG, Dokken KM, Duarte-Gardea M, Peralta-Videa JR, Gardea-Torresdey JL (2009) Accumulation, speciation, and coordination of arsenic in an inbred line and a wild type cultivar of the desert plant species *Chilopsis linearis* (Desert willow). Phytochemistry 70:540-545

Castillo-Michel H, Hernandez-Viezcas J, Dokken KM, Marcus MA, Peralta-Videa JR, Gardea-Torresdey JL (2011) Localization and speciation of arsenic in soil and desert plant *Parkinsonia florida* using μXRF and μXANES. Environ Sci Technol 45:7848-7854

Castillo-Michel H, Hernandez-Viezcas JA, Servin A, Peralta-Videa JR, Gardea-Torresdey JL (2012) Arsenic localization and speciation in the root-soil interface of the desert plant *Prosopis juliflora-velutina*. Appl Spectrosc 66:719-727

Catalano J, Park C, Fenter P, Zhang Z (2008) Simultaneous inner- and outer-sphere arsenate adsorption on corundum and hematite. Geochim Cosmochim Acta 72:1986-2004

Catalano JG, Huhmann BL, Luo Y, Mitnick EH, Slavney A, Giammar DE (2012) Metal release and speciation changes during wet aging of coal fly ashes. Environ Sci Technol 46:11804-11812

Caumette G, Koch I, Estrada E, Reimer KJ (2011) Arsenic speciation in plankton organisms from contaminated lakes: transformations at the base of the freshwater food chain. Environ Sci Technol 45:9917-9923

Caumette G, Koch I, Estrada E, Reimer KJ (2012a) Arsenobetaine formation in plankton: A review of studies from the base of the food chain. J Environ Monitor 11:2841-2853

Caumette G, Koch I, Moriarty M, Reimer KJ (2012b) Arsenic distribution and speciation in *Daphnia pulex*. Sci Total Environ 432:243-250

Chakraborty S, Bardelli F, Mullet M, Greneche J-M, Varma S, Ehrhardt J-J, Banerjee D, Charlet L (2011) Spectroscopic studies of arsenic retention onto biotite. Chem Geol 281:83-92

Charnock JM, Polya DA, Gault AG, Wogelius RA (2007) Direct EXAFS evidence for incorporation of As^{5+} in the tetrahedral site of natural andraditic garnet. Am Mineral 92:1856-1861

Choi S, O'Day PA, Hering JG (2009) Natural attenuation of arsenic by sediment sorption and oxidation. Environ Sci Technol 43:4253-4259

Cismasu C, Ona-Nguema G, Bonnin D, Menguy N, Brown GE Jr (2008) Zinc and arsenic immobilization and magnetite formation upon reduction of maghemite by *Shewanella putrefaciens* ATCC 8071. Geochim Cosmochim Acta 72:A165-A165

Corriveau MC, Jamieson HE, Parsons MB, Campbell JL, Lanzirotti A (2011a) Direct characterization of airborne particles associated with arsenic-rich mine tailings: particle size, mineralogy and texture. Appl Geochem 26:1639-1648

Corriveau MC, Jamieson HE, Parsons MB, Hall GEM (2011b) Mineralogical characterization of arsenic in gold mine tailings from three sites in Nova Scotia. Geochem Explor Environ Anal 11:179-192

Couture RM, Rose J, Kumar N, Mitchell K, Wallschlaeger D, Van Cappellen P (2013) Sorption of arsenite, arsenate, and thioarsenates to iron oxides and iron sulfides: a kinetic and spectroscopic investigation. Environ Sci Technol 47:5652-5659

Covey AK, Furbish DJ, Savage KS (2010) Earthworms as agents for arsenic transport and transformation in roxarsone-impacted soil mesocosms: A µXANES and modeling study. Geoderma 156:99-111

Cramer SP, Siskin M, Brown LD, George GN (1988) Characterization of arsenic in oil-shale and oil-shale derivatives by X-ray absorption spectroscopy. Energy Fuels 2:175-180

Craw D, Bowell RJ (2014) The characterization of arsenic in mine waste. Rev Mineral Geochem 79:473-506

Craw D, Falconer D, Youngson JH (2003) Environmental arsenopyrite stability and dissolution: theory, experiment, and field observations. Chem Geol 199:71-82

Cullen WR, Reimer KJ (1989) Arsenic speciation in the environment. Chem Rev 89:713-764

Cutler JN, Jiang DT, Remple G (2001) Chemical speciation of arsenic in uranium mine tailings by X-ray absorption spectroscopy. Can J Anal Sci Spectros 46:130-135

Cutler JN, Chen N, Jiang DT, Demopoulos GP, Jia Y, Rowson JW (2003) The nature of arsenic in uranium mill tailings by X-ray absorption spectroscopy. J Phys IV 107:337-340

Dagan R, Dubey B, Bitton G, Townsend T (2007) Aquatic toxicity of leachates generated from electronic devices. Arch Environ Contam Toxicol 53:168-173

Datta S, Mailloux B, Jung HB, Hoque MA, Stute M, Ahmed KM, Zheng Y (2009) Redox trapping of arsenic during groundwater discharge in sediments from the Meghna riverbank in Bangladesh. Proc Natl Acad Sci USA 106:16930-16935

De la Rosa G, Parsons JG, Martinez-Martinez A, Peralta-Videa JR, Gardea-Torresdey JL (2006) Spectroscopic study of the impact of arsenic speciation on arsenic/phosphorus uptake and plant growth in tumbleweed (*Salsola kali*). Environ Sci Technol 40:1991-1996

Deditius AP, Utsunomiya S, Renock D, Ewing RC, Ramana CV, Becker U, Kesler SE (2008) A proposed new type of arsenian pyrite: Composition, nanostructure and geological significance. Geochim Cosmochim Acta 72:2919-2933

Deditius AP, Utsunomiya S, Reich M, Kesler SE, Ewing RC, Hough R, Walshe J (2011) Trace metal nanoparticles in pyrite. Ore Geol Rev 42:32-46

Del Razo LM, Styblo M, Cullen WR, Thomas DJ (2001) Determination of trivalent methylated arsenicals in biological matrices. Toxicol Appl Pharmacol 174

Deschamps E, Ciminelli VST, Weidler PG, Ramos AY (2003) Arsenic sorption onto soils enriched in Mn and Fe minerals. Clays Clay Miner 51:197-204

DeSisto SL, Jamieson HE, Parsons MB (2011) Influence of hardpan layers on arsenic mobility in historical gold mine tailings. Appl Geochem 26:2004-2018

Dewan S (2008) Tennessee Ash Flood Larger Than Initial Estimate. The New York Times: A10

Dillon CT (2012) Synchrotron radiation spectroscopic techniques as tools for the medicinal chemist: microprobe X-ray fluorescence imaging, X-ray absorption spectroscopy, and infrared microspectroscopy. Aust J Chem 65:204-217

Dowling CB, Poreda RJ, Basu AR, Peters SL, Aggarwal PK (2002) Geochemical study of arsenic release mechanisms in the Bengal Basin groundwater. Water Resour Res 38:1173

Downward L, Booth CH, Lukens WW, Bridges F (2006) A variation of the F-test for determining statistical relevance of particular parameters in EXAFS fits. Proceedings of the 13th International Conference on X-ray Absorption Fine Structure, Palo Alto. *http://www.slac.stanford.edu/econf/C060709/proceedings.htm* (accessed July 2014)

Drahota P, Filippi M (2009) Secondary arsenic minerals in the environment: a review. Environ Int 35:1243-1255

Duarte G, Ciminelli VST, Dantas MSS, Duarte HA, Vasconcelos IF, Oliveira AF, Osseo-Asare K (2012) As(III) immobilization on gibbsite: Investigation of the complexation mechanism by combining EXAFS analyses and DFT calculations. Geochim Cosmochim Acta 83:205-216

Durham PJ (1988) Theory of XANES. *In:* X-ray Absorption: Principles, Applications, Techniques of EXAFS, SEXAFS, and XANES. Koningsberger DC, Prins R (eds) John Wiley and Sons, New York, p 90

Eiche E, Kramar U, Berg M, Berner Z, Norra S, Neumann T (2010) Geochemical changes in individual sediment grains during sequential arsenic extractions. Water Res 44:5545-5555

Essilfie-Dughan J, Hendry MJ, Warner J, Kotzer T (2012) Microscale mineralogical characterization of As, Fe, and Ni in uranium mine tailings. Geochim Cosmochim Acta 96:336-352

Essilfie-Dughan J, Hendry MJ, Warner J, Kotzer T (2013) Arsenic and iron speciation in uranium mine tailings using X-ray absorption spectroscopy. Appl Geochem 28:11-18

Essington ME (1988) Estimation of the standard free energy of formation of metal arsenates, selenates, and selenites. Soil Sci Soc Am J 52:1574-1579

Etschmann BE, Ryan CG, Brugger J, Kirkham R, Hough RM, Moorhead G, Siddons DP, De Geronimo G, Kuczewski A, Dunn P, Paterson D, de Jonge MD, Howard DL, Davey P, Jensen M (2010) Reduced As components in highly oxidized environments: Evidence from full spectral XANES imaging using the Maia massively parallel detector. Am Mineral 95:884-887

Farquhar ML, Charnock JM, Livens FR, Vaughan DJ (2002) Mechanisms of arsenic uptake from aqueous solution by interaction with goethite, lepidocrocite, mackinawite, and pyrite: An X-ray absorption spectroscopy study. Environ Sci Technol 36:1757-1762

Fawcett SE, Jamieson HE (2011) The distinction between ore processing and post-depositional transformation on the speciation of arsenic and antimony in mine waste and sediment. Chem Geol 283:109-118

Fendorf S, Eick MJ, Grossl P, Sparks PDL (1997) Arsenate and chromate retention mechanisms on goethite. 1. Surface structure. Environ Sci Technol 31:315-326

Fendorf S, Herbel MJ, Tufano KJ, Kocar BD (2007) Biogeochemical processes controlling the cycling of arsenic in soils and sediments. *In:* Biophysico-Chemical Processes of Heavy Metals and Metalloids in Soil Environments. Violante A, Huang PM, Gadd GM (eds) John Wiley and Sons, p 313-338

Fendorf S, Kocar BD (2009) Biogeochemical processes controlling the fate and transport of arsenic: implications for South and Southeast Asia. Adv Agron 104:137-164

Fendorf S, Nico PS, Kocar BD, Masue Y, Tufano KJ (2010) Arsenic chemistry in soils and sediments. *In:* Synchrotron-Based Techniques in Soils and Sediments. Singh B, Grafe M (eds) Elsevier Publishing, p 357-378

Fitzmaurice AG, Bilgin AA, O'Day PA, Illera V, Burris DR, Reisinger HJ, Hering JG (2009) Geochemical and hydrologic controls on the mobilization of arsenic derived from herbicide application. Appl Geochem 24:2152-2162

Foster AL (1999) Partitioning and transformation of arsenic and selenium in natural and laboratory systems. Ph.D. Thesis, Stanford University, Palo Alto, CA

Foster AL (2003) Spectroscopic investigations of arsenic species in solid phases. *In:* Arsenic in Groundwater: Geochemistry and Occurrence. Welch AH, Stollenwerk KG (eds) Kulwer Academic Publishers, Norwell, p 27-65

Foster AL, Ashley RP (2002) Characterization of arsenic species in microbial mats from an inactive gold mine mine. Geochemistry: Explor Environ Anal 2:253-261

Foster AL, Brown GE Jr, Parks GA, Voigt DE, Brantley SL (1995) X-ray absorption near edge structure (XANES) spectroscopic analysis of As-contaminated soils and As-bearing mine tailings. Geol Society Am Abstr Progr 27:163

Foster AL, Brown GE Jr, Parks GA (1998a) X-ray absorption fine-structure spectroscopy study of photocatalyzed, heterogeneous As(III) oxidation on kaolin and anatase. Environ Sci Technol 32:1444-1452

Foster AL, Brown GE Jr, Tingle T, Parks GA (1998b) Quantitative arsenic speciation in mine tailings using X-ray absorption spectroscopy. Am Mineral 83:553-568

Foster AL, Ashley RP, Rytuba JJ (2000a) Direct and *in situ* speciation of arsenic in microbian mats using X-ray absorption spectroscopy. Metal Ions Biol Med 6:62-64

Foster AL, Breit GN, Welch AH, Whitney JW, Yount JC, Islam MS, Alam MM, Islam MK, Islam MN (2000b) *In situ* identification of arsenic species in soil and aquifer sediment from Ramrail, Brahmanbaria, Bangladesh. Am Geophys Union Fall Meeting Suppl 81:H21D-01

Foster AL, Brown GE Jr, Parks GA (2003) X-ray abosrption fine structure study of As(V) and Se(IV) sorption complexes on hydrous Mn oxides. Geochim Cosmochim Acta 67:1937-1953

Foster AL, Breit GN, Yount JC, Whitney JW, Welch AH, Lamothe PJ, Sanzalone RF, Perkins RB, Uddin MdN, Muneem AA, Alam MM (2005) Spectromicroscopy and microdiffraction studies of arsenic-rich iron oxide bands in near-surface sediments of eastern Bangladesh. Abstracts Papers Am Chem Soc 230:U1754-U1755

Foster AL, Berry K, Jacobson EA, Rytuba JJ (2009a) Arsenic species in scute (shell plate) and lung tissue of desert tortoises. EOS Trans AGU Fall Meet Suppl 90:B32B-04

Foster AL, Ona-Nguema G, Tufano K, White R III (2009b) Temporal chemical data for sediment, water, and biological samples from the Lava Cap Mine Superfund Site, Nevada County, California-2006-2008. U.S. Geological Survey Open File Report 2009-1268

Foster AL, Lowers HA, Breit GN, Whitney J, Yount J, Uddin MN, Muneem AA (2010) Arsenic association with secondary iron phases on ferroan micas: Implications for ground water quality in South Asia. Geochim Cosmochim Acta 74:A300-A300

Foster AL, Ashley RP, Rytuba JJ (2011) Arsenic species in weathering mine tailings and biogenic solids at the Lava Cap Mine Superfund Site, Nevada City, CA. Geochem Trans 12:1

Foster AL, Alpers CN, Burlak T, Blum AE, Petersen EU, Basta NT, Whitacre S, Casteel SW, Kim CS, Brown AL (2014) Arsenic chemistry, mineralogy, speciation, and bioavailability/bioaccessibility in soils and mine waste from the Empire Mine, CA, USA. Goldschmidt 2014 Abstracts #726

Frierdich AJ, Catalano JG (2012) Distribution and speciation of trace elements in iron and manganese oxide cave deposits. Geochim Cosmochim Acta 91:240-253

Gao X, Root RA, Farrell J, Ela W, Chorover J (2013) Effect of silicic acid on arsenate and arsenite retention mechanisms on 6-L ferrihydrite: A spectroscopic and batch adsorption approach. Appl Geochem 38:110-120

Gault AG, Rowland HAL, Charnock JM, Wogelius RA, Gomez-Morilla I, Vong S, Leng M, Samreth S, Sampson ML, Polya DA (2008) Arsenic in hair and nails of individuals exposed to arsenic-rich groundwaters in Kandal province, Cambodia. Sci Total Environ 393:168-176

Gaw S, McBride G (2010) Sheep Dip Factsheet No.1. New Zealand Ministry for the Environment

George GG (2000) EXAFSPAK: A Suite of Computer Programs for Analysis of X-Ray Absorption Spectra. *http://www-ssrl.slac.stanford.edu/exafspak.html* (accessed July 2014)

George GN, Pickering IJ (2007) X-ray absorption spectroscopy in biology and chemistry. *In:* Brilliant Light in Life and Materials Sciences. Tsakanov V, Wiedemann H (eds) Springer, Dordrecht, p 97-119

George GN, Prince RC, Singh SP, Pickering IJ (2009) Arsenic K-edge X-ray absorption spectroscopy of arsenic in seafood. Mol Nutr Food Res 53:552-557

Godelitsas A, Nastos P, Mertzimekis TJ, Toli K, Simon R, Goettlicher J (2011) A microscopic and synchrotron-based characterization of urban particulate matter (PM_{10}-$PM_{2.5}$ and $PM_{2.5}$) from Athens atmosphere, Greece. Nuclear Instr Meth Phys Res Sect B 269:3077-3081

Gomez MA, Becze L, Blyth RIR, Cutler JN, Demopoulos GP (2010) Molecular and structural investigation of yukonite (synthetic, natural) and its relation to arseniosiderite. Geochim Cosmochim Acta 74:5835-5851

Grafe M, Sparks DL (2005) Kinetics of zinc and arsenate co-sorption at the goethite-water interface. Geochim Cosmochim Acta 69:4573-4595

Grafe M, Beattie DA, Smith E, Skinner WM, Singh B (2008a) Copper and arsenate co-sorption at the mineral-water interfaces of goethite and jarosite. J Colloid Interface Sci 322:399-413

Grafe M, Tappero RV, Marcus MA, Sparks DL (2008b) Arsenic speciation in multiple metal environments: I. Bulk-XAFS spectroscopy of model and mixed compounds. J Colloid Interface Sci 320:383-399

Grafe M, Tappero RV, Marcus MA, Sparks DL (2008c) Arsenic speciation in multiple metal environments - II. Micro-spectroscopic investigation of a CCA contaminated soil. J Colloid Interface Sci 321:1-20

Guilbert JM, Park CF Jr (1985) The Geology of Ore Deposits. W. H. Freeman and Co., New York

Guo H, Zhang B, Li Y, Berner Z, Tang X, Norra S, Stueben D (2011a) Hydrogeological and biogeochemical constrains of arsenic mobilization in shallow aquifers from the Hetao basin, Inner Mongolia. Environ Pollut 159:876-883

Guo H, Zhang B, Zhang Y (2011b) Control of organic and iron colloids on arsenic partition and transport in high arsenic groundwaters in the Hetao basin, Inner Mongolia. Appl Geochem 26:360-370

Guo H, Ren Y, Liu Q, Zhao K, Li Y (2013) Enhancement of arsenic adsorption during mineral transformation from siderite to goethite: mechanism and application. Environ Sci Technol 47:1009-1016

Hammersley A (2004) Fit2D. *http://www.esrf.eu/computing/scientific/FIT2D/* (accessed July 2104)

Hattori K, Takahashi Y, Guillot S, Johanson B (2005) Occurrence of arsenic (V) in forearc mantle serpentinites based on X-ray absorption spectroscopy study. Geochim Cosmochim Acta 69:5585-5596

Hawthorne FC (1976) The hydrogen postitions in scorodite. Acta Cryst B B32:2891-2992

Heiney P (2002) Datasqueeze. *http://www.datasqueezesoftware.com/* (accessed July 2014)

Helmhart M, O'Day PA, Garcia-Guinea J, Serrano S, Garrido F (2012) Arsenic, copper, and zinc leaching through preferential flow in mining-impacted soils. Soil Sci Soc Am J 76:449-462

Helz G, Tossel JA, Charnock JM, Pattrick RA, Vaughan DV, Garner CD (1995) Oligomerization in As(III) sulfide solutions: theoretical constraints and spectroscopic evidence. Geochim Cosmochim Acta 59:4591-4604

Henderson GS, de Groot FMF, Moulton BJA (2014) X-ray absorption near-edge structure (XANES) Spectroscopy. Rev Mineral Geochem 78:75-138

Hohmann C, Morin G, Ona-Nguema G, Guigner J-M, Brown GE Jr, Kappler A (2011) Molecular-level modes of As binding to Fe(III) (oxyhydr)oxides precipitated by the anaerobic nitrate-reducing Fe(II)-oxidizing *Acidovorax* sp strain BoFeN1. Geochim Cosmochim Acta 75:4699-4712

Hopenhayn C (2006) Arsenic in drinking water: Impact on human health. Elements 2:103-107

Hopp L, Nico PS, Marcus MA, Peiffer S (2008) Arsenic and chromium partitioning in a podzolic soil contaminated by chromated copper arsenate. Environ Sci Technol 42:6481-6486

Hornberger B, de Jonge MD, Feser M, et al. (2008) Differential phase contrast with a segmented detector in a scanning X-ray microprobe. J Synch Rad 15:355-362

Huang Z-C, Chen T-B, Lei M, Hu T-D (2004) Direct determination of arsenic species in arsenic hyperaccumulator *Pteris vittata* by EXAFS. Acta Botanica Sinica 46:46-50

Hudson-Edwards KA, Banerjee DM, Ravenscroft P, McArthur JM, Carter A, Mishra R, Pirohit R, Chatterjee A, Talukder A, Houghton S (2004) A sedimentary framework for arsenic-contaminated groundwater in West Bengal. Geochim Cosmochim Acta 68:A515-A515

Hudson-Edwards KA, Jamieson HE, Charnock JM, Macklin MG (2005) Arsenic speciation in waters and sediment of ephemeral floodplain pools, Rios Agrio-Guadiamar, Aznalcollar, Spain. Chem Geol 219:175-192

Hudson-Edwards KA, Jamieson HE, Lottermoser BG (2011) Mine wastes: past, present, future. Elements 7:375-379

Huffman GP (1993) Nondestructive determination of trace-element speciation in coal and coal ash by XAFS spectroscopy. Energy Fuels 7:482-489

Huffman GP, Huggins FE, Shah N, Zhao JM (1994) Speciation of arsenic and chromium in coal and combustion ash by XAFS spectroscopy. Fuel Process Technol 39:47-62

Huffman GP, Huggins FE, Shah N, Huggins R, Linak WP, Miller CA, Pugmire RJ, Meuzelaar HLC, Seehra MS, Manivannan A (2000) Characterization of fine particulate matter produced by combustion of residual fuel oil. J Air Waste Manage Assoc 50:1106-1114

Huggins FE, Shah N, Zhao JM, Lu FL, Huffman GP (1993) Nondestructive determination of trace-element speciation in coal and coal ash by XAFS spectroscopy. Ener Fuels 7:482-489

Huggins FE, Goodarzi F, Lafferty CJ (1996) Mode of occurrence of arsenic in subbituminous coals. Energy Fuels 10:1001-1004

Huggins FE, Huffman GP, Kolker A, Mroczkowski SJ, Palmer CA, Finkelman RB (2002) Combined application of XAFS spectroscopy and sequential leaching for determination of arsenic speciation in coal. Energy Fuels 16:1167-1172

Huggins FE, Senior CL, Chu P, Ladwig K, Huffman GP (2007) Selenium and arsenic speciation in fly ash from full-scale coal-burning utility plants. Environ Sci Technol 41:3284-3289

Huggins FE, Seidu LBA, Shah N, Huffman GP, Honaker RQ, Kyger JR, Higgins BL, Robertson JD, Pal S, Seehra MS (2009) Elemental modes of occurrence in an Illinois #6 coal and fractions prepared by physical separation techniques at a coal preparation plant. Int J Coal Geol 78:65-76

Ilgen AG, Rychagov SN, Trainor TP (2011) Arsenic speciation and transport associated with the release of spent geothermal fluids in Mutnovsky field (Kamchatka, Russia). Chem Geol 288:115-132

Ilgen AG, Foster AL, Trainor TP (2012) Role of structural Fe in nontronite NAu-1 and dissolved Fe(II) in redox transformations of arsenic and antimony. Geochim Cosmochim Acta 94:128-145

Impellitteri CA (2005) Effects of pH and phosphate on metal distribution with emphasis on As speciation and mobilization in soils from a lead smelting site. Sci Total Environ 345:175-190

Itai T, Masuda H, Takahashi Y, Mitamura M, Kusakabe M (2006) Determination of As-III/As-V ratio in alluvial sediments of the Bengal Basin using X-ray absorption near-edge structure. Chem Lett 35:866-867

Itai T, Takahashi Y, Seddique AA, Maruoka T, Mitamura M (2010) Variations in the redox state of As and Fe measured by X-ray absorption spectroscopy in aquifers of Bangladesh and their effect on As adsorption. Appl Geochem 25:34-47

IXS Standards and Criteria Committee (2000) Error Reporting Reccomendations: A Report of the Standards and Criteria Committee. *http://ixs.csrri.iit.edu/IXS/subcommittee_reports/sc/SC00report.pdf* (accessed July 2014)

Jackson BP, Taylor VF, Karagas MR, Punshon T, Cottingham KL (2012) Arsenic, organic foods, and brown rice syrup. Environ Health Perspect 120:623-626

Jambeck JR, Townsend TG, Solo-Gabriele HM (2008) Landfill disposal of CCA-treated wood with construction and demolition (C&D) debris: arsenic, chromium, and copper concentrations in leachate. Environ Sci Technol 42:5740-5745

James-Smith J, Cauzid J, Testemale D, Liu W, Hazemann J-L, Proux O, Etschmann B, Philippot P, Banks D, Williams P, Brugger J (2010) Arsenic speciation in fluid inclusions using micro-beam X-ray absorption spectroscopy. Am Mineral 95:921-932

Jamieson HE (2011) Geochemistry and mineralogy of solid mine waste: essential knowledge for predicting environmental impact. Elements 7:381-386

Jamieson HE (2014) The legacy of arsenic contamination from mining and processing refractory gold ore at Giant Mine, Yellowknife, Northwest Territories, Canada. Rev Mineral Geochem 79:533-551

Jamieson HE, Walker SR, Parsons MB, Hall GEM (2008) Characterization of multiple secondary minerals in arsenic-rich gold mine tailings. Geochim Cosmochim Acta 72:A424-A424

Jamieson HE, Walker SR, Andrade CF, Wrye LA, Rasmussen PE, Lanzirotti A, Parsons MB (2011) Identification and characterization of arsenic and metal compounds in contaminated soil, mine tailings, and house dust using synchrotron-based microanalysis. Hum Ecol Risk Assess 17:1292-1309

Jean J-S, Lee M-K (2006) Potential etiological agents of Blackfoot disease caused by drinking water in Taiwan: an overview. GSA Fall Annual Meeting, # 67-4

Johnston SG, Burton ED, Keene AF, Planer-Friedrich B, Voegelin A, Blackford MG, Lumpkin GR (2012) Arsenic mobilization and iron transformations during sulfidization of As(V)-bearing jarosite. Chem Geol 334:9-24

Kachenko AG, Grafe M, Singh B, Heald SM (2010) Arsenic speciation in tissues of the hyperaccumulator *P. calomelanos* var. austroamericana using X-ray absorption spectroscopy. Environ Sci Technol 44:4735-4740

Kakoulli I, Prikhodko SV, Fischer C, Cilluffo M, Uribe M, Bechtel HA, Fakra SC, Marcus MA (2014) Distribution and chemical speciation of arsenic in ancient human hair using synchrotron radiation. Anal Chem 86:521-526

Kappen P, Webb J (2013) An EXAFS study of arsenic bonding on amorphous aluminium hydroxide. Appl Geochem 31:79-83

Kashiwabara T, Mitsuo S, Hokura A, Kitajima N, Abe T, Nakai I (2010) *In vivo* micro X-ray analysis utilizing synchrotron radiation of the gametophytes of three arsenic accumulating ferns, *Pteris vittata* L., *Pteris cretica* L. and *Athyrium yokoscense,* in different growth stages. Metallomics 2:261-270

Katz SA, Salem H (2005) Chemistry and toxicology of building timbers pressure-treated with chromated copper arsenate: a review. J Appl Toxicol 25:1-7

Kelly SD, Hesterberg D, Ravel B (eds) (2008) Analysis of Soils and Minerals Using X-ray Absorption Spectroscopy. Soil Science Society of America, Madison, WI

Kim CS, Chi C, Miller SR, Rosales RA, Sugihara ES, Akau J, Rytuba JJ, Webb SM (2013) (Micro)spectroscopic analyses of particle size dependence on arsenic distribution and speciation in mine wastes. Environ Sci Technol 47:8164-8171

Kneebone PE, O'Day PA, Jones N, Hering JG (2002) Deposition and fate of arsenic in iron- and arsenic-enriched reservoir sediments. Environ Sci Technol 36:381-386

Kocar BD, Garrott RA, Inskeep WP (2004) Elk exposure to arsenic in geothermal watersheds of Yellowstone National Park, USA. Environ Toxicol Chem 23:982-989

Kocar BD, Tufano K, Masui Y, Stewart B, Herbel M, Fendorf S (2005) Arsenic mobilization influenced by iron reduction and sulfidogenesis. Geochim Cosmochim Acta 69:A466-A466

Kocar BD, Herbel MJ, Tufano KJ, Fendorf S (2006) Contrasting effects of dissimilatory iron(III) and arsenic(V) reduction on arsenic retention and transport. Environ Sci Technol 40:6715-6721

Kocar BD, Polizzotto ML, Benner SG, et al. (2008) Integrated biogeochemical and hydrologic processes driving arsenic release from shallow sediments to groundwaters of the Mekong delta. Appl Geochem 23:3059-3071

Kocar BD, Borch T, Fendorf S (2010) Arsenic repartitioning during biogenic sulfidization and transformation of ferrihydrite. Geochim Cosmochim Acta 74:980-994

Koch I, Moriarty M, House K, Sui J, Cullen WR, Saper RB, Reimer KJ (2011) Bioaccessibility of lead and arsenic in traditional Indian medicines. Sci Total Environ 409:4545-4552

Kolker A, Huggins FE (2007) Progressive oxidation of pyrite in five bituminous coal samples: An As XANES and 57Fe Mössbauer spectroscopic study. Appl Geochem 22:778-787

Kolker A, Nordstrom DK (2001) Occurrence and micro-distribution of arsenic in pyrite. U.S. Geological Survey Workshop on As in the Environment

Kolker A, Huggins FE, Palmer CA, Shah N, Crowley SS, Huffman GP, Finkelman RB (2000) Mode of occurrence of arsenic in four US coals. Fuel Process Technol 63:167-178

Koningsberger DC, Prins R (1988) X-ray Absorption: Principles, Applications, Techniques of EXAFS, SEXAFS, and XANES. John Wiley and Sons

Kossoff D, Hudson-Edwards KA, Dubbin WE, Alfredsson M, Geraki T (2012) Cycling of As, P, Pb and Sb during weathering of mine tailings: implications for fluvial environments. Mineral Mag 76:1209-1228

Kreidie N, Armiento G, Cibin G, Cinque G, Crovato C, Nardi E, Pacifico R, Cremisini C, Mottana A (2011) An integrated geochemical and mineralogical approach for the evaluation of arsenic mobility in mining soils. J Soils Sed 11:37-52

La Force MJ, Hansel CM, Fendorf S (2000) Arsenic speciation, seasonal transformations, and co-distribution with iron in a mine waste-influenced palustrine emergent wetland. Environ Sci Technol 34:3937-3943

Ladeira ACQ, Ciminelli VST, Duarte HA, Alves MCM, Ramos AY (2001) Mechanism of anion retention from EXAFS and density functional calculations: arsenic (V) adsorbed on gibbsite. Geochim Cosmochim Acta 65:1211-1217

Lande J, Webb S (2007) The Area Diffraction Machine. *https://github.com/joshualande/AreaDiffractionMachine* (accessed July 2014)

Landrot G, Tappero R, Webb SM, Sparks DL (2012) Arsenic and chromium speciation in an urban contaminated soil. Chemosphere 88:1196-1201

Langdon CJ, Winters C, Sturzenbaum SR, Morgan AJ, Charnock JM, Meharg AA, Piearce TG, Lee PH, Semple KT (2005) Ligand arsenic complexation and immunoperoxidase detection of metallothionein in the earthworm *Lumbricus rubellus* inhabiting arsenic-rich soil. Environ Sci Technol 39:2042-2048

Langmuir D, Mahoney J, Rowson J (2006) Solubility products of amorphous ferric arsenate and crystalline scorodite (FeAsO$_4$·2H$_2$O) and their application to arsenic behavior in buried mine tailings. Geochim Cosmochim Acta 70:2942-2956

Langner P, Mikutta C, Kretzschmar R (2012a) Arsenic sequestration by organic sulphur in peat. Nature Geoscience 5:66-73

Langner P, Mikutta C, Kretzschmar R (2012b) Synchrotron-based spectroscopy reveals first evidence for organic sulfur-coordinated arsenic in peat. Chimia 66:877-877

Langner P, Mikutta C, Suess E, Marcus MA, Kretzschmar R (2013) Spatial distribution and speciation of arsenic in peat studied with Microfocused X-ray fluorescence spectrometry and X-ray absorption spectroscopy. Environ Sci Technol 47:9706-9714

Lehner S, Savage K (2008) The effect of As, Co, and Ni impurities on pyrite oxidation kinetics: Batch and flow-through reactor experiments with synthetic pyrite. Geochim Cosmochim Acta 72:1788-1800

Lehner S, Savage K, Ciobanu M, Cliffel DE (2007) The effect of As, Co, and Ni impurities on pyrite oxidation kinetics: An electrochemical study of synthetic pyrite. Geochim Cosmochim Acta 71:2491-2509

Lerotic M, Jacobsen C, Schäfer T, Vogt S (2004) Cluster analysis of soft X-ray spectromicroscopy data. Ultramicroscopy 100:35-57

Leybourne MI, Johannesson KH, Asfaw A (2014) Measuring arsenic speciation in environmental media: sampling, preservation, and analysis. Rev Mineral Geochem 79:371-390

Lim SR, Schoenung JM (2010) Toxicity potentials from waste cellular phones, and a waste management policy integrating consumer, corporate, and government responsibilities. Waste Management 30:1653-1660

Lin J, Chen N, Pan Y (2013a) Arsenic incorporation in synthetic struvite (NH$_4$MgPO$_4$ 6H$_2$O): a synchrotron XAS and single-crystal EPR study. Environ Sci Technol 47:12728-12735

Lin J, Chen N, Nilges MJ, Pan Y (2013b) Arsenic speciation in synthetic gypsum (CaSO$_4$·2H$_2$O): A synchrotron XAS, single-crystal EPR, and pulsed ENDOR study. Geochim Cosmochim Acta 106:524-540

Liu G, Cai Y (2007) Arsenic speciation in soils: An analytical challenge for understanding arsenic biogeochemistry. *In:* Developments in Environmental Science. Vol 5. Sarkar D, Datta R, Hannigan R, (eds). Elsevier, p 685-708

Liu S, Jing C, Meng X (2008) Arsenic re-mobilization in water treatment adsorbents, under reducing conditions: Part II. XAS and modeling study. Sci Total Environ 392:137-144

Liu T-K, Chen K-Y, Yang TF, Chen Y-G, Chen W-F, Kang S-C, Lee C-P (2009) Origin of methane in high-arsenic groundwater of Taiwan - Evidence from stable isotope analyses and radiocarbon dating. J Asian Earth Sci 36:364-370

Liu C, Wright CG, McAdam KG, Taebunpakul S, Heroult J, Braybrook J, Goenaga-Infante H (2012) Arsenic Speciation in Tobacco and Cigarette Smoke. Beitr Tabak Int 25:375-380

Lloyd JR, Oremland RS (2006) Microbial transformations of arsenic in the environment: from soda lakes to aquifers. Elements 2:85-90

Lottermoser BG (2010) Mine Wastes: Characterization, Treatment, Environmental Impacts. Springer-Verlag, Berlin

Lovley DR (2000) Fe(III) and Mn(IV) reduction. *In:* Environmental Microbe-Metal interactions. Lovley DR (ed) ASM Press, Washington DC, p 3-30

Lowers HA, Breit GN, Foster AL, Whitney J, Yount J, Uddin MN, Muneem AA (2007) Arsenic incorporation into authigenic pyrite, Bengal Basin sediment, Bangladesh. Geochim Cosmochim Acta 71:2699-2717

Luo L, Zhang S, Shan X-Q, Jiang W, Zhu Y-G, Liu T, Xie Y-N, McLaren RG (2006) Arsenate sorption on two Chinese red soils evaluated with macroscopic measurements and extended X-ray absorption fine-structure spectroscopy. Environ Toxicol Chem 25:3118-3124

Luo Y, Giammar DE, Huhmann BL, Catalano JG (2011) Speciation of selenium, arsenic, and zinc in class C fly ash. Energy Fuels 25:2980-2987

MacLean LCW, Beauchemin S, Rasmussen PE (2010) Application of synchrotron x-ray yechniques for the determination of metal speciation in (house) dust particles. *In:* Urban Airborne Particulate Matter. Zereini F, Wiseman CLS (eds) Springer, p 193-216

MacPhee MJ, Novak JT, Mutter RN, Cornwell DA (2003) Disposal of wastes resulting from arsenic removal processes. *In:* Arsenic Exposure and Health Effects V. Chappell WR, Abernathy CO, Calderon RL, Thomas DJ (eds) Elsevier, New York, p 483-489

Maher K, Chamberlain CP (2014) Hydrologic regulation of chemical weathering and the geologic carbon cycle. Science 343:1502-1504

Maillot F, Morin G, Juillot F, Bruneel O, Casiot C, Ona-Nguema G, Wang Y, Lebrun S, Aubry E, Vlaic G, Brown GE Jr (2013) Structure and reactivity of As(III)- and As(V)-rich schwertmannites and amorphous ferric arsenate sulfate from the Carnoules acid mine drainage, France: comparison with biotic and abiotic model compounds and implications for As remediation. Geochim Cosmochim Acta 104:310-329

Majzlan J, Drahota P, Filippi M (2014) Parageneses and crystal chemistry of arsenic minerals. Rev Mineral Geochem 79:17-184

Makris KC, Sarkar D, Parsons JG, Datta R, Gardea-Torresdey JL (2007) Surface arsenic speciation of a drinking-water treatment residual using X-ray absorption spectroscopy. J Colloid Interface Sci 311:544-550

Makris KC, Sarkar D, Parsons JG, Datta R, Gardea-Torresdey JL (2009) X-ray absorption spectroscopy as a tool investigating arsenic(III) and arsenic(V) sorption by an aluminum-based drinking-water treatment residual. J Hazard Mater 171:980-986

Malinowski ER (2002) Factor Analysis in Chemistry. John Wiley and Sons, Inc, New York

Manceau A (1995) The mechanism of anion adsorption on iron oxides: Evidence for the binding of arsenate tetrahedra on free $Fe(O,OH)_6$ edges. Geochim Cosmochim Acta 59:3647-3653

Manceau A, Marcus MA, Tamura N (2002) Quantitative speciation of heavy metals in soils and sediments by synchrotron X-ray techniques. Rev Mineral Geochem 49:341-428

Manceau A, Lanson M, Geoffroy N (2007) Natural speciation of Ni, Zn, Ba, and As in ferromanganese coatings on quartz using X-ray fluorescence, absorption, and diffraction. Geochim Cosmochim Acta 71:95-128

Mancini I, Guella G, Frostin M, Hnawia E, Laurent D, Debitus C, Pietra F (2006) On the first polyarsenic organic compound from nature: Arsenicin a from the New Caledonian marine sponge *Echinochalina bargibanti*. Chemistry 12 8989-8994

Mandal BK, Suzuki KT (2002) Arsenic round the world: a review. Talanta 58:201-235

Mandaliev PN, Mikutta C, Barmettler K, Kotsev T, Kretzschmar R (2014) Arsenic species formed from arsenopyrite weathering along a contamination gradient in circumneutral river floodplain soils. Environ Sci Technol 48:208-217

Manning B (2005) Arsenic speciation in As(III)- and As(V)-treated soil using XANES spectroscopy. Microchim Acta 151:181-188

Manning BA, Fendorf SE, Bostick B, Suarez DL (2002) Arsenic(III) oxidation and arsenic (V) adsorption reactions on synthetic birnessite. Environ Sci Technol 36:976-981

Manning BA, Fendorf SE, Goldberg S (1998) Surface structures and stability of arsenic(III) on goethite: Spectroscopic evidence for inner-sphere complexes. Environ Sci Technol 32:2383-2388

Manning BA, Fendorf SE, Suarez DL (2003) Arsenic(III) complexation and oxidation reactions on soil. *In:* Biogeochemistry of Environmentally Important Trace Elements. ACS Symposium Series, Vol. 835. Cai Y, Braids OC (eds) American Chemical Society, p 57-69

Masue-Slowey Y, Kocar BD, Bea Jofré SA, Mayer KU, Fendorf S (2011) Transport implications resulting from internal redistribution of arsenic and iron within constructed soil aggregates. Environ Sci Technol 45:582-588

Masue-Slowey Y, Ying SC, Kocar BD, Pallud CE, Fendorf S (2013) Dependence of arsenic fate and transport on biogeochemical heterogeneity arising from the physical structure of soils and sediments. J Environ Qual 42:1119-1129

Mayhew LE, Templeton AS (2011) Microscale imaging and identification of Fe speciation and distribution during fluid-mineral reactions under highly reducing conditions. Env Sci Technol 45:4468-4474

Melamed D (2005) Monitoring arsenic in the environment: a review of science and technologies with the potential for field measurements. Anal Chim Acta 532:1-13

Meunier L, Walker SR, Wragg J, Parsons MB, Koch I, Jamieson HE, Reimer KJ (2010) Effects of soil composition and mineralogy on the bioaccessibility of arsenic from tailings and soil in gold mine districts of Nova Scotia. Environ Sci Technol 44:2667-2674

Meunier L, Koch I, Reimer KJ (2011a) Effects of organic matter and aging on the bioaccessibility of arsenic. Environ Pollut 159:2530-2536

Meunier L, Koch I, Reimer KJ (2011b) Effect of particle size on arsenic bioaccessibility in gold mine tailings of Nova Scotia. Sci Total Environ 409:2233-2243

Mikutta C, Mandaliev PN, Kretzschmar R (2013a) New clues to the local atomic structure of short-range ordered ferric arsenate from extended X-ray absorption fine structure spectroscopy. Environ Sci Technol 47:3122-3131

Mikutta C, Mandaliev PN, Kretzschmar R (2013b) Response to Comment on "New Clues to the Local Atomic Structure of Short-Range Ordered Ferric Arsenate from Extended X-ray Absorption Fine Structure Spectroscopy". Environ Sci Technol 47:13201-13202

Miot J, Morin G, Skouri-Panet F, Ferard C, Bubry E, Briand J, Wang Y, Ona-Nguema G, Guyot F, Brown GE Jr (2008) XAS study of arsenic coordination in *Euglena gracilis* exposed to arsenite. Environ Sci Technol 42:5342-5347

Miot J, Morin G, Skouri-Panet F, Ferard C, Poitevin A, Aubry E, Ona-Nguema G, Juillot F, Guyot F, Brown GE Jr (2009) Speciation of arsenic in *Euglena gracilis* cells exposed to As(V). Environ Sci Technol 43:3315-3321

Mitchell VL (2014) Health risks associated with chronic exposures to arsenic in the environment. Rev Mineral Geochem 79:435-449

Mogren CL, Webb SM, Walton WE, Trumble JT (2013) Micro x-ray absorption spectroscopic analysis of arsenic localization and biotransformation in *Chironomus riparius* Meigen (*Diptera: Chironomidae*) and *Culex tarsalis* Coquillett (*Culicidae*). Environ Pollut 180:78-83

Moriarty MM, Koch I, Gordon RA, Reimer KJ (2009) Arsenic speciation of terrestrial invertebrates. Environ Sci Technol 43:4818-4823

Moriarty MM, Lai VWM, Koch I, Cui L, Combs C, Krupp EM, Feldmann J, Cullen WR, Reimer KJ (2014) Speciation and toxicity of arsenic in mining-affected lake sediments in the Quinsam watershed, British Columbia. Sci Total Environ 466:90-99

Morin G, Calas G (2006) Arsenic in soils, mine tailings, and former industrial sites. Elements 2:97-101

Morin G, Juillot F, Casiot C, Bruneel O, Personne J-C, Elbaz-Poulichet F, Leblanc M, Ildefons P, Calas G (2003) Bacterial formation of tooeleite and mixed Arsenic(III) or Arsenic(V)-Iron(III) Gels in the Carnoules acid mine drainage, France: A XANES, XRD, and SEM study. Environ Sci Tech 37:1705-1712

Morin G, Ona-Nguema G, Wang Y, Menguy N, Juillot F, Proux O, Guyot F, Calas G, Brown GE Jr (2008) Extended X-ray absorption fine structure analysis of arsenite and arsenate adsorption on maghemite. Environ Sci Technol 42:2361-2366

Morin G, Ona-Nguema G, Wang Y, Menguy N, Juillot F, Calas G, Brown GE Jr (2009) Arsenic(III) polymerization upon sorption on iron(II,III)-(hydr)oxides surfaces: implications for arsenic mobility under reducing conditions. Geochim Cosmochim Acta 73:A906-A906

Morton WE, Dunnette DA (1994) Health effects of environmental arsenic. *In:* Arsenic in the Environment Part 2: Human Health and Ecosystem Effects. Vol 27. Nriagu JO (ed) John Wiley and Sons, Inc., New York, p 293

Munro KL, Mariana A, Klavins AI, Foster AJ, Lai B, Vogt S, Cai Z, Harris HH, Dillon CT (2008) Microprobe XRF mapping and XAS investigations of the intracellular metabolism of arsenic for understanding arsenic-induced toxicity. Chem Res Toxicol 21:1760-1769

Myneni SCB (2002) Soft X-ray spectroscopy and spectromicroscopy studies of organic molecules in the environment. Rev Mineral Geochem 49:485-579

Myneni SCB, Traina SJ, Logan TJ, Waychunas GA (1997) Oxyanion behavior in alkaline environments: sorption and desorption of arsenate in ettringite. Environ Sci Technol 31:1761-1768

Nachman KE, Baron PA, Raber G, Francesconi KA, Navas-Acien A, Love DC (2013) Roxarsone, inorganic arsenic, and other arsenic species in chicken: A U.S.-based market basket sample. Environ Health Perspect, doi: 10.1289/ehp.1206245

Neumann T, Scholz F, Kramar U, Ostermaier M, Rausch N, Berner Z (2013) Arsenic in framboidal pyrite from recent sediments of a shallow water lagoon of the Baltic Sea. Sedimentology 60:1389-1404

Newville M (2014) Fundamentals of XAFS. Rev Mineral Geochem 78:33-74

Nico PS, Fendorf SE, Lowney YW, Holm SE, Ruby MV (2004) Chemical structure of arsenic and chromium in CCA-treated wood: implications of environmental weathering. Environ Sci Technol 38:5253-5260

Nico PS, Ruby MV, Lowney YW, Holm SE (2006) Chemical speciation and bioaccessibility of arsenic and chromium in chromated copper arsenate-treated wood and soils. Environ Sci Technol 40:402-408

Nordstrom DK (2011) Mine waters: acidic to circumneutral. Elements 7:393-398

Nordstrom DK, Archer DG (2003) Arsenic thermodynamic data and environmental geochemistry: an evaluation of thermodynamic data for modeling the aqueous environmental geochemistry of arsenic. *In:* Arsenic in Ground Water. Welch AH, Stollenwerk KG (eds) Kluwer Academic Publishers, Boston, p 1-25

Nriagu JO, Bhattacharya P, Mukherjee AB, Bundschuh J, Zevenhoven R, Loeppert RH (2007) Arsenic in soil and groundwater: an overview. *In:* Trace Metals and other Contaminants in the Environment. Vol Volume 9. Bhattacharya P, Mukherjee AB, Bundschuh J, Zevenhoven R, Loeppert RH (eds) Elsevier, p 3-60

O'Day P (2006) Chemistry and mineralogy of arsenic. Elements 2:77-83

O'Day PA, Vlassopoulos D, Root R, Rivera N (2004) The influence of sulfur and iron on dissolved arsenic concentrations in the shallow subsurface under changing redox conditions. Proc Natl Acad Sci 101:13703-13708

Ona-Nguema G, Morin G, Juillot F, Calas G, Brown GE (2005) EXAFS analysis of arsenite adsorption onto two-line ferrihydrite, hematite, goethite, and lepidocrocite. Environ Sci Technol 39:9147-9155

Ona-Nguema G, Morin G, Wang Y, et al. (2009) Arsenite sequestration at the surface of nano-Fe(OH)2, ferrous-carbonate hydroxide, and green-rust after bioreduction of arsenic-sorbed lepidocrocite by *Shewanella putrefaciens.* Geochim Cosmochim Acta 73:1359-1381

Ona-Nguema G, Morin G, Wang Y, Foster AL, Juillot F, Calas G, Brown GE Jr (2010) XANES evidence for rapid arsenic(III) oxidation at magnetite and ferrihydrite surfaces by dissolved O_2 via Fe^{2+}-mediated reactions. Environ Sci Technol 44:5416-5422

Osanna A, Jacobsen C (2000) Principal component analysis for soft X-ray spectromicroscopy. *In:* X-ray Microscopy. American Institute of Physics, Melville, p 350-355

Paikaray S (2012) Environmental hazards of arsenic associated with black shales: a review on geochemistry, enrichment and leaching mechanism. Rev Environ Sci Bio-Tech 11:289-303

Paikaray S, Hendry MJ, Essilfie-Dughan J (2013) Controls on arsenate, molybdate, and selenate uptake by hydrotalcite-like layered double hydroxides. Chem Geol 345:130-138

Paikaray S, Essilfie-Dughan J, Goettlicher J, Pollok K, Peiffer S (2014) Redox stability of As(III) on schwertmannite surfaces. J Hazard Mater 265:208-216

Paktunc D (2008a) Speciation of arsenic in an anaerobic treatment system at a Pb-Zn smelter site, gold roaster products, Cu smelter stack dust and impacted soil. Proc 9th Intl Congr Appl Mineral: p 343-348

Paktunc D (2008b) Speciation of arsenic in pyrite by micro-X-ray absorption fine-structure spectroscopy (XAFS). Proc 9th Intl Congr Appl Mineral, p 155-158

Paktunc D (2013) Mobilization of arsenic from mine tailings through reductive dissolution of goethite influenced by organic cover. Appl Geochem 36:49-56

Paktunc D, Bruggeman K (2010) Solubility of nanocrystalline scorodite and amorphous ferric arsenate: implications for stabilization of arsenic in mine wastes. Appl Geochem 25:674-683

Paktunc D, Dutrizac JE (2003) Characterization of arsenate-for-sulfate substitution in synthetic jarosite using x-ray diffraction and x-ray absorption spectroscopy. Can Mineral 41:905-919

Paktunc D, Manceau A (2013) Comment on "New Clues to the Local Atomic Structure of Short-Range Ordered Ferric Arsenate from Extended X-ray Absorption Fine Structure Spectroscopy". Environ Sci Technol 47:13199-13200

Paktunc D, Foster A, Laflamme G (2003) Speciation and characterization of arenic in Ketza River mine tailings using X-ray absorption spectroscopy. Env Sci Tech 37:2067-2074

Paktunc D, Foster A, Heald S, Laflamme G (2004) Speciation and characterization of arsenic in gold ores and cyanidation tailings using X-ray absorption spectroscopy. Geochim Cosmochim Acta 68:969-983

Paktunc D, Kingston D, Pratt A, McMullen J (2006) Distribution of gold in pyrite and in products of its transformation resulting from roasting of refractory gold ore. Can Mineral 44:213-227

Paktunc D, Dutrizac J, Gertsman V (2008) Synthesis and phase transformations involving scorodite, ferric arsenate and arsenical ferrihydrite: implications for arsenic mobility. Geochim Cosmochim Acta 72:2649-2672

Paktunc D, Majzlan J, Palatinus L, Dutrizac J, Klementova M, Poirier G (2013) Characterization of ferric arsenate-sulfate compounds: implications for arsenic control in refractory gold processing residues. Am Mineral 98:554-565

Parsons JG, Lopez ML, Castillo-Michel H, Peralta-Videa JR, Gardea-Torresdey JL (2009) Arsenic speciation in biological samples using XAS and mixed oxidation state calibration standards of inorganic arsenic. Appl Spectrosc 63:961-970

Parviainen A, Lindsay MBJ, Perez-Lopez R, Gibson BD, Ptacek CJ, Blowes DW, Loukola-Ruskeeniemi K (2012) Arsenic attenuation in tailings at a former Cu-W-As mine, SW Finland. Appl Geochem 27:2289-2299

Patel PC, Goulhen F, Boothman C, Gault AG, Charnock JM, Kalia K, Lloyd JR (2007) Arsenate detoxification in a *Pseudomonad* hypertolerant to arsenic. Arch Microbiol 187:171–183

Pearce DC, Dowling K, Gerson AR, Sim MR, Sutton SR, Newville M, Russell R, McOrist G (2010) Arsenic microdistribution and speciation in toenail clippings of children living in a historic gold mining area. Sci Total Environ 408:2590-2599

Pichler T, Veizer J, Hall G (1999) Natural input of arsenic into a coral-reef ecosystem by hyrothermal fluids and its removal by Fe(III) oxyhydroxides. Environ Sci Technol 33:1373-1378

Pickering IJ, Gumaelius L, Harris HH, Prince RC, Hirsch G, Banks JA, Salt DE, George GN (2006) Localizing the biochemical transformations of arsenate in a hyperaccumulating fern. Environ Sci Technol 40:5010-5014

Pickering IJ, Prince RC, George MJ, Smith RD, George GN, Salt DE (2000) Reduction and coordination of arsenic in Indian Mustard. Plant Physiol 122:1171-1177

Plumlee GS, Morman SA (2011) Mine wastes and human health. Elements 7:399-404

Plumlee GS, Ziegler TL (1999) The Medical Geochemistry of Dusts, Soils, and Other Earth Materials. *In:* Environmental Geochemistry. Treatise on Geochemistry, Vol 9. Sherwood Lollar B (ed), Elsevier, p 263-310

Polizzotto ML, Harvey CF, Sutton SR, Fendorf S (2005) Processes conducive to the release and transport of arsenic into aquifers of Bangladesh. Proc Natl Acad Sci USA 102:18819-18823

Polizzotto ML, Harvey CF, Li G, Badruzzman B, Ali A, Newville M, Sutton S, Fendorf S (2006) Solid-phases and desorption processes of arsenic within Bangladesh sediments. Chem Geol 228:97-111

Polizzotto ML, Kocar BD, Benner SG, Sampson M, Fendorf S (2008) Near-surface wetland sediments as a source of arsenic release to ground water in Asia. Nature 454:505-U505

Price RE, London J, Wallschlaeger D, Ruiz-Chancho MJ, Pichler T (2013a) Enhanced bioaccumulation and biotransformation of As in coral reef organisms surrounding a marine shallow-water hydrothermal vent system. Chem Geol 348:48-55

Price RE, Savov I, Planer-Friedrich B, Buehring SI, Amend J, Pichler T (2013b) Processes influencing extreme As enrichment in shallow-sea hydrothermal fluids of Milos Island, Greece. Chem Geol 348:15-26

Ravel B, Newville M (2005) Athena, Artemis, Hephaestus: data analysis for X-ray absorption spectroscopy using IFEFFFIT. J Synchrotron Radiat 12:537-541

Reeder RJ, Schoonen MAA, Lanzirotti A (2006) Metal speciation and its role in bioaccessibility and bioavailability. Rev Mineral Geochem 64:59-113

Rehr JJ, Kas JJ, Vila FD, Prange MP, Jorissen K (2010) Parameter-free calculations of x-ray spectra with FEFF9. Phys Chem Chem Phys 12:5503-5513

Ressler T (1998) WinXAS: A program for x-ray absorption spectroscopy data analysis under MS-Windows. J Synchrotron Rad 5:118-122

Ressler T, Wong J, Roos J, Smith IL (2000) Quantitative speciation of Mn-bearing particulates emitted from autos burning (methylcyclopentadienyl)manganese tricarbonyl-added gasolines using XANES Spectroscopy. Environ Sci Technol 34:950-958

Reynolds JG, Naylor DV, Fendorf SE (1999) Arsenic sorption in phosphate-amended soils during flooding and subsequent aeration. Soil Sci Soc Am J 63:1149-1156

Reza AHMS, Jean J-S, Yang H-J, Lee M-K, Woodall B, Liu C-C, Lee J-F, Luo S-D (2010) Occurrence of arsenic in core sediments and groundwater in the Chapai-Nawabganj District, northwestern Bangladesh. Water Res 44:2021-2037

Ritchie VJ, Ilgen AG, Mueller SH, Trainor TP, Goldfarb RJ (2013) Mobility and chemical fate of antimony and arsenic in historic mining environments of the Kantishna Hills district, Denali National Park and Preserve, Alaska. Chem Geol 335:172-188

Robins RG (1981) The solubility of metal arsenates. Metall Trans B 12:103-109

Root RA, Tixit S, Campbell KM, Jew AD, Hering JG, O'Day PA (2007) Arsenic sequestration by sorption processes in high-iron sediments. Geochim Cosmochim Acta 71:5782–5803

Root RA, Vlassopoulos D, Rivera NA, Rafferty MT, Andrews C, O'Day PA (2009) Speciation and natural attenuation of arsenic and iron in a tidally influenced shallow aquifer. Geochim Cosmochim Acta 73:5528-5553

Root RA, Fathordoobadi S, Alday F, Ela W, Chorover J (2013) Microscale speciation of arsenic and iron in ferric-based sorbents subjected to simulated landfill conditions. Environ Sci Technol 47:12992-13000

Rowland HAL, Gault AG, Charnock JM, Polya DA (2005) Preservation and XANES determination of the oxidation state of solid-phase arsenic in shallow sedimentary aquifers in Bengal and Cambodia. Mineral Mag 69:825-839

Ruiz-Chancho MJ, Pichler T, Price RE (2013) Arsenic occurrence and speciation in *Cyclope neritea*, a gastropod inhabiting the arsenic-rich marine shallow-water hydrothermal system off Milos Island, Greece. Chem Geol 348:56-64

Sadiq M (1997) Arsenic chemistry in soils: an overview of thermodynamic predictions and field observations. Water Air Soil Pollut 93:117-136

Sahai N, Lee YJ, Xu HF, Ciardelli M, Gaillard JF (2007) Role of Fe(II) and phosphate in arsenic uptake by coprecipitation. Geochim Cosmochim Acta 71:3193-3210

Salt DE, Prince RC, Pickering IJ (2002) Chemical speciation of accumulated metals in plants: evidence from X-ray absorption spectroscopy. Microchem J 71:255-259

Sarret G, Smits EAHP, Michel HC, Isaure MP, Zhao FJ, Tappero R (2013) Use of synchrotron-based techniques to elucidate metal uptake and metabolism in plants. Adv Agron 119:1-82

Savage KS, Bird DK, Ashley RP (2000a) Legacy of the California gold rush: environmental geochemistry of arsenic in the southern Mother Lode gold district. Int Geol Rev 42:385-415

Savage KS, Tingle TN, O'Day PA, Waychunas GA, Bird DK (2000b) Arsenic speciation in pyrite and secondary weathering phases, Mother Lode Gold District, Tuolumne County, California. Appl Geochem 15:1219-1244

Savage KS, Bird DK, O'Day PA (2005) Arsenic speciation in synthetic jarosite. Chem Geol 215:473-498

Savage KS, Ashley RP, Bird DK (2009) Geochemical evolution of a high arsenic, alkaline pit-lake in the Mother Lode Gold District, California. Econ Geol 104:1171-1211

Sayers DE, Bunker BA (1988) Data Analysis. *In:* X-ray Absorption: Principles, Applications, Techniques of EXAFS, SEXAFS, and XANES. Koningsberger DC, Prins R (eds) John Wiley and Sons, New York, p 211-253

Sayers DE, Stern EA, Lytle FW (1971) New technique for investigating noncrystalline structures: Fourier analysis of the extended x-ray-absorption fine structure. Phys Rev Lett 27:1204-1207

Schwer DR III, McNear DH (2011) Chromated copper arsenate-treated fence posts in the agronomic landscape: soil properties controlling arsenic speciation and spatial distribution. J Environ Qual 40:1172-1181

Scott MJ, Morgan JJ (1995) Reactions at oxide surfaces. 1. Oxidation of As(III) by synthetic birnessite. Environ Sci Technol 29:1898-1905

Seddique AA, Masuda H, Mitamura M, Shinoda K, Yamanaka T, Nakaya S, Ahmed KM (2011) Mineralogy and geochemistry of shallow sediments of Sonargaon, Bangladesh and implications for arsenic dynamics: focusing on the role of organic matter. Appl Geochem 26:587-599

Seyfferth AL, Webb SM, Andrews JC, Fendorf S (2010) Arsenic localization, speciation, and co-occurrence with iron on rice (*Oryza sativa* L.) roots having variable Fe coatings. Environ Sci Technol 44:8108-8113

Seyfferth AL, Webb SM, Andrews JC, Fendorf S (2011) Defining the distribution of arsenic species and plant nutrients in rice (*Oryza sativa* L.) from the root to the grain. Geochim Cosmochim Acta 75:6655-6671

Sherman DM, Randall SR (2003) Surface complexation of arsenic(V) to iron(III) (hydr)oxides: Structural mechanism from *ab initio* molecular geometries and EXAFS spectroscopy. Geochim Cosmochim Acta 67:4223-4230

Shoji T, Huggins FE, Huffman GP, Linak WP, Miller CA (2002) XAFS spectroscopy analysis of selected elements in fine particulate matter derived from coal combustion. Energy Fuels 16:325-329

Singer DM, Fox PM, Guo H, Marcus MA, Davis JA (2013) Sorption and redox reactions of As(III) and As(V) within secondary mineral coatings on aquifer sediment grains. Environ Sci Technol 47:11569-11576

Slejkovec Z, Bajc Z, Doganoc DZ (2004) Arsenic speciation patterns in freshwater fish. Talanta 62:931-936

Slowey AJ, Johnson SB, Newville M, Brown GE Jr (2007) Speciation and colloid transport of arsenic from mine tailings. Appl Geochem 22:1884-1898

Smedley PL (2003) Arsenic in groundwater-south and east Asia. *In:* Arsenic in groundwater: Geochemistry and Occurrence. Welch AH, Stollenwerk KG (eds) Kulwer Academic, Norwell, MA, p 179-209

Smedley PL, Kinniburgh DG, Huq I, Zhen-dong L, Nicolli HB (2001) International perspective on naturally occurring arsenic problems in groundwater. *In:* Arsenic Exposure and Health Effects IV. Chappell WR, Abernathy CO, Calderon RL (eds) Elsevier, New York, p 9-26

Smith PG, Koch I, Gordon RA, Mandoli DF, Chapman BD, Reimer KJ (2005) X-ray absorption near-edge structure analysis of arsenic species for application to biological environmental samples. Environ Sci Technol 39:248-254

Smith E, Smith J, Naidu R (2006) Distribution and nature of arsenic along former railway corridors of South Australia. Sci Total Environ 363:175-182

Smith PG, Koch I, Reimer KJ (2007) Arsenic speciation analysis of cultivated white button mushrooms (*Agaricus bisporus*) using high-performance liquid chromatography - Inductively coupled plasma mass spectrometry, and X-ray absorption Spectroscopy. Environ Sci Technol 41:6947-6954

Smith PG, Koch I, Reimer KJ (2008a) Uptake, transport and transformation of arsenate in radishes (*Raphanus sativus*). Sci Total Environ 390:188-197

Smith PG, Koch I, Reimer KJ (2008b) An investigation of arsenic compounds in fur and feathers using X-ray absorption spectroscopy speciation and imaging. Sci Total Environ 390:198-204

Smith E, Kempson I, Juhasz AL, Weber J, Skinner WM, Graefe M (2009) Localization and speciation of arsenic and trace elements in rice tissues. Chemosphere 76:529-535

Sposito G (1984) The Surface Chemistry of Soils. Oxford University Press, New York

Stollenwerk KG, Breit GN, Welch AH, Yount JC, Whitney JW, Foster AL, Uddin MN, Majumder RK, Ahmed N (2007) Arsenic attenuation by oxidized aquifer sediments in Bangladesh. Sci Total Environ 379:133-150

Strawn D, Doner H, Zavarin M, McHugo S (2002) Microscale investigation into the geochemistry of arsenic, selenium, and iron in soil developed in pyritic shale materials. Geoderma 108:237-257

Su SM, Zeng XB, Li LF, Duan R, Bai LY, Li AG, Wang J, Jiang S (2012) Arsenate reduction and methylation in the cells of *Trichoderma asperellum* SM-12F1, *Penicillium janthinellum* SM-12F4, and *Fusarium oxysporum* CZ-8F1 investigated with X-ray absorption near edge structure. J Hazard Mater 243:364-367

Suess E, Wallschläger D, Planer-Friedrich B (2011) Stabilization of thioarsenates in iron-rich waters. Chemosphere 83 1524–1531

Sun X, Doner H (1998) Adsorption and oxidation of arsenite on goethite. Soil Science 163:278-287

Sun X, Doner HE, Zavarin M (1999) Spectroscopy study of arsenite [As(III)] oxidation on Mn-substituted goethite. Clays Clay Miner 47:474-480

Sutton SR, Bertsch PM, Newville M, Rivers M, Lanzirotti A, Eng P (2002) Microfluorescence and microtomography analyses of heterogeneous earth and environmental materials. Rev Mineral Geochem 49:429-483

Takahashi Y, Ohtaku N, Mitsunobu S, Yuita K, Nomura M (2003) Determination of the As(III)/As(V) ratio in soil by X-ray absorption near-edge structure (XANES) and its application to the arsenic distribution between soil and water. Anal Sci 19:891-896

Taleshi MS, Jensen KB, Raber G, Edmonds JS, Gunnlaugsdottir H, Francesconi KA (2008) Arsenic-containing hydrocarbons: natural compounds in oil from the fish capelin, Mallotus villosus. Chem Commun 2008(39):4706-4707

Tanaka M, Togo YS, Yamaguchi N, Takahashi Y (2014) An EXAFS study on the adsorption structure of phenyl-substituted organoarsenic compounds on ferrihydrite. J Colloid Interface Sci 415:13-17

Tazaki K, Rafiqul IA, Nagai K, Kurihara T (2003) FeAs$_2$ biomineralization on encrusted bacteria in hot springs: An ecological role of symbiotic bacteria. Can J Earth Sci/Rev Can Sci Terre 40:1725-1738

Teo BK (1986) EXAFS: Basic Principles and Data Analysis. Springer-Verlag, Berlin

Uryu T, Yoshinaga J, Yanagisawa Y (2003) Environmental Fate of Gallium Arsenide Semiconductor Disposal. J Ind Ecol 7:103-112

USEPA (2002) The Arsenic Rule: Water Treatment Plant Residuals. US Environmental Protection Agency. *http://www.epa.gov/safewater/arsenic/pdfs/arsenic_training_2002/train2006-residuals.pdf*   Accessed July 2014

USEPA (2006) Reregistration Eligibility Decision for MSMA, DSMA, CAMA, and Cacodylic Acid. US Environmental Protection Agency Report 738-R-06-02

USEPA (2013a) Coal Combustion Residuals. US Environmental Protection Agency. *http://www.epa.gov/osw/nonhaz/industrial/special/fossil/coalashletter.htm* (accessed July 2014)

USEPA (2013b) Frequent Questions: Coal Combustion Residues (CCR) -- Proposed Rule. US Environmental Protection Agency. *http://www.epa.gov/wastes/nonhaz/industrial/special/fossil/ccr-rule/ccrfaq.htm* (accessed July 2014)

USEPA (2013c) Organic Arsenicals. US Environmental Protection Agency. *http://www.epa.gov/oppsrrd1/reregistration/organic_arsenicals_fs.html* (accessed July 2014)

USFDA (2013) FDA response to citizen petition on arsenic-based animal drugs. US Food and Drug Administration   *http://www.fda.gov/AnimalVeterinary/SafetyHealth/ProductSafetyInformation/ucm370568.htm* (accessed July 2014)

Valentim B, Guedes A, Flores D, Ward CR, Hower JC (2009) Variations in fly ash composition with sampling location: case study from a Portuguese power plant. Coal Comb Gas Prod 1:14-24

Veselska V, Majzlan J, Hiller E, Petkova K, Jurkovic L, Durza O, Volekova-Lalinska B (2013) Geochemical characterization of arsenic-rich coal-combustion ashes buried under agricultural soils and the release of arsenic. Appl Geochem 33:153-164

Voegelin A, Weber F-A, Kretzschmar R (2007) Distribution and speciation of arsenic around roots in a contaminated riparian floodplain soil: micro-XRF element mapping and EXAFS spectroscopy. Geochim Cosmochim Acta 71:5804-5820

Voigt DE, Brantley SL, Hennet RJC (1996) Chemical fixation of arsenic in contaminated soils. Appl Geochem 11:633-643

Walker SR, Jamieson HE, Lanzirotti A, Andrade CF, Hall GEM (2005) The speciation of arsenic in iron oxides in mine wastes from the Giant Gold Mine, N.W.T: application of synchrotron micro-XRD and micro-XANES at the grain scale. Can Mineral 43:1205-1224

Walker SR, Parsons MB, Jamieson HE, Lanzirotti A (2009) Arsenic mineralogy of near-surface tailings and soils: influences on arsenic mobility and bioaccessibility in the Nova Scotia Gold Mining Districts. Can Mineral 47:533-556

Walker SR, Jamieson HE, Rasmussen PE (2011) Application of synchrotron microprobe methods to solid-phase speciation of metals and metalloids in house dust. Env Sci Technol 45:8233-8240

Wang HC, Wang PH, Peng CY, Liu SH, Wang YW (2001) Speciation of As in the blackfoot disease endemic area. J Synch Rad 8:961-962

Wang Y, Morin G, Ona-Nguema G, Menguy N, Juillot F, Aubry E, Guyot F, Calas G, Brown GE Jr (2008) Arsenite sorption and the magneite-water interface during aqueous precipitation of magnetite: EXAFS evidence for a new arsenite surface complex. Geochim Cosmochim Acta 72:2573-2586

Wang Y, Morin G, Ona-Nguema G, Juillot F, Guyot F, Calas G, Brown GE Jr (2010) Evidence for different surface speciation of arsenite and arsenate on green rust: an EXAFS and XANES study. Environ Sci Technol 44:109-115

Wang Y, Morin G, Ona-Nguema G, Juillot F, Calas G, Brown GE Jr (2011) Distinctive arsenic(V) trapping modes by magnetite manoparticles induced by different sorption processes. Environ Sci Technol 45:7258-7266

Wang Y, Jiao JJ, Zhu S, Li Y (2013) Arsenic K-edge X-ray absorption near-edge spectroscopy to determine oxidation states of arsenic of a coastal aquifer-aquitard system. Environ Pollut 179:160-166

Waychunas GA, Rea BA, Fuller CC, Davis JA (1993) Surface chemistry of ferrihydrite: part1. EXAFS studies of the geometry of coprecipitated and adsorbed arsenate. Geochim Cosmochim Acta 57:2251-2269

Waychunas GA, Davis JA, Fuller CC (1995) Geometry of sorbed arsenate on ferrihydrite and crystalline FeOOH: re-evaluation of EXAFS results and topological factors in predicting sorbate geometry, and evidence for monodentate complexes. Geochim Cosmochim Acta 59:3655-3661

Waychunas GA (1996) Wide angle X-ray scattering (WAXS) study of "two-line" ferrihydrite structure; effect of arsenate sorption and counterion variation and comparison with EXAFS results. Geochim Cosmochim Acta 60:1765-1781

Webb SM (2005) SixPack Phys Scr T115:1011-1014 (current version: 1.01) *http://home.comcast.net/~sam_webb/sixpack.html* (accessed July 2014)

Webb SM (2006) Sam's Microprobe Analysis Kit (SMAK; current version: 0.52). *http://home.comcast.net/~sam_webb/smak.html* (accessed July 2014)

Webb SM, Galliard J-F, Ma LQ, Tu C (2003) XAS speciation of arsenic in a hyper-accumulating fern. Environ Sci Technol 37:754-760

Welch AH, Lico MS, Hughes JL (1988) Arsenic in ground water of the western United States. Ground Water 26:333-347

Whaley-Martin KJ, Koch I, Moriarty M, Reimer KJ (2012) Arsenic speciation in blue mussels (*Mytilus edulis*) along a highly contaminated arsenic gradient. Environ Sci Technol 46:3110-3118

Whaley-Martin KJ, Koch I, Reimer KJ (2013) Determination of arsenic species in edible periwinkles (Littorina littorea) by HPLC-ICPMS and XAS along a contamination gradient. Sci Total Environ 456:148-153

Wikimedia Commons (2014) h*ttp://en.wikipedia.org/wiki/Wikimedia_Commons* (accessed July 2014)

Wilkie JA, Hering JG (1998) Rapid oxidation of geothermal arsenic(III) in streamwaters of the eastern Sierra Nevada. Environ Sci Technol 32:657-662

Wilkin RT, Ford RG (2006) Arsenic solid-phase partitioning in reducing sediments of a contaminated wetland. Chem Geol 228:156-174

Williams GP (2001) Section 1.1 Electron Binding Energies. *In:* X-ray Data Booklet. Thompson AC, Vaughan D (eds) Lawrence Berkeley National Laboratory, Berkeley, p 161

Winkel LHE, Casentini B, Bardelli F, Voegelin A, Nikolaidis NP, Charlet L (2013) Speciation of arsenic in Greek travertines: co-precipitation of arsenate with calcite. Geochim Cosmochim Acta 106:99-110

Wogelius RA (2013) Adsorption and co-precipitation reactions at the mineral-fluid interface: natural and anthropogenic processes. Cryst Res Technol 48:877-902

Wolf RE, Morman SA, Hageman PL, Hoefen TM, Plumlee GS (2011) Simultaneous speciation of arsenic, selenium, and chromium: species stability, sample preservation, and analysis of ash and soil leachates. Anal Bioanal Chem 401:2733-2745

Wolfe-Simon F, Blum JS, Kulp TR, Gordon GW, Hoeft SE, Pett-Ridge J, Stolz JF, Webb SM, Weber PK, Davies PWC, Anbar AD, Oremland RS (2010) A bacterium that can grow by using arsenic instead of phosphorus. Science 332:1163-1166

Wolthers M, Charlet L, Van der Weijden CH, Van der Linde PR, Rickard D (2005) Arsenic mobility in the ambient sulfidic environment: sorption of arsenic(V) and arsenic(III) onto disordered mackinawite. Geochim Cosmochim Acta 69:3483-3492

Wu X, Tan X, Yang S, Wen T, Guo H, Wang X, Xu A (2013) Coexistence of adsorption and coagulation processes of both arsenate and NOM from contaminated groundwater by nanocrystallined Mg/Al layered double hydroxides. Water Res 47:4159-4168

Xu L, Zhao Z, Wang S, Pan R, Jia Y (2011) Transformation of arsenic in offshore sediment under the impact of anaerobic microbial activities. Water Res 45:6781-6788

Yamauchi H, Fowler BA (1994) Toxicity and metabolism of inorganic and methylated arsenicals. *In:* Arsenic in the Environment. Vol 1. Nriagu JO (ed) John Wiley and Sons, New York, p. 35-53

Yan W, Vasic R, Frenkel AI, Koel BE (2012) Intraparticle reduction of arsenite (As(III)) by nanoscale zerovalent iron (nZVI) investigated with *in situ* X-ray absorption spectroscopy. Environ Sci Technol 46:7018-7026

Yang L, Donahoe RJ (2007) The form, distribution and mobility of arsenic in soils contaminated by arsenic trioxide, at sites in southeast USA. Appl Geochem 22:320-341

Ying SC, Kocar BD, Fendorf S (2012) Oxidation and competitive retention of arsenic between iron- and manganese oxides. Geochim Cosmochim Acta 96:294-303

Ying SC, Kocar BD, Griffis SD, Fendorf S (2011) Competitive microbially and Mn oxide mediated redox processes controlling arsenic speciation and partitioning. Environ Sci Technol 45:5572-5579

Ying SC, Masue-Slowey Y, Kocar BD, Griffis SD, Webb S, Marcus MA, Francis CA, Fendorf S (2013) Distributed microbially- and chemically-mediated redox processes controlling arsenic dynamics within Mn-/Fe-oxide constructed aggregates. Geochim Cosmochim Acta 104:29-41

Yu C, Lavergren U, Peltola P, Drake H, Bergback B, Astrom ME (2014) Retention and transport of arsenic, uranium and nickel in a black shale setting revealed by a long-term humidity cell test and sequential chemical extractions. Chem Geol 363:134-144

Zevenhoven R, Mukherjee AB, Bhattacharya P (2007) Arsenic flows in the environment of the European Union: a synoptic review. *In:* Trace Metals and other Contaminants in the Environment. Vol 9. Bhattacharya P, Mukherjee AB, Bundschuh J, Zevenhoven R, Loeppert RH (eds) Elsevier, New York, p 527-547

Zhao FH, Ren DY, Zheng BS, Hu TD, Liu T (1998) Modes of occurrence of arsenic in high-arsenic coal by extended X-ray absorption fine structure spectroscopy. Chin Sci Bull 43:1660-1663

Zhao K, Guo H (2014) Behavior and mechanism of arsenate adsorption on activated natural siderite: evidences from FTIR and XANES analysis. Environ Sci Pollution Res 21:1944-1953

Zhao Z, Jia Y, Xu L, Zhao S (2011) Adsorption and heterogeneous oxidation of As(III) on ferrihydrite. Water Res 45:6496-6504

Zielinski RA, Foster AL, Meeker GP, Brownfield IK (2007) Mode of occurrence of arsenic in feed coal and its derivative fly ash, Black Warrior Basin, Alabama. Fuel 86:560-572

APPENDIX 1

XAS studies of As sorption in model systems. Standard reporting error for LC-LSF: ±10%. Standard reporting detection limit: <5-10%. Typical reported error for EXAFS-derived interatomic distances (R) is ±0.02 Å, and for coordination number (N) is ±20%. IS = inner-sphere; OS = outer sphere; MS = multiple scattering. See Figure 12 for explanation of sorption complex notation.

Adsorbate/Sorbent	Collection/ Analysis	Arsenic Solid-Phase Species
γ-Al$_2$O$_3$: As^{3+} and As^{5+} (Arai et al. 2001)	B-XAS RT, LT C L (XANES)	As^{3+}: ^{2}C, R_{As-Al} = 3.21 Å As^{5+}: ^{2}C , R_{As-Al} = 3.11 Å Used XANES LC-LSF to show 66% IS and 34% OS
γ-Al(OH)$_3$ (gibbsite): As^{5+} 1. (Ladeira et al. 2001) 2. (Foster 1999, 2003) 3. (Grafe et al. 2008b)	B-XAS LT and RT C	1. ^{2}C ,R_{As-Al} = 3.19 Å; ^{2}E complex not energetically favorable 2. ^{2}C, R_{As-Al} = 3.19-3.21 Å; As^{5+} coppt with am-Al(OH)$_3$: ^{2}C, R_{As-Al} : 3.22 Å 3. ^{2}C R_{As-Al} = 3.17 Å, N_{Al} = 1.6 > ^{2}E, R_{As-Al} = 2.72 Å, N_{Al} = 0.4;
γ-Al(OH)$_3$ (gibbsite): As^{3+} 1. (Duarte et al. 2012) 2. (Ladeira et al. 2001; Foster 1999, Foster 2003) 3. (Grafe et al. 2008b)	B-XAS LT and RT C	1. As^{3+}: ^{2}C, R_{As-Al} = 3.21 Å > ^{1}V R_{As-Al} =3.49 Å 2. As^{3+}: ^{2}C, R_{As-Al} = 3.19-3.21 Å; As^{3+} coppt with am-Al(OH)$_3$: not ^{2}C , R_{As-Al}: 3.50 Å , N = 3.4 3. ^{2}C R_{As-Al} = 3.17 Å, N_{Al} = 1.6 > ^{2}E, R_{As-Al} = 2.72 Å, N_{Al} = 0.4
γ-Al(OH)$_3$ (gibbsite; As^{5+}, Cu^{2+}$_{(aq)}$, and Zn^{2+}$_{(aq)}$); also homogeneous coprecipitation of As^{5+}, Cu^{2+}, and Zn^{2+}) (Grafe et al. 2008b)	B, XAS RT C	+ Cu: ^{2}C R_{As-Cu} = 3.23 Å, N_{Cu} = 1.6 > ^{2}E, R_{As-Cu} = 2.82 Å, N_{Cu} = 0.3 + Zn only = ^{2}C $R_{As-Al\ or\ Zn}$ = 3.13 Å, N_{Zn} = 1.8 > ^{2}E, $R_{As-Al\ or\ Zn}$ = 2.75 Å, $N_{Al\ or\ Zn}$ = 0.8 + Cu and Zn: ^{2}E, R_{As-Cu} = 2.82 Å, N_{Cu} = 0.3; ^{2}C (dominant), $R_{As-Cu\ or\ Zn}$ = 3.25 Å, N = 1.6. As^{5+} co-precipitates: As^{5+}+ Zn $\cong$ koettigite; As^{5+}+ Cu $\cong$ clinoclase [Cu$_3$AsO$_4$(OH)$_3$]
Kaolinite KGal-b [Al$_2$Si2O$_5$(OH)$_4$], **Smectite SWy-2:** Na$_{0.33}$(Al,Mg)$_2$(Si$_4$O$_{10}$) (OH)$_2$·nH$_2$O **Nontronite NAu-1:** M$^+$[Si,Al]$_8$[Al,Fe,Mg]$_4$O$_{20}$(OH)$_4$. 1. (Arai 2010) 2. (Foster 1999, 2003) 3. (Ilgen et al. 2012)	B-XAS LT, RT F, L, C	1. As^{5+}- kaolinite + Ca$_{(aq)}$: OS+ IS [^{2}C, R_{As-Al} =3.11 Å] no ternary species 2. As^{3+}- and As^{5+}- KGal-b: OS+ IS [^{2}C, R_{As-Al}=3.22 Å and 3.19 Å, respectively] Photocatalytic oxidation of As^{3+} by TiO$_2$ in KGal-b. As^{3+}, As^{5+}-smectite: OS adsorption 3. (XANES only) Structural Fe in NAu-1 does not reduce As^{5+}, but Fe(II) added to KGa-1b or NAu-1 does
Biotite K(Mg,Fe)$_3$AlSi$_3$O$_{10}$(F,OH)$_2$ (Chakraborty et al. 2011)	B-XAS LT, no O$_2$ C	As^{3+}: ^{2}C, R_{As-Fe} = 3.37 Å > ^{2}E, R_{As-Fe} = 3.00 Å
Ferrihydrite (FH) Fe$_2$O$_3$·0.5H$_2$O		
FH: As^{5+} adsorbed + coprecipitated 1. (Waychunas et al. 1993, 1996) 2. (Manceau 1995) 3. (Sherman and Randall 2003) 4. (Gao et al. 2013) 5. (Hohmann et al. 2011)	B-XAS RT/LT F, L, C	1. ^{2}C, R_{As-Fe} = 3.28 Å (adsorbed); 3.25 Å (coprecip) > ^{1}V R_{As-Fe} = 3.60 Å (both sorption modes) 2. Adsorbed; ^{2}C, R_{As-Fe} = 3.26 Å > ^{2}E R_{As-Fe} = 2.83 Å 3. Adsorbed; ^{2}C, (2 Fe atoms fit separately): R_{As-Fe} = 3.16 Å + 3.30 Å (adsorbed) and 3.27 Å + 3.38 Å(coprecipitate). 4. Adsorbed; (-) silicic acid (SA): As^{5+}: ^{2}C, R_{As-Fe} = 3.27 Å; As^{3+}: ^{2}C, R_{As-Fe} = 3.43 Å > ^{2}E, R_{As-Fe} = 2.93 Å; (+) SA: same complex 5. Adsorbed = coprecip: ^{2}C, R_{As-Fe} = 3.30 Å > R_{MS} = 3.08 Å

Adsorbate/Sorbent	Collection/ Analysis	Arsenic Solid-Phase Species
FH: As^{3+} adsorbed + coprecipitated 1. (Ona-Nguema et al. 2005, 2010) 2. (Zhao et al. 2011) 3. (Gao et al. 2013) 4. (Hohmann et al. 2011)	B-XAS No O$_2$, LT F, C	1. Adsorbed ^{2}C, R_{As-Fe} = 3.35 Å $\cong$ ^{1}E, R_{As-Fe} = 2.90 Å, same at low and high coverage; not oxidized by ferrihydrite unless Fe^{2+} added 2. Adsorbed partial oxidation in presence of light; ^{2}C complex predominant 3. Adsorbed ^{2}C, R_{As-Fe} = 3.43 Å > ^{2}E, R_{As-Fe} = 2.93 Å; (+) SA: same complex 4. adsorbed = coprecip. ^{2}C, R_{As-Fe} = 3.42 Å $\cong$ ^{2}E, R_{As-Fe} = 2.94 Å, R_{MS} = 3.20 Å (polymers may increase relative abundance of ^{2}E over inorganic systems)
FH: Phenyl-substituted organoarsenic (V) compounds (Tanaka et al. 2014)	B-XAS RT? F, C	Mono-phenyl-substituted (PAA) + Di-phenyl-substituted (DPAA): identical complexes, ^{2}C, R_{As-Fe} = 3.26 Å > ^{1}V R_{As-Fe} = 3.45 Å, pH 4
FH: Mono- and Tetra-thioarsenate (MTAs^{5+} and TTAs^{5+}) (Couture et al. 2013)	B-XAS LT+ move sample F, L, C	~ 20% reduction to inorganic As^{3+} (both species; by XANES w/ LC-LSF). TTAs^{5+} decomposed to MTAs^{5+} (monitored by N$_s$ and N$_O$ in 1st coordination shell) Adsorbed MTAs^{5+}: ^{1}V, R_{As-Fe} = 3.30-3.35 Å, N$_{Fe}$ 0.7-0.9
Goethite (GH) α-FeOOH		
GH: As^{5+} 1. (Waychunas et al. 1993) 2. (Fendorf et al. 1997) 3. (Farquhar et al. 2002) 4. (Sherman and Randall 2003) 5. (Grafe et al. 2008a)	B-XAS RT, LT? F, C	1. ^{2}C, R_{As-Fe} = 3.30 Å, N = 2.1-2.4 2. ^{2}C, R_{As-Fe} = 3.24 Å, N = 0.9-1.6 > ^{2}E, R_{As-Fe} = 2.85 Å, N = 0.6-1.8 $\cong$ ^{1}V R_{As-Fe} = 3.60 Å, N = 0.4-1.0 3. ^{2}C, R_{As-Fe} = 3.30 Å, N = 1.0; > ^{2}E, R_{As-Fe} = 2.85 Å, N = 0.5 4. ^{2}C, R_{As-Fe} = 3.30 Å, N = 2.0 (2 Fe atoms fit separately) 5. ^{2}C R_{As-Fe} = 3.28 Å, N = 1.2-1.7 > ^{2}E, R_{As-Fe} = 2.87 Å, N = 0.2-0.3
GH: As^{3+} 1. (Manning et al. 1998) 2. (Farquhar et al. 2002) 3. (Ona-Nguema et al. 2005) 4. (Bhandari et al. 2012)	B 1,3	1. ^{2}C, R_{As-Fe} = 3.38 Å 2. ^{2}C, R_{As-Fe} = 3.31 Å, N = 2.0 3. ^{2}C, R_{As-Fe} = 3.3-3.4 Å > 1V, R_{As-Fe} = 3.5-3.6 Å 4. photocatalytic oxidation of As^{3+} catalyzed by goethite increases sorption (as As^{5+}, ^{2}C)
GH: adsorbed As^{5+} and Zn^{2+} 1. (Grafe and Sparks 2005) **GH: adsorbed As^{5+} Cu^{2+}, and/ or Zn^{2+}** 2. (Grafe et al. 2008a)	B 3	1.) 100 and 10 ppm As$_{(aq)}$: surface precipitates with structure similar to known Zn hydroxyarsenates 2. (+ Cu) =(+Zn) = (Cu +Zn): ^{2}E, R_{As-Fe} = 2.84-2.87 Å, N = 0.3-0.4; ^{2}C (dominant), R_{As-Fe} = 3.28 Å, N = 1.6-2.1; no complexation with Cu^{2+} or Zn^{2+}
GH: adsorbed Mono- and Tetra- thioarsenate (MTAs^{5+} and TTAs^{5+}) (Couture et al. 2013)	B-XAS RT? F, L C	About 20% reduction to inorganic As^{3+} (both species; XANES w/ LC-LSF); TTAs^{5+} decomposed to MTAs^{5+} (monitored by N$_s$ and N$_O$ in 1st coordination shell) Adsorbed complex of MTAs^{5+}: ^{1}V, N_{Fe} 1.2-1.7 R_{As-Fe} = 3.32 Å
Lepidocrocite (LP) γ-FeOOH		
LP: As^{5+} 1. (Waychunas et al. 1993) 2. (Farquhar et al. 2002) 3. (Sherman and Randall 2003) 4. (Ona-Nguema et al. 2005)	B 1,3	1. ^{2}C, R_{As-Fe} = 3.29 Å, N = 2.3 2. ^{2}C, R_{As-Fe} = 3.31 Å, N = 2.0; 3. (pH 4) As^{5+} : ^{2}C, R_{As-Fe} = 3.29 Å, N = 2.3 (2 Fe atoms fit separately) 4. No O$_2$; ^{2}C, R_{As-Fe} = 3.3-3.4 Å > ^{1}V, R_{As-Fe} = 3.5-3.6 Å; no ^{2}E at high surface coverages

Adsorbate/Sorbent	Collection/ Analysis	Arsenic Solid-Phase Species
LP: As^{3+} 1. (Farquhar et al. 2002) 2. (Ona-Nguema et al. 2005)	B-XAS LT, no O2 F, C	1. As^{3+}: R_{As-Fe} = 2.97 Å, N = 0.5, R_{As-Fe} = 3.41 Å, N = 1.0 2. ^{2}C, R_{As-Fe} = 3.3-3.4 Å > ^{1}V, R_{As-Fe} = 3.5-3.6 Å; no ^{2}E at high surface coverage
Schwertmannite **[(Fe$_{16}$O$_{16}$(OH)$_{12}$(SO$_4$)$_2$]: As^{3+},** **As^{5+}** 1. (Maillot et al. 2013) 2. (Paikaray et al. 2014)	B-XAS C RT	1. ^{2}C and ^{2}E complexes invariant with surface loading. As^{3+}: R_{MS} = 3.17-3.28 Å, ^{2}C: R_{As-Fe} = 3.41 Å, N = 1.8-2.7 > ^{2}E R_{As-Fe} = 2.91-2.97 Å, N = 0.3-0.8; As^{5+}: R_{MS1} = 3.12-3.16 Å, R_{MS2} ~3.20 Å, ^{2}C, R_{As-Fe} = 3.27-3.30 Å, N = 1.8-2.7, R_{MS3} ~4.4 Å 2. ^{2}C for As^{5+} and As^{3+}; partial oxidation of As^{3+}

Hematite α-Fe$_2$O$_3$

Adsorbate/Sorbent	Collection/ Analysis	Arsenic Solid-Phase Species
Hematite: adsorbed As^{5+} 1. (Sherman and Randall 2003) 2. (Bolanz et al. 2013)	B-XAS F, C RT ?	1. ^{2}C, R_{As-Fe} = 3.24 Å, 3.35 Å (2 Fe fit separately) 2. No sorption observed; epitaxially-controlled precipitation of angelellite [Fe$^{3+}_4$(AsO$_4$)$_2$O$_3$]-like phase
Hematite: adsorbed As^{3+} (Ona-Nguema et al. 2005)	B-XAS F, C LT, no O$_2$	^{2}C, R_{As-Fe} = 3.36 Å, N = 0.7 ≅ ^{1}E R_{As-Fe} = 2.89 Å, N = 0.4

Mixed-Valent Fe Oxides, hydroxides and green rust phases

Adsorbate/Sorbent	Collection/ Analysis	Arsenic Solid-Phase Species
Maghemite (γ-Fe$_2$O$_3$), As^{3+} and **As^{5+} vacancy–ordered, synthetic** **"Q" type vs. "C" (cubic)–type** **produced by oxidation of** **biogenic magnetite** (Morin et al. 2008)	B-XAS F, C LT, no O$_2$	Sorption on "C" = "Q" types; type and relative abundance of complex not dependent on surface coverage As^{5+}: ^{2}C, R_{As-Fe} = 3.34-3.35 Å; N = 1.1-1.5; R_{MS} = 3.09 Å As^{3+}: ^{2}E, R_{As-Fe} = 2.90 Å, N = 0.3-0.4 and ^{2}C + ^{1}V (mixed) R_{As-Fe} = 3.45-3.47 Å, N = 0.7-0.8; R_{MS} = 3.21 Å
Magnetite Fe$_3$O$_4$: adsorbed As^{3+} 1. (Ona-Nguema et al. 2010) **As^{3+} coprecipitated with** **magnetite** 2. (Wang et al. 2008)	B-XAS F, C LT, ± O$_2$	(1) Oxidation rapid at neutral pH + O$_2$ (2) No O$_2$; unique ^{3}C complex in vacant tetrahedral site R_{As-Fe} = 3.53 Å; and R_{As-Fe} = 3.30 Å; at 2 highest loadings, an amorphous As^{3+}-Fe^{2+}/Fe^{3+} precipitate with local geometry = ^{2}E, R_{As-Fe} = 3.53 Å
Magnetite Fe$_3$O$_4$: adsorbed and **coprecipitated As^{5+}** (Wang et al. 2008, 2011)		Adsorbed: ^{2}C complex and OS complex at higher sorption densities As in clusters with a magnetite-like local structure
Green Rust (Cl, OH, SO$_4$ **forms): adsorbed As^{3+} and As^{5+}** (Wang et al. 2010) [Fe$^{2+}_{(1-x)}$ Fe$^{3+}_x$(OH)$_2$]$^{x+}$(CO$_3$,Cl,SO$_4$)$^{x-}$	B-XAS F, C LT, no O$_2$	As^{5+}: R_{MS} = 3.07-3.09 Å, ^{2}C, R_{As-Fe} = 3.32-3.35 Å; N = 1.2-1.6 > ^{1}V R_{As-Fe} = 3.48-3.49 Å, N = 1.0-1.3 As^{3+}: R_{MS} = 3.21-3.22 Å, low coverage: ^{2}C, R_{As-Fe} = 3.31 Å; N = 1.2-1.6+ dimer sorbed R_{As-As} = 3.32 Å with ^{1}V R_{As-Fe} = 3.50-3.51 Å, N = 1.0-1.3; high coverage: dimer R_{As-As} = 3.32 Å; N = 1.2 with ^{1}V R_{As-Fe} = 3.50 Å, N = 0.9 and R_{As-Fe} = 4.72 Å, N = 0.3

Manganese (hydr)oxides

Adsorbate/Sorbent	Collection/ Analysis	Arsenic Solid-Phase Species
Birnessite, vernadite As^{3+} and **As^{5+}** 1. (Manning et al. 2002) 2. (Foster et al. 2003)	B-XAS F, C RT	1. ^{2}C, R_{As-Mn} = 3.22 Å; N = 2.0 2. ^{2}C, R_{As-Mn} = 3.16-3.18 Å; N = 1.7-2.2

Adsorbate/Sorbent	Collection/ Analysis	Arsenic Solid-Phase Species
Sulfide minerals		
Troilite (FeS) and Pyrite (FeS₂): As³⁺ 1. (Bostick and Fendorf 2003) 2. (Couture et al. 2013)	B-XAS F, C RT/LT, no O₂	1. Reduction of As³⁺ to As⁻¹ and surface precipitation of FeAsS-like species: R_{As-S} = 2.35 Å, R_{As-Fe} = 2.40 Å 2. (LT) no reduction. On FeS: R_{As-S} = 3.12 Å, N =1.7 + R_{As-Fe} = 2.35 Å, N = 0.1+ R_{As-Fe} = 3.42 Å, N = 2.1. On FeS₂: R_{As-S} = 2.07 Å, N =0.3; R_{As-Fe} = 2.31 Å, N =0.2; R_{As-S} = 3.05 Å, N =0.1 R_{As-Fe} = 3.31 Å, N = 1.1.
Galena (PbS) and Sphalerite (ZnS): As³ (Bostick et al. 2003)	B-XAS F, C RT, no O₂ Flow cell expt.	Sorption by exchange of O- for S- in As³⁺O₃ and formation of adsorbed polynuclear species with proposed stoichiometry As₃S₃(SH)₃. *PbS*: R_{As-S}: 2.51 Å, N = 2.8-3.4; R_{As-As}: 3.65-3.68 Å, N = 2 (fixed). *ZnS*: R_{As-S}: 2.23 Å, N = 2.3; R_{As-As}: 3.35, N = 1.6. Photooxidation on PbS, but not on ZnS.
Mackinawite (Mk) and Pyrite (Py): As³⁺ and As⁵⁺ (pH 5.5-6.5) MK formula: (Fe,Ni)₁₊ₓS, x = 0-0.11; Py formula: FeS₂ (Farquhar et al. 2002)	B-XAS F, C RT, no O₂	All (?) OS complexes give S @~3.10 Å and Fe @~3.4 Å Mk, low As⁵⁺: R_{As-S} 3.11Å, N = 1; R_{As-Fe} = 3.51, N = 1; Mk, high As⁵⁺: R_{As-S} 3.09 Å, N = 1, R_{As-Fe} = 3.38, N = 1 Py, high As⁵⁺: R_{As-Fe} = 3.35 Å, N = 2, no S shell Mk low As³⁺: R_{As-S} = 3.09 Å, N = 1; R_{As-Fe} = 3.40 Å, N = 1; Mk high As³⁺:$R_{As-S/Fe}$ 2.87 Å, N = 1, R_{As-Fe} = 3.32 Å, N = 1 Py high As³⁺: R_{As-S} = 3.02 Å, N = 1, R_{As-Fe} = 3.41 Å, N = 1
FeS and FeS₂: Mono- and Tetra- thioarsenate (MTAs⁵⁺ and TTAs⁵⁺) adsorption (Couture et al. 2013)	B-XAS F, C LT, no O₂	Adsorbed fraction: ¹V or OS (indeterminant), R_{As-S} = 3.08-3.13 Å and R_{As-Fe} = 3.42 Å. Partial reduction of MTAs⁵⁺ and TTAs⁵⁺ on Fe produces isomorphous substitution of As³⁺ for S (~10%): R_{As-Fe} = 2.35 Å, R_{As-S} = 3.12 Å. Less decomposition of TTAs⁺ on sulfides relative to oxides
Carbonates, Phosphates, and Sulfates		
Calcite (CaCO₃): As⁵⁺ (Alexandratos et al. 2007)	B-XAS/μ- XRF F, C RT	Adsorption ≅ coprecipitation: 2 separate ¹V complexes or 1distorted ²C, complex (indeterminant) with R_{As-Ca} = 3.44 Å, N = 1.1-1.9, and R_{As-Ca} = 3.62 Å, N = 2.9. Coordination *not* equivalent to As coordination in Ca-arsenates
Struvite NH₄MgPO₄·6H₂O As⁵⁺ and As³⁺ (Lin et al. 2013a)	B-XAS/ μ-XRF LT	Isomorphous substitution for PO₄ As³⁺: oxidation to As⁵⁺ during synthesis (in air). No As-P or As-Mg neighbors in EXAFS FT due to low scattering amplitude of P and Mg. Instead, MS paths at 2.90-3.47 Å important in fit.
Gypsum CaSO₄·2H₂O: coprecipitated with As⁵⁺ and As³⁺ (Lin et al. 2013b)	B-XAS F, C RT	Isomorphous substitution for SO₄; As⁵⁺ and As³⁺ present in precipitating solutions. Theoretical calculations of XANES. Only 1ˢᵗ shell distances fit, confirming As³⁺ and As⁵⁺ in structure.
Ettringite: As⁵⁺ sorption and coprecipitation (Myneni et al. 1997)	B-XAS F, C RT	Adsorption (low As_tot): R_{As-Al} = 3.21 Å, R_{As-Ca} = 3.57 Å + 4.22 Å (high As_tot): no Al, R_{As-Ca} = 3.20 Å + 3.47-3.5 Å + 3.70-3.73 Å, and R_{As-As} = 4.46 Å Coppt: As_tot: R_{As-Ca1} = 3.18-3.21 Å, R_{As-Ca2} = 3.5-3.6 Å, R_{As-Ca3} = 4.1 Å, R_{As-As} = 5.1 Å
Jarosite: As⁵⁺ adsorption or coprecipitation 1. (Paktunc and Dutrizac 2003) 2. (Savage et al. 2005) 3. (Grafe et al. 2008a) (± Cu_(aq))	B-XAS F, C FT Also XRD	1. Co-precipitation: at least 9.9% As can be incorporated. R_{As-Fe} = 3.26 Å 2. Co-precipitation: up to 20% As accommodated in structure; R_{Fe-As} =3.30-3.37 Å (from Fe EXAFS) 3. Adsorption: (– Cu): standard ²C complex; (+ Cu): surface precipitate similar to euchroite (Cu₂[AsO₄](OH)·3H₂O)

Adsorbate/Sorbent	Collection/ Analysis	Arsenic Solid-Phase Species
Natural Soils		
3 CA soils: As^{3+} and As^{5+} sorption (Manning 2005)	B-XANES L	As^{3+}: significant oxidation upon sorption to all three. As^{3+}-Fe^{3+} hydroxide main species; main As^{5+} species varied (illite or Fe^{3+}hydroxide) As^{5+}: 2 of 3 soils : 50/50 As^{5+} sorbed to Al-Si clay/Al hydroxide and Fe^{3+}hydroxide; 3rd : As^{5+} sorbed to Al-Si clay/Al hydroxide
Mn and Fe-rich soils: As^{5+} **sorption** (Deschamps et al. 2003)	B-XANES L	Evidence for oxidation of As^{3+} and As^{5+} sorption to Fe oxides, but also appreciable As^{3+} and As^{5+} on Mn oxides suggests passivation via precipitation of (potentially Fe-rich) phases
Phosphate-amended soils: As^{5+} sorption (Reynolds et al. 1999)	B-XANES L	Soils amended with $As^{5+}O_4$ and PO_4, then subjected to aeration cycles: ($-O_2$): arsenopyrite formation. ($+O_2$): Fe precipitation + As sorption,+ Mn precipitation. Arsenopyrite destroyed, As^{3+} accumulated in soil.
"Red Soils" from PR China: As^{5+} sorption (Luo et al. 2006)	B-EXAFS C	As^{5+} sorbed on B horizons of Ultisol and Oxisols with high Al and Fe oxide content. Fit with 2C complexes : $R_{As-Al} = 3.18$ Å; $R_{As-Fe} = 3.28$ Å

B = bulk ; U = microbeam; XAS = XANES + EXAFS; Analysis type: fingerprinting (F), linear combination (L), principal component analysis (P), and curve fitting with theoretical standard (C), as discussed in the text. LT = cryogenic < 77K temperature; RT = ambient temperature; No O_2 = anaerobic

APPENDIX 2

As-XAS studies of naturally-occurring As-bearing biogenic minerals and laboratory studies of microbial oxidation/reduction of As-bearing biogenic minerals or synthetic analogs. Typical reported error for EXAFS-derived interatomic distances (R) is ±0.02 Å, and for coordination number (N) is ±20%. FRB, ARB, and SRB = iron-, arsenic-, and sulfate-reducing bacteria, respectively. Standard reporting error for LC-LSF: ±10%. Standard reporting detection limit for LC-LSF: <5-10%.

System	Collection/ Analysis	Arsenic Solid-Phase Species
Microbial Fe/As Oxidation		
As$^{3+,5+}$-Ferrihydrite precipitation by *Acidovorax* (anaerobic Fe^{2+}-oxidizer) (Hohmann et al. 2011)	B-EXAFS (F,C) RT(?) not specified also Fe EXAFS	Low As/Fe: As-goethite (Gt); high As/Fe: As-Ferrihydrite (Fh) (both oxyanions). Higher loading of As^{5+} on Fh but equivalent loading of As^{3+}on Fh or Gt. Exopolymers may increase the relative abundance of As sorbed to ^{2}E sites due to competition for the preferred ^{2}C site
Microbial Fe^{3+}/As^{3+}/SO$_4$ Reduction		
As$^{3+,5+}$-lepidocrocite +FRB/ ARB (*S. putrefaciens*) (Ona-Nguema et al. 2009)	B-XAFS (F,C) LT vacuum dried	As^{3+}-lepidocrocite → hydroxycarbonate green rust (GR-CO$_3$; and ferrous-carbonate hydroxide (FCH) As^{5+}-lepidocrocite → FCH + Fe^{2+}(OH)$_2$ phase with As^{5+} → As^{3+}. Polymerized As^{3+}O$_3$ on Fe^{2+}(OH)$_2$ gives R_{As-As} = 3.3 Å 3.32 Å and R_{As-Fe} = 3.50 Å.
As$^{5+,3+}$ferrihydrite-coated sand + FRB+ ARB (*S. putrefaciens* or *S. barnesii*) or + ARB (*B. benzoevorans*) 1. (Kocar et al. 2006) or **in constructed soil aggregates** 2. (Masue-Slowey et al. 2011, 2013)	B-XAFS (L) RT Air dried (probably in anaerobic chamber) Also: Fe XAFS	1. ARB only: As^{5+} → As^{3+} no sorption = high mobility; FRB/ARB: As^{5+} → As^{3} with sorption on mixed valent Fe phases or Fe^{2+} phases (see Appendix 1) 2. As-Fe hydroxide coated sand made into aggregate. FRB/ARB, + O$_2$, flow system. Anoxia in particle interior (As^{3+}-magnetite), but more accumulates in oxidized particle exterior or in preferential flow fractures as As^{5+}-Fe^{3+} hydroxide).
As^{5+}-ferrihydrite and natural FRB microbial consortium from soil (Burnol et al. 2007)	B-XANES (L) RT	1st month: Fe release only; PO$_4$-rich growth medium leads to vivianite precipitation. No As^{3+} production until Fe^{3+} reduction ceases. CO$_3$ from microbial respiration competes with As^{3+} for sorption sites, leading to increased dissolved As
As^{5+} adsorbed to birnessite + ferrihydrite in constructed soil aggregates (+ agarose) with *S. putrefaciens* ANA-3 (FRB+ARB) (Ying et al. 2013) or **As^{3+}(aq) injected in Donnan reactor** (Ying et al. 2011)	B-XANES (F, L) Air-dried U-XRF (thin sections anaerobically fixed) also: Fe-XAFS	Soil aggregate studies: As^{3+}, As^{5+} cycling is restricted to outside (3 mm) of aggregate (approx. 10 mm radius); the inside remains reducing even in aerated water. Donnan reactor studies: As^{3+} oxidation by birnessite limited by surface passivation due to precipitation of Mn^{2+}CO$_3$ under reducing conditions. Both studies: As^{3+} oxidation by birnessite is limited in the absence of O$_2$.

System	Collection/ Analysis	Arsenic Solid-Phase Species
As^{5+}-ferrihydrite and *D. vulgaris* (SRB) cultures (Kocar et al. 2010) **(+ humic acid)** (Burton et al. 2013)	B-EXAFS (C) RT? Air dried (in anaerobic chamber?) Also: Fe, S XAFS	Low As-loading: FeS in reducing zones and magnetite in intermediate zone. high loading (50% of sites), green rust instead of magnetite. No As uptake in green rust or FeS, no formation of As sulfides due to Fe^{2+} scavenging (+ humic acid): less sorption of As due to formation of thioarsenate species
As^{5+}-jarosite sulfidization (abiotic) (Johnston et al. 2012)	B, XRD, 2	Jarosite $\rightarrow$ Mackinawite (FeS$_{am}$): As^{5+} OS sorption (no reduction)
As^{5+}-ferrihydrite and anaerobic digester sludge microbial community (Root et al. 2013)	B-EXAFS, U-XRF LT (EXAFS); RT (μ-XRF) Hydrated (bulk) Dried no O$_2$ (μ-XRF) also: Fe EXAFS	Low and high SO$_4$ conditions tested. As^{5+} reduced to As^{3+} in both. Low-SO$_4$ endpoint solid phases = As-ferrihydrite> As-siderite > As-green rust (GR). High-SO$_4$ endpoint phases: amorphous As sulfide (realgar-like local structure)

B = bulk ; U = microbeam; XAS = XANES + EXAFS; Analysis type: fingerprinting (F), linear combination (L), principal component analysis (P), and curve fitting with theoretical standard (C), as discussed in the text. LT = cryogenic < 77K temperature; RT = ambient temperature; No O$_2$ = anaerobic

APPENDIX 3

XAS studies of arsenic speciation in rocks, related soils, and rock-forming fluids. Standard reporting error for LC-LSF: ± 10%. Standard reporting detection limit: < 5-10%. Typical reported error for EXAFS-derived interatomic distances (R) is ±0.02 Å, and for coordination number (N) is ±20%.

Rock or Soil	Collection/ Analysis	Arsenic Solid-Phase species
Igneous and/or hydrothermal rocks (Savage et al. 2000b, Kolker and Nordstrom 2001, Paktunc 2008b) non-XAS: (Deditius et al. 2008, 2011)	Bulk XAS (C) U-XAS RT and LT Also Fe EXAFS (also several non-XAS techniques)	Arsenian pyrite is a major host for As across several rock types. As^{-1} pyrite (As → S) is nearly ubiquitous; As^{3+} pyrite (As → Fe), is rare. EXAFS studies of As^{-1} pyrite suggest As clustering. Non-XAS studies have documented the formation of crystalline As-rich nanoparticles of other phases (5-100 nm) in pyrite from epithermal deposits
Fluid inclusions from 3 types of mineral deposits (James-Smith et al. 2010)	U-XRF U1, U3	Reduced As species in all samples (As^{3-}, As^0, and/or As^{-1}); $As^{3+}O_3$ in the high As sample; Beam-induced photo-oxidation precluded analysis of low-As samples
Marine hydrothermal plume particles, East Pacific Rise (Breier et al. 2012)	U-XRF RT	As correlates positively with P, V, Cr, and Fe and is found in both HFO and sulfide particles
Geothermal field siliceous sinter deposit El Tatio, Chile (Alsina et al. 2008, 2013)	B-XAS (F,C); 2008 U-XAS (F, L); 2013 Sinter sections store at 8 °C prior to bulk Fe EXAFS; sections thinned to 280 μm ambient conditions	2013: 63% As^{5+}-HFO and 37% loellingite ($FeAs_2$) (micronodules) Loellingite formation suggests highly anoxic formation conditions 2008: As^{5+}-HFO in ^{2}C complex: R_{As-Fe} = 3.23 Å
Metamorphic Rock: Tso Morari eclogites, NW Himalaya (Forearc mantle serpentinites,) As = 6-275 ppm (Hattori et al. 2005)	B-XAS F, C RT	As mineral hosts: coarse grained magnetite, antigorite (by EMPA), but also minute + rare Ni- sulfides. Speciation: Primarily As^{5+}; much less common As^{3-} and As^{-1}. As is introduced into mantle wedge at depths shallower than 25 km
Skarn deposits with As-zoned garnets (Charnock et al. 2007)	B-XAS	$As^{5+}O_4$ substitutes for SiO_4 in garnet
Fe-Mn exhalative deposit sample (oxidized), Switzerland (Etschmann et al. 2010)	U-XRF/XAS Thin section	$Sr,Ce)Al_3(AsO_4)(SO_4)(OH)_6$ (kemmlizite) is as "hot spots"; also Fluorapatite hosts up to 9.3 wt% As^{5+} and As^{3+}. Hematite hosts low amounts of As^{5+}. Full-spectrum XANES mapping over 3 × 4 mm² area + quantitative XRF data
Moolart Well, laterite-hosted Au deposit Australia (Etschmann et al. 2010)	U-XRF, U-XAS, Epoxy-embedded, thinned, polished soil section MAIA detector	Oxidized pisolitic regolith (800 ppm As). (3 m depth), As concentrated in Fe-rich concretions. As^{5+} predominant, but one concretion showed core of As^{3+} + full-spectrum XANES mapping over 8 x 8 mm² + quantitative XRF data

Rock or Soil	Collection/ Analysis	Arsenic Solid-Phase species
Fe- and Mn- oxide cave deposits, Illinois, USA and ferromanganese coatings on quartz 1. (Frierdich and Catalano 2012 2. (Manceau et al. 2007)	BL (air dried, ground), U-XRF (air dried, intact flakes or edge-oriented nodule/ coated pebbles)	1. As^{5+} only; bound to both Mn and Fe oxides (more enriched in Fe oxide phase 2. Fe/Mn coatings form alternating layers of ferrihydrite and vernadite; no more than 20% of As^{5+} sorbed to phyllomanganate in the ferromanganese coatings (Manceau et al. 2007)
Travertine deposits, Greece (Winkel et al. 2013; Bardelli et al. 2011)	B, U-XRF, U-XAS ambient conditions	Travertines bearing ~900 mg kg^{-1} As; As^{5+} coprecip. with calcite (not adsorbed, not Ca-arsenate); immobilizes up to 25% of As in high As geothermal area
Naturally-weathered and burnt black shale, SW Sweden (Yu et al. 2014)	B-As XANES F, L RT Dried	As^{-1} in "fresh" shale; spectrum consistent with arsenian pyrite. Humidity test cycles (artificial weathering) transformed > 88% of As^{-1} to As^{5+} with no evidence of As^{3+} intermediate. As^{5+} is bound to schwertmannite (Fe hydroxyl-sulfate) and jarosite formed during humidity tests. Simulated seawater interaction liberated As from these phases by pH-dependent desorption.
Coal from eastern and western USA used in power plants (Huggins et al. 1993, 2009; Huffman et al. 1994; Kolker et al. 2000; Zielinski et al. 2007)	B-XANES F, L RT Dried physically fractionated	Bituminous (mostly eastern US): As^{-1}-pyrite Lignite (low-rank, mostly western US): As^{3+}-organic Both oxidize if stored under ambient conditions; wetted, As^{-1}-pyrite in bituminous coal oxidizes fastest in 17-month studies (Kolker and Huggins 2007).
As-rich coal from Guizhou Province, PR China used for domestic heating (Zhao et al. 1998)	B-EXAFS C RT	6 coal samples from Xingren and Xingyi. No inorganic As detected. Theoretical fits indicate that As^{5+} tetrahedral oxyanion is dominant species in 5 of the samples, with As^{3+} trigonal pyramidal oxyanion in the remaining sample. Inference: As associated with organic macerals in all samples.
Oil Shale and derived kerogen concentrate (Cramer et al. 1988)	B-XAS F, L	Oil shales from several sites in US (including Green River), and one site in Brazil (Irati). SEM/EDS: Ni-Co-Fe-arsenides present. XANES fit: As^{-1} (~60%) and As^{5+}(~40%) in Green River, but ratio is variable: Irati contains ~100% As^{-1} and Montana Phosphoria contains only As^{5+}. Kerogen concentrate contains 100% As^{-1} but FT of EXAFS not identical to skutterudite Ni-Co-Fe-arsenide (fingerprinting only).
Soil developed on pyritic shale, CA, USA (Strawn et al. 2002)	U-XRF, U-EXAFS (Fe), U-XANES (As, Se) Ambient dried	As^{-1}- pyrite (presumed original As source) is absent, due to high rate of oxidative dissolution and flushing. As^{5+} predominant in Fe-rich soil aggregates; preferred association with HFO-rich aggregates rather than jarosite-rich aggregates

B = bulk ; U = microbeam; XAS = XANES + EXAFS; Analysis type: fingerprinting (F), linear combination (L), principal component analysis (P), and curve fitting with theoretical standard (C), as discussed in the text. LT = cryogenic < 77K temperature; RT = ambient temperature; No O_2 = anaerobic

APPENDIX 4

XAS studies of As speciation in ore, mine wastes, and mining-impacted sites. Standard reporting error for LC-LSF: $\pm10\%$. Standard reporting detection limit: $<5\text{-}10\%$. Typical reported error for EXAFS-derived interatomic distances (R) is ±0.02 Å, and for coordination number (N) is $\pm20\%$.

Site Description/ Deposit Type	Collection/ Analysis	Major Findings
Soil with calciner residue Cornwall, UK (Polymetallic) (Arcon et al. 2005)	B-XAS RT F, C	Primary minerals: FeAsS and FeAs$_2$ $\rightarrow$ As$_2$O$_3$ in purpose-built calciners. Soil @ 25-30 cm depth: As^{5+} dominant, $R_{\text{As-Al}}$ = 2.54 Å, $R_{\text{As-Fe}}$ = 3.34 Å (N ~1.6 for both). Interpretation: As^{5+} sorbed to Al- or Al/Si phases (^{2}E complex), and amorphous Fe^{3+}As^{5+}O$_4$ precipitate
Cu-W-As mine tailings, SW Finland (Parviainen et al. 2012)	U-XAS (L, C), U-XRF	As^{5+} dominant, sorbed to ferrihydrite, as scorodite or kankite
Stack sample from Cu smelter (Paktunc 2008a)	B-XANES, U-XAS L RT + LT	As^{3+} and As^{5+} mixed, only found on outer surfaces of spherical Cu particles, w/in secondary phases. Hypothesis: condensation of volatilized As during cooling. Soils collected near smelter: As^{5+} (inorganic) and organic As (MMA, TMAO)
Soil contaminated by Anaconda smelter, Tacoma, WA, USA (Beaulieu and Savage 2005)	B-EXAFS C RT	Vashon Island soil that received airborne smelter waste. As primarily associated with clinochlore, the main phyllosilicate. Multiple scattering very important to include for good fit
Gold mine tailings and airborne dusts near historical Au mining districts, Nova Scotia, Canada (Corriveau et al. 2011a,b)	U-XANES (F, L), U-XRF	Airborne: As^{5+} dominant; hydrous ferric arsenate, sorption to hydrous ferric oxide, scorodite Mine tailings: As^{5+} dominant; arsenopyrite, scorodite, amorphous Fe arsenate, Ca-Fe arsenate phase coprecipitated with Fe-oxyhydroxides
Soil contaminated by historical Au roaster waste, Giant Mine, Yellowknife, NW Territories, Canada (Jamieson et al. 2011; Fawcett and Jamieson 2011)	B/U-XANES (L, C) Also: Sb-XANES, μ-XRD	As^{5+} in calcines, As^{3+} in roaster dust; both bonded to O; As$_2$O$_3$ identified in contaminated soil
Gold mine waste, downstream sediments, Denali National Park, Alaska, USA (Ritchie et al. 2013)	B-XAS (L), Also: Sb-XAS	As^{3+} in mine waste materials, As^{5+} in downstream sediments sorbed to Fe-hydroxides
Au mine tailings, Southern Mother Lode, CA, USA (Savage et al. 2000b)	B-EXAFS (F, C) Also: Fe-EXAFS	Pyrite: As^{-1} substitution for S; tailings: As^{5+} sorption as ^{2}C complexes to goethite, substitution for SO$_4$ in jarosite
Au mine tailings, soils, sediments, Giant Gold Mine, Northwest Territories, Canada (Walker et al. 2005)	B/U-XANES (L, P) Also: U-XRD	B-XANES: As^{5+}>As^{3+}, minor As^{-1}; U-XANES: association with maghemite particles

Site Description/ Deposit Type	Collection/ Analysis	Major Findings
Au mine tailings, Nova Scotia, Canada (Walker et al. 2009)	U-XAS (L, P) Also: U-XRD, U-XRF	Scorodite, amorphous hydrous ferric arsenate most common
Au mine tailings, Northern Mother Lode, CA USA 1. Argonaut mine (Foster et al. 1998b) 2. Lava Cap Mine (Foster et al. 2000a, 2011; Foster and Ashley 2002) 3. Empire Mine (Foster et al. 2014)	B-EXAFS F, P, L, C + Fe EXAFS	1. Weathered sulfide concentrate: arsenopyrite >> scorodite (C-type fit) 2. Ore types (arsenopyrite rich>> pyrite-rich) → As^{5+}-ferrihydrite + arseniosiderite in weathered tailings. Microbial samples: As^{5+}-ferrihydrite + unidentified As^{3+} species (latter dominates in filtrate from historic tailings impoundment. Arsenopyrite in subaqueous tailings relatively stable 3. Mining-impacted sediments ("average speciation"): As^{5+}-ferrihydrite/goethite >> jarosite > Ca-Fe + Ca-arsenate minerals > As^{5+} and As^{3+} sorbed to Al-hydroxide > pyrite/ arsenopyrite
Au mine wastes and impacted soils, Empire Mine State Historic Park USA (*N. Mother Lode, CA, USA*) (Burlak et al. 2010; Alpers et al. 2012; Burlak 2012)	U-XRF, U-XAS mineral thin sections + redox mapping	Ore hand samples (weathered): oxidized As^{5+} (minor As^{3+}) on rims of As^{-1} pyrite, arsenopyrite, and cobaltite [(Co, Fe) AsS]. Weathering produces a range of As^{5+}-Fe^{3+}-mineral associations. General trend: As^{5+}-Fe^{3+} hydroxide on rims of As^{-1} pyrite and hydrous ferric arsenate precipitate on rims of arsenopyrite. Some samples contained Ca-Fe arsenate minerals with stoichiometries ≠ arseniosiderite /yukonite
Argonaut Au Mine, Spenceville Au Mine (Mother Lode, CA) and Ruth Au Mine (Mojave Desert, CA) (Foster et al. 1998b)	B-EXAFS (F, C), theory calculations	Argonaut: weathering sulfide concentrate: arsenopyrite + scorodite Spenceville (roaster product): indeterminant (As^{5+}) Ruth: As^{5+} in ^{2}C sorption to Fe^{3+}, Al^{3+} hydroxides (no MS)
Au/Ag mines, Mojave Desert, CA, USA (Kim et al. 2013)	B-EXAFS U-XRF/XANES L, P, C size fractions	Randsburg district. As^{5+}-ferrihydrite/goethite, Ca and Ca-Fe arsenates, amorphous Fe-arsenate, and Na-arsenate used in fits. Greater proportion of As^{5+}-ferrihydrite/goethite in background samples. Unique As/Fe trends in bivariate plots (μ-XRF) are more pronounced in smaller size fractions.
Marcasite (Fe) mine tailings and sediments, northern Latium, Italy (Kreidie et al. 2011)	B-XANES (F)	As^{5+} and As^{3+} in tailings, converts to As^{5+} with increasing distance
Lead-zinc mine-impacted sediment, Aznalcóllar, Spain (Hudson-Edwards et al. 2005)	B-XAS (L, C)	Mix of As^{5+} and As^0, as As-bearing amorphous Fe-hydroxide, As-jarosite, and arsenopyrite
Carnoules Pb-Zn-Sb Mine, France Acid (pH < 4) water downstream of seepage under dam (Maillot et al. 2013)	B-EXAFS (C) LT, air dried Fe-EXAFS	As^{5+} and As^{3+} -rich Fe^{3+} hydroxide/hydroxysulfate suspended particulate material: ^{2}C complexes on schwertmannite (R_{As-Fe} = 3.3 Å; As^{5+} sorbs more than As^{3+}

Site Description/ Deposit Type	Collection/ Analysis	Major Findings
Carnoules Pb-Zn-Sb Mine, France Microbial community in acid (pH < 4) Reigous Creek (Morin et al. 2003)	B-XANES (F, L) RT Air-dried	3 major phases: As^{5+}-Fe^{3+} oxyhydroxide, As^{3+}-Fe^{3+} oxyhydroxide (suspended sediment), and nanocrystalline tooleite $Fe_6(AsO_3)_4(SO_4)(OH)_4 4H_2O$ (stromatolites). Wet season dominated by microbes that can oxidize Fe(II) but not As^{3+} (e.g., *A. ferroxidans*); these force tooleite formation. Dry season dominated by microbes that oxidize As and Fe (e.g., *Thiomonas spp.*)
Pb smelter soils, East Helena, Montana, USA (Impellitteri 2005)	B-XAS (L)	As^{5+} dominates with variable As^{3+} and As^0; scorodite dominant with rare As^0, arsenolite, realgar
Hg mine tailings, Sulphur Bank, CA, USA (Slowey et al. 2007)	B-XAS (L)	As^{5+} adsorbed to Fe (hydr)oxides or coprecipitated with jarosite
Ag mine tailings wetland, Ontario, Canada (Beauchemin and Kwong 2006)	B-XAS (L)	As^{5+} dominant, reducible to As^{3+}, As^{-1} but easily reoxidized; As_2O_5, As_2O_3, scorodite dominant
Weathered Ag mine tailings, Potosí, Bolivia (Kossoff et al. 2012)	U-XANES (L), U-XRF Also: Sb-XANES	Arsenopyrite and As-bearing pyrite weather to scorodite
W mining-impacted stream sediments, Madrid, Spain (Helmhart et al. 2012)	B-XAS (L, C) Also: Fe-XAS	As^{5+} sorbed to Fe-oxide phases
U mine tailings, Saskatchewan, Canada (Cutler et al. 2001, 2003)	B-XAS (F, L)	Amorphous ferric arsenates, adsorbed arsenates and other poorly ordered arsenates
Uranium mine tailings, Saskatchewan, Canada (Essilfie-Dughan et al. 2012, 2013)	B/U-XAS (F, L, C) Also: Fe-XAS	As^{5+} dominant with variable As^{3+} in weathering rinds around particles, sorbed to ferrihydrite as bidentate surface complexes

B = bulk ; U = microbeam; XAS = XANES + EXAFS; Analysis type: fingerprinting (F), linear combination (L), principal component analysis (P), and curve fitting with theoretical standard (C), as discussed in the text. LT = cryogenic < 77K temperature; RT = ambient temperature; No O_2 = anaerobic

APPENDIX 5

Arsenic speciation in floodplain and aquatic sediments, and in organic-rich deposits. Standard reporting error for LC-LSF: ±10%. Standard reporting detection limit: <5-10%. Typical reported error for EXAFS-derived interatomic distances (R) is ±0.02 Å, and for coordination number (N) is ±20%.

System	Data types(s) and analysis	Major findings
River and Floodplain sediment		
Riparian floodplain on Mulde River, Germany (Voegelin et al. 2007)	B/U, As, Fe EXAFS	As^{5+}-Fe^{3+} oxyhydroxide. Inclusion of 3- and 4-leg MS paths within the $As^{5+}O_4$ tetrahedron allowed discrimination between As^{5+}-Fe^{3+} oxyhydroxide and As^{5+}-Al^{3+} oxyhydroxide
Ichinokawa River sediment, Japan (Asaoka et al. 2012)	B-XANES F, RT	As^{5+} only in sediments. Other data suggest sorption to ferric oxyhydroxides dominates speciation
Amous River, Gard, France (Adra et al. 2013)	B3, As-, Fe EXAFS	As^{5+} sorbs to natural Al-rich ferrihydrite (25-30 mol% Al) in ^2C geometry; same for synthetic Al-rich ferrihydrite
Ogosta River, Bulgaria (Mandaliev et al. 2014)	B-EXAFS L, P, LT +Fe-EXAFS	Size-fractionated river floodplain soils (3 depths, 2 sites) impacted by FeAsS mining). Dominant species: As^{5+}-ferrihydrite, As^{5+}-HFO, and FeAsS (latter only in silt and bulk fractions)
As associated with Al and Fe rich phases in geothermal fluids (Ilgen et al. 2011)	B-EXAFS C	As^{5+} and in As^{3+} in < 65 µm river bottom sediments receiving As from geothermal spent fluids. ^2C complexes predominant sorption geometry for both oxyanions on Fe and Al oxyhydroxides.
Riverbed and subsurface soil sediment, Madrid Spain (Helmhart et al. 2012)	B-EXAFS C, LT +Fe-EXAFS	Preferential flowpaths transfer As and Fe from riverbed to subsoil in mining impacted area (acidic water). However, precipitation of As^{5+}-ferrihydrite along flowpath leads to strong localization of As and Fe enrichment and limited transport into larger soil matrix
Floodplain sediments + active or buried paleo-channels Bangladesh 1. (Foster et al. 2000b; Breit et al. 2000, 2004) 2. (Polizzotto et al. 2005, 2006) 3. Datta et al. 2009)	B-XANES (1-3) U-XRF/XAS (2) L +Fe EXAFS (1)	1. Meghna and Titas rivers. As^{5+}-Fe^{3+} hydroxide predominates above water table. $As^{5+} \rightarrow As^{3+}$ and $Fe^{3+} \rightarrow Fe^{2+}$ correlated with tan $\rightarrow$ gray color change in shallow sediment (<3 m depth). Formation of sub-horizontal band(s) rich in As^{5+}-ferrihydrite result from climate and hydrologic conditions. 2. Central Bangladesh, Ganges R. As^{5+}-Fe^{3+} hydroxide $\rightarrow As^{3+}$ in shallow sediment; sulfide grains present in shallow sediment appear detrital. 3. Active Meghna river channel and bank deposits. During dry season, discharge of reduced As^{3+} and Fe^{2+}-bearing water into hyporheic zone leads to oxidation and formation of As^{5+}-ferrihydrite
Lake/Reservoir sediment		
Haiwee Reservoir CA, USA (Kneebone et al. 2002; Root et al. 2007)	B-EXAFS (C) LT hydrated + Fe EXAFS	Precipitate is amorphous (HFO), similar to ferrihydrite; As is adsorbed and/or coprecipitated. As^{5+} is completely reduced to As^{3+} within ~4 cm; products of HFO reduction not observed until 10 cm depth. Amorphous green rust (GR) phase is the main product. As and Fe release is highest in zone of GR formation.

System	Data types(s) and analysis	Major findings
Harvard Pit Lake, California (Savage et al. 2009)	B-XANES RT, F	Oxidation of As^{-1}-pyrite on pit walls produces secondary minerals that incorporate As^{5+} via sorption and/or coprecipitation
Don Pedro Reservoir, California (Savage et al. 2000a)	B-EXAFS C	Mine tailings exposed on shoreline during dry months. Mass wasting of pyrite weathering products (As^{5+}-Fe^{3+}oxyhydroxides and As^{5+} in secondary sulfates (e.g., jarosite) add As to reservoir
Lost Lake, CA, USA (Foster et al. 2011)	B-EXAFS L, P RT and LT hydrated	Lake is a >70 yr. old retention structure for arsenopyrite-rich, fine-grained gold mine tailings. Tailings kept submerged since deposition have > 60% As in arsenopyrite, but tailings that were exposed to variable O_2 on shoreline for 3 yrs had lost ½ that arsenopyrite, converting it to As^{5+}
Lakes in Quinsam Watershed, British Columbia, Canada (Moriarty et al. 2014)	B-XANES LT (Wet); RT freeze-dried, air-dried	Freshwater lakes in coal-mining area examined for As speciation. As^{3+} and As^{5+} quantified by XANES fitting; further speciation detail was not obtained due to low bulk As concentration
Freshwater wetlands/peat deposits		
Coeur d'Alene freshwater wetlands, Idaho, USA (La Force et al. 2000)	B-XANES F, L	impacted by Pb/Zn mining As^{5+}, As^{3+}-O, and As^{3+}-S (orpiment-like) all present w/ seasonal and location variation. Winter rains bring substantial oxidation , but by spring, reductive processes have converted most of the As^{5+} to As^{3+}-S (orpiment-like); this partially oxidizes over the summer
Farr Creek wetland, Ontario, Canada (Beauchemin and Kwong 2006)	B-EXAFS (L, P) RT collection	Column studies with wetland sediment (w/Ag mine tailings) subjected to reduction/oxidation cycles to simulate seasonal changes. As^{5+} reduced to As^{3+} in 30 day expts with sediment from top, intermediate, and bottom layers. Safflorite-like phase $[(Co,Fe)As_2]$ formed in glucose (+) incubations. Re-oxidation rapid in top and bottom, but not in intermediate. PCA coupled to target transformation was not effective for selecting model compounds. LCF fit uniqueness tests = not unique
Delta/Wetland sediment Cambodia (Kocar et al. 2008; Polizzotto et al. 2008)	U-XRF/XAS L	Kandal province/ Mekong *R.* As^{5+}-Fe^{3+} hydroxide → As^{3+} in shallow sediment; high organic matter drives formation of As^{-1} sulfides, but not pervasive enough to sequester all the As. Mobilization of As is most pronounced in wetlands (oxbow lakes)
Groundwater-fed Constructed wetland Massachusetts, USA (Wilkin and Ford 2006)	B-XANES (F, C) RT Magnetic separate	impacted by industrial contamination. As in gregite; abundant reactive Fe limits formation of As-sulfide phases. Gregite is transformation product from mackinawite. Pyrite accounts for < 20% of As after 30 years of sediment deposition
freshwater peatland, Gola di Lago, Switzerland (Langner et al. 2012a,b, 2013)	B/U F, L, C RT/LT	Near surface samples: As bound in α-As_4S_4 as 10-50 um size particles. Deeper: As^{3+} complexed by particulate NOM sulfhydryl groups ; lesser quantity of < 25 μm FeAsS and $(FeAs_xS_{2-x})$ particles

System	Data types(s) and analysis	Major findings
Estuary/Marine sediment		
Marine sediment Jinzhou Bay, NE China (Xu et al. 2011)	B-XANES L +S XANES	Location contaminated by Zn smelter waste. As^{5+} in initial sediment changes to As_2S_3-like phase after 60 days incubated anaerobically
Achterwasser lagoon sediment (estuary), Oder *R.* **sea** (Neumann et al. 2013)	U-XRF Mineral separation heavy liquids	Outlet to Baltic Sea. As^{-1} in framboidal pyrite accounts for 9-55% of total As, but near surface also contain As^{3+} and As^{5+} associated with pyrite. These species might be due to oxidation of pyrite during sample preparation. Developed method to quantify As in pyrite based on μ-XRF signal, correcting for spherical framboid shape and pyrite matrix composition

B = bulk ; U = microbeam; XAS = XANES + EXAFS; Analysis type: fingerprinting (F), linear combination (L), principal component analysis (P), and curve fitting with theoretical standard (C), as discussed in the text. LT = cryogenic <77 K temperature; RT = ambient temperature; No O_2 = anaerobic

APPENDIX 6

Arsenic speciation in bedrock and sedimentary aquifers. Standard reporting error for LC-LSF: ±10%. Standard reporting detection limit: <5-10%. Typical reported error for EXAFS-derived interatomic distances (R) is ±0.02 Å, and for coordination number (N) is ±20%.

Aquifer System	Data types(s) and analysis	Major findings
Taiwan (SW): Blackfoot disease endemic area (Wang et al. 2001)	B-EXAFS C	$As^{5+}O_4$ in soil and well water (R_{As-O} 1.74 Å, N =4)
P.R. China, Pearl River. Delta (Wang et al. 2013)	BXANES F, L Freeze-dried sediment	Coastal sedimentary aquifer/aquitard system As^{5+} predominant (range 30-67%); sand/alluvial terrestrial gravels of basal aquifer contains more than silty clay marine layers (aquitard). Silty layers contain 20-40% As^{-1}. Little variance in As^{3+} (30-40%) among sediment types
PR China, Inner Mongolia (Guo et al. 2011b)	μ-XRF	Hetao basin, an inland restricted, increasingly saline sedimentary aquifer system. Colloidal As is associated with smaller NOM colloids 5-10 kDa (As^{5+} > As^{3+}) than with larger Fe-rich colloids
Cambodia Mekong Delta (Polizzotto et al. 2008; Kocar et al. 2008)	μ-XRF, μ-XAS F (edge position) RT Tape smear	Kocar (2008): Delta and/or wetland sediment adjacent to oxbow pond: 10 cm depth: As^{5+}. 4 m depth: As^{5+} and As^{3+} with "hot spots" of sulfide-As^{-1} (no quant, FeAsS used for As^{-1} model) Conclusion: mobilization and reduction of As occurs near water table as result of reducing conditions.
Cambodia Kien Svay (south of Phonm Penh) (Rowland et al. 2005)	B-XANES F, L LT (77K) Preservation method tests size fractionation	As^{3+}-O, As^{5+}-O; presence of As^{-1}-S (arsenopyrite model) fits was an artifact of their linear combination fitting routine (proof from analysis of 2-component mixtures lacking As^{-1}-S). see other comments above
Bangladesh (E-central) (Foster et al. 2000b; Breit et al. 2000, 2004)	B-XANES F, L LT (77K)	Gray Holocene-age sediment (gray): no As^{5+}- Fe^{3+}-oxyhydroxides. As^{3+} associated with secondary Fe phases on mica phases and/or sorbed to mica or Al^{3+}-oxyhydroxides. As in sulfides is rare to absent. Pleistocene-age Fe-coated sands ca. 24 m depth (reddish-brown): As^{5+}-Fe^{3+}-oxyhydroxides persists due to oxidation by Mn oxides. As^{5+} and As^{3+} present sulfides rare to absent
Bangladesh (central) Active floodplain of Ganges River (Polizzotto et al. 2005, 2006)	U-XRF/XANES L	Holocene-age and Pleistocene-age aquifer sediment: no As^{5+}- Fe^{3+}-oxyhydroxides. Instead, As in sulfides (orpiment-like and Fe-sulfides) common, and as As^{3+} (unidentified). >100 um grains = detrital As-bearing sulfides; evidence against reductive dissolution of As^{5+}-HFO at depth
Bangladesh confluence of Meghna-Old Brahmaputra floodplain (Itai et al. 2006, 2010; Seddique et al. 2011)	B-XANES F, L	As^{3+} and Fe^{2+} dominant in grey aquifer sediment
Bangladesh (NW) Chapai-Nawabganj (Reza et al. 2010)	B-XANES F, L	As^{5+} and As^{3+}; XANES spectral quality did not allow further interpretation. Sequential extraction data suggest As^{5+}-Fe^{3+}-oxyhydroxides

Aquifer System	Data types(s) and analysis	Major findings
Bangladesh (S-central) Coastal floodplain (Lowers et al. 2007)	B-XANES U-XRF/XAS F, L	Authigenic pyrite sequesters arsenic in deep coastal aquifer sediment (gray), but As^{5+} and As^{3+} bound to Fe^{3+}-oxyhydroxides, secondary Fe phases, and mica in shallow aquifer (gray)
	Extended XANES	
India, West Bengal, Chakdaha block (Rowland et al. 2005)	B-XANES F, L	As^{3+}, As^{5+}; As^{3+}-sulfide fits were an artifact of their fitting routine; centrifuging samples in to different size fractions simplifies speciation in each fraction; they also did substantial oxidation tests and validated least squares fitting by prepared mixtures
USA, ID: Western Snake River Plain (Busbee et al. 2009)	B-XANES F, L	Semi-arid sedimentary aquifer; As^{5+} associated with Fe^{3+}-oxyhydroxide-stained sediments is mobilized by irrigation water downward into aquifer
USA, MA: Ft. Devens Reserve (Choi et al. 2009)	B-XANES F, L LT + Mn XANES	Native sediment contained only As^{5+}; limited oxidation of As^{3+} observed when added to sediment
USA, MA: Cape Cod (Singer et al. 2013)	U-XRF/XANES L	Natural Mn-rich goethite coatings on quartz sand oxidize As^{3+} in lab experiments, but reaction was not complete due to passivation or too little Mn relative to As^{3+}.
USA, CA: Southern San Francisco Bay (O'Day et al. 2004; Root et al. 2009)	B-EXAFS F, L, C LT	Upper zone (reduced, brackish): As in sulfides (realgar, orpiment, pyrite); middle zone (transitional): no As-sulfides, As^{3+}-O, As^{5+}-O; deeper oxidized zone (oxic, freshwater): only As^{5+}-O on Fe(III) secondary phases

B = bulk ; U = microbeam; XAS = XANES + EXAFS; Analysis type: fingerprinting (F), linear combination (L), principal component analysis (P), and curve fitting with theoretical standard (C), as discussed in the text. LT = cryogenic < 77K temperature; RT = ambient temperature; No O_2 = anaerobic

APPENDIX 7

XAS studies of arsenic-rich agricultural, industrial, and municipal wastes. Standard reporting error for LC-LSF: ±10%. Standard reporting detection limit: <5-10%. Typical reported error for EXAFS-derived interatomic distances (*R*) is ±0.02 Å, and for coordination number (*N*) is ±20%.

System	Data types(s) and analysis	Major findings
Poultry litter and litter –amended soils (Arai et al. 2003)	μ-XRF/XANES	As^{5+} in roxarsone converts to As^{3+} (orpiment-like) and As^{5+} (fit as roxarsone)
Roxarsone-contaminated soil (Covey et al. 2010)	μ-XRF/XANES F, L	Lab microcosms with earthworms and As^{5+} in roxarsone converted to Me-As species in soil w/in 30 days; Worm tissue: As-glutathione complex
As_2O_3 application, Tyndall AFB, Florida (Fitzmaurice et al. 2009)	B-EXAFS C	No trace of As_2O_3 by As-EXAFS. As^{5+} fitted as complex with Al-oxides or aluminosilicates, but inconsistent with the measured low-mobility of As; not associated with HFO or HAO.
As_2O_3 ("Anaconda" herbicide) application, Gulf Coast, Southern US (Yang and Donahoe 2007)	μ-XANES, μ-XRD, RT F Particles on tape	No trace of As_2O_3 by As-EXAFS. Sandy loam soils contain As^{5+}. Rare Ca-arsenate phase Identified in clay by μ-XANES + μ-XRD. Primary As^{5+} association with Al and/or Fe on clay-sized particles inferred from sequential extractions
Pb- and Al-arsenate stockpiles Massif Central, near Auzon, France (Cancès et al. 2008)	B-XAFS, L, C	35% of As in topsoil (0-5 cm) is arseniosiderite $(Ca_2Fe^{3+}_3(As^{5+}O_4)O_2\ 3H_2O)$; rest is As^{5+}-Fe^{3+}oxyhydroxide. Below 15 cm, no arseniosiderite, only As^{5+}-Fe^{3+}oxyhydroxide. No evidence for original compounds
Pb Arsenate stockpile, northeast US (Arai et al. 2006)	B-XAS, μ-XRF/XANES/ XRD RT	Stagnant water sediments : As^{5+}on Fe^{3+}oxyhydroxides + As^{5+} in calcite/gypsum/ welite $(CaHAsO_4)$ + As^{3+}-S (orpiment-like) Intertidal zone sediments : As^{5+}-Fe^{3+}oxyhydroxides and residual schultenite $(PbHAsO_4)$
As_2O_3 stockpile, San Francisco Bay California (Foster 1999)	B-EXAFS C	No trace of As_2O_3 by As-EXAFS. Surface soil: As^{5+} in hoernesite $Mg_3(AsO_4)_2·8H_2O$ formed by reaction with Bay water (source of Mg). Follow on study to (Voigt et al. 1996)
Industrial waste		
Spent geothermal fluid (Ilgen et al. 2011)	B-EXAFS	Spent fluid: reducing, high temperature, 9 mg L^{-1} As introduced into river. River sediment: As^{5+} sorbed to Al, Fe oxyhydroxides, ^2C geometry dominant. Substantial lowering of dissolved As via this process.
Bauxite ore processing residue spill and leachate neutralization tests samples, Hungary (Burke et al. 2012, 2013)	B As-EXAFS RT C Also Cr, V	Common name = red mud. No second-neighbor atoms were detected around $As^{5+}O_4$ in red mud samples, suggesting As^{5+} is in outer-sphere (highly labile) sorption complex. Red mud *leachate* (alkaline fluid), post neutralization speciation tests: (a) by HCl: As^{5+} likely coprecipitates with Al oxyhydroxides (dominant mineralogy); (b) by gypsum: As^{5+} in calcite, As-EXAFS show no second-neighbor Ca, but As is not PO_4-extractable; (c) by seawater: As^{5+} in hydrotalcite-type phase + sorption to oxyhydroxide (about ½ of the As is PO_4-extractable).

System	Data types(s) and analysis	Major findings
"Class C" fly ash 1. (Shoji et al. 2002) 2. (Huggins et al. 2007) 3. (Zielinski et al. 2007) 4. (Luo et al. 2011) 5. (Catalano et al. 2012) 6. (Veselska et al. 2013)	B-XAS (1-5) U-XAS (6) L, C (varies) RT	Lignite- or sub-bituminous coal → Class C ashes (alkaline pH, Ca-rich). Ca pyroarsenate-like phase identified in C-type fly ashes from USA (4, 5) and from Turkey (2). Similar XANES spectra presented in earlier studies (1, 2). Freshly lagooned class C ash (Slovakia) showed variable speciation at 25 μm scale: As^{5+} dominant in Ca-Al rich glass, but (some samples) had ~60% As^{3+} in high Ca- and vesicular non-glass particles (6).
"Class F" fly ash 1. (Zielinski et al. 2007) 2. (Catalano et al. 2012)	B-XAS L (1); C (2) RT	Bituminous coal → "Class F" fly ash (acidic waters). 1. As^{5+} in jarosite ≅ goethite > uncharacterized FeOOH (LC-LSF fits). 2. As^{5+} in $FeAsO_4$ or angelellite [$Fe_4O_3.(AsO_4)_2$]. Latter speciation is more likely at high temperature and more analogous to Class C findings.
Wet aging or Burial of C- and F-type fly ashes 1. (Catalano et al. 2012) 2. (Veselska et al. 2013)	B-XAS (Catalano) m-XAS (Veleska) RT Also Zn	1. No observed changes in As speciation in C or F types from USA. 2. No significant changes in leaching or speciation of As after 45 years of burial under agricultural field.
Poisoned oil shale catalyst (Cramer et al. 1988)	B-XAS F	The presence of As poisons the Ni-Mo hydrotreating catalyst used for removal of organic N from raw shale oil. Fingerprinting comparison to the NiAs FT shows similarity, but not identical speciation. No As^{3+}-O or As^{5+}-O peaks were observed.
Burnt fuel oil particulates (Huffman et al. 2000)	B-XANES RT	Particulates contain carbonaceous cenospheres (mostly graphitic C) and vesicular particles (dominantly sulfate minerals by XRD and other techniques). Other metals (V, Ni, Fe, Cu and Zn) appear sequestered in sulfates; As^{5+} speciation was not determined (might also be substituting in sulfates)
Solids from Municipal waste treatment		
Fe^{3+} and Al -based Water Treatment Residuals (Fe^{3+}WTR) 1. (Makris et al. 2007, 2009) 2. (Makris et al. 2009)	B-EXAFS C +/- MS	1. As^{3+}, As^{5+} both sorb in ^2E coordination (+ MS only one set of paths); strong sorption insensitive to time, high PO_4 2. As^{3+}, As^{5+} both sorb in ^2E, ^2C, and ^1V coordination (no MS included); strong sorption insensitive to time, high PO_4
Fe-based WTR simulated landfill conditions (Root et al. 2013)		See Appendix 2
Spent adsorbent: hydrous Al oxide, Ti dioxide, and hydrous Fe oxide (Liu et al. 2008)	B-EXAFS C	2 yr anaerobic incubation (no microbes): As^{5+} → As^{3+}: 35% (activated alumina); ~75% (TiO_2); > 80% (hydrous ferric oxide). On HFO: As^{3+} sorbs in ^2C geometry; Fe EXAFS = some pyrite formation. No evidence for As^{3+}-S
Spent adsorbent: amorphous Al hydroxide sludge (Kappen and Webb 2013)	B-XAS F, C LT	^2C complex; some evidence for ordering beyond adsorption, but As-Al distances not equal to that in mansfeldite ($AlAsO_4 \cdot 2H_2O$)

System	Data types(s) and analysis	Major findings
Spent adsorbent: synthetic and natural Siderite (FeCO$_3$) (Guo et al. 2013; Zhao and Guo 2014)	B-XAS F, C RT	(+ O$_2$): goethite forms, enhances As^{3+}adsorption and oxidation to As^{5+}. ^{2}C and ^{1}V complexes for both As^{3+} and As^{5+}. (– O$_2$): As^{3+} forms ^{2}C complexes $R_{\text{As-Fe2+}}$ (siderite) and $R_{\text{As-Fe3+}}$ goethite. Activated natural siderite is partially oxidized, no reduction of As^{5+} upon sorption
Spent adsorbent: Zero valent iron (Yan et al. 2012)	B-EXAFS	As^{3+} sorbs to ferrihydrite outer rind, diffuses inward to Fe^0 core. Forms Fe^0-As^0 (intermetallic) phase (no evidence for As-As bonds or FeAs$_2$)
Water-treatment residual: Ti-based (Ti-WTR) (Liu et al. 2008)	B-XAS L, C	Simulated disposal of WTRs under anaerobic conditions (2 year). Partial reduction of As^{5+} in most adsorbents; complete reduction to As^{3+} in granular ferric oxide adsorbent; sorption in ^{2}C geometry
Hydrotalcite-type layered double hydroxides (LDH) 1. (Paikaray et al. 2013) 2. (Wu et al. 2013)	B-EXAFS C RT	Potential water treatment residual. 1. Fe-rich LDH: As^{5+} binds to Fe-sites (^{1}E geometry) unlike Se^{6+} and Mo^{6+} (outer-sphere sorption in interlayer) 2. Mg-Al LDH: As^{5+} outer-sphere sorption in interlayer
Copper-Chromate-Arsenate-treated wood (fresh and aged) and wood "residue" 1. (Bull et al. 2000) 2. (Nico et al. 2004) 3. (Nico et al. 2006)	B-EXAFS C RT	1. CCA-treated pine wood: As^{5+} bound to Cr in a CrAsO$_4$ • nH$_2$O-like phase; same species in CCA-sludge. 2. Cr-dimer binds to cellulose; As sorbs in ^{2}C geometry to dimer $R_{\text{As-Cr}}$ distance = 3.25 Å. 3. CCA residue from human hand contact is equivalent to that removed by wire brush
Soil contaminated by CCA-treated wood 1. (Nico et al. 2006) 2. (Grafe et al. 2008c) 3. (Hopp et al. 2008) 4. (Schwer and McNear 2011)	1, 3, 4: B-XAS L, C 2, 3, 4 U-XRF/XAS L	1. As in soil was not dissolved by human sweat (suggests low dermal absorption). 2. As^{5+} correlated with Fe, Mn, Cu, and Zn (μ-XRF). Complex speciation with of Cu, Fe arsenates and As^{5+} sorption on goethite and gibbsite suggested. 3. All soil depths: As^{5+} dominant. Surface soil: CCA residues dominant. Soil Bs horizon: As^{5+} sequesterd by proto-imogolite allophane. 4. Most of the As in soil around CCA-fence posts is immobile, but with the large volumes of fencing used, large As loads to soil can result.
Urban particulate matter, Athens Greece (Godelitsas et al. 2011)	U-XRF/XANES F	As concentrated in PM$_{2.5}$ (respirable); As$^{5+} \cong$ As^{3+}
House dust, Ottawa, Canada (Walker et al. 2011)	U-XRF/XRD vacuumed dust samples also: Cu, Cr, Pb	As enriched in living room (LR) dust, not bedrooms (N =2). Strong As-Cr correlation and ID of wood cellulose in dust using μXRD points to CCA-wood of unknown source outside the home as the origin of the As in LR dust.
Tobacco and cigarette smoke (Liu et al. 2012)	B-XANES F, L LT	Smoke: As^{3+}-O + As^{5+}-O > As^{3+}-S >> organoAs. Ash and tobacco: As^{5+}-O. Smoke particulate: As^{3+} > As^{5+}. Cryotrapping required for accurate speciation of smoke and particulates.
Urban contaminated soil, Baltimore, MD, USA (Landrot et al. 2012)	B-EXAFS, U-XRF/XAS RT Sections	Association of As^{5+} with Al and small fraction removed by leach tests suggests formation of a mansfeldite-like phase (AlAsO$_4$·2H$_2$O)

B = bulk ; U = microbeam; XAS = XANES + EXAFS; Analysis type: fingerprinting (F), linear combination (L), principal component analysis (P), and curve fitting with theoretical standard (C), as discussed in the text. LT = cryogenic < 77K temperature; RT = ambient temperature; No O$_2$ = anaerobic

APPENDIX 8

As-XAS studies to improve sequential extraction, *in vitro* bioaccessibility (IVBA), and *in vivo* relative bioavailability (RBA) tests for arsenic. Standard reporting error for LC-LSF: ±10%. Standard reporting detection limit: <5-10%. Typical reported error for EXAFS-derived interatomic distances (R) is ±0.02 Å, and for coordination number (N) is ±20%.

System	Data types(s) and analysis	Major Findings
As^{5+} sorbed to ferrihydrite (Beak et al. 2006)	B-EXAFS C RT	IVBA = 0-5%; no change in ^{2}C sorption geometry post-extraction (see Fig. 12). As released during intestinal step was variable and a function of total As concentration and adsorption maxima. Readsorption of As on neo-precipitated Fe phases during intestinal step is possible.
Residential soils contaminated by mining/smelting (Bradham et al. 2011)	B-EXAFS RT L	Simple linear regression used (multivariate did not improve results). Correlation of individual As species with IVBA/RBA (mouse) was low ($R^2 < 0.2$) and insignificant ($p \geq 0.10$) for all species except arsenopyrite ($R^2 = 0.28$, $p = 0.09$ for RBA). Log (Fe+Al) gave best correlation with IVBA and RBA was best: 0.8 and 0.62, respectively, $p < 0.01$)
Core sediment from Red River Delta, Vietnam (Eiche et al. 2010)	U-XRF RT Epoxy sections	Initial speciation: As mainly with Fe-oxyhydroxides. PO$_4$ leach: 34-66% of As removed, As-Fe correlations in maps disappear + and higher Fe/As ratio on most grains. Most grain coatings dissolved. As-Fe correlations remained in K-bearing grains; interpreted as unreactive silicates
As-contaminated soils mixed with peat moss (Meunier et al. 2011a)	B-XANES/U RT bulk < 150 mm; XRF polished section	Peat moss added to 29 sediments (mine wastes, soils: residential, mine-impacted, and reference), + aging. Significant ($p =0.032$) increase in IVBA in ½ of tested soils, specifically those with native $C_{org} > 25\%$ and > 30% IVBA. No significant correlation between As speciation by XANES and IVBA.
Au mine tailings and impacted sediment, Nova Scotia, Canada (Meunier et al. 2011b)	U RT	Tested IVBA (2 methods) on < 250, < 150, and <45 µm size fractions; Highest % IVB: samples with As^{5+} (species unidentified), potentially adsorbed. Lowest % IVBA: samples with encapsulated As-rich grains (arsenopyrite and/or As^{5+} mineral particles).
Mining-impacted soils, CA, USA (Au mines) (Foster et al. 2014)	B-EXAFS L RT Also: Fe EXAFS	As speciation on 19 samples comparison with IVBA (all 19) and RBA (juvenile swine; 12 of the 19 samples). Significant (> 0.05) Pearson correlations: As-RBA (none), IVBA: As^{5+}-gibbsite, As^{5+} jarosite (> +0.8), and arsenopyrite (-0.99). Also significant correlations with several physical, chemical, and mineralogical parameters.
Traditional Indian medicines (Koch et al. 2011)	B-XANES; L RT	Physiologically-based extraction applied to medicines; 29% had bioaccessible As, which was positively correlated to Σ (As^{3+}-O + As^{5+}-O) and negatively correlated to Σ (As-S).
Bituminous and Lignite (low-rank) Coals (Huggins et al. 2002)	B-XANES; L RT	4-step USGS coal-leaching method (NH$_4$ acetate, HCL, HF, HNO$_3$). Both coals: NH$_4$ acetate → weakly-bound As^{5+} (minor species). HCl → As^{5+} from pyrite oxidation during storage. HNO$_3$ → dominant As species (both coals). Bituminous = As^{-1}-pyrite; Lignite = As^{3+}-O (putative organoAs) + As^{-1}-pyrite. HF released no As.

B = bulk ; U = microbeam; XAS = XANES + EXAFS; Analysis type: fingerprinting (F), linear combination (L), principal component analysis (P), and curve fitting with theoretical standard (C), as discussed in the text. LT = cryogenic < 77K temperature; RT = ambient temperature; No O$_2$ = anaerobic

APPENDIX 9

XAS studies of arsenic speciation in biota. Standard reporting error for LC-LSF: ± 10%. Standard reporting detection limit: < 5-10%. Typical reported error for EXAFS-derived interatomic distances (R) is ± 0.02 Å, and for coordination number (N) is ± 20%. See Figure 1 for explanation of abbreviations.

Organism	Data types(s) and analysis	Major findings
Halomonas spp. (Wolfe-Simon et al. 2010)	U-EXAFS; F, C RT (air-dried)	Intracellular: As^{5+}-O
Pseudomonas spp. (Patel et al. 2007)	U-EXAFS, XANES F, L, C; LT	65% As^{3+}-O; 30% As^{5+}-O; 5%: As^{3+}-S
Euglena gracilis (Miot et al. 2008, 2009)	B-XAS; L, C LT (10°K) dried cells + cell fractions + move between scans	As^{3+}-S (complex with glutathione or phytochelatin) concentrated in cytoplasm when exposed to As^{3+} or As^{5+} ≤100 mg L^{-1} Exposed to >100 mg L^{-1} As^{5+}: same As^{3+} species present but As^{5+} partitions into positively charged fraction of nuclear membrane
Seaweed (George et al. 2009)	B-XANES; L LT (10°K)	(1) R_3As^{5+}O, (2) R_3As^{3+}, (3) As^{5+}-O. Species (1) dominant in red algae *Porphyra. perforate* and commercial nori; species (2) dominant in *Ulva lactuca* and *Nereocystis luetkeana*. Species (3) found only in *Ulva* and *Nereocystis*
Phytoplankton and zooplankton (Caumette et al. 2011)	B-XANES; F, L RT	As^{5+}-O > As^{3+}-S (*tris*-glutathione) >> organoAs (only detected with HPLC); zooplankton: freeze-dried, then ground; phytoplankton: analyzed directly on filter
Trichoderma asperellum, *Penicillium janthinellum,* and *Fusarium oxysporum* (fungi) (Su et al. 2012)	U-XANES; L mycelia on SiN window	*T. asperellum and P. janthinellum*: As^{3+}-O > $DMAs^{5+}$, >> $MMAs^{5+}$. *F oxysporum*: As^{3+}-O > $MMAs^{5+}$ >As^{5+}-O Methylation rate-limiting for arsenic efflux
Agaricus bisporus (white button mushrooms) (Smith et al. 2007)	U-XANES; F LT (-20°C)	AB ≅ As^{3+}-O > $DMAs^{5+}$>> As^{5+}-O ≅ TMAO (HPLC-ICP-MS data) AB localized in stalk and cap, used to maintain turgor pressure
Pteris vittata L. 1. (Webb et al. 2003) 2. (Pickering et al. 2006)	B-XAS (1, 2); U-XRF (2) L (XANES), C (EXAFS) RT (continuous scanning mode)	(1) Fresh young and mature leaves grown in 1000 mg kg^{-1} As: As^{3+}-O; As^{3+}-S detected in leaves grown in 10,000 mg kg^{-1} As. As^{5+} → As^{3+} in root (2) Dried leaves and spores: As^{5+}-O dominant (20% As^{3+} in spore, 0% in dry leaves); As^{5+} not reduced in root
Pityrogramma calomelanos (Kachenko et al. 2010)	U-XRF/XANES; F LT (-20°C) Flash frozen samples	Roots: As^{5+}-O → As^{3+}-O, which is transported to above ground biomass. Leaves: similar results as for *Pteris vittata L.* studies above
Pteris vittata L., Pteris cretica L., (hyperaccumulators) and, *Athyrium yokoscense* (Kashiwabara et al. 2010)	U-XRF/XAS *In vivo* analysis	All species' gametophytes: As^{5+}-O → As^{3+}-O. As accumulates in reproductive area of *P. cretica* (asexual reproduction), but not in *P. vittata* (sexual reproduction)

Organism	Data types(s) and analysis	Major findings
Salsola kali (tumbleweed) (De la Rosa et al. 2006)	B-XAS; F, C LT freeze dried	Plants exposed to aqueous As^{3+}-O or As^{5+}-O. Both treatments, same speciation: As^{3+}-S in roots, stems, leaves
Chilopsis linearis (desert willow) (Castillo-Michel et al. 2009)	B-XAS; F RT freeze dried	Wild-type and cultivated plants exposed to aqueous As^{5+}-O. $As^{5+} \rightarrow As^{3+}$-S (incomplete) leaves > root > stem. Wild type: more As^{3+}-S, and more As tolerance
Parkinsonia florida (Castillo-Michel et al. 2011)	U-XRF/XANES LT (-20°C) fresh root sections	*P. florida* accumulates in roots; no translocation above ground As^{3+}-S >> As^{3+}-O $\cong As^{5+}$-O
Prosopis juliflora-velutina (mesquite) 1. (Aldrich et al. 2007) 2. (Castillo-Michel et al. 2012)	1. B-XAS (RT) 2. U-XRF/XANES (LT) freeze dried (both) ground (1) or sectioned (2)	1. Exposed to aqueous As^{3+}-O and As^{5+}-O (simultaneously): preferential uptake of the latter. As^{5+}-O $\rightarrow As^{3+}$-S in roots; only As^{3+}-S in stems, leaves 2. Little translocation to stems, leaves; same speciation as (1)
Raphanus sativus (radish) (Smith et al. 2008a)	B-XANES, U-XRF/ XAS L B: LT, U: RT	Stem vascular bundles + leaves: alternating As^{3+}-O and As^{5+}-O-rich zones. Radish root: As^{3+}-S >> As^{5+}-O (localized in outer red layer) *In vivo* (leaf, stem); frozen (root)
Pisum sativum (pea) (Castillo-Michel et al. 2007)	B-XANES F, L RT (lyophilized)	Exposed to aqueous As^{3+}-O or As^{5+}-O. Both treatments: As accumulation only in roots. Higher phytotoxicity to As^{3+}-O. In roots: $As^{5+} \rightarrow As^{3+}$; As^{3+}-S >> As^{3+}-O
Oryza sativa **Quest** (rice) (Smith et al. 2009)	B-XAS, U-XAS/XRF L RT freeze dried	Exposed to aqueous As^{5+}-O. As present in all tissues; correlated with Fe in roots, correlated with Cu in leaves. As^{5+}-O and As^{3+}-O, with As^{3+}-O dominant in areal portion
Oryza sativa L. (Calrose rice) (Seyfferth et al. 2010, 2011)	U-XAS/XRF, U-tomography L RT *in vivo*	Exposed to aqueous As^{5+}-O. Young roots (no Fe plaque): As^{5+}-O, As^{3+}-O >> DMA >> As^{3+}-S. Older roots (some Fe plaque): As^{5+}-O $\cong As^{3+}$-O. Root cross sections: xylem As^{5+}-O and DMA. Vacuole: As^{3+}-O and As^{3+}-S. Bran layers: As^{5+}-O >> DMA. Germ: As^{5+}-O, As^{3+}-O; no As in endosperm
Brassica juncea (Indian mustard) (Pickering et al. 2000):	B-XAS F, L, C LT	Exposed to aqueous As^{5+}-O. Uptake primarily in roots, little translocation. Main species: As^{3+}-S (*tris*-Glutathione model; 3 S, $R_{As-S} = 2.25$Å)
Mytilus edulis (blue mussels) (Whaley-Martin et al. 2012) *Littorina littorea* (periwinkles, sea snails) (Whaley-Martin et al. 2013)	B-XAS, U-XRF/ XANES L LT (both) Ground or sectioned	1. As concentrated in digestive gland, as As^{3+}-S and As^{3+}-O; >> organoAs (by HPLC); more inorganic As in mussels from contaminated area 2. Contaminated area: As^{3+}-O >> As^{3+}-S > organoAs >> As^{5+}-O. Pristine site: AB > arsenosugar>> As^{3+}-O $\cong As^{3+}$-S
Mamestra configurata (Bertha armyworm moth) (Andrahennadi and Pickering 2008)	B-XAS (LT); U-XRF/XAS (*in vivo;* N_2 used to anesthetize) L	As^{3+}-S (As^{3+}-*tris*-glutathione model) is dominant in larval, pupal, and adult stages. Arsenic is concentrated in gut. Foregut also contains ~ 20% As^{5+}-O.

Organism	Data types(s) and analysis	Major findings
Lumbricus rubellus (earthworm) 1. (Covey et al. 2010) 2. (Button et al. 2011)	1. U-XRF/XANES; 2. B-XANES 1. *in vivo* 2. Dried @ 50 °C; ground	1. As^{3+}-S (As^{3+}-*tris*-glutathione model) is dominant species 2. As^{5+}-O > As^{3+}-O. Field-collected organisms: more organoAs than lab-grown, might be due to absence of natural gut microbes in latter
Grasshoppers, ants, spiders, caterpillars, moths, dragonflies, slugs, flies (Moriarty et al. 2009)	B-XANES, U-XRF/ XANES LT	All species: mix of As^{5+}-O and As^{3+}-S (low or no organoAs; only MMA and DMA detected). Caterpillars, ants, dragonflies, mosquitoes: As^{5+} > As^{3+}. Moths, slugs, spiders: As^{3+} > As^{5+}
Dogfish (George et al. 2009)	B-XAS; L; LT	skeletal muscle: aliphatic arsonium (fit using AB)
Gopherus agasszzi (Mojave desert tortoise) (Foster et al. 2010)	U-XRF/XAS F, L RT (formalin)	Scute: metabolized As distribution follows growth lamellae; XANES best fit: As^{3+}-O; Exogenous As easily distinguished from metabolized As. Lung tissue: particulates containing As^{5+}-O, As^{3+}-S, and arsenide
Rodent fur and **bird feathers** (Smith et al. 2008b)	U-XRF/XAS RT	As^{3+}-S > As^{5+}-O = As^{3+}-O; Low MMA, DMA. Keratin reduces As^{5+} applied as solution. Exogenous As^{5+} difficult to distinguish
Human nail clippings 1. (Pearce et al. 2010) 2. (Gault et al. 2008)	1. U-XRF/XAS, 2: B-XAS Both: L RT	1. As distribution follows nail growth pattern. As^{3+}, possibly S- and CH_3-coordination 2. As^{3+}-O >> As^{3+}-S (authors note calibration issues)
Human hair 1. (Gault et al. 2008) 2. (Kakoulli et al. 2014)	U-XRF/XAS F, L RT	1. As^{3+} (indeterminate, but see comments above for nail clippings) 2. As^{3+} throughout; increasing As^{5+} with hair age suggests oxidation
Human hepatocyte cells (HepG2) 1. (Munro et al. 2008) 2. (Bacquart et al. 2010)	1, 2: U-XRF/XAS, F, L 1, 2. LT	HepG2 is a cell model for cancer; study effect of Trisenox (As_2O_3), which is used to treat some types of leukemia 1. accumulation in cell nucleus as As^{3+}-S (*tris*-glutathione model) 2. accumulation in cytosol, nucleus, and mitochondria as As^{3+}-O; limited oxidation to As^{5+}-O in nucleus observed in 1 of 9 cells

B = bulk ; U = microbeam; XAS = XANES + EXAFS; Analysis type: fingerprinting (F), linear combination (L), principal component analysis (P), and curve fitting with theoretical standard (C), as discussed in the text. LT = cryogenic < 77K temperature; RT = ambient temperature; No O_2 = anaerobic

Reviews in Mineralogy & Geochemistry
Vol. 79 pp. 371-390, 2014
Copyright © Mineralogical Society of America

Measuring Arsenic Speciation in Environmental Media: Sampling, Preservation, and Analysis

Matthew I. Leybourne

ALS Geochemistry
2103 Dollarton Hwy, North Vancouver
British Columbia V7H 0A7, Canada

Now at: Department of Earth Sciences, Laurentian University,
Sudbury, Ontario P3E 2C6, Canada

mleybourne@laurentian.ca

Karen H. Johannesson

Department of Earth and Environmental Sciences
Tulane University
New Orleans, LA 70118-5698, U.S.A.

Alemayehu Asfaw

ALS Geochemistry
2103 Dollarton Hwy, North Vancouver
British Columbia V7H 0A7, Canada

INTRODUCTION

There is an extensive literature relating to As contents of various geological and biological media, driven in large part because of As-related diseases, which include for example, melanosis, leucomelanosis, keratosis, hyper-keratosis, oedema, gangrene, skin cancer and extensive liver damage. As an example, using the keywords "arsenic" and "environmental" in the academic search engine Scopus yields >12,300 references (title, abstract, and keywords). However, there is also increasing recognition that the toxicology of As is controlled by the form (speciation) of As (Scopus "Arsenic" and "speciation" yields >2900 references). It is commonly held that inorganic As(III) is the most toxic form, followed by arsenate (As(V)), with the various methylated forms generally having much less toxicity, although the epidemiology of the methylated forms has not been as well studied (e.g., Bacquart et al. 2010; Kobayashi 2010; Quazi et al. 2011; Whaley-Martin et al. 2013). However, the effect of inorganic speciation on human metabolism is debatable because As(V) is rapidly reduced after ingestion. Recent studies have shown that cellular biomethylation can result in the production of trivalent methylated As species, which can be more toxic than inorganic As forms (e.g., Styblo et al. 2000; Mass et al. 2001; Dopp et al. 2010; Rahman and Hassler 2014). For a recent review of As toxicology, see Mitchell (2014, this volume). Because in most terrestrial waters As(III) occurs as a neutral species ($H_3AsO_3^\circ$), arsenite is more difficult to remove from solution in terms of water treatment, without first undergoing an oxidation step (e.g., Hu et al. 2012).

There are a large variety of techniques for the measurement of total As and arsenic species, both in the field and in the laboratory. For the former, the main challenges are analytical time, complexity of sample treatment and detection limits, whereas for laboratory analyses of As speciation, the primary challenge is the ability to preserve the As species unchanged from the

1529-6466/14/0079-0006$05.00
http://dx.doi.org/10.2138/rmg.2014.79.6

time of sampling to the time of analysis. In this paper, we review the various techniques for As analysis and speciation and discuss the advantages and pitfalls of different methodologies. Speciation analysis here refers to the analytical protocols that are used to quantify the individual chemical species of As in a sample. In reviewing the literature, an interesting point is that studies investigating collection and preservation methods for subsequent laboratory analyses indicate that stabilization of inorganic As species is possible, and that they can be stable for weeks to months if collected properly (McCleskey et al. 2004). Conversely, studies advocating field based speciation methods highlight the difficulties of species preservation (Voice et al. 2011).

Arsenic species

There are more than fifty As species that have been identified, with a large number of these mainly observed in marine environments (Francesconi and Kuehnelt 2004). The most important of these in terms of geochemical studies are the inorganic species arsenate and arsenite, along with methylated As species, monomethylarsenite (MMA(III)), monomethylarsenate (MMA(V)), dimethylarsenite (DMA(III)), and dimethylarsenate (DMA(V)) (Table 1). In marine animals the dominant species is arsenobetaine (AsB) and arsenocholine (AsC) and arsenosugars are common in some organisms (Dahl et al. 2010; Santos et al. 2013). Inorganic forms of As typically represent < 1% of total As in fish (Dahl et al. 2010).

A more thorough treatment of As speciation is given in Campbell and Nordstrom (2014, this volume). The form that As takes in waters, sediments, biological materials or in gaseous phases is primarily controlled by Eh and pH, organic matter content, microbial activity, and salinity (Smedley and Kinniburgh 2002; Campbell and Nordstrom 2014, this volume). Arsenic has a variety of oxidation states including As(V), As(III), As(0) and As(−III) (Leermakers et

Table 1. Common arsenic species in aqueous, geological and biological samples. Modified from Gong et al. (2002), Wilkin et al. (2003), and Suess et al. (2011).

Name	Abbreviation	Chemical formula
Arsenite (arsenous acid)	As(III)	H_3AsO_3
Arsenate (arsenic acid)	As(V)	H_3AsO_4
Monomethylarsonic acid	MMA(V)	$CH_3AsO(OH)_2$
Monomethylarsonous acid	MMA(III)	$CH_3As(OH)_2$
Dimethylarsinic acid	DMA(V)	$(CH_3)_2AsO(OH)$
Dimethylarsinous acid	DMA(III)	$(CH_3)_2AsOH$
Dimethylarsinoyl ethanol	DMAE	$(CH_3)_2AsOCH_2CH_2OH$
Trimethylarsine oxide	TMAO	$(CH_3)_3AsO$
Monothioarsenate	MTAs(V)	$H_xAsO_3S^{3-x}$
Dithioarsenate	DTAs(V)	$H_xAsO_2S_2^{3-x}$
Trithioarsenate	TTAs(V)	$H_xAsOS_3^{3-x}$
Tetrathioarsenate	TRAs(V)	$H_xAsS_4^{3-x}$
Monothioarsnite	MTAs(III)	$H_xAsO_2S^{3-x}$
Dithioarsenite	DTAs(III)	$H_xAsOS_2^{3-x}$
Trithioarsenite	TTAs(III)	$H_xAs_3^{3-x}$
Tetramethylarsonium ion	Me_4As^+	$(CH_3)_4As^+$
Arsenobetaine	AsB	$(CH_3)_3As^+CH_2COO^-$
Arsenocholine	AsC	$(CH_3)_3As^+CH_2CH_2OH$
Arsines	AsH_3, $MeAsH_2$, Me_2AsH	$(CH_3)_xAsH_{3-x}$; $(x = 0\text{-}3)$

al. 2006); however, the predominant dissolved species of arsenic are arsenate and arsenite. In most natural waters, for example, the MMA(V) and DMA(V) forms of As typically represent at most a few μg L^{-1} (Smedley and Kinniburgh 2002; Kumar and Riyazuddin 2010), although in reducing environments, methylated As can become significant (Kumar and Riyazuddin 2010). Thioarsenates and methylthioarsenates have also been measured in some geothermal systems, and thioarsenites are increasingly being recognized in sulfidic systems (Hollibaugh et al. 2005; Stauder et al. 2005; Planer-Friedrich et al. 2007; Suess et al. 2009) (Table 1). In this review, we focus primarily on the most important As species for geochemical studies. As noted above, these are the most toxic in terms of impact on humans and aquatic life (Leermakers et al. 2006).

One of the challenges for reliable determinations of As species is the preservation during collections, storage and analysis (i.e., to prevent conversion from one species to another); changes in pH, Eh, microbial activity and As losses by volatilization or adsorption to particulates or bottle walls must be avoided, and the species must be quantitatively transferred for analysis. There is, therefore, a challenge for As speciation studies; it is easier to determine inorganic As species by measuring these *in situ* rather than relying on preservation back to the lab, although for organic forms field analysis is too complicated. However, lab-based analytical methods are more precise, robust, and can be integrated with automation technologies to provide more rapid analytical turnaround.

FIELD COLLECTION AND ANALYSIS

Solid samples

Sampling of solid samples for As speciation includes rocks, sediments, and soils, as well as biological samples. Ko et al. (2009) used a standard field test kit for aqueous As (Econo-Quick) to investigate field measurements of As contamination in stream sediments. They found good agreement between field 1 N HCl extractions compared to laboratory HG-ICP-AES analyses. In the field, As was extracted from air-dried and sieved stream sediments (sieved to 75 μm and 150 μm) using 1 N HCl, as per regulatory requirements in South Korea. The resultant solution was then filtered and reacted with KI to reduce all of the As(V) to As(III), with subsequent reaction with sodium borohydride (NaBH$_4$) to produce H$_2$ and arsine gas, the latter was reacted with the test trips and the resultant color compared to standards.

Over the last decade, significant improvements have been made to field portable X-ray fluorescence instrumentation (typically abbreviated as pXRF or FP-XRF) (Potts et al. 2002; Peinado et al. 2010; Weindorf et al. 2013). As an example, Parsons et al. (2013) studied the applicability of pXRF for determining the As contents of soils, and found detection limits in the low ppm (6.8 mg kg^{-1}) with precision of 14.4% relative standard deviation (RSD). Hall et al. (2014) used five different pXRF models (two as benchtop models) to measure a large number of geologically relevant certified reference materials, and found that As was one of the trace elements for which the pXRF performed well, again with detection limits close to 10 mg kg^{-1}. For a more complete discussion on the use of pXRF in geological matrices, the reader is referred to a recent special issue (Hall et al. 2014). Although pXRF is not a speciation tool, for field use it could be used to rapidly quantify As contents, and serve as a screening mechanism for more detailed *in situ* As speciation testing.

In situ measurement of As species for solid samples is constrained by the need for more apparatus and greater time requirements compared to aqueous samples (see below). Huang and Ilgen (2006) performed a detailed study on the effects of sample collection and storage for soils and plant tissues, after noting that problems associated with solid sample collection for As had previously been largely ignored. Prior to this study, freeze drying was the most commonly employed method for preparing and storing sediment and biological samples prior to laboratory analyses (Huang and Ilgen 2006). Huang and Ilgen (2006) collected mineral soils, forest floor materials, wetland soils, moss, and tree needles and subjected them to a variety of processing and storage protocols. As speciation was determined by HPLC-ICP-MS after extraction by a mixture of methanol-water (20% v/v; Fig. 1); this form of extraction was designed to release only the most labile forms of As from these samples. This study revealed large changes in both extractable As abundance and As speciation. Freeze-drying of samples commonly resulted in lower As yields and increases in As(III/V) compared to wet samples, perhaps as a result of changes in organic matter structure and arrangement and/ or to mineralogical changes that occur during freeze-drying (Krachler and Emons 2000; Huang and Ilgen 2006). Conversely, As contents increased on air-drying of samples at 60 °C, consistent with breakdown of organic matter and therefore greater extractability of As at high temperature (Huang and Ilgen 2006). The As(III/V) value decreased in the moss and needles after drying at 60 °C, but increased in the organic-rich soils, most likely as a result of organic matter oxidation during the O_2-rich drying process.

However, other studies have found different results. For example, Dahl et al. (2010) showed that both the As contents and As speciation appeared to be largely unaffected by freezing and storage of fish samples in the laboratory. A recent study of marine/estuarine sediments also employed homogenization and freeze-drying of the sediments after collection for transport to the laboratory, where they were analyzed by HPLC-ICP-MS following extraction with H_3PO_4 and microwave heating (at 120 °C) (Mamindy-Pajany et al. 2013). Despite studies noting the problems with freeze-drying of wet soils and sediments, Mamindy-Pajany et al. (2013) did not comment on possible changes in speciation. Of significance is that this study also measured dissolved sulfide and ferrous Fe concentrations in the pore fluids of the marine sediments and documented the presence of abundant thioarsenates where pore fluid sulfide concentrations were in excess of 1 mg L^{-1}, with thioarsenates up to in excess of 1 mg L^{-1} in extreme cases. Thioarsenates have previously been identified in geothermal and hydrothermal fluids (Planer-Friedrich et al. 2007, 2009; Haertig and Planer-Friedrich 2012; Maher et al. 2013), although they are now increasingly being recognized in contaminated estuarine and landfill environments (Mamindy-Pajany et al. 2013; Zhang et al. 2014).

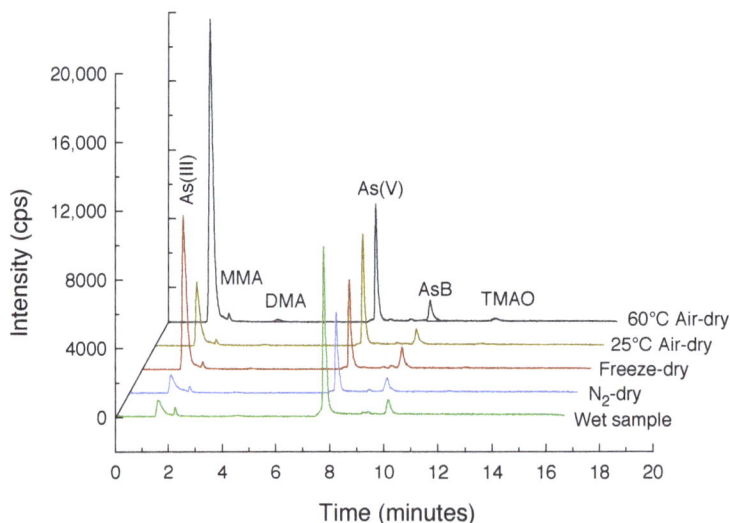

Figure 1. Chromatograms from HPLC-ICP-MS speciation experiments on wetland soils, showing the effects of different drying techniques and temperatures and the resulting As species conversions. Modified from (Huang and Ilgen 2006).

Aqueous samples

Total analyses of As. For the determination of total As, sample preservation for laboratory analysis is relatively straightforward. Most studies of surface and groundwater chemistry use ultrapure HNO_3 acid to keep metals in solution, and this approach works equally well for total As (Hall et al. 1999). The most common method of As analysis in the laboratory is by ICP-MS, and the temperature of the plasma is such that all As is converted to As^+ in the plasma, regardless of the speciation (see below). In marine studies, sample preservation is typically done with ultrapure HCl, and this acidification technique is also appropriate for total As, although complications arise during analysis by IC-MS, as described below.

Owing to the increase in As contamination in drinking water supplies in a number of countries (e.g., >10 million wells need to be tested for As in Bangladesh alone; Kinniburgh and Kosmus 2002; Rahman et al. 2002; Arora et al. 2009), there has been a number of relatively simple field test kits developed for *in situ* total As determination. Commercial kits are available from a number of manufacturers including Hach and Merck (Arora et al. 2009). However, there are a number of serious concerns with these commercially available test kits, as recently reviewed by Arora et al. (2009). Issues include: 1) poor quality of analyses, in particular at concentrations below 50-100 μg L^{-1}, compared to laboratory results; 2) many of the kits are based on hydride generation (i.e., the Gutzeit method; Sanger 1908; Kinniburgh and Kosmus 2002), and as a result generate potentially toxic arsine gas, and 3) make use of concentrated HCl, complicating the method.

Arsenic species. As noted by several research groups (e.g., Hu et al. 2012; Sorg et al. 2014), although there have been a large number of studies on the source, mobilization and transport of As^T in ground and surface waters, there have been relatively few well-characterized studies of terrestrial waters where full As speciation has been determined, owing to sampling complexity, cost, and on-site preparation and analysis time. Most of the studies on As speciation on natural waters report a relatively small number of samples (e.g., Korte 1991; McNeill and Edwards 1997; Hinkle and Polette 1999; Hering and Chiu 2000; Kim et al. 2002, 2003; Kent and Fox 2004; Haque and Johannesson 2006a,b; Haque et al. 2008; Peters and Burkert 2008; Munk et al. 2011).

There are three approaches to performing As speciation analyses on water samples: 1) In-situ separation and analysis; 2) *In situ* separation with laboratory analysis; and 3) preservation of As species followed by analysis in the laboratory.

Sorg et al. (2014) present the largest *in situ* separation study, with ~800 analyses from 65 sites in 28 states in the United States. In this study, As speciation was determined by first applying chromatographic separation of As(III) from As(V) in the field, the latter readily adsorbed to the column matrix, with As(III) eluted and collected for analysis. A similar approach was taken by Haque and Johannesson (2006a), Yalcin and Le (2001), and Wilkie and Hering (1998).

As noted above, there are a number of technical and safety issues with many As kits. To get around the issue of arsine gas production and to improve on the reliability of field As determinations, as well as to allow speciation determination, Hu et al. (2012) recently modified the molybdenum blue method (e.g., Stauffer 1980). This method involves colorimetric determination of As(III) after oxidation with potassium permanganate ($KMnO_4$); this oxidant has the advantage of having minimal interferences, lack of pH dependence, and rapid kinetics (Hu et al. 2012). To determine the As(V), a suitable reductant was investigated, and these authors determined that the most suitable in terms of reaction kinetics, efficiency and least production of harmful by-products was thiourea (CH_4N_2S). The subsequent field method involves acidifying a sample with 1% HCl along with addition of 10 μmol L^{-1} phosphate and subdividing the sample into three 10 ml aliquots. The $KMnO_4$ is added to one aliquot (to oxidize As(III)), the CH_4N_2S to the second (to reduce As(V)) and the third is left untreated. The samples are then left for 30 minutes to react, and a color agent (molybdenum blue) added before measurement by spectrophotometer at 880 nm. The study demonstrated good agreement between the field tests and analyses of As speciation by HPLC-atomic fluorescence spectrometry (Hu et al. 2012). Although the technique shows promise in terms of linear range of analyses (four orders of magnitude), minimization of hazardous by-products and "real-time" analyses, the length of time for an analysis is problematic for large-scale application.

For collection and preservation of As species for subsequent laboratory analysis, there is a large body of literature, although much of it is contradictory, in particular in terms of using different mineral acids to preserve the As species (Hall et al. 1999; McCleskey et al. 2004; Kumar and Riyazuddin 2010). For example, some researchers maintain that field separation is necessary to ensure that no changes occur to As speciation. By contrast, McCleskey et al. (2004) show that field collection strategies that include filtration to remove microorganisms, includes a reagent to prevent oxidation of dissolved Fe and Mn (and subsequent precipitation), and prevents exposure to sunlight should maintain the As(III/V) ratio, for samples without significant dissolved sulfide present.

Despite this disagreement, some general themes are evident from an analysis of the literature. Acidification of the sample is in most cases essential to prevent changes in As speciation before analyses can be performed in the laboratory. A number of reagents have been proposed for acidification including HCl, HNO_3, H_2SO_4, EDTA, and phosphoric acid, among others. Some studies have investigated the utility of different acids and collection methods by using distilled or deionized water, spiked with appropriate As species. Although these studies have their uses, typically the results from synthetic waters are quite different compared to studies that use natural water samples, with generally more rapid redox transformations (Kumar and Riyazuddin 2010). Note also that acidification of sulfidic waters will lead to precipitation of arsenic-sulfur phases via transformation of thioarsenates (Smieja and Wilkin 2003; Planer-Friedrich and Wallschläger 2009).

Filtration of a water sample is also an important consideration (Fig. 2). Except for waters with elevated Fe concentrations, there is typically minimal difference between filtered and unfiltered total As concentrations (e.g., Leybourne and Cameron 2008; Dahl et al. 2010; Sorg et al. 2014). However, filtration also removes much of the microbial content of a water sample (McCleskey et al. 2004), which along with acidification helps to minimize As speciation changes catalyzed by microbes. The choice of pore size is another consideration. Typically in aqueous geochemical studies, filters of 0.45 μm (as dictated by the United States Environmental Protection Agency) or 0.2 μm pore size are used, and many studies suggest that this size separates the particulates from the dissolved form. However, this is entirely operationally defined (Hall et al. 1996a), as many colloids and some organic species are much smaller than 0.45 μm. Some studies of As speciation have used 0.2 or 0.1 μm filters, and noted presence of As associated with natural organic matter and Fe-oxyhydroxides (Neubauer et al. 2013a). Note that although 0.2 μm are typically used to filter microbes, some microbial species have been shown to pass through even 0.1 μm filters (e.g., Wang et al. 2007b; Lee et al. 2010). Although there have been some studies that have used ultrafiltration to investigate trace metal partitioning on fine colloids and associated with DOM, studies that have combined ultrafiltration with As speciation are rare (Neubauer et al. 2013b). This might well be a fruitful avenue for further research. Figure 3 summarizes the changes that can occur for aqueous samples from collection to analysis.

Another consideration is exposure of collected samples to light. Although As(III) is apparently not photochemically reactive, Fe(III) is, so that changes in Fe oxidation and associated formation or dissolution of Fe-oxyhydroxide can promote As(III) oxidation (McCleskey et al. 2004) (Fig. 4). Therefore, most studies now advocate minimization of

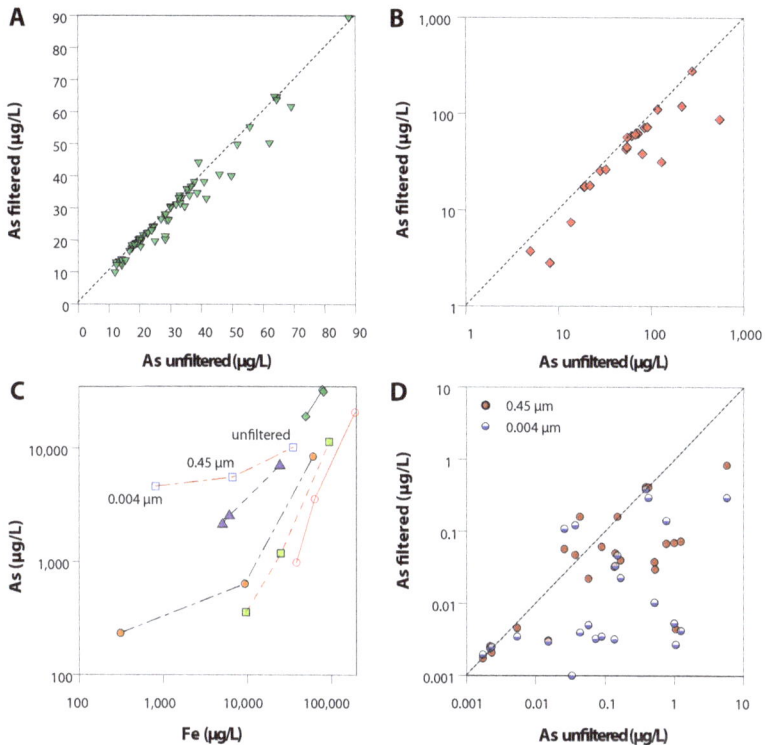

Figure 2. Plots showing the effects of filtration on As concentrations on different water types. A) Total versus filtered (0.45 μm) As concentrations from water supply facilities in the continental United States (Sorg et al., 2014). For most facilities, the amount of As associated with particulates is low. B) Total versus filtered (0.45 μm) As concentrations for groundwaters around the Spence porphyry Cu deposit, Chile. Here, "dissolved" As concentrations are generally similar to the unfiltered aliquots, although there is some As associated with Fe- and Mn-oxyhydroxides (Leybourne and Cameron 2008). C) Plot of As versus Fe for groundwater samples from a tailings pile associated with a volcanogenic massive sulfide deposit (Leybourne et al. 2000, data not previously published). The samples were analyzed for total As and Fe concentrations, < 0.45 μm filtered, and < 0.004 μm ultrafiltered (for details, see Leybourne et al. 2000). Samples are connected by tie lines, with both As and Fe showing decreasing concentrations with extent of filtration. D) Comparison of unfiltered As concentrations with < 0.45 μm filtered, and < 0.004 μm ultrafiltered concentrations for a series of groundwaters from the Bathurst Mining Camp, Canada (Leybourne et al. 2000, data not previously published).

exposure to light, either through storage in the dark (Hall et al. 1999; Leermakers et al. 2006; Kumar and Riyazuddin 2010), or collection in amber sample bottles (McCleskey et al. 2004), with the caveat that high-density polypropylene bottles are the preferred choice for collection for trace metal/metalloid analyses (Hall 1998).

Gaseous samples

Here we will only briefly discuss biovolatilization of As, and methods for sampling and analysis, as this has been the subject of an excellent recent review by Mestrot et al. (2013). However, as noted by these authors, volatilization of As is an understudied yet likely significant part of the As cycle, with implications not just for human and ecosystem health (Mestrot et al. 2013; Wang et al. 2014), but perhaps also in the understanding of migration of metals and metalloids from deeply buried mineralization to the surface environment (e.g., Cameron et al.

Step	Factors that influence As stability and speciation
Sample collection	Changes in pH, oxidation state, temperature, precipiation of new mineral phases (e.g., Fe-, Mn-oxyhydroxides, sulfides), adsorption/ desorption to colloid surfaces
Sample storage	Adsorption to bottle walls, changes induced by microbial activity, exposure to light, precipitation reactions, changes in As speciation
Sample treatment/ preservation	Filtered vs unfiltered, filter size (0.45 μm, 0.2 μm, 0.1 μm, ultrafiltration), acidification/ preservation (HCl, HNO$_3$, H$_2$SO$_4$ etc), As species conversion, formation of sulfides by acidification, temperature of storage
Analytical	Incomplete analyte recovery, changes in oxidation state in chromatography columns, conditions during hydride generation

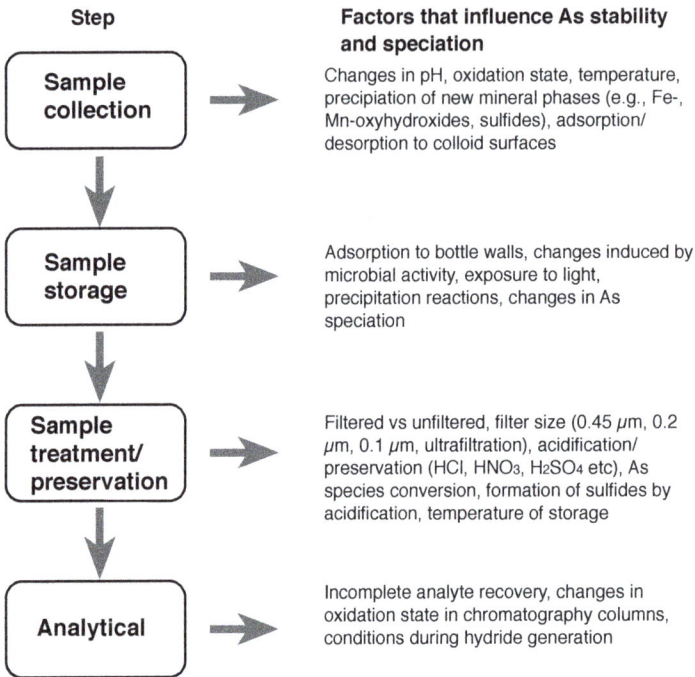

Figure 3. Potential causes of arsenic species interconversion from sampling to analysis. Modified from Kumar and Riyazuddin (2010).

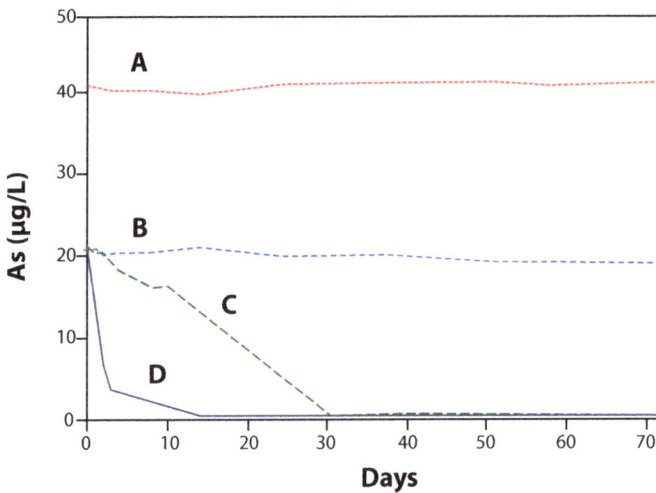

Figure 4. As(III) and As(total) as a function of time for a variety of conditions. A) Total As, with As(III) and As(V) at 20 μg L^{-1} each, and B) As(III) at 20 μg L^{-1} in the presence of 1 mg L^{-1} Fe(III) and 100 mg L^{-1} SO$_4$. Samples were acidified to pH 1.3 with HCl and stored in the dark. C) As(III) at 20 μg L^{-1} with 1 mg L^{-1} Fe(III), 10 mg L^{-1} Fe(II) and 1000 mg L^{-1} SO$_4$. D) As(III) at 20 μg L^{-1} with 10 mg L^{-1} Fe(III), 10 mg L^{-1} Fe(II). Curves C and D represent samples acidified with HCl to pH 1.3, but exposed to light. Modified from McCleskey et al. (2004).

2004). Volatile arsenic species have boiling points below 150 °C, and can be produced either biotically or abiotically (Wang et al. 2014). The primary volatile species are arsine (AsH_3) and the methylated forms ($MeAsH_2$, Me_2AsH, and Me_3As), although they have received relatively little research in terms of As cycling in the environment compared to solid and aqueous forms (Mestrot et al. 2011). Most studies have focussed on these four volatile As species, although ethyl-forms have been observed in landfills (Ilgen and Huang 2013) and chloromethylated arsines as well as a dimethylarsenomercaptane in geothermal areas (Planer-Friedrich et al. 2006). One of the drivers for the renewed interest in volatile forms of arsenic is the recognition that volatile As is more stable than previously thought (Jakob et al. 2010; Mestrot et al. 2011).

Mestrot et al. (2011) used Tedlar (polyvinyl fluoride) bags to trap gaseous As and determine the stability of arsine and the methylated forms under daytime (UV light) and night-time (dark) conditions. Volatile arsenic species after 7 days in the dark showed recoveries of 85%, 69%, 60%, and 25% for AsH_3, $MeAsH_2$, Me_2AsH, and Me_3As, respectively; the more methyl groups, the less stable the volatile species. Under UV light, the methylated forms were unstable, with a half-life on the order of 7 hours (Mestrot et al. 2011). By contrast, AsH_3 was relatively stable, even under UV conditions.

Another area where gaseous forms of As are important is in mineral exploration. Arsenic is a common pathfinder element associated with a variety of mineral deposit types, including epithermal Au, porphyry Cu, and volcanogenic massive sulfide, among others. There have been various attempts over the last couple of decades to trap volatile species in soils overlying mineral deposits using activated carbon traps (Klusman 2009). This method is only applicable to determination of total volatile As, however.

LABORATORY ANALYSIS

Solid samples

Direct analysis of solid samples. There are a number of methods that permit *in situ* analysis of As and As speciation using X-rays, which fall under the broad term X-ray Absorption Spectroscopy (XAS). These methods include X-ray Absorption Near Edge Structure (XANES) and Extended X-Ray Absorption Fine Structure (EXAFS). XAS analysis of As speciation is discussed in detail in Foster and Kim (2014). Other related methods include Nuclear Magnetic Resonance (NMR), Electron Spin Resonance (ESR), X-ray Photoelectron Spectroscopy (XPS), and Mössbauer spectroscopy.

Total extracts. There are a number of issues related to the analysis of As speciation in solid samples, including geological and biological media, including a dearth of certified standards with certified mass fractions values of different As species and lack of standard methods, which leads to considerable interlaboratory differences (Santos et al. 2013). A small number of certified reference materials have recently been introduced including Hijiki CRM 7405-a (Narukawa et al. 2012), tuna fish BCR-627 (Karadjova et al. 2007), and rice NIST SRM 1568 (Juskelis et al. 2013). A number of studies have used microwave-assisted extraction to measure total As in biological tissues, but speciation studies are less common or are focused on differentiating total inorganic arsenic (iAs) from methylated forms, in particular in plants and seafood (Rasmussen et al. 2012). Santos et al. (2013) used both microwave and ultrasonic digestion, and found that the former was more effective in extracting all available As. Critical to preserving As species using this method is a mild extraction (either H_2O or methanol-H_2O) in order to avoid analyte loss and interspecies conversions, while at the same time extracting all available As (Santos et al. 2013). In this study, changes in As speciation was monitored by using oyster certified reference materials and spiking these with known amounts of the main inorganic and methylated As species. Although these authors found little change in As speciation through the use of microwave digestion, it is unclear the extent to which spiking certified materials mimics the natural variations and potential changes in As species compared to the unspiked natural samples tested (mollusk shells). For example, a recent study by (Amaral et al. 2013) investigated a variety of extraction methods, including microwave, and found in some case, in particular where H_2O only or HNO_3 preservation was used, As(III) was converted in part to As(V); methylated forms are more stable under the same conditions (Francesconi and Kuehnelt 2004). Although the methylated forms may be more stable with respect to interconversion of As species, previous studies have also shown that the methylated forms of As can be degraded when exposed to mineral acids like HNO_3 (Ali and Jain 2004). Some studies have shown that As species can change during chromatographic separations (Francesconi and Kuehnelt 2004; Niazi et al. 2011). The conversion of As(III) to As(V) during acid digestion is used to advantage in studies that differentiate iAs from AsB and MMA/DMA. In this case, in addition to acid digestion, H_2O_2 is added to promote the oxidation of arsenite to arsenate (Rasmussen et al. 2012).

Sequential extraction. Sequential extractions have long been used in attempts to understand the partitioning of various species into different fractions of sediments/soils (e.g., Tessier et al. 1979; Gruebel et al. 1988; Hall et al. 1996b; Cardoso Fonseca and Ferreira da Silva 1998; Hall et al. 2003) (Fig. 5). There are, however, limitations associated with this method for understanding As speciation, including limited selectivity (and lack of understanding as to how selective extractions really are) and redistribution/reprecipitation of extracted elements to other mineral phases (e.g., Kim et al. 2014)(Fig. 5). For example, although some studies refer to sequential/partial extractions as "selective", the extent to which a particular extractant is able to quantitatively target a specific phase or phases is largely unknown. In most cases, the selectivity of the extraction is inferred. Although different studies employ different leaches, the basic idea is to quantify the nonspecifically and specifically adsorbed As, As incorporated within amorphous and crystalline Fe and/or Mn oxides and oxyhydroxides, As within sulfide minerals and/or organic matter, and residual As (e.g., Creed et al. 2005; Goh and Lim 2005; Rubinos et al. 2005; Bok Jung and Zheng 2006; Liu and Cai 2007; Mir et al. 2007; Paul et al. 2009; Huang and Kretzschmar 2010; Niazi et al. 2011; Kim et al. 2014).

Sequential extraction procedures are useful for understanding the partitioning of As (and other species) between different portions of geological media, but typically the redox form of As is not preserved or measured (Niazi et al. 2011). However, an increasing number of studies attempt to also determine As speciation associated with each extraction step (Georgiadis et al. 2006; Liu and Cai 2007). For example, Creed et al. (2005) used a three-step sequential extraction procedure (10 mM $MgCl_2$ @ pH 8, 10 mM NaH_2PO_4 @ pH 7 and 10 mM $(NH_4)_2C_2O_4$ @ pH 3) performed as continuous flow in-line via IC-ICP-MS. The IC permitted separation of As(III) from As(V). They found that the initial extraction step was dominated by As(III), whereas As(V) dominated the subsequent extractions (Creed et al. 2005), consistent with the greater lability of As(III) compared to As(V).

Aqueous samples

Here we discuss the main analytical methods for the quantification of As in solution, either aqueous samples or solid samples from which As has been digested/extracted. Instrumentation and hyphenation methodologies have previously been reviewed (Francesconi and Kuehnelt 2004; Terlecka 2005). There a large number of methods to quantify As including atomic absorption spectroscopy (AAS), atomic fluorescence spectroscopy (AFS), inductively-coupled plasma atomic (or optical) emission spectroscopy (ICP-AES or ICP-OES), liquid chromatography (LC) with conductivity or UV detection, time-of-flight spectroscopy, and ICP-mass spectrometry (ICP-MS) among others. It is beyond the scope of this review to go through all of these methods in detail. Instead, we focus on some of the more important developments since the reviews noted above were published.

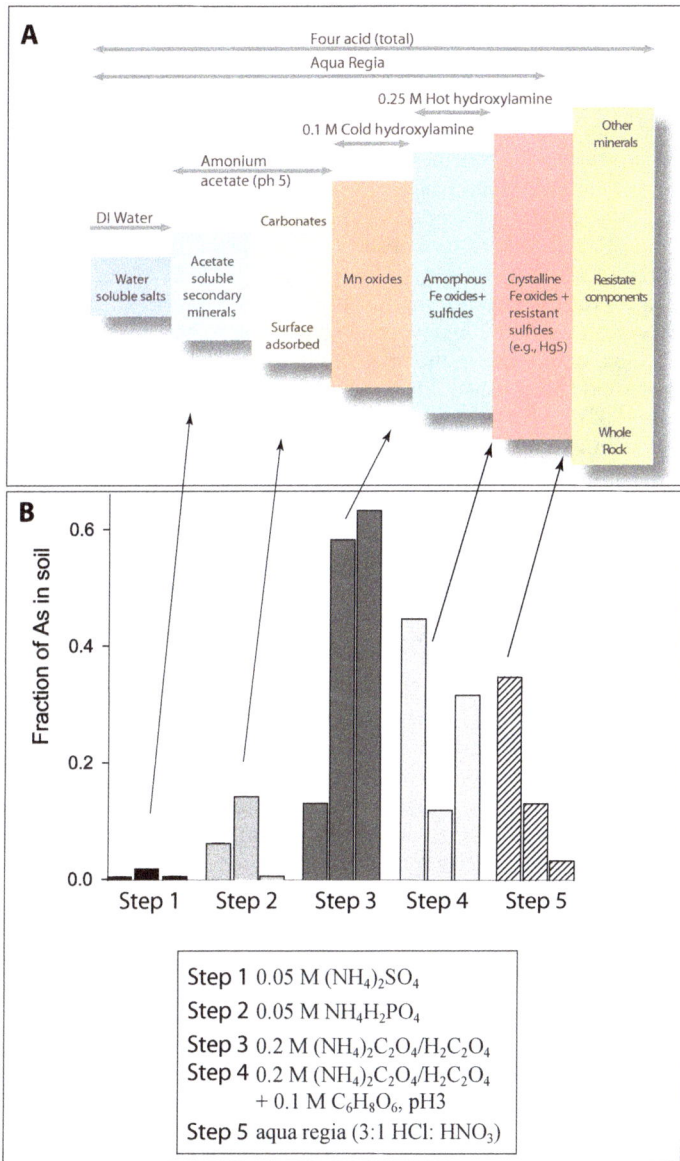

Figure 5. A) Sequential extractions are designed to liberate As associated with different parts of the solid sample; with increasing reagent leach strength, a greater portion of the sample is digested. Total digestion of the sample requires a four-acid digestion. B) Kim et al. (2014) used a five step extraction scheme to account for As associated with different portions of soil samples recovered from two smelter sites and a mine site. The arrows indicate the area of diagram A where each leachate extracted As.

ICP-MS. ICP-MS is the main analytical tool of choice in As detection. One of the great advantages of the ICP plasma is that the temperatures are high enough (up to 10000 K) that all molecular and organic forms of As are broken down to produce an ion stream of As^+, detected at mass 75. ICP-MS instruments are available with different types of mass filter, of which the quadrupole is by far the most common (Q-ICP-MS), owing its relative low cost compared to magnetic-sector instruments (high resolution and multi-collector; HR- and MC-ICP-MS, respectively). Although ICP-MS instruments have very low instrumental detection limits, for As low detection limits are hampered by the formation of polyatomic interferences within the ICP plasma. For many analytes, using a different isotope can minimize the effects of these interferences. Unfortunately, As is monoisotopic; although there are a number of potential interfering species, the most important are $^{40}Ar^{35}Cl^+$ and, especially in geological matrices, $^{40}Ca^{35}Cl^+$. There have been a number of ways suggested to get around these interferences, including use of HR-ICP-MS (in medium to high-resolution mode), reaction/collision cell ICP-MS, electrothermal vaporization (ETV-ICP-MS), hydride generation (HG-ICP-MS) and LC-ICP-MS (or ion chromatography IC-ICP-MS) (Ferguson et al. 2007). As a result, As detection limits can be lowered from low ppb to low ppt in some cases (Ferguson et al. 2007). For HR-ICP-MS, the differences in the mass of the As^+ peak (mass ~ 74.92 amu) and $ArCl^+$ (~74.93 amu) can be distinguished when the resolution of the instrument is set to 10,000 (Ferguson et al. 2007). Q-ICP-MS instruments have resolutions of 300-400 and so are not capable of resolving these two peaks. In a collision cell (CC) or dynamic reaction cell (DRC), an additional quadrupole or octopole is added before the primary mass filter quadrupole. A collision gas (commonly H_2 or He, or a mixture of the two) or reaction gas (e.g., CH_4, NH_4, O_2) is added at low pressures resulting in either kinetic breakdown of the interfering polyatomic ions or charge transfer reactions occur (e.g., Polya et al. 2003; Layton-Matthews et al. 2006; Planer-Friedrich et al. 2007; D'Ilio et al. 2008; Minnich et al. 2008). Reaction cells are increasingly being used for As speciation studies, such as the study by Wang et al. (2007a) who used DRC-IC-ICP-MS for As and Se speciation.

Hyphenated techniques – HPLC/IC. The real power of the low detection capabilities of ICP-MS is evident when combined with an in-line species separation method, or hyphenation. As with elemental analysis of As, there are a large number of separation methods that can be hyphenated with an ICP-MS (or others, such as AAS, AFS, ICP-AES). As well as the ability to separate a variety of As species, one potential advantage of using ion exchange columns in hyphenated systems with ICP-MS is that the problem of $ArCl^+$ is minimized as Cl is eluted at different times compared to the As species. However, different columns behave differently, and elution times can be highly matrix dependent (Sloth et al. 2003). Polya et al. (2003) observed that Cl and As(III) can co-elute. Further, care must be taken to calibrate the columns correctly with known standards. For example, some studies have shown that AsB and As(III) can be co-eluted using anion exchange columns, in which case cation exchange columns are preferable where analysis of AsB is important (Amaral et al. 2013).

A commonly used approach is that of Garbarino et al. (2002), whereby the inorganic arsenic species, As(III) and As(V), as well as the anionic methylated species MMA(V) and DMA(V) are separated, on-line, by HPLC coupled to an ICP-MS. This study used a 100 μL sample injection loop, HNO_3 in 0.5% methanol as the mobile phase, a Dionex AS7 anion-exchange column (4 mm × 250 mm), and a Dionex AG7 guard column (4 mm × 50 mm). The gradient program that is employed to separate the As species uses a mobile phase flow rate of 1 mL min^{-1}, where the initial (i.e., the first 3 minutes of the chromatograph) mobile phase composition consists of 0.0025 N HNO_3 in 0.5% methanol, the following 3 minutes uses 0.05 N HNO_3 in 0.5% methanol, and the remaining ~2 minutes employs 0.0025 N HNO_3 in 0.5% methanol (Garbarino et al. 2002). In this study, the various As species are subsequently directly introduced to the plasma of a HR-ICP-MS detector by coupling the HPLC to the ICP-MS. Detection limits of the HPLC-HR-ICP-MS for the various As species are estimated as low as 75 ng L^{-1} As, although these can be lowered, if necessary, by increasing the injection loop volume (Garbarino et al. 2002). Other studies have also coupled HPLC to quadrupole ICP-MS instruments (e.g., Londesborough et al. 1999; Le et al. 2004). HPLC-ICP-MS (quadrupole) has also been used to determine the reduced methylated arsenic forms, including monomethylarsenite (MMA(III)) and trimethylarsenite (TMA(III)) (Qin et al. 2009). For determination of thioarsenates, chromatographic separations are performed using NaOH as the eluent at alkaline pH conditions to prevent loss of thioaresenates by precipitation (Planer-Friedrich et al. 2007).

Hyphenated techniques – hydride generation. Arsenic forms stable volatile hydrides by reaction with sodium or potassium tetrahydroborate in an acidic medium. Efficient introduction of hydrides of As (Arsine) to atomic spectrometry (AAS, ICP-OES and ICP-MS), which involves efficient volatile hydride formation, separation of the gaseous phase from the liquid phase and transportation of the gases to the detector, provides low detection limits by improving the sample introduction efficiency and separating the analytes from the potential matrix and spectral interferences (Sturgeon and Mester 2002; Pohl 2004). The efficiency of hydride generation depends on the oxidation state of As; because low sensitivity is obtained for higher oxidation states, pre-reduction agents are normally used to reduce As(V) to As(III). L-cysteine is useful in pre-reduction of As and preventing interference from transition metals such as iron, which are found in higher concentrations in environmental samples (Chen et al. 1992). In any case, due to its superior advantage, hydride generation is the most popular sample derivatization method used for inorganic arsenic detection in water samples (Hung et al. 2004).

By varying the hydride generation reaction conditions, the technique can be used for non-chromatographic speciation of arsenic i.e., selective hydride generation (Kumar and Riyazuddin 2007; Pitzalis et al. 2007, 2014; Matoušek et al. 2013; Wang and Tyson 2014). Recently, Pitzalis et al. (2014) developed a HG-AAS method using different reductants such as borane-ammonia, borane-tert-butylamine, and sodium tetrahydridoborate in HCl and $HClO_4$ media, in the presence or absence of L-cysteine(Cys) for the selective determination of inorganic arsenic, monomethylarsonic acid (MMA(V)) and dimethylarsinic acid (DMA(V)). Unique experimental conditions were developed for selective generation of a given species (arsane, arsine, monomethylarsane, and dimethylarsane) in the presence of others species. For example, in this work DMA(V) and MMA(V) could be selectively determined in 0.5 and 10 M $HClO_4$ solutions, respectively, in the presence of Cys, with borane–ammonia as the reducing agent (Pitzalis et al. 2014). Previously, work by the same authors showed that As(III) was selectively determined in 0.005 M CH_3COOH in the presence of Cys (Pitzalis et al. 2007). Selective hydride generation of Arsine with/without the use L- cysteine with preconcentration by cryotrapping and detection by ICP-MS was used for non-chromatographic speciation analysis of inorganic As(III) and As(V) and methylated species in river water (NRC SLRS-4 and SLRS-5) and seawater (NRC CASS-4, CASS-5 and NASS-5) reference materials (Matoušek et al. 2013). In this work, As(III) was selectively determined by HG without L-cysteine and the sum of As (III) and As (V) with L-cysteine; the methylated species were resolved on the basis of thermal desorption of the formed methyl-substituted arsines after collection at −196 °C (Matoušek et al. 2013). Non-chromatographic hydride generation atomic spectrometric techniques for speciation analysis of Arsenic and other analytes was reviewed by Kumar and Riyazuddin (2007).

Hydride generation can also be used for pre-column or post-column derivatization. As discussed for non-chromatographic speciation analysis, pre-column derivitization is based on the formation of volatile arsines, which are cryogenically trapped, and sequentially desorbed into the gas chromatograph (Wuilloud et al. 2004; Asfaw and Beauchemin 2010). In post-column, hydride generation is used after the separation of the arsenic species on a HPLC system (Gómez-Ariza et al. 2000), improving the detection of the species sequentially eluted from HPLC by atomic spectrometry. For example, Gómez-Ariza et al. (2000) reported that with the HPLC-HG-ICP-MS system, ultra-trace levels of As (detection limit 0.1-0.3 μg L^{-1}) can be determined in environmental samples.

Musil et al. (2014) presented a new method for the rapid determination of inorganic forms of As in rice and seafood samples by hydride generation ICP-MS; analytical conditions were optimized so that organic forms (MMA(V) and DMA(V)) did not form hydrides and were sent to waste. The purpose of this study was to provide a rapid screening tool as it is generally thought that only the inorganic forms of As are toxic to human and aquatic life (Musil et al. 2014). As noted previously, trivalent methylated forms of As are also toxic (see Mitchell 2014, this volume). One of the interesting aspects of this study is that it is among the small number of studies that have utilized a new technology in ICP-MS, the triple-quadrupole (refered to as ICP-QQQ; Fernández et al. 2012). In this instrument there are two analytical quadrupoles with a reaction cell in between. Unlike the conventional reaction cell where the cell reaction by-products could be a potential interference (e.g., the formation of AsO^+ m/z 91 could interfere with Zr^+ m/z 91, for samples with high Zr contents), in the triple-quad arrangement the first quadrupole (Q), before the Octopole Reaction System cell, selectively rejects ions from all other masses except the ions from target mass enabling interference-free analysis. This configuration permits greater discrimination and elimination of interfering species on many analytes, in particular As.

Hyphenated techniques – gas chromatography. Using GC as a sample introduction for ICP-MS is powerful technique for speciation analysis because it provides high chromatographic resolution and low background from the carrier gas, improves sample introduction efficiency, and separates the analytes of interest from the matrix (Wuilloud et al. 2004; Asfaw and Beauchemin 2010). However, the capability of GC is limited to volatile and thermally stable species and it requires a derivatization step to transform the analytes to the volatile form (Asfaw and Beauchemin 2010). The volatile species of arsenic commonly analyzed by GC are AsH_3, $MeAsH_2$, Me_2AsH, t-BuAs, Me_3As, Ph_3As and Et_3As (Wuilloud et al. 2004). Volatile species are cryotrapped in a liquid-nitrogen-cooled trap or a NaOH-filled CO_2 trap. For non-volatile species of arsenic, derivatization of the arsenic species into the corresponding hydrides is performed by using sodium borohydride in an acidic medium (Wuilloud et al. 2004). The volatile or the derivatized volatile species are then thermally released into the GC for analysis (Wuilloud et al. 2004; Asfaw and Beauchemin 2010).

Gaseous samples

As noted above, details on the analysis of gaseous As species has recently been reviewed by Mestrot et al. (2013). They show that there are currently three primary means of sampling and analysing the volatile As species. These include: 1) direct sampling with analysis by GC-MS or GC-ICP-MS; 2) cryotrapping (CT) with GC-MS, GC-ICP-MS, or GC-AFS (atomic fluorescence spectroscopy); and 3) chemotrapping with HPLC-ICP-MS or HPLC-AFS. Direct sampling involves collecting air samples via syringe, Tedlar bags, or by solid-phase microextraction (SPME). These methods are best suited to point-source analyses owing to the low volumes of volatile As released from soils (i.e., non-point source). SPME is useful for qualitative assessment of As species (Planer-Friedrich et al. 2006), but issues relating to matrix effects, unknown gas partitioning coefficients and lack of standard reference materials limit the utility of this method (Mestrot et al. 2013). Cryotrapping is also best suited for point-source studies, as the method typically requires the use of Tedlar bags for gas sample collection. This method also suffers from issues related to CO_2 and H_2O in the gas phase. However, CT hyphenated with ICP-MS is capable of low detection limits (Ilgen and Huang 2013). Finally, chemotrapping involves the reaction of volatile As species with liquid or solid phase chemicals such as H_2O_2, NaOCl, $Hg(NO_3)_2$, and activated charcoal, among others (Mestrot et al. 2013). In some cases, quantification of the different species is possible, e.g., where the gas samples are oxidized to their methylarsenate forms, although most studies report only total volatile As concentrations (Mestrot et al. 2013). However, this method can be used for more species-specific analyses. Uroic et al. (2009) used silica gel tubes impregnated with silver nitrate to trap CH_3As from natural gas, and found concentrations ranging from 0.2 to 1800 µg m^{-3}.

Ilgen and Huang (2013) presented a new method for the analysis of volatile As species. They used a crytorapping-cryofocussing (CT-CF) system hyphenated to a GC with detection by both ICP-MS and EI-MS (electron ionization). They found that the use of silanized glass beads (2 mm) within the cryotrap greatly enhanced the efficiency of the cryotrap, in particular for the methylarsenates. The CF was used to reduce peak broadening prior to entering the GC. The use of an ICP-MS as the detector permitted detection limits as low as 0.12-0.41 pg As, with low RSDs of 0.76-1.29%. The EI-MS was used to confirm the form of the various As species. This study also addressed one of the main issues with CT analyses for volatile As species; matrix effects where samples have significant CO_2 or H_2O in the sample volume (Mestrot et al. 2013). By using NaOH in-line prior to the CT, and by using liquid Ar ($-186\,°C$) rather than the more typically used liquid N_2 ($-196\,°C$), Ilgen and Huang (2013) were able to minimize the effects of CO_2 or H_2O, although they note that for samples with very low As, requiring much larger sample volumes, CO_2 or H_2O are likely to be problematic.

SUMMARY

In this paper we have shown that there are a large variety of techniques for field and laboratory measurement of total As and arsenic speciation. Indeed, there is a rich literature on all aspects of As speciation, with increasing recognition that the number of species important in As geochemical cycling and in epidemiological studies is much larger than previously considered. In the laboratory, increasingly sophisticated analytical techniques and instrumentation are pushing ever-lower detection limits and allowing separation of more As species. The main challenge for laboratory techniques, however, is in preserving the sample from the time and location of collection until the sample can be analyzed, preventing any change in As speciation. Conversely, separations and speciation analyses for inorganic arsenic that are performed *in situ* (i.e., in the field) have the advantage that analyses can be performed before interspecies conversion can take place. However, field protocols need to be carefully considered; large-scale studies using field test kits are not as accurate as required to properly assess the impact of As on human and biological health. The researcher therefore has the complicated decision for the optimal protocols for sample collection, sample preservation, and field and laboratory techniques. These decisions ultimately require a "fit-for-purpose" approach.

ACKNOWLEDGMENTS

We thank Britta Planer-Friedrich and an anonymous reviewer for constructive comments that greatly improved the manuscript. Unpublished As data was collected with funding from the Geological Survey of Canada as part of the EXTECH-II program. We also thank Rob Bowell, Kirk Nordstrom, Heather Jamieson, Charlie Alpers and Juraj Mazlan for their efforts in organizing this volume and the short course as part of Goldschmidt 2014.

REFERENCES

Ali I, Jain CK (2004) Advances in arsenic speciation techniques. Int J Env Anal Chem 84(12):947-964, doi: 10.1080/03067310410001729637
Amaral CDB, Dionísio AGG, Santos MC, Donati GL, Nóbrega JA, Nogueira ARA (2013) Evaluation of sample preparation procedures and krypton as an interference standard probe for arsenic speciation by HPLC-ICP-QMS. J Anal At Spectrom 28(8):1303, doi: 10.1039/c3ja50099c
Arora M, Megharaj M, Naidu R (2009) Arsenic testing field kits: some considerations and recommendations. Environ Geochem Health 31 Suppl 1:45-48, doi: 10.1007/s10653-008-9231-4
Asfaw A, Beauchemin D (2010) Sample introduction in ICP-MS: Gas Chromatography. Encyclopedia of Mass Spectrometry 5:58-73

Bacquart T, Devès G, Ortega R (2010) Direct speciation analysis of arsenic in sub-cellular compartments using micro-X-ray absorption spectroscopy. Environ Res 110(5):413-416

Bok Jung H, Zheng Y (2006) Enhanced recovery of arsenite sorbed onto synthetic oxides by l-ascorbic acid addition to phosphate solution: calibrating a sequential leaching method for the speciation analysis of arsenic in natural samples. Water Res 40(11):2168-2180

Cameron EM, Hamilton SM, Leybourne MI, Hall GEM, McClenaghan B (2004) Finding deeply-buried deposits using geochemistry. Geochem: Expl Env Anal 4:7-32

Campbell KM, Nordstrom DK (2014) Arsenic speciation and sorption in natural environments. Rev Mineral Geochem 79:185-216

Cardoso Fonseca E, Ferreira da Silva E (1998) Application of selective extraction techniques in metal-bearing phases identification: a South European Case Study. J Geochem Explor 61:230-212

Chen H, Brindle ID, Le XC (1992) Prereduction of arsenic(V) to arsenic(III), enhancement of the signal, and reduction of interferences by L-cysteine in the determination of arsenic by hydride generation. Anal Chem 64(6):667-672

Creed PA, Schwegel CA, Creed JT (2005) Investigation of arsenic speciation on drinking water treatment media utilizing automated sequential continuous flow extraction with IC-ICP-MS detection. J Environ Monit 7(11):1079-1084

D'Ilio S, Petrucci F, D'Amato M, Di Gregorio M, Senofonte O, Violante N (2008) Method validation for determination of arsenic, cadmium, chromium and lead in milk by means of dynamic reaction cell inductively coupled plasma mass spectrometry. Anal Chim Acta 624(1):59-67, doi: 10.1016/j.aca.2008.06.024

Dahl L, Molin M, Amlund H, Meltzer HM, Julshamn K, Alexander J, Sloth JJ (2010) Stability of arsenic compounds in seafood samples during processing and storage by freezing. Food Chem 123(3):720-727, doi: 10.1016/j.foodchem.2010.05.041

Dopp E, von Recklinghausen U, Diaz-Bone R, Hirner AV, Rettenmeier AW (2010) Cellular uptake, subcellular distribution and toxicity of arsenic compounds in methylating and non-methylating cells. Environ Res 110(5):435-442

Ferguson MA, Fernandez DP, Hering JG (2007) Lowering the detection limit for arsenic: Implications for a future practical quantitation limit. J Am Water Works Assn 99(8):92-98+12

Fernández SD, Sugishama N, Encinar JR, Sanz-Medel A (2012) Triple quad ICPMS (ICPQQQ) as a new tool for absolute quantitative proteomics and phosphoproteomics. Anal Chem 84(14):5851-5857

Foster AL, Kim CS (2014) Arsenic speciation in solids using X-ray absorption spectroscopy. Rev Mineral Geochem 79:257-369

Francesconi KA, Kuehnelt D (2004) Determination of arsenic species: a critical review of methods and applications, 2000-2003. Analyst 129(5):373-395, doi: 10.1039/b401321m

Garbarino JR, Snyder-Conn E, Leiker TJ, Hoffman GL (2002) Contaminants in arctic snow collected over northwest Alaskan sea ice. Water Air Soil Pollut 139(1-4):183-214

Georgiadis M, Cai Y, Solo-Gabriele HM (2006) Extraction of arsenate and arsenite species from soils and sediments. Environ Pollut 141(1):22-29

Goh KH, Lim TT (2005) Arsenic fractionation in a fine soil fraction and influence of various anions on its mobility in the subsurface environment. Appl Geochem 20(2):229-239

Gómez-Ariza JL, Sánchez-Rodas D, Giráldez I, Morales E (2000) Comparison of biota sample pretreatments for arsenic speciation with coupled HPLC-HG-ICP-MS. Analyst 125(3):401-407

Gong Z, Lu X, Ma M, Watt C & Le XC (2002) Arsenic speciation analysis. Talanta 58:77-96

Gruebel KA, Davis JA, Leckie JO (1988) Feasibility of using sequential extraction techniques for arsenic and selenium in soils and sediments. Soil Sci Soc Am J 52(2):390-397

Haertig C, Planer-Friedrich B (2012) Thioarsenate transformation by filamentous microbial mats thriving in an alkaline, sulfidic hot Spring. Environ Sci Technol 46(8):4348-4356

Hall GEM, Bonham-Carter GF, Horowitz AJ, Lum K, Lemieux C, Quemerais B, Garbarino JR (1996a) The effect of using different 0.45 μm filter membranes on 'dissolved' element concentrations in natural waters. Appl Geochem 11:243-249

Hall GEM, Vaive JE, Beer R, Hoashi M (1996b) Selective leaches revisited, with emphasis on the amorphous Fe oxyhydroxide phase extraction. J Geochem Explor 56:59-78

Hall GEM (1998) Cost-effective protocols for the collection, filtration and preservation of surface waters for detection of metals and metalloids at ppb (μg L^{-1}) and ppt (ng L^{-1}) levels. Aquatic Effects Technology Evaluation Program (Task Force on Water Quality Issues), AETE Project 313, Canmet, Ottawa see *http://www.nrcan-rncan.gc.ca/mms/canmet-mtb/mmsl-lmsm/enviro/reports/313.pdf*

Hall GEM, Pelchat JC, Gauthier G (1999) Stability of inorganic arsenic(III) and arsenic(V) in water samples. J Anal At Spectrom 14(2):205-213

Hall GEM, Parkhill MA, Bonham-Carter GF (2003) Conventional and selective leach geochemical exploration methods applied to humus and B horizon soil overlying the Restigouche VMS deposit, Bathurst Mining Camp, New Brunswick. *In*: Massive Sulphide Deposits of the Bathurst Mining Camp, New Brunswick, and Northern Maine. Economic Geology Monograph, 11. Goodfellow WD, McCutcheon SR, Peter JM (eds), Society of Economic Geologists, Inc., Littleton, CO, p 763-782

Hall GEM, Bonham-Carter GF, Buchar A (2014) Evaluation of portable X-ray fluorescence (pXRF) in exploration and mining: Phase 1, control reference materials. Geochem: Expl Env Analy, doi: 10.1144/geochem2013-241

Haque S, Johannesson KH (2006a) Arsenic concentrations and speciation along a groundwater flow path: The Carrizo Sand aquifer, Texas, USA. Chem Geol 228(1-3 SPEC. ISS.):57-71

Haque S, Johannesson KH (2006b) Concentrations and speciation along a groundwater flow path in the upper Floridan aquifer, Florida, USA. Environ Geol 50(3):218-228

Haque S, Ji J, Johannesson KH (2008) Evaluating mobilization and transport of arsenic in sediments and groundwaters of Aquia aquifer, Maryland, USA. J Contam Hydrol 99(1-4):68-84

Hering JG, Chiu VQ (2000) Arsenic occurrence and speciation in municipal ground-water-based supply system. J Environ Eng 126(5):471-474

Hinkle SR, Polette DJ (1999) Arsenic in ground water of the Willamette Basin, Oregon. Water-Resources Investigations Report 98-4205

Hollibaugh JT, Carini S, Gürleyük H, Jellison R, Joye SB, LeCleir G, Meile C, Vasquez L, Wallschläger D (2005) Arsenic speciation in Mono Lake, California: Response to seasonal stratification and anoxia. Geochim Cosmochim Acta 69(8):1925-1937

Hu S, Lu J, Jing C (2012) A novel colorimetric method for field arsenic speciation analysis. J Environ Sci 24(7):1341-1346, doi: 10.1016/s1001-0742(11)60922-4

Huang J-H, Ilgen G (2006) Factors affecting arsenic speciation in environmental samples: sample drying and storage. Int J Env Analy Chem 86(5):347-358, doi: 10.1080/03067310500227878

Huang JH, Kretzschmar R (2010) Sequential extraction method for speciation of arsenate and arsenite in mineral soils. Anal Chem 82(13):5534-5540

Hung DQ, Nekrassova O, Compton RG (2004) Analytical methods for inorganic arsenic in water: A review. Talanta 64(2):269-277

Ilgen G, Huang J-H (2013) An automatic cryotrapping and cryofocussing system for parallel ICP-MS and EI-MS detection of volatile arsenic compounds in gaseous samples. J Anal At Spectrom 28(2):293, doi: 10.1039/c2ja30251a

Jakob R, Roth A, Haas K, Krupp EM, Raab A, Smichowski P, Gómez D, Feldmann J (2010) Atmospheric stability of arsines and the determination of their oxidative products in atmospheric aerosols (PM10): Evidence of the widespread phenomena of biovolatilization of arsenic. J Environ Monit 12(2):409-416

Juskelis R, Li W, Nelson J, Cappozzo JC (2013) Arsenic speciation in rice cereals for infants. J Agric Food Chem 61(45):10670-10676

Karadjova IB, Petrov PK, Serafimovski I, Stafilov T, Tsalev DL (2007) Arsenic in marine tissues - The challenging problems to electrothermal and hydride generation atomic absorption spectrometry. Spectrochim Acta, Part B 62(3 SPEC. ISS.):258-268

Kent DB, Fox PM (2004) The influence of groundwater chemistry on arsenic concentrations and speciation in a quartz sand and gravel aquifer. Geochem Trans 5(1):1-12

Kim EJ, Yoo JC, Baek K (2014) Arsenic speciation and bioaccessibility in arsenic-contaminated soils: sequential extraction and mineralogical investigation. Environ Pollut 186:29-35, doi: 10.1016/j.envpol.2013.11.032

Kim MJ, Nriagu J, Haack S (2002) Arsenic species and chemistry in groundwater of southeast Michigan. Environ Pollut 120(2):379-390

Kim MJ, Nriagu J, Haack S (2003) Arsenic behavior in newly drilled wells. Chemosphere 52(3):623-633

Kinniburgh DG, Kosmus W (2002) Arsenic contamination in groundwater: Some analytical considerations. Talanta 58(1):165-180

Klusman RW (2009) Transport of ultratrace reduced gases and particulate, near-surface oxidation, metal deposition and adsorption. Geochem: Expl Env Anal 9(3):203-213, doi: 10.1144/1467-7873/09-192

Ko I, Kang SY, Kim K-W, Lee CH (2009) Application of arsenic field test kit to stream sediment: effect of fine particles and chemical extraction. Chem Speciation Bioavail 21(1):49-57, doi: 10.3184/095422909x419880

Kobayashi Y (2010) Elucidation of the metabolic pathways of selenium and arsenic by analytical toxicology. J Health Sci 56(2):154-160

Korte N (1991) Naturally occurring arsenic in groundwaters of the midwestern United States. Env Geol Wat Sci 18(2):137-141

Krachler M, Emons H (2000) Extraction of antimony and arsenic from fresh and freeze-dried plant samples as determined by HG-AAS. Fresenius J Anal Chem 368(7):702-707, doi: 10.1007/s002160000578

Kumar AR, Riyazuddin P (2007) Non-chromatographic hydride generation atomic spectrometric techniques for the speciation analysis of arsenic, antimony, selenium, and tellurium in water samples - A review. Int J Env Analy Chem 87(7):469-500

Kumar AR, Riyazuddin P (2010) Preservation of inorganic arsenic species in environmental water samples for reliable speciation analysis. TrAC, Trends Anal Chem 29(10):1212-1223, doi: 10.1016/j.trac.2010.07.009

Layton-Matthews D, Leybourne MI, Peter JM, Scott SD (2006) Determination of selenium isotopic ratios by continuous-hydride-generation dynamic-reaction-cell inductively coupled plasma-mass spectrometry. J Anal At Spectrom 21(1):41-49, doi: 10.1039/b501704a

Le XC, Lu X, Li XF (2004) Arsenic Speciation. Anal Chem 76(1):26A-33A

Lee A, McVey J, Faustino P, Lute S, Sweeney N, Pawar V, Khan M, Brorson K, Hussong D (2010) Use of hydrogenophaga pseudoflava penetration to quantitatively assess the impact of filtration parameters for 0.2-micrometer-pore-size filters. Appl Environ Microbiol 76(3):695-700

Leermakers M, Baeyens W, De Gieter M, Smedts B, Meert C, De Bisschop HC, Morabito R, Quevauviller P (2006) Toxic arsenic compounds in environmental samples: Speciation and validation. TrAC, Trends Anal Chem 25(1):1-10, doi: 10.1016/j.trac.2005.06.004

Leybourne MI, Goodfellow WD, Boyle DR, Hall GEM (2000) Form and distribution of anthropogenic gold mobilized in surface waters and sediments from a gossan tailings pile, Murray Brook massive sulfide deposit, New Brunswick, Canada. Appl Geochem 15:629-646

Leybourne MI, Cameron EM (2008) Source, transport, and fate of rhenium, selenium, molybdenum, arsenic, and copper in groundwater associated with porphyry-Cu deposits, Atacama Desert, Chile. Chem Geol 247:208-228, doi: 10.1016/j.chemgeo.2007.10.017

Liu G, Cai Y (2007) Chapter 31 Arsenic speciation in soils: an analytical challenge for understanding arsenic biogeochemistry. vol 5, p 685-708

Londesborough S, Mattusch J, Wennrich R (1999) Separation of organic and inorganic arsenic species by HPLC-ICP-MS. Fresenius J Anal Chem 363(5-6):577-581

Maher WA, Foster S, Krikowa F, Duncan E, St John A, Hug K, Moreau JW (2013) Thio arsenic species measurements in marine organisms and geothermal waters. Microchem J 111:82-90

Mamindy-Pajany Y, Bataillard P, Séby F, Crouzet C, Moulin A, Guezennec A-G, Hurel C, Marmier N, Battaglia-Brunet F (2013) Arsenic in Marina Sediments from the Mediterranean Coast: Speciation in the Solid Phase and Occurrence of Thioarsenates. Soil Sediment Contam 22(8):984-1002, doi: 10.1080/15320383.2013.770441

Mass MJ, Tennant A, Roop BC, Cullen WR, Styblo M, Thomas DJ, Kligerman AD (2001) Methylated trivalent arsenic species are genotoxic. Chem Res Toxicol 14(4):355-361, doi: 10.1021/tx000251l

Matoušek T, Currier JM, Trojánková N, Saunders RJ, Ishida MC, González-Horta C, Musil S, Mester Z, Stýblo M, Dědina J (2013) Selective hydride generation-cryotrapping-ICP-MS for arsenic speciation analysis at picogram levels: Analysis of river and sea water reference materials and human bladder epithelial cells. J Anal At Spectrom 28(9):1456-1465

McCleskey RB, Nordstrom DK, Maest AS (2004) Preservation of water samples for arsenic(III/V) determinations: an evaluation of the literature and new analytical results. Appl Geochem 19(7):995-1009, doi: 10.1016/j.apgeochem.2004.01.003

McNeill LS, Edwards M (1997) Predicting as removal during metal hydroxide precipitation. J Am Water Works Assn 89(1):75-86

Mestrot A, Merle JK, Broglia A, Feldmann J, Krupp EM (2011) Atmospheric stability of arsine and methylarsines. Environ Sci Technol 45(9):4010-4015, doi: 10.1021/es2004649

Mestrot A, Planer-Friedrich B, Feldmann J (2013) Biovolatilisation: A poorly studied pathway of the arsenic biogeochemical cycle. Environ Sci Processes Impacts 15(9):1639-1651

Minnich MG, Miller DC, Parsons PJ (2008) Deterinination of As, Cd, Pb, and Hg in urine using inductively coupled plasma mass spectrometry with the direct injection high efficiency nebulizer. Spectrochim Acta, Part B 63(3):389-395, doi: 10.1016/j.sab.2007.11.033

Mir KA, Rutter A, Koch I, Smith P, Reimer KJ, Poland JS (2007) Extraction and speciation of arsenic in plants grown on arsenic contaminated soils. Talanta 72(4):1507-1518

Mitchell VL (2014) Health risks associated with chronic exposures to arsenic in the environment. Rev Mineral Geochem 79:435-449

Munk L, Hagedorn B, Sjostrom D (2011) Seasonal fluctuations and mobility of arsenic in groundwater resources, Anchorage, Alaska. Appl Geochem 26(11):1811-1817

Musil S, Petursdottir AH, Raab A, Gunnlaugsdottir H, Krupp E, Feldmann J (2014) Speciation without chromatography using selective hydride generation: inorganic arsenic in rice and samples of marine origin. Anal Chem 86(2):993-999, doi: 10.1021/ac403438c

Narukawa T, Inagaki K, Zhu Y, Kuroiwa T, Narushima I, Chiba K, Hioki A (2012) Preparation and certification of Hijiki reference material, NMIJ CRM 7405-a, from the edible marine algae hijiki (Hizikia fusiforme). Anal BioanalChem 402(4):1713-1722

Neubauer E, Kohler SJ, von der Kammer F, Laudon H, Hofmann T (2013a) Effect of pH and stream order on iron and arsenic speciation in boreal catchments. Environ Sci Technol 47(13):7120-7128, doi: 10.1021/es401193j

Neubauer E, v d Kammer F, Hofmann T (2013b) Using FLOWFFF and HPSEC to determine trace metal-colloid associations in wetland runoff. Water Res 47(8):2757-2769, doi: 10.1016/j.watres.2013.02.030

Niazi NK, Singh B, Shah P (2011) Arsenic speciation and phytoavailability in contaminated soils using a sequential extraction procedure and XANES spectroscopy. Environ Sci Technol 45(17):7135-7142, doi: 10.1021/es201677z

Parsons C, Margui Grabulosa E, Pili E, Floor GH, Roman-Ross G, Charlet L (2013) Quantification of trace arsenic in soils by field-portable X-ray fluorescence spectrometry: considerations for sample preparation and measurement conditions. J Hazard Mater 262:1213-1222, doi: 10.1016/j.jhazmat.2012.07.001

Paul CJ, Ford RG, Wilkin RT (2009) Assessing the selectivity of extractant solutions for recovering labile arsenic associated with iron (hydr)oxides and sulfides in sediments. Geoderma 152(1-2):137-144

Peinado FM, Ruano SM, González MGB, Molina CE (2010) A rapid field procedure for screening trace elements in polluted soil using portable X-ray fluorescence (PXRF). Geoderma 159(1-2):76-82

Peters SC, Burkert L (2008) The occurrence and geochemistry of arsenic in groundwaters of the Newark basin of Pennsylvania. Appl Geochem 23(1):85-98

Pitzalis E, Ajala D, Onor M, Zamboni R, D'Ulivo A (2007) Chemical vapor generation of arsane in the presence of L-cysteine. Mechanistic studies and their analytical feedback. Anal Chem 79(16):6324-6333

Pitzalis E, Onor M, Mascherpa MC, Pacchi G, Mester Z, D'Ulivo A (2014) Chemical generation of arsane and methylarsanes with amine boranes. potentialities for nonchromatographic speciation of arsenic. Anal Chem 86(3):1599-1607

Planer-Friedrich B, Lehr C, Matschullat J, Merkel BJ, Nordstrom DK, Sandstrom MW (2006) Speciation of volatile arsenic at geothermal features in Yellowstone National Park. Geochim Cosmochim Acta 70(10):2480-2491

Planer-Friedrich B, London J, McCleskey RB, Nordstrom DK, Wallschläger D (2007) Thioarsenates in geothermal waters of yellowstone National Park: Determination, preservation, and geochemical importance. Environ Sci Technol 41(15):5245-5251

Planer-Friedrich B, Fisher JC, Hollibaugh JT, Elke S, Wallschläger D (2009) Oxidative transformation of trithioarsenate along alkaline geothermal drainages-abiotic versus microbially mediated processes. Geomicrobiol J 26(5):339-350

Planer-Friedrich B, Wallschläger D (2009) A critical investigation of hydride generation-based arsenic speciation in sulfidic waters. Environ Sci Technol 43(13):5007-5013

Pohl P (2004) Hydride generation - Recent advances in atomic emission spectrometry. TrAC, Trends Anal Chem 23(2):87-101

Polya DA, Lythgoe PR, Abou-Shakra F, Gault AG, Brydie JR, Webster JG, Brown KL, Nimfopoulos MK, Michailidis KM (2003) IC-ICP-MS and IC-ICP-HEX-MS determination of arsenic speciation in surface and groundwaters: preservation and analytical issues. Mineral Mag 67(2):247-261, doi: 10.1180/0026461036720098

Potts PJ, Ramsey MH, Carlisle J (2002) Portable X-ray fluorescence in the characterisation of arsenic contamination associated with industrial buildings at a heritage arsenic works site near Redruth, Cornwall, UK. J Environ Monit 4(6):1017-1024

Qin J, Lehr CR, Yuan C, Le XC, McDermott TR, Rosen BP (2009) Biotransformation of arsenic by a Yellowstone thermoacidophilic eukaryotic alga. PNAS 106(13):5213-5217, doi: 10.1073/pnas.0900238106

Quazi S, Sarkar D, Datta R (2011) Changes in arsenic fractionation, bioaccessibility and speciation in organo-arsenical pesticide amended soils as a function of soil aging. Chemosphere 84(11):1563-1571

Rahman MA, Hassler C (2014) Is arsenic biotransformation a detoxification mechanism for microorganisms? Aquatic Toxicology 146:212-219

Rahman MM, Mukherjee D, Sengupta MK, Chowdhury UK, Lodh D, Chanda CR, Roy S, Selim M, Quamruzzaman Q, Milton AH, Shahidullah SM, Tofizur RM, Chakraborti D (2002) Effectiveness and reliability of arsenic field testing kits: Are the million dollar screening projects effective or not? Environ Sci Technol 36(24):5385-5394

Rasmussen RR, Hedegaard RV, Larsen EH, Sloth JJ (2012) Development and validation of an SPE HG-AAS method for determination of inorganic arsenic in samples of marine origin. Anal Bioanal Chem 403(10):2825-2834, doi: 10.1007/s00216-012-6006-7

Rubinos DA, Arias M, Díaz-Fierros F, Barral MT (2005) Speciation of adsorbed arsenic(V) on red mud using a sequential extraction procedure. Mineral Mag 69(5):591-600

Sanger CR (1908) The quantitative determination of arsenic by the gutzeit method. J Am Chem Soc 30(6):1041-1042

Santos CMM, Nunes MAG, Barbosa IS, Santos GL, Peso-Aguiar MC, Korn MGA, Flores EMM, Dressler VL (2013) Evaluation of microwave and ultrasound extraction procedures for arsenic speciation in bivalve mollusks by liquid chromatography–inductively coupled plasma-mass spectrometry. Spectrochim Acta, Part B 86:108-114, doi: 10.1016/j.sab.2013.05.029

Sloth JJ, Larsen EH, Julshamn K (2003) Determination of organoarsenic species in marine samples using gradient elution cation exchange HPLC-ICP-MS. J Anal At Spectrom 18(5):452-459, doi: 10.1039/B300508A

Smedley PL, Kinniburgh DG (2002) A review of the source, behaviour and distribution of arsenic in natural waters. Appl Geochem 17(5):517-568

Smieja JA, Wilkin RT (2003) Preservation of sulfidic waters containing dissolved AS(III). J Environ Monit 5(6):913-916

Sorg TJ, Chen AS, Wang L (2014) Arsenic species in drinking water wells in the USA with high arsenic concentrations. Water Res 48:156-169, doi: 10.1016/j.watres.2013.09.016

Stauder S, Raue B, Sacher F (2005) Thioarsenates in sulfidic waters. Environ Sci Technol 39(16):5933-5939

Stauffer RE (1980) Molybdenum blue applied to arsenic and phosphorus determinations in fluoride- and silica-rich geothermal waters. Environ Sci Technol 14:1475-1481

Sturgeon RE, Mester Z (2002) Analytical applications of volatile metal derivatives. Appl Spectrosc 56(8):202A-231A

Styblo M, Del Razo LM, Vega L, Germolec DR, LeCluyse EL, Hamilton GA, Reed W, Wang C, Cullen WR, Thomas DJ (2000) Comparative toxicity of trivalent and pentavalent inorganic and methylated arsenicals in rat and human cells. Arch Toxicol 74(6):289-299

Suess E, Scheinost AC, Bostick BC, Merkel BJ, Wallschlaeger D, Planer-Friedrich B (2009) Discrimination of thioarsenites and thioarsenates by X-ray absorption spectroscopy. Anal Chem 81(20):8318-8326

Suess E, Wallschläger D, Planer-Friedrich B (2011) Stabilization of thioarsenates in iron-rich waters. Chemosphere 83(11):1524-1531

Terlecka E (2005) Arsenic speciation analysis in water samples: a review of the hyphenated techniques. Environ Monit Assess 107(1-3):259-284, doi: 10.1007/s10661-005-3109-z

Tessier A, Campbell PGC, Bisson M (1979) Sequential extraction procedure for the speciation of particulate trace metals. Anal Chem 51:844-851

Uroic MK, Krupp EM, Johnson C, Feldmann J (2009) Chemotrapping-atomic fluorescence spectrometric method as a field method for volatile arsenic in natural gas. J Environ Monit 11(12):2222-2230, doi: 10.1039/b913322d

Voice TC, Flores Del Pino LV, Havezov I, Long DT (2011) Field deployable method for arsenic speciation in water. Phys Chem Earth (2002) 36(9-11):436-441, doi: 10.1016/j.pce.2010.03.027

Wang N, Tyson J (2014) Non-chromatographic speciation of inorganic arsenic by atomic fluorescence spectrometry with flow injection hydride generation with a tetrahydroborate-form anion-exchanger. J Anal At Spectrom 29(4):665-673, doi: 10.1039/C3JA50376C

Wang RY, Hsu YL, Chang LF, Jiang SJ (2007a) Speciation analysis of arsenic and selenium compounds in environmental and biological samples by ion chromatography-inductively coupled plasma dynamic reaction cell mass spectrometer. Anal Chim Acta 590(2):239-244, doi: 10.1016/j.aca.2007.03.045

Wang Y, Hammes F, Boon N, Egli T (2007b) Quantification of the filterability of freshwater bacteria through 0.45, 0.22, and 0.1 μm pore size filters and shape-dependent enrichment of filterable bacterial communities. Environ Sci Technol 41(20):7080-7086

Wang P, Sun G, Jia Y, Meharg AA, Zhu Y (2014) A review on completing arsenic biogeochemical cycle: Microbial volatilization of arsines in environment. J Environ Sci 26(2):371-381, doi: 10.1016/s1001-0742(13)60432-5

Weindorf DC, Paulette L, Man T (2013) In-situ assessment of metal contamination via portable X-ray fluorescence spectroscopy: Zlatna, Romania. Environ Pollut 182:92-100

Whaley-Martin KJ, Koch I, Reimer KJ (2013) Determination of arsenic species in edible periwinkles (Littorina littorea) by HPLC-ICPMS and XAS along a contamination gradient. Sci Total Environ 456-457:148-153

Wilkie JA, Hering JG (1998) Rapid oxidation of geothermal arsenic(III) in streamwaters of the eastern Sierra Nevada. Environ Sci Technol 32(5):657-662

Wilkin RT, Wallschläger D, Ford RG (2003) Speciation of arsenic in sulfidic waters. Geochem Trans 4:1-7

Wuilloud JCA, Wuilloud RG, Vonderheide AP, Caruso JA (2004) Gas chromatography/plasma spectrometry - An important analytical tool for elemental speciation studies. Spectrochim Acta, Part B 59(6):755-792

Yalcin S, Le XC (2001) Speciation of arsenic using solid phase extraction cartridges. J Environ Monit 3(1):81-85, doi: 10.1039/B007598L

Zhang J, Kim H, Townsend T (2014) Methodology for assessing thioarsenic formation potential in sulfidic landfill environments. Chemosphere 107:311-8, doi: 10.1016/j.chemosphere.2013.12.075

Reviews in Mineralogy & Geochemistry
Vol. 79 pp. 391-433, 2014
Copyright © Mineralogical Society of America

Microbial Arsenic Metabolism and Reaction Energetics

Jan P. Amend

Departments of Earth Sciences and Biological Sciences
University of Southern California
Los Angeles, California 90089, U.S.A.

janamend@usc.edu

Chad Saltikov

Department of Microbiology and Environmental Toxicology
University of California, Santa Cruz
Santa Cruz, California 95064, U.S.A.

saltikov@ucsc.edu

Guang-Sin Lu

Department of Earth Sciences
University of Southern California
Los Angeles, California 90089, U.S.A.

guangsil@usc.edu

Jaime Hernandez

Department of Microbiology and Environmental Toxicology
University of California, Santa Cruz
Santa Cruz, California 95064, U.S.A.

jhernan4@ucsc.edu

INTRODUCTION

Reviews on the geochemistry, biochemistry, or microbial ecology of arsenic—and there are many—commonly start with statements about the toxicity of this metalloid (Newman et al. 1998; Rosen 2002; Smedley and Kinniburgh 2002; Oremland and Stolz 2003; Oremland et al. 2004, 2009; Silver and Phung 2005; Lloyd and Oremland 2006; Stolz et al. 2006, 2010; Bhattacharjee and Rosen 2007; Paez-Espino et al. 2009; Tsai et al. 2009; Slyemi and Bonnefoy 2012; Cavalca et al. 2013b; Kruger et al. 2013; van Lis et al. 2013; Watanabe and Hirano 2013; Zhu et al. 2014). These introductions are sometimes followed by famous anecdotes of foul play (e.g., was Napoleon I poisoned by his British captors?) and reminders that arsenic was used as a popular medicine, tonic, and aphrodisiac since the 18th century. Recall that the 1908 Nobel Prize in medicine was awarded to Paul Ehrlich, in part, for the discovery of an organoarsenical (Salvarsan) as a treatment for syphilis—this was arguably also the first documented application of what would later become known as "chemotherapy." Readers are then often reminded that arsenic is still used today in pesticides and herbicides, in animal feed, as a wood preservative, in electronic devices, and in specialized medical treatments.

Arsenic is toxic in both of its common oxidation states, the oxidized arsenate, As(V), and the reduced arsenite, As(III). As a molecular analog of phosphate, arsenate uses a phosphate transport system to enter the cell and there inhibits the phosphorylation of ADP and thereby

1529-6466/14/0079-0007$05.00   http://dx.doi.org/10.2138/rmg.2014.79.7

the synthesis of ATP. Arsenate can also substitute for phosphate in various biomolecules, thus disrupting key pathways, including glycolysis. Arsenite is even more toxic than arsenate and enters the cell much like glycerol molecules via aqua-glyceroporins (Cullen and Reimer 1989; Meng et al. 2004; Ehrlich and Newman 2008). It binds to the –SH group of cysteine residues, thereby inactivating proteins and otherwise affecting various receptors and transcription factors (Bhattacharjee and Rosen 2007; Stolz et al. 2010; Slyemi and Bonnefoy 2012). In addition, arsenite can inhibit redox homeostasis and protein folding, among other intracellular processes.

Arsenite is more toxic, more soluble, and more bioavailable than arsenate. As part of their detoxification mechanism, many organisms reduce As(V) inside the cytoplasm, followed by transport of As(III) across the membrane to the extracellular environment by efflux pumps (Mukhopadhyay et al. 2002; Rosen 2002; Silver and Phung 2005). Another detoxification process, though less well understood, especially in microorganisms, involves the methylation and demethylation of arsenicals (Bentley and Chasteen 2002; Stolz et al. 2006; Qin et al. 2009).

In the mid-1990's, another role was ascribed to arsenic, that of reactant in energy-yielding microbial metabolism. The first evidence was provided by the ε-Proteobacterium *Sulfurospirillum* carrying out anaerobic respiration with arsenate as the terminal electron acceptor and organic matter as electron donor (Ahmann et al. 1994; Laverman et al. 1995). More recently, Bacteria that conserve energy by arsenite oxidation have also been described (Muller et al. 2003; Santini and vanden Hoven 2004), and a few years ago, anoxygenic photosynthesis was added to the list of microbially-catalyzed arsenic metabolisms (Budinoff and Hollibaugh 2008; Kulp et al. 2008).

In this review, we provide an account of the isolated and at least partially characterized microorganisms that can transform arsenite or arsenate for the purposes of energy conservation or detoxification. The genetic underpinnings of these enzymatically controlled As-redox processes are then summarized, followed by a brief discussion on the apparent antiquity but minimal modern ecological role of Archaea (relative to Bacteria) in arsenic transformations. We conclude with an assessment on the energy yields as a function of temperature and chemical composition from 15 As-redox reactions that may be available to microorganisms in a wide range of environments.

ARSENIC METABOLISMS IN BACTERIA AND ARCHAEA

The first documented account of microbially-mediated arsenic transformation goes back nearly a century (Green 1918). To date, upwards of 200 strains of bacteria and archaea (241 at last count) are known that can oxidize As(III) or reduce As(V) for energy gain or detoxification (see the Appendix). Based on 16S ribosomal RNA gene sequences, these strains can be assigned to 11 phyla, with only 1 phylum, however, in the Archaea (Crenarchaeota) (Fig. 1). Within the Bacteria, they identify as Aquificae, Deinococcus-Thermus, Chloroflexus, Firmicutes, Actinobacteria, Deferribacteres, Chrysiogenetes, Cyanobacteria, Proteobacteria (including members of the α, β, γ, δ, and ε subgroups), and Bacteroidetes. By far the most isolates are among the γ-Proteobacteria (75), followed by the Firmicutes (52), β-Proteobacteria (40), and α-Proteobacteria (31), which together account for >80% of the total.

These isolates stem from a wide variety of unpolluted, polluted, and engineered sites, including hot springs, lakes, groundwater, soils, marine sediments, mine drainage, and waste streams (see the Appendix). Consequently, their growth temperatures and pH ranges are similarly varied. While most strains are mesophiles (T_{opt} 20-40 °C), 16 thermophiles (T_{opt} >50 °C) have been identified, mostly among the Crenarchaeota, Aquificae, Deferribacteres, and Deinococcus-Thermus. For example, several species of *Pyrobaculum* (Nos. 1-6 in the Appendix) isolated from hydrothermal environments can grow at temperatures up to and

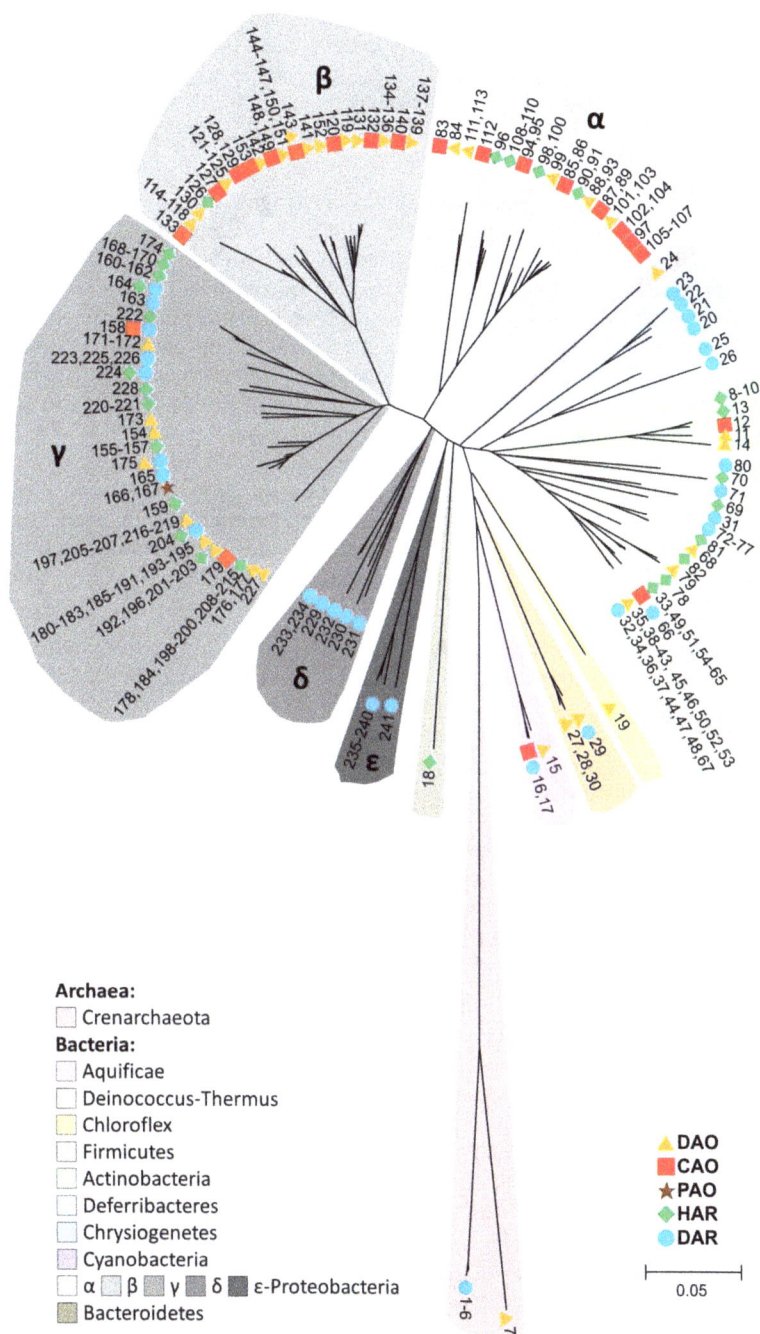

Figure 1. (*for color see Plate 6*) Phylogenetic tree based on 16S rRNA gene sequences of cultivable Archaea and Bacteria that are known to transform of arsenic. Detoxifying arsenite oxidizers (DAOs) are indicated by gold triangles, chemolithotrophic arsenite oxidizers (CAOs) by red squares, phototrophic arsenite oxidizers (PAO) by brown stars, heterotrophic arsenate reducers (HARs) by green diamonds, and dissimilatory arsenate reducers (DARPs) by blue circles. The scale bar represents the substitution of 5% divergence.

even beyond 100 °C. Obligate psychrophiles (T_{opt} <15 °C) are not known among the As-metabolizing organisms, but *Polaromonas* sp. str. GM1 (No. 141) is an arsenite oxidizer capable of growth below 10 °C though its T_{opt} is 20 °C. The pH range for the vast majority of the organisms in the Appendix is circumneutral, but some thrive in soda lakes at pH up to 10-11 (e.g., *Alkalilimnicola ehrlichii*; No. 159), while others are acidophiles living at pH 2-3 (e.g., *Acidicaldus* sp. str. AO5; No. 83).

Arsenic-transforming microorganisms are commonly assigned to one of four groups based on their metabolisms (Oremland and Stolz 2003, 2005): 1) dissimilatory arsenate-reducing prokaryotes (DARPs), 2) arsenate-resistant microorgansims (ARMs), 3) chemoautotrophic arsenite oxidizers (CAOs), and 4) heterotrophic arsenite oxidizers (HAOs). The key factors in these assignments are whether the organisms are using arsenic in their energy metabolisms or only transforming it in detoxification processes, and whether the organisms are heterotrophic or lithotrophic. DARPs (see the Appendix) are predominantly heterotrophic, strictly anaerobic organisms that conserve metabolic energy from the reduction of As(V); in laboratory growth experiments, the electron donors are commonly complex organic compounds (e.g., yeast extract, peptone, tryptone), but a number of DARPs are also capable of using low molecular weight organic acids (e.g., acetate, lactate, pyruvate), glucose, or, in a few cases, inorganic compounds such as hydrogen or reduced sulfur. DARPs are represented in most of the phyla shown in the phylogenetic tree in Figure 1. They are the only representatives among the Deferribacteres, Chrysiogenetes, and δ- and ε-Proteobacteria, but they are not found in the α- and β-Proteobacteria.

Organisms are identified as ARMs because of their putative capability to reduce arsenate for detoxification, not for energy conservation. In many cases, it is the mere presence of a specific protein (ArsC), or more typically the corresponding gene (*arsC*), and not experimentally confirmed arsenate reduction that is used to classify organisms as ARMs. However, some organisms (e.g., *Bacillus* sp. str. 3.9; No. 50) have this particular *arsC* gene but do not appear to reduce As(V), and other organisms (e.g., *Arthrobacter*; Nos. 8-9) do not have the gene but are still capable of arsenate reduction. Hence, here we use the term heterotrophic arsenate reducers (HARs) to represent all organisms where arsenate reduction has been verified, but metabolic energy gain is not coupled to this process (see the Appendix). The HARs are found only in the Bacteria, predominantly in the Firmicutes, Actinobacteria, and α- and γ-Proteobacteria; they are also the only arsenic metabolic group in the Bacteroidetes (Fig. 1).

The CAOs are strictly chemolithoautotrophic organisms that gain metabolic energy from the oxidation of arsenite to arsenate (see the Appendix). All known CAOs are Bacteria, and the vast majority are strict aerobes, using oxygen as the terminal electron acceptor. However, a few species, including strains of *Sinorhizobium* (No. 102), *Azoarcus* (No. 128), and *Alkalilimnicola* (No. 159) have now been identified that can use nitrate as their terminal electron acceptors in the oxidation of As(III) (Oremland et al. 2002; Rhine et al. 2006; Hoeft et al. 2007; Zargar et al. 2010; Rodriguez-Freire et al. 2012). And even selenate (in *Bacillus*; No. 66) and chlorate (in *Azospira* and *Dechloromonas*; Nos. 129, 133) can apparently be used for this purpose. The CAOs dominate the α- and β-Proteobacteria branches in the phylogenetic tree (Fig. 1), with a few isolates in the other bacterial phyla.

A large fraction of the organisms in the Appendix catalyze arsenite oxidation for detoxification. All currently known strains are strictly aerobic, and the majority are heterotrophic—hence the name "heterotrophic arsenite oxidizers" mentioned above. However, a few known species of Bacteria (members of *Hydrogenobaculum* (No. 15), *Chloroflexus* (No. 19), and *Thiobacillus* (Nos. 111, 113)), and even one Archaeon (*Sulfolobus acidocaldarius*; No. 7) are chemolithotrophic, not heterotrophic. They appear to conserve metabolic energy from the aerobic oxidation of hydrogen, ferrous iron, or intermediate oxidation state sulfur compounds. Hence, here we use the more general label of detoxifying arsenite oxidizers (DAOs) to include all organisms that oxidize As(III) without coupling this to a gain in metabolic energy. Most of the DAOs belong to the Firmicutes, and α-, β-, and γ-Proteobacteria; they are also the only known arsenic-transforming organisms among the Chloroflexi and Cyanobacteria (see Fig. 1).

All of the organisms discussed so far are chemotrophs, relying on thermodynamically favorable redox reactions—though not necessarily As-redox reactions—for metabolic energy. Kulp et al. (2008) described two strains of *Ectothiorhodospira* (Nos. 166, 167) from hot spring biofilms in alkaline Mono Lake (California, USA) that use As(III) as the electron donor in anoxygenic photosynthesis (see the Appendix). These organisms grow optimally at the moderately high temperature of 45 °C and pH 9.3, fixing carbon dioxide into biomass with energy from sunlight, not from chemical disequilibrium. For these, we suggest the metabolic group name phototrophic arsenite oxidizers (PAOs); they occupy one branch among the γ-Proteobacteria in the phylogenetic tree (Fig. 1).

With few exceptions, every strain in the Appendix falls into only one of these five categories, with 97 DAOs, 69 HARs, 56 DARPs, 34 CAOs, and 2 PAOs. A few organisms, however, are a bit more versatile in their metabolic capabilities and group in more than one category. For example, *Sulfurihydrogenibium azorense* (No. 16) can gain metabolic energy from the aerobic oxidation of arsenite (making it a CAO) and from the reduction of arsenate with reduced sulfur (making it a DARP), and *Thermus* sp. str. HR13 (No. 29) is a facultative anaerobe that can oxidize arsenite in a detoxification process (making it a DAO) or reduce arsenate for energy with lactate as the electron donor (making it a DARP). More details can be found in a collection of papers that specifically address arsenite metabolism (Santini and Ward 2012). The aforementioned arsenic metabolisms are, of course, catalyzed by specific enzymes that are encoded by specific genes—these are discussed below.

GENES THAT ENCODE FOR ARSENIC ENZYMES

Based on large-scale sequencing projects, our current perspective is that genes related to arsenic resistance are found in almost every microbial genome. However, these projects have also revealed genes for arsenate respiration and arsenite oxidation in microorganisms that apparently do not metabolize arsenic. Four main genes or gene groups have been identified that confer arsenic-based metabolism on a microbial host: *arr* and *ars* for arsenate reduction, and *aio* and *arx* for arsenite oxidation. The major proteins responsible for these arsenic transformations, including the cross-membrane transport, as well as their location in the cytoplasm or periplasm are depicted in Figure 2.

Respiratory arsenate reduction via *arr*

DARPs gain metabolic energy from the reduction of As(V) coupled to the oxidation of organic or inorganic electron donors. The key enzyme, a reductase, responsible for the arsenic transformation is Arr (consisting of subunits A and B), which is encoded by the *arrAB* gene cluster. The first Arr protein to be purified and characterized was from the anaerobic Bacterium *Chrysiogenes arsenatis* (No. 20) (Kraft and Macy 1998). *Bacillus selenitireducens* MLS-10 (No. 48), a gram-positive bacterium isolated from Mono Lake is another DARP in which the *arr* gene has been identified. Interestingly, the respiratory arsenate reductase in this organism has a much higher specificity for As(V) than *C. arsenatis* (Afkar et al. 2003). Perhaps most importantly, the facultative anaerobe *Shewanella* sp. str. ANA-3 (No. 224) was developed into a genetic system to investigate arsenate respiration (Saltikov and Newman 2003). Based on genetic manipulation experiments, it is now firmly established that ArrAB is essential to arsenate respiration—it facilitates electron transfer from cytochromes in the electron transport chain (see Fig. 2) and also short circuits other anaerobic pathways (Saltikov and Newman 2003; Croal et al. 2004; Murphy and Saltikov 2007). It should be pointed out that the

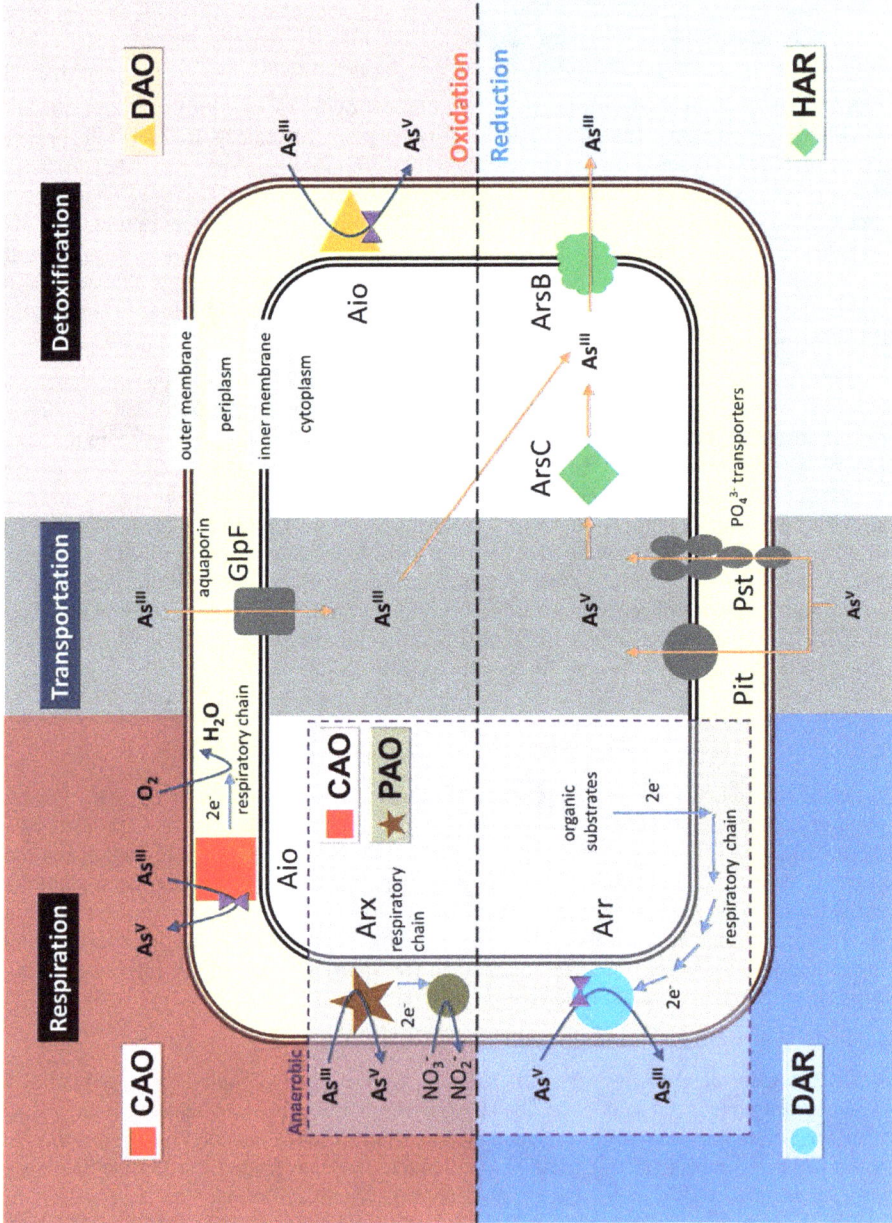

Figure 2. (*for color see Plate 7*) Schematic of inorganic arsenic respiration (*left*), transportation (*center*), and detoxification (*right*) in a composite cell of CAOs, DAOs, HARs, DARs, and PAOs, with oxidative processes in red tones (*top*) and reductive processes in blue tones (*bottom*). The major enzymes involved in arsenic transformation (Aio, Ars, Arr, and Arx) and the key membrane transport proteins (GlpF, Pit, and Pst) are also depicted. Modified from Paez-Espino et al. (2009).

ArrAB protein has not yet been crystallized, and hence questions remain about the structural differences or similarities to proteins involved in aerobic oxidation of arsenite (AioAB, see below), the electrochemical properties of ArrAB, and how these enzymes interact with the electron transport chain (Macy et al. 1996; Kraft and Macy 1998).

Detoxifying arsenate reduction via *ars*

Arsenic-tolerant microorganisms that do not gain metabolic energy from arsenate reduction typically have the *ars* gene cluster that encodes a detoxification mechanism. This involves the reduction and subsequent transport of arsenic (see Fig. 2). Arsenate is first reduced in the cytoplasm with the ArsC protein, a pathway that was first characterized in *E. coli* where *arsC* is located in a plasmid and glutaredoxin and glutathione are needed to reduce the As(V). The resulting As(III), which is far more soluble, is then generally pumped out of the cell with the help of the ArsB protein. In some cases, the efflux of As(III) can also be coupled to the electrochemical proton gradient, where chemical energy in the form of ATP is used to pump As(III) with the help of ArsA (an ATPase) (Rosen et al. 1999; Zhou et al. 2000; Meng et al. 2004).

Aerobic arsenite oxidation via *aio*

The vast majority of CAOs are aerobes that can extract metabolic energy from arsenite oxidation using oxygen as the terminal electron acceptor. In that case, a proton motive force is generated to conserve the energy, which can then be used to support cellular growth. The group of organisms referred to in the Appendix as DAOs are predominantly heterotrophs that oxidize As(III) with oxygen for detoxification, not energy gain. In both groups, the key enzyme is the arsenite oxidase recently named Aio (Fig. 2) (Lett et al. 2012), formerly referred to in the literature as Aro, Aox, or Aso (Silver and Phung 2005). The first detailed characterization of an arsenite oxidase was reported for the heterotrophic Bacterium *Alcaligenes faecalis* (No. 121) (Anderson et al. 1992), followed a few years later by the first purification and identification of AioAB in a CAO, *Rhizobium* sp. str. NT-26 (No. 85) (Santini et al. 2000; vanden Hoven and Santini 2004). Though homologous, these two arsenite oxidases show major differences in their crystal structures, especially within the "B" subunits, which may help explain physiological differences with respect to arsenite oxidation in the energy-conserving CAOs and the detoxifying DAOs (Warelow et al. 2013). The differences may affect how electrons flow through the enzyme and how AioB is "wired" into the electron transport chain.

Anaerobic arsenite oxidation via *arx*

Two dramatically different groups of arsenite oxidizing microorganisms—those that conserve energy by anaerobic respiration and those that harvest light energy in anoxygenic photosynthesis—both appear to rely on the same arsenite oxidase, ArxAB (see Fig. 2). Whether the Arx proteins can also be used in detoxification of As(III) is not known. In 2002, members of the US Geological Survey called out the quest to isolate a microorganism capable of anaerobic arsenite oxidation after this process was observed in anoxic bottom waters of Mono Lake (Hoeft et al. 2002). *Alkalilimnicola ehrlichii* strain MLHE-1 (No. 159) was the first isolate to fit into this metabolic category (Oremland et al. 2002). It is an haloalkaliphilic chemolithoautotroph that gains energy by reducing nitrate with As(III) as the electron donor. By disturbing the *arx* gene in *A. ehrlichii*, the role of Arx as the arsenite oxidase was then elucidated (Zargar et al. 2010). It has been shown *in vitro* that Arx from *A. ehrlichii* has both arsenate reductase and arsenite oxidase enzymatic properties, but physiologically it appears that Arx can only catalyze arsenite oxidation (Richey et al. 2009). It is now clear that Arx is a novel enzyme that clearly diverges from both Arr and Aio (Zargar et al. 2010, 2012); perhaps somewhat surprisingly, its protein sequence is more similar to that of the arsenate reductase (Arr) than to that of the other arsenite oxidase (Aio). Richey et al. (2009) demonstrated that Arx (previously referred to as Arr) can catalyze arsenite oxidation. They also showed that the arsenate respiratory reductase, Arr, may have reverse catalytic activity *in vitro*.

A few years ago, Mono Lake yielded yet another surprise in arsenic microbiology—a purple sulfur Bacterium was identified in a red biofilm that can carry out anoxygenic photosynthesis with As(III) as the sole electron donor (Kulp et al. 2008). This organism appears to transform light energy into chemical energy using a different electron transport chain than *A. ehrlichii*, though this new mechanism has yet to be fully elucidated. Photosynthetic arsenite oxidation has now been described in two haloalkaliphilic *Ectothiorhodospira* strains, the original PHS-1 (No. 167) and MLP2 (No. 166). Photoarsenotrophy was attributed to *arx* genes in those strains as they lack *aio* and the PHS-1 *arx* genes are induced in the presence of arsenite and light (see Fig. 2) (Zargar et al. 2012). Conversely, genome evidence suggests that other photosynthetic Bacteria like *Chloroflexus* may oxidize arsenite through Aio—this has yet to be demonstrated experimentally.

ARSENIC REDOX IN ARCHAEA

Of the 241 microorganisms in the Appendix capable of arsenic metabolism, only 7 belong to the Archaea. This proportion may well reflect a sampling bias as relatively few culturing studies have targeted ecosystems where Archaea are common or even dominant, ecosystems such as hot springs and hydrothermal vents, marine environments, and the subsurface. It has been proposed, however, that arsenic-based metabolism may be quite ancient, perhaps dating back to the earliest evolution of life (van Lis et al. 2013; Zhu et al. 2014). One then naturally looks to the Archaea for signs of its antiquity, particularly because arsenotrophy (i.e., the conservation of metabolic energy from arsenic reduction or oxidation) in Archaea may have predated its evolution in Bacteria. Within the Archaea, "true" arsenotrophy has only been demonstrated in the genus *Pyrobaculum* (Nos. 1-6), though several other strains can respire arsenate with organic or inorganic electron donors, placing them in the DARP group (see the Appendix). The mechanism of arsenate respiration in *Pyrobaculum* remains enigmatic, however, because the genome sequences do not have the typical *arr* genes (Cozen et al. 2009). Based on a transcriptome analysis, it has been hypothesized that a highly expressed gene cluster encoding for a molybdenum enzyme and a FeS subunit, both of which are analogous (but not homologous) to ArrAB, may represent a "novel" type of arsenate respiratory reductase (Cozen et al. 2009). Interestingly, ArrA proteins also appear to have their ancestral roots in molybdenum enzymes, which are present in many Archaea (Hedderich et al. 1999). So, where did ArrA originate? Evidence now points to the divergence with the Archaea, perhaps from a tetrathionate reductase, with the *Pyrobaculum* arsenate reductase as some intermediate form.

What about the other arsenic genes—*ars*, *arx* and *aio*—in the Archaea? *ars* genes are found in many Archaea, and the *ars* pathway is likely to be ancient. Note that genetic studies with *Halobacterium* established *ars* as the primary arsenic detoxification pathway (Wang et al. 2004). For the *arx* genes (and the Arx proteins), it is perhaps too early for evolutionary speculations; there are neither enough sequences, nor cultured reference strains. The origins of *aioA* appear to be within the genes for nitrate reductase or formate dehydrogenase (McEwan et al. 2002; Rhine et al. 2007). The Aio clade is closest to nitrate reductase followed by formate dehydrogenase. Note, for example, that using phylogenetics, AioA was found in several Archaea, including *Pyrobaculum calidifontis*, *Sulfolobus tokodai*, and *Aeropyrum pernix* (van Lis et al. 2013).

ENERGETICS OF ARSENIC REDOX REACTIONS

As noted above, numerous Bacteria and Archaea belonging to the DARPs and CAOs can obtain metabolic energy by catalyzing As-redox reactions. These reactions include aerobic and anaerobic arsenite oxidation, as well as arsenate reduction with organic and inorganic electron donors. In Table 1, we list 15 As-redox reactions, some of which are known to support microbial growth, while others are investigated here as to their *potential* to support arsenotrophs. We included 9 arsenate reduction reactions, with hydrogen (H_2), hydrogen sulfide (H_2S), elemental sulfur (S^0), methane (CH_4), and acetate (CH_3COO^-) as electron donors; we also considered 6 arsenite oxidation reactions, with oxygen (O_2), nitrate (NO_3^-), nitrite (NO_2^-), ferrihydrite ($FeOOH$), pyrolusite (MnO_2), and chlorate (ClO_3^-) as terminal electron acceptors. Of course, all As(V)-reduction reactions can represent As(III)-oxidation reactions when written in reverse, and vice versa. The direction selected in Table 1 is based on whether the reaction is generally energy-yielding or energy-consuming (see below).

Table 1. Redox reactions with arsenate and arsenite.

Arsenate (As^V) reduction	
A	$H_2AsO_4^- + H_2 + H^+ \rightleftharpoons H_3AsO_3 + H_2O$
B	$H_2AsO_4^- + 0.25H_2S + 0.5H^+ \rightleftharpoons H_3AsO_3 + 0.25SO_4^{2-}$
C	$H_2AsO_4^- + 0.33H_2S + 0.33H^+ \rightleftharpoons H_3AsO_3 + 0.33SO_3^{2-}$
D	$H_2AsO_4^- + 0.5H_2S + 0.5H^+ \rightleftharpoons H_3AsO_3 + 0.25S_2O_3^{2-} + 0.25H_2O$
E	$H_2AsO_4^- + H_2S + H^+ \rightleftharpoons H_3AsO_3 + S^0 + H_2O$
F	$H_2AsO_4^- + 0.33S^0 + 0.33H^+ + 0.33H_2O \rightleftharpoons H_3AsO_3 + 0.33SO_4^{2-}$
G	$H_2AsO_4^- + 0.25CH_4 + H^+ \rightleftharpoons H_3AsO_3 + 0.25CO_2 + 0.5H_2O$
H	$H_2AsO_4^- + 0.33CH_4 + H^+ \rightleftharpoons H_3AsO_3 + 0.33CO + 0.67H_2O$
I	$H_2AsO_4^- + 0.25CH_3COO^- + 1.25H^+ \rightleftharpoons H_3AsO_3 + 0.5CO_2 + 0.5H_2O$
Arsenite (As^{III}) oxidation	
J	$H_3AsO_3 + 0.5O_2 \rightleftharpoons H_2AsO_4^- + H^+$
K	$H_3AsO_3 + 0.25NO_3^- + 0.25H_2O \rightleftharpoons H_2AsO_4^- + 0.25NH_4^+ + 0.5H^+$
L	$H_3AsO_3 + 0.33NO_2^- + 0.33H_2O \rightleftharpoons H_2AsO_4^- + 0.33NH_4^+ + 0.33H^+$
M	$H_3AsO_3 + 2FeOOH + 3H^+ \rightleftharpoons H_2AsO_4^- + 2Fe^{2+} + 3H_2O$
N	$H_3AsO_3 + MnO_2 + H^+ \rightleftharpoons H_2AsO_4^- + Mn^{2+} + H_2O$
O	$H_3AsO_3 + 0.33ClO_3^- \rightleftharpoons H_2AsO_4^- + 0.33Cl^- + H+$

Metabolic energy yields can be evaluated from the sign and magnitude of the Gibbs energy of reaction (ΔG_r) in accord with

$$\Delta G_r = \Delta G_r^\circ + RT \ln Q_r \qquad (1)$$

where ΔG_r° denotes the *standard* Gibbs energy of reaction at any temperature and pressure, R and T stand for the gas constant and temperature (in Kelvin), respectively, and Q_r represents the activity product. Values of ΔG_r° and Q_r are calculated with the relations

$$\Delta G_r^\circ = \sum_i v_{i,r} \Delta G_i^\circ \qquad (2)$$

and

$$Q_r = \prod a_i^{v_{i,r}} \qquad (3)$$

respectively, where $v_{i,r}$, ΔG_i°, and a_i represent the stoichiometric reaction coefficient, *standard* Gibbs energy of formation, and activity of the ith species in reaction r.

Because energy yield is, in part, a function of temperature[*] and chemical composition, we first review values of $\Delta G_r°$ at 0-200 °C for the 15 reactions in Table 1, and then calculate the corresponding values of ΔG_r over broad, but geochemically realistic compositional space. We chose this temperature range to reach and presumably exceed the temperature maximum of life on Earth. Values of $\Delta G_i°$ for all of the chemical species used in the reactions in Table 1 are given in Table 2, and corresponding values of $\Delta G_r°$ as a function of temperature are listed in Table 3 and plotted as black curves in Figures 3 and 4. There, it can clearly be seen that the values of $\Delta G_r°$ for all 15 reactions (in the directions as written) are negative over the entire temperature range. With increasing temperature from 0 to 200 °C, values of $\Delta G_r°$ for the 9 As(V) reduction reactions (Fig. 3) become more negative by ~10-45 kJ mol^{-1} As, and those for the 6 As(III) oxidation reactions (Fig. 4) become less negative by ~30-50 kJ mol^{-1} As. All calculations of $\Delta G_i°$ and $\Delta G_r°$ were carried out with the software package SUPCRT92 (Johnson et al. 1992), with additional thermodynamic data for pyrolusite and ferrihydrite from LaRowe and Amend (2014), Robie and Hemingway (1985), and Snow et al. (2013).

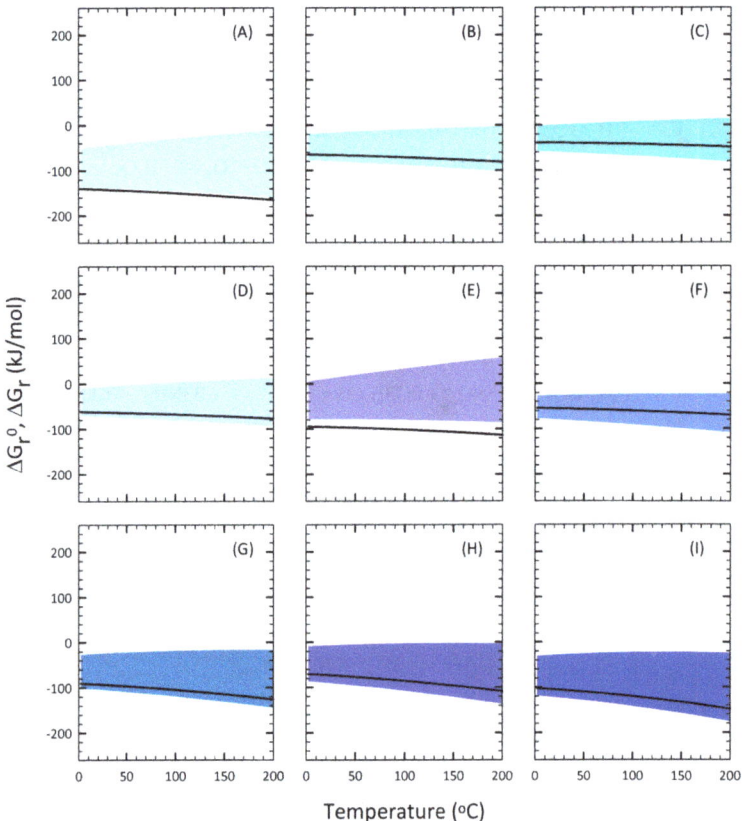

Figure 3. Values of $\Delta G_r°$ (black curves, data from Table 3) and ΔG_r (colored swaths) in kJ mol^{-1} as a function of temperature at P_{SAT} for the 9 arsenate reduction reactions given in Table 1. Values of ΔG_r are calculated with activities given in Table 4.

---

[*] $\Delta G_r°$ is also a function of pressure, but these effects, at least for pressures up to a few hundred bar, are relatively small and certainly much smaller than the effect of temperature from 0 to 100 °C. Therefore, all energy calculations presented here were carried out at the saturation pressure for H_2O (P_{SAT}), in other words the P-T boiling curve for water.

Table 2. Values of ΔG^0 (kJ mol^{-1}) at P_{SAT} as a function of temperature. All compounds are aqueous, except for FeOOH (2-line ferrihydrite) and MnO$_2$ (pyrolusite).

Compound \ T (°C)	2	18	25	37	45	55	70	85	100	150	200
H_3AsO_3	-635.4	-638.5	-639.9	-642.2	-643.8	-645.9	-649.0	-652.2	-655.4	-666.6	-678.5
$H_2AsO_4^-$	-750.5	-752.4	-753.2	-754.6	-755.5	-756.7	-758.5	-760.2	-762.0	-767.8	-773.3
O_2	18.8	17.3	16.5	15.2	14.2	13.0	11.0	8.8	6.6	-1.7	-11.2
H_2	18.9	18.1	17.7	17.0	16.5	15.8	14.6	13.4	12.1	7.1	1.3
NO_3^-	-107.5	-109.9	-110.9	-112.7	-113.8	-115.2	-117.4	-119.4	-121.5	-128.1	-134.3
NO_2^-	-29.3	-31.4	-32.2	-33.7	-34.6	-35.8	-37.5	-39.2	-40.8	-45.8	-50.3
NH_4^+	-77.0	-78.7	-79.5	-80.8	-81.7	-82.9	-84.7	-86.5	-88.4	-94.9	-101.7
FeOOH	-463.7	-464.8	-465.3	-466.1	-466.7	-467.5	-468.7	-469.9	-471.2	-475.9	-481.2
Fe^{2+}	-93.9	-92.2	-91.5	-90.2	-89.4	-88.3	-86.7	-85.0	-83.4	-77.7	-71.6
MnO_2	-463.7	-464.5	-464.8	-465.5	-465.9	-466.5	-467.4	-468.3	-469.3	-472.8	-476.6
Mn^{2+}	-232.1	-231.0	-230.5	-229.7	-229.2	-228.5	-227.5	-226.4	-225.4	-221.8	-217.9
SO_4^{2-}	-743.7	-744.3	-744.5	-744.6	-744.7	-744.7	-744.6	-744.3	-743.9	-741.7	-737.8
SO_3^{2-}	-487.0	-486.8	-486.6	-486.2	-485.8	-485.4	-484.5	-483.5	-482.3	-477.4	-470.5
$S_2O_3^{2-}$	-520.8	-522.1	-522.6	-523.3	-523.8	-524.3	-524.9	-525.4	-525.8	-526.2	-524.9
S^0	0.7	0.2	0	-0.4	-0.7	-1.0	-1.5	-2.0	-2.6	-4.7	-7.1
H_2S	-25.2	-27.1	-27.9	-29.5	-30.6	-31.9	-34.1	36.4	-38.8	-47.3	-56.7
ClO_3^-	-4.2	-6.8	-8.0	-9.9	-11.2	-12.8	-15.1	-17.5	-19.8	-27.4	-34.5
Cl^-	-129.9	-130.9	-131.3	-131.9	-132.4	-132.8	-133.5	-134.1	-134.6	-135.8	-136.0
CO_2	-383.5	-385.2	-386.0	-387.4	-388.5	-389.8	-392.0	-394.3	-396.7	-405.4	-415.2
CO	-117.9	-119.3	-120.0	-121.3	-122.2	-123.5	-125.4	-127.5	-129.8	-138.1	-147.7
CH_4	-32.7	-33.9	-34.5	-35.6	-36.4	-37.5	-39.2	-41.1	-43.2	-50.9	-59.8
CH_3COO^-	-367.4	-368.7	-369.3	-370.4	-371.1	-372.0	-373.3	-374.7	-376.1	-380.8	-385.4
H_2O	-235.6	-236.7	-237.2	-238.0	-238.6	-239.4	-240.6	-241.8	-243.1	-247.7	-252.7
H^+	0	0	0	0	0	0	0	0	0	0	0

Table 3. Values of ΔG_r° (kJ mol^{-1}) at P_{SAT} as a function of temperature for the reactions given in Table 1.

Reaction \ T (°C)	2	18	25	37	45	55	70	85	100	150	200
A	−139.5	−140.9	−141.5	−142.7	−143.5	−144.6	−146.2	−148.0	−149.9	−156.7	−164.7
B	−64.6	−65.4	−65.7	−66.4	−66.9	−67.6	−68.7	−69.8	−71.0	−75.6	−81.1
C	−38.7	−39.1	−39.4	−39.7	−40.0	−40.4	−41.0	−41.7	−42.4	−45.2	−48.6
D	−61.4	−62.2	−62.6	−63.3	−63.7	−64.4	−65.4	−66.4	−67.6	−71.9	−76.9
E	−94.6	−95.5	−95.9	−96.6	−97.2	−97.9	−99.0	−100.3	−101.7	−107.0	−113.7
F	−54.4	−55.2	−55.5	−56.2	−56.7	−57.3	−58.4	−59.5	−60.6	−65.0	−70.0
G	−90.4	−92.2	−93.1	−94.6	−95.8	−97.2	−99.5	−102.0	−104.6	−114.4	−125.9
H	−70.5	−72.4	−73.3	−75.0	−76.2	−77.7	−80.2	−82.9	−85.7	−96.3	−108.6
I	−102.6	−104.8	−105.9	−107.8	−109.2	−111.0	−114.0	−117.1	−120.5	−133.2	−148.1
J	−124.5	−122.6	−121.7	−119.9	−118.7	−117.1	−114.4	−111.6	−108.6	−97.12	−83.7
K	−48.4	−46.9	−46.1	−44.8	−43.8	−42.5	−40.5	−38.4	−36.1	−27.7	−17.9
L	−52.3	−50.7	−50.0	−48.6	−47.7	−46.4	−44.4	−42.3	−40.1	−31.8	−22.1
M	−82.3	−78.9	−77.3	−74.7	−72.9	−70.6	−67.2	−63.7	−60.1	−47.6	−33.7
N	−119.1	−117.2	−116.3	−114.6	−113.5	−112.0	−109.6	−107.1	−104.5	−94.7	−83.2
O	−156.9	−155.3	−154.5	−153.0	−151.9	−150.6	−148.4	−146.0	−143.5	−134.1	−123.1

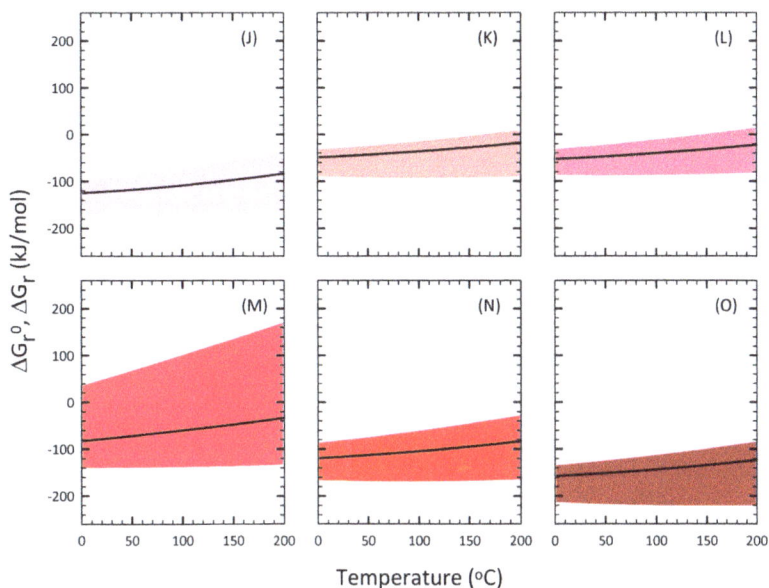

Figure 4. Values of ΔG_r° (black curves, data from Table 3) and ΔG_r (colored swaths) in kJ mol^{-1} as a function of temperature at P_{SAT} for the 6 arsenite oxidation reactions given in Table 1. Values of ΔG_r are calculated with activities given in Table 4.

The value of ΔG_r° by itself does not, however, inform on the favorable direction of a reaction or the magnitude of the energy yield. For that, we need to calculate ΔG_r in accord with Equation (1). In this communication, we do not report energy yields for a specific environment—a good example of that approach was carried out by Inskeep et al. (2005), who evaluated ΔG_r for several As-redox reactions (among others) in a few acidic and circumneutral hot springs of Yellowstone National Park at their *in situ* temperatures. Instead, here we selected broad ranges of activities (Table 4) that cover many, but certainly not all, chemical conditions that may be encountered in natural or engineered systems. For example, the chosen pH range (2-8) covers acidic to slightly alkaline conditions; very high pH systems, including Searles Lake sediments (pH ~10), are outside the range of conditions considered in our calculations. As another example, the activities (a_i) of aqueous $H_2AsO_4^-$ and H_3AsO_3 (10^{-7}-10^{-4}) are representative of systems with moderate- to high-As content, equivalent to ~0.1 μM (12.6-14.1 ppb) at the low end and ~100 μM (12.6-14.1 ppm) at the high end. For context, note that total As in some groundwaters of Bangladesh, India, Vietnam, and Thailand can exceed 25-50 μM (Nordstrom 2002); in shallow-sea hydrothermal vent fluids, it can top 50 μM (Douville et al. 1999; Price et al. 2013); some Yellowstone hot springs contain >100 μM As (Webster and Nordstrom 2003); and sediment porewaters in alkaline lakes can approach 4 mM As (Oremland et al. 2005).

Values of ΔG_r over the range of activities at 0-200 °C are depicted in Figure 3 for the 9 As(V) reduction reactions and in Figure 4 for the 6 As(III) oxidation reactions. The colored swaths (As(V) reduction in Fig. 3; As(III) oxidation in Fig. 4), representing ΔG_r, are plotted alongside values of ΔG_r° (black curves) for comparison. Observe that values of ΔG_r for all of the reactions change demonstrably with changes in temperature and chemical composition. Note further that for all but two reactions, values of ΔG_r° plot within the colored swaths representing ΔG_r. In other words, the chemical composition (and hence the Q-term in Eqn. 1) can contribute either negatively or positively to the energy yields. Only for two reactions were the values of ΔG_r° more negative than all values of ΔG_r at the conditions considered here.

It should be pointed out that the changes in chemical composition (Table 2) can correspond to rather large differences in energy yields: ~50-85 kJ mol^{-1} As at the low temperature end and ~90-150 kJ mol^{-1} As at the high temperature end for As(V) reduction, and ~55-175 kJ mol^{-1} As at the low temperature end and ~100-300 kJ mol^{-1} As at the high temperature end for As(III) oxidation. The energy yields from As(V) reduction (blue shades) vary somewhat from reaction to reaction, but not dramatically so; the reduction of As(V) coupled to the oxidation of H_2 (Reaction A), CH_4 (Reaction G), or acetate (Reaction I) are slightly more exergonic than the other As(V) reduction reactions. However, for the As(III) oxidation reactions (red shades), the energy yields vary much more so from reaction to reaction, with that coupled to chlorate reduction the most exergonic, followed by O_2 reduction.

CONCLUDING REMARKS

Arsenic resistance genes are widespread, perhaps nearly universal, in microorganisms. Even energy metabolism based on arsenic is carried out by a wide variety of predominantly bacterial phyla, with the number of isolates increasing steadily. The near absence of Archaea among the known arsenic metabolizing organisms seems curious and likely represents a sampling bias. Note that thermophiles and marine organisms also appear in very low numbers among the organisms. Perhaps a focused cultivation effort with samples from shallow-sea hydrothermal systems would change the current status. Shallow-sea vents, including those off Ambitle Island (Papua New Guinea, Milos Island (Greece), and Dominica, to name but a few, are characterized by warm (up to ~100 °C), saline fluids and often high levels of arsenite, with nearby mineral precipitates containing high levels of arsenate. The energy maps as a function of temperature and composition provided here can also be used to aid in the culturing efforts. They can be used to identify target environments for arsenotrophy and to design media that are optimized for the growth of novel arsenic-transforming Archaea.

Table 4. Range of activities (a_i) of aqueous species used in reactions in Table 1.

Species (aq)	Activity range
$H_2AsO_4^-$	1×10^{-4} - 1×10^{-7}
H_3AsO_3	1×10^{-4} - 1×10^{-7}
H_2	2×10^{-2} - 1×10^{-8}
H^+	1×10^{-2} - 1×10^{-8}
H_2S	1×10^{-4} - 1×10^{-8}
SO_4^{2-}	6×10^{-2} - 1×10^{-6}
SO_3^{2-}	1×10^{-4} - 1×10^{-8}
$S_2O_3^{2-}$	1×10^{-4} - 1×10^{-8}
CH_4	2×10^{-2} - 1×10^{-6}
CO_2	2×10^{-2} - 1×10^{-6}
CO	1×10^{-4} - 1×10^{-8}
CH_3COO^-	1×10^{-2} - 1×10^{-6}
O_2	max* - 1×10^{-8}
NO_3^-	1×10^{-3} - 1×10^{-7}
NO_2^-	1×10^{-4} - 1×10^{-8}
NH_4^+	1×10^{-3} - 1×10^{-7}
Fe^{2+}	5×10^{-3} - 1×10^{-7}
Mn^{2+}	1×10^{-5} - 1×10^{-8}
ClO_3^-	1×10^{-6} - 1×10^{-9}
Cl^-	0.55 - 1×10^{-6}

* Maximum oxygen activities are equilibrium solubility values with air at the temperature of interest, ranging from 4.4×10^{-4} at 2 °C to 1.7×10^{-4} at 70 °C to 2.8×10^{-4} at 200 °C.

REFERENCES

Abdrashitova SA, Abdullina GG, Ilialetdinov AN (1985) Glucose consumption and dehydrogenase activity of the cells of the arsenite-oxidizing bacterium *Pseudomonas putida*. Mikrobiologiia 54:679-681

Afkar E, Lisak J, Saltikov C, Basu P, Oremland RS, Stolz JF (2003) The respiratory arsenate reductase from *Bacillus selenitireducens* strain MLS10. FEMS Microbiol Lett 226:107-112

Aguiar P, Beveridge TJ, Reysenbach AL (2004) *Sulfurihydrogenibium azorense*, sp nov., a thermophilic hydrogen-oxidizing microaerophile from terrestrial hot springs in the Azores. Int J Syst Evol Microbiol 54:33-39

Ahmann D, Roberts AL, Krumholz LR, Morel FMM (1994) Microbe grows by reducing arsenic. Nature 371:750-750

Amo T, Paje MLF, Inagaki A, Ezaki S, Atomi H, Imanaka T (2002) *Pyrobaculum calidifontis* sp. nov., a novel hyperthermophilic archaeon that grows in atmospheric air. Archaea 1:113-121

Anderson CR, Cook GM (2004) Isolation and characterization of arsenate-reducing bacteria from arsenic-contaminated sites in New Zealand. Curr Microbiol 48:341-347

Anderson GL, Williams J, Hille R (1992) The purification and characterization of arsenite oxidase from *Alcaligenes faecalis*, a molybdenum-containing hydroxylase. J Biol Chem 267:23674-23682

Anderson GL, Ellis PJ, Kuhn P, Hille R (2002) Oxidation of arsenite by *Alcaligenes faecalis*. In: Environmental Chemistry of Arsenic. Frankenberger WT (ed) Marcel Dekker, Inc., New York, p 343-361

Bachate SP, Khapare RM, Kodam KM (2012) Oxidation of arsenite by two beta-proteobacteria isolated from soil. Appl Microbiol Biot 93:2135-2145

Baesman SM, Stolz JF, Kulp TR, Oremland RS (2009) Enrichment and isolation of *Bacillus beveridgei* sp nov., a facultative anaerobic haloalkaliphile from Mono Lake, California, that respires oxyanions of tellurium, selenium, and arsenic. Extremophiles 13:695-705

Bahar MM, Megharaj M, Naidu R (2012) Arsenic bioremediation potential of a new arsenite-oxidizing bacterium *Stenotrophomonas* sp MM-7 isolated from soil. Biodegradation 23:803-812

Battaglia-Brunet F, Dictor MC, Garrido F, Crouzet C, Morin D, Dekeyser K, Clarens M, Baranger P (2002) An arsenic(III)-oxidizing bacterial population: selection, characterization, and performance in reactors. J Appl Microbiol 93:656-667

Battaglia-Brunet F, Joulian C, Garrido F, Dictor MC, Morin D, Coupland K, Johnson DB, Hallberg KB, Baranger P (2006) Oxidation of arsenite by *Thiomonas* strains and characterization of Thiomonas arsenivorans sp nov. Anton Leeuw Int J G 89:99-108

Bentley R, Chasteen TG (2002) Microbial methylation of metalloids: arsenic, antimony, and bismuth. Microbiol Mol Biol R 66:250-271

Bhattacharjee H, Rosen BP (2007) Arsenic metabolism in prokaryotic and eukaryotic microbes. In: Molecular Microbiology of Heavy Metals. Nies DH, Silver S (eds). Springer-Verlag, Heidelberg, p 372-406

Blum JS, Bindi AB, Buzzelli J, Stolz JF, Oremland RS (1998) *Bacillus arsenicoselenatis*, sp nov, and *Bacillus selenitireducens*, sp nov: two haloalkaliphiles from Mono Lake, California that respire oxyanions of selenium and arsenic. Arch Microbiol 171:19-30

Blum JS, Han S, Lanoil B, Saltikov C, Witte B, Tabita FR, Langley S, Beveridge TJ, Jahnke L, Oremland RS (2009) Ecophysiology of "*Halarsenatibacter silvermanii*" strain SLAS-1(T), gen. nov., sp. nov., a facultative chemoautotrophic arsenate respirer from salt-saturated Searles Lake, California. Appl Environ Microb 75:5437-5437

Blum JS, Kulp TR, Han S, Lanoil B, Saltikov CW, Stolz JF, Miller LG, Oremland RS (2012) *Desulfohalophilus alkaliarsenatis* gen. nov., sp nov., an extremely halophilic sulfate- and arsenate-respiring bacterium from Searles Lake, California. Extremophiles 16:727-742

Bouchard B, Beaudet R, Villemur R, McSween G, Lepine F, Bisaillon JG (1996) Isolation and characterization of *Desulfitobacterium frappieri* sp nov, an anaerobic bacterium which reductively dechlorinates pentachlorophenol to 3-chlorophenol. Int J Syst Evol Microbiol 46:1010-1015

Branco R, Chung AP, Morais PV (2008) Sequencing and expression of two arsenic resistance operons with different functions in the highly arsenic-resistant strain *Ochrobactrum tritici* SCII24(T). BMC Microbiol 8:95-106

Branco R, Francisco R, Chung AP, Morais PV (2009) Identification of an *aox* system that requires cytochrome c in the highly arsenic-resistant bacterium *ochrobactrum tritici* SCII24. Appl Environ Microb 75:5141-5147

Brock TD, Freeze H (1969) *Thermus aquaticus* gen. n. and sp. n., a nonsporulating extreme thermophile. J Bacteriol 98:289-297

Bruneel O, Personne JC, Casiot C, Leblanc M, Elbaz-Poulichet F, Mahler BJ, Le Fleche A, Grimont PAD (2003) Mediation of arsenic oxidation by *Thiomonas* sp in acid-mine drainage (Carnoules, France). J Appl Microbiol 95:492-499

Budinoff CR, Hollibaugh JT (2008) Arsenite-dependent photoautotrophy by an Ectothiorhodospira-dominated consortium. ISME J 2:340-343

Cai L, Liu GH, Rensing C, Wang GJ (2009a) Genes involved in arsenic transformation and resistance associated with different levels of arsenic-contaminated soils. BMC Microbiol 9:4-14

Cai L, Rensing C, Li XY, Wang GJ (2009b) Novel gene clusters involved in arsenite oxidation and resistance in two arsenite oxidizers: *Achromobacter* sp SY8 and *Pseudomonas* sp TS44. Appl Microbiol Biot 83:715-725

Campos VL, Valenzuela C, Yarza P, Kampfer P, Vidal R, Zaror C, Mondaca MA, Lopez-Lopez A, Rossello-Mora R (2010) *Pseudomonas arsenicoxydans* sp nov., an arsenite-oxidizing strain isolated from the Atacama desert. Syst Appl Microbiol 33:193-197

Carapito C, Muller D, Tarlin E, Koechler S, Danchin A, Van Dorsselaer A, Leize-Wagner E, Bertin PN, Lett MC (2006) Identification of genes and proteins involved in the pleiotropic response to arsenic stress in *Caenibacter arsenoxydans*, a metalloresistant beta- proteobacterium with an unsequenced genome. Biochimie 88:595-606

Cavalca L, Corsini A, Bachate SP, Andreoni V (2013a) Rhizosphere colonization and arsenic translocation in sunflower (*Helianthus annuus* L.) by arsenate reducing *Alcaligenes* sp strain Dhal-L. World J Microb Biot 29:1931-1940

Cavalca L, Corsini A, Zaccheo P, Andreoni V, Muyzer G (2013b) Microbial transformations of arsenic: perspectives for biological removal of arsenic from water. Future Microbiol 8:753-768

Chang JS, Kim YH, Kim KW (2008) The *ars* genotype characterization of arsenic-resistant bacteria from arsenic-contaminated gold-silver mines in the Republic of Korea. Appl Microbiol Biot 80:155-165

Chang JS, Yoon IH, Lee JH, Kim KR, An J, Kim KW (2010) Arsenic detoxification potential of *aox* genes in arsenite-oxidizing bacteria isolated from natural and constructed wetlands in the Republic of Korea. Environ Geochem Health 32:95-105

Chitpirom K, Tanasupawat S, Akaracharanya A, Leepepatpiboon N, Prange A, Kim KW, Lee KC, Lee JS (2012) *Comamonas terrae* sp nov., an arsenite-oxidizing bacterium isolated from agricultural soil in Thailand. J Gen Appl Microbiol 58:245-251

Chovanec P, Stolz JF, Basu P (2010) A proteome investigation of roxarsone degradation by *Alkaliphilus oremlandii* strain OhILAs. Metallomics 2:133-139

Christiansen N, Ahring BK (1996) *Desulfitobacterium hafniense* sp nov, an anaerobic, reductively dechlorinating bacterium. Int J Syst Bacteriol 46:442-448

Clingenpeel SR, D'Imperio S, Oduro H, Druschel GK, McDermott TR (2009) Cloning and In situ expression studies of the *Hydrogenobaculum* arsenite oxidase genes. Appl Environ Microb 75:3362-3365

Connon SA, Koski AK, Neal AL, Wood SA, Magnuson TS (2008) Ecophysiology and geochemistry of microbial arsenic oxidation within a high arsenic, circumneutral hot spring system of the Alvord Desert. FEMS Microbiol Ecol 64:117-128

Corsini A, Cavalca L, Crippa L, Zaccheo P, Andreoni V (2010) Impact of glucose on microbial community of a soil containing pyrite cinders: Role of bacteria in arsenic mobilization under submerged condition. Soil Biol Biochem 42:699-707

Cozen AE, Weirauch MT, Pollard KS, Bernick DL, Stuart JM, Lowe TM (2009) Transcriptional map of respiratory versatility in the hyperthermophilic Crenarchaeon *Pyrobaculum aerophilum*. J Bacteriol 191:782-794

Croal LR, Gralnick JA, Malasarn D, Newman DK (2004) The genetics of geochemistry. Annu Rev Genet 38:175-202

Cullen WR, Reimer KJ (1989) Arsenic speciation in the environment. Chem Rev 89:713-765

D'Imperio S, Lehr CR, Breary M, McDermott TR (2007) Autecology of an arsenite chemolithotroph: Sulfide constraints on function and distribution in a geothermal spring. Appl Environ Microb 73:7067-7074

Dastidar A, Wang YT (2009) Arsenite oxidation by batch cultures of *Thiomonas arsenivorans* strain b6. J Environ Eng 135:708-715

Dastidar A, Wang YT (2010) Kinetics of arsenite oxidation by chemoautotrophic *Thiomonas arsenivorans* strain b6 in a continuous Stirred Tank Reacto. J Environ Eng 136:1119-1121

Delavat F, Lett MC, Lievremont D (2012) Novel and unexpected bacterial diversity in an arsenic-rich ecosystem revealed by culture-dependent approaches. Biol Direct 7:28-41

Donahoe-Christiansen J, D'Imperio S, Jackson CR, Inskeep WP, McDermott TR (2004) Arsenite-oxidizing *Hydrogenobaculum* strain isolated from an acid-sulfate-chloride geothermal spring in Yellowstone National Park. Appl Environ Microb 70:1865-1868

Douville E, Charlou J, Donval J, Hureau D, Appriou P (1999) As and Sb behaviour in fluids from various deep-sea hydrothermal systems. Earth Planet Sci Lett 328:97-104

Drewniak L, Matlakowska R, Sklodowska A (2008) Arsenite and arsenate metabolism of *Sinorhizobium* sp M14 living in the extreme environment of the Zloty Stok gold mine. Geomicrobiol J 25:363-370

Duquesne K, Lieutaud A, Ratouchniak J, Muller D, Lett MC, Bonnefoy V (2008) Arsenite oxidation by a chemoautotrophic moderately acidophilic *Thiomonas* sp.: from the strain isolation to the gene study. Environ Microbiol 10:228-237.

Ehrlich HL, Newman DK (2008) Geomicrobiology. CRC Press.

Fan H, Su C, Wang Y, Yao J, Zhao K, Wang G (2008) Sedimentary arsenite-oxidizing and arsenate-reducing bacteria associated with high arsenic groundwater from Shanyin, Northwestern China. J Appl Microb 105:529-539

Fisher E (2006) Microbial transformation of arsenic and the characterization of *Clostridium* sp. strain OhILAs. Department of Biological Sciences. Duquesne University, p. 110

Fisher JC, Hollibaugh JT (2008) Selenate-dependent anaerobic arsenite oxidation by a bacterium from Mono Lake, California. Appl Environ Microb 74:2588-2594

Fisher E, Dawson AM, Polshyna G, Lisak J, Crable B, Perera E, Ranganathan M, Thangavelu M, Basu P, Stolz JF (2008) Transformation of inorganic and organic arsenic by *Alkaliphilus oremlandii* sp. nov. strain OhILAs. Ann N Y Acad Sci 1125:230-41

Freikowski D, Winter J, Gallert C (2010) Hydrogen formation by an arsenate-reducing *Pseudomonas putida*, isolated from arsenic-contaminated groundwater in West Bengal, India. Appl Microbiol Biot 88:1636-1371

Fu HL, Meng YL, Ordonez E, Villadangos AF, Bhattacharjee H, Gil JA, Mateos LM, Rosen BP (2009) Properties of Arsenite Efflux Permeases (Acr3) from *Alkaliphilus metalliredigens* and *Corynebacterium glutamicum*. J Biol Chem 284:19887-19895

Fu HL, Rosen BP, Bhattacharjee H (2010) Biochemical characterization of a novel ArsA ATPase complex from *Alkaliphilus metalliredigens* QYMF. FEMS Lett 584:3089-3094

Fujita M, Ike M, Nishimoto S, Takahashi K, Kashiwa M (1997) Isolation and characterization of a novel selenate-reducing bacterium, *Bacillus* sp. SF-1. J Ferment Bioeng 83:517-522

Garcia-Dominguez E, Mumford A, Rhine ED, Paschal A, Young LY (2008) Novel autotrophic arsenite-oxidizing bacteria isolated from soil and sediments. FEMS Microbiol Ecol 66:401-410.

Gihring TM, Banfield JF (2001) Arsenite oxidation and arsenate respiration by a new *Thermus* isolate. FEMS Microbiol Lett 204:335-340

Gihring TM, Druschel GK, McCleskey RB, Hamers RJ, Banfield JF (2001) Rapid arsenite oxidation by *Thermus aquaticus* and *Thermus thermophilus*: Field and laboratory investigations. Environ Sci Technol 35:3857-3862

Giloteaux L, Holmes DE, Williams KH, Wrighton KC, Wilkins MJ, Montgomery AP, Smith JA, Orellana R, Thompson CA, Roper TJ, Long PE, Lovley DR (2013) Characterization and transcription of arsenic respiration and resistance genes during *in situ* uranium bioremediation. ISME J 7:370-383

Green HH (1918) Description of a bacterium which oxidizes arsenite to arsenate, and of one which reduces arsenate to arsenite, isolated from a cattle-dipping tank. S Afr J Sci 14:465-467

Hamamura N, Fukushima K, Itai T (2013) Identification of antimony- and arsenic-oxidizing bacteria associated with antimony mine tailing. Microbes Environ 28:257-263

Handley KM, Hery M, Lloyd JR (2009a) *Marinobacter santoriniensis* sp nov., an arsenate- respiring and arsenite-oxidizing bacterium isolated from hydrothermal sediment. Int J Syst Evol Microbiol 59:886-892

Handley KM, Hery M, Lloyd JR (2009b) Redox cycling of arsenic by the hydrothermal marine bacterium *Marinobacter santoriniensis*. Environ Microbiol 11:1601-1611

Hao XL, Lin YB, Johnstone L, Liu GH, Wang GJ, Wei GH, McDermott TR, Rensing C (2012) Genome Sequence of the Arsenite-Oxidizing Strain *Agrobacterium tumefaciens* 5A. J Bacteriol 194:903-903

Hedderich R, Klimmek O, Kroger A, Dirmeier R, Keller M, Stetter KO (1999) Anaerobic respiration with elemental sulfur and with disulfides. FEMS Microbiol Rev 22:353-381

Herbel MJ, Blum JS, Hoeft SE, Cohen SM, Arnold LL, Lisak J, Stolz JF, Oremland RS (2002) Dissimilatory arsenate reductase activity and arsenate-respiring bacteria in bovine rumen fluid, hamster feces, and the termite hindgut. FEMS Microbiol Ecol 41:59-67

Hoeft SE, Lucas FO, Hollibaugh JT, Oremland RS (2002) Characterization of microbial arsenic reduction in the anoxic bottom waters of Mono Lake, California. Geomicrobiol J 19:23-40

Hoeft SE, Kulp TR, Stolz JF, Hollibaugh JT, Oremland RS (2004) Dissimilatory arsenate reduction with sulfide as electron donor: Experiments with mono lake water and isolation of strain MLMS-1, a chemoautotrophic arsenate respirer. Appl Environ Microb 70:2741-2747

Hoeft SE, Blum JS, Stolz JF, Tabita FR, Witte B, King GM, Santini JM, Oremland RS (2007) *Alkalilimnicola ehrlichii* sp. nov., a novel, arsenite-oxidizing haloalkaliphilic gammaproteobacterium capable of chemoautotrophic or heterotrophic growth with nitrate or oxygen as the electron acceptor. Int J Syst Evol Microbiol 57:504-512

Hoeft SE, Kulp TR, Han S, Lanoil B, Oremland RS (2010) Coupled arsenotrophy in a hot spring photosynthetic biofilm at Mono Lake, California. Appl Environ Microb 76:4633-4639

Hollibaugh JT, Budinoff C, Hollibaugh RA, Ransom B, Bano N (2006) Sulfide oxidation coupled to arsenate reduction by a diverse microbial community in a Soda Lake. Appl Environ Microb 72:2043-2049

Holliger C, Wohlfarth G, Diekert G (1998) Reductive dechlorination in the energy metabolism of anaerobic bacteria. FEMS Microbiol Rev 22:383-398

Hsu HY (2007) Isolation and characterization of arsenite-oxidizing bacteria *Bosea* sp. str. L7506 from arsenic-contaminated groundwater in blackfoot disease region in Taiwan, Department of Bioenvironmental Systems Engineering. National Taiwan University, Taipei, p. 64

Huang YY, Li H, Rensing C, Zhao K, Johnstone L, Wang GJ (2012) Genome sequence of the facultative anaerobic arsenite-oxidizing and nitrate-reducing bacterium *Acidovorax* sp Strain NO1. J Bacteriol 194:1635-1636

Huber R, Kristjansson JK, Stetter KO (1987) *Pyrobaculum* gen. nov., a new genus of neutrophilic, rod-shaped archaebacteria from continental solfataras growing optimally at 100 °C. Arch Microbiol 149:95-101

Huber R, Sacher M, Vollmann A, Huber H, Rose D (2000) Respiration of arsenate and selenate by hyperthermophilic archaea. Syst Appl Microbiol 23:305-314

Ilyaletdinov AN, Abdrashitova SA (1981) Autotrophic arsenic oxidation by a *Pseudomonas arsenitoxidans* culture. Mikrobiologiia 50:135-140

Inskeep WP, Ackerman GG, Taylor WP, Kozubal M, Korf S, Macur RE (2005) On the energetics of chemolithotrophy in nonequilibrium systems: case studies of geothermal springs in Yellowstone National Park. Geobiology 3:297-317

Ito A, Miura JI, Ishikawa N, Umita T (2012) Biological oxidation of arsenite in synthetic groundwater using immobilised bacteria. Water Res 46:4825-4831

Jareonmit P, Mehta M, Sadowsky MJ, Sajjaphan K (2012) Phylogenetic and phenotypic analyses of arsenic-reducing bacteria isolated from an old tin mine area in Thailand. World J Microb Biot 28:2287-2292

Jensen A, Finster K (2005) Isolation and characterization of *Sulfurospirillum carboxydovorans* sp nov., a new microaerophilic carbon monoxide oxidizing epsilon Proteobacterium. Anton Leeuw Int J G 87:339-353

Johnson JW, Oelkers EH, Helgeson HC (1992) SUPCRT92: A software package for calculating the standard molal properties of minerals, gases, aqueous species, and reactions from 1 to 5000 bar and 0 to 1000 °C. Comp Geosci 18:899-947

Jones CA, Langner HW, Anderson K, McDermott TR, Inskeep WP (2000) Rates of microbially mediated arsenate reduction and solubilization. Soil Sci Soc Am J 64:600-608

Kao AC, Chu YJ, Hsu FL, Liao VHC (2013) Removal of arsenic from groundwater by using a native isolated arsenite-oxidizing bacterium. J Contam Hydrol 155:1-8

Kashefi K, Lovely DR (2000) Reduction of Fe(III), Mn(IV), and toxic metals at 100 degrees C by *Pyrobaculum islandicum*. Appl Environ Microb 66:1050-1056

Kraft T, Macy JM (1998) Purification and characterization of the respiratory arsenate reductase of *Chryosiogenes arsenatis*. Europ J Biochem 255:647-653

Kruger MC, Bertin PN, Heipieper HJ, Arsene-Ploetze F (2013) Bacterial metabolism of environmental arsenic-mechanisms and biotechnological applications. Appl Microbiol Biot 97:3827-3841

Kulp TR, Hoeft SE, Asao M, Madigan MT, Hollibaugh JT, Fisher JC, Stolz JF, Culbertson CW, Miller LG, Oremland RS (2008) Arsenic(III) fuels anoxygenic photosynthesis in hot spring biofilms from Mono Lake, California. Science 321:967-970

LaRowe DE, Amend JP (2014) Energetic constraints on life in marine deep sediments. *In*: Microbial Life of the Deep Biosphere. Kallmeyer J, Wagner D (eds). DeGruyter, Berlin, p 279-302.

Laverman AM, Blum JS, Schaefer JK, Phillips EJP, Lovley DR, Oremland RS (1995) Growth of strain SES-3 with arsenate and other diverse electron acceptors. Appl Environ Microb 61:3556-3561

Lebrun E, Brugna M, Baymann F, Muller D, Lievremont D, Lett MC, Nitschke W (2003) Arsenite oxidase, an ancient bioenergetic enzyme. Mol Biol Evol 20:686-693

Lebuhn M, Achouak W, Schloter M, Berge O, Meier H, Barakat M, Hartmann A, Heulin T (2000) Taxonomic characterization of *Ochrobactrum* sp isolates from soil samples and wheat roots, and description of *Ochrobactrum tritici* sp nov and *Ochrobactrum grignonense* sp nov. Int J Syst Evol Microbiol 50:2207-2223

Ledbetter RN, Connon SA, Neal AL, Dohnalkova A, Magnuson TS (2007) Biogenic mineral production by a novel arsenic-metabolizing thermophilic bacterium from the Alvord Basin, Oregon. Appl Environ Microb 73:5928-5936

Legge JW, Turner AW (1954) Bacterial oxidation of arsenite 3: cell-free arsenite dehydrogenase. Aust J Biol Sci 7:496-503

Lett MC, Muller D, Lievremont D, Silver S, Santini J (2012) Unified nomenclature for genes involved in prokaryotic aerobic arsenite oxidation. J Bacteriol 194:207-208

Liao VHC, Chu YJ, Su YC, Hsiao SY, Wei CC, Liu CW, Liao CM, Shen WC, Chang FJ (2011) Arsenite-oxidizing and arsenate-reducing bacteria associated with arsenic-rich groundwater in Taiwan. J Contam Hydrol 123:20-29

Lin PC (2010) Characterization of an arsenate-reducing bacterium *citrobacter* sp. strain L2, Department of Bioenvironmental Systems Engineering. National Taiwan University, Taipei, p. 63

Lin YB, Fan HX, Hao X, Johnstone L, Hu Y, Wei GH, Alwathnani HA, Wang GJ, Rensing C (2012) Draft Genome Sequence of *Halomonas* sp Strain HAL1, a Moderately Halophilic Arsenite-Oxidizing Bacterium Isolated from Gold-Mine. Soil. J Bacteriol 194:199-200

Liu A, Garcia-Dominguez E, Rhine ED, Young LY (2004) A novel arsenate respiring isolate that can utilize aromatic substrates. FEMS Microbiol Ecol 48:323-332

Lloyd JR, Oremland RS (2006) Microbial transformations of arsenic in the environment: From soda lakes to aquifers. Elements 2:85-90

London J (1963) *Thiobacillus intermedius* nov. sp., a novel type of facultative autotroph. Arch Mikrobiol 46:329-337

Lonergan DJ, Jenter HL, Coates JD, Phillips EJP, Schmidt TM, Lovley DR (1996) Phylogenetic analysis of dissimilatory Fe(III)-reducing bacteria. J Bacteriol 178:2402-2408

Luijten M, de Weert J, Smidt H, Boschker HTS, de Vos WM, Schraa G, Stams AJM (2003) Description of *Sulfurospirillum halorespirans* sp nov., an anaerobic, tetrachloroethene-respiring bacterium, and transfer of *Dehalospirillum multivorans* to the genus *Sulfurospirillum* as *Sulfurospirillum multivorans* comb. nov. Int J Syst Evol Microbiol 53:787-793

Luijten M, Weelink SAB, Godschalk B, Langenhoff AAM, van Eekert MHA, Schraa G, Stams AJM (2004) Anaerobic reduction and oxidation of quinone moieties and the reduction of oxidized metals by halorespiring and related organisms. FEMS Microbiol Ecol 49:145-150

Macur RE, Wheeler JT, McDermott TR, Inskeep WP (2001) Microbial populations associated with the reduction and enhanced mobilization of arsenic in mine tailings. Environ Sci Technol 35:3676-3682

Macur RE, Jackson CR, Botero LM, McDermott TR, Inskeep WP (2004) Bacterial populations associated with the oxidation and reduction of arsenic in an unsaturated soil. Environ Sci Technol 38:104-111

Macur RE, Jay ZJ, Taylor WP, Kozubal MA, Kocar BD, Inskeep WP (2013) Microbial community structure and sulfur biogeochemistry in mildly-acidic sulfidic geothermal springs in Yellowstone National Park. Geobiology 11:86-99

Macy JM, Santini JM (2002) Unique modes of arsenate respiration by *Chrysiogenes arsenatis* and *Desulfomicrobium* sp str. Ben-RB. *In*: Environmental Chemistry of Arsenic. Frankenberger WT (ed). Marcel Dekker, New York, p 297-312

Macy JM, Nunan K, Hagen KD, Dixon DR, Harbour PJ, Cahill M, Sly LI (1996) *Chrysiogenes arsenatis* gen. nov., sp. nov., a new arsenate-respiring bacterium isolated from gold mine wastewater. Int J Syst Bacteriol 46:1153-1157

Macy JM, Santini JM, Pauling BV, O'Neill AH, Sly LI (2000) Two new arsenate/sulfate-reducing bacteria: mechanisms of arsenate reduction. Arch Microbiol 173:49-57

Majumder A, Bhattacharyya K, Bhattacharyya S, Kole SC (2013) Arsenic-tolerant, arsenite-oxidising bacterial strains in the contaminated soils of West Bengal, India Sci Total Environ 463:1006-1014

Malasarn D, Saltikov W, Campbell KM, Santini JM, Hering JG, Newman DK (2004) *arrA* is a reliable marker for As(V) respiration. Science 306:455-455

Malasarn D, Keeffe JR, Newman DK (2008) Characterization of the arsenate respiratory reductase from *Shewanella* sp strain ANA-3. J Bacteriol 190:135-142

Marsh RM, Norris PR (1983) The isolation of some thermophilic, autotrophic, iron- and sulphur-oxidizing bacteria. FEMS Microbiol Lett 17:311-315

McEwan A, Ridge J, McDevitt C, Hugenholtz P (2002) The DMSO reductase family of microbial molybdenum enzymes: Molecular properties and role in the dissimilatory reduction of toxic elements. Geomicrobiol J 19:3-21

Meng YL, Liu Z, Rosen BP (2004) As(III) and Sb(III) uptake by GlpF and efflux by ArsB in *Escherichia coli*. J Biol Chem 279:18334-18341

Moreira D, Amils R (1997) Phylogeny of *Thiobacillus cuprinus* and other mixotrophic thiobacilli: Proposal for *Thiomonas* gen nov. Int J Syst Bacteriol 47:522-528

Mukhopadhyay BP, Rosen BP, Phung LT, Silver S (2002) Microbial arsenic: from geocycles to genes to enzymes. FEMS Microbiol Rev 26:311-325

Muller D, Lievremont D, Simeonova DD, Hubert JC, Lett MC (2003) Arsenite oxidase *aox* genes from a metal-resistant beta-proteobacterium. J Bacteriol 185:135-141

Muller D, Simeonova DD, Riegel P, Mangenot S, Koechler S, Lievremont D, Bertin PN, Lett MC (2006) *Herminiimonas arsenicoxydans* sp nov., a metalloresistant bacterium. Int Journal Syst Evol Microbiol 56:1765-1769

Murphy JN, Saltikov CW (2007) The *cymA* gene, encoding a tetraheme c-type cytochrome, is required for arsenate respiration in *Shewanella* species. J Bacteriol 189:2283-2290

Murphy JN, Saltikov CW (2009) The ArsR repressor mediates arsenite-dependent regulation of arsenate respiration and detoxification operons of *Shewanella* sp strain ANA-3. J Bacteriol 191:6722-6731

Nagvenkar GS, Ramaiah N (2010) Arsenite tolerance and biotransformation potential in estuarine bacteria. Ecotoxicology 19:604-613

Newman DK, Ahmann D, Morel FMM (1998) A brief review of microbial arsenate respiration. Geomicrobiol J 15:255-268

Newman DK, Beveridge TJ, Morel FMM (1997a) Precipitation of arsenic trisulfide by *Desulfotomaculum auripigmentum*. Appl Environ Microb 63:2022-2028

Newman DK, Kennedy EK, Coates JD, Ahmann D, Ellis DJ, Lovley DR, Morel FMM (1997b) Dissimilatory arsenate and sulfate reduction in *Desulfotomaculum auripigmentum* sp. nov. Arch Microbiol 168:380-388

Niggemyer A, Spring S, Stackebrandt E, Rosenzweig RF (2001) Isolation and characterization of a novel As(V)-reducing bacterium: Implications for arsenic mobilization and the genus *Desulfitobacterium*. Appl Environ Microb 67:5568-5580

Nordstrom DK (2002) Worldwide occurrences of arsenic in ground water. Science 296:2143-2145

Oremland RS, Stolz JF (2003) The ecology of arsenic. Science 300:939-944

Oremland RS, Stolz JF (2005) Arsenic, microbes, and contaminated aquifers. Trends Microbiol 13:45-49

Oremland RS, Blum JS, Culbertson CW, Visscher PT, Miller LG, Dowdle P, Strohmaier FE (1994) Isolation, growth, and metabolism of an obligately anaerobic, selenate respiring bacterium, strain SES-3. Appl Environ Microb 60:3011-3019

Oremland RS, Hoeft SE, Santini JA, Bano N, Hollibaugh RA, Hollibaugh JT (2002) Anaerobic oxidation of arsenite in Mono Lake water and by facultative, arsenite-oxidizing chemoautotroph, strain MLHE-1. Appl Environ Microb 68:4795-4802

Oremland RS, Stolz JF, Hollibaugh JT (2004) The microbial arsenic cycle in Mono Lake, California. FEMS Microbiol Ecol 48:15-27

Oremland RS, Kulp TR, Blum JS, Hoeft SE, Baesman S, Miller LG, Stolz JF (2005) A microbial arsenic cycle in a salt-saturated, extreme environment. Science 308:1305-1308

Oremland RS, Wolfe-Simon F, Saltikov CW, Stolz JF (2009) Arsenic in the evolution of earth and extraterrestrial ecosystems. Geomicrobiol J 26:522-536

Osborne FH, Ehrlich HL (1976) Oxidation of arsenite by a soil isolate of *Alcaligenes*. J Appl Bacteriol 41:295-305

Osborne TH, Jamieson HE, Hudson-Edwards KA, Nordstrom DK, Walker SR, Ward SA, Santini JM (2010) Microbial oxidation of arsenite in a subarctic environment: diversity of arsenite oxidase genes and identification of a psychrotolerant arsenite oxidiser. BMC Microbiol 10:205-212

Páez-Espino D, Tamames J, de Lorenzo V, Canovas D (2009) Microbial responses to environmental arsenic. Biometals 22:117-130

Phillips SE, Taylor ML (1976) Oxidation of arsenite to arsenate by *Alcaligenes faecalis*. Appl Environ Microb 32:392-399

Price RE, Savov I, Planer-Friedrich B, Bühring S, Amend JP, Pichler T (2013) Processes influencing extreme As enrichment in shallow-sea hydrothermal fluids of Milos Island, Greece. Chem Geol 348:15-26

Qin J, Lehr CR, Yuan C, Le XC, McDermott TR, Rosen BP (2009) Biotransformation of arsenic by a Yellowstone thermoacidophilic eukaryotic alga. Proc Natl Acad Sci 106:5213-5217

Rauschenbach I, Narasingarao P, Häggblom MM (2011) *Desulfurispirillum indicum* sp. nov., a selenate- and selenite-respiring bacterium isolated from an estuarine canal. Int Journal of Syst Evol Microbiol 61:654-658

Rauschenbach I, Bini E, Haeggblom MM, Yee N (2012) Physiological response of *Desulfurispirillum indicum* S5 to arsenate and nitrate as terminal electron acceptors. FEMS Microbiol Ecol 81:156-162

Rehman A, Butt SA, Hasnain S (2010) Isolation and characterization of arsenite oxidizing *Pseudomonas lubricans* and its potential use in bioremediation of wastewater. Afr J Biotechnol 9:1493-1498

Rhine ED, Phelps CD, Young LY (2006) Anaerobic arsenite oxidation by novel denitrifying isolates. Environ Microbiol 8:899-908

Rhine ED, Ni Chadhain SM, Zylstra GJ, Young LY (2007) The arsenite oxidase genes (*aroAB*) in novel chemoautotrophic arsenite oxidizers. Biochem Biophys Res Commun 354:662-667

Rhine ED, Onesios KM, Serfes ME, Reinfelder JR, Young LY (2008) Arsenic transformation and mobilization from minerals by the arsenite oxidizing strain WAO. Environ Sci Technol 42:1423-1429

Richey C, Chovanec P, Hoeft SE, Oremland RS, Basu P, Stolz JF (2009) Respiratory arsenate reductase as a bidirectional enzyme. Biochem Biophys Res Commun 382:298-302

Robie RA, Hemingway BS (1985) Low-temperature molar heat capacities and entropies of MnO_2 (pyrolusite), Mn_3O_4 (hausmanite) and Mn_2O_3 (bixbyite). J Chem Thermo 17:165-181

Rodriguez-Freire L, Sun W, Sierra-Alvarez R, Field JA (2012) Flexible bacterial strains that oxidize arsenite in anoxic or aerobic conditions and utilize hydrogen or acetate as alternative electron donors. Biodegradation 23:133-143

Rosen BP (2002) Biochemistry of arsenic detoxification. FEMS Lett. 529:86-92

Rosen BP, Bhattacharjee H, Zhou T, Walmsley AR (1999) Mechanism of the ArsA ATPase. Biochim Biophys Acta 1461:207-215

Salmassi TM, Venkateswaren K, Satomi M, Nealson KH, Newman DK, Hering JG (2002) Oxidation of arsenite by *Agrobacterium albertimagni*, AOL15, sp nov, isolated from Hot Creek, California. Geomicrobiol J 19:53-66

Salmassi TM, Walker JJ, Newman DK, Leadbetter JR, Pace NR, Hering JG (2006) Community and cultivation analysis of arsenite oxidizing biofilms at Hot Creek. Environ Microbiol 8:50-59

Saltikov CW, Newman DK (2003) Genetic identification of a respiratory arsenate reductase. Proc Natl Acad Sci 100:10983-10988

Saltikov CW, Cifuentes A, Venkateswaran K, Newman DK (2003) The *ars* detoxificaton system is advantageous but not required for As(V) respiration by the genetically tractable *Shewanella* species strain ANA-3. Appl Environ Microb 69:2800-2809

Saltikov CW, Wildman RA, Newman DK (2005) Expression dynamics of arsenic respiration and detoxification in *Shewanella* sp strain ANA-3. J Bacteriol 187:7390-7396

Santini JM, Ward SA (eds) (2012) The Metabolism of Arsenite. CRC Press/Balkema, Leiden, The Netherlands

Santini JM, Sly LI, Schnagl RD, Macy JM (2000) A new chemolithoautotrophic arsenite-oxidizing bacerium isolated from a gold mine: phylogenetic, physiological, and preliminary biochemical studies. Appl Environ Microb 66:92-97

Santini JM, Sly LI, Wen AM, Comrie D, De Wulf-Durand P, Macy JM (2002a) New arsenite- oxidizing bacteria isolated from Australian gold mining environments – Phylogenetic relationships. Geomicrobiol J 19:67-76

Santini JM, Stolz JF, Macy JM (2002b) Isolation of a new arsenate-respiring bacterium-physiological and phylogenetic studies. Geomicrobiol J 19:41-52

Santini JM, vanden Hoven RN, Macy JM (2002c) Characteristics of newly discovered arsenite-oxidizing bacteria. *In*: Environmental Chemistry of Arsenic. Frankenberger WT (ed) Marcel Dekker, New York, p 329-342

Santini JM, vanden Hoven RN (2004) Molybdenum-containing arsenite oxidase of the chemolithoautotrophic arsenite oxidizer NT-26. J Bacteriol 186:1614-1619

Santini JM, Kappler U, Ward SA, Honeychurch MJ, vanden Hoven RN, Bernhardt PV (2007) The NT-26 cytochrome c(552) and its role in arsenite oxidation. Biochim Biophys Acta 1767:189-196

Scholzmuramatsu H, Neumann A, Messmer M, Moore E, Diekert G (1995) Isolation and characterization of *Dehalospirillum multivorans* gen. nov., sp. nov., a tetrachloroethene-utilizing, strictly anaerobic bacterium. Arch Microbiol 163:48-56

Schumacher W, Kroneck PMH, Pfennig N (1992) Comparative systematic study on "*Spirillum* 5175", *Campylobacter* and *Wolinella* species. Arch of Microbiol 158:287-293

Sehlin HM, Lindstrom EB (1992) Oxidation and reduction of arsenic by *sulfolobus-acidocaldarius* strain BC. FEMS Microbiol Lett 93:87-92

Shakoori FR, Aziz I, Rehman A, Shakoori AR (2010) Isolation and characterization of arsenic reducing bacteria from industrial effluents and their potential use in bioremediation of wastewater. Pak J Zool 42:331-338

Shelobolina ES, Vrionis HA, Findlay RH, Lovley DR (2008) *Geobacter uraniireducens* sp nov., isolated from subsurface sediment undergoing uranium bioremediation. Int J Syst Evol Microbiol 58:1075-1078

Silver S, Phung LT (2005) Genes and enzymes involved in bacterial oxidation and reduction of inorganic arsenic. Appl Environ Microb 71:599-608

Simeonova DD (2004) Arsenic oxidation of *Cenibacterium Arsenoxidans*: Potential application in bioremediation of arsenic contaminated water, Department of Biological Sciences. Louis Pasteur University, p 318

Slyemi D, Bonnefoy V (2012) How prokaryotes deal with arsenic (dagger). Environ Microbiol Rep 4:571-586

Smedley PL, Kinniburgh DG (2002) A review of the source, behavior and distribution of arsenic in natural waters. Appl Geochem 17:517-568

Snow CL, Lilova KI, Radha AV, Shi Q, Smith S, Navrotsky A, Boerio-Goates J, Woodfield BF (2013) Heat capacity and thermodynamics of a synthetic two-line ferrihydrite, FeOOH·0.027H2O. J Chem Thermo 58:307-314

Song WF, Deng Q, Bin LY, Wang W, Wu C (2012) Arsenite oxidation characteristics and molecular identification of arsenic-oxidizing bacteria isolated from soil. Appl Mechanics Materials 188:313-318

Sorokin DY, Muyzer G (2010) *Desulfurispira natronophila* gen. nov sp nov.: an obligately anaerobic dissimilatory sulfur-reducing bacterium from soda lakes. Extremophiles 14:349-355

Sorokin ID, Kravchenko IK, Tourova TP, Kolganova TaV, Boulygina ES, Sorokin DY (2008) *Bacillus alkalidiazotrophicus* sp nov., a diazotrophic, low salt-tolerant alkallphile isolated from Mongolian soda soil. Int J Syst Evol Microbiol 58:2459-2464

Sorokin DY, Tourova TP, Sukhacheva MV, Muyzer G (2012) *Desulfuribacillus alkaliarsenatis* gen. nov sp nov., a deep-lineage, obligately anaerobic, dissimilatory sulfur and arsenate- reducing, haloalkaliphilic representative of the order Bacillales from soda lakes. Extremophiles 16:597-605

Stackebrandt E, Schumann P, Schuler E, Hippe H (2003) Reclassification of *Delsulfotomaculum auripigmentum* as *Desulfosporosinus auripigmenti* corrig., comb. nov. Int J Syst Evol Microbiol 53:1439-1443

Stolz JF, Oremland RS (1999) Bacterial respiration of arsenic and selenium. FEMS Microbiol Rev 23:615-627

Stolz JF, Gugliuzza T, Blum JS, Oremland R, Murillo FM (1997) Differential cytochrome content and reductase activity in *Geospirillum barnesii* strain SeS3. Arch Microbiol 167:1-5

Stolz JF, Basu P, Santini JM, Oremland RS (2006) Arsenic and selenium in microbial metabolism. Annu Rev Microbiol 60:107-130

Stolz JF, Perera E, Kilonzo B, Kail B, Crable B, Fisher E, Ranganathan M, Wormer L, Basu P (2007) Biotransformation of 3-nitro-4-hydroxybenzene arsonic acid (roxarsone) and release of inorganic arsenic by *Clostridium* species. Environ Sci Technol 41:818-823

Stolz JF, Basu P, Oremland RS (2010) Microbial arsenic metabolism: New twists on an old poison. Microbe 5:53-59

Sultana M, Hartig C, Planer-Friedrich B, Seifert J, Schlomann M (2011) Bacterial communities in Bangladesh aquifers differing in aqueous arsenic concentration. Geomicrobiol J 28:198-211

Sun WJ, Sierra-Alvarez R, Milner L, Field JA (2010) Anaerobic oxidation of arsenite linked to chlorate reduction. Appl Environ Microb 76:6804-6811

Sung Y, Fletcher KF, Ritalaliti KM, Apkarian RP, Ramos-Hernandez N, Sanford RA, Mesbah NM, Loffler FE (2006) *Geobacter lovleyi* sp nov strain SZ, a novel metal-reducing and tetrachloroethene-dechlorinating bacterium. Appl Environ Microb 72:2775-2782

Suttigarn A, Wang YT (2005) Arsenite oxidation by *Alcaligenes faecalis* strain O1201. J Environ Eng 131:1293-1301

Suzuki K, Wakao N, Sakurai Y, Kimura T, Sakka K, Ohmiya K (1997) Transformation of *Escherichia coli* with a large plasmid of *Acidiphilium multivorum* AIU 301 encoding arsenic resistance. Appl Environ Microb 63:2089-2091

Takai K, Hirayama H, Sakihama Y, Inagaki F, Yamato Y, Horikoshi K (2002) Isolation and metabolic characteristics of previously uncultured members of the order Aquificales in a subsurface gold mine. Appl Environ Microb 68:3046-3054

Takai K, Kobayashi H, Nealson KH, Horikoshi K (2003a) *Deferribacter desulfuricans* sp nov., a novel sulfur-, nitrate- and arsenate-reducing thermophile isolated from a deep-sea hydrothermal vent. Int J Syst Evol Microbiol 53:839-846

Takai K, Kobayashi H, Nealson KH, Horikoshi K (2003b) *Sulfurihydrogenibium subterraneum* gen. nov., sp nov., from a subsurface hot aquifer. Int J Syst Evol Microbiol 53:823-827

Tsai SL, Singh S, Chen W (2009) Arsenic metabolism by microbes in nature and the impact on arsenic remediation. Curr Opin Biotechnol 20:659-667

Turner AW (1949) Bacterial oxidation of arsenite. Nature 164:76-77

Turner AW (1954) Bacterial oxidation of arsenite 1: Description of bacteria isolated from arsenical cattle-dipping fluids. Aust J Biol Sci 7:452-478

Turner AW, Legge JW (1954) Bacterial oxidation of arsenite 2: The activity of washed suspensions. Aust J Biol Sci 7:479-495

Valenzuela C, Campos VL, Yanez J, Zaror CA, Mondaca MA (2009) Isolation of arsenite- oxidizing bacteria from arsenic-enriched sediments from Camarones River, Northern Chile. Bull Environ Contam Toxicol 82:593-596

van Lis R, Nitschke W, Duval S, Schoepp-Cothenet B (2013) Arsenics as bioenergetic substrates. Biochim Biophys Acta (Bioenergetics) 1827:176-188

vanden Hoven RN, Santini JM (2004) Arsenite oxidation by the heterotroph *Hydrogenophaga* sp. str. NT-14: the arsenite oxidase and its physiological electron acceptor. Biochim Biophys Acta (Bioenergetics) 1656:148-155

Volkl P, Huber R, Drobner E, Rachel R, Burggraf S (1993) *Pyrobaculum aerophilum* sp. nov., a novel nitrate-reducing hyperthermophilic archaeum. Appl and Environ Microbiol 59:2918-2926

Wakao N, Koyatsu H, Komai Y, Shimokawara H, Sakurai Y, Shiota H (1988) Microbial oxidation of arsenite and occurrence of arsenite-oxidizing bacteria in acid mine. Geomicrobiol J 6:11-24

Wakao N, Nagasawa N, Matsuura T, Matsukura H, Matsumoto T, Hiraishi A, Sakurai Y, Shiota H (1994) *Acidiphilium multivorum* sp. nov., an acidophilic chemoorganotrophic bacterium from pyritic acid mine drainage. J Gen Appl Microbiol 40:143-159

Wang G, Kennedy SP, Fasiludeen S, Rensing C, DasSarma S (2004) Arsenic resistance in *Halobacterium* sp. strain NRC-1 examined by using an improved gene knockout system. J Bacteriol 186:3187-3194

Wang YT, Suttigarn A (2007) Arsenite oxidation by *Alcaligenes faecalis* strain O1201 in a continuous-flow bioreactor. J Environ Eng 133:471-476

Warelow TP, Oke M, Schoepp-Cothenet B, Dahl JU, Bruselat N, Sivalingam GN, Leimkuhler S, Thalassinos K, Kappler U, Naismith JH, Santini JM (2013) The respiratory arsenite oxidase: structure and the role of residues surrounding the rieske cluster. PLos ONE 8:e72535.

Watanabe T, Hirano S (2013) Metabolism of arsenic and its toxicological relevance. Arch Toxicol 87:969-979

Webster JG, Nordstrom DK (2003) Geothermal Arsenic. *In*: Arsenic in Ground Water. Welch AH, Stollenwerk KG (eds). Kluwer Academic Publishers, Boston, p 101-125

Weeger W, Lievremont D, Perret M, Lagarde F, Hubert JC, Leroy M, Lett MC (1999) Oxidation of arsenite to arsenate by a bacterium isolated from an aquatic environment. Biometals 12:141-149

Wei CC (2008) Isolation and molecular characterization of arsenate respiration and resistance bacterial strain W2 from arsenic-contaminated groundwater in Taiwan, Department of Bioenvironmental Systems Engineering. National Taiwan University, Taipei

Yamamura S, Watanabe K, Suda W, Tsuboi S, Watanabe M (2014) Effect of antibiotics on redox transformations of arsenic and diversity of arsenite-oxidizing bacteria in sediment microbial communities. Environ Sci Technol 48:350-357

Yamamura S, Yamashita M, Fujimoto N, Kuroda M, Kashiwa M, Sei K, Fujita M, Ike M (2007) *Bacillus selenatarsenatis* sp nov., a selenate- and arsenate-reducing bacterium isolated from the effluent drain of a glass-manufacturing plant. Int J Syst Evol Microbiol 57:1060-1064

Ye Q, Roh Y, Carroll SL, Blair B, Zhou JZ, Zhang CL, Fields MW (2004) Alkaline anaerobic respiration: Isolation and characterization of a novel alkaliphilic and metal-reducing bacterium. Appl Environ Microb 70:5595-5602

Yoon IH, Chang JS, Lee JH, Kim KW (2009) Arsenite oxidation by *Alcaligenes* sp strain RS-19 isolated from arsenic-contaminated mines in the Republic of Korea. Environ Geochem Health 31:109-117

Zargar K, Hoeft S, Oremland R, Saltikov CW (2010) Identification of a novel arsenite oxidase gene, *arxA*, in the haloalkaliphilic, arsenite-oxidizing bacterium *Alkalilimnicola ehrlichii* strain MLHE-1. J Bacteriol 192:3755-3762

Zargar K, Conrad A, Bernick DL, Lowe TM, Stolc V, Hoeft S, Oremland RS, Stolz J, Saltikov CW (2012) ArxA, a new clade of arsenite oxidase within the DMSO reductase family of molybdenum oxidoreductases. Environ Microbiol 14:1635-1645

Zavarzina DG, Tourova TP, Kolganova TV, Boulygina ES, Zhilina TN (2009) Description of *Anaerobacillus alkalilacustre* gen. nov., sp nov.-Strictly anaerobic diazotrophic bacillus isolated from soda lake and transfer of *Bacillus arseniciselenatis*, *Bacillus macyae*, and *Bacillus alkalidiazotrophicus* to *Anaerobacillus* as the new combinations *A. arseniciselenatis* comb. nov., *A. macyae* comb. nov., and *A. alkalidiazotrophicus* comb. nov. Microbiol 78:723-731

Zhang SY, Rensing C, Zhu YG (2014) Cyanobacteria-mediated arsenic redox dynamics is regulated by phosphate in aquatic environments. Environ Sci Technol 48:994-1000

Zhou T, Radaev S, Rosen BP, Gatti DL (2000) Structure of the ArsA ATPase: the catalytic subunit of a heavy metal resistance pump. EMBO J 19:4838-4845

Zhu Y-G, Yoshinaga M, Zhao F-J, Rosen BP (2014) Earth abides arsenic transformations. Ann Rev Earth Planet Sci 42:443-467

APPENDIX

Summary of archaeal and bacterial isolates capable of metabolizing arsenic.

No	Strain name	Source site (ambient arsenic conc.)
Archaea / Crenoarchaeota		
1	*Pyrobaculum aerophilum* str. IM2 (DSM 7523)	Boiling marine water hole, Maronti Beach, Ischia, Italy
2	*Pyrobaculum arsenaticum* str. PZ6 (DSM 13514)	Hot spring, Pisciarelli solfatara, Naples, Italy (2.67-5.33 mM)
3	*Pyrobaculum calidifontis* str. VA1 (JCM 11548)	Hot spring, Los Baños and Calamba, Philippines
4	*Pyrobaculum islandicum* str. GEO3 (DSM 4184)	Geothermal power plant, Krafla, Iceland
5	*Pyrobaculum* sp. str. WIJ3	Sediment, White Island Volcano, New Zealand
6	*Pyrobaculum* sp. str. WP30	Joseph's coat spring, Yellowstone National Park (135 µM)
7	*Sulfolobus acidocaldarius* str. BC	Hot spring, southwestern Iceland
Bacteria / Actinobacteria		
8	*Arthrobacter aurescens* str. 51	Soil, Madison River Valley, Montana, USA (2.6 µM)
9	*Arthrobacter* sp. str. 53	Soil, Madison River Valley, Montana, USA (2.6 µM)
10	*Arthrobacter* str. GW14	Water and sediments, groundwater wells, Shanxi, China (2.67-350 µM)
11	*Corynebacterium* sp. str. FW4	Water and sediments, estuarine of the Zuari River, Goa, India (0.001 – 0.14 µM)
12	*Microbacterium oxydans* str. W702	Aquifers, Titas subdistrct, Comilla District, Bangladesh (70 µM)
13	*Microbacterium* sp. str. 46	Soil, Madison River Valley, Montana, USA (2.6 µM)
14	*Micrococcus* sp. str. ES9	Water and sediments, estuarine of the Zuari River, Goa, India (0.001 – 0.14 µM)
Aquificae		
15	*Hydrogenobaculum acidophilus* str. H55	Geothermal spring, Norris Geyser Basin, Yellowstone National Park, USA(35 µM)
16	*Sulfurihydrogenibium azorense* str. Az-Fu1	Hot springs, Furnas, São Miguel Island, Azores, Portugal
17	*Sulfurihydrogenibium subterraneum* str. HGMK1	Subsurface hot aquifer water, Kishikari gold mine, Kagoshima Prefecture, Japan (9.2-10.7 µM)
Bacteroidetes		
18	*Flavobacterium* sp. str. 36	Soil, Madison River Valley, Montana, USA (2.6 µM)
Chloroflex		
19	*Chloroflexus aurantiacus* str. H-10-f1	NR
Chrysiogenetes		
20	*Chrysiogenes arsenatis* str. BAL-1	Gold mine wastewater , Ballarat Goldfields in Ballarat, Victoria, Australia
21	*Desulfurispirillum indicum* str. S5(DSM 22839, ATCC BAA-1389)	Sediments, estuarine channel, Chepauk, Chennai, southern India

O_2 affinity[1]	Arsenic metabolism (conc)[2,3]	e- donor/ e- acceptor/ C Source[4]	Growth T (°C) (range) [optimal][5,6]	Growth pH (range) [optimal][6]	Arsenic related gene[6]	Ref.[7]
FAN	DAR (10 mM)	H_2/ As^V/ CO_2 YE/ As^V/ YE	(74-104) [100]	(5-9) [7]	no *arr*	[1-3]
AN	DAR (10 mM)	H_2/ As^V/ CO_2	(68-100) [95]	6.5-7	no *arr*	[1, 3]
FAN	DAR (10 mM)	YE/ As^V/ YE	(75-100) [90-95]	(5.5-8) [7]	*arr*	[3, 4]
AN	DAR(W) (10 mM)	YE/ As^V/ YE	(74-102) [100]	(5-7) [6]	NR	[3, 5, 6]
AN	DAR (10 mM)	H_2/ As^V/ CO_2	90	6.5-7	NR	[1]
AN	DAR (5 mM)	YE/ As^V/ YE	(60-94) [75]	(3.6-9.0) [4.6-6.6]	NR	[7]
A	DAO (1 mM)	Fe^{2+}, S^0, $S_4O_6^{2-}$/ O_2/ CO_2	65	2	NR	[8, 9]
A	HAR (0.25 mM)	Glucose/ O_2/ Glucose	RT	7	no *arsC*	[10]
A	HAR (0.25 mM)	Glucose/ O_2/ Glucose	RT	7	no *arsC*	[10]
A	HAR (1.33 mM)	Lactate/ O_2/ Lactate	27	7	NR	[11]
A	DAO (13.33 mM)	YE, peptone/ O_2/ YE, peptone	27	7.4	NR	[12]
A	CAO (5 mM)	As^{III}/ O_2/ HCO_3^-	NR	7-8	NR	[13]
A	HAR (0.25 mM)	Glucose/ O_2/ Glucose	RT	7	no *arsC*	[10]
A	DAO (13.33 mM)	YE, peptone/ O_2/ YE, peptone	27	7.4	NR	[12]
MA	DAO (0.05 mM)	H_2/ O_2/ CO_2	[55-60]	[3]	*aoxAB*	[14, 15]
FAN-AN	CAO (2 mM) DAR (W) (2 mM)	As^{III}/ O_2/ CO_2 H_2,S^0,$S_2O_3^{2-}$/ As^V/ CO_2	(50-73) [68]	(5.5-7) [6]	NR	[16]
FAN	CAO (W) (2 mM) DAR (2 mM)	As^{III}/ O_2/ CO_2 H_2,S^0,$S_2O_3^{2-}$/ As^V/ CO_2	(40-70) [60-65]	(6.4-8.8) [7.5]	NR	[16-18]
A	HAR (0.25 mM)	Glucose/ O_2/ Glucose	RT	7	no *arsC*	[10]
NR	DAO	NR	NR	NR	*aoxB*	[19]
AN	DAR (3-5 mM)	Acetate, Pyruvate, Lactate, Succinate, Malate, Fumarate/ As^V/ Acetate, Pyruvate, Lactate, Succinate, Malate, Fumarate	[25-30]	7.4-7.8	*arrAB*	[20, 21]
AN	DAR	Pyruvate/ As^V/ Pyruvate	(4-37) [28]	(6.8-7.6) [7.4]	NR	[22, 23]

No	Strain name	Source site (ambient arsenic conc.)
22	*Desulfurispira natronophila* str. AHT11 (DSM 22071)	Sediments, soda lakes in Kulunda Steppe, Altai, Russia
23	*Desulfurispira natronophila* str. AHT19	Sediments, soda lakes in Kulunda Steppe, Altai, Russia
Cyanobacteria		
24	*Synechocystis*	NR
Deferribacteres		
25	*Deferribacter desulfuricans* str. SSM1	Deep-sea hydrothermal vent chimney, Suiyo Seamount, Izu-Bonin Arc, Japan
26	*Denitrovibrio acetiphilus* str. N2460T	NR
Deinococcus-Thermus		
27	*Thermus aquaticus* str. YT1 (DSM 625)	Hot springs, Yellowstone National Park, USA (0.6-13 µM)
28	*Thermus* sp. str. AO3C	Hot springs, Alvord Desert, Oregon, USA (60µM)
29	*Thermus* sp. str. HR13	Growler hot Spring, California, USA (0.12 mM)
30	*Thermus thermophilus* str. HB8 (DSM 579)	Hot springs at mine, Shizuoka Prefecture, Japan
Firmicutes		
31	*Alkaliphilus metalliredigens* str. QYMF	Leachate ponds, Borax Company, USA
32	*Bacillus alkalidiazotrophicus* str. MS 6, *Anaerobacillus alkalidiazotrophicus* str. MS 6 (NCCB100213,UNIQEM U377)	Mongolian soda soil
33	*Bacillus anthracis* str. SA1	Wastewater samples, Sheikhupura, Pakistan
34	*Bacillus arseniciselenatis* str. E1H, *Anaerobacillus arseniciselenatis* str. E1H	Bottom sediments, Mono Lake, California, USA (0.2 mM)
35	*Bacillus arsenoxydans*	Arsenical cattle-dipping fluids, south Africa
36	*Bacillus benzoevorans.* str. HT-1	Hamster feces
37	*Bacillus beveridgei* str. MLTeJB	Sediments of shallow littoral region, Mono Lake, California, USA (0.2 mM)
38	*Bacillus* flexus str. ADP-25	Soil, rice-growing areas, Nadia district, India (0.11-0.18 mM)
39	*Bacillus* flexus str. AGH-29	Soil, rice-growing areas, Nadia district, India (0.11-0.18 mM)
40	*Bacillus* flexus str. AGO-05	Soil, rice-growing areas, Nadia district, India (0.11-0.18 mM)
41	*Bacillus* flexus str. AGO-S5	Soil, rice-growing areas, Nadia district, India (0.11-0.18 mM)
42	*Bacillus* flexus str. AMO-09	Soil, rice-growing areas, Nadia district, India (0.11-0.18 mM)
43	*Bacillus* flexus str. AMO-7A	Soil, rice-growing areas, Nadia district, India (0.11-0.18 mM)
44	*Bacillus macyae* str. JMM-4 *Anaerobacillus macyae* str. JMM-4 (DSM 16346)	Arsenic-contaminated muds, Central Deborah gold mine, Bendigo, Australia (0.16 mM)
45	*Bacillus megaterium* str. AMO-10	Soil, rice-growing areas, Nadia district, India (0.11-0.18 mM)
46	*Bacillus pumilus* str. AMO-27	Soil, rice-growing areas, Nadia district, India (0.11-0.18 mM)
47	*Bacillus selenatarsenatis* str. SF-1(JCM14380, (DSM18680)	Se-containing discharge, glass processing factory, Japan
48	*Bacillus selenitireducens* str. MLS10	Bottom sediments, Mono Lake, California, USA (0.2 mM)

O₂ affinity[1]	Arsenic metabolism (conc)[2,3]	e- donor/ e- acceptor/ C Source[4]	Growth T (°C) (range) [optimal][5,6]	Growth pH (range) [optimal][6]	Arsenic related gene[6]	Ref.[7]
AN	DAR (5 mM)	Acetate, Propionate/ As^V/ Acetate, Propionate	28	(8.5-10.9) [10.2]	NR	[24]
AN	DAR (5 mM)	Acetate, Propionate/ As^V/ Acetate, Propionate	28	(8.2-10.5) [9.8]	NR	[24]
A	DAO (10 µM)	NR	20-25	NR	NR	[25]
AN	DAR (2mM)	Acetate + H_2/ As^V/ Acetate, CO_2	(40-70) [60-65]	(5-7.5) [6.5]	NR	[26]
NR	DAR	NR/ As^V/ NR	NR	NR	arrA	[27]
A	DAO (1 mM)	YE/ O_2/ YE	(40-79) [70-72]	(6-9.5) [7.5-7.8]	NR	[28, 29]
A	DAO (1-2 mM)	YE/ O_2/ YE	65	7	NR	[30]
FAN	DAO(1 mM) DAR(1 mM)	YE/ O_2/ YE Lactate/ As^V/ Lactate	[75]	[7.5]	aroAB	[31, 32]
FAN	DAO (1 mM)	YE/ O_2/ YE	(47-85) [65-72]	(5.1-9.6) [7-7.5]	aroAB	[29, 32]
AN	DAR	Acetate, Lactate/ As^V/ Acetate, Lactate	(4-45) [35]	(7-11) [9.5]	ACR3 arsA	[33-35]
AN	DAR (10 mM)	Lactate/ As^V/ Lactate	(15-43) [33-35]	(7.8-10.6) [9.5]	NR	[36, 37]
A	HAR (1.33 mM)	Acetate / O_2/ Acetate	(25-42) [37]	(5-9) [7]	NR	[38]
AN	DAR (10 mM)	Lactate/ As^V/ Lactate	20	(8.5-10) [8.5]	ArrA	[37, 39, 40]
A	DAO (13.33 mM)	NR	NR	NR	NR	[41, 42]
AN	DAR (5 mM)	Acetate/ As^V/YE	30-37	NR	NR	[43]
FAN	DAR (5 mM)	Lactate/ As^V/ Lactate	(5-65) [37]	(7.5-10) [8.5]	NR	[44]
A	DAO (1 mM)	LB/ O_2/ LB	30	NR	aoxB	[45]
A	DAO (1 mM)	LB/ O_2/ LB	30	NR	aoxB	[45]
A	DAO (1 mM)	LB/ O_2/ LB	30	NR	aoxB	[45]
A	DAO (1 mM)	LB/ O_2/ LB	30	NR	aoxB	[45]
A	DAO (1 mM)	LB/ O_2/ LB	30	NR	aoxB	[45]
A	DAO (1 mM)	LB/ O_2/ LB	30	NR	aoxB	[45]
AN	DAR (5 mM)	Acetate, Pyruvate, Lactate, Succinate, Glutamate, Malate, H_2/ As^V/ Acetate, Pyruvate, Lactate, Succinate, Glutamate, Malate	28	(7-8.4) [7.8]	NR	[37, 46]
A	DAO (1 mM)	LB/ O_2/ LB	30	NR	aoxB	[45]
A	DAO (1 mM)	LB/ O_2/ LB	30	NR	aoxB	[45]
FAN	DAR (1 mM)	Lactate/ As^V/ Lactate	(25-40) [40]	(7.5-9) [8]	NR	[47, 48]
MA-AN	DAR (10 mM)	Lactate/ As^V/ Lactate	20	(8.5-10) [9.8]	arrA	[39, 40]

No	Strain name	Source site (ambient arsenic conc.)
49	*Bacillus* sp. str. 3.2	Polluted area of a chemical factory, Torviscosa, Udine, Italy (2.27 μM)
50	*Bacillus* sp. str. 3.9	Polluted area of a chemical factory, Torviscosa, Udine, Italy (2.27 μM)
51	*Bacillus* sp. str. 5.8	Polluted area of a chemical factory, Torviscosa, Udine, Italy (2.27 μM)
52	*Bacillus* sp. str. AGH-21	Soil, rice-growing areas, Nadia district, India (0.11-0.18 mM)
53	*Bacillus* sp. str. AGH-31	Soil, rice-growing areas, Nadia district, India (0.11-0.18 mM)
54	*Bacillus* sp. str. AR-10	Groundwater, Choushui River alluvial fan, Taiwan (1.6 μM)
55	*Bacillus* sp. str. AR-9	Groundwater, Choushui River alluvial fan, Taiwan (1.6 μM)
56	*Bacillus* sp. str. MC013	Old tin mine area, Thailand
57	*Bacillus* sp. str. MC120	Old tin mine area, Thailand
58	*Bacillus* sp. str. MC123	Old tin mine area, Thailand
59	*Bacillus* sp. str. MC169	Old tin mine area, Thailand
60	*Bacillus* sp. str. MC194	Old tin mine area, Thailand
61	*Bacillus* sp. str. MC196	Old tin mine area, Thailand
62	*Bacillus* sp. str. MC202	Old tin mine area, Thailand
63	*Bacillus* sp. str. MC203	Old tin mine area, Thailand
64	*Bacillus* sp. str. MC205	Old tin mine area, Thailand
65	*Bacillus* sp. str. MC265	Old tin mine area, Thailand
66	*Bacillus* sp. str. ML-SRAO	Mono Lake, California, USA (0.2 mM)
67	*Bacillus* sp. str. Z-0521	Bottom sediments, low-mineralized soda lake Khadyn, Tuva, upper Yenisey region, Russia
68	*Bacterium* sp. str. AGH-03	Soil, rice-growing areas, Nadia district, India (0.11-0.18 mM)
69	*Caloramator* sp. str. YeAs	Hydrothermal Murky plot, Alvord Basin, Oregon, USA
70	*Clostridium* sp. str. CN-8	Typic Calciaquoll soil (0.02-0.84 mM)
71	*Clostridium* sp. str. OhILAs *Alkaliphilus oremlandii* str. OhILAs	Anoxic sediments from Ohio River, Pittsburgh, PA, USA
72	*Desulfitobacterium frappieri* str. PCP-1	Methanogenic consortium
73	*Desulfitobacterium hafiniese* str. DCB-2	Municipal sludge
74	*Desulfitobacterium hafiniese* str.GBFH	Arsenic-enriched sediments of Lake Coeur d'Alene, Idaho, USA
75	*Desulfosporosinus* sp. str. Y5	Sediments, Onondaga Lake, New York, USA
76	*Desulfotomaculum auripigmentum* str. OREX-4 *Desulfosporosinus auripigmenti* str. OREX-4 (DSM 13351, ATCC700205)	Surface sediments, Upper Mystic Lake, Massachusetts, USA
77	*Desulfuribacillus alkaliarsenatis* str. AHT28T (DSM24608T,UNIQEM U855T)	Sediments, soda lake

O$_2$ affinity[1]	Arsenic metabolism (conc)[2,3]	e- donor/ e- acceptor/ C Source[4]	Growth T (°C) (range) [optimal][5,6]	Growth pH (range) [optimal][6]	Arsenic related gene[6]	Ref.[7]
A	HAR (10 mM)	Glucose / O$_2$/ Glucose	30	NR	arsC no arsB	[49]
A	DAO (3 mM)	Glucose / O$_2$/ Glucose	30	NR	arsC no arsB	[49]
A	HAR (10 mM)	Glucose / O$_2$/ Glucose	30	NR	arsC no arsB	[49]
A	DAO (1 mM)	LB/ O$_2$/ LB	30	NR	aoxB	[45]
A	DAO (1 mM)	LB/ O$_2$/ LB	30	NR	aoxB	[45]
FAN	HAR (10 mM)	Lactate/ O$_2$/ Lactate	25	7	arsC no arrA, arsB, aoxB	[50]
FAN	HAR (10 mM)	Lactate/ O$_2$/ Lactate	25	7	arsC no arrA, arsB, aoxB	[50]
A	HAR (1-20 mM)	Lactate/ O$_2$/ Lactate	30	7	NR	[51]
A	HAR (1-20 mM)	Lactate/ O$_2$/ Lactate	30	7	NR	[51]
A	HAR (1-20 mM)	Lactate/ O$_2$/ Lactate	30	7	NR	[51]
A	HAR (1-20 mM)	Lactate/ O$_2$/ Lactate	30	7	NR	[51]
A	HAR (1-20 mM)	Lactate/ O$_2$/ Lactate	30	7	NR	[51]
A	HAR (1-20 mM)	Lactate/ O$_2$/ Lactate	30	7	NR	[51]
A	HAR (1-20 mM)	Lactate/ O$_2$/ Lactate	30	7	NR	[51]
A	HAR (1-20 mM)	Lactate/ O$_2$/ Lactate	30	7	NR	[51]
A	HAR (1-20 mM)	Lactate/ O$_2$/ Lactate	30	7	NR	[51]
AN	CAO (2-5 mM) DAR(10 mM)	As^{III} /HSeO$_4$/ Acetate, Lactate Lactate/ As^V/ Lactate	NR	9	arrA	[52]
AN	DAR (5 mM)	Mannite/ As^V/ Mannite	(18-40) [30-35]	(8.5-10.7) [9.7]	NR	[37]
A	DAO (1 mM)	LB/ O$_2$/ LB	30	NR	aoxB	[45]
AN	HAR (2.5 mM)	YE/ O$_2$/ YE	(37-75) [55]	(6-8) [7-7.5]	no arr ars	[53]
AN	HAR (6-600 µM)	Glucose/ O$_2$/ Glucose	25	6.5	NR	[54]
AN	DAR (1-40 mM)	Acetate, Lactate, Pyruvate, Formate, H$_2$/ As^V/ Acetate, Lactate, Pyruvate, Formate	(22-50) [37]	(7.5-8.75) [8.5]	arrAB	[40, 55-58]
AN	DAR (10 mM)	Lactate/ As^V/ Lactate	(15-45) [38]	(6-9) [7.5]	NR	[59, 60]
AN	DAR (10 mM)	Lactate/ As^V/ Lactate	(19-40) [37]	[7]	NR	[59, 61]
AN	DAR (10 mM)	Lactate/ As^V/ Lactate	(15-42) [37]	(6-8) [7.5]	NR	[59]
AN	DAR (10 mM)	Lactate/ As^V/ Lactate H$_2$/ As^V/ CO$_2$	30	7-7.2	NR	[62]
AN	DAR (5-10 mM)	H$_2$+Acetate / As^V/ Acetate	[25-30]	[6.4-7]	NR	[63-65]
AN	DAR (5 mM)	Formate/ As^V/ Formate	(NR-43) [35]	(8.5-10.6) [10]	NR	[66]

No	Strain name	Source site (ambient arsenic conc.)
78	*Exiguobacterium* sp. str. WK6	Lake Ohakuri, Otago, New Zealand (0.013-1.3.33 mM)
79	*Geobacillus stearothermophilus* sp. str. AGH-02	Soil, rice-growing areas, Nadia district, India (0.11-0.18 mM)
80	*Halarsenatibacter silvermanii* str. SLAS-1T	Sediment, Searles Lake, California, USA (3mM)
81	*Paenibacillus* sp. str. 3.1	Polluted area of a chemical factory, Torviscosa, Udine, Italy (2.27 µM)
82	*Staphylococcus xylosus* str. 5.2	Polluted area of a chemical factory, Torviscosa, Udine, Italy (2.27 µM)
α–Proteobacteria		
83	*Acidicaldus* sp. str. AO5	Hot springs, Norris Geyser Basin, Yellowstone National Park, USA (35 µM)
84	*Acidiphilium multivorum* str. AIU301 (JCM 8867)	Acid mine drainage, Matsuo sulfur-pyrite mine area, Iwate Prefecture, Japan
85	*Agrobacterium / Rhizobium*-subbranch sp. str. NT-26	Gold mine, Northern Territory, Australia
86	*Agrobacterium / Rhizobium*-subbranch sp. str. NT-25	Gold mine, Northern Territory, Australia
87	*Agrobacterium / Rhizobium*-subbranch sp. str. Ben-5	Water, Central Deborah gold mine in Bendigo, Australia (24 µM)
88	*Agrobacterium albertimagni* str. AOL15 (ATCC BAA-24)	Macrophytes surface, Hot Creek, California, USA (2.7 µM)
89	*Agrobacterium* sp. str. Q	Soil sample, Dinghy Mountain Natural Reserve, China (6.66 mM)
90	*Agrobacterium tumefaciens* str. 42	Soil, Madison River Valley, Montana, USA (2.6 µM)
91	*Agrobacterium tumefaciens* str. GW4	Water and sediments, groundwater wells, Shanxi, China (2.67-350µM)
92	*Agrobacterium tumefaciens* str. 52(5A)	Soil, Madison River Valley, Montana, USA (2.6 µM)
93	*Ancylobacter* sp. str. OL-1	Sediments, Onondaga Lake, New York, USA
94	*Bosea* sp. str. AR-11	Groundwater, Choushui River alluvial fan, Taiwan (1.6 µM)
95	*Bosea* sp. str. WAO	Groundwater, Newark Basin, New Jersey, USA (0.13-2.86 µM)
96	*Caulobacter leidyi*	Mine, tailings, Anaconda, Montana, USA (16.7-340 nM)
97	*Ensifer adhaerens* sp. str. C-1	Conventional sewage treatment plant, Iwate Prefecture, Japan
98	*Ochrobactrum* sp. str. MC197	Old tin mine area, Thailand
99	*Ochrobactrum tritici* str. SCII24(LMG 18957, DSM 13340)	Rhizoplane of wheat (cultivar Lloyd) , Grignon soil, France
100	*Rhizobium loti* str. RM1-2001	Mine, tailings, Anaconda, Montana, USA (16.7-340 nM)
101	*Sinorhizobium* sp. str. A2	Soil, Ichinokawa mine tailing areas, Ehime, Japan (16.53mM)
102	*Sinorhizobium* sp. str. DAO10	Arsenic-contaminated soil, New Jersey, USA (0.13-2.86 µM)
103	*Sinorhizobium* sp. str. GW3	Water and sediments, groundwater wells, Shanxi, China (2.67-350µM)

O_2 affinity[1]	Arsenic metabolism (conc)[2,3]	e- donor/ e- acceptor/ C Source[4]	Growth T (°C) (range) [optimal][5,6]	Growth pH (range) [optimal][6]	Arsenic related gene[6]	Ref.[7]
AN	HAR (1-60 mM)	TYEG / O_2/ TYEG	28	(5-9) [8-9]	*arsB* no *arsC*	[67]
A	DAO (1 mM)	LB/ O_2/ LB	30	NR	*aoxB*	[45]
AN	DAR (5-7 mM)	Glucose, Pyruvate / As^V/ Glucose, Pyruvate HS⁻/ As^V/ HCO_3^-	(28-55) [44]	(8.7-9.8) [9.4]	*arrABD*	[68, 69]
A	HAR (10 mM)	Glucose / O_2/ Glucose	30	NR	no *arsC* no *arsB*	[49]
A	HAR (10 mM)	Glucose / O_2/ Glucose	30	NR	*arsB* no *arsC*	[49]
MA	CAO (1 mM)	As^{III}/ O_2/ CO_2	(50-70) [55-60]	3	NR	[70]
A	DAO (0.67 mM)	Trypticase soy/ O_2/ Trypticase soy	(17-42) [27-35]	(1.9-5.6) [3.5]	NR	[71, 72]
A	CAO (5-10 mM)	As^{III} / O_2/ HCO_3^-	28	[5.5]	*aroAB*	[32, 73-76]
A	CAO (5-10 mM)	As^{III} / O_2/ HCO_3^-	28	8	NR	[73, 74]
A	CAO (5-10 mM)	As^{III} / O_2/ HCO_3^-	28	6.5-8	NR	[73, 74]
A	DAO (0.002-5 mM)	Citrate, Tryptone / O_2/ Citrate, Tryptone	(4-42) [30]	[7-8]	NR	[77]
A	CAO (1.6 mM)	As^{III} / O_2/ HCO_3^-	(20-45) [30]	(4-9) [7]	NR	[78]
A	HAR (0.25 mM)	Glucose/ O_2/ Glucose	RT	7	*arsC*	[10]
A	HAR (1.33 mM)	Lactate/ O_2/ Lactate	27	7	*aoxB*	[11]
A	DAO (75 µM)	Glucose/ O_2/ Glucose	RT	7	*arsC* *aoxAB* *aio* operon	[10, 79]
A	CAO (10 mM)	As^{III} / O_2/ HCO_3^-	30	8	*aroAB*	[32, 80]
FAN	CAO DAO (3 mM)	As^{III} / O_2/ HCO_3^- Lactate/ O_2/ Lactate	(20-45) [37]	[7.4-8.4]	*aroA* *aoxB*	[50, 81]
A	CAO (5 mM)	As^{III} / O_2/ HCO_3^-	30	7.2	*aroAB*	[32, 82]
A	HAR (13 µM)	YEPG/ O_2/ YEPG	NR	NR	NR	[83]
A	CAO (1.46 mM)	As^{III} / O_2/ HCO_3^-	25(10-30)	7(5-8)	NR	[84]
A	HAR (1-20 mM)	Lactate/ O_2/ Lactate	30	7	NR	[51]
A	DAO (5 mM)	YE, As^{III} / O_2/ YE,	30	7.2	*ars* *aoxB*	[85-87]
A	HAR (13 µM)	YEPG/ O_2/ YEPG	NR	NR	NR	[83]
A	DAO (10 mM)	Lactate/ O_2/ Lactate	25	NR	*aioA*	[88]
FA	CAO (5 mM)	As^{III} / O_2/ HCO_3^- As^{III} / NO_3^-/ HCO_3^-	30	7.2	*aroAB*	[32, 89]
A	DAO (1.33 mM)	Lactate/ O_2/ Lactate	27	7	*aoxB*	[11]

No	Strain name	Source site (ambient arsenic conc.)
104	*Sinorhizobium* sp. str. M14	Bottom sediments, Zloty Stok gold mine in Lower Silesia, SW Poland
105	*Sinorhizobium* sp. str. NT-2	Gold mine, Northern Territory, Australia
106	*Sinorhizobium* sp. str. NT-3	Gold mine, Northern Territory, Australia
107	*Sinorhizobium* sp. str. NT-4	Gold mine, Northern Territory, Australia
108	*Sphingomonas echinoides* str. RM2-2001	Mine, tailings, Anaconda, Montana, USA (16.7-340 nM)
109	*Sphingomonas parapaucimobilis* str. GW5	Water and sediments, groundwater wells, Shanxi, China (2.67-350 µM)
110	*Sphingomonas yanoikuyae*	Mine, tailings, Anaconda, Montana, USA (16.7-340 nM)
111	*Thiobacillus ferrooxidans* str. Fe1	Acid mine drainage, Matsuo sulfur-pyrite mine, Iwate Prefecture, Japan (26.7-173 µM)
112	*Thiobacillus* sp. str. S-1	Sediments, Onondaga Lake, New York, USA
113	*Thiobacillus thiooxidans* str. S3	Acid mine drainage, Matsuo sulfur-pyrite mine, Iwate Prefecture, Japan (26.7-173 µM)
β-Proteobacteria		
114	*Achromobacter / Bordetella* (*Alcaligenes*) sp. str. Ben-4	Water, Central Deborah gold mine in Bendigo, Australia (24 µM)
115	*Achromobacter / Bordetella*-subbranch sp. str. NT-10	Gold mine, Northern Territory, Australia
116	*Achromobacter* sp. str. GW1	Water and sediments, groundwater wells, Shanxi, China (2.67-350 µM)
117	*Achromobacter* sp. str. SPB-31	Gardon soil, campus of university of Pune, Pune, India
118	*Achromobacter* sp. str. SY8	Asenic-contaminated soils, Lianyungang, China
119	*Acidovorax* sp. str. GW2	Water and sediments, groundwater wells, Shanxi, China (2.67-350 µM)
120	*Acidovorax* sp. str. NO1	Arsenic contaminated soil, gold mine, Daye, Hubei Provinxe, China
121	*Alcaligenes faecalis* str. NCIB8687	NR
122	*Alcaligenes faecalis* str. YE56	Sewage
123	*Alcaligenes faecalis* str. HLE (str. ANA)	Soil, courtyard, New York, USA
124	*Alcaligenes faecalis* str. O1201	Soil, South Plant Superfund Site in South Point, Ohio, USA
125	*Alcaligenes faecalis* str. RS-19	Duckum abandoned mines, Korea (0.49-1.05 mM)
126	*Alcaligenes* sp. str. Dhal-L	Agricultural soil (2.85 mM)
127	*Alcaligenes* sp. str. H	Soil, Dinghy Mountain Natural Reserve, China (6.66 mM)
128	*Azoarcus* sp. str. DAO1	Arsenic-contaminated soil, New Jersey, USA (0.13-2.86 µM)
129	*Azospira* sp. str. ECC1-pb2	Municipal wastewater treatment plant, Arizona, USA
130	*Bordetella* sp. str. SPB-24	Gardon soil, campus of university of Pune, Pune, India
131	*Comamonas terrae* str. A3-3 (KCTC 22606,NBRC, 106524, PCU 324, TISTR 1906T)	Agricultural soil, Taoland (<0.66 mM)

O_2 affinity[1]	Arsenic metabolism (conc)[2, 3]	e- donor/ e- acceptor/ C Source[4]	Growth T (°C) (range) [optimal][5, 6]	Growth pH (range) [optimal][6]	Arsenic related gene[6]	Ref.[7]
A	CAO(5 mM) HAR	As^{III} / O_2/ CO_2, HCO3- LB/ O_2/ LB	(10-37) [22]	(4-9) [7]	*ars* *aox*	[90]
A	CAO (5-10 mM)	As^{III} / O_2/ HCO_3^-	28	6.5-8	NR	[73, 74]
A	CAO (5-10 mM)	As^{III} / O_2/ HCO_3^-	28	6.5-8	NR	[73, 74]
A	CAO (5-10 mM)	As^{III} / O_2/ HCO_3^-	28	6.5-8	NR	[73, 74]
A	HAR (13 µM)	YEPG/ O_2/ YEPG	NR	NR	NR	[83]
A	HAR (1.33 mM)	Lactate/ O_2/ Lactate	27	7	NR	[11]
A	HAR (13 µM)	YEPG/ O_2/ YEPG	NR	NR	NR	[83]
A	DAO (0.67 mM)	Fe^{2+}, S^0/ O_2/ CO_2	30	3	NR	[91]
A	CAO (10 mM)	As^{III} / O_2/ HCO_3^-	30	8	*aroAB*	[32, 80]
A	DAO (0.67 mM)	Fe^{2+}, S^0/ O_2/ CO_2	30	3	NR	[91]
A	DAO (5-10 mM)	YE/ O_2/ YE	28	5.5-8	NR	[73, 74]
A	DAO (5-10 mM)	YE/ O_2/ YE	28	5.5-8	NR	[73, 74]
A	DAO (1.33 mM)	Lactate/ O_2/ Lactate	27	7	*aoxB*	[11]
A	DAO (2-10 mM)	Acetate/ O_2/ Acetate	(8-50 [42])	(4-10) [6]	NR	[92]
A	DAO (0.8 mM)	LB/ O_2/LB	28	7	*aox X SRA BCD arsR*	[93, 94]
A	DAO (1.33 mM)	Lactate/ O_2/ Lactate	27	7	*aoxB*	[11]
FAN	CAO (20-200 mM)	As^{III} / O_2/ HCO_3^- As^{III} / NO_3^-/ HCO_3^-	NR	NR	*Aio* operon *Ars* operon	[95]
A	DAO (1.33 mM)	LB/ O_2/LB	30, 32-37	7.4	*asoAB*	[96]
FAN	DAO	LB/ O_2/LB	(25-42)	7.35	NR	[97, 98]
A	DAO	NB/ O_2/NB	(25-37)	6.5-8	NR	[73, 74, 97, 99]
A	DAO (0.13-13.33 mM)	NB/ O_2/NB	(10-40) [30]	(4-9) [7]	NR	[100, 101]
A	DAO (1-25 mM)	Glucose/ O_2/ Glucose	30	7	NR	[102]
A	HAR	LB/ O_2/LB	30	NR	*ACR3* *arsC*	[103]
A	CAO (1.6 mM)	As^{III} / O_2/ HCO_3^-	(20-45) [30]	(4-9) [7]	NR	[78]
FA	CAO (5 mM)	As^{III} / O_2/ HCO_3^- As^{III} / NO_3^-/ HCO_3^-	30	7.2	no *aroAB*	[32, 89]
AN	CAO (0.5 mM)	As^{III} / ClO_3^-/ HCO_3^-	28-32	7-7.2	NR	[104]
A	DAO (2-10 mM)	Acetate/ O_2/ Acetate	(8-50 [42])	(4-10) [6]	NR	[92]
A	DAO (NR)	TYEG/ O_2/ TYEG	4-50	5-11	NR	[105]

No	Strain name	Source site (ambient arsenic conc.)
132	*Comamonas testosteroni* str. W30W1a	Aquifers, Titas subdistrct, Comilla District, Bangladesh (70 µM)
133	*Dechloromonas* sp. str. ECC1-pb1	Municipal wastewater treatment plant, Arizona, USA
134	*Hydrogenophaga* sp. str. NT-14	Gold mine, Northern Territory, Australia
135	*Hydrogenophaga* sp. str. NT-5	Gold mine, Northern Territory, Australia
136	*Hydrogenophaga* sp. str. NT-6	Gold mine, Northern Territory, Australia
137	*Hydrogenophaga* sp. str. YED1-18	Aquatic macrophytes, Hot Creek, California, USA (2.7-13 µM)
138	*Hydrogenophaga* sp. str. YED6-21	Aquatic macrophytes, Hot Creek, California, USA (2.7-13 µM)
139	*Hydrogenophaga* sp. str. YED6-4	Aquatic macrophytes, Hot Creek, California, USA (2.7-13 µM)
140	*Hydrogenophaga* sp str. CL-3	Sediments, Onondaga Lake, New York, USA
141	*Polaromonas* sp str. GM1	Giant Mine, Yellowknife, Northwest Territories, Canada (50 mM)
142	*Ralstonia picketii* str. B3	Cheni gold mine, Limousin, France
143	*Thiomonas Arsenivorans* str. B6	Cheni gold mine, Limousin, France
144	*Thiomonas intermedia* str. ATCC15466	A freshwater stream
145	*Thiomonas* sp. str. B1	Carnoule`s mine, Ale`s, France (4.67 mM)
146	*Thiomonas* sp. str. B2	Carnoule`s mine, Ale`s, France (4.67 mM)
147	*Thiomonas* sp. str. B3	Carnoule`s mine, Ale`s, France (4.67 mM)
148	*Thiomonas* sp. str. 3As	Acid mine drainage, Carnoulès, Gard, France (2.67 mM)
149	*Thiomonas* sp. str. NO 115	Mining site, Norway
150	*Thiomonas* sp. str. WJ 68	Mining site, Wales, UK
151	*Thiomonas* sp. str. X19	Acid mine drainages, Carnoulès, Gard, France
152	*Variovorax* sp. str. 34	Soil, Madison River Valley, Montana, USA (2.6 µM)
153	*Zoogloea* sp. str. ULPAs1 *Caenibacter arsenoxydans* str. ULPAs1 *Herminiimonas arsenicoxidans* str. ULPAs1	Wastewater treatment plant contaminated with arsenic, Germany (0.47 mM)
γ-Proteobacteria		
154	*Acinetobacter* sp. str. EW6	Water and sediments, estuarine of the Zuari River, Goa, India (0.001 – 0.14 µM)
155	*Acinetobacter* sp. str. GW13 *Serratia* sp. str. GW13	Water and sediments, groundwater wells, Shanxi, China (2.67-350 µM)
156	*Acinetobacter* sp. str. GW7	Water and sediments, groundwater wells, Shanxi, China (2.67-350 µM)
157	*Acinetobacter* sp. str. GW8	Water and sediments, groundwater wells, Shanxi, China (2.67-350 µM)
158	*Aeromonas* sp. str. CA1	Lake Ohakuri, Otago, New Zealand (0.013-1.3.33 mM)
159	*Alkalilimnicola ehrlichii* str. MLHE1	Mono Lake, California, USA (0.2 mM)
160	*Citrobacter freundii* str. SB2	Wastewater samples, Sheikhupura, Pakistan
161	*Citrobacter* sp. str. AR-7	Groundwater, Choushui River alluvial fan, Taiwan (1.6 µM)
162	*Citrobacter* sp. str. L2	Groundwater from Yunlin County, Taiwan

O_2 affinity[1]	Arsenic metabolism (conc)[2, 3]	e- donor/ e- acceptor/ C Source[4]	Growth T (°C) (range) [optimal][5, 6]	Growth pH (range) [optimal][6]	Arsenic related gene[6]	Ref.[7]
A	CAO (5 mM)	As^{III} / O_2/ HCO_3^-	NR	7-8	NR	[13]
AN	CAO (0.5 mM)	As^{III} / ClO_3^-/ HCO_3^-	28-32	7-7.2	NR	[104]
A	DAO (5-10 mM)	YE/ O_2/ YE	28	5.5-8	aroAB	[73, 74]
A	DAO (5-10 mM)	YE/ O_2/ YE	28	5.5-8	NR	[73, 74]
A	DAO (5-10 mM)	YE/ O_2/ YE	28	5.5-8	NR	[73, 74]
A	DAO (0.1 mM)	YE/ O_2/ YE	21	8	NR	[106]
A	DAO (0.1 mM)	YE/ O_2/ YE	21	8	NR	[106]
A	DAO (0.1 mM)	YE/ O_2/ YE	21	8	NR	[106]
A	CAO (10 mM)	As^{III} / O_2/ HCO_3^-	30	8	aroAB	[32, 80]
A	DAO (4 mM)	YE/ O_2/ YE	(4-20)[20]	5.5	aroA	[107]
A	DAO (1.33 mM)	Lactate/ O_2/ Lactate	25	7	NR	[108]
A	DAO (1.33 mM) CAO (1.33 mM)	Lactate/ O_2/ Lactate As^{III} / O_2/ HCO_3^-	(4-30) [20-30]	(2.5-8) [5.5]	NR	[108-111]
A	DAO (1.33 mM)	YE,$S_2O_3^{2-}$ / O_2/ YE	[30-37]	(1.9-7.0) [5-7]	NR	[109, 112, 113]
A	DAO (2.67 mM)	YE/ O_2/ YE	30	7-7.2	NR	[114]
A	DAO (2.67 mM)	YE/ O_2/ YE	30	7-7.2	NR	[114]
A	DAO (2.67 mM)	YE/ O_2/ YE	30	7-7.2	NR	[114]
A	CAO (2-10 mM)	As^{III} / O_2/ HCO_3^-	[30]	(3-7.5) [6]	aoxB	[115]
A	CAO (1.33 mM)	As^{III} / O_2/ HCO_3^-	25	7	NR	[109]
A	DAO (1.33 mM)	YE,$S_2O_3^{2-}$ / O_2/ YE	25	7	NR	[109]
A	DAO (1.33 mM)	LB / O_2/LB	NR	3.5, 5.5	aioA	[116]
A	DAO (75 µM)	Glucose/ O_2/ Glucose	RT	7	no arsC	[10]
A	CAO (13.3 mM)	As^{III} / O_2/ HCO_3^-	(4-30) [25]	[7-8.5]	acr3 aoxAB	[117-121]
A	DAO (13.33 mM)	YE, peptone/ O_2/ YE, peptone	27	7.4	NR	[12]
A	HAR (1.33 mM)	Lactate/ O_2/ Lactate	27	7	NR	[11]
A	HAR (1.33 mM)	Lactate/ O_2/ Lactate	27	7	NR	[11]
A	HAR (1.33 mM)	Lactate/ O_2/ Lactate	27	7	NR	[11]
AN	HAR (1-60 mM)	TYEG / O_2/ TYEG	28	(5-9) [8-9]	no arsBC	[67]
FAN	CAO (5 mM)	As^{III} / NO_3^-/ HCO_3^-	(13-40) [30]	(7.3-10) [9.3]	arxA arrA no aroAB	[32, 122-124]
A	HAR (1.33 mM)	Acetate / O_2/ Acetate	(25-42) [37]	(5-9) [5]	NR	[38]
FAN	HAR (10 mM)	Lactate/ O_2/ Lactate	25	7	arsBC no arrA, aoxB	[50]
AN	HAR (0.2 mM)	YE/ O_2/ YE	37	7.2	arsC	[125]

No	Strain name	Source site (ambient arsenic conc.)
163	*Citrobacter* sp. str. TSA-1	Termite hindgut
164	*Citrobacter* sp. str. W2	Groundwater from Yunlin County, Taiwan
165	*Ectothiorhodospira* sp. str. MLCB	Mono Lake, California, USA (0.2 mM)
166	*Ectothiorhodospira* sp. str. MLP2	Hot spring biofilms, Mono Lake, California, USA (0.2 mM)
167	*Ectothiorhodospira* sp. str. PHS-1	Hot spring biofilms, Mono Lake, California, USA (0.2 mM)
168	*Enterobacter* sp. str. AR-8	Groundwater, Choushui River alluvial fan, Taiwan (1.6 μM)
169	*Enterobacter* sp. str. MC010	Old tin mine area, Thailand
170	*Enterobacter* sp. str. MC204	Old tin mine area, Thailand
171	*Enterobacteriaceae* sp. str. FW1	Water and sediments, estuarine of the Zuari River, Goa, India (0.001 – 0.14 μM)
172	*Enterobacteriaceae* sp. str. FW3	Water and sediments, estuarine of the Zuari River, Goa, India (0.001 – 0.14 μM)
173	*Halomonas* sp. Str. HAL1	Gold mine, Daye County, Hubei Province, Central China
174	*Klebsiella oxytoca* sp. str. SB1	Wastewater samples, Sheikhupura, Pakistan
175	*Marinobacter santoriniensis* str. NKSG1	Shallow marine hydrothermal sediments, Santorini, Greece (5.3 mM)
176	*Pesudomonas arsenicoxydans* str VC-1	Camarones Valley, Atacama Desert, Chile
177	*Pesudomonas arsenoxydans Xanthomonas arsenoxydans Achromobacter arsenoxydans*	Arsenical cattle-dipping fluids, Queensland, Australia
178	*Pseudomonas aeruginosa*	Mine, tailings, Anaconda, Montana, USA (16.7-340 nM)
179	*Pseudomonas arsenitoxidans*	Mine waters
180	*Pseudomonas fluorescens* str. FN-13	Sediments, Camarones River, Chile (14.66 μM)
181	*Pseudomonas fluorescens* str. FN-15	Sediments, Camarones River, Chile (14.66 μM)
182	*Pseudomonas fluorescens* str. FN-70	Sediments, Camarones River, Chile (14.66 μM)
183	*Pseudomonas fluorescens* str. FN-71	Sediments, Camarones River, Chile (14.66 μM)
184	*Pseudomonas fluorescens* str. RM3-2001	Mine, tailings, Anaconda, Montana, USA (16.7-340 nM)
185	*Pseudomonas fulva* str. OS-10	Duckum abandoned mines, Korea (0.5mM)
186	*Pseudomonas lubricans*	Heavy metal laden industrial wastewater, Sheikhupura, Pakistan
187	*Pseudomonas marginalis* str. FN-41	Sediments, Camarones River, Chile (14.66 μM)
188	*Pseudomonas putida* str. 18	Gold-arsenic deposits
189	*Pseudomonas putida* str. FN-57	Sediments, Camarones River, Chile (14.66 μM)
190	*Pseudomonas putida* str. FN-58	Sediments, Camarones River, Chile (14.66 μM)
191	*Pseudomonas putida* str. FN-66	Sediments, Camarones River, Chile (14.66 μM)

O_2 affinity[1]	Arsenic metabolism (conc)[2,3]	e- donor/ e- acceptor/ C Source[4]	Growth T (°C) (range) [optimal][5,6]	Growth pH (range) [optimal][6]	Arsenic related gene[6]	Ref.[7]
AN	DAR (5 mM)	H_2/ As^V/ Acetate Acetate/ As^V/YE	30-37	NR	NR	[43]
FAN	DAR(2 mM) HAR (2 mM)	Acetate / As^V/ Acetate YE/ O_2/ YE	[37]	[7-8.5]	*arsR* *PhsA*	[126]
AN	DAR (5 mM)	Acetate / As^V/ Acetate	25	9.3	*arrA*	[127]
AN	PAO (2-4 mM)	Light/ H_2O/ CO_2	45	9.3	NR	[128]
AN	PAO (2-4 mM)	Light/ H_2O/ CO_2	45	9.3	*arr* no *aoxB*	[128, 129]
FAN	HAR (10 mM)	Lactate/ O_2/ Lactate	25	7	*arsBC* no *arrA*, *aoxB*	[50]
A	HAR (1-20 mM)	Lactate/ O_2/ Lactate	30	7	NR	[51]
A	HAR (1-20 mM)	Lactate/ O_2/ Lactate	30	7	NR	[51]
A	DAO (13.33 mM)	YE, peptone/ O_2/ YE, peptone	27	7.4	NR	[12]
A	DAO (13.33 mM)	YE, peptone/ O_2/ YE, peptone	27	7.4	NR	[12]
A	DAO	LB / O_2/ LB	NR	NR	*arsC* *arsH* *ACR3* *arsR* *aioA* *aioB*	[130]
A	HAR (1.33 mM)	Acetate / O_2/ Acetate	(25-42) [30]	(5-9) [7]	NR	[38]
FAN	DAO (1-8 mM) DAR (0.5-12 mM)	Acetate, Lactate / O_2/ Acetate, Lactate Acetate, Lactate / As^V/ Acetate, Lactate	[35-40]	(5.5-9) [7-8]	*aoxB* no *arrA* no *arsC*	[131, 132]
A	DAO (5 mM)	Lactate/ O_2/ Lactate	(4-37) [30]	(6.5-10) [7.5-8]	*aox*	[133]
A	DAO (20-100 mM)	YE/ O_2/ YE	(20-40) [40]	(6-6.7) [6.4]	NR	[41, 134-136]
A	HAR (13 µM)	YEPG/ O_2/ YEPG	NR	NR	NR	[83]
A	CAO (13.3-26.7 mM)	As^{III} / O_2/ HCO_3^-	28-35	7.5	NR	[137]
A	DAO (75 µM)	Lactate/ O_2/ Lactate	25	NR	*aroAB*	[138]
A	DAO (75 µM)	Lactate/ O_2/ Lactate	25	NR	*aroAB*	[138]
A	DAO (75 µM)	Lactate/ O_2/ Lactate	25	NR	*aroAB*	[138]
A	DAO (75 µM)	Lactate/ O_2/ Lactate	25	NR	*aroAB*	[138]
A	HAR (13 µM)	YEPG/ O_2/ YEPG	NR	NR	NR	[83]
A	DAO		30	7	*arsB*	[139]
A	DAO (40 mM)	LB / O_2/ LB	(28-35) [30]	(6.4-8) [7]	NR	[140]
A	DAO (75 µM)	Lactate/ O_2/ Lactate	25	NR	*aroAB*	[138]
A	DAO	Glucose/ O_2/ Glucose	(4-28)	(6-9)	NR	[141]
A	DAO (75 µM)	Lactate/ O_2/ Lactate	25	NR	*aroAB*	[138]
A	DAO (75 µM)	Lactate/ O_2/ Lactate	25	NR	*aroAB*	[138]
A	DAO (75 µM)	Lactate/ O_2/ Lactate	25	NR	*aroAB*	[138]

No	Strain name	Source site (ambient arsenic conc.)
192	*Pseudomonas putida* str. OS-1	Myoung-bong arsenic-contaminated gold-silver mines, Korea (13.33 µM)
193	*Pseudomonas putida* str. OS-3	Myoung-bong arsenic-contaminated gold-silver mines, Korea (13.33 µM)
194	*Pseudomonas putida* str. OS-17	Duckum arsenic-contaminated gold-silver mines, Korea (0.5 mM)
195	*Pseudomonas putida* str. OS-18	Duckum arsenic-contaminated gold-silver mines, Korea (0.5 mM)
196	*Pseudomonas putida* str. OW-16	Duckum arsenic-contaminated gold-silver mines, Korea (0.5 mM)
197	*Pseudomonas putida* str. OW-27	Duckum arsenic-contaminated gold-silver mines, Korea (0.5mM)
198	*Pseudomonas putida* str. RS-4	Myoung-bong arsenic-contaminated gold-silver mines, Korea (13.33 µM)
199	*Pseudomonas putida* str. RS-5	Myoung-bong arsenic-contaminated gold-silver mines, Korea (13.33 µM)
200	*Pseudomonas putida* str. RS-17	Duckum arsenic-contaminated gold-silver mines, Korea (0.5 mM)
201	*Pseudomonas putida* str. RW-20	Duckum arsenic-contaminated gold-silver mines, Korea (0.5 mM)
202	*Pseudomonas putida* str. RW-26	Duckum arsenic-contaminated gold-silver mines, Korea (0.5 mM)
203	*Pseudomonas putida* str. RW-27	Duckum arsenic-contaminated gold-silver mines, Korea (0.5 mM)
204	*Pseudomonas putida* str. WB	Groundwater collected from Sahispur, Nadia, India (3.8 µM)
205	*Pseudomonas rhizosphaerae* str. OW-2	Myoung-bong arsenic-contaminated gold-silver mines, Korea (13.33 µM)
206	*Pseudomonas* sp. str. 31	Soil, Madison River Valley, Montana, USA (2.6 µM)
207	*Pseudomonas* sp. str. FW2	Water and sediments, estuarine of the Zuari River, Goa, India (0.001 – 0.14 µM)
208	*Pseudomonas* sp. str. GW10	Water and sediments, groundwater wells, Shanxi, China (2.67-350 µM)
209	*Pseudomonas* sp. str. GW11	Water and sediments, groundwater wells, Shanxi, China (2.67-350 µM)
210	*Pseudomonas* sp. str. GW12	Water and sediments, groundwater wells, Shanxi, China (2.67-350 µM)
211	*Pseudomonas* sp. str. GW6	Water and sediments, groundwater wells, Shanxi, China (2.67-350µM)
212	*Pseudomonas* sp. str. GW9	Water and sediments, groundwater wells, Shanxi, China (2.67-350µM)

O$_2$ affinity[1]	Arsenic metabolism (conc)[2,3]	e- donor/ e- acceptor/ C Source[4]	Growth T (°C) (range) [optimal][5,6]	Growth pH (range) [optimal][6]	Arsenic related gene[6]	Ref.[7]
A	DAO (0-26 mM) HAR (0-66.7 M)	YE/ O$_2$/ YE	30	7	*arsD*	[139]
A	DAO (0-26 mM)	YE/ O$_2$/ YE	30	7	*arsB*	[139]
A	DAO (0-26 mM)	YE/ O$_2$/ YE	30	7	*arsB*	[139]
A	DAO (0-26 mM)	YE/ O$_2$/ YE	30	7	*arsB*	[139]
A	DAO (0-26 mM) HAR (0-66.7 M)	YE/ O$_2$/ YE	30	7	*arsAB* *arsC* *arsH*	[139]
A	DAO (0-26 mM)	YE/ O$_2$/ YE	30	7	*arsB* *arsAB* *arsD*	[139]
A	HAR (0-66.7 M)	YE/ O$_2$/ YE	30	7	*arsB* *arsAB*	[139]
A	HAR (0-66.7 M)	YE/ O$_2$/ YE	30	7	*arsH*	[139]
A	HAR (0-66.7 M)	YE/ O$_2$/ YE	30	7	*arsB* *arsAB* *arsD* *arsH*	[139]
A	DAO (0-26 mM) HAR (0-66.7 M)	YE/ O$_2$/ YE	30	7	*arsB* *arsD* *arsH*	[139]
A	DAO (0-26 mM) HAR (0-66.7 M)	YE/ O$_2$/ YE	30	7	*arsB* *arsAB* *arsD*	[139]
A	DAO (0-26 mM) HAR (0-66.7 M)	YE/ O$_2$/ YE	30	7	*arsB* *arsAB* *arsD*	[139]
FAN	HAR (5 mM) DAR (5 mM)	Acetate, Lactate / O$_2$/ Acetate, Lactate Acetate, Lactate / As^V/ Acetate, Lactate	27	7	*arsC* *arrA*	[142]
A	DAO (0-26 mM)	YE/ O$_2$/ YE	30	7	*arsB* *arsAB* *arsH*	[139]
A	DAO (75 μM)	Glucose/ O$_2$/ Glucose	RT	7	no *arsC*	[10]
A	DAO (13.33 mM)	YE, peptone/ O$_2$/ YE, peptone	27	7.4	NR	[12]
A	HAR (1.33 mM)	Lactate/ O$_2$/ Lactate	27	7	NR	[11]
A	HAR (1.33 mM)	Lactate/ O$_2$/ Lactate	27	7	NR	[11]
A	HAR (1.33 mM)	Lactate/ O$_2$/ Lactate	27	7	NR	[11]
A	HAR (1.33 mM)	Lactate/ O$_2$/ Lactate	27	7	NR	[11]
A	HAR (1.33 mM)	Lactate/ O$_2$/ Lactate	27	7	NR	[11]

No	Strain name	Source site (ambient arsenic conc.)
213	*Pseudomonas* sp. str. AR-1	Groundwater, Choushui River alluvial fan, Taiwan (1.6 µM)
214	*Pseudomonas* sp. str. AR-2	Groundwater, Choushui River alluvial fan, Taiwan (1.6 µM)
215	*Pseudomonas* sp. str. AR-3	Groundwater, Choushui River alluvial fan, Taiwan (1.6 µM)
216	*Pseudomonas* sp. str. As7325	Southern Zhuoshui River alluvial fan, Taiwan (1.6 µM)
217	*Pseudomonas stutzeri* sp. str. TS44	Arsenic-contaminated soil, Hubei, China
218	*Pseudomonas stutzeri* str. GIST-BDan2	Bottom sediment, Damyang wetlands, Korea
219	*Pseudomonas vancouverensis* str. FN-48	Sediments, Camarones River, Chile (14.66 µM)
220	*Psychrobacter* sp. str.AR-4	Groundwater, Choushui River alluvial fan, Taiwan (1.6 µM)
221	*Psychrobacter* sp. str.AR-5	Groundwater, Choushui River alluvial fan, Taiwan (1.6 µM)
222	*Serratia marcescens* str. RS-2	Myoung-bong arsenic-contaminated gold-silver mines, Korea (13.33 µM)
223	*Shewanella putrefaciens* str. CN-32	Subsurface core sample, New Mexico, USA
224	*Shewanella* sp. str. ANA-3 *Shewanella trabarsenatis* *Shewanella arsenana*	Arsenic-treated wooden pier, Massachusetts, USA
225	*Shewanella* sp. str. HAR-4	Fe-As-rich sediments, Haiwee reservoir, California, USA
226	*Shewanella* sp. str. W3-18-1	Marine sediments, Pacific Ocean
227	*Stenotrophomonas* sp. str. MM-7	Soil, Lead Smelter Plant, Port Pirie, South Australia
228	*Vibrio* sp str. AR-6	Groundwater, Choushui River alluvial fan, Taiwan (1.6 µM)

δ –Proteobacteria

No	Strain name	Source site (ambient arsenic conc.)
229	δ-Proteobacteria str. MLMS-1	Bottom water from Mono Lake, California, USA (0.2 mM)
230	*Desulfomicrobium* sp. str. Ben-RB	Black mud, arsenic-contaminated reed bed, Bendigo, Australia
231	*Desulfovibrio* sp. str. Ben-RA	Black mud, arsenic-contaminated reed bed, Bendigo, Australia
232	*Desulfohalophilus alkaliarsenatis* str. SLSR-1	Searles Lake, California, USA
233	*Geobacter lovleyi* str. SZ	Sediments, Su-Zi Creek, Seoul, South Korea
234	*Geobacter uraniireducens* str. Rf4	Subsurface sediments with uranium bioremediation

O$_2$ affinity[1]	Arsenic metabolism (conc)[2,3]	e- donor/ e- acceptor/ C Source[4]	Growth T (°C) (range) [optimal][5,6]	Growth pH (range) [optimal][6]	Arsenic related gene[6]	Ref.[7]
FAN	HAR (10 mM)	Lactate/ O$_2$/ Lactate	25	7	arsBC no arrA, aoxB	[50]
FAN	HAR (10 mM)	Lactate/ O$_2$/ Lactate	25	7	no arrA, arsBC, aoxB	[50]
FAN	HAR (10 mM)	Lactate/ O$_2$/ Lactate	25	7	no arrA, arsBC, aoxB	[50]
A	DAO (10 mM)	Acetate, Succinate, Lactate/ O$_2$/ Acetate, Succinate, Lactate	24-26	7	NR	[143]
A	DAO (0.8 mM)	LB/ O$_2$/LB	28	7	aoxAB arsDAHCR ACR3	[93, 94]
A	DAO (1 mM)	Glucose/ O$_2$/ Glucose	30	7.3	aoxBR	[144]
A	DAO (75 µM)	Lactate/ O$_2$/ Lactate	25	NR	aroAB	[138]
FAN	HAR (10 mM)	Lactate/ O$_2$/ Lactate	25	7	arsC no arrA, arsB, aoxB	[50]
FAN	HAR (10 mM)	Lactate/ O$_2$/ Lactate	25	7	arsB no arrA, arsC, aoxB	[50]
A	HAR (0-66.7 mM)	YE/ O$_2$/ YE	30	7	arsH	[139]
AN	DAR (10 mM)	Lactate / As^V/ Lactate	30	7	arr	[145]
FAN	DAR (10 mM) HAR	Lactate / As^V/ Lactate	25	7	arr arsDABC	[145-149]
AN	DAR (10 mM)	Lactate / As^V/ Lactate	30	7	arr	[150]
AN	DAR (10 mM)	Lactate / As^V/ Lactate	30	7	arr	[145]
A	DAO (1 mM)	Glucose/ O$_2$/ Glucose	25	(5-10)	aoxB	[151]
FAN	HAR (10 mM)	Lactate/ O$_2$/ Lactate	25	7	arsC no arrA, arsB, aoxB	[50]
AN	DAR (10 mM)	HS$^-$/ As^V/ HCO$_3^-$	20	9.8	NR	[152]
AN	DAR (10-30 mM)	Lactate / As^V/ Lactate	[25-30]	6.5	no ars	[153, 154]
AN	DAR (10-30 mM)	Lactate / As^V/ Lactate	28	6.5	arsC	[153]
AN	DAR (5-7mM)	Lactate / As^V/ Lactate S$_2^-$ / As^V/ HCO$_3^-$	(20-50) [44]	(7.75–9.75) [9.25]	arrA	[155]
AN	DAR (5mM)	Acetate / As^V/ Acetate	(10-40) [35]	[6.5-7.2]	arrA	[156, 157]
AN	DAR	Acetate / As^V/ Acetate	[32]	[6.5-7]	arrA	[158]

No	Strain name	Source site (ambient arsenic conc.)
ε-Proteobacteria		
235	*Sulfurospirillum halorespirans* str. PCM-E2	Soil, a polluted site, Netherlands
236	*Dehalospirillum multivorans Sulfurospirillum multivorans*	Activated sludge
237	*Geospirillum barnesii* str. SES-3 *Sulfurospirillum barnesii* str. SES-3	Selenate-contaminated freshwater marsh, Nevada, USA
238	*Geospirillum arsenophilus* str. MIT-13 *Sulfurospirillum arsenophilum* str. MIT-13	Arsenic-contaminated watershed sediments, Massachusetts, USA
239	*Sulfurospirillum carboxydovorans* str. MV	Sediments, Kvitebjoern in North Sea
240	*Spirillum* sp. str. 5175 *Sulfurospirillum deleyianum* str. 5175	Anoxic mud of a forest pond, Braunschweig, Germany
241	*Wolinella succinogenes* str. BRA-1	Bovine rumen fluid

[1]Abbreviation for oxygen affinity: **A**, aerobes; **AN**, anaerobes; **FAN**, facultative anaerobes; **MA**, microaerobes.
[2]Abbreviation for arsenic metabolism: **DAO**, detoxifying arsenite oxidizer; **CAO**, chemolithotrophic arsenite oxidizer; **PAO**, phototrophic arsenite oxidizer; **HAR**, heterotrophic arsenate reducer; **DAR**, dissimilatory arsenate reducer.
[3]**YE**, yeast extract; **TYEG**, tryptone, yeast extract and glucose; **YEPG**, yeast extract, peptone and glucose; **NB**, nutrient broth
[4]**W**, weak growth;
[5]**RT**, room temperature;
[6]**NR**, not reported;
[7]**References:**

[1] Huber et al. (2000) [2] Volkl et al. (1993) [3] Cozen et al. (2009) [4] Amo et al. (2002) [5] Huber et al. (1987) [6] Kashefi and Lovely (2000)[7] Macur et al. (2013) [8] Marsh and Norris (1983) [9] Sehlin and Lindstrom (1992) [10] Macur et al. (2004) [11] Fan et al. (2008) [12] Nagvenkar and Ramaiah (2010) [13] Sultana et al. (2011) [14] Clingenpeel et al. (2009) [15] Donahoe-Christiansen et al. (2004) [16] Aguiar et al. (2004) [17] Takai et al. (2003b) [18] Takai et al.(2002) [19] Lebrun et al. (2003) [20] Krafft and Macy (1998) [21] Macy et al. (1996) [22] Rauschenbach et al. (2012) [23] Rauschenbach et al. (2011) [24] Sorokin and Muyzer G (2010) [25] Zhang et al. (2014) [26] Takai et al. (2003a) [27] Yamamura et al. (2014) [28] Brock and Freeze (1969) [29] Gihring et al. (2001) [30] Connon et al.(2008) [31] Gihring and Banfield (2001) [32] Rhine et al. (2007) [33] Ye et al. (2004) [34] Fu et al. (2009) [35] Fu et al. (2010) [36] Sorokin et al. (2008) [37] Zavarzina et al. (2009) [38] Shakoori et al. (2010) [39] Blum et al. (1998) [40] Richey et al. (2009) [41] Turner (1949) [42] Green (1918) [43] Herbel et al. (2002) [44] Baesman et al. (2009) [45] Majumder et al. (2013) [46] Santini et al. (2002b) [47] Fujita et al. (1997) [48] Yamamura et al. (2007) [49] Corsini et al. (2010) [50] Liao et al. (2011) [51] Jareonmit et al.(2012) [52] Fisher and Hollibaugh (2008) [53] Ledbetter et al. (2007) [54] Jones et al. (2000) [55] Chovanec et al. (2010) [56] Fisher (2006) [57] Stolz et al. (2007) [58] Fisher et al. (2008) [59] Niggemyer et al. (2001) [60] Bouchard et al. (1996) [61] Christiansen and Ahring (1996) [62] Liu et al.(2004) [63] Newman et al. (1997a) [64] Newman et al. (1997b) [65] Stackebrandt et al. (2003) [66] Sorokin et al. (2012) [67] Anderson and Cook (2004) [68] Blum et al. (2009) [69] Oremland et al. (2005)

O_2 affinity[1]	Arsenic metabolism (conc)[2,3]	e- donor/ e- acceptor/ C Source[4]	Growth T (°C) (range) [optimal][5,6]	Growth pH (range) [optimal][6]	Arsenic related gene[6]	Ref.[7]
MA-AN	DAR (10-25 mM)	Acetate / As^V/ Acetate	[25-30]	NR	NR	[159, 160]
MA-AN	DAR (10-25 mM)	Acetate / As^V/ Acetate	(15-33) [30]	[7.3-7.6]	NR	[159-162]
MA-AN	DAR (5-10 mM)	Acetate / As^V/ Acetate	(25-33) [33]	[7.5]	*arrA*	[159, 160, 163-167]
MA-AN	DAR (10 mM)	Acetate / As^V/ Acetate	(20-30) [20]	[7.5]	NR	[159, 160, 167, 168]
MA-AN	DAR (10 mM)	Acetate / As^V/ Acetate	(9.9-31.9) [30.2]	(5.9-8.5)[7.3]	NR	[169]
MA-AN	DAR (10 mM)	Acetate / As^V/ Acetate	(20-36) [30]	7.2	NR	[160, 170]
AN	DAR (5 mM)	Acetate/ As^V/YE	30-37	NR	*arrA*	[43]

[70] D'Imperio et al. (2007) [71] Suzuki et al. (1997) [72] Wakao et al. (1994) [73] Santini et al. (2002b) [74] Santini et al. (2002c) [75] Santini et al. (2007) [76] Santini and vanden Hoven (2004) [77] Salmassi et al. (2002) [78] Song et al. (2012) [79] Hao et al. (2012) [80] Garcia-Dominguez et al. (2008) [81] Hsu (2007) [82] Rhine et al. (2008) [83] Macur et al. (2001) [84] Ito et al. (2012) [85] Branco et al. (2009) [86] Branco et al. (2008) [87] Lebuhn et al. (2000) [88] Hamamura et al. (2013) [89] Rhine et al. (2006) [90] Drewniak et al. (2008) [91] Wakao et al. (1988) [92] Bachate et al. (2012) [93] Cai et al. (2009a) [94] Cai et al. (2009b) [95] Huang et al. (2012) [96] Anderson et al. (1992) [97] Anderson et al. (2002) [98] Phillips and Taylor (1976) [99] Osborne and Ehrlich (1976) [100] Suttigarn and Wang YT (2005) [101] Wang and Suttigarn (2007) [102] Yoon et al. (2009) [103] Cavalca et al. (2013a) [104] Sun et al. (2010) [105] Chitpirom et al. (2012) [106] Salmassi et al. (2006) [107] Osborne et al. (2010) [108] Battaglia-Brunet et al. (2002) [109] Battaglia-Brunet et al. (2006) [110] Dastidar and Wang (2010) [111] Dastidar and Wang (2009) [112] London (1963) [113] Moreira and Amils (1997) [114] Bruneel et al. (2003) [115] Duquesne et al. (2008) [116] Delavat et al. (2012) [117] Simeonova (2004) [118] Carapito et al. (2006) [119] Muller et al. (2003) [120] Muller et al. (2006) [121] Weeger et al. (1999) [122] Hoeft et al. (2007) [123] Oremland et al. (2002) [124] Zargar et al. (2010) [125] Lin (2010) [126] Wei (2008) [127] Hollibaugh et al. (2006) [128] Kulp et al. (2008) [129] Hoeft et al. (2010) [130] Lin et al. (2012) [131] Handley et al. (2009a) [132] Handley et al. (2009b) [133] Campos et al. (2010) [134] Turner and Legge (1954) [135] Legge and Turner (1954) [136] Turner (1954) [137] Ilyaletdinov and Abdrashitova (1981) [138] Valenzuela et al. (2009) [139] Chang et al. (2008) [140] Rehman et al. (2010) [141] Abdrashitova et al. (1985) [142] Freikowski et al. (2010) [143] Kao et al. (2013) [144] Chang et al. (2010) [145] Murphy and Saltikov (2007) [146] Saltikov et al. (2003) [147] Saltikov, C.W., et al. (2005) [148] Malasarn et al. (2008) [149] Murphy and Saltikov (2009) [150] Malasarn et al. (2004) [151] Bahar et al. (2012) [152] Hoeft et al. (2004) [153] Macy et al. (2000) [154] Macy and Santini (2002) [155] Blum et al. (2012) [156] Sung et al. (2006) [157] Giloteaux et al. (2013) [158] Shelobolina et al. (2008) [159] Luijten, M., et al. (2003) [160] Luijten, M., et al. (2004) [161] Holliger et al. (1998) [162] Scholzmuramatsu, H., et al. (1995) [163] Laverman, et al. (1995) [164] Lonergan et al. (1996) [165] Oremland et al. (1994) [166] Stolz et al. (1997) [167] Stolz and Oremland (1999) [168] Ahmann et al. (1994) [169] Jensen and Finster (2005) [170] Schumacher et al. (1992)

Reviews in Mineralogy & Geochemistry
Vol. 79 pp. 435-449, 2014
Copyright © Mineralogical Society of America

Health Risks Associated with Chronic Exposures to Arsenic in the Environment

Valerie L. Mitchell

Department of Toxic Substances Control
California Environmental Protection Agency
Sacramento, California 95826-3200, U.S.A.
Valerie.Mitchell@dtsc.ca.gov

INTRODUCTION

Arsenic (As) is a naturally occurring toxic metalloid that is ubiquitous in the environment. It is found in water, soil, and air and as such is also found in the food supply. Millions of people are exposed to As at concentrations in their drinking water that exceed health-based standards worldwide. The World Health Organization (WHO) has listed As as one of its ten chemicals of major public health concern (WHO 2010). Inorganic As (iAs) is listed as the number one concern on the Priority List of Hazardous Substances by the Agency for Toxic Substances and Disease Registry (ATSDR 2014). This list is prepared by ASTDR and the United States Environmental Protection Agency (USEPA) and ranks the substances that present the greatest risk to public health. The list is based on a number of factors including prevalence, toxicity, and the potential for human exposure. Chronic exposure to high levels of As has proven to cause a variety of cancers, cardiovascular disease, and neurologic impairments in exposed populations (ATSDR 2007).

PREVALENCE OF ARSENIC IN THE ENVIRONMENT

Water

The natural background concentration of As in water is 1 to 2 μg L^{-1} (Hindmarsh and McCurdy 1986; NRC 1999), yet elevated levels of iAs are present in the groundwater worldwide (Fig. 1). Elevated levels of As in groundwater can occur due to dissolution and weathering of As-rich ore deposits (Welch et al. 1999, 2000). This process can be accelerated in geothermal waters (Lord et al. 2012; Bundschuh et al. 2013), leading to contamination of surface and groundwater. For example, in the geothermal springs of Yellowstone National Park in Wyoming, As is known to exceed 1000 μg L^{-1} (Stauffer and Thompson 1984; Ball et al. 1998). These geothermal waters discharge into surface waters resulting in measured concentrations as high has 360 μg L^{-1} in the Madison River at the boundary of the park (Nimick et al. 1998). Mining operations can accelerate the natural rate of dissolution of As-rich deposits through an increase in surface area and weathering. Arsenic concentrations in water associated with the Iron Mountain Mine in California, an area with massive sulfide deposits and some of the most acidic mine waters in the world, been measured as high as 340,000 μg L^{-1} (Nordstrom and Alpers 1999). Mining activities are a major contributor to elevated-As concentrations in groundwater that lead to human exposure through drinking water (Mok et al. 1988; Choprapawon 1998; Pawlak et al. 2008; Chakraborti 2013; Ghosh 2013).

1529-6466/14/0079-0008$05.00

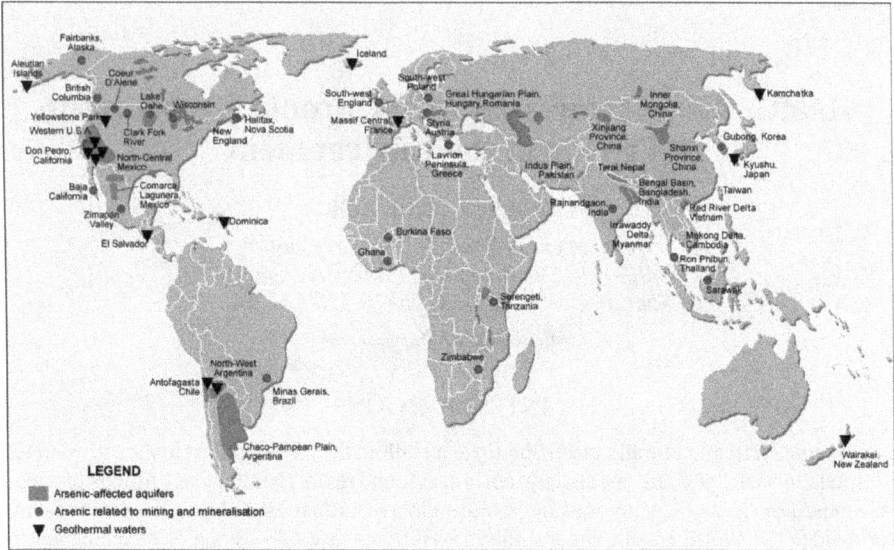

Figure 1. Worldwide Distribution of Arsenic in Groundwater. This information was produced by the British Geological Survey and the Department of Public Health Engineering (BGS/DPHE 2001).

Both the WHO and USEPA use the standard of 10 μg L⁻¹ for As in drinking water (USEPA 2001; WHO 2011). This value is the maximum contaminant level (MCL) for tap water in the United States and is meant to be protective of human health. It is estimated that approximately 10% of regulated public water supplies in the United States exceed this standard (Welch et al. 1999). Approximately 57 million people in Bangladesh alone are chronically exposed to As at concentrations exceeding the WHO standard of 10 μg L⁻¹ through drinking water obtained from tube wells (BGS/DPHE 2001). Argentina, Chile, China, India (West Bengal), Mexico, and Taiwan also have well-documented elevated levels of arsenic in groundwater (BGS/DPHE 2001; WHO 2010).

Soils

The concentration of naturally occurring As in soils depends greatly on the mineralogy of the soils and the rock from which it was derived. Arsenic is a known component of at least 245 mineral species. It is present in high concentrations within sulfide deposits, most commonly as arsenopyrite. Naturally occurring As in soils ranges from undetectable to 40 mg kg⁻¹ with an average of around 5 mg kg⁻¹. Soils overlying sulfide ore deposits may contain much higher concentrations, ranging from hundreds to thousands mg kg⁻¹ As (NAS 1977). Volcanic eruptions can also lead to deposition of As in soils (Liu et al. 2013). Mining of As, lead, zinc, and gold can all bring As in the form of waste rock to the surface (USEPA 1986). Weathering of waste rock and erosion can lead to elevated levels of As in the soil, in respirable dust as well as sediment and surface water through run-off. Other anthropogenic sources of As include smelting operations (Lee-Feldstein 1983; Enterline et al. 1995; Erraguntla et al. 2012) and historical use of arsenicals as pesticides and herbicides (Senesi et al. 1999; Tsuji et al. 2005). Most of the As currently in use in the United States is used in chromated copper arsenate (Reese 1998), which is a common preservative used to pressure treat wood products used as poles, fences, and decks. Weathering and leaching of these treated products can release As into the surrounding soils (Chirenje et al. 2003).

Air

Arsenic is present in the atmosphere at varying levels. The natural, non-anthropogenic, concentration of As in air ranges from 1 to 3 ng m^{-3} and may be as high as 100 ng m^{-3} in urban areas (ATSDR 2007). The largest source of inorganic As in the atmosphere is combustion and high-temperature processes such as wood and fossil fuel burning in fireplaces, combustion engines, and geothermal steam releases (California Air Resources Board 1990). Smelters have been responsible for the highest detected concentrations of As in the atmosphere, reaching concentrations as high as 2,500 ng m^{-3} (Schroeder et al. 1987). Arsenic is also found in cigarette smoke, with an estimated 40-120 ng in mainstream smoke per cigarette (Hughes et al. 1995).

Food supply

A significant source of As exposure in the human population is dietary. The foods most commonly associated with As exposure include rice and fish. The FDA conducts a market basket study yearly to monitor the concentrations of toxic contaminants in the food supply (USFDA 2008). Table 1 summarizes select As content results from the 2006-2008 market basket study. The highest concentrations of As in the food supply is consistently found in fish. One study documented As concentrations ranging from 160 µg L^{-1} in freshwater fish to 2,360 µg L^{-1} in saltwater fish (Schoof et al. 1999). Arsenic in fish and shellfish is primarily organic, in the forms arsenobetaine and arsenocholine, which have very little associated toxicity (Sabbioni et al. 1991). Rice is the major source of inorganic As in the food supply, mostly due to the large amount of water used in its cultivation. Inorganic As accounts for approximately 80 to 91% of the As in rice in south and southeast Asia (Rahman and Hasegawa 2011). An evaluation of rice from 204 commercial rice suppliers worldwide found As in the range of 5 to 710 µg L^{-1} (Zavala and Duxbury 2008). It was further documented that the mean concentration of As in rice in both the United States and the European Union was 198 µg L^{-1}. Arsenic concentrations in rice are further elevated if high-As content water is used for irrigation and/or cooking (Rahman and Hasegawa 2011). Organic As is an approved animal dietary supplement and is found in specifically approved drugs added to poultry and other animal feeds (Lasky 2013). In 2012 it was estimated that 88% of chickens raised for human consumption in the United States are treated with drugs containing As in order to combat parasitic infections and increase weight gain (Nachman et al. 2012). An evaluation of cooked chicken purchased in

Table 1. Arsenic concentrations detected in select food items[1].

Food Description	Mean (µg L^{-1})	Std Dev (µg L^{-1})	Min (µg L^{-1})	Max (µg L^{-1})
Fish Sticks	527	150	247	738
Rice, white, cooked	65	15	41	84
Crisped rice cereal	135	41	81	201
Shrimp, boiled	265	105	136	481
Mushrooms, raw	73	41	27	137
Tuna Noodle Casserole	164	76	78	321
Clam chowder, canned	128	31	59	169
Salmon, steaks, baked	288	49	222	373
Tuna, canned in water, drained	1000	445	378	1850
Cake, chocolate with icing	13	45	0	155

[1]Select results from the US Food and Drug Administration Market Baskets 2006-2008 (USFDA 2008)

the United States compared the inorganic As content of traditionally raised chickens with organically raised chickens (Nachman et al. 2013) and found three times as much inorganic As in the meat of traditionally raised chickens (1.8 µg L^{-1} versus 0.6 µg L^{-1}, respectively).

ARSENIC TOXICITY

Paracelsus, a 16th century physician, first said, "dose makes the poison," when he described the fundamental principal of toxicology. The toxicity of a substance depends on the concentration required to elicit a negative biological response. So what is it about As that makes it such a toxic substance? As discussed previously, As is ubiquitous in the environment and exposures can occur through ingestion of contaminated groundwater, incidental ingestion of contaminated soils, inhalation of respirable dusts, and ingestion of food containing As. Upon ingestion or inhalation, the substance must be absorbed by the body (through the gastrointestinal tract or the lining of the lung) in order to elicit an effect. Bioavailability is the term used to describe how readily the body can absorb a substance thus making it available to interact with various biological systems. Bioavailability is an important metric when evaluating the significance of exposure, as only the bioavailable fraction of the available As can elicit a toxic response. Absorption is greatly influenced by the solubility of a substance; the higher the solubility, the greater the absorption (Marafante and Vahter 1987).

Arsenic in groundwater is readily absorbed and its uptake has been estimated at approximately 95% following ingestion in healthy subjects (Zheng et al. 2002). The bioavailability of As ingested following exposure to contaminated soils is significantly influenced by the mineralogical associations of As within the soil matrix and has a reported range of 3 to 100% (Freeman et al. 1995; Rodriguez et al. 1999; Juhasz et al. 2007; Bradham et al. 2011).

Following absorption, As is chemically transformed during metabolism, which plays an important role in its toxicity. The metabolism of As, as diagrammed in Figure 2, has been extensively studied in both humans and animals and is reviewed elsewhere (Aposhian et al. 2000; Thomas et al. 2001; Hughes 2002). The oxidation state of As (−III, 0, +III, and +V) directly influences its speciation and toxicity. Although As exists in both organic and inorganic compounds, of the greatest risk of toxic exposures and threat to human health is associated with iAs. Inorganic arsenic is commonly found in one of two forms: arsenite (As(III)) and arsenate (As(V)). The prevalence of a particular state of As is determined by the pH and redox condition of the matrix in which it exists. As(III), the reduced trivalent form of As, is the more potent toxin (Thomas et al. 2001; Hughes 2002; Singh et al. 2011). The oxidation state of iAs impacts how the metalloid is metabolized in the body. Arsenite is the preferred substrate for methylation reactions, a key step in As metabolism; arsenate must be reduced to arsenite prior to methylation. Monomethylated arsenicals (MMAs) and dimethylated arsenicals (DMAs) are the major methylated metabolites found in urine and are often used as biomarkers along with As(III) and As(V) to evaluate exposure to iAs (Smith et al. 1977; Thomas et al. 2001; Byun et al. 2013). As with iAs, MMA and DMA can exist in both a +3 and +5 oxidation state. Comparative analysis of the *in vitro* effects of various arsenicals on both rat and human cells have demonstrated that trivalent MMA (MMA(III)) is the most cytotoxic arsenical followed by As(III) (Petrick et al. 2000; Styblo et al. 2000). MMA(III) is highly unstable in urine and as such, efforts to use it as a biomarker in exposed populations have proven difficult (Kalman et al. 2013). Total MMA (MMA(III) + MMA(V)) as a proportion of total arsenic excreted in urine has been shown in a number of epidemiological studies to be associated with increased risk of As related heart disease, skin lesions, skin cancer, bladder cancer, and lung cancer (Smith and Steinmaus 2009; Chen et al. 2013; Melak et al. 2014). These studies support the hypothesis that As methylation is an activation step that increases toxicity and negative health outcomes following environmental exposure.

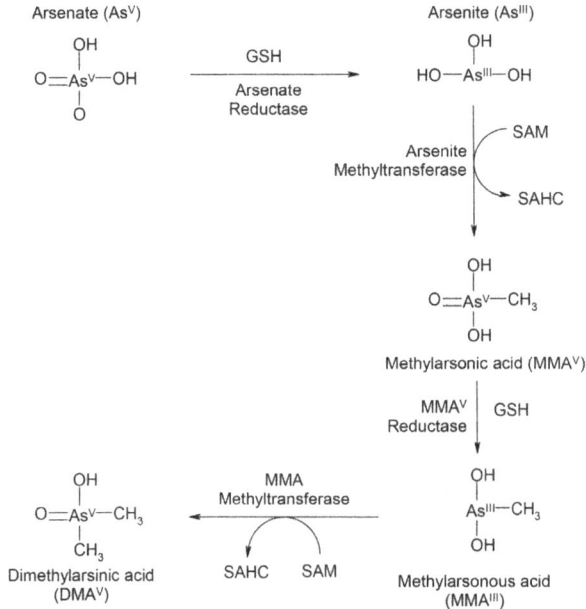

Figure 2. Arsenic Metabolism. The metabolism of As via the methylation pathway (ATSDR 2007). GSH = Glutathione; SAHC = S-adenosylhomocysteine; SAM = S-adenosylmethionine.

The discussion of As toxicity would not be complete without a discussion of the potential mechanisms of action involved. The elucidation of these mechanisms of action for As is an ongoing field of study that has been reviewed extensively by others (Thomas et al. 2001; Hughes 2002; Schoen et al. 2004; Kitchin and Wallace 2008; Druwe and Vaillancourt 2010; Hughes et al. 2011). Ongoing hypotheses include biochemical disruption, oxidative stress, alteration of signal transduction pathways, increased cell proliferation, genotoxicity, DNA methylation, and tumor promotion.

CHRONIC ARSENIC EXPOSURE AND ASSOCIATED HEALTH EFFECTS BY REGION

Taiwan

The effects of environmental exposures to iAs have been long recognized. A large amount of data regarding arseniasis, or chronic arsenic poisoning, comes from a cohort in Southwestern Taiwan where exposure to As-contaminated groundwater was widespread. Due to high salinity in shallow groundwater in this area, residences began using artesian wells as a water source in the 1920s (Chen et al. 1985). Following the adoption of this source of drinking water, a peripheral vascular disease known as Blackfoot Disease (BFD) became endemic in this part of Taiwan. BFD is characterized by coldness and numbness in the extremities that eventually leads to difficulty in walking and onset of gangrene (pictured in Fig. 3), resulting in spontaneous or surgical amputation and an increase in premature death (Tseng 2005). Epidemiological studies subsequently revealed that BFD was associated with the consumption of water from the artesian wells (Tseng 2002). The groundwater in this region of Taiwan contained elevated levels of naturally occurring As as high as 1,820 µg L^{-1} (Tseng et al. 1968) and well water had concentrations ranging from 350 to 1,100 µg L^{-1} (Wu et al. 1989). Tseng et al. surveyed a total of 40,421 people and placed them into 3 different exposure groups: low (0-290 μg L^{-1}), medium (300-590 μg L^{-1}), and high (600 μg L^{-1} or greater) based on the well data from their reported area of residence. The skin cancer rate for the population was 10.6 in 1000. The incidence of cancer increased with age and was three times higher in males than females. The results of these studies form the basis for the drinking water standard of 10 μg L^{-1} set by WHO and USEPA (USEPA 2001; WHO 2011). In addition to skin cancer, there was a high rate (9 in 1000) of BFD in this population (Tseng 1977).

Both skin cancer and BFD incidences coincided with increasing concentrations of As in the groundwater (Tseng 1977). The Taiwanese population was thus further studied to examine the dose response relationship between As concentrations in well water and various cancers (Wu et al. 1989; Chen et al. 1992). Mortality rates of cancers including bladder, kidney, skin, lung, liver, prostate, leukemia, nasopharynx, esophagus, stomach, colon, and uterine cervix in 42 villages were determined based on information provided on death certificates from 1973-1986. Mortality rates were age-adjusted and compared to median arsenic levels in wells. The study examined the same exposure groups used by Tseng et al. (1968) so that the studies could be compared. A significant dose response relationship was observed for bladder, kidney, skin, and lung cancer in both sexes and additionally for liver and prostate cancer in males. Peripheral vascular disease and cardiovascular disease deaths in both sexes also demonstrated a significant dose response relationship with median arsenic levels in well water (Wu et al. 1989).

Chile

Northern Chile is another region where exposure to As in the groundwater has been studied extensively. More than 250,000 people were exposed to high levels of As over a 13-year period when river water, high in naturally occurring As, was diverted in 1957 and used for drinking water. This area in the Andes Mountains is one of the driest inhabited places on earth (McKay et al. 2003) and as such, public water supplies are the only source of water most of the population. This provides investigators the unique opportunity of having relative certainty regarding individual exposure based simply on their residential history. The average concentration of As in drinking water during this period in the most populous city in the area, Antofagasta, was 860 μg L^{-1} (Steinmaus et al. 2013). The installation of a water treatment plant in 1970 reduced these concentrations to below the WHO standard of 10 μg L^{-1}.

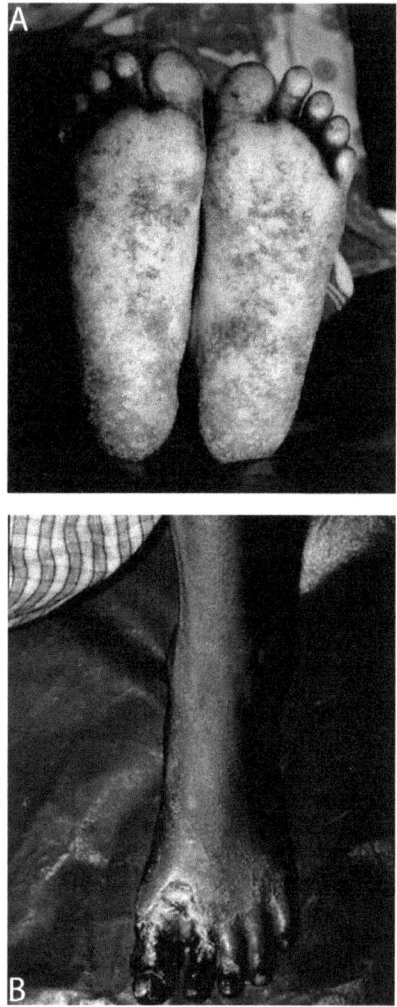

Figure 3. Arsenic-induced Lesions Following Chronic Exposure to Contaminated Drinking Water. (A) Keratosis (B) Gangrene. Images courtesy of the Arsenic Foundation. *http://users.physics.harvard.edu/~wilson/arsenic/Arsenic%20Foudation.html.*

Mortality ratios for bladder, lung, skin, liver, and kidney cancer were evaluated in the exposed population in this region of Chile and standardized against age-matched mortality rates observed in rest of the country for the same cancers (Smith et al. 1998). The bladder cancer standardized mortality ratio (SMR) in exposed women was 8 times greater than that of the rest of the country (SMR=8.2, p<0.001). Similarly, bladder cancer mortality rates in exposed men were increased 6 fold (p<0.001). Skin cancer mortality rates were also increased for both women (SMR=3.2, p=0.012) and men (SMR=7.7, p<0.001), as were lung cancer SMRs (3.8 and 4.1, respectively, p<0.001). Smaller SMRs were seen for kidney cancer in both women (2.7, p<0.001) and men (1.6, p=0.016). This study further estimated that As could be accountable for approximately 7% of all deaths in persons 30 years or older in the exposed population.

The study by Smith et al. (1998) is an ecological study design that evaluated the impact of elevated As in drinking water on mortality rates in the exposed population when compared to a non-exposed population. Another type of study design, known as a case-control study, was carried out by Steinmaus et al. (2013). The case-control study identified individuals with contrasting outcomes (cancer vs. no cancer), determined individual exposures to As in drinking water based on residential history, and then evaluated the association of cancer with varying levels of As exposure. Because data is available on an individual basis, it is possible to evaluate a dose response relationship between As exposure and cancer incidence. A total of 306 bladder and 232 lung cancer cases first diagnosed between 2007 and 2010 were compared to 640 gender and age-matched controls. Odds ratios were calculated based on quartiles of the average As concentration in drinking water prior to 1971 (<11, 11–90, 91–335, and >335 $\mu g\ L^{-1}$). For bladder cancer the corresponding odds ratios were 1.00, 1.36 (95% confidence interval, 0.78 to 2.37), 3.87 (2.25 to 6.64), and 6.50 (3.69 to 11.43), respectively. Corresponding lung cancer odds ratios were 1.00, 1.27 (0.81 to 1.98), 2.00 (1.24 to 3.24), and 4.32 (2.60 to 7.17). For those exposed to the highest levels of As (average 860 $\mu g\ L^{-1}$) during that 13-year period, the odds ratios were 6.88 (3.84 to 12.32) for bladder cancer and 4.35 (2.57 to 7.36) for lung cancer. In other words the odds ratios in those individuals were 7 times higher for bladder cancer and 4 times higher for lung cancer than those exposed to As in drinking water at concentrations below the WHO standards during the corresponding time period. This study demonstrated a clear dose response relationship between As concentrations in drinking water and odds ratios associated with both lung and bladder cancer. Furthermore, this study is unique in that it indicates increased odds ratios for both types of cancer 40 years after the cessation of exposure to As, demonstrating a period of latency. In another study by the same group of investigators, odds ratios for kidney cancer were evaluated in a similar fashion using the same population-based controls (Ferreccio et al. 2013). A total of 122 cancer cases were evaluated including 76 renal cell, 24 transitional cell renal pelvis and ureter, and 22 other kidney cancers. For renal pelvis and ureter cancers, the adjusted odds ratios based on median water concentrations of 60, 300, and 860 $\mu g\ L^{-1}$ were 1.00, 5.71 (1.65 to 19.82), and 11.09 (3.60 to 34.16), respectively. The odds ratios were not increased for renal cell cancers, however. Although these findings are based on a relatively small number of cases (24), they provide evidence that exposure to As in drinking water cause some forms of kidney cancer.

The exposure scenario in Chile has provided a unique opportunity to look at the effects of early life exposures to elevated concentrations of As. Mortality rates in Antofagasta were evaluated for those born just prior to (1950-1957) and during the peak in As concentration in the water supply (1958-1970) (Smith et al. 2006). Mortality rates for 30-49 year olds were evaluated from death certificates for the period from 1989-2000 and compared to rates in the rest of Chile. Those exposed as young children (born 1950-1957) had an SMR for lung cancer 7 times that of age-matched controls (p<0.001). In addition to lung cancer, mortality rates for the non-malignant lung disease, bronchiectasis, were also evaluated and had an SMR of 12.4 (p<0.001). The cohort born between 1950 and 1957, who was assumed to be exposed *in utero*, had a similar SMR for lung cancer as those exposed as small children (SMR=6.1, p<0.001). The evaluation of bronchiectasis mortality rates amongst those exposed *in utero* revealed an SMR 46.2 times greater (p<0.001) than those seen in rest of the Chile. This was the first study to demonstrate that early-life exposures to As could lead to pathogenesis in adulthood. In another study by the same group of investigators the relative risk for childhood mortality (death prior to 20 years of age) due to liver cancer for those born between 1950 and 1957 was 10.6 (95% CI 2.9–39.2; p<0.001) compared to the reference population (Liaw et al. 2008).

Bangladesh

While exposures to As in the water supply are largely historical in Taiwan and Chile, there are ongoing exposures to high As content in drinking water in Bangladesh. Arsenic contamination in the groundwater in Bangladesh is widespread and affects most of the country's inhabitants. In the 1960s and 1970s the United Nations Children's Emergency Fund (UNICEF), along with the Bangladeshi government, began installation of shallow tubewells as a source of pathogen-free drinking water (Ahsan et al. 2006a). An estimated 6-10 million tubewells have been installed with most exploiting groundwater at a depth of 10-50 meters (BGS/DPHE 2001). A survey of these well waters by the British Geological Survey (BGS) revealed that nearly half are contaminated with As exceeding the WHO drinking water standard of 10 μg L^{-1}; 25% exceed the Bangladeshi water standard of 50 μg L^{-1}, 9% of the wells have As content that exceeds 200 μg L^{-1} and nearly 2% exceed 500 μg L^{-1} As in water. The high-As content of these waters was not discovered until the 1990s when an epidemic of previously-unseen skin lesions (such as those pictured in Fig. 3) surfaced (Ahsan et al. 2006a). Due to the ongoing nature of these exposures, a unique opportunity exists to evaluate health effects on an individual basis.

The Health Effects of Arsenic Longitudinal Study (HEALS) was established in 2000 by a group of investigators at Columbia University to evaluate the effects of the full range of As concentrations on various health outcomes in a cohort of nearly 12,000 men and women (Ahsan et al. 2006a). Individual As exposures were evaluated using well data and reported water consumption from participants, 76% of whom drink water from wells containing As in excess of the WHO standard of 10 μg L^{-1} (Ahsan et al. 2006a). Consistent dose response effects of As were observed in the prevalence odd ratios for premalignant skin lesions (Ahsan et al. 2006b). These lesions follow a typical pattern of development starting with hyperpigmentation followed by a thickening of palms and the soles of the feet often accompanied by nodule protrusions known as keratosis (Alain et al. 1993, Fig. 3). Skin lesions were 4 times more common in men than women. Odds ratios as compared to those exposed to As in drinking water below 8 μg L^{-1} ranged from 1.9 (95% CI 1.26, 2.89) in those exposed to drinking water containing between 8.1 and 40.0 μg L^{-1} As to 5.39 (95% CI 3.69, 7.86) for those exposed to drinking water with As between 175.1 and 864.0 μg L^{-1}. These findings are particularly important as they demonstrate significant pathology at concentrations below the current Bangladesh water standard of 50 μg L^{-1}.

A study of As effects on pregnancy outcomes was performed in an area of Bangladesh where the average concentration of As in drinking water was 240 μg L^{-1} (Ahmad et al. 2001). Arsenic at these concentrations had a significant negative impact on pregnancy outcomes in exposed women as compared to women exposed to drinking water with As concentrations less than 20 μg L^{-1}. Exposed women had nearly three times (2.9) as many spontaneous abortions (miscarriages) and stillbirth rates were 2.24 times higher than in the comparison group. Premature birth rates in exposed women also increased 2.54 fold. All of these differences were statistically significant.

Intellectual function was also evaluated in children of some of the HEALS participants. Evaluations were made separately for children aged six and ten years old. A total of 201 ten year olds were evaluated using a subset of intellectual tests to measure Verbal and Performance

Intellectual Quotients (IQ) (Wasserman et al. 2004). Arsenic concentrations were measured in tubewells at each child's home, as well as in their urine. Blood samples were additionally collected from 107 participants to measure blood lead, which is known to negatively affect IQ in children. Intellectual function was significantly reduced in children exposed to As at concentrations greater than 50 µg L^{-1} as compared to those exposed to less than 5.5 µg L^{-1} in drinking water. Significance reduction in intellectual function remained when results were adjusted for socioeconomic background and magnesium in drinking water. A similar study was conducted for 301 children 6 years of age (Wasserman et al. 2007). This study involving the younger children also showed a negative association between As content in drinking water and intellectual function although to a lesser degree. Negative associations between IQ and As exposure were also demonstrated by Nahar et al. for both 9-10 year olds (2013a) and 14-15 year olds (2013b). Reduction of intelligence was associated with As concentration measured in the subjects urine in a dose dependent manner.

United States of America

It is estimated that greater than 10% of the drinking water wells in the United States have concentrations of As exceeding the USEPA drinking water standard of 10 µg L^{-1} (Welch et al. 1999). Less than 1% exceeds the previous standard of 50 µg L^{-1}. Elevated As concentrations in groundwater are more common in the Western United States than the East or the South. A cohort mortality study was done in Millard County, Utah where As concentrations in drinking water were as high as 680 µg L^{-1} with the median As content between 20 and 200 µg L^{-1} in the seven cities evaluated (Lewis et al. 1999). Statistically significant increases were seen in men for SMRs associated with hypertensive heart disease (2.2, 95% CI 1.36-3.36), nephrosis and nephritis (1.72, 95% CI 1.13-2.50), and prostate cancer (1.45, 95% CI 1.07-1.91), as well as for kidney cancer in the medium and high exposure groups (1.75, CI 95% 0.8-3.32). In women, statistically significant increases in mortality rates were associated with hypertensive heart disease (1.73, 95%CI 1.11-2.58) and all other heart disease (1.43, 95%CI 1.11-1.80). Additional increases in mortality rates were seen for benign neoplasms, arteriosclerosis, diabetes, melanoma, and all other malignant neoplasms but none of these increases were statistically significant. Epidemiological studies in the US have been unable to show increases in the mortality rates for the common cancers (skin, lung, bladder) associated with exposures to elevated As in groundwater in other countries, likely due to the fact that the maximum concentrations seen in US waters are similar to the minimum concentrations seen in other countries (Morton et al. 1976; Lewis et al. 1999; Dauphine et al. 2013).

While the bulk of epidemiological data on exposures to As and resulting health effects involve the ingestion of contaminated water, other significant exposures to As do exist. Mining and smelting operations are associated with inhalation exposures to As, primarily in the form of As trioxide (As$_2$O$_3$). Mortality rates due to respiratory cancer were greatly increased in workers at the Anaconda copper smelter in Montana in a dose-dependent manner with those exposed to the highest concentrations (500-5000 micrograms per cubic meter) having a 7 fold increase compared to non-exposed age-matched controls (Lee-Feldstein 1983; Welch et al. 1982). Mortality rate data from the ASARCO copper smelter in Tacoma, Washington show similar results. In addition to increased mortality rates due to respiratory cancer, significant increases were seen for rates associated with cancers of the large intestine and bone in exposed workers (Enterline et al. 1995). Air pollution from the ASARCO copper smelter contaminated over 1000 square miles of soils. Studies have identified elevated concentrations of As and As metabolites in the urine of residents living in close proximity to the smelter (Polissar et al. 1990). Children aged 0-6 who lived within one-half mile of the site saw the greatest increase in urinary As. These increases were attributed to increased hand-to-mouth activities of children, which resulted in the ingestion of contaminated soils. There is ongoing remediation aimed at reducing the concentration of As and other contaminants in these soils.

ENVIRONMENTAL REGULATION OF ARSENIC
IN THE UNITED STATES OF AMERICA

In addition to smelter emissions, elevated concentrations of As in soils are also associated with mining operations and the historical use of arsenical pesticides. Remediation of said soils is an ongoing effort throughout the US by the USEPA as well as individual state environmental agencies. Just as the USEPA set the standard to 10 μg L^{-1} for drinking water, it has established guidance for concentrations of As in soil that are protective of human health. These concentrations vary according to land use (residential or commercial/industrial), but are universally lower than typical background concentrations found in soil. As previously mentioned, background concentrations of As in soil range from 0.1-40 mg kg^{-1}. The health-based residential screening level for As in soils is less than 1 mg kg^{-1}. The fact that health-based screening levels are lower than the naturally-occurring background concentrations of As in soil creates a unique problem for site clean-up and exposure reduction. Exposure to contaminated soils can occur in the form of incidental ingestion, inhalation of contaminated dusts, and through dermal contact. As discussed above, with regards to the smelter site, children are often at greatest risk for exposures to contaminated soils due to hand-to-mouth activities. An important factor to consider in terms of exposures to As in soil is bioavailability. Current risk assessments assume that the As in soils is 100% bioavailable although studies show that it can range from 3 to 100% depending upon speciation and mineral association (Freeman et al. 1995; Rodriguez et al. 1999; Juhasz et al. 2007; Bradham et al. 2011). Methods used to determine the bioavailability of As in soils as a basis for making remediation decisions are discussed in detail by Basta and Juhasz (2014, this volume).

Both the drinking water standard and the soil screening values recommended by the USEPA were calculated using what is known as toxicity criteria. Toxicity criteria are used in human health risk assessments to estimate potential cancer and non-cancer risks to individuals exposed to contaminated media (groundwater, soil, respirable soil dust, etc.). Toxicity criteria for exposure to carcinogens are presented as a Slope Factor (SF) or an Inhalation Unit Risk (IUR), which is an upper-bound estimate of the probability of a person developing cancer after a lifetime exposure to a unit intake (e.g. 1 mg kg^{-1} day^{-1} for ingestion of soils, or 1 μg m^{-3}, for inhalation). Non-cancer toxicity criteria are presented as Reference Doses (RfD) or Concentrations (RfC) for ingestion and inhalation, respectively, and represent the dose/concentration at which no adverse effects are expected even with long-term exposure in sensitive populations.

At the time of publication the USEPA is currently evaluating the As toxicity criteria (USEPA 2014). The USEPA last updated the toxicity criteria for iAs in 1988 (USEPA 1988). The current USEPA oral Slope Factor (SFo) is 1.5 mg kg^{-1} day^{-1}. This slope factor was derived by evaluating the increased risk of skin cancer in the Taiwanese cohort that was exposed to high levels of As in drinking water, as discussed previously (Tseng 1968, 1977). In 1999, it was recommended by the National Research Council (NRC) that the USEPA update their cancer risk assessment to consider the wealth of new information regarding the toxicity of As, such as the epidemiology studies from Chile and Bangladesh. This reassessment began in 2003 and is ongoing. The California Environmental Protection Agency (CaEPA) updated their SFo in 2009 to 9.5 mg kg^{-1} day^{-1}. The California slope factor is higher (indicative of greater cancer potency) primarily due to additional safety factors to account for increased cancer rates due to early life exposures (CaEPA 2009). The RfD for arsenic is 3×10^{-4} mg kg^{-1} day^{-1} and is also based on the original cohort studies from Taiwan (Tseng 1968, 1977). The reference dose is expected to decrease (indicative of a lower exposure concentration resulting in adverse non-cancer effect) following the re-evaluation of the toxicity criteria. The updates to the toxicity criteria for both cancer and non-cancer endpoints are important steps towards ensuring that contaminated sites are cleaned up to levels needed to protect human health.

SUMMARY

Exposures to As concentrations that are harmful to human health occur on a daily basis worldwide. Epidemiological studies have implicated As in skin, lung, bladder, liver, and kidney cancers as well as vascular disease, pre-malignant skin lesions, reduced IQ, poor pregnancy outcomes, diabetes, neuropathy, and overall increased mortality (Tseng 1977; Wu et al. 1989; Smith et al 1998; Ahmad 2001; Wasserman et al. 2004; ATSDR 2007; Vahidnia et al. 2007; Christoforidou et al. 2013; James et al. 2013). Moreover, recent findings in Chile suggest a 40-year latency period for some of these effects to occur (Steinmus et al. 2013). Continued remediation of As contaminated drinking water supplies and soils is necessary to reduce exposure and subsequent negative health outcomes.

REFERENCES

Ahmad SA, Sayed MH, Barua S, Khan MH, Faruquee MH, Jalil A, Hadi SA, Talukder HK (2001) Arsenic in drinking water and pregnancy outcomes. Environ Health Perspect 109(6):629-31

Ahsan H, Chen Y, Parvez F, Argos M, Hussain AI, Momotaj H, Levy D, van Geen A, Howe G, Graziano J (2006a) Health effects of arsenic longitudinal study (HEALS): description of a multidisciplinary epidemiologic investigation. J Exposure Sci Environ Epidemiol 16(2):191-205, doi: 10.1038/sj.jea.7500449

Ahsan H, Chen Y, Parvez F, Zablotska L, Argos M, Hussain I, Momotaj H, Levy D, Cheng Z, Slavkovich V, van Geen A, Howe GR, Graziano JH (2006b) Arsenic exposure from drinking water and risk of premalignant skin lesions in Bangladesh: baseline results from the Health Effects of Arsenic Longitudinal Study. Am J Epidemiol 163(12):1138-1148, doi: 10.1093/aje/kwj154

Alain G, Tousignant J, Rozenfarb E (1993) Chronic arsenic toxicity. Int J Dermatol 32(12):899-901

Aposhian HV, Zheng B, Aposhian MM, Le XC, Cebrian ME, Cullen W, Zakharyan RA, Ma M, Dart RC, Cheng Z, Andrewes P, Yip L, O'Malley GF, Maiorino RM, Van Voorhies W, Healy SM, Titcomb A. (2000) DMPS-arsenic challenge test. II. Modulation of arsenic species, including monomethylarsonous acid (MMAs(III)), excreted in human urine. Toxicol Appl Pharmacol 165(1):74-83

ATSDR (2007) Toxicological Profile for Arsenic. Agency for Toxic Substances and Disease Registry, *http://www.atsdr.cdc.gov/toxprofiles/tp2.pdf* (accessed July 2014)

ATSDR (2014) Support Document to the 2013 Priority List of Hazardous Substances that will be the Subject of Toxicological Profiles. Agency for Toxic Substances and Disease Registry, *http://www.atsdr.cdc.gov/SPL/resources/ATSDR_2013_SPL_Support_Document.pdf* (accessed July 2014)

Ball JW, Nordstrom DK, Jenne EA, Vivit DV (1998) Chemical analyses of hot springs, pools, geysers, and surface waters from Yellowstone National Park, Wyoming, and vicinity, 1974-1975. USGS Open-File Report 98-182

Basta NT, Juhasz A (2014) Using in vivo bioavailability and/or in vitro gastrointestinal bioaccessibility testing to adjust human exposure to arsenic from soil ingestion. Rev Mineral Geochem 79:451-472

BGS/DPHE (2001) Arsenic contamination of groundwater in Bangladesh. British Geological Survey Technical Report WC/00/19. Kinniburgh DG, Smedley PL (eds) British Geological Survey: Keyworth.

Bradham KD, Scheckel KG, Nelson CM, Seales PE, Lee GE, Hughes MF, Miller BW, Yeow A, Gilmore T, Harper S, Thomas DJ (2011) Relative bioavailability and bioaccessibility and speciation of arsenic in contaminated soils. Environ Health Perspect 119:1629-1634

Bundschuh J, Maity JP, Nath B, Baba A, Gunduz O, Kulp TR, Jean JS, Kar S, Yang HJ, Tseng YJ, Bhattacharya P, Chen CY (2013) Naturally occurring arsenic in terrestrial geothermal systems of western Anatolia, Turkey: potential role in contamination of freshwater resources. J Hazard Mater 262:951-9, doi: 10.1016/j.jhazmat.2013.01.039

Byun K, Won YL, Hwang YI, Koh DH, Im H, Kim EA (2013) Assessment of arsenic exposure by measurement of urinary speciated inorganic arsenic metabolites in workers in a semiconductor manufacturing plant. Ann Occup Environ Med 25(1):21, doi: 10.1186/2052-4374-25-21

CaEPA (2009) Technical Support Document for Cancer Potency Factors: Methodologies for derivtion, listing of available values, and adjustments to allow for early life stage exposures. California Environmental Protection Agency, *http://oehha.ca.gov/air/hot_spots/2009/TSDCancerPotency.pdf* (accessed July 2014)

California Air Resources Board (1990) Proposed Identification of Inorganic Arsenic as a Toxic Air Contaminant. State of California Air Resources Board Report, *http://oehha.ca.gov/air/toxic_contaminants/pdf1/inorganic%20arsenic.pdf* (accessed July 2014)

Chakraborti D, Rahman MM, Murrill M, Das R, Siddayya, Patil SG, Sarkar A, Dadapeer HJ, Yendigeri S, Ahmed R, Das KK (2013) Environmental arsenic contamination and its health effects in a historic gold mining area of the Mangalur greenstone belt of Northeastern Karnataka, India. J Hazard Mater 262:1048-55, doi: 10.1016/j.jhazmat.2012.10.002

Chen CJ, Chen CW, Wu MM, Kuo TL (1992) Cancer potential in liver, lung, bladder and kidney due to ingested inorganic arsenic in drinking water. Brit J Cancer 66(5):888-892

Chen CJ, Chuang YC, Lin TM, Wu HY (1985) Malignant neoplasms among residents of a blackfoot disease-endemic area in Taiwan: high-arsenic artesian well water and cancers. Cancer Res 45(11):5895-5899

Chen Y, Wu F, Graziano JH, Parvez F, Liu M, Paul RR, Shaheen I, Sarwar G, Ahmed A, Islam T, Slavkovich V, Rundek T, Demmer RT, Desvarieux M, Ahsan H (2013) Arsenic exposure from drinking water, arsenic methylation capacity, and carotid intima-media thickness in Bangladesh. Am J Epidemiol 178(3):372-381 doi: 10.1093/aje/kwt001

Chirenje T, Ma LQ, Clark C, Reeves M (2003). Cu, Cr and As distribution in soils adjacent to pressure-treated decks, fences and poles. Environ Pollut 124(3):407-417.

Christoforidou EP, Riza E, Kales SN, Hadjistavrou K, Stoltidi M, Kastania AN, Linos A (2013) Bladder cancer and arsenic through drinking water: a systematic review of epidemiologic evidence. J Environ Sci Health Part A Toxic/Hazard Subst Environ Eng 48(14):1764-1775 doi: 10.1080/10934529.2013.823329

Dauphine DC, Smith AH, Yuan Y, Balmes JR, Bates MN, Steinmaus C (2013) Case-control study of arsenic in drinking water and lung cancer in California and Nevada. Int J Environ Res Public Health 10:3310-24, doi:10.3390/ijerph10083310

Druwe IL, Vaillancourt RR (2010) Influence of arsenate and arsenite on signal transduction pathways: an update. Arch Toxicol 84(8):585-596, doi: 10.1007/s00204-010-0554-4

Enterline PE, Day R, Marsh GM (1995) Cancers related to exposure to arsenic at a copper smelter. Occup Environ Med 52(1):28-32

Erraguntla NK, Sielken RL, Jr., Valdez-Flores C, Grant RL (2012) An updated inhalation unit risk factor for arsenic and inorganic arsenic compounds based on a combined analysis of epidemiology studies. Regul Toxicol Pharm 64(2):329-341 doi: 10.1016/j.yrtph.2012.07.001

Ferreccio C, Smith AH, Duran V, Barlaro T, Benitez H, Valdes R, Aguirre JJ, Moore LE, Acevedo J, Vasquez MI, Perez L, Yuan Y, Liaw J, Cantor KP, Steinmaus C (2013) Case-control study of arsenic in drinking water and kidney cancer in uniquely exposed Northern Chile. Am J Epidemiol 178(5):813-818 doi: 10.1093/aje/kwt05

Freeman GB, Schoof RA, Ruby MV, Davis AO, Dill JA, Liao SC, Lapin CA, Bergstrom PD (1995) Bioavailability of arsenic in soil and house dust impacted by smelter activities following oral administration in cynomolgus monkeys. Fundam Appl Toxicol 28(2):215-222

Ghosh A (2013) Evaluation of chronic arsenic poisoning due to consumption of contaminated ground water in West Bengal, India. Int J Prevent Med 4(8):976-979

Hindmarsh JT and McCurdy RF (1986) Clinical and environmental aspects of arsenic toxicity. Crit Rev Clin Lab Sci 23(4):315-47

Hughes K, Meek ME, Newhook R, Chan PK (1995) Speciation in health risk assessments of metals: evaluation of effects associated with forms present in the environment. Regul Toxicol Pharm 22(3):213-220 doi:10.1006/rtph.1995.0003

Hughes MF (2002) Arsenic toxicity and potential mechanisms of action. Toxicol Lett 133(1):1-16

Hughes MF, Beck BD, Chen Y, Lewis AS, Thomas DJ (2011) Arsenic exposure and toxicology: a historical perspective. Toxicol Sci 123(2):305-332 doi: 10.1093/toxsci/kfr184

James KA, Marshall JA, Hokanson JE, Meliker JR, Zerbe GO, Byers TE (2013) A case-cohort study examining lifetime exposure to inorganic arsenic in drinking water and diabetes mellitus. Environ Res 123:33-38, doi: 10.1016/j.envres.2013.02.005

Juhasz AL, Smith E, Weber J, Rees M, Rofe A, Kuchel T, Sansom L, Naidu R (2007) Comparison of in vivo and in vitro methodologies for the assessment of arsenic bioavailability in contaminated soils. Chemosphere 69:961-966

Kalman DA, Dills RL, Steinmaus C, Yunus M, Khan AF, Prodhan MM, Yuan Y, Smith AH (2013) Occurrence of trivalent monomethyl arsenic and other urinary arsenic species in a highly exposed juvenile population in Bangladesh. J Expo Sci Environ Epidemiol 24(2):113-20, doi: 10.1038/jes.2013.14

Kitchin KT, Wallace K (2008) The role of protein binding of trivalent arsenicals in arsenic carcinogenesis and toxicity. J Inorg Biochem 102(3):532-9, doi: 10.1016/j.jinorgbio.2007.10.021

Lasky T (2013) Arsenic levels in chicken. Environ Health Perspect 121(9):A267, doi: 10.1289/ehp.1307083.

Lee-Feldstein A (1983) Arsenic and respiratory cancer in humans: follow-up of copper smelter employees in Montana. J Natl Cancer Inst 70(4):601-10

Lewis DR, Southwick JW, Ouellet-Hellstrom R, Rench J, Calderon RL (1999) Drinking water arsenic in Utah: A cohort mortality study. Environ Health Perspect 107(5):359-365

Liaw J, Marshall G, Yuan Y, Ferreccio C, Steinmaus C, Smith AH (2008) Increased childhood liver cancer mortality and arsenic in drinking water in northern Chile. Cancer Epidemiol Biomarkers Prevent 17(8):1982-1987, doi: 10.1158/1055-9965.epi-07-2816

Liu CC, Kar S, Jean JS, Wang CH, Lee YC, Sracek O, Li Z, Bundschuh J, Yang HJ, Chen CY (2013) Linking geochemical processes in mud volcanoes with arsenic mobilization driven by organic matter. J Hazard Mater 262:980-988, doi: 10.1016/j.jhazmat.2012.06.050

Lord G, Kim N, Ward NI (2012) Arsenic speciation of geothermal waters in New Zealand. J Environ Monit 14(12):3192-3201, doi: 10.1039/c2em30486d

Marafante E, Vahter M (1987) Solubility, retention, and metabolism of intratracheally and orally administered inorganic arsenic compounds in the hamster. Environ Res 42(1):72-82

McKay CP, Friedmann EI, Gomez-Silva B, Caceres-Villanueva L, Andersen DT, Landheim R (2003) Temperature and moisture conditions for life in the extreme arid region of the Atacama desert: four years of observations including the El Nino of 1997-1998. Astrobiol 3(2):393-406 doi: 10.1089/153110703769016460

Melak D, Ferreccio C, Kalman D, Parra R, Acevedo J, Perez L, Cortes S, Smith AH, Yuan Y, Liaw J, Steinmaus C (2014) Arsenic methylation and lung and bladder cancer in a case-control study in northern Chile. Toxicol Appl Pharmacol 274(2):225-231, doi: 10.1016/j.taap.2013.11.014

Mok WM, Riley JA, Wai CM (1988) Arsenic speciation and quality of groundwater in a lead-zinc mine, Idaho. Water Res 22(6):769-774

Morton W, Starr G, Pohl D, Stoner J, Wagner S, Weswig P (1976) Skin cancer and water arsenic in Lane County Oregon. Cancer 37:2523-2532

Nachman KE, Baron PA, Raber G, Francesconi KA, Navas-Acien A, Love DC (2013) Roxarsone, inorganic arsenic, and other arsenic species in chicken: a U.S.-based market basket sample. Environ Health Perspect 121(7):818-824, doi: 10.1289/ehp.1206245

Nachman KE, Raber G, Francesconi KA, Navas-Acien A, Love DC (2012) Arsenic species in poultry feather meal. Sci Total Environ 417-418:183-188 doi:10.1016/j.scitotenv.2011.12.022

Nahar MN, Inaoka T, Fujimura M (2014a) A consecutive study on arsenic exposure and intelligence quotient (IQ) of children in Bangladesh. Environ Health Prevent Med 19(3):194-199, doi: 10.1007/s12199-013-0374-2

Nahar MN, Inaoka T, Fujimura M, Watanabe C, Shimizu H, Tasnim S, Sultana N (2014b) Arsenic contamination in groundwater and its effects on adolescent intelligence and social competence in Bangladesh with special reference to daily drinking/cooking water intake. Environ Health Prevent Med 19(2):151-158, doi: 10.1007/s12199-013-0369-z

NAS (1977) Arsenic. Medical and Biological Effects of Environmental Pollutants. National Academy of Sciences, Washington DC

Nimick DA, Moore JN, Dalby CE, Savka MW (1998) The fate of geothermal arsenic in the Madison and Missouri Rivers, Montana and Wyoming. Water Resour Res 34:3051-3067

Nordstrom DK, Alpers CN (1999) Negative pH, efflorescent mineralogy, and consequences for environmental restoration at the Iron Mountain Superfund site, California. Proc Natl Acad Sci USA 96(7):3455-62

NRC (1999) Arsenic in Drinking Water. National Research Council, National Academy Press, Washington, DC

Pawlak Z, Rauckyte T, Zak S, Praveen P (2008) Study of arsenic content in mine groundwater commonly used for human consumption in Utah. Environ Technol 29(2):217-224, doi: 10.1080/09593330802028956

Petrick JS, Ayala-Fierro F, Cullen WR, Carter DE, Vasken Aposhian H (2000) Monomethylarsonous acid (MMA(III)) is more toxic than arsenite in Chang human hepatocytes. Toxicol Appl Pharmacol 163(2):203-7

Polissar L, Lowry-Coble K, Kalman DA, Huhges JP, Belle G, Covert DS, Burbacher TM, Bolgiano D, Mottet NK (1990) Pathways of human exposure to arsenic in a communit surrounding a copper smelter. Environ Res 53(1):29-47

Rahman MA, Hasegawa H (2011) High levels of inorganic arsenic in rice in areas where arsenic-contaminated water is used for irrigation and cooking. Sci Total Environ 409(22):4645-4655, doi: 10.1016/j.scitotenv.2011.07.068

Reese RG Jr (1998) Arsenic. In: United States Geological Survey Minerals Yearbook,1998, Fairfax, VA

Rodriguez RR, Basta NT, Casteel SW, Pace LW (1999) An in vitro gastrointestinal method to estimate bioavailable arsenic in contaminated soils and solid media. Environ Sci Technol 33:642-649

Sabbioni E, Fischbach M, Pozzi G, Pietra R, Gallorini M, Piette JL (1991) Cellular retention, toxicity and carcinogenic potential of seafood arsenic. I. Lack of cytotoxicity and transforming activity of arsenobetaine in the BALB/3T3 cell line. Carcinogenesis 12(7):1287-1291

Schoen A, Beck B, Sharma R, Dube E (2004) Arsenic toxicity at low doses: epidemiological and mode of action considerations. Toxicol Appl Pharmacol 198(3):253-267, doi: 10.1016/j.taap.2003.10.011

Schoof RA, Yost LJ, Eickhoff J, Crecelius EA, Cragin DW, Meacher DM, Menzel DB (1999) A market basket survey of inorganic arsenic in food. Food Chem Toxicol 37(8):839-846

Schroeder WH, Dobson M, Kane DM, Johnson ND (1987) Toxic trace elements associated with airborne particulate matter: A review. J Air Pollut Control Assoc 37(11):1267-1285

Senesi GS, Baldassarre G, Senesi N, Radina B (1999) Trace element inputs into soils by anthropogenic activities and implications for human health. Chemosphere 39(2):343-77

Singh AP, Goel RK, Kaur T (2011) Mechanisms pertaining to arsenic toxicity. Toxicol Int 18(2):87-93, doi: 10.4103/0971-6580.84258

Smith AH, Goycolea M, Haque R, Biggs ML (1998) Marked increase in bladder and lung cancer mortality in a region of Northern Chile due to arsenic in drinking water. Am J Epidemiol 147(7):660-669

Smith AH, Marshall G, Yuan Y, Ferreccio C, Liaw J, von Ehrenstein O, Steinmaus C, Bates MN, Selvin S (2006) Increased mortality from lung cancer and bronchiectasis in young adults after exposure to arsenic in utero and in early childhood. Environ Health Perspect 114(8):1293-1296

Smith AH, Steinmaus CM (2009) Health effects of arsenic and chromium in drinking water: recent human findings. Annu Rev Public Health 30:107-122

Smith TJ, Crecelius EA, Reading JC (1977) Airborne arsenic exposure and excretion of methylated arsenic compounds. Environ Health Perspect 19:89-93

Stauffer RE, Thompson JM (1984) Arsenic and antimony in geothermal waters of Yellowstone National Park, Wyoming, USA. Geochim Cosmochim Acta 48:2547-2561

Steinmaus CM, Ferreccio C, Romo JA, Yuan Y, Cortes S, Marshall G, Moore LE, Balmes JR, Liaw J, Golden T, Smith AH (2013) Drinking water arsenic in northern chile: high cancer risks 40 years after exposure cessation. Cancer Epidemiol Biomarkers Prev 22(4):623-630, doi: 10.1158/1055-9965.epi-12-1190

Styblo M, Del Razo LM, Vega L, Germolec DR, LeCluyse EL, Hamilton GA, Reed W, Wang C, Cullen WR, Thomas DJ (2000) Comparative toxicity of trivalent and pentavalent inorganic and methylated arsenicals in rat and human cells. Arch Toxicol 74(6):289-299, doi: 10.1007/s002040000134

Thomas DJ, Styblo M, Lin S (2001) The cellular metabolism and systemic toxicity of arsenic. Toxicol Appl Pharmacol 176(2):127-144, doi: 10.1006/taap.2001.9258

Tseng CH (2002) An overview on peripheral vascular disease in blackfoot disease-hyperendemic villages in Taiwan. Angiology 53(5):529-537

Tseng CH (2005) Blackfoot disease and arsenic: a never-ending story. J Environ Sci Health Part C Environ Carcinogen Ecotoxicol Rev 23(1):55-74, doi: 10.1081/gnc-200051860

Tseng WP (1977) Effects and dose--response relationships of skin cancer and blackfoot disease with arsenic. Environ Health Perspect 19:109-119

Tseng WP, Chu HM, How SW, Fong JM, Lin CS, Yeh S (1968) Prevalence of skin cancer in an endemic area of chronic arsenicism in Taiwan. J Natl Cancer Inst 40(3):453-463

Tsuji JS, Van Kerkhove MD, Kaetzel RS, Scrafford CG, Mink PJ, Barraj LM, Crecelius EA, Goodman M (2005) Evaluation of exposure to arsenic in residential soil. Environ Health Perspect 113(12):1735-1740

USEPA (1986) Inorganic Arsenic Risk Assessment for Primary and Secondary Zinc Smelters, Primary Lead Smelters, Zinc Oxide Plants, Cotton Gins, and Arsenic Chemical Plants. US Environmental Protection Agency, EPA-450/5-85-002

USEPA (1988) Special Report on Ingested Inorganic Arsenic: Skin Cancer; Nutritional Essentiality. US Environmental Protection Agency, EPA-625/3-87-013

USEPA (2001) Fact Sheet about the January 2001 arsenic rule. US Environmental Protection Agency, EPA 815-F-00-015, *http://water.epa.gov/lawsregs/rulesregs/sdwa/arsenic/regulations_techfactsheet.cfm* (accessed July 2014)

USEPA (2014) Inorganic Arsenic -- Toxicological Review for IRIS. US Environmental Protection Agency, *http://yosemite.epa.gov/sab/sabproduct.nsf/fedrgstr_activites/Rev%20Tox%20Review%20Inorg%20Ars enic!OpenDocument&TableRow=2.0#2*, (accessed July 1, 2014)

USFDA (2008) Total Diet Study. U.S. Food and Drug Administration *http://www.fda.gov/Food/ FoodScienceResearch/TotalDietStudy/default.htm* (accessed July 1, 2014)

Vahidnia A, van der Voet GB, de Wolff FA (2007) Arsenic neurotoxicity--a review. Human Exp Toxicol 26(10):823-832, doi: 10.1177/0960327107084539

Wasserman GA, Liu X, Parvez F, Ahsan H, Factor-Litvak P, Kline J, van Geen A, Slavkovich V, Loiacono NJ, Levy D, Cheng Z, Graziano JH (2007) Water arsenic exposure and intellectual function in 6-year-old children in Araihazar, Bangladesh. Environ Health Perspect 115(2):285-289, doi: 10.1289/ehp.9501

Wasserman GA, Liu X, Parvez F, Ahsan H, Factor-Litvak P, van Geen A, Slavkovich V, LoIacono NJ, Cheng Z, Hussain I, Momotaj H, Graziano JH (2004) Water arsenic exposure and children's intellectual function in Araihazar, Bangladesh. Environ Health Perspect 112(13):1329-1333

Welch AH, Helsel DR, Focazio MJ, Watkins SA (1999) Arsenic in ground water supplies of the United States, In: Arsenic Exposure and Health Effects. Chappell WR, Abernathy CO, Calderon RL (eds) Elsevier Science, New York, p 9-17

Welch AH, Westjohn DB, Helsel DR, Wanty RB (2000) Arsenic in ground water of the United States-- occurrence and geochemistry: Ground Water 38(4):589-604

Welch K, Higgins J, Oh M, Burchfiel C (1982) Arsenic exposure, smoking, and respiratory cancer in copper smelter workers. Arch Environ Health 37(6):325-35

WHO (2010) Exposure to Arsenic: A Major Public Health Concern. World Health Organization *http://www.who.int/ipcs/features/arsenic.pdf* (accessed July 2014)

WHO (2011) Guidelines for Drinking-Water Quality. 4th Edition. World Health Organization. *http://www.who.int/water_sanitation_health/publications/2011/dwq_guidelines/en/* (accessed July 2014)

Wu MM, Kuo TL, Hwang YH, Chen CJ (1989) Dose-response relation between arsenic concentration in well water and mortality from cancers and vascular diseases. Am J Epidemiol 130(6):1123-1132

Zavala YJ, Duxbury JM (2008) Arsenic in rice: I. Estimating normal levels of total arsenic in rice grain. Environ Sci Technol 42(10):3856-3860

Zheng Y, Wu J, Ng JC, Wang G, Lian W (2002) The absorption and excretion of fluoride and arsenic in humans. Toxicol Lett 133(1):77-82

Reviews in Mineralogy & Geochemistry
Vol. 79 pp. 451-472, 2014
Copyright © Mineralogical Society of America

Using *In Vivo* Bioavailability and/or *In Vitro* Gastrointestinal Bioaccessibility Testing to Adjust Human Exposure to Arsenic from Soil Ingestion

Nicholas T. Basta

School of the Environment and Natural Resources
The Ohio State University
Columbus, Ohio 43210, U.S.A.

basta.4@osu.edu

Albert Juhasz

Centre for Environmental Risk Assessment and Remediation (CERAR)
University of South Australia
Adelaide, South Australia 5095, Australia

Albert.Juhasz@unisa.edu.au

INTRODUCTION

Remedial investigations (RI) conducted on hazardous waste sites should determine (1) the nature and extent of contamination that exists and (2) the extent to which some level of cleanup must be performed to be protective of human health and the environment. The typical RI includes the collection and chemical analyses of site media, including surface and subsurface soils, surface and groundwater, sediment, and biota (plant and animal species). In some instances, air monitoring may be conducted to determine airborne concentrations of contaminants. An integral component of the RI is the development of the Human Health Baseline Risk Assessment. The risk assessment is the foundation upon which site remediation goals are determined and is developed following two fundamental assessments: a toxicity assessment and an exposure assessment to quantify human intake of contaminated media. Subsequently, by measuring the concentration of chemicals detected in site media, the chemical intake dose can then be quantified to complete the exposure assessment.

Contamination of soil with arsenic (As), and its potential impact on human and environmental health, is a global issue. Although As occurs naturally in soil, enrichment of soil-As may occur as a result of a variety of anthropogenic processes including, but not limited to, pesticide/herbicide manufacture and use, mining, smelting, and wood preservation. Arsenic has been ranked the most common inorganic contaminant found in the National Priority List of Sites in the United States (ATSDR 2011). Numerous health effects are associated with As exposure (Lien et al. 1999; Mandal and Suzuki 2002; ATSDR 2011). For example, acute inorganic As poisoning consists of burning/dryness of the oral and nasal cavities, gastrointestinal distress, and muscle spasms. Chronic As exposure results in depression, fatigue, disruption of red cell production, and various forms of cancer.

Arsenic exposure pathways of concern include consumption of contaminated food and water, inhalation of dust and incidental ingestion of soil. In some well-documented regions (e.g., Bangladesh, West Bengal India), consumption of As-contaminated drinking water and rice represents the major exposure sources and pathways. However, at many As-contaminated sites, drinking water is not the risk driver especially when the source water for drinking pur-

1529-6466/14/0079-0009$05.00

poses is hydrologically disconnected from the contaminated site. In addition, the soil-plant barrier (Chaney and Ryan 1994; Basta et al. 2005) often limits plant uptake of As, which minimizes significant exposure via food. At these sites, incidental ingestion of contaminated soil and dust can comprise a significant risk to animals and humans. This exposure pathway may also be considered important for Cd, Ni, F, Hg, Pb, and possibly other elements (Bradham et al. 2014). Many times, incidental soil ingestion by children is an important pathway in assessing public health risks associated with exposure to As-contaminated soils and is often the risk-driver (Chaney and Ryan 1994; Dudka and Miller 1999). Children are considered the most sensitive receptor due to the propensity of hand-to-mouth contact, which may facilitate increased incidental soil and dust ingestion.

Exposure must be quantified considering the magnitude, frequency, and duration of exposure for the receptors and pathways selected for quantitative evaluation. For incidental ingestion, the following formula (Eqn. 1) can be used to quantify average daily chemical intake of As (USEPA 1989):

$$CDI = \frac{(CS)(IR)(CF)(FI)(EF)(ED)}{(BW)(AT)} \tag{1}$$

where CDI = chemical daily intake of arsenic (mg kg d^{-1}), CS = total arsenic concentration in soil (mg kg^{-1}), IR = ingestion rate of soil (mg soil d^{-1}), CF = conversion factor (10^{-6} kg mg^{-1}), FI = fraction ingestion from contaminated source (unitless), EF = exposure frequency (d yr^{-1}), ED = exposure duration (yr), BW = body weight (kg), and AT = averaging time (period over which exposure is averaged – days).

The underlying assumption in quantifying contaminant intake is that all of the As measured by the total metal analysis is related to the absorbed dose. However, there is an inherent problem with the above assumption. For an adverse health effect to be realized, the chemical toxicant (in this case, As) must be dissolved for absorption to occur from the gastrointestinal (GI) tract across the intestinal epithelium into the systemic circulation (e.g., blood). The underlying assumption in quantifying As intake by the above formula is that all of the As measured by the total As analysis is quantified as the absorbed dose. Several forms of As in soil are not soluble under physiochemical conditions associated with the human GI tract. The combination of various chemical species with different soil/solid matrices of As produces a wide range of As solubility and speciation. For example, the solubility of As_2S_3 in water is 0.005 g L^{-1}, while the solubility of As_2O_3 is 37 g L^{-1}. These differences will have a significant impact on the As dose absorbed from ingestion of contaminated soil. Also, other elements (e.g., Fe, P) or soil chemical factors (organic carbon and clay content) may influence As dissolution and absorption from the GI tract across the intestinal epithelium and into the blood.

As a consequence, most risk from As is associated with the soluble forms of As in the GI tract that are biologically available for absorption across the intestinal epithelium, or "bioavailable" to humans. Bioavailable As is the portion of As dose that enters the systemic circulation from the GI tract (i.e., absorbed dose) from an administered dose. It may be expressed as relative bioavailability i.e., the ratio of the absorbed fraction from the exposure medium (i.e., As-contaminated soil) to the absorbed fraction from the dosing medium used in the critical toxicity study (i.e., sodium arsenate). For incidental ingestion, the above CDI exposure calculation can be modified by incorporating relative bioavailability (RBA) (USEPA 1989, Appendix A) as follows:

$$CDI = \frac{(CS)(IR)(CF)(FI)(EF)(ED)(RBA)}{(BW)(AT)} \tag{2}$$

where RBA = relative bioavailability (unitless, 0.0 to 1.0).

Arsenic *RBA* has been reported to vary from 3 to 100%, but typical values are well below the default values of 100% (Rodriguez et al. 1999; Juhasz et al. 2007; Bradham et al. 2011; Denys et al. 2012). As detailed by the aforementioned authors, most As-contaminated soils have an As *RBA* of <50%. Risk assessment is determined directly from the calculated As *CDI*. Cancer risk can be expressed by the following equation (USEPA 1989, Appendix A):

$$Risk = CDI \times SF \tag{3}$$

where *CDI* = the chemical daily intake, and *SF* = the cancer slope factor.

Non-cancer risk can be calculated as (USEPA 1989, Appendix A):

$$Risk \ (hazard \ quotient) = \frac{CDI}{RfD} \tag{4}$$

where *RfD* = the reference dose.

Thus, As *RBA* *must* be considered when determining *CDI* in an exposure assessment. Otherwise, the *CDI* equation, not modified for As *RBA*, would use 100%, likely overestimate *CDI* and risk, and would result in overly conservative and/or unnecessary remedial soil *cleanup targets*. Recently, USEPA (2012a) has provided guidance allowing the use of a default As *RBA* value of 60%, acknowledging the overly conservative approach by not incorporating As *RBA* (i.e., assuming 100% As *RBA*).

Determining As *RBA* requires knowledge of the amount of soil-As dose that is absorbed into the systemic circulation. To date, the most suitable methods for evaluating As *RBA* involve the measurement of As in urine, blood, and/or feces following the administration of contaminated soil to animals such as swine, monkeys, and mice (Roberts et al. 2002; Juhasz et al. 2007; Bradham et al. 2011; USEPA 2012b). However, *in vivo* assessment of As *RBA* requires ethical considerations. In addition, these methodologies are complicated, expensive, and time consuming.

In order to overcome the difficulty and expense associated with *in vivo* trials, research efforts have been directed toward the development of *in vitro* methods to simulate human GI conditions. In this approach, the amount of As dissolved by the *in vitro* GI-extraction method is determined and termed "bioaccessible" As. *In vivo* vs. *in vitro* regression equations are determined from studies that measure As *RBA* and As *in vivo* bioaccessibility (As *IVBA*). These equations are then used to predict As *RBA* from solely laboratory *in vitro* extractions (i.e., *IVBA*). Regardless of the *in vitro* method utilized, As *IVBA* must be well correlated with As *RBA* in order to provide accurate As *RBA* predictions for refining human health exposure assessment.

In this chapter, we review the state of the science of *in vivo* and *in vitro* methods for measuring and predicting soil As *RBA*. The importance and implications of including As *RBA* for refining exposure and characterizing risk at As-contaminated sites will also be discussed.

METHODS FOR DETERMINING BIOAVAILABILITY OF As FROM SOIL INGESTION: *IN VIVO* MODELS

A number of *in vivo* models have been utilized for the assessment of As *RBA* for refining human exposure to contaminated soil via the incidental ingestion pathway (Table 1). Although dogs, rabbits, and rats may be utilized, the majority of As *RBA* studies have been conducted using swine, monkey, and mouse models. Due to their use as a model for humans in preclinical pharmacokinetic studies (Bergeron et al. 2000), monkeys (e.g., *Cebus*, *Cynomolgus*) have been utilized for the assessment of As *RBA* in contaminated soils. Although urinary and fecal

Table 1. Animal models utilized for the assessment of As
relative bioavailability in contaminated soil.

Animal model	Biomarkers of As exposure	Reference
Dogs (Beagles)	Single dose urinary excretion	Groen et al. 1994
Rabbits (New Zealand White)	Single dose urinary excretion	Freeman et al. 1993
Rats (Wistar)	Single dose urinary excretion	Ng et al. 1998; Ellickson et al. 2001
Monkeys (*Cebus, Cynomolgus*)	Single dose urinary excretion	Freeman et al. 1995; Roberts et al. 2002, 2007; USEPA 2009
Swine (Large White)	Steady state urinary excretion	Rodriguez et al. 1999; Casteel et al. 2005, 2009a,b,c, 2010a,b,c; Basta et al. 2007; Denys et al. 2012; Brattin and Casteel 2013
	Single dose blood *AUC*	USEPA 1996; Juhasz et al. 2007, 2008
Mice (C57BL/6)	Steady state urinary excretion	Bradham et al. 2011

excretion of sodium arsenate following oral administration in these species is consistent with observations in humans (Bettley and O'Shea 1975; Pomroy et al. 1980), the prohibitive costs associated with these animal models limit their use. While rats have been used in a variety of medical and environmental applications, their use as an animal model for the assessment of As *RBA* has been limited due to differences in their As-distribution patterns compared to humans (Kelley et al. 2002), differences in intestinal morphology (Patterson et al. 2008), and issues associated with coprophagy. Coprophagy has also been identified as an issue associated with the use of rabbits, in addition to the animal being a hind-gut fermenter, while physiological differences between dog and human—such as gastric pH, gastric emptying, and intestinal transit time (Parrott et al. 2009)—have limited their use for contaminant *RBA* assessment. Although the metabolism of As differs in mice and humans, it has been suggested that similarities are sufficient to create physiologically-based pharmacokinetic models that can be adopted for humans (El-Masri and Kenyon 2008; Evans et al. 2008). Mice offer advantages over other animal models as they are cost effective (i.e., low purchase and husbandry costs), easy to handle, and have the potential to be utilized in numerous laboratories (Bradham et al. 2011). In addition, due to their low cost, larger assay sample sizes can be accommodated that may increase the robustness of As-*RBA* data. To date, swine are the preferred animal model for the assessment of As *RBA* due to their similarity to humans in terms of metabolism and excretion of As, bone development, and mineral metabolism (Weis and La Velle 1991; Patterson et al. 2008). Immature swine share a number of similarities to young children, including gastrointestinal absorption, body weight, and physiologic age. In addition, swine can be trained to ingest contaminated soil, repeat blood samples can be collected, and cross-over studies are possible thereby reducing the variability and the number of animals required (Rees et al. 2009).

When determining As *RBA*, two common approaches have been utilized as the biomarker for As exposure irrespective of animal model adopted: assessment of As urinary excretion or blood As concentration. Similarly, irrespective of the biomarker of As exposure, As *RBA* is calculated by comparing As urinary excretion or area under the blood As concentration time curve (*AUC*) of the contaminated soil to that of sodium arsenate (reference material).

When utilizing As-urinary excretion as the biomarker of As exposure, the urinary-excretion fraction is calculated by dividing the amount of As excreted in the urine by the dose of As administered (Eqn. 5). This may be calculated for single-dose administration (Roberts et al. 2002, 2007; USEPA 2009) or following repeated As doses to achieve steady-state urinary excretion (Rodriguez et al. 1999; Casteel et al. 2005, 2009a,b,c, 2010a,b,c; Basta et al. 2007; Bradham et al. 2011; Denys et al. 2012; Brattin and Casteel 2013). For example, Roberts et al. (2002) collected urine for up to 4 days following the administration of a single oral dose of As-contaminated soil or sodium arsenate to *Cebus* monkeys. In contrast, Bradham et al. (2011) utilized a steady-state dosing approach whereby mice were exposed to As-contaminated soil or sodium arsenate incorporated in feed for a period of 9 days with a 10-day urine collection period. Arsenic *RBA* is then calculated as the ratio of urinary excretion factors when As was administered in the test material (contaminated soil) vs. the reference material (sodium arsenate) (Eqn. 6):

$$UEF = \frac{U_{As}}{D_{As}} \tag{5}$$

where UEF = urinary excretion factor, U_{As} = amount of As excreted in the urine, and D_{As} = dose or As administered.

$$BA = \frac{UEF_{TM}}{UEF_{RM}} \tag{6}$$

where UEF_{TM} = urinary excretion factor for the test material (contaminated soil), and UEF_{RM} = urinary excretion factor for the reference material (sodium arsenate).

When estimating As *RBA* following single-dose administration and blood As analysis, area under the blood As-concentration time curve (*AUC*) is calculated (estimated using a geometric approximation such as the trapezoid rule) for both test and reference materials. Arsenic *RBA* is calculated as the ratio of *AUC* when As was administered in the test material (contaminated soil) versus the reference material (sodium arsenate) following dose normalization (Eqn. 7):

$$RBA = \frac{AUC_{TM}}{AUC_{RM}} \times \frac{D_{RM}}{D_{TM}} \tag{7}$$

where AUC_{TM} = area under the As concentration time curve for the test material (contaminated soil), AUC_{RM} = area under the As concentration time curve for the reference material (sodium arsenate), D_{RM} = dose of As administered in the reference material, and D_{TM} = dose of As administered in the test material.

In addition to the aforementioned, As *RBA* may be determined following administration of multiple test and reference material doses. When monitoring urinary As concentrations, urinary excretion factors are estimated as the regression slope of the relationship between As excreted and dose (Denys et al. 2012). Similarly, when assessing As *RBA* using the *AUC* approach, the *AUC* divided by dose ratio is estimated as the regression slope of the relationship between the blood *AUC* and dose.

Both approaches for estimating As *RBA* offer advantages and disadvantages. Monitoring urinary excretion of As is less sample intensive and does not require surgical procedures to be undertaken on the animal. However, the determination of urinary-excretion factors will underestimate As absorption, as a proportion of absorbed As is excreted in the feces (via billiary excretion) and compartmentalized into keratin-rich tissue (i.e., hair, nails) (Csanaky and Gregus 2002). In addition, for larger-animal models, metabolic cages are unable to separate urine and fecal material, and as a consequence, As excretion may be overestimated through fecal contamination. In contrast, an advantage of the *AUC* approach is that it determines the amount of As that is absorbed into the systemic circulation. However, in order to accomplish this, larger animals (i.e., with a sufficient blood volume for multiple samples) require a surgical procedure to insert jugular catheters for routine blood sampling (see Rees et al. 2009).

For animal models with low blood volume (i.e., mice), multiple animals may be required at each time point in order to construct a pseudo-*AUC* that may introduce additional variability into *AUC* calculations due to intra-species differences in As absorption. In addition, multiple samples are required to generate data for *AUC* determination, which increases the analytical component of As *RBA* calculations.

Irrespective of the methodology utilized for the assessment of As *RBA*, there are limitations and uncertainties associated with the prediction of As *RBA* for humans. Indeed to date, there have been no quantitative comparisons of As *RBA* between humans and either animal model, although it has been suggested that large differences between mammalian species is unlikely (USEPA 2012b). Similarly, limited data are available regarding comparative As *RBA* studies involving different animal models, different biomarkers of As exposure, and single dose versus steady state models. Further details regarding this uncertainty are discussed later in the text (see 'Comparison of animal models for the determination of As relative bioavailability'). In addition, most As *RBA* studies administer As-contaminated soil at dose levels (up to 3500, 2970, and 1650 mg soil per kg body weight (bw) for swine, monkeys, and mice, respectively) that are much higher than typical ingestion rates in children (100 mg d^{-1}) or adults (50 mg d^{-1}) (USEPA 2008). In some cases (i.e., single-dose experiments), elevated soil doses are required due to constraints associated with detection limits during the analysis of As in blood and urine. However, the implications of administering elevated doses (i.e., impact of dose dependency on As *RBA*) and the subsequent extrapolation of As *RBA* data from *in vivo* model to humans are yet to be investigated.

Measurement of As relative bioavailability in contaminated soils

Table 2 and Figure 1 provide a summary of As *RBA* data following the *in vivo* assessment of As-contaminated soils. For comparative purposes, As *RBA* data were subdivided into animal model utilized (swine, monkey, mouse), biomarker of exposure (urine, blood), and source of As contamination (mine-, pesticide-impacted). As detailed in Table 2, a large proportion of studies undertaken to date have assessed As *RBA* in mine-impacted soils (encompassing mining and

Table 2. Summary of studies detailing the assessment of As relative bioavailability in mine site and pesticide-impacted soils.

Animal model	Biomarker of exposure	Source material	# of soils	Soil As concentration (mg kg^{-1}) Min.	Med.	Max	As relative bioavailability (%) Min.	Med.	Max.	Ref.
Swine	Urine	Mine site	59	20	367	25000	3.0	33.0	100	[1]
		Pesticide	5	290	364	388	31.0	47.0	53.0	[2]
	Blood	Mine site	6	88	692	10100	6.9	28.6	78.0	[3]
		Pesticide	16	42	262	1221	6.8	17.3	74.7	[4]
Monkey	Urine	Mine site	6	300	530	1492	5.0	17.5	20.1	[5]
		Pesticide	18	101	334	1412	5.0	19.5	38.0	[6]
Mouse	Urine	Mine site	9	183	829	4495	11.2	42.1	51.6	[7]
		Pesticide	6	322	392	769	20.9	29.8	35.2	[8]

References: [1] Rodriguez et al. (1999); Casteel et al. (2005, 2009b, 2010a,b,c); Basta et al. (2007); USEPA (2010); Denys et al. (2012); [2] Casteel et al. (2009a,c, 2010b); USEPA (2010); [3] USEPA (1996); Juhasz et al. (2007); [4] Juhasz et al. (2007); Juhasz and Smith (2013); [5] Freeman et al. (1995); Roberts et al. (2007); [6] Roberts et al. (2002, 2007); USEPA (2009); [7] Bradham et al. (2011); [8] USEPA (2012b)

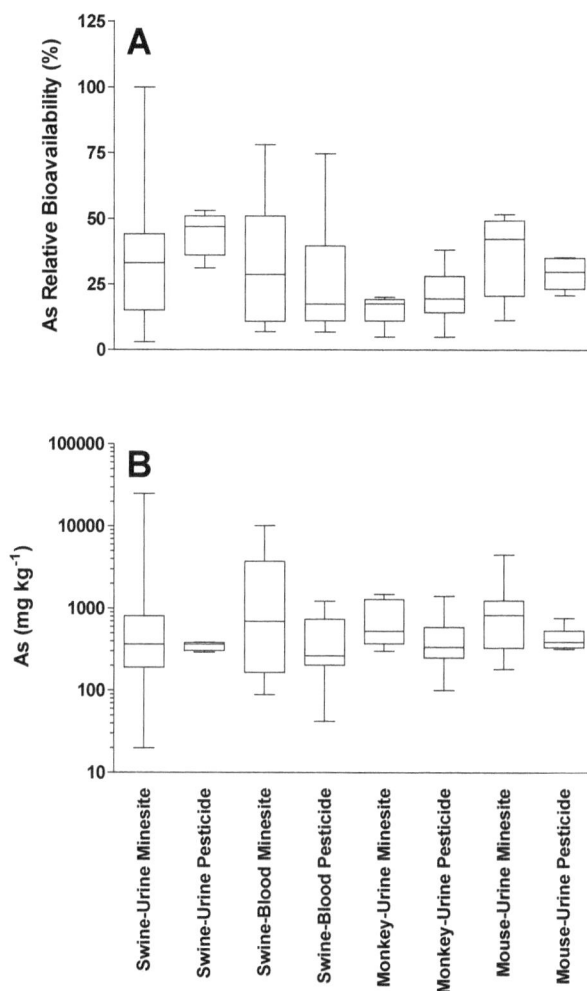

Figure 1. Box plot of As *RBA* data (A) and corresponding soil As concentrations utilized (B). For comparative purposes, As *RBA* data was subdivided into animal model utilized (swine, monkey, mouse), biomarker of exposure (urine, blood), and source of As contamination (mine-, pesticide-impacted). Each box represents the lower and upper quartiles, while the band within the box represents the median As *RBA* value or soil As concentration for the respective subdivision. Whiskers represent minimum and maximum values for each subdivision.

smelting contaminated soils), while fewer As *RBA* studies have focused on pesticide-impacted soils (encompassing orchard, railway corridor, and pesticide manufacturing soils).

Arsenic *RBA* varies considerably irrespective of animal model, biomarker of exposure, and contaminant source. Lower As *RBA* values have been measured for mine- and pesticide-impacted soils using monkey (maximum = 20.1 and 38%, respectively) and mouse models (maximum = 51.6 and 35.2%, respectively), which may be reflective of the source materials' soil properties and As mineralogy that influence As absorption. For example, as depicted in Figure 1A, As *RBA* for mine-impacted soil determined using swine urinary As excretion ranged from 3 to 100% with a median value of 33% (n = 59). Similarly, using swine and *AUC* estimates, As *RBA* ranged from 6.9 to 78% (median = 28.6%; n = 6) and 6.8 to 74.7% (median = 17.3%; n = 16) for mine- and pesticide-impacted soils, respectively (Table 2; Fig. 1A).

Comparison of animal models for the determination of As relative bioavailability

Although As *RBA* may be determined using a number of animal models, limited information is available regarding comparative determinations of the same sample using multiple *in vivo* assays. As detailed in USEPA (2012b), As *RBA* in four pesticide-impacted soils (290-388 mg As kg^{-1}) were compared following administration to swine, monkey, and mouse and monitoring using steady state As urinary excretion. For soil MS1, As-*RBA* data were congruent (P > 0.05) between the three animal models (Fig. 2). However, for the remaining soils, there were significant differences (P < 0.05) between measured As-*RBA* values for swine and monkey (MS4), swine and mouse (MS5), and swine and monkey / mouse (Fig. 2). In three of the four soils analyzed, the swine model provided the most conservative measurement of As *RBA*; while for three of the four soils, there was no significant difference (P > 0.05) in As-*RBA* values derived using monkey and mouse models. For a larger data set, comparative values are available for swine and mouse As-*RBA* estimates (USEPA 2012b; n = 11 including the 4 pesticide-impacted soils detailed above). For 5 of the 11 soils (including 2 NIST reference materials), the 95% confidence limits overlapped; however, for the remaining 6 soils, assessment using the swine assay provided As-*RBA* values that were 1.4- to 2.3-fold higher compared to the mouse assay.

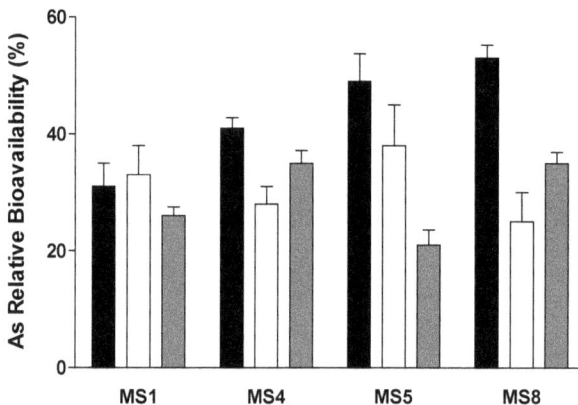

Figure 2. Comparison of As *RBA* values (mean and standard error) for four pesticide-impacted soils determined using swine (■), monkey (□), and mouse (■) *in vivo* models and steady state As urinary excretion (data from USEPA 2012).

USING *IN VITRO* GASTROINTESTINAL BIOACCESSIBILITY METHODS TO PREDICT BIOAVAILABILITY

To overcome the difficulty and expense associated with *in vivo* trials, research efforts over the past two decades have been directed toward the development of *in vitro* methods to predict *RBA* for As and other contaminants of concern (e.g., Pb). As the GI tract is an extremely complex system, *in vitro* methods do not replicate GI tract conditions but aim to mimic important biochemical parameters known to influence the release of contaminants from the soil matrix. As a consequence, these assays determine contaminant concentrations that are solubilized following GI extraction and are therefore potentially available for absorption into the systemic circulation (termed bioaccessibility). *In vitro* bioaccessibility methodologies (*IVBA*) were initially developed to predict Fe absorption and to evaluate Fe nutrition of food (Crews et al. 1983). These *IVBA* methods were then adapted to evaluate Pb bioaccessibility by Ruby et al. (1992) with subsequent development for other contaminants of concern. *IVBA* methodologies may include gastric extraction alone or sequential gastric to intestinal extraction with each phase considering a number of key GI tract physiological factors including:

(1) *pH*: The GI tract pH has been shown to be especially important in its effect on contaminant dissolution (Meunier 2011). In humans, stomach pH conditions range from 1.5 to 2.5 under fasting conditions and increase to pH 4 under fed conditions (Malagelada et al. 1976). In addition, small intestine pH conditions vary from 5.5 to 6.5 in the duodenum and jejunum, respectively, and from 6.5 to 7.5 in the ileum. A number of *in vitro* studies have illustrated the importance of gastric phase pH in influencing As *IVBA* results. *IVBA* methodologies have chosen a gastric phase pH value that represents a worst case scenario (fasted state) for young children. Low-pH stomach values are particularly prudent for the assessment of As *IVBA* (and other inorganic contaminants) as the pH will drive the dissolution of As and As mineral phases, thereby controlling the fraction that is potentially available for uptake. As detailed in Table 3, commonly used As *IVBA* methodologies employ a gastric pH of 1.5-2.5.

(2) *Chemical composition of GI solutions*: The chemical composition of *IVBA* solutions range from simple gastric phase systems that contain limited constituents (e.g., 0.4 M glycine) to highly complex solutions that contain multiple organic and inorganic components (see Table 3). Pepsin and glycine are base constituents in gastric phase solutions, while bile and pancreatin are often added when modifying solution conditions from the gastric to the intestinal phase.

(3) *Presence of food*: Food may be added to *IVBA* assays for comparison of As bioaccessibility between fed and fasted states. Due to the influence of food addition on stomach pH values, As *IVBA* values may decrease as a result of these amendments (Ruby et al. 1996). However, phosphate also *increases* bioaccessible As via desorption from soil Fe, Al, and Mn oxides (Rodriguez et al. 2003), which would promote increased As absorption and bioavailability. Human diets are rich in phosphate (Basta et al. 2007); thus food will affect As *RBA*. *In vivo* research is needed to determine whether dietary phosphate increases or decreases As *RBA*. The Oklahoma State University *in vitro* gastrointestinal extraction (*IVG*) method incorporated the animal feed dosing vehicle used in juvenile swine As *RBA* studies. Other *IVBA* methods, such as the Ohio State University *in vitro* method (OSU-*IVG*) removed the dosing vehicle (i.e., dough ball) for As *IVBA* measurement (Basta et al. 2007).

(4) *Rate of emptying of the stomach and transit time in the small intestine*: Small variations in extraction times are seen between *IVBA* methodologies (Table 3). However, gastric and GI extraction time frames are reflective of nutrition studies that have demonstrated that stomach emptying occurs after 1-2 h, while between 3 and 5 h is required for constituents to pass from the small to the large intestine.

(5) *Soil/solid:gastro(intestinal) liquid ratio*: The soil:liquid ratio has the potential to influence *IVBA* results due to its impact on dissolution kinetics. At small soil:liquid ratios, As *IVBA* may be underestimated as a result of solubility issues arising from diffusion-limited dissolution kinetics (Ruby et al. 1992). However, as demonstrated by Hamel et al. (1998), metal *IVBA* values may not vary significantly when soil:liquid ratios vary from 1:100 to 1:5000 (g mL^{-1}).

(6) *Redox*: The solubility of As can be greatly influencing by solution redox. The stomach and small intestine have aerobic to mildly reducing (>300 mV) conditions. Microbial activity in the colon has strongly sulfate reducing conditions. Absorption of arsenic from the GI tract across the gut epithelium into system circulation (i.e., bioavailability) occurs in the small intestine not the colon. Therefore, redox conditions influencing bioavailability (i.e., stomach, small intestine) are aerobic to mild reducing.

Table 3. Key physiochemical conditions of selected As *IVBA* methods.

Method	Gastric conditions				Intestinal conditions				Ref.
	S:S ratio	Constituents (g L⁻¹)	Time (h)	pH	S:S ratio	Constituents (g L⁻¹)	pH	Time (h)	
OSU *IVG*	1:150	10 g pepsin, 8.77 g NaCl	1	1.8	1:150	0.56 g bile, 056 g pancreatin	6.1	2	[1]
SBRC	1:100	30.03 g glycine	1	1.5	1:100	1.75 g bile, 0.5 g pancreatin	7.0	4	[2]
USEPA 9200	1:100	30.03 g glycine	1	1.5	N/A	—	—	—	[3]
RBALP	1:100	30.03 g glycine	1	1.5	N/A	—	—	—	[4]
PBET	1:100	1.25 g pepsin, 0.5 g sodium malate, 0.5 g sodium citrate, 420 µl lactic acid, 500 µL acetic acid	1	2.5	1:100	1.75 g bile, 0.5 g pancreatin	7.0	4	[5]
DIN	1:50	1 g pepsin, 3 g mucin, 2.9 g NaCl, 0.7 g KCl, 0.27 g KH_2PO_4	2	2.0	1:100	9.0 g bile, 9.0 g pancreatin, 0.3 g trypsin, 0.3 g urea, 0.3 g KCl, 0.5 g $CaCl_2$, 0.2 g $MgCl_2$	7.5	6	[6]
UBM	1:37.5	Saliva (pH 6.5±0.5): 0.896 g KCl, 0.888 g NaH_2PO_4, 0.2 g KSCN, 0.57 g Na_2SO_4, 0.298 g NaCl, 1.8 mL of 1 M NaOH, 0.2 g urea, 0.145 g amylase, 0.05 g mucin, 0.015 g uric acid Gastric Phase (pH 0.9-1.0): 2.752 g NaCl, 0.266 g NaH_2PO_4, 0.824 g KCl, 0.4 g $CaCl_2$, 0.306 g NH_4Cl, 8.3 mL of 37% HCl, 0.65 g glucose, 0.02 mg glucuronic acid, 0.085 g urea, 0.33 g glucosaminehydrochloride, 1 g bovine serum albumin, 3 g mucin, 1 g pepsin	1	1.2	1:100	Duodenal phase (pH 7.4±0.2): 7.012 g NaCl, 5.607 g $NaHCO_3$, 0.08 g KH_2PO_4, 0.564 mg KCl, 0.05 g MgCl2, 0.18 mL of 37% HCl, 0.1 g urea, 0.2 g $CaCl_2$, 1 g bovine serum albumin, 3 g pancreatin, 0.5 g lipase Bile phase (pH8.0±0.2): 5.259 g NaCl, 5.785 g $NaHCO_3$, 0.376 g KCl, 0.18 mL of 37% HCl, 0.25 g urea, 0.222 g $CaCl_2$, 1.8 g bovine serum albumin, 6 g bile	6.5	4	[7]

References: [1] Basta et al. (2007); [2] Kelley et al. (2002); [3] Bradham et al. (2011); [4] Brattin et al. (2013); [5] Ruby et al. (1996); [6] Din (2000); [7] Wragg et al. (2011)

Factors influencing As bioaccessibility and relative bioavailability

The bioavailability of As in soil can be simply divided into two kinetic steps: (1) As dissolution in GI fluids and (2) absorption across the GI epithelium into the blood stream (Fig. 3). Therefore, combining the variability of geochemical forms of As in soil and solid wastes with dissolution chemistry and biological absorption processes in the GI tract results in an extremely complex system.

Factors that affect the rate of As dissolution in the GI tract, the first reaction in Figure 3, will affect measured *IVBA*. As shown in Figure 4, As *IVBA* values have been reported to range from <1 to >90%. The reason for the wide range in As *IVBA* values include differences in As speciation and mineralogy in addition to variability in other soil constituents that may influence As dissolution in the gastric phase or As precipitation in the intestinal phase. Meunier et al. (2010) reported a wide range in As *IVBA* values for seven As minerals. High-As *IVBA* was associated with Ca-Fe arsenate minerals, whereas low As *IVBA* was associated with scorodite or arsenopyrite (Meunier et al. 2010; Meunier 2011). Yang et al. (2002), Rodriguez et al. (2003), Meunier et al. (2010), Meunier (2011), and Whitacre (2013) reported As associated with reactive Fe-oxide minerals had low bioaccessibility. Beak et al. (2006) found Fe-oxide surfaces in ferrihydrite greatly reduced As *IVBA* (to <5%) as a result of strong binuclear-bidentate bonding with the Fe-oxide surface, determined using extended X-ray absorption fine structure X-ray absorption near-edge spectroscopy. The presence of soluble phosphate can also increase As *IVBA*

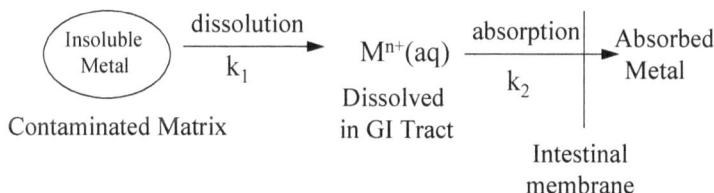

Figure 3. Simplified kinetic rate-limiting steps in the gastrointestinal trace affecting contaminant bioavailability.

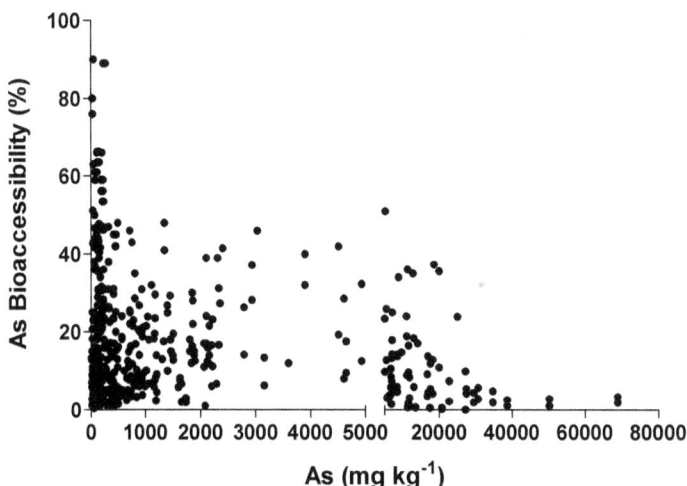

Figure 4. Range in As *IVBA* values following the *in vitro* assessment of As-contaminated soil.

presumably through desorption of As from Fe oxide surfaces (Rodriguez et al. 2003; Basta et al. 2007). However, the presence of phosphate may alter arsenate absorption due to competition between arsenate and phosphate for sodium-coupled phosphate transporters in the gastrointestinal barrier (Eto et al. 2006; Villa-Bellosta and Sorribas 2009). As a result, increased phosphate should reduce As absorption and *RBA*, which is contradictory to *IVBA* results.

Reducing conditions can greatly affect As dissolution and As *IVBA*. Ascorbic acid, a strong reducing agent, is excreted by the human stomach lining at 0.42 to 1.65 mg h⁻¹ (O'Connor et al. 1989). Whitacre et al. (2014) reported that addition of physiological rate of 100 mM ascorbic acid lowers the redox of the *in vitro* GI solutions and can double the concentration of As and As *IVBA* compared to the OSU *IVG* and SBRC methods *IVBA* without ascorbic acid addition. Addition of ascorbic acid lowered the redox of the OSU *IVG in vitro* solution from 350-390 mV to 290-320 mV. These redox values are in the vicinity of the Fe and As redox couples so a small decrease in redox resulted in reduction of As(V) to As(III) and Fe(III) to Fe (II) with a significant increase in dissolved and As *IVBA* (Whitacre et al. 2014).

Due to differences in *IVBA* operational parameters (e.g., pH, solution constituents), analysis of the same As-contaminated soil with different *IVBA* methods may result in different As *IVBA* values. As illustrated in Figure 5, As *IVBA* values can vary within the methodology depending on the *in vitro* phase utilized in addition to between methodologies. Although a number of researchers have observed within and between assay differences when assessing As *IVBA* (Rodriguez et al. 1999; Oomen et al. 2002; Van de Wiele et al. 2007; Juhasz et al. 2009), the suitability of *IVBA* methods to act as a surrogate for As *RBA* lies in its ability to predict *in vivo* values (i.e., the goodness of fit of the As *RBA-IVBA* correlation).

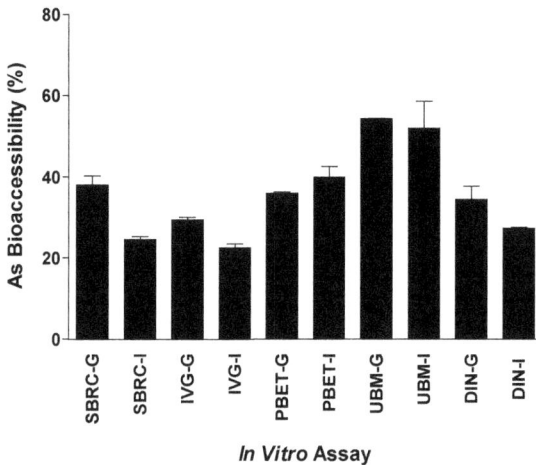

Figure 5. Difference in As *IVBA* values following assessment of an As-contaminated soil (herbicide impacts) with five *IVBA* methodologies (data from Juhasz et al. 2007, 2009).

Correlation of As relative bioavailability and As bioaccessibility

In order for *in vitro* assays to act as a surrogate measurement of As *RBA*, the relationship between As *RBA* and As *IVBA* needs to be established. According to EPA Guidance (USEPA 2007a):

"In the case that a 'validated' in vitro method is used to estimate bioavailability, it is recommended that the protocol specified in the methodology be followed for making the

extrapolation from in vitro data to in vivo values. That is, there is no a priori assumption that all validated in vitro methods must yield results that are identical to in vivo values. Rather, it is assumed that a mathematical equation will exist such that the in vitro result (entered as input) will yield an estimate of the in vivo value (as output)."

For *in vivo–in vitro* comparisons (IVIVC), As *RBA* is expressed as a function of As *IVBA* which is expressed as a percentage as follows:

$$\text{As } IVBA \ (\%) \ = \ \frac{bioaccessible \text{ As } (mg \ kg^{-1})}{total \text{ As } content \ (mg \ kg^{-1})} \times 100 \tag{8}$$

A successful IVIVC, as defined by USEPA (2007a) is the OSU-*IVG* method of Rodriguez et al. (1999) replotted with data from Basta et al. (2007) (Fig. 6). Arsenic *RBA* can be determined using the mathematically-simple linear regression in Figure 6 and meet the requirement of USEPA (2007a) where "the *in vitro* result (entered as input) will yield an estimate of the *in vivo* value (as output)." Several other *in vitro* methods have also met and exceeded the USEPA (2007a) guidance requirements of a successful IVIVC. These include the Solubility Bioaccessibility Research Consortium (SBRC) method (Juhasz et al. 2007, 2009, 2011), USEPA method 9200 (Bradham et al. 2011), the UBM method (Wragg et al. 2011; Denys et al. 2012), and the *RBA*LP method (Brattin et al. 2013). The most important criteria to select an *in vitro* method is a successful *in vivo–in vitro* comparison (IVIVC). *In vitro* methods have not been evaluated against *in vivo* data are not acceptable for human risk assessment.

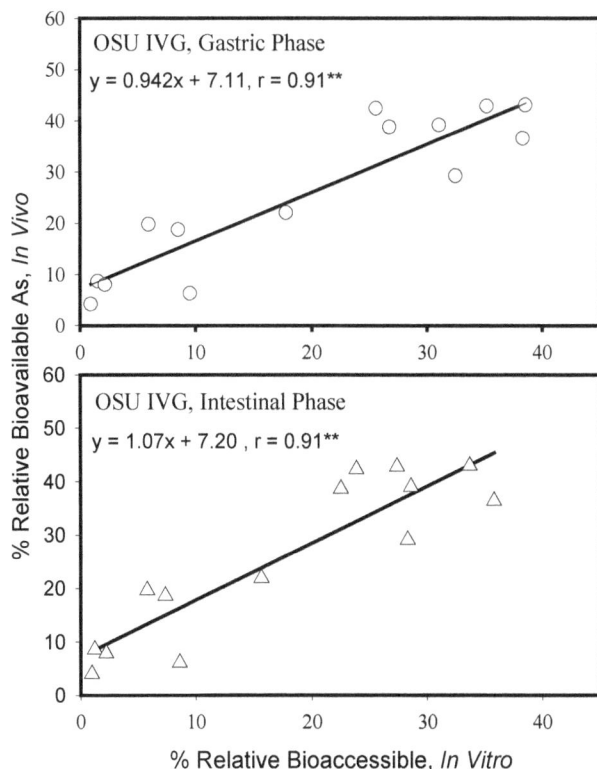

Figure 6. *In vivo in vitro* correlation (IVIVC) for juvenile swine As *RBA* vs. OSU-*IVG* As *IVBA* (Basta et al. 2007).

USEPA (2007a) does not define the criteria for the "goodness of fit" parameters for the IVIVC regression. However, Wragg et al. (2011) reported goodness of fit parameters adopted from guidance developed by the U.S. Department of Health and Human Services Food and Drug Administration (1997). The criteria below may be applied to IVIVC and validation studies. Criteria include:

- a linear relationship between *in vivo* and *in vitro* data with a correlation coefficient of (r) >0.8 and a slope >0.8 and <1.2 (for the initial correlation and subsequent validation data sets, respectively);

- a within-laboratory repeatability of ≤10% RSD (for *in vivo* and *in vitro* assays); and

- a between-laboratory reproducibility of ≤20% RSD (for *in vivo* and *in vitro* assays).

A number of studies have detailed As IVIVC. The OSU-*IVG* method was the first to report that an inexpensive *IVBA* method can be used to predict As *RBA* and risk to children from ingestion of As-contaminated soil (Rodriguez et al. 1999). This method incorporated the animal feed dosing vehicle used in the juvenile swine As *RBA*. OSU-*IVG* removed the dosing vehicle and made slight changes to conditions of the original OSU-*IVG* (Basta et al. 2007). Juhasz et al. (2007) reported the SBRC method, described by Kelley et al. (2002), was strongly correlated with As *RBA*. The physiochemical conditions of this method are the basis for the Relative Bioavailability Leaching Procedure (RBALP) method of Drexler and Brattin (2007). Likewise, Pb *IVBA* measured by RBALP was strongly correlated with Pb *RBA* determined using the juvenile swine model (Drexler and Brattin 2007). The RBALP has gained guidance from USEPA to be used for adjusting the oral bioavailability of Pb in soil for use in human health risk assessment (USEPA 2007b). Bradham et al. (2011) reported the USEPA Method 9200, the same method as SBRC, was strongly correlated with As *RBA*. Similarly, Brattin et al. (2013) reported that the RBALP method, which has the same physiochemical conditions as SBRC, was strongly correlated with As *RBA*. The Unified Bioaccessibility Method (UBM), developed by the Bioaccessibility Research Group of Europe (BARGE), is a more complex *IVBA* method capable of predicting As *RBA* in addition to Pb and Cd *RBA* (Wragg et al. 2011; Denys et al. 2012).

As detailed in Table 4, the aforementioned studies have developed IVIVC utilizing data generated from swine, monkey, or mouse assays and a variety of bioaccessibility methodologies with the goodness of fit (r^2) varying from poor (0.14) to excellent (0.99). For some studies where multiple *in vitro* assays were utilized (e.g., Juhasz et al. 2009), comparison of *in vivo* and *in vitro* results indicated that the correlation between As *RBA* and As *IVBA* was variable depending on the methodology and phase utilized. In the study of Juhasz et al. (2009), the slope of the IVIVC varied significantly, particularly for SBRC and PBET assays when extraction conditions were modified from the gastric to the intestinal phase. For example, for the SBRC assay, the slope of the relationship increased from 0.99 to 1.64; while for the PBET assay, the slope increased from 1.16 to 1.76. A common misconception in the absence of the IVIVC is that the relationship (or slope of the line) between As *RBA* and As *IVBA* is 1. For some linear regression models this is true (see Table 4); however, for others (notably SBRC-I and PBET-I from Juhasz et al. 2009), this assumption would result in an underestimation of As *RBA*. Conversely, assuming a linear regression slope of 1 would result in a conservative prediction of As *RBA* based on the *in vivo* (monkey)–*in vitro* (SBRC-I) correlation of Brattin et al. (2013). As a consequence, an understanding of the IVIVC and how it was derived is critical when utilizing *in vitro* data as a surrogate for As *RBA* for refining human health exposure assessment. In addition to the slope of the IVIVC, another parameter to consider is the *y*-intercept. For some linear regression models, the *y*-intercept was small (Rodriguez et al. 1999; Basta et al. 2007; Juhasz et al. 2007, 2009; Denys et al. 2012); however, for others, large *y*-intercepts were calculated (Juhasz et al. 2009; Brattin et al. 2013), suggesting issues with predictive capabilities at low As *RBA* values. Similar problems with large *y*-intercepts were reported for soils contaminated by gold mining activities (Basta et al. 2014). The As *RBA* was

Table 4. Linear regression models for predicting As relative bioavailability using *in vitro* assays.

Study	# of soils	Source	As (mg kg⁻¹)	Animal model	Biomarker	In vitro assay	Phase	Slope	y-intercept	r²
Rodriguez et al. (1999)	14	Mine	233–17500	Swine	SSUE[a]	IVG	Gastric	0.91	7.37	0.85[kl]
						IVG	Intestinal	1.02	7.55	0.82[kl]
Basta et al. (2007)	9	Mine	405–17500	Swine	SSUE[a]	IVG	Gastric	1.09	9.48	0.85[l]
						IVG	Intestinal	0.99	7.79	0.92[l]
Juhasz et al. (2007)	12	Mine, pesticide	42–1114	Swine	AUC[b]	SBRC	Gastric	0.99	1.69	0.75
Juhasz et al. (2009)						SBRC	Intestinal	1.64	5.63	0.65
						IVG	Gastric	0.85	14.32	0.57
						IVG	Intestinal	1.11	13.97	0.57
						PBET	Gastric	1.16	10.10	0.64
						PBET	Intestinal	1.76	5.68	0.67
						DIN	Gastric	1.77	5.73	0.55
						DIN	Intestinal	1.46	9.20	0.53
Juhasz et al. (2011)						UBM[d]	Gastric	0.99	0.80	0.52
						UBM[d]	Intestinal	1.08	–3.73	0.59
						UBM (1:37.5)[e]	Gastric	1.05	6.29	0.52
						UBM 1:100[f]	Gastric	0.93	5.07	0.48
Bradham et al. (2011)	11	Mine, pesticide	173–6899	Mouse	SSUE	SBRC	Gastric	0.72	5.64	0.92
Denys et al. (2012)	16	Mine	18–25000	Swine	SSUE	UBM	Gastric	1.03	–1.51	0.99
						UBM	Intestinal	1.10	–0.51	0.99
Brattin et al. (2013)	20	Mine, pesticide	181–3957	Swine	SSUE	SBRC[g]	Gastric	0.62	19.68	0.72
						SBRC[h]	Intestinal	0.31	35.45	0.14
						SBRC (+ P)[i]	Intestinal	0.35	32.55	0.18
	19	Mine, pesticide	123–1492	Monkey	SDUE[c]	SBRC[g]	Gastric	0.32	11.07	0.34
						SBRC[h]	Intestinal	0.43	17.10	0.71
						SBRC (+ P)[i]	Intestinal	0.58	14.26	0.76

NOTES: [a]Steady state urinary excretion. [b]Area under the blood As concentration time curve. [c]Single dose urinary excretion. [d]The UBM assay also incorporates an initial saliva phase. [e]The saliva phase was omitted; however, the soil:solution ratio was maintained at the prescribed ratio (1:37.5). [f]The saliva phase was omitted; however, the soil:solution ratio was increased to 1:100. [g]The extraction fluid was not identified as the SBRC assay; however, its composition (0.4 M glycine adjusted to pH 1.5) was the same as detailed by Kelley et al. (2002). [h]The extraction fluid was composed of 0.4 M glycine adjusted to pH 7.0. [i]The extraction fluid was composed of 0.4 M glycine (pH 7.0) with the addition of 0.05 M phosphate and 0.1 M hydroxylamine. [j]Predicted As relative bioavailability = slope *x* measured *in vitro* value ± *y* intercept. [k]Data replotted as *RBA* vs. As *IVBA* for 14 soils in Rodriguez et al. (1999) updated using *RBA* data and no dosing vehicle conditions described in Basta et al. (2007). [l]Data replotted as *RBA* vs. As *IVBA* using data in Basta et al. (2007).

low ranging from 4.0 to 23.7 % for the 12 study soils. The soils contained significant levels of hydrous ferric oxide (i.e., goethite, ferrihydrite) and scorodite. Arsenic associated with these Fe minerals are known to have low solubility of associated As and As *IVBA* (Meunier et al. 2010). IVIVC plots have *y*-intercept values that were up to most of the measured As *RBA* resulting in underestimation of As *RBA* by *in vitro* methods. Whitacre et al. (2014) reported that addition of 100 mM ascorbic acid to the OSU-*IVG* method significantly improved agreement between As *RBA* determined by the Juvenile Swine model. A modified OSU-*IVG* method, which contained 100 mM ascorbic acid, increased As *IVBA* and accurately predicted As *RBA* for the gold mining soils (Basta et al. 2014; Whitacre et al. 2014).

Validation of As relative bioavailability-bioaccessibility relationships

The data depicted in Table 4 illustrate the research efforts undertaken to develop linear regression models for the prediction of As *RBA* from *in vitro* data. By utilizing simple, rapid, and inexpensive *IVBA* assays for As *RBA* predictions, issues associated with the use of *in vivo* assays—namely ethical considerations, lengthy timeframes of bioassays, and cost (especially for larger animal models)—may be overcome. While a number of researchers have demonstrated a strong correlation between As *RBA* and As *IVBA* for some *in vitro* assays (see Table 4), these relationships are yet to be validated using a set of soils outside of the empirical data set used in the correlation study (Juhasz et al. 2013). Performance validation is an important next step for As *RBA-IVBA* research in order to ensure robust linear regression models for predicting As *RBA* for the refinement of human health exposure. By using an independent data set as opposed to internal validation methods (e.g., split sample, cross validation, and bootstrapping methods), the predictive performance of the linear regression models can truly be evaluated (Emami 2006).

As detailed by Juhasz et al. (2013), a number of parameters and performance criteria need to be considered during As IVIVC and validation studies. During both the initial IVIVC phase and subsequent validation experiments, parameters associated with the As-contaminated soil sample set need to be taken into account to meet suitability and robustness criteria. These parameters include:

- The number of soils (data points) required to develop the initial As IVIVC, in addition to the number of independent soils to validate the relationship.

- Soil properties and source of As contamination associated with both primary and secondary soil sample sets. If linear regression models are developed for utilization across all As-contaminated environments, *in vivo* and *in vitro* data for model construction needs to be generated from soils with diverse physicochemical properties encompassing a wide range of As contaminant sources. However, if linear regression models are developed for site- or region-specific use (e.g., mining impacted environments), sample sets should reflect the pertinent physicochemical properties across the site/region.

- Arsenic concentration in the soil sample set used for As IVIVC assessment. As shown in Figure 1B, the concentration of As in soil used in As *RBA* studies has ranged up to 25,000 mg kg^{-1}—several orders of magnitude higher than relevant health investigation levels. For relevance to human health risk assessment outcomes, it may be appropriate to utilize soil with As concentrations that range up to 10 times higher than health investigation levels.

In addition to the above soil-sample parameters, some consideration is required regarding IVIVC goodness of fit parameters as detailed above (Wragg et al. 2011). As suggested by Juhasz et al. (2013), an additional criterion should also be applied to determine the performance of the independent data set (i.e., validation data) compared to data used to construct the linear regression models (correlation data). This may involve statistical analysis to determine whether the slope and y intercepts of the original vs. independent *in vivo–in vitro* relationships differ significantly or are in agreement (e.g., P < 0.05). Although validation studies are yet to be published, research efforts are being undertaken in order to generate robust predictive models for estimating As *RBA* for the refinement of human health exposure.

APPLYING As RELATIVE BIOAVAILABILITY / BIOACCESSIBILITY TO THE CHARACTERIZATION OF As RISK IN SOILS

Application of site-specific As-*RBA* and/or *IVBA* data is discussed in Basta et al. (2001). Risk to human receptors from exposure to As-contaminated soils is evaluated for carcinogenic effects by multiplying the quantified chemical daily intake by the cancer slope factor (see Eqn. 3). The slope factor (SF) converts estimated daily intakes averaged over a lifetime of exposure directly to incremental risk of an individual developing cancer (USEPA 1989).

The potential for noncarcinogenic effects (hazard quotient) can be evaluated by comparing an exposure level over a specified period (typically, a lifetime) with a reference dose (RfD; as detailed in the Integrated Risk Information System [IRIS]) derived for a similar exposure period (see Eqn. 4) (USEPA 1989). Site-specific As-*RBA* data obtained by direct measurement (*in vivo* assay) or predicted through the use of regression equations derived from IVIVC studies (*in vitro* methods) may be used in risk assessment. Using measured or predicted As-*RBA* values, adjustments to toxicity values can be made as follows:

$$RfD_{adjusted} = \frac{RfD_{IRIS}}{RBA} \tag{9}$$

$$SF_{adjusted} = SF_{IRIS} \times RBA \tag{10}$$

The importance of site-specific As-*RBA* data and its impact on cleanup levels is demonstrated by the Anaconda Superfund Site, Montana. Arsenic *RBA* was evaluated to determine the effect specific data had on cleanup levels as compared to using USEPA defaults (Walker and Griffin 1998). Site-specific soil As *RBA* was determined using *cynomolgus* monkeys and was found to be 25% for indoor dust and 18.3% for soil (Freeman et al. 1995). Walker and Griffin (1998) estimated that using the USEPA default value (100% *RBA*) rather than site-specific *RBA* values resulted in a 200% overestimation of As exposure. They reported that risk from exposure to As-contaminated soils was reduced from 1.7×10^{-4} to 4.0×10^{-5} using site-specific bioavailability instead of the standard USEPA default.

The usefulness and/or ability to adjust *CDI* for *RBA* depend on many issues, including (1) the contaminant concentration in the soil and (2) the chemical properties of the soil/geomedia, and on regulatory acceptance. Because animal models and *IVBA* methods have inherent uncertainty in measuring and predicting As *RBA*, *CDI* adjustments will be less likely used on highly contaminated (>500 mg kg^{-1} As) than moderately contaminated soils (<200 mg kg^{-1} As) (Chaney et al. 2008). Often, the highly contaminated area is localized and considerably smaller than the moderately contaminated area on a site. Excavation and replacement of the highly contaminated area may be feasible but is less feasible for large areas of moderately contaminated soil. Thus, *RBA* adjustments are needed for moderately contaminated soils. Cleanup of many sites are often considered at <50 mg As kg^{-1} soil. It is difficult to obtain As *RBA* values using animal models for contaminated soils that contain <100 mg As kg^{-1}. Arsenic in urine or blood, used to determine As *RBA* from animal diets in dosing studies, may be at the limits of detection in these biological matrices. A strong advantage of *IVBA* methods is the ability to estimate As *RBA* at low soil As concentrations, including background levels of <10 mg As kg^{-1} (Chaney et al. 2008). In addition, *IVBA* methods are not limited by background As from food as *in vivo* animal models.

The use of site-specific As *RBA* data in risk assessment may dramatically lower cleanup costs. The demand for site-specific soil As *RBA* data will likely increase because As *RBA* and *IVBA* methods are inexpensive compared to site cleanup based on overly conservative soil As *RBA*. Site-specific As *RBA* information will lower the degree of uncertainty in risk assessment and provide scientifically derived data to aid in the selection of appropriate remedies that are cost effective and protective of human health and the environment.

FUTURE DIRECTIONS, KNOWLEDGE GAPS, AND RESEARCH CHALLENGES

Research on As *RBA* over the past two decades clearly shows values can be much lower than 100% and even lower than the USEPA default value of 60%. Site-specific assessment incorporating As *RBA* would prevent unnecessary expensive remedial action of soils with low *RBA* As while protecting public health and the environment. Guidance exists for the use of appropriate animal-model dosing trials to determine site-specific *RBA*. However, disadvantages in conducting animal studies include expense, specialized facilities and personnel, and time to conduct the dosing trials. In addition to cost, time, facilities, and personnel needed, animal studies will not be able to meet the demand for site-specific *RBA*. Federal agencies in the US view the reduction, refinement, and replacement of animal use in toxicity testing as a priority. Both the US National Institute of Environmental Health Sciences (NIEHS) National Toxicology Program Roadmap for the 21st Century and the National Research Council's Vision and Strategy of Toxicity Testing in the 21st Century emphasize the goal of ensuring that new and improved test methods reduce, refine, or replace the use of animals where scientifically feasible (*http://iccvam.niehs.nih.gov/*).

The science of contaminant bioaccessibility, using *in vitro* gastro(intestinal) methods, has matured in the last two decades. Use of bioaccessibility methods to predict Pb *RBA* in site-specific human health risk assessment has gained regulatory acceptance (USEPA 2007b). Arsenic is the next contaminant likely to gain regulatory acceptance for the use of bioaccessibility methods to predict As *RBA*. Several *in vitro* methods have reported accurate prediction of As *RBA* determined using surrogate animal models for limited numbers of contaminated soils and geomedia (Rodriguez et al. 1999; Basta et al. 2007; Juhasz et al. 2007; 2009; 2011; Bradham et al. 2011; Denys et al. 2012; Brattin et al. 2013). Research is needed to demonstrate the type of sites (i.e., soil, mineralogy, contaminant source) where specific As *in vitro* methods are accurate predictors of contaminant. Validation of *in vitro* methods for a wide range of soils and contaminated media is the most pressing knowledge and research gap that is slowing regulatory acceptance (Juhasz et al. 2013). Other research needs are:

- Soil or dietary parameters that influence As *RBA* and As *IVBA*, e.g., *in vivo* research is needed to determine whether dietary phosphate increases or decreases As *RBA*.

- Comparison of As *RBA* between different animal models and endpoints—some of this research has been initiated; however, a much larger data set is needed.

- Round robin studies to determine the between-laboratory variability in As *RBA* values—round robin studies have been undertaken to determine within-laboratory reproducibility and between-laboratory variability for bioaccessibility assays, but such studies have not been undertaken (presumably due to cost limitations) for *in vivo* assays.

- Assessment of the effect of amendments to minimize As *RBA*—does a reduction in As *RBA* (determined using an *in vivo* approach) also correspond to a decrease in As *IVBA*?

REFERENCES

ATSDR (2011) Priority list of hazardous substances. Agency for Toxic Substances and Disease Registry, *http://www.atsdr.cdc.gov/spl/* (accessed July 2014)

Basta NT, Rodriguez RR, Casteel SW (2001) Bioavailability and risk of arsenic exposure by the soil ingestion pathway. *In*: Environmental Chemistry of Arsenic. Frankenberger WT (ed) Marcel Dekker, Inc., New York

Basta NT, Ryan JA, Chaney RL (2005) Trace element chemistry in residual-treated soil: Key concepts and metal bioavailability. J Environ Qual 34:49-63

Basta NT, Foster JN, Dayton EA, Rodriguez RR, Casteel SW (2007) The effect of dosing vehicle on arsenic bioaccessibility in smelter-contaminated soil. J Environ Health Sci Part A 42:1275-1281

Basta NT, Whitacre SW, Meyers P, Mitchell VL, Alpers CN, Foster AL, Casteel SW, Kim CS (2014) Using *in vitro* gastrointestinal and sequential extraction methods to characterize site-specific arsenic bioavailability. Goldschmidt 2014 Conference, *http://goldschmidt.info/2014/uploads/abstracts/finalPDFs/134.pdf* (accessed July 2014)

Beak DG, Basta NT, Scheckel KG, Traina SJ (2006) Bioaccessibility of arsenic (V) bound to ferrihydrite using a simulated gastrointestinal system. Environ Sci Technol 40:1364-1370

Bergeron RJ, Merriman RL, Olson SG, Wiegand J, Bender J, Streiff RR, Weimar WR (2000) Metabolism and pharmacokinetics of N^1,N^{11}-diethylnorspermine in a Cebus apella primate model. Cancer Res 60:4433-4439

Bettley FR, O'Shea JA (1975) The absorption of arsenic and it relation to carcinoma. Br J Dermatol 92:563-568

Bradham KD, Scheckel KG, Nelson CM, Seales PE, Lee GE, Hughes MF, Miller BW, Yeow A, Gilmore T, Harper S, Thomas DJ (2011) Relative bioavailability and bioaccessibility and speciation of arsenic in contaminated soils. Environ Health Perspect 119:1629-1634

Bradham KD, Laird BD, Rasmussen PE, Siciliano SD, Hughes MF (2014) Assessing the bioavailability and risk from metal contaminated soils and dust. Hum Ecol Risk Assess 20:272-286

Brattin W, Casteel S (2013) Measurement of arsenic relative bioavailability in swine. J Toxicol Environ Health Part A 76:449-457

Brattin W, Drexler J, Lowney Y, Griffin S, Diamond G, Woodbury L (2013) An *in vitro* method for estimation of arsenic relative bioavailability in soil. Toxicol Environ Health Part A 76:458-478

Casteel SW, Fent G, Tessman R, Brattin WJ, Wahlquist AM (2005) Relative Bioavailability of Arsenic and Vanadium in Soil from a Superfund Site in Palestine, Texas. Prepared by University of Missouri, Columbia and SRC. Prepared for U.S. Environmental Protection Agency, Office of Superfund Remediation Technology Innovation. *http://www.epa.gov/superfund/bioavailability/pdfs/PTX%20As-V%20RBA%20 Report_Final%202005_508.pdf* (accessed July 2014)

Casteel SW, Weis CP, Henningsen GW, Brattin WJ (2006) Estimation of relative bioavailability of lead in soil and soil-like materials using young swine. Environ Health Perspect 114:1162-1171

Casteel SW, Fent G, Myoungheon L, Brattin WJ, Hunter P (2009a) Relative Bioavailability of Arsenic in Barber Orchard Soils. Prepared for U.S. Environmental Protection Agency, Office of Superfund Remediation Technology Innovation. Prepared by University of Missouri, Columbia and SRC. Report # SRC TR-09-245. *http://www.epa.gov/superfund/bioavailability/pdfs/Barber%20Orchard%20Swine%20RBA%20 9-18-09_508.pdf* (accessed July 2014)

Casteel SW, Fent G, Myoungheon L, Brattin WJ, Wahlquist AM (2009b) Relative Bioavailability of Arsenic in NIST SRM 2710 (Montana Soil). Prepared for U.S. Environmental Protection Agency, Office of Superfund Remediation Technology Innovation. Prepared by University of Missouri, Columbia and SRC *http://www.epa.gov/superfund/bioavailability/pdfs/NIST1_As%20RBA%20Report%20Final%203-13-09_508.pdf* (accessed July 2014)

Casteel SW, Fent G, Myoungheon L, Brattin WJ, Hunter P (2009c) Relative Bioavailability of Arsenic in a Mohr Orchard Soil. Prepared for U.S. Environmental Protection Agency, Office of Superfund Remediation Technology Innovation. Prepared by University of Missouri, Columbia and SRC. Report # SRC TR-09-254. *http://www.epa.gov/superfund/bioavailability/pdfs/Mohr%20Orchard%20RBA%20 Report_10-28-09_SRC_508.pdf* (accessed July 2014)

Casteel SW, Fent G, Myoungheon L, Brattin WJ, Hunter P (2010a) Relative Bioavailability of Arsenic in an Iron King Soil. Prepared for U.S. Environmental Protection Agency, Office of Superfund Remediation Technology Innovation. Prepared by University of Missouri, Columbia and SRC. Report # SRC-09-041. *http://www.epa.gov/superfund/bioavailability/pdfs/Iron%20King%20Swine%20RBA%2002-25-10_SRC_508.pdf* (accessed July 2014)

Casteel SW, Fent G, Knight LE, Brattin WJ, Hunter P (2010b) Relative Bioavailability of Arsenic in an ASARCO and Hawaiian soil. Prepared for U.S. Environmental Protection Agency, Office of Superfund Remediation Technology Innovation. Prepared by University of Missouri, Columbia and SRC. Report # SRC TR-10-108. *http://www.epa.gov/superfund/bioavailability/pdfs/AS-HI%20RBA%20Revised%20 FINAL%20Report%2003-07-12.pdf* (accessed July 2014)

Casteel SW, Fent G, Myoungheon L, Brattin WJ, Hunter P (2010c) Relative Bioavailability of Arsenic in NIST SRM 2710a (Montana Soil). Prepared for U.S. Environmental Protection Agency, Office of Superfund Remediation Technology Innovation. Prepared by University of Missouri, Columbia and SRC. Report # SRC TR-09-0951

Chaney RL, Ryan JA (1994) Risk Based Standards for Arsenic, Lead and Cadmium in Urban Soils. (ISBN 3-926959-63-0) DECHEMA, Frankfurt. 130 pp

Chaney, RL, Basta NT, Ryan JA (2008) Element bioavailability and bioaccessibility in soils: what is known now, and what are the significant data gaps? *In*: SERDP and ESTCP Expert Panel Workshop on Research and Development Needs for Understanding and Assessing the Bioavailability of Contaminants in Soils and Sediments, SERDP and ESTCP, Arlington, VA, November, 2008, p B-36–B-72, *http://docs.serdp-estcp.org/content/download/8049/99405/version/1/file/Bioavailability_Wkshp_Nov_2008.pdf* (accessed July 2014)

Crews, HM, Burrell, JA, McWeeney. DJ (1983) Preliminary enzymolysis studies on trace element extractability from food. J Sci Food Agric 34:997-1014

Csanaky I, Gregus Z (2002) Species variations in the biliary and urinary excretion of arsenate, arsenite and their metabolites. Comp Biochem Physiol Part C, Toxicol Pharmacol 131:355-365

Denys S, Caboche J, Tack K, Rychen G, Wragg J, Cave M, Jondreville C, Feidt C (2012) *In vivo* validation of the unified BARGE method to assess the bioaccessibility of arsenic, antimony, cadmium, and lead in soils. Environ Sci Technol 46:6252-6260

DIN (2000) Soil quality-absorption availability of organic and inorganic pollutants from contaminated soil material. Standard DIN E19738. Berlin: Deutsches Institut fur Normung e.V.

Drexler JW, Brattin WJ (2007) An *in vitro* procedure for estimation of lead relative bioavailability: With validation. Hum Ecol Risk Assess 13:383-401

Dudka S, Miller WP (1999) Permissible concentrations of arsenic and lead in soils based on risk assessment. Water Air Soil Pollut 113:127-132

Ellickson KM, Meeker RJ, Gallo MA, Buckley BT, Lioy PJ (2001) Oral bioavailability of lead and arsenic from a NIST standard reference soil material. Arch Environ Toxicol 40:128-135

El-Masri HA, Kenyon EM (2008) Development of a human physiologically based pharmacokinetic (PBPK) model for inorganic arsenic and its mono- and di-methylated metabolites. J Pharmacokinet Pharmacodyn 35:31-68

Emami J (2006) *In vitro-in vivo* correlation: from theory to application. J Pharm Pharmaceut Sci 9:169-189

Eto N, Tomita M, Hayashi M (2006) NaPi-mediated transcellular permeation is the dominant route in intestinal inorganic phosphate absorption in rats. Drug Metab Pharmacokinet 21:217-221

Evans MV, Dowd SM, Kenyon EM, Hughes MF, El-Masri HA (2008) A physiologically based pharmacokinetic model for intravenous and ingested dimethylarsinic acid in mice. Toxicol Sci 104:250-260

Freeman GB, Johnson JD, Killinger JM, Liao SC, Davis AO, Ruby MV, Chaney RL, Lovre SC, Bergstrom PD (1993) Bioavailability of arsenic in soil impacted by smelter activities following oral administration in rabbits. Fundam Appl Toxicol 21:83-88

Freeman GB, Schoof RA, Ruby MV, Davis AO, Dill SC, Liao SC, Lapin CA, Bergstrom PD (1995) Bioavailability of arsenic in soil and house dust impacted by smelter activities following oral administration in cynomolgus monkeys. Fundam Appl Toxicol 28:215-222

Groen K, Vaessen H, Kliest JJG, de Boer JLM, Ooik TV, Timmerman A, Vlug FF (1994) Bioavailability of inorganic arsenic from bog ore-containing soil in the dog. Environ Health Perspect 102:182-184

Hamel SC, Buckley B, Lioy PJ (1998) Bioaccessibility of metals in soils for different liquid to solid ratios in synthetic gastric fluid. Environ Sci Technol 32(3):358-362

Juhasz AL, Smith E (2013) Quantifying arsenic relative bioavailability in Hawai'i soil. University of South Australia report to the Hawai'i Department of Health

Juhasz AL, Smith E, Weber J, Rees M, Rofe A, Kuchel T, Sansom L, Naidu R (2007) Comparison of *in vivo* and *in vitro* methodologies for the assessment of arsenic bioavailability in contaminated soils. Chemosphere 69:961-966

Juhasz AL, Smith E, Weber J, Naidu R, Rees M, Rofe A, Kuchel T, Sansom L (2008) Effect of soil ageing on *in vivo* arsenic bioavailability in two dissimilar soils. Chemosphere 71:2180-2186

Juhasz AL, Smith E, Weber J, Naidu R, Rees M, Rofe A, Kuchel T, Sansom L (2009) Assessment of four commonly employed *in vitro* arsenic bioaccessibility assays for predicting *in vivo* arsenic relative bioavailability in contaminated soils. Environ Sci Technol 43:9887-9894

Juhasz AL, Weber J Smith E (2011) Influence of saliva, gastric and intestinal phases on the prediction of As relative bioavailability using the unified Bioaccessibility Research Group of Europe method (UBM). J Hazard Materials 197: 161-168

Juhasz AL, Basta NT, Smith E (2013) What is required for the validation of *in vitro* assays for predicting contaminant relative bioavailability? Considerations and criteria. Environ Pollut 180:372-375

Kelley ME, Brauning SE, Schoof RA, Ruby MV (2002) Assessing Oral Bioavailability of Metals in Soil. Battelle Press, Columbus, OH

Lien HC, Tsai TF, Lee YY, Hsiao CH (1999) Merkel cell carcinoma and chronic arsenicism. J Am Acad Dermatol 41:641-643

Magagelada JR, Lonstreth GG, Summerskill WHJ, Go VLW (1976) Measurement of gastric functions during digestion of ordinary solid meals in man. Gastroenterology 70:203-210

Mandal BK, Suzuki KT (2002) Arsenic round the world: A review. Talanta 58:201-235

Menunier L (2011) Physico-chemical parameters influencing the bioaccessibility of arsenic from tailings and soils. Ph.D. Dissertation. Royal Military College of Canada, Kingston, ON, Canada

Meunier L, Walker SR, Wragg J, Parsons MB, Koch I, Jamieson HE, Reimer KJ (2010) Effects of soil composition and mineralogy on the bioaccessibility of arsenic from tailings and soil in gold mine districts of Nova Scotia. Environ Sci Technol 44:2267-2674

Ng JC, Kratzmann SM, Qi L, Crawley H, Chiswell B, Moore MR (1998) Speciation and absolute bioavailability: risk assessment of arsenic-contaminated sites in a residential suburb in Canberra. Analyst 123:889-892

O'Connor HJ, Schorah CJ, Habibzedah N, Axon ATR (1989) Vitamin C in the human stomach: relation to gastric pH, gastroduodenal disease, and possible sources. Gut 30:436-442

Oomen AG, Hack A, Minekus M, Zeijdner E, Cornelis C, Schoeters G, Verstraete W, Van de Wiele T, Wragg J, Rompelberg CJM, Sips, AJAM, Van Wijnen JH (2002) Comparison of five *in vitro* digestion models to study the bioaccessibility of soil contaminants. Environ Sci Technol 36:3326-3334

Parrott N, Lukacova V, Fraczkiewicz G, Bolger MB (2009) Predicting pharmacokinetics of drugs using physiologically based modelling – application to food effects. Am Assoc Pharm Sci 11:45-53

Patterson JK, Lei, XG, Miller DD (2008) The pig as an experimental model for elucidating the mechanisms governing dietary influence on mineral absorption. Exp Biol Med 233:651-664

Pomroy C, Charbonneau SM, McCullough RS, Tam GKH (1980) Human retention studies with ^{74}As. Toxicol Appl Pharmacol 53:505-556

Rees M, Sansom L, Rofe A, Juhasz AL, Smith E, Weber J, Naidu R, Kuchel T (2009) Principles and application of an *in vivo* swine assay for the determination of arsenic bioavailability in contaminated matrices. Environ Geochem Health 31:167-177

Roberts SM, Weimar WR, Vinson JRT, Munson JW, Bergeron RJ (2002) Measurement of arsenic bioavailability in soil using a primate model. Toxicol Sci 67:303-310

Roberts SM, Munson JW, Lowney YW, Ruby MV (2007) Relative oral bioavailability of arsenic from contaminated soils measured in the Cynomolgus monkey. Toxicol Sci 95:281-288

Rodriguez RR, Basta NT, Casteel SW, Pace LW (1999) An *in vitro* gastrointestinal method to estimate bioavailable arsenic in contaminated soils and solid media. Environ Sci Technol 33:642-649

Rodriguez RR, Basta NT, Casteel SW, Armstrong FP, Ward DC (2003) Chemical extraction methods to assess bioavailable As in contaminated soil and solid media. J Environ Qual 32:876-884

Ruby MV, Davis A, Kempton JH, Drexler JW, Bergstrom PD (1992) Lead bioavailability: Dissolution kinetics under simulated gastric conditions. Environ Sci Technol 26:1242-1248

Ruby MV, Davis A, Schoof R, Eberle S, Sellstone CM (1996) Estimation of lead and arsenic bioavailability using a physiologically based extraction test. Environ Sci Technol 30:422-430

U.S. Department of Health and Human Services Food and Drug Administration (1997) Guidance for industry extended release oral dosage forms: Development, evaluation and application of *in vitro* / *in vivo* correlations. U.S. Department of Health and Human Services Food and Drug Administration, September 1997

USEPA (1989) Risk Assessment Guidance for Superfund (RAGS), Volume I, Human Health Evaluation Manual (Part A). U.S. Environmental Protection Agency, Washington, DC: Office of Emergency and Remedial Response, EPA/540/1-89/002

USEPA (1996) Bioavailability of Arsenic and Lead in Environmental Substrates. U.S. Environmental Protection Agency, Region 10: Seattle, WA. EPA910/R-96-002

USEPA (2007a) Guidance for Evaluating the Oral Bioavailability of Metals in Soils for Use in Human Health Risk Assessment. U.S. Environmental Protection Agency, OSWER 9285.7-80

USEPA (2007b) Estimation of relative bioavailability of lead in soil and soil-like materials using *in vivo* and *in vitro* methods. U.S. Environmental Protection Agency, OSWER 9285.7-77, May 2007

USEPA (2008) Child-Specific Exposure Factors Handbook. U.S. Environmental Protection Agency, National Center for Environmental Assessment, Office of Research and Development: Washington, DC. EPA/600/R-06/096F

USEPA (2009) Relative Bioavailability of Arsenic from Soil Barber Orchard Superfund Site Waynesville, North Carolina. Prepared for U.S. Environmental Protection Agency, Region 4 by Center for Environmental & Human Toxicology, University of Florida. (available through U.S. EPA Region 4 Administrative Record Index for the Barber Orchard (Explanation of Significant Differences) NCSDN0406989)

USEPA (2010) Relative bioavailability of arsenic in soil at 11 superfund sites using an *in vivo* juvenile swine method. U.S. Environmental Protection Agency. EPA OSWER Directive #9200.0-76. *http://epa.gov/superfund/bioavailability/pdfs/as_in_vivo_rba_main.pdf* (accessed July 2014)

USEPA (2012a) Recommendation for default value for relative bioavailability of arsenic in soil. U.S. Environmental Protection Agency, OSWER 9200.1-113. *http://www.epa.gov/superfund/bioavailability/pdfs/Arsenic%20Bioavailability%20POLICY%20Memorandum%2012-20-12.pdf* (accessed July 2014)

USEPA (2012b) Compilation and review of data on relative bioavailability of arsenic in soil. U.S. Environmental Protection Agency, OSWER 9200.1-113. *http://www.epa.gov/reg3hwmd/risk/human/rb-concentration_table/documents/ArsenicBioavailability.pdf* (accessed July 2014)

Van de Wiele, TR, Oomen AG, Wragg J, Cave M, Minekus M, Hack A, Cornelis C, Rompelberg CJM, de Zwart LL, Klinck B, van Wijen J, Verstraete,W, Sips, AJAM (2007) Comparison of five *in vitro* digestion models to *in vivo* experimental results: lead bioaccessibility in the human gastrointestinal tract. J Environ Sci Health Part A 42:1203-1211

Villa-Bellosta R, Sorribas V (2009) Role of rat sodium/phosphate cotransporters in the cell membrane transport of arsenate. Toxicol. Appl. Pharmacol. 232:125-134

Walker S, Griffin S (1998) Site-specific data confirm arsenic exposure predicted by the U.S. Environmental Protection Agency. Environ Health Perspect 106:133-139

Weis CP, LaVelle JM (1991) Characteristics to consider when choosing an animal model for the study of lead bioavailability. Chem Speciation Bioavailability 3:113-119

Whitacre SD, Basta NT, Dayton EA (2013) Soil controls on bioaccessible arsenic fractions. J Environ Health Sci Part A 48(6):620-628

Whitacre SW, Basta NT, Casteel SW, Foster AL, Meyers P, Mitchell VL. (2014) Bioavailability measures for arsenic in California gold mine tailings. Goldschmidt 2014 Conference, *http://goldschmidt.info/2014/uploads/abstracts/finalPDFs/2695.pdf*

Wragg J, Cave M, Basta N, Brandon E, Casteel S, Denys S, Gron C, Oomen A, Reimer K, Tack K, Van de Wiele T (2011) An inter-laboratory trial of the unified BARGE bioaccessibility method for arsenic, cadmium and lead in soil. Sci Total Environ 409:4016-4030

Yang JK, Barnett MO, Jardine PM, Basta NT, Casteel SW (2002) Adsorption, sequestration, and bioaccessibility of As (V) in soils. Environ Sci Technol 36:4562-4569

Reviews in Mineralogy & Geochemistry
Vol. 79 pp. 473-505, 2014
Copyright © Mineralogical Society of America

The Characterization of Arsenic in Mine Waste

Dave Craw

Department of Geology
University of Otago
Dunedin, New Zealand 9054

dave.craw@otago.ac.nz

Robert J. Bowell

SRK Consulting
17 Churchill Way
Cardiff, CF10 2HH, Wales, United Kingdom

rbowell@srk.co.uk

INTRODUCTION

Arsenic is dispersed widely in nature and is the 47th most abundant element among the 88 known natural elements. The average crustal abundance is 1.5 ppm, with higher concentrations in reduced shales and coals. It is concentrated in many metal-bearing mineral deposits being a chalcophile element. It occurs in many metallic deposits including those of Cu, Ag, Au, Zn, Hg, U, Sn, Pb, Mo, W, Ni, Co and PGE. Arsenic is often more dispersed than ore minerals in those deposits and as such is a useful indicator in geochemical exploration (Boyle and Jonasson 1973; Hale 1981; Cohen and Bowell 2014). Consequently, elevated concentrations of As are common in mine waste and process waste from metal-bearing ores (Bowell et al. 1994, 2013; Thornton 1994; Craw and Pacheco 2002; Lazareva et al. 2002). In particular, As is the main element of environmental concern in most hardrock mines. In this chapter, we outline the principal occurrences of As at mine sites, and their environmental significance. As an example of the detailed controls on As geochemistry in mine waste we focus on the gold mines in New Zealand.

ARSENIC IN DIFFERENT MINERAL DEPOSIT TYPES

Many mines are significant point sources for As in the environment, and some mine sites have high concentrations of As locally, often exceeding 1% in ore or waste. For almost all of these mines, the As is a natural but undesirable component of the ore, and therefore the As is discarded with the rest of the mine wastes. Hence, mine wastes, especially mine tailings, are major repositories of As and have to be managed carefully.

Arsenic is an especially common constituent of sulfide-bearing mineral deposits, where As typically occurs either as separate As minerals (Table 1) or in solid solution in the dominant sulfide

Table 1. Major primary mineral hosts for As in mine waste.

Mineral	Formula
Arsenopyrite	FeAsS
Arsenian pyrite or Marcasite	$Fe(S,As)_2$
Enargite-Luzonite (dimorphs)	Cu_3AsS_4
Orpiment	As_2S_3
Realgar	AsS
Tennantite	$(Cu,Fe)_{12}As_4S_{13}$

1529-6466/14/0079-0010$05.00
http://dx.doi.org/10.2138/rmg.2014.79.10

minerals, and commonly as both. Arsenic-bearing pyrite is particularly common in most deposit types. The amount of As in pyrite can range from a few ppm to several % (Bowell et al. 1999; Reich et al. 2005). Since most of these As-bearing minerals readily oxidize in the surficial environment, the contained As can become available for formation of secondary minerals or may dissolve in contact waters.

SECONDARY ARSENIC MINERALS IN MINE WASTES

Arsenic in mine waters is primarily dispersed by the oxidation of sulfide minerals, the principal one being arsenian pyrite with the main As sulfides being arsenopyrite and realgar (Table 1, 2; Bowell 1994b; Craw et al. 1999, 2003; Bowell and Parshley 2003; O'Day 2006; Yu et al. 2007; McKibben et al. 2008; Corkhill and Vaughan 2009; Lengke et al. 2009). However, another important source of As is dispersion from secondary processed materials, very often As-oxides or arsenates or arsenides (Ettler et al. 2009; Dey et al. 2010; Jamieson 2014, this volume). At the Tsumeb smelter in Namibia, slag material, a waste product of smelting, contains up to 75,870 mg kg^{-1} of As (Dey et al. 2010). Detailed mineralogy of this material has identified As as highly complex Cu-Pb-Ca arsenates, As-oxides and as post-smelting formed Cu-Zn-Ge arsenides (Day et al. 2000).

Oxidation of primary sulfides results in the release of As, sulfate and metals. The more common secondary As minerals are given in Table 3. Consequently, weathering leads to the formation of secondary As minerals (Bowell 1994a; Majzlan et al 2012a,b). The most common phase in mine waste is scorodite ($FeAsO_4 \cdot 2H_2O$) but there are more than 394 other As-bearing oxide or oxyanion minerals (Mandarino and Back 2004; IMA 2014) including five arsenites that are largely only known from one locality, Tsumeb (Keller 1977; Gebhard 1999; Ettler et al. 2009; Bowell 2014, this volume). Occurrences of complex arsenate mineral assemblages are generally most common in base metal deposits due to the abundance of potential cations for precipitation of arsenate minerals. The most well-known of these deposits is Tsumeb in Namibia (Gebhard 1999). Here, over forty arsenates and five arsenite minerals occur, of which fifteen only occur in this deposit (Bowell 2014, this volume).

Other complex arsenate assemblages include Penberthy Croft in Cornwall (Betterton 2000) where a complex assemblage of base metal arsenates are present in quartz veins hosted in shale at the edge of a granite batholith. The locality is particularly famous for the abundance of bayldonite. In the Czech Republic at Kutná Hora, the ancient mine dumps described by Agricola in his De Re Metallica (Agricola 1556) boast a complex assemblage of Fe-arsenate minerals and is the type locality for kaňkite and bukovskýite (Majzlan et al. 2012 a,b). In the United States in Nevada, there are two important mineral deposits for secondary As mineralogy and these are the Getchell mine where oxidation of realgar and orpiment in carbonate host rocks has generated a comprehensive assemblage of calcium-magnesium arsenates due to the limited Fe and base metal content of the ore (Stolberg and Dunning 1985). The second is the Majuba porphyry deposit, Pershing County where copper arsenates, particularly olivenite, clinoclase, arthurite, cornubite and cornwallite are present along with abundant scorodite, pharmacosiderite and zeunerite in oxidized arsenopyrite-rich assemblage (Jensen 1985).

At the Gold Hill mine in Utah, the intrusion of quartz monzonite into Paleozoic limestones produced a mineralized skarn that through oxidation has generated a complex secondary mineral assemblage with characteristic metal signatures in different parts of the oxide deposit. This includes a Zn assemblage formed where oxidized sphalerite and arsenopyrite occur in limestone. This resulted in an assemblage dominated by adamite and austenite replacing earlier arseniosiderite, symplesite and kottigite (Kokinos and Wise 1993). A colorful Fe and Pb assemblage was formed in the main gossan with oxidation of a galena and arsenopyrite rich zone resulting in the formation of carminite, arseniosiderite, scorodite, mimetite and

Table 2. Summary of common mineral deposit types, with their typical As minerals and potential for environmental acid rock drainage (ARD) generation.

Deposit type	As content	Typical As minerals	Alteration scale	Fe-S mineral	Dominant other metals	ARD potential
Orogenic Au	High	arsenopyrite	Minor, 0.1-10 m	Pyrite	Sb	Low
High sulfidation epithermal Au	Moderate	luzonite, enargite realgar, orpiment	Large halo, 100's of m	Jarosite, pyrite	Cu, Pb, Zn	Moderate
Low sulfidation epithermal Au	Moderate-low	arsenopyrite	Large halo, 100's of m	Pyrite abundant	Cu, Pb, Zn, Hg	High
Carlin Au	High	arsenian pyrite realgar orpiment native arsenic	Large to minor 100's of m to >1 km	Variable pyrite	Sb, Hg, Tl	Low
Witwatersrand	Low to Moderate	arsenian pyrite arsenopyrite	Minor, 0.1-10 m	pyrite	U, Ti, S	Moderate
Porphyry Cu	Moderate-low	enargite	Large halo, 100's of m	Pyrite abundant	Cu, Pb, Zn, Mo	High
Granitoid Sn	Moderate-high	arsenopyrite	Large halo, 100's of m	Pyrite, pyrrhotite	Cu, Pb, Zn, W	Moderate
Magmatic sulfide	Low-moderate	arsenopyrite	Minor	Pyrrhotite, pyrite	Cu, Ni	High
Massive sulfide	Low		Minor	Pyrite, pyrrhotite	Cu, Pb, Zn	High

Table 3. Common secondary minerals
host to As in mine waste.

Mineral	Formula
Adamite-Olivenite	$Zn_2(AsO_4)(OH) - Cu_2(AsO_4)(OH)$
Adelite	$Ca,Mg(AsO_4)(OH)$
Annabergite	$Ni_3(AsO_4)_2 \cdot 8H_2O$
Arsenic	As
Arsendescloizite	$Pb,Zn(AsO_4)(OH)$
Arseniosiderite	$Ca_2Fe^{3+}_3(AsO_4)_3O_2 \cdot 3H_2O$
Arsenoclasite	$Mn^{2+}_5(AsO_4)_2(OH)_4$
Arsenogorceixite	$HBaAl_3(AsO_4)_2(OH)_6$
Arsenolite	As_2O_3
Arthurite	$CuFe^{3+}_2(AsO_4)_2(OH)_2 \cdot 4H_2O$
Austinite	$Ca,Zn(AsO_4)(OH)$
Bayldonite	$Pb,Cu_3(AsO_4)_2(OH)_2$
Bukovskyite	$Fe^{3+}_2(AsO_4)(SO_4)(OH) \cdot 9H_2O$
Clinoclase	$Cu_3(AsO_4)(OH)_3$
Cornwallite/Cornubite	$Cu(AsO_4)_2(OH)_4$
Conichalcite	$CaCu(AsO_4)(OH)$
Duftite	$PbCu(AsO_4)(OH)$
Erythrite	$Co_3(AsO_4)_2 \cdot 8H_2O$
Kankite	$FeAsO_4 \cdot 3.5H_2O$
Kottigite	$Zn_3(AsO_4)_2 \cdot 8H_2O$
Lavendulan/Lemanskiite (dimorph)	$NaCaCu_5(AsO_4)_4Cl \cdot 5H_2O$
Lotharmeyerite	$Ca(Zn,Mn^{3+})_2(AsO_4)_2(OH,H_2O)_2$
Mansfieldite	$AlAsO_4 \cdot 2H_2O$
Mimetite	$Pb_5(AsO_4)_3Cl$
Parnauite	$Cu_9(AsO_4)_2(SO_4)(OH)_{10} \cdot 7H_2O$
Pharmacosiderite	$KFe^{3+}(AsO_4)_3(OH)_4 \cdot 6\text{-}7H_2O$
Pharmacolite	$CaHAsO_4 \cdot 2H_2O$
Picropharmacolite	$H_2Ca_4M(AsO_4)_4 \cdot 11H_2O$
Scorodite	$FeAsO_4 \cdot 2H_2O$
Seelite	$MgUO_2(As^{3+}O_3)_{1.4}(As^{5+}O_4)_{0.6} \cdot 7H_2O$
Strashimirite	$Cu_8(AsO_4)_4(OH)_4 \cdot 5H_2O$
Symplesite	$Fe^{2+}_3(AsO_4)_2 \cdot 8H_2O$
Tyrolite	$Ca_2Cu_2(AsO_4)_4(CO_3)(OH)_8 \cdot 11H_2O$
Tooeleite	$Fe^{3+}_6(As^{3+}O_3)_4(SO_4)(OH)_4 \cdot 4H_2O$
Uranospinite	$Ca(UO_2)_2(AsO_4)_2 \cdot 10H_2O$
Weilite	$CaAsO_3OH$
Yukonite	$Ca_2Fe^{3+}_2(AsO_4)_4(OH) \cdot 12H_2O$
Zeunerite	$Cu^{2+}(UO_2)_2(AsO_4)_2 \cdot 10\text{-}16H_2O$
Zykaite	$Fe_3(AsO_4)_3(SO_4)(OH) \cdot 15H_2O$

pharmacosiderite amongst others. The most vivid assemblage, however, is the complex assemblage formed is the copper arsenate one after massive tennantite and chalcopyrite which resulted in the formation of abundant conichalcite, olivenite along with cornwallite, clinoclase, lavendulan, parnauite, strashimirite, tyrolite and others.

Where the system is Fe limited, Ca arsenates can form in the supergene zone and arsenolite can be stabilized even in intensely oxidized rocks, such as at the Getchell mine, Nevada (Stolberg and Dunning 1985). Here picopharmacolite, adelite and pharmacolite occur as fracture coatings and as replacements of realgar and orpiment in carbonaceous limestone.

At the Ashanti mine in Ghana, boxworks were produced by the oxidation of arsenopyrite (Fig. 1). The replacement initially is along fractures and grain boundaries with hematite, goethite and arsenolite. As oxidation becomes more intensive goethite, bukovskýite and kaňkite are formed and these are later replaced by scorodite that is the most abundant alteration product in the saprock zone (Bowell 1994b). In the oxide zone, amorphous hydrous ferric oxide and pitticite (a poorly ordered hydrated ferric arsenate) form granular crusts presumably reflecting hydrogeochemical dispersion of As. In the oxide zone, "clinker-like" boxworks of goethite, hematite and scorodite are present and gold-goethite intergrowths are also observed as pseudomorphs after arsenopyrite crystals. Goethite in the gossans hosts up to 5 wt% As by electron microprobe analysis.

In addition to these minerals, As can also occur as a trace element in goethite (FeOOH) and jarosite ($KFe_3(SO_4)_2(OH)_6$) as well as poorly crystalline and amorphous Fe-aluminum precipitates (Scott 1987; Waychunas et al. 1993; Foster et al. 1998; Lenain and Courtin-Nomade 2003; Petrunic et al. 2006; Slowey et al. 2007; Courtin-Nomade et al. 2009). The variable solubility of these As minerals will influence chemical dissolution, mobility and

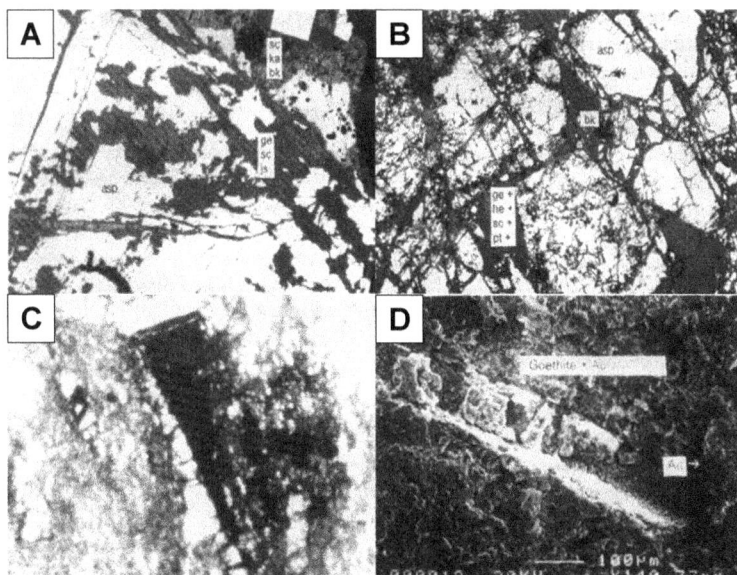

Figure 1. Photomicrographs of arsenopyrite oxidation in natural gossans and weathered zones, Justice mine, Ghana (Bowell 1994b). A. Partial oxidation of arsenopyrite to secondary arsenates and goethite, reflected light, field of view 1 mm; B. Intense alteration of arsenopyrite leading to extensive replacement, reflected light, field of view 1 mm; C. Pseudomorph of arsenopyrite by goethite; D. Gold grains associated with goethite pseudomorph. Bk-bukovskýite ge=goethite; Au=gold; he=hematite; js=jarosite; ka=kaňkite; pt=pitticite; sc=scorodite.

attenuation of As and ultimately the bioavailability from mine waste and impacted sediments and soils (Courtin-Nomade et al. 2005; Craw et al. 2007; Haffert and Craw 2008a,b; Walker et al. 2009). Fe-oxyhydroxides also provide an effective attenuation mechanism for As. This has been shown for natural Fe-oxyhydroxides where As concentrations in lakes receiving mine drainage are similar to unimpacted lakes but in which the Fe-oxyhydroxides in the sediments contain anomalous As, for example Moira Lake in Ontario (Azcue and Nriagu 1995) and in mine drainage from Mout Bischoff tin mine in western Tasmania (Gault et al. 2005).

Arsenic solubility in mine waste

Unlike cations of metallic elements, As can be mobile at extremely acidic and at neutral to highly alkaline pH conditions as well as in highly oxidized and even reduced waters (Bowell et al. 1994; Clara and Maghales 2002; Roddick-Lanzilotta et al. 2002; O'Day 2006;). Arsenic exists in natural waters in primarily two oxidation states, As(III) (arsenite, $H_nAsO_3^{3-n}$) and As(V) (arsenate, $H_nAsO_4^{3-n}$). In oxygenated environments, arsenate is predominant and occupies a large part of the same stability field as ferric hydroxide on a Pourbaix diagram, indicating the two can coexist (Fig. 2).

The more reduced arsenite species is generally the more mobile in mine waters (Bowell 2003) and is the more toxic species (Yamauchi and Fowler 1994). The oxidation of arsenite to arsenate is relatively rapid at higher pH and salinity. Oxidation can also be catalyzed by particular bacteria (Casiot et al. 2003, 2004). In addition, methylation of As can occur in some mine waters leading to formation of monomethyl As and dimethyl arsinic acids (Bowell et al. 1994).

As can be pertained from Figure 2, the oxidation state of Fe and the formation of hydrous ferric oxides can strongly control As mobility, particularly for arsenate (Bednar et al. 2005; Paktunc et al. 2008). In an oxidizing environment with a pH greater than 3, hydrous ferric oxide forms quickly but goethite is thermodynamically more stable. The strong influence of adsorption onto this phase over a pH range of approximately 3-6 can attenuate or limit arsenate mobility significantly (Bowell 1994a; Clara and Magalhaes 2002; Herbal and Fendorf 2006; O'Day 2006). The formation of hydrous ferric oxides is an efficient removal mechanism for As

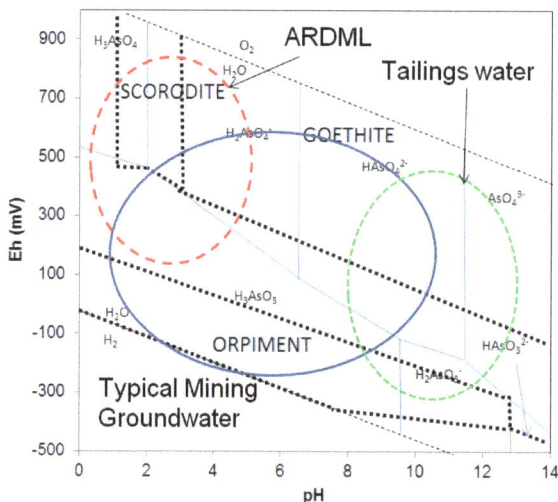

Figure 2. Pourbaix (Eh-pH) diagram showing the stability fields of Fe phases and aqueous species, overprinted on the predominance fields for aqueous As species (light dashed lines show stability field of water). Activities: Fe = 10^{-3} M; As, S = 10^{-6} M; HCO_3 = 10^{-4} M,

from mine waters (Bowell and Craw 2014, this volume). The formation of schwertmannite and jarosite also has the potential to co-precipitate or adsorb As as well (Foster et al. 1998; Clara and Magalhaes 2002; Courtin-Nomade et al. 2005; Asta et al. 2009; Lazareva et al. 2013). In some cases, As release from mine waste can occur as fine grained mineral particles or colloids with As occurring as a sorbed species to fine grained jarosite or hydrous ferric oxide (Slowey et al. 2007).

Arsenic solubility can also be limited by other controls with the formation of secondary arsenate minerals due to saturation with respect to components ions such as those shown in Table 3. Arsenate can also substitute for sulfate in alunite-jarosite minerals and gypsum and for carbonate in calcite and other carbonate bearing minerals (Foster et al. 1998; Savage et al. 2000a,b; Giere et al. 2003; Roman-Ross et al. 2006; Haffert and Craw 2008a).

Arsenic solubility can also arise from dissolution of secondary minerals, particularly those only stable in extreme environments, such as arsenolite from roasted sulfide ore that is stable in alkaline, reduced environments, however exceptions exist such as the Giant mine in Canada (Jamieson 2014). Scorodite solubility is also controlled by pH, being particularly soluble at very low pH and at near neutral to alkaline pH above pH 5 (Dove and Rimstidt 1985; Majzlan 2011). Other potential redissolution mechanisms include the crystallization of amorphous As phases to more crystalline products. For example, goethite and hematite have a lower number of sorption sites for arsenate than ferrihydrite and so recrystallization of amorphous ferric hydroxide will result in release of adsorbed arsenate (Waychunas et al. 1993; Paktunc et al. 2003; Majzlan 2011). Reduction of ferric to ferrous Fe can also lead to desorption of adsorbed As species and if reduction of arsenate to arsenite occurs then increased solubility occurs as well (O'Day 2006; Drahota et al. 2013). However, in highly reducing environments, the presence of sulfide will limit As mobility through the attenuation of As as As sulfide (pararealgar) that has been identified in basal sediments from mine pit lakes (Bowell and Parshley 2005).

Schafer et al (2006) observed As associated with secondary calcite in precipitates formed during evaporation of synthetic pit lake water at a mine in Nevada. Based on electron microscopy analysis, As had higher affinity for secondary calcite than for aragonite or amorphous iron-aluminum-sulfate precipitate (Fig. 3).

Environmental issues associated with arsenic in mine wastes

Since the uses of As are limited and it occurs often in complex forms, As is rarely recovered and therefore considered a waste material. Stringent guidelines exist in monitoring As concentrations in the environment with typical maximum permissible receiving As concentrations in surface water being around 0.01-0.05 mg L^{-1}. A comparison of monitoring levels globally for As in different materials is given in Table 4. Despite these limits, national and international standards are not protective.

The WHO guideline value for As in drinking water (0.10 mg L^{-1}) is based on the consumption of 2 liters of water per day. On average, in the As-affected districts of Bangladesh and West Bengal, adults actually consume 3-4 liters water per day. People working in agriculture or other occupations involving heavy labor are likely to be drinking considerably more, up to 4-6 liters of water per day (Chakraborti et al. 2010).

Bioavailability/bioaccessibility of arsenic in mine waste

In this chapter the authors use the term bioavailability to define an element that is chemically available for sorption or utilization by a biological organism. Bioaccessibility is taken in this chapter as being the actual amount of the element dissolved in the *in vitro* tests. This is perhaps more a geochemist's interpretation, whereas toxicologists may have variation on this.

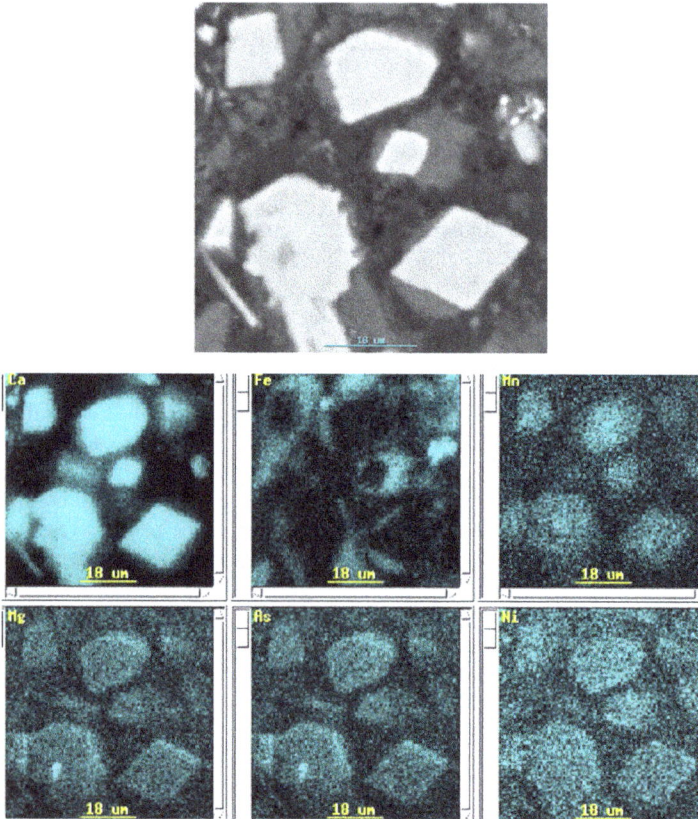

Figure 3. Electron micrograph of secondary calcite grains formed by experimental evaporation of synthetic pit water. From the EDX imaging it can be observed that As is associated with the calcite grains along with Fe without any other metal(loid). From Schafer et al. (2006). Reproduced with permission of the authors.

Mildly reducing environments, such as saturated mine tailings, can lead to the reduction of both ferric Fe and arsenate. Reduction increases the mobility of both elements and is often enhanced by the activity of microorganisms (LeBlanc et al. 1996; Macur et al. 2001; Herbal and Fendorf 2006; Babechuk et al. 2009; Casiot et al. 2009). Consequently, seepage waters from tailings repositories can contain elevated Fe and As and this can result in impacts to the environment (Azcue and Nriagu 1995; Grimes et al. 1995; Stichbury et al. 2000; Craw et al. 2002b).

Where high sulfide contents exist in mine waste, iron- and sulfur-oxidizing bacteria such as *Leptospirillum ferrooxidans* can assist in the oxidation of arsenopyrite. This has been confirmed in experimental studies (Corkhill et al. 2008; McKibbens et al. 2008). In the Matsuo pyrite mine in Iwate Prefecture, Japan, arsenite-oxidizing bacteria have been isolated (Wakao et al. 1988). Here, As in acidic mine drainage (pH 2-2.4) contains up to 13 mg L^{-1} total As. Arsenite-resistant bacteria were observed with up to 27 colonies/mL in mine water samples (Wakao et al. 1988). Six strains of arsenite-oxidizing bacteria have been described as acidophilic gram-negative aerobic rod-shaped bacteria. In experimental studies, these bacteria were shown to be highly specialized in that they could not oxidize ferrous Fe or elemental sulfur as a sole energy source (Wakao et al. 1988). Arsenite-oxidizing bacteria have been characterized even in subarctic environments and were active at 4 °C in biofilms collected at the Giant mine, Canada (Osborne et al. 2010).

Table 4. Comparison of maximum permissible concentrations for As in surface discharge waters applicable to mining and processing operations.

Regulatory Authority/ Country	Value (mg L^{-1})	Class of water	Reference
Australia	0.01	Receiving water standard	Australian and New Zealand Environment and Conservation Council (2000)
Canada	0.01 0.5 0.005	Receiving water Discharge water Aquatic life	Angus Environmental (1991); Canadian Government (2002)
Chile	0.5	Effluent standard	Chilean Ministry of Environment (2010)
DRC	0.4	Effluent standard	SADC (2012)
European Union	0.05	Discharge to surface water	Flanagan (1999)
Ghana	0.01	Discharge to surface water	SADC (2012)
Namibia	0.5	Surface water discharge	SADC (2012)
New Zealand	0.1	Surface water discharge	Australian and New Zealand Environment and Conservation Council (2000)
Russia	0.05	Discharge to fresh water (fisheries)	Ministry of Environment (2002)
South Africa	0.5 0.05	Discharge to surface water Receiving water	SADC (2012)
Turkey	0.1 0.02	Polluted water Good quality water	Flanagan (1999) Flanagan (1999)
USA	0.01 0.005	Surface water general standard Aquatic life standard	USEPA (2012)
Zambia	0.15 0.05	Effluent standard Receiving surface water	SADC (2012)

Analysis of ferric oxyhydroxide-microbial mats from the Lava Cap lode gold mine in California found the mats enriched up to 1000-fold relative to contact water (Foster and Ashley 2002). The predominant organism in the cyanobacterial mats was proposed to be *Leptothrix ochracea* which have sheaths that are commonly 10 μm thick. From trace element analyses of the mats, As in the bacteria appears to be associated with exopolysaccharide matrix secreted by the cyanobacteria. In *Leptothrix* dominated ferric oxyhydroxides, the XAFS (X-ray absorption fine structure) analyses clearly indicate that arsenate is the predominant As species and occurs as an adsorbed and/or co-precipitated complex (Foster and Ashley 2002). The fate of As with the microbial mats is not well understood. In the dry summer season in this part of California, a build-up of partially to completely desiccated microbial mats can occur and these are then flushed by rain in the fall and winter, dispersing the As downstream as a result in the geochemical changes in the environment. The main parameter identified by Foster and Ashley (2002) for the dispersion of As are changes in the redox state that would lead to mobilization of As as ferric oxyhydroxides formed by inorganic and biogenic processes are reductively dissolved, due to the burial or reaction with organic matter. This mechanism would lead to secondary dispersion identified in other studies in historic gold mining districts (Leblanc et al. 1996; Craw and Pacheco 2002; Casiot et al. 2003). A critical factor in this potential dispersion is the physical properties of the ferric oxyhydroxide in terms of particle size and inclusions or rinds of grains (Walker et al. 2009). In a study of As mineralogy and bioaccessibility in tailings in Nova Scotia, a correlation between particle porosity, permeability, surface area and degree of crystallinity were cited as important factors in determining particle dissolution and As release (Walker et al. 2009). Mineralogy in this study was seen as an indicator of bioaccessibility and bioavailability and also provides an insight into the role of biological activity in the dispersion of As. Bioavailability of As was proposed to be highest in sulfide fractions, based on the result of selective extraction for pyrite-rich tails in Almagrera, Spain (Alvarez-Valero et al. 2009). Here, more than 50% of the As is associated with the sulfide fraction.

Arsenic-tolerant flora species have been identified at similar massive sulfide deposits, such as at São Domingo mining area of central Portugal (García et al. 2012). Here, *Erica andevalensis* (an endemic heather) and *Erica australis* have been sampled and found to contain up to 25 mg kg^{-1} and 11.6 mg kg^{-1} (dry mass) As, respectively. These plants are growing on As-rich soils with concentrations up to 7924 mg As kg^{-1} (García et al. 2012). The plants were identified as having very different tolerance mechanisms as *E. australis* accumulates mainly arsenite while in *E. andevalensis,* arsenate is the dominant As species (García et al. 2012). A critical feature in the speciation and preference for As and As species by plants may be the role of ferrous and ferric Fe and soil phosphate levels that would influence co-precipitation in soil or promote plant uptake of dissolved species (Sahai et al. 2007). This study focused on plant As uptake and found that chemically-mixed ferric oxyhydroxide-phosphate-arsenate phases are present in the plant and are similar to those found in natural oxic soils at circumneutral or slightly acidic pH. Similar observations to this experimental work have been made in a regional study in the semi-arid mining district of Zimapan in Mexico. Here, As concentrations in *Prosopis laevigata* (mesquite) and *Acacia farnesiana* (huizache) occur with up to 83 mg kg^{-1} As in *P. laevigata* and up to 225 mg kg^{-1} in *A. farnesiana* (Armienta et al. 2008). Despite variations of several orders of magnitude in surface substrate As concentrations, such variation was not reflected in plant As concentration indicating these species are As-tolerant plants rather than hyperaccumulators and therefore potentially suitable for land stabilization on eroded mine waste.

Toxicity of arsenic in mine waste

Risk to health from As in mine waste has been assessed in several studies (Adriano 2001; Meunier et al. 2010; Plumlee and Morman 2011). Symptoms of chronic As poisoning have been reported in populations reliant on water in several mining areas (Smedley et al. 1996; Smith et al. 1998). Epidemiological evidence of adverse effects at lower exposure levels than the traditional criteria of 0.05 mg L^{-1} has prompted a lower water quality criteria of 0.01 mg L^{-1} for drinking water (Van Leeuwen 1993).

Results of a survey in Korea around the abandoned Myungbong gold mine found that potential human health threats were greatly increased through As uptake in consumed rice grown on As-impacted soils on mine waste (Lee et al. 2008). Against a reference dose, the non-cancer health hazard index showed that the toxic risk due to As was 7.8 times greater and that the increased risk of cancer from As for exposed individuals increased to 1 in 1000, above the regulatory 1 in 10,000 risk. Thus the daily intake of locally produced rice was shown to present a potential health threat due to long term As exposure.

The death of a moose calf in 1994 in the Kenai Fjords National Park in Alaska promoted a study of potential source/media samples. The results indicated that As concentrations within an uncontained historic tailings pond exceeded human health risk-based limits (Lockwood et al. 1997). As a result of this study, historic tailings in the park were stabilized in a concrete mixture and removed for off-site disposal.

Arsenic speciation in mine waste is considered an important control on bioavailability and several studies have applied mineralogy to interpret Physiologically Based Extractive Tests (PBET) and As bioavailability (Walker et al. 2009; Meunier et al. 2010; Bowell et al. 2013). Gastric to intestinal extraction tests indicate that in general As bioavailability decreases via digestion for calcium iron arsenate, lead arsenate, arsenic trioxide > amorphous iron arsenate, As-bearing iron-(oxy)hydroxides > As-rich pyrite and simple As sulfides (such as realgar) > arsenopyrite, scorodite (Meunier et al. 2010; Plumlee and Morman 2011).

Rapid evaluation of As-bearing wastes in Cornwall in a World Heritage site identified the potential for human health impacts from waste residues associated with former As calciners (Bowell et al. 2013). Use of PBET demonstrated that by contrast to natural rock materials, the calciner waste contained appreciable bioavailable As. There are several limitations to the PBET test that need to be recognized; firstly the lack of knowledge regarding how close the PBET values relate to human bioaccessibility; and secondly the lack of soil reference materials for which the As human bioavailability data are known, and that can be used to check *in vitro* tests such as PBET (Cave et al. 2003). The testwork does not generally indicate one particular trend (Table 5). Some samples, such as those from the Wheal Fortune (WFS) mine indicate that As leaching is higher in the first two steps that simulate stomach-acid conditions and decrease in the alkaline tests that simulate intestine conditions. A common observation in these tests is fouling of the test by precipitation of a brown precipitate, possibly a ferric oxyhydroxide phase, supporting this mechanism. However in some samples, noticeably from the Tolgus Calciner and the Higher Condurrow soils, the As concentration increases from the initial digestion to the later digestion stages, possibly reflecting presence of scorodite or similar arsenates thar are less stable in the higher pH intestinal fluids. Clearly, other mechanisms are active and for these materials, the role of unspent lime or charcoal in the calcine solids or of soil phosphate or organic acids in the case of the Condurrow soil may also be influencing the behavior of the ferric oxyhydroxide phase surface chemistry. In addition, the reagents utilized in the test contain organic acids, malic acid, sodium citrate, and bile salts that contain phosphorus. These also play an important role in inhibiting As sorption onto the ferric oxyhydroxide phases.

Further, in the context of the study area, it is unlikely that children aged 2-3 years old would be allowed to become exposed to contaminated material and the PBET does not include the potential bioaccessibility to adults. Consequently, a cautious approach is adopted in interpreting the results of the PBET analysis in the context of this study. The PBET enables the user exposure to be evaluated more rigorously as the test simulates the leaching of sediment in the human gastrointestinal tract, and determines the bioaccessibility of a particular element (i.e., the total fraction that is available for adsorption during passage through the small intestine).

The comparison of the PBET tests to the leach and selective extraction results indicate that much of the As bioavailability can be explained by the presence of adsorbed As, highly soluble As(V) minerals, amorphous iron As(V) minerals and arsenolite as indicated by the higher bioavailability of As in the overburden. The more As-rich waste in the sulfide mine dump and overburden at Wheal Fortune also has higher predicted bioavailability, most likely due to the higher solubility of secondary minerals developed in the oxidized material. This can be observed in the correlation of predicted bioavailable As to leachable As in the leach results (Fig. 4).

Table 5. Results of PBET testwork, Higher Condurrow Mine, Cornwall (from Bowell et al. 2013).

	Total As (mg kg⁻¹)	Phase 1 (mg kg⁻¹)	Phase 2 (mg kg⁻¹)	Phase 3 (mg kg⁻¹)	Phase 4 (mg kg⁻¹)	Phase 5 (mg kg⁻¹)	Bioaccessibility
HCSS01	434	105	115	112	25	10	26%
HCSS02	8088	375	415	506	577	558	7%

Key to leach steps:
Phase 1: Stomach Bio-accessibility 20 minutes. **Phase 2:** Stomach Bio-accessibility 40 minutes. **Phase 3:** Stomach Bio-accessibility 60 minutes. **Phase 4:** Intestinal Bio-accessibility 1 hour. **Phase 5:** Intestinal Bio-accessibility 3 hours.

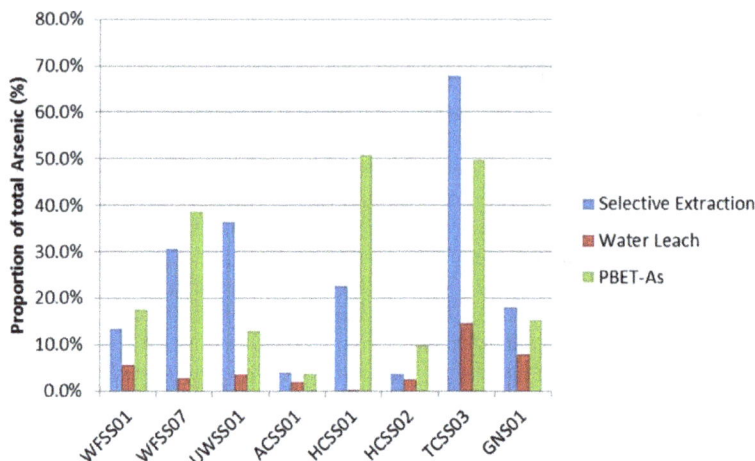

Figure 4. Comparison of leachable As in "reactive fraction" from selective extraction tests, a water leach and PBET-stomach leach tests (taken from Bowell et al. 2013).

The concentration of predicted bioavailable As can be determined from results of the soluble As fraction in selective extraction tests where the sum of As in the "soluble fractions" is taken as the water soluble, ionic, adsorbed and carbonate fractions (Fig. 4). For the Wheal Fortune overburden material, the predicted bioavailable As is higher than the sum of the soluble fractions. This discrepancy can be accounted for by including the amorphous ferric oxyhydroxide fraction and probably also includes a portion of the crystalline ferric oxyhydroxide and sulfide fractions as well. This reflects the solubility of at least a portion of the secondary As(V) minerals on these mine dumps. Against the control site, all sites show anomalous levels of bioavailable As.

Where water leachate and selective extraction results indicated low leachable or soluble As, the PBET results also reflected low potential for bioavailability, confirming that any As present is not in a reactive form with respect to human digestion. Bioavailability is not uniform even within a single site, for example the results from the Higher Condurrow Shaft (shown in Table 5). Overburden sample HCSS01 displays a high portion of As in amorphous or AVS forms in the selective extraction results and up to 400 mg kg^{-1} leached in the BS EN12457 water leach. This has resulted in a potential bioavailability of up to 26% of the total As. However, sample HCSS02 (also overburden) contains 8,000 mg kg^{-1} of As with a high portion of As reporting to the sulfide fraction (in this sample as tennantite and enargite inclusions in chalcopyrite) contains less than 7% bioavailable As demonstrating mineralogical controls on bioavailability.

GEOLOGICAL CONTROLS ON ARSENIC IN MINE WASTE

The processes that govern the mobilization and attenuation of As can be characterized and classified (Bowell et al. 2000). Equally, despite individual peculiarities, mineral deposits can also be classified according to mineralogical and geologic characteristics and by extension, environmental geochemistry characteristics can be added to such a classification (Plumlee 1994; du Bray 1995). The release of As is primarily dependent on the ore and gangue mineralogy and host rock lithology that in turn controls the composition of the mine waters (Plumlee 1994). The important controls in terms of As leaching are: As mineralogy; presence of Fe sulfide, oxide, hydroxide or silicate minerals or siderite (provide an excess of Fe over As);

susceptibility of these minerals to weathering which in turn is a function of porosity and grain size; the geometry of the ore body, and exposure of ore and waste components to air and water; presence, metabolic activity and speciation of microbes; buffering capacity and potential for attenuation of As on Fe oxyhydroxides, clays or zeolites; physical-chemical conditions of the environment.

As an example of geological controls, the As chemistry of water draining from mine waste and tailings from different gold mines is shown in Figure 5 and in Figure 6 for pH against Fe/As ratio. A high As and low pH signature is observed for high-sulfidation epithermal systems, massive sulfide and banded Fe formation deposits. By comparison, gold-bearing porphyries also plot as acidic pH water but with much lower As content. In part, this is due to the much higher Fe/As ratio in these systems in water. Most of the orogenic gold deposits from Canada, Australia, Ghana, Zimbabwe and New Zealand occupy the center of the diagram, indicating in the generally neutral pH and highly variable As concentrations in mine water (Fig. 5). In part, this reflects the As chemistry of the deposits and in some cases the variable Fe contents of the waters (Fig. 6). Several deposits appear to overlap, including skarn type gold deposits, Carlin-type deposits and low-sulfidation epithermal systems show well-buffered water quality in which the pH is circumneutral to alkaline and As chemistry is variable. For Carlin-type deposits, the very high As content and low-Fe content of mine water results in extremely high As at neutral pH for the Getchell deposit in Nevada. Despite precursor geology due to the high lime content of gold process tailings most of them plot as highly alkaline waters with little or no Fe giving extremely high ratios of As (Fig. 6).

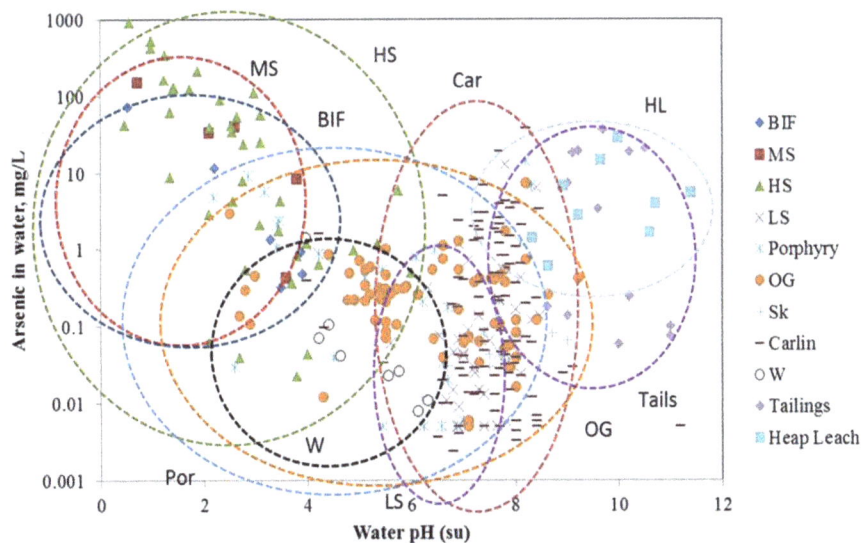

Figure 5. (*for color see Plate 8*) Arsenic concentration versus pH in gold mine waters from published studies. Key: BIF – banded Fe formation; MS – massive sulfide; W – Witwatersrand; Carlin - Carlin type; HS – high-sulfidation epithermal; LS – low-sulfidation epithermal; POR – gold-rich porphyries; Sk – skarn-hosted gold deposits; OG – orogenic or shear zone hosted gold deposits; TCN – cyanide tailings or heap leach facilities (data sourced from; Bowell et al. 1994; Azcue and Nriagu 1995; Grimes et al. 1995; Schwartz 1995; Leblanc et al. 1996; Smedley et al. 1996; Odor et al. 1998; Bennett and Tempel 2000; Bowell et al. 2000, 2003; Savage et al. 2000a,b; Stichbury et al. 2000; Welch et al. 2000; Bowell 2001, 2002; Williams 2001; Lazareva et al. 2002; Wong et al. 2002; Loredo et al. 2003; Craw et al. 2004; Oyarzun et al. 2004; Bowell and Parshley 2005; Haffert and Craw 2008b; Tutu et al. 2008; Bowell et al. 2009; Cidu et al. 2009; Coetzee et al. 2010; Warrender et al. 2012; Drahota et al. 2013).

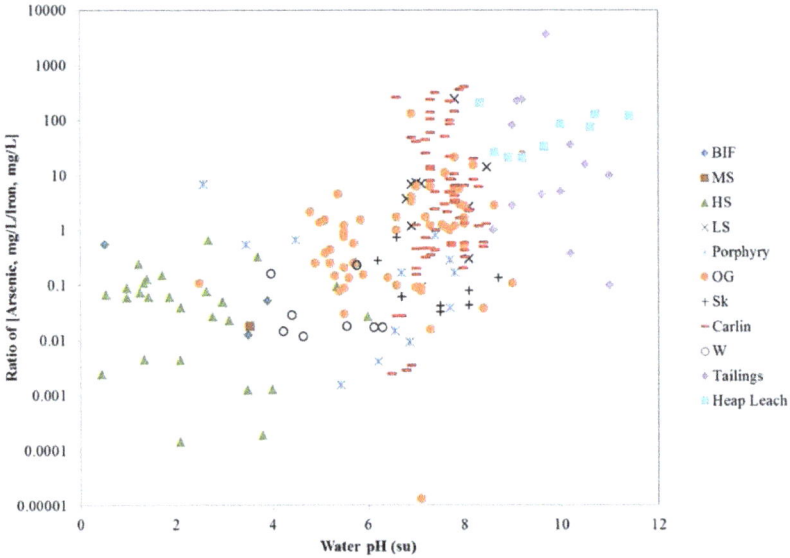

Figure 6. (*for color see Plate 9*) Correlation of pH to As/Fe ratio in mine waters. Same data sources and key as Figure 5.

GOLD MINING AND PROCESSING

Gold mines are major sources of As in the environment, because most such mines have As concentrations that are substantially greater than the gold that is mined (typically <30 mg kg^{-1}). This is especially true for orogenic gold mines (Table 2), where As contents up to 5 wt% (50,000 mg kg^{-1}) can occur throughout the ore. In addition, orogenic gold mines typically contain abundant carbonate minerals, in alteration zones and host rocks, so that acid rock drainage is rare and mine waters have mostly circumneutral pH. As-bearing minerals can be soluble in circumneutral to alkaline pH waters. Management of As in these waters is a higher priority at orogenic gold mines than at other mine sites where more complex metal suites and acid rock drainage (ARD) prevail. Consequently, this section focuses on the relatively simple chemistry of As at orogenic mine sites, with some comparison to epithermal gold mines.

Arsenopyrite is the most common As mineral in orogenic gold mines (Tables 1 and 2), so the environmental mobility of As around these mines is strongly controlled by processes that affect arsenopyrite. Surface dissolution of arsenopyrite involves formation of both As(III) and As(V), so the relative importance of these species is also documented where appropriate, as well as the nature and rates of transformation of arsenopyrite to As(III), and then to As(V).

The high As content of gold deposits is a consequence of natural processes that result in similar mobility of Au and As in hydrothermal fluids that concentrate these elements. These hydrothermal fluids deposit gold and As under the same physicochemical conditions, and so the gold is intimately intergrown with As minerals, particularly arsenopyrite in orogenic gold deposits. Gold can occur in solid solution in arsenopyrite and pyrite, as micron-scale grains enclosed in these sulfides, and as free grains of gold that occur, with sulfides, enclosed in quartz veins. Hence, gold extraction in a mine involves processing of large volumes of As-bearing rock to remove the gold that is essentially a minor or trace element.

The process system at a gold mine typically involves most or all of the stages listed in Table 6. Most of these stages involve interaction between freshly-exposed As sulfide mineral

Table 6. Summary of gold mine processes relevant to orogenic and epithermal deposits, and their potential effects on environmental availability of As.

Processing stage	Product	Effects on As minerals	Environmental significance
Mine excavation (open pits; underground tunnels)	Bare rock walls	Exposure of surfaces to oxidation and dissolution in water	Water runoff with elevated As accumulates in pits and tunnels, and discharges when full
Waste rock deposition	Large piles of oxidised rock with low As	Exposure of surfaces to oxidation and dissolution in water	Water discharge from surface and beneath may have elevated As, acidification
Primary ore crushing & grinding	Gravel, sand, silt ore feed	Finer grain size, exposed sulfide mineral surfaces, e.g., arsenopyrite, arsenian pyrite	As mobilization in mine waters; potential acidification of tailings
Sulfide concentration by flotation or gravity	Concentrate sand, silt	Enhanced proportion of As-bearing sulfides, e.g., arsenopyrite, arsenian pyrite	As mobilization in mine waters; oxidation & acidification of tailings
Roasting, calcining, pressure oxidation	Oxide- & sulfate-rich residue	Oxidation to As-trioxide, As-hematite, arsenates (especially scorodite)	Mobilization of As from residues; dissolution of As minerals, e.g. arsenolite, scorodite;
Heap leaching by weak cyanide-lime solution	Similar residue to mined ore	Exposure of surfaces to oxidation and dissolution in water	Alkaline water draindown solutions may contain As in addition to CN. Long term may be acidic
Cyanidation	Tailings gravel, sand, silt with pH 10-11	Dissolution of As during process agitation	High-As tailings water; high pH facilitates dissolution of As from scorodite, As-ferrihydrite in tailings
Tailings deposition	Gravel, sand, silt in long-term impoundment	Exposure to dissolution by tailings waters or rain	Drainage waters potentially have long-term high dissolved As

grains and oxygenated waters. The enhanced exposure of surfaces of the As minerals is the key process that results in elevated As contents of mine waters. This process occurs naturally during geological uplift, weathering, and erosion, but at a much slower rate than in the mine setting. If gold is contained within sulfide minerals, then these minerals are concentrated for further processing. Typically they are finely milled (to exposure the gold bearing minerals that are typically less than 0.1 mm in size). With a finer grain size more sulfide surface is exposed to oxygen and water and greater oxidation and possible leaching of As could occur. By contrast in a heap leach, the particle sizes are a lot coarser (generally run of mine or crushed to -300 mm) and if sulfides are present they may be only partially exposed to the atmosphere, such that sulfide oxidation is much less intense. In addition, lime is used in a heap leach or cyanidation circuit to prevent volatilization of cyanide (at pH 6.5 as HCN). Excess lime is usually present to buffer any acid although As concentration in solution may still be high.

Oxidation of sulfide minerals is commonly deliberately accelerated in the mine process system, so that solid solution Au and particulate gold are liberated from within the sulfide mineral grains. This stage of the process exposes gold to the transporting slurry in the processing plant, and therefore facilitates dissolution of Au by cyanide or similar chemical reagents. The most common modern process for acceleration of oxidation to release gold is pressure-oxidation, which is done at elevated temperatures (>200 °C) in an oxygen atmosphere. Similar results are obtained by roasting the ore under normal atmospheric conditions at even higher temperatures. The residue of these oxidation processes is normally various arsenite and arsenate minerals, As-bearing hematite, arsenolite, amorphous As-bearing Fe oxyhydroxides or oxides, or Ca-arsenates.

Apart from the few grams of gold/tonne of rock that is extracted, the rest of the mined ore is deposited as mine tailings, which are normally sand or finer grained sediments (Table 6). These tailings contain all the As minerals, now in a fine-grained form, that are available for dissolution in surface waters. Hence, tailings management is one of the largest issues around gold mines, particularly with respect to As in orogenic gold mines. Similarly, waste rocks removed from the mine to expose ore grade material are also disposed of in the vicinity of the mine. These waste rock piles typically have lower As contents than the ore (Table 2). However, epithermal gold deposits have more extensive alteration haloes, and As-bearing sulfides may extend for hundreds of meters beyond the ore grade zones (Table 2), and waste rock at such mines can generate As-bearing and/or acid runoff waters.

CASE STUDIES: NEW ZEALAND GOLD MINES

Orogenic gold mines in New Zealand provide case studies for this chapter that involve a range of mine types, ore compositions, processing systems, and climatic settings (Table 7). Arsenic mobility is a key feature of all these mine sites, with a range of As minerals involved. The following case studies provide a summary of As geochemical processes that occur at these sites. The orogenic gold observations are contrasted with observations from low-sulfidation epithermal gold mines, where As-related issues are distinctly less significant.

The geological setting of the New Zealand gold mines is summarized in Figure 7 (after Craw 2001), showing that there is a clear spatial and tectonic separation between orogenic and epithermal gold deposits. The orogenic deposits were formed in metagreywacke basement during collisional tectonic processes between Paleozoic and Cenozoic, and similar deposits are currently forming near the active transpressive plate boundary, the Alpine Fault (Fig. 7). In contrast, epithermal deposits formed in the North Island in the Cenozoic, and are still forming locally, associated with subduction-related volcanism (Fig. 7). Hence, it is possible to predict environmental issues associated with gold mines, including As issues, on a regional scale, based on tectonic setting of deposit formation.

Table 7. Summary of New Zealand As-bearing gold mine wastes mentioned in the text, with principal primary (1°) and secondary As minerals. Locations are shown on Figure 6.

Deposit type	Location	Age	Climate (rainfall)	Setting of site	Region pH	Local pH of As waste	Metallic ore minerals (1°)	Secondary As minerals	Principal As reactions
Orogenic	Waiuta	Historic, closed 1951	Wet temperate (2300 mm/yr)	Fe-poor, subaerial roaster residue	5-7	3-5	Arsenopyrite (minor pyrite)	Arsenolite waste, oxidized to scorodite cement	Scorodite dissolution; arsenolite oxidation
Orogenic	Waiuta	Historic, closed 1951	Wet temperate (2300 mm/yr)	Fe-poor, saturated roaster residue	5-7	5-7	Arsenopyrite (minor pyrite)	Arsenolite tailings	Arsenolite dissolution, As(III) oxidation
Orogenic	Globe	Historic, closed 1920	Wet temperate (2300 mm/yr)	Fe-rich, underground mine drainage	5-7	7-8	Arsenopyrite, As-pyrite	Pharmacosiderite, amorphous As-HFO precipitates	Pyrite, arsenopyrite oxidation to ferrihydrite, adsorption of As(V)
Orogenic	Bullendale	Historic, closed 1907	Subalpine (900 mm/yr)	Fe-rich, roaster residue	7-8	2.5-3.5	Arsenopyrite, pyrite	Scorodite, kaňkite, zýkaite, As-ferrihydrite	Fe arsenate dissolution & acidification
Orogenic	Macraes	1990-1999	Cool, arid (600 mm/yr)	Fe-rich, sulfide concentrate	7	2-10 cyanidation tailings	Arsenopyrite, pyrite	Scorodite cement in acidified tailings crust	Scorodite deposition from acidification, desiccation
Orogenic	Macraes	Active since 1999	Cool, arid (600 mm/yr)	Fe-rich pressure oxidation residue	7	6-10 cyanidation tailings	Arsenopyrite, pyrite	Scorodite, As-HFO process residue	As(V) adsorption to ferrihydrite
Epithermal (low sulfidation)	Martha	Active since 1989	Warm moist, maritime (2200 mm/yr)	Fe-rich, abundant pyrite impregnation of ore	3-7	7-8 cyanidation tailings	Pyrite, chalcopyrite, sphalerite, galena, arsenopyrite,	As-HFO as sulfide oxidation residue during & after processing	Oxidation of sulfides; As(V) adsorption to ferrihydrite
Epithermal (low sulfidation)	Thames area	Many small historic mines, closed by 1940	Warm moist, maritime (2000 mm/yr)	Fe-rich, abundant acid rock drainage with Cu, Pb, Zn, Cd; As minor	2-7	3-6	Pyrite, chalcopyrite, sphalerite, galena; As in solid solution	As-HFO as sulfide oxidation residue after deposition	Oxidation of sulfides; As(V) adsorption to ferrihydrite

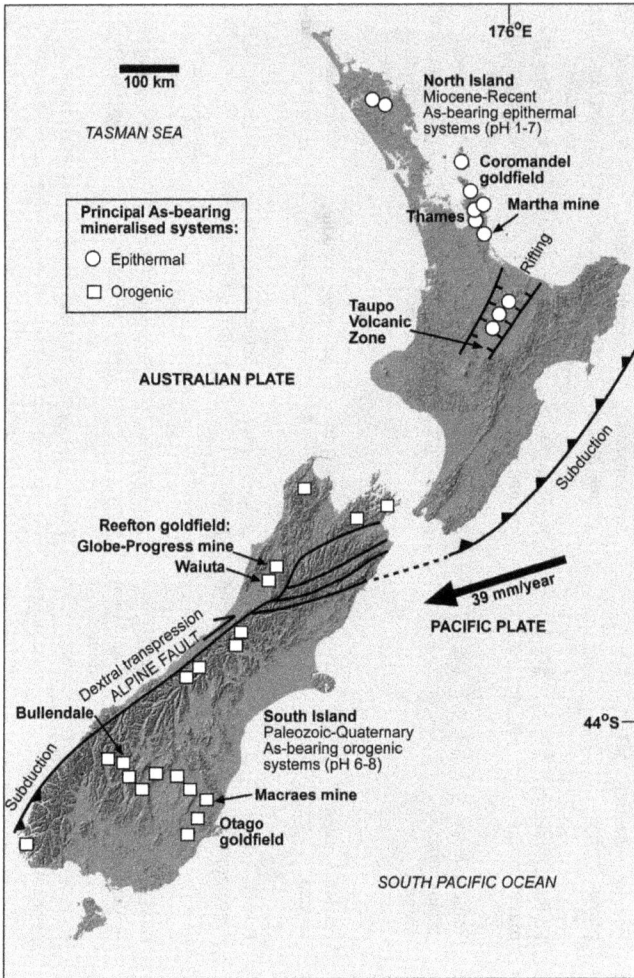

Figure 7. Map of New Zealand showing the locations of mines mentioned in the text and Table 7. Active tectonic setting is shown, as this is currently controlling formation of orogenic gold deposits in the South Island and epithermal gold deposits in the North Island. Active and ancient orogenic gold deposits in the South Island contain abundant As that is readily dissolved in the ambient circumneutral pH waters. Active and ancient epithermal gold deposits in the North Island contain some As, subordinate to base metals, and this As has limited solubility in the associated acid mine drainage.

Arsenopyrite oxidation in orogenic gold mines

Arsenopyrite is the principal As-bearing mineral in New Zealand orogenic gold deposits, so environmental issues regarding As at these mines initially revolve around oxidation of arsenopyrite. Arsenopyrite is stable under relatively reducing conditions over a broad pH range (Fig. 8A). Arsenopyrite remains stable when host rocks are saturated by water, especially slow-moving groundwater, to within a few meters of the surface. Groundwater in contact with arsenopyrite can contain dissolved As up to 0.3 mg L^{-1} in mineralized rocks before mining commences, and similar water As concentrations can persist in these rocks below the active excavations. Dissolved As contents of these waters increase where oxidation has been enhanced by mining activity, and concentrations near to 1 mg L^{-1} are common.

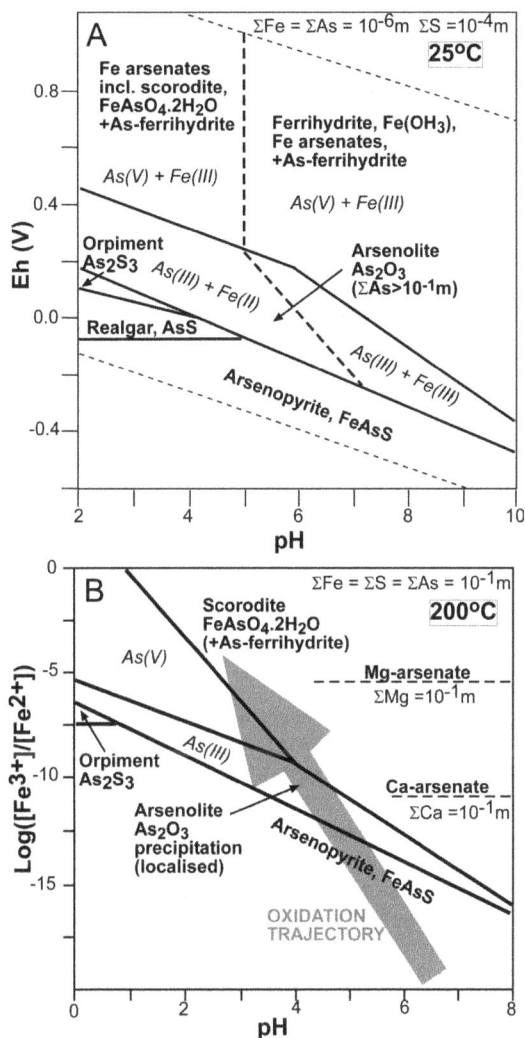

Figure 8. Summary pH-redox diagrams for arsenopyrite in orogenic gold deposits (after Craw et al. 2003; Craw 2006). A. Environmental oxidation of arsenopyrite results in a range of As-bearing phases around South Island gold mines, predominantly As-bearing HFO compounds, and scorodite which dissolves in circumneutral and high pH waters. B. Oxidation of arsenopyrite during roasting to release contained gold at a mine. This oxidation is accompanied by acidification, and can yield precipitates of arsenolite, scorodite, and As-HFO in processed tailings.

Initial oxidation of arsenopyrite yields dissolved As(III), which is highly soluble so that arsenolite will precipitate only when dissolved As reaches extremely high levels (Fig. 8A). Accompanying Fe from dissolution of arsenopyrite and pyrite is also soluble as Fe(II) under acid and circumneutral conditions, but at higher pH, Fe(III) precipitates (Fig. 8B). With further oxidation, As(III) oxidizes to As(V), and Fe(III) forms precipitates under most pH conditions, except in strongly acid or alkaline conditions. Oxidation of As(III) in natural waters from arsenopyrite-bearing rocks occurs on a time scale of days (Fig. 9), and slow-moving groundwater seepages discharge into the environment with essentially 100% As(V).

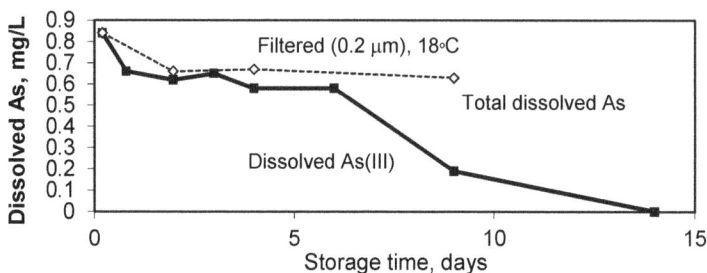

Figure 9. Time scale for oxidation of dissolved As in groundwater from arsenopyrite-bearing rock, discharged from an underground tunnel, Macraes orogenic gold mine (Table 3). Water was collected with essentially 100% As(III), and was re-analyzed repeatedly over 2 weeks (after Craw and Pacheco 2002).

Oxidation of arsenopyrite at high temperature during gold mine processing results in similar changes to those described for surficial conditions (Fig. 8A,B). However, the relatively small, almost closed, systems of these processes can result in locally high As concentrations, and arsenolite is a common precipitate. Some historic roasting processes deliberately volatilized the As so that arsenolite precipitated in flues, from which it was extracted as a by-product. In a modern pressure oxidation system, arsenolite can precipitate in the reaction vessel, but most discharging material is fully oxidized to As(V). Precipitates of this As(V) include scorodite and As-bearing HFO (Fig. 8B).

Iron-poor orogenic gold mine wastes

Ore rich in arsenopyrite but with relatively little pyrite results in relatively Fe-poor mine wastes, in which the geochemical processes are dominated by As minerals. The relatively low Fe content of mine wastes can be further depleted by the processing system that selectively removes Fe precipitates. Both these situations have contributed to Fe-poor mine wastes at the historic Waiuta mining area (Fig. 7; Table 7). The quartz vein system that was mined contained abundant arsenopyrite and only minor pyrite, and this As-rich ore was roasted to save As as a by-product. The roasting process produced arsenolite and hematite from arsenopyrite, and residues from this process remain at the site. Similar roasting processes at the orogenic gold mine at Bullendale (Fig. 7; Table 7) produced mine wastes that are now dominated by Fe arsenates, with no arsenolite. Geochemical processes that have occurred in these mine wastes are summarized in Fig. 10A-C.

Some of the original arsenolite discarded on the ground inside and outside the historic buildings at Waiuta has remained unaffected by the subsequent 60 years of damp climate, where it has been protected from direct rainfall incursion. However, much of the arsenolite has dissolved, and the As(III) has oxidized to As(V) that reacted with dispersed hematite to form scorodite cement in the mine wastes (Fig. 10A). This scorodite cement also undergoes dissolution in rainstorms, yielding runoff waters with dissolved As up to 30 mg L^{-1}. A similar site at Bullendale (Fig. 7, Table 7; Haffert and Craw 2010), but without preserved arsenolite, is dominated by scorodite and the closely-related Fe arsenates, kaňkite and zýkaite.

A notable feature of these low-Fe mine wastes is that the sites are locally acidic, with pH down to ~3. This is distinctly different from the pH of other mine wastes in the area, and the ambient pH of the region, which is either circumneutral or alkaline. Likewise, most other orogenic gold sites in New Zealand have circumneutral or alkaline pH, apart from local sites with abundant As and low Fe contents (Table 7). The acidification of the arsenolite-bearing Waiuta sites has resulted from two As reactions: As(III) oxidation, and scorodite dissolution (Fig. 10A). The As(III) oxidation occurs as arsenolite dissolves (Fig. 7A) via the reaction:

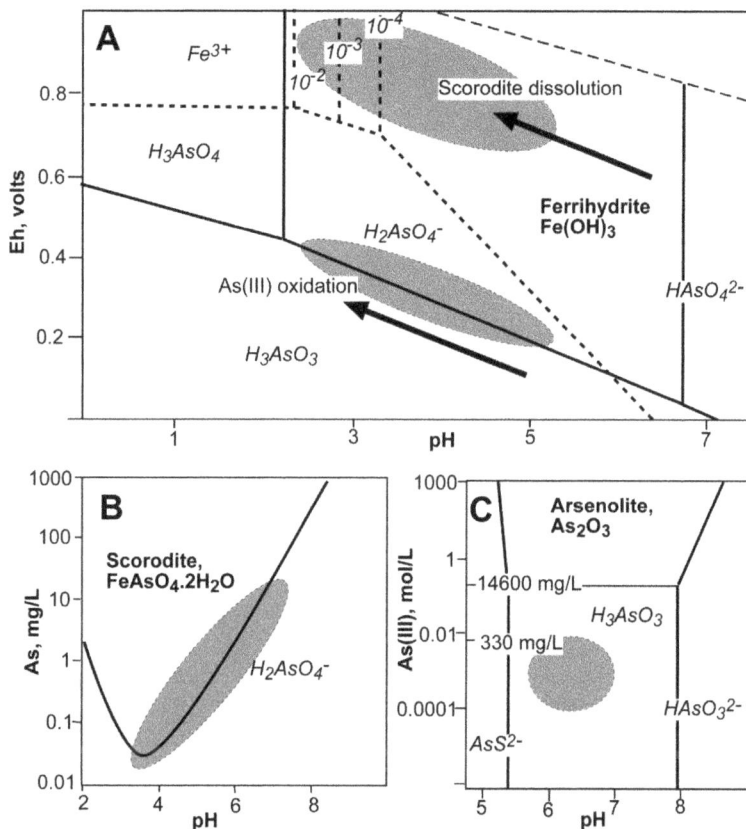

Figure 10. Geochemical diagrams for Fe-poor orogenic gold mine sites, Waiuta and Bullendale (Fig. 4, Table 7; after Haffert and Craw 2008a,b, 2010; Haffert et al. 2010), showing the environmental stability of As minerals and environmental mobility of dissolved As. Grey ellipses show ranges of water parameters at the mine sites. A. Eh-pH diagram showing dissolved As reactions that cause environmental acidification in mine tailings: oxidation of arsenolite (lower arrow and grey ellipse) and dissolution of scorodite (upper arrow and grey ellipse). B. Waters running off mine tailings with abundant scorodite and/or kaňkite have pH that evolves towards the low dissolved As point of the scorodite dissolution curve of Krause and Ettel (1988). C. Water-saturated arsenolite-rich tailings at Waiuta have extremely high dissolved As(III) at circumneutral pH, but observed dissolved As is still well below saturation with As(III).

$$As_2O_3 \text{ (arsenolite)} + 3H_2O \rightleftharpoons 2H_3AsO_3 \text{ (aq)} \quad (1)$$

and this is followed by:

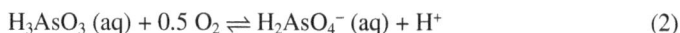

$$H_3AsO_3 \text{ (aq)} + 0.5\,O_2 \rightleftharpoons H_2AsO_4^- \text{ (aq)} + H^+ \quad (2)$$

Reaction (2), the oxidation of arsenite to arsenate ions, generates acidity, and this is the process that causes acidification at the Waiuta site. Likewise, scorodite (or kaňkite) dissolves incongruently via the reaction:

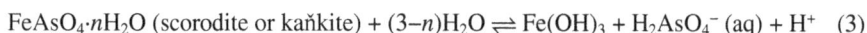

$$FeAsO_4 \cdot nH_2O \text{ (scorodite or kaňkite)} + (3-n)H_2O \rightleftharpoons Fe(OH)_3 + H_2AsO_4^- \text{ (aq)} + H^+ \quad (3)$$

Reaction (3) also generates acidity, although the greater solubility of arsenolite at Waiuta means that scorodite dissolution is only a minor part of the acidification process which is reversed when scorodite precipitates *in situ* (Fig. 10A). However, scorodite and kaňkite dissolution is responsible for acidification of the Bullendale site (Fig. 10A). There is a broad geochemical limit to these acidification processes, defined by the arsenate species predominance boundary between $H_2AsO_4^-$ (aq) and H_3AsO_4 (aq), at pH near 2 (Fig. 10A; Haffert et al. 2010).

Scorodite becomes progressively less soluble with decreasing pH from circumneutral conditions, towards a minimum at pH 3-4 (Fig. 10B). Hence, the acidification processes encourage scorodite precipitation or re-precipitation, and limit the amount of dissolved As(V) in waters emanating from the sites (Fig. 10B). Under circumneutral conditions, dissolved As(V) can exceed 10 mg L^{-1}, but runoff from the acidic mine wastes has dissolved As(V) concentrations <0.1 mg L^{-1} (Fig. 10B). The generation of hydrogen ions by Reactions (2) and (3) decreases towards this pH limit (Haffert et al. 2010). Therefore, this acidification process results in natural limitation of dissolved As contents in discharging mine waters in contact with scorodite.

Water-saturated tailings at Waiuta are relatively reduced, and As(III) predominates. Arsenolite is preserved, without any transformation to scorodite in these reduced tailings after 60 years of abandonment. Dissolved As(III) is locally very high in the tailings pore waters, up to 330 mg L^{-1} (Fig. 10C). Even this very high dissolved As is not saturated with respect to arsenolite, and higher dissolved As(III) is theoretically possible (Fig. 10C).

Iron-rich orogenic gold mine wastes

Orogenic gold deposits with abundant pyrite are more common than the Fe-poor sites described in the previous section. Oxidation of pyrite generates sulfuric acid, which reacts with abundant carbonate minerals in host rocks and alteration zones. The carbonate content generally greatly exceeds the acid generation potential of the pyrite on all but the centimeter scale, and the ambient pH of mine waters remains circumneutral or weakly alkaline. The pyrite oxidation and neutralization processes cause precipitation of abundant hydrous ferric oxide (HFO) that is typically X-ray amorphous but has mineralogical affinities to ferrihydrite, $Fe(OH)_3$.

The ferrihydrite precipitates are extremely effective at adsorbing As(V) from associated mine waters because of the very high surface area of the nanoparticles. This results in extraction of dissolved As that has been derived from arsenopyrite oxidation at the same time as pyrite oxidation occurs, coupled with precipitation of As-rich ferrihydrite. At the ambient circumneutral pH of these mines, scorodite deposition is unlikely unless dissolved As(V) reaches very high levels, >100 mg L^{-1} (Fig. 11B). Adsorption of As prevents these high As levels being attained in the discharge waters.

The Globe orogenic gold mine occurs in the same goldfield as the Waiuta mines (Fig. 6; Table 7) but has abundant pyrite coexisting with arsenopyrite, and therefore the environmental response to oxidation of mine rocks is distinctly different at these mine sites (Fig. 11A,B). Waters emanating from historic Globe mine underground tunnels have precipitated As-HFO deposits that are locally several meters thick. The As contents of these precipitates can be as high as 30 wt% locally, and bulk deposits can have As contents of >10 wt% (Fig. 12). Some of the initial precipitates have been remobilized into veins (Fig. 9B), with better-defined crystal structures, including pharmacosiderite.

The amount of As adsorbed by the ferrihydrite precipitates at the Globe mine varies widely, as does the amount of dissolved As in the associated mine waters (Fig. 12). The ratio of adsorbed to dissolved As, or K_d, typically ranges between 10^3 and 10^5, and can be as high as 10^6 (Fig. 12). Laboratory experiments on adsorption of As from mine waters from Macraes orogenic gold mine (Fig. 6; Table 7) yield a predicted K_d of 10^5, similar to the higher values observed in the discharges from historic Globe mine (Fig. 12). Hence, precipitation of ferrihydrite at these Fe-rich mine sites can be an effective natural As-extraction and remediation process. However, despite these adsorption processes, dissolved As(V) concentrations of up to 50 mg L^{-1} have drained from the site, albeit in small volumes. This dissolved As(V) is much higher than that emanating from the self-acidified scorodite-rich tailings in the Fe-poor systems described above.

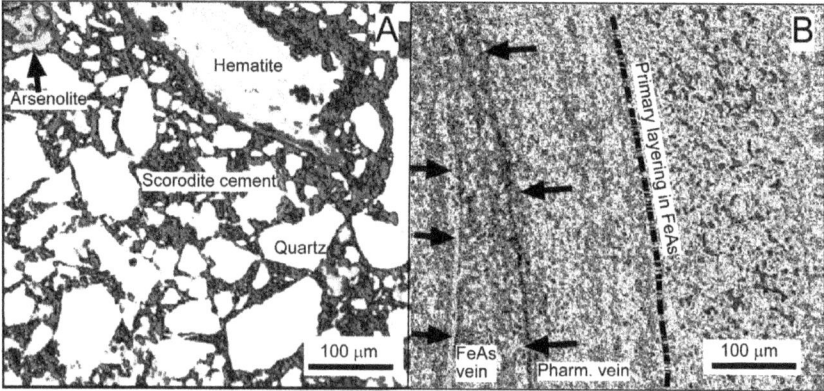

Figure 11. Microphotographs from As-rich mine wastes at Waiuta and Globe historic orogenic gold mine sites (after Haffert and Craw 2008a). A. Scorodite has formed a cement in tailings from a roaster at the Fe-poor Waiuta site. The scorodite has formed as a secondary mineral from arsenolite residue (remnants labeled in top left) at the roaster, with Fe provided by dissolution of minor hematite impurities in the tailings. B. Electron microprobe As map of As-bearing Fe oxyhydroxide precipitate at the Fe-rich Globe mine site. Lighter shades represent areas richer in As. Primary precipitate layering is near-vertical in this image, and is crosscut by subparallel veins of pharmacosiderite (Pharm) and amorphous As-rich Fe oxyhydroxide.

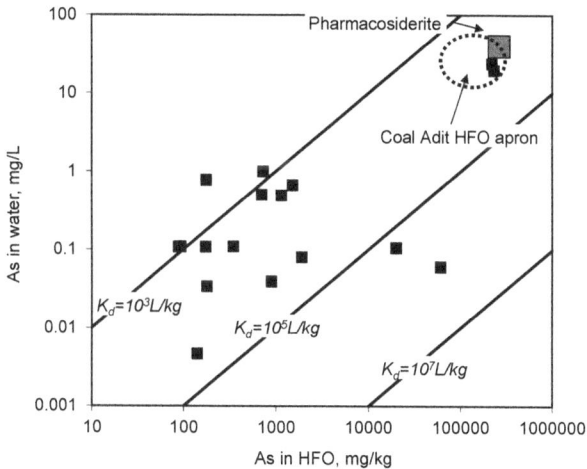

Figure 12. Arsenic contents of mine waters discharging from historic underground mine tunnels at the Fe-rich (pyritic) Globe orogenic gold mine (Fig. 6, Table 7), and of coexisting Fe oxyhydroxide precipitates which have adsorbed As. Diagonal lines are theoretical bulk K_d for As in solids and water. Maximum observed K_d is similar to laboratory experimental results of $\sim 10^{-5}$. The highest solid and water As compositions are associated with precipitates outside a deep drainage tunnel, the Coal Adit (dotted circle, top right), in which pharmacosiderite veins have precipitated. Data after Roddick-Lanzilotta et al. (2002) and Craw et al. (2004).

Modern Macraes orogenic gold mine

The Macraes gold mine (Fig. 7; Table 7) that opened in 1990 is a world-class deposit with >9 Moz resource. The ore is Fe-rich, with abundant pyrite as well as arsenopyrite. Development of the mine involved several changes to the processing system, leading to variations in the geochemical setting of As in the mine wastes. Initial mining involved cyanidation of a sulfide concentrate, with disposal of tailings in a dedicated impoundment (Fig. 13). Then for several years, these concentrate tailings were remixed with the bulk tailings in the main tailings dam. Subsequently, pressure-oxidation was introduced to oxidize the sulfide concentrate. The sulfide concentrate tailings have since been reprocessed through a more efficient system for their contained residual gold.

In the initial stages of mining, when arsenopyrite went right through the process stream, arsenopyrite became oxidized and partially dissolved as it was crushed and agitated in the ore slurry. Most such dissolution of As occurred in the cyanidation process, where arsenopyrite was fine grained and the solution pH was alkaline (pH 10-11). The ore slurry was oxygenated by air during agitation, and this oxidized alkaline solution was capable of dissolving As up to several hundred mg L^{-1} (Fig. 14).

Concentrations of >400 mg L^{-1} have been recorded in this process stream. Almost all of the dissolved As was As(V), although some As(III) occurred locally. The very high levels of As(V) in the cyanidation process were probably mitigated by adsorption to ferrihydrite that was generated at the same time, from both arsenopyrite and pyrite.

The process discharge waters remained alkaline in the concentrate tailings impoundment, although they evolved with time towards pH 8 (Fig. 13). The high pH and high As contents means that the dissolved As(V) concentrations are limited by scorodite solubility (Fig. 15), and secondary scorodite is widespread in the tailings. The mine is in a semiarid environment, with abundant evaporation, so evaporative concentration of mine waters has occurred in the shallow tailings at times, resulting in scorodite crust formation, and localized cementation of tailings by scorodite. Oxidation of pyrite in the surface crusts (1-5 cm thick) resulted in acidification, which also promotes scorodite deposition as scorodite solubility decreases with pH (Fig. 10B).

The high dissolved As of tailings waters was attenuated in the tailings impoundment, and seepages from the impoundment had lower, but significant, dissolved As. This dissolved As,

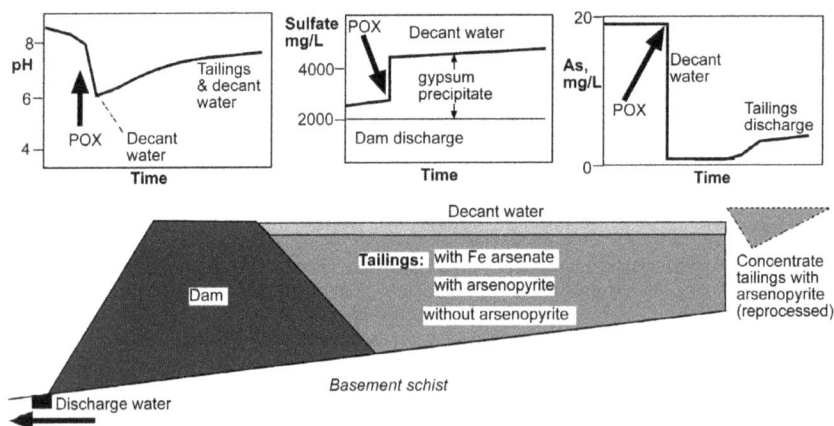

Figure 13. Sketch of the mine tailings system at the active Macraes orogenic gold mine (after Craw 2003). Early mining (starting 1990) involved separate deposition of arsenopyrite-pyrite rich concentrate, and the main tailings dam contained little arsenopyrite. The concentrate tailings have since been reprocessed, and all arsenopyrite and pyrite were deposited in the main tailings dam until 1999, when pressure-oxidation processing (POX) yielded As-rich ferrihydrite and scorodite to the tailings. Geochemical changes in the tailings over the change to pressure oxidation are indicated in sketch graphs at top. Dissolved As in discharging mine waters (currently recycled) is lower after introduction of pressure-oxidation because of enhanced adsorption to Fe oxyhydroxides.

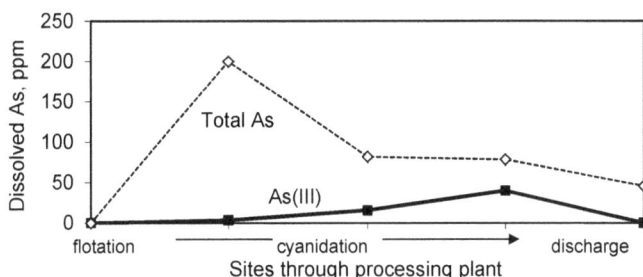

Figure 14. Total As and As(III) analyses through the processing system at Macraes orogenic gold mine, as arsenopyrite is separated with pyrite by flotation from the silicates, reground to ~15 microns, passed through cyanidation tanks (pH 10-11), and then discharged to tailings. Data after Craw and Pacheco (2002).

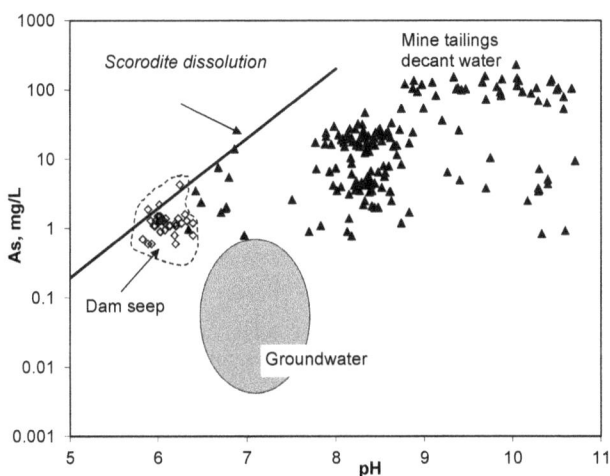

Figure 15. Mine water compositions at the Macraes orogenic gold mine, showing the high As contents of high-pH waters in the sulfide concentrate tailings (see Fig. 13) that were discharged from the cyanidation plant with pH of 10.5 (Table 7). Evaporative drying at the tailings surface caused As saturation as pH was neutralized, with localized precipitation of scorodite. Seeps from the tailings dam have lower pH, probably because of pyrite oxidation, and also had As concentrations that were saturated with respect to scorodite and were controlled by scorodite deposition. Background groundwater As contents in arsenopyrite-bearing rocks unaffected by mining are indicated with the grey ellipse.

all As(V), was typically 1-10 mg L^{-1} (Fig. 15) but eventually evolved as high as 20 mg L^{-1}. Attenuation of As in the tailings impoundment was facilitated by adsorption of As to ferrihydrite that was formed by oxidation of pyrite. In addition, this pyrite oxidation generated acid that contributed to a general lowering of discharge water pH (Fig. 15). The lower pH also limited dissolved As concentrations by further precipitation of scorodite (Fig. 15).

Most dissolved As in discharged sulfide concentrate tailings was As(V) as indicated in Figure 15. However, the tailings contained abundant arsenopyrite and pyrite, and the water-saturated parts consequently maintain low redox *in situ*. The pore waters soon became dominated by As(III), on a time scale of weeks or months (Fig. 16). However, an oxidation front progressively moved down through the tailings, focused on the most permeable layers, on a time scale of 5-10 years, resulting in progressive oxidation of As(III) to As(V), with associated deposition of scorodite cement (Fig. 16). Scorodite formed cements adjacent to desiccation cracks (centimeter scale) in old tailings, but this scorodite was rapidly redissolved

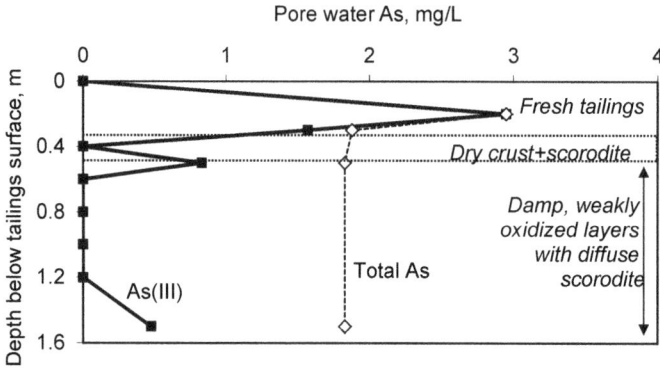

Figure 16. Relative distribution of As(III) and total As in pore waters in a section dug through the surface zone of the concentrate tailings impoundment, Macraes orogenic gold mine. This section includes old tailings (~5-10 years) at the base, with a dry crust, on which fresh tailings (1-2 months old) had been deposited. Fresh tailings waters largely contain As(III), and old tailings are dominated scorodite and dissolved As(V). Data after Craw and Pacheco (2002).

when new tailings were deposited in the desiccation cracks. The generally low permeability of the tailings ensured that chemical changes occurred by diffusion, rather than advection of water through the tailings. Experimental neutralization of the scorodite-bearing acid crust on the tailings with wet limestone resulted in diffusive neutralization and scorodite dissolution on the 10 cm scale in less than a year (Craw et al. 2002a).

The above-described As mobility at Macraes mine changed completely with the introduction of pressure oxidation to the process system. The sulfide concentrate is oxidized, producing amorphous Fe arsenate, As-ferrihydrite, and scorodite. Small amounts of arsenolite are precipitated in the reactor vessel, but this mineral does not persist through the reactor to the discharge point. The resultant oxidized solution is highly acidic, and has to be neutralized before cyanidation. Some acidity persists through to the tailings in the form of unoxidized Fe(II) that hydrolyses in the tailings impoundment to lower the overall tailings pH (Fig. 16). Tailings waters are so rich in sulfate that gypsum precipitates within the tailings system (Fig. 16). Dissolved As in tailings discharge waters dropped dramatically with the onset of pressure oxidation (Fig. 16), presumably because of the abundance of new ferrihydrite surfaces for adsorption of As.

Epithermal gold mines

In contrast to the orogenic gold mine sites described above, epithermal mine sites in the North Island of New Zealand (Fig. 7) typically involve some degree of acid rock drainage development. This acidification is driven by the abundant pyrite in the deposits themselves, and in the widespread alteration haloes surrounding the deposits (Craw 2001). Some epithermal vein systems contain abundant carbonate minerals, and these can be sufficient to neutralize acidity produced by pyrite oxidation acid for small volumes of rock. However, this is minor compared to extensive acid rock drainage that can develop from the adjacent waste rocks. Cyanidation mine tailings commonly have circumneutral pH because the high pH that cyanide requires (Table 6) mitigates acid generated by pyrite oxidation. However, these tailings typically require storage under water to prevent further pyrite oxidation overcoming their neutral pH status. The process of acidification around epithermal deposits is a natural one, and mineralized rocks, without mines, can also contribute acid waters with pH down to 3 to nearby streams.

In addition to the ARD issues, epithermal deposits commonly contain a wide variety of minor and trace metals, and many of these, such as Cu, Pb, and Zn, are generally more abundant than As. This distinction is shown for Cu and As for some typical North Island (Coromandel area, Fig. 7) epithermal mine rocks, compared to some orogenic gold mine rocks from the South Island (Fig. 17). This distinction is exacerbated by the common acid mine waters around epithermal mines, which readily dissolve and transport the bivalent cationic metals, especially Cu, Pb, Zn and Cd (Webster-Brown and Craw 2005). In contrast, As is relatively less soluble in acid solutions (Fig. 8B). Consequently, most waters associated with these mineralized rocks have dissolved As at or below 0.01 mg L^{-1} (Fig. 17).

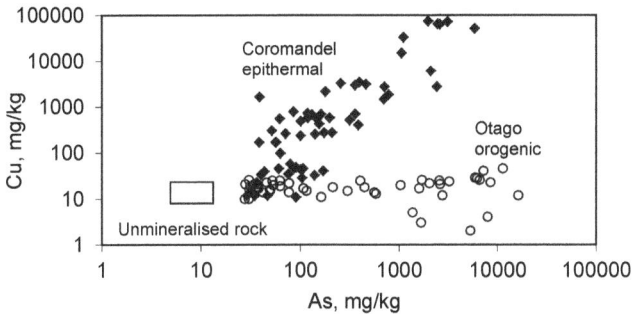

Figure 17. Comparison of mineralized rock compositions in epithermal gold deposits (Coromandel) and Otago orogenic gold deposits (Fig. 6, Table 7). The relative significance of base metals in epithermal deposits is indicated by the relatively high Cu contents. Waters derived from these epithermal deposits typically have low pH and strongly elevated dissolved base metal contents (Cu, Pb, Zn, Cd), but relatively low As contents. Oxidation of abundant epithermal pyrite causes this acidification and provides Fe oxyhydroxide precipitates for adsorption of As from mine waters. Data from Craw (2001).

Epithermal/geothermal spring deposits

Active geothermal systems are associated with the subduction-driven volcanism in the central North Island, and these are surface expressions of deeper epithermal gold-forming systems such as those that have become mines in the Coromandel area (Fig. 7). These active geothermal systems bring gold and also abundant dissolved As to the surface, where As-bearing precipitates can form. The precipitates include amorphous As, Sb, Tl sulfides whose deposition is at least partially facilitated by biological processes. The geochemistry of As-bearing waters draining from As-rich springs is also affected by biological processes, some of them on a daily basis as photosynthesis waxes and wanes and changes the geochemical environment, especially the dissolved oxygen content (Pope et al. 2004). The daytime increase in dissolved oxygen causes an increase in dissolved As, whereas lower dissolved oxygen at night facilitates deposition of As sulfide precipitates (Pope et al. 2004).

Some geothermal spring systems have elevated Hg accompanying the As, and these are most abundant in the far north of the North Island (Fig. 7). Both active and young fossil springs occur in this area, and some of these sites have been mined for mercury and prospected for gold. There is a crude positive relationship between As and Hg in mineralized rocks and spring precipitates in these geothermal systems, and As typically exceeds Hg (Fig. 18A). Most mercury occurs as cinnabar in localized concentrations, and some of the cinnabar has As in solid solution. The cinnabar is accompanied by abundant pyrite and/or marcasite, which also have As in solid solution. Oxidation of the Fe sulfides in the surficial environment results in extensive acid rock drainage around mineralized sites, with pH commonly ~3. Leaching experiments show that both Hg and As can be readily mobilized in acid solutions representative

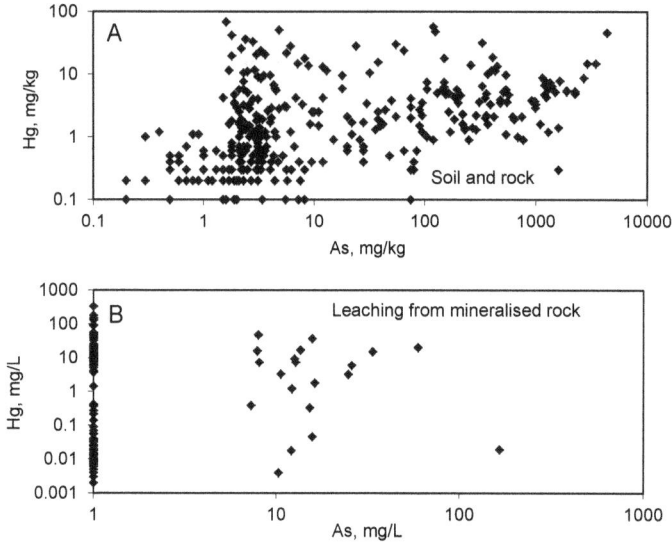

Figure 18. Concentrations of As and Hg in rocks and soils (A), and in leachates from mineralized rocks (B) in epithermal/geothermal spring deposits in the far north of the North Island of New Zealand (Fig. 7). Arsenic exceeds Hg in many rocks, soils, and leachates but dissolved As is very low in nearby streams. Data after Craw et al. (2002b) and references therein.

of acid rock drainage settings (Fig. 18B). However, dilution of these leachates occurs in the environment, and dissolved As in streams is generally low (~0.01 mg L^{-1}).

CONCLUSIONS

This review demonstrates that As is an important minor or trace element in many mineral deposits, particularly those of gold. In contrast to cations of metals, As is present in mine waters over a range of reduction-oxidation potential and varying pH conditions reflecting the complex, multi-valence chemistry of an oxyanion rather than cation metal behavior.

Although potentially mobile, several natural attenuation mechanisms exist for As oxyanions. Co-precipitation can occur with sulfide in low reduction-oxidation potential environments to form As sulfides and some cation metals to form arsenides. Co-precipitation with a range of metals is possible for arsenate, and even in restricted environments, for arsenite. Adsorption can occur onto a range of minerals, particularly Fe oxyhydroxides, providing a local sink for As. However, adsorbed As can also constitute a secondary As source if changes in redox potential and/or pH facilitate As desorption and release into the environment. Arsenic shows an affinity for biological utilization, despite its potential toxicity, and arsenite can be utilized by some bacteria as an energy source. Arsenic uptake in plants is complex and some species are sufficiently evolved to selectively uptake different species of As and restrict toxicity through co-precipitation in the plant structure.

As an example of the application of the characterization the As geochemistry of contrasting gold mineral deposits and mines in New Zealand reflects that even in a small area with similar geology a wide variation in As dispersion in the environment can occur. The examples presented are representative of similar gold deposits elsewhere in the world and underlines the importance of site-specific case studies in order to better understand the geochemistry of As in mine wastes and waters.

REFERENCES

Adriano DC (2001) Trace Elements in the Terrestrial Environment. 2nd Edition, Springer-Verlag

Agricola G (1556) De Re Metallica. Translated by HC Hoover and LH Hoover. Dover Publications

Alvarez-Valero AM, Saex R, Perez-Lopez R, Delgado J, Nieto JM (2009) Evaluation of heavy metal bioavailability from Almagrera pyrite-tailings dam. J Geochem Explor 102:87-94

Angus Environmental (1991) Review of Environmental Quality Standards. Inland Water Directorate, Ottawa, Science Series

Armienta MA, Ongley LK, Rodriguez R, Cruz O, Mango H, Villasenor G (2008) Arsenic distribution in mesquite (*Prosopis laevigata*) and huizache (*Acacia farnesiana*) in the Zimapan mining area, Mexico. Geochem Explor Environ Anal 8:191-197

Asta MP, Cama J, Martinez M Gimenez J (2009) Arsenic removal by goethite and jarosite in acidic conditions and its environmental implications. J Hazard Mater 171:965-972

Australian and New Zealand Environment and Conservation Council (2000) National Water Quality Management Strategy: Paper No. 4. Australian and New Zealand Guidelines for Fresh and Marine Water Quality. Volume 1. October 2000, Canberra

Azcue JM, Nriagu JO (1995) Impact of abandoned mine tailings on the arsenic concentrations in Moira Lake, Ontario. J Geochem Explor 52:81-89

Babechuk MG, Weisener CG, Fryer BJ, Paktunc D, Maunders C (2009) Microbial reduction of ferrous arsenate: biogeochemical implications for arsenic mobilization. Appl Geochem 24:2332-2341

Bednar AJ, Garbarino JR, Ranville JF, Wildeman TR (2005) Effects of iron on arsenic speciation and redox chemistry in acid mine waters. J Geochem Explor 85:55-62

Bennett JC, Tempel RN (2000) Geochemical controls on water quality in groundwaters at the Getchell Mine. *In:* Geology and Ore Deposits 2000: The Great Basin and Beyond. Cluer JK, Price JG, Struhsacker EM, Hardyman RF, Morris CL (eds) Geological Society of Nevada Symposium Proceedings, Reno/Sparks, May 2000. p 783-798

Betterton J (2000) Famous mineral localities: Penberthy Croft mine, St Hilary, Cornwall, England. UK J Mines Miner 20:7-37

Bowell RJ (1994a) Arsenic sorption by iron oxyhydroxides and oxides. Appl Geochem 9:279-286

Bowell RJ (1994b) Sulfide oxidation and the Production of Gossans, Ashanti Mine, Ghana. Int Geol Rev 36:732-752

Bowell RJ (2001) Hydrochemistry of the Getchell underground mine. Part 1: mine water chemistry. Mine Water Environ 20:81-97

Bowell RJ (2002) Hydrogeochemical dynamics of Pit Lakes. *In:* Mine Water Hydrogeology and Geochemistry. Younger PL, Robins N (eds) Geol Soc London Special Pub 97:181-212

Bowell RJ (2014) Hydrogeochemistry of the Tsumeb deposit: implications for arsenate mineral stability. Rev Mineral Geochem 79:589-627

Bowell RJ, Craw D (2014) The management of arsenic in the mining industry. Rev Mineral Geochem 79:507-532

Bowell RJ, Parshley JV (2003) Arsenic cycling in the mining environment. *In:* USEPA Workshop on Managing arsenic Risks to the Environment: Characterization of Waste, Chemistry and Treatment and Disposal. Proceedings and Summary Report, Ohio, USEPA/625/R-03/010, p 10-12

Bowell RJ, Parshley JV (2005) Control of pit lake water chemistry by secondary minerals, Summer camp Pit, Getchell mine, Nevada. Chem Geol 215:373-385

Bowell RJ, Morley NH, Din VK (1994) Arsenic speciation in porewaters, Ashanti, Ghana. Appl Geochem 9:15-22

Bowell RJ, Baumann M, Gingrich M, Tretbar D, Perkins WF, Fisher PC (1999) The occurrence of gold at the Getchell Mine, Nevada. J Geochem Explor 67:127-144

Bowell RJ, Rees SB, Parshley JV (2000) Geochemical predictions of metal leaching and acid generation: geologic controls and baseline assessment. *In:* Geology and ore Deposits 2000: The Great Basin and Beyond. Cluer JK, Price JG, Struhsacker EM, Hardyman RF, Morris CL (eds) Geological Society of Nevada Symposium Proceedings, Reno/Sparks, May 2000, p. 799-823

Bowell RJ, Kabellega D, Knol R, McIveen G, Rees SB, Shoo K, Stephen R (2003) Establishing a geochemical baseline for environmental assessment at the Geita mine, Tanzania. *In:* Proceedings of the 8th International Congress on Mine water and the Environment, Johannesburg, South Africa. Armstrong D, deVillers A, Kleinman R, Norton P, McCarthy T (eds) p 403-415

Bowell RJ, Parshley JV, McClelland G,Upton CB, Zhan JG (2009) Geochemical evaluation of heap rinsing of the Gold Acres Heap, Cortez joint venture, Nevada Miner Eng 22:479-489

Bowell, RJ, Rees SB, Barnes A, Prestia A, Warrender R, Dey BM (2013) Geochemical assessment of arsenic toxicity in mine sites along the proposed Mineral Tramway Project, Camborne, Cornwall. Geochem Explor Environ Anal 13:145-158

Boyle RW, Jonasson IR (1973) The geochemistry of arsenic and its use as an indicator element in geochemical prospecting. J Geochem Explor 2:251-296

Canadian Government (2002) Metal mine effluent Regulations SOR/2002-222. Ottawa. March 2, 2012. 62 p

Casiot C, Morin G, Juillot F, Bruneel O, Personne JC, LeBlanc M, Dusqesne K, Bonnefoy B, Elbaz-Pouchlichet F (2003) Bacterial immobilization and oxidation of arsenic in acid mine drainage (Carnoules Creek, France). Water Res 37:2929-2936

Casiot C, Bruneel O, Personne JC, LeBlanc M, Elbaz-Pouchlichet F (2004) Arsenic oxidation and bioaccumulation by the acidophilic protozoan *Euglena mutablis*, in acid mine drainage (Carnoules Creek, France). Sci Total Environ 320:259-267

Casiot C, Egal M, Elbaz-Pouchet F, Bruneel O, Bancon-Montigny C, Cordier MA, Gomez E, Aliaume C (2009) Hydrological and geochemical control of metals and arsenic in a Mediterranean river contaminated by acid mine drainage; preliminary assessment of impacts on fish (*Leuciscus cephalus*). Appl Geochem 24:787-799

Cave MR, Wragg J, Palumbo B, Klinck BA (2003) Measurement of Bioaccessibility of Arsenic in UK soils. P5-062/TR1. Environment Agency

Chakraborti D, Mahmudur M, Bhaskar Das, Murrill M, Dey S, Mukherjee S, Dhar RK, Biswas BK, Chowdhury U, Roy S, Sorif S, Selim S, Rahman M, Quamruzzaman Q (2010) Status of groundwater arsenic contamination in Bangladesh: A 14-year study report. Water Res 44:5789-5802

Chilean Ministry of the Environment (2010) Effluent standards applied to industrial and mining effluents. Publication 10/6/1.

Cidu R, Biddau R, Fanfani L (2009) Impact of past mining activity on the quality of groundwater in SW Sardinia (Italy). J Geochem Explor 100:125-132

Clara M, Magalhaes F (2002) Arsenic. An environmental problem limited by solubility. Pure Appl Chem 74:1843-1850

Coetzee H, Hobbs PJ, Burgess JE, Thomas A, Keet M (2010) Mine Water Management In The Witwatersrand Gold Fields With Special Emphasis On Acid Mine Drainage. Report to the inter-ministerial committee on acid mine drainage. Council for Geosciences, Pretoria. 146p

Cohen DR, Bowell RJ (2014) Exploration Geochemistry. *In:* Treatise on Geochemistry. Holland HD, Turekian KK (eds) Elsevier, p 623-650

Corkhill CL, Vaughan DJ (2009) Arsenopyrite oxidation. Appl Geochem 24:2342-2361

Corkhill CL, Wincott PL, Lloyd JR, Vaughan DJ (2008) The oxidative dissolution of arsenopyrite (FeAsS) and enargite (Cu_3AsS_4) by *Leptospirillum ferrooxidans*. Geochim Cosmochim Acta 72:5616-5633

Courtin-Nomade A, Grosbois C, Bril H, Roussel C (2005) Spatial variability of arsenic in some iron-rich deposits generated by acid mine drainage. Appl Geochem 20:383-396

Courtin-Nomade A, Grosbois C, Liu B, Beckett P, Fakra SC, Beny JM, Foster AL (2009) The weathering of a sulfide orebody: speciation and fate of some potential contaminants. Can Mineral 47:493-508

Craw D (2001) Tectonic controls on gold deposits and their environmental impact, New Zealand. J Geochem Explor 73:43-56

Craw D (2003) Geochemical changes in mine tailings during a transition to pressure-oxidation process discharge, Macraes mine, New Zealand. J Geochem Explor 80:81-94

Craw D (2006) Pressure-oxidation autoclave as an analogue for acid-sulphate alteration in epithermal systems. Mineral Dep 41:357-368

Craw D, Pacheco D (2002) Mobilization and bioavailability of arsenic around mesothermal gold deposits in a semiarid environment, Otago, New Zealand. Sci World J 2:308-319

Craw D, Chappell D, Nelson M, Walrond M (1999) Consolidation and incipient oxidation of alkaline arsenopyrite-bearing mine tailings, Macraes mine, New Zealand. Appl Geochem 14:485-498

Craw D, Koons PO, Chappell DA (2002a) Arsenic distribution during formation and capping of an oxidized sulphidic mine soil, Macraes mine, New Zealand. J Geochem Explor 76:13-29

Craw D, Chappell D, Black A (2002b) Surface run-off from mineralized road aggregate, Puhipuhi, Northland, New Zealand. New Zealand J Marine Freshwater Res 36:105-116

Craw D, Falconer D, Youngson JH (2003) Environmental arsenopyrite stability and dissolution: theory, experiment and field observations Chem Geol, 199:71-82

Craw D, Wilson N, Ashley PM (2004) Geochemical controls on the environmental mobility of Sb and As at mesothermal antimony and gold deposits. Trans Inst Min Metall 113:B3-B10

Craw D, Rafaut C, Haffert L, Paterson L (2007) Plant colonization and arsenic uptake on high arsenic mine wastes, New Zealand. Water Air Soil Pollut 179:351-364

Dey BM, Bowell RJ, Hutton-Ashkenny MJ, Grogan J (2010) Characterization and reprocessing of lead reduction furnace slag, Tsumeb, Namibia. *In:* Proceedings of Zinc'10, Cape Town, RSA. Wills B (ed) p 10-18

Dove PM, Rimstidt JD (1985) The solubility and stability of scorodite, $FeAsO_4 \cdot 2H_2O$. Am Mineral 70:838-844

Drahota P, Novakova B, Matousek T, Mihaljevic M, Rohovec J (2013) Diel variation of arsenic, molybdenum and antimony in a stream draining natural As geochemical anomaly. Appl Geochem 31:84-93

du Bray EA (ed) (1995) Preliminary compilation of Descriptive Geoenvironmental Mineral Deposit models, U.S.Geological Survey Open File Report 95-831

Ettler V, Johan Z, Kribek B, Sebek O, Mihaljevic M (2009) Mineralogy and environmental stability of slags from the Tsumeb smelter, Namibia. Appl Geochem 24:1-15

Flanagan PJ (1990) Parameters of water quality: Interpretation and standards. Environmental Resources Unit. Environmental Protection Agency, Ireland.

Foster AL, Ashley RP (2002) Characterization of arsenic species in microbial mats from an inactive gold mine. Geochem Explor Environ Anal 2:253-261

Foster AL, Brown GE Jr, Tingle TN, Parks GA (1998) Quantitative arsenic speciation in mine tailings using X-ray absorption spectroscopy. Am Mineral 83:553-568

García BM, Perez-Lopez R, Ruiz-Chancho MJ, Lopez-Sanchez J, Rubio R, Abreu MM, Nieto J, Cordoba F (2012) Arsenic speciation in soils and *Erica andevalensis* Cabezudo & Rivera and *Erica australis* from Sao Domingos Mine area, Portugal. J Geochem Explor 119-120:51-59

Gault AG, Cooke DR, Townsend AT, Charnock JM, Polya D (2005) Mechanisms of arsenic attenuation in acid mine drainage from Mount Bischoff, western Tasmania. Sci Total Environ 345:219-228

Gebhard G (1999) Tsumeb. A unique mineral locality. GG Publishing, Grossenseifen, Germany

Giere R, Sidenko NV, Lavareva EV (2003) The role of secondary minerals in controlling the migration of arsenic and metals from high-sulfide wastes (Berikul gold mine, Siberia). Appl Geochem 18:1347-1359

Grimes DJ, Ficklin WH, Meier AL, McHugh JB (1995) Anomalous gold, antimony, arsenic and tungsten in groundwater and alluvium around disseminated gold deposits along the Getchell Trend, Humboldt County, Nevada. J Geochem Explor 52:351-371

Haffert L, Craw D (2008a) Mineralogical controls on environmental mobility of arsenic from historic mine processing residues, New Zealand. Appl Geochem 23:1467-1483

Haffert L, Craw D (2008b) Processes of attenuation of dissolved arsenic downstream from historic gold mine sites, New Zealand. Sci Total Environ 405:286-300

Haffert L, Craw D (2010) Geochemical processes influencing arsenic mobility at Bullendale historic gold mine, Otago, New Zealand. New Zealand J Geol Geophys 53:129-142

Haffert L, Sander SG, Hunter KA, Craw D (2010) Evidence for arsenic-driven redox chemistry in a wetland system: a field voltammetric study. Environ Chem 7:386-397

Hale M (1981) Pathfinder application of arsenic, antimony and bismuth in geochemical exploration. J Geochem Explor 15:307-323

Herbel M, Ferndorf S (2006) Biogeochemical processes controlling the speciation and transport of arsenic within iron coated sands. Chem Geol 228:16-32

IMA (2014) The new IMA list of minerals. A work in progress. Updated: March 2014. *http://pubsites.uws.edu. au/ima-cnmnc/IMA_Master_List_(2014-03).pdf* (accessed June 2014)

Jamieson HE (2014) The legacy of arsenic contamination from mining and processing refractory gold ore at Giant Mine, Yellowknife, Northwest Territories, Canada. Rev Mineral Geochem 79:533-551

Jensen M (1985) The Majuba Hill Mine, Pershing County. Mineral Record 16:57-72

Keller P (1977) Paragensis: assemblages, sequences, associations. *In:* Tsumeb [Nambia]! Wilson W (ed) Mineral Record 8:38-47

Kim M-J (2010) Effects of pH, adsorbate/adsorbent ratio, temperature and ionic strength on the adsorption of arsenate onto soil. Geochem Explor Environ Anal 10:407-412

Kokinos M, Wise WS (1993) Famous mineral localities: the Gold Hill mine, Tooele County, Utah. Mineral Record 24:11-22

Krause E, Ettel VA (1988) Solubility and stability of scorodite, $FeAsO_4 \cdot 2H_2O$: new data and further discussion. Am Mineral 73:850-854

Lazareva EV, Shuvaeva O, Tsimbalist VG (2002) Arsenic speciation in the tailings impoundment of a gold recovery plant in Siberia. Geochem Explor Environd Anal 2:263-268

Leblanc M, Achard B, Othman DB, Luck JM, Bertrand-Sarfait J, Personne JC (1996) Accumulation of arsenic from acid mine waters by ferruginous bacterial accretions (stromatolites). Appl Geochem 11:541-554

Lee J-S, Lee SW, Chon W, Kim K-W (2008) Evaluation of human exposure to arsenic due to rice ingestion in the vicinity of abandoned Myungbong Au-Ag mine site, Korea. J Geochem Explor 96:231-235

Lenain J-F, Courtin-Normade A (2003) Visual-Statistical classification of As-Fe rich products of alteration of tailings from the Enguiales tungsten mine, France. Can Mineral 41:1135-1146

Lengke MF, Sanpawanitchakit C, Tempel RN (2009) The oxidation and dissolution of arsenic-bearing sulfides. Can Mineral 47:593-613

Lockwood M, Ferndorf S, Stromquist L (1997) Isolation of arsenic-bearing tailings, Beauty Bay mine, Kenai Fjords, Alaska. *In:* Tailings and Mine Waste. Balkema Rotterdam, p 701-709

Loredo J, Ordóñez A, Baldo C, García Iglesias J (2003) Arsenic mobilization from waste piles of the El Terronal mine, Asturias, Spain. Geochem Explor Environ Anal 3:1-9

Macur RE, Wheeler JT, Mcdermott TR, Inskeep WP (2001) Microbial populations associated with the reduction and enhanced mobilization of arsenic in mine tailings. Environ Sci Technol 35:3676-3682

Majzlan J (2011) Thermodynamic stabilization of hydrous ferric oxide by adsorption of phosphate and arsenate. Environ Sci Technol 45:4726-4732

Majzlan J, Drahota P, Filippi M, Grevel K-D, Kahl W-A, Plášil J, Woodfield BF, Boerio-Goates J (2012a) Thermodynamic properties of scorodite and parascorodite ($FeAsO_4 \cdot 2H_2O$), kaňkite ($FeAsO_4 \cdot 3.5H_2O$), and $FeAsO_4$. Hydrometallurgy 117-118:47-56

Majzlan J, Lazic B, Armbruster T, Johnson MB, White MA, Fisher RA, Plášil J, Loun J, Škoda R, Novák M (2012b) Crystal structure, thermodynamic properties, and paragenesis of bukovskýite, $Fe_2(AsO_4)(SO_4)$ (OH)$\cdot 9H_2O$. J Mineral Petrol Sci (Japan) 107:33-148

Mandarino JA, Back ME (2004) Fleischer's Glossary of Mineral Species. The Mineralogical Record, Tucson. 307p

McKibben MA, Tallbant BA, del Angel JK (2008) Kinetics of inorganic arsenopyrite oxidation in acidic aqueous solutions. Appl Geochem 23:121-135

Meunier L, Walker SR, Wragg J, Parsons MB, Koch I, Jamieson HE, Reimer KJ (2010) Effects of soil composition and mineralogy on the bioaccessibility of arsenic from tailings and soil in gold mine districts of Nova Scotia. Environ Sci Technol 44:2667-2674

Ministry of the Environment (2012) Water quality limits. Moscow. 12p

O'Day PA (2006) Chemistry and mineralogy of arsenic. Elements 2:77-83

Odor L, Wanty RB, Horvath I, Fugedi U (1998) Mobilization and attenuation of metals downstream from a base-metal mining site in the Matra Mountains, northeastern Hungary. J Geochem Explor 65:47-60

Osbourne TH, Jamieson HE, Hudson-Edwards KA, Nordstrom DK, Walker SR, Ward SA, Santini JM (2010) Microbial oxidation of arsenite in a subarctic environment: diversity of arsenite oxidase genes and identification of a psychrotolerant arsenite oxidizer. BMC Microbiology 10, 205

Oyarzun R, Lillo J, Higueras P, Oyarcun J, Maturana H (2004) Strong arsenic enrichment in sediments from the Elqui watershed, Northern Chile: industrial (gold mining) vs geologic processes. J Geochem Explor 84:53-64

Paktunc D, Foster A, Laflamme G (2003) Speciation and characterization of arsenic in Ketza River mine tailings using X-ray absorption spectroscopy. Environ Sci Technol 37:2067-2074

Paktunc D, Dutrizac J, Gertsman V (2008) Synthesis and phase transformations involving scorodite, ferric arsenate and arsenical ferrihydrite: Implications for arsenic mobility. Geochim Cosmochim Acta 72:2649-2672

Petrunic BM, Al TA, Weaver L (2006) A transmission electron microscopy analysis of secondary minerals formed in tungsten-mine tailings with an emphasis on arsenopyrite oxidation. Appl Geochem 24:2222-2233

Plumlee G (1994) Environmental Geology models of Mineral Deposits: SEG Newsletter, v.16, p.5-6

Plumlee G, Morman SA (2011) Mine Wastes and Human Health. Elements 7:399-404

Pope JG, McConchie DM, Clark MD, Brown KL (2004) Diurnal variations in the chemistry of geothermal fluids after discharge, Champagne Pool, Waiotapu, New Zealand. Chem Geol 203:253-272

Reich M, Kesler SE, Utsunomiya S, Palenik CS, Chryssoulis SL, Ewing RC (2005) Solubility of gold in arsenian pyrite. Geochim Cosmochim Acta 69:2781-2796

Roddick-Lanzilotta AJ, McQuillan AJ, Craw D (2002) Infrared spectroscopic characterisation of arsenate (V) ion adsorption from mine waters, Macraes Mine, New Zealand. Appl Geochem 17:445-454

Roman-Ross G, Cuello GJ, Turillas X, Fernandez-Martinez A, Charlet L (2006) Arsenite sorption and co-precipitation with calcite. Chem Geol 233:328-336

SADC (2012) Environmental Legislation Handbook. Third Edition. Development Bank of South Africa

Sahai N, Lee YJ, Xy H, Ciardelli M, Gaillard J-F (2007) Role of Fe(II) and phosphate in arsenic uptake by coprecipitation. Geochim Cosmochim Acta 71:3193-3210

Savage KS, Bird DK, Ashley RP (2000a) Legacy of the California Gold Rush: Environmental Geochemistry of Arsenic in the Southern Mother Lode Gold District. Int Geol Rev 42:385-415

Savage KS, Tingle TN, O'Day PA, Waychunas GA, Bird DK (2000b) Arsenic speciation in pyrite and secondary weathering phases, Mother Lode Gold District, Tuolumne County, California. Appl Geochem 15:1219-1244

Schafer WS, Logsdon M, Zhan G, Espell R (2006) Post-Betze pit lake water quality prediction. Proceedings of the 7th ICARD symposium, St Louis, May 1-3 2006. ASMR, Lexington

Schwartz MO (1995) Arsenic in porphyry copper deposits: economic geology of a polluting element. Int Geol Rev 37:9-25

Scott KM (1987) Solid solution in, and classification of, gossan-derived members of the alunite-jarosite family, Queensland, Australia. Am Mineral 72:178-187

Slowey AJ, Johnson SB, Newville M, Brown GE Jr (2007) speciation and colloid transport of arsenic from mine tailings. Appl Geochem 22:1884-1898

Smedley PL, Edmunds WM, Pelig-Ba KB (1996) Mobility of arsenic in groundwater in the Obuasi area of Ghana. *In:* Environmental Geochemistry and Health Special Publication, 113. Appleton JD, Fuge R, McCall GJH (eds) Geological Society, London, p 163-181

Smith AH, Goycolea M, Haque R, Biggs ML (1998) Marked increase in bladder and lung cancer morality in a region of northern Chile due to arsenic in drinking water. Am J Epidemiology 147:660-669

Stichbury ML, Bain JG, Blowes DW, Douglas-Gould W (2000) Microbially mediated reductive dissolution of arsenic bearing minerals in a gold mine tailings impoundment. ICARD 2000 Proceedings 5th International Conference Acid Rock Drainage, p 97-103

Stolberg CS, Dunning GE (1985) Getchell Mine. Mineral Record 16:15-24

Thornton I (1994) Sources and pathways of arsenic in the geochemical environment. *In:* Environmental Geochemistry and Health Special Publication, 113. Appleton JD, Fuge R, McCall GJH (eds) Geological Society, London, p 153-162

Tutu H, McCarthy TS, Cukrowska EM (2008) The chemical characteristics of acid mine drainage with particular reference to sources, distribution and remediation: the Witwatersrand Basin, South Africa, as a case study. Appl Geochem 23:3666-3684

United States Environmental Protection Agency (2012) National Water Quality Criteria. *http://water.epa.gov/ scitech/swguidance/standards/index.cfm* (Accessed Apr 2014)

Van Leeuwen FXR (1993) The new WHO guideline value for arsenic in drinking water. *In:* Proceedings First International Conference on Arsenic Exposure and Health Effects, New Orleans, 1993. Society of Environmental Geochemistry and Health, p 30-32

Wakao N, Koyatsu H, Komai Y, Shimokawara H, Sakurai Y, Shiota H (1988) Microbial oxidation of arsenite and occurrence of arsenite-oxidizing bacteria in acid mine water from a sulfur-pyrite mine. Geomicrobiol J 6:11-24

Walker SR, Parsons MB, Jamieson HE, Lanzirotti A (2009) Arsenic mineralogy of near-surface tailings and soils: Influences on arsenic mobility and bioaccessibility in the Nova Scotia gold mining districts. Can Mineral 47:533-556

Warrender R, Bowell RJ, Prestia A, Barnes A, Mansanares W, Miller M (2012) The application of predictive geochemical modelling to determine backfill requirements at Turquoise ridge Joint Venture, Nevada. Geochem Explor Environ Anal 12:339-347

Waychunas GA, Rea BA, Fuller CC, Davis JA (1993) Surface chemistry of ferrihydrite: Part 1: EXAFS studies of the geometry of coprecipitated and adsorbed arsenate. Geochim Cosmochim Acta 57:2251-2269

Webster-Brown J, Craw D (2005) Examples of trace metal mobility around historic and modern metal mines. *In:* Metal Contaminants in New Zealand. Moore TA, Black A, Centeno JA, Harding JS, Trumm DA (eds), Resolutionz Press, Christchurch NZ, p. 213-230

Welch AH, Westjohn DB, Helsel DR, Wanty RB (2000) Arsenic in groundwater of the United States: occurrence and geochemistry. Ground Water 38:589-604

Williams TM (2001) Arsenic in mine waters: an international study. Environ Geol 40:267-278

Wong HKT, Gauthier A, Beauchamp S, Todon R (2002) Impact of toxic metals and metalloids from the Caribou gold-mining areas in Nova Scotia, Canada. Geochem Explor Environ Anal 2:235-241

Yamauchi H, Fowler BA (1994) Toxicity and metabolism of inorganic and methylated arsenicals. *In:* Arsenic in the Environment. Part 2. Nriagu JO (ed) Wiley, New York, p 35-53

Yu Y, Zhu Y, Gao Z, Gammons CH, Li D (2007) Rates of arsenopyrite oxidation by oxygen and Fe(III) at pH 1.8-12.6 and 15-45 °C. Environ Sci Technol 41:6460-6464

Reviews in Mineralogy & Geochemistry
Vol. 79 pp. 507-532, 2014
Copyright © Mineralogical Society of America

The Management of Arsenic in the Mining Industry

Robert J. Bowell

*SRK Consulting
17 Churchill Way
Cardiff, CF10 2HH, Wales, United Kingdom*

rbowell@srk.co.uk

Dave Craw

*Department of Geology
University of Otago
PO Box 56
Dunedin 9054, New Zealand*

dave.craw@otago.ac.nz

INTRODUCTION

Arsenic contamination of mine and metallurgical waters has long been recognized as a global problem. More stringent guidelines, based on demonstration of potential toxicity to humans and ecological receptors, have motivated regulators and operators to address both legacy sites and existing or future operational discharges to mitigate potential impacts. The safe disposal of material considered to be hazardous is a natural part of good housekeeping for any industrial development. This is particularly so for the mining industry, which historically was not always well managed in this aspect and as a result, has a high-political profile today.

Arsenic can occur in several oxidation states in natural waters although the trivalent arsenite (As(III)) or pentavalent arsenate (As(V)) are the most common (Smedley and Kinniburgh 2002). The most thermodynamically stable species over the natural range of groundwater redox conditions (150-500 mV, Bass Becking et al. 1960) and pH (4-7, Baas Becking et al. 1960) are $H_2AsO_4^-$, $HAsO_4^-$, and in acid rock drainage waters (pH below 5) $H_2AsO_4^-$. In more reduced waters, $As(OH)_3$ is the most common species. Thioarsenic species may also be present but in general are not observed in natural waters. The kinetics of arsenic reduction-oxidation (redox) reactions is not rapid, so the predicted proportions of arsenic species based on thermodynamic calculations do not always correspond to analytical results (O'Neil 1990). An Eh-pH diagram showing the thermodynamically stable regions for arsenic species is shown in Figure 1.

Because of arsenic toxicity, the World Health Organization placed a guideline maximum allowable concentration of arsenic in drinking water of 10 µg L^{-1} (WHO 1998). The USEPA reduced the drinking water standard from 50 to 10 µg L^{-1} in 2002 (USEPA 2001). Arsenite is considered to be more toxic than arsenate. Inorganic As(III) is a known and recognized human carcinogen with a poorly understood mechanism. As(III) can disrupt cell division. Multiple skin cancer lesions are common. In addition, there is a link between arsenic exposure and internal cancers of the bladder, liver, lungs, and kidneys (Hopenhayn 2006). Initial symptoms of arsenic exposure are the formation of hyperpigmentation on the skin, which usually appears as fine freckles distributed symmetrically, and keratosis (formation of calluses).

1529-6466/14/0079-0011$05.00

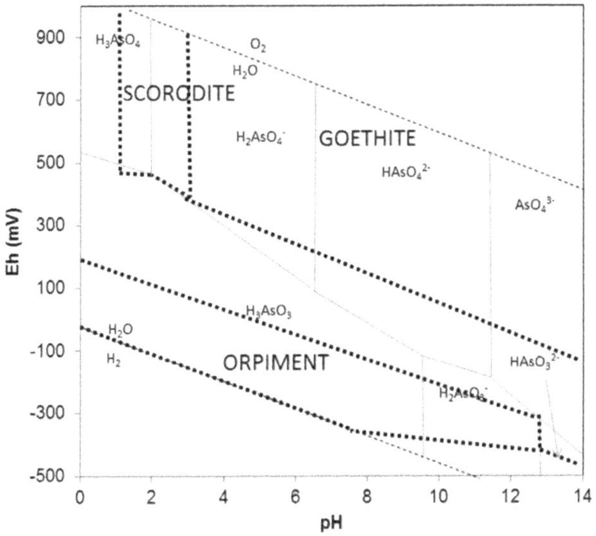

Figure 1. Eh-pH diagram of arsenic species (aqueous and solid) commonly found in natural waters. 298 K, 1 atm, $a_{As} = 10^{-3}$; $a_{Fe} = 10^{-5}$; $a_{Ca} = 10^{-2}$.

Oxidation of As-bearing sulfide minerals (most importantly As-bearing or arsenian pyrite) and reduction of arsenate adsorbed to iron oxyhydroxides exposed by earth movement and water table changes are the largest contributing factor to As mobilization (Bowell 2001). This can occur due to natural processes or a legacy of industrial activity such as mining. A general scheme showing the cycling of As in the environment of an Orogenic gold deposit is shown in Figure 2.

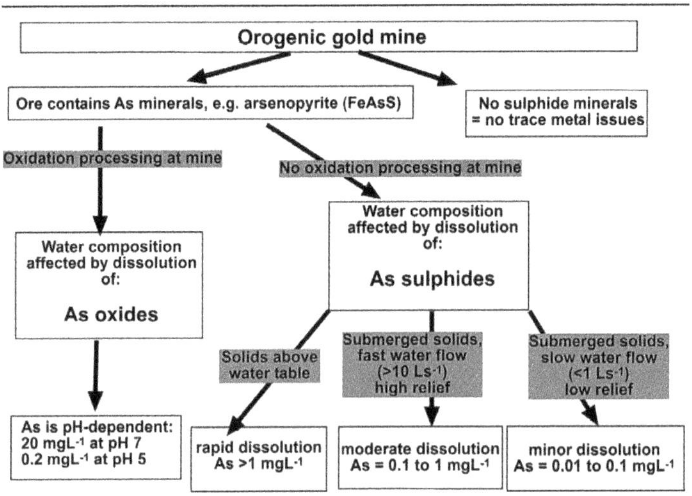

Figure 2. Generalized estimation of As contents of discharge waters that emanate from mine tailings stored under a range of conditions. The estimates are based on observations of As contents from As-rich orogenic gold mines (Cavanagh et al. 2014). The As concentrations are indicative only, and are controlled mainly by mine process systems, relief, and rainfall.

APPROACH TO ARSENIC MANAGEMENT

The most common methods for As management are containment or removal. Containment of As is more applicable where As is present as a solid waste such as dust or process waste. Removal of As through precipitation, sorption, or membrane methods are more common where As is present in water or a highly leachable form. Table 1 provides a summary of the As management strategies discussed in this chapter.

No single solution fits all scenarios, so the advantages and disadvantages need to be evaluated on a case-by-case basis to identify the most cost-effective solutions. A common issue with all methods is the need for long-term disposal of the high-As residues and spent treatment media. In the case of membrane treatment, this is a highly concentrated but small-volume precipitate. With iron or calcium co-precipitation, larger volumes of residues may be produced. The methods shown in Table 1 are reviewed in this chapter and the advantages and disadvantages of each expanded upon.

The types of mining waste discussed for the purposes of this chapter are defined below (Bowell et al. 1998).

Waste rock

Waste rock is material excavated as overburden or rock outside the ore body that is removed to gain access to the ore but would normally not contain any ore material of significance. The rock would be placed in specific sites on the surface around the mining operation. Some waste rock may be taken underground to provide backfill and roof support. Waste rock has also been used as a source of construction material if suitable. At some mine

Table 1. Summary of Arsenic Management Strategies

Method	Advantages	Disadvantages
Physical Containment	Relatively low cost	Arsenic still chemically active Can leak or seep from structure
Chemical Containment	High cost Chemically stable Proven technology	Requires high chemical input Can require chemically stable environment or isolation

Precipitation	Uses common chemicals
	Can be relatively low cost
		Generates a high volume waste that
		  can be chemically active
		Pre-oxidation and pH adjustment
		  required
		High requirement for chemical
		  consumables

Sorption	Lower sludge production Fewer chemicals required	Periodic media replacement Requires stable chemical environment Requires careful monitoring
Membrane Removal	Well defined performance High removal Low space demand Removes other contaminants Small volume of waste produced	Very high costs – capital and operating Concentrated waste Not stable with high oxidizers
Phytoremediation	Low cost, "green solution" Visually acceptable No emissions or monitoring needed	Unproven technology Uncertain long-term stability Issues with food web transfer

sites, As concentrations in waste rock are not sufficiently high to require more than a suitable cover or physical containment.

Low-grade ore stock piles

These are similar to waste rock piles but are typically kept separate for possible ore processing at a later stage, usually towards the end of the mining operation. Therefore, they represent a potential source of contaminants due to prolonged exposure to chemical weathering.

Tailings impoundment

Tailings and slimes dams represent the waste product remaining after crushing, grinding, processing of the ore, and extraction of the concentrate. Deposition can be in a dry form or hydraulically emplaced in the impoundment. Traditionally these "dams" are a major source of potential problems from escape of contaminants into the environment. For example, a poly-metallic-sulfide ore can generate elevated anions, cations, and metals together with chemical reagents from the process.

Contaminated water

Water contaminated with As by mining may come from a number of sources (Fig. 3) (Bowell et al. 1998).

Underground or open-pit dewatering. Water typically has high total dissolved solids (TDS) in excess of 1000 mg L^{-1}. Typically in metal mines and some coal mines, sulfate is in excess of 1000 mg L^{-1} (Lottermoser 2010). Some of this water may be highly acidic and contain high levels of dissolved metals, particularly Fe, such as in the Berkeley pit lake, Montana (Gammons and Duaime 2006). At this locality, the lake contains in excess of 600 mg L^{-1} Fe(III) and almost 800 mg L^{-1} Fe(II) in addition to 150-200 mg L^{-1} Cu and approximately 10,000 mg L^{-1} sulfate at a pH less than 2.6.

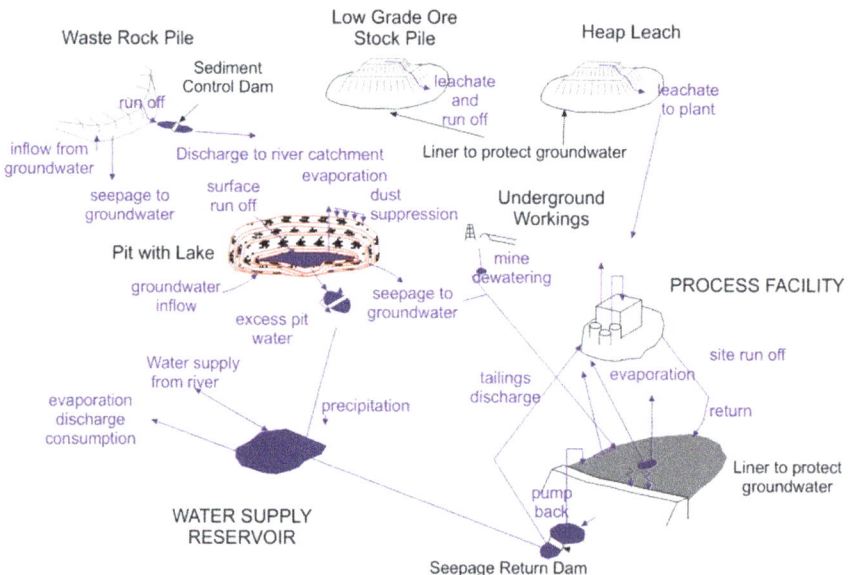

Figure 3. Schematic diagram showing potential routes of water use, consumption and effluent discharge at a mining site (modified after Smith and Mudder 1991).

Pit lakes, developed in open pits are influenced by exchange with groundwater and can also impact local and regional groundwater. Surface runoff and reaction with soluble salts, precipitation, evaporation, as well as the chemistry of any water discharged directly into the pit affect the chemistry of the lake. By contrast underground mines tend to reflect the characteristics of regional groundwater rather than open pits once flooded. For example, the flooded underground mines of Butte under the Berkeley pit range in pH from 4.5 to 7 and shows highly variable metal contents with Cu varying from <0.016 mg L^{-1} to in excess of 1000 mg L^{-1} and sulfate from 68 mg L^{-1} to more than 20,000 mg L^{-1} (Gammons et al. 2006). Despite having similar geology to the Berkeley pit, the underground workings have been flooded by groundwater and so oxygen is limited. Metal and salt release reflects flushing of partially flooded upper workings blended with regional groundwater. There is little influence from evaporation or sulfide oxidation on most of the mine waters and hence the wide range of observed chemistry.

Tailings effluent waters. These may be highly alkaline or highly acidic (Bodenan et al. 2004). For example, tailings waters from gold cyanide leach operations are typically pH > 10 with high levels of sulfate but low levels of metals, although oxyanions such as arsenate, antimonate, and molybdenate are high (Smith and Mudder 1991). Dispersion of historic tailings can lead to release and secondary accumulation of As away from the mining areas. For example at Moira Lake, Ontario, the accumulation of As in lake sediments is attributed to tailings erosion and release over 160 years of mining (Azcue and Nriagu 1995).

Water contaminated by natural interaction with minerals. For example, at the Clara mineral deposit in Germany, oxidation of As-bearing sulfide minerals in barite-fluorite veins has led to the release of As into streams around the area up to concentrations of 0.15 mg L^{-1} (Zhu et al. 2003). Evidence of natural contamination from mineral interactions can be preserved in sediments distal to mineralized zones, even in areas where no mining has occurred. For example, in the Elqui watershed in Chile, early Holocene sequences show evidence of As enrichment related to erosion of As-rich epithermal systems (Oyarzun et al. 2004).

Saline groundwater. This may be released into the mine from regional groundwater storage. For example, water with high chloride, Mg, and sulfate from an evolved mix of seawater and mine water was observed at the Levant mine in Cornwall (Bowell and Bruce 1995).

Chemical reagents

Most forms of mineral processing utilize inorganic and organic reagents in the separation of economic elements and minerals from waste rock. Additionally, drilling fluids and petroleum products may be major waste components at a mine site. Part of the chemical containment required at a mine site will involve collection, treatment, and disposal of waste chemicals. These are very often the most hazardous materials on site (Lottermoser 2010).

ARSENIC CONTAINMENT

In the management of mine waste there are three broad strategies that can be applied with respect to arsenic control (Bowell et al. 1998): (1) limit source release through control of generating processes such as limiting sulfide oxidation, (2) restrict the interaction between As-bearing phases and the environment such as in an engineered waste facility, and (3) control the overall release of As by collection and treatment of any effluents.

Physical containment

Physical containment is the most common method of limiting the interaction of mine waste with the environment (Bentley 1996). Such methods include placement of mine waste in an engineered repository, such as a lined and covered tailings impoundment, or placement underground in a cemented backfill. Old tailings impoundments allowed drainage of water to the subsurface. More modern impoundments incorporate engineered low-permeability linings.

Mechanical removal of sulfides can be achieved by traditional or innovative mineral processing techniques. However, this still presents the problem of the disposal of the sulfides in a different manner elsewhere and is not applicable where sulfides are present in large quantities underground.

A more practical approach is to exclude the sulfides by precipitation of an insoluble, non-reactive precipitate thereby isolating the sulfides from oxidants. Ferric phosphates and oxy-hydroxides have been proposed and the rate of sulfide oxidation has been observed to decrease in samples amended with phosphate (Achmed 1991).

Exclusion of bacteria can also significantly reduce sulfide oxidation; this may involve the use of bactericides either as an intimate mixture with tailings or back fill material, or applied directly onto sulfide surfaces (Kleinmann 1998; Sand et al. 2007). However, such treatment requires continual reapplication based on currently available bactericides and is only suitable as a short-term option, although slow-release pellets may help to provide long-term bacterial inhibition (Kleinmann 1998).

Limiting oxygen entry into tailings impoundments or underground workings can greatly impede oxidation by placing a diffusion barrier between reactive sulfides and the atmosphere. Such a barrier could include a geomembrane or geofabric cover which actively consumes moisture and/or oxygen (Fig. 4).

In underground mine workings, flooding will impede the access of oxygen to sulfide mineral surfaces, but there may be consequences to the environment depending on the local hydrology (Gammons et al. 2006).

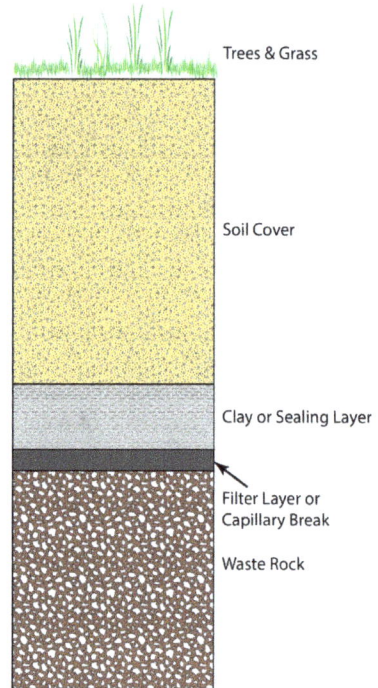

Figure 4. Schematic diagram of the use of a geofabric to restrict oxidation of sulfides and release of As.

With tailings material, water is the main cover option during operations but on closure and rehabilitation a similar cover can be applied to restrict oxygen ingress to sulfide tailings (Romano et al. 2003). As an option, organic carbon has been added to such covers in order to promote activity of sulfate reducing bacteria (Lindsey et al. 2009). A restriction on these covers is obviously climate and the ability of the environment to sustain a water cover. Where the tailings are oxidized such covers are also limited due to the presence of metals as water soluble salts. Here a barrier layer such as peat is applied to prevent mobilization of metals (Lottermoser 2010). In most environments, dry covers tend to be the more common for closure and rehabilitation. The design and configuration of dry covers are complex and can include single or multiple layers of materials to restrict air or water ingress to the tailings. A geofabric can be applied as a sealant or a capillary break or filter layer to capture and remove any water preventing seepage into reactive tailings (Patterson et al. 2006).

Fine-grained tailings covers rely on their moisture-retaining characteristics to maintain high moisture contents above the water table. Naturally formed covers can be encouraged by intentional formation of "hard pans" (Blowes et al. 1991). The zone acts as a barrier for water and oxygen movement thus limiting sulfide oxidation.

Placement of As-bearing waste in a neutralized sludge under a water cover has been demonstrated to be an effective management strategy (Beauchemin et al. 2004). However such strategies require that the site of physical containment be stable and not subject to slope instability or physical changes to the landscape (Zinck 2000).

Placement in a contained paste backfill in underground workings is another common method of containment (Lottermoser 2010). The backfill is essentially a mix of As-bearing waste, Portland cement (as a binder), and fly ash or waste rock (for strength). The stabilization of such material requires that there is no fluctuation in water or physical movement of the area, otherwise chemical or physical degradation of the backfill can occur (Ouellet et al. 1998). At the Sappes mine in Greece, backfill of As-rich mine waste and process waste has occurred in a shallow underground operation (Bowell et al. 1998; Bowell 2014). The mineralogy of the backfill demonstrates that the particles of tailings are cemented rather than being chemically altered (Fig. 5).

The stability of residues is impacted by the presence of sulfide minerals. If the residues oxidize with sulfides, such as pyrite, sulfuric acid can result (Nordstrom 1982). This causes the dissolution of the Portland cement, destabilizing the backfill, and resulting in the formation of expansive sulfate minerals such as ettringite and gypsum (Fig. 3B) by a reaction such as;

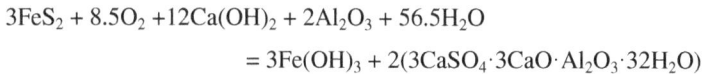

$$3FeS_2 + 8.5O_2 + 12Ca(OH)_2 + 2Al_2O_3 + 56.5H_2O$$

$$= 3Fe(OH)_3 + 2(3CaSO_4 \cdot 3CaO \cdot Al_2O_3 \cdot 32H_2O)$$

Such reactions can generate high-crystalline pressure (up to 180 MPa; Hughes 2008). Thus, the durability of the mortar is compromised and structural integrity is lost. Arsenic in the basic, oxidizing environment (pH of approximately 10.5-12; Eh in the range 80-170 mV) forms AsO_4^{3-} that is highly mobile and can be leached at high concentrations (Fig. 6).

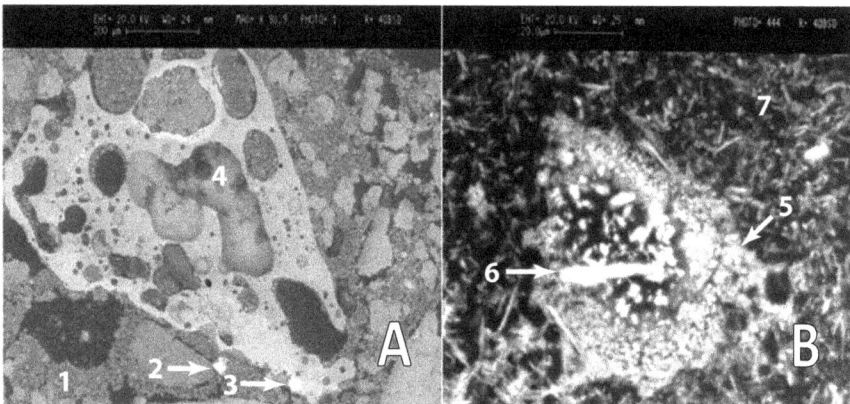

Figure 5. Photomicrographs of experimental-paste backfill material, Sappes gold project, Greece. (A) Typical backfill composition with tailings particles cemented in lime and flyash (taken from Bowell 2014). Essentially sulfides and gangue phases remain unchanged. 1 = fly ash, 2 = enargite, 3 = pyrite, 4 = fragments of tailing encapsulated in cement. (B) Sulfide oxidation in backfill leads to decomposition of lime and formation of gypsum, clinoclase, cornubite, and arseniosiderite in the alteration crust indicating limited arsenic mobilization (Bowell 2014). Within the altered gypsum zone, a non-crystalline Ca-Al-hydrated silicate forms that appears to act as a barrier to further reaction of the sulfide particles. 5 = cornubite, clinoclase, arseniosiderite alteration crust, 6 = residual enargite after partial oxidation, 7 = gypsum.

Where the paste backfill remains intact, leaching of the fill is controlled not by solubility or mass flux, but by diffusion of As from the backfill, with the outer rind of the fill acting as a diffusion barrier. In this instance, leaching concentration of As is low even over a prolonged period of time. Initially in diffusion tests, the leached As concentration can be high and is related to flushing of adsorbed As on Fe hydroxides and oxidation of exposed enargite. However once a reaction rind of gypsum and Ca-Al hydrated silicate is formed, As concentration in the leachate rapidly decreases, and thereafter As concentration can be below that in nearby groundwater in a few days (Fig. 7). In addition, pH decreases from approximately pH 11 to 10 over the same period.

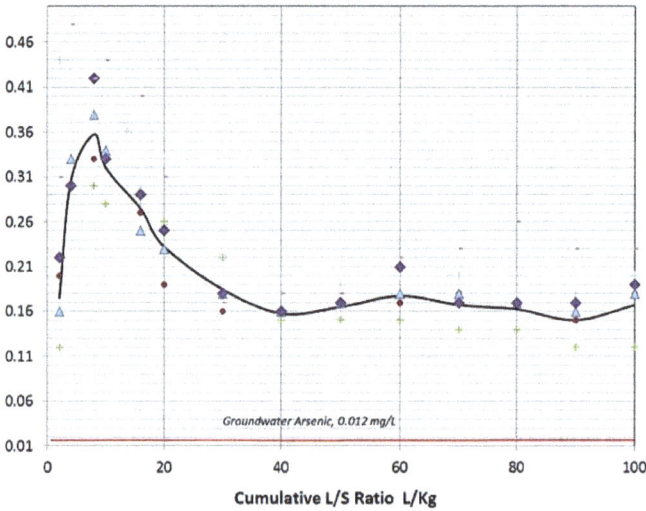

Figure 6. Leaching of As from paste-backfill material as a function of water:fill ratio (Bowell 2014).

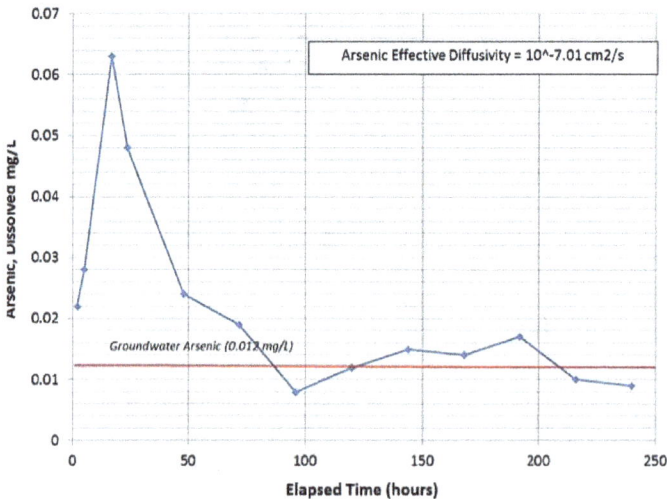

Figure 7. Leaching of As from experimental paste backfill over time (Bowell 2014).

Chemical containment

Chemical containment of As-bearing materials is increasingly becoming an established practice because of the limitations of long-term storage of As-bearing wastes and increasingly stringent standards (Machingawuta and Broadbent 1994; Nolan et al. 2012). Chemical containment includes the precipitation of As with Ca or Fe in a process stream prior to discharge, incorporation of As into silicate slag or clay, use of silicate polymer to encapsulate the As, or formation of briquettes with As incorporated in the structure (Harris and Krause 1993; Machingawuta and Broadbent 1994; Carter et al. 1995; Swash and Monhemius 1995; Wang et al. 2000a; Zinck 2000; Luganov and Sajin 2002; Mohan and Pittman 2007).

Most metals can be precipitated as their respective hydroxides or hydrated metal oxides. During neutralization, slurry can be aerated to transform reduced metals to oxidized forms to improve recovery of metals as more stable hydroxides. In practice, the most frequently applied method for the treatment of Fe-rich waters is oxidation of iron and simultaneous pH adjustment and hydroxide precipitation. Typical process configurations for hydroxide precipitation comprise a neutralization-precipitation stage, and a settling or clarification stage.

The pH adjustment is most commonly achieved by $Ca(OH)_2$ (added as a slurry). Sodium-based reagents preclude the formation of gypsum but result in higher total dissolved metals levels for the treated discharge water stream. The neutralization-precipitation stage is typically completed in a single process step, complemented with pre-aeration or direct aeration. Final clarification is achieved by flocculation and settling of the Fe-hydroxide precipitates.

The use of hydrothermal precipitation with Fe to remove As, particularly as ferric arsenate, has been proposed in recent years (Swash and Monhemius 1995, 1996). This disposal option is based on the known stabilities of these minerals. For example, more than 60% of known As-minerals are arsenates, of which the most common is scorodite, $FeAsO_4 \cdot 2H_2O$. A number of elaborate schemes have now been proposed, including MIRO's Scorodite process (Swash and Monhemius 1995). At high-temperature conditions (around 300 °C) precipitated As is chemically incorporated into the structure of the crystalline material. This is much more stable than As which has been "stabilized" through surface adsorption with ferrihydrite (Harris and Krause 1993).

Ferrihydrite is a disordered ferric hydroxide typically with high surface area, formed by ferrolysis (oxidation and hydrolysis of ferrous iron) at ambient temperatures. With time, ferrihydrite becomes unstable and converts to goethite (Waychunas et al. 1993). Arsenical ferrihydrite is produced by the neutralization of effluents which contain As together with ferric iron. Low solubilities of As are associated with products with an Fe:As molar ratio of >3:1. Lower Fe:As ratios tend to produce a highly-soluble product (Harris and Krause 1993). Adsorption of oxyanions in mildly acidic conditions is an effective method of attenuating some elements such as As (Bowell 1994). However some environmental changes, such as reduction or pH increase, will lead to liberation of the contained As species.

Fixation of As can also be achieved within a slag matrix (Machingawuta and Broadbent 1994; Kontopoulos et al. 1996). Up to 10 wt% As can be stabilized in slags, and solidification and stabilization are used for many hazardous materials. This involves the mixing of sludges with a cement binder to produce a solid that is structurally sound and relatively impermeable (Emmett and Khoe 1994). For example, it has been shown experimentally that up to 70% As_2O_3 (as Ca-arsenates) can be incorporated into cement, but high-As loadings reduce structural strength and solidification of the cement. In cement-stabilized materials, leachability is reduced as porosity and permeability are reduced. The high levels of lime added cause increases in Ca contents and pH of test solutions, potentially leading to erroneous measurements of leachability. Cements can deteriorate rapidly and the overall characteristics of the waste need to be considered before this disposal option can be chosen. The cement requires a dry, low-humidity, CO_2-free environment at constant temperature for long-term stability (Malhotra 1991).

Understanding the environmental stability for any product created is essential in assessing a particular treatment option. Protocols for the geochemical assessment of material have been published elsewhere and cover determination of acid-generating capacity, and leachable- and total-metal concentrations through a variety of procedures (Price 1997). These tests are used as the major criteria for determining safety of long-term disposal of waste products and highlight the need to take into account geochemical processes that will modify the physical and chemical characteristics of the compounds. Table 2 summarizes the long-term stability considerations for As-rich waste.

Table 2. Geochemical considerations in the long-term stability of contaminant waste products.

Product	Consideration
Ferrihydrite	Dehydration leads to instability, recrystallization to goethite Precipitate at ambient temperature. Biogeochemical reduction of Fe (III) to Fe (II)
Ca-precipitated cements	High intrinsic solubility Ca-salts converted to $CaCO_3$. Release of contaminants (influence of CO_2) High CaO levels, high pH (11-12) CaO converts to $CaCO_3$
Crystalline products (e.g. scorodite)	Low solubility Geological stability High production cost
Slags	Recrystallize glass-devitrification Unknown long-term stability Quenched slags, low solubility Low efficiency, high volume High production cost; can theoretically get energy back

Encapsulation in polysilicates has demonstrated potential to limit As solubility from industrial waste (Carter et al. 1995). Arsenic is bound through a covalent-bonded matrix through the action of the silicates and cement. Curing the polysilicate prior to final disposal was found to be a key issue in the treatment of such materials (Dey 1997).

ARSENIC REMOVAL BY ACTIVE TREATMENT

There are a number of different technologies for the removal of As from natural waters which can be grouped by type of treatment process: chemical precipitation, adsorption, membrane filtration, *in situ* treatment, and biological remediation (Fields et al. 2000; Wang et al. 2000a,b; Bowell 2003; Shih 2005; Hutton-Ashkenny et al. 2011; Nolan et al. 2012). Chemical precipitation, coagulation, and filtration are standard methods for treating As-bearing waters. These treatment facilities require the use of a reagent to precipitate As flocculants that are then removed by filtration. This method produces toxic sludge that requires disposal and requires very efficient pre-oxidation of As(III) to As(V) to ensure attenuation of As (Malik et al. 2009).

Chemical precipitation

Water treatment facilities using chemical precipitation and sedimentation of As are common. A number of cities in India have facilities that utilize this method (Mohan and

Pittman 2007). A plant in Kolkata oxidizes arsenic using 2 mg L^{-1} bleach powder before coagulation and precipitation with 40 mg L^{-1} of alum. Technical problems can arise related to the continuous addition of chemicals with fluctuating water inflows. Dosing should therefore be automated as a function of the water inflow (Gupta et al. 2009).

For most water-treatment processes, an important criterion for chemical precipitation of As is the oxidation of arsenite to arsenate (Krause and Ettel 1985, 1989; Wang et al. 2000b; Bowell 2003). Without this step, arsenite is difficult to precipitate regardless of pH or the concentration of Fe or other considerations (Bowell 2003). Ozone, hydrogen peroxide, potassium permanganate, and sulfur dioxide have all been shown to be effective in the oxidation of arsenite (Wang et al. 2000b). The benefit of using oxidizing agents was tested on Fe(II)-As(III) and Fe(III)-As(III) solutions to consider As oxidation and precipitation (Bowell 2003). Hydrogen peroxide oxidation, without pH control, is rapid in alkaline solutions but much slower below pH 6. The results of titrating hydrogen peroxide at a pH = 2.0 ± 0.5 with As(III) solutions is shown in Figure 8.

An iron precipitation and filtration facility is operated at the Getchell mine, Nevada (Bowell 2003). This facility can operate at flow rates up to 230 L h^{-1}. It was found that, at the desired flow rate through the filter, an iron dose of 25 mg L^{-1} achieved lowering of As from approximately 8 mg L^{-1} to less than 0.01 mg L^{-1}. Increased Fe doses and higher filtration rates had a linear effect to decrease filter life-time before backwashing was required. In addition, maintaining a pH > 5 was important in stabilizing ferric arsenate, FeAsO$_4$·2-4H$_2$O, rather than precipitation of ferric hydroxide that subsequently adsorbs arsenic. However, increased filtration rate appeared to have little effect on effluent-As concentration at high-Fe doses. In practice, all of these factors exhort a strong control on As removal and as such the flow rate, Fe dosing and pH design parameter must be suited to the precipitate formed and the required filtration rate.

An extension of the precipitation of amorphous ferric arsenate is the production of the more stable crystalline phase, scorodite, FeAsO$_4$·2H$_2$O (Robbins and Glastras 1987; Ugarte and Monhemius 1992; Demopoulos et al. 1995; Swash and Monhemius 1995; Papassiopi et al. 1996; Doušová et al. 2005; Fujita et al. 2008a). Conditions of large crystalline grains of

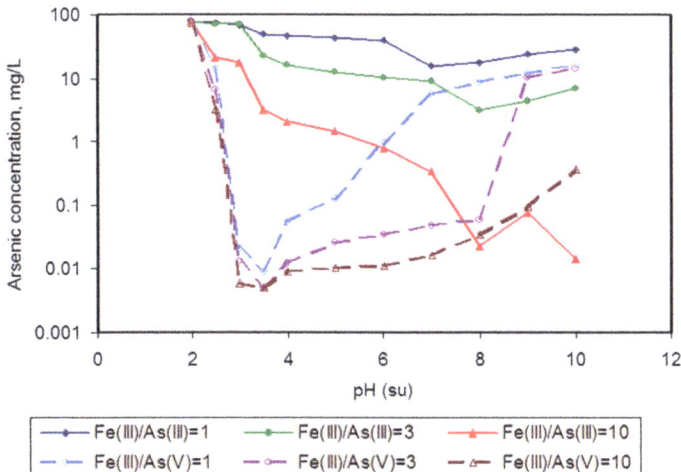

Figure 8. Arsenic (III) removal through oxidation (analysis for As(III)) by hydrogen peroxide over time (figure from Bowell 2003).

scorodite have been reported for contrasting different conditions from ambient temperature (Papassiopi et al. 1996; Doušová et al. 2005) through temperatures of 80 to 90 °C (Demopoulos et al. 1995; Fujita et al. 2008a,b) to hydrothermal temperatures in autoclaves of 250-400 °C (Ugarte and Monhemius 1992; Swash and Monhemius 1995). Solution chemistry is an important factor; high chloride and sulfate or high concentrations of sodium inhibit the formation of scorodite even at elevated temperatures. In contrast, Cu and Zn do not appear to cause an issue, although precipitation of mixed cation arsenates does occur in scorodite precipitation due to trace element incorporation into the scorodite lattice (Demopoulos et al. 1995; Fujita et al. 2008c). Arsenic concentrations up to 50,000 mg L^{-1} can be treated to produce scorodite and reduce effluent arsenic to less than 0.02 mg L^{-1} by this method (Fujita et al. 2008a).

Even in highly contaminated mine water it is possible to reduce solution As although a study by Doušová et al. (2005) showed that a complex mineralogy of ferrous-, ferric-, zinc-, and other metal-arsenates and arsenate-sulfate minerals form. Similar testing on mixed arsenate-sulfate minerals such as bukovskýite (Fe$^{3+}_2$(AsO$_4$)(SO$_4$)(OH)·9H$_2$O) demonstrate that such phases can be considerably more stable under ambient leaching conditions (Ugarte and Monhemius 1992; Swash and Monhemius 1995; Bowell 2003) than ferric arsenate (Table 3).

In addition, longer curing times are considered advantageous to producing more stable products (Akhter et al. 1997). As way of an example, the solubility of various arsenate salts in the Toxic Characteristic Leach Test (TCLP) is shown in Table 3. This test utilizes pH-adjusted ethanoic acid (pH 2.88 or pH 4.93) as a simulant of an environmental leach solution and a standard USEPA test procedure, test 1311 (USEPA 1992). As can be observed, scorodite may precipitate.

The critical factors for precipitation of a stable arsenic solid are: the chemistry of the As-containing species, Fe:As or Ca:As ratio, the size of grains, crystallinity, hydration of precipitated solids, and presence of other oxyanions in the precipitated solid lattice such as sulfate. Scorodite-type solids, particularly those precipitated at higher temperature, show the lowest solubility in the TCLP test (Table 3).

Calcium precipitation of As as sparingly soluble Ca arsenates has been used as a control, however such salts typically fail environmental leaching tests and so are limited in application (Zinck 2000). Sulfate control in mine waters can be effectively undertaken with addition of barium salts (Maree and Bosman 1989; Maree et al. 2004). The formation of Ba$_3$(AsO$_4$)$_2$ should produce a more insoluble salt than addition of Ca salts, although no experimental work on this has been published.

Zero-valent iron and organic carbon have also been demonstrated as suitable chemical precipitation methods for As (Mohan and Pittman 2007). These solids are particularly useful as reducing agents (electron donors) in permeable reactive barriers (PRBs) where the aim is to treat groundwater chemistry. At a gold mine in Canada, a pilot scale (13 m^2) PRB using a reactive mixture of zero-valent iron and organic carbon (wood chips) removed As as arsenic sulfide with the coincidental reduction of sulfate as well (Bain et al. 2009). The pilot scale project showed up to 99% removal of inflow As within the reactive mixture although over time this was blinded by precipitates (Bain et al. 2009).

Adsorption

As pH increases in water, ferric hydroxide (or hydrous ferric oxide, HFO) solubility decreases with a minimum at around pH 6. At low pH, precipitated HFO tends to scavenge negatively charged oxyanions as the surface of the HFO is positively charged in the Helmholtz layer (Deng and Stumm 1994). In low-pH environments these HFO particles are usually colloidal sized and have a high reactivity proportional to surface area (Fig. 9). As the pH increases and colloid particles aggregate as Fe-OH bonds become longer and more rigid, reflecting an increase in the excess of hydroxyl ions, the net surface charge of the particles becomes nega-

Table 3. Comparison of leachability in TCLP tests for arsenic-bearing precipitates

Precipitate Type	As concentration (mg L^{-1})	Reference
Arsenic adsorbed to natural goethite Fe:As 5:1, pH 4.5	11.7	Bowell and Parshley 2003
Arsenic adsorbed to synthetic ferrihydrite Fe:As 8:1, 20 °C, pH 5	0.97	Bowell and Parshley 2003
Synthetic scorodite 90 °C, pH 1.5	<0.01	Bowell and Parshley 2003
Synthetic kaňkite 110 °C, pH 3	0.37	Bowell and Parshley 2003
Synethetic bukovskýite 90 °C, pH 2.5	<0.01	Bowell and Parshley 2003
Ca$_3$(AsO$_4$)$_2$ 25 °C, pH 11	380	Bowell and Parshley 2003
Adelite CaMg(AsO$_4$)(OH)	4.2	Bowell and Parshley 2003
Synthetic scorodite 150 °C, pH <1	<0.8	Swash and Monhemius 1995
Fe-arsenate Fe:As 2.3:1 (molar ratio), 200 °C pH <1	<0.34	Swash and Monhemius 1995
FeAsO$_4$, 225 °C pH <1	11.9	Swash and Monhemius 1995
Ferrihydrite Fe:As 9:1, 20 °C, pH5	0.4	Swash and Monhemius 1995
Ferrihydrite Fe:As 2.3:1, 20 °C, pH5	1.2	Swash and Monhemius 1995
Ferrihydrite Fe:As 1.5:1, 20 °C, pH5	50.2	Swash and Monhemius 1995
Precipitated calcium arsenate pH > 11; low CO$_2$	>1500	Swash and Monhemius 1995
White cement, 28 days	3.6	Akhter et al. 1997
White cement, 1 year	10	Akhter et al. 1997
Ordinary Portland cement (OPC) 1:1 OPC: Flyash (FA), 28 days	7.9	Akhter et al. 1997
Ordinary Portland cement (OPC) 1:1 OPC: Flyash (FA), 1 year	94	Akhter et al. 1997
Ordinary Portland cement (OPC) 1:1 OPC: Flyash (FA), 1 year	130	Akhter et al. 1997

tive. In the case of goethite, this occurs at a pH between 6 and 7 (Parfitt 1980; Hiemstra and van Riemsdijk 1996). The pH at which this occurs is termed the point of zero charge (Deng and Stumm 1994). As pH increases the surface of the HFO particles attracts metal cations and releases oxyanions (Fig. 9).

Aluminum and manganese hydroxides and clay minerals can act in a similar fashion, although in general they show a lower efficiency for As adsorption than HFO phases (Wang and Mulligan 2006). Adsorption on clay minerals is enhanced by the presence of Fe in the clay structure (Lin and Puls 2000).

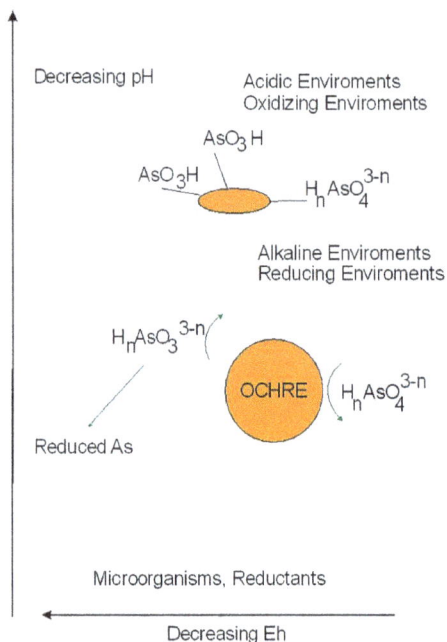

Figure 9. Schematic diagram showing implications of HFO chemistry and geochemical environment on solute adsorption at different pH (modified after a similar scheme by Deng and Stumm 1994).

In circumneutral to alkaline, oxic environments, As typically forms species such as $H_nAsO_4^{-(3-n)}$. Such ions tend to form soluble products and, as they are not strongly adsorbed, their dissolved concentrations can increase from continued release and evaporation. Consequently pH plays a critical role in the control of natural As attenuation.

Adsorption can occur naturally (Wang and Mulligan 2006; Asta et al. 2009) and can also occur in engineered facilities on synthetic clays and hydroxides or on precipitates from natural mine drainage (Rait et al. 2010). With natural HFO materials, adsorption onto schwertmannite is rapid and effective although sulfate competes with arsenate for adsorption sites (Wang and Mulligan 2006). Where mineral transformations occur, As desorption can occur. For example, when schwertmannite is transformed to goethite and jarosite, As species can be desorbed because of the lower sorption capacity of goethite and jarosite (Asta et al. 2009).

Batch removal tests have been undertaken on zeolite minerals (Payne and Abdel-Fattah 2005). Up to 50% of available arsenate and 30% of arsenite were removed in tests on natural chabazite and up to 40% and 20%, respectively, on clinoptilolite (Payne and Abdel-Fattah 2005). A more complex evaluation was undertaken on zeolite-rich basalt by Elizalde-González et al. (2001). They observed higher loadings of $H_2AsO_4^-$ than H_3AsO_4 at equivalent starting arsenic concentrations. Addition of iron in these tests did not significantly improve removal efficiency.

Lipps et al. (2010) published a study for the USEPA on the use of adsorption columns to treat contaminated well water to the WHO guideline of 10 µg L^{-1}. The pilot study was able to assess the cost effectiveness of the system, the technical capabilities of two media types, and the reliability of the water treatment system design. The feed well water to the plant contained 34.6 to 50.2 µg L^{-1} of As, predominantly as As (III). An oxidative media was applied to treatment coupled with an absorptive media. Over a 2-year pilot plant monitoring period minimal engineering difficulties were observed reflecting the simple design that worked on passive principles. The only electronic systems were the well pump and a booster pump to send water to a pressure tank feeding the columns. After installation and commissioning, there was no requirement for skilled operators to maintain the plant, apart from 7-monthly media replacement and monthly sediment filter inspections.

ARSENIC REMOVAL BY PASSIVE TREATMENT

Passive treatment involves relatively simple engineered systems that are set up at a site and largely left to operate with minimal input apart from minor maintenance and monitoring effectiveness. These systems are relatively inexpensive compared to active treatment systems (next section). However, the longevity of passive treatment systems has not been tested for more than about 20 years (Munshower 1994), and long-term treatment operations of these systems may require periodic reconstruction. Most passive treatment systems for As extraction involve adsorption of As(V) to iron oxyhydroxide precipitate material (HFO) that is produced at the site (Roddick-Lanzilotta et al. 2002; Milham and Craw 2009). However, importation of HFO, that is a waste product elsewhere, is also an option (Rait et al. 2010).

The adsorption capacity of the HFO is dependent on its surface area, and most natural precipitates of HFO have very-high surface area, and are ideal for As extraction (Roddick-Lanzilotta et al. 2002). Adsorption of As also depends on the pH of the discharging waters. Adsorption of As is most effective at low pH (~pH 3), and decreases steadily to alkaline pH where adsorption is largely ineffective at pH ~10 (Fig. 10). The effectiveness of As adsorption from a solution can be measured by comparing the amount of As adsorbed versus the amount of As remaining in solution in equilibrium with the adsorbing HFO. This ratio is the bulk distribution constant:

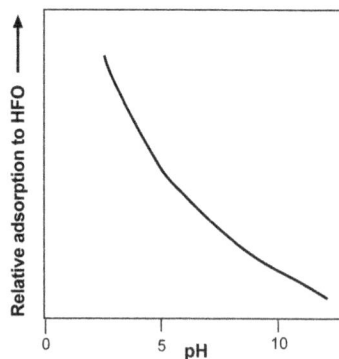

Figure 10. Relative adsorption of As(V) to natural HFO precipitate formed during oxidation of pyrite at an orogenic gold mine (after Roddick-Lanzilotta et al. 2002). Vertical axis represents infrared observations of surface properties of HFO; relative adsorption of arsenate at 830 cm[-1].

$$K_d \ (\mathrm{L \ kg^{-1}}) = \frac{\mathrm{As \ concentration \ in \ solid \ (mg \ kg^{-1})}}{\mathrm{As \ concentration \ in \ water \ (mg \ L^{-1})}}$$

K_d is a useful indicator of the theoretical amount of As that can be extracted from water. The experiments of Roddick-Lanzilotta et al. (2002), as indicated in Figure 10, suggest that K_d for waters with pH near 7 is ~10^5. This is confirmed with observations of naturally-developed systems at an old mine site, at which pyrite oxidation has led to precipitation of abundant HFO, that has adsorbed dissolved As from discharging waters (Fig. 11). These HFO deposits have variable As contents, but the data imply K_d ranging almost to 10^6 (Fig. 11). Clearly, HFO can extract large amounts of As from highly As-rich solutions. The effectiveness is generally more limited at low-As concentrations, but this process is an important component of many active arsenic extraction processes (as described above). For example, a modern mine developed at the same site as the naturally-developed system in Figure 11 uses an active treatment system for the same As-rich waters (Fig. 12). The active treatment system makes HFO from addition of ferric chloride, and the resultant HFO lowers dissolved As contents by two orders of magnitude (Fig. 12).

Membrane filtration

Several methods of membrane filtration have been successfully applied to As management. Jessica et al. (2006) evaluated cost effectiveness of different As removal techniques in the USA. Reverse Osmosis (RO) was found to be the most cost effective solution followed by activated alumina.

Other methods of membrane separation that could be used for As removal includes Molecular Recognition Technology (Bruening et al. 1991; Amos et al 2000; Bradshaw et al. 2000; Rahman et al. 2013). This process allows for rapid separation of As from effluents by use of a ligand with a high affinity for As that benefits from the high electronegativity of As and polarization. To date, however, the process has yet to be commercially applied due to high costs. Solvent extraction has also been proposed but as yet untested for As removal (Iberhan and Wisniewski 2003).

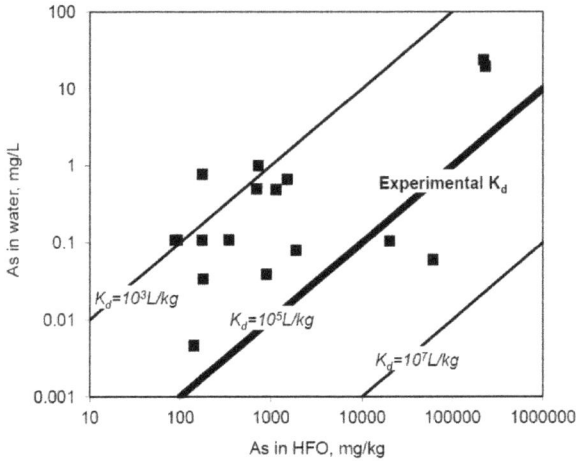

Figure 11. Arsenic content of mine waters discharging from historic underground mine tunnels and of coexisting iron oxyhydroxide precipitates which have adsorbed As, at the Fe-rich (pyritic) Globe orogenic gold mine, New Zealand (see Craw and Bowell 2014, this volume). Diagonal lines are theoretical bulk K_d for As in solids and water. Observed K_d is similar to laboratory experimental results of ~10^5 (heavy line; Roddick-Lanzilotta et al. 2002).

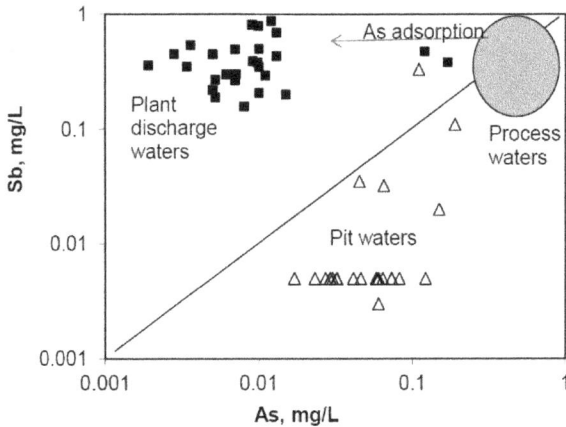

Figure 12. Arsenic and Sb contents of waters at the modern Globe orogenic gold mine, New Zealand. Mine waters and process waters typically have As/Sb ratio near to 1, reflecting the ore composition. The water treatment system at the processing plant uses ferric chloride to make HFO for As adsorption, and this treatment plant discharges waters (black squares) with ~ two orders of magnitude lower As than the processing waters (after Milham and Craw 2009).

Wang et al. (2011) completed a 10-month pilot-scale reverse osmosis (RO) treatment facility for potable water treatment at a school. The feed to the unit contained 18.2 μg L^{-1} of As, predominantly as As (V) and required filtering using a 5 μm filter prior to treatment over a 10-month period. During this time, the feed pump required replacement, indicating the technical nature of this treatment option. After installation and commissioning there was no requirement for skilled operators to maintain the facility, apart from the pump replacement.

In situ treatment

A field trial for enhanced natural attenuation was completed. In the Carson Valley, Nevada, USA, field trials of a passive remediation system for As in high-alkalinity (high-pH) and low-Fe concentration groundwaters has been published (Welch et al. 2000). The scenario greatly limits the number of attenuation sites on the limited availability Fe oxyhydroxides. The high-alkaline pH limits As oxyanion adsorption due to the abundance of free hydroxyl ions (Fig. 9). Arsenic concentrations in the aquifer were in the range of 30-36 μg L^{-1} and showed different ratios of As species (III and V). Injection cycles of dissolved oxygen, Fe(II) and hydrochloric acid were used to lower the pH and promote iron oxide formation in the aquifer. From aquifers containing As(V), the extracted waters had As concentrations below 0.05 mg L^{-1}, although elevated Fe and Mn concentrations were observed in some waters due to oxidation of As(III) to As(V).

In 2005, a permeable reactive barrier (PRB) of zero-valent iron (ZVI) was installed at an old smelting site in Montana (Wilkin et al. 2008). This solution has successfully treated a point source of As contamination in groundwater from an up-gradient concentration of over 25,000 μg L^{-1} to <10 μg L^{-1} down-gradient. The PRB had been operating successfully for 2 years at the time of publication. The major problem with the PRB is that As contamination extends below the depth of the barrier. The major advantage of PRBs is that As and other toxic substances can be stabilized in the subsurface in a location where exposure to organisms is minimized. Unfortunately cost estimations were not included in the report by Wilkin et al. (2008). An important conclusion of the study was a recommendation for thorough site characterization prior to implementation of the treatement scheme. This should include detailed hydrogeological and hydrogeochemical understanding of As dispersion in the groundwater and relation to permeable zones within the aquifer.

Manna et al. (2010) set up a membrane distillation facility in Bangladesh where water vapor passes through a membrane, condenses on the other side, and liquid water remains on the feed side of the membrane. A solar powered system was tested with a feed water containing 200 μg L^{-1} As, and this was treated to less than 10 μg L^{-1} As. The rate of the distillation process was 85 kg m^{-2} h^{-1} as the feed concentration increased to 600 μg L^{-1}. The design of the system is not suited to large flows but use of such a scheme may overcome the shortcomings of the membrane plant discussed above. Fouling would also be less of a problem in this system so very fine 5 μm prefilters are not required.

Biological remediation

Bacteria can be used to enhance the rate of As(III) oxidation by coupling the reaction to the reduction of oxygen or nitrate for energy (Wang and Zhao 2009). This process reduces As toxicity in water and forms a useful pre-treatment prior to precipitation of As(V). In Malgara, northern Greece, a water treatment plant has begun using bio-oxidation as a pretreatment method to oxidize As, ammonium, and manganese (Katsoyiannis et al. 2008). The feed concentration of As is 20 μg L^{-1} (70% as As(III)) and Fe concentrations are particularly low at 165 μg L^{-1}. Final As concentrations sent to the consumer are below 10 μg L^{-1}. Initial plant operation showed limitations in the number of favorable Fe-adsorption sites for As removal. Therefore, in order to remove arsenic, the biologically oxidized As(V) is coagulated onto HFO via hydrolysis of FeClSO$_4$ which are subsequently removed by filtration.

ENVIRONMENTAL ATTENUATION

Natural attenuation through the sorption of As onto naturally occurring iron oxyhydroxides is widely applied as a management control particularly where low cost solutions are required (Mohan and Pittman 2007; Wang and Mulligan 2007). In addition, natural biological processes can also limit the release of arsenic from mine waste. There are three main approaches to usage of plants for treatment of mine wastes: *phytoextraction*, which deliberately uses plants to extract As from the substrate; *phytostabilization*, which uses plants to stabilize the exposed surfaces of As-bearing wastes to prevent wind and water erosion; and *phytoisolation*, which uses plant cover to form a barrier between As-rich wastes and the rest of the environment. All these methods have the initial drawback of the need for establishment of plants on the mine wastes. Mine wastes, especially mine tailings, can be hostile environments for plant introduction and growth (Munshower 1994; Tordoff et al. 2000; Wong 2003; Mains et al. 2006a).

Environmental attenuation on minerals

Mineral removal of As is generally through adsorption (as discussed above). The most useful materials are iron oxyhydroxides and oxides or hydrous ferric oxides. These have higher removal efficiency than other minerals (Wang and Mulligan 2006). Aluminum and manganese hydroxides and oxides may also be useful. Clay minerals, unless high in iron, have low adsorption efficiency for As and other components in mine waters such as organic acids can deactivate the clays, further lowering adsorption efficiency.

Frequently, it is a combination of minerals that proves the best candidate for sorption of As species (Elizalde-González et al. 2001). In the Zimapan mining district in Mexico, As concentrations often exceed 1 mg L^{-1} (Ongley et al. 2001). Reaction with the calcareous shales of the Soyatal formation, that are rich in chlorite, kaolinite, illite and calcite, in a rock:water ratio of 1:10, demonstrated removal of As from initial concentrations of 0.5-1 mg L^{-1} to less than 0.03 mg L^{-1}. In field trials in South West England, the addition of vermiculite as fuller's earth to As contaminated soils reduced the bioavailable As in the soil from 20% to less than 2%. Other clays such as montmorillonite, sepiolite, and bentonite had less efficiency in sorbing As (Jones and Mitchell 1993).

Phytoextraction

Phytoextraction makes use of certain plants natural propensity to accumulate As to anomalously-high concentrations. A pilot plant in New Mexico (Elless et al. 2005) successfully produced As-free water at a maximum flow rate of 1,900 L per day over a three-month period without any breakthrough. However, it was necessary to rotate the plants through nutrient solutions during treatment. Water loss was only about 5% by evapotranspiration. The major drawbacks of the technology are lack of knowledge regarding plant stress factors that could hinder growth of a crop at a large scale exposed to high-As concentrations, the disposal of the As-bearing biomass, and the large area required for high flow rates. In addition, residence time for this design solution can be in the region of 5.5 h compared with 0.3 h for ion exchange methods of As removal.

Phytostabilization

Large volumes of As-bearing mine wastes are impractical as targets for phytoextraction, which only affects the upper few meters of substrate. In particular, As-bearing mine tailings impoundments are a priority for physical stability and environmental reasons, as the sandy tailings are easily mobilized by wind and rain. However, the tailings surface can be a hostile environment for plant establishment, as they are biologically sterile when they are deposited. The tailings also allow rapid water drainage and evaporation, and low plant nutrient availability, and commonly have sub-optimal pH for plant growth (Munshower 1994; Tordoff et al. 2000; Wong 2003; Mains et al. 2006a). The tailings commonly have elevated levels of toxic metals and other chemical constituents that can inhibit plant growth (Bruce et al. 2003; Baroni et al. 2004). All these issues have to be addressed for successful plant establishment. Irrespective of the effectiveness of plant cover, As-bearing leachate will almost certainly emanate from the mine wastes after stabilization, and these leachates have to be managed, and established plants may contain elevated As that is available to other parts of the ecosystem.

In addition to vegetation establishment issues, plants can incorporated As into their structures (see previous section), so the plant species to be established have to be selected with the final land use in mind. Hence, the most effective species combine tenacious growth and effective-As exclusion. Lottermoser (2010) suggests gorse shrubs (*Europaeus*) as a suitable plant for this process on As-rich mine wastes in Cornwall, UK. Even this hardy plant struggles on dry sandy tailings, and dryland grasses are most effective in this setting. Rye, corn, and, barley have been shown to be useful phytostabilizers in the near-desert conditions of semi-arid tailings in New Zealand, with rye and corn being the most effective (Mains et al. 2006a,b). These grasses have only limited-As uptake by their shoots, although their roots take up to 400 ppm As (Fig. 13).

Establishment success of phytostabilization plantings can be enhanced by addition of surface-fertilizer amendments, which overcomes some plant-nutrient issues. This increases the expense of the operation, and care has to be taken to ensure that amendments do not enhance As uptake of plants, or increase dissolved As contents of leachates from the stabilized tailings. For example, elevated As in grass shoots in some of the trial plots in tailings in Figure 13 arose after amendment with a biosolid fertilizer (Mains et al. 2006b). Phosphorus is one of the principal nutrients lacking in mine wastes, and addition of fertilizers such as superphosphate greatly enhances plant establishment success (Mains et al. 2006b). One concern with phosphorus addition is that, because phosphate ions behave chemically in a similar manner to arsenate ions, As mobility may be enhanced by competition for sorption sites. This can occur on a small-experimental scale, but on the scale of 1 m thick tailings, this has been shown to be negligible (Fig. 14).

Phytoisolation

The aim of phytoisolation is to produce a complete vegetation cover on the mine wastes, preferably with abundant plant litter and a viable ecosystem that progressively builds in thickness and is self-sustaining. The processes of initial establishment of plants on mine wastes described in the phytostabilization section apply to the phytoisolation system as well. Also, there is the ongoing issue of seepage of As-bearing water and this has to be managed separately, as for phytostabilization, either by active or passive treatment processes (above).

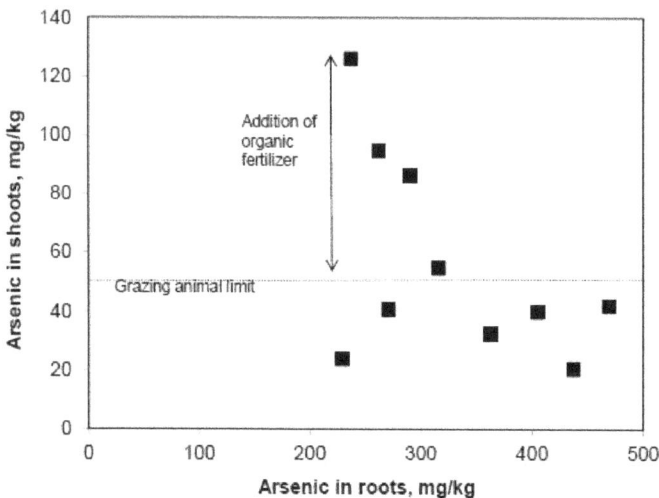

Figure 13. Arsenic contents of roots and shoots in grasses grown in trial plots of mine tailings (~1500 mg kg^{-1} As) (after Mains et al. 2006b). General animal grazing As limits in grasses (NRC 1980) are indicated. Shoots with As >50 mg kg^{-1} were grown with addition of organic fertilizer.

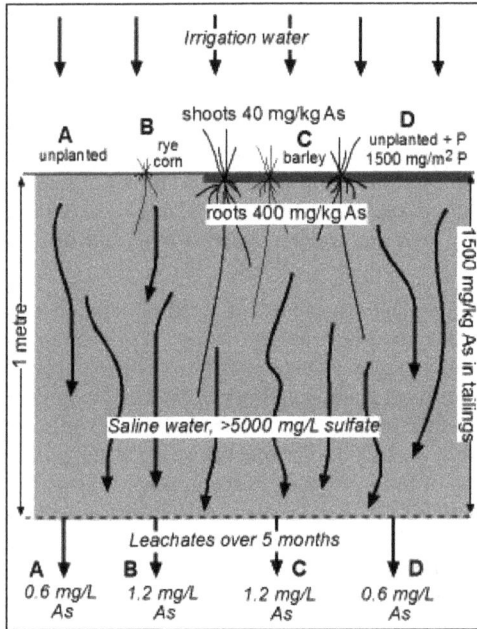

Figure 14. Schematic diagram of the upper meter of tailings, summarizing the effects on dissolved As in leachates from variably vegetated mine tailings with ~1500 ppm As (after Mains et al. 2006b). Amended tailings are indicated with heavy grey line at surface. Irrigation water (negligible As) was added from above, and was sampled as leachate at the base. Different water pathways are shown as A-D. Tailings with plants growing in them released higher concentrations of As (pathway B,C) than unplanted tailings (pathway A). Addition of low-P amendment (pathways C and D) did not alter the release of As.

Selection of appropriate plants for initiation and development of phytoisolation is critical to success, and this is a local decision appropriate to specific sites. The most effective way of plant selection is to note which local species will grow on a site naturally, or on nearby abandoned sites. Natural colonization can occur rapidly with appropriate species if there is a local seed bank. Plants can be remarkably tolerant of high-As contents in solids and associated water runoff (Figs. 14, 15). In general, mosses can form an excellent first stage of revegetation provided there is sufficient moisture, although mosses can also take up abundant As into their structures (Fig. 16). Grasses are also effective at providing rapid cover (see phytostabilization section, above), but can also take up significant quantities of As (Fig. 15). Shrubs and trees form the most effective long-term covers, and take up the least As (Fig. 15). Establishment of shrubs may require an early moss cover, in which shrubs can germinate and develop a root system (Craw et al. 2007).

CONCLUSIONS

Arsenic requires management in the mining effluents and this can occur at the source or site of generation, captured along the flow path, or treated prior to discharge. Issues of As chemistry and engineering need to be accounted for in the development and application of suitable methods.

Arsenic removal methods applied to natural waters focus on chemical precipitation, filtration of particulates, or biological or chemical oxidation of As(III) to As(V) and subsequent precipitation. It is often difficult to impossible to infer true performance in the field based soley on laboratory data. Two examples of As treatment that holds potential long term benefits are membrane distillation (membrane filtration technology) and biosorption (adsorption biotechnology).

Biosorption through bacterial uptake and As removal from water or soils involves the direct adsorption of As followed by co-precipitation generally with Fe and to a lesser extent

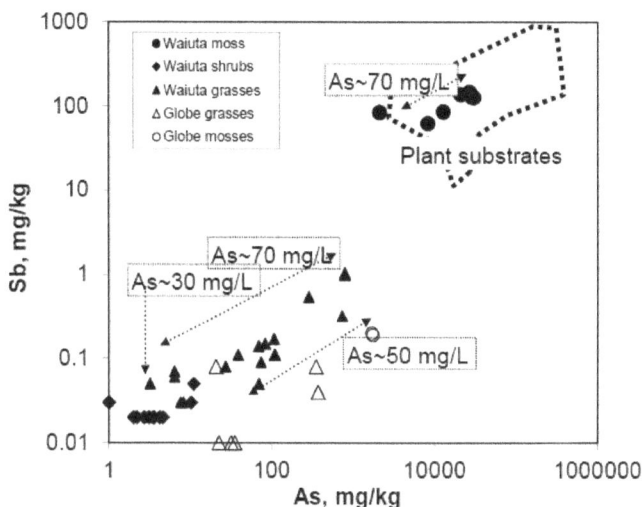

Figure 15. Metalloid uptake into plants growing on wastes at a New Zealand As-rich gold processing site (after Craw et al. 2007) at the low-Fe Waiuta mine, and at the nearby high-Fe Globe mine site (Craw and Bowell 2014, this volume). Antimony occurs in solid solution in arsenopyrite and has been mobilized in the environment after arsenopyrite oxidation. Compositions of mine wastes at both sites fall into the heavy dotted field (top right). Dissolved As contents of waters that run off these sites are indicated in boxes and with dotted arrows. Shrubs take up the least As, and grasses and mosses are highly variable in As uptake. The highest As uptake is in moss (*Pohlia wahlenbergii*) growing in water-saturated tailings containing arsenolite (see Fig. 16).

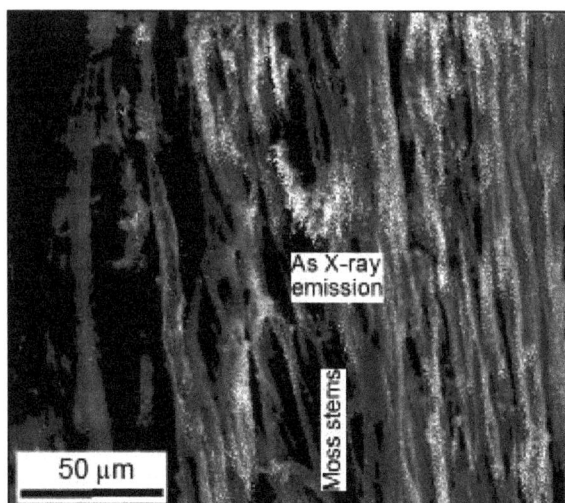

Figure 16. Arsenic X-ray image (white dots) from electron microprobe scanning of moss *Pohlia wahlen-bergii* that contains up to 3 wt% As (after Craw et al. 2007), growing on substrate with >10% As that is saturated with water with >50 mg L^{-1} dissolved As.

Mn (Wang and Zhao 2009). These processes are limited by the rate of precipitation or sorption of the As species onto the solid and by oxidation of reduced As to suitable oxidized As(V) species. Genetically modified *E. Coli* produced cells which accumulated 5× as much As(III) and 60× as much As(V) as other cultures and have demonstrated removal rates up to 98% of As in laboratory tests from contaminated water (50 μg L^{-1} As) within 1 h (Kostal et al. 2004). The application of high-rate, high-capacity adsorbents in a reliable and relatively simple column using the adsorption mechanism could possibly provide a sustainable and cost effective, as well as efficient, approach to As treatment particularly in rural communities (Mench et al. 2010).

Mine water impacts are usually limited to a local area but characterized by high concentrations. For this type of contamination, *in situ* treatments have potential based on the ability to efficiently handle groundwater and treat. This is due to the ability to efficiently handle point sources in groundwater and a wider area of contamination without costly infrastructure.

REFERENCES

Achmed SM (1991) Electrochemical and surface chemical methods for the prevention of atmospheric oxidation of sulfide tailings and acid generation. *In:* Proceedings Second Int. Conf. Abatement Acidic Drainage, MEND Secretariat, Ottawa, Ontario. 2:305-319

Akhter H, Cartledge FK, Roy A, Tittlebaum ME (1997) Solidifcation/stabilization of arsenic salts: Effects of long cure times. J Hazard Mater 52:247-264

Amos G, Hopkins W, Izatt SR, Bruening RL, Dale JB, Krakowiak KE (2000) Extraction, recovery, and recycling of metals from effluents, electrolytes, and product streams using molecular recognition technology. In: Environmental Improvements in Mineral Processing and Extractive Metallurgy. Sánchez MA, Vergara F, Castro SH (eds) University of Concepción, Chile, May 9-13 2000 2:1-16

Asta MP, Cama J, Martinez M, Gimenez J (2009) Arsenic removal by goethite and jarosite in acidic conditions and its environmental implications. J Hazard Mater 171:965-972

Azcue JM, Nriagu JO (1995) Impact of abandoned mine tailings on the arsenic concentrations in Moira Lake, Ontario. J Geochem Explor 52:81-89

Baas Becking LGM, Kaplan IR, Moore D (1960) Limits of the natural environment in terms of pH and oxidation-reduction-potentials. J Geol 6:243-284

Bain J, Blowes D, Wilkens JA (2009) Evaluation of the treatment of groundwater arsenic at mining and industrial sites using ZVI and BOFS permeable reactive barriers. *In:* Proceedings of Securing the Future and 8th ICARD, June 22-26 2009, Skelleftea, Sweden. T3-1, 4 p

Baroni F, Boscagli A, Di Lella LA, Protano G, Riccobono F (2004) Arsenic in soil and vegetation of contaminated areas in southern Tuscany (Italy). J Geochem Explor 81:1-14

Beauchemin S, Fiset JF, MacKinnon T, Hesterberg D (2004) Impact of a water cover on the stability of arsenic in a neutralization sludge. *In:* Advances in Mineral Resource Management, Environmental Geotechnology, Hania 2004, Greece. Agioutantis Z (ed) Heliotopos Conferences, p 519-524

Bentley SP (ed) (1996) Engineering Geology of Waste Disposal. Geological Society Engineering Geology Special Publication, 11. The Geological Society, London

Blowes DW, Ptacek CJ, Jambor L (1994) Remediation of low quality drainage in tailings impoundments. *In:* Environmental Geochemistry of Sulphide Mine Waste. Jambor JL, Blowes DW (eds) Mineralogical Association of Canada Short Course 22:365-379

Blowes DW, Reardon EJ, Jambor JL, Cherry JA (1991) The formation and potential importance of cemented layers in inactive mine tailings. Geochim Cosmochim Acta 55:965-978

Bodenan F, Baranger P, Lassin A, Azaroual M, Gaucher E, Braibant G (2004) Arsenic behaviour in gold-ore mill tailings, Massif Central, France: hydrogeochemical study and investigation of *in situ* redox signatures. Appl Geochem 19:1785-1800

Bowell RJ (1994) Arsenic sorption by iron oxyhydroxides and oxides. Appl Geochem 9:279-286

Bowell RJ (2003) The influence of speciation in the removal of arsenic from mine waters. Land Contam Reclam 11:231-238

Bowell RJ (2014) Mineralogical and chemical assessment of paste backfill stability and aging. Minerals Engineering, accepted

Bowell RJ, Bruce I (1995) Geochemistry of ochre and mine waters, Levant mine, Cornwall. Appl Geochem 10:237-250

Bowell, RJ, Parshley JV (2003) Arsenic cycling in the mining environment. *In*: USEPA Workshop on Managing Arsenic Risks to the Environment: Characterization of Waste, Chemistry and Treatment and Disposal. Proceedings and Summary Report, Ohio, USEPA/625/R-03/010, 10-12

Bowell, RJ, Williams KP, Connelly RJ (1998) Chemical containment of mining waste. *In*: Chemical Containment in the Geosphere. Geological Society Special Publication 157. Metcalf R, Rochelle CR (eds) Geological Society of London, p 213-241

Bradshaw JS, Izatt RM, Savage PB, Bruening RL, Krakowiak KE (2000) The design of ion selective macrocycles and the solid-phase extraction of ions using molecular recognition technology: a synopsis. Supramolecular Chem 12:23-26

Bruce SL, Noller BN, Grigg AH, Mullen BF, Mulligan DR, Ritchie PJ, Currey N, Ng JC (2003) A field study conducted at Kidston gold mine to evaluate the impact of arsenic and zinc from mine tailing to grazing cattle. Toxicol Lett 137:23-34

Bruening ML, Mitchell DM, Bradshaw JS, Izatt RM, Bruening RL (1991) Effect of organic solvent and anion type on cation binding constants with silica gel bound macrocycles and their use in designing selective concentrator columns. Anal Chem 63:21-24

Carter M, Baker N, Burford RP (1995) Polymer encapsulation of arsenic-containing wastes. J Appl Polymer Sci 58: 2039-2046

Cavanagh JE, Pope J, Harding JS, Trumm D, Craw D, Rait R, Greig H, Niyogi D, Buxton R, Champeau O, Clemens A (2014) A framework for predicting and managing water quality impacts of mining on streams: a user's guide. *http://www.crl.co.nz/downloads/geology/FrameworkUsersGuideOct2010.pdf* (accessed July 2014)

Craw D, Bowell RJ (2014) The characterization of arsenic in mine waste. Rev Mineral Geochem 79:473-506

Craw D, Rufaut CG, Haffert L, Paterson LA (2007) Plant colonization and arsenic uptake on high arsenic mine wastes, New Zealand. Water Air Soil Pollut 179:351-364

Demopoulos GP, Droppert DJ, van Weert G (1995) Precipitation of crystalline scorodite from chloride solutions. Hydrometallurgy 38:245-261

Deng Y, Stumm W (1994) Reactivity of aquatic iron (III) oxyhydroxides- implications for redox cycling of iron in natural waters. Appl Geochem 9:23-36

Dey BM (1997) Treatment of Acid Mine Drainage. PhD thesis, University College of Wales. 180p.

Doušová B, Kolousek D, Kovanda F, Novotna M (2005) Removal of As(V) species from extremely contaminated mining water. Appl Clay Sci 28:31-42

Elizalde-González MP, Mattusch J, Eincke D, Wennrich R (2001) Sorption on natural solids for arsenic removal. Chem Eng J 81:187-195

Elless MP, Poynton CY, Willms CA, Doyle MP, Lopez AC, Sokkary DA, Ferguson BW, Blaylock MJ (2005) Pilot-scale demonstration of phytofiltration for treatment of arsenic in New Mexico drinking water. Water Res 39:3863-3872

Emmet MA, Khoe G (1994) Environmental stability of As-iron hydroxides. Proceedings EPD Congress, TMS Annual Conference. Warren GW (ed) p 153-166

Fields K, Chen A, Wang L (2000) Arsenic removal from drinking water by iron removal plants, Ohio, EPA/600/R-00/086

Fujita T, Taguch, R, Abumiya M, Matsumot, M, Shibata E, Nakamura T (2008a) Novel atmospheric scorodite synthesis by oxidation of ferrous sulfate solution. Part 1. Hydrometallurgy 90:92-102

Fujita T, Taguchi R, Abumiya M, Matsumoto M, Shibata E, Nakamura T (2008b) Novel atmospheric scorodite synthesis by oxidation of ferrous sulfate solution. Part 2. Effect of temperature and air. Hydrometallurgy 90:85-91

Fujita T, Taguchi R, Abumiya M, Matsumoto M, Shibata E, Nakamura T (2008c) Effects of zinc, copper and sodium ions on ferric arsenate precipitation in a novel atmospheric scorodite process. Hydrometallurgy 93:30-38

Gammons GH, Duaime TE (2006) Long term changes in the limnology and geochemistry of the Berkeley Pit Lake, Butte, Montana. Mine Water Environ 25:76-85

Gammons GH, Metesh JJ, Snyder DM (2006) A survey of the geochemistry of flooded mine shaft water in Butte, Montana. Mine Water Environ 25:100-107

Gupta BS, Chatterjee S, Rott U, Kauffman H, Bandopadhyay A, DeGroot W, Nag NK, Carbonell-Barrachina AA, Mukherjee S (2009) A simple chemical free arsenic removal method for community water supply - A case study from West Bengal, India. Environ Pollut 157:3351 – 3353

Harris, GB, Krause E (1993) The disposal of arsenic from Metallurgical processes: its status regarding ferric arsenate. *In*: Extractive Metallurgy of Nickel, Cobalt, and Associated Metals. Volume 1. TMS Annual Meeting, Denver, February 21-25. Abstracts volume

Hiemstra P, van Riemsdijk D (1996) Calculated values for point of zero charge for some common minerals. J Colloid Interface Sci 179:488-508

Hopenhayn C (2006) Arsenic in drinking water: impact on human health. Elements 2:103-107

Hughes P (2008) Performance of composite paste fill fences. CIM AGM- Edmonton, Alberta, May 2008. Electronic Proceedings, p 5-7

Hutton-Ashkenny M, Bowell RJ, Dey BM (2011) Engineering Solutions for the Removal of Arsenic from Aqueous Solution: A Review. In: Mine Water – Managing the Challenges. Rüde C, Freund A, Wolkersdorfer C (eds) IMWA 2011 Aachen, Germany, 353-358

Iberhan L, Wisniewski M (2003) Removal of arsenic (III) and arsenic (V) from sulfuric acid solution by liquid-liquid extraction. J Chem Technol Biotechnol 78:659-665

Jessica S-M, Kevin JB, Smith AE (2006) Cost effective arsenic reductions in private well water in Maine. J Am Water Resour Assoc 42:1237–1245

Jones K, Mitchell PB (1993) Monitoring and remediation of arsenic contamination of agricultural soils in south west England. Land Contam Rem 1:46-54

Katsoyiannis AI, Zikoudib A, Huga SJ (2008) Arsenic removal from groundwaters containing iron, ammonium, manganese and phosphate: A case study from a treatment unit in northern Greece. Desalination 224:330-339

Kleinmann RLP (1998) Bacterial control of acidic drainage. In: Coal Mine Drainage Prediction and Pollution Prevention in Pennsylvania. The Pennsylvania Department of Environmental Protection, Pittsburgh 15:1-6

Kontopoulos A, Komnitsa K, Xendis A (1996) Environmental characterisation of lead smelter slags in Laviron. In: Proceedings of the International Conference on Minerals, Metals & the Environment, Prague, September 3-6 1996. Institution of Mining and Metallurgy, 405-420

Kostal J, Yang R, Wu CH, Mulchandani A, Chen W (2004) Enhanced arsenic attenuation in engineered bacterial cells expressing ArsR. Appl Environ Microbio 70:4582-4287

Krause E, Ettel VA (1985) Ferric arsenates-are they safe? CIM annual meeting, Vancouver, Canada, 1-20

Krause E, Ettel VA (1989) Solubilities and stabilities of ferric arsenate compounds. Hydrometallurgy 22:311-357

Lin Z, Puls RW (2000) Adsorption, desorption and oxidation of arsenic affected by clay minerals and aging process. Environ Geol 39:753-759

Lindsay MBJ, Condon PD, Jambor JL, Lear KG, Blowes DW, Ptacek CJ (2009) Mineralogical geochemical and microbial investigation of a sulfide-rich tailings deposit characterized by neutral drainage. Appl Geochem 24:2212-2221

Lipps JP, Chen ASC, Wang L, Wang A, McCall SE (2010) Arsenic removal from drinking water by absorptive media. US EPA Demonstration Project at Spring Brook Mobile Home Park in Wales, ME. Final Performance Evaluation Report. EPA/600/R-10/012

Lottermoser B (2010) Mine Waste. Springer-Verlag, Berlin. 400p

Luganov VA, Sajin EN (2002) Detoxification of Arsenic bearing wastes. In: Green Conference Proceedings, Cairns, Queensland. May 2002. p 375-378

Machingwuta NC, Broadbent CP (1994) Incorporation of As into silicate slags. Trans Inst Min Metall 103:C1-8

Mains D, Craw D, Rufaut CG, Smith CMS (2006a) Phytostabilisation of gold mine tailings, New Zealand. Part 1: Plant establishment in an alkaline substrate. Int J Phytorem 8:131-147

Mains D, Craw D, Rufaut CG, Smith CMS (2006b) Phytostabilisation of gold mine tailings, New Zealand. Part 2: Experimental evaluation of arsenic mobilization during revegetation. Int J Phytorem 8:163-183

Malhotra VM (1991) Fibre enforced high volume fly ash shotcrete for controlling aggressive leachates from exposed rock surfaces and mine tailings. In: Proceedings Second Int. Conf. Abatement Acidic Drainage, 2, 27-41. MEND Secretariat, Ottawa, Ontario

Malik AH, Khan ZM, Mahmood Q, Nasreen S, Bhatti ZA (2009) Perspectives of low cost arsenic remediation of drinking water in Pakistan and other countries. J Hazard Mater 168:1-12

Manna AK, Sen M, Martin AR, Pal P (2010) Removal of arsenic from contaminated groundwater by solar-driven membrane distillation. Environ Pollut 158:805-811

Maree JP, Bosman DJ (1989) Chemical removal of sulfate, calcium and metals from mining and power station effluents. Water Sewage and Effluent 9:10-25

Maree JP, Strobos G, Greben H, Netshidaulu E, Steyn E, Christie A, Gunther P, Waanders FB (2004) Treatment of acid leachate from coal discard using calcium carbonate and biological sulfate removal. Mine Water Environ 23:144-151

Mench M, Lepp N, Bert V., Schwitzguebel J, Gawronski SW, Schroeder P, Vangronsveld J (2010) Successes and limitations of phytotechnologies at field scale: outcomes, assessment and outlook from COST Action 859. J Soils Sediments 10:1039-1070

Milham L, Craw D (2009) Antimony mobilization through two contrasting gold ore processing systems, New Zealand. Mine Water Environ 28:136-145

Mohan D, Pittman U (2007) Arsenic removal from water/wastewater using adsorbents – A critical review. J Hazard Mater 142:1-53

Munshower FF (1994) Disturbed Land Revegetation. CRC Press, Inc., U.S.A.

Nolan AL, McKay D, Lunsmann F, Fensom M, Wood D (2012) Chemical immobilisation of complex arsenic co-contaminated industrial waste. *In*: The 4th International Congress on Arsenic in the Environment – Understanding the Geological-Medical Interface of Arsenic, Cairns, Australia, July 22-27, 2012. CRC Press, 443-447

Nordstrom DK (1982) Aqueous pyrite oxidation and the consequent formation of secondary minerals. *In*: Acid Sulfate Weathering. Kittrick JA (ed) Soil Science Society of America p 37-56

NRC (1980) Mineral Tolerance of Domestic Animals. National Research Council, Washington DC, p 122-124

O'Neil P (1990) Arsenic. In: Heavy Metals in Soils. Alloway BJ (ed) Springer, Berlin, p 105-121

Ongley LK, Armienta MA, Heggerman K, Lathrop AS, Mango H, Miller W, Pickelner S (2001) Arsenic removal from contaminated water by the Soyatal Formation, Zimapan Mining District, Mexico - a potential low-cost low-technology remediation system. Geochem Explor Environ Anal 1:25-31

Ouellet J, Benzaazoua M, Servant S (1998) Mechanical, mineralogical and chemical characterization of a paste backfill. *In*: Tailings and Mine Waste'98. Balkema, Rotterdam 139-146

Oyarzun R, Lillo J, Higueras P, Oyarzun J, Maturana H (2004) Strong arsenic enrichments in sediments from the elqui watershed, Northern Chile: industrial (gold mining at El Indio-Tambo districts) vs. geologic processes. J Geochem Explor 84:53-64

Papassiopi N, Vimkova E, Nenov V, Molnar L (1996) Removal and fixation of arsenic in the form of ferric arsenate. Three parallel experimental studies. Hydrometallurgy 41:243-253

Parfitt RL (1980) Chemical properties of variable charge soils. *In*: Soils with Variable Charge. Theng BKG (ed) Sp. Pub. New Zealand Society of Soil Science, 167-194

Patterson BM, Robertson BS, Woodbury RJ, Talbot B, Davis GB (2006) Long term evaluation of a composite cover overlaying a sulfidic tailings facility. Mine Water and the Environment 25:137-145

Payne KB, Abdel-Fattah TM (2005) Adsorption of arsenate and arsenite by iron-treated activated carbon and zeolites: effects of pH, temperature, and ionic strength. J Environ Sci Health 40:723-749

Price WA (1997) Guidelines for the Prediction of Acid Rock Drainage and Metal Leaching for Mines in British Columbia: Part I. In: Proc. 4th International Conference on Acid Rock Drainage, Vancouver, BC, p 15-30

Rait R, Trumm D, Pope J, Craw D, Newman N, MacKenzie H (2010) Adsorption of arsenic by iron rich precipitates from two coal mine drainage sites on the West Coast of New Zealand. NZ J Geol Geophys 53:177-193

Rahman IM, Begum ZA, Furusho Y Mizutani S, Maki T, Hasegawa H (2013) Selective separation of tri- and pentavalent arsenic in aqueous matrix with a macrocycle-immobilized solid-phase extraction system. Water Air Soil Pollut 224(5):1-11

Robbins RG, Glastras MV (1987) The precipitation of arsenic from aqueous solution in relation to disposal from hydrometallurgical processes. *In*: Proceedings of the AusIMM, Extractive Metallurgy, May 1987, p 223-229

Roddick-Lanzilotta AJ, McQuillan AJ, Craw D (2002) Infrared spectroscopic characterisation of arsenate(V) ion adsorption from mine waters, Macraes Mine, New Zealand. Appl Geochem 17:445-454

Romano CG, Mayer KU, Jones DR, Ellerbroek DA, Blowes DW (2003) Effectiveness of various cover scenarios on the rate of sulfide oxidation in mine tailings. J Hydrol 271:171-187

Sand W, Jozsa PG, Kovacs ZM, Săşăran N, Schippers A (2007) Long term evaluation of acid rock drainage mitigation measures in large lysimeters. J Geochem Explor 92:205-211

Shih M (2005) An overview of arsenic removal by pressure driven membrane processes. Desalination 172:85-97

Smedley PL, Kinniburgh DG (2002) A review of the source, behaviour and distribution of arsenic in natural waters. Appl Geochem 17:517-568

Smith A, Mudder T (1991) Chemistry and Treatment of Cyanidation Wastes. Mining Journal Press, London

Swash PM, Monhemius AJ (1995) Hydrothermal precipitation, characterization and solubility testing in the Fe-Ca-AsO$_4$ system. *In*: Sudbury '95--Mining in the Environment, CANMET, p 17-28

Swash PM, Monhemius AJ (1996) The characteristics of calcium arsenate compounds and their relevance to industrial waste disposal. *In*: Minerals, Metals and the Environment II, Institute of Mining and Metallurgy, Prague, September 1996, p 353-362

Tordoff GM, Baker AJM, Willis AJ (2000) Current approaches to the revegetation and reclamation of metalliferous mine wastes. Chemosphere 41:219-228

Ugarte FJG, Monhemius AJ (1992) Characterization of high temperature arsenic containing residues from hydrometallurgical processes. Hydrometallurgy 30:69-86

USEPA (1992) EPA Test Method 1311 - TCLP, Toxicity Characteristic Leaching Procedure. US Environmental Protection Agency, Cincinnati Ohio, 38 p. *http://www.ehso.com/cssepa/TCLP_from%20EHSOcom_Method_1311.pdf* (accessed July 2014)

USEPA (2001) Arsenic and Clarifications to Compliance and New Source Monitoring Rule: A Quick Reference Guide, EPA 816-F-01-004 January 2001, 2 p. *http://water.epa.gov/drink/info/arsenic/upload/2005_11_10_arsenic_quickguide.pdf* (accessed July 2014)

Wang L, Chen A, Fields K (2000a) Arsenic Removal from Drinking Water by Ion Exchange and Activated Alumina Plants. US Environmental Protection Agency EPA/600/R-00/088

Wang L, Lewis GM, Chen ASC (2011) Arsenic and Antimony Removal From Drinking Water by Point-Of-Entry Reverse Osmosis Coupled with Dual Plumbing Distribution. US Environmental Protection Agency EPA/600/R-11/026

Wang Q, Nishimura T, Umetsu Y (2000b) Oxidative precipitation for arsenic removal in Effluent treatment. *In*: Minor Elements 2000. Young C (ed) Society for Mining, Metallurgy and Exploration, Colorado, p 39-52

Wang S, Mulligan CN (2006) Natural attenuation processes for remediation of arsenic contaminated soils in groundwater. J Hazard Mater B 138:459-470

Wang S, Zhao X (2009) On the potential of biological treatment for arsenic contaminated soils and groundwater. J Environ Manage 90:2367-2376

Waychunas GA, Rea BA, Fuller CC, Davis JA (1993) Surface chemistry of ferrihydrite: Part 1: EXAFS studies of the geometry of coprecipitated and adsorbed arsenate. Geochim Cosmochim Acta 57:2251-2269

Welch AH, Westjohn DB, Helsel DR, Wanty RB (2000) Arsenic in ground water of the United States: occurrence andgeochemistry. Ground Water 38:589-604

WHO (2011) Arsenic in Drinking-water; Background document for development of WHO Guidelines for Drinking-water Quality WHO/SDE/WSH/03.04/75/Rev/1. *http://www.who.int/water_sanitation_health/dwq/chemicals/arsenic.pdf* (accessed July 2014)

Wilkin RT, Acree SD, Beak DG, Ross RR, Lee TR, Paul CJ (2008) Field Application of a Permeable Reactive Barrier for Treatment of Arsenic in Ground Water. US Environmental Protection Agency, EPA 600/R-08/093

Wong MH (2003) Ecological restoration of mine degraded soils, with emphasis on metal contaminated soils. Chemosphere 50:775-780

Zhu Y, Merkel BJ, Stober I, Bucher K (2003) The Hydrogeochemistry of Arsenic in the Clara mine, Germany. Mine Water Environ 22:110-117

Zinck JM (2000) The abundance, behaviour and stability of As, Cd, Pb and Se in Lime treatment sludges. *In*: Minor Elements 2000. Young C (ed) Society for Mining, Metallurgy and Exploration, Colorado, p 213-223

Reviews in Mineralogy & Geochemistry
Vol. 79 pp. 533-551, 2014
Copyright © Mineralogical Society of America

12

The Legacy of Arsenic Contamination from Mining and Processing Refractory Gold Ore at Giant Mine, Yellowknife, Northwest Territories, Canada

Heather E. Jamieson

*Department of Geological Sciences & Geological Engineering
Miller Hall, Queen's University
Kingston, Ontario, Canada, K7L 3N6*

jamieson@queensu.ca

INTRODUCTION

The case of the Giant mine illustrates how a large, long-lived Au mine has resulted in a complex regional legacy of As contamination and an estimated remediation cost of almost one billion Canadian dollars (AANDC 2012). The mine, located a few km north of the city of Yellowknife on the shore of Great Slave Lake (Figs. 1, 2) produced more than 7 million troy ounces of Au (approximately 220 tonnes) from a largely underground operation. Giant mine was the largest producer in the Yellowknife greenstone belt, which produced more than 12 million troy ounces (~370 tonnes) in total (Bullen and Robb 2006). Arsenopyrite-bearing Au ore was roasted from 1949 to 1999 as a pretreatment for cyanidation (Fig. 3a). Poor or nonexistent emission controls in the early years resulted in the release of an estimated 20,000 tonnes of roaster-generated As_2O_3 to the surrounding environment through stack emissions (CPHA 1977; Wrye 2008). Over the lifetime of the mine, however, most of the As_2O_3 (237,000 tonnes) was stored in underground chambers (Fig. 3b) and is a now an ongoing source of As to groundwater and surface water (INAC 2007; Jamieson et al. 2013). Other roaster products include As-bearing maghemite and hematite (calcine) were deposited with tailings and re-mobilized into creek and lake sediments. Under reducing conditions, post-depositional remobilization of As associated with roaster-generated Fe oxides results in release of As to sediment pore water and reprecipitation of some As as a sulfide phase (Fawcett and Jamieson 2011). However, As(III) in maghemite and As_2O_3 persists in the oxidizing conditions of near-surface tailings and soils (Walker et al. 2005; Jamieson et al. 2013).

Ore roasting increases the solubility, toxicity, and bioaccessibility of As by converting sulfide-hosted As to oxide-hosted As. At Giant, understanding the processing history is critical to characterizing mine waste, assessing the risk to human and ecosystem health, and predicting long-term stability and optimal management. Roasting is still used worldwide today for improving recovery of refractory Au ores and for removing As from copper ores, but in modern operations As-bearing roaster products are managed as hazardous waste or sold for use in various industrial applications. Other Au mines in the Yellowknife area also roasted arsenopyrite in the early years of mining and undoubtedly have contributed to As contamination in the region. At the Con mine, the largest other Au mine in the region, managers were able to cease roasting when they encountered less refractory ores, and then converted to pressure oxidation in the late 1980's, resulting in an Fe arsenate waste product considered to be less problematic than As_2O_3 (Walker et al. 2014). Giant mine is considered the largest industrial source of bioavailable arsenic in the region (MVRB 2013).

1529-6466/14/0079-0012$05.00

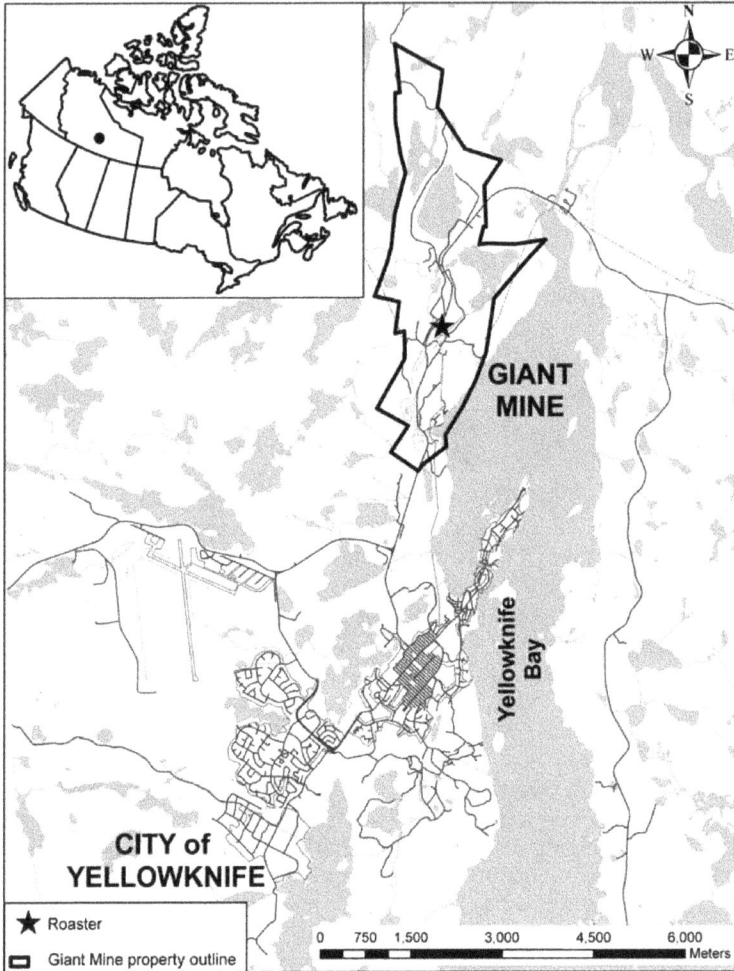

Figure 1. Map showing location of Giant mine and the City of Yellowknife, with inset showing location of Yellowknife, NWT, Canada. Solid square indicates the sample site for the data shown in Figure 4. Map is modified from Bromstad and Jamieson (2012).

The history of Giant mine is also a remarkable example of how relations between mining companies, local population, and government authorities have evolved over the last 75 years. Local Aboriginal communities were not consulted when mines were first established in the 1930's and 1940's yet were affected by removal of access to their traditional land and exposed to arsenic contamination along with other local residents (Sandlos and Keeling 2012). A Dene child died of arsenic poisoning in 1951 in an area near the mine. Pollution controls were put in place following this incident but emissions and spills of As persisted. Giant was operated by a series of private mining companies for fifty years but was placed into receivership in April 1999, and after no purchasers were identified by the receiver, ownership of the mine reverted to the Crown (federal government). Giant mine is now considered an abandoned mine and the environmental liability is public. Modern mining operations in the Northwest Territories require multi-stage permitting and extensive community consultation. The City of Yellowknife, the capital city and home for half the population of the Northwest Territories, developed as a

Figure 2. (*for color see Plate 10*) Air photo of the Giant Mine property ca. 2000. Data show total As concentration in outcrop soil samples from Bromstad (2011). The industrial remediation guideline for Giant mine NWT is 340 mg kg⁻¹.

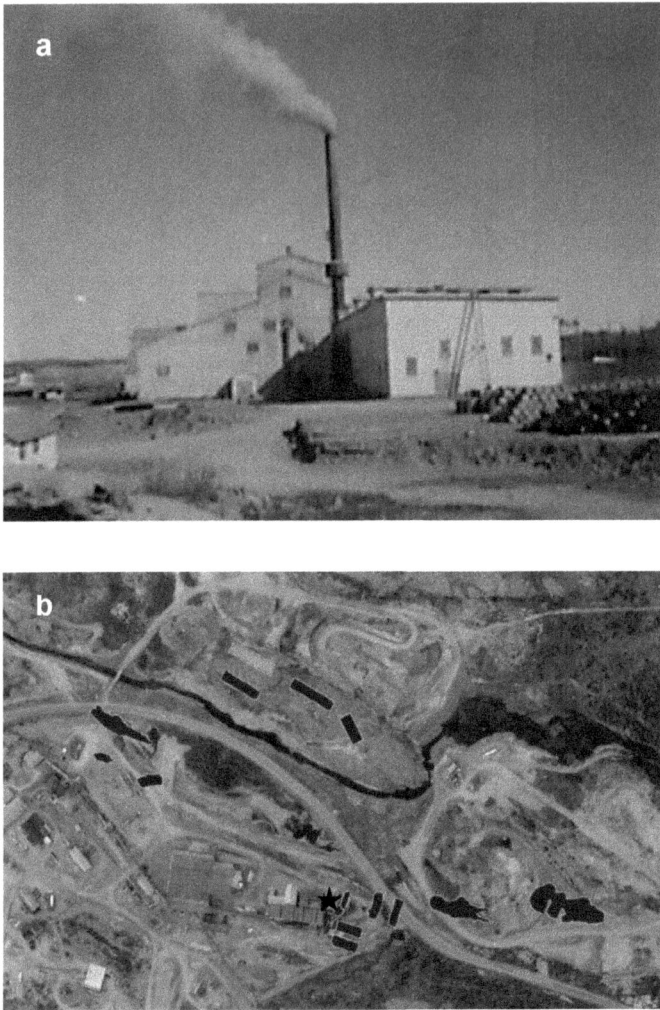

Figure 3. (a) Ore roaster at Giant mine operating in 1950's. Courtesy GNWT Archives (b) Solid black areas show location of underground storage chambers for As$_2$O$_3$ dust. Roaster location indicated by star. [Used with permission of the Giant Mine Remediation Team from MVRB (2013).]

center of Au mining. In the last fifteen years, the economy of the region has been fuelled by diamond mining, a sector operating with environmental controls and community participation that would have been unimaginable in the early days of the Giant mine.

The objective of this chapter is to report the lessons learned from research on the history and nature of As contamination at the Giant mine site. The focus is on the geochemical and mineralogical speciation of As in tailings, soils, sediments and pore water, and the long-term stability of these materials. Previous work has documented the concentration and distribution of As in these media (INAC 2007; Bromstad and Jamieson 2012). Factors that resulted in this widespread and complex contamination are discussed in the hope that modern mining will avoid such potentially damaging and costly problems and that risk assessment of other abandoned sites will benefit from this example.

FACTORS INFLUENCING THE LEGACY OF
ARSENIC CONTAMINATION AT GIANT MINE

Nature of mineralization

The predominance of As and Sb as elements of environmental concern, the absence of acidic drainage, and the requirement of oxidative pretreatment of the ore are all factors resulting from the nature of the Au ore at Giant, and thus may be applicable to other deposits of this type. Gold mineralization at Giant mine occurs within shear zones as disseminated sulfides in broad, silicified zones or quartz–carbonate veins bounded by muscovite or chlorite schist (Boyle 1961; Siddorn et al. 2006). The deposit lies within the Yellowknife Supergroup, part of the Slave Structural Province of the Canadian Shield. The Yellowknife Supergroup consists of Archaean metavolcanic and metasedimentary rocks intruded by younger granitoids. Several early Proterozoic gabbro and diabase dikes crosscut the area, and several faults divide the volcanic and granitoid rock units (Siddorn et al. 2006). The eastern half of the Giant Mine property lies principally over the Kam Group of the Yellowknife Bay Formation, which consists of variolitic pillowed and massive flows dominated by basalts and metamorphosed to greenschist facies. Ore shoots typically contain less than 10% sulfide and commonly less than 5% (Coleman 1957). The sulfide and sulfosalt mineralogy at the Giant mine includes (in decreasing order of abundance): pyrite (average As content 0.68 wt%), arsenopyrite, sphalerite, chalcopyrite, sulfosalts (jamesonite, berthierite, bournonite and tetrahedrite), pyrrhotite, and galena. However, the actual abundance is variable within individual ore shoots and different parts of the mine (Coleman 1957; Canam 2006; Hubbard et al. 2006; Walker et al. 2014). A study of flotation concentrate in 1990 (originating from stopes being mined at the time) indicated that pyrite and arsenopyrite combined account for 95% of the sulfides and that marcasite, chalcopyrite, sphalerite, acanthite, boulangerite, tetrahedrite, berthierite, gudmundite, stibnite were present in trace amounts (Walker et al. 2014).

Giant falls into the deposit type known as orogenic Au deposits, or lode Au deposits, which are typically associated with quartz-carbonate veins and have relatively low sulfide content (Groves et al. 1998; Seal and Hammarstrom 2003; Siddorn et al. 2006). Most of these ores have associated As and Sb in sufficiently high concentrations that these elements are enriched in tailings and waste rock and are potentially leached into groundwater and surface water. Other elements such as Cu, Pb, and Zn are present in flotation tailings at Giant, as would be predicted from the ore mineralogy described above. As shown in Figure 4, these elements have been remobilized into sediments (Fawcett 2009; Andrade et al. 2010). Although the concentrations of these other elements may, in some cases, exceed guidelines for sediment and soil quality, the focus of the risk assessment and remediation at Giant has been on As (MVRB 2013).

The concentration of As in tailings at Giant (1,000 to 5,000 mg kg^{-1}) is governed by the nature of the mineralization and modifications related to processing methods. Pyrite and arsenopyrite are the dominant sulfide minerals but overall sulfide concentrations in tailings are relatively low (total S typically <1%; INAC 2007). Furthermore, much of the As was diverted from the tailings by ore roasting, and is now present as As_2O_3 in underground storage chambers and as airfall deposits in soils and sediments in the area (Wrye 2008; Bromstad 2011). In comparison, historic tailings from orogenic Au deposits in Nova Scotia contain a wider range of total As concentrations (10 mg kg^{-1} to 31 wt%; median 2,550 mg kg^{-1}; Parsons et al. 2012), reflecting the dominance of arsenopyrite and the highly variable sulfide content in the ore veins (0.1 to 5%, locally 50 to 75%; Kontak and Jackson 1999). The extremely high As content of some of the Nova Scotia tailings is associated with the disposal of arsenopyrite-rich gravity concentrates (Walker et al. 2009). Overall, this type of Au deposit is generally low in total sulfide content with arsenopyrite and/or pyrite as the dominant sulfide minerals (Groves et al. 1998).

a

210Pb Dates
(yr)

Yellowknife Bay Aug 2003 (YKBS-03)

As, Pb, Sb, Zn, and Cu (mg/kg)

b

Yellowknife Bay Aug 03 (YKBS-03) and Apr 2004 (YKBW-04)

Dissolved As (µg/L)

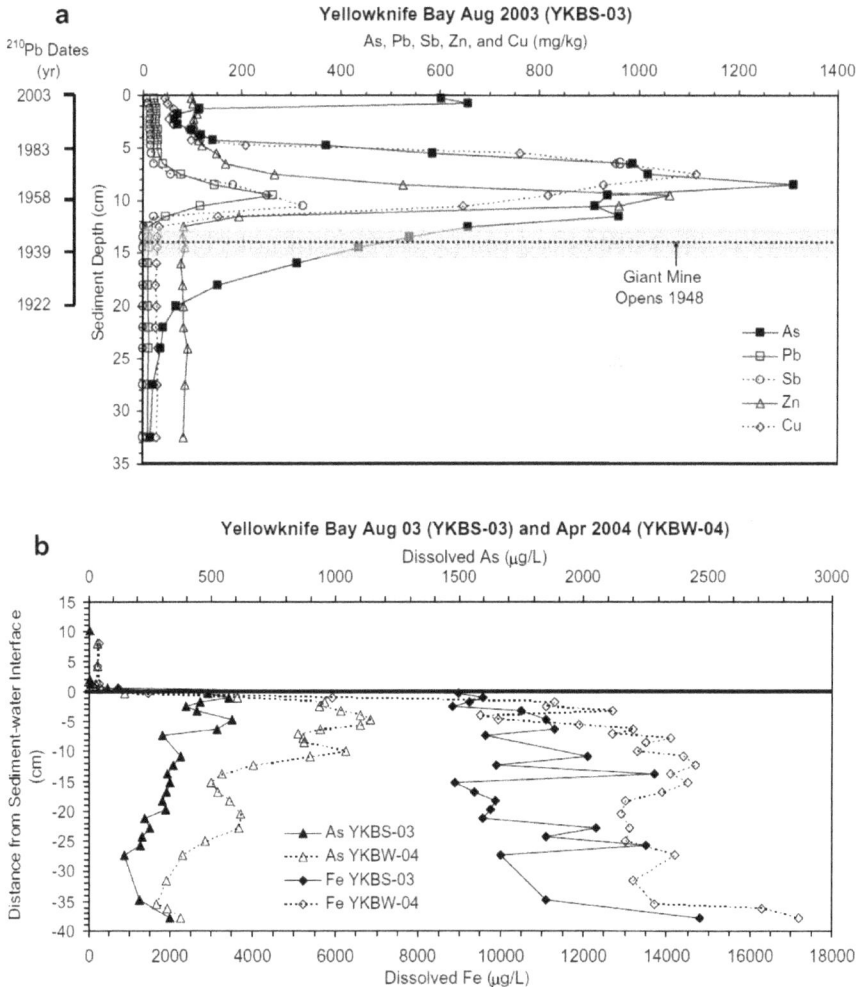

Figure 4. (a) Solid phase metal(loid) concentrations from sediment core with associated timeline. (b) Dissolved As and Fe in sediment pore waters. Location of the sample site in Yellowknife Bay is shown by solid square in Figure 1. Water depth at this location is 13 m. {Reprinted with permission of Elsevier from Andrade et al. (2010), *App Geochem*, Vol 25, Fig, 5 p. 204.]

Critical to the environmental legacy at any mine is the ratio of neutralizing carbonate minerals to acid-generating iron-sulfide minerals because this governs the tendency of tailings and waste rock to produce acid leachate (Bowell et al. 2000; Seal and Hammarstrom 2003). Groves et al. (1998) described orogenic Au deposits as typically having 3-5% sulfide minerals, mainly Fe-sulfides, and 5-15% carbonate minerals. Kontak and Jackson (1999) report generally subequal modal amounts of sulfide and carbonate in the auriferous veins from Nova Scotia. This suggests that acid rock drainage may or may not be associated with mine waste from these deposits. Measured pH values associated with streams in the Nova Scotia Au mining districts are typically greater than 4 and often near-neutral, suggesting effective carbonate buffering of pH (Parsons et al. 2012). However, acidic pore waters are found in the hardpans produced from weathering of arsenopyrite-rich gravity concentrates (DeSisto et al. 2011).

At Giant mine, a significant excess of carbonate, mostly ferroan dolomite, relative to Fe sulfide (pyrite and arsenopyrite) in the tailings precludes the formation of acidic drainage. Acid-base accounting tests demonstrated high neutralization potential (111 to 234 tonnes $CaCO_3$ equivalent per 1000 tonnes of tailings) and low acid generation potential (2.5 to 32.8 tonnes $CaCO_3$ equivalent per 1000 tonnes of tailings) (INAC 2007). Tailings pore waters and streams on the mine site have pH values consistently between 5.5 and 8.5 (INAC 2007; Walker 2006; Fawcett et al. 2014). If the balance between acid-generating and acid-neutralizing minerals in the tailings had been such that acid leachate was produced, it is likely that the drainage would have contained higher concentration of Cu, Zn, Pb, Ni and other metals present in sulfide and sulfosalt minerals vulnerable to acidic dissolution. Given that As and Sb are mobile as oxyanion species at near-neutral pH and tend to adsorb at low pH (Wilson et al. 2010), it is possible that lower pH leachate would have produced lower concentrations of these metalloids. Remnant pyrite and arsenopyrite grains in the Giant mine tailings and waste rock exhibit rims of As-bearing secondary Fe oxyhydroxides, suggesting that sulfide oxidation occurred during weathering. The net neutralizing nature of the mine waste is inherited from the original, carbonate-rich mineralogy and enhanced by sulfide removal during processing. The relatively low ore-roasting temperatures used at Giant (ca. 500 °C) destroyed much of the arsenopyrite but did not affect the ferroan dolomite or calcite (Walker et al. 2014).

The third influence of the nature of the mineralization on the environmental legacy at Giant is that it dictated the choice for ore processing. Some free-milling Au, which is usually defined as ore where >95% Au recovery can be achieved with simple cyanidation or gravity separation (Marsden and House 2006), was present in parts of the ore body exploited early in mining (Tait 1961), but it was soon realized that most of the Au was refractory, incorporated submicroscopically within arsenopyrite, and thus required pre-treatment (Halverson 1990; Stefanski and Halverson 1992; Walker et al. 2014). The presence of Sb in the ore complicated the roasting process because Sb concentrations exceeding 0.75% may cause clinkering (formation of glassy phases not amenable to cyanidation) of the roaster bed (Marsden and House 2006; Fawcett and Jamieson 2011). The remoteness of the site precluded shipping a concentrate to another facility. Ore roasting was likely the obvious choice for Au recovery from the Giant mine ore in 1949. Coupled with the lack of emission controls and regulations, roasting led to the dispersion of large amounts of As_2O_3 through stack emissions and the accumulation of even larger amounts of As_2O_3 in underground storage chambers.

Ore roasting

The influence of ore roasting on the environmental legacy at Giant is a direct consequence of the nature of the As-hosting solid species in the tailings, stored roaster waste, soils, and sediments in the area. Processing converted most of the original arsenopyrite to As-bearing roaster products: Fe oxides (maghemite and hematite) and As_2O_3 (arsenolite). These phases are potentially more soluble and more bioaccessible than the original arsenopyrite, and are vulnerable to reductive dissolution.

The As gases produced during roasting of arsenopyrite condensed to As_2O_3 with a crystallographic structure similar to arsenolite (Wrye 2008; Bromstad 2011; Walker et al. 2014). Because the As_2O_3 at Giant is an industrial product, in contrast to that produced by natural weathering of As-rich sulfides (Nordstrom and Archer 2003; Drahota and Filippi 2009), it is referred to as As_2O_3. The solubility of arsenic trioxide-bearing dust from the Giant mine site has been reported as 11,000 to 15,000 mg L^{-1}, based on laboratory measurements of samples from Giant (Riveros et al. 2000); the As solubility may be limited by Sb content, but this is not well documented. Seepage from underground drillholes and fractures near the storage chambers contain up to 4,000 mg As L^{-1} in pH-neutral water (INAC 2007; Jamieson et al. 2013). Approximately 70% of the As is As(III). Near these seepage points, and on stope walls where low-flow drips from old drill holes and fractures produce As(III)-rich water, biofilms develop (Fig. 5). These contain psychrophilic As(III)-oxidizing bacteria (Osborn et al. 2010). As discussed by Bowell and Craw (2014, this volume), As(III) oxidation to As(V) is an acid-generating reaction. In this case, carbonate in the wallrock and concrete in the bulkheads of the storage chambers neutralizes the water, and yukonite precipitates as a product of the reaction (Fig. 5).

Figure 5. (*for color see Plate 11*) Biofilm developed on underground stope wall where As-rich water drips from an drillhole. Mineral precipitate is yukonite $Ca_7Fe_{12}(AsO_4)_{10}(OH)_{20} \cdot 15H_2O$.

Walker et al. (2005, 2014) have described the As-hosting roaster-generated Fe oxides in detail. They are nanocrystalline composite grains, predominantly maghemite (γ-Fe_2O_3) with a smaller amount of hematite (Fig. 6). Arsenic is incorporated during roasting in a mixed oxidation state with the ratio of As(III)/As(V) determined by the partial pressure of O_2 at the reacting surface during combustion of the sulfide (i.e. lower P_{O2}, higher As(III)). Gas phase As chemisorbs to the maghemite surface. The As concentration of maghemite grains in the tailings range from <0.5 to 7 wt% As. Where further transformation of maghemite to hematite (α-Fe_2O_3) occurs, the transformation is accompanied by rapid crystallite growth, a decrease in surface area, and concurrent decrease in the As content (<2 wt%).

Two types of roasted material were collected and treated with cyanide for Au recovery, electrostatic precipitator (ESP) dust, and calcine (roaster bed product). Fawcett and Jamieson (2011) have shown that in the ESP waste stream, As(III) and Sb(III) are dominant in the bulk sample and in the roaster-generated iron oxides that host these elements, but As(V) and Sb (V) are dominant in the bulk calcine sample and in the roaster Fe-oxide grains therein. As explained in the next section, both ESP and calcine waste streams were combined with flotation tails, resulting in a mixture of oxidation states and mineralogical hosts of As in tailings deposits.

The most important consequence of ore roasting at Giant, in terms of the environmental legacy, is the production of approximately 300,000 tonnes of As_2O_3 as waste during a time of few emission controls, few regulations, and limited options regarding re-use of the As_2O_3. In many cases, mining operations eliminated or reduced the amount of As_2O_3 they need to store

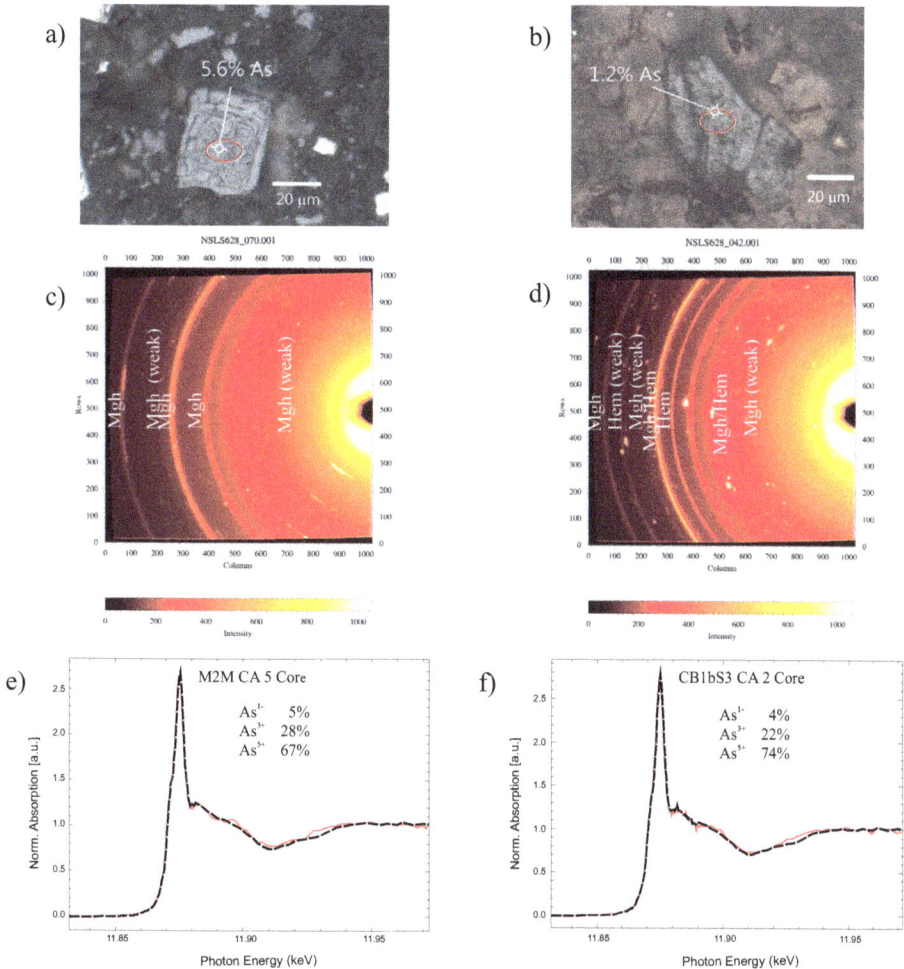

Figure 6. (*for color see Plate 13*) Selected analyses of two grains of roaster iron oxides. a) Reflected-light photomicrograph of square concentric roaster iron oxide from calcine sample (M2M). Total As established by EPMA (designated in white). b) Transmitted- and reflected light photomicrograph of target grain from shore-line tailings sample (CB1bS3). Total As by EPMA as indicated. c) micro-XRD image of target indicated by ellipse in (a). Pattern corresponds to maghemite. Three arcs in lower right-hand corner are chlorite reflections. d) Micro-XRD image of target indicated by ellipse in (b). Pattern is a mixture of maghemite and hematite. e) Micro-XANES analysis of target in (a). Sample spectrum is lighter undashed line, dashed line is best-fit linear combination for result shown. f) Micro-XANES analysis of target in (b). [Reprinted with permission of the Mineralogical Association of Canada, from Walker et al. (2005) *Can Mineral*, Vol. 43, Fig. 8. p. 1218.]

by selling a high-purity version for use in herbicides, insecticides, or the arsenic acid used in the formulation of chromated copper arsenate (CCA) preservatives for the pressure treating of lumber, all products used during most of the years that the Giant mine operated. Other operations, however, also stored arsenic-rich roaster waste. Across town, the Con mine stored their As_2O_3, considerably less in volume, on surface. The Campbell mine in NW Ontario, which, like Giant, has operated continuously since 1949, has 20,000 tonnes of As_2O_3 stored underground.

The remote location of Giant mine precluded effective marketing and sale of As_2O_3. Attempts were made to develop markets for As_2O_3 in the USA and South America although re-mining the storage chambers would have been challenging. As late as 2002, one of the alternative solutions for the As_2O_3 stored underground at Giant was extraction, treatment, and sale of high-purity As_2O_3 to American wood preservative manufacturers. This alternative was subsequently dropped from consideration because changes to U.S. regulations severely restricted the use of arsenic as a wood preservative, and it was considered unlikely that a future market would exist for the Giant mine As_2O_3 (INAC 2002). Thus, a potential resource became an environmental liability.

Modern alternatives for oxidative pre-treatment include hydrometallurgical processes such as low- or high-pressure oxidation, nitric acid treatment, chlorination, and bio-oxidation (Marsden and House 2006). Roasters are still effective in terms of recovery and cost in many cases, and are still used. However appropriate emission controls limit dispersion of As_2O_3 and regulations require disposal in an environmentally acceptable manner.

Waste disposal practices

The distribution of As-bearing waste material is an important factor regarding the environmental impact of Giant mine operations. In the first two years of operation, mill tailings were discharged directly onto the shore of Yellowknife Bay (Fig. 2). These shoreline tailings were stabilized with rip-rap in 2002, but much of the material had already been eroded and dispersed into the bay (Andrade et al. 2010).

Beginning in 1951, tailings were deposited into several former lakes on mine property. Later, tailings dams and several impoundments were constructed (Fig. 2). The processing produced three solid waste streams including flotation tails, cyanided calcine residue, and cyanided ESP dust. Table 1 shows that the lower tonnage streams (calcine and ESP) carry high concentrations of As as fine-grained particles, mostly roaster-generated maghemite and hematite, and possibly some As_2O_3 (Walker et al. 2005, 2014). The total As concentration in tailings ranges from 1,100 to 5,000 mg kg^{-1}, with an average approximately 2,700 mg kg^{-1}. Petrographic examination and electron microprobe examination of the tailings indicate the presence of unroasted arsenopyrite in addition to roaster-generated iron oxide (Walker 2006). Pyrite exhibits oxidation rims of Ca-Fe arsenate (yukonite). The combination of sulfide-hosted and oxide-hosted arsenic requires a carefully-designed cover that will maintain the stability of both. If a thick cover containing organic material is used, there is a risk that oxide-hosted As,

Table 1. Arsenic contributions from the main solid effluent streams at Giant.

Tailings Stream	Year†	Approximate Discharge Rate (tpd)	Arsenic Concentration (wt%)	Arsenic Loading (tpd)
Flotation (70-80% <0.075 mm)	1999	1000	0.09	0.9
	1963	794	0.28	2.2*
Calcine Residue (90% <0.045 mm)	1999	170	2.5	4.2
	1963	122	1.2	1.5*
ESP Residue (90% <0.014 mm)	1999	9	4.4	0.4
	1963	9	6.2	0.6

tpd = dry tonnes per day.

† For 1999, discharge rate extrapolated from Royal Oak Mines (1992) and As concentration from Walker et al. (2014). For 1963, data from Grainge (1963).

* Estimated loading. Actual loading to tailings is unknown, since some tailings were cycloned for mine-backfill from 1957 to 1976 removing approximately 50% of the coarsest material and an unknown proportion of As.

particularly roaster-generated Fe oxides, could destabilize and release As, as described from other Au-mining areas (e.g., Martin and Pederson 2002).

Prior to sedimentation control, a significant amount of fine tailings, including ESP dust and calcine, flowed over the ice-covered tailings dams each spring and was deposited at the upstream end of Baker Pond, a natural water body located along Baker Creek (Bérubé et al. 1974). Some fine-grained material travelled down Baker Creek into Yellowknife Bay (Mudroch et al. 1989; Fawcett and Jamieson 2011). A breakwater that extends across most of the mouth of Baker Creek was built in 1964 (Fig. 2). Air photos show that this resulted in the capture of fine sediment behind the breakwater, which is covered by dense aquatic vegetation (*Equisetum fluviatile*), and the establishment of a narrow creek channel (Fawcett and Jamieson 2011). Baker Creek was also impacted by numerous tailings spills originating from faulty tailings pipelines (Fawcett 2009).

Giant mine tailings and fine-grained tailings fractions are now present in both subaerial and subaqueous settings, as both near-surface and buried sediment. Arsenic is mostly in the form of As(III) and As(V) associated with maghemite and lesser amounts of hematite. In some cases, the As appears to be hosted in stable form, but in other cases the As-hosting phases are unstable or soluble, and are likely to release As to groundwater and surface water. Walker et al. (2005) used microXANES to show that the ratio of As(III)/As(V) is similar in maghemite grains collected directly from the roaster in 1999 and grains exposed subaerially in shoreline tailings for 60 years. The lack of oxidation of the As(III) over time in these phases requires an explanation, because surface-sorbed $(As^{III}O_3)^{-3}$ might be expected to oxidize to $(As^{V}O_4)^{-3}$. More work is needed to understand the kinetics of As oxidation in this environment. Walker et al. (2014) suggested that the most plausible reason is that some of the As(III) is incorporated within the maghemite framework associated with structural vacancies. Andrade (2006) sampled shoreline tailings that had been eroded and redeposited at 1 m depth in Yellowknife Bay. The mixed oxidation state of As in the roaster-generated iron oxides was similar to that on the onshore tailings, indicating that these materials are stable in the oxygenated, high-energy, organic-poor, near-shore environment (Andrade 2006).

There is clear evidence of the release of As from buried tailings in sediment in Baker Pond, in the vegetated area behind the Baker Creek breakwater, and at depth in Yellowknife Bay (Andrade 2006; Fawcett et al. 2014). Figure 4 shows post-depositional mobility and upward migration of pore-water As from the mid-core enrichment zone from Yellowknife Bay sediment, which is related to high-As releases in the early years of operation, likely due to spills, tailings pond decanting, and possibly elevated stack emissions. These results are from the deep-water (13 m) site, approximately 1 km from the mouth of Baker Creek (Fig. 2). Andrade et al. (2010) described this release of As from contaminated sediments, which are suboxic, relatively rich in organic material, and contain arsenate- and sulfate-reducing microbes. Some of the upwardly-migrating pore-water As is recaptured at the sediment-water interface by a thin layer of Fe-Mn oxyhydroxides. Post-depositional mobilization of As is also apparent in sediments from Baker Pond and Baker Creek channel. Roaster-generated maghemite exhibits lower ratios of As(III)/As(V) (and Sb(III)/Sb(V)) in sediments at 10 to 20 cm depth compared to near-surface material, suggesting either *in situ* reduction of the metalloid or preferential release of As(V). At depth, approximately 10 to 20% of the arsenic released to pore water in these locations is attenuated via precipitation as sulfide (realgar) or sorption to sulfide surfaces (Fawcett and Jamieson 2011).

Stack emissions of As_2O_3 have left an important environmental legacy at Giant. Table 2 shows the daily emissions rate, which decreased from more than 7 to 0.1 tonnes per day from 1949 to 1990, as control measures such as ESPs were installed. By the time the emission-control process was completely refined in 1963, 86% of the total airborne As emissions at Giant had been released into the surrounding area. Stack emissions consisted of SO_2 and

As-bearing gases, both produced by roasting of arsenopyrite. The As gases condensed into a dust with As-hosting phases consisting mostly of As_2O_3 in the mineral structure of arsenolite, with some As-bearing, roaster-generated Fe oxides in the form of maghemite (Bromstad and Jamieson 2012). Wrye (2008) and Bromstad (2011) have shown that particles of aerially distributed As_2O_3 persist in the near-surface soils over much of the Giant mine property. Figure 7 shows examples of As_2O_3 and the rarer roaster-generated Fe oxide particles, which are clearly visible in samples examined by scanning electron microscopy and identified by synchrotron-based microXRD.

There is little visual evidence of dissolution or instability of the As_2O_3 particles in the soils near Giant mine. Arsenic from stack emissions appears to have concentrated in soils that are found in the small pockets of soil downwind of the roaster on the large outcrops that cover approximately 30% of the mine property (Fig. 2). Outcrop depressions are known sinks for runoff after rainfall and freshet at Giant mine, trapping water and promoting evaporation (Spence and Woo 2002), a feature that may be important at other smelter and roaster sites. Most of the outcrop soil pockets were not included in the delineation of soils destined for removal in the remediation plan, although some of these areas may become publicly accessible. Total As concentrations in soils sampled by Wrye (2008) and Bromstad (2011) range up to 5,760 mg kg^{-1}. The highest values are similar to concentrations in the tailings, demonstrating that the ore roaster effectively diverted As from the tailings to the soil and water in the surrounding environment, particularly in the early days of operation. Based on As-Sb-Au soil concentrations and documented improvements in recovering Au and Sb from roaster emissions,

Table 2. Estimates of Aerial Emissions of As_2O_3 from the Giant Mine Roaster between 1949 and 1999. (tpd = tonnes per day).

Year	As_2O_3 Emissions † (tpd)	As_2O_3 Emissions‡ (tonnes)
1949-1951*	7.3	7963.6
1952*	5.5	1989.3
1953*	5.5	1989.3
1954	5.5	1990.9
1955	2.9	1061.8
1956	2.7	995.5
1957	3.0	1078.4
1958*	1.5	547.5
1959	0.1	19.1
1960	0.1	27.4
1961-1963*	0.2	163.3
1964	0.3	114.5
1965	0.0	0.0
1966	0.2	88.8
1967	0.1	47.3
1968	0.2	83.0
1969	0.3	109.5
1970	0.2	80.5
1971	0.9	320.2
1972	0.4	145.2
1973	0.4	147.7
1974	0.2	80.5
1975-1992*	0.3	1698.9
1991-1993*	0.1	54.8
1994	0.01	3.6
1995	0.01	3.5
1996	0.01	3.5
1997	0.02	7.6
1998	0.015	5.3
1999	0.01	3.7
Total As_2O_3 Emissions (tonnes)		**20,824**

† As_2O_3 emissions estimates were measured from the top of the roasting stack and were reported by following sources; CPHA 1977, EnviroCan 2007, GNWT 1993, INAC 2007, Tait 1961.

* Emissions for this year or period of years have been estimated based on information from the above sources and known changes to roaster operations.

‡ As_2O_3 emissions for the year or period of years assuming roasting occurred for 365 days. This is an overestimate of emissions, however the dates when the roaster was not in operation are unknown and the majority of As_2O_3 generated was during the first 10 years of roasting.

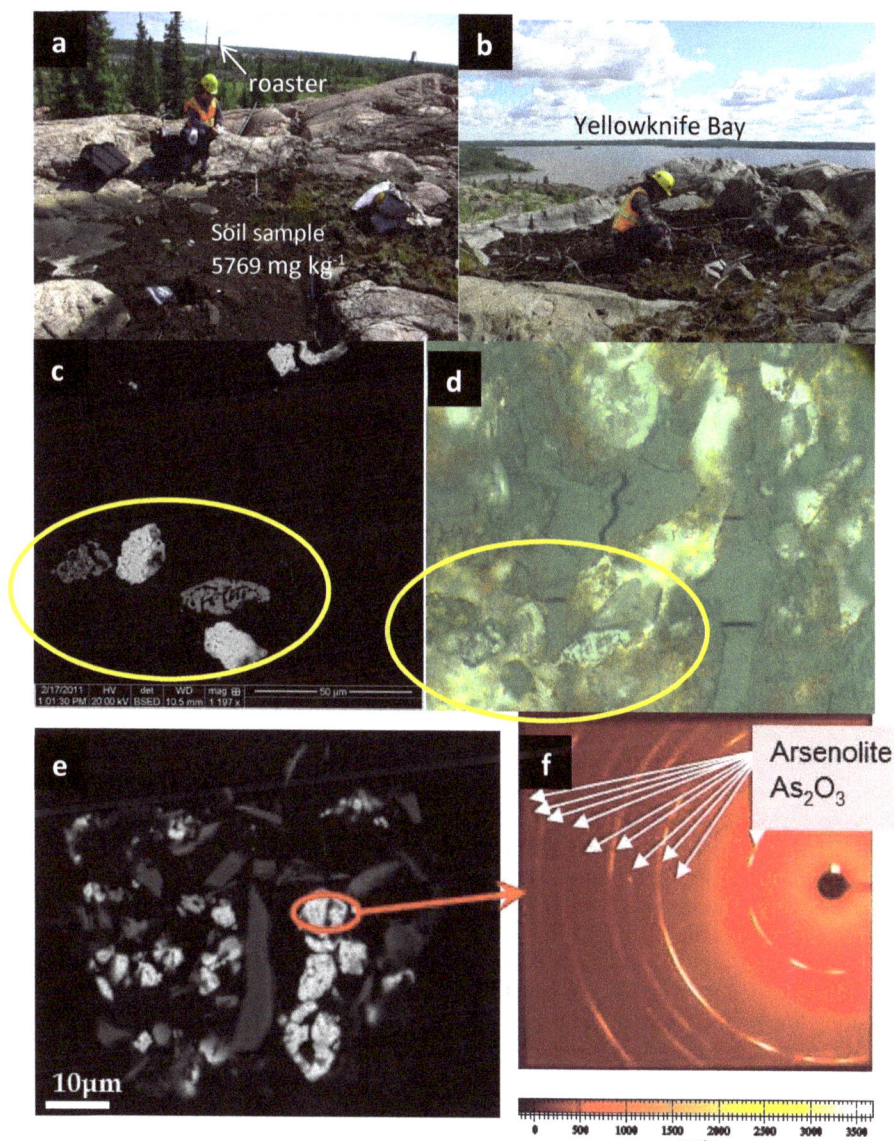

Figure 7. (*for color see Plate 14*) Roaster-derived As-bearing particles in soils near Giant mine. (a) and (b) show typical soil pockets on large outcrops, the sites that contained the highest concentration of roaster-derived soil. (c) back-scattered electron image of a thin section of soil showing light (As_2O_3) and dark (roaster-generated Fe oxide) particles. (d) petrographic image of the same area as (c), combined transmitted and reflected light. (e) As_2O_3 particles in soil (f) microXRD of one of these particles indicating the pattern corresponding to arsenolite. From Wrye (2008) and Bromstad (2011).

Bromstad (2011) concluded that most of the arsenic trioxide in the soils is more than 50 years old, persisting from early stack emissions despite its high solubility and the expectation that arsenic trioxide would have dissolved after years of soil exposure. Haffert and Craw (2008) document reaction rims of scorodite on particles of roaster-generated As_2O_3 probably caused by exposure to acidic, Fe-bearing fluids. In contrast, the As_2O_3 particles in soil near Giant do not have reactions rims, likely because they were never exposed to acid mine drainage, and exhibit only subtle textural evidence of dissolution (Bromstad 2011).

Suction lysimeters were used to sample pore water in soil pockets immediately following summer rainfalls. Soil pore-water concentrations are as high as 2 mg L^{-1} of As, considerably less than the underground seepage waters, but still more than two orders of magnitude higher than the World Health Organization drinking water guideline of 0.01 mg L^{-1}. The pore waters are pH-neutral and the major ions are Ca^{2+}, Na^+, and SO_4^{2-} (Bromstad 2011; Jamieson et al. 2013).

Giant mine has been a managed site during and since mine closure with limited public access except for the area around the outlet of Baker Creek. However, post-remediation plans call for some areas to be unfenced, and human exposure to As_2O_3-contaminated soil may be increased. The risk associated with possible ingestion of soil can be evaluated using bioaccessibility testing designed to evaluate the solubility of As in body fluids (see Basta and Jurasz 2014, this volume). Plumlee and Morman (2011) have ranked gastric-intestinal bioaccessibility of various As compounds, and include arsenic trioxide as one of the most bioaccessible forms and arsenopyrite as one of the least.

Bioaccessible As in three soil samples from a single soil pocket was measured at the US Geological Survey laboratories in Denver using *in vitro* techniques (Bromstad 2011). A simulated gastric fluid was used for the size fraction less than 250 μm and indicated that 29 to 42% of the total As (238 to 4,760 mg kg^{-1}) would be bioaccessible if ingested. A second series of tests using a simulated lung fluid to leach the size fractions less than 20 μm indicated that 15 to 20% would be bioaccessible if inhaled. In both cases, there is no correlation between the percentage of bioaccessible As and the total concentration of As, suggesting homogeneity in As host properties (Bromstad 2011).

In summary, three key factors—the nature of the mineralization, the ore-roasting procedures, and the waste-disposal practices—were influential in creating the complicated environmental problems near the Giant mine. The nature of the mineralization resulted in high concentrations of As in various wastes, abundant neutralizing capacity limiting acid drainage and most importantly, arsenopyrite-hosted refractory Au, which required oxidative pre-treatment for Au recovery. The choice of ore roasting created oxide-hosted arsenic (As_2O_3 and roaster-generated maghemite and hematite) in mill tailings and other waste emissions, materials which are more soluble, more bioaccessible, and vulnerable to reductive dissolution compared with sulfide-hosted As which is present at other mine-waste sites. Finally, the practice of co-depositing fine-grained roaster waste with flotation tailings, and the poorly-controlled release of As_2O_3 from stack emissions resulted in widespread distribution of As in soils, sediment, and water. Much of the roaster-generated As(III) has persisted under oxidizing conditions but shows evidence of post-depositional transformation under reducing conditions.

Modern mining and processing practices can limit As contamination by understanding that this type of Au deposit is typically accompanied by metalloids (As and Sb) that are mobile in neutral-pH water, employing alternative methods of oxidative pre-treatment of refractory ore that do not produce As_2O_3, or using modern ore roasting with efficient emission controls, and employing appropriate tailings and processing waste methods and monitoring. The story of Giant mine also provides some lessons for remediation design of older and abandoned As-contaminated mine sites, namely the importance of understanding processing and disposal history, distinguishing between natural and anthropogenic As, and recognition of post-depositional As mobility.

REMEDIATION PLAN

A comprehensive remediation plan, based on approximately five years of work, was filed for environmental assessment in 2007 (INAC 2007). The "developer" of the site is identified as the Giant Mine Remediation Directorate, which involves both the federal and territorial governments. As a result, the entire submission, including approximately 50 consultants' reports, is available to researchers. The Mackenzie Valley Environmental Impact Review Board, which is the regulatory body in the Northwest Territories, responded in June 2013 to the proposed remediation plan (MVRB 2013). The complexity of the site, the amount of technical information gathered, the requirement to involve other parties, notably the local First Nations communities and NGOs (Non-Governmental Organizations), and the nature of the environmental assessment process have all contributed to the long timeline and large volume of documentation. At the time of writing of this chapter, the Minister of Aboriginal Affairs and Northern Development Canada has provided preliminary response to the regulator's suggested modifications to the remediation plan. Very limited remediation has taken place to date, although some buildings, including the roaster, are being dismantled, and a highway bypass designed to avoid the area overlying the arsenic trioxide storage chambers has been constructed.

Relevant to the consideration of other sites where roaster waste has accumulated are the alternatives considered for the 237,000 tonnes of As_2O_3 stored in underground chambers. More than 50 possible techniques for reducing the risk associated with these chambers were considered, with the top two being *in situ* freezing or removal and encapsulation in cement or bitumen. The selected alternative is the frozen-block method which involves pumping coolant underground and then maintaining frozen ground using thermosyphons, filled with compressed carbon dioxide, as passive heat pumps (INAC 2007). This technology is used commonly in the Canadian north for ice core dams and other applications. Yellowknife is located in an area of discontinuous permafrost, but the large underground opening and subsequent partial flooding of the underground workings has all but eliminated permafrost in the mine area. The remediation plan considered the effects of global warming and concluded that, even after 100 years of sustained temperature increases based on estimates from the ICCP (International Climate Change Partnership), the thermosyphon installation should be adequate. There has been considerable discussion about other aspects of climate change, including the risk of flooding and potential effects on both the frozen chambers and other features on site.

The remediation plan calls for removal or covering of approximately 960,000 m^3 of contaminated soils to industrial site-specific standard of 340 mg kg^{-1} total As which was established by the City of Yellowknife in 2003 (GNWT 2003). A residential guideline of 160 mg kg^{-1} was established at the same time. These criteria are much higher than the Canadian Soil Quality Guideline of 12 mg kg^{-1} (CCME 1997), and were based on the expectation that exposure would be limited by the cold climate and that the "average natural background concentration of arsenic in and around Yellowknife was determined to be 150 ppm." The industrial guideline was also based on the expectation of "little or no public access." Given more recent evidence that the roaster stack emissions have introduced widespread contamination by As_2O_3, and the construction of a highway bypass adjacent to some of the highly contaminated outcrop soil pockets, the evaluation of natural background and future public access may need to be re-examined.

The remediation plan indicates that monitoring, maintenance, and management is designed for perpetuity. This perpetual aspect has significance for funding, the emergence of new technology that might be better than the frozen-block method, climate change (including global warming or glaciation), possible seismic activity, and community anxiety. Given that the funding source for perpetual care would be the Canadian federal government, the source of funding for perpetuity is likely more reliable than at privately-owned sites. Kempton et al. (2010) have described perpetual environmental management at hardrock mines as a global dilemma and noted that, while perpetual treatment is usually less expensive than permanent stabilization, it results in increased social and financial complexity. An analysis of perpetual care at contaminated sites by Kuyek (2011) highlighted the need for "public engagement and activism in order to create the political will for the enormous funding that is required" to maintain funding.

The remediation plan proposes collection and removal of As from surface and groundwater that would reduce the loading of As to Yellowknife Bay, part of Great Slave Lake. The contaminated water would be treated on site and released to Yellowknife Bay via a diffuser system. This would allow year-around release of treated water instead of the current situation which requires winter storage of water in tailings ponds, increasing the possibility of flooding and uncontrolled release of untreated water.

One of the important lessons learned from Giant mine concerns the limitations of remediation. The As released from the site will decrease from the current level of approximately 900 kilograms per year to less than 610 kg per year (MVRB 2013). This may be disappointing to community members in Yellowknife, and in other areas impacted by historic mining activities, who expect that remediation means a complete "clean-up" of contamination. The Giant plan is essentially a control measure because without any remediation, various failures of infrastructure and natural events could results in releases in thousands of kg of As per year.

Although the estimated cost of the proposed remediation is very high ($CA 903,535,080, as reported by AANDC 2012), it is exceeded by the monetary value of the Au extracted over the lifetime of Giant Mine. Bullen and Robb (2006) have determined that the cumulative revenue for Giant and two other nearby mines exceeded $CA 5.5 billion, with the contribution from Giant estimated at $CA 2.75 billion, recalculated in terms of 2006 dollars. In their analysis, Bullen and Robb (2006) described the impact of the revenues on the Yellowknife economy as money flowing to employees and local businesses. Unlike many mining districts, the pattern of revenue flow to the NWT gross domestic product did not follow a boom-and-bust cycle because of the staggered timing of the two large mines (Con and Giant) and the historical change in Au price. Thus the Au mines provided a sustained economic benefit that launched the city of Yellowknife, led to its establishment as the capital of the NWT in 1967, and lasted until the diamond mines provided a new source of mineral wealth.

Local First Nations were largely excluded from these direct economic benefits during the operating years of the Au mines, although a few individuals found employment in the mining industry. Indirect benefits to First Nations included improved access to health care and transportation infrastructure (Bullen and Robb 2006; Sandlos and Keeling 2012). In general, First Nations have opposed Au mining in the Yellowknife area (Bullen and Robb 2006) and have been impacted historically by the environmental degradation of the region and direct exposure to arsenic contamination, partially because of the proximity of their community and the lack of resources to defend their interests in the past (MVRB 2013).

The real cost of Giant mine to the Yellowknife community, in terms of health impacts and environmental damage, cannot be easily evaluated, nor can the ongoing value of the site as a source of future business opportunities and employment. For better or for worse, a significant portion of the economic history and future of Yellowknife is associated with Giant mine.

CONCLUSIONS

Ore and host-rock mineralogy and geochemistry have a direct influence on the environmental legacy at any mine. In the case of Au deposits, where significant processing occurs on the mine site (such as roasting, pressure oxidation, heap leach, mercury amalgamation, and/or cyanidation), solid waste to be managed includes processing waste in addition to waste rock and mill tailings. Orogenic Au deposits such as Giant have refractory ore that requires destruction of the host arsenopyrite to enable cyanidation to recover Au. Ore roasting is an effective pre-treatment for refractory Au but has the unfortunate consequence of producing significant quantities of As_2O_3. In the case of Giant mine, much of the mining and processing took place before containment technology for this hazardous waste was developed, and the remoteness of the location meant that the As_2O_3 could not be sold, as was the case for other roaster operations of the same vintage. The environmental legacy involves both surface and subsurface As-rich material present as sulfide minerals (primary arsenopyrite and secondary realgar) and oxide minerals (roaster-generated Fe oxides and As_2O_3, weathering products of arsenopyrite such as yukonite and As-bearing goethite). Although no acidic drainage is present at Giant mine due to a generous supply of carbonate minerals, the sensitivity of As phases to changes in redox results in post-depositional mobility.

Conversion of arsenopyrite to As_2O_3 has significantly increased As bioaccessibility in potentially ingested dusts and soils at Giant mine. More importantly, this conversion also increased the concentration of dissolved As in groundwater and surface waters, and thus, the general mobility and bioavailability of As. The proposed remediation plan involves a series of measures that would reduce but not eliminate the release of this dissolved As to Great Slave Lake, one of the largest lakes in Canada.

For the current residents of Yellowknife, the mine that helped to launch the city during the gold rush of the 1930's and 1940's is now, for many, an ongoing source of concern as well as an alarming expense, albeit one shared with other Canadian taxpayers. The voice of the original First Nations residents of the area, barely heard in the early years of mine development, is now a critical part in site management decisions. Remediation has replaced mining as the focus of employment associated with Giant mine. Diamond mines and other potential mineral developments in the Canadian North routinely incorporate community consultation.

For researchers outside Yellowknife, the case of Giant mine provides lessons regarding the particular complexities that mining and processing As-bearing Au ores involve. The remoteness of the location and the severity of the climate have influenced the environmental legacy, but more important factors are the nature of ore and host-rock mineralization, and the history of processing and waste management, factors that influence mine waste anywhere on the globe.

REFERENCES

AANDC (Aboriginal Affairs and Northern Development Canada) (2012) Internal Audit Report: Value for Money Audit of the Giant Mine Remediation Project. Audit and Assurances Survey Branch Project 12-32, *http://www.aadnc-aandc.gc.ca/eng/1366814305245/1366814424097* (accessed Apr 2014)

Andrade CF (2006) Arsenic cycling and speciation in mining-impacted sediments and pore-waters from Yellowknife Bay, Great Slave Lake, NWT. MSc Thesis, Queen's University, Kingston, Ontario, Canada

Andrade CF, Jamieson HE, Praharaj T, Fortin D, Kyser TK (2010) Biogeochemical cycling of arsenic in mine-impacted sediments and co-existing pore waters. Appl Geochem 25:199-211

Basta NT, Juhasz A (2014) Using in vivo bioavailability and/or in vitro gastrointestinal bioaccessibility testing to adjust human exposure to arsenic from soil ingestion. Rev Mineral Geochem 79:451-472

Bérubé Y, Frenette M, Gilbert R, Anctil C (1974) Studies of mine waste containment at two mines near Yellowknife, N.W.T. Indian Affairs and Northern Development Canada, QS–3038–000–EE–A1, ALUR 72–73–32

Bowell RJ, Craw D (2014) The management of arsenic in the mining industry. Rev Mineral Geochem 79:507-532

Bowell RJ, Rees SB, Parshley JV (2000) Geochemical prediction of metal leaching and acid generation: geologic controls and baseline assessment. *In:* Geology and Ore Deposits 2000: The Great Basin and Beyond, Geol Soc Nevada Symp Proc. Cluer JK, Price JG, Struhsacker EM, Hardyman RF, Morris CL (eds) p 799-823

Boyle RW (1961) The geology, geochemistry, and origin of the gold deposits of the Yellowknife district. Geol Surv Can Mem 310

Bromstad MJ (2011) The characterization, persistence, and bioaccessibility of roaster-derived arsenic in surface soils at giant mine, Yellowknife, NWT. MSc thesis, Queen's Univ. Kingston, Ontario, Canada

Bromstad MJ, Jamieson HE (2012) Giant Mine, Yellowknife, Canada: Arsenite waste as the legacy of gold mining and processing. *In:* The Metabolism of Arsenite. Arsenic in the Environment 5. Santini JA, Ward SA (eds) CRC Press, p 25-42

Bullen W, Robb M (2006) Economic Contribution of Gold Mining in the Yellowknife Mining District. *In:* Gold in the Yellowknife Greenstone Belt, Northwest Territories: Results of the EXTECH III Multidisciplinary Research Project. Anglin CD, Falck H, Wright DF, Ambrose EJ (eds) Geol Assoc Can, Min Dep Div Spec Pub 3:38-49

Canam TW (2006) Discover, mine production, and geology of the Giant Mine. *In:* Gold in the Yellowknife Greenstone Belt, Northwest Territories: Results of the EXTECH III Multidisciplinary Research Project. Anglin CD, Falck H, Wright DF, Ambrose EJ (eds) Geol Assoc Can, Min Dep Div Spec Pub 3:188-196

CCME (Canadian Council of Ministers of the Environment) (1997) Canadian Soil Quality Guideline for the Protection of Environmental and Human Health. Arsenic (Inorganic). *http://ceqg-rcqe.ccme.ca/download/en/257/* (accessed April 2014)

Coleman LC (1957) Mineralogy of the Giant Yellowknife Gold Mine, Yellowknife, NWT. Econ Geol 52:400-425

CPHA (Canadian Public Health Association) (1977) Final Report- Canadian Public Health Association Task Force of Arsenic, Yellowknife, Northwest Territories. Canadian Public Health Association, Ottawa, December 1977

DeSisto SD, Jamieson HE, Parsons MB (2011) Influence of hardpan layers on arsenic mobility in historical gold mine tailings. Appl Geochem 26:2004-2018

Drahota P, Filippi M (2009) Secondary arsenic minerals in the environment: A review. Environ Int 35:1243-1255

EnviroCan (2007) National Pollutant Release Inventory (NPRI). Environment Canada. Online data search 1994-1999 Facility on-site releases, Royal Oak Mines Inc., Giant Mine, *http://www.ec.gc.ca/pdb/querysite/query_e.cfm* (accessed July 2014)

Fawcett SE (2009) Speciation and mobility of antimony and arsenic in mine waste and the aqueous environment in the region of the Giant mine, Yellowknife, Canada. PhD thesis, Queen's University, Kingston, Ontario, Canada

Fawcett SE, Jamieson HE (2011) The distinction between ore processing and post-depositional transformation on the speciation of arsenic and antimony in mine waste and sediment. Chem Geol 283:109-118

Fawcett SE, Jamieson HE, Nordstrom DK, McCleskey RB (2014) Recognizing the distinct geochemical nature of As and Sb associated with mine-waste, sediment, and water at the Giant Mine, Yellowknife. Appl Geochem (in press)

GNWT (Government of the Northwest Territories) (2003) Environmental Guideline for Contaminated Site Remediation, 35 p, *http://www.enr.gov.nt.ca/_live/documents/content/siteremediation.pdf* (accessed Apr 2014)

Grainge JW (1963) Water Pollution, Yellowknife Bay, Northwest Territories. Canadian Department of National Health & Welfare, Public Health Engineering Division, Edmonton, Alberta, Dec 1963

Groves DR, Goldfarb RJ, Gebre-Mariam M, Robert F (1998) Orogenic gold deposits: a proposed classification in the context of their crustal distribution and relationship to other gold deposit types. Ore Geol Rev 13:7-27

Haffert L, Craw D (2008) Mineralogical controls on environmental mobility of arsenic from historic mine processing residues, New Zealand. Appl Geochem 23:1467-1483

Halverson GB (1990) Fluosolids roasting practice at Giant Yellowknife Mines Ltd. *In:* Proceedings of the 96th Annual Northwest Mining Association Convention, Spokane Washington, Dec 5-7, 1990

Hubbard LJ, Marshall DD, Anglin CD, Thorkelson D, Robinson MH (2006) Giant Mine: Alteration, mineralization, and ore-zone structures with an emphasis of the Supercrest zone. *In:* Gold in the Yellowknife Greenstone Belt, Northwest Territories: Results of the EXTECH III Multidisciplinary Research Project. Anglin CD, Falck H, Wright DF, Ambrose EJ (eds) Geol Assoc Can, Min Dep Div Spec Pub 3:197-212

INAC (Indian and Northern Affairs Canada) (2007) Giant Mine remediation plan. Report of the Giant Mine remediation team. Department of Indian Affairs and Northern Development as submitted to the Mackenzie Valley Land and Water Board, Yellowknife, Canada, 2007, 260 p, *http://reviewboard.ca/upload/project_document/EA0809-001_Giant_Mine_Remediation_Plan_1328900464.pdf* (accessed Apr 2014)

Jamieson HE, Bromstad ML, Nordstrom DK (2013) Extremely arsenic-rich, pH-neutral waters from the Giant Mine, Canada. Proceedings of the First International Conference on Mine Water Solutions in Extreme Environments. Mine Water Solutions, p 82-84

Kempton H, Bloomfield TA, Hanson JL, Limerick P (2010) Policy guidance for identifying and effectively managing perpetual environmental impacts from new hardrock mines. Environ Sci Policy 13:558-566

Kontak DJ, Jackson SJ (1999) Documentation of variable trace- and rare-earth-element abundances in carbonates from auriferous quartz veins in Meguma lode-gold deposits, Nova Scotia. Can Mineral 37:469-488

Kuyek J (2011) Theory and practice of perpetual care of contaminated sites: a review of the literature and some case studies. Submitted to Mackenzie Valley Environmental Impact Board, July 2011, *http:// www.miningwatch.ca/sites/www.miningwatch.ca/files/Kuyek-theory%20and%20Practice%20final%20 (July%202011)-1.pdf* (accessed Apr 2014)

Marsden JO, House CI (2006) The Chemistry of Gold Extraction, 2nd edition. Society for Mining, Metallurgy, and Exploration, Inc., Littleton, Colorado, 651 p

Martin AJ, Pederson TF (2002) Seasonal and interannual mobility of arsenic in a lake impacted by metal mining. Environ Sci Technol 36:1516-1523

Mudroch A, Joshi SR, Sutherland D, Mudroch P, Dickson KM (1989) Geochemistry of sediments in the Back Bay and Yellowknife Bay of Great Slave Lake. Environ Geol Water Sci 14:35-42

MVRB (Mackenzie Valley Review Board) (2013) Report of Environmental Assessment and Reasons for Decision, Giant Mine Remediation Project. EA0809-001, 245 p, *http://reviewboard.ca/upload/project_ document/EA0809-001_Giant_Report_of_Environmental_Assessment_June_20_2013.PDF* (accessed Apr 2014)

Nordstrom DK, Archer DG (2003) Arsenic thermodynamic data and environmental geochemistry. *In:* Arsenic in Ground Water: Geochemistry and Occurrence. Welch AH, Stollenwerk KG (eds) Kluwer Academic Publishers, Boston, p 1-25

Osborne TH, Jamieson HE, Hudson-Edwards KA, Nordstrom DK, Walker SR, Ward SA, Santini JM (2010) Microbial oxidation of arsenite in a subarctic environment: diversity of arsenite oxidase genes and identification of a psychotolerant arsenite oxidizer. BMC Microbiol 10:205, doi: 10.1186/1471-2180-10-205

Parsons MB, LeBlanc KWG, Hall GEM, Sangster, AL, Vaive JE, Pelchat P (2012) Environmental geochemistry of tailings, sediments and surface waters collected from 14 historical gold mining districts in Nova Scotia. Geol Surv Can, Open File 7150, 321 p, doi:10.4095/291923

Plumlee GS, Morman SA (2011) Mine wastes and human health. Elements 7:399-404

Riveros PA, Dutrizac JE, Chen TT (2000) Recovery of marketable arsenic trioxide from arsenic rich roaster dust. *In* Environmental Improvements in Mineral Processing and Extractive Metallurgy: Proceedings of the V International Conference on Clean Technologies for the Mining Industry, Santiago, Chile, May 9-13, 2000. Sánchez MA, Vergara F, Castro SH (eds) Univ Concepcion, Chile 2:135-149

Royal Oak Mines (1992) The Giant Mine, Water License N1L3-0043, Submission in support of water license renewal, September 1992. Unpublished report, Royal Oak Mines Inc, Yellowknife Division.

Sandlos J, Keeling A (2012) Giant Mine: Historical Summary. Memorial Univ, St. John's, Newfoundland, Canada, 20 p, *http://www.reviewboard.ca/registry/project.php?project_id=69* (accessed Apr 2014)

Seal RR, Hammarstrom JM (2003) Geoenvironmental models of mineral deposits: Examples from massive sulfide and gold deposits. *In:* Environmental Aspects of Mine Wastes. Jambor JL, Blowes DW, Ritchie AIM (eds) Mineral Assoc Can Short Course 31:11-50

Siddorn JP, Cruden AR, Hauser RL, Armstrong JP, Kirkham G (2006) The Giant-Con Gold Deposits: Preliminary Integrated Structural and Mineralization History. *In:* Gold in the Yellowknife Greenstone Belt, Northwest Territories: Results of the EXTECH III Multidisciplinary Research Project. Anglin CD, Falck H, Wright DF, Ambrose EJ (eds), Geol Assoc Can, Min Dep Div Spec Pub 3:212-231

Spence C, Woo M (2002) Hydrology of subarctic Canadian Shield: bedrock upland. J Hydrol 232:111-127

Stefanski MJ, Halverson GB (1992) Gold recovery improvement investigations at Giant Yellowknife mine. *In:* Project Summaries, Canada-Northwest Territories Mineral Development Subsidiary Agreement, 1987-1991. Richardson DG, Irving M (eds), Geol Surv Can Open File 2484:217-219

Tait RJC (1961) Recent progress in milling and gold extraction at Giant Yellowknife Gold Mines Limited. Can Inst Mining Metall Trans 64:204-216

Walker SR (2006) The solid phase speciation of arsenic in roasted and weathered sulfides at the Giant gold mine, Yellowknife, NWT. PhD thesis, Queen's University, Kingston, Ontario, Canada

Walker SR, Jamieson HE, Lanzirotti A, Andrade CF, Hall GEM (2005) The speciation of arsenic in iron oxides in mine wastes from the Giant Gold Mine, N.W.T.: Application of synchrotron micro-XRD and micro-XANES at the grain scale. Can Mineral 43:1205-1224

Walker SR, Parsons MB, Jamieson HE, Lanzirotti A (2009) Arsenic mineralogy of near-surface tailings and soils: Influences on arsenic mobility and bioaccessibility in the Nova Scotia gold mining districts. Can Mineral 47:533-556

Walker SR, Jamieson HE, Lanzirotti A, Hall GEM, Peterson RC (2014) The effect of ore roasting on arsenic oxidation state and solid phase speciation in gold mine tailings. Geochem Explor Environ Anal (in press)

Wilson SC, Lockwood PV, Ashley PM, Tighe M (2010) The chemistry and behaviour of antimony in the soil environment with comparisons to arsenic: A critical review. Environ Pollut 158:1169-1181

Wrye L (2008) Distinguishing between natural and anthropogenic sources of arsenic in soils from the Giant Mine, Northwest Territories, and the North Brookfield Mine, Nova Scotia. MSc Thesis, Queen's University, Kingston, Ontario, Canada

Reviews in Mineralogy & Geochemistry
Vol. 79 pp. 553-587, 2014
Copyright © Mineralogical Society of America

13

Arsenic Associated with Historical Gold Mining in the Sierra Nevada Foothills: Case Study and Field Trip Guide for Empire Mine State Historic Park, California

Charles N. Alpers

U.S. Geological Survey, California Water Science Center
Sacramento, California, U.S.A.

cnalpers@usgs.gov

Perry A. Myers

California Department of Toxic Substances Control
Sacramento, California, U.S.A.

Daniel Millsap

California Department of Parks and Recreation
Sacramento, California, U.S.A.

Tamsen Burlak Regnier

Clean Harbors
San Diego, California, U.S.A.

INTRODUCTION

The Empire Mine, together with other mines in the Grass Valley mining district, produced at least 21.3 million troy ounces (663 tonnes) of gold (Au) during the 1850s through the 1950s, making it the most productive hardrock Au mining district in California history (Clark 1970). The Empire Mine State Historic Park (Empire Mine SHP or EMSHP), established in 1975, provides the public with an opportunity to see many well-preserved features of the historic mining and mineral processing operations (CDPR 2014a,b,c).

A legacy of Au mining at Empire Mine and elsewhere is contamination of mine wastes and associated soils, surface waters, and groundwaters with arsenic (As), mercury (Hg), lead (Pb), and other metals. At EMSHP, As has been the principal contaminant of concern and the focus of extensive remediation efforts over the past several years by the State of California, Department of Parks and Recreation (DPR) and Newmont USA, Ltd. In addition, the site is the main focus of a multidisciplinary research project on As bioavailability and bioaccessibility led by the California Department of Toxic Substances Control (DTSC) and funded by the U.S. Environmental Protection Agency's (USEPA's) Brownfields Program.

This chapter was prepared as a guide for a field trip to EMSHP held on June 14, 2014, in conjunction with a short course on "Environmental Geochemistry, Mineralogy, and Microbiology of Arsenic" held in Nevada City, California on June 15-16, 2014. This guide contains background information on geological setting, mining history, and environmental history at EMSHP and other historical Au mining districts in the Sierra Nevada, followed by descriptions of the field trip stops.

1529-6466/14/0079-0013$05.00   http://dx.doi.org/10.2138/rmg.2014.79.13

GEOLOGICAL SETTING

Regional geology

Empire Mine SHP is located in the Grass Valley mining district on the western-sloping foothills of the Sierra Nevada (Fig. 1). The Sierra Nevada Foothills (SNFH) orogenic gold province stretches 150 miles (240 km) from the town of Mariposa to northern Sierra County (Fig. 1a). The SNFH province is subdivided into four zones, or belts (Clark 1970): (1) the Mother Lode, which refers to Au-quartz vein deposits in close proximity to the Melones Fault Zone, from the Mariposa to Auburn (Fig. 1b), (2) the West Belt, located to the west of the Mother Lode, in and near the Bear Mountains Fault Zone (Fig. 1b), (3) the East Belt, located to the east of the Mother Lode in the vicinity of the Calaveras-Shoo Fly Thrust (Figs. 1b,c), and (4) the "Northern Sierra Nevada" or "Northern Mines" area, which includes the Grass Valley, Alleghany, and Washington mining districts (Fig. 1b,c). Deposits in the SNFH orogenic gold province consist of low-sulfide Au-quartz veins (Lindgren 1895; Clark 1970; Böhlke and Kistler 1986; Böhlke 1999; Ashley 2002) and mineralized schist, greenstone, and granitic rocks. Including placer deposits exploited by hydraulic mining, drift mining, and dredging, the SNFH orogenic gold province produced more than 100 million troy ounces (3,100 tonnes) of Au from 1848 to 1960 (Clark 1970).

The deposits of the SNFH orogenic gold province were emplaced during the Cordilleran orogen and reside within a series of accreted terranes that were deformed after docking with the Pacific margin of the United States (McCuaig and Kerrich 1998). Volcanic arcs and accreted terranes were deposited on the Pacific margin from the Early Triassic through the Late Jurassic, with a magmatic arc developing in the Late Jurassic (Schweickert 1981; Schweickert et al. 1984; Goldfarb et al. 1998; Ernst et al. 2008a,b). From 155 to 123 Ma, deformation and metamorphism occurred, followed by emplacement of plutons and the Sierra Nevada batholith from 151 to 80 Ma (Goldfarb et al. 1998). Most of the deposits are associated with steeply-dipping thrust faults (Tuminas 1983; Ernst et al. 2008a,b) where extensive Au mineralization took place between about 144 and 108 Ma (Böhlke and Kistler 1986; Goldfarb et al. 1998; Böhlke 1999).

Carbon-dioxide–rich fluids resulting from metamorphism of an accretion complex were injected into major fracture zones within the metamorphic accretion complex (Johnston 1940; Böhlke 1989). The resulting Au-quartz veins are generally low in sulfide content, hence their classification by the U.S. Geological Survey as "low-sulfide Au-quartz veins" (Berger 1986; Ashley 2002) or "orogenic (metamorphic-hosted) Au-quartz veins" (Böhlke 1982). Other names for this Au deposit type include greenstone-hosted, mesothermal, metamorphic rock-hosted, orogenic, shear-zone hosted, slate-belt, and turbidite-hosted (Groves et al. 1998).

Typical sulfide mineralogy includes pyrite, galena, sphalerite, chalcopyrite, arsenopyrite ± pyrrhotite. Additional accessory and gangue minerals may include tellurides such as hessite and altaite (Johnston 1940), scheelite, bismuth, tetrahedrite-tennantite, stibnite, molybdenite, calcite, dolomite, ankerite, siderite, and fluorite (Berger 1986) ± nickeline ($NiAs$) and cobaltite ($CoAsS$) (Savage et al. 2000b; Burlak 2012). Pyrite and arsenopyrite tend to be abundant in alteration zones around veins and in disseminated ore (Ashley 2002). The pyrite associated with these alteration zones, from which waste rock typically is generated, tends to be As-bearing (i.e., arsenian pyrite), with up to 6 wt% As (Savage et al. 2000a,b; Reich and Becker 2006; Burlak 2012). The pyrite may also contain Ni and (or) Co (Savage et al. 2000b; Burlak 2012).

Carbonate minerals including calcite, dolomite, ankerite, magnesite, and siderite are commonly associated with the Au-quartz veins and alteration halos (Lindgren 1895, 1896a; Böhlke 1988, 1989). The carbonate minerals provide acid-neutralizing capacity that typically leads to near-neutral rather than acidic drainage from tunnels and waste piles associated with abandoned Au mines in the Sierra Nevada Foothills region (Ashley 2002). An exception to this is areas where piles of sulfide concentrate are deposited, such as one at the Argonaut mine

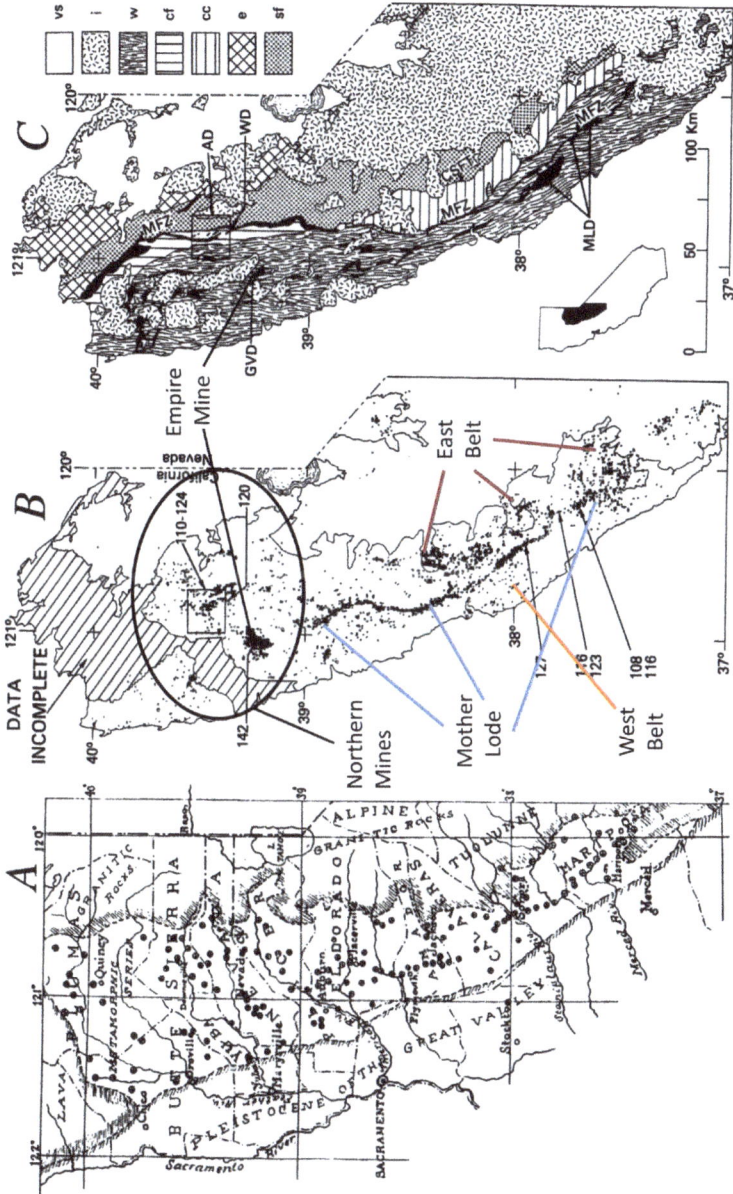

Figure 1. Distribution of gold-quartz veins in the Sierra Nevada foothills metamorphic belt (modified from Böhlke 1999). (a) Gold quartz veins of the "central gold belt" of California (from Lindgren 1895; Plate 1). (b) Lode gold deposits of the Sierra Nevada foothills region from the U.S. Geological Survey Computer Resources Information Bank (through 1984) with radiometric ages of hydrothermal minerals (in Ma; from Böhlke and Kistler 1986). (c) Selected geologic features (from Böhlke and Kistler 1986): black areas—ultramafic rocks; vs—unmetamorphosed volcanic and sedimentary rocks (post-Cretaceous); i—intrusive rocks (mostly Mesozoic); w—metavolcanic and metasedimentary rocks of the western foothills; cf—Calaveras Formation; cc—Calaveras Complex; e—eastern Paleozoic and Mesozoic melange and arc rocks; sf—Shoo Fly Complex; MFZ—Melones fault zone; CSFT—Calaveras–Shoo Fly thrust; MLD—Mother Lode district; GVD—Grass Valley district; AD—Alleghany district; WD—Washington district. [Used by permission of the Geological Society of America, from Böhlke 1999, *Geol Soc Am Spec Paper 338*, Fig. 1, p 58.]

tailings in the Jackson mining district (Ashley 2002); these sulfide concentrates were derived from ore brought from the Plymouth mine (R. Ashley, written commun., 2014).

Geology of the Grass Valley area

The Grass Valley Fault Zone and the Wolf Creek Fault Zone near Grass Valley (Tuminas 1983; Mayfield and Day 2000) allowed CO_2-rich hydrothermal fluids to migrate upward through the metamorphic belt, remobilizing and depositing silica and other constituents at temperatures ranging from about 230 °C to 370 °C (Böhlke and Kistler 1986; Weir and Kerrich 1987). Fractures filled with multiple generations of quartz, carbonates, various metallic sulfides, and Au while being reactivated by faulting (Knaebel 1931; Johnston 1940; Böhlke 1989; Goldfarb et al. 1998).

The Grass Valley mining district lies within a granodiorite pluton of Mid- to Late-Jurassic age (162-153 Ma) based on high-precision Ar-Ar dating of biotite and hornblende (Taylor et al. 2011). Hornblende from the Grass Valley pluton gave a Cretaceous K-Ar age of 127 Ma (Böhlke and Kistler 1986), which may represent a later metamorphic event related to Au mineralization. The pluton is bordered by Jurassic-Triassic-age serpentinite and ophiolite, Upper Jurassic accretionary sequence volcanic rocks, slate, and greenschist, and Carboniferous- to Triassic-age Calaveras Complex chert and fine grained argillite (Fig. 1c) (Knaebel 1931; Johnston 1940; Mayfield et al. 2000; Ernst et al. 2008a). The Calaveras Complex, Jurassic-Triassic arc, and the Upper Jurassic accretionary sequence are part of the Western Metamorphic Belt (Ernst et al. 2008a). This belt was deformed and metamorphosed during the Nevadan orogeny (Ernst et al. 2008a,b) and is intruded by various igneous rocks, mostly granitoids (Mayfield and Day 2000). Other common igneous and metamorphic rocks in the area include: amphibolite, schist, serpentine, gabbro, diorite, quartz porphyry, carbonates, limestone, dolomite, and intrusive dikes of various compositions (Lindgren 1896a,b; Knaebel 1931; Johnston 1940).

Rock types within Empire Mine SHP are predominantly metavolcanic rock (mapped by Lindgren (1986a) as pophyrite), diabase, granodiorite, and andesite (Fig. 2). The geology of the Grass Valley mining district is somewhat distinct from the rest of the Mother Lode, especially the other mining districts to the south. One important difference is the relative lack of slate formations (e.g., Mariposa Slate) in the Grass Valley area (Bowen and Crippen 1948). Structurally, the most productive veins in the Grass Valley district tend to be shallow dipping (about 35°, mostly west but some east; Johnston 1940; Clark 1970), whereas productive veins tend to dip more steeply in the southern and central parts of the Mother Lode (about 70° east or west; Bowen and Crippen 1948; Clark 1970).

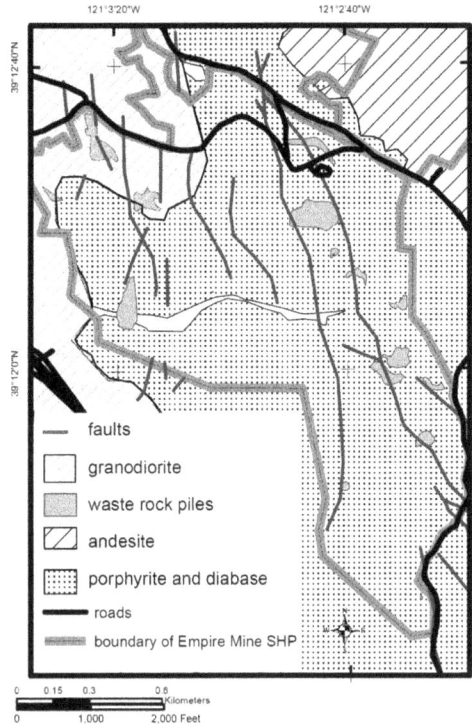

Figure 2. Geology of a portion of the Grass Valley mining district, including most of Empire Mine State Historic Park (modified from Lindgren 1896a).

MINING HISTORY

The discovery in January 1848 of Au nuggets in the tail race of a lumber mill by James Marshall on the South Fork American River at Coloma, California, led to the well-known Gold Rush in 1849 (Clark 1970). The massive influx of fortune seekers created many mining camps with thousands to tens of thousands of miners and merchants throughout the Sierra Nevada. The production of Au in California (Fig. 3) was dominated by hydraulic mining of placer deposits from the 1850s until 1884, when the decision by Judge Lorenzo Sawyer in the case Woodruff vs. North Bloomfield Mining and Gravel Company largely shut down the practice (Kelley 1959). In 1893, the Caminetti Act created the California Debris Commission (a federal agency) which required all mines (hydraulic and later, hardrock) to impound their tailings (Averill 1946; Hagwood 1981; Kelley 1984). Improvements in lode mining technology led to increased production from underground mining during the 1890s (Clark 1970). Dredging of placer deposits became an important source of Au from the early 1900's up to World War I (1914-1917) (Averill 1946).

Gold production in California declined during the 1920's and then picked up greatly during the Great Depression of the 1930's (Fig. 3), spurred by the increase of the price of Au from $20.67 to $35 per troy ounce. The increase in production during the 1930's came from large-scale dredges and large lode mines, including those in the Grass Valley mining district. Most Au production halted in late 1942 because of World War II, with Limitation Order L-208 by the War Production Board. There was a small resurgence after World War II, but statewide production fell during the 1950s and early 1960s (Fig. 3) because of increased costs, with the price of Au still fixed at $35 per troy ounce (Clark 1970; Craig and Rimstidt 1998).

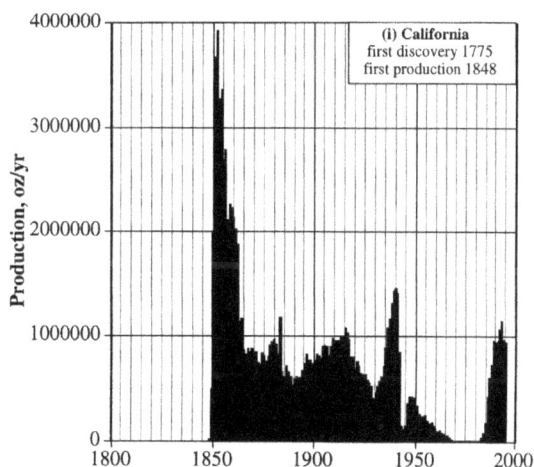

Figure 3. History of gold production in California from 1848 to 1995. From Craig and Rimstidt (1998). [Used by permission of Elsevier, from Craig and Rimstidt (1998), *Ore Geology Reviews*, Vol. 13, Fig. 4i, p. 426.]

Major developments in the mining history at Empire Mine are outlined in Table 1, starting with discovery of Au at the site in 1850. The 30-stamp Steamboat Mill on Ophir Hill was completed in 1865 (Bohakel 1980). Prior to 1879, mine workings were relatively shallow, within 1,200 feet (370 m) of the surface. The early history of lode mining at the Betsy, Daisy Hill, Heuston Hill, Prescott Hill, and Sebastopol mines is described by Bean (1867), Selverston (2008), and Hilton and Selverston (2013).

In 1879, William Bourne Jr. took over management of Empire mine and several shafts were sunk to greater depth. Under the direction of George W. Starr as superintendent (1887-1893 and 1898-1927), the mine flourished and became the largest Au producer in California (McQuiston 1986). The main inclined shaft at Empire Mine was deepened from 2,800 to 3,500 feet starting in 1902 (Mining and Scientific Press 1902). Newmont Mining Company purchased the Empire mine and the North Star mine in 1929 and created the Empire-Star Mines Company Ltd., which produced more than 100,000 troy ounces (3.1 tonnes) Au per year

Table 1. Mining history at Empire Mine State Historic Park

Date	Event
June 1850	Discovery of gold in Grass Valley area by George McKnight
Oct. 1850	Discovery of gold at Empire mine site by George Roberts
1851-1852	Ophir Hill mine consolidated claims of small-scale miners
1852	Ophir Hill mine purchased by John Rush, name changed to Empire Quartz Hill Company
1852-1869	Several changes in mine ownership; surface structures and processing plants modernized
1869	William Bourne gains controlling interest in Empire mine
1876	Death of William Bourne
1876-1879	Slump in production; mining to 1,200 feet (366 m) below surface
1879	William Bourne Jr. takes over management of Empire mine
1879-1884	Several shafts pushed deeper than 1,200 feet (366 m)
1884	Mine regains profitability
1886	Empire mill expanded to 40 stamps
1887-1893	George W. Starr works as superintendent of Empire mine
1898-1927	George W. Starr returns as superintendent of Empire mine
1900	Empire mine incline deeper than 3,000 feet (914 m)
1910	Cyanide plant opens at Empire mine, capacity 150 tons per day
1911-1915	Pennsylvania and WYOD (Work Your Own Diggins) mines acquired by Empire mine
1929	Newmont Mining Company purchases Empire mine and North Star mine
1930's	Empire-Star Mines Co. produces more than 100,000 troy ounces (3.1 tonnes) Au per year
1942	War Production Board halts gold mining (L-208)
1945	Mine reopens at end of WWII
July 5, 1956	Miners go on strike
Jan. 1957	Removal of underground equipment begins
May 28, 1957	Last pump shut down, mine closed
Sept. 1959	Equipment sold at auction
1975	State of California purchases Empire-Star mine properties for $US 1,250,000

Sources: McQuiston (1986), CDPR (2014b,c)

during the 1930's. Limitation Order L-208 by the War Production Board classified all mines that produced more than 1,200 tons during 1941 as non-essential and required the Empire-Star mines to close in 1942. After World War II, the mine resumed production from 1945 to 1956. Total Au production from the Empire and North Star mines, based on fairly complete records, is estimated to have been 5,855,000 troy ounces (182 tonnes) (McQuiston 1986) from about 10,346,000 short tons (9,386,000 tonnes) of ore (Long et al. 1998), so the average Au grade was 0.56 troy ounces per short ton (~19 mg kg^{-1}).

The combined Empire and North Star was the most productive historic Au mine in California (Clark 1970). Total Au production in the Grass Valley mining district was at least 21.3 million troy ounces (663 tonnes) of Au (Goldfarb et al. 2005). Total Au production in California through 1995 was about 115 million troy ounces (3,600 tonnes), of which about 75% came from veins and placer deposits in the metamorphic complexes of the western Sierra Nevada foothills (Clark 1970, 1985; USBM 1983-1995; Craig and Rimstidt 1998; Böhlke

1999). About 18% of California's historic Au production came from the Grass Valley mining district, and much of that was from the Empire and North Star mines.

ENVIRONMENTAL STUDIES AT OTHER SIERRA NEVADA GOLD MINES

Arsenic contamination has been studied at several mining districts in the SNFH orogenic gold province. This section summarizes the work done at these other locations, to put the Empire Mine in a regional context. The locations of mines and mining districts discussed in this section are shown in Figure 1c.

Relationship of arsenic to gold mineralization

Samples of gold ore, wall rock, and mine waste analyzed from several mining districts in the SNFH region indicate a wide range of As concentrations, from less than 10 to more than 10,000 mg kg^{-1} (Fig. 4). Arsenic is typically higher in the gold ore and wall rock from deposits in the northern Sierra Nevada (e.g., Alleghany district) than in deposits from the southern part of the region (e.g., Hodson district). In the north, arsenopyrite is the main host of As in unweathered deposits (e.g., Lindgren 1896a; Ferguson and Gannett 1932), with minor arsenian pyrite (e.g., Burlak 2012), whereas in the south, As is hosted primarily by arsenian pyrite with rare arsenopyrite (Savage et al. 2000a,b).

Harvard Pit, Jamestown Mining District

The Jamestown mining district is located in western Tuolumne County, California (Fig. 5). Mineralization occurs in low-sulfide gold-quartz veins developed in the vicinity of the Melones Fault Zone (Dohms et al. 1985; Allgood 1990). Production of Au in the Jamestown mining

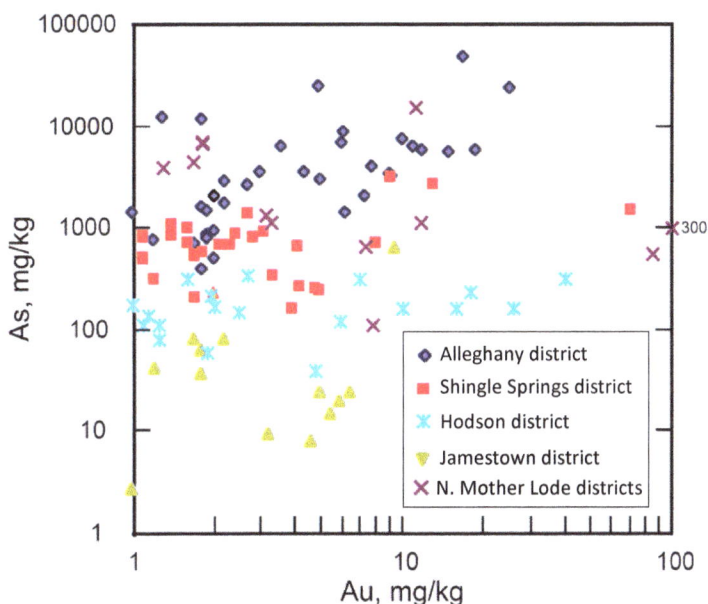

Figure 4. Scatter plot showing concentrations of gold and arsenic in samples of ore, wall rock and waste rock in several mining districts of the Mother Lode Region. Data sources: Alleghany (Böhlke 1986); Shingle Springs (Big Canyon mine, Nash 1988); Hodson (Chaffee and Sutley 1994); Jamestown (Harvard mine, Ashley 2002); N. Mother Lode, including Sutter Creek, Nashville, Georgetown, Spanish Flat, Kelsey, and Nevada City districts (Ashley 2002). Modified from Savage et al. (2000a) and Ashley (2002).

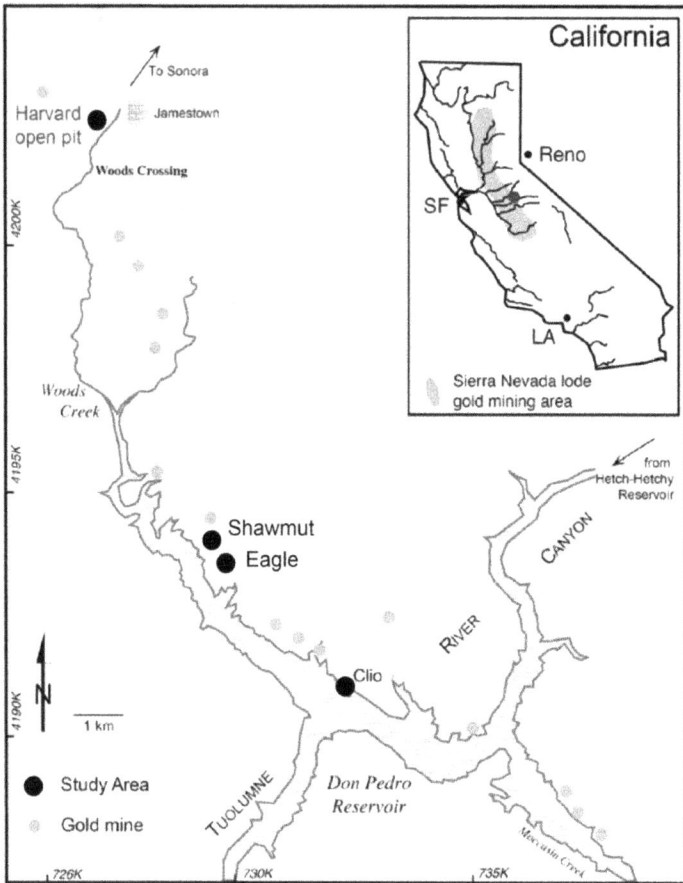

Figure 5. Map showing gold mines in the vicinity of Jamestown, California and Don Pedro Reservoir (modified from Savage et al. 2009) [Used by permission of the Society of Economic Geologists, from Savage et al. (2009), *Econ Geol*, Vol. 104, Fig. 1, p 1172.]

district during 1850-1994 was approximately 2,300,000 troy ounces (72 tonnes) (Long et al. 1998). The Harvard pit was developed by the Sonora Mining Corporation during 1986-1994. The pit produced about 660,000 troy ounces of Au from about 17,000,000 short tons of ore with an overall stripping ratio of about 4.5:1 (Ashley and Savage 2001). Two smaller nearby pits, the Crystalline pit and South Crystalline pit, were backfilled to prevent the accumulation of water. Dimensions of the Harvard pit are approximately 830 m in length, 460 m in width, and 185 m in depth; elevation at pit bottom is 265 m above sea level (Ashley and Savage 2001). The Harvard pit began to fill with water soon after mining stopped in 1994. The water level had reached about 357 m in late 2000 (Ashley and Savage 2001) and 384 m in late 2007.

Detailed studies of water chemistry and mineralogy in the Harvard pit (Savage et al. 2000a, 2009) documented a monomictic lake that has developed in the pit, with alkaline water of the Ca-Mg-HCO$_3$-SO$_4$ type, containing As concentrations up to 1.2 mg L^{-1} and total dissolved solids (TDS) up to 2,000 mg L^{-1}. Pit walls are comprised of two major geologic units in fault contact: the hanging wall is composed of interlayered slate, metavolcanic and metavolcaniclastic rocks, and schists and the footwall rocks are chlorite-actinolite and talctremolite schists generated by metasomatism of greenschist facies mafic and ultramafic igneous rocks (Savage et al. 2009). Primary sulfide minerals include arsenian pyrite (up to 6 wt% As) in the ore zone and hanging wall, and gersdorffite, niccolite, and cobaltite, mainly in the footwall with rare arsenopyrite (Savage et al. 2009). Weathering products of arsenian pyrite include goethite and jarosite on pit walls and in joints, and efflorescent copiapite group minerals (mostly magnesiocopiapite) and Mg-sulfates (most commonly hexahydrite) that accumulate on wall rock faces during the dry season; weilite, a Ca-arsenate, was observed in one sample (Savage et al. 2009). In weathered rock samples from the ultramafic foot wall and ore zone, the three highest As concentrations (722-1,260 mg kg^{-1}) had relatively low Al and relatively high Mg, Ni, Cr, and Co (Savage et al. 2009). Weathered rock samples from the hanging wall (slate and schist) had As up to 385 mg kg^{-1} with higher As in samples with magnesiocopiapite and jarosite compared to samples with gypsum and hexahydrite (Savage et al. 2009).

Covariation of As and bicarbonate was observed in the hypolimnion of the Harvard pit lake; both rose in concentration as stratification proceeded in the summer and declined during winter rains (Savage et al. 2009). A schematic cross-section through the pit (Savage et al. 2009) illustrates some of its geologic and hydrologic features (Fig. 6). Geochemical modeling with PHREEQC (Parkhurst and Appelo 1999) suggests that pit lake characteristics are related to evasion of CO_2 to the atmosphere, interaction with pit walls including dissolution of efflorescent salts during the first flush and seasonal rainfall, and As sorption (Savage et al. 2009).

Eagle-Shawmut and Clio Mines, Jacksonville Mining District

The Eagle-Shawmut and Clio mines are located in the Jacksonville mining district, adjacent to Don Pedro Reservoir, near Jamestown, California (Fig. 5). Production of Au from the Eagle-Shawmut mine was approximately 290,000 troy ounces (9.0 tonnes) during 1896-

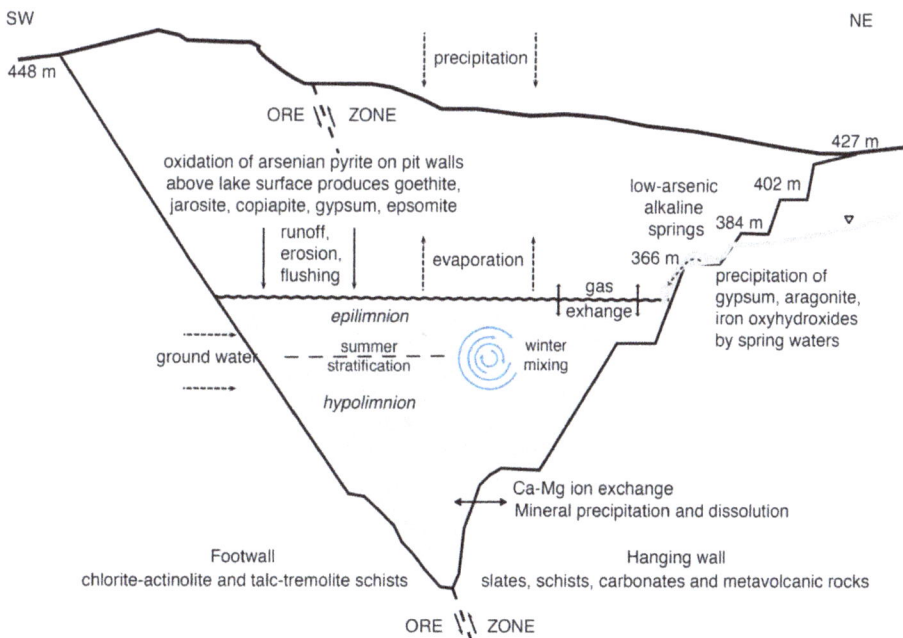

Figure 6. Schematic cross-section through Harvard Pit, showing geologic and hydrologic features [Used by permission of the Society of Economic Geologists, from Savage et al. (2009), *Econ Geol*, Vol. 104, Fig. 14, p 1191.]

1946 (Knopf 1929; Long et al. 1998). During the early 1960s, many reservoirs were built along the western slope of the Sierra Nevada. An example of this is Don Pedro Reservoir, which submerged mill tailings from the Eagle-Shawmut and Clio mines (Fig. 5).

Arsenopyrite is rare at the Eagle-Shawmut and Clio mines, where the primary sulfide mineralogy is dominated by arsenian pyrite, as is typical in the southern Mother Lode (Savage et al. 2000b). The As content of pyrite was documented to decrease with distance from the Melones Fault Zone, the locus of Mother Lode low-sulfide Au-quartz vein mineralization (Fig. 7) based on electron microprobe measurements (Savage et al. 2000b). Near the fault zone, the As in pyrite ranged from 0 to 2.1 wt%; approximately 0.5 km away, the maximum As in pyrite was less than 1.0 wt%, and from 1.5 to 3 km from the fault, the maximum As was less than 0.5 wt% (Fig. 7). It is well known in the field of geochemical exploration that the As content of soil and stream sediment can be an effective exploration guide in this and similar geologic settings (e.g., Bowell 1994; Thornton and Farago 1997; Mueller et al. 2007).

Weathering of arsenian pyrite leads to release of As, which is attenuated by adsorption on secondary ferric iron minerals such as goethite, and by coprecipitation with jarosite (Savage et al. 2000a; Ashley and Savage 2001). The secondary iron mineralogy in mine waste at the Eagle-Shawmut and Clio mines is controlled by the acid-buffering capacity of the pyrite-bearing materials, with jarosite forming at lower pH in environments with lower acid-buffering capacity (Savage et al. 2000a).

Mesa de Oro and Central Eureka Mine, Amador County

Mesa de Oro is a housing subdivision that was built on As-contaminated mill tailings of the Central Eureka mine, located in the town of Sutter Creek in Amador County, California

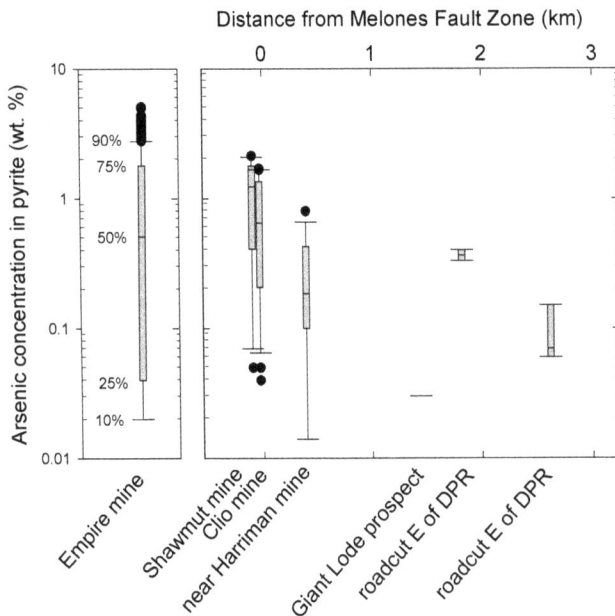

Figure 7. Box plots showing the concentration of arsenic in pyrite (logarithmic scale) from waste rock at Empire Mine (from Burlak 2012) compared with mines, prospects, and roadcuts in the vicinity of Don Pedro Reservoir (DPR) (modified from Savage et al. 2000a). The data from Burlak (2012) are individual electron microprobe analyses, whereas the data from Savage et al. (2000a) represent averages of 3 to 9 electron microprobe analyses of individual pyrite grains, plotted as a function of distance from the Melones fault zone. "Zero" on the horizontal axis marks the approximate center of the Melones fault zone where samples from the Clio and Shawmut mines are plotted (locations in Fig. 5). The distance +0.5 km denotes the eastern edge of the fault zone: the samples are from a mineralized outcrop near the Harriman mine (Savage et al. 2000a).

(Fig. 1). The Central Eureka mine operated from the 1890's to the 1960's and produced 456,000 troy ounces of Au (14.2 tonnes) (Long et al. 1998). Concentrations of As in ten samples of soil containing mill tailings ranged from 115 to 1,320 mg kg^{-1} with an average of 374 mg kg^{-1} (Salocks et al. 1996), compared to a background As value of 22 mg kg^{-1}; yard soils in the subdivision, derived mostly from tailings, had As from 90 to 1,142 mg kg^{-1}; vacuum bags from homeowners in the area had As up to 889 mg kg^{-1} (Alexander Law Group 2014). Arsenic problems at Mesa de Oro were well publicized (Greenwald 1995), and resulted in a substantial remediation program (USEPA 1998; DTSC 2005).

Sequential extraction studies on the Mesa del Oro subdivision soils by Salocks et al. (1996) used distilled water and 1.0 M magnesium chloride, each of which solubilized less than 1% of total sample As. Extraction with 0.1 M phosphate buffer (pH 8) solubilized 1 to 14% of total sample As, suggesting a significant pool of surface-bound, soluble As. Final extraction with 0.25 M hydroxylamine hydrochloride in HCl solubilized 4 to 35% of total sample As, indicating As associated with Fe oxides that likely developed from oxidation of pyrite. A separate experiment with 0.1 M HCl (pH 2) dissolved from 24 to 34% of total sample As. Salocks et al. (1996) concluded that up to 50% of total As in the samples from Mesa de Oro was soluble in acidic solutions similar to stomach acid and therefore potentially bioavailable.

An *in vivo* study with mice used a sample of mine tailings from Mesa de Oro with an As concentration of 691 mg kg^{-1} soil (Golub et al. 1999). Because the study was done with a large dose (5 mg As kg^{-1} body weight) relative to other studies and a single time-point estimate (one hour after ingestion by lavage), a measure of percent bioavailability was not possible; nevertheless the level of bioavailability was considered relatively high based on bioaccumulation in various tissues relative to a lower-As control (Golub et al. 1999).

Argonaut Mine, Amador County

The Argonaut mine, located in the town of Jackson in Amador County, California (Fig. 1), produced 940,000 troy oz. of Au (29 tonnes) during 1893-1942 (Long et al. 1998). The speciation and bioaccessibility of As in a sample of mill tailings from the Argonaut mine with an As concentration of 260 mg kg^{-1} was studied by various leach tests (Borch et al. 1994a,b) and by X-ray absorption spectroscopy (XAS) (Foster et al. 1998). The leach tests consisted of a four-step sequential extraction procedure (Tessier et al. 1979); each sample was successively reacted with 0.25M KCl, 0.1 M K$_2$HPO$_4$, 1 M NaOAc, and citrate-dithionite solutions. The extraction data for the Argonaut mine tailings sample suggested that ~6% of the total As is easily removed, ~2% is associated with carbonates, and only ~2% is associated with ferric hydroxides. Borch et al. (1994b) suggested that the remaining As is present in low-solubility phases.

Results of XAS analysis of the Argonaut tailings (Foster et al. 1998) indicate that ~20% of As was bound in arsenopyrite and arsenian pyrite and ~80% of As was in a precipitate such as scorodite; however, no precipitate was detected by X-ray diffraction (XRD) or electron microprobe analysis (EMPA), suggesting that the phase is poorly crystalline and/or has low abundance.

Lava Cap Mine, Nevada County

The Lava Cap and Banner mines exploited a vein system in the southeastern part of the Nevada City mining district (Fig. 8). Total Au production was 520,000 troy ounces (16.2 tonnes) (Long et al. 1998). A major rain storm in January 1997 caused the failure of a tailings dam at the Lava Cap mine, in Nevada County, California (Fig. 8). Samples of Fe-oxyhydroxide-bearing bacterial mats from the Lava Cap mine area were highly enriched in As (7,960 mg kg^{-1}; n = 3) (Foster and Ashley 2002; Foster et al. 2011). In comparison, other materials collected in the area had the following As concentrations: ore samples, 949 mg kg^{-1} (n = 4), mill tailings collected by tube core in 1999-2000 from the tailings remaining behind the failed log dam, 1,043 mg kg^{-1} (n = 6); tailings-rich Lost Lake bottom sediment, 1,543 mg kg^{-1} (n = 3), and subaerially-exposed tailings at the Lost Lake water-line, 530 mg kg^{-1} (n = 2) (Foster et al. 2011). The primary

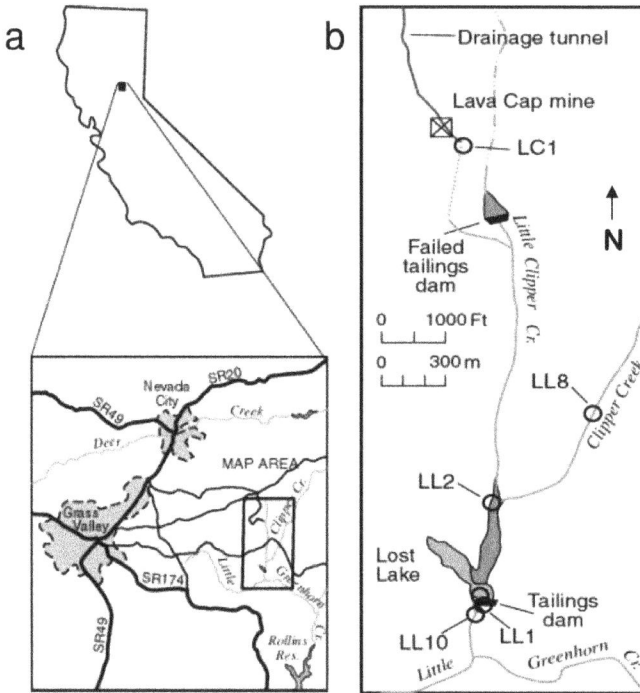

Figure 8. (a) Regional map of California with location of Lava Cap mine; (b) simplified map of Lava Cap mine and Lost Lake area, with major drainages and sampling points indicated. [Used by permission of The Geological Society of London from Foster and Ashley (2002) *Geochem Explor Environ Anal,* Vol. 2, Fig. 1, p. 254.]

host for As in Lava Cap tailings was arsenopyrite; pyrite was rare to absent because it was mostly removed in the flotation process used in milling to concentrate selected sulfide minerals. XAS analysis showed that tailings submerged in Lost Lake for >50 years were not substantially oxidized, whereas tailings exposed to wet-dry cycles at the water line during ~3 years following the dam failure were ~30% oxidized (Foster et al. 2011). The predominant oxidation products identified by X-ray Absorption Fine Structure (XAFS) were As-bearing ferric oxyhydroxides and a Ca-Fe arsenate (possibly arseniosiderite; Foster et al. 2011).

ENVIRONMENTAL CHARACTERIZATION AND REMEDIATION AT EMPIRE MINE STATE HISTORIC PARK

Mining, milling, and other mineral processing activities associated with Empire Star Mining Company and its predecessors resulted in accumulations of waste rock and mill tailings at various locations within Empire Mine SHP. South of the former cyanide plant, areas known as the "Conveyance Corridor" and "Sand Dam" conveyed mill tailings and contain stockpiled mill tailings, respectively (Fig. 9). After the DPR acquired the Empire Mine SHP in 1974-75 (Selverston and Hilton 2013; CDPPR 2014c), an extensive trail system was established. The trails were used by joggers, equestrians, hikers, and bikers (DTSC 2006). In some areas, portions of the trails were found to contain historic mine waste or mill-related materials with high concentrations of As, which led to further action, described below.

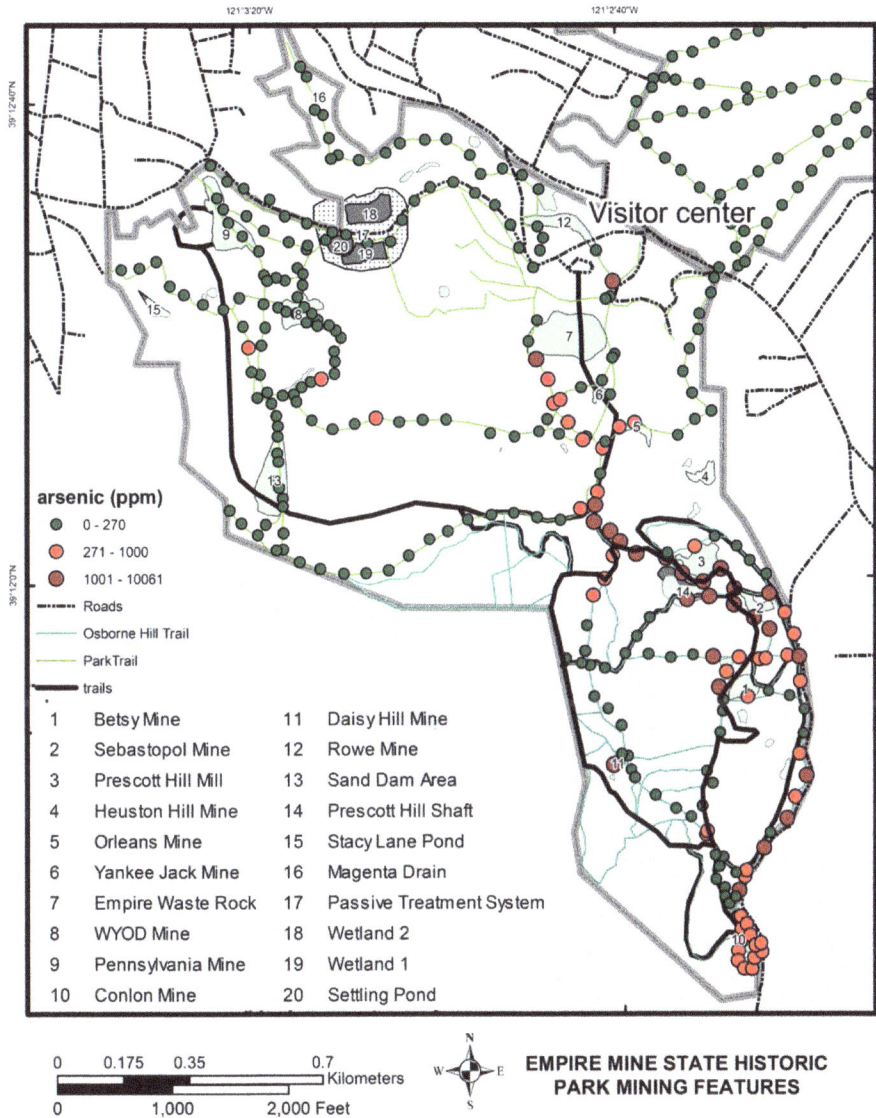

Figure 9. (*for color see Plate 15*) Map of Empire Mine State Historic Park showing mining features. Waste-rock piles shown in light grey (green in the digitial version). Data for arsenic concentration on trails prior to remediation determined by field X-ray fluorescence. Modified from MFG (2008b).

Various state regulatory and park personnel, and contractors to the State of California and Newmont USA, Ltd. conducted site investigations at EMSHP from 1977 to present. A Preliminary Endangerment Assessment (PEA) determined that the mill tailings contained elevated levels of metals (DTSC 1992). The PEA recommended to conduct a full Remedial Investigation/Feasibility Study (RI/FS) and to close selected trails in the area. The DTSC has taken the lead on the RI/FS process, which is ongoing under the federal Comprehensive Environmental Response, Compensation and Liability Act (CERCLA, or Superfund).

Environmental restoration work at the site has been divided into 11 operable units, or sub-areas (Table 2). Documents related to environmental restoration at EMSHP are organized by sub-area (DTSC 2014).

Osborne Hill Area trails

During 2006, 81 samples of trail soil were sampled, sieved to <0.25 mm, and analyzed for 17 metals (MFG 2008b). The most prevalent contaminant of concern was As, however Pb and Cd were elevated locally. In 2006, the California DTSC recommended a remediation goal of 270 mg kg^{-1} based on estimated exposure to runners (Klein 2006). Field X-ray fluorescence (XRF) was used to determine *in situ* As concentration at more than 400 trail locations (MFG 2008b). Several trails were identified as having As concentrations above the remediation goal (Fig. 9). Interim actions taken in 2006-07 to manage the As-contaminated trail surfaces included covering or closing some trail sections (DTSC 2008). A more comprehensive remediation of trails was completed in 2013 (DTSC 2014).

Red Dirt Pile

A remnant sulfide-rich stockpile of tailings or low-grade Au ore, known locally as the Red Dirt Pile (Table 2, Fig. 10), was one of the first areas to be remediated at Empire Mine SHP. The majority of the material (approximately 46,000 short tons, or 41,700 tonnes) was removed to the McLaughlin Mine (Lake County, California) for Au recovery in the late 1980s. Some of the sulfide-rich material was left behind and the soil underneath the stockpile area was impacted by acidic leachate. The soil was found to have an average pH of 3.6 and concentrations of Pb and As as high as 6,000 mg kg^{-1} and 2,460 mg kg^{-1}, respectively (CVRWQCB 2012). During rain events, storm runoff from the Red Dirt Pile impacted the local receiving water, Little Wolf Creek. In 2007, additional contaminated material was removed, and the area was covered with a geosynthetic clay liner and combination of vegetated soil cover and asphalt (coarse base) cover; the asphalt portion is now a parking lot (DTSC 2014).

Characterization of waste-rock piles

Thirteen waste-rock piles at the EMSHP were inventoried and characterized during 2007 (Table 3) (MFG 2008a, MFG 2009, WME 2013a). These waste piles ranged in volume from 1,000 to 66,500 m^3 and were located in the central and southern areas within the EMSHP (Table 3, Fig. 9). The 13 waste-rock piles evaluated represent the largest of 138 waste dumps identified within the EMSHP (Selverston and Hilton 2013). A screening-level assessment of 17 metals to identify potential constituents of concern to human health (MFG 2009) in the waste-rock piles identified As, Cd, and Pb (Table 3) based on comparison to the California Human Health Screening Levels (CHHSLs) for residential and commercial/industrial use for As and Cd (CEPA 2005) and the USEPA Region 9 industrial soil Regional Screening Level (RSL) for Pb (USEPA 2009). The concentration of As in individual waste-rock samples ranged from 10.1 to 15,300 mg kg^{-1} (MFG 2009). The waste-rock piles with the highest mean As concentrations (> 6,000 mg kg^{-1}) were at the Betsy, Prescott Hill, and Woodbury mine sites (Table 3).

Samples of soil and waste rock were collected from 20 locations within the EMSHP in September 2009 as part of a study on As bioavailability and bioaccessibility (Mitchell et al. 2010, 2011, 2012; Alpers et al. 2012; DTSC 2014). Samples were taken at depths ranging from 0 to 1.3 m from trenches dug with a backhoe or by hand. Samples were screened in the field to <4.76 mm (#4 mesh) and then sieved in the laboratory at Ohio State University to <0.25 mm (#60 mesh). Polished epoxy grain mounts were prepared for the <0.25 mm fraction, and polished thin sections were prepared from co-collected hand samples of weathered waste rock.

The mineralogy and composition of primary sulfides and secondary weathering products in hand samples from Empire Mine SHP waste-rock piles (Table 4) were determined by Burlak

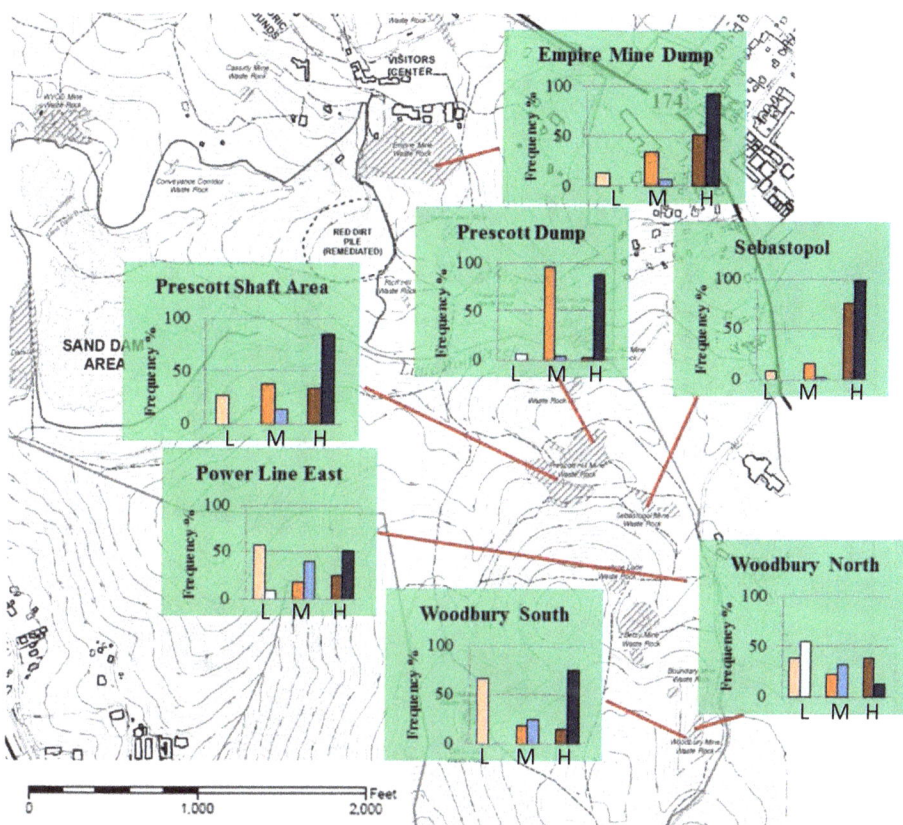

Figure 10. Arsenic concentration in pyrite (left bar) and hydrous ferric oxides (right bar) in waste-rock material from Empire Mine State Historic Park, from electron microprobe analysis. L, low: less than 0.15 wt%; M, medium: 0.15 to 1.0 wt%; H, high: greater than 1.0 wt%. Modified from Burlak (2012).

(2012) using EMPA. The As content of arsenopyrite averaged 42.09 wt%, corresponding to a formula of $Fe(As_{0.874}S_{1.105})$, based on 217 EMPA observations. This is similar to the composition of arsenopyrite from the Oriental mine in the Alleghany district (Fig. 1) and is consistent with formation below 400°C (Sharp et al. 1985). The As content of pyrite in waste-rock samples from Empire Mine SHP ranged from <0.04 to 5.11 wt%, with a median value of 0.51 wt% based on 524 EMPA observations (Burlak 2012; Table 4). The geographic distribution of the data (Fig. 10) shows that As in pyrite was higher in the central waste-rock piles (Empire, Prescott Hill, and Sebastopol) and lower in the southern area (Power Line East, Woodbury North, and Woodbury South).

The composition of secondary weathering products observed in Empire Mine SHP waste-rock piles using EMPA (Burlak 2012) is shown in Figures 11 and 12 and is summarized in Table 4. The most abundant secondary minerals were hydrous ferric oxides (HFO; ferrihydrite and goethite) with variable amounts of As. Scorodite was found in several samples and kaňkite in two samples. The presence of goethite, ferrihydrite, and scorodite in the sieved samples (<0.25 mm) was confirmed by powder XRD (A. Blum, U.S. Geological Survey, written commun.); ferrihydrite was also confirmed by bulk Fe X-ray absorption spectroscopy (XAS) (Foster et al. 2014; see also Foster and Kim 2014, this volume). There is an apparent mixing trend between ferrihydrite and scorodite compositions (Fig. 11), similar in some respects to the trend found in

Table 2. Operable Units for environmental restoration at Empire Mine State Historic Park

Operable Unit	Description	Status of restoration
Conveyance Corridor	This OU extends from the Cyanide Plant area to the historic Hard Rock Trail at the Sand Dam. It was used to hydraulically transport tailing from the former stamp mill and Cyanide Plant to the Sand Dam Tailing area.	Characterization data have been collected. The cleanup is deferred until work on OU's upgradient is completed to prevent re-contamination.
Cyanide Plant	This area includes the cyanide plant ruins and a portion of the conveyance corridor leading to the Sand Dam	The selected remedy was implemented in 2012. Currently stormwater runoff is being studied to determine if additional measures are necessary to protect water quality.
Historic Grounds	This area includes the lawns, gardens and natural areas surrounding the Empire Cottage and other historic buildings	A Remedial Action Assessment was completed in 2013 that determined concentrations of metals present in soils do not pose an unacceptable risk to human health and the environment.
Historic Mine and Mill Sites	This operable unit covers historic mine and mill sites identified for characterization within the Park. The sites are grouped into three areas; Central Area, Osborne Hill Area, and Union Hill Area.	Characterization work has been completed. Concentrations of metals in waste rock and soils were determined not to pose an unacceptable risk to human health in 2013. Evaluation of the risk to ecological receptors is ongoing.
Magenta Drain	The Magenta Drain is a mine tunnel that was constructed to drain water from the underground workings of the Empire Mine. Water flowing from the Magenta Drain portal drains to a surface channel which then flows to an unnamed creek in Woodpecker Ravine, passing through Memorial Park in the City of Grass Valley. Since November 2011, the water is pumped to a wetland system where iron, manganese, and arsenic are removed by passive treatment.	A passive treatment system for water discharged from the drain was completed in 2011. An NPDES permit is in effect for this discharge. Bed and bank sediments in the channel of the unnamed creek in Woodpecker Ravine will be characterized after mining-derived sediment has been removed.
Mine Yard and Stamp Mill	The historic mine yard includes the Visitors Center, several historic buildings and the open yard connecting them. It is located adjacent to the historic grounds.	Characterization data have been collected. A remedy selection document is in preparation.

Table 2 (continued). Operable Units for environmental restoration at Empire Mine State Historic Park

Operable Unit	Description	Status of restoration
Red Dirt Pile	The RDP is a remnant sulfide material stockpile associated with historic milling activities that was removed and covered with an engineered cover system and partially with asphalt. It is located south and slightly west of the historic mine yard and adjacent to the cyanide plant foundation.	The RDP was covered with an engineered cover system and is currently in the operations and maintenance phase.
Residences	Inside the Park are eight residences available for use by California Dept. of Parks and Recreation staff. The eight residences consist of six houses, constructed in the early-to-mid 1930's, and two trailer homes.	Characterization of metals concentrations in soils is completed and a remedy selection document is in preparation.
Sand Dam	This area is located in the southwest portion of the Park. The Sand Dam embankment was built in 1917 to contain mine tailings from the stamp mill and cyanide plant. The area contains a varied landscape of vegetated areas, perennial and seasonal ponds, and bare ground. Approximately 4.8 million tons of tailings were discharged to the Sand Dam area between 1917 and 1956 (McQuiston 1986)	Characterization data have been collected. This OU is down-gradient from the Conveyance Corridor and cleanup will commence after work on the Conveyance Corridor to prevent re-contamination.
Stacy Lane Pond	A swale-fill tailing deposit retained by a waste rock embankment bisected by the Stacy Lane Trail.	Characterization data have been collected. Remedy selection for this OU will proceed concurrently with the Sand Dam as they are similar.
Trails	The trail system traverses most of the 856 acres inside the Park boundaries and consists of approximately 12 miles of trails, in total. The trails are used by pedestrians, equestrians, and bicyclists.	A remedy was selected and implemented. A construction completion report is scheduled for submission in 2014.

Table 3. Waste rock piles at Empire Mine State Historic Park

	Estimated Area (ha)	Estimated Max. Height (m)	Estimated Volume (m³)	Rock Types*	Total Metals Exceeding Criteria**	As Exposure Point Concentration*** (mg kg⁻¹)
Betsy Mine	0.28	8.5	9,480	db, ox	As	8,600
Conlon Mine	0.40	12.2	47,326	db	As	4,250
Daisy Hill	0.09	7.6	3,716	db	As	5,810
Empire Mine	1.90	1.5	14,450	db, gd	As, Cd, Pb	927
Heuston Hill Mine	0.22	7.6	13,609	db, ox	As	5,810
Josephine Lode	0.07	3.0	1,391	db, ox	As	3,850
Orleans Mine	0.20	6.1	1,032	db	As	268
Pennsylvania Mine	1.09	6.1	66,516	db, gd	Cd, Pb	24
Prescott Hill Mine	1.17	12.2	23,548	db	As, Cd, Pb	10,250
Rowe Mine - A	0.18	4.6	8,334	db	N/A	N/A
Rowe Mine - B	0.55	9.1	50,766	db, gd	As, Cd	226
Sebastopol Mine	0.22	12.2	6,598	db, ox	As	2,460
Woodbury Mine	0.13	4.6	1,254	db, ox	As	8,210
WYOD	0.65	9.1	60,017	db, gd, ox	As	224

Data from MFG (2009), WME (2013a)

* Rock types: db, diabase; gd, granodiorite; ox, iron oxides and clays

** Criteria for As and Cd: California Human Health Screening Levels (CEPA 2005); Criterion for Pb: USEPA Regional Screening Level (USEPA 2009)

*** Exposure Point Concentration = 95 percentile for locations with n>6; maximum for n<6   N/A - data not available

Table 4. Arsenic-bearing minerals at Empire Mine State Historic Park (data from Burlak 2012)

Mineral	Alternative name	Formula	Arsenic (wt%)		
			range	average	median
Primary minerals					
Arsenopyrite		$FeAs_{(1-x)}S_{(1+x)}$	40.1 to 4.2	42.1, n = 217	
Pyrite	Arsenian pyrite	$Fe(S,As)_2$	<0.04 to 5.1	1.28, n = 413*	0.51, n = 524
Cobaltite		$(Co, Fe)AsS$	44.0 to 44.8	44.4, n = 18	
Secondary minerals					
Scorodite		$FeAsO_4 \cdot 2H_2O$	(stoichiometric)	32.46	
Kaňkite		$FeAsO_4 \cdot 3.5H_2O$	(stoichiometric)	29.06	
Goethite and ferrihydrite	Hydrous ferric oxide (HFO)	$FeOOH$ and $5Fe_2O_3 \cdot 9H_2O$	<0.05 to 17.9	6.49, n = 678*	5.41, n = 726
Nanoscale mixture (?) of ferrihydrite and scorodite	Hydrous ferric arsenate (HFA)	$[xFeAsO_4 \cdot 4\text{-}7H_2O + y5Fe_2O_3 \cdot 9H_2O]$	18.0 to 48.1	32.9, n = 316	32.5, n = 316
Jarosite		$KFe_3(SO_4,HAsO_4)(OH)_6$	0.3 to 0.4	0.35	
Unknown Ca-Fe-arsenate (1)	G1 (Fig. 13)	$Ca_3Fe_{24}(AsO_4)_{10}(OH)_{48} \cdot 9H_2O$	18.8 to 22.7	20.6, n = 13	
Unknown Ca-Fe-arsenate (2)	P2 (Fig. 13)	$Ca_3Fe_7(AsO_4)_6(OH)_9 \cdot 7H_2O$		30.2, n = 1	

n = number of electron microprobe observations

* data points with no detectable arsenic not included in averages;

arsenic detection limits: pyrite, <0.04 wt% (111 points): HFO, <0.15 wt% (48 points)

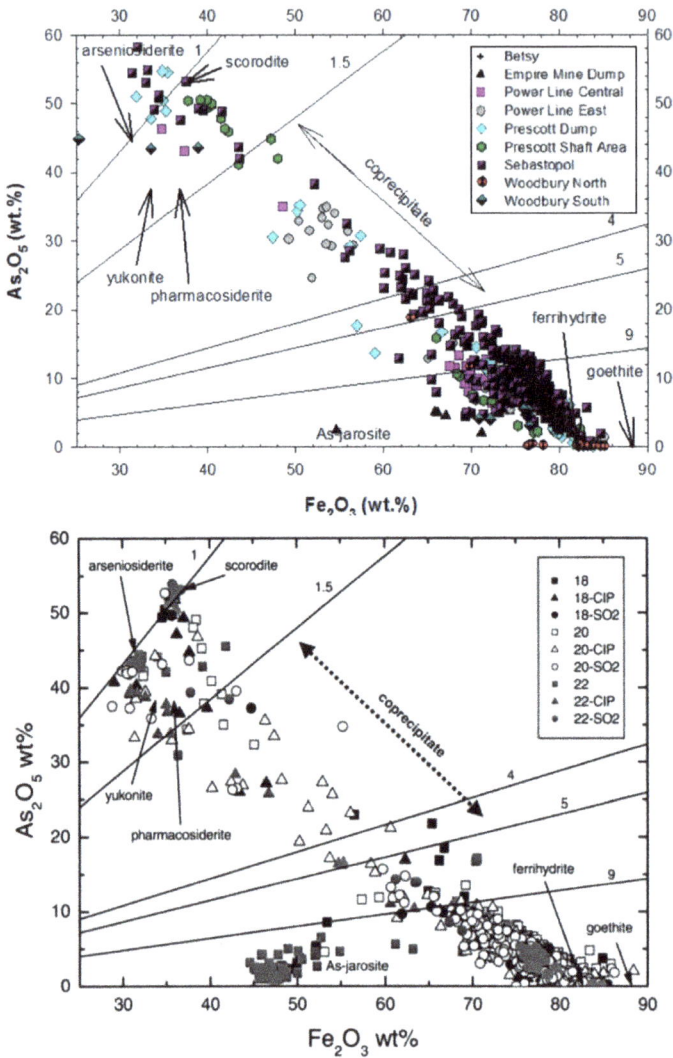

Figure 11. Plots of Fe_2O_3 vs. As_2O_5 data from EMPA of hydrous iron oxide and arsenate weathering products of sulfide oxidation. (a) data from Empire Mine SHP (Burlak 2012), (b) data from Ketza River mine, Canada (Paktunc et al. 2004). Diagonal lines are Fe/As molar ratios; arrows point to end-member mineral compositions. [Used by permission of Elsevier, from Paktunc et al. (2004), Geochim Cosmochim Acta, Vol. 68, Fig. 3, p. 976.]

weathered, As-rich mill tailings from the Au deposit at Ketza River, Yukon, Canada (Paktunc et al. 2004). Material along this compositional trend with values of the Fe:As molar ratio greater than 3 is considered to be ferrihydrite with less than about 18 wt% As, and materials with molar Fe:As less than 3 are considered to be hydrous ferric arsenate (HFA) with about 18 to 48 wt% As, following the terminology of Walker et al. (2009). Paktunc et al. (2008) suggested that HFA is not a well-defined mineral but rather may consist of a nano-scale mixture of ferrihydrite and amorphous or poorly crystalline scorodite.

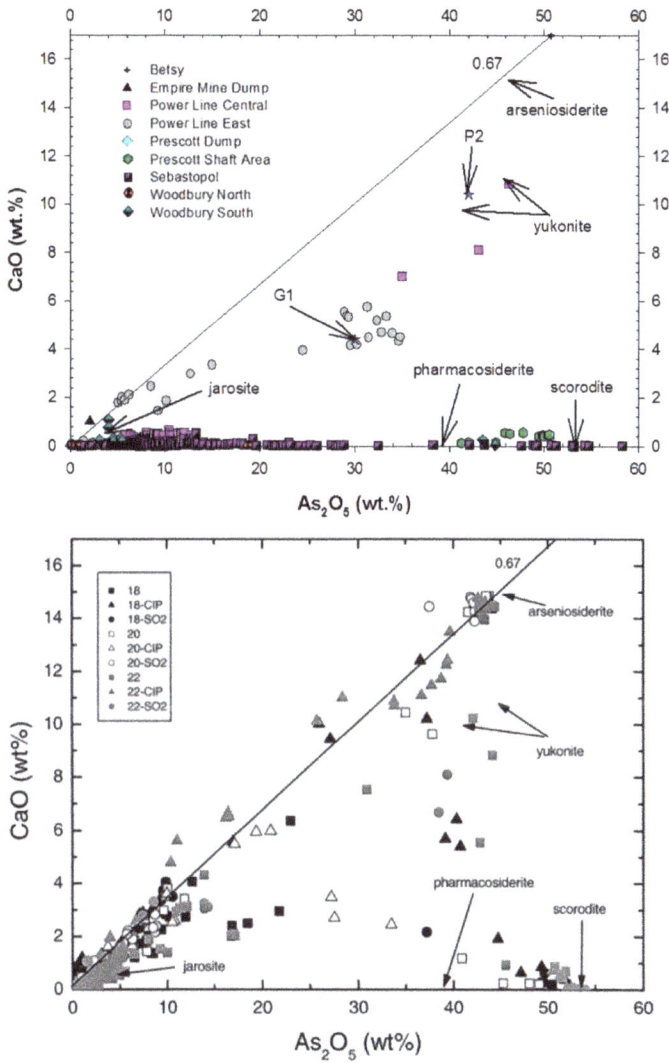

Figure 12. Plots of As$_2$O$_5$ vs. CaO data from EMPA of hydrous iron oxide and arsenate weathering products of sulfide oxidation. (a) data from Empire Mine SHP (Burlak 2012), (b) data from Ketza River mine, Canada (Paktunc et al. 2004). Yukonite has several proposed formulas (Garavelli et al. 2009); yukonite arrows point to two of those compositions. The 0.67 diagonal line is the Ca/As molar ratio for arseniosiderite. G1 and P2 are possible unknown Ca-Fe-arsenates (see text and Burlak 2012). Other arrows point to end-member mineral compositions. [Used by permission of Elsevier, from Paktunc et al. (2004), Geochim Cosmochim Acta, Vol. 68, Fig. 4, p. 976.]

Previous work at other locations (e.g., Nova Scotia, Canada) has shown that Ca-arsenates, Ca-Fe-arsenates (such as yukonite and arseniosiderite), and K-Fe-arsenate (e.g., pharmaco-siderite) are more bioavailable than other As-bearing Fe oxides found in mine waste (Walker et al. 2009; Meunier et al. 2010; DeSisto et al. 2011). At Empire Mine SHP, neither yukonite, arseniosiderite, nor pharmacosiderite were found using EMPA, however two possible unknown hydrous Ca-Fe-arsenates were described (Fig. 12, Table 4; Burlak 2012). Preliminary data for bulk As XAFS required the addition of synthetic Ca-arsenate or Ca-Fe-arsenate (yukonite or arseniosiderite) to fit spectra for 6 of 19 samples, but the modeled relative abundance of these Ca-As-bearing phases was low (6 to 14%; A. Foster, U.S. Geological Survey, written commun., 2014).

Secondary weathering products (HFO and HFA) were observed most commonly on the rims of the primary sulfide minerals, arsenian pyrite and arsenopyrite (Fig. 13). In some cases, the secondary minerals were observed replacing the cores of sulfide grains (Figs. 13b,d). The molar As:Fe of HFA was typically 0.5 to 1.0, less than or equal to the As:Fe of its arsenopyrite precursor. The molar As:Fe of HFO was typically higher than the As:Fe of the arsenian pyrite with which it was spatially associated. A linear least-squares regression (albeit with low n) for paired data from the same mineral grain indicates that As:Fe in HFO was a factor of two higher than As:Fe in pyrite (Fig. 14), whereas the median concentration of As in HFO (5.41 wt%, Table 4) was about 10-fold higher than the median As in pyrite (0.51 wt%, Table 4). The higher concentrations of As in HFO relative to precursor arsenian pyrite may be caused by multiple factors, including preferential weathering of As-rich regions in zoned arsenian pyrite, or weathering of nearby arsenopyrite and incorporation of arsenate in HFO and HFA by adsorption or co-precipitation.

A QEMSCAN® (Quantitative Evaluation of Minerals by SCANning electron microscopy) image of weathered waste rock from Empire Mine SHP (Fig. 15) shows arsenopyrite grains (red) with associated weathering products (light green; "Element_Map-As" in legend) corresponding to HFA. Pyrite (yellow) and its weathering product, HFO (dark brown; "Fe Oxides(G07)" in legend) are also shown. The QEMSCAN® unit was programmed to distinguish between HFA and HFO based on As counts corresponding to about 6 wt% As (Burlak 2012).

Microbeam synchrotron XAS (μXAS; see Foster and Kim 2014, this volume) images of the same thin-section of waste rock as in Figure 15 are shown in Figure 16. Masking different regions of the field of view based on relative counts of As vs. Fe shows the distribution of high-As weathering products (HFA) rimming grains of arsenopyrite. Arsenian pyrite and its weathering products (HFO) are grouped together in Figure 16 because the As:Fe ratio is not sufficiently different; however these phases can be distinguished readily in the QEMSCAN® image (Fig. 15) based on S counts (Burlak 2012).

Bioavailability and bioaccessibilty of As in soils and waste rock from Empire Mine SHP

Bioavailability of As in waste rock and soils from Empire Mine SHP were evaluated by *in vivo* testing using juvenile swine (Casteel and Naught 2011), which are considered to be a good physiological model for gastrointestinal absorption in children (Casteel et al. 1996). Mine waste embedded in a ball of dough consisting of moistened feed (typically about 5 g) was fed to the swine twice daily for 14 days. A soluble form of As (Na arsenate) was used for comparison, and some animals received feed with no added As as a control. Six EMSHP samples tested during 2011 ranged in total As from 302 to 12,041 mg kg^{-1} (Casteel and Naught 2011). Results for these six samples for all days tested indicate that relative bioavailability (RBA) of As ranged from 4.0% (90% confidence interval 3.3 to 4.6%) to 23.7% (10.9 to 36.5%). Results for another six samples tested during 2012-13 are preliminary (S. Casteel, unpub. data).

Bioaccessibility was evaluated in 25 samples of soil and waste rock by using several *in vitro* gastrointestinal (IVG) methods, including two versions of the Ohio State University IVG method and the Relative Bioavailability Leaching Procedure (Basta et al. 2014). Based on bulk Fe and As K-edge XAS (see Foster and Kim 2014, this volume), the predominant As mineralogy in the soils and wastes was As(V)-ferrihydrite, and As(V and III) associated with Al oxyhydroxide and gibbsite, with lesser amounts of arsenopyrite, arsenian pyrite, scorodite, and As-bearing jarosite (Foster et al. 2014). Soil As was fractionated into 5 pools using a sequential extraction procedure (SEP): F1, non-specifically sorbed; F2, specifically sorbed; F3, amor-

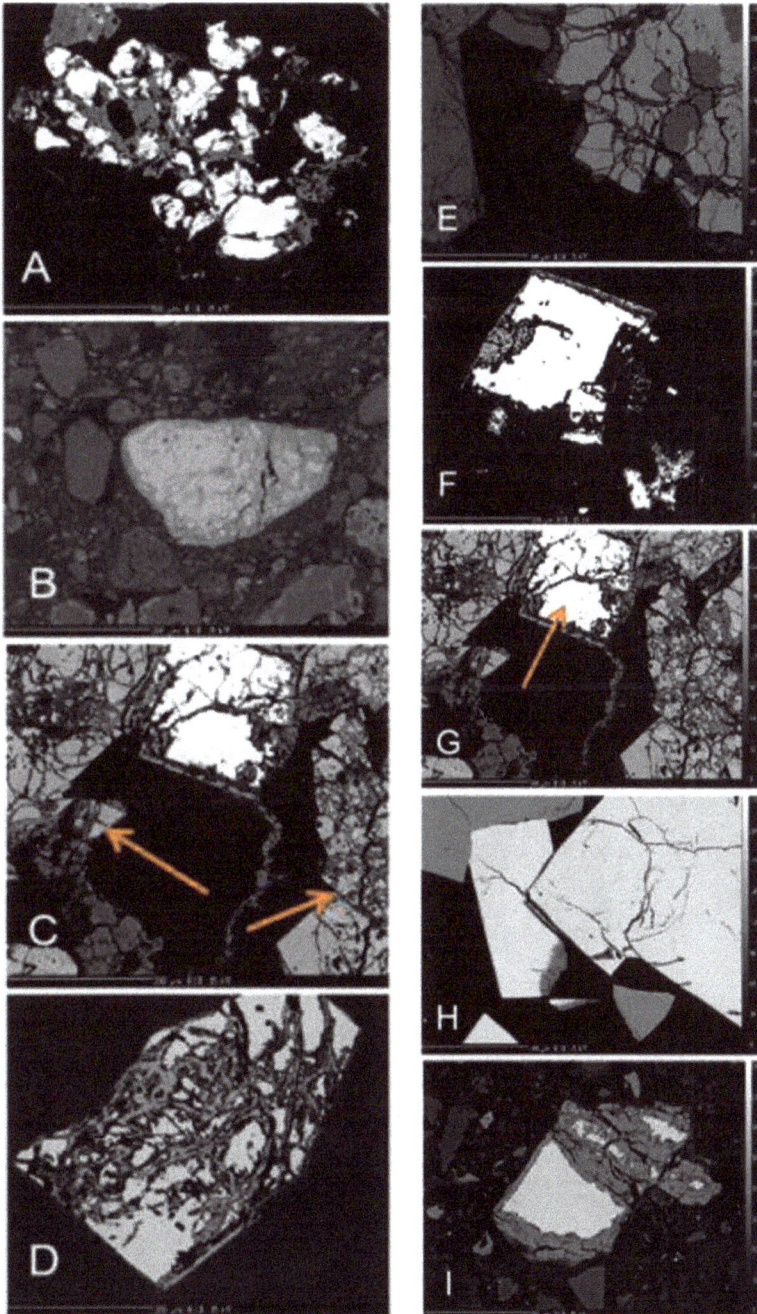

Figure 13. Backscatter electron (BSE) images from Burlak (2012) of polished sections from waste rock at Empire Mine SHP. Pyrite/HFO grains: (A) Sebastopol, (B) Empire Mine Dump , (C) Prescott Shaft area, (D) Empire Mine Dump soil; Arsenopyrite/HFA grains: (E) Power Line East, (F) and (G) Prescott Shaft area, (H) Betsy, (I) Prescott Shaft area soil. All BSE images illustrate rim weathering textures; images (B) and (D) also represent core weathering.

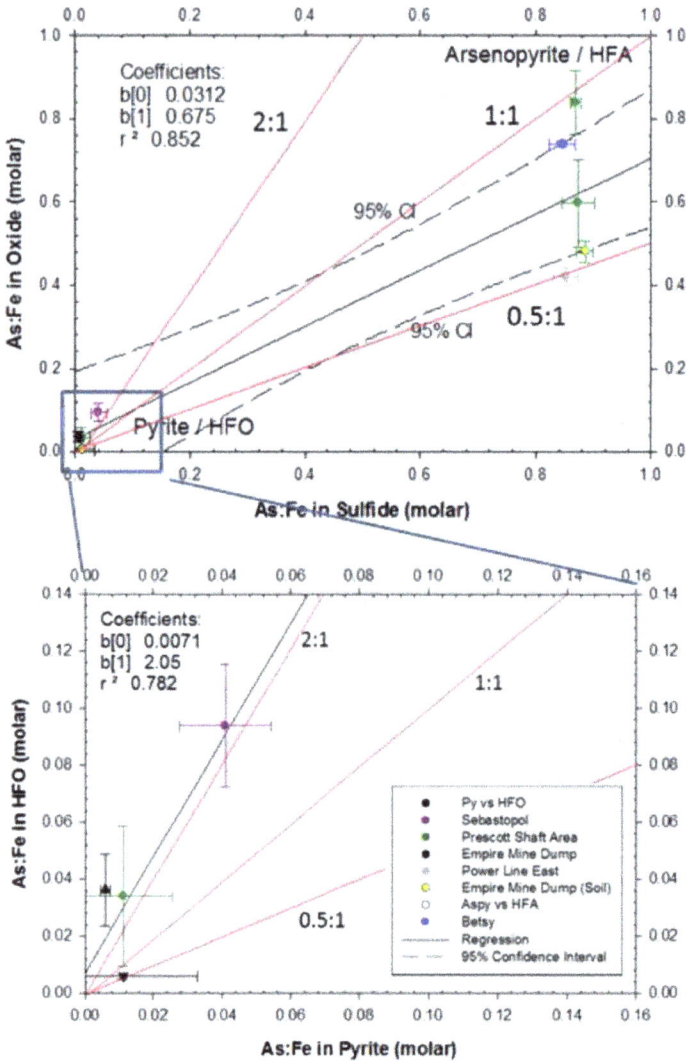

Figure 14. Molar bidirectional plots (from Burlak 2012) of paired EMPA data for primary sulfide minerals and weathering products: pyrite/HFO and arsenopyrite/HFA. Diagonal lines (pink in color version) indicate ratio of As:Fe in oxides relative to As:Fe in sulfides. Solid black lines represent linear least-squares regression fits. Dashed lines in top plot represent 95% confidence interval. Regression coefficients: b[0] is intercept, b[1] is slope. Triangles represent core weathering of grains and circles represent rimming. Regression line including all data (top) has slope of 0.675. Regression line including only pyrite/HFO data has steeper slope of 2.05.

phous and poorly crystalline oxides of Fe and Al; F4, well-crystallized oxides of Fe and Al; and F5, residual (Basta et al. 2014). Most As was associated with reactive Fe and Al oxide fractions (F3). Strong linear relationships were found between bioaccessible As (IVG) and RBA As (*in vivo*, n=12). However, all *in vitro* methods except one under-predicted RBA As for several high-Fe samples of soil and waste rock from EMSHP. Comparison of *in vivo* RBA As and SEP results show F1 + F2 < RBA As < F3. The sum of SEP F1+F2+F3 provided a conservative

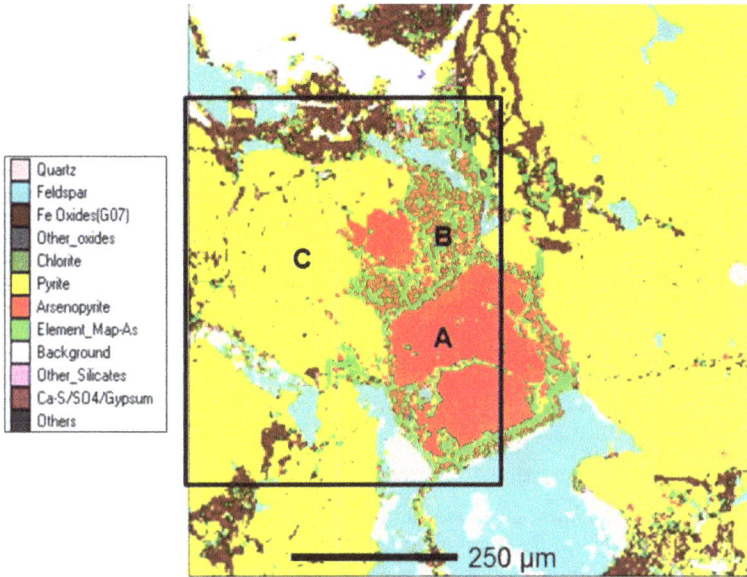

Figure 15. (*for color see Plate 12*) QEMSCAN® image of a polished thin section of waste-rock from the Prescott Shaft area, Empire Mine SHP. (A) arsenopyrite; (B) hydrous ferric arsenate (HFA); (C) arsenian pyrite and hydrous ferric oxide (HFO). From Burlak (2012).

Figure 16. (*for color see Plate 16*) μXAS images of a thin section of waste rock from Prescott Shaft (same view as QEMSCAN® image in Fig. 16). Linear trends are non-quantitative point counts for As vs. Fe collected from Beamline 2-3 at the Stanford Synchrotron Radiation Laboratory. Slopes are for the upper and lower bound of each encircled linear trend. The slopes for pyrite and HFO could not be completely separated, so they are combined. From Burlak (2012).

estimate of RBA As (Basta et al. 2014). Additional IVG tests and analysis of total As on eleven size fractions ranging from ≥2,830 μm to ≤20 μm revealed an inverse correlation between particle size and mass of As released, a relationship best explained by normalization to both initial (<2,830 μm) As concentration and surface area (Buckendorf and Kim 2014; Kim et al. 2014)

Significant ($p < 0.05$) negative correlations (Pearson) between the abundance of arsenopyrite and/or arsenian pyrite and *in vitro* (n = 25) and *in vivo* (n = 12) datasets were found for XRD, QEMSCAN®, and bulk XAS datasets (both As and Fe) (Foster et al. 2014). Significant positive correlations with *in vivo* and *in vitro* datasets were found for the concentration of As in ferrihydrite (by EMPA) and the abundance of HFO (by powder XRD). Positive correlations (significance not yet determined) were also found with *in vivo* and *in vitro* datasets and the abundance of Fe (hydr)oxides (by bulk Fe XAS), and relative abundance of As(V and III) associated with Al oxyhydroxide, gibbsite, or kaolinite (by bulk As XAS). The quantity of ferrihydrite (FH) and/or As concentration in FH are two lab-measureable sample parameters that correlate strongly with *in vitro* bioaccessibility and/or *in vivo* bioavailability data from Empire Mine SHP samples (Foster et al. 2014).

Passive water treatment at Magenta Drain

The closure of the mines and cessation of pumping water from the underground mine workings in 1957 has led to flooding of most of the underground mine workings. The mine workings can be considered to comprise a massive mine pool with an identified underground "spill point" that contributes in maintaining a relatively constant mine pool surface elevation (Gusek et al. 2011). The Magenta Drain is a drainage adit that is connected to the mine workings and it discharges near-neutral pH mining influenced water (MIW) with dissolved concentrations of Fe, As, and Mn that are the principal contaminants of environmental concern (MFG 2007). A passive water treatment system was designed to treat the effluent from the Magenta Drain (Gusek et al. 2011).

The passive water-treatment consists of a settling pond and two aerobic wetlands (Figs. 17, 18). Water is pumped uphill from the Magenta Drain to the pond. The treatment system

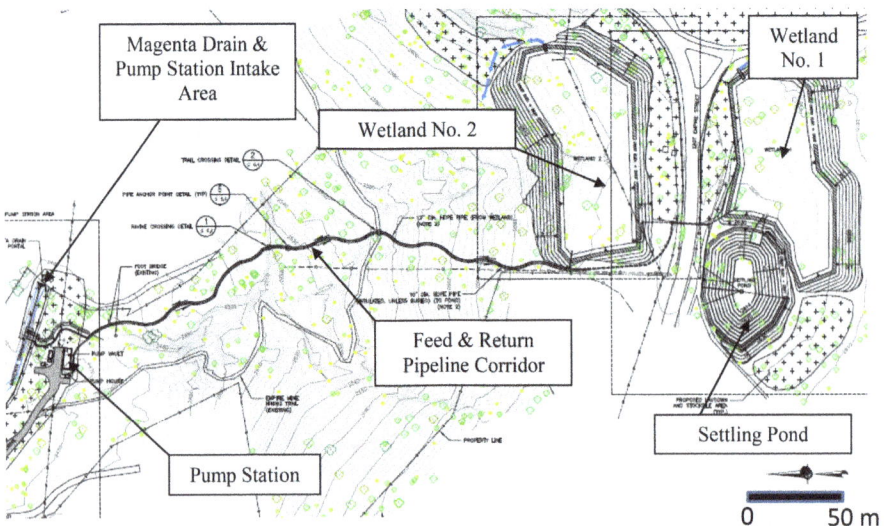

Figure 17. Site layout plan for passive wetland treatment system for Magenta drain effluent. Modified from Golder Construction Services (2011). [Used by permission of the American Society of Mining and Reclamation, from Gusek et al. (2011), *Proceedings, Natl Mtg Am Soc Min Reclam 2011*, Fig. 2, p. 235.]

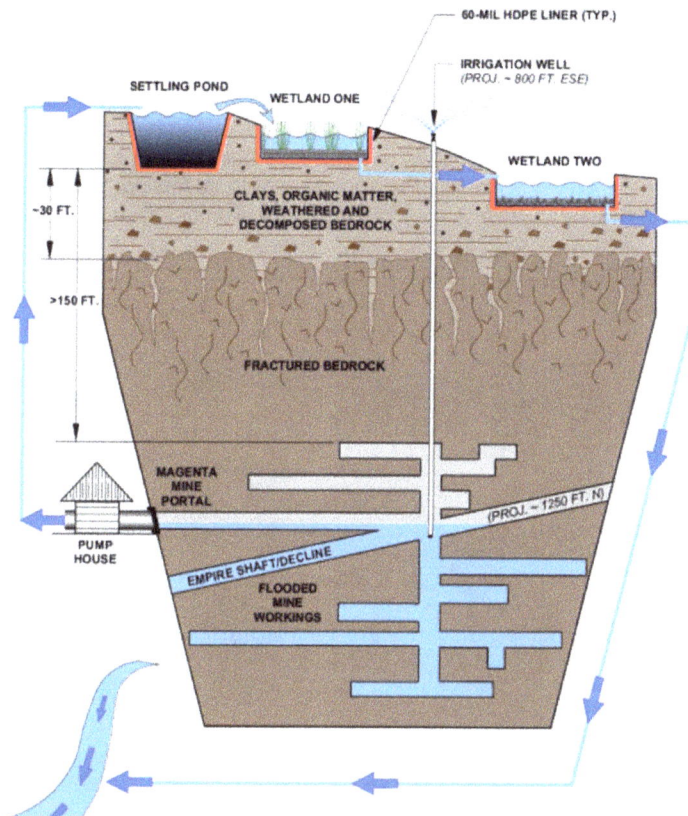

Figure 18. Schematic cross-section of Magenta Drain passive mine water treatment system. [Used by permission of the American Society of Mining and Reclamation, from Gusek et al. (2011), *Proc Natl Mtg Am Soc Min Reclam 2011*, Fig. 4, p. 243.]

was constructed at higher elevation than the effluent because limitations on available land within the park boundary precluded a gravity-fed approach (Gusek et al. 2011). The goal of the water treatment system is to meet regulatory limits for several constituents: total suspended solids, settleable solids, pH, turbidity, color, and total recoverable Al, Sb, As, Ba, Cd, Cr, Co, Cu, Fe, Pb, Mn, Hg, Ni, Th, V, and Zn (Gusek et al. 2011). The system was designed for a peak flow of 3.7 m^3 min^{-1} (1,200 gal·min^{-1}), which corresponds to the flow peak for a 25-year recurrence interval storm event (Gusek et al. 2011).

The Magenta Drain treatment system is based on an aerobic wetland design. During testing, the ratio Fe:As in Magenta Drain MIW ranged from 88:1 to 184:1. These ratios greatly exceed the minimum Fe:As of 3:1 to 5:1 recommended by Langmuir et al. (1999) to ensure As removal by adsorption to hydrous ferric oxides (Gusek et al. 2011) that form in the wetlands by oxidation of Fe(II) to Fe(III). The final design includes a wetland water surface of 8,225 m^2 (88,000 ft^2). The system is designed to optimize removal of Fe and As in the Settling Pond and Wetland 1, and Mn in Wetland 2.

The passive water-treatment system was constructed beginning in April 2011 and began operation in November 2011. The passive water treatment system is currently in the second year of a three-year maturation period, during which interim effluent limitations are required by the Regional Water Quality Control Board through a Time Schedule Order. After the three-year maturation period, final effluent limits will apply. Box plots showing concentrations of Fe, As, and Mn at various stages of treatment during October 2012 through December 2013 are shown in Figure 19, along with the final effluent limits (dashed lines). The plots indicate that there is substantial reduction in the concentration of As and Fe both within the Pond and in Wetland 1. Manganese (Mn) is effectively removed in Wetlands 1 and 2. With the exception of a few outliers (about 10% of Fe data and about 20% of Mn), all three contaminants of concern were being treated during the period shown in Figure 19 to within the final regulatory criteria.

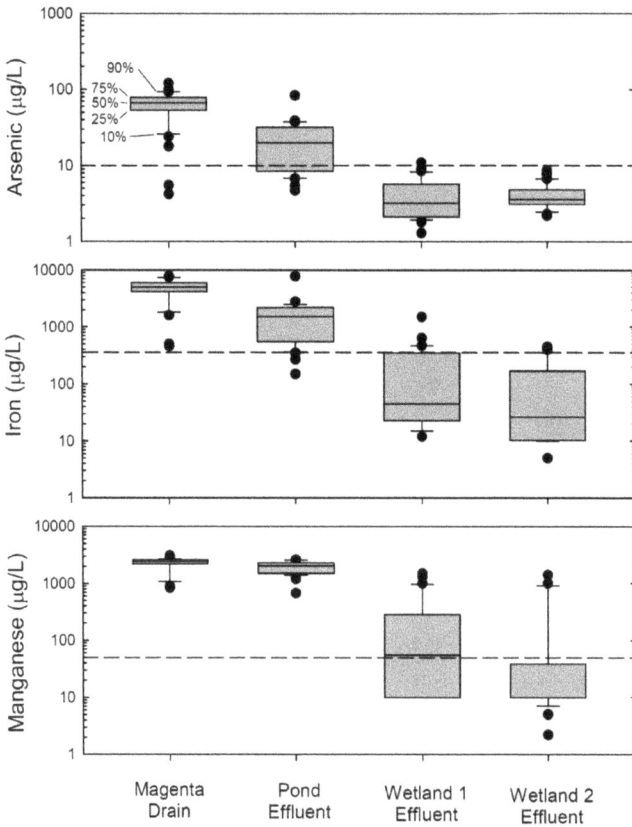

Figure 19. Box plots (logarithmic scale) showing concentrations of arsenic, iron, and manganese in effluents from Magenta Drain, Pond, Wetland 1, and Wetland 2 during October 2012 through December 2013. Dashed lines represent final effluent limits.

FIELD TRIP STOPS

1. Stamp mill and mine yard

The Mine Yard at Empire Mine State Historic Park contains many well-preserved features from when the mine was active (prior to 1957). The machine shop and carpentry shops are preserved much as they were during active operations. There are many examples of machinery (pumps, etc.) preserved. The foundation of the stamp mill is preserved. An intact 10-stamp mill (from another location) is on display.

An exhibit allows park visitors to observe and descend a short distance in the main inclined shaft. At the time of mine closure in 1956, the greatest depth of 11,007 feet (3,355 m) on the incline was reached in the North Star shaft, almost one mile (1.6 km) vertically below the surface (McQuiston 1986).

2. Empire Mine waste dump

The Empire mine waste rock pile is the largest in area in the State Historic Park, covering nearly 2 hectares, although its estimated volume (14,450 m^3) is only the sixth largest (Table 3). A variety of rock types are present, including metamorphic rocks, diabase, and granodiorite. Rocks with visible iron stain typically contain sulfide minerals including arsenopyrite and arsenian pyrite.

3. Cyanide plant

The cyanide plant was constructed in 1910, with a capacity of 150 tons per day (McQuiston 1986). There were numerous expansions and modifications that occurred over time. Following closure of the Empire Mine in 1956, process tanks and equipment were removed and the cyanide plant building was demolished; however, the concrete foundations remain. Some historical photographs and a map of the preserved foundation are shown in Figure 20.

Residual deposits of tailings, other mine and mill related materials, and demolition debris that were present within the foundations were removed during 2013 (WME 2013b).

4. Prescott shaft and waste pile

After an initial discovery in 1855, the Prescott Hill quartz mine was developed in 1885 by John L. Smith (Hartwell 1885). At that time, the claim included hoisting works, inclined shaft, a stone powder house, a tunnel, and 16 shafts. The incline shaft was 60 m (200 ft) deep. It intersected the Orleans tunnel at a depth of 30 m (100 ft). The Orleans tunnel continues on to the Betsey (sic) Quartz Mine (Hartwell 1885).

The Prescott Hill waste-rock pile is one of three in the Empire Mine SHP that are 12.2 m in height (Table 3). A sample of fine-grained material taken from the waste rock had an As concentration of more than 10,000 mg kg^{-1} (Table 3). Data from EMPA indicate that nearly all of the pyrite had As concentration of 0.15 to 1.0 wt% and nearly all of the HFO had As of more than 1.0 wt% (Fig. 10). This is the location of the hand sample used by Burlak (2012) to demonstrate the use of QEMSCAN® and μ-EXAFS to document the texture of oxidation products of arsenopyrite (HFO) and arsenian pyrite (HFA) (Figs. 15, 16).

5. Visitor Center

A highlight of the Visitor Center is the 3D scale wire model of the underground mine workings beneath the Empire Mine SHP (and other parts of Grass Valley). In all, there are 592 km (367 miles) of underground mine workings as part of the Empire-Star Mines Co. properties. The model was constructed for use by the mining company during operations.

6. Superintendent's office and other rooms

Upstairs in the building adjacent to the parking lot, there are several office rooms preserved in condition similar to how they were during mine operations. There are the Superintendent's office, the office of the Chief Geologist, a map room, and an assay room. The map was considered highly confidential during mine operations and access to the map room was carefully controlled.

7. Magenta Drain constructed wetlands

It is a short walk down East Empire Street to the Magenta Drain passive water-treatment system. A plan view (Fig. 17) shows the location of the Settling Pond, Wetland 1, and Wetland 2. A schematic cross-section (Fig. 18) shows the relative elevation of the Magenta mine

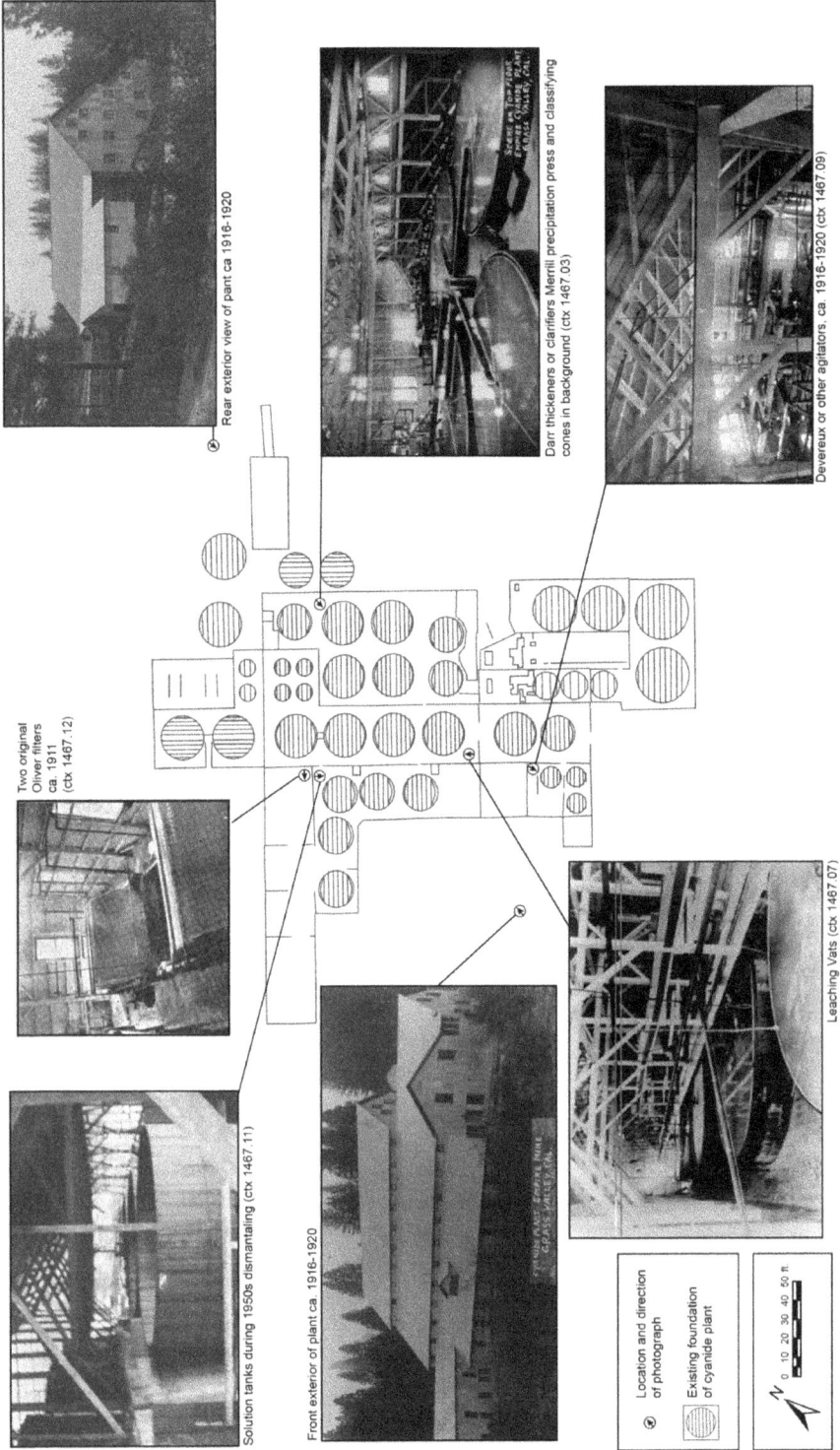

Rear exterior view of pant ca 1916-1920

Darr thickeners or clarifiers Merrill precipitation press and classifying cones in background (ctx 1467.03)

Devereux or other agitators, ca. 1916-1920 (ctx 1467.09)

Two original Oliver filters ca. 1911 (ctx 1467.12)

Solution tanks during 1950s dismantaling (ctx 1467.11)

Front exterior of plant ca. 1916-1920

Leaching Vats (ctx 1467.07)

Location and direction of photograph

Existing foundation of cyanide plant

N   0   10   20   30   40   50 ft

Figure 20. Historical views of cyanide plant at Empire Mine State Historic Park, with map of preserved foundation. Ctx numbers refer to State Park archive.

portal and other mine workings. Box plots showing the performance of the system during October 2012 through December 2013 (Fig. 19) indicate that there is substantial reduction in the concentration of As and Fe both within the Settling Pond and in Wetland 1, whereas Mn is effectively removed in Wetlands 1 and 2.

CONCLUDING REMARKS

Empire Mine State Historic Park (EMSHP) includes many historical features and cultural resources that are accessible to the public. These resources are used, in part, to educate visitors about EMSHP's role in California's rich mining history. In addition, the park's trails are open and accessible to the public for recreational purposes. The legacy of As contamination at EMSHP has resulted in the need for remediation in several areas to reduce risk of As exposure to the public and to park workers. Ongoing research at EMSHP is making Empire Mine a useful laboratory for improving the understanding of As mineralogy, bioavailability, and bioaccessibility in mining environments.

ACKNOWLEDGMENTS

The authors wish to thank the U.S. Environmental Protection Agency for funding through the Brownfields Program. Rick Fears (California DTSC) assisted with GIS and drafting of maps. Review comments by Roger Ashley, J.K. Böhlke, Andrea Foster, Charles Kaehler, Keith Long, Kaye Savage, and Kelly Sexsmith improved the manuscript. Reference to any commercial product is for information only and does not constitute endorsement by the U.S. Geological Survey.

REFERENCES

Alexander Law Group, LLC (2014) Environmental pollution and property damage: overcoming the defenses of consent and custom and practice – anatomy of a $2,000,000 property damage settlement. Part ii. *http://www.alexanderinjury.com/library-toxic-27b/* (accessed July 2014)

Allgood GM (1990) Geology and operations at the Jamestown Mine, Sonora Mining Corporation, California. *In*: Yosemite and the Mother Lode Gold Belt: Geology, Tectonics, and the Evolution of Hydrothermal Fluids in the Sierra Nevada of California. Landefeld LA, Snow GG (eds) AAPG Pacific Section, p 147-154

Alpers CN, Burlak TL, Foster AL, Basta NT, Mitchell VL (2012) Arsenic and old gold mines: mineralogy, speciation, and bioaccessibility. Goldschmidt 2012 Abstracts. Mineral Mag 76(6):1418

Ashley RP (2002) Geoenvironmental model for low-sulfide gold-quartz vein deposits. US Geol Survey Open-file Report 02-195K, p 176-195, *http://pubs.usgs.gov/of/2002/of02-195/OF02-195K.pdf* (accessed July 2014)

Ashley RP, Savage KS (2001) Analytical data for waters of the Harvard open pit, Jamestown Mine, Tuolumne County, California, March 1998-September 1999. US Geol Survey Open-File Report OF-01-74, 13 p. plus XLS file.

Averill CV (1946) Placer mining for gold in California. California State Division of Mines and Geology Bulletin 135, 336 p

Basta NT, Whitacre S, Meyers P, Mitchell VL, Alpers CN, Foster AL, Casteel SW, Kim CS (2014) Using in vitro gastrointestinal and sequential extraction methods to characterize site-specific arsenic bioavailability. Goldschmidt 2014 Abstracts, #2172

Bean EF (compiler) (1867) History and Directory of Nevada County. Daily Gazette Book and Job Office, Nevada (City, California) 424 p

Berger BR (1986) Descriptive model of low-sulfide Au-quartz veins. *In*: Mineral deposit models. Cox DP, Singer DA (eds) US Geol Survey Bulletin 1693:239

Bohakel CA (1980) A brief history of the Empire Mine at Grass Valley. Empire Mine Park Association, Grass Valley, California, 34 p

Böhlke JK (1982) Orogenic (metamorphic-hosted) gold-quartz viens. *In*: Characteristics of Mineral Deposit Occurrences. Erickson RL (ed) US Geol Surv Open-File Report 82-795:70-76

Böhlke JKFP (1986) Local wall rock control of alteration and mineralization reactions along discordant gold quartz veins, Alleghany, California. PhD thesis, Univ Calif Berkeley, 308 p

Böhlke JK (1988) Carbonate-sulfide equilibria and "stratabound" disseminated epigenetic gold mineralization: a proposal based on examples from Alleghany, California, U.S.A. Appl Geochem 3:499-516

Böhlke JK (1989) Comparison of metasomatic reactions between a common CO_2-rich vein fluid and diverse wall rocks: Intensive variables, mass transfers, and Au mineralization at Alleghany, California. Econ Geol 84:291-327

Böhlke JK (1999) Mother Lode gold. *In*: Classic Cordilleran Concepts: A View from California. Moores EM, Sloan D, Stout DL (eds) Geol Soc Am Spec Paper 338:55-67

Böhlke JK, Kistler RW (1986) Rb-Sr, K-Ar, and stable isotope evidence for the ages and sources of fluid components of gold-bearing quartz veins in the northern Sierra Nevada foothills metamorphic belt, California. Econ Geol 81:296-322

Borch RS, Hastings LL, Tingle TN (1994a) A simple gastrointestinal extraction procedure for evaluating the human health risks posed by ingestion of mine waste and contaminated soil. Geol Soc Am Abstracts with Programs 26:433

Borch RS, Hastings LL, Tingle TN, Verosub KL (1994b) Speciation and in vitro gastrointestinal extractability of arsenic in two California gold mine tailings. Eos 75:190

Bowell RJ (1994) Sorption of arsenic by iron oxides and oxyhydroxides in soils. Appl Geochem 9:279-286

Bowen OE, Crippen RA Jr. (1948) Geologic maps and notes along Highway 49. *In*: The Mother Lode County (centennial edition), Geologic Guidebook along Highway 49 – Sierran Gold Belt, California. Jenkins OP (ed) Calif Div Mines Bull 141:35-86

Buckendorf L, Kim CS (2014) Relationships between particle size, arsenic concentration, surface area, and bioaccessibility of mine tailings from the Empire Mine, CA. Goldschmidt 2014 Abstracts, #3209

Burlak T (2012) The mineralogical fate of arsenic during weathering of sulfides in gold-quartz veins: a microbeam analytical study. MSc thesis, California State University, Sacramento, 142 p *https://www.yumpu.com/en/document/view/16475309/the-mineralogical-fate-of-arsenic-during-weathering-* (accessed July 2014)

Casteel SW, Cowart RP, Weis CP, Henningsen GM, Hoffman E, Brattin WJ, Starost MF, Payne JT, Stockham SL, Becker SV, Turk JR (1996) A swine model for determining the bioavailability of lead from contaminated media. *In*: Advances in Swine in Biomedical Research, Vol. 2. Tumbleson LB, Schook ME (eds) Plenum Press, New York, p 637-646

Casteel SW, Naught L (2011) Relative bioavailability of arsenic for California DTSC soil study. Univ. of Missouri, Prepared for Calif Dept Toxic Substances Control, Aug 25, 2011. *http://www.envirostor.dtsc. ca.gov/public/profile_report.asp?global_id=29100003* (accessed July 2014)

CEPA (2005) Use of California Human Health Screening Levels (CHHSLs) in Evaluation of Contaminated Properties. California Environmental Protection Agency January 2005, *http://www.calepa.ca.gov/ Brownfields/documents/2005/CHHSLsGuide.pdf* (accessed July 2014)

CDPR (2014a) Empire Mine State Historic Park. California Department of Parks and Recreation, *http://parks. ca.gov/?page_id=499* (accessed July 2014)

CDPR (2014b) About the Park. California Department of Parks and Recreation, *http://parks.ca.gov/?page_ id=21647* (accessed July 2014)

CDPR (2014c) The Empire's Success. California Department of Parks and Recreation, *http://parks. ca.gov/?page_id=21648* (accessed July 2014)

Chaffee MA, Sutley SJ (1994) Analytical results, mineralogical data, and distributions of anomalies for elements and minerals in three Mother Lode-type gold deposits, Hodson mining district, Calaveras County, California. US Geol Surv Open-File Report 94-640, 216 p

Clark WB (1970) Gold districts of California. Calif Div Mines Geol Bull 193, 186 p

Clark WB (1985) Gold districts of California: An update. Calif Geol 38:3-4

Craig JR, Rimstidt JD (1998) Gold production history of the United States. Ore Geol Rev 13:407-464

CVRWQCB (2012) Mining — Region 5 Success Stories: Empire Mine State Historic Park, Red Dirt Pile, Nevada County. Central Valley Regional Water Quality Control Board, *http://www.waterboards.ca.gov/ centralvalley/water_issues/mining/region5_success_stories/empire_mine_state_park/index.shtml.* (accessed July 2014)

DeSisto SL, Jamieson HE, Parsons MB (2011) Influence of hardpan layers on arsenic mobility in historical gold mine tailings. Appl Geochem 26: 2004-2018

Dohms PH, Hoagland RD, Allgood GM (1985) Geology of the Jamestown mine area, Mother Lode gold belt, Tuolumne County, California. *In*: Geologic Cross Section across the Southern Mother Lode Belt: Harvard Mine, Royal-Mountain King Mine, Gold Cliff Mine. Slavik G (ed) Geol Soc Nevada Spec Pub 2, p 38-56

DTSC (1992) Preliminary Endangerment Assessment Report, Empire Mine State Historic Park. Dept. of Toxic Substances Control. *http://www.envirostor.dtsc.ca.gov/public/profile_report.asp?global_id=29100003* (accessed July 2014)

DTSC (2005) Sutter Creek approves special building and land use controls at and around Mesa De Oro. Dept. of Toxic Substances Control Fact Sheet, September 2005, *https://dtsc.ca.gov/SiteCleanup/upload/ MesaDeOro_FS_LandUse_0905.pdf* (accessed July 2014)

DTSC (2006) Actions Addressing Mining Waste to Begin at Empire Mine State Historic Park. Dept. of Toxic Substances Control Fact Sheet, July 2006, 6 p, *http://www.parks.ca.gov/pages/23071/files/ empireminefactsheetjuly212006.pdf* (accessed July 2014)

DTSC (2008) Cleanup Plan for the Osborne Hill Area Trails at the Empire Mine State Historic Park is Available for Review and Comment. Dept. of Toxic Substances Control Fact Sheet, November 2008, 4 p, *https://dtsc. ca.gov/SiteCleanup/upload/EmpireMine_FS_dRAW_1108.pdf* (accessed July 2014)

DTSC (2014) Empire Mine State Park (29100003). Dept. of Toxic Substances Control, ENVIROSTOR database, *http://www.envirostor.dtsc.ca.gov/public/profile_report.asp?global_id=29100003* (accessed July 2014)

Ernst WG, Snow CA, Scherer HH (2008a) Contrasting early and late Mesozoic petrotectonic evolution of northern California. Geol Soc Am Bull 120:179-194

Ernst WG, Snow CA, Scherer HH (2008b) Mesozoic transpression, transtension, subduction, and metallogenesis in northern and central California. Terra Nova 20:394-413

Ferguson HG, Gannett RW (1932) Gold quartz veins of the Alleghany district, California. US Geol Surv Prof Paper 172, 139 p

Foster AL, Ashley RP (2002) Characterization of arsenic species in microbial mats from an inactive gold mine. Geochem Explor Environ Anal 2:253-261

Foster AL, Kim CS (2014) Arsenic speciation in solids using X-ray absorption spectroscopy. Rev Mineral Geochem 79:257-369

Foster AL, Alpers, CN, Burlak T, Blum AE, Petersen EU, Basta NT, Whitacre S, Casteel SW, Kim CS, Brown AL (2014) Arsenic chemistry, mineralogy, speciation, and bioavailability/bioaccessibilty in soils and mine waste from the Empire Mine, CA, USA. Goldschmidt 2014 Abstracts, #726

Foster AL, Ashley RP, Rytuba JJ (2011) Arsenic species in weathering mine tailings and biogenic solids at the Lava Cap Mine Superfund site, Nevada City, CA. Geochem Trans 12:1, doi: 10.1186/1467-4866-12-1 *http://www.geochemicaltransactions.com/content/12/1/1*

Foster AL, Brown Jr GE, Tingle TN, Parks GA (1998) Quantitative arsenic speciation in mine tailings using X-ray absorption spectroscopy. Am Mineral 83:553-568

Garavelli A, Pinto D, Vurro F, Mellini M, Viti C, Balić-Žunić T, Ventura GD (2009) Yukonite from the Grotta Della Monaca cave, Sant'Agata Di Esaro, Italy: characterization and comparison with cotype material from the Daulton Mine, Yukon, Canada. Can Mineral 47:39-51

Golder Construction Services (2011) Final Report – Magenta Drain waste water passive system design basis. Prepared for California Dept of Parks and Recreation, Empire Mine State Historical Park, Grass Valley, CA, Agreement C07e0017, Work Order 17-014709-003, March 2011, 51 p

Goldfarb RJ, Baker T, Dube, B, Groves DI, Hart CJR, Gosselin P (2005) Distribution, character, and genesis of gold deposits in metamorphic terranes. *In*: Economic Geology 100th Anniversary Volume, 1905-2005. Hedenquist JW, Thompson JFH, Goldfarb RJ, Richards JP (eds) Society of Economic Geologists, Littleton, Colorado, p 407-450

Goldfarb RJ, Phillips GN, Nokleberg WJ (1998) Tectonic setting of synorogenic gold deposits of the Pacific Rim. Ore Geol Rev 13:185-218

Golub MS, Keen CL, Commisso JF, Salocks CB, Hathaway TR (1999) Arsenic tissue concentration of immature mice one hour after oral exposure to gold mine tailings. Environ Geochem Health 21:199-209

Greenwald J (1995) Arsenic and old mines: As Montanans battle a new gold rush, Californians are dealing with the poisonous legacy of the past. Time Magazine, Sept 25, 1995

Groves DI, Goldfarb RJ, Gebre-Mariam M, Hagemann SG, Robert F (1998) Orogenic gold deposits—A proposed classification in the context of their crustal distribution and relationship to other gold deposit types. Ore Geol Rev 13:7-27

Gusek J, Josselyn L, Agster W, Lofholm S, Millsap D (2011) Process selection & design of a passive treatment system for the Empire Mine State Historic Park, *In*: Proceedings 2011 Natl Mtg Amer Soc Mining Reclam, Barnhisel RI (ed) ASMR, Lexington, Kentucky, p 232-253, *http://www.asmr.us/Publications/ Conference%20Proceedings/2011/0232-Gusek-CO-1.pdf* (accessed July 2014)

Hagwood JJ Jr (1981) The California Debris Commission: A history of the hydraulic mining industry in the western Sierra Nevada of California, and of the governmental agency charged with its regulation. U.S. Army Corps of Engineers, 102 p

Hartwell JG (1885) U.S. Deputy Mineral Surveyor's Report submitted with plat and notes of Mineral Survey No. 2366 (Prescott Hill Quartz Mine). On file at the U.S. BLM office, Sacramento, Calif.

Hilton S, Selverston M (2013) The Empire Mine California Historic District revisited: Evolution of a golden landscape. Proceedings of the Society for California Archaeology, 47th Annual Meeting, 27:205-216, *http://scahome.org/wp-content/uploads/2013/10/Proceedings.27Hilton.pdf* (accessed July 2014)

Johnston WD Jr (1940) The gold quartz veins of Grass Valley, California. U. S. Geol Survey Prof Paper 194:1-101

Kelley R (1959) Gold vs. Grain: The Hydraulic Mining Controversy in California's Sacramento Valley. Arthur H. Clarke, Glendale, California, 327 p

Kelley R (1984) Review of: The California Debris Commission: A history of the hydraulic mining industry in the western Sierra Nevada of California, and of the governmental agency charged with its regulation, by J.J. Hagwood Jr. Public Historian 6(2):106-108

Kim CS, Anthony TL, Buckendorf L, O'Connor KP, Rytuba JJ (2014) Transport, bioaccessibility and risk assessment of fine-grained arsenic-bearing mine tailings. Goldschmidt 2014 Abstracts, #4645

Klein K (2006) Remediation goals for the recreational runner on trails, Empire Mine State Historical Park, Grass Valley, PCA: 11045, Site: 100235-00, memorandum to P. Myers, Sept. 29, 2006, *http://www.envirostor. dtsc.ca.gov/public/final_documents2.asp?global_id=29100003&doc_id=6018901* (accessed July 2014)

Knaebel JB (1931) The veins and crossings of the Grass Valley District, California. Econ Geol 26:375-398

Knopf A (1929) The Mother Lode system of California: US Geol Surv Prof Paper 157, 88 p

Langmuir D, Mahoney J, MacDonald A, Rowson J (1999) Predicting arsenic concentrations in the pore waters of buried uranium mill tailings. Geochim Cosmochim Acta 63:2279-3394

Lindgren W (1895) Characteristic features of California gold-quartz veins. Bull Geol Soc Am 6:221-240

Lindgren W (1896a) The gold quartz veins of Nevada City and Grass Valley districts, California. US Geol Survey Annual Report 17:13-262

Lindgren W (1896b) Nevada City Special Folio. US Geol Survey, Geol Atlas 29

Long KR, DeYoung JH Jr, Ludington SD (1998) Database of significant deposits of gold, silver, copper, lead, and zinc in the United States. US Geol Survey Open-File Report 98-206A, 33 p, *http://geopubs.wr.usgs. gov/open-file/of98-206/of98-206a.pdf* (accessed July 2014)

Mayfield JD, Day H (2000) Ultramafic rocks in the Feather River Belt, northern Sierra Nevada, California. *In*: Field Guide to the Geology and Tectonics of the Northern Sierra Nevada. Brooks ER, Dida LT (eds) Calif Dept Conserv, Div Mines Geol, Spec Pub 122:1-15

McCuaig CT, Kerrich R (1998) P-T-t-deformation-fluid characteristics of lode gold deposits: Evidence from alteration systematics. Ore Geol Rev 12:381-453

McQuiston FW (1986) Gold: The saga of the Empire Mine, 1950–1956. Empire Mine Park Association, Blue Dolphin Press, Inc, Nevada City, Calif. 95 p

Meunier L, Walker SR, Wragg J, Parsons MB, Koch I, Jamieson HE, Reimer KJ (2010) Effects of soil composition and mineralogy on the bioaccessibility of arsenic from tailings and soil in gold mine districts of Nova Scotia. Environ Sci Tech 44:2667-2674

MFG (2007) Final Magenta Drain Assessment, 2007 Work Plan, Empire Mine State Historic Park, July 2007. MFG, Inc, Fort Collins, Colo. Prepared for California Dept. of Parks and Recreation and Newmont USA, Ltd. *http://www.envirostor.dtsc.ca.gov/public/profile_report.asp?global_id=29100003* (accessed July 2014)

MFG (2008a) Historic mine and mill site survey, Empire Mine State Historic Park, March 2008. MFG, Inc, Fort Collins, Colo. Prepared for California Dept. of Parks and Recreation and Newmont USA, Ltd. *http://www. envirostor.dtsc.ca.gov/public/profile_report.asp?global_id=29100003* (accessed July 2014)

MFG (2008b) Osborne Hill Area trails data evaluation and remedial options analysis report, Empire Mine State Historic Park, July 2008. MFG, Inc, Fort Collins, Colo. Prepared for California Dept of Parks and Recreation and Newmont USA Ltd. *http://www.envirostor.dtsc.ca.gov/public/profile_report.asp?global_ id=29100003* (accessed July 2014)

MFG (2009) Data transmittal and evaluation report for Historic Mine and Mill Sites, 2008 Work Plan, Empire Mine State Historic Park, October 2009. MFG, Inc, Fort Collins, Colo. Prepared for Newmont USA, Ltd. *http://www.envirostor.dtsc.ca.gov/public/profile_report.asp?global_id=29100003* (accessed July 2014)

Mining and Scientific Press (1902) Mining Summary: Nevada County. Aug. 16, 1902: 93

Mitchell VL, Alpers CN, Basta NT, Berry DL, Christopher JP, Eberl DD, Kim CS, Fears RL, Foster AE, Myers PA, Parsons B (2010) Identifying predictors for bioavailability of arsenic in soil at mining sites. Toxicologist 114:412

Mitchell VL, Alpers CN, Basta NT, Burlak T, Casteel SW, Fears RL, Foster AL, Kim CS, Myers PA, Petersen E (2011) The role of iron in the reduced bioavailability of arsenic in soil. Toxicologist 120 (Supp 2):415

Mitchell VL, Alpers CN, Basta NT, Casteel SW, Foster AL, Kim CS, Naught L, Myers PA (2012) Alternative methods for the prediction of bioavailability of arsenic in mining soils. Toxicologist 126:321

Mueller SH, Goldfarb RJ, Verplanck PL, Trainor TP, Sanzolone RF, Adams M (2007) Sediment geochemistry of epizonal and shear-hosted mineral deposits in the Tintina gold province—Arsenic and antimony distribution and mobility. *In*: Recent U.S. Geological Survey Studies in the Tintina Gold Province, Alaska, United States, and Yukon, Canada—Results of a 5-Year Project. Gough LP, Day WC (eds) US Geol Surv Sci Invest Rept 2007-5289-G, 9 p

Nash JT (1988) Geology and geochemistry of gold deposits of the Big Canyon area, El Dorado County, California. US Geol Surv Bull 1854, 40 p

Paktunc D, Dutrizac J, Gertsman V (2008) Synthesis and phase transformations involving scorodite, ferric arsenate and arsenical ferrihydrite: Implications for arsenic mobility. Geochim Cosmochim Acta 72:2649-2672

Paktunc D, Foster A, Heald S, Laflamme G (2004) Speciation and characterization of arsenic in gold ores and cyanidation tailings using X-ray absorption spectroscopy. Geochim Cosmochim Acta 68:969-983

Parkhurst DL, Appelo CAJ (1999) User's guide to PHREEQC (version 2): A computer program for speciation, batch-reaction, one-dimensional transport, and inverse geochemical calculations. US Geol Surv Water-Res Investig Report 99-4259, 312 p

Reich M, Becker U (2006) First-principles calculations of the thermodynamic mixing properties of arsenic incorporation into pyrite and marcasite. Chem Geol 225:278-290

Salocks C, Hathaway T, Ziarkowski D, Walker W(1996) Physical characterisation, solubility, and potential bioavailability of arsenic in tailings from a former gold mine. Toxicologist 16:48

Savage KS, Ashley RP, Bird DK (2009) Geochemical evolution of a high-arsenic, alkaline pit lake in the Mother Lode gold district, California. Econ Geol 104:1171-1211

Savage KS, Bird DK, Ashley RP (2000a) Legacy of the California Gold Rush: Environmental geochemistry of arsenic in the southern Mother Lode Gold District. Int Geol Rev 42:385-415

Savage KS, Tingle TN, O'Day PA, Waychunas GA, Bird DK (2000b) Arsenic speciation in pyrite and secondary weathering phases, Mother Lode Gold District, Tuolumne County, California. Appl Geochem 15:1219-1244

Schweickert RA (1981) Tectonic evolution of the Sierra Nevada Range. *In*: The geotectonic development of California (Rubey Volume I). Ernst WG (ed) Englewood Cliffs, New Jersey, Prentice-Hall, Inc., p 87-131

Schweickert RA, Bogen NI, Girty GH, Hanson RE, Merguerian C (1984) Timing and structural expression of the Nevadan orogeny, Sierra Nevada, California. Geol Soc Am Bull 95:967-979

Selverston MD (2008) Historic context for Empire Mine Historic District, Nevada County, California. Prepared for California Dept Parks and Recreation, Anthropological Studies Center, Sonoma State Univ, July 2008, 53 p, *http://www.parks.ca.gov/pages/980/files/appendix%20f-1%20-%20historical%20context(1).pdf* (accessed July 2014)

Selverston MD, Hilton SM (2013) Mine remediation and historical archaeology: A gold mine's tale. Proceedings of the Society for California Archaeology, 47th Annual Meeting, Berkeley, Calif, March 7-10, 2013, 27:193-204, *http://scahome.org/wp-content/uploads/2013/10/Proceedings.27Selverston.pdf* (accessed July 2014)

Sharp ZD, Essene EJ, Kelly WC (1985) A re-examination of the arsenopyrite geothermometer: Pressure considerations and applications to natural assemblages. Can Mineral 23:517-534

Taylor RD, Goldfarb RJ, Lee JP (2011) Age constraints on the emplacement of the Grass Valley Granodiorite, California, and relation to lode gold formation. Geol Soc Amr Abstracts with Programs 43(5):470

Tessier A, Campbell PGC, Bisson M (1979) Sequential extraction procedure for the speciation of particulate trace metals. Anal Chem 51:844-851

Thornton I, Farago M (1997) The geochemistry of arsenic. *In*: Arsenic: Exposure and Health Effects, Abernathy CO, Calderon RL, Chappell WR (eds) Chapman & Hall, ISBN 978-94-011-5864-0 (eBook)

Tuminas A (1983) Structural and stratigraphic relations in the Grass Valley-Colfax area of the northern Sierra Nevada foothills, California. Ph.D. thesis, Univ Calif Davis, 415 p

USBM (1983-1995) U.S. Bureau of Mines, Minerals Yearbook: Washington, D.C., U.S. Government Printing Office

USEPA (1998) Central Eureka Mine: Work completed in residential areas of former mining site. U.S. Environmental Protection Agency Fact Sheet, *http://yosemite.epa.gov/r9/sfund/r9sfdocw.nsf/95831d90484 434d7882574260072fadf/a6d83ec20cf4efc6882570070063c34c/$FILE/CentEur.pdf* (accessed July 2014)

USEPA (2009) Regional Screening Levels (RSL) for Chemical Contaminants at Superfund Sites. U.S. Environmental Protection Agency RSL Table Update, April 2009, *http://www.epa.gov/region09/superfund/ prg/* (accessed July 2014)

Walker SR, Parsons MB, Jamieson HE, Lanzirotti A (2009) Arsenic mineralogy of near-surface tailings and soils: influences on arsenic mobility and bioaccessibility in the Nova Scotia gold mining districts. Can Mineral 47:533-556

Weir RH, Kerrick DM (1987) Mineralogic, fluid inclusion, and stable isotope studies of several gold mines in the Mother Lode, Tuolumne and Mariposa Counties, California. Econ Geol 82:328-344

WME (2013a) Remedial action assessment, Historic Mine and Mill Sites, Empire Mine State Historic Park, Grass Valley, California. Worthington Miller Environmental, LLC. Prepared for Newmont USA Ltd., Jan 2013, 49 p, *http://www.envirostor.dtsc.ca.gov/public/profile_report.asp?global_id=29100003* (accessed July 2014)

WME (2013b) Cyanide plant and adit project area: Remedial action implementation plan – completion report, Empire Mine State Historic Park, Grass Valley, California. Worthington Miller Environmental, LLC. Prepared for Newmont USA Ltd., Feb 2013, revised July 2013, 27 p plus figures, tables, and appendices, *http://www.envirostor.dtsc.ca.gov/public/profile_report.asp?global_id=29100003* (accessed July 2014)

Reviews in Mineralogy & Geochemistry
Vol. 79 pp. 589-627, 2014
Copyright © Mineralogical Society of America

Hydrogeochemistry of the Tsumeb Deposit: Implications for Arsenate Mineral Stability

Robert J. Bowell

SRK Consulting
Churchill House, Churchill Way
Cardiff CF10 2HH, Wales, United Kingdom
rbowell@srk.co.uk

INTRODUCTION

The Tsumeb base-metal deposit contained one of the most diverse examples of mineralogical paragenesis ever observed within a single mineral deposit (Keller 1977). The deposit hosted approximately 307 minerals and 232 of those minerals are most likely formed in the oxidation zone. Of the total number, 69 minerals were first described from the deposit. Arsenic minerals show the greatest diversity in the Tsumeb deposit: 63 arsenates, 6 arsenites, and 7 arseno-sulfate minerals (see Appendix 1). Typically, As content was around 1% in the ore zone, and was intermittently produced as a by-product (white As oxide).

Mineralization is hosted in the Otavi dolomite. The main ore body is a pipe that comprises of massive peripheral ores, manto-style ores, and disseminated and stringer ores. These ores were subjected to extensive oxidation not just from surficial surface weathering but also along deep-seated permeable faults that developed complex secondary mineral assemblages at depth.

Due to the karstic nature of the host dolomite, there has been considerable water flow through the deposit and during operations into the mine workings, even during early mining. As such, water chemistry within the mine has a varied composition reflecting the different areas of the mine, water source, and geochemical reactions with host rock and the ore. In addition to water, which has been locally enriched from sulfide oxidation, saline and dilute water can be observed in the mine.

With such a complex mineralogy and paragenesis, it is possible to describe the geochemical conditions that influenced the mineral evolution of the deposit and predict interactions with groundwater. The extent to which current mine water reflects mineral paragenesis and the observed As-mineral assemblage in the mine is reviewed and used to provide an understanding of the formation of the oxide zone and the geochemical conditions at the time of formation compared to the operational mining environment.

LOCATION OF THE TSUMEB MINE

The Tsumeb deposit is located in northeast Namibia, Africa, approximately 500 km to the north of the capital, Windhoek (Fig. 1). The rugged topography of the area is formed by weathered carbonate rocks with the highest mountain elevation at 1677 m above sea level. The valleys are characterized by a cover of calcrete (caliche) overgrown by dense thorn bush and scrubby vegetation. The mountain slopes contain even more resilient species of shrubs and cacti, the latter attesting to annual temperatures that commonly exceed 30 °C. The rainfall in the Otavi Mountain Land is the highest in Namibia, with Tsumeb recording an average of 520 mm per annum (King 1994).

1529-6466/14/0079-0014$05.00

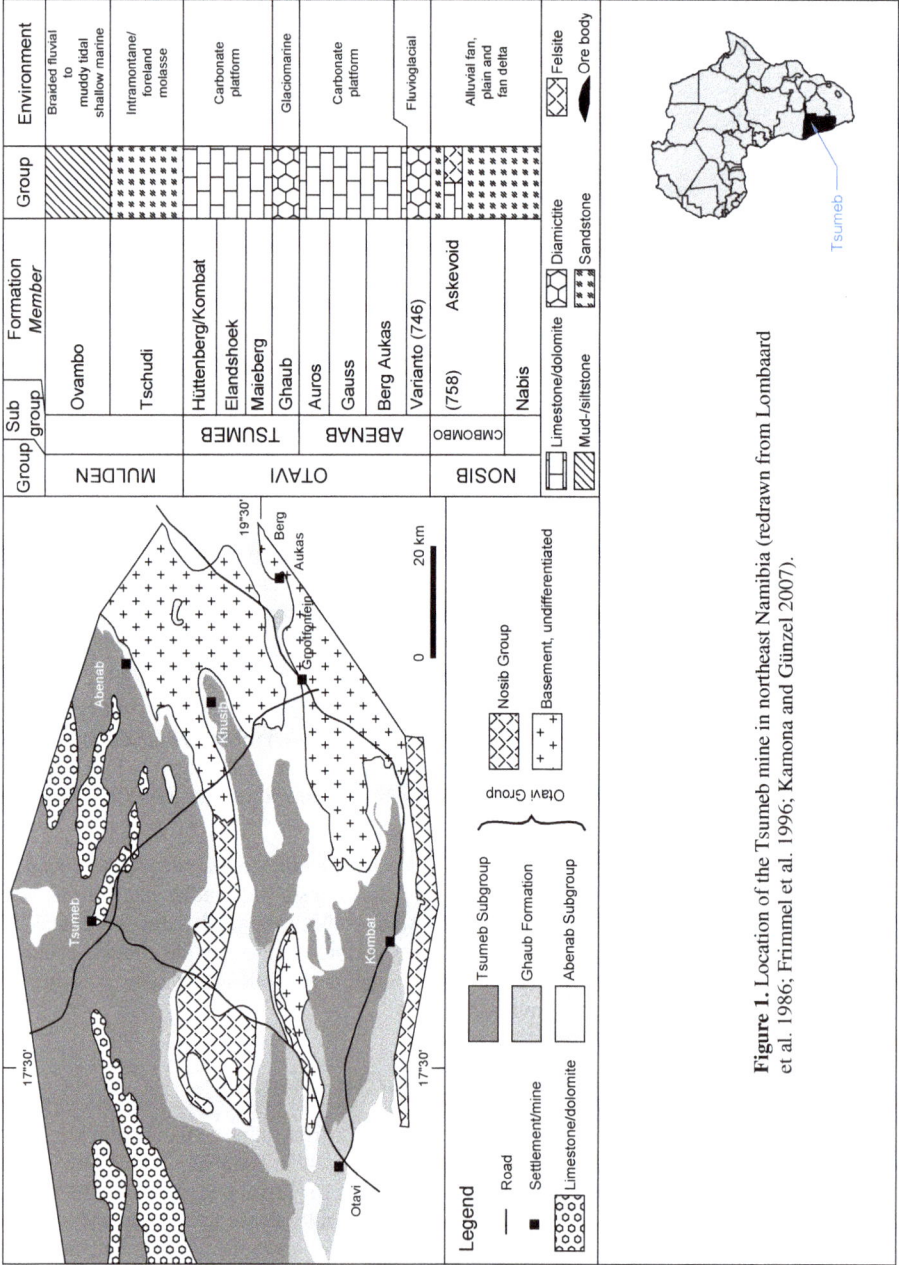

Figure 1. Location of the Tsumeb mine in northeast Namibia (redrawn from Lombaard et al. 1986; Frimmel et al. 1996; Kamona and Günzel 2007).

HISTORY OF THE TSUMEB MINE

The name Tsumeb comes from the Nama language and most likely translates as "place of the moss." The name reflecting the hill of outcropping green, oxidized-copper ore (Söhnge 1967). The first European account of copper occurred in 1857 by missionaries. Ovambo traders provided the first rich samples of ore from the Otavi Mountains. The Bushmen named the place Tsomsoub, meaning "to dig a hole in loose ground that keeps collapsing," a reference to the karst nature of the local dolomite (Clark 1957). European adventurers obtained mineral rights to the area over the next 50 years, however it was not until 1893 that the first organized mine by the South West Africa Company, occurred (Gebhard 1999).

The first geologist, Mathew Rogers, surveyed the deposit in 1903. The town was founded in 1905 by the German colonial power to support the newly formed mine and the company was purchased by the Otavi-Minen-und-Eisenbahngesellschaft (O.M.E.G.) company in 1906. Soon after, a waterpipe line and rail link to Otavi to the south was constructed. However, the mine was short lived and World War 1 caused a downturn in activities at Tsumeb with the the cessation of operations fro 1914-1921. The 1920s brought in renewed mining activity, with deeper levels being exploited. The mine's output peaked in 1930 but the Great Depression led to temporary closure.

The next phase came with Tsumeb Corporation Ltd. (T.C.L., a subsidiary of Newmont Mining Corporation) from 1946 to 1987. This was later purchased by Goldfields Namibia Ltd. The post-war period was the most productive for Tsumeb.

The 1990s witnessed a drop in base-metal prices with all-time low-metal values that forced many base-metal mines throughout the world to close. The mine lingered on until August 1996, when strikers prevented essential services. Within a few days, the famous De Wet Shaft had flooded, and the mine was forced to close. Current water level oscillates around a depth of 300 m below the surface (Level 5 in the mine).

The pipe was famous for high-grade ore. A good percentage of the ore (called "direct smelting ore") was so rich that it was sent straight to the smelter situated near the town without first having to be processed through the mineral enrichment plant. The deposit was mined by a large open pit and by several shafts. The district was once the foremost producer of Pb in Africa and, over its lifetime, yielded 6 Mt Cu, 11 Mt Pb, 2 Mt Zn, and 2600 t of Ag along with a small amount of by-product Cd, Mo, and Ge from 1906 to closure in 1996 (Frimmel et al. 1996; Chetty and Frimmel 2000).

The Tsumeb mine is also renowned among mineral collectors, particularly for the secondary mineral assemblages; many of the rarest minerals are As-bearing. There is still oxidized ore waiting to be recovered in the old upper levels of the mine; however, it is highly unlikely that the deepest levels will ever be reopened.

GEOLOGY OF THE TSUMEB DEPOSIT

The deposit lies in the upper part of the Otavi group, which consists of limestone and dolomite of Neoproterozoic age (Miller 2008). The pipe is infilled by feldspar-bearing sandstone of the overlying Mulden group. It is this sandstone that is the host for the pipe, which indicates that karst formation took place soon after deposition and lithification of the Otavi dolomite (Miller 2008). The mineralization is characterized by large-scale alteration (calcification, silicification, and limited argillization of the host rocks) and common hydrothermal carbonate veins. The deposit contained a great diversity of ore minerals of Pb, Cu, Zn, Ag, As, Sb, Cd, Co, Ge, Ga, Au, Fe, Hg, Mo, Ni, Sn, and W, as well as V, containing about 12% Pb, 4.3% Cu, 3.5% Zn, 100 ppm Ag, and 50 ppm Ge based on run-of-mine ore for the period 1964 to 1974 (Lombaard et al. 1986). There is an extensive literature on the geology and mineralogy of the mine (Schneiderhöhn 1929; Clark 1931; Frondel and Ito 1957 Pinch and Wilson 1977; Weber and Wilson 1977; Emslie 1979; Lombaard et al. 1986; Hughes 1987; Pirajno and Joubert 1993; King 1994; Frimmel et al. 1996; Gebhard 1999; von Bezing et al. 2007; Kamona and Günzel 2007).

The ore body is a pipe that is approximately 120×15 m in cross section, steeply dipping and extending from the surface to at least 1000 m in depth (Fig. 2). Controls for ore emplacement include interrupted circular fracturing, core breccia, and an internal mass of feldspathic sandstone altered to a pseudo-aplite (the Tsumeb Pipe). Local rocks include mainly Late Proterozoic sedimentary terrane. The Tsumeb pipe comprises large lenses, smaller veins, and pods of high-grade massive ore, mostly emplaced along the marginal and arcuate fractures in brecciated and foliated zones, and also disseminations in altered rock types forming large tonnages of low-grade ore. It can be classified into manto and massive sulfide ores, disseminated sulfide and vein ores, and secondary sulfide and oxide ores (Lombaard et al. 1986). The Tsumeb deposit appears to have been formed after the peak events of the Damara orogeny at approximately 530 ± 40 Ma (Allsopp and Ferguson 1970; Frimmel et al. 1996; Kamona and Günzel 2007).

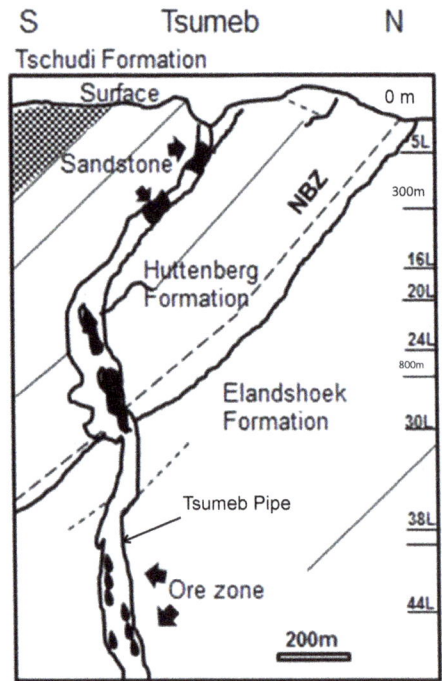

Figure 2. Cross-section of the Tsumeb deposit (based on a similar figure by Weber and Wilson 1977). The stratigraphic units of limestone are indicated by name. The Tsumeb pipe is shown as the central anastomosing structure comprising of sandstone (or in earlier descriptions, aplite) as well as ore zones (shown in black). The nomenclature on the right shows depth below ground surface and the level number of the mine that increases from surface to depth.

The massive peripheral ores comprise up to 40% total metal (Pb+Cu+Zn) content, which occur next to and within the feldspathic sandstone, and are prominent down to Level 20, below which they become thin and eventually pinch out at Level 34 (Fig. 2). The medium- to coarse-grained sulfide assemblage of the massive peripheral ore consists of galena and sphalerite together with tennantite, enargite, bornite, and chalcocite. Characteristic for this ore type is the presence of angular fragments of chert and dolomite, which are interpreted as relics of the wall rock. In places, the massive peripheral ore progressively grades into less mineralized feldspathic sandstone; remnants of sandstone can also be found within the massive ore, indicating the replacement character of the ore. The complex hypogene ore consists of the following minerals in order of abundance: galena, tennantite, sphalerite, chalcocite, enargite, and bornite together with lesser chalcopyrite, germanite, renierite, and pyrite (Hayes 1984). Supergene chalcocite, djurleite, digenite, and covellite are important in an upper and a lower oxidation zone.

The Cu-bearing hypogene sulfide mineralization resulted from the interaction of hot circulating Cu-rich saline solutions and host dolomites (Haynes 1984; Frimmel et al. 1996). Hydrothermal fluids that formed the main sulfide deposit have predicted formation temperatures of 210-280 °C and were moderately saline (6-12 wt% NaCl equivalent) based on fluid inclusion data (Haynes 1984). Phase relations for the copper sulfides indicate a paragenesis of chalcopyrite followed by bornite, that was contemporary to and replaced by chalcocite, that was partially replaced by enargite and tennantite, with subsequent later overprint by covellite and chalcocite. The interpretation of this paragenesis of a single stage Cu-rich solution reacting with dolomite host rock precipitating the sulfides. Concomitantly dolomite dissolution buffered pH provided an increase in pH over time resulting in late stage calcite precipitation (Haynes 1984).

Manto ores form concordant to semi-concordant extensions of the massive peripheral ore into the wall rock dolomite. They occur between at a depth of 780 to 930 m below the surface (Levels 26 and 30), with the most pronounced formation as a breccia on at a depth of 840 m (Level 28) (Fig. 3). They typically consist of bornite, chalcocite, djurleite, tennantite, galena, and sphalerite. They were formed by replacement, like the massive peripheral ores. Throughout the pipe, tennantite is the most persistent primary copper mineral. Chalcocite and bornite are locally important in the section at a depth of 720 m to 930 m (between Levels 24 and 30), mainly in the massive ore bodies. The primary source of As is tennantite, with lesser enargite in the upper levels.

Feldspathic sandstone, dolomite, and dolomite breccias are the main host for the stringer veins and the disseminated ore types throughout the mine. Bornite, chalcocite, and tennantite are the predominant ore minerals, accompanied by lesser amounts of galena and sphalerite. In the feldspathic sandstone, the equigranular ore minerals are evenly scattered and have mainly replaced feldspar. In the dolomite adjoining the massive peripheral ores, ore minerals form irregular blebs and discontinuous veins. The dolomite breccia contains finely scattered sulfides in the matrix, as well as sulfides on minute fractures.

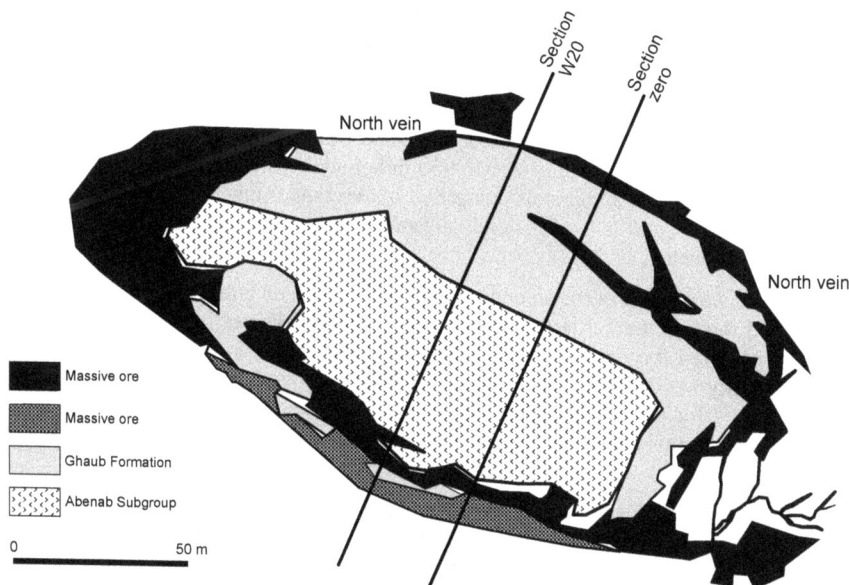

Figure 3. Geological plan of the Tsumeb pipe deposit on Level 28 (depth of 840 m below surface) showing development of the Manto massive ore and the calcitized breccia-style ores (based on mine plans).

Three zones of oxidation have been identified in the deposit; these contain secondary mineral assemblages derived by supergene alteration of primary sulfides. Supergene minerals predominate from the surface to Level 12 (360 m). From Level 12 to Level 25 (760 m), the ore body consists of almost unaltered sulfides. Below Level 25, a second oxidation zone is encountered, containing a profusion of perfectly developed secondary minerals. The oxidation zone persists to the base of the mine at a depth of 1410 m (Level 47). The lower or third oxidation zone is a mixed sulfide-oxide ore and is unique in that in hosts only partially oxidized As-minerals (arsenites), such as reinerite and schneiderhöhnite (Fig. 4).

Figure 4. (*for color see Plate 17*) Examples of arsenites from Tsumeb. (*left*) Schneiderhöhnite with minor ludlockite from Level 42 Stope 0W, Tsumeb mine. Scale bar = 5 mm. (*right*) Reinerite, probably the finest ever collected. ~5 cm in height. Crystal Classics specimen, photograph by Ed Loye. Reproduced with permission of Crystal Classics.

The most intense effects of sulfide oxidation are evident at 840 to 870 m (Levels 28 and 29) where the permeable North Break Zone intersects the pipe (e.g., Fig. 5). The main As-bearing oxide minerals comprise bayldonite, mimetite, adamite, olivenite, duftite, conichalcite, austenite, and beudantite. Bayldonite is more abundant in the first or upper oxidation zone than in the lower ones (Bartelke 1976; Keller 1977).

The secondary ores have economic significance from the surface to a depth of 300 m (Level 11) and at a depth interval of 750 to 930 m below the surface (between Levels 25 and 30) (Fig. 2). The main oxide minerals comprised of cerussite, mimetite, wulfenite, malachite, native copper, cuprite, duftite, conichalcite, olivenite, smithsonite, and willemite (Keller 1977; Gebhard 1999; Bowell et al. 2014).

Geological weathering of the orebody started with uplift and erosion in the late Cretaceous and continued through the Tertiary to the present day. The climate in northern Namibia has changed over the last forty million years with increasing aridity, from humid conditions with dense forest to dry woodland approximately 25 million years ago to the grassland environment that has persisted for the last 4 million years (McCarthy and Rubidge 2005). This climatic change influenced groundwater chemistry to some extent, and thus the nature of the oxidation zone at Tsumeb.

The geochemical ratios of Cu, Pb, and Zn varied throughout the deposit, with Pb being the dominant metal. Vertical metal zoning was evident in the deeper levels, and the overall metal content decreased with depth below 900 m (Level 30). From the surface to a depth of 180 m (Level 6), high-Cu values were encountered, probably due to supergene enrichment, whereas the corresponding low-Zn content may have been caused by leaching of the near-surface ore.

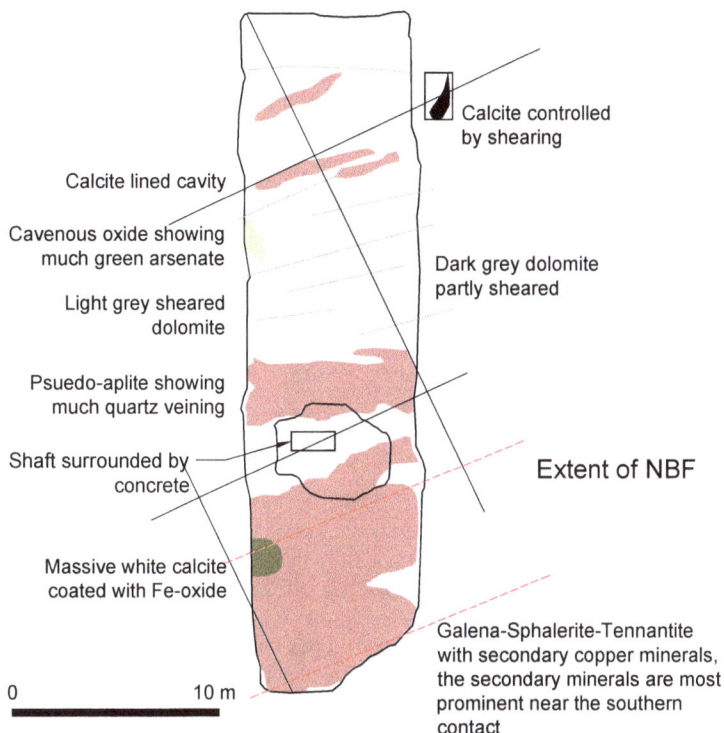

Figure 5. Detailed geological plan, from underground mapping, of Level 30 Stope 0W, showing the impact of the North Break zone on the Tsumeb mineralization. The North Break fault (NBF) is indicated by dashed lines.

Copper is about as abundant as Pb from surface to a depth of 300 m (Level 10). However below this to a depth of 600 m (Level 20), the ratio changes with Cu:Pb:Zn ratio of roughly 2:5:3 (Lombaard et al. 1986). Copper concentrations continued to increase relative to Pb from there to a depth of 810 m (Level 27) reaching peak concentrations at a depth of 870 m (Level 29), where Cu equaled Pb in content due to high grade ore zones on the southern side of the pipe. Deeper in the mine, the pipe structure narrowed and was generally Pb-rich, with a metal ratio of Cu:Pb:Zn of approximately 3:5:2 (Söhnge 1967). The Cu grade was appreciably lower at Level 44 (1320 m depth below surface)but improved with depth (Level 47, 1410 m below surface).

HYDROGEOCHEMISTRY OF THE CURRENT MINE

Two studies have published groundwater geochemistry data from the Tsumeb mine (Marchant 1980; Ingwersen 1990). Data from these studies have been compiled into a geochemical database in order to evaluate the hydrogeochemical characteristics of the deposit (Appendix 2). In an evaluation of the hydrogeochemical exploration methods, Marchant (1980) used the Na/Cl ratios and sulfate concentrations to determine that there were several sources of water in the Tsumeb mine (Table 1).

To distinguish the water types, the classification of Marchant (1980) is adopted here, with shallow referring to water in the upper part of the mine (surface to a depth of 660 m or Level 22). Some of the water types are characterized by only a few samples and so caution must be

Table 1. Classification of Tsumeb mine waters (compiled from data in Marchant 1980).

Water type	SO$_4$ (mg L^{-1})	Cl (mg L^{-1})	Na (mg L^{-1})	Na/Cl	SO$_4$/Cl	Flow (L min^{-1})
Stope water	433-1056	100-164	94-140	0.84-0.94	2.83-10.56	1-33
Dilute water shallow stope	42	3	3	1	14	1
Shallow undiluted waters	104-560	104-556	54-109	1.07-1.15	1.37-8.22	3-22
North Break fault	422-717	89-294	53-176	0.6-0.76	1.84-3.25	2-264
Intermediate waters	194-477	103-182	97-155	0.81-0.94	1.88-2.64	2->>150
Deep sodium water	204-216	77-86	83-86	1.02-1.08	2.37-2.81	1-2
Deep dilute water	21-26	6-9	10-16	1.67-1.78	2.33-4.33	5-39
Deep chloride water	52	136	28	0.21	0.38	2

applied, but the study essentially demonstrated that significant differences are encountered in the mine and that the nature of the water type is controlled by hydrological processes, such that different water types can be observed schematically in different portions of the mine (Fig. 6). Each water type shows different compositions that will influence the stability of As minerals and helps to explain the stability of certain minerals or mineral assemblages within parts of the Tsumeb mine. The pH of water for all water types tend to be high due to the widespread application of lime-rich shotcrete for geotechnical stability underground. This manmade material distorts the natural hydrogeochemical conditions and needs be kept in context when reviewing the data (Table 2). The complete database is provided in Appendix 2.

Shallow stope waters

Shallow stope waters are generally sulfate-rich, although sulfate chemistry is highly variable in these waters and have relatively high concentrations of Na and Cl (Fig. 7). The waters tend to show a universal low-Mg content. Due to the presence of active sulfide oxidation in the upper workings, there are relatively high levels of trace elements, including up to 2.6 mg L^{-1} As (Fig. 8). Trace metals such as Cu, Zn, and Pb tend to be low compared to deeper waters in the mine. In the Ficklin plot (Fig. 9), it can be seen that the shallow stope waters occupy the low-metal near-neutral segment of the plot and only the intermediate and deeper waters occupy higher-metal segments of the plot. As can be expected, the more oxidized zones of the mine required more cementation for stabilization and therefore, waters in those areas have a tendency towards high pH, as can be observed in a depth vs. pH plot particularly in the area of the North Break Fault waters (location shown in Fig. 2 and water chemistry in Fig. 10). Arsenic shows a negative or negligible correlation to other parameters except Fe (Table 3). This is interpreted as reflecting the adsorption of As onto Fe oxyhydroxides or the formation of scordite..

The stope waters tend to be associated with the relatively dry upper levels of the mine on the northern side of the orebody (Marchant 1980). There was no visible accumulation of floor water in this area and only a few boreholes that actively produced water hence the few samples reported by Marchant (1980) or Ingwersen (1990).

Dilute upper stope waters

In the upper levels of the mine, a few analyzed waters are highly dilute and are interpreted as reflecting surface effects in the shallow workings. The water is largely meteoric and shows low overall solute levels (Table 2). These waters show a Na-Mg-HCO$_3$ characteristic in macrochemistry (Fig. 7).

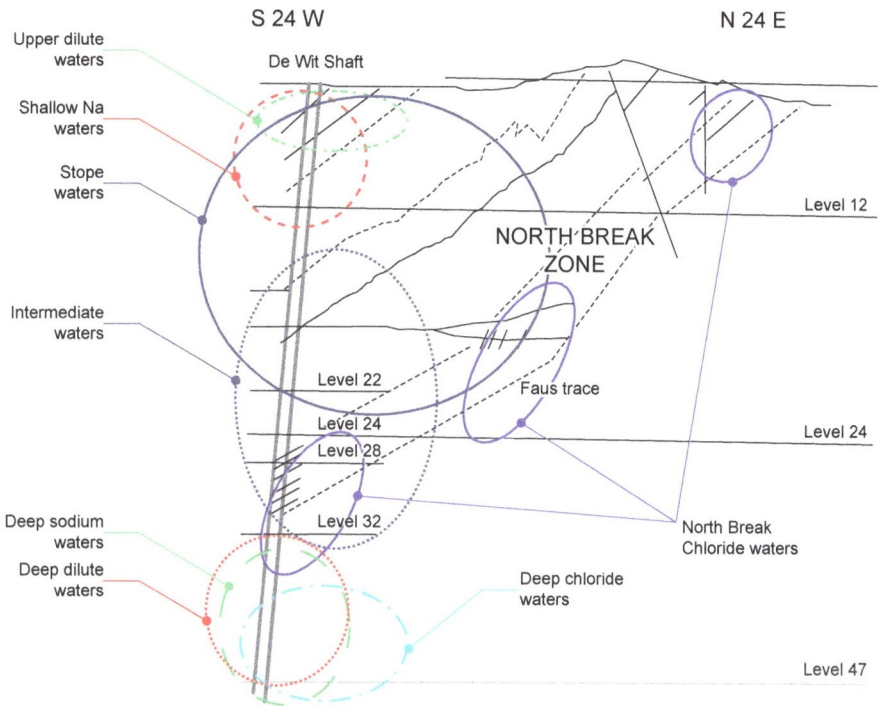

Figure 6. Schematic representation of the zone of influence of different water types at Tsumeb.

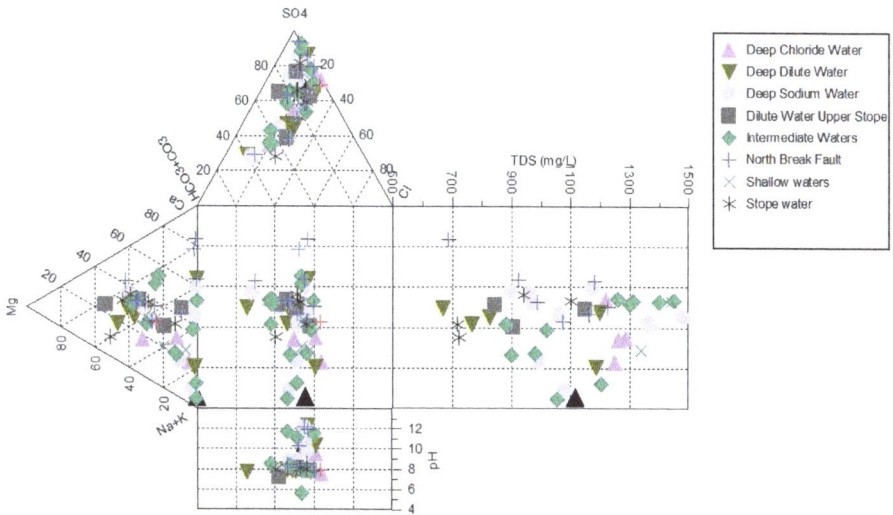

Figure 7. Durov plot for Tsumeb groundwaters.

Table 2. Summary of hydrogeochemistry of Tsumeb waters. Concentrations are listed in mg L^{-1} except for pH and Eh (mV).

Water type	pH		Eh		Cl		SO$_4$		Na		Mg		Ca	
	Min	Max	Min	Max	Min	Max	Min	Max	Min	Max	Min	Max	Min	Max
Deep chloride water	6.4	9.5	55	135	137	181	330	665	119	245	10.2	92	62.8	141
Deep dilute water	7.7	12.3	35	331	36	171	125	1250	55.5	310	0.96	67.9	43.2	570.4
Deep sodium water	7	12.2	7	308	57	134	204	835	83	204.9	1.14	84	20.9	282.6
Dilute water upper stope	7.2	8.1	120	280	48	212	384	597	42.5	151.7	18.5	89	42.2	158
Intermediate waters	5.6	11.7	27	432	61	172	247	1875	75.2	265	0.14	99	15.5	270.8
North Break fault	6.4	12.2	26	331	37	294	230	1367	38.4	176	0.75	115	104.6	409.1
Shallow waters	7.8	8.3	270	314	149	232	455	652	143	239	15.9	95	100	154
Stope water	7.2	8.3	123	312	79	164	141	1460	89.8	250	24.3	125.3	21.2	272.7

Water type	As		Fe		Mn		Cu		Pb		Zn		Cd	
	Min	Max	Min	Max	Min	Max	Min	Max	Min	Max	Min	Max	Min	Max
Deep chloride water	0.01	7.6	0.03	0.08	0.028	0.06	0.01	0.17	0.0048	0.31	0.02	0.21	0.0007	0.0007
Deep dilute water	1.5	7.4	0.08	0.13	0.01	0.04	0.04	0.22	0.19	23	0.02	0.18	0.01	0.01
Deep sodium water	0.001	120.5	0.03	0.14	0.01	0.11	0.01	1.47	0.0019	9.28	0.02	2.21	0.0004	0.06
Dilute water upper stope	0.018	3.7	0.03	0.09	0.02	0.03	0.02	0.49	0.0108	0.48	0.02	0.16	0.0017	0.01
Intermediate waters	0.001	37.3	0.03	0.14	0.01	2.07	0.01	131.2	0.0045	15.69	0.02	196.13	0.0007	1.84
North Break fault	0.004	37.1	0.03	0.43	0.01	0.09	0.01	3.61	0.0097	11.05	0.02	0.69	0.0006	0.08
Shallow waters	0.01	25.9	0.03	0.1	0.02	0.09	0.01	0.2	0.0121	0.19	0.02	0.33	0.0046	0.02
Stope water	0.017	7.8	0.03	0.1	0.02	0.03	0.01	0.22	0.0133	0.4	0.06	1.8	0.03	0.32

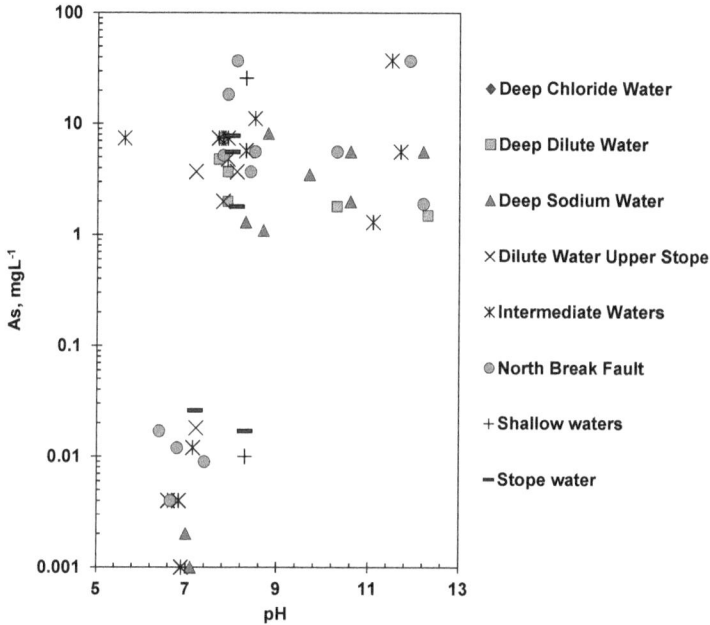

Figure 8. Plot of As vs. pH.

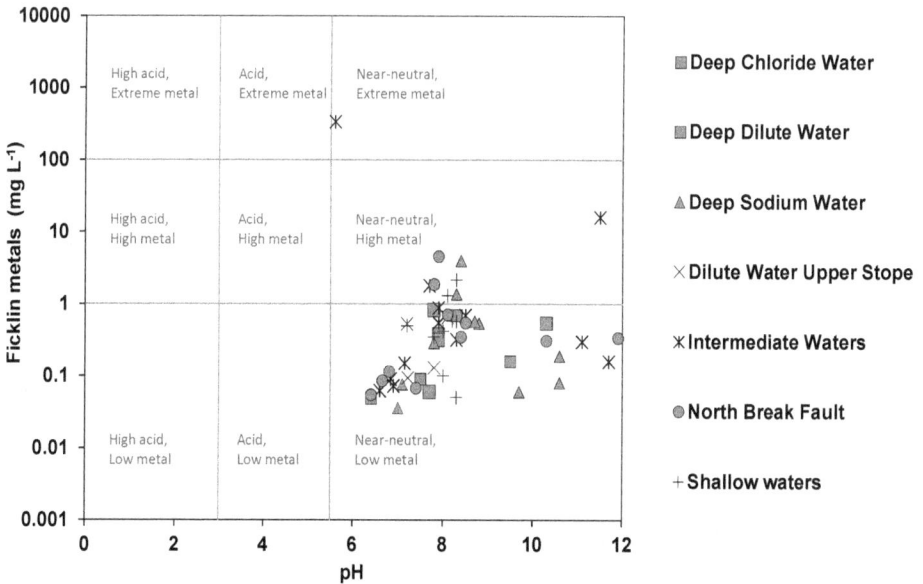

Figure 9. Ficklin plot for Tsumeb groundwaters. Concentration of
Ficklin metals (Cu, Co, Cd, Pb, Ni, and Zn) plotted vs. pH.

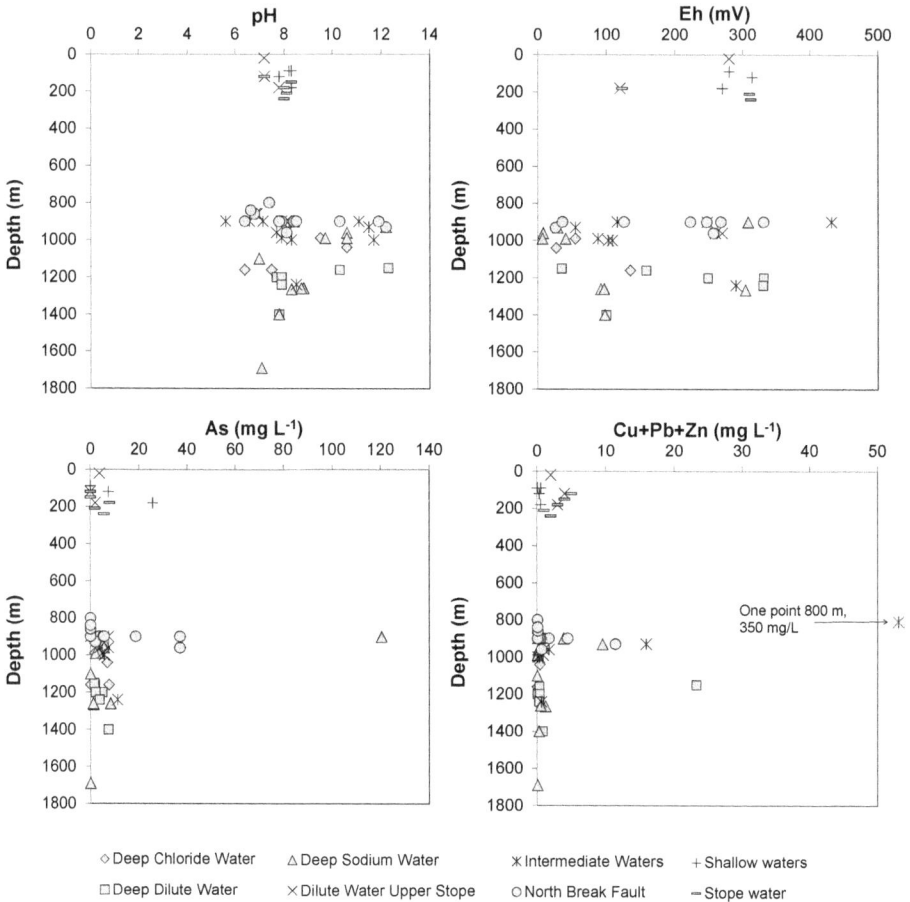

Figure 10. Depth profiles of pH, Eh (mV), Cu+Pb+Zn (mg L⁻¹),
and As (mg L⁻¹) for the Tsumeb mine, Namibia.

Within the waters there is a crude correlation of As with Cu, Zn, Pb, and SO_4 indicating the influence of sulfide oxidation in the upper stopes (Table 4). The correlation factors are given only for those elements with sufficient number of analyses above detection to allow for calculation of meaningful coefficients. However there is a poor correlation to pH, possibly reflecting the use of shotcrete modifying local-borehole water chemistry. These waters are considered to be modified from natural groundwater and represent contact with exposed rocks in the open-pit upper workings and from seepage into karstic fractures in the vicinity of the waste management area of the mining complex (Marchant 1980).

Shallow undilute waters

This group of waters are distinct from the dilute shallow waters in that they typically contain higher levels of SO_4 and Cl (Fig. 7), and higher As and trace metal contents (Figs. 8-9). These waters are considered representative of natural groundwater and are similar to groundwater elsewhere in the Tsumeb dolomite rocks in northern Namibia (Marchant 1980). Water in these stopes are derived from boreholes associated with mineralized fractures and are often observed associated with metal-rich secondary precipitates, for example on Level 7

Table 3. Correlation coefficients for stope water in the Tsumeb mine.

	pH	As	Cl	SO$_4$	Mg	Cu	Fe	Mn	HCO$_3$	Zn	Ca	K	Na
pH	1.00												
As	0.23	1.00											
Cl	-0.061	-0.73	1.00										
SO$_4$	0.084	-0.60	0.92	1.00									
Mg	-0.20	-0.74	0.96	0.77	1.00								
Cu	-0.068	-0.074	0.44	0.71	0.21	1.00							
Fe	0.34	0.73	-0.067	0.076	-0.17	0.40	1.00						
Mn	-0.37	-0.79	0.30	0.022	0.48	-0.51	-0.89	1.00					
HCO$_3$	-0.067	0.21	0.062	-0.25	0.28	-0.55	0.26	0.15	1.00				
Zn	0.51	-0.67	0.50	0.40	0.49	-0.26	-0.45	0.55	0.00025	1.00			
Ca	0.13	-0.67	0.83	0.96	0.64	0.71	-0.091	0.074	-0.49	0.47	1.00		
K	0.29	0.15	0.097	0.47	-0.19	0.87	0.41	-0.68	-0.75	-0.19	0.56	1.00	
Na	0.22	-0.18	0.54	0.82	0.28	0.94	0.35	-0.46	-0.59	0.046	0.84	0.89	1.00

Table 4. Correlation coefficients for dilute upper stope water in the Tsumeb mine.

	pH	As	Cl	SO$_4$	Mg	Cu	Fe	Mn	HCO$_3$	Zn	Ca	K	Na
pH	1.00												
As	0.99	1.00											
Cl	-1.00	-0.99	1.00										
SO$_4$	0.64	0.74	-0.64	1.00									
Mg	-0.99	-0.97	0.99	-0.55	1.00								
Cu	0.82	0.89	-0.82	0.97	-0.75	1.00							
Fe	0.98	1.00	-0.98	0.77	-0.96	0.91	1.00						
Mn	-0.94	-0.89	0.94	-0.35	0.97	-0.58	-0.87	1.00					
HCO$_3$	-0.92	-0.86	0.93	-0.30	0.96	-0.54	-0.84	1.00	1.00				
Zn	0.78	0.86	-0.77	0.98	-0.70	1.00	0.88	-0.52	-0.48	1.00			
Ca	-0.37	-0.24	0.38	0.47	0.47	0.23	-0.20	0.66	0.70	0.29	1.00		
K	0.95	0.98	-0.94	0.86	-0.90	0.96	0.99	-0.78	-0.75	0.94	-0.048	1.00	
Na	-0.50	-0.61	0.50	-0.99	0.40	-0.91	-0.65	0.18	0.13	-0.93	-0.62	-0.76	1.00

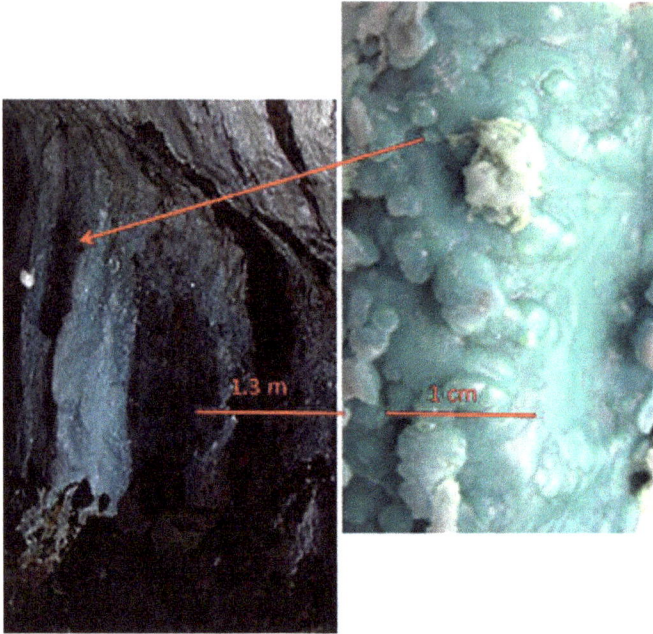

Figure 11. (*for color see Plate 18*) (*left*) Aragonite associated with seepage zone in Level 7 Stope 9E, Tsumeb mine. These precipitates (*right*) are associated with bayldonite, such as the one shown.

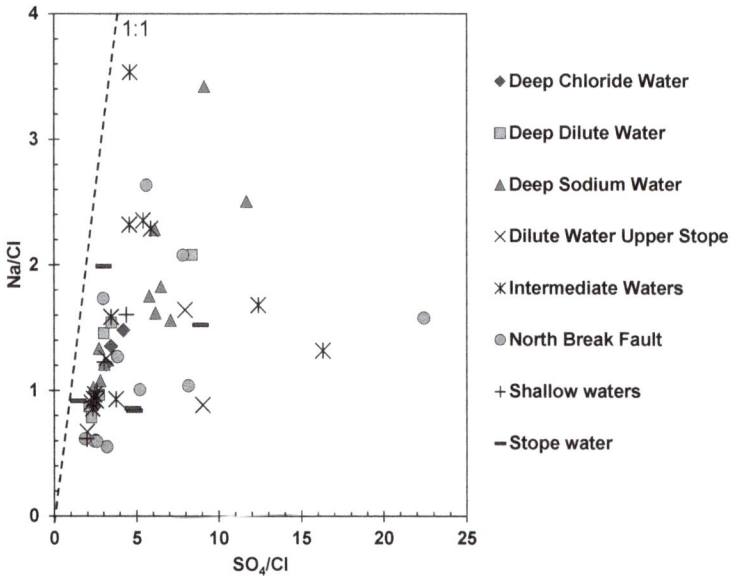

Figure 12. Na/Cl vs. SO₄/Cl for Tsumeb mine waters.

Stope E9, Cu- and Pb-rich aragonites are associated with a water in-flow that was associated with microcrystalline bayldonite (Fig. 11). These waters tend to have a Cl-HCO$_3$ dominated composition with variable cation chemistry (Fig. 6).

Within these waters As shows a correlation with high pH, Ca, and Mg but not SO$_4$ (Table 5). This may reflect the occurrence of secondary minerals assemblages such as the one in Figure 11 that is a product of high pH promoting As mobilization in highly-alkaline (Ca-Mg rich) waters. These waters tend to be well buffered although due to cementation it is not always possible to distinguish background pH and its anthropogenic modification.

Intermediate waters

The defining point between the shallow and deeper waters at Tsumeb was proposed reasonably by Marchant (1980) as being at the upper surface of the cone of dewatering at Level 22 in the mine (Fig. 5). Water in this upper level was classified as being intermediate based on Na/Cl and SO$_4$/Cl ratios (Fig. 12). These samples showed characteristics of both the shallow and the deeper waters of the mine. These waters show a wide variation in salinity but actually report the highest salinity and show a mixed chemistry with no definite affinity for macrochemicals. Within these waters, metal and As levels can be highly variable due to water originating from boreholes in the oxidized mineral lodes in the second oxidation zone (Fig. 10).

Within these waters, the well mixed area results in poor correlation of most ions due to the complex nature of water sources (Table 6). Arsenic shows a positive correlation to Pb, Zn, Cl, pH, and Ca, the latter reflecting the secondary mineralogy possibly dominated by mimetite. The influence of saline oxidizing water from the North Break fault on the intermediate waters is apparent from the data (Figs. 7-10, 12) with higher sulfate and trace metals in some of the intermediate waters. Consequently there is considerable overlap between the two groups.

North Break Fault

The groundwater that characterizes water in the North Break Fault (NBF) is typically highly variable tending to have a lower overall TDS than deeper waters (Fig. 6) but with variable levels of Cl reflecting the source of evolved shallow oxygenated water than observed deeper in the mine (Fig. 6). Chloride is slightly lower in the NBF than other groundwaters below a depth of 660 m (Level 22) reflecting the source as largely meteoric water recharging along the deep seated fault structure but higher than the shallow waters (Fig. 5). The chemical affinity of the water tends to be Ca-Cl.

In places sulfate is also a major component but only where the fault zone cuts mineralized fractures. Here higher metal levels and sulfate are observed (Fig. 10). Although this does not always correlate to high As that shows two populations, one high in As (above 1 mg L^{-1}) and the other low (below 0.01 mg L^{-1}) presumably related to pH control (Fig. 7). The higher SO$_4$/Cl ratios are higher for these waters compared to water from elsewhere in the NBF and other water types in the mine (Fig. 12). Arsenic shows distinct correlation to SO$_4$ and Cu possibly reflecting tennantite but not to chloride, as in the intermediate waters. Zinc is also positively correlated but Ca is not again possibly reflecting a strong influence of the tennantite source to these elements (Table 7).

Deep waters

The deep mine waters at Tsumeb show considerable overlap in geochemistry, with similar ranges in pH and trace metals (Figs. 7-8). The limited number of samples of mine water below Level 38 (1200 m below ground surface) reflects the effects of cementation used to retain geotechnical stability despite high-hydrostatic pressure. They are sub-divided based on Na and Cl chemistry. The sub-classes, as defined by Marchant (1980), are:

(1) *Deep sodium water.* High Na and sulfate concentrations with slightly higher trace element concentrations of As and metals, possibly due to close proximity to

Table 5. Correlation coefficients for shallow undiluted water in the Tsumeb mine.

	pH	As	Cl	SO₄	Mg	Cu	Fe	Mn	HCO₃	Zn	Ca	K	Na
pH	1.00												
As	0.24	1.00											
Cl	-0.37	-0.99	1.00										
SO₄	-0.52	0.71	-0.60	1.00									
Mg	0.33	-0.83	0.75	-0.98	1.00								
Cu	-0.41	0.78	-0.69	0.99	-1.00	1.00							
Fe	-0.72	0.50	-0.37	0.97	-0.89	0.93	1.00						
Mn	0.50	-0.72	0.62	-1.00	0.98	-1.00	-0.96	1.00					
HCO₃	0.38	-0.80	0.71	-0.99	1.00	-1.00	-0.92	0.99	1.00				
Zn	-0.43	0.77	-0.67	1.00	-0.99	1.00	0.94	-1.00	-1.00	1.00			
Ca	-0.28	-1.00	1.00	-0.68	0.81	-0.76	-0.46	0.69	0.78	-0.74	1.00		
K	0.47	0.97	-0.99	0.51	-0.67	0.61	0.27	-0.53	-0.63	0.59	-0.98	1.00	
Na	-0.41	0.79	-0.69	0.99	-1.00	1.00	0.93	-1.00	-1.00	1.00	-0.76	0.61	1.00

Table 6. Correlation coefficients for intermediate water in the Tsumeb mine.

	pH	As	Cl	SO₄	Mg	Cu	Fe	Mn	HCO₃	Zn	Ca	K	Pb	Na
pH	1.00													
As	0.56	1.00												
Cl	-0.17	0.20	1.00											
SO₄	0.17	0.36	-0.066	1.00										
Mg	-0.79	-0.50	0.34	-0.15	1.00									
Cu	0.054	0.099	-0.28	-0.26	-0.42	1.00								
Fe	0.66	0.29	-0.48	0.32	-0.62	0.34	1.00							
Mn	-0.44	-0.30	0.41	-0.22	0.56	-0.35	-0.82	1.00						
HCO₃	-0.57	-0.53	0.43	-0.58	0.84	-0.41	-0.64	0.60	1.00					
Zn	0.072	0.47	0.14	-0.11	-0.32	0.82	0.15	-0.26	-0.35	1.00				
Ca	-0.33	0.31	0.40	0.69	0.46	-0.47	-0.25	0.19	0.050	-0.10	1.00			
K	0.65	0.14	-0.49	0.19	-0.83	0.49	0.63	-0.51	-0.77	0.29	-0.51	1.00		
Pb	0.44	0.95	0.47	0.15	-0.47	-0.03	0.09	-0.24	-0.44	0.33	0.31	0.11	1.00	
Na	0.76	-0.0029	-0.12	-0.22	-0.61	-0.090	0.36	-0.16	-0.20	-0.26	-0.66	0.57	0.09	0.09

Table 7. Correlation coefficients for the North Break Fault waters in the Tsumeb mine.

	pH	As	Cl	SO$_4$	Mg	Cu	Fe	Mn	HCO$_3$	Zn	Ca	K	Na
pH	1.00												
As	0.26	1.00											
Cl	-0.63	-0.31	1.00										
SO$_4$	0.63	0.11	-0.0040	1.00									
Mg	-0.85	-0.59	0.84	-0.41	1.00								
Cu	0.049	0.39	0.25	0.15	-0.045	1.00							
Fe	0.24	0.45	0.091	0.25	-0.22	0.98	1.00						
Mn	-0.80	-0.60	0.59	-0.34	0.79	-0.37	-0.55	1.00					
HCO$_3$	-0.82	-0.59	0.43	-0.55	0.75	-0.44	-0.61	0.96	1.00				
Zn	0.14	0.37	0.15	0.11	-0.089	0.97	0.98	-0.49	-0.52	1.00			
Ca	0.59	-0.13	-0.15	0.90	-0.40	-0.16	-0.075	-0.11	-0.28	-0.20	1.00		
K	0.31	0.74	-0.026	0.32	-0.41	0.88	0.93	-0.63	-0.70	0.86	-0.011	1.00	
Na	-0.25	0.035	0.74	0.13	0.47	0.11	0.049	0.11	-0.042	0.074	-0.15	0.065	1.00

Table 8. Correlation coefficients for the deep sodium waters in the Tsumeb mine.

	pH	As	Cl	SO$_4$	Mg	Cu	Fe	Mn	HCO$_3$	Zn	Ca	K	Na
pH	1.00												
As	-0.082	1.00											
Cl	0.66	0.39	1.00										
SO$_4$	0.32	0.29	0.13	1.00									
Mg	-0.81	0.31	-0.16	-0.37	1.00								
Cu	-0.12	0.87	0.10	0.45	0.20	1.00							
Fe	0.72	0.28	0.63	0.57	-0.43	0.41	1.00						
Mn	-0.66	-0.23	-0.21	-0.75	0.74	-0.45	-0.74	1.00					
HCO$_3$	-0.65	-0.12	-0.16	-0.80	0.77	-0.33	-0.68	0.98	1.00				
Zn	-0.11	1.00	0.40	0.27	0.35	0.87	0.29	-0.20	-0.094	1.00			
Ca	0.67	-0.021	0.83	-0.014	-0.17	-0.27	0.50	-0.055	-0.070	-0.0046	1.00		
K	0.46	-0.36	-0.27	0.55	-0.81	-0.024	0.32	-0.72	-0.75	-0.40	-0.26	1.00	
Na	0.59	0.44	0.27	0.74	-0.62	0.62	0.72	-0.95	-0.88	0.40	-0.012	0.64	1.00

Table 9. Correlation coefficients for the deep chloride waters in the Tsumeb mine.

	pH	As	Cl	SO$_4$	HCO$_3$	Mg	Cu	Mn	Fe	Zn	Ca	K	Na
pH	1.00												
As	0.86	1.00											
Cl	-0.23	-0.68	1.00										
SO$_4$	0.70	0.24	0.48	1.00									
HCO$_3$	-0.92	-0.59	-0.13	-0.92	1.00								
Mg	-0.98	-0.89	0.36	-0.63	0.88	1.00							
Cu	0.38	0.100	0.54	0.51	-0.44	-0.19	1.00						
Mn	0.92	-0.16	0.71	0.89	-0.79	-0.65	-0.180	1.00					
Fe	0.85	0.55	0.24	0.82	-0.89	-0.74	0.79	-0.12	1.00				
Zn	-0.036	-0.28	0.69	0.25	-0.075	0.23	0.91	-0.18	0.48	1.00			
Ca	-0.75	-0.86	0.43	-0.19	0.48	0.69	-0.51	0.041	-0.68	-0.22	1.00		
K	0.89	0.94	-0.60	0.41	-0.71	-0.96	-0.067	-0.77	0.52	-0.47	-0.65	1.00	
Na	0.99	0.77	-0.074	0.80	-0.97	-0.95	0.45	0.92	0.90	0.055	-0.68	0.82	1.00

Table 10. Correlation coefficients for the deep dilute waters in the Tsumeb mine.

	pH	As	Cl	SO$_4$	Mg	Cu	Fe	Mn	HCO$_3$	Zn	Ca	K	Na
pH	1.00												
As	-0.55	1.00											
Cl	0.54	0.045	1.00										
SO$_4$	0.80	-0.12	0.92	1.00									
Mg	-0.94	0.51	-0.29	-0.60	1.00								
Cu	-0.40	0.26	0.44	0.12	0.69	1.00							
Fe	-0.31	0.77	0.56	0.27	0.43	0.61	1.00						
Mn	0.92	-0.28	0.68	0.91	-0.79	-0.20	-0.094	1.00					
HCO$_3$	-0.88	0.72	-0.60	-0.76	0.69	0.032	0.33	-0.82	1.00				
Zn	-0.22	-0.32	0.26	-0.010	0.46	0.77	0.18	-0.28	-0.21	1.00			
Ca	-0.088	0.59	0.67	0.42	0.19	0.47	0.94	0.034	0.15	0.19	1.00		
K	0.95	-0.43	0.68	0.85	-0.88	-0.31	-0.065	0.85	-0.80	-0.13	0.21	1.00	
Na	0.93	-0.33	0.81	0.96	-0.79	-0.12	0.038	0.94	-0.85	-0.088	0.24	0.96	1.00

mineralized zones for these boreholes. The waters tend to Na-SO$_4$ dominance in macrochemistry (Fig. 7).

(2) *Deep chloride water.* Generally the deepest samples had the highest chloride in the mine (levels 42-44, below 2000 m depth). These tended to have low concentrations of Na as well with the principal cations being Ca and K producing a K-Ca-Cl macrochemistry (Fig. 7).

(3) *Deep dilute water.* These show lower solute concentrations of all three elements and are essentially Ca-Mg-CO$_3$ waters (Fig. 7).

Correlation coefficients for trace elements vary in the three water types reflecting the different concentrations (Tables 8-10).

Below a depth of 720 m (Level 24 in the mine), country rocks are saturated with water and flow rates generally increase with depth. These increases are not regular due to variations in the permeability of rocks, the degree of cementation, and the presence of mineralized fractures. The general dewatering rate was on the order of 20,000 m^3 per day indicating a dynamic hydrological and connected system.

The most significant distinction is the strong correlation of As to pH in the dilute deep stope waters in contrast to the insignificant correlation in the other two water types (Tables 8-10). Presumably this reflects the stronger control of redox in these waters and/or the limited range of pH observed in the waters (Fig. 9).

ARSENIC HYDROGEOCHEMISTRY

Redox potential (Eh) and pH are the most important factors controlling As speciation in natural waters (Smedley et al. 2002). Under oxidizing conditions H$_2$AsO$_4^-$ (dihydrogen arsenate) is predominant in acidic waters (less than pH 6.9), while in alkaline waters HAsO$_4^{2-}$ (hydrogen arsenate) is predominant (Fig. 13). The majority of Tsumeb water samples are

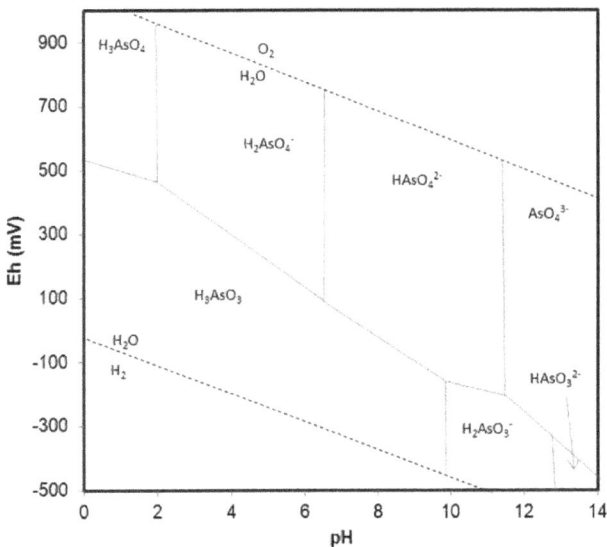

Figure 13. Eh-pH diagram for As speciation at 298 K, 1 bar, and $a_{As} = 10^{-3}$.

$HAsO_4^{2-}$ dominant, which is derived from the weathering of secondary As minerals at near neutral pH. However in a sample from the North Break Fault, $H_2AsO_4^-$ is predicted to be the dominant species in the mildly oxidizing and weakly acidic conditions. Strong adsorption or precipitation of As under these conditions may explain the apparently low As concentration compared to metals in the NBF waters (Fig. 8). In addition two discrete groups can be observed in the relationship of As to SO_4/Cl ratio in the waters with higher As where sulfate content is higher, possibly reflecting a direct source of sulfide oxidation (Fig. 14). The intermediate waters are an exception to this and show no correlation to sulfate/chloride ratio (Fig. 14).

In the deeper mine waters, intermediate and stope waters with higher pH, AsO_4^{3-} (arsenate) is predicted to dominate As speciation, reflecting the extremely alkaline conditions (above pH 10), which may explain the higher mobility of As in some of these samples. There is generally a significant positive correlation when compared with divalent metal cations, as reported in Tables 3 to 10. This can be observed also in Figure 15 where, in general, high As is accompanied by high divalent cations. Exceptions exist, particularly in the stope waters and deep mine waters that are characterized by high pH and As mobility.

The mobilization of As at high pH is also reflected in the positive correlation of As with high calcium at high pH (Fig. 16). The general positive correlation of As to Ca reflects more the presence of Ca from lime in the high pH waters than any natural geochemical relationship and is distinct from the relationship to Mg. For some of the intermediate waters, exceptions to this are perhaps due to higher Mg in these waters (Table 2).

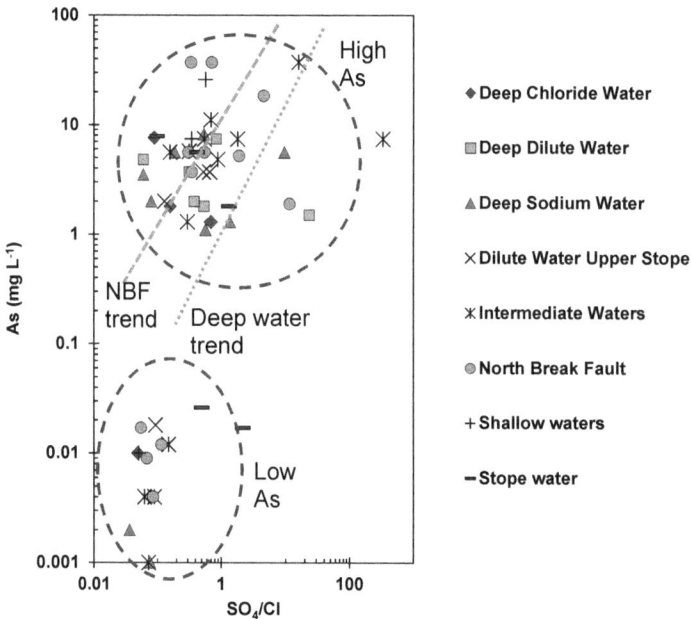

Figure 14. Variation of As vs. SO_4/Cl ratio. The plot shows a significant low-As zone and a high-As zone. In the high-As samples (mineralized) NBF and deep waters the correlation of As to sulfate can be observed. The low-As region indicates unmineralized or background waters and the high-As region water in contact with mineralized zones. Trend lines indicate approximate trend in data.

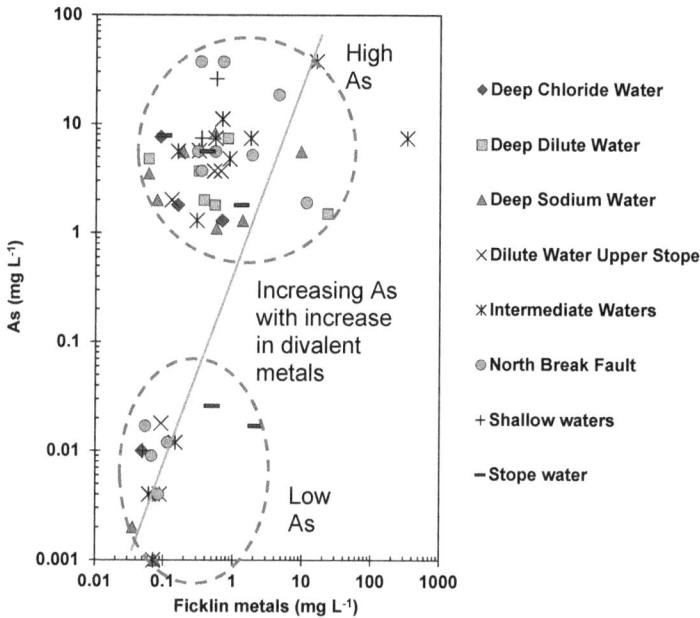

Figure 15. Variation in As with sum of divalent cations or Ficklin metals (Cu, Co, Ni, Pb, Zn). The low-As region indicates unmineralized or background waters and the high-As region water in contact with mineralized zones. General overall trend observed and illustrated by trend line.

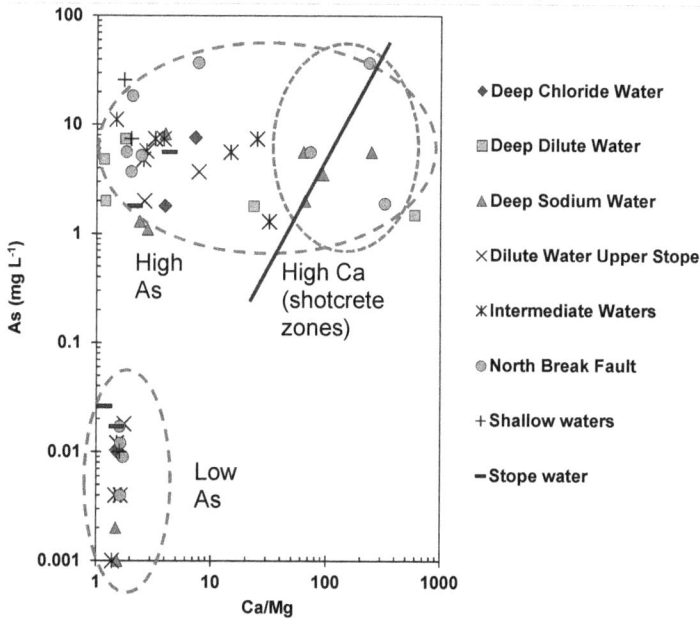

Figure 16. Comparison of As in water to Ca/Mg ratio. The low-As region indicates unmineralized or background waters and the high-As region water in contact with mineralized zones. The zone of waters in contact with cemented or shotcrete zones is also shown. For these waters a trend can be observed of increasing As with increasing Ca/Mg ratio.

ARSENATE SECONDARY MINERALOGY

The comprehensive assemblage of arsenate minerals at Tsumeb, preserved with excellent crystallinity, allows the systematic study of the progressive formation and replacement of secondary minerals revealing a complex paragenesis (Keller 1977; Keller and Bartelke 1982). Many of the arsenate minerals occur in discrete zones, particularly on the northern periphery of the deposit (Söhnge 1967).

Within the three oxidation zones present in the Tsumeb deposit, it can be generally be observed that (1) bayldonite, mimetite, arsentsumebite, and olivenite are the dominant arsenate in the first oxidation zone, (2) duftite, mimetite, conichalcite, and adamite-olivenite dominate the second oxidation zone, and (3) the arsenate mineral assemblages of the third oxidation zone are characterized by discrete pods of distinct chemistry. Examples of the latter include the legrandite-rich Zn pocket found on Level 42 of the mine (Gebhard 1999) and unusual pockets of arsenite minerals that occur in only one or two pockets in the mine and occur within a sulfide mineral host (Gebhard 1999). Although arsenates comprise only a small portion of the total mass of secondary oxidation zone minerals at Tsumeb, they are nonetheless distinct and in evaluating their paragenesis, it is essential to assess other non-arsenate minerals as well. However, such a study cannot be comprehensive due to the varied and numerous assemblages present in the mine (Fig. 17).

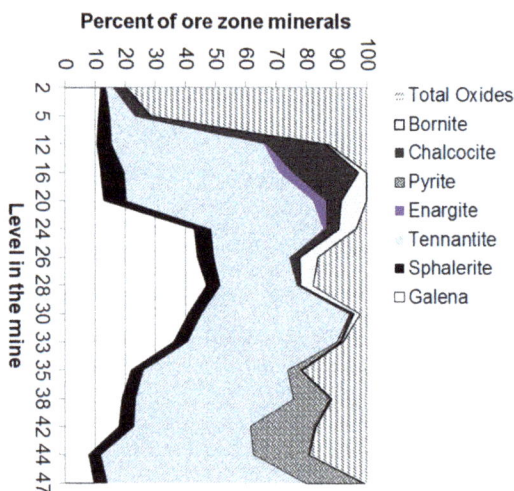

Figure 17. Approximate proportion of oxide zone minerals at different depths in the Tsumeb deposit (based on data presented in Söhnge 1967; Bartelke 1976; Lombaard et al. 1986).

The most plentiful secondary minerals in the upper or first oxidation zone include cerussite, smithsonite, malachite, azurite, anglesite, mimetite, mottramite, bayldonite, and olivenite. In the second oxidation zone, cerussite, smithsonite, malachite, duftite, mottramite, mimetite, willemite, wulfenite, azurite, olivenite-adamite, and dioptase are the more important mineral phases. Examples of the arsenate minerals are shown in Figure 18.

A striking feature of mineral specimens from Tsumeb is the different assemblages that are unique to particular levels of the mine. For example, bayldonite can occur with azurite in several parts of the mine but only with arsentsumebite and malachite in the shallow levels of the first oxidation zone. Duftite is present at least twice in the paragenesis and is often associated with malachite and dioptase, indicative of higher-pH conditions. In these zones, duftite can be observed both to have been replaced by cerussite and also to replace cerussite. By contrast, only one generation of mimetite appears to occur in the secondary mineral assemblage, despite its widespread presence, and as such, is a useful indicator of the paragenesis of the secondary mineral assemblage at Tsumeb (Fig. 19).

Secondary minerals formed in response to oxidation of sulfides, dissolution of gangue carbonate minerals, and release of component ions into groundwater. As concentrations of component ions evolved and/or a change occurs in the physical-chemical conditions of the

Figure 18. *(for color see Plate 19)* Examples of secondary arsenate mineral assemblages from Tsumeb. (a) Bayl-donite and arsentsumebite on azurite, Level 8 "Easter Pocket," first oxidation zone. Scale bar = 1 cm. (b) Olivenite on cuprian adamite, Level 10 Stope 15W, first oxidation zone. Scale bar = 2.5 cm. (c) Balydonite psuedomorphs of mimetite, first oxidation zone. Scale bar = 1.5 cm. (d) Keyite, second oxidation zone, Level 28 Stope 10E. Scale bar = 0.3 mm. (e) Olivenite with gartrellite, second oxidation zone. Scale bar = 1 cm. (f) Conichalcite, molbdofornacite with dioptase, second oxidation zone. Level 30 Stope 10W. Scale bar = 1 cm. (g) Legrandite on matrix, third oxidation zone. Scale bar = 1 mm. (h) Mimetite, "Gem pocket," second oxidation zone, Level 34. Scale bar = 1 cm.

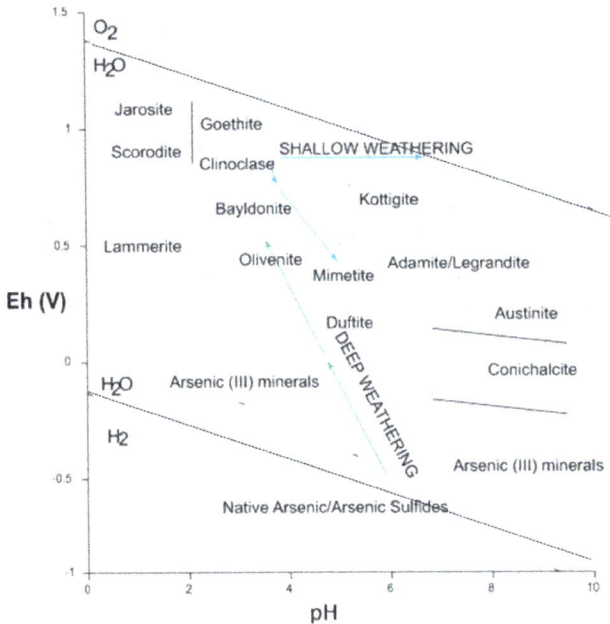

Figure 19. Schematic Eh-pH diagram showing relative stability fields for various significant Tsumeb arsenate and arsenite mineral species. Data based from Garrels (1953, 1954, 1957), Baas Becking et al. (1960), Sato (1960), Williams (1963), Garrels and Christ (1965), Mann and Deutscher (1977, 1980), Dove and Rimstidt (1980), Pourbaix and Yang (1981), Wagman et al.(1982), Robie et al. (1984), Magalhães et al. (1986, 1988), Brookins (1988), Woods and Garrels (1986, 1987), Krause and Etter (1988), Inegbenebor et al. (1989), Ingwersen (1990), and Marini and Accornero (2007).

environment, typically viewed as redox potential (Eh) and degree of acidity (pH), then different secondary minerals were formed. Secondary-As minerals are generally sensitive to changes in both Eh and pH. A crude schematic representation of the stability of the major arsenates is given in Figure 19.

A complex paragenetic model was proposed by Keller (1977) to explain the distribution of all of the secondary minerals at Tsumeb and updated by Keller and Bartelke (1982) to account for newly characterized arsenates. A simplified version of this scheme is shown in Figure 20, with only the secondary arsenate minerals included.

Based on this paragenesis, it can be observed that the initial arsenate minerals formed were beudantite, helmutwinklerite, and stranskiite. These are then partially altered to carminite, scorodite, warikahnite, olivenite, conichalcite, and keyite. These mineralogical observations are congruent with the reported hydrogeochemistry of the mine.

In the near-surface neutral to mildly acidic pH environment, bayldonite, arsentsumebite, mimetite, and olivenite are the most commonly observed arsenates. Often psuedomorphs are observed, for example bayldonite after mimetite (Fig. 18c). Increasing pH leads to saturation and formation of carbonate minerals, so psuedomorphs after adamite by smithsonite and after mimetite by cerussite occur. These were formed alongside minerals such as austinite, conichalcite, and duftite. The excess arsenate from these replacements went into solution and hence higher-As concentration is reported in the deeper mine waters.

In general, only one or two arsenate minerals occur in any particular assemblage. However, two important exceptions to this occur in the second oxidation zone in a massive

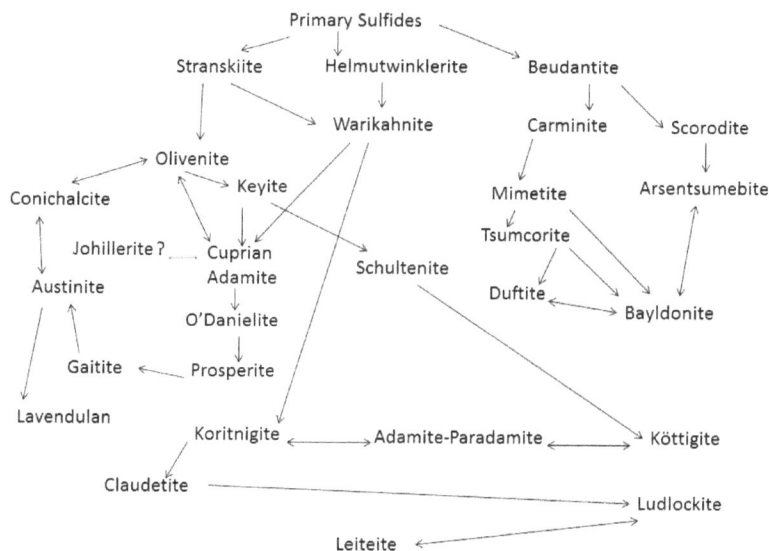

Figure 20. Schematic paragenesis of arsenate minerals in the Tsumeb oxidation zone (simplified from schemes by Keller 1977; Keller and Bartelke 1982).

tennantite-germanite ore on Level 31 in pillar E9 (Keller and Bartelke 1982). Here, the earliest arsenates are emerald-green cuprian adamite and conichalcite. They occur with keyite (e.g., Fig. 18d) as well as schultenite, koritnigite, warikahnite, helmutwinklerite, prosperite, gaitite, o'danielite, and johillerite. In addition, remnants of stranskiite, largely replaced by warikahnite, and olivenite are observed in specimens from this locality as well as an overprint of claudetite, austinite, leiteite and ludlockite. The formation conditions of these minerals are most likely mildly acidic (around pH 5) with a limited oxidation potential. The latter reflects the stabilization of As (III) in the assemblage in the form of the rare mineral, claudetite. The pH constraint is based on the stabilization of koritnigite, o'danielite, and johillerite in this assemblage (Zettler et al. 1979).

COMPARISON OF MINERALOGY TO HYDROGEOCHEMISTRY

The thermodynamic data for secondary minerals from Tsumeb has been collated and evaluated by Ingwersen (1990). The data presented in that thesis is used here to evaluate the observed textural relationships of secondary As minerals in the Tsumeb deposit. Additional information on mineral solubility and stability constants was taken from Garrels (1953, 1954, 1957), Baas Becking et al. (1960), Sato (1960), Williams (1963), Garrels and Christ (1965), Rickard (1970), Mann and Deutscher (1977, 1980), Dove and Rimstidt (1980), Pourbaix and Yang (1981), Wagman et al. (1982), Robie et al. (1984), Magalhães et al. (1986, 1988), Brookins (1988), Woods and Garrels (1986, 1987), Krause and Ettler (1988), Inegbenebor et al. (1989), and Marini and Accornero (2007).

In this section, the major arsenate mineralogy of Cu, Pb, Fe, and Zn arsenates is discussed in reference to the existing groundwater chemistry within the mine. The quantification of thermodynamic numerical values for the arsenates is perhaps the biggest limitation in undertaking this study. Gibbs free energy of formation for common arsenates was largely taken from Ingwersen (1990) and supplemented with information from other sources (Appendix 3). The assessment of this database and evaluation of the quality of data is presented in Ingwersen

(1990) and is not repeated here. Suffice to say that all data is reported at 298 K, 1 atmospheric pressure and for all solutions, dilute conditions were assumed. Activity diagrams were constructed using Ingwersen's data compilation in Geochemists Workbench.

The stability field of As species in natural waters is shown in Figure 13. The thermodynamic data within Geochemists Workbench, modified after Wagman et al (1982), were used as a basis for the speciation calculations. This section will summarize the stability of different arsenate minerals with respect to each other and groundwater chemistry at Tsumeb to demonstrate the observed mineralogical paragenesis.

Copper-zinc arsenate system

The stability fields of phases in the Cu-Zn-As-CO_3 system are shown in Figure 21. The chemistry of the system was generated from the mine water chemistry reported above.

End-members and solid solution mixed members from the adamite-olivenite series occur throughout the oxide zone at Tsumeb. One of the more intriguing examples is a zoned crystal that shows alternating bands of adamite (Zn) and olivenite (Cu) in a single crystal collected from the second oxidation zone on Level 30, Stope 10W (Fig. 22). For this example (Fig. 22), the solutions must have been in equilibrium between the two end-member compositions or rather for Cu and Zn ion activity in solution (Ingwersen 1990):

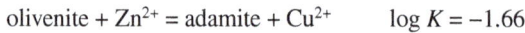

$$\text{olivenite} + Zn^{2+} = \text{adamite} + Cu^{2+} \qquad \log K = -1.66$$

Where Cu concentrations exceed an activity of 10^{-4} at low pH, olivenite would be expected to precipitate in favor of adamite. Interestingly, no modern Tsumeb mine waters plot in the olivenite field, indicating that under current conditions, adamite is the more stable mineral phase. However, olivenite is more common than adamite at Tsumeb, which perhaps indicates that Cu

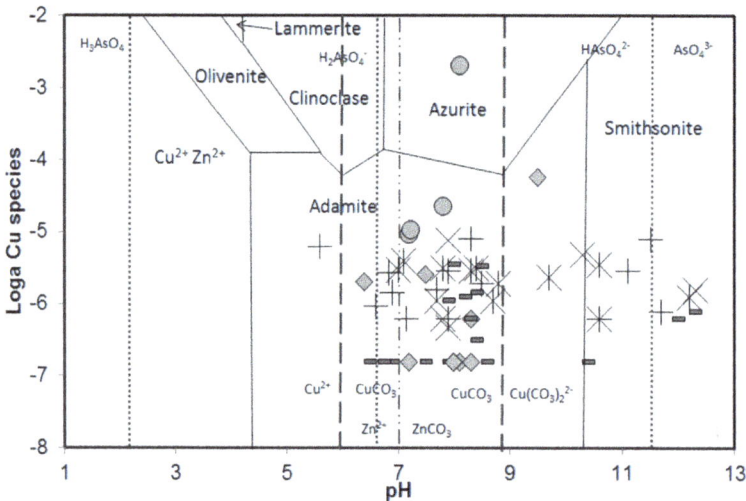

Figure 21. Stability fields of minerals in the copper-zinc-arsenate-carbonate system at 298 K and 1 bar, with $P_{CO2} = 10^{-2}$, $a_{As} = 10^{-3}$, and $a_{Zn} = 10^{-2}$. Stability fields for minerals shown as solid black lines. Aqueous species shown as chemical formula. Dashed lines: ·····As species ----Cu species -·-·-Zn species

◇ Deep Chloride Water
□ Deep Dilute Water
△ Deep Sodium Water
✕ Dilute Water Upper Stope
✹ Intermediate Waters
○ North Break Fault
+ Shallow waters
▬ Stope water

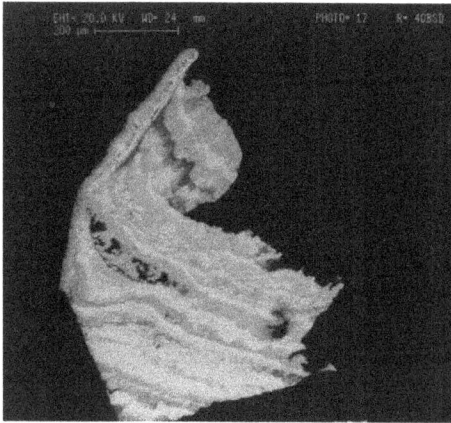

Figure 22. Cyclic precipitation of olivenite-adamite crystal. Lighter areas are zinc-rich.

activity was higher or pH lower during the initiation of ambient weathering and oxidation in the late Cretaceous period than is currently observed. However, the former is unlikely given that Zn is generally more soluble and active in aqueous solution than Cu and that at higher Cu activity, the stability field of adamite is reduced, but to the benefit of azurite and smithsonite, not olivenite (Fig. 23). Consequently, it can be postulated that pH was lower during the formation of these minerals, which seems reasonable if sulfide oxidation was taking place.

Azurite and smithsonite are more common at Tsumeb in the shallower areas of the mine than adamite or olivenite and dominate later mineral assemblages in the paragenesis, so it can be deduced that current weathering conditions and mine water chemistry are similar to the later stages of secondary mineral formation but not the early stages. This indicates that groundwater chemistry was buffered, probably by host dolomite, from the effects of low pH from sulfide oxidation.

Under atmospheric conditions, malachite replaces azurite and clinoclase is predicted to be more stable. Despite the common occurrence of olivenite at Tsumeb, clinoclase is a compara-

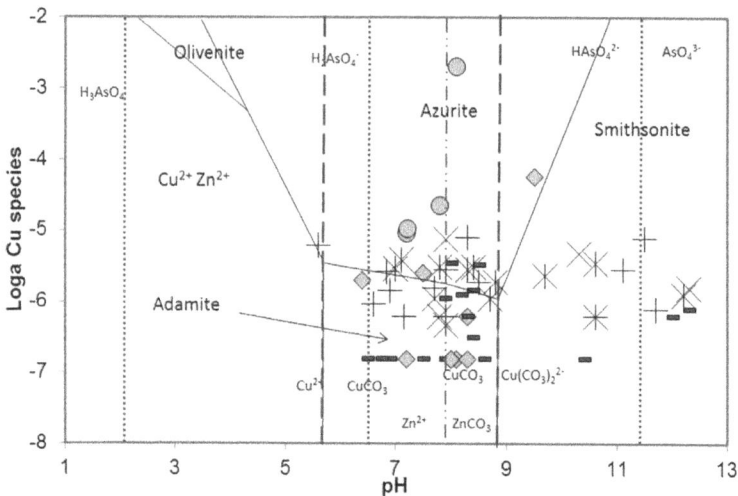

Figure 23. Stability fields of minerals in the copper-zinc-arsenate-carbonate system at 298 K and 1 bar, with $P_{CO2} = 10^{-2}$, $a_{As} = 10^{-6}$, and $a_{Zn} = 10^{-4}$. Stability fields for minerals shown as solid black lines. Aqueous species shown as chemical formula. Dashed lines: ·····As species ----Cu species ----Zn species

◇ Deep Chloride Water
□ Deep Dilute Water
△ Deep Sodium Water
× Dilute Water Upper Stope
✶ Intermediate Waters
○ North Break Fault
+ Shallow waters
▪ Stope water

tively rare mineral having only been identified once in the 19th century in early specimens from Tsumeb (Gebhard 1999). Presumably this indicates that during initial formation of the secondary minerals, sulfide oxidation buffered pH sufficiently to maintain dominance of olivenite over clinoclase and this is reflected in the common observation of olivenite and azurite occurring together in the secondary oxidation zone, such as the example shown in Figure 24. Thus it can be postulated that the content of CO_2 in groundwater during the formation of Tsumeb was lower than in current mine waters.

Figure 24. (*for color see Plate 20*) Azurite and olivenite. First oxidation zone. Scale bar is 1 cm.

The addition of Ca to this system greatly modifies the mineralogy and reflects the stabilization of conichalcite and calcite at higher pH (Fig. 25). Calcite and conichalcite are observed occurring together at Tsumeb (Fig. 26). The common occurrence indicates the pH of the solution.

A common occurrence at Tsumeb is conichalcite with malachite associated with calcite and cerussite (Fig. 24). If the minerals in this assemblage are coeval, then the pH of the precipitating solution can be deduced. In addition, it can be postulated that the Cu ion activity was restricted ($a_{Cu} < 10^{-2}$) and CO_2 pressure was $< 10^{-2}$ because malachite is present and not azurite,

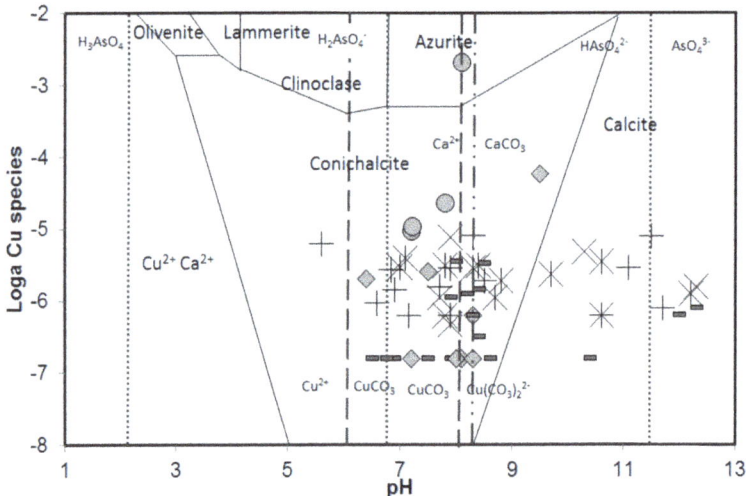

Figure 25. Stability fields of minerals in the copper-zinc-arsenate-carbonate system 298 K and 1 bar, with $P_{CO2} = 10^{-2}$, $a_{As} = 10^{-3}$, and $a_{Ca} = 10^{-2}$. Stability fields for minerals shown as solid black lines. Aqueous species shown as chemical formula. Dashed lines: ⋯⋯As species ----Cu species -·-·-Ca species

◇ Deep Chloride Water
□ Deep Dilute Water
△ Deep Sodium Water
× Dilute Water Upper Stope
✳ Intermediate Waters
○ North Break Fault
+ Shallow waters
= Stope water

Figure 26. (*for color see Plate 21*) (a) Conichalcite on cerussite, malachite and calcite. Second oxidation zone, Level 28 Stope 0W. Scale bar = 10 mm. (b) Conichalcite with malachite as inclusions in calcite. Second oxidation zone, Level 30. Scale bar = 8 mm.

and because cerussite occurs and not mimetite, that As activity was low (a_{As} <10^{-4}) and Pb activity was minimal as PbO was not observed (a_{Pb} <10^{-3}).

Iron-arsenate system

On the whole, Fe concentration is low in the Tsumeb ore compared to other base metal deposits in the vicinity or indeed globally, with the average Fe concentration in ore reporting to the Tsumeb smelter for the period 1964-1974 being only 1.5% (Lombaard et al. 1986). Beaudantite is the earliest identified iron-bearing arsenate in the secondary mineralogy (Fig. 20).

Although pyrite is restricted to the lowest levels of the mine, goethite is present along with hematite in the oxide ore throughout the mine, much of it formed by dissolution of abundant siderite and ankerite in the carbonate gangue of the mineral lodes as well as leaching of clinochlore. Against current mine water chemistry, scorodite is predicted to be unstable (Fig. 27). Claudetite is reported from Tsumeb as a rare mineral in the second oxidation zone and some of the water from the deeper parts of the mine would indicate that it is stable in contact with the minerals. Scorodite is also considered rare at Tsumeb, occurring in two localities in the second oxidation zone on Levels 28 and 30 and then at the base of the third oxidation zone on Level 45-47. Despite that, scorodite crystals are unusually large (up to 5 cm) and are considered some of the world's finest crystals (von Bezing et al. 2007) (Fig. 28). Despite this, the mineral is no longer predicted to be in equilibrium with current mine waters.

The discrepancies between mineral stability and current water chemistry indicates that the groundwater was more acidic in the past, in at least part of the mine, causing scorodite to be stable. The presence of pyrite and its oxidation in the lower portions of the mine (Fig. 17) supports the potential of acidic conditions, which would favor stabilization of scorodite over goethite.

Lead-arsenate system

Keller (1984) states that duftite is the most widely spread Cu- or Pb-arsenate mineral in the Tsumeb deposit and that only a small amount of Pb is required to form duftite and bayldonite in place of clinoclase or olivenite. If this is true, then the widespread occurrence of duftite in the mine and the lack of clinoclase provide an important indication that the activity of Pb species was greater than 10^{-6} in most solutions (Ingwersen 1990) and was on the boundary of this activity in the shallow parts of the mine where olivenite and bayldonite coexist.

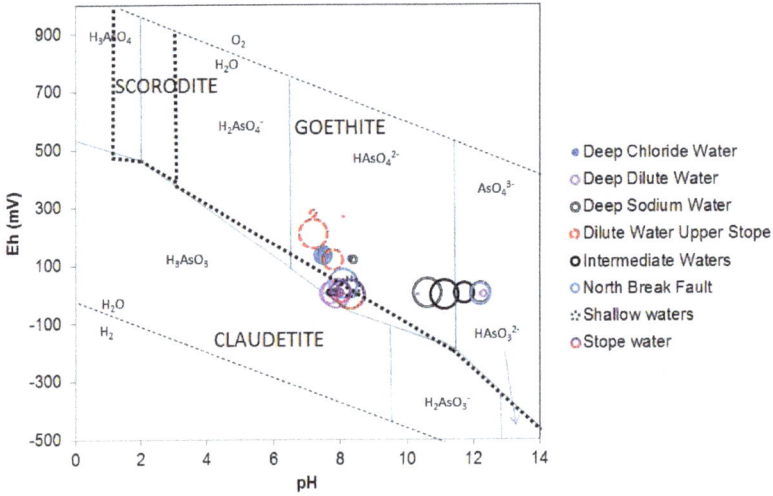

Figure 27. Stability fields of minerals in the iron-arsenate-carbonate system 298 K and 1 bar, with $P_{CO2} = 10^{-2}$, $a_{As} = 10^{-3}$, $a_{Fe} = 10^{-5}$, and $a_{Ca} = 10^{-2}$. Aqueous species shown as chemical formula. Minerals in upper case.

The presence of Pb in mine waters generates large stability fields for not just bayldonite and duftite but also for schultenite. Nevertheless, bayldonite and duftite are predicted to predominate in the current mine waters (Fig. 29).

Consistent with the previous systems, malachite is more stable than azurite at low partial pressures of CO_2 (Fig. 30). Thus, the widespread occurrence of malachite in the second oxidation zone compared with azurite indicates that sulfide oxidation or secondary mineral formation occurred at lower-CO_2 pressure than during the formation of the first oxidation zone, despite the assemblage being formed deeper in the deposit. Given its vector, the North Break Fault possibly influenced this by recharging meteoric water deep into the deposit.

Figure 28. (*for color see Plate 22*) Scorodite. Second oxidation zone. Tsumeb. Scale bar = 1 cm.

The introduction of Cl in the system, even at relatively low activity (such as 10^{-6}) stabilizes mimetite with respect to duftite and bayldonite at lower Cu activities. This indicates that Cu activities were low at the point in the secondary mineral assemblage that mimetite was formed, but increased during alteration as a common pseudomorph in the first oxidation zone is bayldonite after mimetite (Fig. 18e). Under atmospheric conditions and with low As and Cl activity it can be observed that current mine waters would favor stabilization of mimetite and duftite rather than bayldonite (Fig. 30).

This result is a little surprising given the widespread occurrence of bayldonite in the shallow workings of the mine but once again perhaps reinforces the assessment that P_{CO2} levels have decreased since the formation of the oxide zone at Tsumeb.

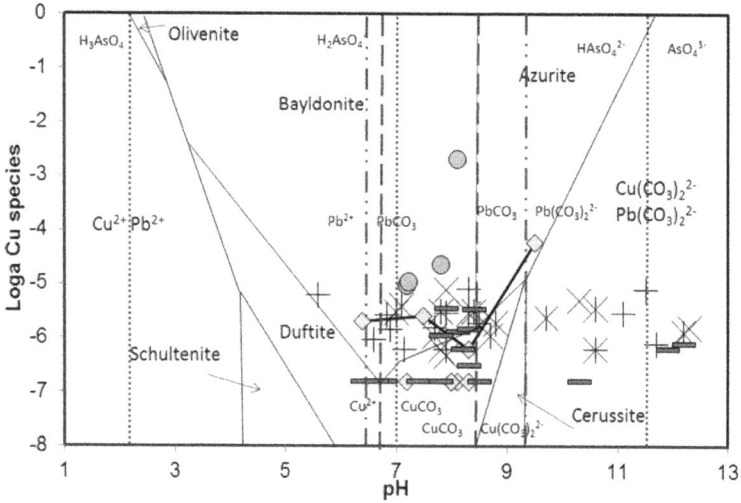

Figure 29. Stability fields of minerals in the copper-lead-arsenate-carbonate system 298 K and 1 bar, with $P_{CO2} = 10^{-2}$, $a_{As} = 10^{-3}$, and $a_{Pb} = 10^{-6}$. Stability fields for minerals shown as solid black lines. Aqueous species shown as chemical formula. Dashed lines: ·····As species ----Cu species -----Pb species.

◇ Deep Chloride Water
☐ Deep Dilute Water
△ Deep Sodium Water
⸜ Dilute Water Upper Stope
✕ Intermediate Waters
○ North Break Fault
+ Shallow waters
▬ Stope water

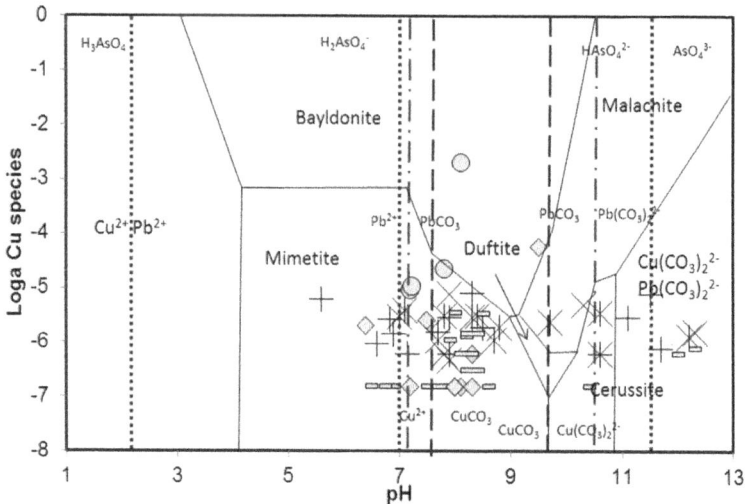

Figure 30. Stability diagram for minerals in the lead-copper-arsenate-chloride-carbonate system at low arsenate and chloride activity at 298 K and 1 bar, with $P_{CO2} = 10^{-2}$, $a_{As} = 10^{-6}$, $a_{Pb} = 10^{-5}$, and $a_{Cl} = 10^{-6}$. Stability fields for minerals shown as solid black lines. Aqueous species shown as chemical formula. Dashed lines: ·····As species ----Cu species -----Pb species

◇ Deep Chloride Water
☐ Deep Dilute Water
△ Deep Sodium Water
⸜ Dilute Water Upper Stope
✕ Intermediate Waters
○ North Break Fault
+ Shallow waters
▬ Stope water

Zinc-arsenate system

The Zn arsenate system at Tsumeb is complex with 23 known Zn-bearing arsenate minerals (Appendix 1). Despite this there is only sufficient thermodynamic data for four of these minerals, adamite, austinite, köttigite, and legrandite (Magalhães et al. 1988). Of these only adamite is particularly widespread in the deposit (Pinch and Wilson 1977).

Köttigite and legrandite have only been reported from the third oxidation zone (at a depth of 1320 m or Level 44, the "zinc pocket"). They are associated with, and partially replacing, adamite and in association with leiteite (von Bezing et al. 2008).

Austinite is more widespread in the mine and forms a solid solution with conichalcite. Despite the stability shown for existing mine waters, adamite is more widespread in the deposit than austinite (Fig. 31). However, a limitation here may be that pure adamite is comparatively rare and most of the adamite contains some Cu as discussed above. This would undoubtedly impact thermodynamic properties but it is not possible to quantify at present. However, it is evident that the current mine waters have a lower Zn activity with respect to Zn arsenates than during formation of the oxide zone at Tsumeb. This is not unreasonable considering that during oxide zone formation, presumably with active sulfide oxidation, that mine water pH was lower. This lower pH would promote divalent metal cation mobility, particularly for elements like Zn that demonstrate highly mobile chemistry over a wide pH range and is very sensitive to pH variations (Day and Bowell 2005).

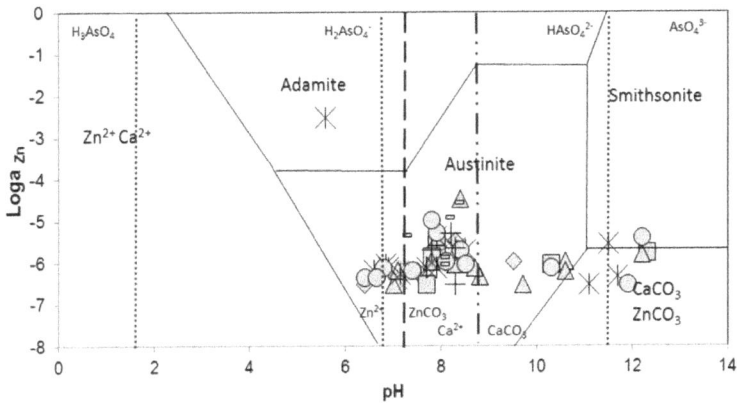

Figure 31. Stability fields of minerals in the zinc-arsenate-carbonate system at 298 K and 1 bar, with $P_{CO2} = 10^{-2}$, $a_{As} = 10^{-3}$, and $a_{Ca} = 10^{-2}$. Stability fields for minerals shown as solid black lines. Aqueous species shown as chemical formula. Dashed lines: ⋯⋯As species ----Zn species ⋯⋯Ca species

◇ Deep Chloride Water
□ Deep Dilute Water
△ Deep Sodium Water
✕ Dilute Water Upper Stope
✻ Intermediate Waters
○ North Break Fault
+ Shallow waters
– Stope water

SUMMARY

There are strong geochemical relationships observed in the current mine water chemistry and in the existing secondary mineral assemblage at Tsumeb.

The As minerals at Tsumeb are complex but by studying the mineral associations a coherent scheme of mineral formation can be deduced. Initial development of the secondary mineral assemblages started with oxidation of sulfide minerals, such as tennantite, that generated not only metal ions but also protons that reduced groundwater pH by a reaction such as,

$$Cu_{12}As_4S_{13} + 30.5O_2 + 7H_2O = 12Cu^{2+} + 13SO_4^{2-} + 4H_3AsO_4 + 2H^+$$

Metal ions and acidity (hydrogen ions, sulfuric, and arsenic acid) were produced by the oxidation of this mineral, which is the most abundant sulfide in the Tsumeb deposit. However, as the deposit is hosted in dolomite, the acidity was quickly neutralized and secondary minerals were deposited in response to changes in solution chemistry. The earliest arsenates in the paragenesis, based on work by Keller (1977), were carminite, beudantite, and scorodite. These were formed by oxidation and dissolution of the sulfides and often occur on partially oxidized sulfide matrix.

Zinc is more mobile than Cu or Pb and was transported in the weathering solutions. Much of the main ore deposit at Tsumeb is characterized by Cu- and Pb-secondary minerals, reflecting the lower mobility of these elements, and the arsenate assemblage is no exception. Formation of minerals such as mimetite appears to have occurred in only mildly acidic solutions; where Cu activity was high, duftite and bayldonite formed in preference to mimetite. Changes in As and Cu activity are probably the cause of more than one generation of duftite.

Assessment of the current mine waters at Tsumeb provides limited insight into the relative stability of the minerals. Modification of natural water chemistry by interaction with cement in shotcrete further compounds the difficulty in applying current mine water chemistry to the conditions of formation in the oxidizing deposit in the Cretaceous-Tertiary period. However, some generalizations can be made about the chemistry that are consistent with the paragenesis proposed by Keller (1977). It appears that metal and As activity have decreased over time in the water contacting the oxidized sulfide deposit. The lower predicted P_{CO2} than would have occurred during oxidation may be explained by interaction with cement; this would offset the $Ca(OH)_2$-$CaCO_3$ balance in the water and thus lower CO_2 partial pressures.

In terms of paleoclimate, the general trend of increasing aridity in the environment since the Cretaceous may explain the higher Cl content of the mine waters than would be predicted from the mineral paragenesis. However, this is by no means certain and the widespread occurrence of mimetite in the oxidation zone appears to indicate that the Cl content has not changed appreciably since that time. In terms of the macro chemistry of the mine waters, general predictions of mineral stability hold true and despite some local modification, mine waters at Tsumeb are a useful indicator of previous water chemistry and can assist in unravelling the complexity and geochemical controls on the mineral paragenesis of the oxide zone in this unique mineral deposit.

REFERENCES

Allsopp HL, Ferguson J (1970) Measurements relating to the genesis of the Tsumeb pipe, south west Africa. Earth Planet Sci Lett 9:448-453

Baas Becking LGM, Kaplan IR, Moore D (1960) Limits of the natural environment in terms of pH and oxidation-reduction-potentials. J Geol 6:243-284

Bartelke W (1976) Die Erzlagerstätte von Tsumeb/Südwest-afrika und ihre Mineralien. Aufschluss 27:393-439

Bowell RJ, Davies AA, Mocke H (2014) Mineralogical evidence of changes in groundwater chemistry during the weathering of the Tsumeb base metal deposit, with reference to the Mineral Collection of the Namibian Geological Survey. Geological Survey of Namibia Special Publication

Brookins DG (1988) Eh-pH Diagrams for Geochemistry. Berlin (Springer)

Chetty D, Frimmel HE (2000) The role of evaporates in the genesis of base sulphide mineralization in the Northern Platform of the Pan-African Damara Belt, Namibia: geochemical and fluid inclusion evidence from carbonate wallrock alteration. Mineral Dep 35:364-376

Clark CW (1931) The ore deposits of the Otavi mountains. Mineral Mag 44:265-272

Clark JD (1957) Pre-European copper workings in south and central Africa. Rhodesian Mining J 29:362-381

Day SJ, Bowell RJ (2005) Atypical and typical zinc geochemistry in a carbonate setting, Sä Dena Hes Mine, Yukon Territory, Canada. Geochem Explor Environ Anal 5:255-266

Dove PM, Rimstidt JD (1980) The solubility and stability of scorodite, $FeAsO_4 \cdot 2H_2O$. Am Mineral 70:838-844

Emslie DP (1979) The mineralogy and geochemistry of the copper, lead and zinc sulphides of the Otavi mountainland, south west Africa. PhD thesis (unpublished), University of the Orange Free State, Bloemfontein, South Africa

Frimmel HE, Deane JG, Chadwick PJ (1996) Pan-African tectonism and the genesis of base metal sulfide deposits in the foreland of the Damara Orogen, Namibia. *In*: Carbonate-Hosted Lead–Zinc Deposits. Special Publication No. 4. Sangster DF (ed) Society of Economic Geologists, Littleton, p 204–217

Frondel C, Ito J (957) Geochemistry of germanium in the oxide zone of the Tsumeb mine, south-west Africa. Am Mineral 42:743-753

Garrels RU (1953) Some thermodynamic relations among the Vanadium oxides, and their relation to the oxidation state of the uranium ores of the Colorado Plateaus. Am Mineral 38:1251-1261

Garrels RU (1954) Mineral species as functions of pH and oxidation -reduction potentials, with special reference to the zone of oxidation and secondary enrichment of sulphide ore deposits. Geochim Cosmochim Acta 5:153-168

Garrels RU (1957) Some free energy values from geologic relations. Am Mineral 42:780-791

Garrels RU, Christ CH (1965) Solutions, Minerals and Equilibria. Freeman & Cooper, San Francisco

Gebhard G (1999) Tsumeb. A Unique Mineral Locality. GG Publishing, Grossenseifen, Germany

Haynes FM (1984) A geochemical model for sulfide paragenesis and zoning in the Cu-Fe-As-S system (Tsumeb, South West Africa/Namibia). Chem Geol 47:183-190

Hughes M (1987) Tsumeb orebody, Namibia, and related dolostone-hosted base metal ore deposits of central Africa. PhD thesis (unpublished). University of the Witwatersrand, Johannesburg,

Inegbenebor AI, Thomas JH, Williams PA (1989) The chemical stability of mimetite and distribution coefficients for pyromorphite-mimetite solid-solution. Mineral Mag 53:363-371

Ingwersen G (1990) Die sekundären Mineralbildungen der Pb-Zn-Cu-Lagerstätte Tsumeb, Namibia. Doktor der Naturwissenschaften, Dissertation, University Stuttgart

Kamona AF, Günzel A (2007) Stratigraphy and base metal mineralization in the Otavi Mountain Land, Northern Namibia—a review and regional interpretation. Gondwana Res 11:396-413

Keller P (1977) Paragensis. *In:* Tsumeb! The World's Greatest Mineral Locality. Wilson WE (ed) Mineral Record 8:38-47

Keller P (1984) Tsumeb/Naimbia - eine der spektakulärsten Mineral-futidsteüen der Erde. Lapis 7B:13-63

Keller P, Bartelke W (1982) Tsumeb! New minerals and their associations. Mineral Record 13:137-147

King CHM (1994) Carbonates and mineral deposits of the Otavi mountainland. *In*: Proterozoic Crustal & Metallogenic Evolution-Excursion. 4. International Conference. McManus MNC (ed) Geological society & Geological Survey of Namibia, 40 p

Krause E, Ettel VA (1988) Solubility and stability of scorodite, $FeAsO_4 \cdot 2H_2O$; New data and further discussion, Am Mineral 73:850-854

Lombaard AF, Gunzel A, Innes J, Kruger TL (1986) the Tsumeb lead-copper-zinc-silver deposit, south west Africa/Namibia. *In*: Mineral deposits of southern Africa. Anhaeusser CR, Maske S (eds) Geological Society of South Africa, Johannesburg, p 1761-1787

Magalhães MCF, Pedrosa de Jesus JD, Williams PA (1986) Stability constants and formation of Cu(II) and Zn(II) phosphate minerals in the oxidized zone of base metal orebodies. Mineral Mag 50:33-39

Magalhães MCF, Pedrosa de Jesus JD, Williams PA (1988) The chemistry of formation of some secondary arsenate minerals of Cu(II), Zn(II) and Pb(II). Mineral Mag 52:679-690

Mann AW, Deutscher RX (1977) Solution geochemistry of copper in water containing carbonate, sulphate and Chloride ions. Chem Geol 19:253-265

Mann AW, Deutscher RX (1980) Solution geochemistry of lead and zinc in water containing carbonate, sulphate and chloride ions. Chem Geol 29:293-311

Marchant JW (1980) Hydrogeochemical Exploration at Tsumeb. PhD thesis, University of Cape Town. 4 volumes 850 p.

Marini L, Accornero M (2007) Prediction of the thermodynamic properties of metal-arsenate and metal-arsenite aqueous complexes to high temperature and pressures and some geological consequences Environ Geol 52:1343-1363

McCarthy T, Rubidge B (2005) The story of Earth and Life. Struik, Cape Town.

Miller R (2008) The Geology of Namibia, 3 volumes. Geological Survey of Namibia. Ministry of Mines and Energy, Windhoek

Naümov GB, Ryzhenko BN, Khodakovsky IX (1974) Handbook of thermodynamic data. NTIS-Report, 373. US Geological Survey, Menlo Park

Pinch WW, Wilson WE (1977) Minerals: Descriptive list. *In:* Tsumeb! The World's Greatest Mineral Locality. Wilson WE (ed) Mineral Record 8:17-37

Pirajno F, Joubert BD (1993) Overview of carbonate-hosted deposits in the Otavi mountain land, Namibia: implications for ore genesis. J African Earth Sci 16:265-272

Pourbaix M, Yang X (1981) Chemical and electrochemical equilibria in the presence of a gaseous phase. 5. oxygen-hydrogen-iron. CEBELCOR Rapt. Tech. 1401

Rickard DT (1970) The chemistry of copper in natural aqueous Solutions. Stockholm Contrib Geol 23:1-64

Robie RA, Hemingway BS, Fisher JR (1984) Thermodynamic Properties of Minerals and related Substances at 298,15 K and 1 bar (10 Pascals) Pressure and at Higher Temperatures. U.S. Geological Survey Bulletin 1452

Sato M (1960) Oxidation of sulfide ore bodies. 1. Geochemical environments in terms of Eh and pH. Econ Geol 55: 928-961

Schneiderhöhn H (1929) Das Otavibergland und seine Erzlagerstätten. Zeit F Prakt Geol 37:87-116

Smedley PL, Nicolli HB, Macdonald DMJ, Barros AJ, Tulli JO (2002) Hydrogeochemistry of As and other inorganic constituents in groundwaters from La Pampa, Argentina. Appl Geochem 17:259-284

Söhnge G (1967) Tsumeb - a historical Sketch. Scientific Research in South West Africa, 92 S. Windhoek

von Bezing L, Bode R, Jahn S (2007) Namibia: Mineralogy and Localities. Edition Schloss Freudenstein. Bode Verlag. Haltern

Wagman DD, Uvans WH, Parkim VB, Schumm HH, Halow I, Bailey SU, Curney K, Nutall RL (1982) The NBS tables of chemical thermodynamic properties: Selected values for inorganic and Ct and C2 organic substances in SI units. J Phys Chem Ref Data 11(Supplement No. 2):392S

Weber D, Wilson WE (1977) Geology. *In:* Tsumeb! The World's Greatest Mineral Locality. Wilson WE (ed) Mineral Record 8:14-16

Williams SA (1963) Stability relations of some arsenates of zinc and cadmium. Econ Geol 58:599-608

Woods TL, Garrels RU (1986) Phase relations of some cupric hydroxy minerals. Econ Geol 81:1989-2007

Woods TL, Garrels RU (1987) Thermodynamic values at low temperature for natural inorganic materials. Oxford University Press, New York

Zettler F, Riffle H, Hess H, Keller P (1979) Cobalthydrogenarsenat-Monohydrat. Darstellung und Kristallstruktur. Z Anorg Allg Chemie 454:134-144

APPENDIX 1

Arsenic-bearing minerals found at Tsumeb. Tsumeb is the type locality for the minerals in bold type. Those with an asterisk are only found at Tsumeb.

Primary Arsenic minerals

Arsenopyrite	FeAsS
Enargite-Luzonite	Cu_3AsS_4
Germanocolusite	$Cu_{13}V(Ge,As)_3S_{16}$
Glaucodote	$(Co,Fe)AsS$
Gratonite	$Pb_9As_4S_{15}$
Realgar	AsS
Renierite	$(Cu,Zn)_{11}(Ge,As)_2Fe_4S_{16}$
Seligmannite	$PbCuAsS_3$
Tennantite	$(Cu,Fe)_{12}As_4S_{13}$

Secondary Arsenic Minerals

Oxides	
Claudetite	As_2O_3

Arsenites, Arsenic Oxides

Gebhardite*	$Pb_8(As_2O_5)_2OCl_6$
Leiteite*	$(Zn,Fe)As_2O_4$
Ludlockite	$(Fe,Pb)As_2O_6$
Reinerite*	$Zn_3(AsO_3)_2$
Schneiderhöhnite	$Fe^{3+}Fe^{2+}_3As_5O_{13}$

Arsenates

Adamite	$Zn_2(AsO_4)(OH)$
Agardite-(Y)	$(Y,La,Ca)Cu_6(AsO_4)_3(OH)_6 \cdot 3H_2O$
Andyrobertsite-(Ca)*	$CaKCu_5(AsO_4)AsO_2(OH,H_2O) \cdot 2H_2O$
Andyrobertsite-(Cd)*	$CdKCu_5(AsO_4)AsO_2(OH,H_2O) \cdot 2H_2O$
Arsenbrackebuschite	$Pb_2(Zn,Fe)(AsO_4)_2(OH) \cdot H_2O$
Arsendesclozite	$PbZn(AsO_4)(OH)$
Arsenohopeite*	$Zn_3(AsO_4)_2 \cdot 4H_2O$
Arseniosiderite	$Ca_2Fe_3O_2(AsO_4)_3 \cdot 3H_2O$
Arsenogoyazite	$(Sr,Ca,Ba)Al_3(AsO_4,PO_4)_2(OH,F)_5 \cdot H_2O$
Arsentsumebite*	$Pb_2Cu(AsO_4)(SO_4)(OH)$
Austinite	$Ca(Zn,Cu)(AsO_4)(OH)$
Bayldonite	$PbCu_3(AsO_4)_2(OH)_2$
Betpakdalite	$HgK_4(H_2O)_{92}Ca_8Fe_{12}As_8Mo_{12}O_{148} \cdot 8H_2O$
Beudantite	$PbFe_3(AsO_4)(SO_4)(OH)_6$
Carminite	$PbFe_2(AsO_4)_2(OH)_2$
Ceruleite	$Cu_2Al_7(AsO_4)_4(OH)_{13} \cdot 12H_2O$
Chalcophyllite	$Cu_{18}Al_2(AsO_4)_3(SO_4)_3(OH)_{27} \cdot 3H_2O$
Chenevixite	$Cu_2Fe_2(AsO_4)_2(OH)_4 \cdot H_2O$
Chudobaite*	$(Mg,Zn)_5H_2(AsO_4)_4 \cdot 10H_2O$
Clinoclase	$Cu_3(AsO_4)(OH)_3$
Conichalcite	$CaCu(AsO_4)(OH)$
Cuprian adamite	$(Cu,Zn)_2(AsO_4)(OH)$
Cobalt-adamite	$(Co,Zn)_2(AsO_4)(OH)$
Davidlloydite*	$Zn_3(AsO_4)_2 \cdot 4H_2O$
Duftite	$Pb,Cu(AsO_4)(OH)$
Erthyrite	$Co_3(AsO_4)_2 \cdot 8H_2O$
Erikapohlite*	$Cu_3(Zn,Cu,Mg)_4Ca_2(AsO_4)_6 \cdot 2H_2O$
Fahleite*	$Zn_5CaFe_2(AsO_4)_6 \cdot 14H_2O$
Feinglosite*	$Pb_2Zn,Fe(AsO_4)(SO_4)(OH)$

Ferrilotharmeyerite*	**Ca(Zn,Cu)(Fe,Zn)(AsO₃OH)₂(OH)₃**
Gaitite*	**Ca₂Zn(AsO₄)₂·2H₂O**
Gallobeudantite*	**PbGa₃(AsO₄)₂(SO₄)₂(OH)₆**
Gartrellite	PbCu(Fe,Cu)(AsO₄)₂(OH,H₂O)₂
Hedyphane	(Ca,Pb)₅(AsO₄)₃Cl
Helmutwinklerite*	**PbZn₂(AsO₄)₂·2H₂O**
Hidalgoite	PbAl₃(AsO₄)(SO₄)(OH)₆
Hörnesite	Mg₃(AsO₄)₂·8H₂O
Ianbruceite	**Zn₂(OH)(H₂O)(H₂O)(AsO₄)](H₂O)₂**
Jamesite	**Pb₂Zn₂Fe₅O₄(AsO₄)₅**
Johillerite*	**Na(Mg,Zn)₃Cu(AsO₄)₃**
Keyite*	**(Cu,Zn,Cd)₃(AsO₄)₂**
Koritnigite	**Zn(AsO₃)(OH)·H₂O**
Köttigite	Zn₃(AsO₄)₂·8H₂O
Lammerite	Cu₃(AsO₄)(PO₄)
Lavendulan	NaCaCu₅(AsO₄)₄Cl·5H₂O
Legrandite	Zn₂(AsO₄)(OH).H₂O
Lukrahnite	**Ca(Cu,Zn)(Fe³⁺,Zn)(AsO₄)₂(OH,2H₂O)**
Mawbyite	Pb(Fe³⁺,Zn)₂(AsO₄)₂·2(OH,H₂O)
Metazeunerite	Cu(UO₂)₂(AsO₄)₂·8H₂O
Mimetite	Pb₅(AsO₄)₃Cl
Mixite	Cu₆Bi(AsO₄)₃(OH)₆·3H₂O
Molybdofornacite	**Pb₂Cu(AsO₄)(PO₄)(MoO₄)(CrO₄)(OH)**
O'Danielite*	**NaZn₃H₂(AsO₄)₃**
Ojuelaite	(Zn,Cu)Fe₂(AsO4)₂(OH)₂·4H₂O
Olivenite	Cu₂(AsO₄)(OH)
Paradamite	Zn₂(AsO₄)(OH)
Parnauite	Cu₉(AsO₄)₂(SO₄)(OH)₁₀·7H₂O
Pharmacosiderite	KFe₄(AsO₄)₃(OH)₄·67H₂O
Philipsbornite	PbAl₃(AsO₄)₂(OH)₅·H₂O
Prosperite	CaZn₂(AsO₄)₂·H₂O
Roselite-beta	Ca₂Co(AsO₄)₂·2H₂O
Schultenite	PbHAsO₄
Scorodite	FeAsO₄·2H₂O
Segnitite	PbFe₃H(AsO₄)₂(OH)₆
Sewardite*	**CaFe³⁺₂(AsO₄)₂(OH)₂**
Stranskiite	**Zn₂Cu(AsO₄)₂**
Thometzekite*	**Pb(Cu,Zn)₂(AsO₄)₂·2H₂O**
Tsumcorite	**Pb(Zn,Fe)₂(AsO₄)₂·2H₂O**
Warikahnite*	**Zn₃(AsO₄)₂·2H₂O**
Wilhelmkleinite*	**Pb(Zn,Fe)₂(AsO₄)₂·2H₂O**
Zincroselite	Ca₂Zn(AsO₄)₂·2H₂O

APPENDIX 2

Composition of Tsumeb mine waters and classification (data compilation from Marchant 1980 and Ingwersen 1990). All concentrations are in mg L^{-1} except pH and Eh (mV).

pH	Eh	Cl	SO$_4$	Na	Mg	Ca	As	Fe	Mn	Cu	Pb	Zn	Cd
Deep Chloride Water													
9.5	55	158	665	234.2	29	116.4	1.8	0.08		0.09		0.07	
8.3	125	158	502	190.9	67	62.8	1.3	0.08		0.17	0.31	0.21	
7.5	135	181	628	245	10.2	74.4	7.6	0.06	0.06	0.05		0.04	
		137	330	119	92	141	0.01	0.03	<0.028	0.01	0.005	0.02	0.001
Deep Dilute Water													
10.3	158	168	507	245	2.95	69	1.8	0.08	0.04	0.08	0.19	0.06	
12.3	35	149	1250	310	0.96	570.4	1.5	0.13	0.02	0.15	23	0.11	0.01
7.7	249	36	125	55.5	48.4	56.1	4.8	0.08	0.01	0.04		0.02	
7.9	331	112	234	97.9	64.3	76.9	2	0.09	0.01	0.2		0.18	
7.9	330	93	254	89.8	67.1	43.2	3.7	0.08	0.02	0.19		0.1	
7.8	100	171	384	134.7	67.9	121.2	7.4	0.13	0.02	0.22	0.52	0.09	
Deep Sodium Water													
8.3	304	68	794	170.4	51	123.1	1.3	0.11	0.11	0.6	0.6	0.06	0.06
8.4	308	117	674	204.9	70	126.1	1.5	0.11	0.04	1.47	0.21	2.21	0.01
12.2	29	134	366	178.8	1.14	282.6	5.6	0.14	0.01	0.18	9.28	0.1	
10.6	8	118	835	184.2	2.81	178.3	5.6	0.11	0.03	0.12		0.07	
8.8	92	57	520	195	5.21	20.9	8.2	0.05	0.01	0.51		0.03	
8.7	97	76	464	173.3	18	50.9	1.1	0.1	0.03	0.5		0.05	
9.7	41	86	558	157.2	1.65	154.6	3.5	0.06	0.01	0.04		0.02	
10.6	7	83	510	134.3	2.23	145.5	2	0.09		0.04		0.04	
7.8	98	116	354	140.3	68.9	121.2		0.06	0.01	0.21		0.07	0.01
		86	204	88	84	126	0.002	0.03	<0.028	0.01	0.002	0.02	0.000
		77	216	83	80	123	0.001	0.03	<0.028	0.01	0.018	0.043	0.001
Dilute Water Upper Stope													
8.1	270	75	597	123.2	18.5	144.6	3.7	0.09	0.03	0.49		0.16	0.01
7.2	280	48	434	42.5	57.3	42.2	3.7	0.05		0.03	0.48	0.02	
7.8	120	121	384	151.7	35.4	93.7	2	0.06	0.02	0.07		0.06	
		212	422	142	89	158	0.018	0.03	<0.028	0.02	0.011	0.057	0.002
Intermediate Waters													
7.9	248	83	1028	75.2	66.9	260.8	7.4	0.08	0.02	0.04	0.4	0.05	
7.7	260	100	540	155	4.47	111.8	7.4	0.06	0.02	0.1	1.59	0.05	
11.7	103	75	345	265	1.05	15.5	5.6	0.08	0.02	0.07		0.03	
11.1	116	110	502	255	1.26	40.2	1.3	0.09	0.01	0.07	0.21	0.02	
8.5	290	71	247	112.6	47.9	71	11.1	0.08	0.03	0.25	0.28	0.13	
5.6	432	115	1875	151.4	82.6	270.8	7.4	0.14	2.07	131.2	2.09	196.13	1.84
7.9	88	138	339	133.9	23.1	60	4.8	0.07	0.04	0.4	0.21	0.27	
8.3	109	61	358	139.5	20.6	56.8	5.7	0.06	0.05	0.18		0.14	
11.5	55	172	648	160	0.14	200	37.3	0.07	0.03	0.06	15.69	0.18	
		147	343	126	90	149	0.004	0.03	<0.028	0.01	0.005	0.043	0.001
		148	352	138	99	138	0.001	0.03	<0.028	0.01	0.006	0.05	0.001
		143	320	129	91	139	0.012	0.05	0.141	0.01	0.094	0.036	0.001
		137	350	126	96	142	0.004	0.067	0.338	0.01	0.005	0.059	0.001
North Break Fault													
7.9	268	242	778	144	67.6	140	18.5	0.43	0.04	3.61	0.5	0.35	0.02
8.4	247	74	414	104.5	60.9	123.5	3.7	0.1	0.03	0.16		0.13	
12.2	26	100	300	127.4	0.75	243.5	1.9	0.1		0.13	11.05	0.26	
8.1	257	77	603	133.3	18.3	141.2	37.1	0.1	0.03	0.31	0.27	0.07	
7.8	331	137	712	137.9	57	141.5	5.2	0.11	0.09	0.69	0.36	0.69	0.08
11.9	36	37	302	38.4	1.16	272.7	37.1	0.07	0.01	0.04	0.28	0.02	
8.5	223	129	498	163.5	57	104.6	5.6	0.08	0.02	0.1	0.36	0.06	
10.3	126	61	1367	96.3	5.64	409.1	5.6	0.1	0.03	0.05	0.2	0.05	0.01
		285	712	172	115	186	0.017	0.03	<0.028	0.01	0.010	0.029	0.002
		294	717	176	115	189	0.012	0.03	<0.028	0.04	0.02	0.05	0.002
		237	436	146	91	158	0.009	0.03	<0.028	0.01	0.010	0.043	0.001
		89	230	53	74	122	0.004	0.03	<0.028	0.01	0.042	0.029	0.001
Shallow waters													
8.2	280	149	652	239	15.9	100		0.09	0.05	0.04	0.19	0.33	0.02
7.8	314	226	588	225	71.2	144.2	7.4	0.1	0.02	0.18		0.13	0.01
8.3	270	192	585	235	65.5	114	25.9	0.08	0.09	0.2	0.19	0.14	
		232	455	143	95	154	0.01	0.03	<0.028	0.01	0.012	0.02	0.005
Stope water													
8.1	310	164	1460	250	125.3	272.7	1.8	0.09	0.03	0.22	0.4	0.78	0.03
8	312	79	239	157.1	24.3	105.5	5.6	0.06	0.02	0.12	0.21	0.09	
8	123	98	141	89.8	68	21.2	7.8	0.1	0.02	0.04		0.06	
		130	631	109	102	155	0.017	0.03	<0.028	0.01	0.016	1.8	0.32
		134	638	115	115	136	0.026	0.03	<0.028	0.091	0.013	0.292	0.1
10.6	27	121	498	250	0.18	17.4	6.9	0.08	0.04	0.08	0.31	0.03	

APPENDIX 3

Compilation of Gibbs free energy for secondary arsenate minerals
(from Ingwersen 1990).

Mineral	ΔG_j° (W mol^{-1})	Source
Adamite $Zn_2(AsO_4)(OH)$	-1252.9	Magalhães et al. (1988);
	-1253.2	Ingwersen (1990)
	-1252.2	
Annabergite $Ni_3(AsO_4)_2 \cdot 8H_2O$	-3482.3	Naümov et al. (1974)
Austinite $CaZn(AsO_4)(OH)$	-1652.5	Magalhães et al. (1988)
Bayldonite $PbCu_3(AsO_4)_2(OH)_2$	-1809.8	Magalhães et al. (1988)
Clinoclase $Cu_3(AsO_4)(OH)_3$	-1211.2	Magalhães et al. (1988)
Conichalcite $CaCu(AsO_4)(OH)$	-1471.7	Magalhães et al. (1988)
Cornubite $Cu_8(AsO_4)_2(OH)_4$	-2057.9	Magalhães et al. (1988)
Duftite $PbCu(AsO_4)(OH)$	-961.1	Magalhães et al. (1988)
Erythrite $Co_3(AsO_4)_2 \cdot 8H_2O$	-3530.5	Naümov et al. (1974)
Euchroite $Cu_2(AsO_4)(OH) \cdot 3H_2O$	-1552.7	Magalhães et al. (1988)
Köttigite $Zn_3(AsO_4)_2 \cdot 8H_2O$	-3795.2	Magalhães et al. (1988)
Lammerite $Cu_3(AsO_4)_2$	-1300.8	Magalhães et al. (1988)
Legrandite $Zn_2(AsO_4)(OH) \cdot H_2O$	-1488.6	Magalhães et al. (1988)
Mansfieldite $Al(AsO_4) \cdot 2H_2O$	-1708.8	Naümov et al. (1974)
Mimetite $Pb_5(AsO_4)_3Cl$	-2674.3	Inegbenebor et al. (1989)
	-2650.9	Ingwersen (1990)
	-2638.9	
Olivenite $Cu_2(AsO_4)(OH)$	-846.4	Magalhães et. al. (1988)
	-845.0	Ingwersen (1990)
	-864.0	
Rooseveltite $Bi(AsO_4)$	-613.6	Naümov et al. (1974)
Schultenite $PbH(AsO_4)$	-809.2	Magalhães et al. (1988)
Scorodite $FeAsO_4 \cdot 2H_2O$	-1279.1	Krause and Ettel (1988)

INDEX

Arsenic minerals (in which As is essential to the crystal structure)

Formulae of selected arsenic minerals are listed in Chapter 2, APPENDIX 1 (p. 174-183).

1529-6466/14/0060-0ind$00.00
DOI: 10.2138/rmg.2014.79.ind

Arsenic-bearing minerals (not essential to the structure) and other minerals

Formulae of selected minerals are listed in Chapter 2, APPENDIX 2 (p. 184).